# Future Petroleum Provinces of the World

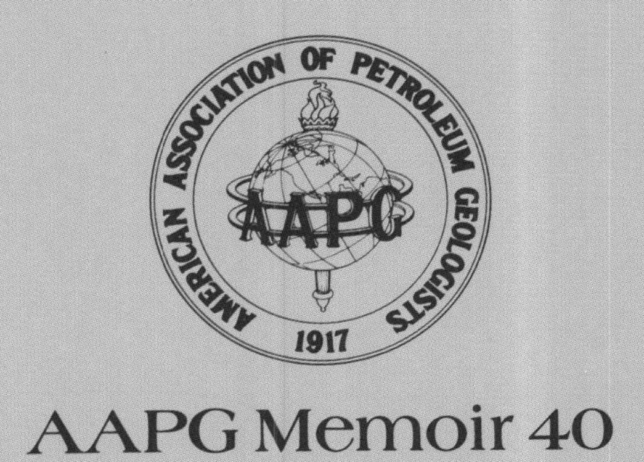


AAPG Memoir 40

# Future Petroleum Provinces of the World

Edited by
Michel T. Halbouty

Proceedings of the
Wallace E. Pratt Memorial Conference
Phoenix, December 1984

Published by
American Association of Petroleum Geologists
Tulsa, Oklahoma 74101, U.S.A.

Published April, 1986

Library of Congress Cataloging-in-Publication Data

Wallace E. Pratt Memorial Conference (1984 : Phoenix, Ariz.)
Future petroleum provinces of the world.

(AAPG memoir ; 40)
Includes bibliographies and index.
1. Petroleum–Geology–Congresses. 2. Gas, Natural –Geology–Congresses. I. Halbouty, Michel Thomas, 1909- . II. Title. III. Series.
TN863.W35 1984 553.2′82 85-30801
ISBN 0-89181-317-9



Association editor: James Helwig
Science Director: Ronald L. Hart
Project editors: Victor Van Beuren, Anne H. Thomas and Thomas L. Warren
Design and Production: S. Wally Powell
Typographers: Eula Matheny, Tricia Kinion, Elaine DiLoreto and Phyllis Kenney

# Dedication

## Wallace Everette Pratt

1885—1981

This volume is dedicated to the memory of Wallace E. Pratt, who believed that the human element—the human factor—is the ultimate tool in the practice of the science of geology.

# Table of Contents

# Foreword

When asked about the philosophy of the Wallace E. Pratt Memorial Conference on Future Petroleum Provinces of the World, which was held in December of 1984 in Phoenix, Arizona, I recalled that Wallace E. Pratt, noted as one of the world's most eminent geologists, recognized early in his prolific career that the most valuable natural resource in the world is the human mind. He concluded one of his most memorable papers with the words, "oil is found in the minds of men."

Pratt's career was one of continually pushing back the frontiers of oil and gas exploration. Changing methodology, technology, and applications of new concepts have led the global petroleum industry to success after success and to ever-higher levels of achievement in the search for oil and gas.

The petroleum geologist's active combination of knowledge, specialized technical skills, intuitiveness and innovation is what leads to petroleum discoveries. In brief, without the primary exploration tool—the mind—all other "tools", from whatever source, are inconsequential.

The idea for the conference was conceived and formulated by Fred Dix, Executive Director of AAPG, and myself shortly after Wallace Pratt's death in December of 1981. With the abovementioned thoughts uppermost in our minds, the organizers of the Wallace E. Pratt Memorial Conference on Future Petroleum Provinces of the World proposed to illustrate Pratt's premise.

With the dedicated efforts of Program Chairman Bill St. John, we worked diligently to obtain a worldwide representation of papers and to coordinate the sessions. The one and a half years spent organizing the conference proved well worth the effort, as evidenced by the quality of authors and subject areas that were selected.

This volume is a compilation of 31 of the papers which were written by authors and numerous co-authors who represented private companies, government agencies, and academia at the conference. These articles address the challenging questions of where the world's explorationists *must* search for the precious commodity on which the world's people are most dependent.

The scope of this volume is truly worldwide, ranging in coverage from philosophical overviews of world oil and gas potential, to known giant fields and areas for future giant discoveries, to basin mechanics, to specific fields and well-known producing basins, to the potential of relatively unexplored remote basins, and even to exploration possibilities in an as yet unexplored and little-known continent, Antarctica.

Michel T. Halbouty
Houston, Texas

# Introduction to the Wallace E. Pratt Memorial Conference on Future Petroleum Provinces of the World[1]

Michel T. Halbouty[2]
*Michel T. Halbouty Energy Co.*
*Houston, Texas*

Wallace E. Pratt is recognized as one of the world's most outstanding petroleum geologists. He combined our science of geology and its application as an art in the search for petroleum. In this regard, he was remarkably successful. His business acumen was superb and comprehensive. Altogether, he was an amazing personality.

I had the privilege of knowing Wallace for more than forty years, and it was indeed a privilege for me to have had the pleasure of not only referring to him as a colleague but as a good friend. During that period, we developed a close bond between us. I was his student and he was the student-teacher. Even after he retired to Arizona, we maintained a relationship until his death. As I occasionally re-read his letters, I again and again feel the presence of this great geoscientist.

Wallace Pratt viewed our science as a way of life, which made him a practical and effective philosopher. He would continually philosophize on the state of our science, the strength and weakness of our government and the nation, and its people as well. He philosophized about continuing education long before our Association became involved in this program. He strongly felt that a geologist should add to his knowledge daily and that it was a must to learn and know the other allied sciences. He thought that the profession was not paying much attention to the rocks in the field, and that too much emphasis was given to the more passive means of learning and knowing. Above all, he philosophized about AAPG and its objectives and often spoke of its strength and weaknesses, but he took great pride in being one of its founders and past presidents.

Wallace Everette Pratt was born in Phillipsburg, Kansas, on March 15, 1885, and died on Christmas Day, December 25, 1981, at the age of 96. He held two bachelor's degrees and one master's degree from the University of Kansas. He began his professional career working for the Mines Bureau of the Government of the Philippines in Manila, then worked in Houston with Texaco from 1915–1918. Humble Oil and Refining Company hired him as its first geologist in 1918, and he was chief geologist until 1923. In 1924 he became a director of the company, and concurrently from 1930 through 1933 he also served as a vice president of Humble. He joined Standard Oil of New Jersey in 1937 and retired in 1946 as a vice president and director of the executive committee.

Many cite his greatest achievement with Humble as leasing the large King Ranch in south Texas for oil and gas rights. The ranch became a notable producer of oil and gas, with hundreds of wells. But to his fellow geoscientists, he is best known for his

[1]Special AAPG volume of papers from the Wallace E. Pratt Memorial Conference on Future Petroleum Provinces of the World, Phoenix, Arizona, December 2–5, 1984.

[2]Chairman of the Board and Chief Executive Officer of Michel T. Halbouty Energy Co., Houston Texas, and Conference General Chairman.

ability to use all of the available data in his search for petroleum. No ideas generated for this cause were ignored. He encouraged far-fetched concepts, and many which he accepted and used are those we use today in our search. He often cited the need for innovative and creative geological thinking. He was a student and a teacher, accepting with enthusiasm each role as the occasion demanded.

When Wallace was presented the first Sidney Powers Memorial Medal in 1945, Everette Lee DeGolyer made the following comments: "Throughout his whole career as an executive and administrator he has demonstrated a superb ability to select and train men and, through the exercise of his genius in this phase of his activity, he has multiplied, made effective, and improved upon his own material and technical skills manyfold. In so doing he has raised the profession of petroleum geology to an eminence and a dignity which it would not otherwise have attained. More than any other man, perhaps, he has brought to the general public an awareness and appreciation of the importance of the science of geology to the arts of prospecting for and producing oil."

Twenty-seven years later, in 1972, Wallace was awarded the first AAPG Human Needs Award and he asked me to be his citationist which, of course, was indeed a privilege and an honor. Several weeks before the presentation, I visited Wallace at his home in Tucson and it was a most memorable experience.

He was 87 at the time. I had always enjoyed our meetings, but this one was indeed special. His student-teacher syndrome was working overtime. For instance, he admonished me for not admonishing the profession for not deliberately searching for the subtle traps. He claimed that our large future reserves would be found in these traps—a belief we jointly shared—and he laid the blame on exploration supervisors who were afraid to go beyond the seismic closures, a belief we also jointly shared. If he were alive today, I would tell him that the search for subtle traps has not changed much since that day in 1972, but I would also tell him that we are coming around, slowly but surely.

He was a firm believer that human beings need encouragement, and that they need to believe they are capable of performing some useful function, no matter how difficult. Wallace spread his optimism around the world for most of his 96 years. He encouraged geologists to try a little harder, hold onto their convictions a little longer, and to think of the unusual instead of the usual. He religiously practiced this philosophy.

His faith and his optimism, teamed with his knowledge and imagination, resulted in the discovery of large petroleum reserves in many parts of the world. His expertise and the petroleum reserves which culminated therefrom were a blessing, and still are a blessing, to the world's people. Wallace was a man whose heart and brain were ever paired. This great scientist loved nature so much that his work was also his pleasure.

It was indeed appropriate that Wallace Pratt received AAPG's first Human Needs Award, as his endeavors benefited mankind in many ways. He stoutly believed that geology is more than a science, and as I stated earlier, he called it "the way of life"! On this premise he felt that man should base his existence firmly on a foundation of geology and its application to human needs.

In his acceptance speech of the Award, which he asked that I present as his doctors forbade him to travel, he said: "Earthly life is the only life in the universe known to man—*so far*. And Mother Earth is the only possible abode for life known to man—*so far*. Life first appeared on the earth billions of years ago, on an earth already billions of years old, but still a starkly barren, cloud-enveloped, totally inorganic planet. From its very inception life faithfully recorded in the inert substance of its habitat, the disintegrating rocks of the land surface, the astounding record of its own evolution, even as it progressed. This marvelous history, incredulous geologists would read with amazement, eons later. And it is this identical record, again, that geology now holds out to us as the authentic way of life."

And then to make his point, he quoted an observation made by another science philosopher, who said: "There are within us the same chemical elements that make up the mountains, the pine trees, the seashore. We are indeed of the earth—brothers of the boulders, cousins of the clouds, and distant kin, by way of chemical tie-up, of the fossil plants and animals."

So it was that Pratt was a living symbol of the science of geology. His creed *was the science*, and he lived as a student and teacher of it. He was a founding member of AAPG, a past president, and a lifelong supporter of the Association. There is no question in my mind that he would beam with pleasure and his eyes would twinkle to know that this special publication bears his name and is dedicated to honor his memory.

The world's petroleum potential is enormous. Our giant fields of today were once the frontiers of yesteryear and, surely, our continually advancing exploration concepts, methods and technology will convert the frontier areas of today into the giants of tomorrow. What we all will learn from this excellent selection of literature will add immeasurably to the wealth of knowledge that men such as Wallace Pratt left as a legacy to us all.

I wish you good reading and good hunting!

# Basins and New Frontiers: An Overview

Michel T. Halbouty
*Michel T. Halbouty Energy Co.*
*Houston, Texas*

Although the global transition to alternative energy sources has begun, the world's chief reliance for energy resources during the coming decades will be on oil and natural gas supplies. Therefore, petroleum exploration must discover the oil and gas that lie untapped in both the known petroleum producing areas of the world and the frontier regions. These frontier areas—deserts, ice-covered lands, deep waters, and remote continental interiors—hold vast hydrocarbon accumulations. These sectors contain the future oil and gas discoveries that could make the difference between a proper energy transition or a global catastrophe.

Explorationists must, therefore, re-evaluate the mature and developing petroleum regions of the world. The vast ocean areas and the remote continental interiors must also be carefully and thoroughly appraised to ascertain their petroleum potential. In conjunction with these investigations, new and better uses of geology, geophysics, petroleum engineering, and technology must enhance not only exploration, but also development and production.

This paper presents an overview of the basins of the world where oil and gas will be found.

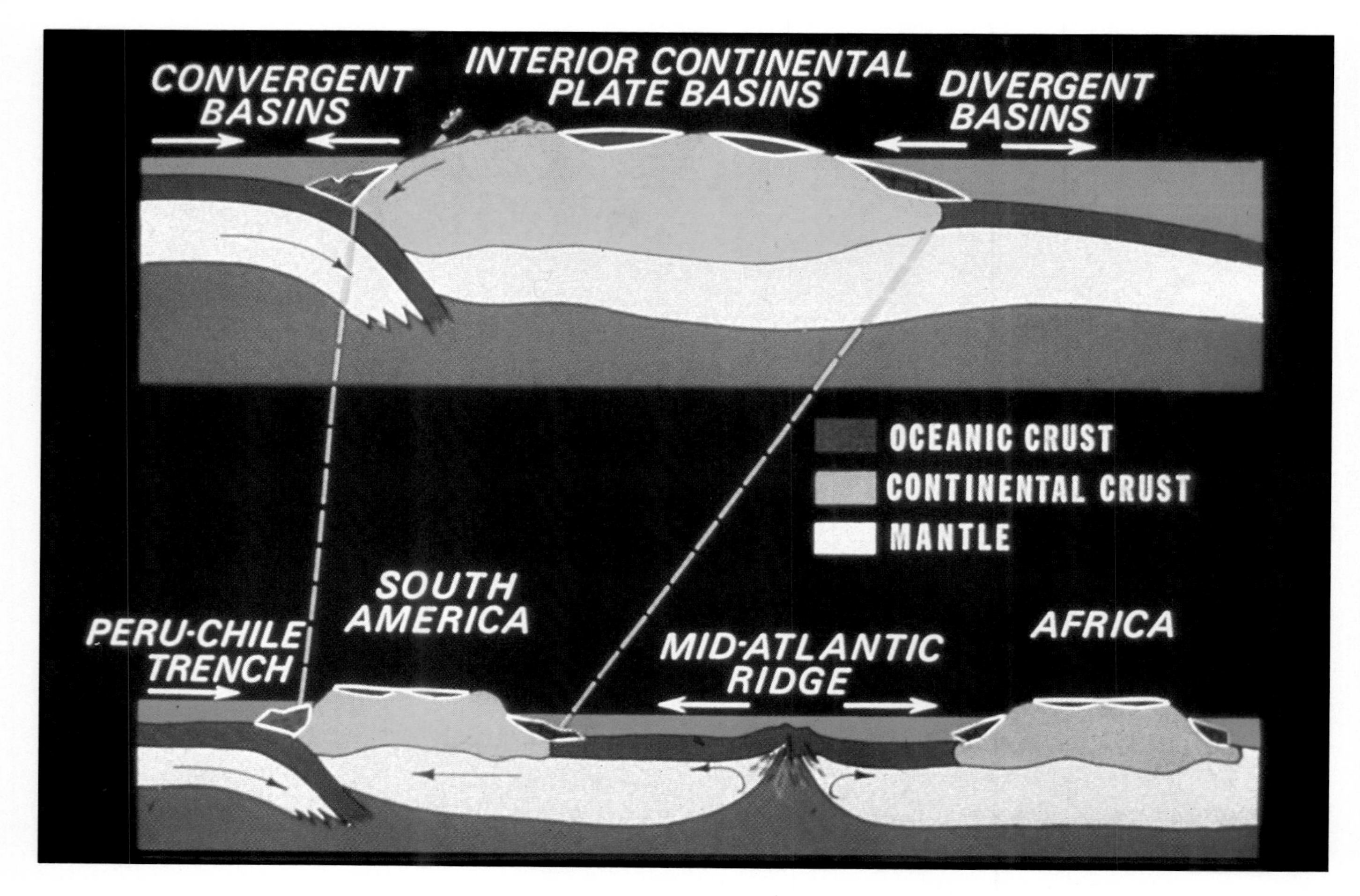


Figure 1. *Generalized cross-section through the American and African plates (Wood, 1979).*

## INTRODUCTION

The world's energy requirements are so interconnected that all countries must jointly consider the development and production of future, consumable energy supplies. In the industrialized, developed countries of the world, science and technology rapidly advance toward developing and producing alternate sources of energy and efficiently exploiting their remaining hydrocarbon reserves and potential. In the developing countries, exploring for commercial oil and gas deposits is intensifying. Yet, even though most of the world has begun a transition from conventional oil and gas energy toward alternative energy sources, world energy survival will depend on oil and natural gas supplies for at least the next four to five decades.

Many possibilities exist for increasing the world's energy supply from petroleum, as well as from other sources, but the lead times for commercial production are very long. The delays come from time needed to demonstrate new technology, obtain regulatory permits, provide environmental safeguards, develop infrastructures in remote locations, construct facilities in harsh physical environments, and, finally, complete financial arrangements.

These problems suggest the need for new solutions, including new technologies for drilling and producing oil and gas. As part of a unified exploration effort, each specialized geoscientific discipline, whether geology, geophysics, or petroleum engineering, must provide the bold and innovative thinking that leads to future discoveries of oil and gas.

The continuing worldwide search for new reserves of petroleum must consider scores of geologically attractive areas, especially the frontier regions that need to be extensively explored to meet the soaring demand for petroleum supplies.

## BASIN CLASSIFICATION

Although the geologic history of sedimentary basins is complex, modern models and concepts of basin development provide a generalized classification that is useful in studying worldwide hydrocarbon distribution, particularly in the areas considered most promising for future discoveries of oil and gas.

Figure 1 shows a generalized cross-section through the American and African plates reflecting a three-fold classification of the world sedimentary basins (convergent, divergent, and interior continental plate basins; Wood, 1979). The divisions are based on concepts of crustal plate movement over long periods of time and show changes in their relative positions. Figure 2

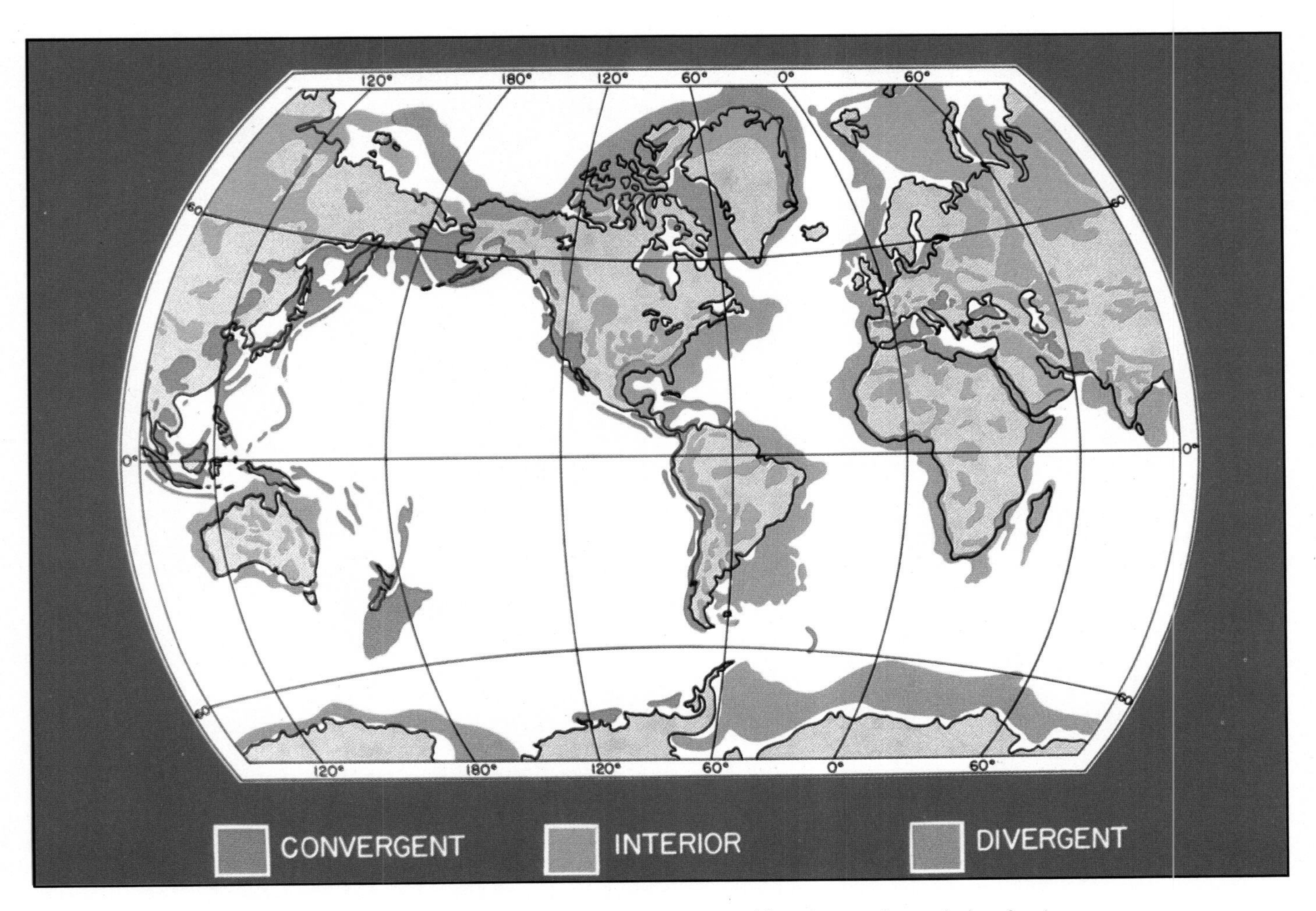


**Figure 2.** *Distribution of convergent, divergent, and interior continental plate basins.*

illustrates the distribution of these three basic classifications of sedimentary basins of the world.

Convergent basins, shown in red, are identified on the margins of plates moving toward one another. Divergent basins, shown in green, are on continental margins moving away from one another. Interior continental plate basins, shown in blue, are those basins formed far from continental margins.

The highest degree of petroleum exploration has been conducted in the world's interior continental plate basins. But exploration in the divergent and convergent basins, many of which are partially or totally offshore regions, is developing rapidly because of the continual advances in drilling and production technology and applying new geological and geophysical concepts.

In the divergent basins of the world, many interesting exploration plays have been identified: for example, offshore Canada, Newfoundland, the U.S. Atlantic coast, and the Brazilian offshore. Exploration of offshore Africa, from the southern to the northern tip of the continent, reveals basins with indications of good-to-excellent prospects for future recoveries of oil and gas. Offshore European basins also show high potential.

Convergent basins are often prolific, containing major oil and gas accumulations. One example is the Sumatra basin of Indonesia, estimated to have over 8 billion barrels of oil equivalent in the North and Central Sumatra basins alone (St. John, 1980). Exploration is still in the early stages in many convergent basin areas, notably offshore Asia and Southeast Asia, and particularly the offshore China basins.

## EXPLORATORY STATUS—BASINS OF THE WORLD

Figure 3 shows the status of exploration in the sedimentary basins of the world. For brevity, many of the basins have been combined. Of the world's prospective sedimentary basins, categorized as intensively explored, moderately explored, partially explored, and essentially unexplored, only 27% currently produce commercial hydrocarbons. Another 40% of the basins have been partially and moderately explored and tested but have not yielded commercial quantities of petroleum. The remaining 33% of the world's basins have had little or essentially no exploration activity.

The world's sedimentary basins contain a prospective area of approximately 29,978,130 sq mi (77,643,000 sq km) (Table 1). About 10,191,167 sq mi (26,395,000 sq km) of this prospective area lies in the world's ocean basins, where large regions of

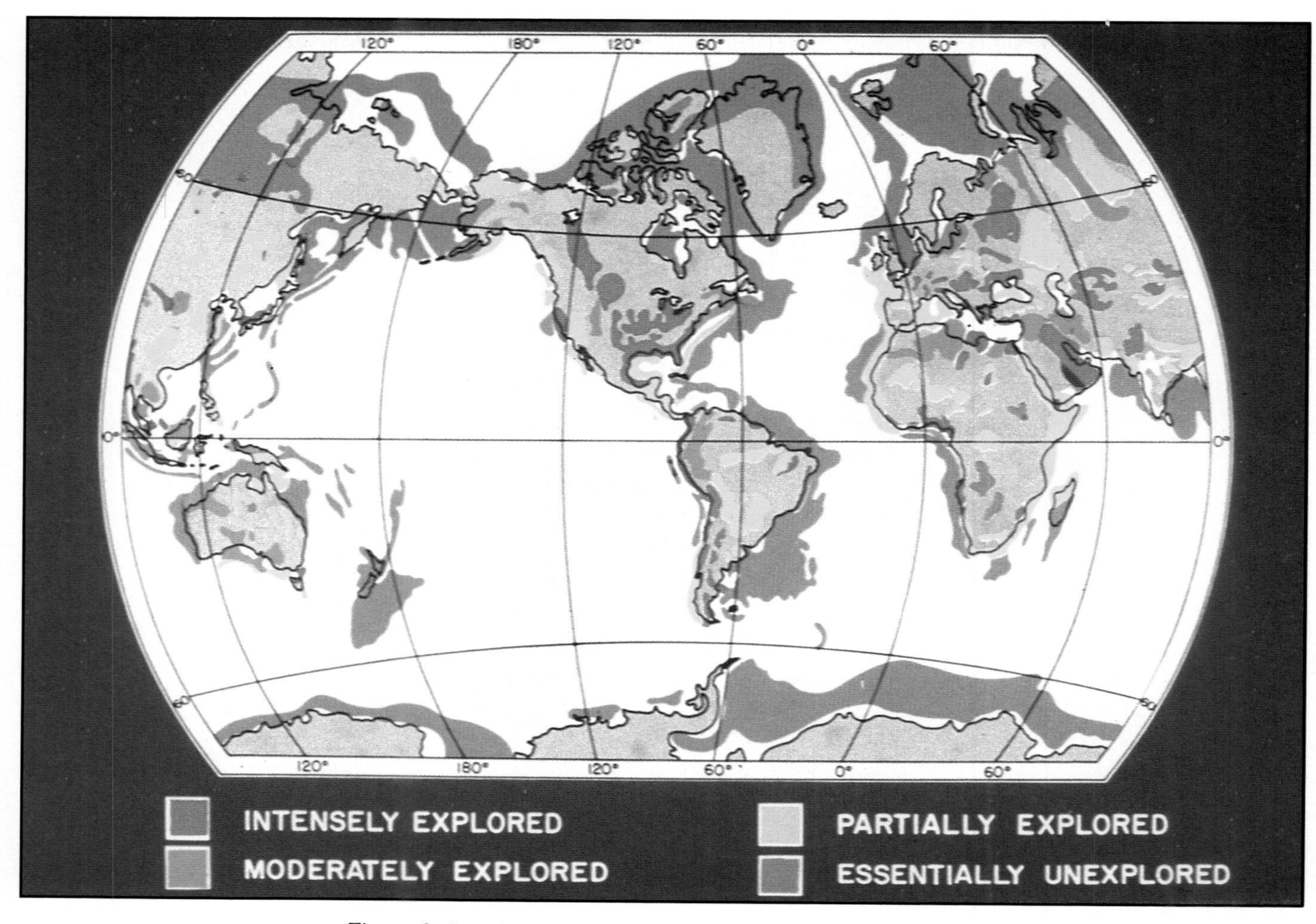


Figure 3. *Petroleum exploratory status of the basins of the world.*

Table 1. *Approximate prospective areas of the sedimentary basins of the world.*

| | Total (1,000 sq km) | Onshore[1] (1,000 sq km) | Offshore[2] (1,000 sq km) |
|---|---|---|---|
| Japan | 644 | 80 | 564 |
| Eastern Europe | 1,015 | 900 | 115 |
| Antarctica | 1,042 | 0 | 1,042 |
| PRC | 2,472 | 1,787 | 685 |
| Middle East | 3,669 | 2,152 | 1,517 |
| Western Europe | 3,848 | 1,944 | 1,904 |
| Canada | 5,167 | 3,084 | 2,083 |
| Australia - NZ | 6,604 | 4,424 | 2,180 |
| Latin America | 7,851 | 4,843 | 3,008 |
| USA | 8,247 | 6,604 | 1,643 |
| S & SE Asia | 8,916 | 3,705 | 5,211 |
| Africa/Madagascar | 13,223 | 11,725 | 1,498 |
| USSR | 14,945 | 10,000 | 4,945 |
| Total | 77,643 | 51,248 | 26,395 |

[1]B. Grossling (1976)
[2]Michel T. Halbouty, independent research (1983)

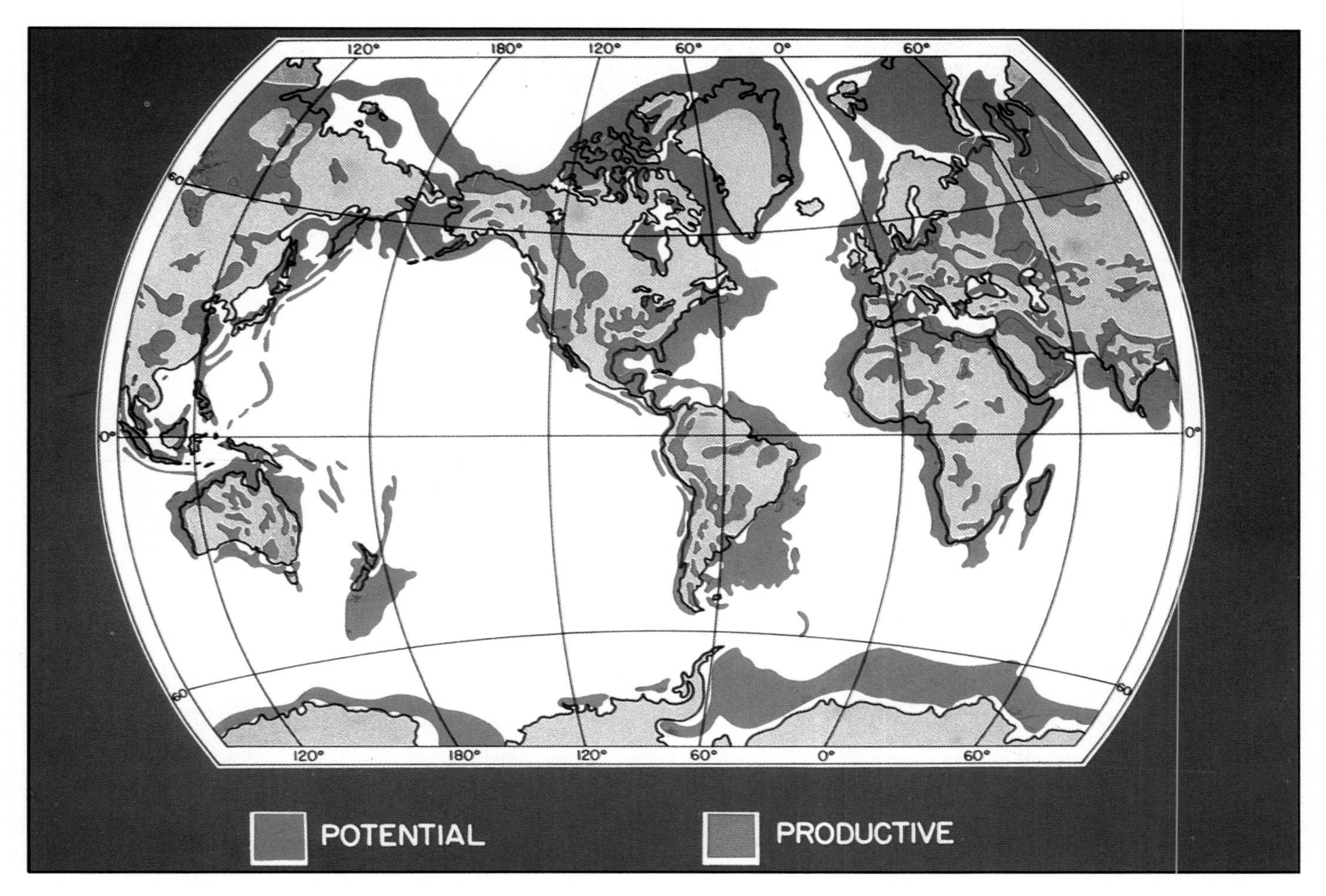


Figure 4. *Sedimentary petroleum basins—currently productive and potentially productive.*

thick sediments of the continental margins extend far beyond the base of the continental slope into the abyssal depths (Halbouty, 1984). These small ocean basins, or restricted seas, form a distinctive region that includes some of the most productive offshore areas, such as the Gulf of Mexico, North Sea, Caribbean Sea, Persian Gulf, Suez graben, Mediterranean Sea, Lake Maracaibo, and the Indonesian ocean basins. There also remain many essentially unexplored ocean basins, such as the Bering Sea, the Norwegian Sea, and the Scotia Sea, which have high potential for future petroleum discoveries (Hedberg, 1981).

Approximately 19,786,963 sq mi (51,248,000 sq km) of the prospective basin area is onshore (Grossling, 1976). Large extents of sedimentary area do not correlate with large volumes of hydrocarbons, nor do small areas correlate with small amounts of hydrocarbons. For example, the Middle East has a relatively small prospective area, 1,416,609 sq mi (3,669,000 sq km), yet it holds the world's largest hydrocarbon reserves.

Onshore basins hold high potential in many parts of the world, especially in the more mature productive areas. Geologists and geophysicists have good opportunities to find additional petroleum reserves by using advanced technology. Three-dimensional seismic investigations, digital recalculations of old seismic data, and the use of remote-sensing images from satellites and spacecraft are some of the methods they use to find clues for discovering "new" reserves in "old" areas.

The majority of the unexplored basins are considered the frontier areas of the world. These basins are located in harsh physical environments such as the Arctic, in deep water, or remote interiors of continents. Many of these frontier regions are restricted because of disputes, political boundaries, or governmental regulations. Others have been bypassed because of a combination of remote location and low geologic potential.

A careful examination of the world's productive sedimentary basins and those areas remaining to be explored shows that many regions remain to be tested (Figure 4). The illustration shows both the currently productive and the potential areas for petroleum exploration. It is interesting to note the purple color that represents the areas of petroleum potential, especially unexplored and untested continental shelves and slopes. These areas are the new frontiers.

We can divide the productive basins of the world into two subjective categories: mature and developing. Many explorationists believe that as much as 30% of the undiscovered commercial potential oil and gas lies in today's mature producing areas. For example, in the Rocky Mountain basins of the western United States, more than 20 wells have been drilled through Precambrian rocks—igneous rocks, mostly granite—into sedimentary sections that are concealed and virtually

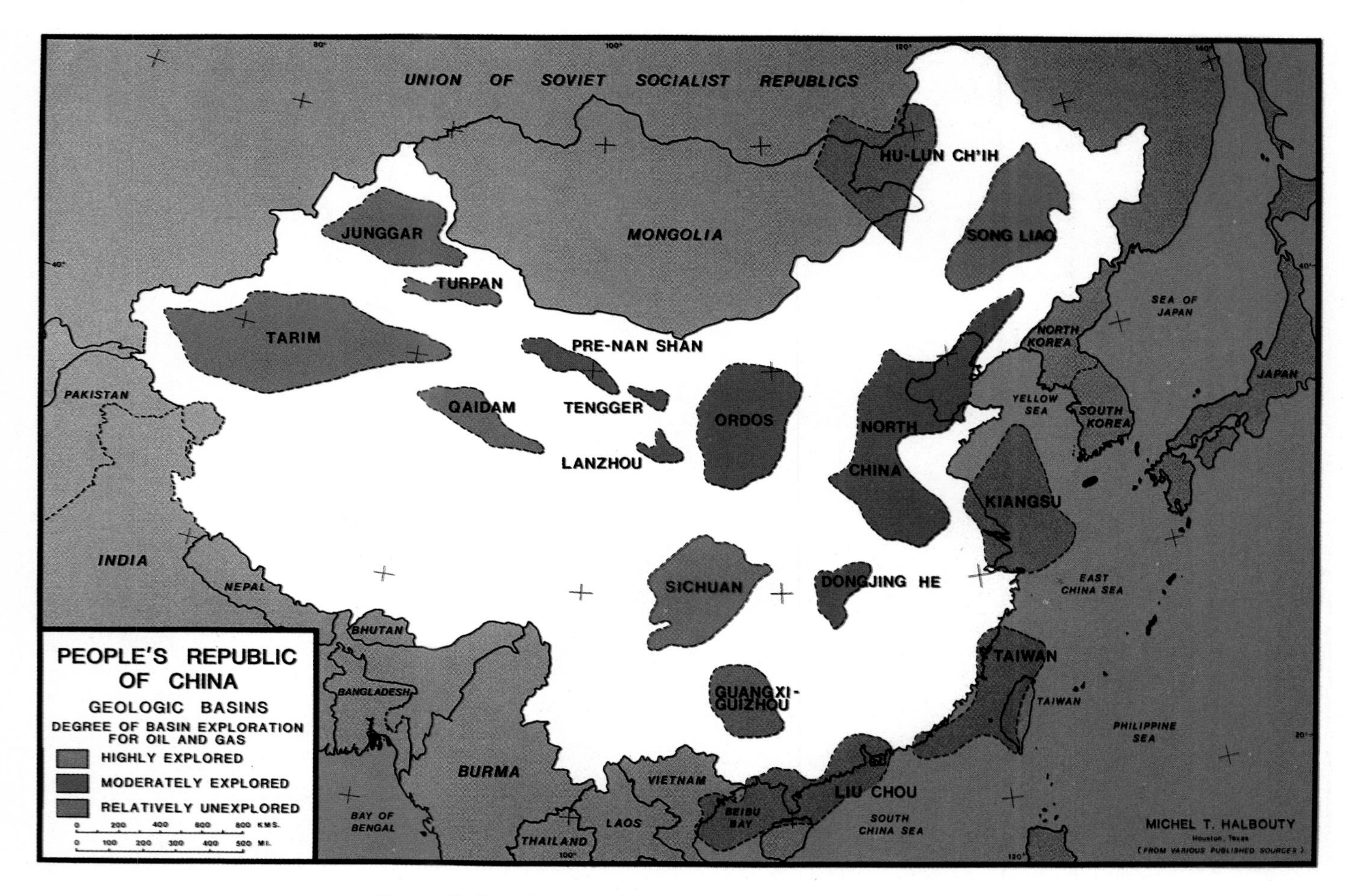


Figure 5. *Petroleum exploration status of the basins of China.*

unexplored beneath thrust sheets.

Also, many of the Middle East sedimentary basins have not been thoroughly evaluated. Recent finds have refuted the previously held belief that the top of the Permian Khuff Formation is an economic "basement", even though thousands of feet of sediment lie below this horizon. The formation is now recognized as a prolific gas reservoir, and many explorationists believe additional oil and gas will be found by deeper drilling (St. John, 1980).

## POTENTIAL PETROLEUM REGIONS

The majority of the world's prospective petroleum-producing area currently is non-productive. Actually, most of these areas essentially are unexplored. Their locations make any hydrocarbon accumulations difficult to find and costly to produce. Frontier areas currently being explored include the U.S. offshore Atlantic and Pacific basins; the offshore regions of Argentina and Brazil; and the east coasts of India, Vietnam, and China.

### China

The U.S., India, Argentina, and China are only a few of the prime examples of countries that have extensive, essentially unexplored outer continental shelf areas. For instance, the offshore Chinese basins shown in Figure 5 represent some of the world's most interesting potential petroleum-producing areas. The meandering coastline of China covers more than 4,000 mi (6,440 km) with the offshore basins offering an estimated 339,770 sq mi (880,000 sq km) of prospective area. Many new basin areas in this coastal region have only recently been offered for petroleum exploration.

Recent exploration success bordering and in the South China Sea has confirmed that the coastal shelves have an excellent petroleum potential. These results prove the viability of a continued, intensive search for hydrocarbons. Only two of the onshore basins in China are in the category of mature production: the Sungliao and Sichuan. Others are either moderately explored, relatively unexplored, or inadequately explored, while China's western basins have had very little exploration.

Also, more than 85% of the new reserves found in China in the past three years have been in the eastern portion of the country. Short-term exploring efforts continue to be centered in the east, near the country's extensive industrial complexes and transportation facilities. Increasing exploring activity, however, is expected in western China where a substantial petroleum potential might exist.

### Africa and South America

The North African basins from Algeria through Libya to the Egyptian Sinai should prove to be an active and fertile territory

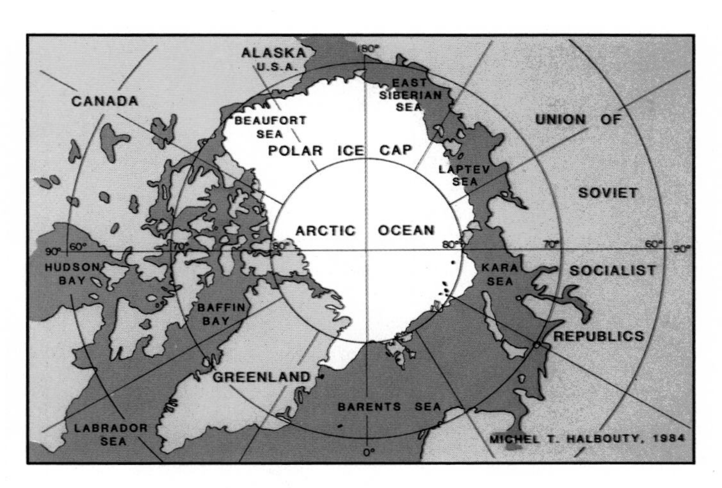


Figure 6. *Circum-Arctic region.*

for intense exploration. Spurred by recent hydrocarbon discoveries off the Ivory Coast, future offshore exploration could identify a major province (Tucker and Timms, 1981).

The significant future onshore fields in South America will likely be in the Oriente basin of southern Ecuador, Lake Maracaibo in the Maracaibo basin of Venezuela, and the Magdalene Delta of the Gulf of Venezuela. Offshore eastern Brazil also shows promise, as does the Magellenes basin of Argentina. In Mexico, a supergiant hydrocarbon province is developing in the Salinas basin and along the Western Campeche Bank (St. John, 1980).

### Pacific Basin and Asian Coast

In the basins of the Pacific Ocean and along Asia's coast, the most favorable area extends from the northwest Australian offshore shelf basins to the offshore Sabah basin. This region, excluding Indonesia, is under-explored. Promising areas in New Zealand include the Great South basin, the east coast of South Island to the Canterbury Bight, the east coast of North Island, and the established Taranaki basin.

### North America

In the U.S., exploration should be successful in the Alaskan North Slope, the Anadarko basin in Oklahoma, the Permian basin of west Texas, and the Gulf Coast area. New and deeper production is likely in the Eastern Overthrust Belt which has received much attention in recent years. Applying improved geophysical processing techniques and greater understanding of tectonics encourages drilling in many areas previously considered unfavorable for hydrocarbon accumulation. Offshore prospects exist in the Georges Bank and the Baltimore Canyon areas off the east coast. There are further offshore prospects in the deeper reaches of the Gulf of Mexico and the Pacific coastal regions from Oregon to California. The recent discoveries in the Santa Maria basin along the southern California coast attest to the potential for giant fields in the area. For example, the Point Arguello field is conservatively estimated to hold as much as 500 million barrels of oil.

Along the United States and Canadian shelves, there are seismic indications of thick layers of sedimentary rock with trapping features that could hold oil and gas accumulations equal to those in fields already found onshore (Harrison, 1979). Figure 6 shows the most promising of the U.S. outer continental shelf lands. Remarkably, fewer than 2% of these offshore areas have ever been leased. In Alaska, for example, only the Prudhoe Bay and Cook Inlet areas have had any appreciable exploration; yet, the Alaskan offshore represents almost 50% of the entire federal offshore reaches of the United States. The basin areas in Alaska where there are future plans for exploration contain four of the most intriguing of the unexplored offshore areas of the world: the St. George, Norton, North Aleutian, and Navarin basins.

In Canada, significant discoveries are predicted off the Labrador coast. In the Beaufort area, recent discoveries, such as Kopanoar, Koahoak, Tarsuit, and Issungak, show rich reserves. Seismic mapping in this region revealed more than 100 structures that could trap substantial amounts of oil and gas. Other promising regions are the Alberta basin, the Beaufort Sea/Mackenzie Delta area of Alaska-Canada, the Canadian Sverdrup basin, and the Sable Island area on the Canadian Scotian Shelf (Geophysics, 1982).

Another harsh, offshore environment that has proven highly successful for oil and gas reserves is the east coast of Canada. In 1979, the giant Hibernia field was discovered off Newfoundland, Canada. Reserves for the field are estimated at 1.8 billion barrels of oil. Although the water depth in the area is 262 ft (80 m), weather conditions and the presence of icebergs will make it a continual challenge to develop and produce (Arthur et al, 1982).

### Circum-Arctic Region

The Circum-Arctic region as a whole covers a vast territory, a part of which Figure 6 illustrates. Many researchers believe that in the combined Arctic basins, including the Arctic areas not shown in Figure 6, there is the potential for accumulations of oil and gas equal to those of the Middle East. Exploring in Arctic basins represents the most costly search ever, but there is great potential for finding the giant accumulations of oil and gas the world needs so crucially.

The Prudhoe Bay and Kuparuk fields in Alaska and the Urengoy field in the U.S.S.R. are prime examples of the giant field potential of the region. In the Bering Sea two fore-arc sedimentary basins are attracting considerable attention: the Anadyr and the previously mentioned Navarin basins which contain thick sedimentary sequences. Onshore, the Anadyr basin underlies the lowlands that are flanked on the north, west, and south by the Koryak Range, U.S.S.R. Offshore U.S.S.R., the East Anadyr trough and Navarin basin underlie the virtually flat Bering Sea shelf and contain thick sedimentary sequences that, in places, exceed 7.5 mi (12 km) (Marlow, Cooper, and Childs 1983).

Over half of the Navarin basin exists in U.S. territorial waters, and the northwest portion of the basin exists in U.S.S.R. waters. The United States has recently offered certain portions of the basin for lease and interest is rather high because of the potential of this large basin. It will be interesting to follow the course of exploration in this high-potential area. It may take the cooperation of both nations, the U.S. and the U.S.S.R., to fully develop whatever petroleum resources are found in the Navarin basin. Such a joint venture would be not only interesting and historical, but also productive for both.

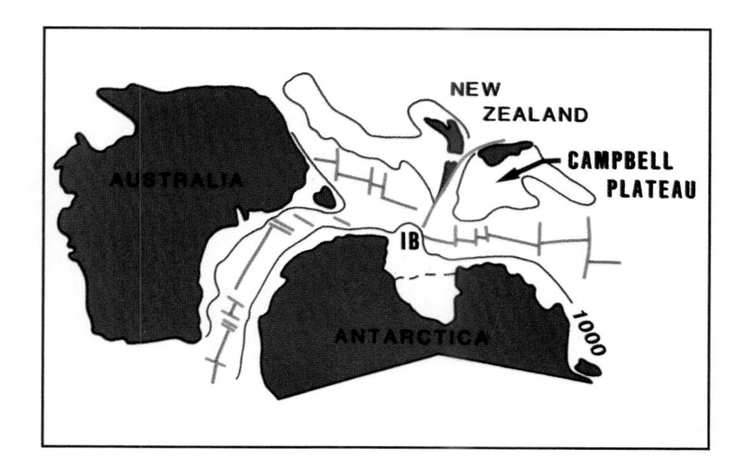


Figure 7. *Possible configuration of Antarctica, Australia, and New Zealand approximately 80 million years ago (modified from Weissel et al, 1977).*

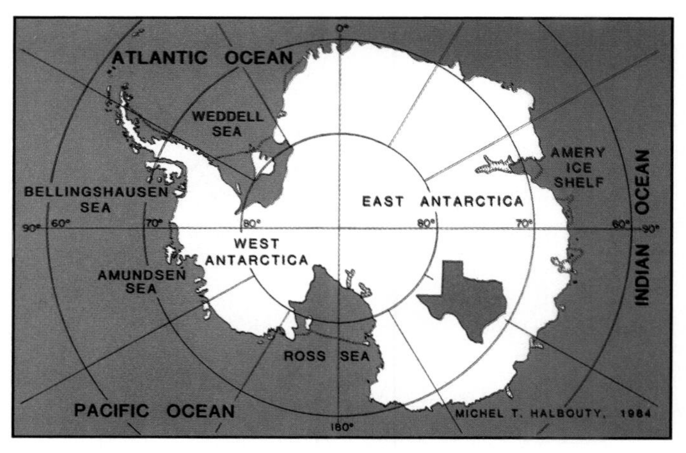


Figure 8. *Size comparison of Antarctica and the state of Texas.*

## Antarctic Region

In considering the world's harsh environments for future petroleum exploration, we must not discount the possibilities that exist in Antarctica, the last true frontier resource area on the earth. Although other regions, such as the Circum-Arctic, parts of South America, and Asia, are sparsely explored, we can at least estimate their potential.

The Antarctic continental margins occupy an area roughly similar to the continental margins of North America. They are composed of sediment wedges many kilometers thick. Although the continent's subsurface land mass may contain a significant, undetermined petroleum potential, geoscientific studies and surveys indicate that sizable reserves exist in the immediate offshore areas.

A portion of the offshore Antarctic margins that are adjacent to the Bass-Gippsland basins of the Bass Strait, Australia, viewed in a Gondwanaland reconstruction, may prove to contain large quantities of hydrocarbons, and could be an analog of the Bass Strait oil fields of Australia. Figure 7 shows the configuration of Antarctica, Australia, and New Zealand as it might have looked some 80 million years ago. The relative movement of Australia and New Zealand from Antarctica is also indicated. About 2.5 billion barrels of proven petroleum reserves are associated with the Cretaceous and Tertiary sediments of the rift basins in the Gippsland basin of Australia, which was apparently formed during or just prior to the breakup of Australia and New Zealand from Antarctica.

The large embayment of the Weddell Sea and Shelf, which adjoins Antarctica and is located south of South America, may be another promising target for exploration, especially where large sediment thicknesses exist. Reconnaissance data for the embayment of the Amory ice shelf region in the Indian Ocean sector of Antarctica show a thick sediment wedge.

The Antarctic continental shelves are deeper than Arctic shelves. Typical shelf depths are 1,312 to 1,640 ft (400 to 500 m). Deep sea drilling has revealed formations that suggest that the seabeds in the Ross, Billingshausen, and Weddell Seas may overlie substantial oil and gas deposits. The potential in the combined Ross, Weddell, and Bellingshausen Seas is 15 to 50 billion barrels of oil equivalent.

Formidable problems involving the technology of exploration, environmental policies, and international politics preclude extracting petroleum for many, many years. But as experience has taught us in the Arctic, with enough economic incentive, we can find solutions to seemingly unsolvable problems (Halbouty, 1983).

Needless to say, Antarctica is a large continent whose margins have just begun to receive attention. To illustrate the immense size of Antarctica, I have superimposed a map of Texas on just a small portion of this vast region of the earth (Figure 8). Antarctica is the epitome of a petroleum frontier, and I predict that if sufficient exploration is conducted, sizable reserves will be found in this intriguing continent.

## Deep-Water Oceans And Hostile Environments

We need to pay more and more attention in coming decades to the world's challenging offshore petroleum frontiers. Using current technological advances in offshore drilling, neither depth nor harsh environmental conditions can be considered barriers. As explorationists are looking to the oceans for major new reserves, especially to deeper waters and hostile environments such as the Arctic regions, engineers are diligently working on designs of new drilling rigs and production platforms. As long as there are inspiration, courage, innovative skills, and financial backing, there is no limit to what the global petroleum industry can accomplish.

Twenty-five years ago, the entire North Sea was considered out of the question for exploration and production because of the extremely hostile environment. Yet, since in 1969 when the first commercial discovery was made, numerous oil and gas finds have made this part of the world practically self-sufficient in petroleum energy.

Only ten years ago, drilling in 200 to 300 ft (60 to 90 m) of water was considered too hazardous and difficult. Production from a conventional platform in over 1,000 ft (305 m) of water has proven feasible. Ten years from now, production from 1,000 ft will be considered common and relatively shallow.

Drilling in water depths beyond 10,000 feet (3,048 m) is rapidly becoming feasible, and could occur by the year 2000, possibly even sooner. Water depth records for offshore exploration keep falling as drilling technology advances. The present

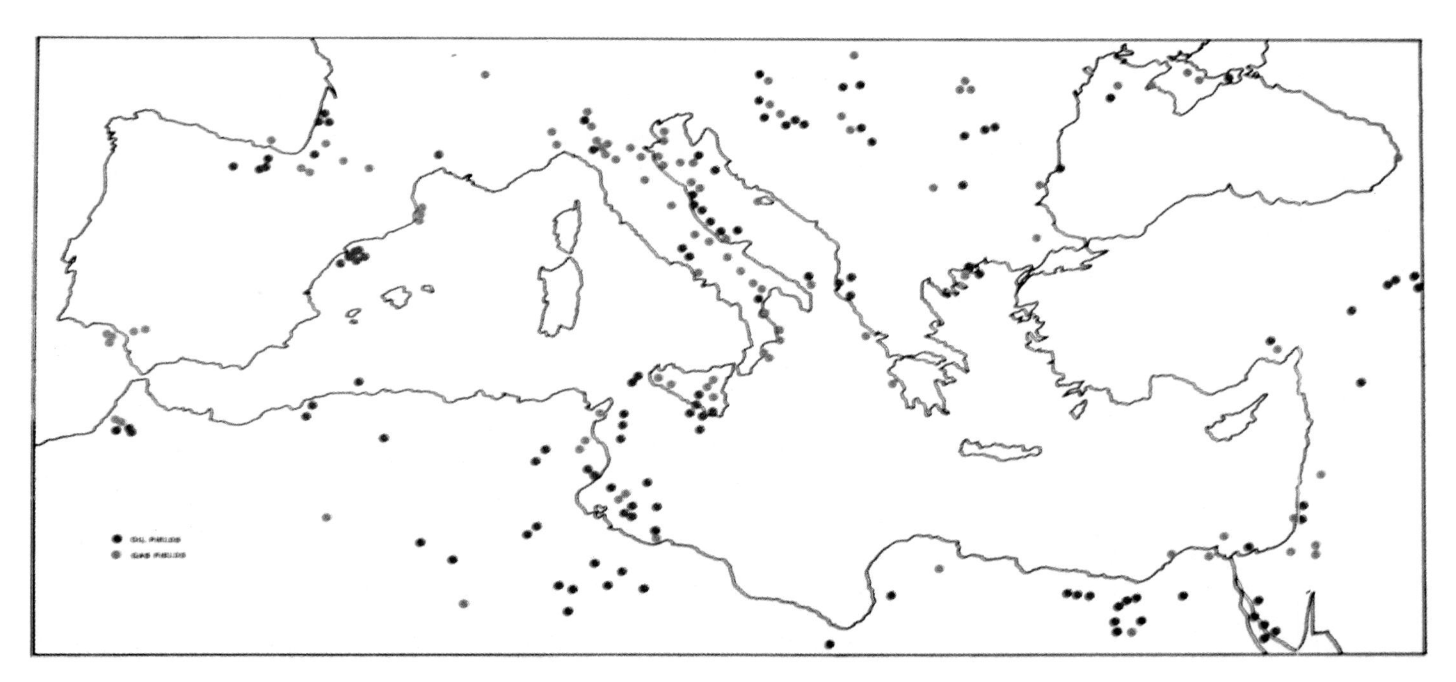

**Figure 9.** *Oil and gas fields of the circum-Mediterranean region.*

record set in late 1983 off the U.S. Atlantic coast is just over 6,400 ft (1,950 m). Engineering and technological breakthroughs are quickly opening vast areas previously considered almost impossible to explore—areas such as the deeper portions of the Gulf of Mexico, the Mediterranean Sea, and the South China Sea, to name only a few.

For example, Figure 9 shows the oil and gas fields of the circum-Mediterranean region. Note the predominance of petroleum deposits in the structural belt that separates the western and eastern segments of the Mediterranean. The region as a whole has not been adequately explored, especially in the eastern portion.

However, the majority of the fields shown in the illustration are relatively new discoveries, attributed to the measure of exploration activity. In other words, the greater the exploration, the greater the number of discoveries, which is a formula for the entire world. New and significant fields in offshore western Spain, onshore and offshore Africa, Sicily, Italy, Egypt, and in other areas, convey the petroleum potential and importance of this entire region to the future world's economy.

New discoveries will occur in the Mediterranean basins that have not yet been adequately explored. These include the basins offshore north Africa and Turkey, and the basins within the central Mediterranean Sea, including the unexplored deltas and continental shelves.

In the last five years more new conceptual ideas have emerged in the geosciences than have been promulgated in the preceding fifty years. During the next few years, geoscientists will contribute more solutions for the current "impossible" problems.

### Subtle Traps

In this regard, explorationists pay too little attention to the deliberate search for reserves that lie in less obvious, subtle traps. We must study the not-so-obvious reservoirs of petroleum, such as those in stratigraphic traps, those lying below unconformities, those associated with buried geomorphological features, and those that may or may not be associated with structure. A substantial number of our large and small petroleum reservoirs which are classified as structural traps are, in fact, partly stratigraphic.

Had the crests of such structures been proven dry by initial exploration, and had drilling not been continued into the flank areas, many such areas possibly would have been abandoned and condemned as dry. Present production on the flanks of such features, being stratigraphically controlled, might still be undiscovered today. There are abandoned dry structures in all the world's petroleum provinces, drilled on top but untested on the flanks.

Stratigraphic accumulations of petroleum may exist on one or more flanks of many of these features. Finding the elusive traps will require highly imaginative thinking combined with every applicable scientific discipline. The geophysicists' computers, combined with new display techniques, are tremendous aids in developing the scientific data needed for finding subtle traps. Great advances have been stimulated by the use of seismic modeling, better data acquisition and data processing, and the extensive use of color seismic display formats. But these kinds of data are just a small part, and must remain just a part, of the effort to search for the subtle traps.

The other and most important part is evaluating the source and the kinds of sediments, facies changes, geologic history, tectonic control of sedimentation, and kinds of environmental conditions present when sediments were deposited. Those are the keys in the search for the stratigraphic and other subtle traps. In other words, we need plain geological thinking, or better, pure geological thinking. There are billions of barrels of oil and trillions of cubic feet of gas accumulated in these kinds of subtle traps, waiting for the drill to find them.

## CONCLUSION

At this point, I would like to make a few comments about the world's future oil and gas supplies. It has been predicted over and over for decades that the world was running out of oil and gas. Those pessimistic doomsayers have consistently been proven wrong. New oil and gas fields are discovered almost daily, all over the world, in both known producing areas and in the frontier regions both onshore and offshore.

As one familiar with the world's geologic potential, I firmly believe that we will find in the future at least as much oil and substantially more gas than we have produced to date. Geoscientists are limited only by their imagination, innovation, and determination. In the coming decades there will be tremendous strides made in petroleum geology, geophysics, and petroleum engineering.

The challenge for all of us, whether we are geologists, geophysicists, engineers, or independent explorationists, is to devise new concepts and skills to explore in areas considered to be too remote or difficult to drill. In my five decades of petroleum exploration, I have observed that whenever substantial reserves of oil and gas have been discovered, they have always been brought to market. In the future we will find the additional reserves because of the tremendous potential under the unexplored low lands, high lands, swamps, ice, deserts, forests, and waters of the seas. As long as there is a petroleum potential any place in the world, the hunt will go on and on and on to discover and produce.

## REFERENCES

Arthur, K. R., D. R. Cole, G.C.L. Henderson and D. W. Kushnit, 1982, Geology of the Hibernia discovery, *in* M. T. Halbouty, ed., The deliberate search for the subtle trap, AAPG Memoir 32: Tulsa, AAPG, p. 181–196.

Geophysics, the Leading edge, 1982, Giants still exist in Canada, v. 1, n. 4 (September) p. 26–47.

Grossling, B., 1976, Window on oil: a survey of world petroleum sources: London, Financial Times Ltd., p. 1–140.

Halbouty, M. T., 1983, Geologists ponder billion bbl question: Offshore (June 20) v. 47, no. 7, p. 40.

Halbouty, M. T., 1984, Reserves of natural gas outside the Communist Block countries, *in* Proceedings of the 11th World Petroleum Congress: London, John Wiley & Sons, Ltd., v. 2, p. 281–292.

Harrison, G. R., 1979, Exploratory drilling, the polar challenge, *in* Proceedings of the 10th World Petroleum Congress: London, Heyden & Sons, v. 2, p. 243.

Hedberg, H. D., 1981, Hydrocarbon resources beneath the oceans, *in* J. J. Mason, ed., Petroleum Geology in China: Tulsa, Pennwell, p. 249–253.

Marlow, M. S., A. K. Cooper, and J. R. Childs, 1983, Tectonic evolution of Gulf of Anadyr and formation of Anadyr and Navarin basins, AAPG Bulletin, v. 67, p. 646–665.

St. John, B., 1980, Sedimentary basins of the world and giant hydrocarbon accumulations: Tulsa, AAPG, 26 p.

Tucker, P. W., and C. J. Timms, 1981, The world outlook for natural gas in the 1980s, Petroleum Review (December), p. 35–39.

Weissel, J. K., D. E. Hayes, and E. M. Herron, 1977, Plate tectonic synthesis: the displacements between Australia, New Zealand, and Antarctica since the Late Cretaceous: Marine Geology, v. 25, p. 231–277.

Wood, P. W. J., 1979, New slant on potential world petroleum reserves: Ocean Industry, v. 14, no. 4, p. 59–64, 66, 68–70.

# Giant Oil and Gas Fields

S. W. Carmalt
*Consultant*
*Geneva, Switzerland*

Bill St. John
*Primary Fuels, Inc.*
*Houston, Texas*

At present, there are 509 known giant oil and gas fields in the world. A giant is defined as having 500 million barrels of recoverable oil or equivalent gas. Less than one-third of this discovered amount has been produced to date; thus, remaining reserves are sufficient, at present production rates, to last well into the 21st century.

Geologically, the giants are found most commonly in provinces that can be classified as having formed in continental crust and having been associated with a plate collision. Basins in these provinces are found across a wide range of geographic areas, many of which remain only moderately or lightly explored. The basins offer, therefore, significant geologic scope for future exploration.

The discovery rate of giant fields has decreased since the late 1960s, indicating that the Hubbert cycle of oil resource exploitation is in a mature phase. Given the basins yet to be explored, this decrease is due to economic, rather than geologic, factors.

## INTRODUCTION

The giant oil and gas fields deserve special study for several reasons. First, they are few in number (509), while containing about two-thirds of the discovered, recoverable reserves. Hence, knowledge of relatively few fields contributes greatly to an understanding of the available supply of hydrocarbons. Second, the geology of the giant fields is important both in exploring for other fields in the same basins, and in the exploration for new giants in other basins. Third, the pattern of discovery of giant fields contributes to our understanding of the magnitude of the undiscovered hydrocarbon resource, with the implications that this has for the future.

## DEFINITIONS

For this discussion, a *giant* oil or gas field is considered to be one for which the estimate of ultimately recoverable oil is 500 million bbl of oil or gas equivalent. *Reserves*, in this paper, refer to this ultimately recoverable amount—including the amounts produced to date. Gas is converted to oil at a ratio of 6,000 cu ft/bbl. Some fields are, therefore, giants only because their combined amounts of oil- and gas-equivalent total at least 500 million bbl, and not because either resource is that great by itself. Nehring (1978) terms these fields *Combination Giants*. Table 1 lists the giant fields in order of the size of their combined recoverable reserves.

Two definition problems are important in compiling data on giant fields. The first is the distinction between a single field and a collection of close, geologically similar fields. Nehring (1978, p. 6–7) has proposed the definition that we follow in this paper.

> ...A field is a producing area containing in the subsurface (1) a single pool uninterrupted by permeability barriers, (2) multiple pools trapped by a common geologic feature, or (3) laterally distinct multiple pools within a common formation and trapped by the same type of geologic feature where the lateral separation between pools does not exceed one-half mile.

The second definition problem concerns recoverable reserves. Uneconomic reserves cannot exist; this is a contradiction because reserves are defined as those hydrocarbons that are economically recoverable. Hence very large accumulations of oil or gas, technically recoverable with existing methods, may nevertheless not be reserves for economic reasons. For example, there are several gas-filled structures in southwestern Iran, containing a minimum of 4.5 trillion cu ft (tcf) or 750 million bbl equivalent each, that have yet to be developed even though more than a decade has elapsed since their discovery; also, to the best of our knowledge there are no development plans at this time. We have taken the position that, at present, fields such as these have no economically recoverable reserves and are, therefore, not giant fields. We recognize that this situation can change.

Three previous, comprehensive lists of giant fields have been published: Halbouty, et al (1970), considered both giant oil and giant gas fields; Nehring (1978) listed giant oil fields; and St. John (1980) tabulated giant oil 'equivalent' accumulations. The Nehring collection specifically did not consider discoveries after the end of 1975. While we have not used a similar cut-off date in compiling Table 1, the more recent discoveries that we list could prove to have been over-estimated. Also, recent discoveries and fields not listed may, in fact, be giants when viewed from a greater historical perspective.

Table 2 is an alphabetical list of the fields listed in Table 1. In addition, Table 2 shows alternate names for the giant fields, providing a cross-reference. The need for a list of cross-reference names arises for many reasons:

(1) Some fields cross frontiers and have different names in each country;
(2) Fields sometimes are known commonly by the name of the discovery well, but receive a different name as a field;
(3) Fields that were once thought to be several different fields may be single fields under the definitions used in this paper;
(4) Alternate names sometimes arise from transliteration or translation difficulties or both; and
(5) Names that are commonly recognized may differ from official names.

Table 2 also includes the names of fields that we do not consider to be giants, but which are commonly named as such. These can either be erstwhile giants, whose production has not been as expected, or potential giants for which our best reserve estimate is still below 500 million bbl equivalent. We have specifically included all the field names found on either Halbouty's or Nehring's lists.

Both Tables 1 and 2 contain geographic coordinates for the fields we name. In some cases these are accurate only to about the nearest degree, but we list the data in order that there will be a minimum of confusion concerning the fields.

## GIANT FIELDS

The 509 giant fields contain a total of 868,115 million bbl of oil and 3,193 tcf of natural gas. Converting gas to oil equivalent results in the giants containing 1,400,343 million barrels of oil equivalent (BOE).

Of these amounts, a minimum of 270,760 million bbl of oil was produced before mid-1984, in addition to 198 tcf of gas. These figures are equivalent to 31% of the oil and 6% of the gas already produced from these fields; the figures may be somewhat low because of unavailable production data in several instances. Nevertheless, approximately 66% of the oil and over 90% of the gas remains to be recovered from the known giants. This conclusion means that 575,000 million bbl of oil and 2,850 tcf of gas remain.

For oil, these reserves are sufficient for 26.3 years of production at 60 million bbl per day. This production rate closely matches the IPE's (1984) average worldwide consumption figures of 61.7, 60.1 and 58.5 million bbl per day for 1980, 1981, and 1982, respectively. We therefore conclude that a resource-caused supply problem is highly unlikely for the remainder of this century, even in the unlikely event that there are no further oil giants discovered anywhere in the world.

For gas, the 1983 worldwide production of 55.1 tcf (IPE, 1984) indicates that the remaining reserves would last 51.7 years, at present consumption rates.

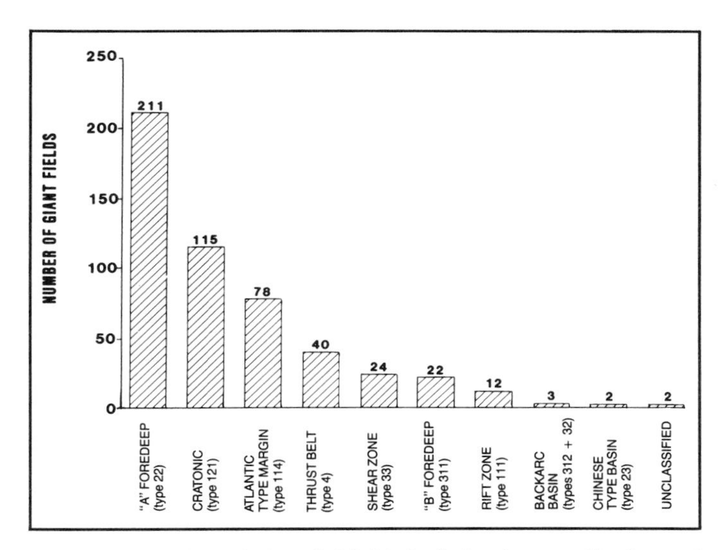


Figure 1. *Number of giant fields by Bally basin type. Horizontal axis: Bally classification; Vertical axis: number of fields.*

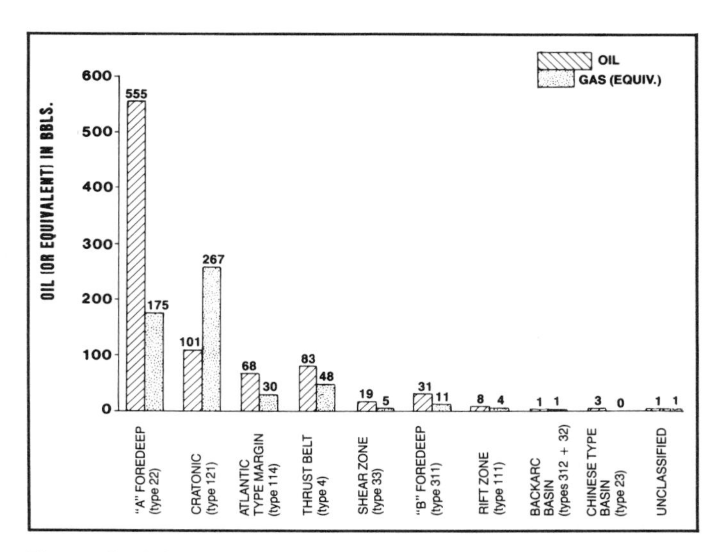


Figure 2. *Oil and gas in giant fields by Bally basin type. Horizontal axis: Bally classification; Vertical axis: oil (billion bbl).*

## GEOLOGY OF GIANT FIELDS

Giant fields are found in a variety of geologic settings. In order to understand better the factors involved in the formation of a giant, the fields have been grouped according to the basin classifications proposed by Klemme (1971) and Bally and Snelson (1980). Both of these classifications are founded on tectonic criteria, and assume that the factors that allow a basin to be classed according to its tectonic history are also significant in the formation of a giant oil field.

Giant fields are given the same basin classification as the petroleum provinces in which they are located. These provinces are classified using the St. John, Klemme, and Bally basin classifications (St. John, Bally, and Klemme, 1984). Where provinces have been listed as a combined type, the first classification listed has been used in the statistics.

The Klemme and the Bally and Snelson classifications are of basins, while St. John's presentation is of petroleum provinces. This arrangement is because some of the provinces are not basins in a strict sense.

## THE BALLY-SNELSON CLASSIFICATION

The Bally and Snelson (1980) basin classification divides basins into three major categories depending upon the tectonic environment. In particular, the major divisions reflect the degree to which a basin is associated with megasutures, which are major compressional plate boundaries. Category 1 basins are formed on the rigid lithosphere, with no association of megasutures. Examples are rifts, oceanic abyssal plains, and 'Atlantic-type' margins. Category 2 basins, termed perisutural, are formed on the rigid lithosphere in association with a megasuture. Examples are deep-sea trenches, foredeeps, and 'Chinese-type' basins. Category 3 basins, termed episutural, are part of a megasuture. Examples are forearc basins, backarc basins, and 'Basin and Range' basins. St. John, Bally and Klemme (1984) have modified and extended the classification with 4th and 5th categories. Category 4 is folded belts, and category 5 is plateau basalts. Table 3 shows the classification, including the detailed type-numbers referred to in Table 1. Figure 1 shows the distribution of fields according to the major basin divisions.

The largest group of fields (211 fields) is found in the "Foredeep and underlying platform sediments division." This category includes the Arabian province (but not the Zagros foldbelt fields). It also encompasses, among others, the Alberta, North Slope, and Permian provinces of North America, much of the Venezuelan production in South America, and several of the provinces in the Soviet Union.

The next largest category (115 fields) is the stable crust ("Cratonic basins on pre-Mesozoic continental crust"). Major provinces in this category include both the North Sea and Western Siberia.

The 'Atlantic margin' category (78 fields) includes the U.S. Gulf Coast, the Campeche region in Mexico, and the Niger delta. The 'Thrust Belts' (40 fields) show as an important category almost solely as the result of the Zagros fields in Iran and Iraq.

Figure 2 illustrates the distribution of hydrocarbons in the giant fields by the Bally and Snelson classification. Gas distribution is normally lower than that of oil. However, more gas than oil occurs in the stable crust, cratonic basins.

Table 3 shows the oil and gas "richness", for the various basin types. This value is obtained by dividing the total oil or gas reserves found in giant fields in the particular basin type by the number of productive provinces of that type. This results in barrels of oil or tcf of gas per productive province. The richness value need not follow the same pattern as the number of fields; however, it does. Hence, the same four basin types that have the most giant fields also prove to be the richest with respect to either oil or gas.

Table 4 also shows the number of provinces of each particular basin classification that do not contain giant fields. Because these are the basin types that have yielded the most giants in the past, and that are the richest in oil and gas on a per-basin measure, they would seem to be the best candidates for future

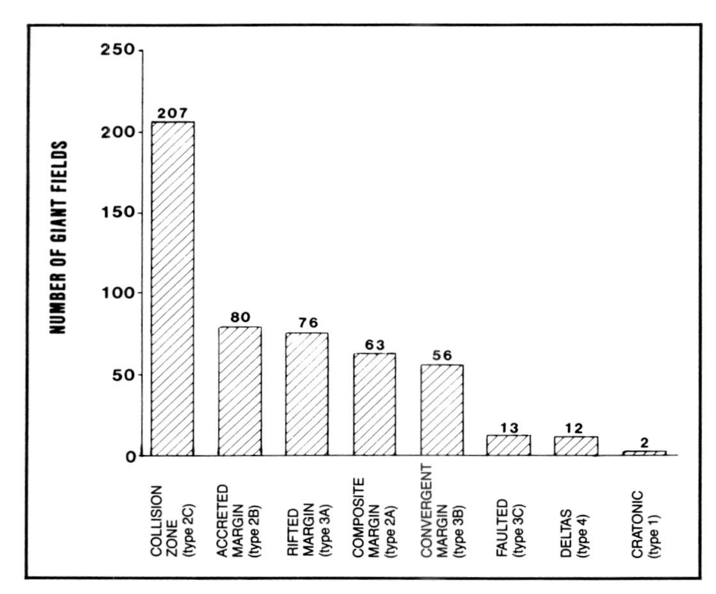


Figure 3. *Number of giant fields by Klemme basin type. Horizontal axis: Klemme classification; Vertical axis: number of fields.*

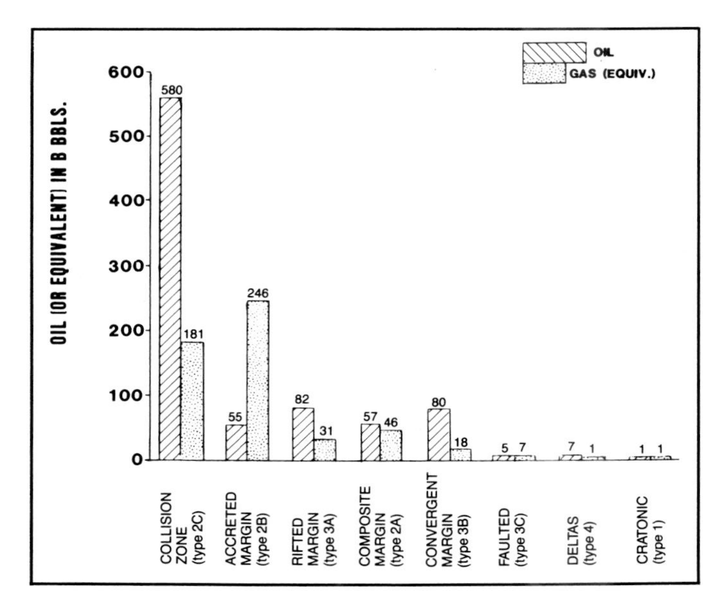


Figure 4. *Oil and gas in giant fields by Klemme basin type. Horizontal axis: Klemme classification; Vertical axis: oil (billion bbl).*

exploration. Their distribution is shown in Table 5, which indicates that they are located primarily in reasonably accessible areas, both politically and geographically, making exploration relatively easy.

St. John, Bally, and Klemme (1984) have shown which provinces are productive from smaller-than-giant fields. Because the size of discoveries in any particular basin tends to decrease with continued exploration (Root and Drew, 1979), there will be some productive basins that can be considered unlikely to contain giants. The Michigan basin in the U.S. is an example.

Nevertheless, there remain a large number of basins available for exploration.

## KLEMME BASIN CLASSIFICATION

The Klemme basin classification (Klemme, 1971) was proposed as an exploration tool based on plate tectonic concepts. The version of the classification used in this paper is that given in St. John, Bally and Klemme (1984). This revised version is significantly different from the original version, including a new numbering scheme. Basins are divided into five major categories according to their tectonic history, with categories 2 and 3 having important sub-divisions. Category 1 is continental interior basins; the Michigan basin would be an example. Category 2 is multi-cycle basins on continental crust, divided into composite margin basins, accreted margin basins, and collision zone basins. Category 3 is various rifted margin basins, divided into cratonic and accreted rift basins, rifted convergent margin basins, and rifted passive margin basins. Category 4 is deltas, and category 5 is forearc basins. The Klemme classification does not cover basins on oceanic crust, which have low oil and gas potential according to most exploration geologists.

Table 6 shows the classification in greater detail, together with the type-numbers used in Table 1. The 799 petroleum provinces listed by St. John, Bally and Klemme (1984) have also been classified by the Klemme basin system.

Figure 3 shows the distribution of giant fields according to the Klemme basin types. The most productive category in the Klemme scheme, by far, is a basin on continental crust formed by a plate collision. There are 207 giant fields in this category, including those in the Arabian Gulf area.

After the plate-collision type, the kinds of basins with the most giants are found along continental margins near plate boundaries. These types are cratonic margin complex, cratonic accreted margin complex, rifted continental margin, and rifted convergent margin.

Giant oil fields and giant gas fields are found, statistics show, in different Klemme basin types. This difference in occurrence is dominated by the collision-zone setting for the oil fields of the Arabian–Persian Gulf, as opposed to the accreted-margin setting for the giant gas fields of Western Siberia. This can be seen in Figure 4.

The oil and gas richness of the Klemme basins with giant fields is shown in Table 7. In oil richness, there is a sharp difference between the most productive type of basin, continental collision, and all the others. Gas richness, similarly, is greatest in accreted complex margin basin type.

The geographic distribution of provinces without giants is shown in Table 8. As with the Bally basin types, there is a wide distribution range, indicating considerable potential for continued exploration.

## GEOLOGIC SETTING OF GIANT FIELDS

The similarity of both the Bally-Snelson and the Klemme analysis of the distribution of giants is that the most prolific provinces are those on the continental crust, but in a mobile zone associated with collision. We do not know whether this occurrence is due entirely to the additional thicknesses of carbonates and clean arenites found in these environments, or whether there are additional factors in these settings (e.g. organic productivity, burial conditions).

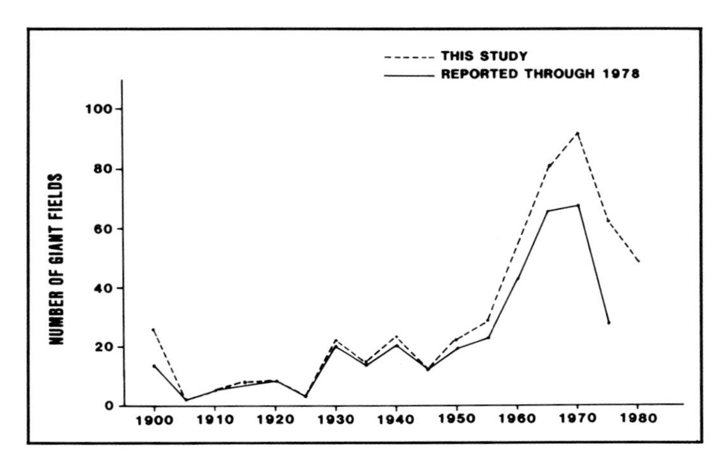


Figure 5a. *Discovery of giant oil and gas fields. Horizontal axis: discovery year; Vertical axis: number of fields.*

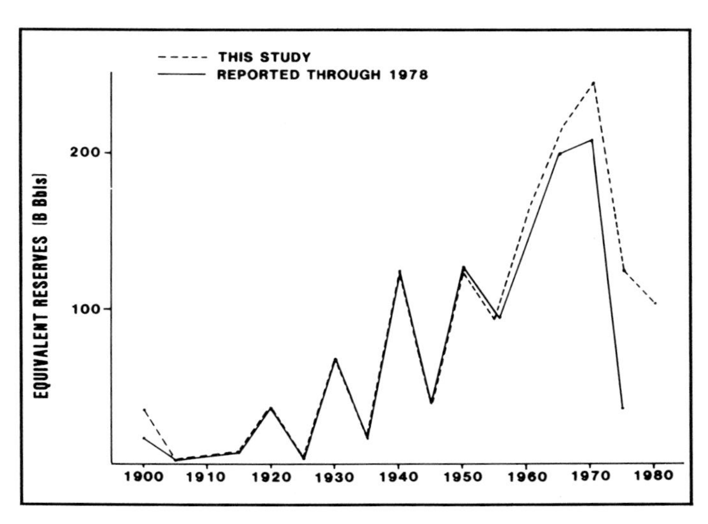


Figure 5b. *Discovery of giant oil and gas fields. Horizontal axis: discovery year; Vertical axis: billion bbl equivalent discovered.*

## DISCOVERY RATES

There is not an immediate resource-based supply problem. Further, there are a large number of favorable basins or provinces in which to explore for more giants. But finding more giant fields requires not only the right geology, but also a favorable economic environment.

Exploration geologists and geophysicists are continuing to discover giant fields, but at a decreasing rate. Furthermore, the giants discovered most recently have been smaller, on average, than those discovered in the past.

Figures 5a and 5b show the discovery rate of giant fields since 1900. The dashed line indicates the number of fields in the present compilation, plotted by the year in which they were discovered. The solid line represents the fields considered giants in this study that were also listed by either Halbouty et al (1970) or Nehring (1978). The curves begin their significant divergence in the early 1960s, which indicates that it may take as much as 20 years for a giant field to be recognized as such.

The rate of discovery shown in Figure 5a decreases to nearly zero during the last five years. This decrease reflects the fact that the most recent discoveries have not yet been properly evaluated. But it is critical that the peaks of the two curves are aligned at dates during the late 1960s. If the sharp declines observed are only the result of the closeness between the discovery date and the publication date for a list of giants, then the peak for the total curve would be offset to the right of the peak for older data. Because it is not offset, we conclude that the decline in giant discoveries is real.

Because the physical potential for additional giant discoveries appears ample, we also conclude that the observed decrease in finding rates is the result of economic, including political, rather than geologic factors.

Figure 6 is taken from Wood (1979), and shows how increased prices have the effect of increasing supply across the entire range of field sizes, including the giants. An increase in price drives the system in the way in which Wood suggests for the short term: it increases the number of economically viable fields. But for the long term, it is the economic return to the explorer that matters. While prices have risen significantly since 1970, and have stimulated production from considerable additional reserves, it is much less clear whether these increased prices have provided a significantly increased return to the oil explorers, primarily the major international oil companies.

## FUTURE OIL SUPPLIES

Present oil supplies, and those for the immediate future, are ample. The concentration of reserves in a single geographic area may be cause for political and economic concern, but this is not a geologic problem. But even the temporary, politically driven supply shortages of the last 12 years have raised the question of how long the "Age of Oil" can last.

Hubbert (1949, 1974) has proposed that resources are exploited in such a way that production plotted against time forms a bell-shaped curve. Hubbert further shows that the peak in production can be predicted from a peak in reserve discoveries. Hubbert's curves describe the resource exploitation cycle for a number of naturally occurring, finite resources.

Giant fields can serve as a proxy for all reserves in this analysis. Except in the U.S., the bulk of the known oil resources occurs in the giant fields (Wood, 1980). Hence, the decline in giant field discoveries observed in Figure 5 is an indicator that the discovery of giant field reserves has passed beyond the halfway or median point. The drop in worldwide production since 1980 may, therefore, be permanent.

The criticism of Hubbert's curves is that they reflect economic conditions. That is also their strength, unless exploration is to be done without regard to economics. Figure 7 shows the production curve for Pennsylvanian anthracite coal (Pyros, 1979, citing Saward, 1934; Saward, 1964; and U.S. Dept. Interior, 1976). Because anthracite occurs as specific horizons in folded sediments, it is relatively easy to calculate the volume of coal in place, based on surface exposures. Allowing for the need to leave some coal unmined, the amount produced is approximately 50% of the known and technologically available reserves as estimated in 1895 (Pyros, 1979). The coal is there; the economics are not.

Similarly, we believe that the oil is in place to be discovered. But the decline in the discovery rate for giant fields is pointing to

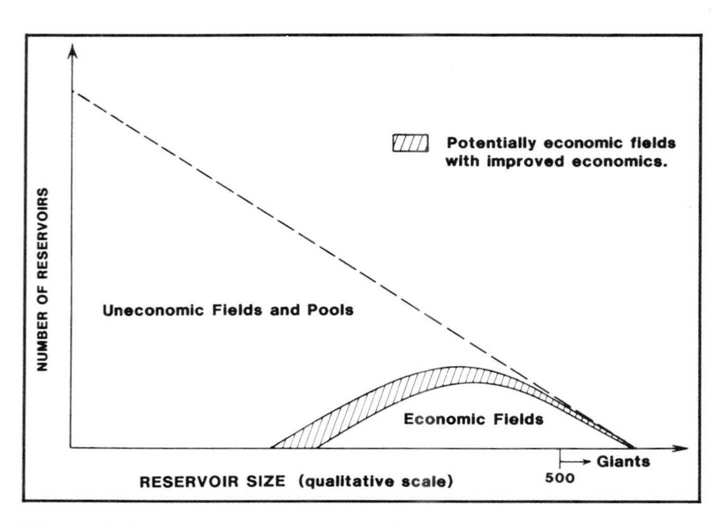


Figure 6. *Increase in reserves with better economics. Horizontal axis: reservoir size (qualitative); Vertical axis: number of reservoirs (qualitative); modified from Wood, 1979.*

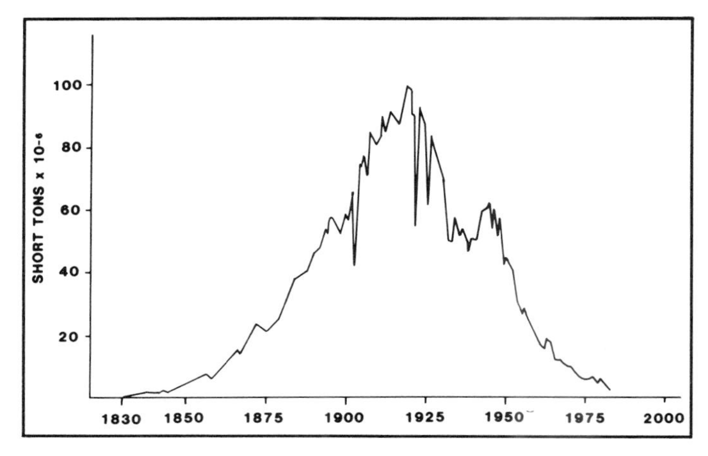


Figure 7. *Pennsylvania anthracite production. Horizontal axis: year; Vertical axis: amount produced.*

the fact that the economic cycle for petroleum is mature, and such additional discoveries may never be made.

## CONCLUSIONS

The known, combined oil and gas resources from giant fields are very large. Without any additional discoveries, a physical supply is available that will last into the next century. This conclusion does not take into account potential shortages caused by political antagonism and interference.

The best geologic settings for giant fields are basins on continental crust that are related in some way with plate collisions. The large number of such basins that do not presently have giant fields are distributed in a variety of political and geographic settings. On solely geologic criteria, there is ample exploration potential.

The discovery rate for giant fields has declined sharply in recent years, mainly because of economic factors. According to Hubbert's discovery–production model (1949, 1974) this decline, if it continues, indicates that the oil resource exploitation cycle is mature.

## ACKNOWLEDGMENTS

The sources used to compile Table 1 were the annual development issues of the AAPG Bulletin, Halbouty et al (1970), Halbouty (1970), Halbouty (1980), International Petroleum Encyclopedia (referred to as IPE) (1984), Nehring (1978, 1981), the Oil and Gas Journal (1975–1984), St. John (1980), and Petroconsultants' computer files (1983). In addition, colleagues too numerous to mention have discussed specific field data with us, which we acknowledge with gratitude.

## REFERENCES CITED

American Association of Petroleum Geologists, AAPG Bulletin [the Developments issue(s) for each year]: Tulsa, Oklahoma.

Bally, A. W., and S. Snelson, 1980, Facts and principles of world petroleum occurrence: realms of subsidence, *in* A. D. Miall, ed., Facts and Principles of World Petroleum Occurrence: Canadian Society of Petroleum Geologists Memoir 6, p. 9–94.

Halbouty, M., T., A. A. Meyerhoff, Robert E. King, Robert H. Dott, Sr., H. Douglas Klemme, and Theodore Shabad, 1970, World's giant oil and gas fields, geologic factors affecting their formation, and basin classification [Part I], *in* M. T. Halbouty, ed., Geology of Giant Petroleum Fields: AAPG Memoir 14, p. 502–528.

——, ed., 1970, Geology of giant petroleum fields: AAPG Memoir 14, 575 p.

——, 1980, Giant oil and gas fields of the decade 1968–1978: AAPG Memoir 30, 596 p.

Hubbert, M. K., 1949, Energy from fossil fuels: Science, 109, p. 103–109.

——, 1975, U.S. Energy Resources, a review as of 1972, U.S. Senate Committee on Interior and Insular Affairs, Background Paper prepared pursuant to Senate Resolution 45 A National Fuels and Energy Policy Study; Serial 93-40 (92-75): Washington, D.C., U.S. Government Printing Office.

International Petroleum Encyclopedia, 1984, International Petroleum Encyclopedia: Tulsa, Oklahoma, PennWell Publishing Co., 410 p.

Klemme, H. D., 1974, The giants and the supergiants: Oil and Gas Journal, 1, 8, and 15 March 1971.

Nehring, R., 1978, Giant oil fields and world oil resources: Santa Monica, California, Rand Corporation Report R-2284-CIA, 162 p.

—— and E. R. Van Driest, II, 1981, The discovery of significant oil and gas fields in the United States: Santa Monica, California, Rand Corporation Report R-2654/1-USGS/DOE, 2 vols.

Oil and Gas Journal, Oil and Gas Journal; Tulsa, Oklahoma, PennWell Publishing Co.

Petroconsultants, 1983, Oil and gas field records: Dublin, Ireland, Petroconsultants Ltd., computer database.

Pyros, J. J., 1979, Anthracite coal in Pennsylvania: unpublished paper, Princeton, New Jersey, Princeton University Geology

*continued on page 52*

Table 1. *Giant oil and gas fields.*

| Field Name (discovery year) | Country Province | Recoverable Reserves | | | Petroleum Province | Classification | | Depth feet (meters) | Trap Type | Geologic Age (lithology) | Latitude Longitude | Remarks |
|---|---|---|---|---|---|---|---|---|---|---|---|---|
| | | Equiv. Oil (mm bbl) | Oil (mm bbl) | Gas (tcf) | | Bally-Type | Klemme-Type | | | | | |
| 1. Ghawer (1948) | Saudi Arabia | 87,500 | 82,000 | 33.00 | Arabian (450) | 221 | 2Ca | 7,500 ( 2,200) | Anticline | Jurassic (carbonate) | N 25°56′ E 49°14′ | 34 API |
| 2. Burgan (1938) | Kuwait | 87,083 | 75,000 | 72.50 | Arabian (450) | 221 | 2Ca | 4,800 (1,400) | Anticline | Cretaceous (sandstone) | N 28°55′ E 47°56′ | 32 API |
| 3. Urengoy (1966) | USSR RSFSR-Tyuman | 47,602 | 2 | 285.59 | West Siberian (774) | 1212 | 2B | 4,000 ( 1,200) | Anticline | Cretaceous (sandstone) | N 66°40′ E 76°45′ | Alias: Vostochno Urengoy; See also separate field: Severo Urengoy (USSR) |
| 4. Safaniya (1951) | Saudi Arabia | 38,066 | 36,100 | 11.79 | Arabian (450) | 221 | 2Ca | 5,250 ( 1,600) | Anticline | Cretaceous (sandstone) | N 28° 2′ E 48°45′ | 27 API; Alias: Khafji (Divided Zone) |
| 5. Bolivar Coastal (1917) | Venezuela | 30,100 | 30,100 | — | Maracaibo (243) | 222 | 3Bc | 3,000 ( 900) | Stratigraphic | Miocene (sandstone) | N 10° 0′ W 71°10′ | Alias: Bachaquero; Cabimas; Ceuta; Lagunillas; Lama; Tia Juana |
| 6. Yamburg (1969) | USSR RSFSR-Tyumen | 27,983 | — | 167.89 | West Siberian (774) | 1212 | 2B | 3,600 ( 1,000) | Broad Arch | Cretaceous (sandstone) | N 68°10′ E 76° 0′ | |
| 7. Bovanenkovo (1971) | USSR RSFSR-Tyumen | 24,416 | — | 146.50 | West Siberian (774) | 1212 | 2B | 4,000 ( 1,200) | Anticline | Cretaceous (sandstone) | N 70°24′ E 68°24′ | |
| 8. Cantarell Complex (1976) | Mexico | 20,000 | 20,000 | — | Campeche (32) | 114 | 2Cc/3C | 5,000 ( 1,500) | Fltd. anticline | Cretaceous (carbonate) | N 19°30′ W 91°57′ | 27 API; Alias: Akal; Nohoch |
| 9. Zakum (1964) | Abu Dhabi | 18,400 | 18,400 | — | Arabian (450) | 221 | 2Ca | 9,100 ( 2,700) | Anticline | Jurassic (carbonate) | N 24°52′ E 53°41′ | |
| 10. Manifa (1957) | Saudi Arabia | 17,800 | 17,000 | 4.79 | Arabian (450) | 221 | 2Ca | 7,600 ( 2,300) | Anticline | Cretaceous (sandstone) | N 27°43′ E 49° 2′ | 29 API; also reserves in Cretaceous and Jurassic carbonates |
| 11. Kirkuk (1927) | Iraq | 17,000 | 17,000 | — | Zagros (464) | 41 | 2Ca | 2,800 ( 800) | Anticline | Oligocene (carbonate) | N 34°45′ E 44°11′ | 36 API; Cretaceous also important |
| 12. Marun (1963) | Iran | 16,195 | 9,500 | 40.16 | Zagros (464) | 41 | 2Ca | 7,450 ( 2,200) | Anticline | Oligocene (carbonate) | N 31°10′ E 49°19′ | 33 API |
| 13. Hassi R'Mel (1956) | Algeria | 16,000 | 1,000 | 90.00 | Sahara (351) | 121 | 2B | 6,900 ( 2,100) | Anticline | Triassic (sandstone) | N 32°56′ E 3°16′ | |
| 14. Zapolyarnoye (1965) | USSR RSFSR-Tyumen | 15,716 | — | 94.29 | West Siberian (774) | 1212 | 2B | 4,000 ( 1,200) | Anticline | Cretaceous (silty sandst) | N 67° 0′ E 79°30′ | |
| 15. Samotlor (1965) | USSR RSFSR-Tyumen | 15,114 | 15,114 | — | West Siberian (774) | 1212 | 2B | 7,300 ( 2,200) | Anticline | Cretaceous (sandstone) | N 61°10′ E 77° 0′ | |
| 16. Romashkinskoye (1948) | USSR RSFSR-Tata | 14,510 | 14,510 | — | Volga-Ural (773) | 221 | 2A | 5,800 ( 1,700) | Anticline | Devonian (sandstone) | N 54°40′ E 52°20′ | |
| 17. Rumaila (1953) | Iraq | 14,000 | 14,000 | — | Arabian (450) | 221 | 2Ca | 10,650 ( 3,200) | Anticline | Cretaceous (sandstone) | N 30°26′ E 47°20′ | |
| 18. Prudhoe Bay (1968) | USA Alaska | 13,783 | 9,450 | 26.00 | North Slope (109) | 221 | 2Cc | 8,500 ( 2,500) | Fltd. anticline | Permian (sandstone) | N 70°20′ W 149° 0′ | Nehring (1981) 1-1 |
| 19. Northwest Dome (1976) | Qatar | 13,333 | — | 80.00 | Arabian (450) | 221 | 2Ca | — ( — ) | Anticline | Permian (carbonate) | N 26°34′ E 51°46′ | Also called North Dome. Reserves could be as much as 300 tcf. |
| 20. Abqaiq (1941) | Saudi Arabia | 13,133 | 12,800 | 2.00 | Arabian (450) | 221 | 2Ca | 6,500 ( 1,900) | Anticline | Jurassic (carbonate) | N 25°57′ E 49°41′ | 37 API |
| 21. Hugoton (1926) | USA Kansas | 12,995 | 1,412 | 69.50 | Anadarko (9) | 221 | 2A | 2,850 ( 800) | Stratigraphic | Permian (dolomite) | N 37° 0′ W 101° 0′ | Alias: Panhandle (Halbouty gas giant no. 11, Halbouty oil giant no. 73, Nehring giant no. 90). Also in Okla. & Texas |

Table 1. *Continued.*

| Field Name (discovery year) | Country Province | Recoverable Reserves | | | Petroleum Province | Classification | | Depth feet (meters) | Trap Type | Geologic Age (lithology) | Latitude Longitude | Remarks |
|---|---|---|---|---|---|---|---|---|---|---|---|---|
| | | Equiv. Oil (mm bbl) | Oil (mm bbl) | Gas (tcf) | | Bally-Type | Klemme-Type | | | | | |
| 22. Berri (1964) | Saudi Arabia | 12,616 | 12,000 | 3.69 | Arabian (450) | 221 | 2Ca | 7,400 ( 2,200) | Anticline | Jurassic (carbonate) | N 26°59′ E 49°37′ | 33 API |
| 23. Groningen (1959) | Netherlands | 12,333 | — | 74.00 | German, Northw (398) | 121 | 2B | 9,650 ( 2,900) | Fltd. uplift | Permian (sandstone) | N 53°11′ E 6°46′ | Alias: Groningen (Germany) |
| 24. Ahwaz (1958) | Iran | 12,160 | 10,160 | 12.00 | Zagros (464) | 41 | 2Ca | 6,500 ( 1,900) | Anticline | Miocene (carbonate) | N 31°20′ E 48°44′ | 32 API |
| 25. Agha Jari (1938) | Iran | 11,993 | 9,000 | 17.95 | Zagros (464) | 41 | 2Ca | 6,500 ( 1,900) | Anticline | Miocene (carbonate) | N 30°50′ E 49°48′ | 34 API |
| 26. Orenburg (1966) | USSR RSFSR - Orenburg | 11,166 | 700 | 62.79 | Volga-Ural (773) | 221 | 2A | 5,300 ( 1,600) | Anticline | Permian (carbonate) | N 51°38′ E 54°17′ | Alias: Krasnoyarskoye; Krasnyy Kholm |
| 27. Raudhatain (1955) | Kuwait | 11,000 | 8,800 | 13.19 | Arabian (450) | 221 | 2Ca | 8,000 ( 2,400) | Anticline | Cretaceous (sandstone) | N 29°52′ E 47°45′ | 35 API |
| 28. Arkticheskoye (1968) | USSR RSFSR - Tyumen | 10,800 | 300 | 63.00 | West Siberian (774) | 1212 | 2B | 7,850 ( 2,300) | Anticline | Cretaceous (sandstone) | N 69°33′ E 70°48′ | |
| 29. Zuluf (1965) | Saudi Arabia | 10,600 | 10,600 | — | Arabian (450) | 221 | 2Ca | 5,800 ( 1,700) | Anticline | Cretaceous (sandstone) | N 28°22′ E 49°10′ | |
| 30. Bermudez (1958) | Mexico | 9,916 | 7,000 | 17.50 | Campeche (32) | 114 | 2Cc/3C | 14,200 ( 4,300) | Fltd. anticline | Cretaceous (dolomite) | N 18° 0′ W 93° 0′ | 29 API: Alias: Antonio J. Bermudez; Cunduacan; Iride; Ojaicaque; Samaria |
| 31. Kyrtaiol'skoye (1970) | USSR RSFSR - Komi | 9,166 | — | 55.00 | West Siberian (774) | 1212 | 2B | 8,200 ( 2,500) | Anticline | Devonian Sandstone | N 57°13′ E 64°55′ | |
| 32. Murban Bu Hasa (1962) | Abu Dhabi | 9,120 | 8,000 | 6.71 | Arabian (450) | 221 | 2Ca | 7,600 ( 2,300) | Anticline reef | Cretaceous (carbonate) | N 23°30′ E 53°10′ | Alias: Bu Hasa |
| 33. Medvezh'ye (1967) | USSR RSFSR - Tyumen | 9,099 | — | 54.59 | West Siberian (774) | 1212 | 2B | 4,250 ( 1,200) | Anticline | Cretaceous (sandstone) | N 67°20′ E 73°50′ | |
| 34. Khurais (1957) | Saudi Arabia | 9,016 | 8,500 | 3.09 | Arabian (450) | 221 | 2Ca | 5,100 ( 1,500) | Anticline | Jurassic (carbonate) | N 25° 5′ E 48° 3′ | 33 API |
| 35. Hassi Messaoud (1956) | Algeria | 9,000 | 9,000 | — | Ghadames (312) | 121 | 2B | 11,000 ( 3,300) | Fltd. anticline | Cambrian (sandstone) | N 31°45′ E 6° 5′ | 46 API |
| 36. Troll (1979) | Norway Blocks 31/ 2,3,5,6 | 8,966 | 1,400 | 45.39 | North Sea, Nor (420) | 1211 | 3A | 4,750 ( 1,400) | Anticline | Jurassic (sandstone) | N 60°46′ E 3°33′ | |
| 37. Gach Saran (1928) | Iran | 8,500 | 8,500 | — | Zagros (464) | 41 | 2Ca | 4,200 ( 1,200) | Anticline | Oligocene (carbonate) | N 30°13′ E 50°49′ | 31 API; production also from Cretaceous, to 7000 ft. |
| 38. B. Structure (1972) | Iran | 8,333 | — | 50.00 | Arabian (450) | 221 | 2Ca | 7,500 ( 2,200) | Anticline | Cretaceous (sandstone) | N 28°24′ E 50° 9′ | |
| 39. Kangan (1973) | Iran | 8,333 | — | 50.00 | Zagros (464) | 41 | 2Ca | — ( — ) | Anticline | Permian (dolomite) | N 27°55′ E 52° 1′ | |
| 40. Dauletabad (1976) | USSR Turkmenia SSR | 8,118 | — | 48.70 | Turkmen (768) | 222 | 2Ca | 9,800 ( 2,900) | Stratigraphic | Cretaceous (sandstone) | N 36°12′ E 61°32′ | Alias: Donmez; Sovietabad; Vostochnyy Daul Full name: Dauletabad-Vostochnyy Daul or Dauletabad-Donmez |
| 41. Moran (1956) | India Assam | 8,042 | 42 | 48.00 | Assam (474) | 221 | 2Cb | 9,200 ( 2,800) | Fltd. anticline | Oligocene (sandstone) | N 27°20′ E 95°10′ | |

Table 1. *Continued.*

| Field Name (discovery year) | Country Province | Recoverable Reserves | | | Petroleum Province | Classification | | Depth feet (meters) | Trap Type | Geologic Age (lithology) | Latitude Longitude | Remarks |
|---|---|---|---|---|---|---|---|---|---|---|---|---|
| | | Equiv. Oil (mm bbl) | Oil (mm bbl) | Gas (tcf) | | Bally-Type | Klemme-Type | | | | | |
| 42. Daqing (1959) | China Heilongjiang | 8,000 | 8,000 | — | Songliao (663) | 3122 | 3A | 4,492 ( 1,300) | Anticline | Cretaceous (sandstone) | N 46°30′ E 125°20′ | Alias: Taching |
| 43. Rumaila North (1961) | Iraq | 8,000 | 8,000 | — | Arabian (450) | 221 | 2Ca | 10,500 ( 3,200) | Anticline | Cretaceous (sandstone) | N 30°30′ E 47°16′ | 34 API |
| 44. Abu Sa'fah (1963) | Saudi Arabia | 7,850 | 7,500 | 2.09 | Arabian (450) | 221 | 2Ca | 6,400 ( 1,900) | Anticline | Jurassic (carbonate) | N 26°55′ E 50°25′ | 30 API |
| 45. Kharasaveyskoye (1974) | USSR RSFSR - Tyumen | 7,449 | — | 44.69 | West Siberian (774) | 1212 | 2B | 5,000 ( 1,500) | Anticline | Cretaceous (sandstone) | N 70°58′ E 67° 0′ | |
| 46. Minas (1944) | Indonesia Sumatra | 7,000 | 4,000 | 18.00 | Sumatra, Centr (670) | 3122 | 3Ba | 2,400 ( 700) | Anticline | Miocene (sandstone) | N 0°45′ E 101°26′ | 350 API; Gas figure may be optimistic. |
| 47. Shaybah (1968) | Saudi Arabia | 7,000 | 7,000 | — | Arabian (450) | 221 | 2Ca | 5,000 ( 1,500) | Anticline | Cretaceous (carbonate) | N 22°33′ E 54° 2′ | Alias: Zarrarah (Saudi Arabia) |
| 48. Majnoon (1976) | Iraq | 7,000 | 7,000 | — | Arabian (450) | 221 | 2Ca | 7,000 ( 2,100) | Anticline | Cretaceous (carbonate) | N 31° 4′ E 47°34′ | |
| 49. Qatif (1945) | Saudi Arabia | 6,800 | 6,000 | 4.79 | Arabian (450) | 221 | 2Ca | 7,700 ( 2,300) | Anticline | Jurassic (carbonate) | N 25°32′ E 49°59′ | 33 API |
| 50. Kruzernshtern | USSR RSFSR - Tyumen | 6,583 | — | 39.50 | West Siberian (774) | 1212 | 2B | — ( — ) | Anticline | Cretaceous (sandstone) | N 70°50′ E 66°55′ | |
| 51. Sarir C (1961) | Libya | 6,500 | 6,500 | — | Sirte (354) | 1211 | 3A | 8,650 ( 2,600) | Fltd. anticline | Cretaceous (sandstone) | N 27°39′ E 22°31′ | 37 API |
| 52. Dorra (1967) | Divided Zone | 6,183 | 350 | 35.00 | Arabian (450) | 221 | 2Ca | 6,000 ( 1,800) | Anticline | Cretaceous (carbonate?) | N 28°57′ E 49° 7′ | |
| 53. Severo Urengoy (1971) | USSR RSFSR - Tyumen | 5,833 | — | 35.00 | West Siberian (774) | 1212 | 2B | 3,800 ( 1,100) | Anticline | Cretaceous (sandstone) | N 67° 7′ E 76° 0′ | See also separate field Urengoy |
| 54. Shatlyskoye (1968) | USSR Turkmenia SSR | 5,783 | 50 | 34.39 | Turkmen (768) | 222 | 2Ca | 11,500 ( 3,500) | Anticline | Cretaceous (sandstone) | N 37°31′ E 61°15′ | Alias, Shatlyk; Sheketli; Shekhitli |
| 55. Fedorovskoye (1971) | USSR RSFSR - Tyumen | 5,633 | 2,500 | 18.79 | West Siberian (774) | 1212 | 2B | 5,900 ( 1,700) | Anticline | Cretaceous (sandstone) | N 63°12′ E 73°34′ | |
| 56. East Texas (1930) | USA Texas (6) | 5,600 | 5,600 | — | East Texas Sal (51) | 1143 | 2Cc | 3,700 ( 1,100) | Stratigraphic | Cretaceous (sandstone) | N 32°30′ W 95° 0′ | Nehring (1981) 9d-1 |
| 57. Umm Shaif (1958) | Abu Dhabi | 5,300 | 5,300 | — | Arabian (450) | 221 | 2Ca | 9,150 ( 2,700) | Anticline | Jurassic (carbonate) | N 25°13′ E 53°13′ | 37 API |
| 58. Shengli (1962) | China Shandong | 5,250 | 5,000 | 1.50 | Huabei (560) | 3122 | 3A | 10,000 ( 3,000) | Anticline | Permian (carbonate) | N 37°32′ E 118°30′ | |
| 59. Pars (1966) | Iran | 5,000 | — | 30.00 | Zagros (464) | 41 | 2Ca | 9,000 ( 2,700) | Anticline | Cretaceous (carbonate) | N 27°58′ E 51°15′ | |
| 60. Sabriya (1958) | Kuwait | 4,893 | 4,000 | 5.35 | Arabian (450) | 221 | 2Ca | 8,000 ( 2,400) | Anticline | Cretaceous (sandstone) | N 29°50′ E 47°47′ | 32 API |
| 61. Marjan (1967) | Saudi Arabia | 4,575 | 4,575 | — | Arabian (450) | 221 | 2Ca | 7,000 ( 2,100) | Anticline | Cretaceous (clastics) | N 28°27′ E 49°43′ | 30 API; Alias: Fereidoon (Iran) |
| 62. Zubair (1948) | Iraq | 4,500 | 4,500 | — | Arabian (450) | 221 | 2Ca | 9,450 ( 2,800) | Anticline | Cretaceous (sandstone) | N 30°25′ E 47°40′ | |
| 63. Gazlinskoye (1956) | USSR Uzbekistan SSR | 4,393 | 60 | 26.00 | Kara-Kum (735) | 222 | 2Ca | 2,550 ( 700) | Anticline | Cretaceous (sandstone) | N 40°21′ E 63°20′ | Some production from carbonates |
| 64. Asab (1966) | Abu Dhabi | 4,333 | 4,000 | 2.00 | Arabian (450) | 221 | 2Ca | 7,750 ( 2,300) | Anticline | Cretaceous (carbonate) | N 23°24′ E 54°16′ | Alias: Abu Jidu |

Table 1. *Continued.*

| Field Name (discovery year) | Country Province | Recoverable Reserves | | | Petroleum Province | Classification | | Depth feet (meters) | Trap Type | Geologic Age (lithology) | Latitude Longitude | Remarks |
|---|---|---|---|---|---|---|---|---|---|---|---|---|
| | | Equiv. Oil (mm bbl) | Oil (mm bbl) | Gas (tcf) | | Bally-Type | Klemme-Type | | | | | |
| 65. Khursaniya (1956) | Saudi Arabia | 4,283 | 4,100 | 1.09 | Arabian (450) | 221 | 2Ca | 6,600 ( 2,000) | Anticline | Jurassic (carbonate) | N 27°14′ E 49°14′ | 31 API |
| 66. Murban Bab (1954) | Abu Dhabi | 4,258 | 3,500 | 4.54 | Arabian (450) | 221 | 2Ca | 8,450 ( 2,500) | Anticline | Cretaceous (carbonate) | N 23°52′ E 53°32′ | 44 API; Alias: Bab |
| 67. Amal (1959) | Libya | 4,250 | 4,250 | — | Sirte (354) | 1211 | 3A | 9,900 ( 3,000) | Anticline | Cretaceous (sandstone) | N 29°27′ E 21° 9′ | Other horizons with different lithologies also produce |
| 68. Sovetskoye (1962) | USSR RSFSR - Tomsk | 4,200 | 4,200 | — | West Siberian (774) | 1212 | 2B | 11,130 ( 3,300) | Anticline | Cretaceous (sandstone) | N 60°50′ E 77° 0′ | Alias: Sosnino Sovetskoye; Medvedev |
| 69. Arlanskoye (1955) | USSR RSFSR - Bashkir | 4,100 | 4,100 | — | Volga-Ural (773) | 221 | 2A | 4,200 ( 1,200) | Anticline | (sandstone) | N 55°55′ E 54°20′ | |
| 70. Rag-e-Safid (1964) | Iran | 4,041 | 2,400 | 9.84 | Zagros (464) | 41 | 2Ca | 6,500 ( 1,900) | Anticline | Miocene (carbonate) | N 32°27′ E 50° 0′ | 27 API; Also Cretaceous to 8300 ft. |
| 71. Vuktyl'skoye (1964) | USSR RSFSR - Komi | 3,840 | 340 | 21.00 | Pechora (755) | 221 | 2B | 8,850 ( 2,600) | Anticline | Permian (carbonate) | N 63°36′ E 57°17′ | Also produces from sandstones |
| 72. Paris (1964) | Iran | 3,826 | 3,000 | 4.95 | Zagros (464) | 41 | 2Ca | 5,000 ( 1,500) | Anticline | Oligocene (carbonate) | N 31° 0′ E 49°56′ | 39 API; Alias: Faris |
| 73. Dukhan (1940) | Qatar | 3,783 | 3,200 | 3.50 | Arabian (450) | 221 | 2Ca | 6,700 ( 2,000) | Anticline | Jurassic (carbonate) | N 25°20′ E 50°47′ | 41 API |
| 74. Pazanan (1961) | Iran | 3,650 | 1,130 | 15.11 | Zagros (464) | 41 | 2Ca | 4,700 ( 1,400) | Anticline | Oligocene (carbonate) | N 32°22′ E 50°11′ | 37 API; Also No. 6 on Halbouty gas list |
| 75. La Brea (1868) | Peru | 3,500 | 1,000 | 15.00 | Talara (279) | 111 | 3Bb | 5,500 ( 1,600) | Fault blocks | Eocene (sandstone) | S 4°40′ W 81°15′ | 33 API; Alias: Parinas; Talara |
| 76. Gialo (1961) | Libya | 3,500 | 3,500 | — | Sirte (354) | 1211 | 3A | 2,200 ( 600) | Anticline | Eocene (carbonate) | N 28°41′ E 21°24′ | 38 API; Sandstones reservoirs also present |
| 77. Statfjord (1974) | Norway Block 33/12 | 3,383 | 3,000 | 2.29 | North Sea, Nor (420) | 1211 | 3A | 8,000 ( 2,400) | Fault & Unconf | Jurassic (sandstone) | N 61°12′ E 1°50′ | 29 API; Alias: Statfjord (United Kingdom) |
| 78. Aghar (1974) | Iran | 3,333 | — | 20.00 | Zagros (464) | 41 | 2Ca | — ( — ) | Anticline | Permian (carbonate) | N 28°35′ E 52°48′ | |
| 79. Elmworth (1976) | Canada Alberta | 3,333 | — | 20.00 | Alberta (4) | 221 | 2A | — ( — ) | Stratigraphic | Cretaceous (sandstone) | N 54°30′ W 118°30′ | Alias: Wapati. Tight formation - reserves potentially 300 tcf. |
| 80. Bibi Hakimeh (1961) | Iran | 3,300 | 2,200 | 6.59 | Zagros (464) | 41 | 2Ca | 4,570 ( 1,300) | Anticline | Miocene (carbonate) | N 30° 3′ E 50°37′ | 30 API |
| 81. Minagish (1959) | Kuwait | 3,275 | 2,400 | 5.25 | Arabian (450) | 221 | 2Ca | 10,000 ( 3,000) | Anticline | Cretaceous (sandstone) | N 29° 5′ E 47°31′ | 34 API |
| 82. Semakovskoye | USSR RSFSR - Tyumen | 3,166 | — | 19.00 | West Siberian (774) | 1212 | 2B | — ( — ) | | | N 69° 0′ E 75° 0′ | |
| 83. Carthage (1936) | USA Texas (6) | 3,150 | 150 | 18.00 | East Texas Sal (51) | 1143 | 2Cc | 6,000 ( 1,800) | Anticline | Cretaceous (limestone) | N 32°30′ W 94°30′ | Alias: Cotton Valley Nehring (1981) 9d-2 |
| 84. Shebelinka (1950) | USSR Ukraine SSR | 3,099 | — | 18.59 | Dnepr-Donets (723) | 222 | 3A | 5,000 ( 1,500) | Salt anticline | Permian (sandstone) | N 49°27′ E 36°37′ | |
| 85. Usinskoye (1962) | USSR RSFSR - Komi | 3,100 | 3,100 | — | Pechora (755) | 221 | 2B | — ( — ) | Anticline | Permian (carbonate) | N 66° 8′ E 57°16′ | |
| 86. Awali (1931) | Bahrain | 3,041 | 1,000 | 12.25 | Arabian (450) | 221 | 2Ca | 1,900 ( 500) | Anticline | Cretaceous (carbonate) | N 26° 2′ E 50°32′ | Alias: Bahrain. Also listed as a giant oil field—no. 96 on Halbouty list and no. 123 on Nehring list. |
| 87. Sredne Botnobin (1970) | USSR RSFSR - Irkutsk | 3,032 | 149 | 17.29 | Vilyuy (772) | 221 | 2A | 6,000 ( 1,800) | Anticline | Proterozoic (sandstone) | N 61°15′ E 111°40′ | Alias: Botnobin. Also produces from Cambrian. |

Table 1. *Continued.*

| Field Name (discovery year) | Country Province | Recoverable Reserves | | | Petroleum Province | Classification | | Depth feet (meters) | Trap Type | Geologic Age (lithology) | Latitude Longitude | Remarks |
|---|---|---|---|---|---|---|---|---|---|---|---|---|
| | | Equiv. Oil (mm bbl) | Oil (mm bbl) | Gas (tcf) | | Bally-Type | Klemme-Type | | | | | |
| 88. Malgobek Voznesenskoye (1915) | USSR RSFSR - Chechen-In | 3,000 | 3,000 | — | Amur (710) | 221 | 3Bc | 1,640 ( 500) | Anticline | Miocene (sandstone) | N 43°36′ E 44°30′ | Alias: Aliyurt; Voznesenka. Also productive from Cretaceous carbonates to 10,900 ft. |
| 89. Duri (1941) | Indonesia Sumatra | 3,000 | 3,000 | — | Sumatra, Centr (670) | 3122 | 3Ba | 770 ( 200) | Anticline | Eocene (sandstone) | N 1°20′ E 101°15′ | 21 API; Recently announced secondary recovery project is responsible for approx. 2500 MMBbls. |
| 90. Jaladi (1978) | Saudi Arabia | 3,000 | 3,000 | — | Arabian (450) | 221 | 2Ca | 9,000 ( 2,700) | Anticline | Jurassic (carbonate) | N 27° 4′ E 48°37′ | |
| 91. Wilmington (1932) | USA California | 2,977 | 2,758 | 1.31 | Los Angeles (89) | 332 | 3Bb | 2,200 ( 600) | Fltd. anticline | Miocene (sandstone) | N 33°45′ W 118°20′ | Nehring (1981) 2c-1 |
| 92. Brent (1971) | United Kingdom Block 211/29 | 2,696 | 2,163 | 3.19 | North Sea, Nor (420) | 1211 | 3A | 8,500 ( 2,500) | Fault blocks | Jurassic (sandstone) | N 61° 6′ E 1°42′ | |
| 93. Komsomolskoye (1966) | USSR RSFSR - Tyumen | 2,666 | — | 16.00 | West Siberian (774) | 1212 | 2B | 2,875 ( 800) | Anticline | Cretaceous (sandstone) | N 64°20′ E 76°50′ | See also separate field Severo Komsomol |
| 94. Yamsovey (1970) | USSR RSFSR - Tyumen | 2,500 | — | 15.00 | West Siberian (774) | 1212 | 2B | 2,900 ( 800) | Anticline | Cretaceous (sandstone) | N 65°36′ E 76° 0′ | |
| 95. Nasser (1959) | Libya | 2,425 | 2,200 | 1.34 | Sirte (354) | 1211 | 3A | 6,000 ( 1,800) | Fltd. anticline | Cretaceous (carbonate) | N 28°54′ E 19°48′ | 38 API; Alias: Zelten |
| 96. Balakhano Sabunchino (1896) | USSR Azerbaijan SSR | 2,400 | 2,400 | — | Kura (740) | 222 | 3Bc | 4,500 ( 1,300) | Anticline | Pliocene (sandstone) | N 40°40′ E 50°20′ | Full name: Balakhano Sabunchino Ramanin. Alias: Ramany; Sabunchi |
| 97. Yubileynoye (1968) | USSR RSFSR - Tyumen | 2,383 | — | 14.29 | West Siberian (774) | 1212 | 2B | 4,000 ( 1,200) | Anticline | Cretaceous (sandstone) | N 64°50′ E 75°57′ | |
| 98. Ekofisk (1969) | Norway Block 2/4 | 2,353 | 1,700 | 3.91 | North Sea, Nor (420) | 1211 | 3A | 10,000 ( 3,000) | Salt dome | Eocene (chalk) | N 56°32′ E 3°12′ | 41 API |
| 99. Messoyakhskoye (1967) | USSR RSFSR - Tyumen | 2,333 | — | 14.00 | West Siberian (774) | 1212 | 2B | 3,000 ( 900) | Anticline | Cretaceous (sandstone) | N 69°20′ E 82°30′ | Jurassic also productive |
| 100. Nar (1975) | Iran | 2,333 | — | 14.00 | Zagros (464) | 41 | 2Ca | 6,900 ( 2,100) | Anticline | Pliocene (carbonate) | N 27°47′ E 52°18′ | |
| 101. Wafra (1953) | Divided Zone | 2,308 | 1,725 | 3.50 | Arabian (450) | 221 | 2Ca | 3,600 ( 1,000) | Anticline | Cretaceous (sandstone) | N 28°36′ E 47°54′ | 20 API; Also produces from Eocene carbonate |
| 102. Ust'bakykskoye (1961) | USSR RSFSR - Tyumen | 2,300 | 2,300 | — | West Siberian (774) | 1212 | 2B | 8,850 ( 2,600) | Anticline | Cretaceous (sandstone) | N 62°24′ E 72°40′ | Alias: Balyk; Ust Balyk |
| 103. Eunice (1929) | USA New Mexico | 2,291 | 941 | 8.09 | Permian (120) | 222 | 2A | 3,700 ( 1,100) | Anticline | Permian (sandstone) | N 32°30′ W 103°15′ | Alias: Jalmat; Monument Nehring (1981) 4a-1 |
| 104. Arun (1971) | Indonesia Sumatra | 2,283 | — | 13.69 | Sumatra, North (671) | 3122 | 3Ba | 10,000 ( 3,000) | Reef | Miocene (limestone) | N 5° 3′ E 97°15′ | |
| 105. Mazunin (1960) | USSR RSFSR - Perm | 2,250 | — | 13.50 | Volga-Ural (773) | 221 | 2A | 5,000 ( 1,500) | Anticline | Carboniferous (carbonate) | N 57°10′ E 56°35′ | |
| 106. Hibernia (1979) | Canada Newfoundland | 2,183 | 1,850 | 2.00 | Grand Banks (62) | 114 | 3C/3A | 9,930 ( 3,000) | Rollover antcl | Cretaceous (sandstone) | N 46°50′ W 58°50′ | |
| 107. Midway Sunset (1894) | USA California | 2,179 | 2,090 | 0.53 | Sacramento San (134) | 332 | 3Bb | 5,000 ( 1,500) | Anticline | Miocene (sandstone) | N 35°10′ W 119°30′ | Alias: Sunset Nehring (1981) 2a-1 |
| 108. Cactus (1972) | Mexico | 2,166 | 1,700 | 2.79 | Campeche (32) | 114 | 2Cc/3C | 13,500 ( 4,100) | Fltd. anticline | Cretaceous (limestone) | N 17°52′ W 93°10′ | 35 API |

Table 1. *Continued.*

| Field Name (discovery year) | Country Province | Recoverable Reserves: Equiv. Oil (mm bbl) | Recoverable Reserves: Oil (mm bbl) | Recoverable Reserves: Gas (tcf) | Petroleum Province | Classification: Bally-Type | Classification: Klemme-Type | Depth feet (meters) | Trap Type | Geologic Age (lithology) | Latitude Longitude | Remarks |
|---|---|---|---|---|---|---|---|---|---|---|---|---|
| 109. Karanj (1963) | Iran | 2,133 | 1,650 | 2.89 | Zagros (464) | 41 | 2Ca | 7,800 (2,300) | Anticline | Oligocene (carbonate) | N 31° 0′ E 49°52′ | 34 API |
| 110. Abu Ghirab (1971) | Iraq | 2,106 | 1,880 | 1.35 | Zagros (464) | 41 | 2Ca | 9,500 (2,800) | Anticline | Miocene (limestone) | N 32°25′ E 47°24′ | 34 API; Alias: Dehluran (Iran) |
| 111. Gubkinskoye (1965) | USSR RSFSR - Tyumen | 2,083 | — | 12.50 | West Siberian (774) | 1212 | 2B | 2,275 (600) | Anticline | Cretaceous (sandstone) | N 65° 0′ E 76°42′ | |
| 112. Seria (1929) | Brunei | 2,063 | 1,730 | 2.00 | Brunei-Sabah (503) | 3122 | 2Cc | 2,500 (700) | Fltd. anticline | Miocene (sandstone) | N 4°37′ E 114°19′ | 34 API |
| 113. Yates (1926) | USA Texas (8) | 2,046 | 2,046 | — | Permian (120) | 222 | 2A | 1,250 (300) | Anticline | Permian (carbonate) | N 31° 0′ W 102°50′ | Nehring (1981) 4c-1 |
| 114. Wasson (1936) | USA Texas (8A) | 2,041 | 1,878 | 0.97 | Permian (120) | 222 | 2A | 6,200 (1,800) | Anticline | Permian (dolomite) | N 33° 5′ W 102°45′ | Nehring (1981) 4d-1 |
| 115. Verkhne Vilyuchanskoye (1975) | USSR RSFSR - Yakutsk | 2,010 | 260 | 10.50 | Vilyuy (772) | 221 | 2A | — (—) | Sandbar | Proterozoic (sandstone) | N 64° 0′ E 122° 0′ | Alias: Vilyuehanskoye |
| 116. Bibieybatskoye (1871) | USSR Azerbaijan SSR | 2,000 | 2,000 | — | Kura (740) | 222 | 3Bc | 5,000 (1,500) | Fltd. anticline | Pliocene (sandstone) | N 40°10′ E 50°20′ | Also spelled: Bibi Eybat |
| 117. Haft Kel (1928) | Iran | 2,000 | 2,000 | — | Zagros (464) | 41 | 2Ca | 3,500 (1,000) | Anticline | Oligocene (carbonate) | N 31°26′ E 49°33′ | 38 API |
| 118. Poza Rica (1930) | Mexico | 2,000 | 2,000 | — | Tampico (156) | 1143 | 2Cc | 7,100 (2,100) | Anticline | Cretaceous (carbonate) | N 20°30′ W 97°27′ | 35 API; Hydrodynamics important in trap |
| 119. Zapadno Surgutskoye (1962) | USSR RSFSR - Tyumen | 2,000 | 2,000 | — | West Siberian (774) | 1212 | 2B | 7,625 (2,300) | Anticline | Cretaceous (sandstone) | N 62°27′ E 73° 6′ | Alias: Surgut; West Surgut |
| 120. Mubarrax (1969) | Abu Dhabi | 2,000 | 2,000 | — | Arabian (450) | 221 | 2Ca | 9,000 (2,700) | Structural | Cretaceous (carbonate) | N 24°27′ E 53°42′ | 38 API |
| 121. Forties (1970) | United Kingdom Block 21/10 | 2,000 | 2,000 | — | North Sea, Nor (420) | 1211 | 3A | 7,000 (2,100) | Anticline | Paleocene (sandstone) | N 57°44′ E 0°58′ | 37 API |
| 122. Harmaliyah (1972) | Saudi Arabia | 2,000 | 2,000 | — | Arabian (450) | 221 | 2Ca | 7,500 (2,200) | Anticline | Jurassic (carbonate) | N 24°32′ E 49°28′ | |
| 123. Renqiu (1975) | China Hebei | 2,000 | 2,000 | — | Huabei (560) | 3122 | 3A | 10,000 (3,000) | Anticline | Cretaceous (carbonate) | N 38°45′ E 116° 0′ | |
| 124. Baghdad, East (1979) | Iraq | 2,000 | 2,000 | — | Arabian (450) | 221 | 2Ca | 10,000 (3,000) | Anticline | Cretaceous (sandstone) | N 33° 8′ E 44°30′ | |
| 125. Kern River (1899) | USA California | 1,949 | 1,949 | — | Sacramento San (134) | 332 | 3Bb | 1,000 (300) | Stratigraphic | Pliocene (sandstone) | N 35°30′ W 119° 0′ | Nehring (1981) 2a-2 |
| 126. Scurry (1948) | USA Texas (8A) | 1,926 | 1,701 | 1.34 | Permian (120) | 222 | 2A | 5,000 (1,500) | Reef | Carboniferous (limestone) | N 32°45′ W 101° 0′ | Alias: Kelly; Snyder; Diamond M. Nehring (1981) 4d-2 |
| 127. Sassan (1967) | Iran | 1,900 | 1,900 | — | Arabian (450) | 221 | 2Ca | 8,250 (2,500) | Anticline | Cretaceous (carbonate) | N 25°31′ E 53° 8′ | Alias: Abu al Bukhoosh (Abu Dhabi) Production also from Jurassic |
| 128. Qishm - Salagh (1961) | Iran | 1,875 | — | 11.25 | Zagros (464) | 41 | 2Ca | 10,255 (3,100) | Anticline | Cretaceous (carbonate) | N 26°42′ E 55°35′ | Alias: Salagh |
| 129. Uzen'skoye (1961) | USSR Kazakhstan SSR | 1,875 | 1,875 | — | Mangyshlak (746) | 222 | 2Ca | 2,667 (800) | Anticline | Jurassic (sandstone) | N 43°15′ E 52°45′ | |

Table 1. *Continued.*

| Field Name (discovery year) | Country Province | Recoverable Reserves: Equiv. Oil (mm bbl) | Recoverable Reserves: Oil (mm bbl) | Recoverable Reserves: Gas (tcf) | Petroleum Province | Classification: Bally-Type | Classification: Klemme-Type | Depth feet (meters) | Trap Type | Geologic Age (lithology) | Latitude Longitude | Remarks |
|---|---|---|---|---|---|---|---|---|---|---|---|---|
| 130. Leman (1966) | United Kingdom Block 49/27 | 1,843 | 10 | 11.00 | North Sea, Sou (421) | 1211 | 2B/3A | 6,600 (2,000) | Fltd. anticline | Permian (sandstone) | N 53° 5′ E 2° 8′ | |
| 131. Abu Hadriya (1940) | Saudi Arabia | 1,840 | 1,840 | — | Arabian (450) | 221 | 2Ca | 9,500 (2,800) | Anticline | Jurassic (carbonate) | N 27°22′ E 49° 0′ | 35 API |
| 132. Ampa, Southwest (1963) | Brunei | 1,833 | 900 | 5.59 | Brunei-Sabah (503) | 3122 | 2Cc | 7,600 (2,300) | Anticline | Miocene (sandstone) | N 4°43′ E 114° 8′ | |
| 133. Pembina (1953) | Canada Alberta | 1,800 | 1,800 | — | Alberta (4) | 221 | 2A | 5,200 (1,500) | Permeability | Cretaceous (sandstone) | N 56° 0′ W 115°30′ | Includes West Pembina |
| 134. Defa (1960) | Libya | 1,800 | 1,800 | — | Sirte (354) | 1211 | 3A | 5,300 (1,600) | Anticline Reef | Paleocene (carbonate) | N 28° 5′ E 19°55′ | |
| 135. Augila (1966) | Liyba | 1,800 | 1,800 | — | Sirte (354) | 1211 | 3A | 8,000 (2,400) | Stratigraphic | Cretaceous (sandstone) | N 29°10′ E 21°29′ | |
| 136. Blanco (1927) | USA New Mexico | 1,788 | 25 | 10.58 | San Juan (139) | 222 | 2B | 6,065 (1,800) | Porosity wedge | Cretaceous (sandstone) | N 36°40′ W 106°30′ | Alias: San Juan Nehring (1981) 3e-1 |
| 137. Vyngapurovskoye (1968) | USSR RSFSR - Tyumen | 1,766 | — | 10.59 | West Siberian (774) | 1212 | 2B | 2,900 (800) | Anticline | Cretaceous (sandstone) | N 63°12′ E 76°42′ | |
| 138. Yuzhno Russkoye (1968) | USSR RSFSR - Tyumen | 1,766 | — | 10.59 | West Siberian (774) | 1212 | 2B | 3,350 (1,000) | Anticline | Cretaceous (sandstone) | N 66° 5′ E 80°25′ | See also separate field Russkoye (USSR) |
| 139. Mamontovskoye (1965) | USSR RSFSR - Tyumen | 1,750 | 1,750 | — | West Siberian (774) | 1212 | 2B | 6,300 (1,900) | Anticline | Cretaceous (sandstone) | N 61°24′ E 72°50′ | |
| 140. Yuzhno Samburg (1978) | USSR RSFSR - Tyumen | 1,750 | — | 10.50 | West Siberian (774) | 1212 | 2B | 10,500 (3,200) | Anticline | Cretaceous (sandstone) | N 66°27′ E 77° 5′ | Alias: Samburg |
| 141. Elk Hills (1919) | USA California | 1,719 | 1,478 | 1.44 | Sacramento San (134) | 332 | 3Bb | 8,000 (2,400) | Fltd. anticline | Miocene (sandstone) | N 35°10′ W 119°20′ | Nehring (1981) 2a-3 |
| 142. Kotur Tepe (1956) | USSR Turkmenia SSR | 1,710 | 1,460 | 1.50 | Caspian, South (720) | 222 | 3Bc | 8,800 (2,600) | Fltd. anticline | Pliocene (sandstone) | N 39°25′ E 53°44′ | Alias: Leninskoye; Tepe |
| 143. Slaughter (1936) | USA Texas (8A) | 1,671 | 1,505 | 1.00 | Permian (120) | 222 | 2A | 6,500 (1,900) | Stratigraphic | Permian (dolomite) | N 33°15′ W 102°30′ | Alias: Levelland Nehring (1981) 4d-3 |
| 144. Hateiba (1963) | Libya | 1,666 | — | 10.00 | Sirte (354) | 1211 | 3A | 8,600 (2,600) | Anticline | Cretaceous (sandstone) | N 29°43′ E 19°41′ | |
| 145. Iris (1977) | Mexico | 1,666 | 1,500 | 1.00 | Salinas (Mex) (136) | 22 | 2Cc | 14,800 (4,500) | Fltd. salt dome | Cretaceous (dolomite) | N 17°43′ W 93°23′ | 33 API; Alias: Giraldas |
| 146. Ardeshir (1961) | Iran | 1,600 | 1,600 | — | Arabian (450) | 221 | 2Ca | 3,000 (900) | Anticline | Miocene (sandstone) | N 29°18′ E 49°32′ | |
| 147. Umm Gudair (1962) | Kuwait | 1,600 | 1,600 | — | Arabian (450) | 221 | 2Ca | 9,200 (2,800) | Anticline | Cretaceous (carbonate) | N 28°50′ E 47°42′ | 25 API; Alias: Umm Gudair (Divided zone) |
| 148. Intisar D (1967) | Libya | 1,583 | 1,500 | — | Sirte (354) | 1211 | 3A | 9,400 (2,800) | Reef | Paleocene (carbonate) | N 28°54′ E 20°58′ | 39 API; Alias: Idris |
| 149. Bagadzhin (1971) | USSR Turkmenia SSR | 1,583 | — | 9.50 | Kara-Kum (735) | 222 | 2Ca | 10,100 (3,000) | Anticline | Cretaceous (carbonate) | N 39°20′ E 62°29′ | |
| 150. Kupal (1965) | Iran | 1,550 | 1,200 | 2.09 | Zagros (464) | 41 | 2Ca | 12,000 (3,600) | Anticline | Miocene (limestone) | N 31°13′ E 49°16′ | 33 API |
| 151. Umm ad Dalkh (1973) | Abu Dhabi | 1,550 | 500 | 6.29 | Arabian (450) | 221 | 2Ca | 9,000 (2,700) | Anticline | Cretaceous (carbonate) | N 24°33′ E 54° 8′ | |
| 152. Mukhanovskoye (1945) | USSR RSFSR - Kuybyshev | 1,530 | 1,530 | — | Volga-Ural (773) | 221 | 2A | 6,900 (2,100) | Anticline | Carboniferous (sandstone) | N 53° 0′ E 50°30′ | Plus large amounts of gas, no estimate available |

Table 1. *Continued.*

| Field Name (discovery year) | Country Province | Recoverable Reserves: Equiv. Oil (mm bbl) | Recoverable Reserves: Oil (mm bbl) | Recoverable Reserves: Gas (tcf) | Petroleum Province | Classification: Bally-Type | Classification: Klemme-Type | Depth feet (meters) | Trap Type | Geologic Age (lithology) | Latitude Longitude | Remarks |
|---|---|---|---|---|---|---|---|---|---|---|---|---|
| 153. Sho-Vel-Tum (1914) | USA Oklahoma | 1,500 | 1,433 | — | Ardmore (14) | 222 | 2A | 1,500 ( 400) | Fltd. anticline | Carboniferous (sandstone) | N 34°30′ W 97°30′ | Nehring (1981) 6b-2 |
| 154. Dammam (1938) | Saudi Arabia | 1,500 | 1,500 | — | Arabian (450) | 221 | 2Ca | 4,700 ( 1,400) | Anticline | Jurassic (carbonate) | N 26°19′ E 50° 8′ | 35 API |
| 155. Mara (1945) | Venezuela | 1,500 | 1,500 | — | Maracaibo (243) | 222 | 3Bc | 5,240 ( 1,500) | | | N 10°52′ W 71°51′ | 22 API |
| 156. Bai Hassan (1953) | Iraq | 1,500 | 1,500 | — | Zagros (464) | 41 | 2Ca | 4,800 ( 1,400) | Anticline | Oligocene (carbonate) | N 35°34′ E 44° 1′ | 34 API; Production also from Cretaceous |
| 157. Karamai (1955) | China Xinjiang | 1,500 | 1,500 | — | Zhungeer (705) | 23 | 3A | 4,000 ( 1,200) | Fault blocks | Triassic (sandstone) | N 45°38′ E 85° 0′ | |
| 158. Lamar (1957) | Venezuela | 1,500 | 1,500 | — | Maracaibo (243) | 222 | 3Bc | 13,000 ( 3,900) | Fltd. anticline | Eocene (sandstone) | N 9°43′ W 71°32′ | |
| 159. Suwaidiyah (1959) | Syria | 1,500 | 1,500 | — | Zagros (464) | 41 | 2Ca | 5,600 ( 1,700) | Fltd. anticline | Cretaceous (limestone) | N 36°56′ E 42° 8′ | Alias: Souedie |
| 160. Pravdinskoye (1964) | USSR RSFSR - Tyumen | 1,500 | 1,500 | — | West Siberian (774) | 1212 | 2B | 7,200 ( 2,100) | Anticline | Cretaceous (sandstone) | N 61°36′ E 71°48′ | |
| 161. El Morgan (1965) | Egypt | 1,500 | 1,500 | — | Suez, Gulf of (358) | 111 | 3A | 5,750 ( 1,700) | Fltd. anticline | Miocene (sandstone) | N 28°14′ E 33°26′ | 30 API |
| 162. Buzurgan (1970) | Iraq | 1,500 | 1,500 | — | Zagros (464) | 41 | 2Ca | 10,000 ( 3,000) | Anticline | Miocene (limestone) | N 32°16′ E 47°14′ | 23 API; Production also from Cretaceous sandstone at 13,000 ft. |
| 163. Messla (1971) | Libya | 1,500 | 1,500 | — | Sirte (354) | 1211 | 3A | 8,800 ( 2,600) | Stratigraphic | Cretaceous (sandstone) | N 27°55′ E 22°24′ | |
| 164. Shurtanskoye (1976) | USSR Uzbekistan SSR | 1,500 | — | 9.00 | Tadzhik (761) | 222 | 2Ca | 9,500 ( 2,800) | Anticline reef | Cretaceous (carbonate) | N 38°58′ E 67° 2′ | |
| 165. Bombay High (1974) | India Gujarat | 1,463 | 1,300 | 0.79 | Cambay (507) | 1141 | 3A | 4,500 ( 1,300) | Fltd. anticline | Miocene (limestone) | N 19°34′ E 71°24′ | |
| 166. Yuzhno Cheremshanskoye (1969) | USSR RSFSR - Tomsk | 1,440 | 1,440 | — | West Siberian (774) | 1212 | 2B | 6,200 ( 1,800) | Anticline | Cretaceous (sandstone) | N 58° 0′ E 78° 0′ | Alias: Cheremshanka; Sorgut |
| 167. Sredne Vilyuy (1963) | USSR RSFSR - Yakutsk | 1,433 | — | 8.59 | Vilyuy (772) | 221 | 2A | 9,000 ( 2,700) | Anticline | Triassic (sandstone) | N 63°48′ E 122°20′ | Alias: Vilyuyskoye. Also produces from Jurassic. |
| 168. Kirpichlin (1972) | USSR Turkmenia SSR | 1,433 | — | 8.59 | Kara-Kum (735) | 222 | 2Ca | 9,900 ( 3,000) | Anticline | Cretaceous (carbonate) | N 40° 0′ E 61° 0′ | |
| 169. Sui (1952) | Pakistan | 1,416 | — | 8.50 | Indus (564) | 221 | 2Cb | 4,750 ( 1,400) | Anticline | Eocene (carbonate) | N 28°39′ E 69°12′ | |
| 170. Bahrah (1956) | Kuwait | 1,400 | 1,400 | — | Arabian (450) | 221 | 2Ca | 8,400 ( 2,500) | Anticline | Cretaceous (sandstone) | N 29°34′ E 47°55′ | |
| 171. Waha (1960) | Libya | 1,400 | 1,400 | — | Sirte ( 354) | 1211 | 3A | 6,600 ( 2,000) | Anticline | Paleocene (carbonate) | N 28°24′ E 19°55′ | 36 API; also Cretaceous reservoirs |
| 172. Rankin, North (1971) | Australia Western Australia | 1,400 | 100 | 7.79 | Dampier (533) | 1141 | 3C | 9,000 ( 2,700) | Fault blocks | Triassic (sandstone) | S 19°36′ E 116° 7′ | |
| 173. Oseberg (1979) | Norway Block 30/6 | 1,400 | 1,000 | 2.39 | North Sea, Nor (420) | 1211 | 3A | 9,000 ( 2,700) | Fault blocks | Jurassic (sandstone) | N 60°33′ E 2°47′ | |

Table 1. *Continued.*

| Field Name (discovery year) | Country Province | Recoverable Reserves Equiv. Oil (mm bbl) | Oil (mm bbl) | Gas (tcf) | Petroleum Province | Classification Bally-Type | Klemme-Type | Depth feet (meters) | Trap Type | Geologic Age (lithology) | Latitude Longitude | Remarks |
|---|---|---|---|---|---|---|---|---|---|---|---|---|
| 174. Sleipner (1974) | Norway Blocks 15/4 & 15/9 | 1,398 | 315 | 6.50 | North Sea, Nor (420) | 1211 | 3A | 11,500 ( 3,500) | Fltd. anticline | Jurassic (sandstone) | N 58°30′ E 1°43′ | |
| 175. Ventura Avenue (1916) | USA California | 1,396 | 991 | 2.43 | Ventura (164) | 222 | 2A | 7,000 ( 2,100) | Fltd. anticline | Pliocene (sandstone) | N 34°20′ W 119°20′ | Alias: Rincon Nehring (1981) 2b-1a |
| 176. Bassin, South (1977) | India Gulf of Cambay | 1,395 | — | 8.36 | Cambay (507) | 1141 | 3A | 6,000 ( 1,800) | Anticline | Miocene (limestone) | N 19°10′ E 72° 6′ | |
| 177. Zarzaitine (1958) | Algeria | 1,366 | 900 | 2.79 | Illizi (316) | 121 | 1 | 4,600 ( 1,400) | Fltd. anticline | Devonian (sandstone) | N 28° 3′ E 9°47′ | 42 API |
| 178. Badak (1972) | Indonesia Kalimantan | 1,356 | 190 | 7.00 | Mahakam (590) | 3122 | 3Ba/4 | 4,500 ( 1,300) | Anticline | Miocene (sandstone) | S 0°20′ E 117°26′ | |
| 179. Lacq (1951) | France | 1,349 | — | 8.09 | Aquitaine (377) | 221 | 3A | 13,300 ( 4,000) | Anticline | Cretaceous (carbonate) | N 43°25′ W 0°43′ | Permeability also important in trap. Frequently listed as Lacq Profond. |
| 180. Zapadno Krestishchenskoy (1968) | USSR Ukraine SSR | 1,349 | — | 8.09 | Dnepr-Donets (723) | 222 | 3A | 9,760 ( 2,900) | Anticline | Permian (sandstone) | N 49°41′ E 35° 5′ | Alias: Krestishchenskoye Also produces from Carboniferous to 12,370 ft. |
| 181. Severo Stavropol (1950) | USSR RSFSR - Stavropol | 1,333 | — | 8.00 | Amur (710) | 221 | 3Bc | 3,000 ( 900) | Anticline | Oligocene (sandstone) | N 45°13′ E 41°57′ | Alias: Pelagiada. Also produces from Eocene and Miocene |
| 182. Naipskoye (1970) | USSR Turkmenia SSR | 1,333 | — | 8.00 | Kara-Kum (735) | 222 | 2Ca | 6,500 ( 1,900) | Anticline | Cretaceous (sandstone) | N 40°57′ E 61°12′ | |
| 183. Gullfaks (1978) | Norway Block 34/10 | 1,331 | 1,225 | 0.63 | North Sea, Nor (420) | 1211 | 3A | 6,750 ( 2,000) | Fault blocks | Jurassic (sandstone) | N 61°11′ E 2°13′ | Alias: Statvik |
| 184. Novo Elkhov (1955) | USSR RSFSR - Tatar | 1,300 | 1,300 | — | Volga-Ural (773) | 221 | 2A | — ( — ) | Anticline | Devonian (sandstone) | N 54°40′ E 51°30′ | Alias: Elkhov |
| 185. Sajaa (1980) | Sharjah | 1,300 | 300 | 6.00 | Arabian (450) | 221 | 2Ca | 14,400 ( 4,300) | | Cretaceous (carbonate) | N 25°20′ E 55°40′ | 48 API |
| 186. Kingfish (1967) | Australia Victoria | 1,291 | 1,250 | — | Gippsland (550) | 111 | 3A | 7,750 ( 2,300) | Fltd. anticline | Eocene (sandstone) | S 38°36′ E 148°13′ | 47 API |
| 187. Zhetybayskoye (1961) | USSR Kazakhstan SSR | 1,283 | 1,100 | 1.09 | Mangyshlak (746) | 222 | 2Ca | 6,300 ( 1,900) | Anticline | Jurassic (sandstone) | N 43°30′ E 52° 0′ | |
| 188. Masjid-i-Suleman (1908) | Iran | 1,266 | 1,125 | 0.84 | Zagros (464) | 41 | 2Ca | 1,600 ( 400) | Anticline | Oligocene (carbonate) | N 32° 0′ E 49°18′ | 39 API; Cretaceous may have considerable additional gas |
| 189. Shkapovskoye (1944) | USSR RSFSR - Bashkir | 1,251 | 1,251 | — | Volga-Ural (773) | 221 | 2A | 6,750 ( 2,000) | Anticline | Devonian (sandstone) | N 54°32′ E 54° 0′ | |
| 190. Cerro Azul (1909) | Mexico | 1,250 | 1,250 | — | Tampico (156) | 1143 | 2Cc | 2,100 ( 600) | Atoll Reef | Cretaceous (carbonate) | N 21°20′ W 97°41′ | 22 API; Alias: Amatlan; Naranjos |
| 191. Kuparuk (1969) | USA Alaska | 1,250 | 1,250 | — | North Slope (109) | 221 | 2Cc | 6,500 ( 1,900) | Stratigraphic | Jurassic (sandstone) | N 70°10′ W 149°30′ | Nehring (1981) 1-4 |
| 192. Bul Hanine (1970) | Qatar | 1,250 | 1,250 | — | Arabian (450) | 221 | 2Ca | 3,750 ( 1,100) | Anticline | Jurassic (limestone) | N 25°26′ E 52°42′ | 35 API |
| 193. Sredneyamal'skoye (1970) | USSR RSFSR - Tyumen | 1,250 | — | 7.50 | West Siberian (774) | 1212 | 2B | 500 ( 100) | Anticline | Cretaceous (sandstone) | N 69.° 8′ E 71°10′ | Alias: Nurm |
| 194. Taking (1973) | China Heilongjiang | 1,250 | 1,250 | — | Songliao (663) | 3122 | 3A | — ( — ) | Anticline | Cretaceous (sandstone) | N 46°37′ E 125°18′ | Alias: New Taching |
| 195. Huntington Beach (1920) | USA California | 1,231 | 1,095 | 0.81 | Los Angeles (89) | 332 | 3Bb | 5,000 ( 1,500) | Fltd. anticline | Miocene (sandstone) | N 33°30′ W 117°50′ | Nehring (1981) 2c-2 |

Table 1. *Continued.*

| Field Name (discovery year) | Country Province | Recoverable Reserves | | | Petroleum Province | Classification | | Depth feet (meters) | Trap Type | Geologic Age (lithology) | Latitude Longitude | Remarks |
|---|---|---|---|---|---|---|---|---|---|---|---|---|
| | | Equiv. Oil (mm bbl) | Oil (mm bbl) | Gas (tcf) | | Bally-Type | Klemme-Type | | | | | |
| 196. Neftyanyye Kamni (1949) | USSR Azerbaijan SSR | 1,225 | 1,225 | — | Mangyshlak (746) | 222 | 2Ca | 5,400 ( 1,600) | Fltd. anticline | Pliocene (sandstone) | N 40°13′ E 50°50′ | Alias: Kamni |
| 197. Sarkhan (1969) | Iran | 1,216 | 800 | 2.50 | Zagros (464) | 41 | 2Ca | — ( — ) | Anticline | Cretaceous (carbonate) | N 33°18′ E 47°45′ | |
| 198. B Structure (1973) | Thailand | 1,216 | — | 7.29 | Thai (685) | 3122 | 3A | 2,600 ( 700) | Fltd. anticline | Miocene (sandstone) | N 8° 5′ E 102°17′ | |
| 199. Monroe (1916) | USA Louisiana | 1,204 | — | 7.22 | Louisiana Salt (90) | 1143 | 2Cc | — ( — ) | Anticline | Cretaceous (limestone) | N 32°30′ W 92° 0′ | Nehring (1981) 9b-1 |
| 200. Koakoak | Canada NWT Franklin | 1,200 | 1,200 | — | Mackenzie (91) | 1142 | 4 | 11,000 ( 3,300) | | Tertiary (sandstone) | N 70°22′ W 134° 7′ | |
| 201. Sarir L (1966) | Libya | 1,200 | 1,200 | — | Sirte (354) | 1211 | 3A | 9,000 ( 2,700) | Anticline | Cretaceous (sandstone) | N 27°48′ E 22°29′ | Alias: L - 65 |
| 202. Intisar A (1967) | Libya | 1,200 | 1,200 | — | Sirte (354) | 1211 | 3A | 9,500 ( 2,800) | Reef | Paleocene (carbonate) | N 29° 2′ E 20°46′ | 45 API; Alias: Idris |
| 203. Bu Attifel (1968) | Libya | 1,200 | 1,200 | — | Sirte (354) | 1211 | 3A | 14,300 ( 4,300) | Anticline | Cretaceous (sandstone) | N 28°51′ E 22° 7′ | 41 API; Alias: A-100 |
| 204. Frigg (1971) | Norway Block 25/1 | 1,183 | — | 7.09 | North Sea, Nor (420) | 1211 | 3A | 6,000 ( 1,800) | Stratigraphic | Paleocene (sandstone) | N 59°53′ E 2° 5′ | Alias: Frigg (United Kingdom) |
| 205. Point Thompson (1975) | USA Alaska | 1,183 | 350 | 5.00 | North Slope (109) | 221 | 2Cc | — ( — ) | | | N 70°20′ W 146° 0′ | Alias: Flaxman Island |
| 206. Sarhun | Iran | 1,166 | — | 7.00 | Zagros (464) | 41 | 2Ca | — ( — ) | Anticline | Cretaceous (carbonate) | N 27°20′ E 56°26′ | |
| 207. Shiyoukou - Tungchi (1955) | China Sichuan | 1,166 | — | 7.00 | Sichuan (658) | 22 | 2A | 5,740 ( 1,700) | Fltd. anticline | Triassic (carbonate) | N 29° 5′ E 106°42′ | Alias: Tungchi |
| 208. Ninian (1974) | United Kingdom Blocks 3/3 & 3/8 | 1,158 | 1,100 | — | North Sea, Nor (420) | 1211 | 3A | 10,430 ( 3,100) | Fault blocks | Jurassic (sandstone) | N 60°48′ E 1°29′ | 35 API |
| 209. Rhourde Nouss (1962) | Algeria | 1,152 | 27 | 6.75 | Ghadames (312) | 121 | 2B | 8,800 ( 2,600) | Anticline | Triassic (sandstone) | N 29°42′ E 6°44′ | Also produces from Cambrian-Ordovician at about 14,000 ft. NGL reserves larger than oil indicated |
| 210. Layavozhskoye (1965) | USSR RSFSR - Komi | 1,133 | 500 | 3.79 | Pechora (755) | 221 | 2B | 6,560 ( 2,000) | Anticline | Permian (carbonate) | N 67°36′ E 55° 5′ | |
| 221. Long Beach (1921) | USA California | 1,118 | 937 | 1.08 | Los Angeles (89) | 332 | 3Bb | 7,500 ( 2,200) | Fltd. anticline | Miocene (sandstone) | N 33°50′ W 118°10′ | Nehring (1981) 2c-3 |
| 212. Lenghu (1958) | China Qinghai | 1,100 | 1,100 | — | Chaidamu (521) | 23 | 3A | 2,340 ( 700) | Anticline | Miocene (sandstone) | N 38°40′ E 93°20′ | |
| 213. Maydan (1963) | Qatar | 1,100 | 1,100 | — | Arabian (450) | 221 | 2Ca | 7,100 ( 2,100) | Anticline | Jurassic (carbonate) | N 25°38′ E 52°31′ | 38 API; Alias: Mahzan |
| 214. Fateh (1966) | Dubai | 1,100 | 1,100 | — | Arabian (450) | 221 | 2Ca | 8,000 ( 2,400) | Anticline | Cretaceous (carbonate) | N 25°36′ E 54°26′ | 32 API |
| 215. Fateh, Southwest (1970) | Dubai | 1,100 | 1,100 | — | Arabian (450) | 221 | 2Ca | 8,000 ( 2,400) | Anticline | Cretaceous (carbonate) | N 25°30′ E 54°18′ | |
| 216. Raguba (1961) | Libya | 1,083 | 1,000 | — | Sirte (354) | 1211 | 3A | 5,400 ( 1,600) | Fltd. anticline | Cretaceous (carbonate) | N 29° 5′ E 19° 5′ | 43 API; also Cambro-Ordovician sandstones |
| 217. Pelyatkinskoye (1969) | USSR RSFSR-Krasnoyar | 1,083 | — | 6.50 | West Siberian (774) | 1212 | 2B | 9,680 ( 2,900) | Anticline | Jurassic (sandstone) | N 69°40′ E 81°40′ | |

Table 1. *Continued.*

| Field Name (discovery year) | Country Province | Recoverable Reserves Equiv. Oil (mm bbl) | Oil (mm bbl) | Gas (tcf) | Petroleum Province | Classification Bally-Type | Klemme-Type | Depth feet (meters) | Trap Type | Geologic Age (lithology) | Latitude Longitude | Remarks |
|---|---|---|---|---|---|---|---|---|---|---|---|---|
| 218. Palyanovo (1972) | USSR RSFSR - Tyumen | 1,083 | — | 6.50 | West Siberian (774) | 1212 | 2B | 5,000 ( 1,500) | Anticline | Cretaceous (sandstone) | N 61°45′ E 67° 1′ | |
| 219. Edda (1969) | Norway Block 2/7 | 1,068 | 68 | 6.00 | North Sea, Nor (420) | 1211 | 3A | — ( — ) | | | N 56°26′ E 3° 8′ | |
| 220. Riachuelo | Brazil Sergipe | 1,066 | — | 6.39 | Sergipe-Alagoa (275) | 1141 | 3C | — ( — ) | | | S 10°45′ W 37°15′ | |
| 221. Maastakh (1967) | USSR RSFSR - Yakutsk | 1,066 | — | 6.39 | Vilyuy (772) | 221 | 2A | 5,850 ( 1,700) | Anticline | Jurassic (sandstone) | N 63° 0′ E 123° 0′ | |
| 222. Loma de la Lata (1977) | Argentina | 1,022 | 289 | 4.39 | Neuquen (247) | 221 | 2B | 7,000 ( 2,100) | | Cretaceous (limestone) | S 38°30′ W 68°38′ | |
| 223. Katy (1934) | USA Texas (3) | 1,020 | 20 | 6.00 | Gulf Coast (66) | 1143 | 2Cc/4 | 6,775 ( 2,000) | Anticline | Eocene (sandstone) | N 29°30′ W 95°50′ | Nehring (1981) 7c-1 |
| 224. Alrar (1960) | Algeria | 1,018 | 235 | 4.69 | Illizi (316) | 121 | 1 | 7,000 ( 2,100) | Anticline | | N 28°30′ E 9°34′ | Alias: D-52 (Libya) |
| 225. Elkfisk (1970) | Norway Block 2/7 | 1,002 | 502 | 3.00 | North Sea, Nor (420) | 1211 | 3A | 9,000 ( 2,700) | Anticline | Paleocene (carbonate) | N 56°26′ E 3°12′ | |
| 226. Quiriquire (1928) | Venezuela | 1,000 | 1,000 | — | Maturin (244) | 221 | 2Ca | 1,800 ( 500) | Stratigraphic | Pliocene (sandstone) | N 10° 0′ W 63°10′ | 16 API |
| 227. Tuymazy (1937) | USSR RSFSR - Bashkir | 1,000 | 1,000 | — | Volga-Ural (773) | 221 | 2A | 4,000 ( 1,200) | Anticline | Devonian (sandstone) | N 54°32′ E 53°35′ | |
| 228. Boscan (1946) | Venezuela | 1,000 | 1,000 | — | Maracaibo (243) | 222 | 3Bc | 7,000 ( 2,100) | Anticline | Eocene (sandstone) | N 10°29′ W 71°59′ | |
| 229. Nahr Umr (1948) | Iraq | 1,000 | 1,000 | — | Arabian (450) | 221 | 2Ca | 9,000 ( 2,700) | Anticline | Cretaceous (sandstone) | N 30°46′ E 47°42′ | |
| 230. Fadhilli (1949) | Saudi Arabia | 1,000 | 1,000 | — | Arabian (450) | 221 | 2Ca | 8,050 ( 2,400) | Anticline | Jurassic (carbonate) | N 26°57′ E 49°13′ | 39 API |
| 231. Jambur (1954) | Iraq | 1,000 | 1,000 | — | Zagros (464) | 41 | 2Ca | 8,000 ( 2,400) | Anticline | Miocene (carbonate) | N 35° 0′ E 44°15′ | |
| 232. Urdaneta (1955) | Venezuela | 1,000 | 1,000 | — | Maracaibo (243) | 222 | 3Bc | 7,000 ( 2,100) | Fault blocks | Miocene (sandstone) | N 10° 6′ W 71°50′ | |
| 233. Centro (1957) | Venezuela | 1,000 | 1,000 | — | Maracaibo (243) | 222 | 3Bc | 10,000 ( 3,000) | Fltd. anticline | Eocene (sandstone) | N 9°50′ W 71°27′ | 36 API |
| 234. Idd-el-Shargi (1960) | Qatar | 1,000 | 1,000 | — | Arabian (450) | 221 | 2Ca | 8,000 ( 2,400) | Anticline | Jurassic (carbonate) | N 25°38′ E 52°27′ | 35 API |
| 235. Khangiran (1968) | Iran | 1,000 | — | 6.00 | Turkmen (768) | 222 | 2Ca | 7,500 ( 2,200) | Anticline | Cretaceous (carbonate) | N 36°34′ E 60°46′ | |
| 236. Vozey (1972) | USSR RSFSR - Komi | 1,000 | 1,000 | — | Pechora (755) | 221 | 2B | — ( — ) | | | N 66°40′ E 57°20′ | |
| 237. Drake Point (1973) | Canada NWT Franklin | 1,000 | — | 6.00 | Sverdrup (151) | 1143 | 2Cc | — ( — ) | | Jurassic (sandstone) | N 76°25′ W 107°50′ | |
| 238. Hamrin (1973) | Iraq | 1,000 | 1,000 | — | Zagros (464) | 41 | 2Ca | 8,000 ( 2,400) | Anticline | Miocene (carbonate) | N 34°30′ E 44° 0′ | |
| 239. Jabal Fauqui (1974) | Iraq | 1,000 | 1,000 | — | Zagros (464) | 41 | 2Ca | 10,000 ( 3,000) | Anticline | Miocene (carbonate) | N 32° 0′ E 47°39′ | Alias: Fakkeh |
| 240. Isis (1976) | Tunisia | 1,000 | 1,000 | — | Pelagian (347) | 1141 | 2Ca | — ( — ) | | | N 33° 0′ E 11°30′ | Alias: Isis (Libya) |

Table 1. *Continued.*

| Field Name (discovery year) | Country Province | Recoverable Reserves Equiv. Oil (mm bbl) | Oil (mm bbl) | Gas (tcf) | Petroleum Province | Classification Bally-Type | Klemme-Type | Depth feet (meters) | Trap Type | Geologic Age (lithology) | Latitude Longitude | Remarks |
|---|---|---|---|---|---|---|---|---|---|---|---|---|
| 241. Ratga (1977) | Kuwait | 1,000 | 1,000 | — | Arabian (450) | 221 | 2Ca | — ( — ) | Anticline | Miocene (sandstone) | N 30° 2′ E 47° 3′ | |
| 242. Kopanoar (1979) | Canada NWT Franklin | 1,000 | 500 | 3.00 | Mackenzie (91) | 1142 | 4 | 10,500 ( 3,200) | | Tertiary (sandstone) | N 70°23′ W 135° 6′ | 28 API |
| 243. Bu Tini (1980) | Abu Dhabi | 1,000 | 1,000 | — | Arabian (450) | 221 | 2Ca | 11,000 ( 3,300) | Anticline | Cretaceous (carbonate) | N 24°34′ E 52°59′ | |
| 244. Espoir (1980) | Ivory Coast | 1,000 | 1,000 | — | Ivory Coast (317) | 1142 | 3C | 6,800 ( 2,000) | | Cretaceous (sandstone) | N 5° 0′ W 4°28′ | 33 API; Alias: Jacqueville |
| 245. Cisco (1981) | Canada NWT Franklin | 1,000 | 1,000 | — | Sverdrup (151) | 1143 | 2Cc | 6,500 ( 1,900) | | Jurassic (sandstone) | N 77°25′ W 106°26′ | |
| 246. Whitney Canyon (1977) | USA Wyoming | 998 | 115 | 5.29 | Western Overth (169) | 41 | 2A | 9,000 ( 2,700) | Thrust Fold | Carboniferous (limestone) | N 41°30′ W 110°40′ | Alias: Carter Creek |
| 247. Caillou Island (1930) | USA Louisiana | 990 | 657 | 2.00 | Gulf Coast (66) | 1143 | 2Cc/4 | 7,500 ( 2,200) | Salt Dome | Miocene (sandstone) | N 29° 5′ W 90°30′ | Note: Bay Marchand and Timbalier Bay are listed separately. Gas production to 1976 // Nehring (1981) 8b-1 |
| 248. Sangachaly Duvannyy (1963) | USSR Azerbaijan SSR | 976 | 860 | 0.69 | Kura (740) | 222 | 3Bc | 11,880 ( 3,600) | Anticline | Pliocene (sandstone) | N 40° 1′ E 49°32′ | Alias: Baku Archipelago; Duvannyy |
| 249. Panuco (1901) | Mexico | 971 | 971 | — | Tampico (156) | 1143 | 2Cc | 1,450 ( 400) | Fltd. anticline | Cretaceous (carbonate) | N 22°10′ W 98° 0′ | Alias: Ebano |
| 250. Old Ocean (1934) | USA Texas (3) | 962 | 129 | 5.00 | Gulf Coast (66) | 1143 | 2Cc/4 | 10,300 ( 3,100) | Anticline | Oligocene (sandstone) | N 28°55′ W 95°45′ | Nehring (1981) 7c-4 |
| 251. Oficina (1937) | Venezuela | 960 | 960 | — | Maturin (244) | 221 | 2Ca | 5,200 ( 1,500) | Fault blocks | Oligocene (sandstone) | N 8°52′ W 64°18′ | 22 API; Alias: Fria; Guico |
| 252. Basin (1947) | USA New Mexico | 960 | — | 5.75 | San Juan (139) | 222 | 2B | 6,500 ( 1,900) | Facies Change | Cretaceous (sandstone) | N 37° 0′ W 107°40′ | Alias: Dakota; San Juan. Nehring (1981) 3e-2 |
| 253. Greta (1934) | USA Texas (2) | 958 | 958 | — | Gulf Coast (66) | 1143 | 2Cc/4 | 5,500 ( 1,600) | Rollover antcl | Oligocene (sandstone) | N 28°15′ W 97°10′ | Alias: Tom O'Connor Nehring (1981) 7b-1 |
| 254. Maui (1969) | New Zealand | 958 | 75 | 5.29 | Taranaki (679) | 1141 | 3Bb/3C | 9,500 ( 2,800) | Fltd. anticline | Eocene (sandstone) | S 39°40′ E 173°19′ | Number on Halbouty list actually 27a |
| 255. Kaybob South (1958) | Canada Alberta | 956 | 540 | 2.50 | Alberta (4) | 221 | 2A | — ( — ) | Reef | Devonian (carbonate) | N 54°20′ W 115°50′ | |
| 256. Gassi Touil (1961) | Algeria | 950 | 500 | 2.69 | Ghadames (312) | 121 | 2B | 6,500 ( 1,900) | Anticline | Triassic | N 30°23′ E 6°31′ | |
| 257. Kettleman Hills North Do (1928) | USA California | 948 | 456 | 2.95 | Sacramento San (134) | 332 | 3Bb | 7,200 ( 2,100) | Anticline | Miocene (sandstone) | N 36° 5′ W 120°15′ | Full name: Kettleman Hills North Dome; Nehring (1981) 2a-4 |
| 258. Achakskoye (1966) | USSR Turkmenia SSR | 933 | — | 5.59 | Kara-Kum (735) | 222 | 2Ca | 5,560 ( 1,600) | Anticline | Cretaceous (sandstone) | N 41° 6′ E 60°58′ | |
| 259. Kandymskoye (1967) | USSR Uzbekistan SSR | 933 | — | 5.59 | Kara-Kum (735) | 222 | 2Ca | 8,200 ( 2,500) | Anticline | Jurassic (carbonate) | N 39°27′ E 63°35′ | Alias: Khadzhiy |
| 260. Swan Hills (1957) | Canada Alberta | 931 | 931 | — | Alberta (4) | 221 | 2A | 8,100 ( 2,400) | Reef | Devonian (carbonate) | N 54°45′ W 115°40′ | 40 API |
| 261. Maleh Kuh (1968) | Iran | 923 | 210 | 4.27 | Zagros (464) | 41 | 2Ca | — ( — ) | Anticline | Cretaceous (carbonate) | N 33°13′ E 47°36′ | Gas field |

Table 1. *Continued.*

| Field Name (discovery year) | Country Province | Recoverable Reserves: Equiv. Oil (mm bbl) | Recoverable Reserves: Oil (mm bbl) | Recoverable Reserves: Gas (tcf) | Petroleum Province | Classification: Bally-Type | Classification: Klemme-Type | Depth feet (meters) | Trap Type | Geologic Age (lithology) | Latitude Longitude | Remarks |
|---|---|---|---|---|---|---|---|---|---|---|---|---|
| 262. Sidi Abderrahman | Tunisia | 916 | — | 5.50 | Pelagian (347) | 1141 | 2Ca | — ( — ) | | | N 37°30′ E 10°30′ | Alias: Cap Bon |
| 263. Kharg (1963) | Iran | 916 | 250 | 4.00 | Zagros (464) | 41 | 2Ca | 10,000 ( 3,000) | Anticline | Cretaceous (carbonate) | N 29°24′ E 50°18′ | |
| 264. Lyantor (1966) | USSR RSFSR - Tyumen | 916 | — | 5.50 | West Siberian (774) | 1212 | 2B | 6,700 ( 2,000) | Anticline | Cretaceous (sandstone) | N 63°30′ E 72° 0′ | |
| 265. Surakhanskoye (1870) | USSR Azerbaijan SSR | 900 | 900 | — | Kura (740) | 222 | 3Bc | 800 ( 200) | Fltd. anticline | Pliocene (sandstone) | N 40°26′ E 50° 2′ | |
| 266. La Paz (1925) | Venezuela | 900 | 900 | — | Maracaibo (243) | 222 | 3Bc | 4,250 ( 1,200) | Fltd. anticline | Cretaceous (carbonate) | N 10°40′ W 72° 0′ | |
| 267. South Sand Belt (1926) | USA Texas (8) | 900 | 900 | — | Permian (120) | 222 | 2A | 3,000 ( 900) | Anticline | Permian (dolomite) | N 31°20′ W 102°45′ | Alias: Sand Belt; Ward Estes. Field extends into New Mexico Nehring (1981) 4c-2 |
| 268. Fahud (1964) | Oman | 900 | 900 | — | Arabian (450) | 221 | 2Ca | 2,000 ( 600) | Fltd. anticline | Cretaceous (carbonate) | N 22°20′ E 56°30′ | 33 API |
| 269. Unity (1980) | Sudan | 900 | 900 | — | Sudan (357) | 111 | 3A | 6,000 ( 1,800) | Stratigraphic | Cretaceous (sandstone) | N 9°28′ E 29°41′ | |
| 270. Salt Creek (1906) | USA Wyoming | 891 | 766 | 0.75 | Powder River (124) | 222 | 2A | 1,000 ( 300) | Anticline | Cretaceous (sandstone) | N 43°30′ W 106°15′ | Nehring (1981) 3h-1 |
| 271. Piper (1973) | United Kingdom Block 15/17 | 887 | 837 | — | North Sea, Nor (420) | 1211 | 3A | 8,200 ( 2,500) | Fault blocks | Jurassic (sandstone) | N 58°28′ E 0°16′ | 37 API |
| 272. Megionskoye (1961) | USSR RSFSR - Tyumen | 885 | 885 | — | West Siberian (774) | 1212 | 2B | 6,900 ( 2,100) | Anticline | Cretaceous (sandstone) | N 60°50′ E 76°26′ | Also Jurassic to 7150 ft. |
| 273. Antipayutin (1978) | USSR RSFSR - Tyumen | 883 | — | 5.29 | West Siberian (774) | 1212 | 2B | 3,300 ( 1,000) | Anticline | Cretaceous (sandstone) | N 69°20′ E 75° 5′ | |
| 274. Naft-i-Safid (1935) | Iran | 878 | 400 | 2.86 | Zagros (464) | 41 | 2Ca | 5,000 ( 1,500) | Anticline | Oligocene (carbonate) | N 31°39′ E 49°15′ | 37 API; Production also from Cretaceous |
| 275. Novoportovskoye (1964) | USSR RSFSR - Tyumen | 876 | 260 | 3.69 | West Siberian (774) | 1212 | 2B | 3,650 ( 1,100) | Anticline | Cretaceous (sandstone) | N 72°23′ E 68° 0′ | Alias: Novyy Port. Also produces from Jurassic to 6,700 ft. |
| 276. Coalinga (1887) | USA California | 867 | 830 | — | Sacramento San (134) | 332 | 3Bb | 2,200 ( 600) | Anticline | Miocene (sandstone) | N 36°10′ W 120°30′ | Nehring (1981) 2a-6 |
| 277. Goldsmith (1934) | USA Texas (8) | 866 | 866 | — | Permian (120) | 222 | 2A | 5,500 ( 1,600) | Anticline | Permian (dolomite) | N 31°45′ W 102°30′ | Alias: Andector Nehring (1981) 4c-3 |
| 278. Hawkins (1940) | USA Texas (6) | 850 | 850 | — | East Texas Sal (51) | 1143 | 2Cc | 4,800 ( 1,400) | Fltd. anticline | Cretaceous (sandstone) | N 32°45′ W 95° 0′ | Nehring (1981) 9d-3 |
| 279. Malongo North & South (1966) | Angola Cabinda | 850 | 850 | — | Cabinda (296) | 1141 | 3C | 8,500 ( 2,500) | Anticline | Cretaceous (sandstone) | S 5°21′ E 12° 5′ | 30 API; Alias: Cabinda B |
| 280. Anschutz Ranch East (1978) | USA Utah | 846 | 180 | 4.00 | Western Overth (169) | 41 | 2A | 6,800 ( 2,000) | Thrust Fold | Jurassic (limestone) | N 41°10′ W 111°15′ | |
| 281. Huangkuanshan (1955) | China Sichuan | 833 | — | 5.00 | Sichuan (658) | 22 | 2A | 3,000 ( 900) | Anticline | Triassic (carbonate) | N 29°12′ E 105°52′ | |
| 282. Sarajeh (1958) | Iran | 833 | — | 5.00 | Zagros (464) | 41 | 2Ca | 5,960 ( 1,800) | Anticline | Oligocene (carbonate) | N 35° 0′ E 51°30′ | |
| 283. Tazovskoye (1962) | USSR RSFSR - Tyumen | 833 | 300 | 3.19 | West Siberian (774) | 1212 | 2B | 3,800 ( 1,100) | Anticline | Cretaceous (sandstone) | N 67°20′ E 78°54′ | Alias: Taz |

Table 1. *Continued.*

| Field Name (discovery year) | Country Province | Recoverable Reserves | | | Petroleum Province | Classification | | Depth feet (meters) | Trap Type | Geologic Age (lithology) | Latitude Longitude | Remarks |
|---|---|---|---|---|---|---|---|---|---|---|---|---|
| | | Equiv. Oil (mm bbl) | Oil (mm bbl) | Gas (tcf) | | Bally-Type | Klemme-Type | | | | | |
| 284. Bubullime (1963) | Albania | 833 | — | 5.00 | Ionian (406) | 41 | 2Cb | — ( — ) | Anticline | | N 40°48′ E 19°42′ | |
| 285. Gidgealpa (1964) | Australia South Australia | 833 | — | 5.00 | Cooper (529) | 1211 | 3A/2B | 6,820 ( 2,000) | Anticline | Permian (sandstone) | S 27°20′ E 139°30′ | |
| 286. Moomba (1964) | Australia Queensland | 833 | — | 5.00 | Cooper (529) | 1211 | 3A/2B | 7,850 ( 2,300) | Anticline | Permian (sandstone) | S 27°20′ E 141°30′ | |
| 287. Barqan (1969) | Saudi Arabia | 833 | 250 | 3.50 | Red Sea, East (461) | 111 | 3A/3C | 6,400 ( 1,900) | Anticline | Miocene (sandstone) | N 27°50′ E 35°15′ | Also no. 182 on Halbouty giant oil field list. |
| 288. Severo Komsomol (1969) | USSR RSFSR - Tyumen | 833 | — | 5.00 | West Siberian (774) | 1212 | 2B | 3,500 ( 1,000) | Anticline | Cretaceous (sandstone) | N 64°40′ E 75°30′ | See also separate field Komso-molskoye |
| 289. Morecambe (1977) | United Kingdom Block 110/2 | 833 | — | 5.00 | Irish (407) | 0 | 2B | 3,100 ( 900) | | | N 53°52′ E 3°38′ | |
| 290. Claresholm (1978) | Canada Alberta | 833 | — | 5.00 | Alberta (4) | 221 | 2A | — ( — ) | Stratigraphic | Cretaceous (sandstone) | N 49°55′ W 113°30′ | |
| 291. Whitefish (1979) | Canada NWT Franklin | 833 | — | 5.00 | Sverdrup (151) | 1143 | 2Cc | 5,500 ( 1,600) | | Jurassic (sandstone) | N 77°12′ W 106°53′ | Pays also at 2900 and 6900 ft |
| 292. Meskala (1980) | Morocco | 833 | — | 5.00 | Essaouira (306) | 1141 | 3C | 11,000 ( 3,300) | | Triassic (sandstone) | N 31°27′ W 9°21′ | Tight |
| 293. Buena Vista (1909) | USA California | 831 | 648 | 1.09 | Sacramento San (134) | 332 | 3Bb | 5,000 ( 1,500) | Anticline | Pliocene (sandstone) | N 35° 5′ W 119°30′ | Also called Buena Vista Hills Nehring (1981) 2a-5 |
| 294. Algyo (1965) | Hungury | 816 | 200 | 3.69 | Pannonian (422) | 321 | 3Bc | 7,500 ( 2,200) | Fltd. anticline | Pliocene (sandstone) | N 46°20′ E 20°15′ | |
| 295. Rangely (1902) | USA Colorado | 804 | 755 | — | Uinta (160) | 222 | 2A | 6,000 ( 1,800) | Anticline | Carboniferous (sandstone) | N 40°10′ W 108°45′ | Nehring (1981) 3b-2 |
| 296. Redwater (1948) | Canada Alberta | 805 | 805 | — | Alberta (4) | 221 | 2A | 3,300 ( 1,000) | Reef Complex | Devonian (carbonate) | N 53°50′ W 113°10′ | 35 API |
| 297. Jones Creek (1967) | Nigeria | 800 | 800 | — | Niger Delta (340) | 1141 | 4 | 6,000 ( 1,800) | Rollover antcl | Miocene (sandstone) | N 5°41′ E 5°23′ | 29 API |
| 298. Molango West (1969) | Angola Cabinda | 800 | 800 | — | Cabinda (296) | 1141 | 3C | 5,000 ( 1,500) | Anticline | Miocene (sandstone) | S 5°25′ E 12° 3′ | 29 API |
| 299. Shushufindi (1969) | Ecuador | 800 | 800 | — | Putumayo (265) | 221 | 2A | 7,500 ( 2,200) | Anticline | Cretaceous (sandstone) | S 0°10′ W 76°38′ | |
| 300. Bolshoye Chernogor (1970) | USSR RSFSR - Tyumen | 800 | 800 | — | West Siberian (774) | 1212 | 2B | 7,500 ( 2,200) | Anticline | Cretaceous (sandstone) | N 71°40′ E 77° 0′ | Alias: Chernogor |
| 301. Pokachev (1970) | USSR RSFSR - Tyumen | 800 | 800 | — | West Siberian (774) | 1212 | 2B | — ( — ) | | | N 61° 0′ E 75°20′ | |
| 302. Vostochno Tarkosalin (1971) | USSR RSFSR - Tyumen | 799 | — | 4.79 | West Siberian (774) | 1212 | 2B | 4,000 ( 1,200) | Anticline | Cretaceous (sandstone) | N 65° 8′ E 78°23′ | Alias: Tarkosalinskoye |
| 303. Handil (1974) | Indonesia Kalimantan | 800 | 800 | — | Mahakam (590) | 3122 | 3B | 6,500 ( 1,900) | Fltd. anticline | Miocene (sandstone) | S 0°51′ E 117°17′ | |
| 304. Ixtoc (1979) | Mexico | 800 | 800 | — | Campeche (32) | 114 | 2Cc/3C | 10,100 ( 3,000) | Fltd. salt dome | Paleocene (dolomite) | N 19°39′ W 92°20′ | Alias: Abkatun; Kanaab; Taratunich |
| 305. El Borma (1964) | Tunisia | 773 | 640 | 0.79 | Ghadames (312) | 121 | 2B | 7,800 ( 2,300) | Fltd. anticline | Triassic (sandstone) | N 31°41′ E 9°12′ | 42 API; Alias: El Borma (Algeria) |
| 306. Bayou Sale (1937) | USA Louisiana | 772 | 189 | 3.50 | Gulf Coast (66) | 1143 | 2Cc/4 | 9,500 ( 2,800) | Salt ridge | Miocene (sandstone) | N 29°30° W 91°30′ | Nehring (1981) 8c-2 |

Table 1. *Continued.*

| Field Name (discovery year) | Country Province | Recoverable Reserves | | | Petroleum Province | Classification | | Depth feet (meters) | Trap Type | Geologic Age (lithology) | Latitude Longitude | Remarks |
|---|---|---|---|---|---|---|---|---|---|---|---|---|
| | | Equiv. Oil (mm bbl) | Oil (mm bbl) | Gas (tcf) | | Bally-Type | Klemme-Type | | | | | |
| 307. Yefremovskoye (1965) | USSR Ukraine SSR | 766 | — | 4.59 | Dnepr-Donets (723) | 222 | 3A | 6,000 ( 1,800) | Salt anticline | Permian (sandstone) | N 49°28′ E 36°15′ | |
| 308. Sitio Grande (1972) | Mexico | 765 | 765 | — | Salinas (Mex) (136) | 22 | 2Cc | 15,000 ( 4,500) | Fault blocks | Cretaceous (dolomite) | N 17°47′ W 93° 7′ | 35 API |
| 309. Satah (1975) | Abu Dhabi | 762 | 200 | 3.37 | Arabian (450) | 221 | 2Ca | 12,000 ( 3,600) | Anticline | Jurassic (carbonate) | N 24°55′ E 52°33′ | |
| 310. Santa Fe Springs (1919) | USA California | 762 | 622 | 0.84 | Los Angeles (89) | 332 | 3Bb | 1,145 ( 300) | Fltd. anticline | Pliocene (sandstone) | N 33°55′ W 118° 0′ | Production also from Miocene, to depths of 9000 ft. Nehring (1981) 2c-4 |
| 311. Hastings (1934) | USA Texas (3) | 760 | 760 | — | Gulf Coast (66) | 1143 | 2Cc/4 | 6,200 ( 1,800) | Fltd. anticline | Oligocene (sandstone) | N 29°35′ W 95°10′ | Nehring (1981) 7c-3 |
| 312. Kuleshovskoye (1958) | USSR RSFSR - Kuybyshev | 759 | 759 | — | Volga-Ural (773) | 221 | 2A | 5,500 ( 1,600) | Anticline | Carboniferous (sandstone) | N 52°40′ E 50°30′ | Carbonates reservoirs also |
| 313. Ekofisk West (1969) | Norway Block 2/4 | 756 | 390 | 2.19 | North Sea, Nor (420) | 1211 | 3A | — ( — ) | | | N 56°33′ E 3° 5′ | |
| 314. Oklahoma City (1928) | USA Oklahoma | 755 | 755 | — | Anadarko (9) | 221 | 2A | 5,000 ( 1,500) | Fltd. anticline | Ordovician (sandstone) | N 35°30′ W 97°30′ | Nehring (1981) 6b-3 |
| 315. Mansuri (1963) | Iran | 755 | 655 | 0.59 | Zagros (464) | 41 | 2Ca | 7,100 ( 2,100) | Anticline | Oligocene (carbonate) | N 30°53′ E 48°50′ | 28 API; Production also from Cretaceous |
| 316. McElroy (1926) | USA Texas (8) | 750 | 750 | — | Permian (120) | 222 | 2A | 2,900 ( 800) | Anticline | Permian (dolomite) | N 31°20′ W 102°15′ | Alias: Dune Nehring (1981) 4c-5 |
| 317. July (1967) | Egypt | 750 | 750 | — | Suez, Gulf of (358) | 111 | 3A | 9,000 ( 2,700) | Fault blocks | Miocene (sandstone) | N 28°15′ E 33°14′ | 32 API |
| 318. Belridge South (1911) | USA California | 741 | 730 | — | Sacramento San (134) | 332 | 3Bb | 1,000 ( 300) | Anticline | Miocene (sandstone) | N 35°30′ W 119°30′ | Nehring (1981) 2a-9 |
| 319. Conroe (1931) | USA Texas (3) | 741 | 741 | — | Gulf Coast (66) | 1143 | 2Cc/4 | 5,200 ( 1,500) | Fltd. anticline | Eocene (sandstone) | N 30°20′ W 95°40′ | Nehring (1981) 7c-2 |
| 320. Bay Marchand (1949) | USA Louisiana | 740 | 615 | 0.75 | Gulf Coast (66) | 1143 | 2Cc/4 | 5,000 ( 1,500) | Salt ridge | Neogene (sandstone) | N 29° 5′ W 90°15′ | This is Bay Marchand Block 2. Note: Caillou Island and Timbalier Bay are listed separately.// Nehring (1981) 8a-1 |
| 321. Indefatigable (1966) | United Kingdom Block 49/18 | 733 | 6 | 4.35 | North Sea, Sou (421) | 1211 | 2B/3A | 8,500 ( 2,500) | Fltd. anticline | Permian (sandstone) | N 53°24′ E 2°31′ | |
| 322. Zhongyuan (1975) | China Henan | 733 | 733 | — | Huabei (560) | 3122 | 3A | 9,300 ( 2,800) | | Paleogene (sandstone) | N 35°15′ E 115°30′ | 37 API |
| 323. Guaricema | Brazil Sergipe | 716 | — | 4.29 | Sergipe-Alagoa (275) | 1141 | 3C | — ( — ) | | | S 11°10′ W 36°45′ | |
| 325. Anastasiyevsko (1953) | USSR RSFSR- Krasnodar | 713 | 247 | 2.79 | Caucasus, Nort (721) | 222 | 2Ca | 7,000 ( 2,100) | Anticline | Miocene (sandstone) | N 45° 5′ E 38° 5′ | Alias: Troiskoye |
| 325. Marlin (1966) | Australia Victoria | 710 | 244 | 2.79 | Gippsland (550) | 111 | 3A | 7,050 ( 2,100) | Anticline | Eocene (sandstone) | S 38°14′ E 148°14′ | Also produces from Paleocene at about 10,000 ft. |
| 326. Mene Grande (1914) | Venezuela | 700 | 700 | — | Maracaibo (243) | 222 | 3Bc | 4,130 ( 1,200) | Anticline | Eocene (sandstone) | N 9°47′ W 70°55′ | |
| 327. Meren (1965) | Nigeria | 700 | 600 | 0.59 | Niger Delta (340) | 1141 | 4 | 5,750 ( 1,700) | Rollover antcl | Miocene (sandstone) | N 5°47′ E 4°53′ | 31 API |

Table 1. *Continued.*

| Field Name (discovery year) | Country Province | Recoverable Reserves Equiv. Oil (mm bbl) | Oil (mm bbl) | Gas (tcf) | Petroleum Province | Classification Bally-Type | Klemme-Type | Depth feet (meters) | Trap Type | Geologic Age (lithology) | Latitude Longitude | Remarks |
|---|---|---|---|---|---|---|---|---|---|---|---|---|
| 328. Halibut (1967) | Australia Victoria | 700 | 700 | — | Gippsland (550) | 111 | 3A | 7,400 ( 2,200) | Anticline | Eocene (sandstone) | S 38°24′ E 148°19′ | 43 API |
| 329. Forcados Yokri (1968) | Nigeria | 700 | 700 | — | Niger Delta (340) | 1141 | 4 | 10,850 ( 3,300) | Rollover antcl | Tertiary (sandstone) | N 5°22′ E 5°18′ | 31 API; Alias: Yokri |
| 330. Gudao (1968) | China Shandong | 700 | 700 | — | Huabei (560) | 3122 | 3A | 3,900 ( 1,100) | Fltd. anticline | Miocene (sandstone) | N 37°52′ E 118°45′ | Drape over horst block |
| 331. Ardjuna B (1969) | Indonesia Java Sea | 700 | 600 | 0.59 | Java, West (570) | 3122 | 3Ba | 2,800 ( 800) | Fltd. anticline | Miocene (sandstone) | S 5°54′ E 107°42′ | 37 API |
| 332. Agave (1976) | Mexico | 700 | 700 | — | Salinas (Mex) (136) | 22 | 2Cc | 15,000 ( 4,500) | Fltd. anticline | Cretaceous (carbonate) | N 17°45′ W 92°53′ | Gas and condensate are probably more important, no estimate |
| 333. Qusahwira (1976) | Abu Dhabi | 700 | 700 | — | Arabian (450) | 221 | 2Ca | 7,000 ( 2,100) | Anticline | Cretaceous (carbonate) | N 22°40′ E 55° 0′ | |
| 334. Block 16/26 (1977) | United Kingdom Block 16/26 | 700 | 200 | 3.00 | North Sea, Nor (420) | 1211 | 3A | 12,000 ( 3,600) | Fault blocks | Cretaceous (sandstone) | N 58° 5′ E 1° 6′ | |
| 335. Bouri (1977) | Libya | 700 | 700 | — | Pelagian (347) | 1141 | 2Ca | 7,500 ( 2,200) | Anticline | Eocene (carbonate) | N 33°54′ E 21°30′ | Alias: B-NC-41 |
| 336. Gydan (1978) | USSR RSFSR - Tyumen | 700 | — | 4.19 | West Siberian (774) | 1212 | 2B | 5,000 ( 1,500) | Anticline | Cretaceous (sandstone) | N 70°40′ E 77°25′ | |
| 337. Myl'dzhino (1964) | USSR RSFSR - Tomsk | 695 | 112 | 3.50 | West Siberian (774) | 1212 | 2B | 8,000 ( 2,400) | Anticline | Jurassic (sandstone) | N 59°25′ E 78°10′ | Also productive from Cretaceous; stratigraphic trapping also important |
| 338. Colomo (1951) | Mexico | 675 | 50 | 3.75 | Salinas (Mex) (136) | 22 | 2Cc | 5,576 ( 1,700) | Fltd. salt dome | Miocene (sandstone) | N 17°58′ W 92°28′ | 48 API; Alias: Chilapilla; Jose Colomo |
| 339. Mazalij (1971) | Saudi Arabia | 675 | 675 | — | Arabian (450) | 221 | 2Ca | — ( — ) | Anticline | Jurassic (carbonate) | N 24°21′ E 48° 8′ | |
| 340. McArthur River (1965) | USA Alaska | 670 | 550 | 0.72 | Cook Inlet (41) | 311 | 3Bb | 5,400 ( 1,600) | Anticline | Miocene (sandstone) | N 60°45′ W 152° 0′ | Nehring (1981) 1-2 |
| 341. Yibal (1963) | Oman | 670 | 670 | — | Arabian (450) | 221 | 2Ca | 8,465 ( 2,500) | Fltd. anticline | Permian (limestone) | N 21°54′ E 56°13′ | 38 API; Alias: Shuaiba |
| 342. Nakhodkinskoye | USSR RSFSR - Tyumen | 666 | — | 4.00 | West Siberian (774) | 1212 | 2B | — ( — ) | | | N 67°30′ E 77°30′ | |
| 343. Peschanyy More (1952) | USSR Azerbaijan SSR | 666 | 600 | — | Kura (740) | 222 | 3Bc | 10,500 ( 3,200) | Shale diapir | Pliocene (sandstone) | N 40°18′ E 50° 0′ | |
| 344. Opon (1956) | Colombia Santander | 666 | — | 4.00 | Magdalena, Mid (240) | 332 | 3Bc | 8,700 ( 2,600) | Fltd. anticline | | N 6°25′ W 73°52′ | |
| 345. Mari (1957) | Pakistan | 666 | — | 4.00 | Indus (564) | 221 | 2Cb | 2,260 ( 600) | Anticline | Eocene (carbonate) | N 27°56′ E 69°45′ | |
| 346. Gugurtlinskoye (1965) | USSR Turkmenia SSR | 666 | — | 4.00 | Kara-Kum (735) | 222 | 2Ca | 3,600 ( 1,000) | Anticline | Cretaceous (limestone) | N 40°13′ E 62°33′ | Also productive from Jurassic sandstones at 6900 ft. |
| 347. Petrel (1969) | Australia Northern Territory | 666 | — | 4.00 | Bonaparte Gulf (497) | 1141 | 3C/2B | 12,000 ( 3,600) | Anticline | Permian (sandstone) | S 12°50′ E 128°28′ | |
| 348. Hecla (1975) | Canada NWT Franklin | 666 | — | 4.00 | Sverdrup (151) | 1143 | 2Cc | — ( — ) | | Jurassic (sandstone) | N 76°21′ W 110°15′ | |
| 349. Neytin (1975) | USSR RSFSR - Tyumen | 666 | — | 4.00 | West Siberian (774) | 1212 | 2B | 2,000 ( 600) | Anticline | Cretaceous (sandstone) | N 69°50′ E 69°45′ | |

Table 1. *Continued.*

| Field Name (discovery year) | Country Province | Recoverable Reserves Equiv. Oil (mm bbl) | Oil (mm bbl) | Gas (tcf) | Petroleum Province | Classification Bally-Type | Klemme-Type | Depth feet (meters) | Trap Type | Geologic Age (lithology) | Latitude Longitude | Remarks |
|---|---|---|---|---|---|---|---|---|---|---|---|---|
| 350. Kharvutin (1976) | USSR RSFSR - Tyumen | 666 | — | 4.00 | West Siberian (774) | 1212 | 2B | 4,000 ( 1,200) | Anticline | Cretaceous (sandstone) | N 67°28′ E 72°43′ | |
| 351. Bradford (1871) | USA Pennsylvania | 658 | 658 | — | Appalachian (12) | 221 | 2A | 2,000 ( 600) | Anticline | Devonian (sandstone) | N 42° 0′ W 78°30′ | Field extends into New York State |
| 352. Kazanskoye (1967) | USSR RSFSR - Tomsk | 658 | 75 | 3.50 | West Siberian (774) | 1212 | 2B | 8,350 ( 2,500) | Anticline | Jurassic (sandstone) | N 57°15′ E 79°30′ | |
| 353. Mocane Laverne (1952) | USA Oklahoma | 653 | 20 | 3.79 | Anadarko (9) | 221 | 2A | 4,250 ( 1,200) | Stratigraphic | Carboniferous (sandstone) | N 35°20′ W 98°10′ | Alias: Laverne Nehring (1981) 6b-4 |
| 354. Starogroznenskoye (1893) | USSR RSFSR - Chechen-In | 650 | 650 | — | Kura (740) | 222 | 3Bc | 13,500 ( 4,100) | Anticline | Cretaceous (sandstone) | N 43°20′ E 45°40′ | Reservoirs in Miocene also |
| 355. Korobkovskoye (1949) | USSR RSFSR - Volgograd | 650 | 150 | 3.00 | Volga-Ural (773) | 221 | 2A | 3,000 ( 900) | Anticline | Carboniferous (sandstone) | N 50°16′ E 45°23′ | Also produces from carbonates. Total productive zone runs from 1600 to 13500 ft; U. Carboniferous to Devonian. |
| 356. Usanovskoye (1963) | USSR RSFSR - Komi | 650 | 650 | — | Pechora (755) | 221 | 2B | 6,500 ( 1,900) | Anticline | Carboniferous (carbonate) | N 66°10′ E 57°40′ | Alias: Usa |
| 357. Zevardin (1968) | USSR Uzbekistan SSR | 650 | — | 3.89 | Turkmen (768) | 222 | 2Ca | 8,700 ( 2,600) | Reef | Jurassic (carbonate) | N 38°51′ E 64°51′ | |
| 358. Sacha (1969) | Ecuador | 650 | 650 | — | Putumayo (265) | 221 | 2A | 7,500 ( 2,200) | Anticline | Cretaceous (sandstone) | S 0°20′ W 76°53′ | |
| 359. Nembe Creek (1973) | Nigeria | 650 | 650 | — | Niger Delta (340) | 1141 | 4 | 7,500 ( 2,200) | Fltd. anticline | Miocene (sandstone) | N 4°26′ E 6°19′ | 35 API |
| 360. Glynsko Rozbyshevskoye (1958) | USSR Ukraine SSR | 643 | 310 | 2.00 | Dnepr-Donets (723) | 222 | 3A | 6,200 ( 1,800) | Anticline | Permian (sandstone) | N 50°32′ E 33°36′ | Alias: Rozbyshev |
| 361. Webster (1937) | USA Texas (3) | 640 | 640 | — | Gulf Coast (66) | 1143 | 2Cc/4 | 6,100 ( 1,800) | Fltd. anticline | Oligocene (sandstone) | N 29°20′ W 95°10′ | Nehring (1981) 7c-5 |
| 362. Cheleken (1965) | USSR Turkmenia SSR | 640 | 640 | — | Caspain, South (720) | 222 | 3Bc | 8,200 ( 2,500) | Fltd. anticline | Paleocene (sandstone) | N 39°20′ E 53°10′ | Pliocene also productive |
| 363. Magnus (1974) | United Kingdom Block 211/12 | 640 | 565 | — | North Sea, Nor (420) | 1211 | 3A | 9,200 ( 2,800) | Fault blocks | Jurassic (sandstone) | N 61°37′ E 1°17′ | |
| 364. Guara East (1942) | Venezuela | 630 | 630 | — | Maturin (244) | 221 | 2Ca | 7,000 ( 2,100) | Fault blocks | Miocene (sandstone) | N 8°57′ W 63°59′ | 26 API |
| 365. Vega East (1982) | Italy | 625 | 625 | — | Caltanisetta (384) | 41 | 2Cb | 7,800 ( 2,300) | Fltd. anticline | Jurassic (carbonate) | N 36°32′ E 14°38′ | |
| 366. Golden Trend (1944) | USA Oklahoma | 624 | 458 | 1.00 | Anadarko (9) | 221 | 2A | 6,600 ( 2,000) | Fltd. anticline | Carboniferous (sandstone) | N 34°40′ W 97°45′ | Nehring (1981) 6b-5 |
| 367. Timbalier Bay (1938) | USA Louisiana | 624 | 507 | 0.69 | Gulf Coast (66) | 1143 | 2Cc/4 | 12,000 ( 3,600) | Salt dome | Miocene (sandstone) | N 29° 5′ W 90°22′ | Note: Bay Marchand and Caillou Island are listed separately. Nehring (1981) 8b-2 |
| 368. Bastian Bay (1941) | USA Lousiana | 621 | 80 | 3.25 | Gulf Coast (66) | 1143 | 2Cc/4 | 12,675 ( 3,800) | Salt ridge | Miocene (sandstone) | N 29°30′ W 89°50′ | Nehring (1981) 8b-3 |
| 369. Malossa (1973) | Italy | 619 | 325 | 1.76 | Po (424) | 221 | 2Cb/4 | 18,000 ( 5,400) | Fltd. anticline | Triassic (dolomite) | N 45°30′ E 9°34′ | 51 API; Jurassic and Cretaceous also productive |
| 370. Shibarghan (1959) | Afghanistan | 616 | 17 | 3.59 | Tadzhik (761) | 222 | 2Ca | 4,000 ( 1,200) | Anticline | Cretaceous (sandstone) | N 37° 0′ E 66° 0′ | Also produces from Jurassic carbonates |

Table 1. *Continued.*

| Field Name (discovery year) | Country Province | Recoverable Reserves Equiv. Oil (mm bbl) | Oil (mm bbl) | Gas (tcf) | Petroleum Province | Classification Bally-Type | Klemme-Type | Depth feet (meters) | Trap Type | Geologic Age (lithology) | Latitude Longitude | Remarks |
|---|---|---|---|---|---|---|---|---|---|---|---|---|
| 371. Luginetskoye (1967) | USSR RSFSR - Tomsk | 613 | 30 | 3.50 | West Siberian (774) | 1212 | 2B | 7,750 ( 2,300) | Anticline | Jurassic (sandstone) | N 60° 9′ E 80° 0′ | |
| 372. Midgard (1983) | Norway Block 6507/11 | 613 | 13 | 3.59 | Helgeland (405) | 1141 | 3C | — ( — ) | | | N 65° 5′ E 7°15′ | |
| 373. Maykop (1909) | USSR RSFSR - Krasnoyar | 612 | 73 | 3.22 | Caucasus, Nort (721) | 222 | 2Ca | 8,500 ( 2,500) | Anticline | Cretaceous (sandstone) | N 44°37′ E 40°15′ | |
| 374. Coalinga Nose (1938) | USA California | 605 | 505 | 0.59 | Sacramento San (134) | 332 | 3Bb | 6,400 ( 1,900) | Stratigraphic | Miocene (sandstone) | N 36°15′ W 120°20′ | Also called Coalinga East Nehring (1981) 2a-7 |
| 375. Tul'skiy (1969) | USSR RSFSR - Krasnodar | 603 | 70 | 3.19 | Amur (710) | 221 | 3Bc | 8,250 ( 2,500) | Anticline | Cretaceous (sandstone) | N 44°30′ E 41°30′ | |
| 376. South Pass Block 24 (1950) | USA Louisiana | 600 | 475 | 0.75 | Gulf Coast (66) | 1143 | 2Cc/4 | 8,000 ( 2,400) | Fltd. anticline | Miocene (sandstone) | N 29° 0′ W 89°15′ | Nehring (1981) 8a-3 Gas production to 1976 |
| 377. Karachukhur Zykhskoye (1928) | USSR Azerbaijan SSR | 600 | 600 | — | Kura (740) | 222 | 3Bc | 5,000 ( 1,500) | Fltd. anticline | Pliocene (sandstone) | N 40°22′ E 50° 2′ | Alias: Zykh |
| 378. Kuang (1940) | Indonesia Sumatra | 600 | 600 | — | Sumatra, South (672) | 3122 | 3Ba | 5,249 ( 1,600) | | | S 4° 0′ E 104° 0′ | 25 API |
| 379. Lung Nussu (1956) | China Sichuan | 600 | 600 | — | Sichuan (658) | 22 | 2A | 4,265 ( 1,300) | Anticline | Jurassic (sandstone) | N 30°32′ E 106° 8′ | |
| 380. Bahi (1958) | Libya | 600 | 600 | — | Sirte (354) | 1211 | 3A | 6,000 ( 1,800) | Anticline | Paleocene (carbonate) | N 29°52′ E 17°35′ | |
| 381. Imo River (1959) | Nigeria | 600 | 600 | — | Niger Delta (340) | 1141 | 4 | 5,800 ( 1,700) | Rollover antcl | Miocene (sandstone) | N 4°58′ E 7°12′ | 32 API |
| 382. Judy Creek (1959) | Canada Alberta | 600 | 600 | — | Alberta (4) | 221 | 2A | — ( — ) | Reef | Devonian (carbonate) | N 54°25′ W 115°40′ | |
| 383. Prilukskoye (1959) | USSR Ukraine SSR | 600 | 600 | — | Dnepr-Donets (723) | 222 | 3A | 5,900 ( 1,700) | Deep salt dome | Carboniferous (sandstone) | N 50°36′ E 32°24′ | Alias: Dnepr |
| 384. Darius (1961) | Iran | 600 | 600 | — | Arabian (450) | 221 | 2Ca | 12,000 ( 3,600) | Anticline | Cretaceous (carbonate) | N 29°26′ E 49°22′ | 32 API; Production also from Oligocene–Miocene; Gas production is from Kharg Field, listed separate |
| 385. Urtabulakskoye (1963) | USSR Uzbekistan SSR | 599 | — | 3.59 | Kara-Kum (735) | 222 | 2Ca | 8,200 ( 2,500) | Anticline | Jurassic (carbonate) | N 39° 2′ E 64°31′ | |
| 386. Agan (1966) | USSR RSFSR - Tyumen | 600 | 600 | — | West Siberian (774) | 1212 | 2B | — ( — ) | | | N 62° 0′ E 75°20′ | |
| 387. Severo Varyegan (1971) | USSR RSFSR - Tyumen | 600 | 600 | — | West Siberian (774) | 1212 | 2B | — ( — ) | | | N 62°50′ E 74°55′ | |
| 388. Rimthan (1974) | Saudi Arabia | 600 | 600 | — | Arabian (450) | 221 | 2Ca | 3,000 ( 900) | Anticline | Jurassic (carbonate) | N 28°29′ E 47° 4′ | |
| 389. Clair (1977) | United Kingdom Block 206/8 | 600 | 600 | — | North Sea, Nor (420) | 1211 | 3A | 6,000 ( 1,800) | Fltd. anticline | Devonian (sandstone) | N 60°41′ W 2°33′ | |
| 390. Venture (1979) | Canada Nova Scotia | 599 | — | 3.59 | Scotia Shelf (143) | 1141 | 3C | 17,650 ( 5,300) | | (sandstone) | N 44° 2′ W 59°37′ | |
| 391. Smackover (1922) | USA Arkansas | 599 | 566 | — | Louisiana Salt (90) | 1143 | 2Cc | 2,500 ( 700) | Fltd. anticline | Cretaceous (sandstone) | N 33°20′ W 92°50′ | Nehring (1981) 9a-1 |

Table 1. *Continued.*

| Field Name (discovery year) | Country Province | Recoverable Reserves Equiv. Oil (mm bbl) | Oil (mm bbl) | Gas (tcf) | Petroleum Province | Classification Bally-Type | Klemme-Type | Depth feet (meters) | Trap Type | Geologic Age (lithology) | Latitude Longitude | Remarks |
|---|---|---|---|---|---|---|---|---|---|---|---|---|
| 392. Spraberry Trend (1949) | USA Texas (8) | 594 | 594 | — | Permian (120) | 222 | 2A | 7,000 ( 2,100) | Anticline | Permian (sandstone) | N 31°45′ W 101°30′ | Nehring (1981) 4c-4 |
| 393. Miranga (1965) | Brazil Bahia | 590 | 590 | — | Reconcavo (266) | 111 | 3A | 3,500 ( 1,000) | Fltd. anticline | Cretaceous (sandstone) | S 12°21′ W 38°12′ | 40 API |
| 394. Elk Basin (1915) | USA Wyoming | 585 | 543 | — | Big Horn (25) | 222 | 2A | 3,900 ( 1,100) | Anticline | Carboniferous (sandstone) | N 45° 0′ W 108°50′ | Nehring (1981) 3h-2 |
| 395. Parsons Lake | Canada NWT Mackenzie | 583 | — | 3.50 | Mackenzie (91) | 1142 | 4 | — ( — ) | | | N 68°40′ W 133°30′ | |
| 396. Rio Vista (1936) | USA California | 583 | — | 3.50 | Sacramento San (134) | 332 | 3Bb | 3,800 ( 1,100) | Faulted Dome | Eocene (sandstone) | N 38°10′ W 121°45′ | Nehring (1981) 2a-8 |
| 397. Matzen (1949) | Austria | 583 | 475 | 0.64 | Vienna (446) | 321 | 3Bc | 4,000 ( 1,200) | Dome | Pliocene (sandstone) | N 48°25′ E 16°43′ | |
| 398. Reynosa (1949) | Mexico | 583 | — | 3.50 | Gulf Coast (66) | 1143 | 2Cc/4 | 6,300 ( 1,900) | Fltd. anticline | Oligocene (sandstone) | N 26° 0′ W 98° 0′ | Alias: Hidalgo (USA); Klump (USA) |
| 399. Samantepinskoye (1964) | USSR Turkmenia SSR | 583 | — | 3.50 | Kara-Kum (735) | 222 | 2Ca | 7,700 ( 2,300) | Anticline | Jurassic (carbonate) | N 39° 5′ E 64°19′ | |
| 400. Mellion - Rousse (1965) | France | 583 | — | 3.50 | Aquitaine (377) | 221 | 3A | 14,900 ( 4,500) | Anticline | Cretaceous (carbonate) | N 43°15′ W 0°30′ | Alias: Rousse. Production also from Jurassic at 16,125 ft. |
| 401. Salzwedel (1965) | Germany East | 583 | — | 3,50 | German, Northw (398) | 121 | 2B | 10,000 ( 3,000) | Salt Dome | Permian (sandstone) | N 52°53′ E 11° 4′ | Alias: Wustrow |
| 402. Hewett (1966) | United Kingdom Block 48/29 | 583 | — | 3.50 | North Sea, Sou (421) | 1211 | 2B/3A | 4,500 ( 1,300) | Fltd. anticline | Triassic (sandstone) | N 53° 4′ E 1°40′ | Permian also produces at about 6500 ft. |
| 403. Dengizkul'khanzak-skoye (1967) | USSR Uzbekistan SSR | 583 | — | 3.50 | Kara-Kum (735) | 222 | 2Ca | 7,600 ( 2,300) | Anticline | Jurassic (carbonate) | N 39°10′ E 64°26′ | |
| 404. Nyda (1967) | USSR RSFSR - Tyumen | 583 | — | 3.50 | West Siberian (774) | 1212 | 2B | 3,750 ( 1,100) | Anticline | Cretaceous (sandstone) | N 66°30′ E 73° 0′ | |
| 405. Bakhar (1968) | USSR Azerbaijan SSR | 583 | — | 3.50 | Volga-Ural (773) | 221 | 2A | 13,100 ( 3,900) | Fltd. anticline | Pliocene (sandstone) | N 40° 9′ E 50° 6′ | |
| 406. Placid (1970) | Netherlands Blocks L/10 & L/11 | 583 | — | 3.50 | North Sea, Sou (421) | 1211 | 2B/3A | 12,000 ( 3,600) | Fault blocks | Permian (sandstone) | N 53°29′ E 4°11′ | |
| 407. Yetypurovskoye (1971) | USSR RSFSR - Tyumen | 583 | — | 3.50 | West Siberian (774) | 1212 | 2B | 2,500 ( 700) | Anticline | Cretaceous (sandstone) | N 64° 0′ E 77°40′ | |
| 408. Zapadno Tarkosalin (1972) | USSR RSFSR - Tyumen | 583 | — | 3.50 | West Siberian (774) | 1212 | 2B | 3,500 ( 1,000) | Anticline | Cretaceous (siltstone) | N 64°53′ E 77°15′ | Alias: Tarkosalin |
| 409. Chuchupa (1973) | Colombia Guajira | 583 | — | 3.50 | Guajira (222) | 332 | 3Bb | 6,000 ( 1,800) | Anticline | Pliocene (sandstone) | N 11°50′ W 72°50′ | Alias: Abllena; Riohacha |
| 410. Yuzhno Tambey (1974) | USSR RSFSR - Tyumen | 583 | — | 3.50 | West Siberian (774) | 1212 | 2B | 8,200 ( 2,500) | Anticline | Cretaceous (sandstone) | N 71°21′ E 71°59′ | Alias: Tambey |
| 411. Verkhnepurpey (1976) | USSR RSFSR - Tyumen | 583 | — | 3.50 | West Siberian (774) | 1212 | 2B | 3,600 ( 1,000) | Anticline | Cretaceous (sandstone) | N 65°10′ E 76°50′ | |
| 412. Serrablo (1979) | Spain | 583 | — | 3.50 | Ebro (394) | 221 | 2Cb | 6,500 ( 1,900) | Anticline | Eocene (sandstone) | N 42°35′ W 0°24′ | |

Table 1. *Continued.*

| Field Name (discovery year) | Country Province | Recoverable Reserves | | | Petroleum Province | Classification | | Depth feet (meters) | Trap Type | Geologic Age (lithology) | Latitude Longitude | Remarks |
|---|---|---|---|---|---|---|---|---|---|---|---|---|
| | | Equiv. Oil (mm bbl) | Oil (mm bbl) | Gas (tcf) | | Bally-Type | Klemme-Type | | | | | |
| 413. Nipa (1945) | Venezuela | 580 | 580 | — | Maturin (244) | 221 | 2Ca | 7,500 ( 2,200) | Faults | Miocene (sandstone) | N 9° 5′ W 64°13′ | 30 API |
| 414. Bateman Lake (1937) | USA Louisiana | 575 | 75 | 3.00 | Gulf Coast (66) | 1143 | 2Cc/4 | 9,750 ( 2,900) | Anticline | Miocene (sandstone) | N 29°40′ W 91°10′ | Nehring (1981) 8c-1 |
| 415. Lab-e-Safid (1968) | Iran | 573 | 550 | — | Zagros (464) | 41 | 2Ca | 4,500 ( 1,300) | Anticline | Miocene (carbonate) | N 32°39′ E 48°41′ | 35 API |
| 416. Albuskjell | Norway Blocks 1/6 & 2/4 | 572 | 172 | 2.39 | North Sea, Nor (420) | 1211 | 3A | — ( — ) | | | N 56°40′ E 3° 5′ | |
| 417. Bonnie Glen | Canada Alberta | 570 | 445 | 0.75 | Alberta (4) | 221 | 2A | — ( — ) | | | N 53° 0′ W 113°50′ | |
| 418. Cowden South (1932) | USA Texas (8) | 568 | 485 | — | Permian (120) | 222 | 2A | 4,700 ( 1,400) | Anticline | Permian (dolomite) | N 32°55′ W 100°35′ | Alias: Foster; Johnson // Nehring (1981) 4c-9. ** Also Permeability/ Porosity Pinchout |
| 419. Central Luconia F-06 | Malaysia Sarawak | 566 | — | 3.39 | Sarawak (649) | 3122 | 2Cc | — ( — ) | Reef | Miocene (carbonate) | N 4°38′ E 112°19′ | Sandstones in section also productive |
| 420. Tang-i-bijar (1965) | Iran | 566 | — | 3.39 | Zagros (464) | 41 | 2Ca | 4,000 ( 1,200) | Anticline | Cretaceous (carbonate) | N 33°42′ E 45°54′ | |
| 421. Fullerton (1942) | USA Texas (8) | 565 | 315 | 1.50 | Permian (120) | 222 | 2A | 8,000 ( 2,400) | Anticline | Permian (dolomite) | N 32°10′ W 102°40′ | Nehring (1981) 4c-10 |
| 422. Ankleshwar (1960) | India Gujarat | 565 | 520 | — | Cambay (507) | 1141 | 3A | 4,000 ( 1,200) | Anticline | Eocene (sandstone) | N 21°36′ E 72°55′ | |
| 423. Keystone (1930) | USA Texas (8) | 564 | 314 | 1.50 | Permian (120) | 222 | 2A | 9,500 ( 2,800) | Fltd. anticline | Ordovician (dolomite) | N 31°55′ W 103° 0′ | Nehring (1981) 4c-11 |
| 424. Goodwyn (1971) | Australia Western Austra-lia | 560 | 60 | 3.00 | Dampier (533) | 1141 | 3C | 9,200 ( 2,800) | Anticline | Triassic (sandstone) | S 19°42′ E 115°54′ | |
| 425. Abu Jifan (1973) | Saudi Arabia | 560 | 560 | — | Arabian (450) | 221 | 2Ca | 5,500 ( 1,600) | Anticline | Jurassic (limestone) | N 24°45′ E 47°48′ | |
| 426. Obagi (1964) | Nigeria | 558 | 450 | 0.64 | Niger Delta (340) | 1141 | 4 | 6,500 ( 1,900) | Rollover antcl | Tertiary (sandstone) | N 5°14′ E 6°40′ | |
| 427. Lawhah (1975) | Saudi Arabia | 552 | 552 | — | Arabian (450) | 221 | 2Ca | 6,750 ( 2,000) | Anticline | Cretaceous (limestone) | N 28°15′ E 49°36′ | |
| 428. Attaka (1970) | Indonesia Kalimantan | 550 | 550 | — | Mahakam (590) | 3122 | 3Ba | 7,500 ( 2,200) | Anticline | Miocene (sandstone) | S 0°10′ E 117°38′ | 43 API |
| 429. Nilam (1974) | Indonesia Kalimantan | 550 | — | 3.29 | Mahakam (590) | 3122 | 3Ba/4 | — ( — ) | | | S 1° 0′ E 117° 0′ | |
| 430. Bassein, North (1976) | India Gulf of Cambay | 550 | 550 | — | Cambay (507) | 1141 | 3A | 5,700 ( 1,700) | Anticline | Miocene (limestone) | N 19°19′ E 72° 0′ | |
| 431. Feni (1980) | Bangladesh | 550 | — | 3.29 | Bengal (488) | 221/ 41/* | 2Ca/* | 100 ( 30) | Anticline | Pliocene (sandstone) | N 23° 0′ E 91°17′ | * Bally: 221/41/42 Klemme: 2Ca/ 2Cb/4 |
| 432. Van (1928) | USA Texas (5) | 546 | 546 | — | East Texas Sal (51) | 1143 | 2Cc | 2,800 ( 800) | Fltd. anticline | Cretaceous (sandstone) | N 32°30′ W 95°30′ | Nehring (1981) 9c-1 |
| 433. San Ardo (1947) | USA California | 545 | 529 | — | Ventura (164) | 332 | 3Bb | 2,200 ( 600) | Anticline | Miocene (sandstone) | N 36° 0′ W 120°50′ | Nehring (1981) 2b-2 |
| 434. West Ranch (1938) | USA Texas (2) | 544 | 378 | 1.00 | Gulf Coast (66) | 1143 | 2Cc/4 | 6,000 ( 1,800) | Anticline | Oligocene (sandstone) | N 29° 0′ W 96°30′ | Nehring (1981) 7b-2 |

Table 1. *Continued.*

| Field Name (discovery year) | Country Province | Recoverable Reserves Equiv. Oil (mm bbl) | Oil (mm bbl) | Gas (tcf) | Petroleum Province | Classification Bally-Type | Klemme-Type | Depth feet (meters) | Trap Type | Geologic Age (lithology) | Latitude Longitude | Remarks |
|---|---|---|---|---|---|---|---|---|---|---|---|---|
| 435. Sahil (1967) | Abu Dhabi | 541 | 500 | — | Arabian (450) | 221 | 2Ca | 9,000 (2,700) | Anticline | Cretaceous (carbonate) | N 23°40′ E 54°22′ | |
| 436. Cushing (1912) | USA Oklahoma | 540 | 490 | — | Anadarko (9) | 221 | 2A | 2,700 (800) | Anticline | Carboniferous (sandstone) | N 36° 0′ W 96°40′ | Nehring (1981) 6b-7 |
| 437. Eugene Island Block 330 (1971) | USA Louisiana | 540 | 290 | 1.50 | Gulf Coast (66) | 1143 | 2Cc/4 | 4,300 (1,300) | Fltd. anticline | Pleistocene (sandstone) | N 28°30′ W 91°30′ | Nehring (1981) 8a-4 |
| 438. Seminole (1936) | USA Texas (8A) | 533 | 460 | — | Permian (120) | 222 | 2A | 5,100 (1,500) | Anticline | Permian (dolomite) | N 32°30′ W 102°40′ | Nehring (1981) 4d-4 |
| 439. Burbank (1920) | USA Oklahoma | 533 | 533 | — | Anadarko (9) | 221 | 2A | 2,850 (800) | Stratigraphic | Carboniferous (sandstone) | N 36°40′ W 96°50′ | Nehring (1981) 6b-8 |
| 440. Cowden North (1930) | USA Texas (8) | 533 | 450 | — | Permian (120) | 222 | 2A | 4,400 (1,300) | Anticline | Permian (dolomite) | N 32° 0′ W 100°25′ | Nehring (1981) 4c-8 |
| 441. Thompson (1931) | USA Texas (3) | 533 | 499 | — | Gulf Coast (66) | 1143 | 2Cc/4 | 5,400 (1,600) | Fltd. salt dome | Miocene (sandstone) | N 29°20′ W 95°30′ | Nehring (1981) 7c-6 |
| 442. Satun (1980) | Thailand | 533 | — | 3.19 | Thai (685) | 3122 | 3A | 5,800 (1,700) | Fltd. anticline | Pliocene (sandstone) | N 91°61′ E 101°24′ | |
| 443. La Gloria (1939) | USA Texas (4) | 531 | 31 | 3.00 | Gulf Coast (66) | 1143 | 2Cc/4 | 6,000 (1,800) | Rollover antcl | Oligocene (sandstone) | N 27°10′ W 98° 5′ | Nehring (1981) 7d-3 |
| 444. Tiger Shoal (1958) | USA Louisiana | 530 | 30 | 3.00 | Gulf Coast (66) | 1143 | 2Cc/4 | 9,000 (2,700) | Fltd. anticline | Miocene (sandstone) | N 29°20′ W 92° 0′ | Nehring (1981) 8a-5 |
| 445. Natih (1960) | Oman | 530 | 480 | — | Arabian (450) | 221 | 2Ca | 7,500 (2,200) | Fltd. anticline | Cretaceous (carbonate) | N 22°26′ E 56°43′ | 32 API |
| 446. Brea (1884) | USA California | 528 | 441 | 0.52 | Los Angeles (89) | 332 | 3Bb | 4,000 (1,200) | Fltd. anticline | Pliocene (sandstone) | N 34° 0′ W 118°30′ | Alias: Olinda; Sansinena Nehring (1981) 2c-5 |
| 447. Grand Isle Block 43 (1956) | USA Louisiana | 524 | 358 | 1.00 | Gulf Coast (66) | 1143 | 2Cc/4 | 5,000 (1,500) | Fltd. anticline | Miocene (sandstone) | N 28°55′ W 90° 0′ | Nehring (1981) 8a-10 |
| 448. Kenai (1959) | USA Alaska | 525 | — | 3.14 | Cook Inlet (41) | 311 | 3Bb | 6,600 (2,000) | Fltd. anticline | Neogene (sandstone) | N 60°20′ W 151°10′ | Nehring (1981) 1-3 |
| 449. Vacuum (1929) | USA New Mexico | 524 | 524 | — | Permian (120) | 222 | 2A | 6,400 (1,900) | Anticline Reef | Permian (dolomite) | N 33° 0′ W 103°15′ | Nehring (1981) 4a-4 |
| 450. Rhourde el Baguel (1962) | Algeria | 521 | 500 | — | Ghadames (312) | 121 | 2B | 7,950 (2,400) | Fltd. anticline | Cambrian (sandstone) | N 31°23′ E 6°57′ | 40 API |
| 451. Bayrmalinskoye (1962) | USSR Turkmenia SSR | 516 | — | 3.09 | Turkmen (768) | 222 | 2Ca | 9,100 (2,700) | | Cretaceous (sandstone) | N 35°57′ E 61° 0′ | |
| 452. West Delta Block 30 (1949) | USA Louisiana | 516 | 424 | 0.54 | Gulf Coast (66) | 1143 | 2Cc/4 | 8,600 (2,600) | Salt Dome | Miocene (sandstone) | N 29°19′ W 89°40′ | Nehring (1981) 8a-7 |
| 453. South Pass Block 27 (1954) | USA Louisiana | 516 | 383 | 0.79 | Gulf Coast (66) | 1143 | 2Cc/4 | 12,000 (3,600) | Fltd. salt dome | Miocene (sandstone) | N 28°58′ W 89°20′ | Gas production to 1976 Nehring (1981) 8a-6 |
| 454. Vermilion Block 39 (1949) | USA Louisiana | 515 | 15 | 3.00 | Gulf Coast (66) | 1143 | 2Cc/4 | 8,700 (2,600) | Fltd. anticline | Miocene (sandstone) | N 29°20′ W 92°15′ | Nehring (1981) 8a-2 |
| 455. Agua Dulce (1928) | USA Texas (4) | 513 | 147 | 2.19 | Gulf Coast (66) | 1143 | 2Cc/4 | — (—) | | | N 27°50′ W 97°30′ | Alias: Stratton Nehring (1981) 7d-2 |
| 456. Sand Hills (1931) | USA Texas (8) | 512 | 246 | 1.59 | Permian (120) | 222 | 2A | 4,000 (1,200) | Anticline | Permian (dolomite) | N 31°30′ W 102°20′ | Nehring (1981) 4c-6 |

Table 1. *Continued.*

| Field Name (discovery year) | Country Province | Recoverable Reserves Equiv. Oil (mm bbl) | Oil (mm bbl) | Gas (tcf) | Petroleum Province | Classification Bally-Type | Klemme-Type | Depth feet (meters) | Trap Type | Geologic Age (lithology) | Latitude Longitude | Remarks |
|---|---|---|---|---|---|---|---|---|---|---|---|---|
| 457. Fyzabad Group (1913) | Trinidad | 510 | 510 | — | Paria (258) | 332 | 2Ca | 8,000 ( 2,400) | Mud diapir | Miocene (sandstone) | N 10° 0′ W 61°35′ | Alias: Coora; Palo Seco; Quarry Trapping mechanism is very complex |
| 458. Yarino Kamennolozhskoye (1954) | USSR RSFSR - Perm | 510 | 510 | — | Volga-Ural (773) | 221 | 2A | 5,400 ( 1,600) | Anticline | Carboniferous (sandstone) | N 58°10′ E 56°20′ | Alias: Kamennyy Log. May have more gas than oil, but no estimate of gas reserves is available. |
| 459. Swan Hills South (1959) | Canada Alberta | 510 | 510 | — | Alberta (4) | 221 | 2A | 8,000 ( 2,400) | Reef | Devonian (carbonate) | N 54°30′ W 115°40′ | |
| 460. Thistle (1973) | United Kingdom Block 211/18 | 510 | 510 | — | North Sea, Nor (420) | 1211 | 3A | 8,500 ( 2,500) | Fault blocks | Jurassic (sandstone) | N 61°22′ E 1°35′ | |
| 461. Blinebry - Drinkard (1935) | USA New Mexico | 506 | 256 | 1.50 | Permian (120) | 222 | 2A | 6,000 ( 1,800) | Anticline | Permian (dolomite) | N 33° 0′ W 103°20′ | Nehring (1981) 4a-3 |
| 462. Borregos (1937) | USA Texas (4) | 503 | 503 | — | Gulf Coast (66) | 1143 | 2Cc/4 | 7,000 ( 2,100) | Fltd. anticline | Oligocene (sandstone) | N 27°15′ W 98° 0′ | Alias: Seeligson; T.C.B. Gas production to 1976 // Nehring (1981) 7d-1 |
| 463. Bangko | Indonesia Sumatra | 500 | 500 | — | Sumatra, Centr (670) | 3122 | 3Ba | — ( — ) | | | N 1°40′ E 101° 0′ | |
| 464. Claymore | United Kingdom Block 14/9 | 500 | 500 | — | North Sea, Nor (420) | 1211 | 3A | — ( — ) | | | N 58°25′ W 0°20′ | |
| 465. Loango | Congo | 500 | 500 | — | Cabinda (296) | 1141 | 3C | — ( — ) | | | S 4°50′ E 11° 0′ | |
| 466. Maharah | Saudi Arabia | 500 | 500 | — | Arabian (450) | 221 | 2Ca | — ( — ) | Anticline | Cretaceous (carbonate) | N 28°23′ E 49°25′ | |
| 467. La Cira (1918) | Colombia Santander | 500 | 500 | — | Magdalena, Mid (240) | 332 | 3Bc | 1,500 ( 400) | Anticline | Eocene (sandstone) | N 6°58′ W 73°45′ | 25 API; also productive from Oligocene and Cretaceous Note: Infantas is a separate field nearby. |
| 468. Pledger (1932) | USA Texas (3) | 500 | — | 3.00 | Gulf Coast (66) | 1143 | 2Cc/4 | 6,800 ( 2,000) | Fltd. anticline | Oligocene (sandstone) | N 29°10′ W 95°55′ | Nehring (1981) 7c-7 |
| 469. Laochunmiao (1938) | China Gansu | 500 | 500 | — | Jiuquan (572) | 32 | 3A | 7,550 ( 2,300) | Anticline | Miocene (sandstone) | N 39°50′ E 97°30′ | Alias: Ya-erh-hsia; Yumen. Also productive from Eocene to 8,860 ft. |
| 470. Leduc Woodbend (1947) | Canada Alberta | 500 | 500 | — | Alberta (4) | 221 | 2A | 5,700 ( 1,700) | Reef | Devonian (carbonate) | N 53° 0′ W 113° 0′ | 40 API; Alias: Woodbend |
| 471. Puckett (1952) | USA Texas (8) | 500 | — | 3.00 | Permian (120) | 222 | 2A | 13,370 ( 4,000) | Anticline | Ordovician (dolomite) | N 30°40′ W 102°30′ | Nehring (1981) 4c-13 |
| 472. Nahorkatiya (1953) | India Assam | 500 | 500 | — | Assam (474) | 221 | 2Cb | 8,000 ( 2,400) | Anticline | Oligocene (sandstone) | N 27°17′ E 95°20′ | |
| 473. Ozeksuatskoye (1953) | USSR RSFSR - Stavropol | 500 | 500 | — | Caucasus, Nort (721) | 222 | 2Ca | 10,400 ( 3,100) | Anticline | Cretaceous (sandstone) | N 44°32′ E 45°20′ | |
| 474. Mata (1954) | Venezuela | 500 | 500 | — | Maturin (244) | 221 | 2Ca | 9,500 ( 2,800) | Fltd. anticline | Miocene (sandstone) | N 9°13′ W 64° 6′ | 34 API; Alias: Jusepin; Mulata; Muri; Pirital; Santa Barbara; Tacat. Jusepin is No. 204 on Nehring giant list |
| 475. Soldado Main (1954) | Trinidad | 500 | 500 | — | Paria (258) | 332 | 2Ca | 4,750 ( 1,400) | Fltd. anticline | Pliocene (sandstone) | N 10°12′ W 61°59′ | 25 API |
| 476. Belayim (1955) | Egypt | 500 | 500 | — | Suez, Gulf of (358) | 111 | 3A | 8,500 ( 2,500) | Anticline | Miocene (sandstone) | N 28°37′ E 33°13′ | 22 API |

Table 1. *Continued.*

| Field Name (discovery year) | Country Province | Recoverable Reserves Equiv. Oil (mm bbl) | Oil (mm bbl) | Gas (tcf) | Petroleum Province | Classification Bally-Type | Klemme-Type | Depth feet (meters) | Trap Type | Geologic Age (lithology) | Latitude Longitude | Remarks |
|---|---|---|---|---|---|---|---|---|---|---|---|---|
| 477. Vermilion Block 14 (1956) | USA Louisiana | 500 | — | 3.00 | Gulf Coast (66) | 1143 | 2Cc/4 | 11,050 (3,300) | Fltd. anticline | Miocene (sandstone) | N 29°30′ W 92°15′ | Nehring (1981) 8a-8 |
| 478. Bomu (1958) | Nigeria | 500 | 500 | — | Niger Delta (340) | 1141 | 4 | 7,200 (2,100) | Rollover antcl | Tertiary (sandstone) | N 4°40′ E 7°17′ | 36 API |
| 479. Dahra Hofra (1959) | Libya | 500 | 500 | — | Sirte (354) | 1211 | 3A | 3,400 (1,000) | Reef | Paleocene (carbonate) | N 29°31′ E 17°46′ | 39 API; Alias: Hofra |
| 480. Mabruk (1959) | Libya | 500 | 500 | — | Sirte (354) | 1211 | 3A | 4,000 (1,200) | Anticline | Eocene (sandstone) | N 29°57′ E 17°17′ | |
| 481. Tuba (1960) | Iraq | 500 | 500 | — | Arabian (450) | 221 | 2Ca | 10,000 (3,000) | Anticline | Cretaceous (carbonate) | N 30°23′ E 47°26′ | |
| 482. Belayim Marine (1961) | Egypt | 500 | 500 | — | Suez, Gulf of (358) | 111 | 3A | 8,000 (2,400) | Fltd. anticline | Miocene (sandstone) | N 28°37′ E 33° 5′ | 30 API |
| 483. Luhais (1961) | Iraq | 500 | 500 | — | Arabian (450) | 221 | 2Ca | 8,000 (2,400) | Anticline | Cretaceous (sandstone) | N 30°18′ E 46°46′ | |
| 484. Uchkyrskoye (1961) | USSR Uzbekistan SSR | 500 | — | 3.00 | Kara-Kum (735) | 222 | 2Ca | 5,000 (1,500) | Anticline | Cretaceous (siltstone) | N 40° 9′ E 63° 3′ | |
| 485. Samah (1962) | Libya | 500 | 500 | — | Sirte (354) | 1211 | 3A | 9,000 (2,700) | Anticline | Cretaceous (carbonate) | N 28°13′ E 19° 8′ | 33 API; also Cambro–Ordovician sandstones |
| 486. Divjake (1963) | Albania | 500 | — | 3.00 | Ionian (406) | 41 | 2Cb | — ( — ) | Anticline | | N 41° 1′ E 19°28′ | Facies change is also important to trapping mechanism |
| 487. Gomez (1963) | USA Texas (8) | 500 | — | 3.00 | Permian (120) | 222 | 2A | 19,871 (6,000) | Fltd. anticline | Ordovician (carbonate) | N 31° 0′ W 103° 0′ | Nehring (1981) 4c-7 |
| 488. Salymskoye (1963) | USSR RSFSR - Tyumen | 500 | 500 | — | West Siberian (774) | 1212 | 2B | 7,875 (2,400) | Anticline | Cretaceous (sandstone) | N 60°30′ E 72° 0′ | |
| 489. Okan (1964) | Nigeria | 500 | 500 | — | Niger Delta (340) | 1141 | 4 | 7,000 (2,100) | Rollover antcl | Miocene (sandstone) | N 5°30′ E 5° 5′ | 36 API |
| 490. West Sole (1965) | United Kingdom Block 48/6 | 500 | — | 3.00 | North Sea, Sou (421) | 1211 | 2B/3A | — ( — ) | | | N 53°45′ E 1° 5′ | Alias: Sole West |
| 491. Jianghan (1966) | China Hebei | 500 | 500 | — | Jianghan (571) | 0 | 3A | 500 (100) | Anticline | Devonian (sandstone) | N 30°28′ E 112°40′ | Alias: Chien Chiang |
| 492. Shah (1966) | Abu Dhabi | 500 | 500 | — | Arabian (450) | 221 | 2Ca | 8,000 (2,400) | Anticline | Cretaceous (carbonate) | N 22°50′ E 54°20′ | |
| 493. Jana (1967) | Saudi Arabia | 500 | 500 | — | Arabian (450) | 221 | 2Ca | 5,000 (1,500) | Anticline | Jurassic (limestone) | N 27°30′ E 48°45′ | |
| 494. Ab-e-Teimur (1968) | Iran | 500 | 500 | — | Zagros (464) | 41 | 2Ca | 10,500 (3,200) | Anticline | Cretaceous (carbonate) | N 31°13′ E 48°30′ | |
| 495. Russkoye (1968) | USSR RSFSR - Tyumen | 500 | 500 | — | West Siberian (774) | 1212 | 2B | 3,300 (1,000) | Anticline | Cretaceous (sandstone) | N 66°40′ E 80°25′ | See also separate field Yuzhno Russkoye (USSR) |
| 496. Snapper (1968) | Australia Victoria | 499 | 100 | 2.39 | Gippsland (550) | 111 | 3A | 10,500 (3,200) | Fault blocks | Paleogene | S 38°12′ E 148° 0′ | |
| 497. Beurdeshik (1969) | USSR Turkmenia SSR | 500 | — | 3.00 | Tadzhik (761) | 222 | 2Ca | 4,920 (1,500) | Anticline | Jurassic (carbonate) | N 39°45′ E 68°46′ | Alias: Buyerdeshik |
| 498. Saath al Raazboot (1969) | Abu Dhabi | 500 | 500 | — | Arabian (450) | 221 | 2Ca | 8,000 (2,400) | Anticline | Jurassic (carbonate) | N 24°45′ E 53° 0′ | |
| 499. Champion (1970) | Brunei | 500 | 500 | — | Brunei-Sabah (503) | 3122 | 2Cc | 4,500 (1,300) | Anticline | Pliocene (sandstone) | N 5°50′ E 114°45′ | 23 API |

Table 1. *Continued.*

| Field Name (discovery year) | Country Province | Recoverable Reserves Equiv. Oil (mm bbl) | Oil (mm bbl) | Gas (tcf) | Petroleum Province | Classification Bally-Type | Klemme-Type | Depth feet (meters) | Trap Type | Geologic Age (lithology) | Latitude Longitude | Remarks |
|---|---|---|---|---|---|---|---|---|---|---|---|---|
| 500. Vat'yegan (1971) | USSR RSFSR - Tyumen | 500 | 500 | — | West Siberian (774) | 1212 | 2B | — ( — ) | | | N 61° 5′ E 76° 0′ | |
| 501. Beryl A (1972) | United Kingdom Block 9/13 | 500 | 500 | — | North Sea, Nor (420) | 1211 | 3A | 9,900 ( 3,000) | Fault blocks | Jurassic (sandstone) | N 59°33′ E 1°32′ | |
| 502. Kholmogor (1973) | USSR RSFSR - Tyumen | 500 | 500 | — | West Siberian (774) | 1212 | 2B | — ( — ) | | | N 62°30′ E 74° 0′ | |
| 503. Rio Nuevo (1975) | Mexico | 500 | 500 | — | Salinas (Mex) (136) | 22 | 2Cc | 14,000 ( 4,200) | Fltd. anticline | Cretaceous (dolomite) | N 17°40′ W 93°20′ | Listed separately, but may be part of Cactus |
| 504. Audrey (1976) | United Kingdom Block 48/6 | 500 | — | 3.00 | North Sea, Sou (421) | 1211 | 2B/3A | 9,000 ( 2,700) | Fltd. anticline | Permian (sandstone) | N 53°42′ E 1° 8′ | |
| 505. Halfayah (1977) | Iraq | 500 | 500 | — | Arabian (450) | 221 | 2Ca | 12,600 ( 3,800) | Anticline | Cretaceous (carbonate) | N 31°47′ E 47°21′ | |
| 506. Char (1979) | Canada NWT Franklin | 500 | — | 3.00 | Sverdrup (151) | 1143 | 2Cc | 5,150 ( 1,500) | | Jurassic (sandstone) | N 77°37′ W 99°31′ | |
| 507. Issungnak (1980) | Canada NWT Franklin | 500 | 500 | — | Mackenzie (91) | 1142 | 4 | 14,000 ( 4,200) | | Jurassic (sandstone) | N 70° 1′ W 134°19′ | 35 API |
| 508. Block 30/3 (1981) | Norway Block 30/3 | 500 | 500 | — | North Sea, Nor (420) | 1211 | 3A | 9,000 ( 2,700) | Fault blocks | Jurassic (sandstone) | N 60°48′ E 2°55′ | |
| 509. Point Arguillo (1981) | USA California | 500 | 500 | — | Ventura (164) | 332 | 3Bb | — ( — ) | | | N 34°30′ W 120°40′ | |

Table 2. *Giant fields cross reference list. Estimates in million barrels (mm bbl).*

| Field Name | Country | Location | | Remarks |
|---|---|---|---|---|
| | | Latitude | Longitude | |
| A-100 | Libya | N 28°51′ | E 22° 7′ | See Bu Attifel (Libya) |
| Ab-e-Teimur | Iran | N 31°13′ | E 48°30′ | Giant field number 494 in Table 1. |
| Abkatun | Mexico | N 19°39′ | W 92°20′ | See Ixtoc (Mexico) |
| Abqaiq | Saudi Arabia | N 25°57′ | E 49°41′ | Giant field number 20 in Table 1. |
| Abu al Bukhoosh | Abu Dhabi | N 25°30′ | E 53° 9′ | See Sassan (Iran) |
| Abu Ghirab | Iraq | N 32°25′ | E 47°24′ | Giant field number 110 in Table 1. |
| Abu Hadriya | Saudi Arabia | N 27°22′ | E 49° 0′ | Giant field number 131 in Table 1. |
| Abu Jidu | Abu Dhabi | N 23°24′ | E 54°16′ | See Asab (Abu Dhabi) |
| Abu Jifan | Saudi Arabia | N 24°45′ | E 47°48′ | Giant field number 425 in Table 1. |
| Abu Sa'fah | Saudi Arabia | N 26°55′ | E 50°25′ | Giant field number 44 in Table 1. |
| Achakskoye | USSR | N 41° 6′ | E 60°58′ | Giant field number 258 in Table 1. |
| Agan | USSR | N 62° 0′ | E 75°20′ | Giant field number 386 in Table 1. |
| Agave | Mexico | N 17°45′ | W 92°53′ | Giant field number 332 in Table 1. |
| Agha Jari | Iran | N 30°50′ | E 49°48′ | Giant field number 25 in Table 1. |
| Aghar | Iran | N 28°35′ | E 52°48′ | Giant field number 78 in Table 1. |
| Agua Dulce | USA | N 27°50′ | W 97°30′ | Giant field number 455 in Table 1. |
| Ahwaz | Iran | N 31°20′ | E 48°44′ | Giant field number 24 in Table 1. |
| Akal | Mexico | N 19°30′ | W 91°57′ | See Cantarell Complex (Mexico) |
| Al Bunduq | Abu Dhabi | N 25° 0′ | E 53° 9′ | Present estimate less than 500 mm bbl equivalent. |
| Al Hout | Divided Zone | N 28°50′ | E 49°15′ | Present estimate less than 500 mm bbl equivalent. |
| Albuskjell | Norway | N 56°40′ | E 3° 5′ | Giant field number 416 in Table 1. |
| Algyo | Hungury | N 46°20′ | E 20°15′ | Giant field number 294 in Table 1. |
| Aliyurt | USSR | N 43°36′ | E 44°30′ | See Malgobek Voznesenskoye (USSR) |
| Alrar | Algeria | N 28°30′ | E 9°34′ | Giant field number 224 in Table 1. |
| Amal | Egypt | N 28° 0′ | E 33°40′ | Present estimate less than 500 mm bbl equivalent. |
| Amal | Libya | N 29°27′ | E 21° 9′ | Giant field number 67 in Table 1. |
| Amatlan | Mexico | N 21°20′ | W 97°41′ | See Cerro Azul (Mexico) |
| Ampa, Southwest | Brunei | N 4°43′ | E 114° 8′ | Giant field number 132 in Table 1. |
| Anastasiyevsko | USSR | N 45° 5′ | E 38° 5′ | Giant field number 324 in Table 1. |
| Andector | USA | N 31°45′ | W 102°30′ | See Goldsmith (USA) |
| Ankleshwar | India | N 21°36′ | E 72°55′ | Giant field number 422 in Table 1. |
| Anschutz Ranch East | USA | N 41°10′ | W 111°15′ | Giant field number 280 in Table 1. |
| Antipayutin | USSR | N 69°20′ | E 75° 5′ | Giant field number 273 in Table 1. |
| Antonio J. Bermudez | Mexico | N 18° 0′ | W 93° 0′ | See Bermudez (Mexico) |
| Ardeshir | Iran | N 29°18′ | E 49°32′ | Giant field number 146 in Table 1. |
| Ardjuna B | Indonesia | S 5°54′ | E 107°42′ | Giant field number 331 in Table 1. |
| Arenque | Mexico | N 21° 0′ | E 97° 0′ | Present estimate less than 500 mm bbl equivalent. |
| Arkticheskoye | USSR | N 69°33′ | E 70°48′ | Giant field number 28 in Table 1. |
| Arlanskoye | USSR | N 55°55′ | E 54°20′ | Giant field number 69 in Table 1. |
| Arun | Indonesia | N 5° 3′ | E 97°15′ | Giant field number 104 in Table 1. |
| Asab | Abu Dhabi | N 23°24′ | E 54°16′ | Giant field number 64 in Table 1. |
| Attaka | Indonesia | S 0°10′ | E 117°38′ | Giant field number 428 in Table 1. |
| Audrey | United Kingdom | N 53°42′ | E 1° 8′ | Giant field number 504 in Table 1. |
| Augila | Libya | N 29°10′ | E 21°29′ | Giant field number 135 in Table 1. |
| Awali | Bahrain | N 26°02′ | E 50°32′ | Giant field number 86 in Table 1. |
| B Structure | Iran | N 28°24′ | E 50° 9′ | Giant field number 38 in Table 1. |
| B Structure | Thailand | N 8° 5′ | E 102°17′ | Giant field number 198 in Table 1. |
| B-NC-41 | Libya | N 33°54′ | E 21°30′ | See Bouri (Libya) |
| Bab | Abu Dhabi | N 23°52′ | E 53°32′ | See Murban Bab (Abu Dhabi) |
| Bachaquero | Venezuela | N 9°55′ | W 71° 7′ | See Bolivar Coastal (Venezuela) |
| Badak | Indonesia | S 0°20′ | E 117°26′ | Giant field number 178 in Table 1. |
| Bagadzhin | USSR | N 39°20′ | E 62°29′ | Giant field number 149 in Table 1. |
| Baghdad, East | Iraq | N 33° 8′ | E 44°30′ | Giant field number 124 in Table 1. |
| Bahi | Libya | N 29°52′ | E 17°35′ | Giant field number 380 in Table 1. |
| Bahrah | Kuwait | N 29°34′ | E 47°55′ | Giant field number 170 in Table 1. |
| Bahrain | Bahrain | N 26° 2′ | E 50°32′ | See Awali (Bahrain) |
| Bai Hassan | Iraq | N 35°34′ | E 44° 1′ | Giant field number 156 in Table 1. |
| Bakhar | USSR | N 40° 9′ | E 50° 6′ | Giant field number 405 in Table 1. |
| Baku Archipelago | USSR | N 40° 1′ | E 49°32′ | See Sangachaly Duvannyy (USSR) |
| Balakhano Sabunchino | USSR | N 40°40′ | E 50°20′ | Giant field number 96 in Table 1. |
| .Ballena | Colombia | N 11°50′ | W 72°50′ | See Chuchupa (Colombia) |
| Balyk | USSR | N 62°24′ | E 72°40′ | See Ust'bakykskoye (USSR) |
| Bangko | Indonesia | N 1°40′ | E 101° 0′ | Giant field number 463 in Table 1. |
| Barqan | Saudi Arabia | N 27°50′ | E 35°15′ | Giant field number 287 in Table 1. |
| Barracouta | Australia | S 38°17′ | E 147°43′ | Present estimate less than 500 mm bbl equivalent. |
| Basin | USA | N 37° 0′ | W 107°40′ | Giant field number 252 in Table 1. |
| Bassein, North | India | N 19°19′ | E 72° 0′ | Giant field number 430 in Table 1. |
| Bassin, South | India | N 19°10′ | E 72° 6′ | Giant field number 176 in Table 1. |
| Bastian Bay | USA | N 29°30′ | W 89°50′ | Giant field number 368 in Table 1. |
| Bateman Lake | USA | N 29°40′ | W 91°10′ | Giant field number 414 in Table 1. |
| Bay Marchand | USA | N 29° 5′ | W 90°15′ | Giant field number 320 in Table 1. |
| Bayou Sale | USA | N 29°30′ | W 91°30′ | Giant field number 306 in Table 1. |
| Bayrmalinskoye | USSR | N 35°57′ | E 61° 0′ | Giant field number 451 in Table 1. |

Table 2. *Continued.*

| Field Name | Country | Location | | Remarks |
|---|---|---|---|---|
| | | Latitude | Longitude | |
| Bekasap | Indonesia | N 1°14′ | E 101°10′ | Present estimate less than 500 mm bbl equivalent. |
| Belayim | Egypt | N 28°37′ | E 33°13′ | Giant field number 476 in Table 1. |
| Belayim Marine | Egypt | N 28°37′ | E 33° 5′ | Giant field number 482 in Table 1. |
| Belridge South | USA | N 35°30′ | W 119°30′ | Giant field number 318 in Table 1. |
| Bergen | Netherlands | N 52°40′ | E 4°30′ | Present estimate less than 500 mm bbl equivalent. |
| Bermudez | Mexico | N 18° 0′ | W 93° 0′ | Giant field number 30 in Table 1. |
| Berri | Saudi Arabia | N 26°59′ | E 49°37′ | Giant field number 22 in Table 1. |
| Beryl A | United Kingdom | N 59°33′ | E 1°32′ | Giant field number 501 in Table 1. |
| Beurdeshik | USSR | N 39°45′ | E 68°46′ | Giant field number 497 in Table 1. |
| Bibi Eybat | USSR | N 40°10′ | E 50°20′ | See Bibieybatskoye (USSR) |
| Bibi Hakimeh | Iran | N 30° 3′ | E 50°37′ | Giant field number 80 in Table 1. |
| Bibieybatskoye | USSR | N 40°10′ | E 50°20′ | Giant field number 116 in Table 1. |
| Binak | Iran | N 29°45′ | E 50°20′ | Present estimate less than 500 mm bbl equivalent. |
| Blanco | USA | N 36°40′ | W 106°30′ | Giant field number 136 in Table 1. |
| Blinebry – Drinkard | USA | N 33° 0′ | W 103°20′ | Giant field number 461 in Table 1. |
| Block 16/26 | United Kingdom | N°58°05′ | E 1° 6′ | Giant field number 334 in Table 1. |
| Block 30/3 | Norway | N 60°48′ | E 2°55′ | Giant field number 508 in Table 1. |
| Bolivar Coastal | Venezuela | N 10° 0′ | W 71°10′ | Giant field number 5 in Table 1. |
| Bolshoye Chernogor | USSR | N 71°40′ | E 77° 0′ | Giant field number 300 in Table 1. |
| Bombay High | India | N 19°34′ | E 71°24′ | Giant field number 165 in Table 1. |
| Bomu | Nigeria | N 4°40′ | E 7°17′ | Giant field number 478 in Table 1. |
| Bonnie Glen | Canada | N 53°00′ | W 113°50′ | Giant field number 417 in Table 1. |
| Borregos | USA | N 27°15′ | W 98° 0′ | Giant field number 462 in Table 1. |
| Boscan | Venezuela | N 10°29′ | W 71°59′ | Giant field number 228 in Table 1. |
| Botnobin | USSR | N 61°15′ | E 111°40′ | See Sredne Botnobin (USSR) |
| Bouri | Libya | N 33°54′ | E 21°30′ | Giant field number 335 in Table 1. |
| Bovanenkovo | USSR | N 70°24′ | E 68°24′ | Giant field number 7 in Table 1. |
| Bradford | USA | N 42° 0′ | W 78°30′ | Giant field number 351 in Table 1. |
| Brae | United Kingdom | N 58°41′ | E 1°17′ | Present estimate less than 500 mm bbl equivalent. |
| Brea | USA | N 34° 0′ | W 118°30′ | Giant field number 446 in Table 1. |
| Brent | United Kingdom | N 61° 6′ | E 1°42′ | Giant field number 92 in Table 1. |
| Bu Attifel | Libya | N 28°51′ | E 22° 7′ | Giant field number 203 in Table 1. |
| Bu Hasa | Abu Dhabi | N 23°30′ | E 53°10′ | See Murban Bu Hasa (Abu Dhabi) |
| Bu Tini | Abu Dhabi | N 24°34′ | E 52°59′ | Giant field number 243 in Table 1. |
| Bubullime | Albania | N 40°48′ | E 19°42′ | Giant field number 284 in Table 1. |
| Buena Vista | USA | N 35° 5′ | W 119°30′ | Giant field number 293 in Table 1. |
| Bul Hanine | Qatar | N 25°26′ | E 52°42′ | Giant field number 192 in Table 1. |
| Burbank | USA | N 36°40′ | W 96°50′ | Giant field number 439 in Table 1. |
| Burgan | Kuwait | N 28°55′ | E 47°56′ | Giant field number 2 in Table 1. |
| Bushgan | Iran | N 29°20′ | E 52° 0′ | Present estimate less than 500 mm bbl equivalent. |
| Buyerdeshik | USSR | N 39°45′ | E 68°46′ | See Beurdeshik (USSR) |
| Buzurgan | Iraq | N 32°16′ | E 47°14′ | Giant field number 162 in Table 1. |
| Cabimas | Venezuela | N 10°22′ | W 71°27′ | See Bolivar Coastal (Venezuela) |
| Cabinda B | Angola | S 5°21′ | E 12° 5′ | See Malongo North & South (Angola) |
| Cactus | Mexico | N 17°52′ | W 93°10′ | Giant field number 108 in Table 1. |
| Caillou Island | USA | N 29° 5′ | W 90°30′ | Giant field number 247 in Table 1. |
| Cantarell Complex | Mexico | N 19°30′ | W 91°57′ | Giant field number 8 in Table 1. |
| Cap Bon | Tunisia | N 37°30′ | E 10°30′ | See Sidi Abderrahman (Tunisia) |
| Carter Creek | USA | N 41°30′ | W 110°40′ | See Whitney Canyon (USA) |
| Carthage | USA | N 32°30′ | W 94°30′ | Giant field number 83 in Table 1. |
| Central Luconia F-06 | Malaysia | N 4°38′ | E 112°19′ | Giant field number 419 in Table 1. |
| Centro | Venezuela | N 9°50′ | W 71°27′ | Giant field number 233 in Table 1. |
| Cerro Azul | Mexico | N 21°20′ | W 97°41′ | Giant field number 190 in Table 1. |
| Ceuta | Venezuela | N 9°47′ | W 71° 5′ | See Bolivar Coastal (Venezuela) |
| Champion | Brunei | N 5°50′ | E 114°45′ | Giant field number 499 in Table 1. |
| Char | Canada | N 77°37′ | W 99°31′ | Giant field number 506 in Table 1. |
| Cheleken | USSR | N 39°20′ | E 53°10′ | Giant field number 362 in Table 1. |
| Cheremshanka | USSR | N 58° 0′ | E 78° 0′ | See Yuzhno Cheremshanskoye (USSR) |
| Chernogor | USSR | N 71°40′ | E 77° 0′ | See Bolshoye Chernogor (USSR) |
| Chien Chiang | China | N 30°27′ | E 112°40′ | See Jianghan (China) |
| Chilapilla | Mexico | N 17°58′ | W 92°28′ | See Colomo (Mexico) |
| Chuchupa | Colombia | N 11°50′ | W 72°50′ | Giant field number 409 in Table 1. |
| Cisco | Canada | N 77°25′ | W 106°26′ | Giant field number 245 in Table 1. |
| Clair | United Kingdom | N 60°41′ | W 2°33′ | Giant field number 389 in Table 1. |
| Claresholm | Canada | N 49°55′ | W 113°30′ | Giant field number 290 in Table 1. |
| Claymore | United Kingdom | N 58°25′ | W 0°20′ | Giant field number 464 in Table 1. |
| Coalinga | USA | N 36°10′ | W 120°30′ | Giant field number 276 in Table 1. |
| Coalinga Nose | USA | N 36°15′ | W 120°20′ | Giant field number 374 in Table 1. |
| Colomo | Mexico | N 17°58′ | W 92°28′ | Giant field number 338 in Table 1. |
| Comodoro Rivadavia | Argentina | S 46° 0′ | W 67°30′ | Present estimate less than 500 mm bbl equivalent. |
| Conroe | USA | N 30°20′ | W 95°40′ | Giant field number 319 in Table 1. |
| Coora | Trinidad | N 10° 0′ | W 61°35′ | See Fyzabad Group (Trinidad) |
| Cormorant North | United Kingdom | N 61°13′ | E 1°10′ | Present estimate less than 500 mm bbl equivalent. |

Table 2. *Continued.*

| Field Name | Country | Location | | Remarks |
|---|---|---|---|---|
| | | Latitude | Longitude | |
| Cotton Valley | USA | N 32°30′ | W 94°30′ | See Carthage (USA) |
| Cowden Complex | USA | N 31°55′ | W 100°30′ | See Cowden North (USA) and Cowden South (USA) |
| Cowden North | USA | N 32° 0′ | W 100°25′ | Giant field number 440 in Table 1. |
| Cowden South | USA | N 32°55′ | W 100°35′ | Giant field number 418 in Table 1. |
| Cunduacan | Mexico | N 18° 0′ | W 93 0′ | See Bermudez (Mexico) |
| Cushing | USA | N 36° 0′ | W 96°40′ | Giant field number 436 in Table 1. |
| D – 52 | Libya | N 28°30′ | E 9°35′ | See Alrar (Algeria) |
| Dahra Hofra | Libya | N 29°31′ | E 17°46′ | Giant field number 479 in Table 1. |
| Dakota | USA | N 37° 0′ | W 107°40′ | See Basin (USA) |
| Dammam | Saudi Arabia | N 26°19′ | E 50° 8′ | Giant field number 154 in Table 1. |
| Daqing | China | N 46°30′ | E 125°20′ | Giant field number 42 in Table 1. |
| Darius | Iran | N 29°26′ | E 49°22′ | Giant field number 384 in Table 1. |
| Dauletabad | USSR | N 36°12′ | E 61°32′ | Giant field number 40 in Table 1. |
| Defa | Libya | N 28° 5′ | E 19°55′ | Giant field number 134 in Table 1. |
| Dehluran | Iran | N 32°30′ | E 47°11′ | See Abu Ghirab (Iraq) |
| Dengizkul'khanzakskoye | USSR | N 39°10′ | E 64°26′ | Giant field number 403 in Table 1. |
| Diamond M | USA | N 32°45′ | W 101° 0′ | See Scurry (USA) |
| Divjake | Albania | N 41° 1′ | E 19°28′ | Giant field number 486 in Table 1. |
| Dnepr | USSR | N 50°36′ | E 32°34′ | See Prilukskoye (USSR) |
| Donmez | USSR | N 36°12′ | E 61°32′ | See Dauletabad (USSR) |
| Dorra | Divided Zone | N 28°57′ | E 49° 7′ | Giant field number 52 in Table 1. |
| Drake Point | Canada | N 76°25′ | W 107°50′ | Giant field number 237 in Table 1. |
| Dukhan | Qatar | N 25°20′ | E 50°47′ | Giant field number 73 in Table 1. |
| Dune | USA | N 31°20′ | W 102°15′ | See McElroy (USA) |
| Dunlin | United Kingdom | N 61°16′ | E 1°36′ | Present estimate less than 500 mm bbl equivalent. |
| Duri | Indonesia | N 1°20′ | E 101°15′ | Giant field number 89 in Table 1. |
| Duvannyy | USSR | N 40° 1′ | E 49°32′ | See Sangachaly Duvannyy (USSR) |
| East Texas | USA | N 32°30′ | W 95° 0′ | Giant field number 56 in Table 1. |
| Ebano | Mexico | N 22°10′ | W 98° 0′ | See Panuco (Mexico) |
| Edda | Norway | N 56°26′ | E 3° 8′ | Giant field number 219 in Table 1. |
| Edofisk | Norway | N 56°32′ | E 3°12′ | Giant field number 98 in Table 1. |
| Ekofisk West | Norway | N 56°33′ | E 3° 5′ | Giant field number 313 in Table 1. |
| El Borma | Algeria | N 31°41′ | E 9°12′ | See El Borma (Tunisia) |
| El Borma | Tunisia | N 31°41′ | E 9°12′ | Giant field number 305 in Table 1. |
| El Morgan | Egypt | N 28°14′ | E 33°26′ | Giant field number 161 in Table 1. |
| Eldfisk | Norway | N 56°26′ | E 3°12′ | Giant field number 225 in Table 1. |
| Elk Basin | USA | N 45° 0′ | W 108°50′ | Giant field number 394 in Table 1. |
| Elk Hills | USA | N 35°10′ | W 119°20′ | Giant field number 141 in Table 1. |
| Elkhov | USSR | N 54°40′ | E 51°30′ | See Novo Elkhov (USSR) |
| Elmworth | Canada | N 54°30′ | W 118°30′ | Giant field number 79 in Table 1. |
| Emeraude Marine | Congo | S 5° 4′ | E 11°50′ | Present estimate less than 500 mm bbl equivalent. |
| Espoir | Ivory Coast | N 5° 0′ | W 4°28′ | Giant field number 244 in Table 1. |
| Eugene Island Block 330 | USA | N 28°30′ | W 91°30′ | Giant field number 437 in Table 1. |
| Eunice | USA | N 32°30′ | W 103°15′ | Giant field number 103 in Table 1. |
| Fadhilli | Saudi Arabia | N 26°57′ | E 49°13′ | Giant field number 230 in Table 1. |
| Fahud | Oman | N 22°20′ | E 56°30′ | Giant field number 268 in Table 1. |
| Fakkeh | Iraq | N 32° 0′ | E 47°39′ | See Jabal Fauqui (Iraq) |
| Faris | Iran | N 31° 0′ | E 49°56′ | See Paris (Iran) |
| Fateh | Dubai | N 25°36′ | E 54°26′ | Giant field number 214 in Table 1. |
| Fateh, Southwest | Dubai | N 25°30′ | E 54°18′ | Giant field number 215 in Table 1. |
| Fedorovskoye | USSR | N 63°12′ | E 73°34′ | Giant field number 55 in Table 1. |
| Feni | Bangladesh | N 23° 0′ | E 91°17′ | Giant fields number 431 in Table 1. |
| Fereidoon | Iran | N 28°32′ | E 49°44′ | See Marjan (Saudi Arabia) |
| Flaxman Island | USA | N 70°20′ | W 146° 0′ | See Point Thompson (USA) |
| Forcados Yokri | Nigeria | N 5°22′ | E 5°18′ | Giant field number 329 in Table 1. |
| Forest | Trinidad | N 10° 0′ | W 61°35′ | See Fyzabad Group (Trinidad) |
| Forties | United Kingdom | N 57°44′ | E 0°58′ | Giant field number 121 in Table 1. |
| Foster | USA | N 32°55′ | W 100°35′ | See Cowden South (USA) |
| Fria | Venezuela | N 8°52′ | W 64°18′ | See Oficina (Venezuela) |
| Frigg | Norway | N 59°53′ | E 2° 5′ | Giant field number 204 in Table 1. |
| Frigg | United Kingdom | N 59°53′ | E 2° 5′ | See Frigg (Norway) |
| Fullerton | USA | N 32°10′ | W 102°40′ | Giant field number 421 in Table 1. |
| Fulmar | United Kingdom | N 56°29′ | E 2° 9′ | Present estimate less than 500 Mmm bbl equivalent. |
| Fyzabad Group | Trinidad | N 10° 0′ | W 61°35′ | Giant field number 457 in Table 1. |
| Gach Saran | Iran | N 30°13′ | E 50°49′ | Giant field number 37 in Table 1. |
| Gassi Touil | Algeria | N 30°23′ | E 6°31′ | Giant field number 256 in Table 1. |
| Gazlinskoye | USSR | N 40°21′ | E 63°20′ | Giant field number 63 in Table 1. |
| Ghasha | Abu Dhabi | N 24°34′ | E 52°55′ | Present estimate less than 500 mm bbl equivalent. |
| Ghawar | Saudi Arabia | N 25°56′ | E 49°14′ | Giant field number 1 in Table 1. |
| Gialo | Libya | N 28°41′ | E 21°24′ | Giant field number 76 in Table 1. |
| Gidgealpa | Australia | S 27°20′ | E 139°30′ | Giant field number 285 in Table 1. |
| Giraldas | Mexico | N 17°43′ | W 93°23′ | See Iris (Mexico) |
| Glynsko Rozbyshevskoye | USSR | N 50°32′ | E 33°36′ | Giant field number 360 in Table 1. |

Table 2. *Continued.*

| Field Name | Country | Location | | Remarks |
|---|---|---|---|---|
| | | Latitude | Longitude | |
| Golden Trend | USA | N 34°40′ | W 97°45′ | Giant field number 366 in Table 1. |
| Goldsmith | USA | N 31°45′ | W 102°30′ | Giant field number 277 in Table 1. |
| Gomez | USA | N 31° 0′ | W 103° 0′ | Giant field number 487 in Table 1. |
| Goodwyn | Australia | S 19°42′ | E 115°54′ | Giant field number 424 in Table 1. |
| Grand Isle Block 43 | USA | N 28°55′ | W 90° 0′ | Giant field number 447 in Table 1. |
| Greta | USA | N 28°15′ | W 97°10′ | Giant field number 253 in Table 1. |
| Grondin | Gabon | S 1°26′ | E 8°40′ | Present estimate less than 500 mm bbl equivalent. |
| Groningen | Germany | N 53°11′ | E 6°46′ | See Groningen (Netherlands) |
| Groningen | Netherlands | N 53°11′ | E 6°46′ | Giant field number 23 in Table 1. |
| Guara East | Venezuela | N 8°57′ | W 63°59′ | Giant field number 364 in Table 1. |
| Guaricema | Brazil | S 11°10′ | W 36°45′ | Giant field number 323 in Table 1. |
| Gubkinskoye | USSR | N 65° 0′ | E 76°42′ | Giant field number 111 in Table 1. |
| Gudao | China | N 37°52′ | E 118°45′ | Giant field number 330 in Table 1. |
| Gugurtlinskoye | USSR | N 40°13′ | E 62°33′ | Giant field number 346 in Table 1. |
| Guico | Venezuela | N 8°52′ | W 64°18′ | See Oficina (Venezuela) |
| Gullfaks | Norway | N 61°11′ | E 2°13′ | Giant field number 183 in Table 1. |
| Gura Ocnitei | Romania | N 45°10′ | E 25°40′ | Present estimate less than 500 mm bbl equivalent. |
| Gydan | USSR | N 70°40′ | E 77°25′ | Giant field number 336 in Table 1. |
| Haft Kel | Iran | N 31°26′ | E 49°33′ | Giant field number 117 in Table 1. |
| Halfayah | Iraq | N 31°47′ | E 47°21′ | Giant field number 505 in Table 1. |
| Halibut | Australia | S 38°24′ | E 148°19′ | Giant field number 328 in Table 1. |
| Hamrin | Iraq | N 34°30′ | E 44° 0′ | Giant field number 238 in Table 1. |
| Handil | Indonesia | S 0°51′ | E 117°17′ | Giant field number 303 in Table 1. |
| Harmaliyah | Saudi Arabia | N 24°32′ | E 49°28′ | Giant field number 122 in Table 1. |
| Hassi Messaoud | Algeria | N 31°45′ | E 6° 5′ | Giant field number 35 in Table 1. |
| Hassi R'Mel | Algeria | N 32°56′ | E 3°16′ | Giant field number 13 in Table 1. |
| Hastings | USA | N 29°35′ | W 95°10′ | Giant field number 311 in Table 1. |
| Hateiba | Libya | N 29°43′ | E 19°41′ | Giant field number 144 in Table 1. |
| Hawkins | USA | N 32°45′ | W 95° 0′ | Giant field number 278 in Table 1. |
| Hecla | Canada | N 76°21′ | W 110°15′ | Giant field number 348 in Table 1. |
| Hewett | United Kingdom | N 53° 4′ | E 1°40′ | Giant field number 402 in Table 1. |
| Hibernia | Canada | N 46°50′ | W 58°50′ | Giant field number 106 in Table 1. |
| Hidalgo | USA | N 26° 0′ | W 98° 0′ | See Reynosa (Mexico) |
| Hofra | Libya | N 29°31′ | E 17°46′ | See Dahra Hofra (Libya) |
| Huangkuanshan | China | N 29°12′ | E 105°52′ | Giant field number 281 in Table 1. |
| Hugoton | USA | N 37° 0′ | W 101° 0′ | Giant field number 21 in Table 1. |
| Huntington Beach | USA | N 33°30′ | W 117°50′ | Giant field number 195 in Table 1. |
| Idd-el-Shargi | Qatar | N 25°38′ | E 52°27′ | Giant field number 234 in Table 1. |
| Idris | Libya | N 29° 0′ | E 20°50′ | See Intisar A (Libya) and Intisar D (Libya) |
| Illinois Old Fields | USA | N 39° 0′ | W 88° 0′ | Present estimate less than 500 mm bbl equivalent. |
| Imo River | Nigeria | N 4°58′ | E 7°12′ | Giant field number 381 in Table 1. |
| Indefatigable | United Kingdom | N 53°24′ | E 2°31′ | Giant field number 321 in Table 1. |
| Infantas | Colombia | N 6°58′ | W 73°45′ | Present estimate less than 500 mm bbl equivalent. |
| Intisar A | Libya | N 29° 2′ | E 20°46′ | Giant field number 202 in Table 1. |
| Intisar D | Libya | N 28°54′ | E 29°58′ | Giant field number 148 in Table 1. |
| Iride | Mexico | N 18° 0′ | W 93° 0′ | See Bermudez (Mexico) |
| Iris | Mexico | N 17°43′ | W 93°23′ | Giant field number 145 in Table 1. |
| Isis | Libya | N 33° 0′ | E 11°30′ | See Isis (Tunisia) |
| Isis | Tunisia | N 33° 0′ | E 11°30′ | Giant field number 240 in Table 1. |
| Issungnak | Canada | N 70° 1′ | W 134°19′ | Giant field number 507 in Table 1. |
| Ixtoc | Mexico | N 19°39′ | W 92°20′ | Giant field number 304 in Table 1. |
| Jabal Fauqui | Iraq | N 32° 0′ | E 47°39′ | Giant field number 239 in Table 1. |
| Jacqueville | Ivory Coast | N 5° 0′ | W 4°28′ | See Espoir (Ivory Coast) |
| Jaladi | Saudi Arabia | N 27° 4′ | E 48°37′ | Giant field number 90 in Table 1. |
| Jalmat | USA | N 32°30′ | W 103°15′ | See Eunice (USA) |
| Jambur | Iraq | N 35° 0′ | E 44°15′ | Giant field number 231 in Table 1. |
| Jana | Saudi Arabia | N 27°30′ | E 48°45′ | Giant field number 493 in Table 1. |
| Jianghan | China | N 30°28′ | E 112°40′ | Giant field number 491 in Table 1. |
| Johnson | USA | N 32°30′ | W 100°35′ | See Cowden South (USA) |
| Jones Creek | Nigeria | N 5°41′ | E 5°23′ | Giant field number 297 in Table 1. |
| Jose Colomo | Mexico | N 17°58′ | W 92°28′ | See Colomo (Mexico) |
| Judy Creek | Canada | N 54°25′ | W 115°40′ | Giant field number 382 in Table 1. |
| July | Egypt | N 28°15′ | E 33°14′ | Giant field number 317 in Table 1. |
| Jusepin | Venezuela | N 9°13′ | W 64° 6′ | See Mata (Venezuela) |
| Kamennyy Log | USSR | N 58°10′ | E 56°20′ | See Yarino Kamennolozhskoye (USSR) |
| Kamni | USSR | N 40°13′ | E 50°50′ | See Neftyanyye Kamni (USSR) |
| Kanaab | Mexico | N 19°39′ | W 92°20′ | See Ixtoc (Mexico) |
| Kandymskoye | USSR | N 39°27′ | E 63°35′ | Giant field number 259 in Table 1. |
| Kangan | Iran | N 27°55′ | E 52° 1′ | Giant field number 39 in Table 1. |
| Karachukhur Zykhskoye | USSR | N 40°22′ | E 50° 2′ | Giant field number 377 in Table 1. |
| Karamai | China | N 45°38′ | E 85° 0′ | Giant field number 157 in Table 1. |
| Karanj | Iran | N 31° 0′ | E 49°52′ | Giant field number 109 in Table 1. |
| Karatchouk | Syria | N 37° 5′ | E 42°10′ | Present estimate less than 500 mm bbl equivalent. |

Table 2. *Continued.*

| Field Name | Country | Location | | Remarks |
|---|---|---|---|---|
| | | Latitude | Longitude | |
| Katy | USA | N 29°30′ | W 95°50′ | Giant field number 223 in Table 1. |
| Kaybob South | Canada | N 54°20′ | W 115°50′ | Giant field number 255 in Table 1. |
| Kazanskoye | USSR | N 57°15′ | E 79°30′ | Giant field number 352 in Table 1. |
| Kelly | USA | N 32°45′ | W 101° 0′ | See Scurry (USA) |
| Kenai | USA | N 60°20′ | W 151°10′ | Giant field number 448 in Table 1. |
| Kern River | USA | N 35°30′ | W 119° 0′ | Giant field number 125 in Table 1. |
| Kettleman Hills North Dome | USA | N 36° 5′ | W 120°15′ | Giant field number 257 in Table 1. |
| Keystone | USA | N 31°55′ | W 103° 0′ | Giant field number 423 in Table 1. |
| Khadzhiy | USSR | N 39°27′ | E 63°35′ | See Kandymskoye (USSR) |
| Khafji | Divided Zone | N 28°25′ | E 48°54′ | See Safaniya (Saudi Arabia) |
| Khangiran | Iran | N 36°34′ | E 60°46′ | Giant field number 235 in Table 1. |
| Kharasaveyskoye | USSR | N 70°58′ | E 67° 0′ | Giant field number 45 in Table 1. |
| Kharg | Iran | N 29°24′ | E 50°18′ | Giant field number 263 in Table 1. |
| Kharvutin | USSR | N 67°28′ | E 72°43′ | Giant field number 350 in Table 1. |
| Kholmogor | USSR | N 62°30′ | E 74° 0′ | Giant field number 502 in Table 1. |
| Khurais | Saudi Arabia | N 25° 5′ | E 48° 3′ | Giant field number 34 in Table 1. |
| Khursaniya | Saudi Arabia | N 27°14′ | E 49°14′ | Giant field number 65 in Table 1. |
| Kingfish | Australia | S 38°36′ | E 148°13′ | Giant field number 186 in Table 1. |
| Kirkuk | Iraq | N 34°45′ | E 44°11′ | Giant field number 11 in Table 1. |
| Kirpichlin | USSR | N 40° 0′ | E 61° 0′ | Giant field number 168 in Table 1. |
| Klump | USA | N 26° 0′ | W 98° 0′ | See Reynosa (Mexico) |
| Koakoak | Canada | N 70°22′ | W 134° 7′ | Giant field number 200 in Table 1. |
| Komsomolskoye | USSR | N 64°20′ | E 76°50′ | Giant field number 93 in Table 1. |
| Kopanoar | Canada | N 70°23′ | W 135° 6′ | Giant field number 242 in Table 1. |
| Korobkovskoye | USSR | N 50°16′ | E 45°23′ | Giant field number 355 in Table 1. |
| Kotur Tepe | USSR | N 39°25′ | E 53°44′ | Giant field number 142 in Table 1. |
| Krasnoyarskoye | USSR | N 51°38′ | E 54°17′ | See Orenburg (USSR) |
| Krasnyy Kholm | USSR | N 51°38′ | E 54°17′ | See Orenburg (USSR) |
| Krestishchenskoye | USSR | N 49°41′ | E 35° 5′ | See Zapadno Krestishchenskoye (USSR) |
| Kruzernshtern | USSR | N 70°50′ | E 66°55′ | Giant field number 50 in Table 1. |
| Kuang | Indonesia | S 4° 0′ | E 104° 0′ | Giant field number 378 in Table 1. |
| Kuleshovskoye | USSR | N 52°40′ | E 50°30′ | Giant field number 312 in Table 1. |
| Kupal | Iran | N 31°13′ | E 49°16′ | Giant field number 150 in Table 1. |
| Kuparuk | USA | N 70°10′ | W 149°30′ | Giant field number 191 in Table 1. |
| Kurten | USA | N 30°50′ | E 96°15′ | Present estimate less than 500 mm bbl equivalent. |
| Kyrtaiol'skoye | USSR | N 57°13′ | E 64°55′ | Giant field number 31 in Table 1. |
| L – 65 | Libya | N 27°48′ | E 22°29′ | See Sarir L (Libya) |
| La Brea | Peru | S 4°40′ | W 81°15′ | Giant field number 75 in Table 1. |
| La Cira | Colombia | N 6°58′ | W 73°45′ | Giant field number 467 in Table 1. |
| La Gloria | USA | N 27°10′ | W 98° 5′ | Giant field number 443 in Table 1. |
| La Paz | Venezuela | N 10°40′ | W 72° 0′ | Giant field number 266 in Table 1. |
| Lab-e-Safid | Iran | N 32°39′ | E 48°41′ | Giant field number 415 in Table 1. |
| Lacq | France | N 43°25′ | W 0°43′ | Giant field number 179 in Table 1. |
| Lagunillas | Venezuela | N 10° 7′ | W 71°16′ | See Bolivar Coastal (Venezuela) |
| Lama | Venezuela | N 9°55′ | W 71°32′ | See Bolivar Coastal (Venezuela) |
| Lamar | Venezuela | N 9°43′ | W 71°32′ | Giant field number 158 in Table 1. |
| Laochunmiao | China | N 39°50′ | E 97°30′ | Giant field number 469 in Table 1. |
| Laverne | USA | N 35°20′ | W 98°10′ | See Mocane Laverne (USA) |
| Lawhah | Saudi Arabia | N 28°15′ | E 49°36′ | Giant field number 427 in Table 1. |
| Layavozhskoye | USSR | N 67°36′ | E 55° 5′ | Giant field number 210 in Table 1. |
| Leduc Woodbend | Canada | N 53° 0′ | W 113° 0′ | Giant field number 470 in Table 1. |
| Leman | United Kingdom | N 53° 5′ | E 2° 8′ | Giant field number 130 in Table 1. |
| Lenghu | China | N 38°40′ | E 93°20′ | Giant field number 212 in Table 1. |
| Leninskoye | USSR | N 39°25′ | E 53°44′ | See Kotur Tepe (USSR) |
| Levelland | USA | N 33°15′ | W 102°30′ | See Slaughter (USA) |
| Lima – Indiana | USA | N 40°55′ | W 84° 5′ | Present estimates less than 500 mm bbl equivalent. |
| Loango | Congo | S 4°50′ | E 11° 0′ | Giant field number 465 in Table 1. |
| Loma de la Lata | Argentina | S 38°30′ | W 68°38′ | Giant field number 222 in Table 1. |
| Long Beach | USA | N 33°50′ | W 118°10′ | Giant field number 211 in Table 1. |
| Luginetskoye | USSR | N 60° 9′ | E 80° 0′ | Giant field number 371 in Table 1. |
| Luhais | Iraq | N 30°18′ | E 46°46′ | Giant field number 483 in Table 1. |
| Lung Nussu | China | N 30°32′ | E 106′ 8′ | Giant field number 379 in Table 1. |
| Lyantor | USSR | N 63°30′ | E 72° 0′ | Giant field number 264 in Table 1. |
| Maastakh | USSR | N 63° | E 123° | Giant field number 221 in Table 1. |
| Mabruk | Libya | N 29°57′ | E 17°17′ | Giant field number 480 in Table 1. |
| Magnus | United Kingdom | N 61°37′ | E 1°17′ | Giant field number 363 in Table 1. |
| Maharah | Saudi Arabia | N 28°23′ | E 49°25′ | Giant field number 466 in Table 1. |
| Mahzan | Qatar | N 25°38′ | E 52°31′ | See Maydan (Qatar) |
| Majnoon | Iraq | N 31° 4′ | E 47°34′ | Giant field number 48 in Table 1. |
| Maleh Kuh | Iran | N 33°13′ | E 47°36′ | Giant field number 261 in Table 1. |
| Malgobek Voznesenskoye | USSR | N 43°36′ | E 44°30′ | Giant field number 88 in Table 1. |
| Malongo North & South | Angola | S 5°21′ | E 12° 5′ | Giant field number 279 in Table 1. |
| Malossa | Italy | N 45°30′ | E 9°34′ | Giant field number 369 in Table 1. |

Table 2. *Continued.*

| Field Name | Country | Location | | Remarks |
|---|---|---|---|---|
| | | Latitude | Longitude | |
| Mamontovskoye | USSR | N 61°24′ | E 72°50′ | Giant field number 139 in Table 1. |
| Manifa | Saudi Arabia | N 27°43′ | E 49° 2′ | Giant field number 10 in Table 1. |
| Mansuri | Iran | N 30°53′ | E 48°50′ | Giant field number 315 in Table 1. |
| Mara | Venezuela | N 10°52′ | W 71°51′ | Giant field number 155 in Table 1. |
| Mari | Pakistan | N 27°56′ | E 69°45′ | Giant field number 345 in Table 1. |
| Marjan | Saudi Arabia | N 28°27′ | E 49°43′ | Giant field number 61 in Table 1. |
| Marlin | Australia | S 38°14′ | E 148°14′ | Giant field number 325 in Table 1. |
| Marun | Iran | N 31°10′ | E 49°19′ | Giant field number 12 in Table 1. |
| Masjid-i-Suleman | Iran | N 32° 0′ | E 49°18′ | Giant field number 188 in Table 1. |
| Mata | Venezuela | N 9°13′ | W 64° 6′ | Giant field number 474 in Table 1. |
| Matzen | Austria | N 48°25′ | E 16°43′ | Giant field number 397 in Table 1. |
| Maui | New Zealand | S 39°40′ | E 173°19′ | Giant field number 254 in Table 1. |
| Maydan | Qatar | N 25°38′ | E 52°31′ | Giant field number 213 in Table 1. |
| Maykop | USSR | N 44°37′ | E 40°15′ | Giant field number 373 in Table 1. |
| Mazalij | Saudi Arabia | N 24°21′ | E 48° 8′ | Giant field number 339 in Table 1. |
| Mazunin | USSR | N 57°10′ | E 56°35′ | Giant field number 105 in Table 1. |
| McArthur River | USA | N 60°45′ | W 152° 0′ | Giant field number 340 in Table 1. |
| McElroy | USA | N 31°20′ | W 102°15′ | Giant field number 316 in Table 1. |
| Medvedev | USSR | N 60°50′ | E 77° 0′ | See Sovetskoye (USSR) |
| Medvezh'ye | USSR | N 67°20′ | E 73°50′ | Giant field number 33 in Table 1. |
| Megionskoye | USSR | N 60°50′ | E 76°26′ | Giant field number 272 in Table 1. |
| Mellion – Rousse | France | N 43°15′ | W 0°30′ | Giant field number 400 in Table 1. |
| Mene Grande | Venezuela | N 9°47′ | W 70°55′ | Giant field number 326 in Table 1. |
| Meren | Nigeria | N 5°47′ | E 4°53′ | Giant field number 327 in Table 1. |
| Mesa Verde | USA | N 36°40′ | W 106°30′ | See Blanco (USA) |
| Meskala | Morocco | N 31°27′ | W 9°21′ | Giant field number 292 in Table 1. |
| Messla | Libya | N 27°55′ | E 22°24′ | Giant field number 163 in Table 1. |
| Messoyakhskoye | USSR | N 69°20′ | E 82°30′ | Giant field number 99 in Table 1. |
| Midale | Canada | N 49°25′ | W 103°30′ | See Weyburn Midale (Canada) |
| Midgard | Norway | N 65° 5′ | E 7°15′ | Giant field number 372 in Table 1. |
| Midway Sunset | USA | N 35°10′ | W 119°30′ | Giant field number 107 in Table 1. |
| Minagish | Kuwait | N 29° 5′ | E 47°31′ | Giant field number 81 in Table 1. |
| Minas | Indonesia | N 0°45′ | E 101°26′ | Giant field number 46 in Table 1. |
| Miranga | Brazil | S 12°21′ | W 38°12′ | Giant field number 393 in Table 1. |
| Mocane Laverne | USA | N 35°20′ | W 98°10′ | Giant field number 353 in Table 1. |
| Molango West | Angola | S 5°25′ | E 12° 3′ | Giant field number 298 in Table 1. |
| Monroe | USA | N 32°30′ | W 92° 0′ | Giant field number 199 in Table 1. |
| Monument | USA | N 32°30′ | W 103°15′ | See Eunice (USA) |
| Moomba | Australia | S 27°20′ | E 141°30′ | Giant field number 286 in Table 1. |
| Moran | India | N 27°20′ | E 95°10′ | Giant field number 41 in Table 1. |
| Morecambe | United Kingdom | N 53°52′ | E 3°38′ | Giant field number 289 in Table 1. |
| Moreni | Romania | N 45°20′ | E 26°40′ | Present estimate less than 500 mm bbl equivalent. |
| Morgan | Egypt | N 28°14′ | E 33°26′ | See: El Morgan (Egypt) |
| Mubarraz | Abu Dhabi | N 24°27′ | E 53°42′ | Giant field number 120 in Table 1. |
| Mukhanovskoye | USSR | N 53° 0′ | E 50°30′ | Giant field number 152 in Table 1. |
| Mulata | Venezuela | N 9°13′ | W 64° 6′ | See Mata (Venezuela) |
| Murban Bab | Abu Dhabi | N 23°52′ | E 53°32′ | Giant field number 66 in Table 1. |
| Murban Bu Hasa | Abu Dhabi | N 23°30′ | E 53°10′ | Giant field number 32 in Table 1. |
| Murchison | Norway | N 61°23′ | E 1°44′ | See Murchison (United Kingdom) |
| Murchison | United Kingdom | N 61°23′ | E 1°44′ | Present estimate less than 500 mm bbl equivalent. |
| Muri | Venezuela | N 9°13′ | W 64° 6′ | See Mata (Venezuela) |
| Myl'dzhino | USSR | N 59°25′ | E 78°10′ | Giant field number 337 in Table 1. |
| Nafoora | Libya | N 29°10′ | E 21°29′ | See Augila (Libya) |
| Naft-i-Safid | Iran | N 31°39′ | E 49°15′ | Giant field number 274 in Table 1. |
| Nahorkatiya | India | N 27°17′ | E 95′20′ | Giant field number 472 in Table 1. |
| Nahr Umr | Iraq | N 30°46′ | E 47°42′ | Giant field number 229 in Table 1. |
| Naipskoye | USSR | N 40°57′ | E 61°12′ | Giant field number 182 in Table 1. |
| Nakhodkinskoye | USSR | N 67°30′ | E 77°30′ | Giant field number 342 in Table 1. |
| Nar | Iran | N 27°47′ | E 52°18′ | Giant field number 100 in Table 1. |
| Naranjos | Mexico | N 21°20′ | W 97°41′ | See Cerro Azul (Mexico) |
| Nasr | Abu Dhabi | N 25°15′ | E 53°30′ | Present estimate less than 500 mm bbl equivalent. |
| Nasser | Libya | N 28°54′ | E 19°48′ | Giant field number 95 in Table 1. |
| Natih | Oman | N 22°26′ | E 56°43′ | Giant field number 445 in Table 1. |
| Neftyanyye Kamni | USSR | N 40°13′ | E 50°50′ | Giant field number 196 in Table 1. |
| Nembe Creek | Nigeria | N 4°26′ | E 6°19′ | Giant field number 359 in Table 1. |
| New Taching | China | N 46°37′ | E 125°18′ | See Taking (China) |
| Neytin | USSR | N 69°50′ | E 69°45′ | Giant field number 349 in Table 1. |
| Nilam | Indonesia | S 1° 0′ | E 117° 0′ | Giant field number 429 in Table 1. |
| Ninian | United Kingdom | N 60°48′ | E 1°29′ | Giant field number 208 in Table 1. |
| Nipa | Venezuela | N 9° 5′ | W 64°13′ | Giant field number 413 in Table 1. |
| Nohoch | Mexico | N 19°30′ | W 91°57′ | See Cantarell Complex (Mexico) |
| Northwest Dome | Qatar | N 26°34′ | E 51°46′ | Giant field number 19 in Table 1. |
| Nove Elkhov | USSR | N 54°40′ | E 51°30′ | Giant field number 184 in Table 1. |

Table 2. *Continued.*

| Field Name | Country | Location | | Remarks |
|---|---|---|---|---|
| | | Latitude | Longitude | |
| Novoportovskoye | USSR | N 72°23′ | E 68° 0′ | Giant field number 275 in Table 1. |
| Novyy Port | USSR | N 72°23′ | E 68° 0′ | See Novoportovskoye (USSR) |
| Nurm | USSR | N 69° 8′ | E 71°10′ | See Sredneyamal'skoye (USSR) |
| Nyda | USSR | N 66°30′ | E 73° 0′ | Giant field number 404 in Table 1. |
| Obagi | Nigeria | N 5°14′ | E 6°40′ | Giant field number 426 in Table 1. |
| Ocnitei | Romania | N 45°10′ | E 25°40′ | See Gura Ocnitei (Rumania) |
| Oficina | Venezuela | N 8°52′ | W 64°18′ | Giant field number 251 in Table 1. |
| Ojaicaque | Mexico | N 18° 0′ | W 93° 0′ | See Bermudez (Mexico) |
| Okan | Nigeria | N 5°30′ | E 5° 5′ | Giant field number 489 in Table 1. |
| Oklahoma City | USA | N 35°30′ | W 97°30′ | Giant field number 314 in Table 1. |
| Oktyabr'skoye | USSR | N 42°40′ | E 46° 0′ | Present estimate less than 500 mm bbl equivalent |
| Old Ocean | USA | N 28°55′ | W 95°45′ | Giant field number 250 in Table 1. |
| Olinda | USA | N 34° 0′ | W 118°30′ | See Brea (USA) |
| Opon | Colombia | N 6°25′ | W 73°52′ | Giant field number 344 in Table 1. |
| Orenburg | USSR | N 51°38′ | E 54°17′ | Giant field number 26 in Table 1. |
| Oseberg | Norway | N 60°33′ | E 2°47′ | Giant field number 173 in Table 1. |
| Ozeksuatskoye | USSR | N 44°32′ | E 45°20′ | Giant field number 473 in Table 1. |
| Palo Seco | Trinidad | N 10° 0′ | W 61°35′ | See Fyzabad Group (Trinidad) |
| Palyanovo | USSR | N 61°45′ | E 67° 1′ | Giant field number 218 in Table 1. |
| Panhandle | USA | N 35°30′ | W 101°15′ | See Hugoton (USA) |
| Panuco | Mexico | N 22°10′ | W 98° 0′ | Giant field number 249 in Table 1. |
| Parinas | Peru | S 4°40′ | W 81°15′ | See La Brea (Peru) |
| Paris | Iran | N 31° 0′ | E 49°56′ | Giant field number 72 in Table 1. |
| Pars | Iran | N 27°58′ | E 51°15′ | Giant field number 59 in Table 1. |
| Parsons Lake | Canada | N 68°40′ | W 133°30′ | Giant field number 395 in Table 1. |
| Pazanan | Iran | N 32°22′ | E 50°11′ | Giant field number 74 in Table 1. |
| Pelagiada | USSR | N 45°13′ | E 41°57′ | See Sevro Stavropol Pelagia (USSR) |
| Pelyatkinskoye | USSR | N 69°40′ | E 81°40′ | Giant field number 217 in Table 1. |
| Pembina | Canada | N 56° 0′ | W 115°30′ | Giant field number 133 in Table 1. |
| Peschanyy More | USSR | N 40°18′ | E 50° 0′ | Giant field number 343 in Table 1. |
| Petrel | Australia | S 12°50′ | E 128°28′ | Giant field number 347 in Table 1. |
| Piper | United Kingdom | N 58°28′ | E 0°16′ | Giant field number 271 in Table 1. |
| Pirital | Venezuela | N 9°13′ | W 64° 6′ | See Mata (Venezuela) |
| Placid | Netherlands | N 53°29′ | E 4°11′ | Giant field number 406 in Table 1. |
| Pledger | USA | N 29°10′ | W 95°55′ | Giant field number 468 in Table 1. |
| Point Arguillo | USA | N 34°30′ | W 120°40′ | Giant field number 509 in Table 1. |
| Point Thompson | USA | N 70°20′ | W 146° 0′ | Giant field number 205 in Table 1. |
| Pokachev | USSR | N 61° 0′ | E 75°20′ | Giant field number 301 in Table 1. |
| Poza Rica | Mexico | N 20°30′ | W 97°27′ | Giant field number 118 in Table 1. |
| Pravdinskoye | USSR | N 61°36′ | E 71°48′ | Giant field number 160 in Table 1. |
| Prilukskoye | USSR | N 50°36′ | E 32°24′ | Giant field number 383 in Table 1. |
| Prudhoe Bay | USA | N 70°20′ | W 149° 0′ | Giant field number 18 in Table 1. |
| Puckett | USA | N 30°40′ | W 102°30′ | Giant field number 471 in Table 1. |
| Punga | USSR | N 62°40′ | E 64°18′ | See Punginskoye (USSR) |
| Punginskoye | USSR | N 62°40′ | E 64°18′ | Present estimate less than 500 mm bbl equivalent. |
| Qatif | Saudi Arabia | N 25°32′ | E 49°59′ | Giant field number 49 in Table 1. |
| Qishm – Salagh | Iran | N 26°42′ | E 55°35′ | Giant field number 128 in Table 1. |
| Quarry | Trinidad | N 10° 0′ | W 61°35′ | See Fazabad Group (Trinidad) |
| Quiriquire | Venezuela | N 10° 0′ | W 63°10′ | Giant field number 226 in Table 1. |
| Qusahwira | Abu Dhabi | N 22°40′ | E 55° 0′ | Giant field number 333 in Table 1. |
| Rag-e-Safid | Iran | N 32°27′ | E 50° 0′ | Giant field number 70 in Table 1. |
| Raguba | Libya | N 29° 5′ | E 19° 5′ | Giant field number 216 in Table 1. |
| Rainbow | Canada | N 57°30′ | W 119° 0′ | Present estimate less than 500 mm bbl equivalent. |
| Ramadan | Egypt | N 28°18′ | E 33°19′ | Present estimate less than 500 mm bbl equivalent. |
| Ramany | USSR | N 40°40′ | E 50°20′ | See Balakhano Sabuchino (USSR) |
| Rangely | USA | N 40°10′ | W 108°45′ | Giant field number 295 in Table 1. |
| Rankin, North | Australia | S 19°36′ | E 116°07′ | Giant field number 172 in Table 1. |
| Ratga | Kuwait | N 30° 2′ | E 47° 3′ | Giant field number 241 in Table 1. |
| Raudhatain | Kuwait | N 29°52′ | E 47°45′ | Giant field number 27 in Table 1. |
| Redwater | Canada | N 53°50′ | W 113°10′ | Giant field number 296 in Table 1. |
| Renqiu | China | N 38°45′ | E 116° 0′ | Giant field number 123 in Table 1. |
| Reynosa | Mexico | N 26° 0′ | W 98° 0′ | Giant field number 398 in Table 1. |
| Rhourde el Baguel | Algeria | N 31°23′ | E 6°57′ | Giant field number 450 in Table 1. |
| Rhourde Nouss | Algeria | N 29°42′ | E 6°44′ | Giant field number 209 in Table 1. |
| Riachuelo | Brazil | S 10°45′ | W 37°15′ | Giant field number 220 in Table 1. |
| Rimthan | Saudi Arabia | N 28°29′ | E 47° 4′ | Giant field number 388 in Table 1. |
| Rincon | USA | N 34°20′ | W 119°20′ | See Ventura Avenue (USA) |
| Rio Nuevo | Mexico | N 17°40′ | W 93°20′ | Giant field number 503 in Table 1. |
| Rio Vista | USA | N 38°10′ | W 121°45′ | Giant field number 396 in Table 1. |
| Riohacha | Colombia | N 11°50′ | W 72°50′ | See Chuchupa (Colombia) |
| Romashkinskoye | USSR | N 54°40′ | E 52°20′ | Giant field number 16 in Table 1. |
| Rostam | Iran | N 25°40′ | E 52°10′ | Present estimate less than 500 mm bbl equivalent. |
| Rousse | France | N 43°15′ | W 0°30′ | See Mellion-Rousse (France) |

Table 2. *Continued.*

| Field Name | Country | Location |  | Remarks |
|---|---|---|---|---|
|  |  | Latitude | Longitude |  |
| Rozbyshev | USSR | N 50°32′ | E 33°36′ | See Glynsko Rozbyshevskoye (USSR) |
| Rumaila | Iraq | N 30°26′ | E 47°20′ | Giant field number 17 in Table 1. |
| Rumaila North | Iraq | N 30°30′ | E 47°16′ | Giant field number 43 in Table 1. |
| Rumaitha | Abu Dhabi | N 23°35′ | E 54°10′ | Present estimate less than 500 mm bbl equivalent. |
| Russkoye | USSR | N 66°40′ | E 80°25′ | Giant field number 495 in Table 1. |
| Saatyh al Raazboot | Abu Dhabi | N 24°45′ | E 53° 0′ | Giant field number 498 in Table 1. |
| Sabriya | Kuwait | N 29°50′ | E 47°47′ | Giant field number 60 in Table 1. |
| Sabunchi | USSR | N 40°40′ | E 50°20′ | See Balakhano Sabuchino (USSR) |
| Sacha | Ecuador | S 0°20′ | W 76°53′ | Giant field number 358 in Table 1. |
| Safaniya | Saudi Arabia | N 28° 2′ | E 48°45′ | Giant field number 4 in Table 1. |
| Sahil | Abu Dhabi | N 23°40′ | E 54°22′ | Giant field number 435 in Table 1. |
| Sajaa | Sharjah | N 25°20′ | E 55°40′ | Giant field number 185 in Table 1. |
| Salagh | Iran | N 26°42′ | E 55°35′ | See Qishm-Salagh (Iran) |
| Salt Creek | USA | N 43°30′ | W 106°15′ | Giant field number 270 in Table 1. |
| Salymskoye | USSR | N 60°30′ | E 72° 0′ | Giant field number 488 in Table 1. |
| Salzwedel | Germany | N 52°53′ | E 11° 4′ | Giant field number 401 in Table 1. |
| Samah | Libya | N 28°13′ | E 19° 8′ | Giant field number 485 in Table 1. |
| Samantepinskoye | USSR | N 39° 5′ | E 64°19′ | Giant field number 399 in Table 1. |
| Samaria | Mexico | N 18° 0′ | W 93° 0′ | See Bermudez (Mexico) |
| Samburg | USSR | N 66°27′ | E 77° 5′ | See Yuzhno Samburg (USSR) |
| Samotlor | USSR | N 61°10′ | E 77° 0′ | Giant field number 15 in Table 1. |
| San Andres | Mexico | N 20°24′ | W 97°11′ | Present estimate less than 500 mm bbl equivalent. |
| San Ardo | USA | N 36° 0′ | W 120°50′ | Giant field number 433 in Table 1. |
| San Juan | USA | N 36°40′ | W 106°30′ | See Basin (USA) and Blanco (USA) |
| Sand Belt | USA | N 31°20′ | E 102°45′ | See South Sand Belt (USA) |
| Sand Hills | USA | N 31°30′ | W 102°20′ | Giant field number 456 in Table 1. |
| Sangachaly Duvannyy | USSR | N 40° 1′ | E 49°32′ | Giant field number 248 in Table 1. |
| Sansinena | USA | N 34° 0′ | W 118°30′ | See Brea (USA) |
| Santa Barbara | Venezuela | N 9°13′ | W 64° 6′ | See Mata (Venezuela) |
| Santa Fe Springs | USA | N 33°55′ | W 118° 0′ | Giant field number 310 in Table 1. |
| Sarajeh | Iran | N 35° 0′ | E 51°30′ | Giant field number 282 in Table 1. |
| Sarir C | Libya | N 27°39′ | E 22°31′ | Giant field number 51 in Table 1. |
| Sarir L | Libya | N 27°48′ | E 22°29′ | Giant field number 201 in Table 1. |
| Sarkhan | Iran | N 33°18′ | E 47°45′ | Giant field number 197 in Table 1. |
| Sarkhun | Iran | N 27°20′ | E 56°26′ | Giant field number 206 in Table 1. |
| Sassan | Iran | N 25°31′ | E 53° 8′ | Giant field number 127 in Table 1. |
| Satah | Abu Dhabi | N 24°55′ | E 52°33′ | Giant field number 309 in Table 1. |
| Satun | Thailand | N 91°61′ | E 101°24′ | Giant field number 442 in Table 1. |
| Scurry | USA | N 32°45′ | W 101° 0′ | Giant field number 126 in Table 1. |
| Seeligson | USA | N 27°15′ | W 98° 0′ | See Borregos (USA) |
| Semakovskoye | USSR | N 69° 0′ | E 75° 0′ | Giant field number 82 in Table 1. |
| Seminole | USA | N 32°30′ | W 102°40′ | Giant field number 438 in Table 1. |
| Seria | Brunei | N 4°37′ | E 114°19′ | Giant field number 112 in Table 1. |
| Serrablo | Spain | N 42°35′ | W 0°24′ | Giant field number 412 in Table 1. |
| Severo Komsomol | USSR | N 64°40′ | E 75°30′ | Giant field number 288 in Table 1. |
| Severo Stavropol | USSR | N 45°13′ | E 41°57′ | Giant field number 181 in Table 1. |
| Severo Urengoy | USSR | N 67° 7′ | E 76° 0′ | Giant field number 53 in Table 1. |
| Severo Varyegan | USSR | N 62°50′ | E 74°55′ | Giant field number 387 in Table 1. |
| Shah | Abu Dhabi | N 22°50′ | E 54°20′ | Giant field number 492 in Table 1. |
| Shatlyk | USSR | N 37°31′ | E 61°15′ | See Shatlyskoye (USSR) |
| Shatlyskoye | USSR | N 37°31′ | E 61°15′ | Giant field number 54 in Table 1. |
| Shaybah | Saudi Arabia | N 22°33′ | E 54° 2′ | Giant field number 47 in Table 1. |
| Shebelinka | USSR | N 49°27′ | E 36°37′ | Giant field number 84 in Table 1. |
| Sheketli | USSR | N 37°31′ | E 61°15′ | See Shatlyskoye (USSR) |
| Shekhitli | USSR | N 37°31′ | E 61°15′ | See Shatlyskoye (USSR) |
| Shengli | China | N 37°32′ | E 118°30′ | Giant field number 58 in Table 1. |
| Shibarghan | Afghanistan | N 37° 0′ | E 66° 0′ | Giant field number 370 in Table 1. |
| Shiyoukou – Tungchi | China | N 29° 5′ | E 106°42′ | Giant field number 207 in Table 1. |
| Shkapovskoye | USSR | N 54°32′ | E 54° 0′ | Giant field number 189 in Table 1. |
| Sho-Vel-Tum | USA | N 34°30′ | W 97°30′ | Giant field number 153 in Table 1. |
| Shuaiba | Oman | N 21°54′ | E 56°13′ | See Yibal (Oman) |
| Shurom | Iran | N 31°12′ | E 51° 0′ | Present estimate less than 500 mm bbl equivalent. |
| Shurtanskoye | USSR | N 38°58′ | E 67° 2′ | Giant field number 164 in Table 1. |
| Shushufindi | Ecuador | S 0°10′ | W 76°38′ | Giant field number 299 in Table 1. |
| Sidi Abderrahman | Tunisia | N 37°30′ | E 10°30′ | Giant field number 262 in Table 1. |
| Sitio Grande | Mexico | N 17°47′ | W 93° 7′ | Giant field number 308 in Table 1. |
| Slaughter | USA | N 33°15′ | W 102°30′ | Giant field number 143 in Table 1. |
| Sleipner | Norway | N 58°30′ | E 1°43′ | Giant field number 174 in Table 1. |
| Smackover | USA | N 33°20′ | W 92°50′ | Giant field number 391 in Table 1. |
| Snapper | Australia | S 38°12′ | E 148° 0′ | Giant field number 496 in Table 1. |
| Snyder | USA | N 32°45′ | W 101° 0′ | See Scurry (USA) |
| Soldado Main | Trinidad | N 10°12′ | W 61°59′ | Giant field number 475 in Table 1. |
| Sole West | United Kingdom | N 53°45′ | E 1° 5′ | See West Sole (United Kingdom) |

Table 2. *Continued.*

| Field Name | Country | Location | | Remarks |
|---|---|---|---|---|
| | | Latitude | Longitude | |
| Solenaya | USSR | N 69° 5′ | E 81°50′ | See Yuzhno Solenaya (USSR) |
| Sorgut | USSR | N 58° 0′ | E 78° 0′ | See Yuzhno Cheremshanskoye (USSR) |
| Sosnino Sovetskoye | USSR | N 60°50′ | E 77° 0′ | See Sovetskoye (USSR) Full name = Sosnino Sovetskoye Medvede |
| Souedie | Syria | N 26°56′ | E 42° 8′ | See Suwaidiyah (Syria) |
| South pass Block 24 | USA | N 29° 0′ | W 89°15′ | Giant field number 376 in Table 1. |
| South Pass Block 27 | USA | N 28°58′ | W 89°20′ | Giant field number 453 in Table 1. |
| South Sand Belt | USA | N 31°20′ | W 102°45′ | Giant field number 267 in Table 1. |
| Sovetskoye | USSR | N 60°50′ | E 77° 0′ | Giant field number 68 in Table 1. |
| Sovietabad | USSR | N 36°12′ | E 61°32′ | See Dauletabad (USSR) |
| Spraberry Trend | USA | N 31°45′ | W 101°30′ | Giant field number 392 in Table 1. |
| Sredne Botnobin | USSR | N 61°15′ | E 111°40′ | Giant field number 87 in Table 1. |
| Sredne Vilyuy | USSR | N 63°48′ | E 122°20′ | Giant field number 167 in Table 1. |
| Sredneyamal'skoye | USSR | N 69° 8′ | E 71°10′ | Giant field number 193 in Table 1. |
| Starogrozneskoye | USSR | N 43°20′ | E 45°40′ | Giant field number 354 in Table 1. |
| Statfjord | Norway | N 61°12′ | E 1°50′ | Giant field number 77 in Table 1. |
| Statfjord | United Kingdom | N 61°12′ | E 1°50′ | See Statfjord (Norway) |
| Statvik | Norway | N 61°11′ | E 2°13′ | See Gullfaks (Norway) |
| Stavropol | USSR | N 45°13′ | E 41°57′ | See Sevro Stavropol (USSR) |
| Sui | Pakistan | N 28°39′ | E 69°12′ | Giant field number 169 in Table 1. |
| Sunset | USA | N 35°10′ | W 119°30′ | See Midway Sunset (USA) |
| Surakhanskoye | USSR | N 40°26′ | E 50° 2′ | Giant field number 265 in Table 1. |
| Surgut | USSR | N 62°27′ | E 73° 6′ | See Zapadno Surgutskoye (USSR) |
| Suwaidiyah | Syria | N 36°56′ | E 42° 8′ | Giant field number 159 in Table 1. |
| Swan Hills | Canada | N 54°45′ | W 115°40′ | Giant field number 260 in Table 1. |
| Swan Hills South | Canada | N 54°30′ | W 115°40′ | Giant field number 459 in Table 1. |
| T.C.B. | USA | N 27°15′ | W 98° 0′ | See Borregos (USA) |
| Tacat | Venezuela | N 9°13′ | W 64° 6′ | See Mata (Venezuela) |
| Taching | China | N 46°30′ | E 125°20′ | See Daqing (China) |
| Taking | China | N 46°37′ | E 125°18′ | Giant field number 194 in Table 1. |
| Talara | Peru | S 4°40′ | W 81°15′ | See La Brea (Peru) |
| Tambey | USSR | N 71°21′ | E 71°59′ | See Yuzhno Tambey (USSR) |
| Tang-i-bijar | Iran | N 33°42′ | E 45°54′ | Giant field number 420 in Table 1. |
| Taratunich | Mexico | N 19°39′ | W 92°20′ | See Ixtoc (Mexico) |
| Tarkosalinskoye | USSR | N 65° 0′ | E 77°45′ | See either Zapadno Tarkosalin (USSR) or Vostochno |
| Taz | USSR | N 67°20′ | E 78°54′ | See Tazovskoye (USSR) |
| Tazovskoye | USSR | N 67°20′ | E 78°54′ | Giant field number 283 in Table 1. |
| Tepe | USSR | N 39°25′ | E 53°44′ | See Kotur Tepe (USSR) |
| Thistle | United Kingdom | N 61°22′ | E 1°35′ | Giant field number 460 in Table 1. |
| Thompson | USA | N 29°20′ | W 95°30′ | Giant field number 441 in Table 1. |
| Tia Juana | Venezuela | N 10°15′ | W 71°25′ | See Bolivar Coastal (Venezuela) |
| Tiger Shoal | USA | N 29°20′ | W 92° 0′ | Giant field number 444 in Table 1. |
| Timbalier Bay | USA | N 29° 5′ | W 90°22′ | Giant field number 367 in Table 1. |
| Tom O'Connor | USA | N 28°15′ | W 97°10′ | See Greta (USA) |
| Troitskoye | USSR | N 45° 5′ | E 38° 5′ | See Anastasiyevsko (USSR) |
| Troll | Norway | N 60°46′ | E 3°33′ | Giant field number 36 in Table 1. |
| Tuba | Iraq | N 30°23′ | E 47°26′ | Giant field number 481 in Table 1. |
| Tul'skiy | USSR | N 44°30′ | E 41°30′ | Giant field number 375 in Table 1. |
| Tungchi | China | N 29° 5′ | E 106°42′ | See Shiyoukou-Tungchi (China) |
| Tuymazy | USSR | N 54°32′ | E 53°35′ | Giant field number 227 in Table 1. |
| 28 April | USSR | N 40°10′ | E 50°52′ | Present estimate less than 500 mm bbl equivalent. |
| Uchkyrskoye | USSR | N 40° 9′ | E 63° 3′ | Giant field number 484 in Table 1. |
| Umm ad Dalkh | Abu Dhabi | N 24°33′ | E 54° 8′ | Giant field number 151 in Table 1. |
| Umm Gudair | Divided Zone | N 28°50′ | E 47°42′ | See Umm Gudair (Kuwait) |
| Umm Gudair | Kuwait | N 28°50′ | E 47°42′ | Giant field number 147 in Table 1. |
| Umm Shaif | Abu Dhabi | N 25°13′ | E 53°13′ | Giant field number 57 in Table 1. |
| Unity | Sudan | N 9°28′ | E 29°41′ | Giant field number 269 in Table 1. |
| Urdaneta | Venezuela | N 10° 6′ | W 71°50′ | Giant field number 232 in Table 1. |
| Urengoy | USSR | N 66°40′ | E 76°45′ | Giant field number 3 in Table 1. |
| Urtabulakskoye | USSR | N 39° 2′ | E 64°31′ | Giant field number 385 in Table 1. |
| Usa | USSR | N 66°10′ | E 57°40′ | See Usanovskoye (USSR) |
| Usanovskoye | USSR | N 66°10′ | E 57°40′ | Giant field number 356 in Table 1. |
| Usinskoye | USSR | N 66° 8′ | E 57°16′ | Giant field number 85 in Table 1. |
| Ust Balyk | USSR | N 62°24′ | E 72°40′ | See Ust'bakykskoye (USSR) |
| Ust'bakykskoye | USSR | N 62°24′ | E 72°40′ | Giant field number 102 in Table 1. |
| Uzen'skoye | USSR | N 43°15′ | E 52°45′ | Giant field number 129 in Table 1. |
| Vacuum | USA | N 33° 0′ | W 103°15′ | Giant field number 449 in Table 1. |
| Valhall | Norway | N 56°16′ | E 3°24′ | Present estimate less than 500 mm bbl equivalent |
| Van | USA | N 32°30′ | W 95°30′ | Giant field number 432 in Table 1. |
| Vat'yegan | USSR | N 61° 5′ | E 76° 0′ | Giant field number 500 in Table 1. |
| Vega East | Italy | N 36°32′ | E 14°38′ | Giant field number 365 in Table 1. |
| Ventura Avenue | USA | N 34°20′ | W 119°20′ | Giant field number 175 in Table 1. |
| Venture | Canada | N 44° 2′ | W 59°37′ | Giant field number 390 in Table 1. |

Table 2. *Continued.*

| Field Name | Country | Location | | Remarks |
|---|---|---|---|---|
| | | Latitude | Longitude | |
| Verkhne Vilyuchanskoye | USSR | N 64° 0′ | E 122° 0′ | Giant field number 115 in Table 1. |
| Verkhnepurpey | USSR | N 65°10′ | E 76°50′ | Giant field number 411 in Table 1. |
| Vermilion Block 14 | USA | N 29°30′ | W 92°15′ | Giant field number 477 in Table 1. |
| Vermilion Block 39 | USA | N 29°20′ | W 92°15′ | Giant field number 454 in Table 1. |
| Viking | United Kingdom | N 53°35′ | E 2° 8′ | Present estimate less than 500 mm bbl equivalent. |
| Vilyuchanskoye | USSR | N 64° 0′ | E 122° 0′ | See Verkhne Vilyuchanskoye (USSR) |
| Vilyuyskoye | USSR | N 63°48′ | E 122°20′ | See Sredne Vilyuy (USSR) |
| Vostochno Tarkosalin | USSR | N 65° 8′ | E 78°23′ | Giant field number 302 in Table 1. |
| Vostochno Urengoy | USSR | N 67° 5′ | E 77°15′ | See Urengoy (USSR) |
| Vostochnyy Daul | USSR | N 36°12′ | E 61°32′ | See Dauletabad (USSR) |
| Vozey | USSR | N 66°40′ | E 57°20′ | Giant field number 236 in Table 1. |
| Voznesenka | USSR | N 43°36′ | E 44°30′ | See Malgobek Voznesenskoye (USSR) |
| Vuktyl'skoye | USSR | N 63°36′ | E 57°17′ | Giant field number 71 in Table 1. |
| Vyngapurovskoye | USSR | N 63°12′ | E 76°42′ | Giant field number 137 in Table 1. |
| Wafra | Divided Zone | N 28°36′ | E 47°54′ | Giant field number 101 in Table 1. |
| Waha | Libya | N 28°24′ | E 19°55′ | Giant field number 171 in Table 1. |
| Wapati | Canada | N 54°30′ | W 118°30′ | See: Elmworth (Canada) |
| Ward Estes | USA | N 31°20′ | W 102°45′ | See South Sand Belt (USA) |
| Wasson | USA | N 33° 5′ | W 102°45′ | Giant field number 114 in Table 1. |
| Webster | USA | N 29°20′ | W 95°10′ | Giant field number 361 in Table 1. |
| West Delta Block 30 | USA | N 29°10′ | W 89°40′ | Giant field number 452 in Table 1. |
| West Delta Block 73 | USA | N 28°55′ | W 89°40′ | Present estimate less than 500 mm bbl equivalent. |
| West Ranch | USA | N 29° 0′ | W 96°30′ | Giant field number 434 in Table 1. |
| West Sole | United Kingdom | N 53°45′ | E 1° 5′ | Giant field number 490 in Table 1. |
| West Surgut | USSR | N 62°27′ | E 72° 6′ | See Zapadno Surgutskoye (USSR) |
| Weyburn Midale | Canada | N 49°25′ | W 103°30′ | Present estimate less than 500 mm bbl equivalent. |
| Whitefish | Canada | N 77°12′ | W 106°53′ | Giant field number 291 in Table 1. |
| Whitney Canyon | USA | N 41°30′ | W 110°40′ | Giant field number 246 in Table 1. |
| Wilmington | USA | N 33°45′ | W 118°20′ | Giant field number 91 in Table 1. |
| Woodbend | Canada | N 53° 0′ | W 113° 0′ | See Leduc Woodbend (Canada) |
| Wustrow | Germany | N 52°53′ | E 11° 0′ | See Salzwedel (Germany) |
| Ya-erh-hsia | China | N 39°50′ | E 97°30′ | See Laochunmio (China) |
| Yamburg | USSR | N 68°10′ | E 76° 0′ | Giant field number 6 in Table 1. |
| Yamsovey | USSR | N 65°36′ | E 76° 0′ | Giant field number 94 in Table 1. |
| Yarino Kamennolozhskoye | USSR | N 58°10′ | E 56°20′ | Giant field number 458 in Table 1. |
| Yates | USA | N 31° 0′ | W 102°50′ | Giant field number 113 in Table 1. |
| Yefremovskoye | USSR | N 49°28′ | E 36°15′ | Giant field number 307 in Table 1. |
| Yetypurovskoye | USSR | N 64° 0′ | E 77°40′ | Giant field number 407 in Table 1. |
| Yibal | Oman | N 21°54′ | E 56°13′ | Giant field number 341 in Table 1. |
| Yokri | Nigeria | N 5°22′ | E 5°18′ | See Forcados Yokri (Nigeria) |
| Yubileynoye | USSR | N 64°50′ | E 75°57′ | Giant field number 97 in Table 1. |
| Yumen | China | N 39°50′ | E 97°30′ | See Laochunmio (China) |
| Yuzhno Cheremshanskoye | USSR | N 58° 0′ | E 78° 0′ | Giant field number 166 in Table 1. |
| Yuzhno Russkoye | USSR | N 66° 5′ | E 80°25′ | Giant field number 138 in Table 1. |
| Yuzhno Samburg | USSR | N 66°27′ | E 77° 5′ | Giant field number 140 in Table 1. |
| Yuzhno Solenaya | USSR | N 69° 5′ | E 81°50′ | Present estimate less than 500 mm bbl equivalent. |
| Yuzhno Tambey | USSR | N 71°21′ | E 71°59′ | Giant field number 410 in Table 1. |
| Zakum | Abu Dhabi | N 24°52′ | E 53°41′ | Giant field number 9 in Table 1. |
| Zapadno Krestishchenskoy | USSR | N 49°41′ | E 35° 5′ | Giant field number 180 in Table 1. |
| Zapadno Surgutskoye | USSR | N 62°27′ | E 73° 6′ | Giant field number 119 in Table 1. |
| Zapadno Tarkosalin | USSR | N 64°53′ | E 77°15′ | Giant field number 408 in Table 1. |
| Zapolyarnoye | USSR | N 67° 0′ | E 79°30′ | Giant field number 14 in Table 1. |
| Zarrarah | Abu Dhabi | N 22°34′ | E 54° 2′ | See Shaybah (Saudi Arabia) |
| Zarzaitine | Algeria | N 28° 3′ | E 9°47′ | Giant field number 177 in Table 1. |
| Zelten | Libya | N 28°54′ | E 19°48′ | See Nasser (Libya) |
| Zevardin | USSR | N 38°51′ | E 64°51′ | Giant field number 357 in Table 1. |
| Zhetybayskoye | USSR | N 43°30′ | E 52° 0′ | Giant field number 187 in Table 1. |
| Zhongyuan | China | N 35°15′ | E 115°30′ | Giant field number 322 in Table 1. |
| Zubair | Iraq | N 30°25′ | E 47°40′ | Giant field number 62 in Table 1. |
| Zuluf | Saudi Arabia | N 28°22′ | E 49°10′ | Giant field number 29 in Table 1. |
| Zykh | USSR | N 40°22′ | E 50° 2′ | See Karachukhur Zykhskoye (USSR) |

Table 3. *Bally and Snelson Basin Classification—modified (after St. John, Bally and Klemme, 1984)*

1. Basins located on the rigid lithosphere, no associated with formation of megasutures
   - 11. Related to formation of oceanic crust
     - 111. Rifts
     - 112. Oceanic transform fault associated basins
     - 113-OC. Oceanic abyssal plains
     - 114. 'Atlantic-type' passive margins (shelf, slope & rise) which straddle continental and oceanic crust
       - 1141. Overlying earlier rift systems
       - 1142. Overlying earlier transform systems
       - 1143. Overlying earlier backarc basins of types 321 and 322
   - 12. Located on pre-Mesozoic continental lithosphere
     - 121. Cratonic basins
       - 1211. Located on earlier rifted grabens
       - 1212. Located on former backarc basins of type 321
2. Perisutural basins on rigid lithosphere associated with formation of compressional megasuture
   - 21-OC. Deep sea trench of moat on oceanic crust adjacent to B-subduction margin
   - 22. Foredeep and underlying platform sediments, or moat on continental crust adjacent to A-subduction margin
     - 221. Ramp with buried grabens, but with little or no block faulting
     - 222. Dominated by block faulting
   - 23. 'Chinese-type' basins associated with distal block faulting related to compressional or megasuture and without associated A-subduction margin
3. Episutural basins located and mostly contained in compressional megasuture
   - 31. Associated with B-subduction zone
     - 311. Forearc basin
     - 312. Circum-Pacific backarc basin
       - 3121-OC. Backarc basins floored by oceanic crust and associated with B-subduction (marginal sea *sensu stricto*)
       - 3122. Backarc basins floored by continental or intermediate crust, associated with B-subduction
   - 32. Backarc basins, associated with continental collision and on concave side of A-subduction arc
     - 321. On continental crust of 'Pannonian-type' basin
     - 322. On transitional and oceanic crust or 'W Mediterranean-type' basins
   - 33. Basins related to episutural megashear systems
     - 331. 'Great basin-type' basin
     - 332. 'California-type' basin
4. Folded belt
   - 41. Related to A-subduction
   - 42. Related to B-subduction
5. Plateau basalts

Table 4. *Distribution of giant fields, Bally-Snelson classification.*

| | "A" Foredeep | Cratonic | Atlantic Type | Thrusts | Shear | "B" Basins | Other Types | All Provinces |
|---|---|---|---|---|---|---|---|---|
| Number of provinces with giant fields | 35 | 9 | 20 | 4 | 7 | 10 | 13 | 98 |
| % provinces with giants | 30 | 11 | 12 | 10 | 7 | 6 | 7 | 12 |
| Number of non-giant provinces | 78 | 69 | 146 | 34 | 85 | 139 | 150 | 701 |
| "Richness" | | | | | | | | |
| Oil (BBbl) per giant province | 15.9 | 11.2 | 3.4 | 20.6 | 2.7 | 3.1 | 1.0 | 8.9 |
| Gas (tcf) per giant province | 30.1 | 171.7 | 9.1 | 71.6 | 4.1 | 6.6 | 2.6 | 32.6 |

Table 5. *Distribution of provinces without giants, Bally-Snelson classification.*

| | "A" Foredeep | Cratonic | Atlantic-Type | Thrusts | Shear | "B" Basins | Others |
|---|---|---|---|---|---|---|---|
| USSR | 13 | 11 | 1 | 1 | 10 | 16 | 7 |
| North America | 27 | 13 | 18 | 8 | 28 | 18 | 38 |
| Middle East | 3 | 0 | 3 | 2 | 11 | 1 | 2 |
| Central and South America | 10 | 6 | 28 | 5 | 16 | 21 | 15 |
| Europe | 10 | 8 | 15 | 1 | 1 | 2 | 20 |
| Africa (sub-Sahara) | 3 | 7 | 24 | 0 | 0 | 0 | 20 |
| North Africa | 4 | 1 | 8 | 1 | 1 | 0 | 5 |
| Australia, New Zealand & Pacific | 2 | 15 | 32 | 8 | 3 | 5 | 11 |
| Far East | 5 | 2 | 9 | 8 | 15 | 69 | 19 |
| Antarctica | 1 | 6 | 7 | 0 | 0 | 4 | 0 |
| Oceans | 0 | 0 | 1 | 0 | 0 | 3 | 13 |
| Total | 78 | 69 | 146 | 34 | 85 | 139 | 150 |

*Additional data for Tables 4 and 7:*

| Table 4 - Bally and Snelson classification | | Table 7 - Klemme classification | |
|---|---|---|---|
| Number | Basin Type | Number | Basin Type |
| 46 | "A" Foredeep | 27 | Collision zone |
| 18 | Cratonic | 12 | Accreted margin |
| 30 | Atlantic type | 29 | Rifted margin |
| 11 | Thrusts | 26 | Composite margin |
| 15 | Shear | 34 | Convergent margin |
| 26 | "B" basins | 38 | Other types |
| 20 | Other types | 166 | All provinces |
| 166 | All provinces | | |

Table 6. *Klemme basin classification—modified (after St. John, Bally, and Klemme, 1984).*

1. Craton interior basins
2. Continental multicycle basins
   A. Craton margin - composite
   B. Craton/accreted margin - complex
   C. Crustal collision zone - convergent plate margin
      a. Closed
      b. Trough
      c. Open
3. Continental rifted basins
   A. Craton and accreted zone rift
   B. Rifted convergent margin - oceanic consumption
      a. Back arc
      b. Transform
      c. Median
   C. Rifted passive margin - divergence
      a. Parallel
      b. Transform
4. Delta basins
   A. Synsedimentary
   B. Structural
5. Forearc basins

Department, 32 p.

Root, D. H., and L. J. Drew, 1979, The pattern of petroleum discovery rates: American Scientist, v. 67, n. 6, p. 648–652.

Saward, F. W., 1934, Saward's Annual—1934: a standard statistical review of the coal trade: New York, Saward's Journal.

Saward, R. B., 1964, Saward's Annual—1964: a standard statistical review of the coal trade: New York, Saward's Journal.

St. John, B., 1980, Sedimentary basins of the world and giant hydrocarbon accumulations (A short text to accompany the map: Sedimentary Basins of the World): AAPG, 24 p.

———, 1982, Sedimentary basins and giant fields, *in* Proceedings of the Southwestern Legal Foundation Exploration and Economics of the Petroleum Industry, Vol. 20: New York, New York, Matthew Bender & Co., p. 1–29.

———, A. W. Bally, and H. D. Klemme, 1984, Sedimentary provinces of the world—hydrocarbon productive and nonproductive: AAPG.

United States Department of the Interior, 1976, Minerals yearbook-1976: Washington, D.C., U.S. Government Printing Office.

Wood, P. W. J., 1979, New slant on potential world petroleum resources: World Oil (April 1979 issue), 9 p.

———, 1980, The future availability of petroleum worldwide, *in* Proceedings of the Southwestern Legal Foundation Exploration and Economics of the Petroleum Industry, Vol. 18: New York, New York, Matthew Bender & Co., p. 1–15.

Table 7. *Distribution of giant fields, Klemme classification.*

| | Collision Zone | Accreted Margin | Rifted Margin | Composite Margin | Convergent Margin | Other Types | All Provinces |
|---|---|---|---|---|---|---|---|
| Number of provinces with giant fields | 27 | 9 | 18 | 13 | 19 | 12 | 98 |
| % provinces with giants | 37 | 18 | 15 | 21 | 10 | 3 | 12 |
| Number of non-giant provinces | 45 | 39 | 100 | 48 | 158 | 311 | 701 |
| "Richness" | | | | | | | |
| Oil (BBbl) per giant province | 21.5 | 6.1 | 4.6 | 4.4 | 4.2 | 1.1 | 8.9 |
| Gas (tcf) per giant province | 40.3 | 164.0 | 10.4 | 21.3 | 5.8 | 4.6 | 32.6 |

Table 8. *Distribution of provinces without giants, Klemme classification.*

| | Collision | Accreted | Rifted | Composite | Convergent | Others |
|---|---|---|---|---|---|---|
| USSR | 2 | 7 | 15 | 6 | 13 | 16 |
| North America | 9 | 7 | 7 | 26 | 37 | 64 |
| Middle East | 5 | 0 | 2 | 0 | 12 | 3 |
| Central and South America | 0 | 4 | 6 | 9 | 30 | 52 |
| Europe | 10 | 8 | 15 | 0 | 6 | 18 |
| Africa (sub-Sahara) | 0 | 7 | 13 | 1 | 0 | 33 |
| North Africa | 7 | 2 | 3 | 2 | 2 | 4 |
| Australia, New Zealand & Pacific | 2 | 2 | 19 | 3 | 12 | 38 |
| Far East | 10 | 1 | 18 | 1 | 46 | 51 |
| Antarctica | 0 | 1 | 2 | 0 | 0 | 15 |
| Oceans | 0 | 0 | 0 | 0 | 0 | 17 |
| Total | 45 | 39 | 100 | 48 | 158 | 311 |

# Antarctica—Geology and Hydrocarbon Potential

Bill St. John
*Primary Fuels, Inc.,*
*Subsidiary of Houston Industries Incorporated*
*Houston, Texas*

Antarctica covers approximately 14 million sq km (5,405,800 sq mi) and hosts an estimated 90% of the world's ice. Glacial ice covers about 98% of the continent, with an average thickness of 2,000 m (6,562 ft).

Based on limited data, geologists have identified 21 sedimentary basins for the Antarctic and immediately adjacent areas. These include 6 onshore, subglacial basins and 15 offshore basins. Excluding 11 basins considered to have little or no potential, the other 10 basins contain an estimated 16.9 million cu km (4.05 million cu mi) of sediment having a potential hydrocarbon yield of 203 billion barrels oil equivalent.

## INTRODUCTION[1]

Antarctica is larger than the United States and Mexico combined, and covers approximately 14 million sq km (5,405,800 sq mi). Some 90% of the world's ice covers 98% of the continent. Snow and falling ice crystals accumulating on the interior ice sheet enable it to flow outward continuously, maintaining a thickness that averages 2,000 m (6,562 ft) and in places exceeds 4,500 m (14,760 ft).

The enormous ice load has depressed the continent by about 600 m (1,968 ft), causing an estimated one-third of the land surface to lie below sea level. Mountain ranges reach over 5,000 m (16,400 ft) above sea level and are surrounded by the ice surface that in general lies between 2,000 and 4,000 m (6,562 and 13,123 ft) above sea level.

The continental shelf of Antarctica averages 30 km (19 mi) in width, compared with the global average of 70 km (43 mi). Depths of 400–600 m (1,312–1,968 ft) at the shelf edge are common and reach 800 m (2,625 ft) in the Ross Sea.

Lines of equal temperature in Antarctica tend to form a symmetrical pattern centered over East Antarctica, with the sharpest temperature changes appearing along the margin of the ice sheet, particularly in the east. January averages vary from slightly below freezing along the coast to below −30°C (−22°F) on the high interior plateau. These same areas, respectively, experience July averages of about −20°C (−4°F) and below −65°C (−85°F). Winter lows in excess of −80°C (−112°F) have been recorded at several stations. The world's record of −88°C (−126.4°F) was measured at Vostok on August 24, 1960. Even in summer, the absolute maximum temperatures do not exceed −15°C (5°F) over a large area of interior Antarctica. Annual absolute temperature ranges of 45°C (113°F) to 65°C (149°F) occur in Antarctica, but are less than in northeast Siberia where ranges of up to 100°C (212°F) between summer and winter are common.

Scientific surveys show that the Antarctic icepack (Figure 1) has different structural and behavioral characteristics than the Arctic pack. Ice in the Antarctic seas originates on or at the edge of a polar landmass and is dispersed by strong winds blowing it northward into the unrestricted expanses of the Atlantic, Indian, and Pacific Oceans, while the Arctic sea ice forms and moves about in the largely landlocked Arctic Ocean basin.

The Antarctic icepack experiences a much greater annual variation in extent than does its Arctic counterpart. The Antarctic pack grows from an average minimum of 2.6 million sq km (1.00 million sq mi) in March—half the minimum size of the Arctic pack—to about 18.8 million sq km (7.25 million sq mi) in September, more than 1.6 times the Arctic maximum and better than a sevenfold increase in area.

Because 85% of the Antarctic icepack melts each year, pack ice in the following year is predominantly first-year replacement ice, with an average thickness of only 1.5 m (5 ft). Antarctic sea ice also has a more uniform thickness than Arctic sea ice, as the divergent forces of Antarctic winds and currents tend to disperse the pack, minimize the formation of pressure ridges, and make the icepack generally more navigable by icebreakers. The same factors, nevertheless, make travel in this region hazardous. The pack moves quickly with the winds—up to 65 km (40 mi) in a single day—and ships easily can be caught in some of the more complex multiyear ice that is trapped by certain coastal configurations of the Ross, Bellingshausen, and Weddell Seas.

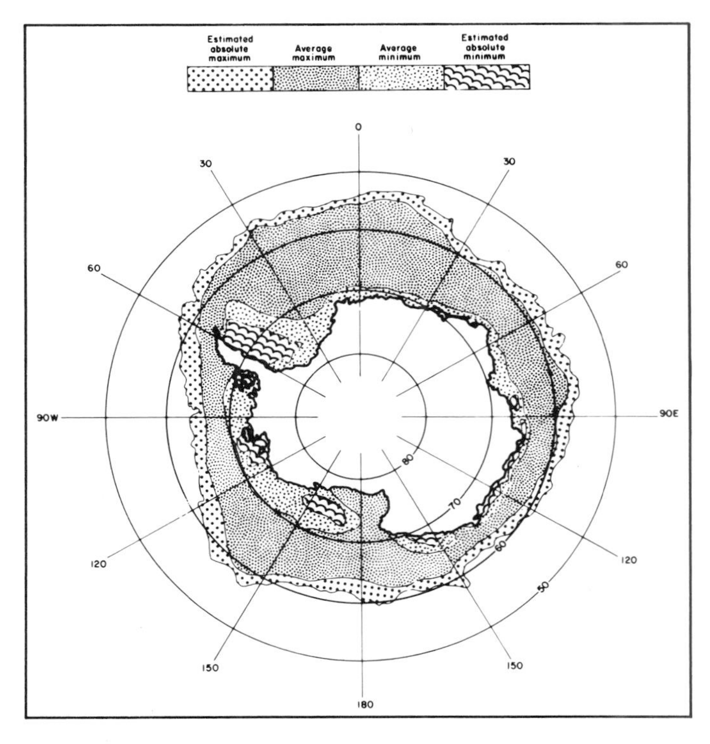


Figure 1. *Extent of sea ice relative to Antarctica, 1971–76.*

Antarctic ice islands, commonly called *tabular icebergs*, are significantly larger and more numerous than their Arctic counterparts; some with observed horizontal dimensions of more than 60 km (37 mi) by 100 km (62 mi) have calved away from the Ross, Ronne, and other ice shelves. These huge tabular floes, towering as high as 70 to 80 m (230 to 262 ft) above the sea surface, have been observed grounded at depths of 500 m (1,640 ft)—a sobering thought to those contemplating offshore petroleum extraction. Antarctic icebergs, however, rarely find their way into Southern Hemisphere shipping lanes.

## TERRITORIAL CLAIMS AND THE INTERNATIONAL GEOPHYSICAL YEAR (IGY)

Claims of 7 nations (Figure 2)—Argentina, Australia, Chile, France, New Zealand, Norway, and the United Kingdom—were not annulled by the 1959 Antarctic Treaty, but were frozen for the duration of the treaty, which expires in 1991.

The United Kingdom was the first nation to claim a part of Antarctica (the "Falkland Islands Dependencies," 1908), and in 1920, it urged New Zealand and Australia to follow suit. Claims were subsequently advanced by them, later by France and Norway, and still later by Argentina and Chile. By the early 1940s, most of the continent had been claimed.

Neither the United States nor the U.S.S.R. has made any claims to Antarctica, and neither recognizes the claims made by others.

[1]The general text of the Introduction, Territorial Claims and the IGY, and the Antarctic Treaty sections is after Central Intelligence Agency, 1978 (reprinted 1979), Polar Regions Atlas.

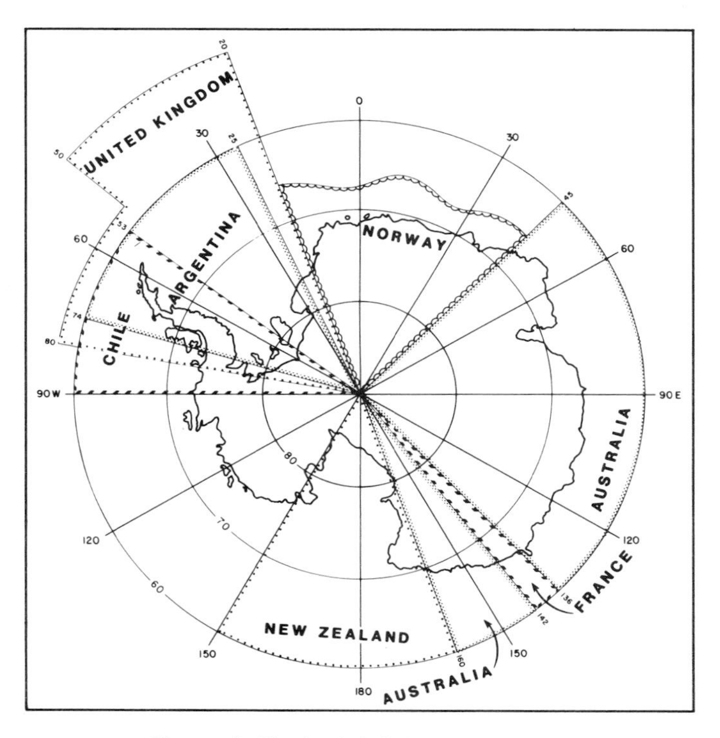


Figure 2. *Territorial claims to Antarctica.*

The International Geophysical Year (IGY), conducted in 1957–58, was the first major world scientific effort that involved Antarctica. Previous cooperative endeavors—the First and Second Polar Years (1882–83 and 1932–33)—stressed the Arctic in their respective studies of geomagnetism, meteorology, and auroral phenomena, barely touching on Antarctica.

Planning and implementing the IGY were successful, international, cooperative efforts. By 1958, over 10,000 scientists and technicians were working at 2,500 IGY stations around the globe. The Antarctic area had 50 stations maintained by 12 countries: Argentina, Australia, Belgium, Chile, France, Japan, New Zealand, Norway, South Africa, the United Kingdom, the United States, and the U.S.S.R..

The IGY activities in Antarctica made a significant contribution to the world's scientific knowledge. Researchers obtained valuable new data from meteorological and seismic observations, studies of upper atmosphere physics, and magnetic measurements. Cores from the continental ice sheet added to glaciological knowledge. Traverses over the inland ice sheet by scientists from the United States, the British Commonwealth countries, the Soviet Union, and France gave new information on ice temperature, density, and thickness; on ice surface elevations; and on magnetic and gravity fields. The Commonwealth Transantarctic Expedition, which in three months trekked from Shackleton Base on the Weddell Sea to Scott Base on the Ross Sea via the South Pole, was the most spectacular traverse and the first surface crossing of Antarctica. A key side benefit of the IGY was the vast improvement in Antarctic geographical knowledge.

## ANTARCTIC TREATY (Table 1)

The Antarctic Treaty, signed in 1959 and going into force on June 23, 1961, established for at least 30 years a legal framework for Antarctica. It formalized and guaranteed the kind of free access and research rights that had spelled success for the International Geophysical Year.

The 12 original signatories to the Treaty undertook mutual obligations in Antarctica for conducting scientific activities. For over 20 years, they have kept the continent demilitarized, cooperated in mutually beneficial scientific exploration, established measures to preserve the environment, and consulted and agreed on measures of common interest.

The treaty stipulates that Antarctica should "forever . . . be used exclusively for peaceful purposes" and "not become the scene or object of international discord." Nuclear explosions and disposal of radioactive wastes are prohibited (making this the first international nuclear test-ban agreement) as are "measures of a military nature." These obligations are supported by provisions for exchange of scientific information and personnel and for observers. The treaty freezes the territorial claims issue by providing that no party use the treaty as a basis to assert, support, deny, or extend existing claims or make new ones for the duration of the treaty. Living resources (mineral resources are not mentioned) are to be preserved. Disputes are to be settled peacefully by the parties involved or, ultimately, by the International Court of Justice. Any UN member state may accede to the treaty but can gain consultative (decision-making) status only after demonstrating its Antarctic interest "by conducting substantial scientific research activity there."

Seven states have acceded since the treaty entered into force. One of these, Poland, became a consultative member in July, 1977—2 years after beginning regular Antarctic research expeditions and 5 months after establishing a permanent scientific station in the South Shetland Islands.

Rules for operating on the continent are set by the treaty states in consultative meetings held outside Antarctica. Close ties with the Scientific Committee on Antarctic Research (SCAR) have helped keep the focus scientific, as most of the 118 resolutions passed at the 9 consultative meetings held so far attest. But new interest in Antarctica, fanned by world energy and protein shortages and spurred by technological advances, goes far beyond pure science to much tougher questions of exploitation and international policies toward the continent.

In recent years, the worldwide search for new resources and the implications of the Law of the Sea negotiations have drawn international attention to the resource potential of the Antarctic. The interest in hydrocarbons, coal, and minerals, however, has not advanced beyond the speculative state. Yet, food sources are already the object of commercial exploitation, especially fish, whales, and krill, a small shrimp-like creature that abounds in Antarctic waters.

Mineral resources pose a difficult problem both physically and politically. Although extraction may be more technically feasible now than when the treaty was drafted, the problem of offshore drilling for petroleum on Antarctica's continental shelf is complicated by the costliness of such an endeavor and by the uncertain status of the claims issue. The implications of Antarctic minerals exploitation have become a key item at recent consultative meetings. The topic was discussed in 1972, but only in terms of possible effects on the Antarctic environment. Three years later, voluntary restraint on mineral exploitation was urged until more information was available. The 1977 meeting reaffirmed the need for voluntary restraint, sought new environmental and technical studies to obtain more knowledge,

*Table 1: Summary of basic provisions of the Antarctic Treaty.*

| ANTARCTIC TREATY | |
|---|---|
| ARTICLE I. | Antarctica shall be used for peaceful purposes only. All military measures, including weapons testing are prohibited. Military personnel and equipment may be used, however, for scientific purposes. |
| ARTICLE II. | Freedom of scientific investigation and cooperation shall continue. |
| ARTICLE III. | Scientific program plans, personnel, observations, and results shall be freely exchanged. |
| ARTICLE IV. | The treaty does not recognize, dispute, or establish territorial claims. No new claims shall be asserted while this treaty is in force. |
| ARTICLE V. | Nuclear explosions and disposal of radioactive wastes are prohibited. |
| ARTICLE VI. | All land and sea shelves below 60° South Latitude are included, but high seas are covered under international law. |
| ARTICLE VII. | Treaty-state observers have free access — including aerial observation — to any area and may inspect all stations, installations, and equipment. Advance notice of all activities and of the introduction of military personnel must be given. |
| ARTICLE VIII. | Observers under Article VII and scientific personnel under Article III are under the jurisdiction of their own states. |
| ARTICLE IX. | Treaty states shall meet periodically to exchange information and take measures to further treaty objectives, including the preservation and conservation of living resources. These consultative meetings shall be open to contracting parties that conduct substantial scientific research in the area. |
| ARTICLE X. | Treaty states will discourage activities by any country in Antarctica that are contrary to the treaty. |
| ARTICLE XI. | Disputes are to be settled peacefully by the parties concerned or, ultimately, by the International Court of Justice. |
| ARTICLE XII. | After the expiration of 30 years from the date the treaty enters into force, any member state may request a conference to review the operation of the treaty. |
| ARTICLE XIII. | The treaty is subject to ratification by signatory states and is open for accession by any state that is a member of the UN or is invited by all the member states. |
| ARTICLE XIV. | The United States is the repository of the treaty and is responsible for providing certified copies to signatories and acceding states. |

and called for legal and political talks to be held on mineral rights (both onshore and offshore) before the scheduled 1979 meeting in Washington.

Even if differences among treaty states on minerals were totally resolved, there might be a problem from interested parties that until now have had little or no involvement in Antarctica. Some Third World spokesmen in the UN General Assembly, in UN specialized agencies, and at regional meetings, have voiced an interest in having Antarctica managed as they would like to see the deep seabed handled—under the common heritage of mankind principle and the jurisdiction of the proposed Seabed Authority. Another potentially difficult problem for treaty members (whose only recourse is "to exert appropriate efforts, consistent with the United National Charter") would be how to deal with Antarctic newcomers who might refuse to accede to or observe treaty rules. These matters must be addressed if the broad terms of the treaty are to remain effective and Antarctica is not to "become the scene or object of international discord."

Until recent years, nations active in Antarctica were interested mainly in issues related to demilitarization of the continent and science, and some, in whaling and fishing in the surrounding areas. Now, one of the main areas of international concern is energy.

The Antarctic treaty is silent on the question of hydrocarbon and mineral exploitation, a recognition that commercial activity was not imminent and that its inclusion in the treaty could have jeopardized an accommodation on the sensitive claims issue. Pressure to settle the resource question began building in the late 1960s and intensified at the time of the 1973 oil embargo. The consultative parties' response has been to discuss the issue at the Treaty Consultative Meetings held in New Zealand in 1972, in Oslo in 1975, and in London in 1977.

The Oslo session and a follow-up special meeting held in Paris in 1976 resulted in serious consideration of the legal and political issues for the first time. A recommendation, passed in Oslo, called for the parties to exercise restraint in mineral-related activities while seeking timely, agreed solutions; to study the environmental implications of these activities; and to promote geological studies. The most significant accomplishments have been an airing of the view of the participating governments and obtaining a commitment to finding a solution that will preserve the viability of the Antarctic Treaty and protect the Antarctic environment.

Continuing negotiations promise to be complex because they relate to the question of territorial claims—both on land and offshore—environmental preservation, and maintaining a political balance among interested countries. Failing a negoti-

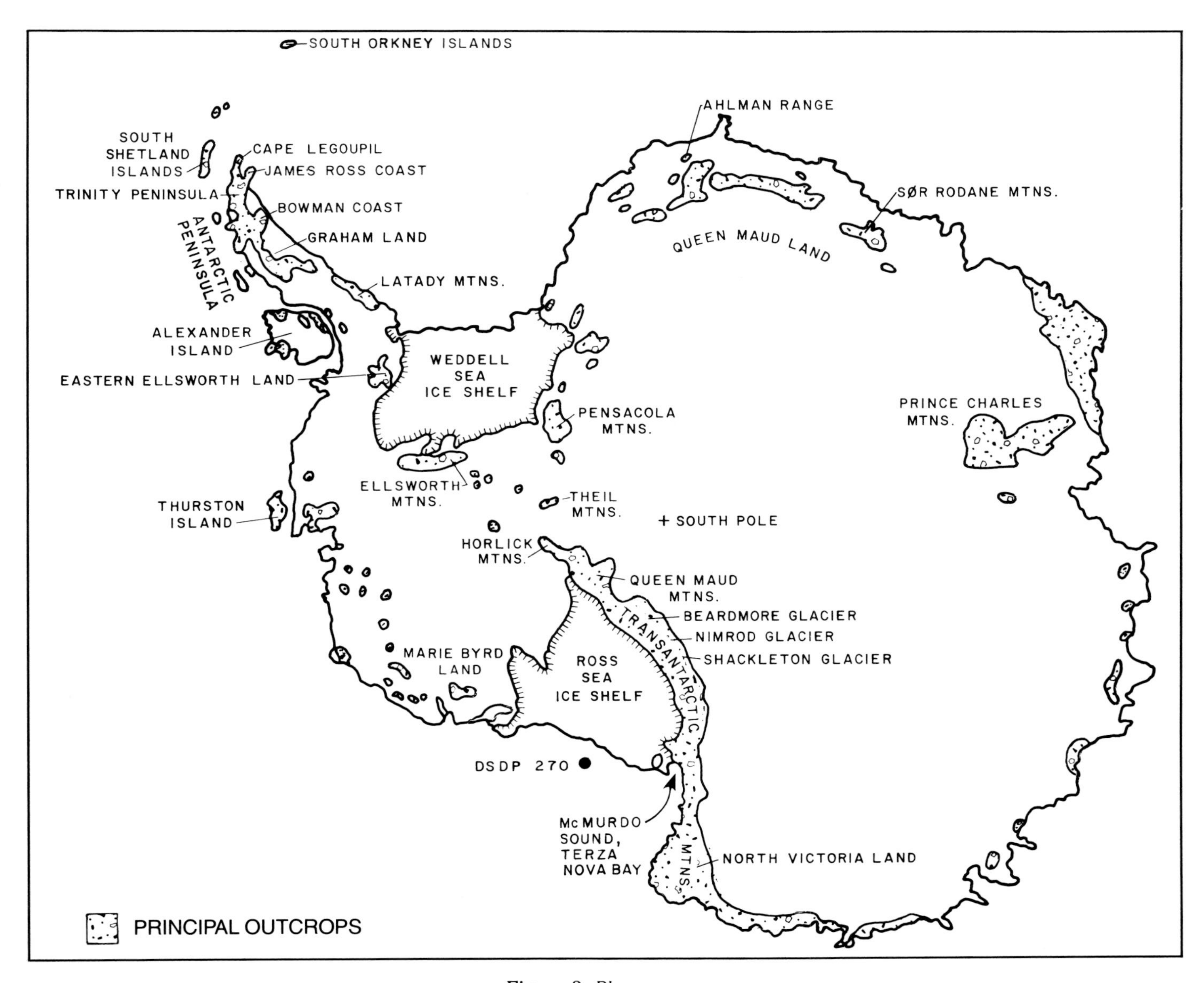


Figure 3. *Place names.*

ated solution, one or more countries could begin exploitation of a resource on the grounds that it is for a peaceful purpose and thus is permitted under the Antarctic Treaty.

## LITHO-PALEOGEOGRAPHIC RECONSTRUCTIONS AND STRATIGRAPHY

### General

The major depositional cycles from Precambrian through Pleistocene are developed from the relative changes of sea level described by Vail et al (1977). Figure 3 indicates the place names used in the text, while Figure 4 shows a stratigraphic chart for Antarctica and adjacent land masses. The litho-paleogeographic reconstructions parallel the major depositional cycles and are assigned geochronologic names covering the major period of that particular time span but which may in fact include some time and rock units slightly older or younger. Paleolatitude data are after Creer (1965, 1968a, 1968b, 1970, 1973); Habicht (1979); Irving (1964); and McElhinny and Luck (1970).

### Cambrian-Early Ordovician (490–600 m.y.B.P.) (Figure 5)

During the time span Cambrian through Early Ordovician, seafloor spreading occurred between the continental masses of North America-Greenland-Eurasia and the ancient Gondwana. Gondwana encompassed the present-day masses of South America, Africa, Madagascar, Arabia, India, Antarctica, and the Indian Ocean Seychelles Island group, plus the continental mass of Australia-Papua-New Guinea.

Gondwana was inverted relative to today's position. At the close of the Early Ordovician, the South Pole was located somewhere near western Morocco, and the tip of southern Africa was situated at approximately lat. 25°S. Antarctica was bisected by the paleoequator.

A central highland mass covered eastern Antarctica, southwest Australia, India, Madagascar, central Africa, and central eastern South America. Separate highs were exposed in the northern Brazilian Shield area, central Argentina-Falkland (Malvinas) Islands, northwest Australia, and that part of northern Australia-Papua now occupied by the Gulf of Carpentaria.

A fringe area of continental lowlands bordered the central, eroding continental high and received river, dune, and other sediments typical of continental deposition.

A transitional zone separated the exposed continental areas from the zone of pure marine deposition. *Transitional zone*, as used in this paper, refers to the part of the continental mass either transgressed or regressed by the sea during the time period represented by each litho-paleogeographic map. Sediments deposited in the transitional zone may be either marine or continental, but overall will be transitional between the two and will normally be represented by interfingering continental and marine beds.

A transitional zone of the Cambrian-Early Ordovician covered most of northern Africa; much of northern South America, including the Amazon River basin; and extended toward Argentina in a narrow band along what is currently the western part of South America. The transitional zone spread over a broad embayment extending from Argentina east to the Karroo basin of South Africa. Across Antarctica, the transitional zone blanketed the area now occupied by the Transantarctic Mountains and connected to eastern Australia. From eastern Australia, the zone extended over much of northern and northwestern Australia to that part of Greater India that has since underthrust the Asian continent.

The Cambrian-Early Ordovician sediments, of what is today West Antarctica and Australia, are indicative of deep water and migrating sialic island arcs. This indication implies a possible landward-migrating spread center and associated subduction zone fringing the continental margins of Antarctica-Australia. As the oceanic crust was being consumed through subduction beneath the continental masses, the deepwater sediments on oceanic crust were probably being scraped off and accreted onto the continental margin.

A subduction zone fringed Antarctica-Australia, but there is considerable question as to its direction of dip at any given time—whether landward or seaward. Deepwater clastics, chert, and basic igneous oceanic rocks are found onshore eastern Australia and are inferred in West Antarctica (Ramsey and Stanley, 1976; Solomon and Griffiths, 1972).

Throughout the Paleozoic-Mesozoic-Cenozoic time, the subduction zone probably reversed its direction of dip more than once. The predominant dip would have been landward, as deepwater sediments on the oceanic crust accumulated along the continental margin. However, reversals resulting in seaward dip and eventual obduction of oceanic crust onto the continent must have occurred, as evidenced by the presence in eastern Australia of ophiolites, chert, tuff, and andesite. The direction the subduction zone dips is indicated as landward on the various enclosed reconstructions.

In many parts of Antarctica, Cambrian sedimentation was an extension of Precambrian sedimentation. Precambrian sediments are known (Figures 3 and 4) from the Ellsworth Mountains, the Transantarctic, the Pensacola, and Thiel Mountains; from Queen Maud Land; and possibly from the coastal Marie Byrd Land/Ross Sea area (DSDP 270). In general, the Precambrian of East Antarctica is composed solely of igneous and metamorphic rocks. There is no known Precambrian in the Antarctic Peninsula or in the interior Marie Byrd Land of West Antarctica.

The Precambrian sediments of the Ahlman Ridge of Queen Maud Land (Craddock et al, 1969–70), are indicative of continental clastic deposition, whereas the Precambrian sediments of the Ellsworth Mountains reflect shallow marine carbonate conditions. The Precambrian of the Transantarctic and Pensacola Mountains consists of extremely thick, tightly folded, turbidite silty sandstones and shales interbedded with volcanic basalt.

Cambrian sediments indicate more shallow marine conditions overall; carbonates dominate Cambrian deposition in Antarctica. Williams et al (1971) report Cambrian (?) coarse-grained marbles interlayered with schist and gneiss in the Northern Transantarctic Mountains. Laird, Mansergh, and Cheppell, (1971) report the Cambrian Byrd Group of the Central Nimrod Glacier area of the Transantarctic Mountains to be composed of limestone/marble, breccia/conglomerate, sandstone/quartzite, and shale. Elliot and Coates (1971) and Yeats et al (1965) describe the Cambrian of the Southern Transantarctic Queen Maud Mountains as coarse-grained marble, clastics, and silicic volcanics.

Minshew (1966) describes the basal sediments of the Wisconsin Range Horlick Mountains as basement of granite and folded metamorphic rocks, phyllite, and schist, overlain by slightly metamorphosed conglomeratic sandstone with limestone and shale containing Middle Cambrian trilobites. In the Pensacola Mountains, Behrendt et al (1974) and Nelson, Schmidt, and Schopf (1967) describes Middle Cambrian limestone overlain by Upper Cambrian transgressive marine felsic flows, pyroclastics, siltstone, and mudstone. Ford and Barrett (1975) describe the basement rock of DSDP Leg 28, site 270 as primarily calcareous metamorphic rock (marble), suggesting a Cambrian age. The Cambrian of the Ellsworth Mountains is described by Webers (1972) as shallow marine clastic in the lower part overlain by Upper Cambrian, nearshore, fossiliferous, crystalline limestone.

Craddock et al (1969–70); Elliot (1975); and Laird, Mansergh, and Cheppell (1971) describe the Cambrian sediments of Antarctica in detail.

In the Argentina-South Africa embayment, the sediments were predominantly marine shales and limestones in the Argentina area (Padula and Mingramm, 1963; Harrington, 1962), and interfingering, continental, massive tillites, sandstones, dolomitic, and black limestones in South Africa (Furon, 1963). early Paleozoic sedimentation along the marine-marginal Antarctica-Australia continental mass apparently decreased markedly toward the end of the Early Ordovician. The close of the Early Ordovician was a time of maximum deformation: the Ross Orogen (Figure 6), in Antarctica, and the corresponding Adelaide Orogen in Australia (Craddock, 1972; Griffiths, 1971a, 1971b; Solomon and Griffiths, 1972). Earlier sediments were uplifted, folded, and intruded by calc-alkalic plutons (Yeats et al, 1965). Oliver (1972) compares the similarities of the South Australian Adelaide geosyncline and subsequent orogen with the comparable Antarctic Ross geosyncline

→

Figure 4. *Stratigraphic chart—Antarctica and adjacent land mass.*

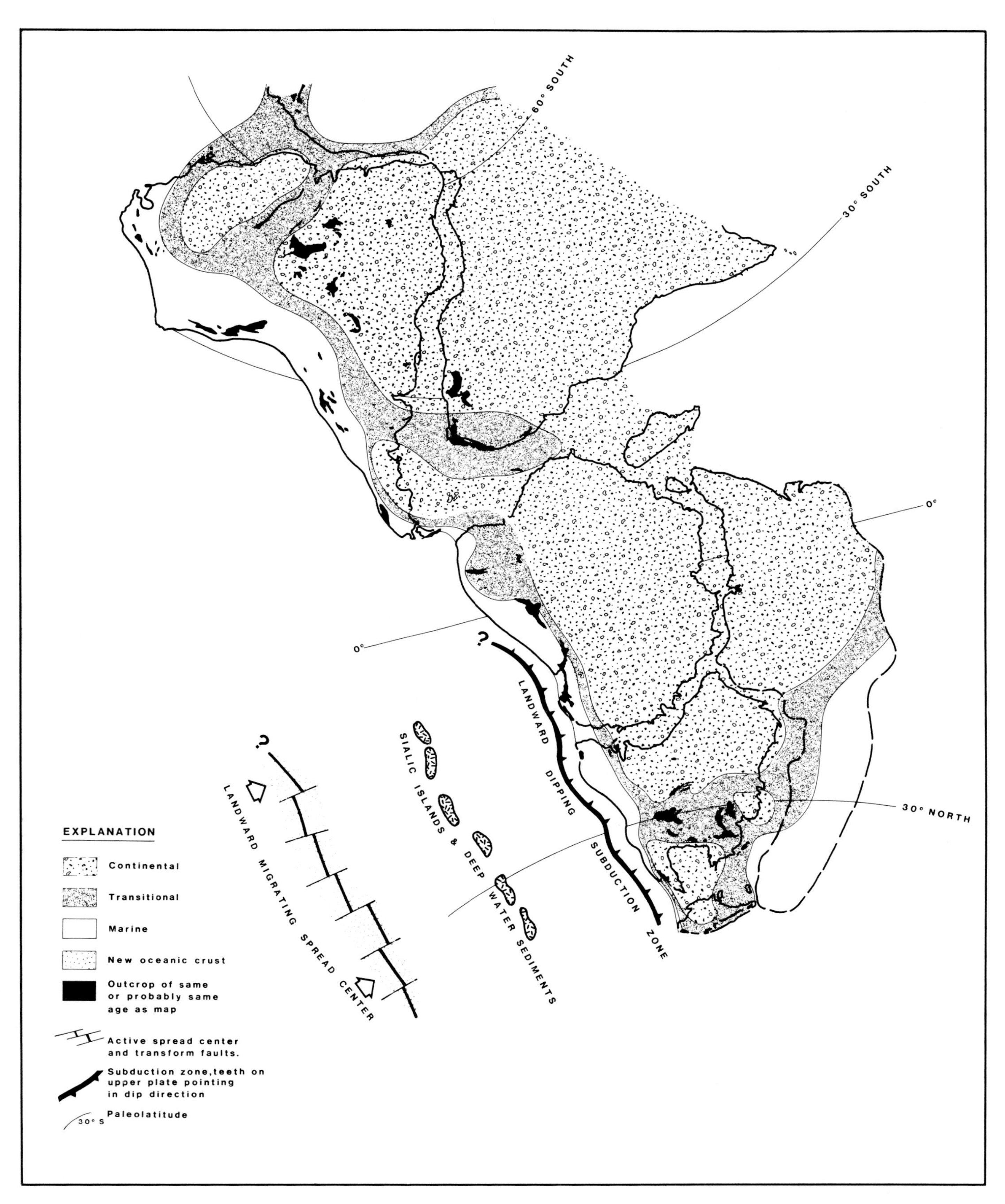


Figure 5. *Cambrian-Early Ordovician reconstruction.*

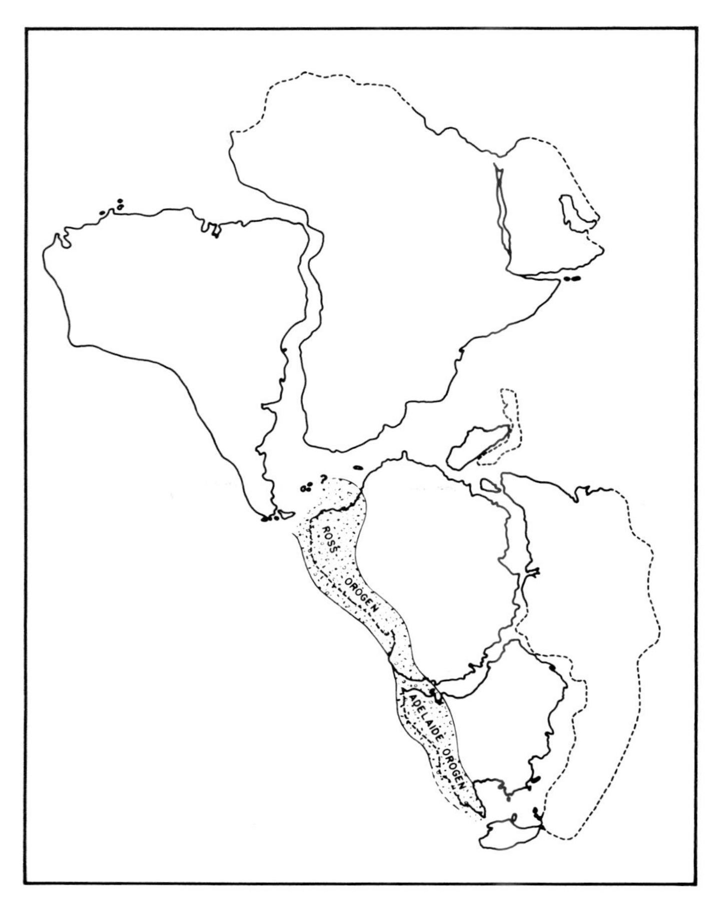


Figure 6. *Early Paleozoic Ross-Adelaide Orogen.*

and orogen, and concludes that they were connected and continuous during Precambrian-Ordovician time. Webby (1978) refers to the event in south Australia and Tasmania as the *Delamerian Orogeny*.

### Middle Ordovician-Silurian (405–490 m.y.B.P.) (Figure 7)

By the close of the Silurian, the South Pole was located in east-central South America and most of South Africa lay between the pole and lat. 60°S. Clastic sediments predominated, and tillite deposits are found over much of South America and Africa today.

In Gondwana, a central highland mass continued through easternmost South America, eastern Antarctica, southwestern Australia, Madagascar and the Seychelles, India, and Central Africa. Separate continental highs were also exposed. Evidence suggests a single high extended from the northern Brazilian Shield area to northwest Africa. A smaller high occurred on the southern Brazilian Shield while Amazon basin drainage was directed westward at this time. A high occupied the Falkland (Malvinas) Plateau-Burdwood Bank region offshore Argentina. The Gulf of Carpentaria of northern Australia and part of northwest Australia were also emergent at this time.

As in the Cambrian-Early Ordovician, depositional lowlands surrounded the continental highs. Areas of purely marine sedimentation were rare, and such occurrences are today known only in eastern Australia, northernmost Africa, and northwestern South America. Transitional-zone sediments covered roughly half of Gondwana. Extensive deposits of transitional sediments presently occur in North Africa and north- and west-central South America.

A marine transgression extended southward across northern Argentina east to the Karroo basin of South Africa. Marine sandstone, shale, and limestone were deposited in northwest Argentina, while a more continental facies of fossiliferous sandstones and non-fossiliferous quartzites and tillites was being laid down in South Africa.

No sediments of proven Middle Ordovician-Silurian age occur in the Antarctic. Craddock (1969–70) and Stump (1973) describe continuous sedimentation from Cambrian to Devonian in the Ellsworth Mountains. The Ellsworth Mountains section consists of thick and conformable clastic sediments of inferred Ordovician-Silurian age, although no fossils were found. Behrendt et al (1974) and Nelson, Schmidt, and Schopf (1967) describe similar sediments in the Pensacola Mountains (Brown Ridge Conglomerate and Elliot Sandstone). Nelson, Schmidt, and Schopf (1967) place them above the Cambrian (?) Wiems Formation and beneath the Devonian (?) Elbow Formation, and they refer to them as being of Paleozoic age. Behrendt et al assign the section an early (?)-middle Paleozoic age. In dating the structures toward the north end of the Transantarctic Mountains, Griffiths (1971b) mentions that the metamorphosed Cambrian Robertson Bay Group contains an infaulted strip of Silurian (?) to Devonian Bowers Group sediment.

Throughout the Transantarctic Mountains, the Middle-Late Ordovician was a time of mountain building, and the Silurian was a period of extensive erosion. This episode is termed the *Borchgrevink Orogen* (Figure 8) in the Northern Transantarctic Mountains correlative with a parallel event in southern Australia. Craddock (1972) refers to the *Tasman Orogen* of Australia while Griffiths (1971a, 1971b) uses the term *Lachlan Orogen*.

Middle Ordovician sediments deposited in central and northwestern Australia were shallow marine sands, shale, evaporites, and carbonates (Sprigg, 1967). During the Late Ordovician-Silurian, central and northwestern Australia were uplifted and continental conditions prevailed. Extensive shallow-marine-shelf carbonates were deposited in the Tasmanian shelf from Middle through Late Ordovician, followed by Early Silurian sandstones, reflecting nearby uplift and renewed erosion.

### Devonian-Early Carboniferous (325–405 m.y.B.P.) (Figure 9)

The proto-Atlantic was no longer in existence by the close of the Early Carboniferous. The southern Gondwana plate had joined the northern Laurasia plate to form the supercontinent Pangaea. The intervening geosynclinal sediments were metamorphosed.

A major exposed land mass remained elevated in eastern Antarctica, southwestern Australia, southern India, and southwestern Arabia, extending from east Africa across central Africa to central east South America. The northern and southern Brazilian Shield rose above the depositional zones and individual highs occurred in western Argentina. A narrow zone of continental sedimentation bounded the exposed highlands; the Australian Gulf of Carpentaria remained exposed.

A major transgression by the surrounding sea during the Devonian resulted in a vast transitional zone. The transgression was almost complete in South America and in northern

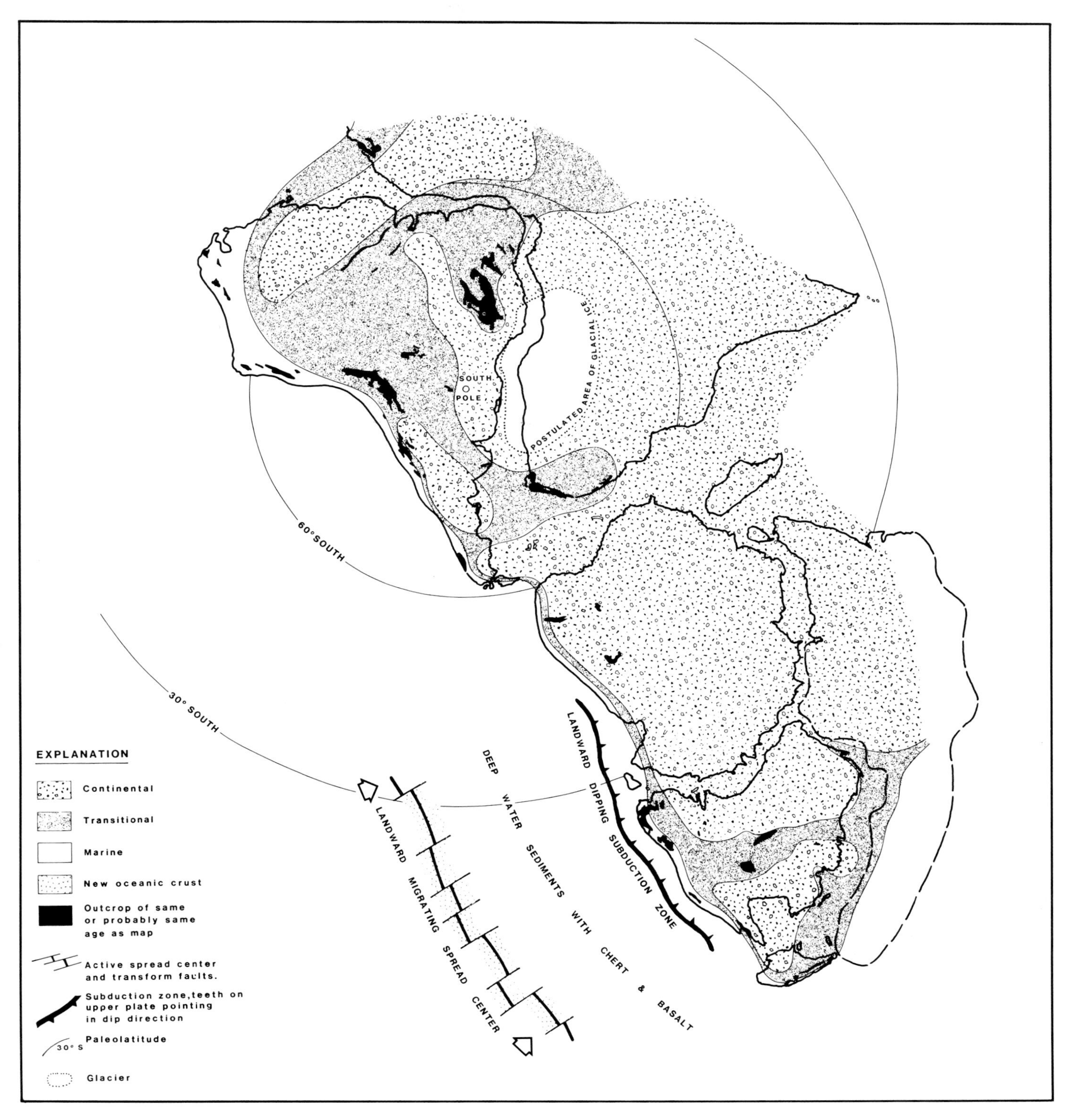


**Figure 7.** *Middle Ordovician-Silurian reconstruction.*

Africa. Much of what is now the central part of Antarctica was covered by the transgression that also covered the margins of Australia. A regressive cycle began in the Early Carboniferous.

The South Pole was situated near the southern tip of South America and mostly clastic sediments, including tillites, were deposited in the Polar zone of South America, Africa, and Antarctica.

Great thicknesses of clastic sediments were deposited in South Africa (Furon, 1963) and Argentina (Roque et al, 1959; Harrington, 1962; Padula and Mingramm, 1963). Reed (1949) reports 3,000 m (9,840 ft) of Devonian clastic sediments in the Falkland (Malvinas) Islands, and Woolard (1970) points to the geological similarities of the Antarctic Peninsula and Patagonia (Argentina).

The Beacon Supergroup in the Transantarctic Mountains of Antarctica is essentially a thick (up to 3,200 m [10,500 ft]), sequence of non-marine sediments, ranging from Devonian to Jurassic (Barrett, 1970). The Supergroup consists of two Groups: The Taylor Group, older and considered to be Devonian and possibly earliest Carboniferous; and the overlying Victoria Group, extending from Early Carboniferous to Early Jurassic.

Near the coast, in northern Victoria Land, or northernmost Northern Transantarctic Mountains, Gair et al (1969–70) describe 270 m (886 ft) of clastic sediments, but they neither refer them to the Beacon Group nor assign any age older than Permian. Further south in the McMurdo Sound area, Harrington (1970) and McKelvey et al (1970) describe 2,300 m (7,546 ft) of Beacon Supergroup rocks. They assign five formation names to the Taylor Group as follows:

| ———————— | unconformity ———————— |
|---|---|
| Aztec Siltstone | 40–138 meters, red and green, siltstone with abundant freshwater fish remains |
| Beacon Heights Orthoquartzite | 300+ meters, cliff-forming |
| Arena Sandstone | 385 meters, buff colored slope forming sandstone |
| Altar Mountain Formation | 162 meters, arkosic sandstone overlain by quartz sandstone and maroon–green siltstone |
| New Mountain Sandstone | 46–220 meters, light colored, crossbedded, quartz sandstone with siltstone and conglomerates |
| ———————— | unconformity ———————— |

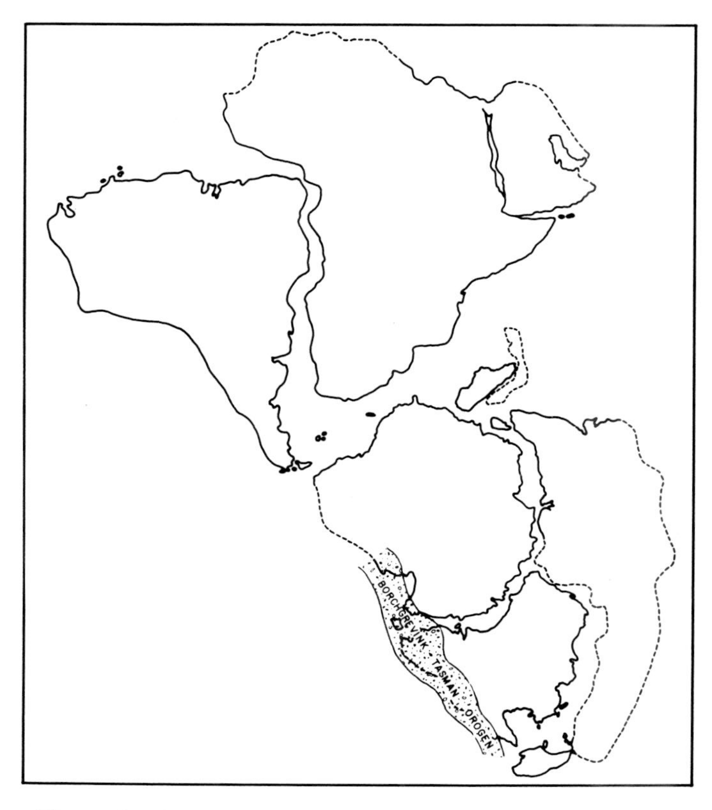


Figure 8. *Middle Paleozoic Brochgrevink-Tasman Orogen.*

Warren (1969–70) assigns 1,000 m (3,280 ft) to the Beacon section in the Terra Nova Bay area of McMurdo Sound. He describes the Taylor group as basal, massive, cross-bedded quartz sandstone containing Devonian plant fragments, overlain by Middle-Late Devonian siltstone containing freshwater fish remains. Warren further believed the section to be conformable with the overlying Permian.

Grindley and Laird (1969–70) measured 2,000+ m (6,560+ ft) of continental Beacon Supergroup sediments along the Shackleton Coast. They describe a 35 m (115 ft) thick, unfossiliferous quartz sandstone. Laird, Mansergh, and Cheppell (1971) described the Alexandra Formation as 700 m (2,297 ft) thick, poorly sorted, cross-bedded nearshore to continental sandstone in the Minrod Glacier area. They assigned the formation name *Castle Crags* to the underlying 123 m (403 ft) thick sandstone and shale sequence, and interpreted it as a fluvial-lacustrine deposit.

In the Beardmore Glacier area, Lindsay (1968) found only the Alexandra formation of the Taylor Group. He described it as 300 m (984 ft) thick, non-marine, well sorted, cross-bedded, quartz sandstone.

Yeats et al (1965) and McGregor and Wade (1969–70) note 1,800+ m (5,905+ ft) of Beacon Supergroup in the Queen Maud Mountains. Others (Yeats et al, 1965) assigned no age but described a basal 61.5 m (202 ft) section called the *Butters Formation* that may be equivalent to the Alexandra. McGregor and Wade believe the oldest sediment to be Permian.

In the Ohio Range at the southern tip of the Transantarctic Mountains, Long (1964) and Mirsky (1969–70) describe the 50 m (164 ft) thick Early Devonian Horlick Formation as a sandstone-siltstone-shale sequence. They stated that the Middle Devonian carboniferous section was missing. Boucot et al (1967) assign the Early Devonian age to marine invertebrates found in a thin, interbedded layer.

Behrendt et al (1974) and Schmidt and Ford (1969–70) describe 2,400+ m (7,870+ ft) of clastic sediments from the Pensacola Thiel Mountains that are younger than Early Ordovician but no younger than Devonian. These are probably Beacon equivalents, but the authors do not say so. The datable unit in the section is the uppermost sandstone with black carbonaceous beds with probable Late Devonian plant fragments. They assign the following names to the section:

| ———————— | unconformity ———————— |
|---|---|
| Dover Sandstone | (Late Devonian) |
| ———————— | unconformity ———————— |
| Heiser Sandstone | |
| Elbow Formation | ? |
| Elliott Sandstone | |
| Brown Ridge Conglomerate | |
| ———————— | unconformity ———————— |

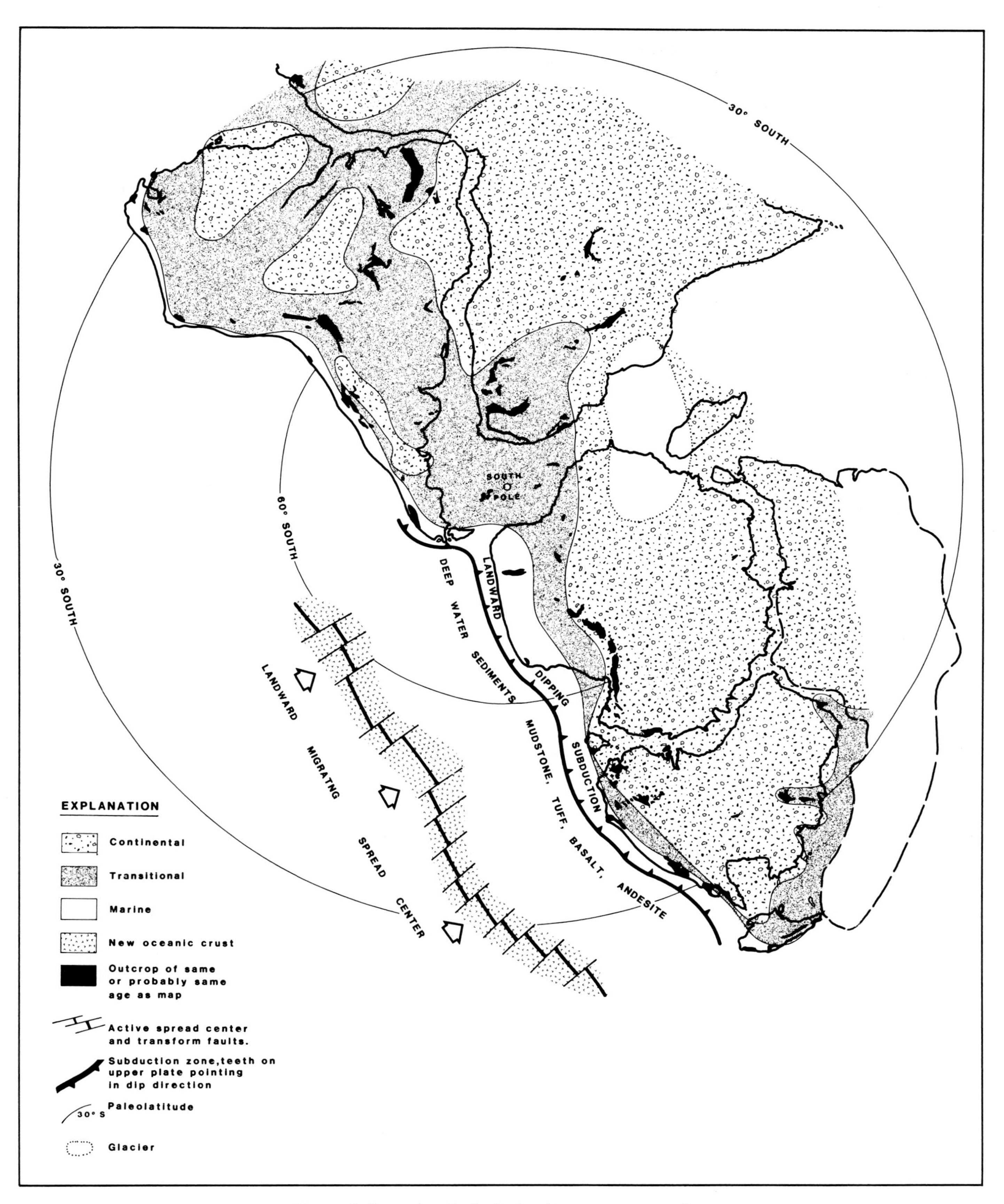


Figure 9. *Devonian-Early Carboniferous reconstruction.*

Craddock (1969–70) describes the Crashsite Quartzite in the Ellsworth Mountains as 3,200+ m (10,500+ ft) of thick and cross-bedded, well sorted, medium to coarse-grain quartzite containing Devonian brachiopods. In addition, Elliot (1975) describes the supposedly Carboniferous-age Trinity Peninsula Series of the Antarctic Peninsula.

A thick graywacke-shale sequence on Alexander Island, the Trinity Peninsula Series on the northern Antarctic Peninsula, the Miers Bluff Formation in the South Shetland Islands, and the Graywacke-Shale Formation of the South Orkney Islands are quartzose graywacke-shale turbidite sequences generally regarded (on rather sparse evidence) as contemporaneous and of late Paleozoic age. Previously, geologists have grouped the quartzose graywackes of the Sandebugten Series on South Georgia with the Trinity Peninsula Series, but structural and petrographic data now suggest that they are much younger.

The Trinity Peninsula Series is the most extensive, estimated to be more than 13,000 m (42,650 ft) thick. Quartzose graywacke and shale form thick, interbedded sequences and predominate markedly over other rock types that include conglomerate, conglomeratic mudstone, arkose, quartzite, siltstone, carbonaceous shale, limestone, chert, and greenschist. Conglomerate clast lithology and sandstone petrology suggest a largely granitic provenance with subordinate sedimentary and metamorphic rocks. The greenschists are likely to have been basic volcanic rocks. Basalts and pillow lavas are abundant in parts of the sequence on Alexander Island, and some of the sandstones have a significant proportion of volcanic detritus. Plant microfossils recovered from the Trinity Peninsula Series at Hope Bay and from the Alexander Island sequence suggest a Carboniferous age. Continental and nearshore marine sedimentation dominated southeastern Australia.

The subduction zone along the Antarctica-eastern Australia margin remained active based on the occurrence of deepwater, oceanic crust type sediments, including mudstone, tuff, basalt, and andesite.

### Late Carboniferous (270–325 m.y.B.P.) (Figure 10)

The supercontinent Pangaea remained intact throughout the Late Carboniferous. Compression continued along the thrust zones of the Appalachian-Hercynian and Mauritanide orogenic belts. In Antarctica the circumpacific orogeny exhibited a peak of tectonic activity (Woolard, 1970).

The landward-dipping subduction zone along the west coasts of North and South America continued to consume Pacific oceanic crust. Destruction of oceanic crust through subduction continued along the coasts of Australia and Antarctica, where Woolard (1970) notes an increase in igneous and tectonic activity. Deepwater sediments found along the Antarctica-east Australia margin contain abundant chert, tuff, and trachyte.

The South Pole was located in central Antarctica and all of Gondwana lay south of the equator.

A continuously exposed continental high extended from Antarctica and southern Australia northward across the Seychelles continental crust fragment, Madagascar and western India, across eastern Africa and western Arabia, through central and northwest Africa to eastern North America. The central highland extended southward from North America and northwest Africa to the northern Brazilian Shield. It also extended southward along the eastern South America and western central Africa margins. Another high occurred in northern Argentina-Chile along the pre-Andean belt. Northernmost Australia also remained emergent.

Transitional zone sediments were deposited in southern Argentina and southern South Africa. Furon (1963) points out that the Karoo basin sediments of this age are predominantly continental, and marine incursions are only observed in the extreme southern and southwestern portions of the basin. Harrington (1962) describes great thicknesses of tillites with thin marine and marine-glacial intercalcations from Argentina.

The Victoria Group is the upper division of the Beacon Supergroup of Antarctica. The Victoria Group is a heterogenous association of glacial beds, alluvial plain deposits, and volcaniclastic sediments ranging in age from Late Carboniferous to Jurassic. It ranges in thickness from only 10 m (33 ft) in northern Victoria Land to as much as 2,400 m (7,870 ft) in the Beardmore Glacier area. Equivalent rocks in the Pensacola Mountains are 715 m (2,346 ft) thick and 2,615 m (8,580 ft) thick in the Ellsworth Mountains.

The lowermost rock unit of the Victoria Group and equivalents is a tillite of indeterminate age that rests disconformably or nonconformably on rocks of Devonian or older age, and lies disconformably below or grades into overlying sediments of Permian age. The very top of the tillite sequence, reflecting the closing stage of glaciation, is Early Permian age (Schopf, cited in Elliot, 1975).

The tillite ranges from Early Carboniferous to Early Permian age and is referred to as the *Pagoda Formation* or *Metschel Tillite* in the Transantarctic Mountains, the *Gale Mudstone* in the Pensacola Mountains, the *Whiteout Conglomerate* in the Ellsworth Mountains, and is equivalent to the *Dwyka* tillite of South Africa. With the possible exception of the Trinity Peninsula Series, geologists consider the only Late Carboniferous age sediments in Antarctica to be tillites.

To understand the glaciation of the Carboniferous-Permian of Antarctica, we must consider the glacial evidence of all Gondwana for this period. Regional syntheses of Gondwana glaciation are reviewed by Crowell and Frakes (1968); Frakes and Crowell (1969,1970); and Frakes, Matthews, and Crowell (1971). In summary, continental glaciers began to form in early to Middle Carboniferous in eastern Antarctica, South America, Madagascar, part of southern Australia, and even peninsular India (Reed, 1949). Glacial debris was deposited along the margins of the active sheet glaciers until sometime during the Early Permian when the climate reversed, warm weather dominated, and plant life flourished.

In many areas of Antarctica, erosion, subsequent to tillite deposition and prior to forming the Permian coal beds, removed all trace of the glacial episode except the polished and striated surface of the underlying rocks. Harrington (1970) uses the term *Maya Erosion Surface* for this unconformity separating the Taylor and Victoria Groups in the Transantarctic Mountains. Matz and Hayes (1967) observe Permian *Glossopteris* and coal beds disconformable on freshwater, fish-bearing sandstone-siltstone-shale of Devonian age in Southern Victoria Land. Lindsay (1968) marks the absence of Carboniferous in the Beardmore Glacier area and records Permian Pagoda Formation disconformable on Devonian Alexandra Formation.

The tillite was also preserved in many areas. McKelvey et al (1970) propose the term *Metschel Tillite* for a 27–28 m (89–92 ft) thick glacial sequence of tillite, varved siltstone, and sandstone in the Olympus and Boomerang Ranges of the McMurdo Sound

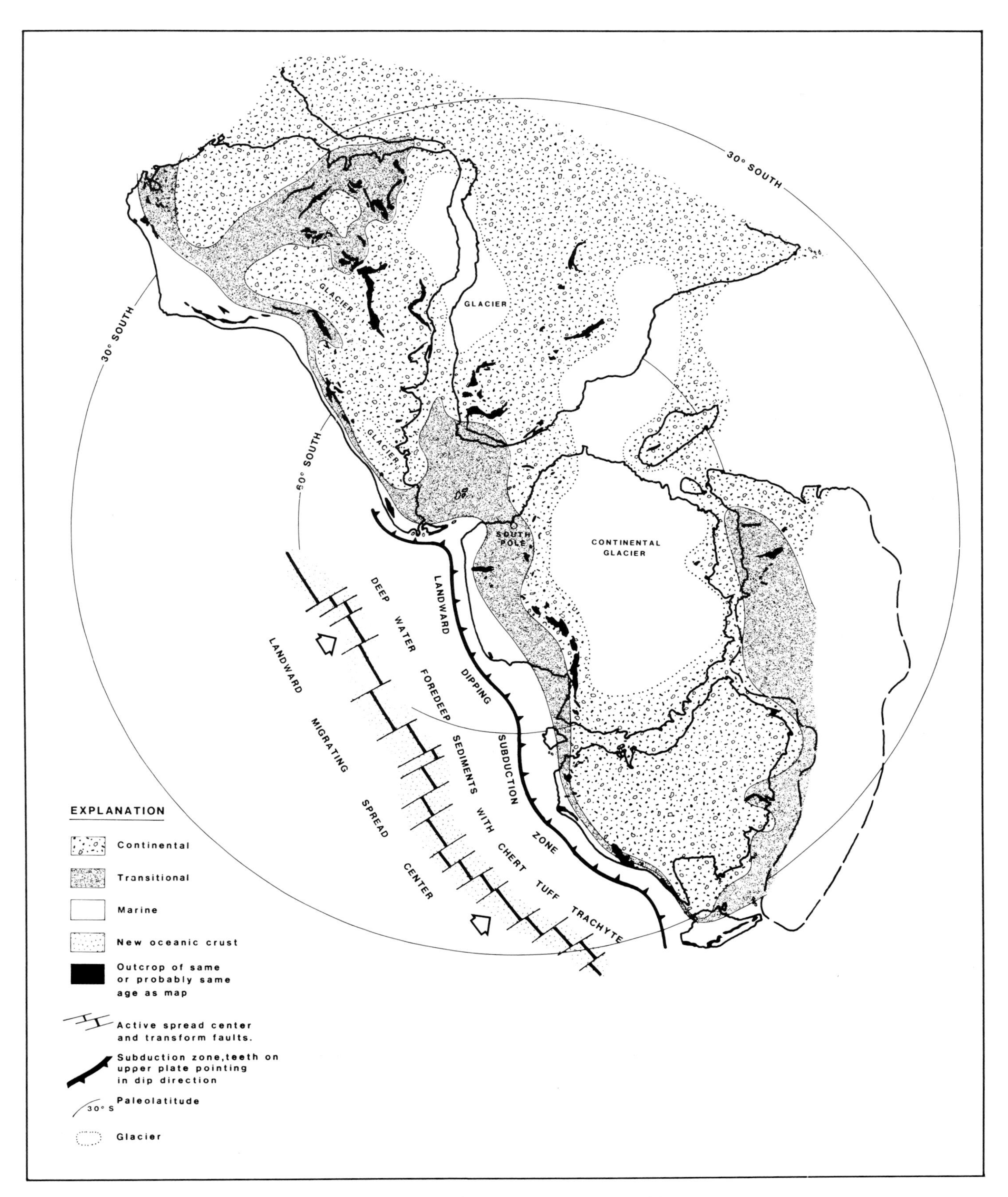


Figure 10. *Late Carboniferous reconstruction.*

area of Victoria Land. They further note that the Metschel lay disconformably on the underlying Devonian Aztec Siltstone and was in disconformable contact with the Permian Weller Coal Measures above. Laird, Mansergh, and Cheppell (1971) record 110 m (361 ft) of Pagoda Tillite unconformable on the Alexandra Formation in the Nimrod Glacier area. In the Queen Maud Mountains, Elliot and Coates (1971) observe tillite 3–150 m (10–492 ft) thick, but consider it Permian age, while Yeats et al (1965) describe the upper 14 m (46 ft) of the Butters Formation as a conglomeratic quartzite, considering it possibly equivalent to the Late Carboniferous Pagoda Formation of others.

In the Ohio Range of the Horlick Mountains, Long (1964) maps Buckeye Tillite disconformable on the Devonian Horlick Formation and nonconformable on the basement complex. Long describes the Buckeye Tillite as 213–275 m (699–902 ft) thick, blue gray boulder clay with thin lenses of Permian (?) age spore-bearing siltstone and sandstone. Minshew (1966) describes Buckeye Tillite from the Wisconsin Range of the Horlick Mountains as 80–140 m (262–459ft) thick, disconformable on basement, and grading upward into the Permian Weaver Formation that *Glossopteris* beds capped.

Behrendt et al (1974) mark the occurrence of the Gale Mudstone, a tillite, in the Pensacola Mountains with 200 m (657 ft) exposed. It lay above the Devonian Dover Sandstone and below the Permian Pecora Formation coal beds.

In the Ellsworth Mountains, Craddock (1969–70) maps 915 m (3,002 ft) of Whiteout Conglomerate, an unsorted bouldery graywacke considered a tillite. The Whiteout Conglomerate lies above the Devonian Crashsite Quartzite and below the Permian Polarstar Formation. Stump (1973) interprets a major ice sheet in Late Carboniferous-Early Permian time in this area to have dumped continental glacial debris over a previous, sediment-filled marine depression.

The thick graywacke sequence of the Trinity Peninsula Series of the Antarctic Peninsula has long been considered both of Carboniferous age and the oldest known rock of the Peninsula. Other similar sections have been correlated with the Trinity Peninsula Series, based on lithology, and assigned a similar age. The fact that a section such as the Sandebugten Series of South Georgia Island is now considered of Mesozoic age casts doubt on the validity of the Paleozoic origin of any part of the Antarctic Peninsula.

**Permian (225–280 m.y.B.P.) (Figure 11)**

Most of Gondwana was continental highland during the Permian, and sedimentation was limited to the continental margins and isolated interior basins.

Continental glaciation, initiated in Early Carboniferous, continued into Early Permian time. Tillite deposits from this glacial episode are found in Antarctica, South America, Africa, Madagascar, and India.

An island-chain moved across the Pacific landward from a spread center. The continental fragments or mass that now form New Zealand, the Campbell Plateau, Lord Howe Rise, Norfolk Ridge, and the Queensland Plateau, though still separate, were rapidly approaching the margin of Antarctica-Australia. The fragments of floating continental crust moved over and destroyed the older Pacific oceanic crust. Eventually, these continental fragments were thrust upon the western margins of North America, South America, Antarctica, and Australia and became part of the individual cratons. The subduction zone remained active along the continental margins facing the Pacific Ocean.

The combined continental masses of Gondwana continued their northern migration, and South America-Africa began to impinge upon the equatorial zone. The South Pole was located in Antarctica, and the Falkland Plateau was situated slightly south of the 60° latitude.

Extensive continental highland existed over northeast South America and north and central Africa. A second large occurrence of highland extended over eastern Antarctica, southern Australia, Madagascar, southern India, and the smaller continental crust segment represented by the Seychelles.

The bulk of the remaining continental mass was dominated by continental lowlands. The transition zone was narrow along western South America and extended into the interior only through northern Argentina into the Parana basin of Paraguay. Only minimal marine deposit occurs along the continental margins.

Thick, continental clastics of Permian age appear in the Beacon Supergroup of Antarctica, as do Permian continental clastics in southeastern and northwestern Australia.

Geologists have identified the uppermost beds of the glacial tillite deposits of Antarctica as of Permian age, and several writers have considered the entire tillite interval to be Permian. All recognize a Permian age for the post-glacial sequence, especially the carbonaceous, coal-bearing unit containing abundant *Glossopteris* flora.

In Victoria Land, Matz and Hayes (1967) note Permian, fluvial arkosic sandstones, and conglomerates with *Glossopteris* unconformably overlying the Devonian fish beds. McKelvey et al (1970), propose the following units as Permian in the McMurdo Sound area:

| | | |
|---|---|---|
| **Triassic** | Lashly Fm | |
| | Fleming Fm | 45 meters coarse, cross-bedded orthoquartzite with siltstone |
| **Permian** | Feather Conglomerate | 160 meters |
| | Weller Coal Measure | 170 meters cross-bedded sandstone/ siltstone with coal seam with *Glossopteris* |
| | ————— unconformity ————— | |
| **Carboniferous** (?) | Metschel Tillite | |

Laird, Mansergh, and Cheppell (1971) presented the following units for the Nimrod Glacier area:

| | | |
|---|---|---|
| **Jurassic** | Ferrar Dolerite/Holyoake Gabbro | |
| | Buckley Coal Measures | 240+ meters cross-bedded yellow to orange fluvial sandstone with coal seams at top |
| **Permian** | Mackellar Fm | 150 meters dark carbonaceous sandstone/ shale sequence |
| | Pagoda Tillite | |
| | ————— unconformity ————— | |

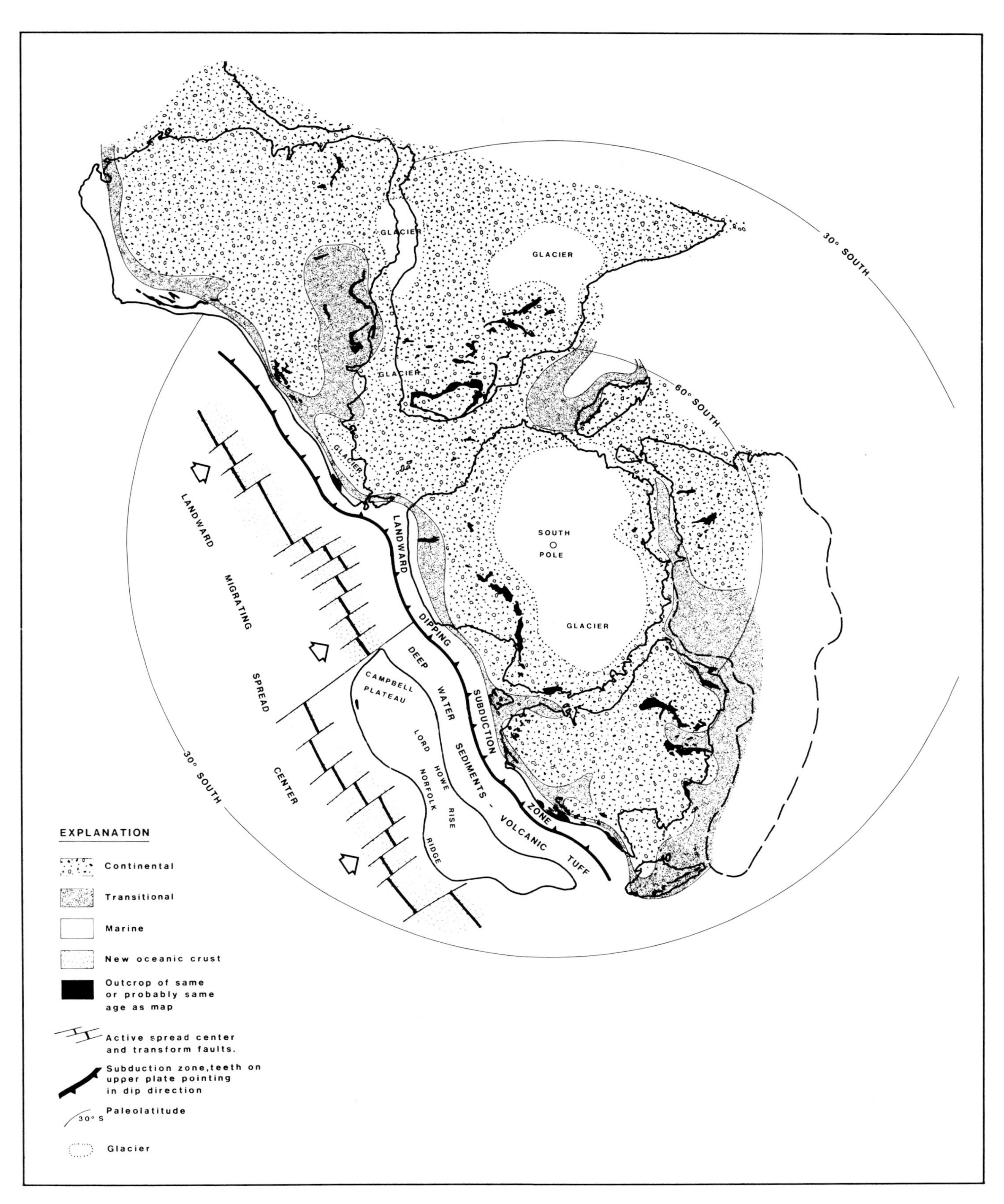


Figure 11. *Permian reconstruction.*

In the Beardmore Glacier area, Barrett's (1968, 1970a, and 1972) followed Lindsay's (1968) descriptions of the Permian:

> The Beacon Supergroup in the Beardmore Glacier area rests unconformably on a Precambrian to lower Paleozoic meta-sedimentary sequence intruded by granite plutons. The 2,600 meter thick sequence, which is intruded and overlain by sills and flows of the tholeiitic Ferrar Group, comprises eight formations. Here the six Beacon Supergroup formations above the glacial Pagoda Formation (Permian) are discussed. The Mackellar Formation, which conformably overlies the Pagoda Formation, consists of 60 to 140 meters of laminated dark shale and light gray fine sandstone. The Fairchild Formation comprises 130–222 meters of massive arkosic sandstone. The Buckley formation is a crudely cyclic sequence about 750 meters thick; sandstone beds rest on erosion surfaces and grade upward into carbonaceous shale. Coal forms as much as 6% of the section. Leaves (mainly *Glossopteris*) and stems are common. Intermediate-acid volcanic detritus, now partly zeolitized, appears 100–300 meters above the base of the formation and dominates the upper part.

Coates (1972) considers the glacial Pagoda Formation of the Queen Maud Mountains to be Permian, and Elliot and Coates (1971) describe the Permian section there as

| | |
|---|---|
| Buckley Fm | arkosic, volcanic sandstone, dark gray shale, and coal seams with *Glossopteris*, 450 meters |
| Fairchild Fm | massive arkosic sandstone, 200 meters |
| Mackellar Fm | medium-dark gray shale with sandstone interbeds, 160 meters |
| Pagoda Tillite | tillite, 3–150 meters |

In the Ohio Range of the Horlick Mountain, Long (1964) describes 2 units above the Buckeye Tillite covered by a Jurassic basalt sill. He considers both units Permian. The lower Discovery Ridge Formation lay unconformably on the Buckeye and consisted of 550 m (1,804 ft) of shale with abundant animal trails and graded upward into the Mount Glossopteris Formation. The latter was approximately 600 m(1,970 ft) thick, a feldspathic sandstone-siltstone-shale sequence with interbedded, semi-anthracite coal containing *Glossopteris* flora and fossil logs, suggesting a fluvial-swamp environment to Long. Minshew (1966) reported that in the Wisconsin Range, the Buckeye Tillite grades upward into the Permian Weaver Formation. The Weaver, 430 m(1,410 ft) thick, has a lower 85 m(279 ft) fissile shale overlain by a 220 m(722 ft) siltstone/shale sequence capped by 125 m (410 ft) of cliff-forming fine-to-medium-grain, well-sorted sandstone with thin black shale at the top containing *Glossopteris*. Minshew believes this sequence to be regressive. Disconformably overlying the Weaver Formation is the Permian Mount Glossopteris Formation, a transgressive, coarse-grain, conglomeratic sandstone, cross-bedded, and 25 m (82 ft) thick.

The youngest sediment observed in the Pensacola Mountains is the Permian Pecora Formation overlying the glacial Gale Mudstone (Behrendt et al, 1974). The Pecora is a well-bedded, light tan, quartzose sandstone-siltstone-shale sequence with interbedded carbonaceous and coaly layers containing *Glossopteris*.

In the Ellsworth Mountains, Craddock (1969–70) and Castle and Craddock (1975) describe the Permian Polarstar Formation as 1,700 m (5,577 ft) thick, siltstone to fine-grain sandstone interbedded with shale and exhibiting cross-bedding, ripple marks, shallow-water organisms, and *Glossopteris* coal beds. They interpret deposits in a prograding delta.

The only sedimentary section in extreme eastern Antarctica is the Permian beds of the Prince Charles Mountains. Mond (1972) describes a detailed, measured section with three subdivisions:

| | |
|---|---|
| Flagstone Bench Formation | Sandstone with iron concretions. Current-bedded, light colored very coarse feldspathic sandstone with pebble horizons interbedded with red-brown to brown feldspathic sandstone and grit grading to sandy siltstone. Thickness 400–600 meters. Source to the west. Probable fluvial deposits into broad valley. |
| Bainmedart Coal Measures | Approximately 65 coal seams from 8–350 centimeters thick interbedded with sandstone, siltstone and shale. About 40% of section is light colored arkosic sandstone, very fine to pebble conglomerate. About 40% of section is gray to brownish red siltstone and gray to brownish red siltstone and shale. Bedding thin laminated to cross-bedded. Erosional channels in coals. Thickness 1,800 meters. Non-marine, sourced from west. Flood plain environment. Upper Permian spores, pollen, and plant leaves. |
| Radok Conglomerate | Green-gray to red-brown to purple fluvial pebble-cobble conglomerate of basement rocks in matrix of quartz and feldspar coarse-grain sandstone. Interbeds of carbonaceous siltstone and shale. |

Balme and Playford (1967) confirm the Permian age of plant microfossils, while Bennett and Taylor (1972) analyze the coals.

The Permian of South Africa consists of sands and clays interbedded with coal seams. It is known for its reptile remains and extensive flora, including complete tree trunks. Reed (1949) reports 3,550 m (11,647 ft) of continental clastic sediments of Permian-Triassic age on the Falkland (Malvinas) Islands. In northern Argentina, continental, marine, glacial, and fluvioglacial sediments represent the Permian period.

### Triassic (189–225 m.y.B.P.) (Figure 12)

The Triassic was much like the Permian in that the continents remained joined and, for the most part, uplifted into con-

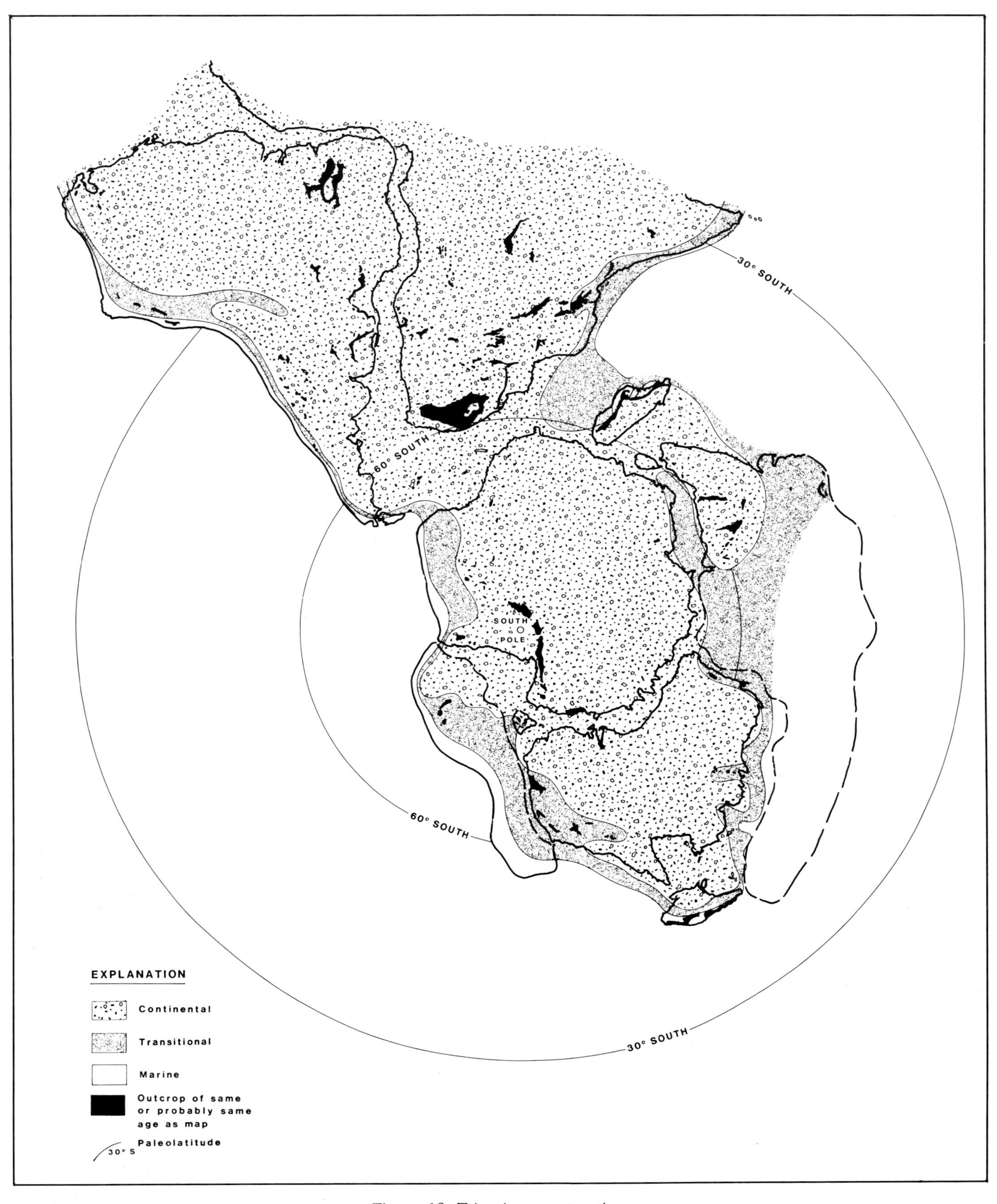


Figure 12. *Triassic reconstruction.*

tinental highlands and lowlands. If anything, regression was more pronounced and areas of lowland deposits more extensive.

The eastward migrating Pacific Island chain moved closer to the west coasts of the continental margins. The New Zealand-Campbell Plateau-Lord Howe Rise-Norfold Ridge-Queensland Plateau mass had reached and accreted onto Antarctica-Australia by the close of the Triassic period, becoming what is termed the New Zealand Complex.

The South Pole lay somewhere south of the Gondwana land mass, and the northern edges of South America-Africa reached and began to cross the equator. Triassic evaporites are common in the sediments of the paleo-equatorial zone.

Continental deposition occurred over southern Africa, southern South America, and parts of western Antarctica. In the Falkland (Malvinas) Islands, continental beds of Permian-Triassic age reportedly reach a thickness of 3,550 m (11,647 ft).

Du Toit (1937) postulates that early Mesozoic deformation was marginal to the East Antarctica shield, calling this event *Gondwanian Orogen* (Figure 13). Ford (1972) uses the term *Weddell Orogeny* for the early Mesozoic deformation in the Pensacola Mountains, while Craddock (1972) uses *Ellsworth Orogeny* for the parallel event in the Ellsworth Mountains. *Cape Orogen* is the equivalent event in South Africa with *Sierra Orogen* of South America considered a correlative. In Australia, the *New England Orogeny* deformed Silurian-Permian age sediments. These various terms apply to a single geologic event I will call the *Gondwanian Orogen,* as originally proposed by Du Toit. This early Mesozoic event was largely restricted to the Triassic period.

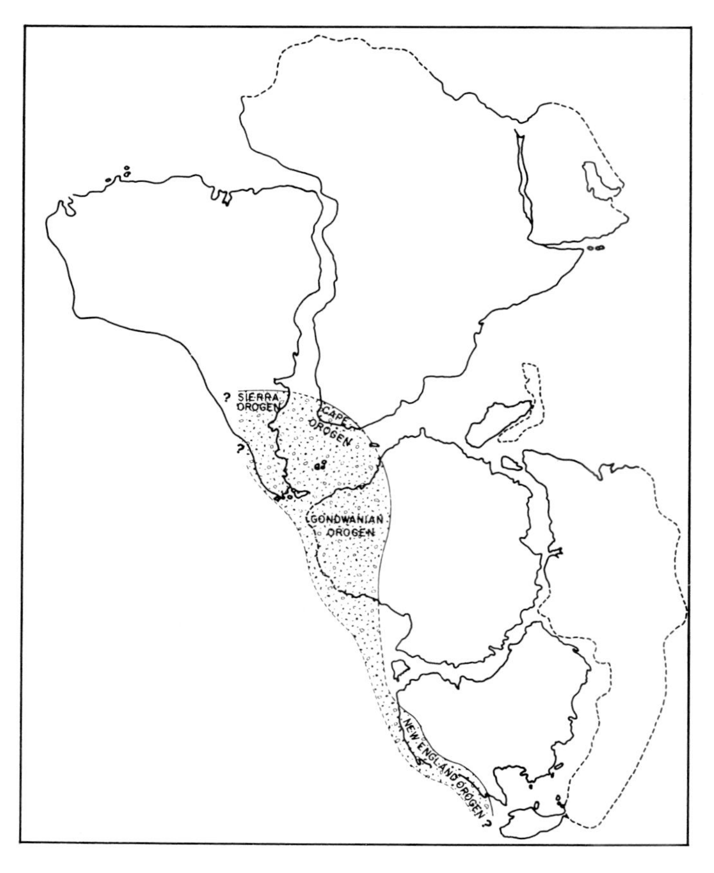


Figure 13. *Early Mesozoic Gondwanian Orogen.*

The stratigraphic record of the Gondwanian Orogen is varied. The Trinity Peninsula Series of supposedly Carboniferous age may actually be Triassic, as are the Legoupil Formation sediments at Cape Legoupil (Thomson, 1975). Early researchers considered the Legoupil Formation Carboniferous as was the Sandebugten Series on South Georgia, which Elliott (1975) now considers early Mesozoic. Triassic age may also apply to (1) the thick graywacke-shale sequence on Alexander Island (Grikurov, 1972); (2) in the South Shetland islands (Dalziel, 1972); and (3) in the South Orkney Islands (Dalziel, 1972). Stump (1973) has determined Triassic dates for some rocks on Thurston Island. The graywacke-shale sequence of the Antarctic Peninsula was deformed by the Gondwanian Orogen and is overlain conformably by Upper Jurassic rocks, thus giving a minimum age to both the deformed sediments and the orogeny.

There are no post-Permian sediments or igneous rocks in the Ellsworth Mountains; however, the early Mesozoic Gondwanian Orogeny deformed the entire Precambrian to Permian sequence. Behrendt et al (1974) report that in the Pensacola and Thiel Mountains, Triassic tectonic activity deformed Permian and older sediments into broad folds that were subsequently intruded by Jurassic dolorite. Long (1964) stated that in the Horlick Mountains, a Jurassic sill terminated Permian sediments and that the Triassic was absent.

Elliot and Coates (1971) and Elliot et al (1970) assign Triassic age to the uppermost Beacon Supergroup sediments in the Queen Maud Mountains. The 3 formations described above the Permian Buckley and below the Jurassic dolerites and basalts were named, from the base up, *Fremouw, Falla,* and *Preble.* The basal Fremouw is an arkosic sandstone and greengray mudstone over 700 m (2,295 ft) thick. The Fremouw Formation is Early Triassic age based on the identification of bones of amphibians and various reptiles, including *Lystrosaurus,* a land-living reptile. Its remains have been found in South Africa and India, further supporting the present reconstruction of Gondwana.

Overlying the Fremouw Formation in the Queen Maud Mountains is the Falla Formation, a 200+ m (656+ ft) sandstone-shale sequence containing *Dicroidium* flora. Pyroclastic deposits of the overlying Preble Formation are at least 460 m (1,510 ft) thick. Barrett (1972) describes similar sediments with the same formation names from the Beardmore Glacier area.

La Prade (1972) describes a 563 m (1,847 ft) thick Mount Kenyon Formation from the Shackleton Glacier area and assigns it a Triassic age, although it is more likely Permian. The unit contains abundant *Glossopteris* flora and is a massive graywacke sandstone with shale interbeds and petrified logs and is considered of fluvial origin.

McKelvey et al (1970) describe, from the Olympus and Boomerang Ranges near McMurdo Sound, the Triassic Lashly Formation. It is a 300 m (984 ft) sequence of sandstone-shale-carbonaceous bands with Middle Triassic microfossils.

Elliot (1975) suggests that a rising geanticline in West Antarctica occurred during the Triassic with graywacke-shale sequences being deposited on the Pacific side and marine and non-marine strata on the continental side. Acidic volcanism began at this time and peaked during the subsequent Jurassic period.

**Jurassic (130.5–189 m.y.B.P.) (Figure 14)**

Gondwana began to break up as oceanic crust started to appear, separating South America-Africa-Madagascar from Antarctica-Australia-India. At the same time, accretion of island arcs was occurring along the Pacific margin. Jurassic oceanic crust also appeared off northwest Australia as sea floor spreading began; part of greater India separated and began its move to become part of southeast Asia.

The first definition of major plates began during the Jurassic. By the close of the Jurassic, four major plates existed in the study area, plus three minor plates within the major Gondwana Plate. Other than the Gondwana Plate, there were the South America-Africa Plate, the Pacific Plate, and the Southeast Asian Plate.

The landward-migrating spread center in the Pacific Ocean was near the continental margins and formed a triple junction. One branch of the spread center paralleled the west coast of South America, a second branch separated South America-Africa-Madagascar from the rest of Gondwana, and the third branch lay offshore Antarctica-Australia. Three island arc bodies, which now compose the Antarctica Peninsula and much of West Antarctica, are depicted as separated by landward-dipping subduction zones but approaching Antarctica *en mass*.

The Paleoequator passed through northern Africa and South America. The Falkland (Malvinas) Plateau lay at approximately lat. 40°S. The South Pole was situated in Antarctica.

Much of the Gondwana continent remained highland. Extensive continental deposition, however, occurred only in Australia, southern Africa, and in east central and southeastern South America. In Antarctica, shallow marine basins typified the Antarctic Peninsula deposits. Extensive volcanic activity marked the end of the Triassic and beginning of the Jurassic. Jurassic volcanic rocks are considered basement for much of southern Argentina and West Antarctica. This designation holds true for the Falkland (Malvinas), Burdwood, and South Georgia basins.

In the eastern Patagonian area of South America, economic "basement" consists of thick quartz porphyries, breccias, and tuffs. These volcanic rocks are considered of Jurassic age and are mostly continental but grade upward into marine tuff and shale (Katz, 1973). The equivalent strata to the west and south are more complex and reflect the eugeosynclinal facies of volcaniclastics, graywacke, and radiolarian chert associated with migrating island arcs and subduction zones.

Suarez (1976) interprets the Triassic (?), graywacke-shale sequence discussed earlier as turbidites and pillow lavas forming a seafloor trench assemblage accreted onto older continental crust. Suarez identifies three major Upper Jurassic-Lower Cretaceous units in the Antarctic Peninsula and Alexander Island that he related to an ensialic volcanic arc, a forearc or intra-arc marine basin, and a back-arc marine basin. Suarez compares the Antarctic Peninsula sequence with a similar sequence in the South American Andes of Tierra del Fuego. The mechanics of the Tierra del Fuego back-arc basin formation and subsequent destruction are illustrated by Bruhn (1979); Bruhn and Dalziel (1977); and Harrison, Barron, and Hay (1979).

The Gondwanian Orogeny possibly extended into the Early Jurassic over the southernmost South American-Antarctic Peninsula-West Antarctica region. Jurassic sediments capped by Upper Jurassic volcanic rocks overlie disconformably the graywacke-shale sequence of probable Triassic age in the Antarctic Peninsula and Eastern Ellsworth Land.

Stevens (1967) identifies Upper Jurassic ammonites and belemnites from marine sediments of James Ross Island, Graham Land, and Eastern Ellsworth Land. He states that they resemble those of South America and South Georgia Island. Thomson (1972) identifies Upper Jurassic-Lower Cretaceous ammonites from Alexander Island clastic sediments.

Williams et al (1972) assign Jurassic (?) age to the Latady Formation in type locality Latady Mountains on the Lassiter Coast. The sediments are tightly folded, black siltstone and shale with minor tan to white graywacke and quartzite with a thickness of 600 m (1,968 ft). The sediments were deposited in shallow water and are overlain at Mount Poster by 1,000 m (3,280 ft) of folded lava flows and ash-flow tuffs of Jurassic age.

Laudon (1972) summarizes the stratigraphy and geologic history of Eastern Ellsworth Land as

| Age | Rock Units | Events |
|---|---|---|
| Recent | | Partial deglaciation |
| Pleistocene | | Continental glaciation<br>Alpine glaciation |
| Pliocene | Olivine-basalt flows | Volcanism |
| Late Tertiary | | Arching and block-faulting of post-Andean peneplain |
| Early Tertiary | | Development of post-Andean peneplain |
| Middle Cretaceous | Andean intrusives | |
| Early Cretaceous | Mafic dikes | |
| | | Orogeny, folding of stratified rocks |
| Jurassic | Clastic sedimentary rocks; felsic volcanic rocks, andesite flows | Geosynclinal deposition<br><br>Volcanism |
| Pre-Jurassic | | Unknown |

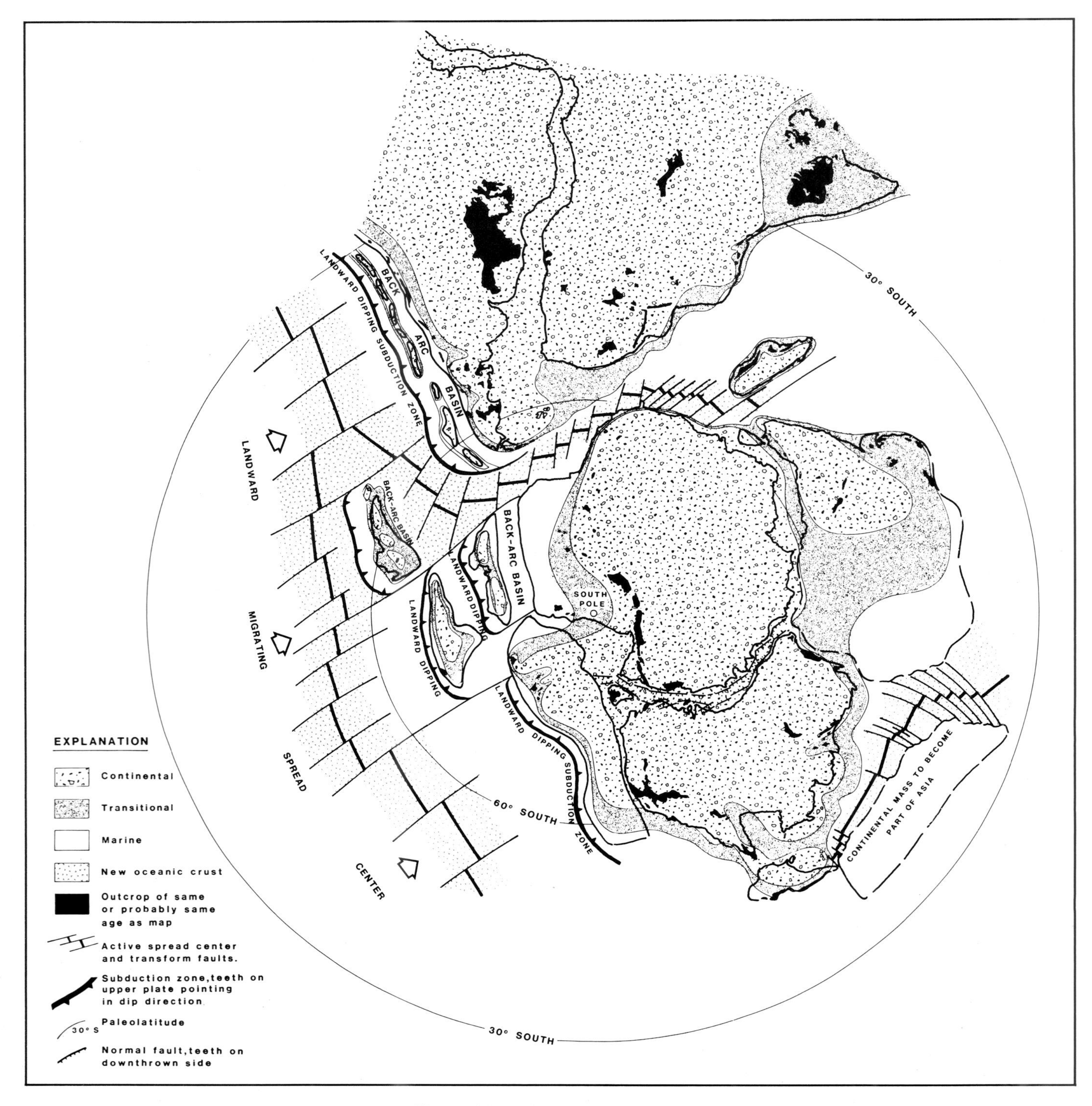


Figure 14. *Jurassic reconstruction.*

The Late Jurassic was a time of widespread calc-alkaline volcanism throughout much of the Peninsula area and, to an apparently limited extent, in western Marie Byrd Land (Stump, 1973). Early to Middle-Jurassic plutonism occurred in the area between the Ellsworth and Pensacola Mountains and also to a limited extent in western Marie Byrd Land and the Antarctic Peninsula. Dolerite sills and tholeiitic flood basalts of this period are found throughout the Transantarctic Mountains. These are interpreted as precursors to the separation of Africa and Antarctica.

Craddock et al (1969–1970) notes either Jurassic dolerite sills (Ferrar) and dikes, or extrusive basalts (Kirkpatrick Basalt and Holyoake Gabbro) in Antarctica at the following locations: Ohio Range of Horlick Mountains, Queen Maud Mountains, Shackleton Glacier area, Terra Nova Bay of McMurdo Sound, North Victoria Land, and in coastal East Antarctica—Ahlman Ridge, Sør Rondane Mountains, and Prince Charles Mountains.

There are Lower Jurassic dolerites in Tasmania (Sprigg, 1967) and Lower-Middle Jurassic, carbonaceous shale-siltstone-sandstone in central Australia as well as carbonaceous shale, coaly sandstone, conglomerate, tuff, and basalt in eastern Australia. Upper Jurassic continental sandstone, carbonaceous siltstone, coal, and tholeiitic basalt occur in the Perth basin. In interior Australia, Upper Jurassic continental sandstones and coal measures are found. Upper Jurassic plateau dolerites are extensive in Tasmania, whereas arkosic and graywacke sandstones with coal and tuff beds occur throughout the Otway, Bass, and Gippsland basins. Similar sediments probably occur in adjacent, offshore Antarctica basins.

Kraus (1967) and Ahmad (1969) believe that the breakup of Gondwana started in the Triassic, and they predicted Triassic oceanic crust in the Indian Ocean. Later data, however, indicate Jurassic initiation of Indian Ocean crust and uplift accompanied by a horst-graben complex between Antarctica and India and between Antarctica and Australia.

Sastri, Raju, and Sinha (1974) describe the evolution of the sedimentary basins along the east coast of India and conclude that they originated in early Mesozoic time. The oldest sediments, unconformable on the Precambrian basement complex, are Permian fluvio-glacial tillites correlated with those of Antarctica and Africa. Jurassic and Lower Cretaceous sediments are also comparable.

### Early Cretaceous (96.5–130.5 m.y.B.P.) (Figure 15)

Elements of sea-floor spreading evident today are evolved from the spreading in existence by the close of the Early Cretaceous.

The South Atlantic spread center met at a triple junction, with one branch extending southeastward into the Indian Ocean and another branch separating South America and Antarctica in the region of today's Weddell sea.

The Early Cretaceous marked the end of Gondwana as a continental entity. The rift between Antarctica-Australia and India began and extended along India's southeast and west coasts to separate it from Africa-Madagascar and Antarctica, and India began its separate migration northward. The Seychelles became isolated in the early Indian Ocean, and Antarctica became widely separated from the Falkland (Malvinas) Plateau-Argentina.

For the first time, South America became a continent separated by sea from other masses of continental crust. South America continued its northward drift, during which time the Falkland (Malvinas) Plateau was at approximately lat. 50°S. The South Pole was situated in the area of the Ross Sea shelf of Antarctica. Subduction persisted along the Pacific margin of Antarctica and southern South America. The Antarctica-Australia-New Zealand complex remained a single land mass, with the ancestral Antarctica Peninsula nearing its present position relative to Antarctica.

The combined Antarctica-Australia continent was dominantly continental high, except for a broad zone of continental lowland sedimentation extending from the Arafura Sea north of Australia and south across eastern Australia. Transitional and marine sediments were deposited only along the outermost shelf areas.

Cretaceous deposition occurred only in the Antarctic Peninsula and West Antarctica. East Antarctica remained a stable craton and the Transantarctic, Pensacola, and Ellsworth Mountains were being uplifted and eroded. Marine sedimentation continued from the Jurassic into Early Cretaceous along the Pacific margin.

An Early Cretaceous trough existed in the Alexander Island area with shallow marine sediments farther offshore. Thomson (1972) and Taylor (1972) identify Early Cretaceous ammonites from Alexander Island.

Elliot (1975) summarizes Cretaceous deposition of the Antarctic Peninsula as follows:

> The James Ross Island area forms part of an apparently extensive Late [sic] Mesozoic-Cenozoic sedimentary basin on the southeast side of Antarctic Peninsula. The Cretaceous sequence has been estimated at 5,000 meters thick. The lower 1,500 meters, which crops out on the northwest side of James Ross Island, includes numerous conglomeratic beds. The upper part, the Snow Hill Island Series, is much more widespread and is generally a finer-grained clastic sequence. Ammonite faunas established a Campanian age. The post-Early Cretaceous Seymour Island Series rests with slight angular unconformity on the Campanian strata.
>
> Isolated outcrops of clastic sediments as far south as the Bowman Coast are assigned a Cretaceous age on their enclosed faunas or by lithologic identification. Conglomeratic strata in a possibly analogous position crop out on the South Orkney Islands; invertebrate and plant fossils suggest a tentative age of Jurassic or Cretaceous.
>
> Cretaceous strata on the Pacific side of the Peninsula crop out on Livingston and Snow Islands in the South Shetlands and on Alexander Island. Volcanic sequences elsewhere in the South Shetlands, generally regarded as Upper Jurassic or Cenozoic, may include Cretaceous strata. Farther south, a thick, shallow-marine clastic sequence, the Fossil Bluff Series, has been assigned a Late [sic] Jurassic-Early Cretaceous age. The sequence consists of conglomerate, sandstone, and mudstone derived from the east and constitutes deltaic and interdeltaic environments of a shoreline facies. Volcanic detritus is abundant in much of the succession.

Continental clastic deposits were laid down in the offshore Otway, Bass, and Gippsland basins of southern Australia (Sprigg, 1967). Marine clastic sediments with glauconite, foraminifera, molluscs, and radiolarians of Early Cretaceous age are found in the Perth and Eucla basins. In interior Australia, fossiliferous marine mudstones occur.

### Late Cretaceous (65–96.5 m.y.B.P.) (Figure 16)

By the close of the Cretaceous, the southern continental masses had separated except for Antarctica-Australia.

Africa, South America, Madagascar, India, Antarctica-Australia, and New Zealand-Campbell Plateau were each isolated by surrounding ocean while Antarctica and Australia remained united. However, the New Zealand Complex separated and began the rotational movement toward its present position.

The landward-dipping subduction zone along the west coast of South America remained active and was offset, via a transform fault, to a position seaward of the New Zealand Complex

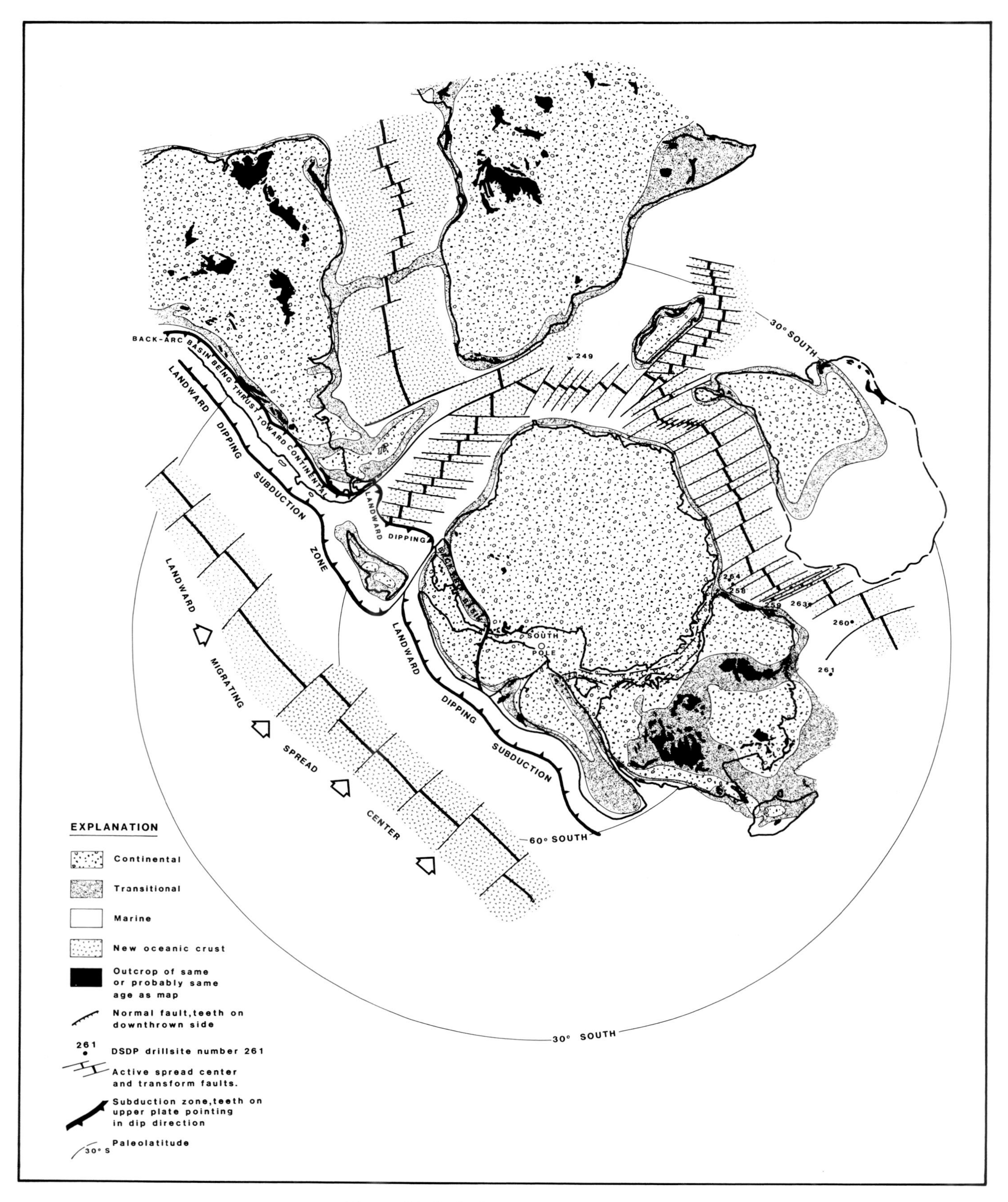


Figure 15. *Early Cretaceous reconstruction.*

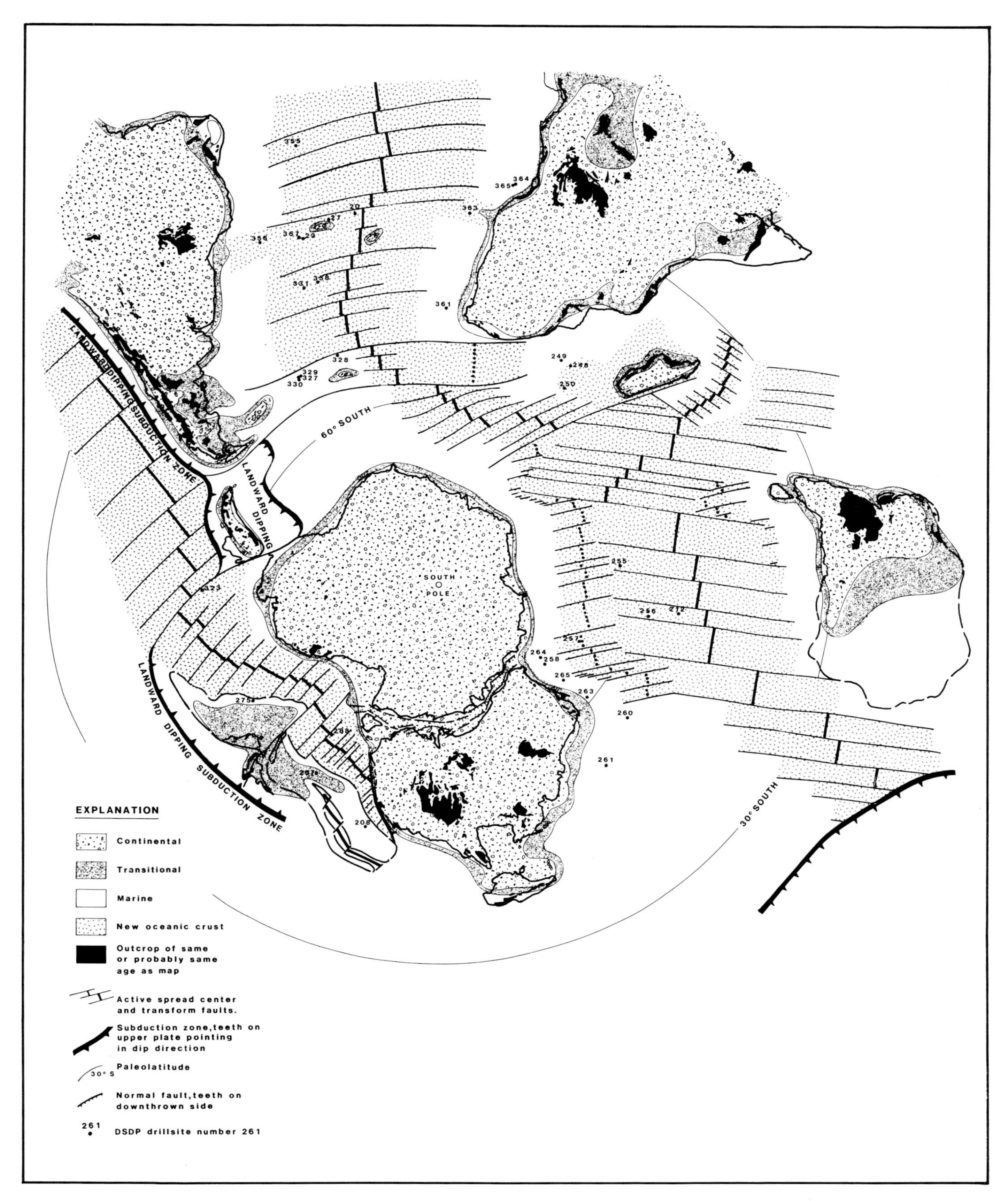


Figure 16. *Late Cretaceous reconstruction.*

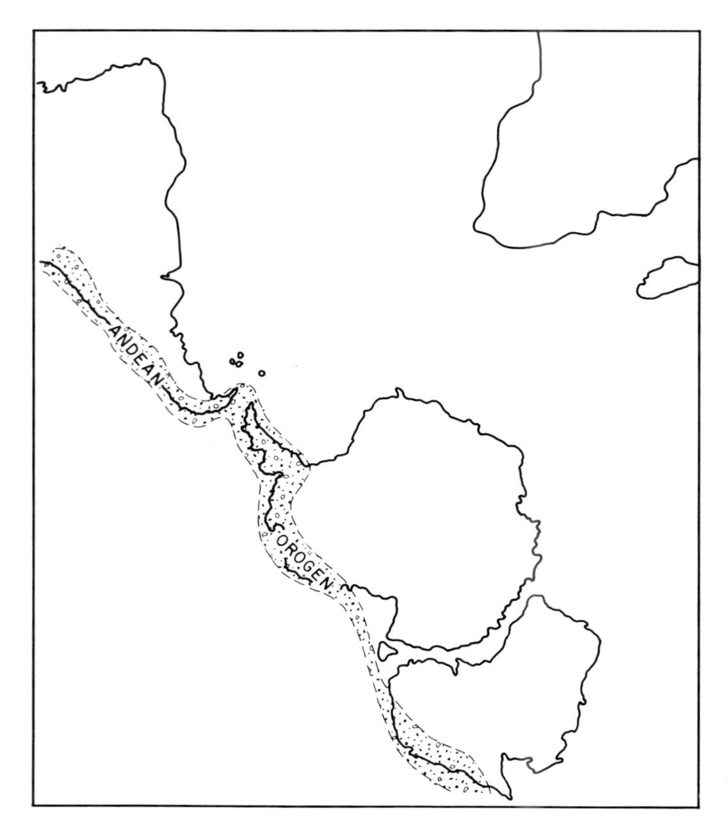


Figure 17. *Late Cretaceous-early Tertiary Andean Orogen.*

landmass. Sea-floor spreading continued in the South Atlantic and extended southeastward into the Indian Ocean sector to the major transform extending from Antarctica to the east coast of Madagascar.

The southwest branch of the South Atlantic triple junction ceased activity near the close of the Early Cretaceous (LaBrecque and Hayes, 1979a, 1979b). No additional separation between the "W" fracture zone (FZ) and Antarctica has occurred after Early Cretaceous.

The spread center separating Antarctica-Australia from India and India from Africa-Madagascar was not joined at this time with the South Atlantic spread center. There is evidence for jumps by the Late Cretaceous center south of Africa (LaBrecque and Hayes, 1979b) and between Antarctica-Australia and India (Johnson, Powell, and Veevers, 1976).

Volcanic extrusion-intrusion accompanied by uplift of the oceanic crust occurred in the southern Indian Ocean to form a sea floor high that was later to become the Kergulen Plateau and Broken Ridge.

The bulk of Antarctica-Australia was emergent during the Late Cretaceous with deposition of continental sediments postulated for depressions in the crust of West Antarctica and northeastern and northern Australia. Central India and Madagascar were high, with fringing lowlands. Much of South America and Africa received continental clastic sediments, while relatively narrow transitional zones occurred along the continental lowlands. The most extensive transitional zone was present in eastern Argentina and over the New Zealand complex.

A shallow-marine basin located at the northern end of the Antarctic Peninsula received some sediment (Stump, 1973). The predominant event of the Late Cretaceous in Antarctica and much of the adjacent land masses, however, was the Andean Orogeny that produced large volumes of granodioritic-granitic plutonic intrusives. These plutons are evident from the South American Andes, through the Antarctic Peninsula and West Antarctica, to eastern Australia.

**Paleocene-Early Eocene (49–65 m.y.B.P.) (Figure 18)**

The southern continents and continental crust island segments were completely separated by the close of the early Eocene. Australia and Antarctica were the last major elements to part.

Additional evolution of plates occurred by the close of the early Eocene. The South America and Africa Plates were relatively unchanged other than by the addition of oceanic crust along the mid-ocean ridge spread centers. The major change was the final breakup of ancestral Gondwana into separate Antarctica and Australia Plates. The spread center between Antarctica and Australia defining the plate boundaries extended into the Indian Ocean, resulting in adjustment of the old Gondwana and Indian Plate boundaries. The Indian Plate split into a western portion, retaining the plate name, and an eastern segment, which is called the Investigator Plate (after the Investigator FZ of that area). The New Zealand and Lord Howe Plates added new oceanic crust, and there was some change in position along the transform between them.

There was insignificant change in the Pacific Plate during the Paleocene-early Eocene. Paired triple junctions in the Southeast Pacific led to the formation of the Aluk Plate and the Bellingshausen (B) Plate. The eastern part of the Pacific became the Farallon Plate by virtue of the division caused by the Southeast Pacific Rise spread center.

Paired triple junctions were also evident in the southeastern Pacific Ocean near the juncture of South America and the Antarctic Peninsula. Subduction continued along the west coast of South America and offshore western Antarctic Peninsula. The southeast Pacific Ocean spread center extended between West Antarctica and the New Zealand Complex into the incipient southeast Indian Ocean between Antarctica and Australia, and from there through the anomalous Kerguelen Plateau-Broken Ridge uplift and on northward. A triple junction of the Indian Ocean spread center occurred north of Kerguelen Plateau-Broken Ridge, with a northward migrating spread center located east of the forming Ninetyeast Ridge.

Antarctica and Australia were predominantly emergent with rare continental clastic sediments in Australia and only postulated deposition in Antarctica crustal depressions. Studies of various DSDP cores (Table 2, after Kennett et al, 1974) indicate temperate climates for this period in Antarctica with any glaciation restricted to high areas. Portions of the New Zealand Complex were emergent during this time.

Transgressive and marine sediments in a narrow band surrounded the continental lowlands with significant and broad, transgressive incursions in eastern Argentina and northeastern Africa. Transgression may have occurred into the Wilkes basin of Antarctica and the Officer basin of southern Australia. Western Australia was covered by marine waters as was much of Papua-New Guinea, so that most of the New Zealand Complex was submerged.

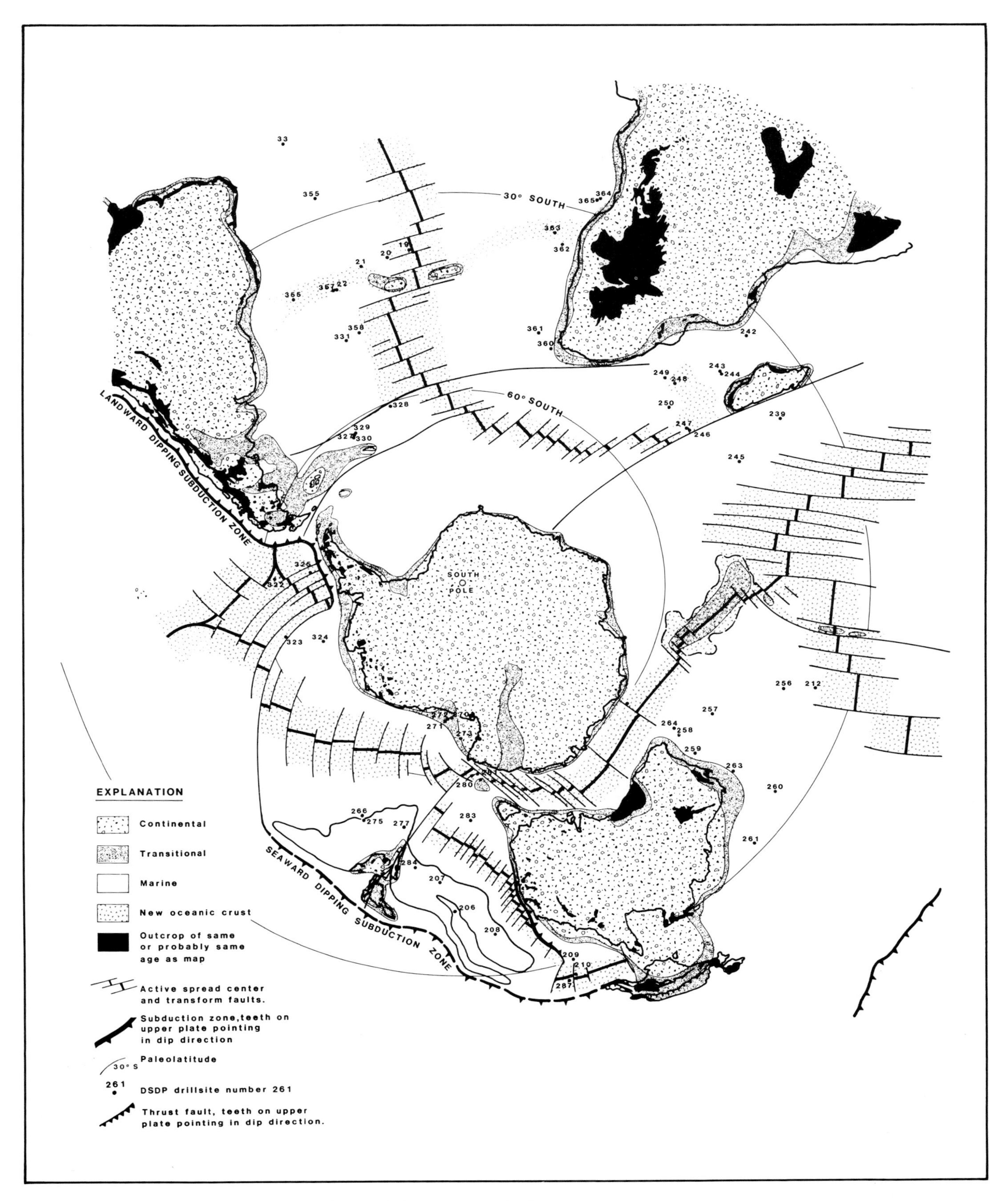


Figure 18. *Paleocene-early Eocene reconstruction.*

*Table 2. Cenozoic sequence of some major Antarctic climatic, glacial, and high southern latitude oceanographic events (after Kennett et al, 1974)*

| Age (myBP) | Epoch | Event |
|---|---|---|
| 0.7 | Late Pleistocene | Further increase in up-welling and biogenic productivity at Antarctic Convergence increase in ice-rafting. |
| 2.6 | Late Pliocene | Development of Northern Hemisphere glaciation. |
| 4.2-5 | Late Miocene - early Pliocene | Development of Antarctic ice sheet much thicker than present: major regression: global marine cooling; development of West Antarctic ice sheet: major increase in upwelling and biogenic productivity at Antarctic Convergence. Increase in ice rafting. |
| 10-13 | Middle Miocene - early late Miocene | Development of major ice cap on East Antarctica. First recorded ice rafting in northern Antarctic waters south of New Zealand (Site 278). |
| 22 | Early Miocene | Initiation of Antarctic Convergence with low degree of upwelling. |
| 25-30 | Late Oligocene | Development of Circum-Antarctic Current and related major changes in deep-sea sediment patterns. |
| 22-38 | Oligocene | Prolonged Antarctic glaciation but no substantial ice sheet formed. Prolonged active bottom currents and deep-sea erosion in many areas. Increased calcareous biogenic productivity and accumulation in central Pacific. |
| 38 | Eocene - Oligocene boundary | Major global cooling; development of Antarctic glaciation at sea level; extensive production of sea ice and Antarctic bottom water; development of present thermohaline oceanic circulation; enhanced deep-sea erosion; some iceberg production; major change in abyssal benthic faunas. Large, abrupt depth increase in calcium-carbonate compensation depth. |
| 38-53 | Eocene | Temperate climates; any Antarctic glaciation restricted to higher elevations. Ice-rafted sediments to southeast Pacific. |

**Middle Eocene-Early Oligocene (30–49 m.y.B.P.) (Figure 19)**

The Bouvet triple junction, still active today, was initiated south of the major Falkland (Malvinas)-Agulhas fracture zone. The southeastern branch extended to the Madagascar fracture zone and the southwestern branch, possibly initiated along a leaky transform, extended westward to the west-dipping subduction zone located east of South Georgia and South Orkney Islands.

Landward or east-dipping subduction continued in the Pacific Ocean along the western South America border and extended southward along the Antarctic Peninsula to the Eltanin fracture zone. The peculiar spread pattern seen earlier in the southeast Pacific Ocean continued with one leg of the eastern triple junction beginning to disappear into the subduction zone off southwestern Chile.

The Pacific Ocean spread center stretched westward, separating the New Zealand Complex from Antarctica, and continued between Antarctica and Australia and, from there, northwestward across the Indian Ocean.

By the close of the early Oligocene time, the seafloor uplift in the Indian Ocean offshore Enderby Land of East Antarctica had separated into two masses: the Kerguelen Plateau and Broken Ridge. Parts of these uplifts extended above sea level and subaerial erosion occurred; carbonates began to form in the shallow water around the emergent volcanic and oceanic basalt islands.

There was major global cooling and development of Antarctic glaciation at sea level and some iceberg and sea ice development (Table 2).

DSDP Site 270 encountered Oligocene sediments above the metamorphic basement on the Ross Sea Shelf. These are described as 25 m (82 ft) of coarse sedimentary breccia over basement, with the top 3 m (10 ft) highly altered and interpreted as a paleosol. The breccia are glacial sediments with molluscs, foraminifera, and diatoms ranging in age from Oligocene to Recent. Because of the current, extensive ice cover and extremely adverse weather conditions, interpretating the environment of sediment deposition on and around Antarctica is largely speculative. Geologists believe that most of East Antarctica and the Transantarctic Mountains formed highlands undergoing erosion, while other source-area highs were located in West Antarctica and the Antarctic Peninsula. Continental lowlands are interpreted east of the Transantarctic Mountains in the Polar and Wilkes basins, and a broad, topographic low extended across West Australia from the Ross Sea embayment through the Marie Byrd basin to the Weddell Sea embayment. Marine transgression may have occurred in the Ross and Weddell Sea embayments and into the Wilkes basin.

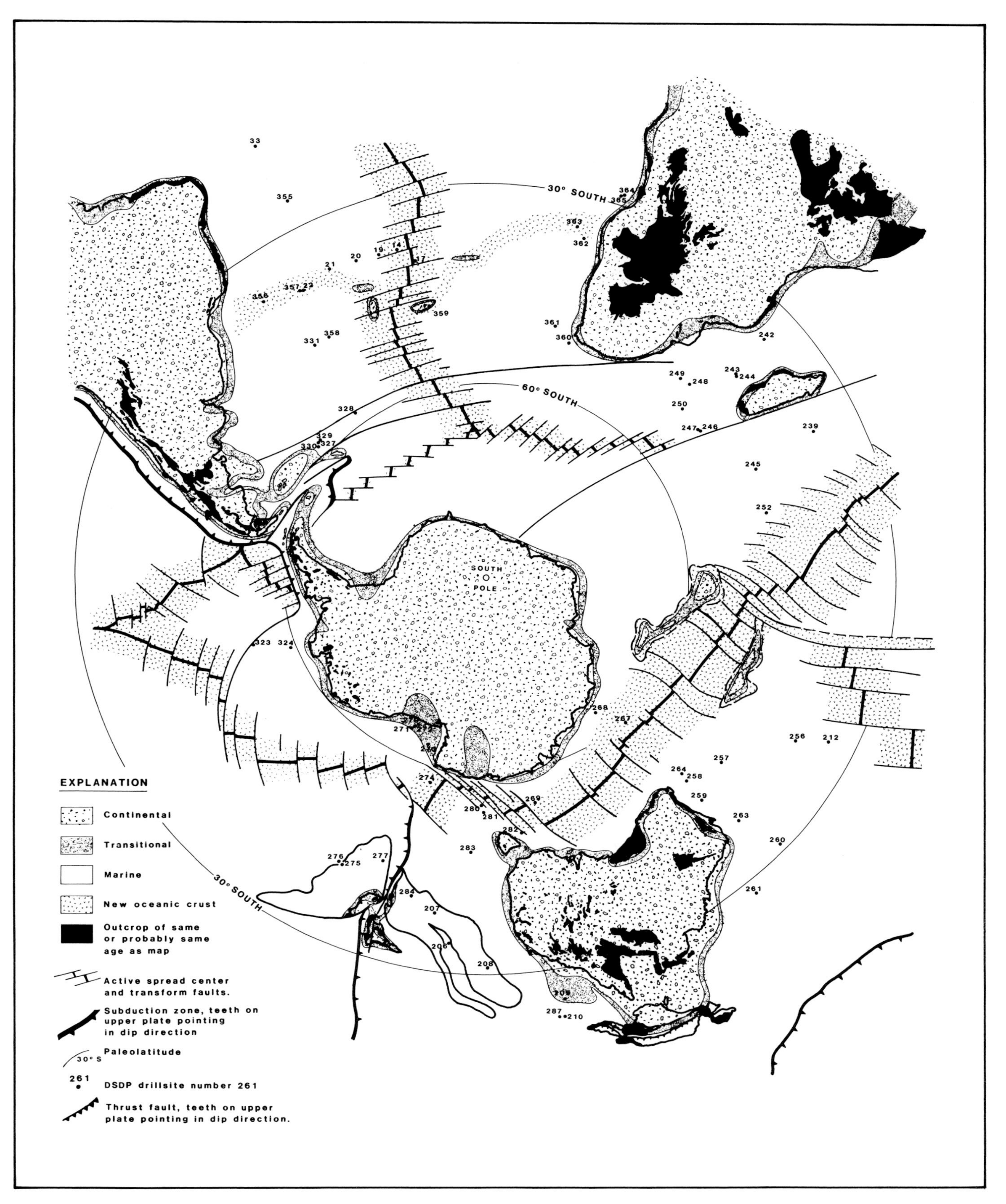


Figure 19. *Middle Eocene-early Oligocene reconstruction.*

Late Oligocene-Middle Miocene (11.5–30 m.y.B.P.) (Figure 20)

The easternmost Pacific unit, the Farallon Plate, broke into the Nazca, Gorda, and Cocos Plates sometime after 26 m.y.B.P. (Handschumacher, 1976). It was also during the late Oligocene-middle Miocene that the Scotia Sea came into being. Four small plates are associated with the creation of the Scotia Sea: (1) Bellingshausen (B), (2) Scotia (S), (3) South Sandwich (SS), and (4) South Orkney (SO).

Behind the west-dipping subduction zone of the South Sandwich Trench, back-arc spreading began in the newly forming Scotia Sea. Development of the volcanic South Sandwich Islands also began at this time. The west-dipping subduction zone extended to the south along the eastern margin of the South Orkney continental block, and that back-arc spreading at this time split the South Orkney block away from the Antarctic Peninsula, thereby creating oceanic crust between them and separating a single, elongate basin into 2 basins: the Bransfield and the South Orkney.

The east-dipping subduction zone remained active off the west South America coast. The easternmost previous triple junction became disjointed, with one leg continuing northward in the Pacific Ocean, a second leg disappearing into the subduction zone offshore Chile, and the third extending from offshore of the west coast of the Antarctic Peninsula into the Scotia Sea. At this time oceanic crust appeared, the Scotia Sea began to form, and the South Georgia block became separated from the Burdwood Bank.

The sea floor spread center of the southwest Pacific Ocean followed its previous pattern and extended between Antarctica and New Zealand into the Indian Ocean between Antarctica and Australia.

The present-day ice cap of Antarctica began to develop in the late Oligocene-middle Miocene (Table 2). Thick and extensive glacial ice in East Australia and smaller glacial ice deposits in West Antarctica affected the isostatic balance of Antarctica, and the continent began to subside relative to sea level. As subsidence increased, the sea invaded the topographic low between the Ross and Weddell Sea embayments, and direct communication was established between the South Atlantic and Pacific Oceans.

Glacial sediments (silty clay with pebbles, of late Oligocene-middle Miocene age) were encountered in DSDP boreholes 270, 272, and 273.

Late Miocene-Pleistocene (0–11.5 m.y.B.P.) (Figure 21)

The diamond-shaped Antarctica plate was bounded by active spread centers. Relatively narrow bands of new oceanic crust developed in the southern extensions of the South Atlantic spread center, compared to the width of new crust in the Pacific, Indian Oceans, and the northern sector of the South Atlantic Ocean.

Minor spread activity was indicated in the Scotia Sea, while the East African Rift remained active. Subduction continued off the west coast of South America, the Antarctic Peninsula, and east of the South Sandwich Islands.

The Antarctica ice sheet was much thicker in the late Miocene than at present (Table 2), and it was then that the extensive West Antarctica ice sheet developed. It was not until the late Pliocene that glaciation developed in the Northern Hemisphere. DSDP 270, 271, and 273 penetrated late Miocene-Pleistocene, glacial, silty, pebbly clay on the Ross Sea shelf of Antarctica.

## SEDIMENTARY BASINS

Based on the limited data, 21 basins (Figure 22) are presented for Antarctica and immediately adjacent areas. These include the Amundsen, Bellingshausen, Bransfield, Dufek, Enderby, Graham, Kerguelen (three basins—East, South and West), Marie Byrd, Polar, Queen Fabiola and Queen Maud basins. Also included are the Ross (three basins—East, West and Ice Shelf), Scotia, Scott, South Orkney, Weddell, and Wilkes basins.

The occurrence and thickness of sediments are from various published data that include isopach maps of small areas, DSDP well sites, seismic sonobuoy, refraction and reflection profiles, depth-to-basement interpretations of magnetic data, and sediment thickness derived from gravity surveys.

The proposed configuration and sediment thicknesses of the basins of Antarctica are highly speculative. However, data do indicate the presence of these basins and suggest their general shape and extent, both laterally and vertically. Published data currently available suggest that the continental shelves adjacent to the Queen Maud, Enderby, and Scott basins are sediment-free basement rocks. As data become available, however, the presence of shelfal sediments will become evident. The present interpretation of the broad and generalized shape of the basins closely parallels the early studies of frontier areas in other parts of the world. We can anticipate that as as geologists acquire additional data, details will emerge. They will redefine the single basins mentioned here to numerous smaller basins as they determine structural boundaries, and basin depths assume different levels.

Additional basins may be present beneath the glacial ice cap. Those suggested here result from only a few traverses across a vast continent. As workers acquire additional data, we can identify previously unknown buried basins and ranges, such as the recently discovered subglacial Gamburtsev Mountains of East Antarctica (Grushinsky and Frolov, 1967).

In addition to the basins, Mond (1972) describes a 2,500 m (8,200 ft) thick section of Permian sediments in the Beaver Lake area of the Prince Charles Mountains.

### Amundsen Basin

Tucholke and Houtz (1976) present a sediment isopach map of the Bellingshausen basin and the adjacent part of the Amundsen basin. DSDP 323 and 324 penetrated fine-grained, deepwater clastic sediments and chert in the east end of the basin.

#### *Regional Geology and Stratigraphy*

The Amundsen basin occupies the continental shelf, slope and abyssal zone of the Amundsen Sea offshore West Antarctica. It extends from Peter Island in the east to the Amundsen Plateau in the west. Water depth ranges from 500 to 5,000 m (1,640 to 16,404 ft). Much of the sediment lies on Late Cretaceous oceanic crust, and the basin is interpreted to be bounded to the south by a pull-apart or divergent margin.

The New Zealand Complex was attached to the Antarctica landmass until the Late Cretaceous, at which time it began its pivotal movement away from Antarctica via the divergent

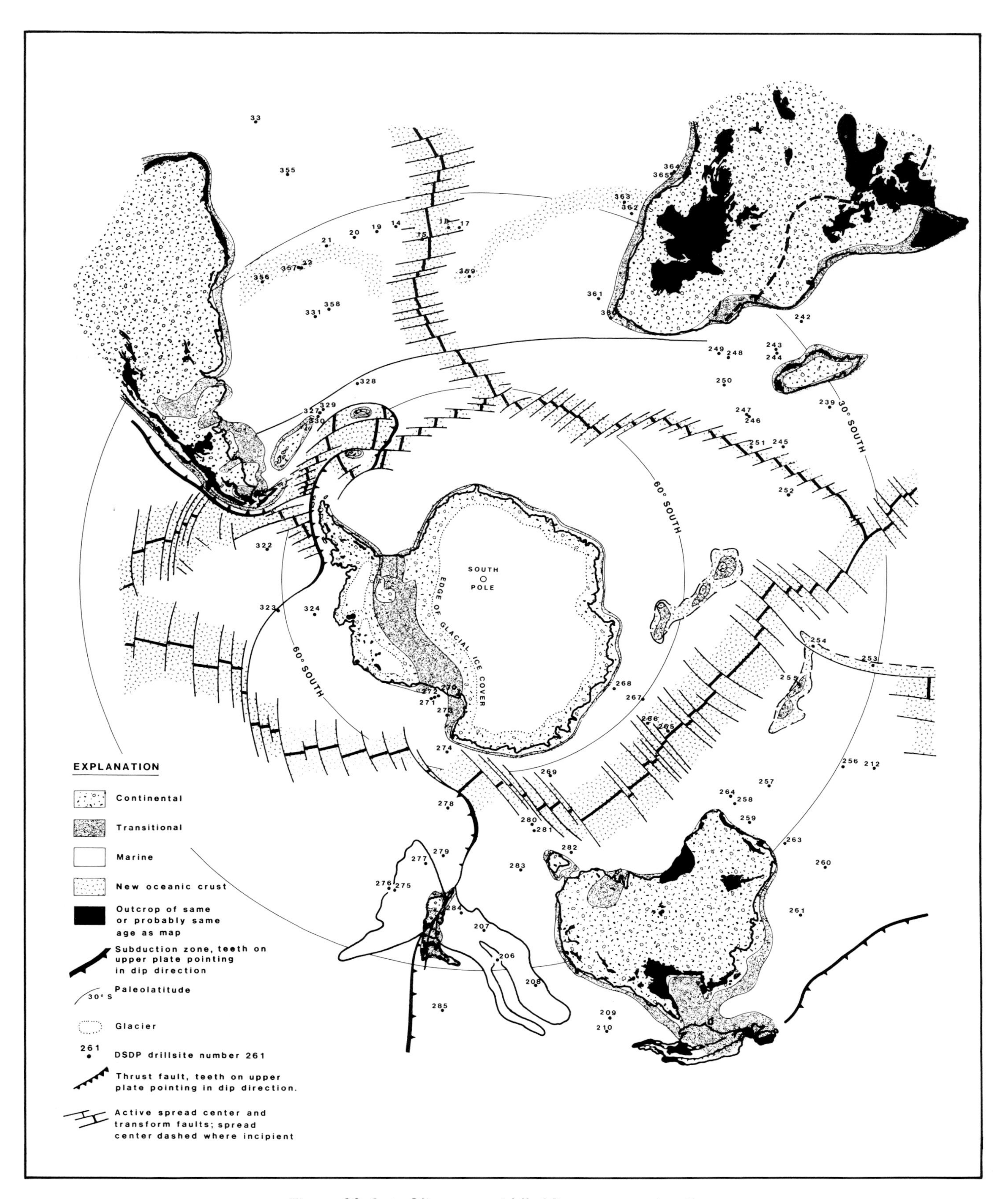


Figure 20. *Late Oligocene-middle Miocene reconstruction.*

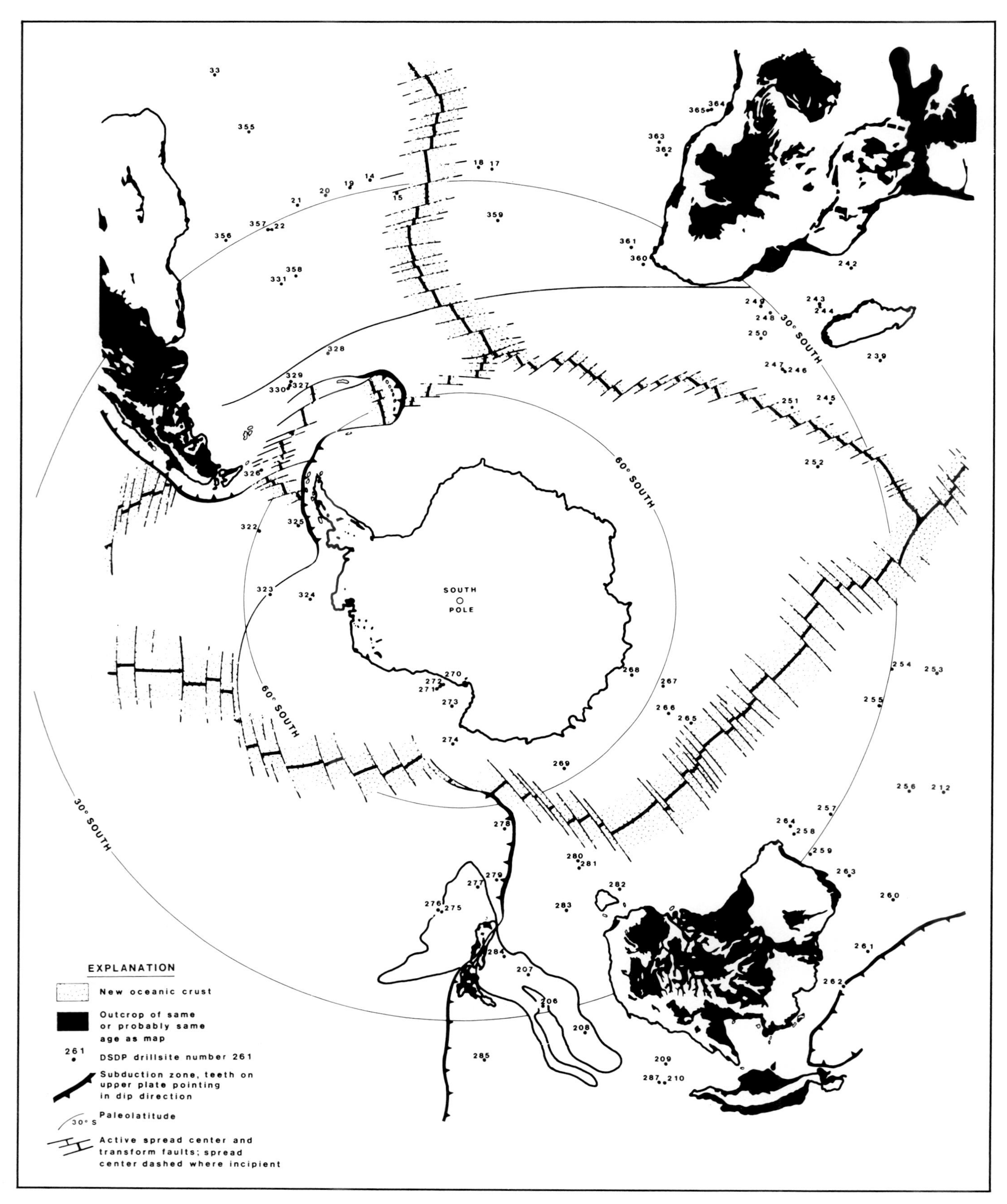


Figure 21. *Late Miocene-Pleistocene reconstruction.*

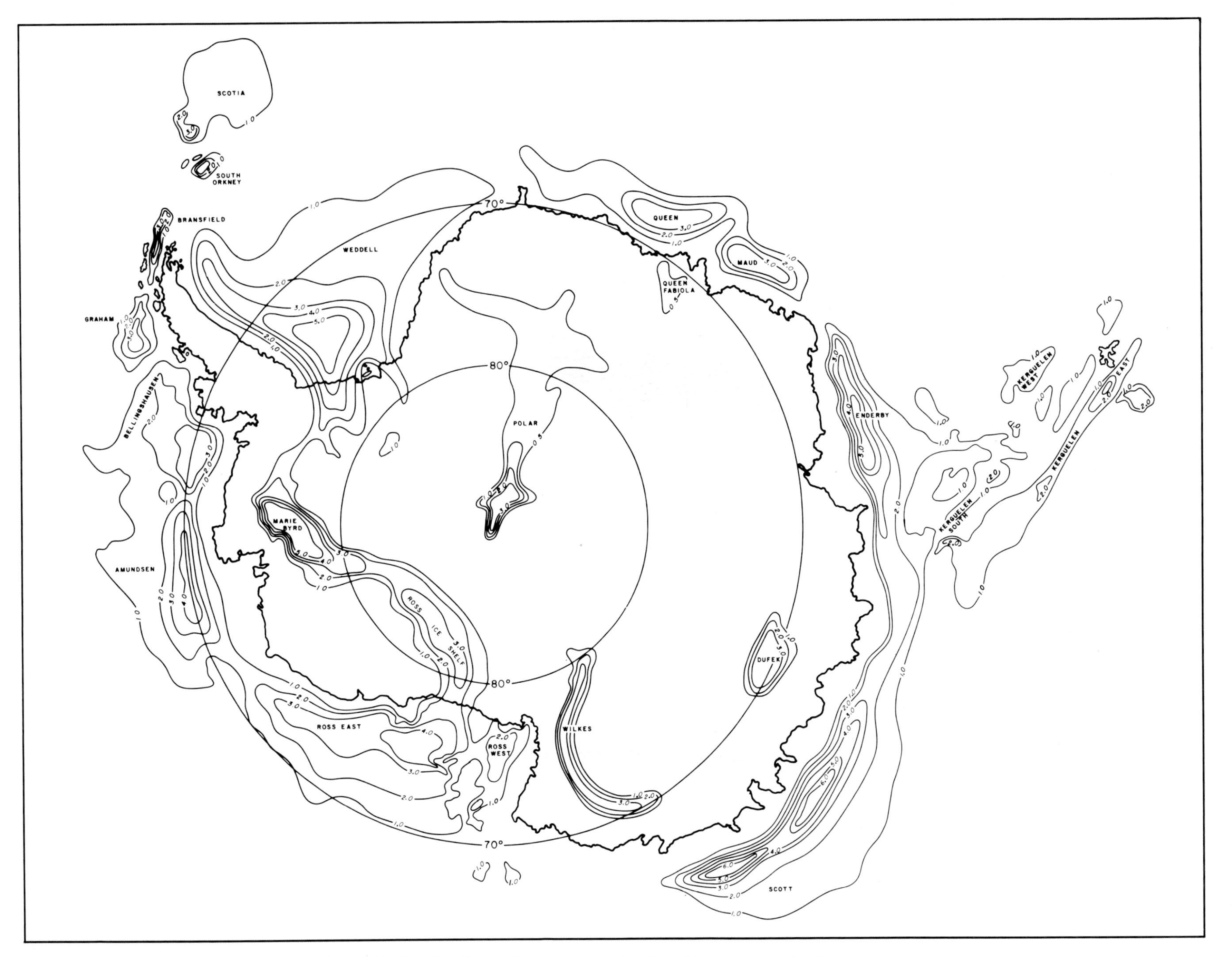


Figure 22. *Total sediment thickness of the basins of Antarctica and adjacent areas.*

process of sea floor spreading. DSDP site 323 penetrated pre-Danian, fine, clastic sediments over Maestrichtian or older oceanic crust. Geologists predict a maximum age of Late Cretaceous for the sediments of the Amundsen basin. The basin may prove comparable to the basins located on the Campbell Plateau southeast of New Zealand, as described by Sanford (1980).

*Hydrocarbon potential*

The Amundsen basin covers approximately 0.783 million sq km (302,340 sq mi) and sediments reach an estimated thickness of 4 km (2 mi). There is an estimated 1.64 million cu km (393,215 cu mi) of sediment, and the calculated possible hydrocarbon yield is 19.7 billion BOE.

| Amundsen Basin Hydrocarbon Parameters | | |
|---|---|---|
| Reservoir | — | quartz sandstones of Late Cretaceous to Miocene age and possible fractured cherts of Oligocene-early Miocene age. |
| Source | — | fine-grained clastic deep water sediments, claystone to siltstone, of Late Cretaceous to Pleistocene age. |
| Seal | — | fine-grained clastic deep water clay, claystone, siltstone, and possibly chert layers. |
| Trap | — | (structural) block-faulted features related to pull apart tectonics. |
| | — | (stratigraphic) permeability pinchouts through lateral facies changes. |
| Timing of migration | — | Tertiary |
| Maturation | — | unknown |

## Bellingshausen Basin

Houtz et al (1973) published a sediment isopach map of the exterior or abyssal portion of the Bellingshausen basin. Their map was intended to supersede all previous efforts. Houtz (1974) presents profile lines across part of the Bellingshausen basin, indicating sediment thickness of 2–3 km (1–2 mi). Tucholke and Houtz (1976) present a generalized sediment isopach map using data from Houtz et al (1973), the Conrad 15 cruise, and data from DSDP Leg 25. DSDP site 325, drilled in the abyssal portion of the basin, penetrated 1.2 km (1 mi) of sediment before reaching basalt basement (Schroeder, 1976). The sediment isopach of the Bellingshausen basin (Figure 22) uses these data plus general forms based on bathymetry and coastal configuration.

*Regional Geology and Stratigraphy*

The Bellingshausen basin, as defined in this report, lies offshore the northwest coast of the Antarctic Peninsula. Water depth ranges from 300 to 500 m (984 to 1,640 ft).

The bulk of the sediments lie on oceanic crust ranging in age from Paleocene to Recent. It is bounded to the south by a subduction zone or convergent margin.

The Antarctic Peninsula and West Antarctica are considered to be composed of what were one or more mobile minor plates during the Mesozoic. Geologists do not expect to find sediments older than Paleocene east of the Eltainin FZ in the Bellingshausen Sea basin, nor older than Cretaceous to the west.

That part of Antarctica adjacent to the Bellingshausen Sea basin is believed to have accreted onto the ancestral Antarctica craton by Paleocene time and to be composed of migrating island arcs and deformed, deep-water sediments. The rocks were subjected to orogeny during the Late Cretaceous-early Tertiary (Andean) periods, and the area is marked by Late Cretaceous-Tertiary intrusive/extrusive igneous rocks.

The Bellingshausen basin sediments have all been deposited in the polar zone, and we expect only clastic sediments. DSDP sites in the area, 322–325, penetrated only clastic sediments, many of which were fine-grained. There were, however, unconsolidated sands, and much of the sand was quartz, which is encouraging from the standpoint of having reservoir rock.

*Hydrocarbon Potential*

The Bellingshausen basin covers approximately 0.53 million sq km (204,649 sq mi), and sediments reach an estimated thickness of 3 km (2 mi). There is an estimated 1.09 million cu km (261,418 cu mi) of sediment and the calculated possible hydrocarbon yield is 13.1 billion BOE (Table 3).

| Bellingshausen Basin Hydrocarbon Parameters | | |
|---|---|---|
| Reservoir | — | quartz sandstones of Late Cretaceous to Miocene age and possible fractured cherts of early Oligocene-Miocene age. |
| Source | — | fine-grained clastic deep water sediments, clay-stone to siltstone, of Late Cretaceous to Pleistocene age. |
| Seal | — | fine-grained clastic, deep-water clay, claystone, siltstone, and possibly chert layers. |
| Trap | — | (structural) possible regional structural reversal into subduction zone east of Eltanin FZ and block-faulted features related to pull apart tectonics west of Eltanin FZ. |
| | — | (stratigraphic) permeability pinchouts through lateral facies changes. |
| Time of migration | — | Tertiary |
| Maturation | — | Unknown |

## Bransfield Basin

The Bransfield basin appears as a tension-induced, back-arc graben basin. Though small, it could be important for its hydrocarbon potential. The isopach contours are speculative as to thickness, but they conform to the probable shape based on existing bathymetry. Piston cores of the sediments contain traces of hydrocarbons (Anderson, 1979a). Tarney, Saunders, and Weaver (1977) suggest the basin formed only 4 m.y.B.P. when active spreading in the Drake Passage ended.

*Table 3. Area, sediment volume, and potential hydrocarbon yield of the sedimentary basins of Antarctica.*

| Basin | Area Million $Km^2$ | Volume Million $Km^3$ | Potential Hydrocarbon Yield In Barrels Oil Equivalent (Billion BOE)* |
|---|---|---|---|
| WEDDELL | 1.768 | 4.05 | 48.6 |
| ROSS** | 1.627 | 3.82 | 45.8 |
| SCOTT | 1.328 | 3.33 | 39.9 |
| AMUNDSEN | 0.783 | 1.64 | 19.7 |
| ENDERBY | 0.655 | 1.62 | 19.5 |
| QUEEN MAUD | 0.613 | 1.23 | 14.8 |
| BELLINGSHAUSEN | 0.530 | 1.09 | 13.1 |
| MARIE BYRD | 0.348 | 1.05 | — |
| WILKES | 0.233 | 0.40 | — |
| DUFEK | 0.133 | 0.32 | — |
| GRAHAM | 0.105 | 0.22 | — |
| POLAR | 0.087 | 0.19 | 0.9 |
| BRANSFIELD | 0.033 | 0.07 | 0.7 |
| SOUTH ORKNEY | 0.024 | 0.06 | — |
| KERGUELEN*** | 0.644 | — | — |
| QUEEN FABIOLA | — | — | — |
| SCOTIA | — | — | — |
| TOTAL | 8.268 | 19.10 | 203.0 |

* Potential hydrocarbon yield based on estimated worldwide basin average of 12,000 barrels oil or oil equivalent (6,000 cubic feet gas equivalent to one barrel oil) per $km^3$ of sediment.
** Includes Ross East, Ross West, and Ross Ice Shelf.
*** Includes Kerguelen East, Kerguelen South, and Kerguelen West.

*Regional Geology and Stratigraphy*

The Bransfield basin lies beneath the strait separating the South Shetland Islands from the northern tip of the Antarctic Peninsula. It appears to be a back-arc, extensional graben feature with a young oceanic crust basement. The feature is probably no more than 3–5 m.y. old and is associated with the subduction-related 5 km (3 mi) deep trench located 120 km (75 mi) northwest of the South Shetland Islands where 15–20 m.y. old oceanic crust is being consumed. Youthful volcanos, Deception and Bridgeman Islands, mark the young spread center.

The South Shetland Islands have a volcanic history extending from the Jurassic. The islands also contain quartzites, gneiss, and, on Livingstone Island, up to 3 km (2 mi) of sediment considered equivalent to the Trinity Peninsula Series of probably Triassic age. The southeast coast of the South Shetland island arc is marked by down-to-the-southeast normal faults.

*Hydrocarbon Potential*

The basin is a graben-like feature, approximately 65 km (40 mi) wide and 400 km (250 mi) long, covering approximately 33,500 sq km (12,935 sq mi). Water depth varies from zero to over 2,000 m (6,560 ft). Sediments reach an estimated thickness of 3 km (2 mi) with the estimated sediment volume at 74,500 cu km (17,985 cu mi). Calculated hydrocarbon potential is 0.9 billion BOE (Table 3).

The Bransfield basin is considered hydrocarbon prospective even though the feature and its contained sediments are geologically young. Anderson (1979a) reports traces of hydrocarbons in piston cores from the basin. Young, relatively shallow, clastic sediments can contain mature hydrocarbon units, given an adequate heat regime.

The Bransfield basin has many similarities to the hydrocarbon productive back-arc Central Sumatra basin.

| Bransfield Basin Hydrocarbon Parameters | | |
|---|---|---|
| Reservoir | — | quartz sandstones of Pliocene age sourced from pre-existing quartzite and silica-rich Mesozoic, igneous rocks. |
| Source | — | fine-grained clastic deep water sediments, claystone to siltstone, of Pliocene to Pleistocene age. |
| Seal | — | fine-grained clastic sediments, claystone to siltstone. |
| Trap | — | (structural) block faults associated with backarc spreading and tilted fault blocks. |
| | — | (stratigraphic) lateral facies changes in clastic sediments. |
| Timing of migration | — | Pliocene to Recent. |
| Maturation | — | possibly mature as abundant heat and relatively high temperatures are to be expected in a backarc spread regime. |

**Dufek Basin**

Heezen, Tharp, and Bentley (1972) infer the extent of the Dufek basin, while Grikurov, Ravich, and Soloviev (1972) propose a slightly different configuration. Both interpretations were based on gravity surveys. The deepest part of the Dufek basin corresponds to the area of overlap of the 2 interpretations.

*Regional Geology and Stratigraphy*

The Dufek basin is located onshore East Antarctica along lat. 70°S and between long. 90°E and 120°E. Ice cover is approximately 2,500 m (8,200 ft) thick. The extent and configuration of the basin are based on data by Grikurov, Ravich, and Soloviev (1972) and Heezen, Tharp, and Bentley (1972).

The basement rock is Precambrian igneous/metamorphic. Data suggest the basin area was part of a continental highland from Cambrian through Early Carboniferous and as covered by continental glaciers during Late Carboniferous and part of the Permian and is believed to have remained continental from Permian to Recent.

Sediments in the Dufek basin should be continental and probably no older than Permian. The basal, or oldest, sediments may be coal-bearing clastics over tillite, equivalent to the Permian section in the Prince Charles Mountains. Continental Mesozoic-early Oligocene clastic sediments may complete the section as the present glacial cover began in the late Oligocene.

*Hydrocarbon Potential*

The Dufek basin covers approximately 133,550 sq km (51,565 sq mi) and sediments should reach a thickness of 3 km (2 mi). Sediment volume is estimated to be 322,000 cu km (77,120 cu mi). Hydrocarbons, if any, will probably be dry gas coming from coal. (Table 3)

| Dufek Basin Hydrocarbon Parameters | | |
|---|---|---|
| Reservoir | — | sandstones of Permian-Triassic age. |
| Source | — | Permian coal beds. |
| Seal | — | fine-grained clastic sediments, clay or shale. |
| Trap | — | (structural) anticlines, faults. |
| | — | (stratigraphic) lateral facies changes, unconformities. |
| Timing of migration | — | Mesozoic to mid-Tertiary |
| Maturation | — | unknown |

## Enderby Basin

The postulated Enderby basin is speculative in large part. Houtz's (1974) seismic profiler line indicates 5–6 km (3–4 mi) of sedimentary section, while Houtz, Hayes and Markl (1977) extend the isopach contours from the Kerguelen Plateau.

*Regional Geology and Stratigraphy*

The Enderby basin lies offshore Antarctica south of the Kerguelen Plateau and between long. 50°E–90°E. Water depth ranges from 500 m (1,640 ft) at the shelf edge to 4,000 m (13,120 ft) in the abyssal plain.

The Enderby basin was formed during the Early Cretaceous when India separated from Antarctica and began its northward migration. Stratigraphically, it should be very similar to the eastern India basins, with pull-apart features expected. Sediments will range in age from Early Cretaceous to Recent, and they should be comparable to the east coast basins of India. Some Gondwana Permian-Triassic, continental clastic sediments, similar to the Talchir Shale of the Palar basin of India, could be preserved in pre-pull-apart lows. There could also be gas-generating Permian coal beds equivalent to those in the nearby Prince Charles Mountains of East Antarctica.

Uplift accompanied by igneous intrusion, prior to pull apart, occurred during the Jurassic and actual breakup was in the Early Cretaceous. Any Jurassic sediments present would be of continental origin, derived from horst blocks and deposited in grabens. Jurassic basic sills, dikes, and possible extrusive rocks are anticipated. Clastic sediments of Early Cretaceous to Recent age deposited in a marine environment should complete the stratigraphic column and form the major percentage of the sediment. No carbonates or evaporites are anticipated.

*Hydrocarbon Potential*

The Enderby basin covers approximately 655,000 sq km (252,915 sq mi) and sediments are estimated to reach at least 5 km (3 mi) thickness. Estimated sediment volume is 1.62 million cu km (388,234 cu mi). Potential hydrocarbon yield is calculated to be 19.5 billion BOE (Table 3).

| Enderby Basin Hydrocarbon Parameters | | |
|---|---|---|
| Reservoir | — | quartz sandstones and/or conglomerates ranging in age from Permian to Tertiary. |
| Source | — | possible gas-generating Permian coal beds overlain by Early Cretaceous to Recent marine, fine-grained clastic sediments — clay, shale siltstone. |
| Seal | — | fine-grained clastic sediments — clay, shale or siltstone. |
| Trap | — | (structural) block faults, tilted block faults. |
| | — | (stratigraphic) unconformities associated with tilted fault blocks, lateral facies changes. |
| Timing of migration | — | Late Cretaceous to Recent, |
| Maturation | — | unknown, but probably more mature in the deeper, older section of sediments near the pull apart margin. |

## Graham Basin

Houtz (1974) published a structure section from a seismic survey of offshore Graham Land, or Palmer Peninsula. The section indicates 2 + km (1 + mi) thickness. DSDP Site 322, located on the abyssal extremity of the basin, penetrated fine-grained, clastic sediments, including sandstone, before drilling into basalt (Craddock and Hollister, 1976).

*Regional Geology and Stratigraphy*

The Graham basin lies offshore the north coast of Graham Land of the Antarctic Peninsula. Water depth varies from 500 to 4,000 m (1,640 to 13,123 ft).

Sediments lie on oceanic crust of late Oligocene to Recent age. The southern edge is believed to be bounded by an inactive subduction zone. No sediments older than late Oligocene are anticipated.

*Hydrocarbon Potential*

The Graham basin covers approximately 0.105 million sq km (40,540 sq mi), and sediments reach an estimated thickness of 3 km (2 mi). There is an estimated 0.22 million cu km (52,869 cu mi). Hydrocarbon potential is poor.

| Graham Basin Hydrocarbon Parameters | | |
|---|---|---|
| Reservoir | — | quartz sandstones of Miocene age and possible fractured cherts of Oligocene-early Miocene age. |
| Source | — | fine-grained, clastic, deep water clay, claystone, siltstone, and possibly chert layers. |
| Trap | — | (structural) possible regional structural reversal into subduction zone. |
| | — | (stratigraphic) permeability pinchouts through lateral facies changes. |
| Timing of migration | — | late Tertiary |
| Maturation | — | unknown |

### Kerguelen—East, South and West Basins

The Kerguelen Plateau is elongated, approximately 1,800 km (1,118 mi) long, and marked by northwest-southeast striking down to the northeast normal faults (Ewing et al, 1971). Houtz, Hayes, and Markl (1977) used sonobouy and seismic reflection profiles to demonstrate that sediments 1–2 km (0.6–1.2 mi) thick cover the Plateau, reaching their greatest thickness at the foot of the fault scarps. Ponding of sediments in the lows created by tilted fault blocks has formed the long, sedimentary wedges denoted as separate basins (Figure 22). Watkins et al (1974) show that the basement rock of the Kerguelen Plateau consists of oceanic and volcanic basalt. There is no indication of continental material present.

The three basins have negligible hydrocarbon potential.

### Marie Byrd Basin

A marine, pre-glacial seaway may extend from the Ross Sea to the Weddell Sea, with a probable sedimentary basin, or basins, located along the intervening depression (NEWS, 1978) . The configuration and thickness based on depth to magnetic basement of Marie Byrd basin and the connection to the Ross Sea basin are after Woolard (1970) and Elliot (1975). Heezen, Tharp, and Bentley (1972) referred to the deep depression as the Byrd Subglacial basin.

*Regional Geology and Stratigraphy*

The Marie Byrd basin is onshore West Antarctica. It lies between long. 75°E and 120°E. and lat. 74°S to 82°S. Ice cover over the subglacial basin varies from less than 1,000 m (3,280 ft) to over 3,500 m (114,830 ft).

Geologists believe that part of West Antarctica underlying the Marie Byrd basin and forming its pre-sediment "basement" consists of migrated volcanic island arcs and deformed, deep-water sediment, accreted onto the ancestral Antarctica craton. The deformed sediments are probably equivalent to the Trinity Series and correlative units seen to the north in the Antarctic Peninsula. Jurassic volcanics are probably also present in the "basement" complex.

Emplacement of the various migrating minor plates that form West Antarctica, including the Antarctic Peninsula, was virtually completed by the close of the Late Cretaceous. The Tertiary was a period of deposition, compression, and intrusive/extrusive igneous activity.

The "basement" should consist of Jurassic and older, deformed deep-water sediments and volcanics. A basal coarse clastic fill of Early Cretaceous age may overlie basement. The overlying Early-Late Cretaceous clastics may be fine-grained and possibly contain hydrocarbon source material. Paleocene-Eocene rocks will include volcanics and volcaniclastics, while the early Oligocene was a period of erosion. The latest Oligocene-middle Miocene was the period when a marine seaway separated East Antarctica from westernmost West Antarctica, and possibly covered the area of the present-day glacier. Only Cenozoic marginal tillites and some volcanics are to be expected, plus some volcanics.

*Hydrocarbon Potential*

The Marie Byrd basin contains sediments as thick as 5 km (3 mi). The area of the basin containing sediments thicker than 1 km (0.6 mi) is 348,000 sq km (134,370 sq mi).

Sedimentary volume is calculated to be 1.05 million cu km (251,630 cu mi). Potential hydrocarbon yield of the basin is considered poor, primarily because of the prohibitive thickness of the glacial cover.

The Marie Byrd basin may be very similar in stratigraphy and development to the hydrocarbon productive Magallanes basin of Argentina.

| Marie Byrd Basin Hydrocarbon Parameters | | |
|---|---|---|
| Reservoir | — | quartz sandstone of Early Cretaceous age. |
| Source | — | fine-grained claystone or shale of Cretaceous age. |
| Seal | — | fine-grained claystone or shale of Cretaceous age. |
| Trap | — | (structural) compressional folds, domes overlying deep seated igneous intrusions. |
| | — | (stratigraphic) facies changes, permeability barriers, unconformities. |
| Timing of migration | — | early Tertiary |
| Maturation | — | predictably mature |

### Polar Basin

Heezen, Tharp, and Bentley (1972) define the general area of the Polar basin, while Grikurov, Ravich, and Soloviev (1972) indicate a possible thick sedimentary section beneath the South Pole.

*Regional Geology and Stratigraphy*

The deepest part of the Polar basin is located near the geographic South Pole; however, the crustal depression extends northward along long. 10°E. Grikurov, Ravich, and Soloviev (1972) and Heezen, Tharp, and Bentley (1972) define the general limits of the basin. Ice cover exceeds 3,500 m (11,480 ft) in places.

The Polar basin is located on the stable East Antarctica craton, and the basement is Precambrian igneous/metamorphic rock. During Cambrian-Early Ordovician time, the basin area was continental and probably highland. Continental-deposit lowland to erosion highland fringed the area from Late Carboniferous through Silurian. Marine transgression from the west may have reached the basin area during the Devonian-Early Carboniferous. Marine regression and growth of a continental glacier occurred during the Late Carboniferous-Permian. A continental lowland environment may have allowed coal to develop during the Permian-Triassic, and Jurassic volcanic rocks may be present in the basin. The Polar basin was continental high during most of the Mesozoic, but it began to develop a troughlike depression toward the end of the Cretaceous. Continental clastics were possibly deposited from Paleocene to early Oligocene.

Glaciation was initiated in the late Oligocene and has continued until today.

### *Hydrocarbon Potential*

The area of the Polar basin containing sediments of 1-km (0.6-mi) thickness or greater covers 87,000 sq km (54,060 mi) and sediment is believed to reach a minimum thickness of 3,000 m (9,840 ft). Sediment volume is calculated to be 193,500 cu km (46,330 cu mi). Potential for hydrocarbons is poor because of the thick glacial cover.

| Polar Basin Basin Hydrocarbon Parameters | | |
|---|---|---|
| Reservoir | — | basal Devonian sandstones, uppermost porous Permian tillite beds, sandstones within Permian-Triassic coal section. |
| Source | — | Permian-Triassic coal beds (dry gas) |
| Seal | — | fine-grained clastics, clay and shale, of Carboniferous-Permian tillite beds or within Permian-Triassic lagoonal series; possibly Jurassic extrusive flow over coal series. |
| Trap | — | (structural) faults, if present. |
| | — | (stratigraphic) facies variations within coal bed sequence, unconformities. |
| Timing of migration | — | Mesozoic |
| Maturation | — | unknown |

## Queen Fabiola Basin

The small, Queen Fabiola basin is inferred only (Heezen, Tharp, and Bentley, 1972), and definitive data are lacking.

### *Regional Geology and Stratigraphy*

Heezen, Tharp, and Bentley (1972) infer the Queen Fabiola basin, located in East Antarctica adjacent to the coast between long. 28°E—36°E. The less than 1 km (0.6 mi) thick sediments of the basin extend over 35,500 sq km (13,700 sq mi).

The basement is Precambrian igneous/metamorphic rock. The area was continental high in Cambrian-Silurian time and covered by or adjacent to a continental glacier from the Devonian through the Early Permian. The area has remained continental until the present.

Any sediments in the Queen Fabiola basin will be continental clastics, possibly with some coal beds of Permian age.

### *Hydrocarbon Potential*

Not prospective.

## Queen Maud Basin

Only limited data are available for the proposed Queen Maud basin. Such a broad expanse of the fringing offshore should not be barren of sediments and portions of three reflection lines (LaBreque and Hayes, 1979b) across the west end of the proposed basin indicate 1 + km (0.6 + mi) thickness of sediment. The presentation of this area (Figure 22) is speculative but possible.

### *Regional Geology and Stratigraphy*

The Queen Maud basin lies offshore Antarctica opposite southern Africa. Water depth reaches 4,000 m (13,123 ft).

The basin resulted from pull apart when Africa/Madagascar began to separate from Antarctica in the Jurassic, and India drifted away in the Early Cretaceous. Paleozoic sediments may have been preserved in pre-pull-apart depressions, but there is no evidence other than the occurrence of such rocks on the fringing edges of the continental margins of Africa/Madagascar/India.

Early Jurassic basic sills and dikes and possibly extrusive flows may begin the post-basement sequence. Late Jurassic sediments, if present, would probably be continental clastics derived from eroding horst blocks and deposited in grabens.

Oceanic crust had appeared between Antarctica and the other continental masses by the close of the Early Cretaceous. Lower Cretaceous sediments should be continental-to-marine clastics, and Upper Cretaceous sediments marine clastics. Marine clastics with ice-rafted material near the upper surface will also represent the Tertiary.

### *Hydrocarbon Potential*

Figure 22 shows that the Queen Maud basin covers 613,000 sq km (236,698 sq mi) and contains 1.23 million cu km (294,752 cu mi) sediment. Potential hydrocarbon yield is calculated to 14.8 billion BOE (Table 3).

The Queen Maud basin should be stratigraphically comparable to the Limpopo and Morondava basins of Mozambique and Madagascar, which are relatively unexplored.

| Queen Maud Basin Hydrocarbon Parameters | | |
|---|---|---|
| Reservoir | — | quartz sandstones of Cretaceous, early Tertiary age sourced from Precambrian igneous/metamorphic shield. |
| Source | — | fine-grained, marine clastics, claystone, or shale; Cretaceous or Tertiary age. |
| Seal | — | fine-grained, marine clastics. |
| Trap | — | (structural) faults, drape over fault blocks. |
| | — | (stratigraphic) lateral facies changes, unconformities. |
| Timing of migration | — | Late Cretaceous-Tertiary |
| Maturation | — | unknown but probably immature due to passive nature of pull apart margin, anticipated low thermal regime, and relatively thin pile of sediment. |

### Ross East, West, and Ice Shelf Basins

The shape and thickness of the Ross Sea basin is fairly well established. Houtz et al (1973) mapped the exterior portion of it in the vicinity of the Balleny Islands. Houtz and Davey (1973) mapped the shelfal portion of the Ross Sea with sonobuoy and reflection profiler. In addition, four DSDP wells (270, 271, 272, and 273) have been drilled within the Ross Sea province. Ford and Barrett (1975) describe the metasedimentary basement rock encountered in DSDP 270. The offshore Ross Sea is divided by a series of basement highs oriented north-south along long. 180°. Hayes and Davey (1975) suggest that the Iselin Plateau, at the shelf edge and on line with the other basement highs, is of continental character. The Ross West basin has approximately 2.5 km (1 mi) sediment thickness, whereas the extent, shape, and isopach of the eastern extremity of the basin are speculative, as is the portrayal of the onshore portion (Figure 22).

*Regional Geology and Stratigraphy*

The Ross basins, as considered in this report, cover a vast area onshore and offshore. Water depth ranges from 500–4,000 m (1,640–13,120 ft). Ice cover averages 500 m (1,640 ft) over the Ross Ice Shelf. Total area of the sedimentary basins, within the 1 km (0.6 mi) isopach, is 1.627 million sq km (628,234 sq mi). Approximately one-third of the area is onshore or beneath the Ross Ice Shelf. The basins extend from approximately long. 125°W through long. 180°E to 165°E. The onshore portion begins at lat. 85°S and extends northward to lat. 72°S. The basins are bordered to the west by the Transantarctic Mountains.

Much of what are now the Ross basins were oceanic during the Paleozoic and early Mesozoic. West Antarctica was the site of an east-dipping, subduction zone from Cambrian through Permian. The subduction zone probably "flipped" and reversed the dip direction one or more times, but the predominant dip direction was toward East Antarctica and, as a result, large numbers of rocks were accreted onto the craton.

The New Zealand Complex was a part of the Gondwana mass in Triassic time. During the Permian-Triassic, the Ross basins were continental, and continental Permian-Triassic sediments occur over the basement somewhere in the basins. Continental-to-transitional conditions prevailed during the Jurassic, plus there was intrusive and extrusive emplacement of basic igneous rocks.

Uplift, faulting, and erosion occurred during the Early Cretaceous, and scattered continental clastics are anticipated.

The Late Cretaceous was the period when the New Zealand Complex separated from Antarctica and oceanic crust was introduced between them. At least the outer margins of the Ross basins were transgressed by marine waters.

The topographic low extending across West Antarctica is believed to have existed by the Paleocene and to have received continental lowland clastic through early Oligocene, although marine conditions prevailed near the ocean verges.

As the ice cover formed and thickened during the late Oligocene, the continental mass began to sink and the transgression over the low ended. A marine link formed over West Antarctica between the Pacific and Atlantic Oceans that probably continued until middle Miocene time, when glacial ice displaced the connecting seaway.

The Ross basins are structurally complex. A series of basement highs along long. 180° separates the area into 2 basins. Hayes and Davey (1975) interpret the highs as marking an ancient rift. In addition, silling is evident in seismic sections across the shelf edge, slope, and rise (Houtz and Davey, 1973). An elevated basement block, apparently faulted along the northern or seaward edge, resulted in a closed basin landward. Seismic profiler data indicate that the Ross West region contains gentle folds that plunge northward and overlie older structures that trend east-northeast.

*Hydrocarbon Potential*

The Ross basins cover 1.627 million sq km (628,234 sq mi) and are estimated to contain 3.82 millions cu km (916,148 cu mi) sediments. The potential hydrocarbon yield is calculated to be 45.8 billion BOE (Table 3).

The Ross West basin may be similar in structure and stratigraphy to the productive Gippsland basin of southern Australia, whereas the Ross East basin may prove comparable to the basins of the Campbell Plateau offshore New Zealand.

Geochemical analysis of cores from DSDP sites 271, 272 and 273A indicate the presence of methane, ethane, and heavier hydrocarbons.

| Ross Basin(s) Hydrocarbon Parameters | | |
|---|---|---|
| Reservoir | — | quartz sandstones ranging in age from Devonian to Tertiary. |
| Source | — | dry gas from Permian coal beds, and oil or gas from fine-grained, marine clastics of latest Jurassic to early Tertiary age. |
| Seal | — | fine-grained clastics, clay, or shale, from Permian to Tertiary age. |
| Trap | — | (structural) compressional anticlines, drape over deep-seated highs, faults. |
| | — | (stratigraphic) unconformities, turbidites, lateral facies changes. |
| Timing of migration | — | Mesozoic to Tertiary |
| Maturation | — | mature |

### Scotia Basin

Refraction data in or around the Scotia Sea are by Allen (1966), Ewing et al (1971), and Harrington, Barker, and Griffins (1970). None of these data pertain to the shelf but instead represent deep water.

The floor of the Scotia basin is oceanic crust, and the sediments are believed to be entirely deep-sea, fine-grain clastic material with little or no hydrocarbon potential.

### Scott Basin

Sufficient data exist for Scott basin located offshore Wilkes Land to map the deepwater sediments with some confidence. This offshore sedimentary area is referred to here as the Scott basin. Payne and Connolly (1972) state that the abyssal plain along the continental margin of this part of Antarctica is covered with terrigenous turbidite sediments up to 5 km (3 mi) thick. Houtz et al (1973) indicate the general extent of the sediments in the abyssal area. Houtz (1974), using seismic profiler and sonobouy data, conclude that the sedimentary thickness is 5–6 km (3–4 mi). Houtz (1974) also proposes that the offshore basins of southern Australia are mirrored by like basins offshore the coastal segments of Australia. Data are lacking for the adjacent shelf area, so the nature of the rocks is not known.

*Regional Geology and Stratigraphy*

The offshore Scott basin is situated off the Antarctica coast, south of Australia in the southeast Indian Ocean. Water depths range from 500 to over 4000 m (1,640 to over 13,120 ft). The basin is estimated to cover 1,328 million sq km (512,780 sq mi). Sediments reach 6 km (4 ft) of thickness and much of the sediment lies on oceanic crust.

Antarctica and Australia remained joined as part of Gondwana until the Paleocene. That part of Gondwana relative to the Scott basin remained continental, probably highland, from Precambrian through the Carboniferous. The ancestral Pacific Ocean may have transgressed the area north of Victoria Land including Tasmania. Continental to transitional sediments of Permian age may be found in the eastern part of the Scott basin.

The Triassic was a period of stability, and continental conditions prevailed. Rifting between Antarctica and Australia began in the Jurassic, continuing through the Cretaceous. Great thicknesses of Jurassic-Cretaceous continental clastics filled the grabens throughout the rift zone. There is a possibility of marine invasion in very latest Cretaceous time.

The continents of Antarctica and Australia separated in the Paleocene, and oceanic crust appeared as the rift continued to grow. Marine conditions prevailed, while the waters of the Pacific and Indian Oceans joined through the developing seaway. Deltaic deposition developed in the early Tertiary along the new marine margins of both continents. Turbidite deposits occur at the distal end of the deltas (Payne and Conolly, 1972). Carbonates developed along the Australia margin in late Tertiary time, but Antarctica remained in the frigid polar zone, and only clastic sediments were laid down. Ice-rafted coarse clastic debris blanketed the basin from late Oligocene to present.

*Hydrocarbon Potential*

The Scott basin covers 1.328 million sq km (512,781 sq mi) and contains 3.33 million cu km (798,488 cu mi) sediment. Potential hydrocarbon yield is calculated to be 39.9 billion BOE (Table 3).

Evolution and stratigraphy of the Scott basin should prove comparable to that of the Otway and Eucla basins of southern Australia.

| Scott Basin Hydrocarbon Parameters | | |
|---|---|---|
| Reservoir | — | Jurassic to Cretaceous quartz sandstones in preserved grabens on continental crust; Tertiary deltaic and turbidite sandstones over faulted continental margin and adjacent oceanic crust. |
| Source | — | possibly marine Permian or Upper Cretaceous in grabens on continental crust; fine-grained marine clastics or oozes in Tertiary section. |
| Seal | — | fine-grained clastics |
| Trap | — | (structural) fault blocks, drape-over faults, reverse roll-over in down-to-the-basin growth faults, anticlinal folds developed during turbidite deposition. |
| | — | (stratigraphic) unconformities, sand pinch-outs, permeability barriers, bar sands. |
| Timing of migration | — | Tertiary |
| Maturation | — | probably immature at shallow depths but mature at depth; low heat flow is to be expected along pull apart margins. |

## South Orkney Basin

The South Orkney basin, across the postulated Tertiary spread center located east of the Bransfield basin, appears to match the Bransfield basin in width and position. Harrington, Barker, and Griffiths, 1970) define the South Orkney basin based on sonobouy refraction and magnetic data. They interpret up to 5 km (3 mi) of Cretaceous-Tertiary sediments above metamorphic basement.

*Regional Geology and Stratigraphy*

The South Orkney basin is located on the submarine plateau that supports the South Orkney Islands. The area lies off the eastern tip of the Antarctic Peninsula along the southern verge of the Scotia Sea. The basin covers 24,000 sq km (9,267 sq mi) with a water depth of 1,000 m (3,281 ft).

Seismic data indicate the presence of up to 5 km (3 mi) of sediments (Harrington, Barker, and Griffiths, 1970). Refraction lines indicate a layer of 2.0–2.5 km/sec (6,560–8,200 ft/sec) sediment over a 3.2–3.7 km/sec (10,500–12,140 ft/sec) layer, which in turn overlies basement rock having 5.0–5.5 km/sec (16,400–18,000 ft/sec) velocity. Harrington, Barker, and Griffiths correlate the 5.0–5.5 km/sec basement complex to the metamorphic rocks cropping out on the South Orkney Islands; the 3.2–3.7 km/sec layer to Cretaceous sediments, including conglomerate beds described from the islands; and the 2.0–2.5 km/sec layer to the Tertiary.

An alternate interpretation would assign the 2.0–2.5 km/sec layer to the Tertiary, but could also place the 3.2–3.7 km/sec layer in the Tertiary. The Bransfield and South Orkney basins may have been joined once and were part of the same structural and depositional basin that was later separated via east-west spreading. Recent back-arc spreading in the Bransfield basin has resulted in volcanos along the rift zone. The presence of a distinctive, linear, magnetic anomaly bordering the southern edge of the South Orkney basin may indicate a similar event.

*Hydrocarbon Potential*

The 24,000 sq km (9,267 sq mi) South Orkney basin contains an estimated 60,000 cu km (14,396 cu mi) sediment. Calculated potential hydrocarbon yield is 0.7 billion BOE (Table 3).

The South Orkney basin may have had a similar development to that of the oil-producing back-arc basins of Sumatra.

| South Orkney Basin Hydrocarbon Parameters | | |
|---|---|---|
| Reservoir | — | quartz sandstone of Cretaceous-Tertiary age. |
| Source | — | marine fine-grained clastics of Cretaceous-Tertiary age. |
| Seal | — | fine-grained clastics. |
| Trap | — | (structural) faults, anticlines. |
| | — | (stratigraphic) facies changes. |
| Timing of migration | — | Tertiary |
| Maturation | — | unknown, possibly mature; magnetic anomaly map represent basalt emplacement with accompanying high heat flow. |

## Weddell Basin

The Weddell basin should to be extensive. Although Wright and Williams, comps. (1974) cite the probable presence of up to

4 km (2 mi) of sediment in the shelf portion of the Weddell Sea, there are no published, hard data available. The extent, configuration, and thickness are speculative. LaBrecque and Hayes (1979b) published 2 seismic reflection lines across part of the northern and eastern extremities of the basin, which indicate up to 1 km (0.6 mi) thickness. Anderson's (1972) dissertation focused on the shallow marine geology of the Weddell Sea.

*Regional Geology and Stratigraphy*

The Weddell basin lies offshore of the east coast of the Antarctic Peninsula. Water depth ranges from zero along the shoreline to 4,000 m (2,485 ft) in the abyssal plain. The basin covers approximately 1.768 million sq km (682,678 sq mi). Sediments probably reach a depth of at least 5 km (3 mi).

The evolution of the Weddell basin is intimately associated with the development of the Antarctic Peninsula that is composed of fragments of migrating island arcs, volcanic arcs, and piled-up volumes of deep-water sediment scraped off oceanic crust during subduction.

The extremely complex history of the Weddell basin begins in the Jurassic. Gondwana remained an entity throughout the Paleozoic, and breakup leading to the evolving of the Weddell basin occurred in Jurassic time.

South America-Africa separated from the rest of Gondwana in the Jurassic, as eastward-migrating spread centers and subduction zones brought the ancestral components of the Antarctic Peninsula toward Antarctica. Africa continued to move in the same direction relative to Antarctica in Early Cretaceous time, but South America's relative direction changed as Africa and South America also began to separate, with a wide belt of oceanic crust growing between them.

A westward-dipping subduction zone then developed along the east coast of the Antarctic Peninsula. The west-dipping subduction zone remained active through the Late Cretaceous. Deformed and upthrust deep-water sediments intruded by igneous rocks should form the bulk of pre-Tertiary sediments. Contemporaneously eroded clastics may have filled depositional lows in the developing Weddell basin. East of the west-dipping subduction zone, deep-water oozes and fine-grained clastics (clay and shale) accumulated.

Marine conditions dominated the Tertiary deposits in the Weddell basin, and quartz sands and feldspar-derived clays were probably deposited from the granitic craton of Antarctica. Future seismic should indicate the presence of deltas and turbidite deposits. The area of the basin remained too close to the pole for carbonates or evaporites to form, and only clastic sediments can be expected.

*Hydrocarbon Potential*

The Weddell basin, as visualized in this report, contains 4.05 millions cu km (973,242 cu mi) sediment. Potential hydrocarbon yield for this vast accumulation of sediment is 48.6 billion BOE (Table 3).

Deuser (1971) points out stratigraphic and structural similarities between the Weddell Sea and Lake Maracaibo and suggests that the Weddell Sea has the greatest hydrocarbon potential in the Antarctic. Certain of Deuser's hypotheses regarding tectonics have proven untrue; however, many of his speculations relative to stratigraphy appear valid.

| Weddell Basin Hydrocarbon Parameters | | |
|---|---|---|
| Reservoir | — | quartz sands of possible Jurassic-Cretaceous age and Tertiary age deposited in deltas and turbidites. |
| Source | — | fine-grained, marine clastics of Cretaceous-Tertiary age. |
| Seal | — | fine-grained clastics of Cretaceous-Tertiary age. |
| Trap | — | (structural) down-to-the-basin growth faults, drape over shale diapirs. |
| | — | (stratigraphic) typical deltaic and turbidite traps plus unconformities. |
| Timing of migration | — | Tertiary |
| Maturation | — | immature at shallow depths, mature at greater depths. |

## Wilkes Basin

The most probable site of a thick-sediment accumulation onshore Eastern Antarctica is the Wilkes basin. Heezen, Tharp, and Bentley (1972) label this area the Wilkes subglacial basin. Grikurov, Ravich, and Soloviev (1972) indicate a much narrower trough within the area outlined by Heezen, Tharp, and Bentley (1972). Grikurov, Ravich, and Soloviev (1972) propose that the depression is filled with Mesozoic-Cenozoic sediments.

*Regional Geology and Stratigraphy*

The Wilkes basin lies on the east flank of the Transantarctic Mountains in East Antarctica. It has been identified by gravity, magnetic, and refraction surveys. The glacial ice cap covers the entire basin and ranges from 500 to 3,500 m (1,640 to 11,483 ft) thick. The Wilkes basin covers approximately 233,000 sq km (89,970 sq mi) and sediments reach 3 km (1.9 mi) thickness.

The history of the Wilkes basin relates to the history of the Transantarctic Mountains. The stratigraphy of the Transantarctic Mountains is known; however, the structure remains an enigma. Previous workers (Hayes and Davey, 1975) consider the mountains to represent a monoclinal fold or an upfaulted block. It is hard to visualize a monoclinal fold extending the length of the Transantarctic Mountains, over 2,000 km (771 mi), so I will adopt the horst block concept.

The basement of the Wilkes basin is projected to be formed of folded Precambrian and Cambrian metamorphics that Ordovician acidic intrusives have injected. The Silurian was a period of uplift and erosion, and it left behind a regional, peneplained, unconformable surface.

Devonian clastics overlie the regional unconformity and are unconformable overlain by Carboniferous-Permian glacial tillite. Permian clastic sediments and coal beds overlie the tillite. Additional clastics, in places containing coal and tuff beds, were deposited during the Triassic. Extensive igneous activity occurred during the Jurassic; basic dikes, and sills penetrated the section, and extrusive flows covered much of the area.

The Wilkes basin was part of the continental high through the Cretaceous, but it may have been invaded by the sea in Paleocene-Oligocene time when breakup occurred between Antarctica and Australia. Growth of glacial ice from late Oligocene on precluded additional sediment accumulation.

*Hydrocarbon Potential*

The Wilkes basin contains 402,000 cu km (96,160 cu mi) sediment. The hydrocarbon potential is considered poor because of the extensive glacial overburden.

The Wilkes basin may have developed in a similar manner as the oil and gas productive basins east of the Rocky Mountains, but earlier in geologic time.

| Wilkes Basin Hydrocarbon Parameters | | |
|---|---|---|
| Reservoir | — | quartz sandstone of Devonian to Jurassic age. |
| Source | — | dry gas from Permian coal beds and possibly hydrocarbons from marine, early Tertiary shales or clays. |
| Seal | — | fine-grained clastics of Devonian to early Tertiary age or extrusive basalt of Jurassic age. |
| Trap | — | (structural) faults, drape over horst blocks. |
| | — | (stratigraphic) facies changes and unconformities. |
| Timing of migration | — | Mesozoic to early Tertiary |
| Maturation | — | presumed mature |

## SUMMARY AND CONCLUSIONS

Ten of the twenty-one basins associated with Antarctica are considered viable hydrocarbon exploration targets at this time. These include the Weddell, Ross West, Ross East, Scott, Amundsen, Enderby, Bellinghausen, Queen Maud, Bransfield, and South Orkney basins. The other 11 are precluded because of glacial cover or poor hydrocarbon potential. The hydrocarbon potential of these 10 basins is estimated to be 203 billion barrels oil equivalent.

The problems of exploration and development will be formidable. First, agreement must be reached among several nations and each will have different political goals and objectives. Second, guarantees of environmental protection must be satisfied. Third, logistical and technical difficulties must be solved.

Many of these problems have been addressed. An international workshop including oil technologists and environmentalists was held in March 1979, at Bellagio, Italy. The firm conclusion (Holdgate and Tinker, 1979) was that technology is already adequate for seismic surveys and exploratory drilling on the Antarctic continental shelf. This conclusion was arrived at based on the experience gained and technology evolved from operations in the Arctic, even considering the extreme water depth, severe climate, presence of pack ice and icebergs, and remoteness of Antarctica.

Current technology will enable us to explore seismically and to drill the offshore area, in spite of the great depth of the water and presence of the bothersome pack ice and icebergs. Technology is not yet available to penetrate the up to 3,500 m (11,480 ft) of glacial ice over the onshore basins, even if we could define drillable targets with seismic.

Special technical problems will be faced during exploratory drilling offshore Antarctica. The continental shelf, as deep as 800 m (2,625 ft), has an average shelf-edge depth of 500 m (1,640 ft) as compared to a 200-m (656-ft) average for the shelves of the rest of the world. Icebergs are frequent, enormous, and are reported to have left scour marks 100 m (328 ft) or more deep in water depths exceeding 500 m (1,640 ft) (Anderson, 1979b).

Offshore, semi-submersible drilling rigs under construction (Figure 23) and scheduled for drilling in the Arctic could also be used in Antarctica. These units are designed to drill in water depths of 460 m (1,509 ft) and to drill to depths of 6,400 m (20,997 ft). These advanced, severe-environment units withstand 100-mile-per-hour winds, 100-foot waves, and operate year-round above the Arctic Circle. The totally enclosed rigs will enable the crew to work in a shirtsleeve environment even in the most severe weather and temperature conditions.

Exploratory drilling will be feasible, although difficult. To be exploitable, extremely large reserves must be proven. This need means that gas will probably not be viable because of the economics involved. Oil development will require giant or supergiant accumulations (500 million barrels or more recoverable).

The requirement for giant accumulations means that only very large prospects will be considered for drilling, which in turn suggests that a relatively coarse grid of seismic lines will be sufficient.

Antarctica does have potential for large hydrocarbon reserves, and the technical expertise, ultimately, is available. The sole hindrance is jurisdiction—from whom do we get a drilling permit?

## ACKNOWLEDGMENTS

This paper is modified from a speculative report (St. John, 1980) written for and published by JEBCO, 68 High Street, Weybridge, Surrey, England. I appreciate John E. Bobbit and Peter Dolan of JEBCO for permission to present this condensed version of the original report. I am grateful to Sonat Offshore Drilling, Inc. for permission to reproduce artist's rendering of their extreme-environment semisubmersibles under construction for work in the Arctic. Without the able assistance of Stevenie Frye, Robin Malone, and Jacqueline Urick, this manuscript and its illustrations would not have been completed. I owe these three my utmost appreciation and gratitude.

## REFERENCES CITED

Adie, R. J., 1972, Recent advances in the geology of the Antarctic Penninsula, *in* R. J. Adie, ed., Antarctic geology and geophysics: Oslo, Universitetsforlaget, p. 121–142.

Ahmad, R., 1969, The origin of the Indian Ocean: Indian National Institute of Science Bulletin, v. 38, n. 1, p. 363–375.

Allen, A., 1966, Seismic refraction investigations in the Scotia Sea: British Antarctic Survey, Science Reports,, n. 55, p. 44.

Anderson, J. B., 1972, The marine geology of the Weddell Sea: Florida State University Ph.D. thesis, 233 p.

——, 1979a, Personal communication: Rice University.

——, 1979b, Geologic observations on the Antarctic continental margin: applications to exploration in high latitude seas: 1979 Offshore Technology Conference Proceedings, Houston, v. 1, p. 643–646.

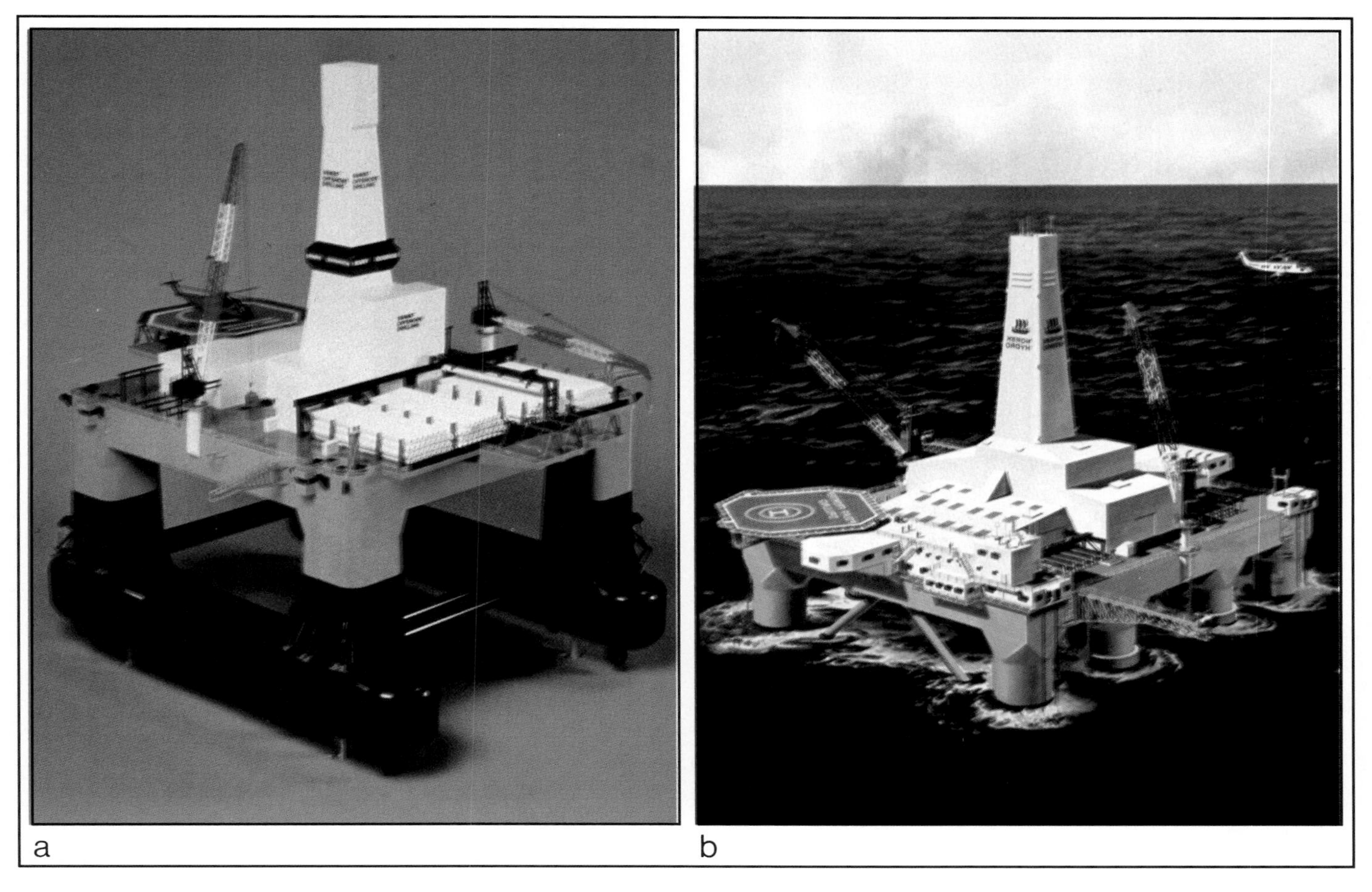

Figure 23. *Severe-environment semisubmersible drilling rigs (a) Model (b) artist's concept.*


Balme, B. E., and G. Playford, 1967, Late Permian plant microfossils from the Prince Charles Mountains, Antarctica: Revue de Micropaleontologie, v. 10, n. 3, p. 179–192.

Barrett, P. J., 1968, The post-glacial Permian and Triassic Beacon rocks in the Beardmore Glacier area, central Transantarctic Mountains, Antarctica: Ohio State University Ph.D. thesis, 535 p.

——, 1970, Stratigraphy and paleogeography of the Beacon Supergroup in the Transantarctic Mountains, Antarctica: International Union of Geological Sciences, 2nd Gondwana Symposium, Proceedings and Papers, South Africa, p. 249–256.

——, 1970b, Paleocurrent analysis of the mainly fluviatile Permian and Triassic Beacon rocks, Beardmore Glacier area, Antarctica: Journal of Sedimentary Petrology, v. 40, n. 1, p. 395–411.

——, 1972, Stratigraphy and petrology of the mainly fluviatile Permian and Triassic part of the Beacon Supergroup, Beardmore Glacier area, *in* R. J. Adie, ed., Antarctic geology and geophysics: Oslo, Universitetsforlaget, p. 365–372.

Behrendt, J. C., et al, 1974, Geophysical investigations of the Pensacola Mountains and adjacent glacerized areas of Antarctica: U.S. Geological Survey Professional Paper, n. 844, 24 p.

Bennett, A. J. R., and G. H. Taylor, 1972, Coals from the vicinity of the Prince Charles Mountains, *in* R. J. Adie, ed., Antarctic geology and geophysics: Oslo, Universitetsforlaget, p. 585–589.

Boucot, A. J., et al, 1967, Devonian of Antarctica: Alberta Society of Petroleum Geologists, International Devonian System Symposium Proceedings, v. 1, p. 639–648.

Brady, H., and H. Martin, 1979, Ross Sea region in the middle Miocene: a glimpse into the past: Science, v. 203, p. 437–438.

Bruhn, R. L., 1979, Rock structures formed during back-arc basin deformation in the Andes of Tierra del Fuego: Geological Society of America Bulletin, pt. 1, v. 90, p. 998–1012.

——, and I. W. D. Dalziel, 1977, Destruction of the Early Cretaceous marginal basin in the Andes of Tierra del Fuego, *in* M. Talwani, and W. C. Pittman, III, eds., Island arcs, deep sea trenches and back-arc basins: American Geophysical Union Maurice Ewing Series I, p. 395–405.

Castle, J. W., and C. Craddock, 1975, Deposition and metamorphism of the Polarstar Formation (Permian), Ellsworth Mountains: Antarctic, Journal, v. 10, n. 5, p. 239–241.

Central Intelligence Agency, 1978 (reprinted 1979), Polar regions Atlas: Produced by National Foreign Assessment Center, CIA, 66 p.

Coates, D. A., 1972, Pagoda formation: evidence of Permian glaciation in the Central Transantarctic Mountains, *in* R. J. Adie, ed., Antarctic geology and geophysics: Oslo, Universitetsforlaget, p. 359–364.

Craddock, C., 1969–70, Geology of the Ellsworth Mountains, *in* C. Craddock, et al, Geologic maps of Antarctica: Antarctica Map Folio Series, New York, American Geographical Society, Folio 12.
——, et al, 1969–70, Geologic maps of Antarctica: Antarctic Map Folio Series, New York, American Geographical Society, Folio 12.
——, 1972, Antarctic tectonics, *in* R. J. Adie, ed., Antarctic geology and geophysics: Oslo, Universitetsforlaget, p. 449–455.
——, and C. D. Hollister, 1976, Geologic evolution of the southeast Pacific basin, *in* C. D. Hollister, et al, Initial reports of the Deep Sea Drilling Project, v. 35: Washington (U.S. Government Printing Office), p. 723–743.
Creer, K. M., 1965, Paleomagnetic data from the Gondwanic continents: Royal Society of London, Philosophical Transactions, v. 258A, p. 27–40.
——, 1968a, Arrangements of the continents during the Paleozoic era: Nature, v. 219, n. 5149, p. 41–44.
——, 1968b, Paleozoic paleomagnetism: Nature, v. 219, p. 246–250.
——, 1970, Review and interpretation of paleomagnetic data from the Gondwanic continents: International Union of Geological Sciences, 2nd Gondwana Symposium, Proceedings and Papers, South Africa, p. 55–72.
——, 1973, A discussion of the arrangement of paleomagnetic poles on the map of Pangaea for epochs in the Phanerozoic, *in* D. H. Tarling, and S. K. Runcorn, eds., Implications of continental drift to the earth sciences: New York, Academic press, v. 1, p. 47–76.
Crowell, J. C., and L. A. Frakes, 1968, Late Paleozoic glacial facies and the origin of the South Atlantic Ocean: 23rd International Geologic Congress Symposium Proceedings, v. 13, p. 291–302.
Dalziel, I. W. D., 1972, Large-scale folding in the Scotia Arc, *in* R. J. Adie, ed., Antarctic geology and geophysics: Oslo, Universitetsforlaget, p. 47–55.
Deuser, W. G., 1971, Lake Maracaibo and Weddell Sea: comparison in petroleum geology: AAPG Bulletin, v. 55, n. 5, p. 705–708.
Du Toit, A. L., 1937, Our wandering continents: Edinburgh, Oliver and Boyd, 366 p.
Elliot, D. H., 1975, Tectonics of Antarctica: a review: American Journal of Science, v. 275-A, p. 45–106.
——, and D. A. Coates, 1971, Geological investigations in the Queen Maud Mountains: Antarctic Journal, v. 3, p. 114–118.
——, et al, 1970, Triassic tetrapods from Antarctica: evidence for continental drift: Science, v. 169, n. 3951, p. 1197–1201.
Ewing, J., et al, 1971, Structure of the Scotia Sea and Falkland Plateau, Journal of Geophysical Research, v. 76, p. 7118–7137.
Ford, A. B., 1972, Weddell orogeny—latest Permian to early Mesozoic deformation at the Weddell Sea margin of the Transantarctic Mountains, *in* R. J. Adie, ed., Antarctic geology and geophysics: Oslo, Universitetsforlaget, p. 419–425.
——, and P. J. Barrett, 1975, Basement rocks of the south-central Ross Sea, Site 270, DSDP Leg 28, *in* D. E. Hayes, et al, Initial reports of the Deep Sea Drilling Project, v. 28: Washington (U.S. Government Printing Office), p. 861–868.
Frakes, L. A., and J. C. Crowell, 1969, Late Paleozoic glaciation: I, South America: Geological Society of America Bulletin, v. 80, p. 1107–1042.
——, and ——, 1970, Late Paleozoic glaciation: II, Africa exclusive of the Karoo basin: Geological Society of America Bulletin, v. 81, p. 2261–2286.
——, J. L. Matthews, and J. C. Crowell, 1971, Late Paleozoic glaciation: III, Antarctica: Geological Society of America Bulletin, v. 82, p. 1581–1604.
Furon R., 1963, The geology of Africa: New York, Hafner Publishing Company, 377 p.
Gair, H. S., et al, 1969–70, The geology of northern Victoria Land, *in* C. Craddock, et al, Geologic maps of Antarctica: Antarctica Map Folio Series, New York, American Geographical Society, Folio 12.
Griffiths, J. R., 1971a, Continental margin tectonics and the evolution of south-east Australia: Australian Petroleum Exploration Association Journal, v. 11, pt. 1, p. 75–79.
——, 1971b, Reconstruction of the south-west Pacific margin of Gondwanaland: Nature, v. 234, n. 5326, p. 203–207.
Grikurov, G. E., 1972, Tectonics of the Antarctandes, *in* R. J. Adie, ed., Antarctic geology and geophysics: Oslo, Universitetsforlaget, p. 163–167.
——, M. G. Ravich, and D. S. Soloviev, 1972, Tectonics of Antarctica, *in* R. J. Adie, ed., Antarctic geology and geophysics: Oslo, Universitetsforlaget, p. 457–468.
Grindley, G. W., and M. G. Laird, 1969–70, Geology of the Shackleton Coast, *in* C. Craddock, et al, Geological maps of Antarctica: Antarctica: Antarctica Map Folio Series, New York, American Geographical Society, Folio 12.
Grushinsky, N. P., and A. I. Frolov, 1967, Some conclusions on the structure of Antarctica: Geological Society of Australia, Journal, v. 14, n. 2, p. 215–223.
Habicht, J. K. A., 1979, Paleoclimate, paleomagnetism, and continental drift: AAPG Studies in Geology n. 9, 31 p.
Handschumacher, D. W., 1976, Post-Eocene plate tectonics of the Eastern Pacific, *in* The Geophysics of the Pacific Ocean basin and Its Margins: American Geophysical Union, Monograph 19, p. 177–202.
Harrington, H. J. J., 1962, Paleographic development of South America: AAPG Bulletin, v. 46, n. 10, p. 1773–1814.
——, 1970, Recent advances in the study of the Beacon Supergroup in the Transantarctic Mountains: International Union of Geological Sciences, 2nd Gondwana Symposium, Proceedings and Papers, p. 257–264.
Harrington, P. K., P. F. Barker, and D. H. Griffiths, 1970, Crustal structure of the South Orkney Islands area from seismic refraction and magnetic measurements: preprint from Scientific Committee on Antarctic Research/International Union of Geological Sciences Symposium on Antarctic Geology and Solid Earth Geophysics, Oslo, Norway.
Harrison, C. G. A., D. J. Barron, and W. W. Hay, 1979, Mesozoic evolution of the Antarctic Peninsula and the Southern Andes: Geology, v. 7, p. 374–378.
Hayes, D. E., and F. J. Davey, 1975, A geophysical study of the Ross Sea, Antarctica, *in* D. E. Hayes, et al, Initial reports of the Deep Sea Drilling Project, v. 28: Washington (U.S. Government Printing Office), p. 887–907.
Heezen, B. C., M. Tharp, and C. R. Bentley, 1972, Morphology of the earth in the Antarctic and Subantarctic: Antarctic Map Folio Service, American Geographical Society, Folio 16, pl. 7.
Holdgate, M. W., and J. Tinker, 1979, Oil and other minerals in

the Antarctic: report of a workshop sponsored by the Rockefeller Foundation, held at the Foundation's Study and Conference Center, Bellagio, Italy, 5–8 March 1979: published by Scientific Committee on Antarctic Research in care of Scott Polar Research Institute, Cambridge, U.K., 51 p.

Houtz, R. E., 1974, Continental margin of Antarctica: Pacific-Indian sectors, *in* C. A. Burk, and C. L. Drake, eds., The geology of continental margins: New York, Spring-Verlag, p. 655–658.

———, and F. J. Davey, 1973, Seismic profiler and sonobuoy measurements in Ross Sea, Antarctica: Journal of Geophysical Research, v. 78, n. 17, p. 3448–3468.

———, D. E. Hayes, and R. G. Markl, 1977, Kerguelen Plateau bathymetry, sediment distribution and crustal structure: Marine Geology, v. 25, p. 95–130.

———, et al, 1973, Plate 5—Sediment isopachs, *in* H. G. Goodell, et al, Marine sediments of the southern oceans: Antarctic Map Folio Service, American Geographical Society, Folio 17.

Irving, E., 1964, Paleomagnetism and its application to geological and geophysical problems: New York, John Wiley and Sons, Inc., 399 p.

Johnson, B. D., C. McA. Powell, and J. J. Veevers, 1976, Spreading history of the eastern Indian Ocean and Greater India's northward flight from Antarctica and Australia: Geological Society of America Bulletin, v. 87, p. 1560–1566.

JEBCO, 1976, Offshore Argentina Petroleum Exploration Reports: 68 High Street, Weybridge, Surrey, England 138 BL, JEBCO Pet. Dev. N. V., 263 p.

Katz, H. R., 1973, Contrasts in tectonic evolution of orogenic belts in the South-East Pacific: Royal Society of New Finland, Journal, v. 3, n. 3, p. 333–361

Kennett, J. P., et al. 1974, Cenozoic paleoceanography in the Southwest Pacific Ocean, Antarctic glaciation, and the development of the circum-Antarctic current, *in* J. P. Kennett, et al, eds., Initial reports of the Deep Sea Drilling Project, v. 29: Washington (U.S. Government Printing Office) p. 1155–1169.

Kent, P. E., 1974, Leg 25 results in relation to east Africa coastal stratigraphy, *in* E. S. W. Simpson, et al, eds., Initial reports of the Deep Sea Drilling Project, v. 25: Washington (U.S. Government Printing Office), p. 679–684.

Kraus, E. C., 1967, Die Bodenstruktur des Indischen Ozeans und dessen Geschichte (Basement structure of the Indian Ocean and its history): Geologische Rundschau, v. 56, n. 2, p. 373–393.

Kyle, P. R., J. Adams, and P. C. Rankin, 1979, Geology and petrology of the McMurdo Volcanic group at Rainbow Ridge, Brown Peninsula, Antarctica: Geological Society of America Bulletin, v. 90, pt. 1, p. 676–688.

LaBreque, J. L., and D. E. Hayes, 1979a, Plate tectonics south of the Agulhas/Falkland fracture zone (abs): American Geophysical Union Transactions, v. 60, n. 18, p. 394.

———, and ———, 1979b, Seafloor spreading history of the Agulhas basin: preprint, 26 p.

Laird, M. G., G. D. Mansergh, and J. M. A. Cheppell, 1971, Geology of the Central Nimrod Glacier area, Antarctica: New Zealand Journal of Geology and Geophysics, v. 14, n. 3, p. 427–468.

La Prade, K. E., 1972, Permian-Triassic Beacon Supergroup of the Shackleton Glacier area, Queen Maud Range, Transantarctic Mountains, *in* R. J. Adie, ed., Antarctic geology and geophysics: Oslo, Universitetsforlaget, p. 373–378.

Laudon, T. S., 1972, Stratigraphy of Eastern Ellsworth Land, *in* R. J. Adie, ed., Antarctic geology and geophysics: Oslo, Universitetsforlaget, p. 215–223.

Lindsay, J. F., 1968, Stratigraphy and sedimentation of the lower Beacon rocks of the Queen Alexandria, Queen Elizabeth, and Holland Ranges, Antarctica, with emphasis on Paleozoic glaciation: Ohio State University Ph.D. thesis, 318 p., abs. only.

Long, W. E., 1964, The stratigraphy of the Ohio Range, Antarctica: Ohio State University Ph.D. thesis, 363 p., abs. only.

Matz, D. B., and M. O. Hayes, 1967, Sedimentary history of Devonian and Permian rocks of Beacon Sandstone in Ross Sea region, South Victoria Land, Antarctica (abs): Geological Society of America Abstracts for 1967 Meetings, p. 277.

McElhinny, M. W., and G. R. Luck, 1970, Paleomagetism and Gondwanaland: Science, v. 168, n. 3933, p. 830–832.

McGregor, V. R., and F. A. Wade, 1969–70, Geology of the Western Queen Maud Mountains, *in* C. Craddock, et al, Geologic maps of Antarctica: Antarctic Map Folio Series, New York, American Geographical Society, Folio 12.

McKelvey, B. C., et al, 1970, Stratigraphy of the Beacon Supergroup between the Olympus and Boomerang Ranges, Victoria Land, Antarctica: Nature, v. 227, p. 1126–1128.

Minshew, V. H., 1966, Stratigraphy of the Wisconsin Range, Horlick Mountains, Antarctica: Science, v. 152, n. 3711, p. 637–638.

Mirsky, A., 1969–70, Geology of the Ohio Range-Liv Glacier area; *in* C. Craddock, et al, Geological maps of Antarctica: Antarctic Map Folio Series, New York, American Geographical Society, Folio 12.

Mond, A., 1972, Permian sediments of the Beaver Lake area, Prince Charles Mountains, *in* R. J. Adie, ed., Antarctic geology and geophysics: Oslo, Universitetsforlaget, p. 585–589.

Nelson, W. H., D. L. Schmidt, and J. M. Schopf, 1967, Structure and stratigraphy of the Pensacola Mountains, Antarctica (abs.): Annual Geological Society of America Cordilleran Section, Seismological Society of America and Paleontological Society Pacific Coast Section Meeting.

NEWS, 1978, RISP/RIGGS bypasses limits of man-made discipline: American Geophysical Union Transactions, v. 59, n. 10, p. 906–908.

Oliver, R. L., 1972, Some aspects of Antarctic-Australian geological relationships, *in* R. J. Adie, ed., Antarctic geology and geophysics: Oslo, Universitetsforlaget, p. 859–864.

Padula, E., and A. Mingramm, 1963, The fundamental geological pattern of the Chaco-Parana basin (Argentina) in relation to its oil possibilities: Sixth World Petroleum Congress, Section 1, Paper 1, p. 293–310.

Payne, R. R., and J. R. Conolly, 1972, Turbidite sedimentation off the Antarctic Continents, *in* D. E. Hayes, ed., Antarctic Oceanology II: the Australian-New Zealand sector: American Geophysical Union Antarctic Research Series, v. 19, p. 349–364.

Ramsey, W. R. H., and J. M. Stanley, 1976, Magnetic anomalies over the western margin of the New England foldbelt, northeast New South Wales: Geological Society of America Bulletin, v. 87, p. 1421–1428.

Reed, F. R. C., 1949 (2nd edition), The geology of the British Empire: London, Edward Arnold and Co., 764 p.

Rogue, P. C., et al, 1959, The sedimentary basins of Argentina: Fifth World Petroleum Congress, Section I, Paper 49, p. 883–899.

Sanford, R. M., 1980, Exploration results off S. New Zealand: Oil Gas Journal, 4 Feb., p. 83–90.

Sastri, V. V., A. T. R. Raju, and R. N. Sinha, 1973, Evolution of the Mesozoic sedimentary basins on the east coast of India: Australian Petroleum Exploration Association Journal, v. 14, pt. 1, p. 29–41.

Schmidt, D. L., and A. B. Ford, 1969–70, Geology of the Pensacola and Thiel Mountains, *in* C. Craddock, et al, Geological maps of Antarctica: Antarctic Map Folio Series, New York, American Geographical Society, Folio 12.

Schroeder, F. W., 1976, A geological survey at Site 325 in the Bellingshausen basin, *in* C. D. Hollister, et al, Initial reports of the Deep Sea Drilling Project, v. 35: Washington (U.S. Government Printing Office), p. 251–261.

Solomon, M., and J. R. Griffiths, 1972, Tectonic evolution of the Tasman Orogenic Zone, Western Australia: Nature (Physical Science), v. 237, p. 3–6.

Sprigg, R. C., 1967, A short geological history of Australia: Australian Petroleum Exploration Association Journal, v. 7, p. 59–82.

Stevens, G. R., 1967, Upper Jurassic fossils from Ellsworth Land, West Antarctica, and notes on Upper Jurassic biogeography of the South Pacific region: New Zealand Journal of Geology and Geophysics, v. 10, n. 2, p. 345–393.

St. John, B., 1980, Geology and hydrocarbon potential of Antarctica: 68 High Street, Weybridge, Surrey, England, JEBCO, 360 p., 52 figs., 6 tbls., 19 pls.

Stump, E., 1973, Earth evolution in the Trans-Antarctic Mountains and West Antarctica, *in* D. H. Tarling, and S. K. Runcorn, eds., Implications of continental drift to the earth sciences: New York, Academic Press, v. 2, p. 909–924.

Suarez, M., 1976, Plate-tectonic model for southern Antarctic Peninsula and its relation to Southern Andes: Geology, v. 4, n. 4, p. 211–214.

Tarney, J., A. D. Saunders, and S. D. Weaver, 1977, Geochemistry of volcanic rocks from the island arcs and marginal basins of the Scotia arc region, *in* M. Talwani and W. C. Pittman, III, eds., 1977, Island arcs, deep sea trenches and back-arc basins: American Geophysical Union Maurice Ewing Series 1, p. 367–377.

Tatsumi, T., and K. Kizaki, 1969–70, Geology of the Lutzow-Holm Bay region and the 'Yamato Mountains' (Queen Fabiola Mountains), *in* C. Craddock, et al, Geologic maps of Antarctica: Antarctic Map Folio Series, New York, American Geographical Society, Folio 12.

Taylor, B. J., 1972, Stratigraphical correlation in South-East Alexander Island, *in* R. J. Adie, ed., Antarctic geology and geophysics: Oslo, Universitetsforlaget, p. 149–153.

Thomson, M.R.A., 1972, Ammonite faunas of south-eastern Alexander Island and their stratigraphical significance, *in* R. J. Adie, ed., Antarctic geology and geophysics: Oslo, Universitetsforlaget, p. 155–160.

——, 1975, First marine Triassic fauna from the antarctic Peninsula: Nature, v. 257, n. 5527, p. 577–578.

Trail, D. S., and I., R. McLeod, 1969–70, Geology of the Lambert Glacier area, *in* C. Craddock, et al, Geological maps of Antarctica: Antarctic Map Folio Series, New York, American Geographical Society, Folio 12.

Tucholke, B. E., and R. E. Houtz, 1976, Sedimentary framework of the Bellingshausen basin from seismic profiler data, *in* C. D. Hollister et al, Initial reports of the Deep Sea Drilling Project, v. 35: Washington (U.S. Government Printing Office), p. 197–227.

Vail, P. R., et al, 1977, Seismic stratigraphy and global changes of sea level, *in* C. E. Payton, ed., Seismic stratigraphy—applications to hydrocarbon exploration: AAPG Memoir 26, p. 49–212.

Wade, F. A., 1969–70, Geology of Marie Byrd Land, *in* C. Craddock, et al, Geological maps of Antarctica: Antarctic Map Folio Series, New York, American Geographical Society Folio 12.

Warren, G., 1969–70, Geology of the Terra-Nova Bay—McMurdo Sound area, Victoria Land, *in* C. Craddock, et al, Geological maps of Antarctica: Antarctic Map Folio Series, New York, American Geographical Society, Folio 12.

Watkins, N. D., et al, 1974, Kerguelen: continental fragment or oceanic island?: Geological Society of America Bulletin, v. 85, p. 201–212.

Webb, P. N., 1974, Micropaleontology, paleoecology, and correlation of the Pecten gravels, Wright Valley, Antarctica, and description of Trochoelphildiella Onyxi N. Gen.: New Zealand Journal, Foraminiferal Research, V. 4, p. 184–199.

Webby, B. D., 1978, History of the Ordovician continental platform shelf margin of Australia: Geological Society of Australia, Journal, v. 25, pts. 1–2 p. 41–63.

Webers, G. F., 1972, Unusual Upper Cambrian fauna from West Antarctica, *in* R. J. Adie, ed., Antarctic geology and geophysics: Oslo, Universitetsforlaget, p. 235–237.

Williams, P. F., et al, 1971, The structural and metamorphic geology of basement rocks in the McMurdo Sound area, Antarctica: Geological Society of Australia, Journal, v. 18, pt. 2, p. 127–142.

Williams, P. L., et al, 1972, Geology of the Lassiter Coast area, Antarctica Penninsula: Preliminary reports, *in* R. J. Adie, ed., Antarctic geology and geophysics: Oslo, Universitetsforlaget, p. 143–148.

Woolard, G. P., 1970, The geology of Antarctica: Hawaii Institute of Geophysics, University of Hawaii, reproduced by National Technical Information Service, Springfield, Va., (22141), 99 p.

Wright, N. A., and P. L. Williams, comps., 1974, Mineral resources of Antarctica: U.S. Geological Survey, Circular 705, 29 p.

Yeats, V. L., et al, 1965, Geology of the central portion of the Queen Maud Range, Transantarctic Mountains: Science, v. 150, p. 1808–1809.

# Circum-Arctic Petroleum Potential

A. R. Green
A. A. Kaplan
R. C. Vierbuchen
*Exxon Production Research Company*

More than 30 sedimentary basins surround the Arctic Ocean. A reconstruction of the ancient structural, depositional, and climatic trends has been made of the more important basins to evaluate the relative hydrocarbon potential of this frontier area.

Although tectonic movement has been active throughout the Phanerozoic, four periods of tectonism were of particular importance in forming and developing the Arctic basins: the Caledonian orogeny during the Devonian, the formation of the Ural Mountains during the Permian–Triassic collision of the Siberian and Baltic Platforms, the opening of the Canada Deep during the Late Jurassic–Early Cretaceous, and sea-floor spreading along the Nansen–Gakkel Ridge during the early Tertiary. These events created a complex paleoenvironmental history for the Arctic basins, resulting in a variety of sources, seals, and reservoirs.

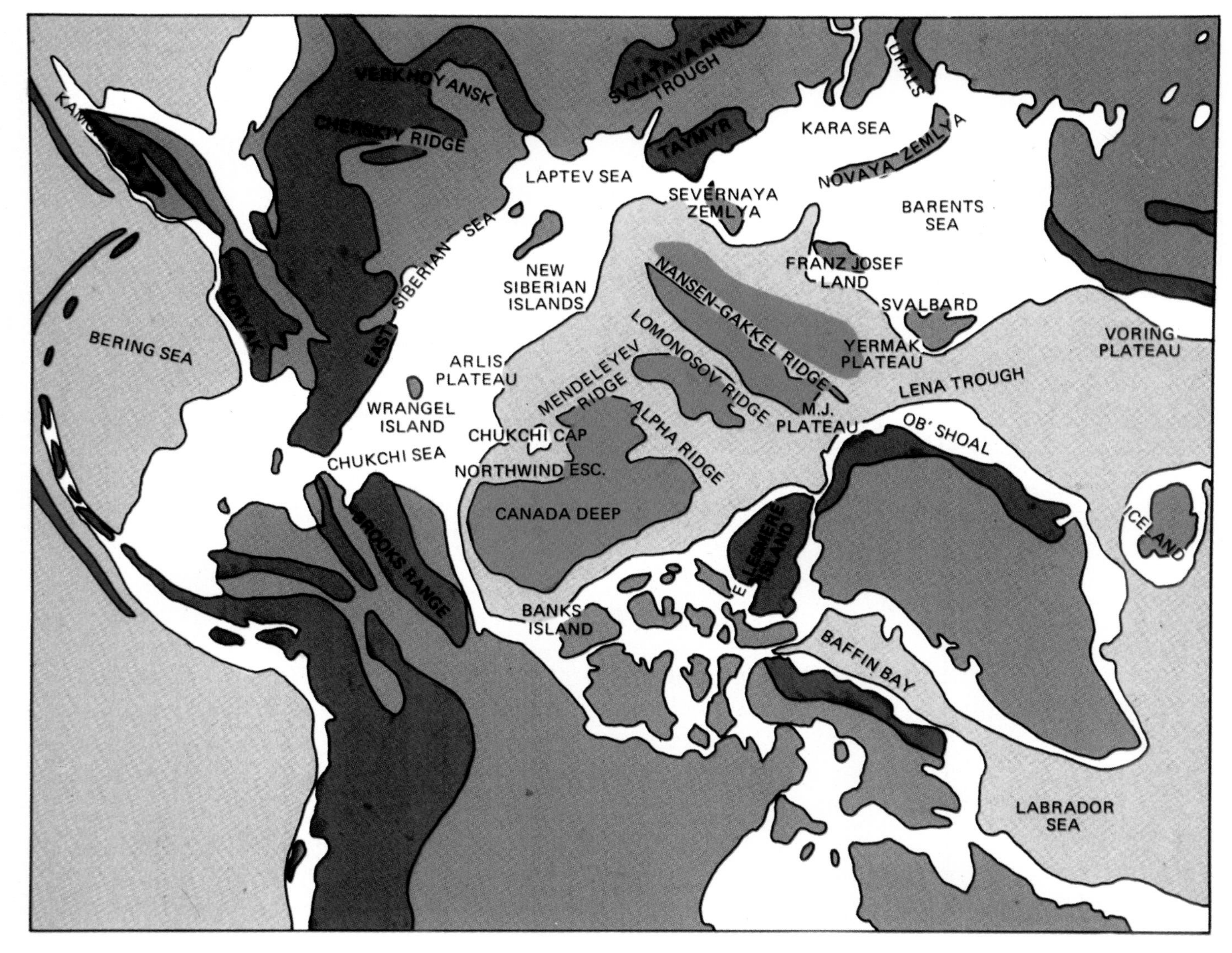


Figure 1. *Arctic Ocean basin.*

## INTRODUCTION

The Arctic Ocean basin, which separates the Eurasian and North American continents, is more than 4 km (2.5 mi) deep, covers more than 13 million sq km (5 million sq mi), and contains many of the world's least understood, major physiographic features (Figure 1). The shelves that surround the deep oceanic basin are some of the widest in the world (Figure 2). Nearly 60% of the Arctic Ocean is less than 1 km (0.6 mi) deep, and over 80% of it is less than 3 km (2 mi) deep (Demenitskaya and Hunkins, 1970). Much of the Arctic shelf is underlain by thick, sedimentary basins containing Precambrian, Paleozoic, Mesozoic, and Tertiary sediments (Figure 3).

We present in this paper a series of tectonic reconstructions that outline the tectonic history of the Arctic from Late Devonian to middle Tertiary. These reconstructions include maps depicting the crustal history of the sedimentary basins, which in turn determined the physical architecture of the basins, their

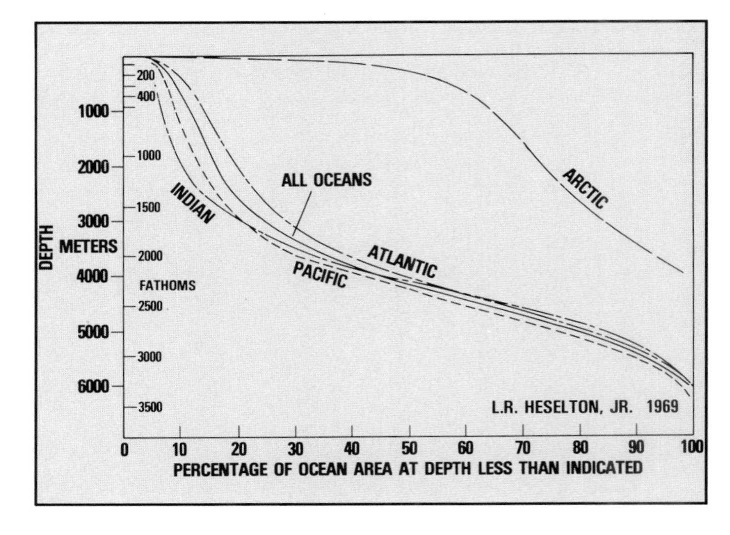


Figure 2. *Hypsometry of world oceans.*

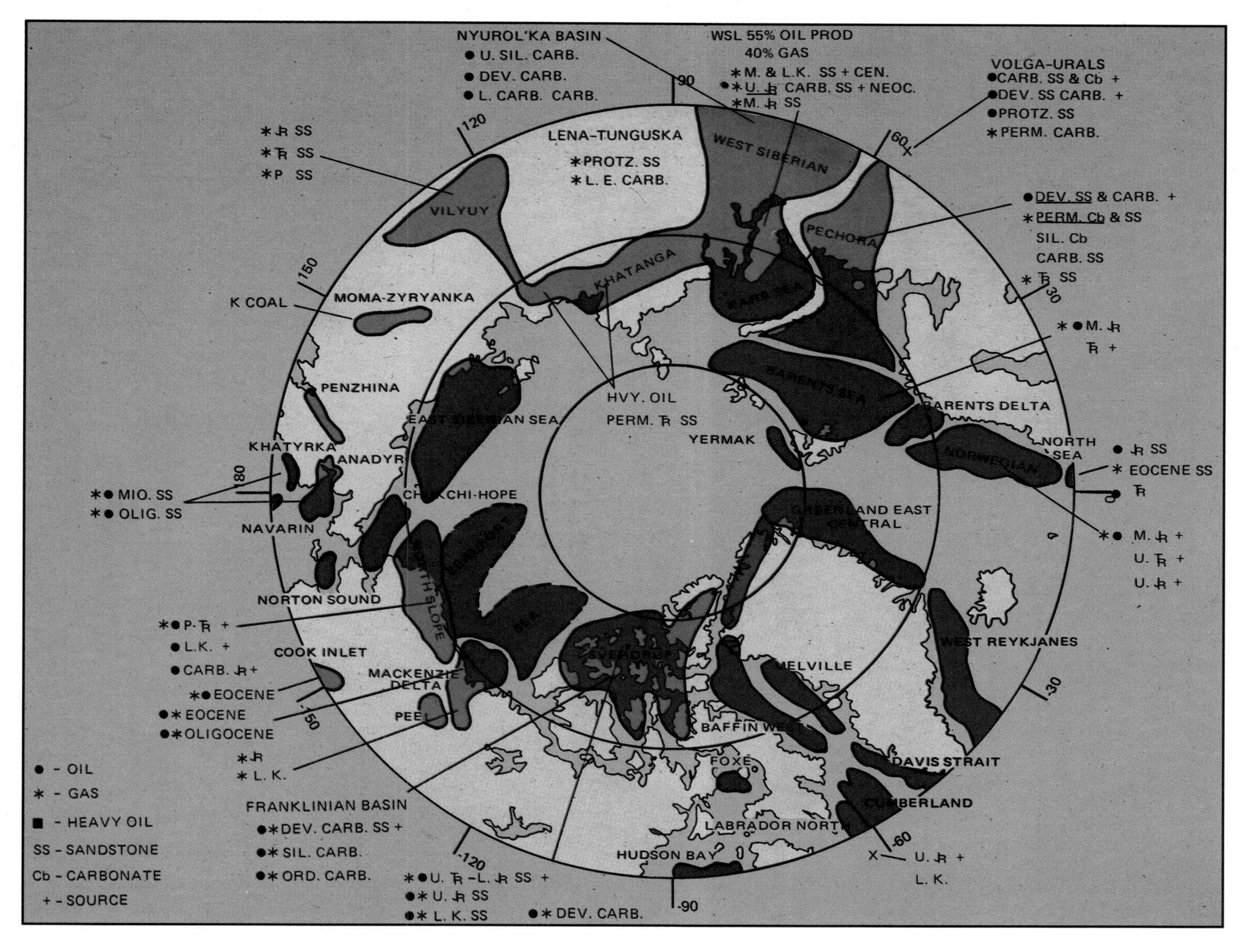


Figure 3. *Sedimentary basins of the Arctic.*

size and shape, basement grain, structural style, heat flow history, and bottom configuration at the time of deposition. The tectonic history that led to the Arctic sedimentary basins also set the stage for the dynamic interactions between the lithosphere and the atmosphere, the hydrosphere, and the biosphere, which created the unique suite of depositional environments for each basin (Green, 1985).

We also present a series of paleogeographic and paleoenvironmental maps to depict basin filling, particularly the deposition of organic-rich source rocks, sealing sediments, and reservoirs for petroleum accumulations.

We conclude by briefly summarizing the petroleum occurrence in the Arctic. The known petroleum accumulations reveal the sites where effective petroleum systems have developed, that is, where source rocks have been deposited, preserved, and matured; and where reservoir rocks and traps have formed effective plumbing and trapping systems for commercial deposits of oil and gas. We combine observations of known petroleum occurrences with maps depicting the geological history of the various basins, to postulate the hydrocarbon potential of the Arctic.

## TECTONIC SETTING

The deep central Arctic Ocean floor has thin oceanic crust (Figures 4 and 5) (Vogt and Kovacs, 1984). In the Canada Deep, north of Alaska, the smooth ocean floor is at 3,600 m (11,800 ft). Approximately 200 km (120 mi) north of the Alaskan coast, sediments are 4–5 km (2–3 mi) thick, and have seismic velocities of 1.8 to 3.6 km/sec (5,900 to 11,800 ft/sec). This sedimentary section overlies oceanic crust approximately 8–10 km (5–6 mi) thick (Mair and Lyons, 1981; Baggeroer and Falconer, 1982). The Eurasian Deep, which is bisected by an active spreading center, is about 4 km (2.5 mi) deep just north of the Morris Jesup Rise. At that point, the sedimentary cover is 1.5–2 km (1–1.4 mi) thick and overlies oceanic crust 4–7 km thick (Duckworth, Baggeroer, and Jackson, 1982; Kristoffersen et al, 1982).

Many significant bathymetric features occur in the central

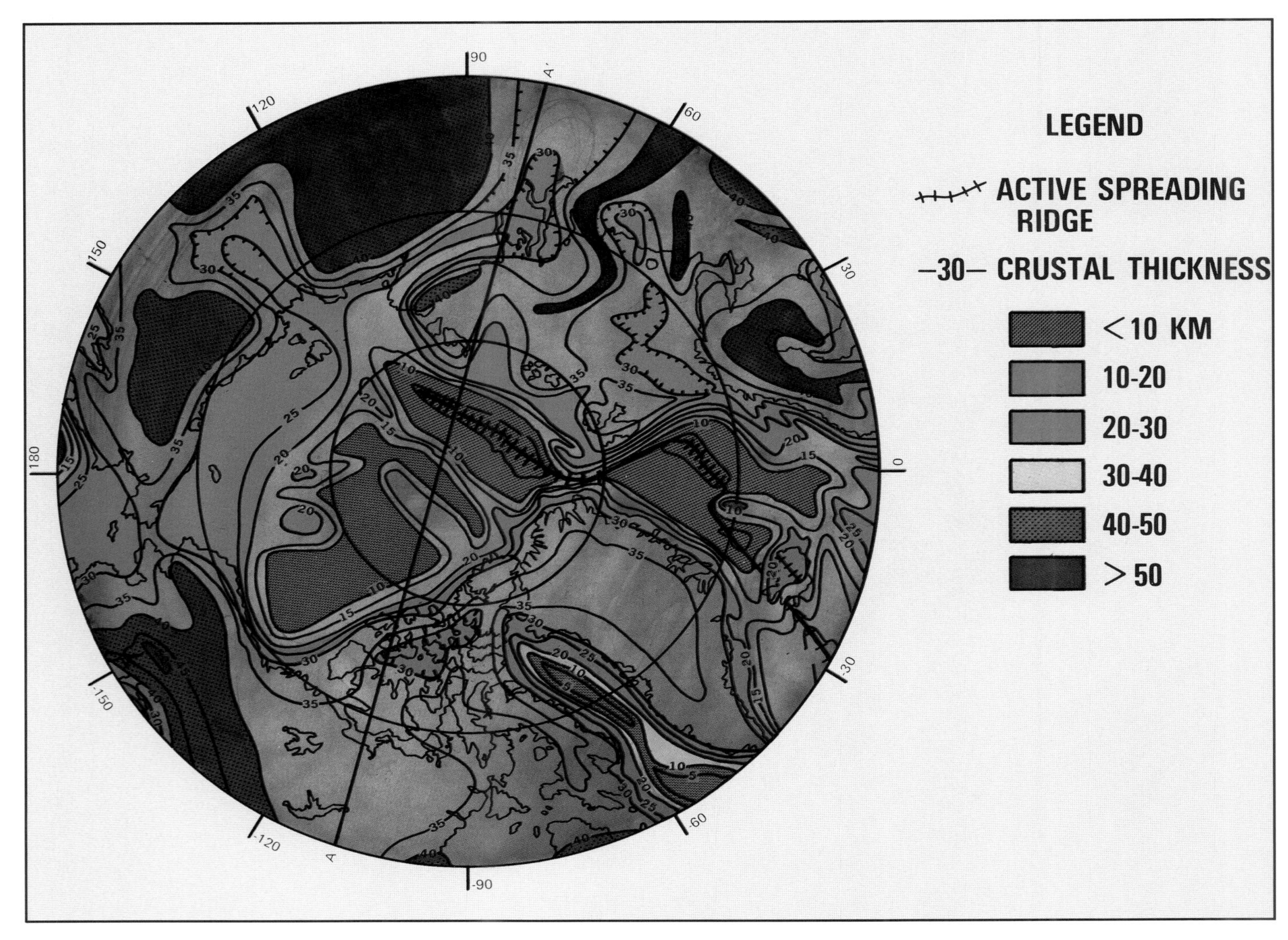


Figure 4. *Crustal thickness of the Arctic.*

Arctic Ocean (Johnson et al, 1978). The Chukchi Cap and Northwind Escarpment may be continental remnants (Forsyth, 1978), as is the Lomonosov Ridge (Weber, 1979). The Alpha Ridge, Mendeleyev Ridge, and Arlis Plateau constitute a complex ridge that may include remnants of thinned, injected, and deformed continental crust. The wide shelves of the Arctic are underlain by continental crust that has been thinned by rifting in areas such as the Barents Sea, the Vøring Plateau, northeastern Greenland, the West Siberian basin, the Kara Sea, the Khatanga basin, the East Siberian Sea, the Sverdrup basin, and Baffin Bay (Gramberg et al, 1984).

During the late Proterozoic and throughout the Phanerozoic, tectonic accretion and the collision of major continental blocks created the complex crust that now surrounds the deep Arctic Ocean (Volk et al, 1984). Figure 6, a generalized suture map, summarizes these events. We review them in chronological order in the discussion of geological evolution.

We have distinguished nine major tectonic blocks within the Arctic portions of the Eurasian and North American continents. These blocks are separated by present or former plate boundaries and are, for the most part, characterized by contrasting geological histories. We could further subdivide most blocks into terrains and composite terrains, but the scale of this analysis makes it impractical.

The nine blocks considered here are:

1. *Baltic Platform*: Extends from the late Paleozoic suture of the Urals in the east, to the Caledonian orogenic belt in Scandinavia. Its northern border lies north of the Barents Sea and is bounded by a passive continental margin from which the Lomonosov ridge was separated at the beginning of the Cenozoic.
2. *Siberian Platform*: Extends from the Urals in the west, to the Cherskiy and Sette–Daban sutures in the east. Its northern edge is bounded by the Uralian Suture in Novaya Zemlya, the Taymyr suture, and by the western end of the Bol'shoy Lyakhov-Anyuy suture.
3. *Omolon-Kolyma Terranes*: Continental block surrounded by accreted, arc-related rocks.
4. *Okhotsk Terranes*: Continental block surrounded by accreted, arc-related rocks.
5. *Koryak Block*: Composed of a series of oceanic terranes,

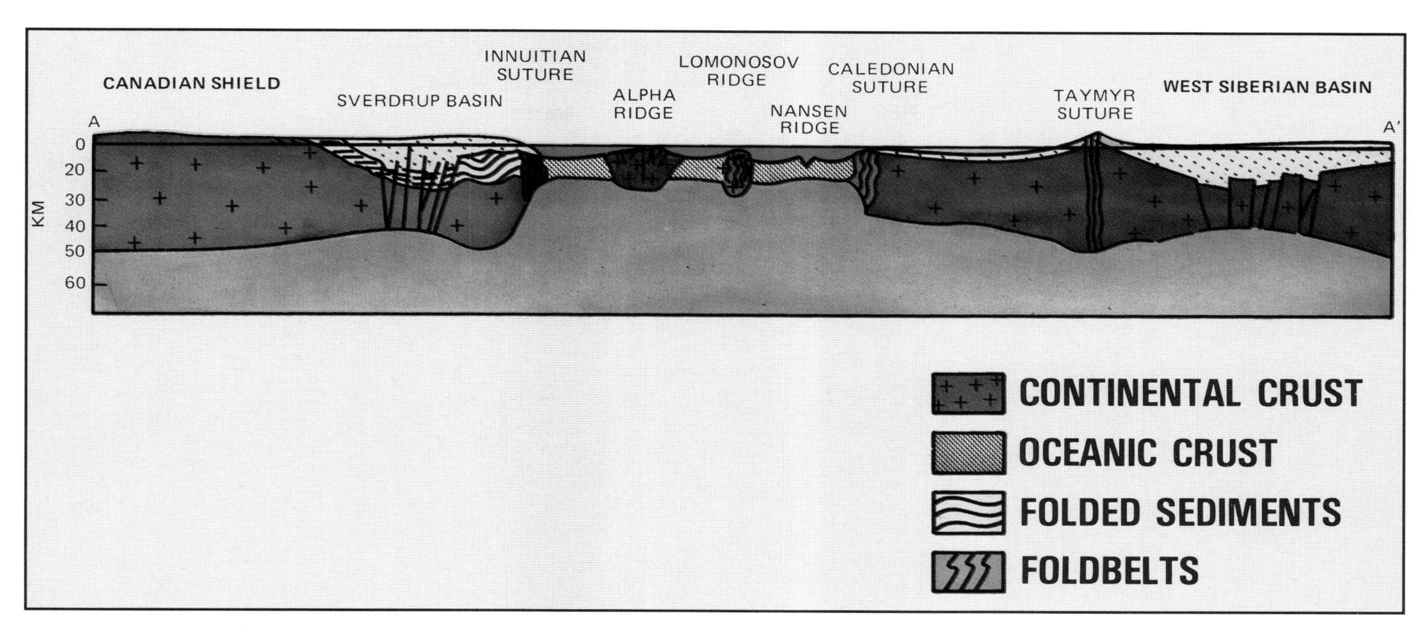


**Figure 5.** *Crustal cross-section of the Arctic.*

accreted to Eurasia during the Cretaceous and early Tertiary. All three blocks lie between the eastern margin of the Siberian Platform and the convergent margins of the northwestern Pacific Ocean.

6. *New Siberian Islands–Chukotka–North Slope Alaska Block*: Bounded by the rifted margin of the Amerasian basin on the north side and by sutures extending from the Brooks Range through the Seward Peninsula and South Anyuy Zone to a western extension in the Bol'shoy Lyakhov Island.
7. *North America*: Bounded by the Cordilleran, Appalachian, and Innuitian orogenic belts. The Appalachian and Innuitian belts were sites of Paleozoic suturing that rifted in the Mesozoic to form, respectively, the Atlantic Ocean and the Amerasian basin of the Arctic Ocean.
8. *Greenland*: East and west margins of Greenland are bounded by Late Cretaceous and early Cenozoic rifts. In east Greenland, rifting occurred along the Caledonian suture between Greenland and Norway. The north coast of Greenland was the site of moderate Innuitian orogenic activity and has been involved in rifting, strike-slip, and minor convergent events associated with the formation of the Arctic Ocean. Greenland drifted away from the North American block in the Late Cretaceous and early Tertiary, creating the Labrador Sea and Baffin Bay.
9. *South and Central Alaska Composite Terrains*: Consist of numerous allochthonous terrains that contain rocks ranging in age from Precambrian to Tertiary, and accreted to the North American block, from the south and west, during the Mesozoic and Tertiary.

## GEOLOGIC EVOLUTION OF THE ARCTIC

### Pre-Devonian

During the middle and late Proterozoic, plate convergence formed foldbelts and sutured continental blocks that later drifted to a north-polar position. Orogenesis occurred in southeastern Canada, where Grenvillian rocks were deformed; in northern Greenland, where Carolinian folding took place; and in northern Finnmark (Stuart, Pringle, and Roberts, 1975) and the U.S.S.R., where Baykalian deformation occurred.

In the latest Proterozoic, rifting and subsequent sea-floor spreading produced a new ocean basin (Iapetus). The continental margins along the various branches of the Iapetus Ocean collected thick accumulations of late Proterozoic and early Paleozoic sediments. In east Greenland these sediments attained thicknesses of 17 km (10 mi) (Henriksen and Higgins, 1976). Lower Paleozoic sediments filled the north Greenland basin (Surlyk and Hurst, 1984).

Closure of Iapetus, which began in the Early Ordovician, culminated as the Caledonian orogeny. This complex orogeny affected a widespread region from the Arctic to northern Europe and eastern North America. By the Middle Silurian, the Iapetus Ocean west of Norway had been consumed and the continental blocks of Greenland and the Baltic Platform had collided. Throughout Late Silurian and Early Devonian, eastward thrust faulting continued, and by Early Devonian, the final welding of the Old Red Continent was complete. When the post-Caledonian history of the Arctic began, southwestern Norway was receiving Old Red deposits in interior basins at equatorial latitudes.

The late Proterozoic and early Paleozoic history of the Arctic region created the crustal framework upon which younger sedimentary basins formed. Our discussion concentrates on the Late Devonian to Tertiary, because most sediments of the Arctic that produce hydrocarbons were deposited during this period. The late Precambrian and early Paleozoic basins, and their petroleum potential, are beyond the scope of this paper.

In the following discussions of the crustal reconstructions and the paleogeographic maps, the compass direction cited refers to the paleo-direction depicted on the respective maps, unless we indicate otherwise.

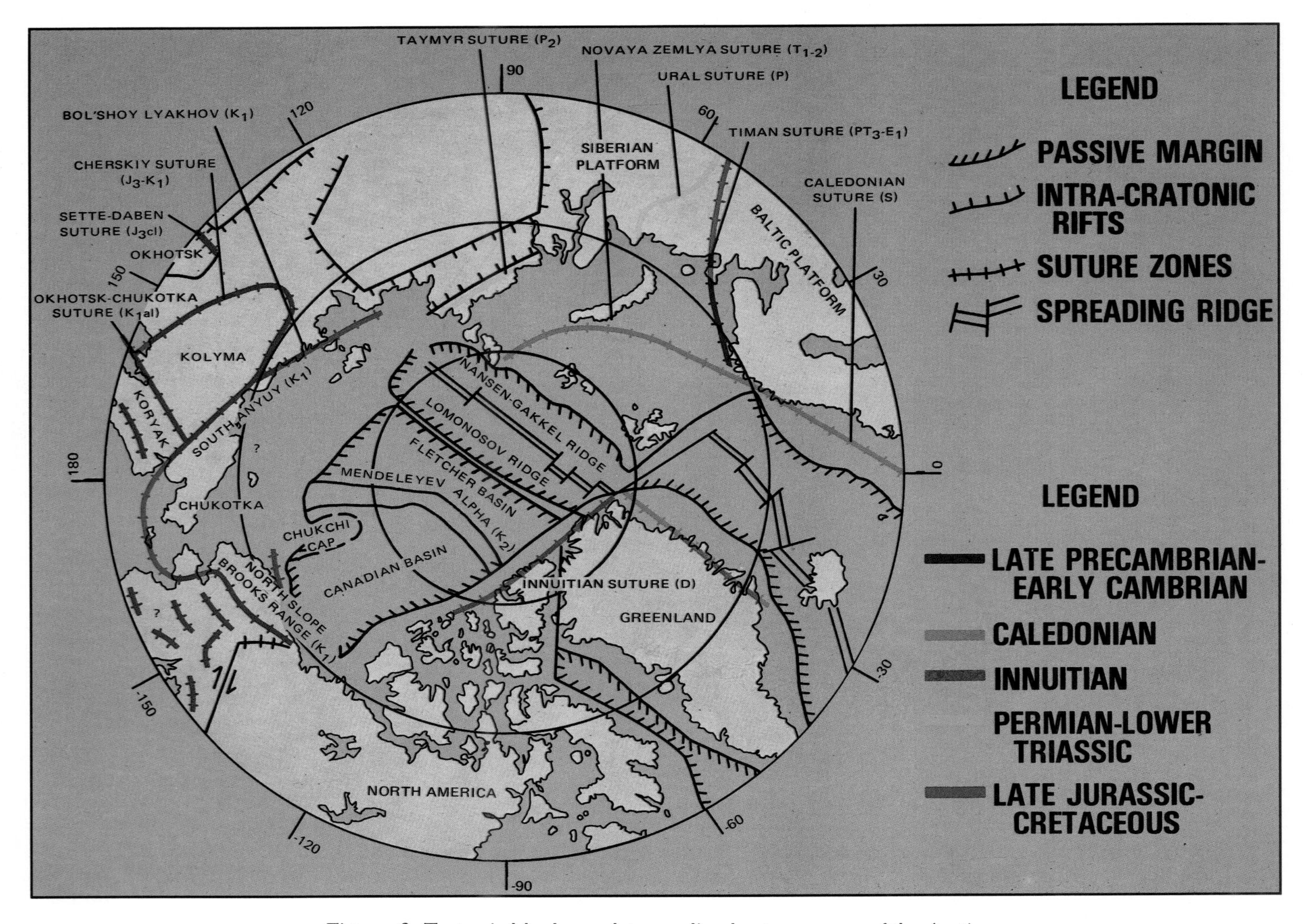


Figure 6. *Tectonic blocks and generalized suture zones of the Arctic.*

Late Devonian (370 Ma)

Late Devonian was a time of major tectonic reorganization of crustal plates (Figures 7 and 8). Most Caledonian convergence between Europe and North America ceased by the Late Devonian, but transform faults (Svalbardian deformation) remained active (Vogt, 1936). Harland (1969) and Faleide, Gudlaugsson, and Jacquart (1984) postulate that, at this time, Europe began a translation of several hundred kilometers to the north with respect to North America and Greenland (Figure 8). Such major strike-slip displacements account for the variation in pre-Devonian rocks of east, central, and west Svalbard and the Devonian basins of southwestern Norway.

During the Middle and Late Devonian, a continental collision occurred between the Siberian Platform and the North American block, forming the Innuitian foldbelt (Christie, 1979; Trettin, Balkwill, and Price, 1979; DOBEX, 1980). The Franklinian geosyncline, which had been collecting sediments throughout the lower Paleozoic, was partially inverted by this compressive event. The Boothia Arch and the Cornwallis foldbelt were also active intermittently throughout the Silurian and into the Late Devonian (Kerr, 1977). The Late Devonian–Early Mississippian Ellesmerian orogeny consisted primarily of compressive, sinistral strike-slip movement. Eisbacher (1983) links this transpressional, sinistral orogeny of the Arctic to the Antler orogeny of Nevada and California. Following the Caledonian and Innuitian deformation, a series of sutures extended from the northern side of the Chukota–New Siberian Islands block, through Svalbard and the Caledonides of eastern Greenland and western Norway to Great Britain.

Late Devonian and Early Carboniferous rifting that followed soon after Innuitian suturing, initiated the proto-Arctic Ocean. The rift system (depicted in the center of Figure 8), occurred deep within the Siberian Platform. As a result, a large part of what was the Siberian Platform remained with northern Canada, namely what is now the North Slope–Chukotka–New Siberian Islands block. At the end of the Devonian, North America and the Siberian Platform faced an ocean that was opening between them (the proto-Arctic Ocean). At the same time, the Kolyvan–Tomsk Sea was closing between Kazakhstan and the Siberian Platform, and the Uralian Sea was closing between Kazakhstan and the Baltic Platform (right center of Figure 8).

This convergence resulted in thrusting and volcanic activity along the margins of the Siberian and Baltic Platforms (Mel'nikova, 1983). Along the edge of the Baltic Platform, a back-arc basin (Pechora basin) was forming behind a subduction zone.

Within and adjacent to the Caledonian foldbelt, a thick accu-

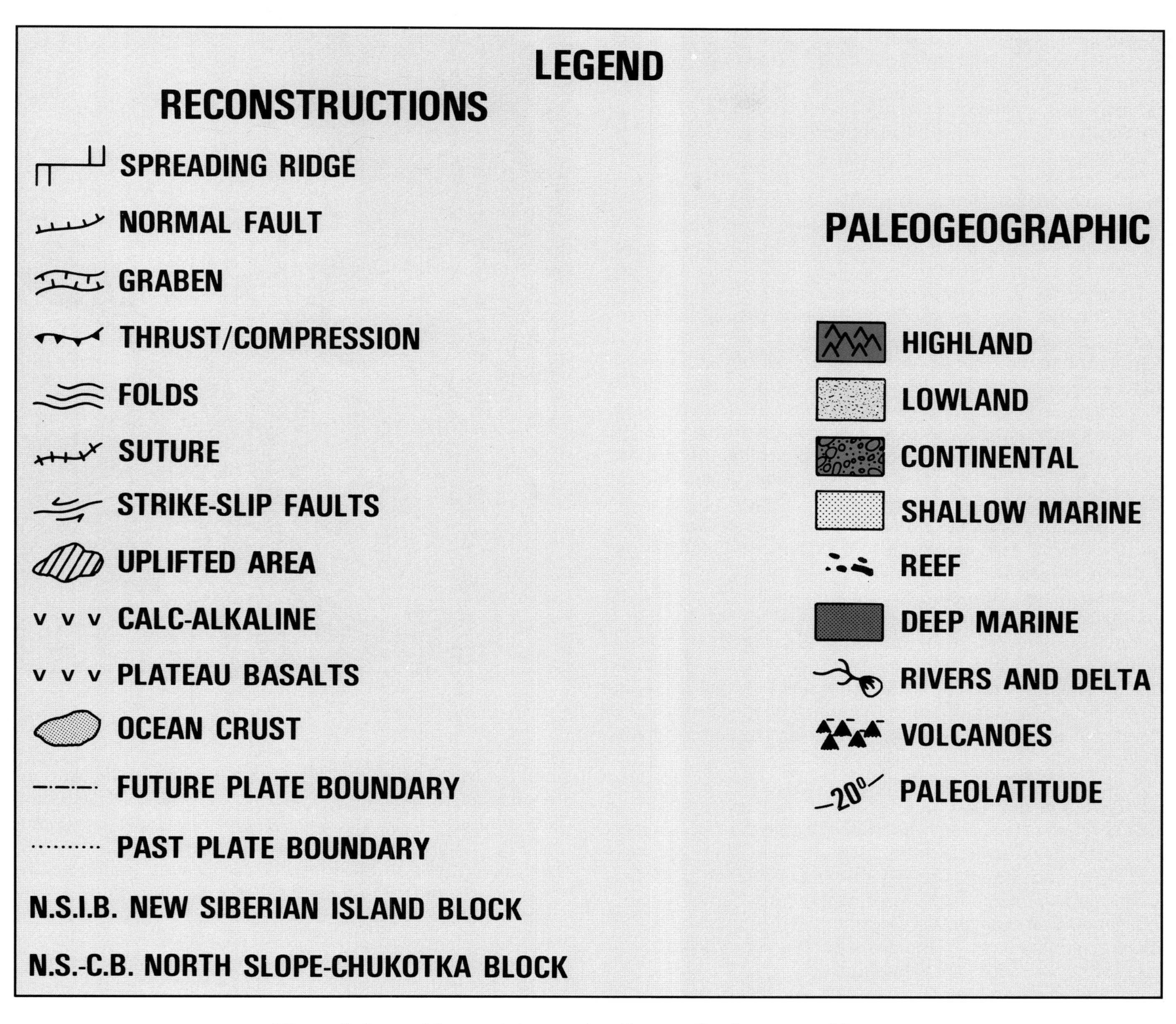


Figure 7. *Legend for crustal reconstructions and paleogeographic maps.*

mulation of synchronous and post-orogenic, dominantly nonmarine sediments were deposited (Figure 9). These rocks (Old Red Sandstone) include the spectacular alluvial fan deposits of Scotland, southwestern Norway, Greenland, and Spitsbergen, as well as the coastal plain deposits of Wales, southern England, the North Sea, and northwestern Germany. The land mass where these rocks were deposited was just north of the equator. The climate was tropical to subtropical in northern Canada and western Greenland and semi-arid on the Baltic Platform. Many of the intermontane basins, in which the Old Red Sandstone was deposited, are linear and steep-sided. The abrupt, straight basin margins, plus the cyclic, coarse-grained alluvial basin-fill are typical of wrench-related basins. Outstanding examples occur along the extension of the Great Glen fault in the Shetland Islands and in southwestern Norway, where long, narrow basins contain great thicknesses (± 8 km or 5 mi) of clastic rocks (Steel, 1976). Organic-rich, lacustrine mudstones were being deposited in northeast Scotland and east Greenland in similar basins.

Red beds, evaporites, and carbonates were deposited upon the Baltic Platform. To the east, in rifts underlying the Pechora basin, subaerial basaltic volcanics are interbedded with deltaic sands and bituminous shales, and overlain mainly by limestones. The total section is more than 3 km (2 mi) thick (Krems, Vasserman, and Matvievskaya, 1974). These shallow water to non-marine sediments grade eastward into reefs and then farther eastward to shales, cherty rocks, and carbonates that were deposited upon a continental slope.

Carbonates and evaporites accumulated upon the paleo-southwestern flanks of the Siberian Platform during Late Devonian (Pogrebitsky, 1971). On the paleo-northern side of the Siberian Platform, widespread rifting occurred (Levashov, 1979). In the Vilyuy Rift, basalts, tuffs, siltstones, sandstones, and more than a kilometer of evaporites were deposited. Along the Verkhoyansk continental margin, adjacent to the Vilyuy Rift, shelf carbonates were deposited.

The Upper Devonian sediments of the North Slope–Chukotka block and New Siberian Islands block are similar. In the New Siberian Islands, rift-related basalts of Frasnian age are interbedded with thick (up to 7.2 km or 4.5 mi) sequences of dominantly marine clastic rocks (Vol'nov, 1975). Platform carbonate facies occur landward of the new continental margin. In Chukotka, marine sandstones and siltstones, 360–400 m (1,200–1,300 ft) thick, were deposited.

Along the Arctic coast of Alaska, the Upper Devonian rocks reflect a major change in paleogeographic setting. Lower Paleozoic, volcanic-rich, basinal shales and sands were folded during the Innuitian event (Dillon et al, 1980), and became a major source of clastic detritus. The direction of sediment transport reversed in the Late Devonian, and sediments began to prograde to the north (present-day south). Thick sandstones, shales, and conglomerates (Kanayut Conglomerate) were shed from the Barrow Arch, in Alaska, and a ridge on the north flank of the Arctic Islands.

During the Late Devonian and through the Early Carboniferous Ellesmerian orogeny, the area that would become the foundation of the Sverdrup basin was first eroded and then covered by shallow-water deltaic deposits. The Sverdrup basin began to subside soon after the end of the Ellesmerian orogeny (Miall, 1976). In the Selwyn basin of the Yukon, restricted and anoxic waters prevailed within a large inland sea connected to the Pacific (Goodfellow and Jonasson, 1984).

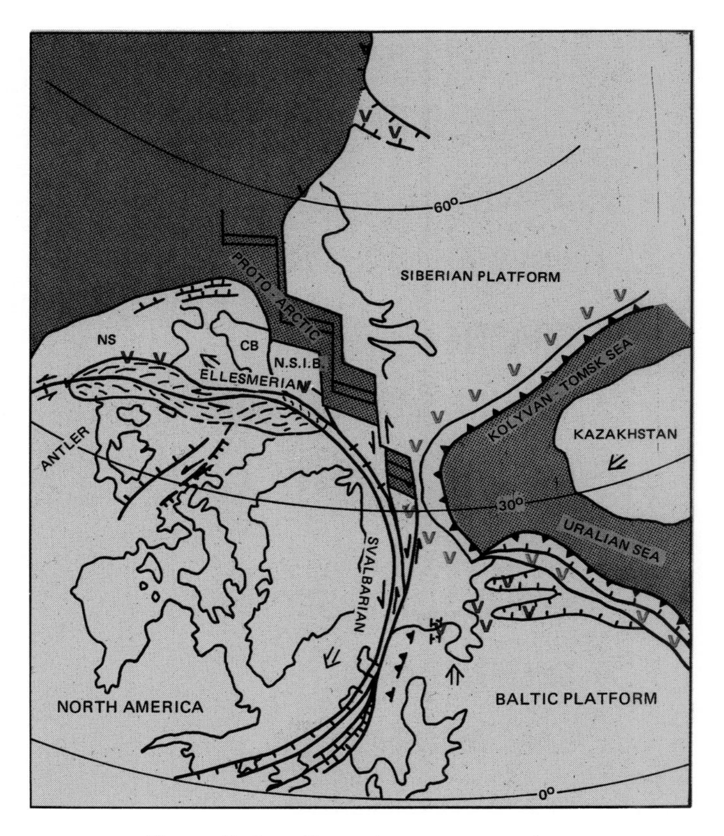


Figure 8. *Late Devonian reconstructions.*

## Late Carboniferous (295 Ma)

Carboniferous tectonism was relatively quiescent when compared with the dramatic events of the Devonian. The Kolyvan–Tomsk Sea had closed along a suture (Figure 10). Svalbardian wrench faults remained active during the Early Carboniferous, juxtaposing Scandinavia and Greenland and connecting to the convergent margin of the Uralian Ocean (Van der Voo and Scotese, 1981). In middle Carboniferous time, the Svalbard Platform was broken by a series of north-northwest to south-southeast oriented asymmetric rift basins (Gjelberg and Steel, 1983). The newly formed proto-Arctic Ocean, between the Siberian Platform and the North American blocks, was continuous with the older Uralian Sea (Figures 10 and 11).

Thick Carboniferous deposits, composed of andesitic volcaniclastic and deep-water (accretionary wedge) clastic sediments, suggest that the eastern side of the Baltic Platform continued to be a convergent, continental margin. Behind the magmatic arc, carbonate, evaporite, and clastic sediments, derived from the west, accumulated on a broad continental shelf. These rocks grade eastward into cherty limestones and then to flysch.

The paleo-southwestern margin of the Siberian Platform received sediments consisting of Lower Carboniferous carbonates, grading upward into sandy limestones, culminating in Upper Carboniferous clastic deposits. On the paleo-northwestern margin of the Siberian Platform, thick carbonates, clastics, and shelf facies were deposited. A major delta formed at the seaward opening of the Vilyuy Rift.

A passive margin occurred along the paleo-northeastern margin of the North American–Siberian blocks, collecting carbonate and clastic sediments. In the New Siberian Islands, Carboniferous rocks resemble those of the Upper Devonian. Platform carbonates graded seaward to relatively deep-water clastic sediments and siliceous carbonates (paleo-north, present-day south). Along the Chukotka segment of the Anyuy suture, a middle to Late Carboniferous fauna is associated with a 400 m (1,300 ft) section of basalts, spilites, diabases, and siliceous and clastic sediments (Til'man, Afitskiy, and Chekhov, 1977). This material is further evidence that along the Anyuy suture, this block faced an ocean basin in post-Devonian time. North Slope Carboniferous sediments consist of shallow-water marine to intertidal carbonates and sandstones in the region of the Barrow Arch. The sediments grade southward (present-day) to deeper water basinal facies (cherts and shales) (Churkin, Nokleberg, and Huie, 1979).

In the Sverdrup basin, a low-relief carbonate barrier formed along the northern basin rim. The barrier silled the basin and a thick sequence of evaporites developed in the center of the basin (Schwerdtner and Osadetz, 1983). Shallow water to nonmarine clastic rocks accumulated along the southern margin of the basin (Meneley, Henao, and Merritt, 1975).

## Permian–Triassic (230–240 Ma)

The Urals and the mountains of the Taymyr Peninsula are the result of a Permian collision of the Siberian and Baltic Platforms, which destroyed the Uralian Sea (Figures 12 and 13). Only in Novaya Zemlya was suturing incomplete at the beginning of this period, as indicated by the presence of thick,

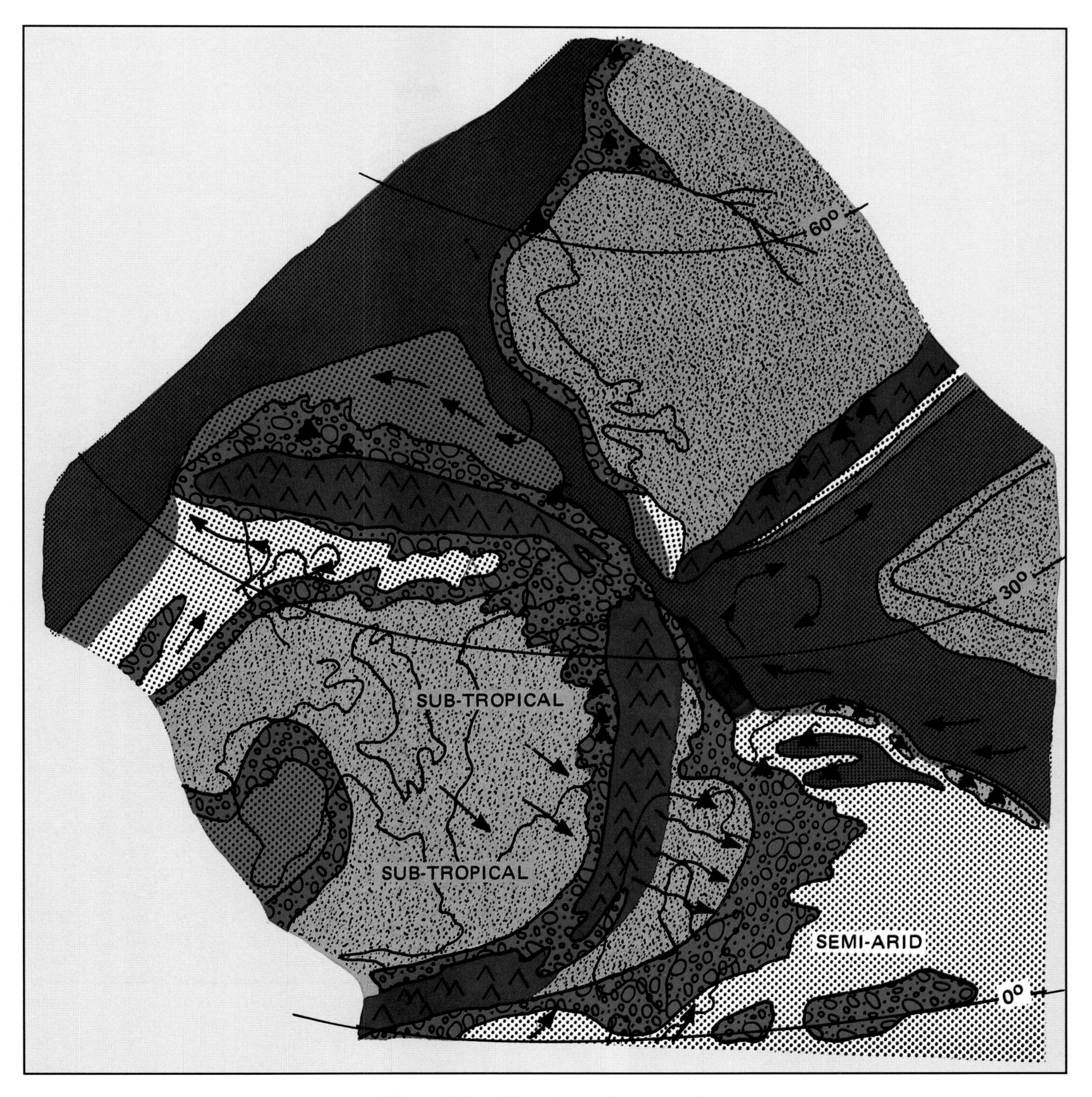


**Figure 9.** *Late Devonian paleogeography.*

marine, Upper Permian shales (Ustritsky, 1977). South of the Urals (present-day west), shales and sandstones were deposited in a foreland trough on the flanks of the Uralian collision zone. As a result of the collision, rift basins underlying the Pechora basin were partially inverted, and large structural traps formed (Milanovskiy, 1981).

The suturing of Novaya Zemlya in Early to Middle Triassic culminated a long series of Paleozoic compressional events that were followed by, and partly synchronous with, Late Permian to Mesozoic extension of the crust. North of the Urals (present-day east), widespread rifting, crustal thinning, and volcanism marked the inception of the West Siberian basin (Milanovskiy, 1978). Extension north (present-day east) of the West Siberian basin caused extrusion of the voluminous Upper Permian to Middle Triassic Tunguska, or Siberian plateau basalts. The present area of these basalts and associated sills exceeds $10^6$ sq

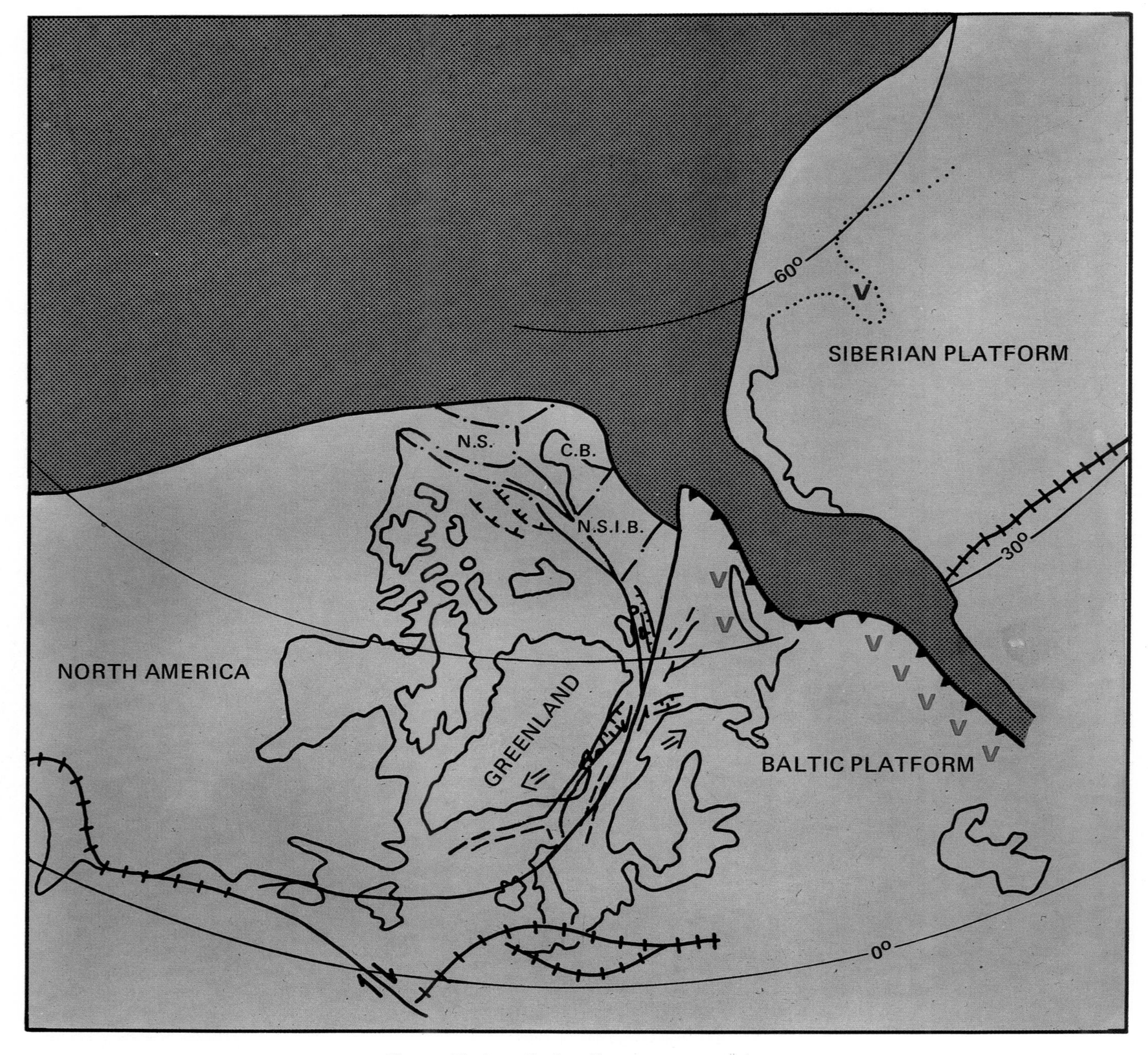


Figure 10. *Late Carboniferous reconstruction.*

km (380,000 sq mi). Basalts of this age also occur in the Vilyuy and Khatanga Rifts. Rifting and basaltic volcanism covered the Siberian Platform from just south (present-day west) of the Urals, across the West Siberian basin and Tunguska to the Vilyuy Rift, a distance of nearly 4,000 km (2,500 mi).

Rates of sedimentation increased in the Vilyuy and Khatanga Rifts. Rapid deposition along the Verkhoyansk passive margin was contemporaneous with extrusion and uplift of the Siberian plateau basalts.

A major rift system extended from northern South America through eastern North America, western Europe, and up through the North Sea, as far as east Greenland and the Barents Sea. Rift basins in this system formed between Late Permian and Late Triassic. In some of these basins, faulting and subsidence continued into the Late Cretaceous and early Tertiary. Triassic sediments, deposited in rift basins between east central Greenland and Norway, include thick sections of evaporites and record rapid deposition in alternating shallow-marine and non-marine environments. The Upper Permian, organic-rich, Posidonia shales and the overlying Triassic sands were deposited in the restricted basins of east Greenland (Surlyk, 1983).

The Permian section, in the Trømso and Hammerfest basins of the Barents Sea, is thought to be composed primarily of evaporites and carbonates. Some salt diapirs in this area have radii

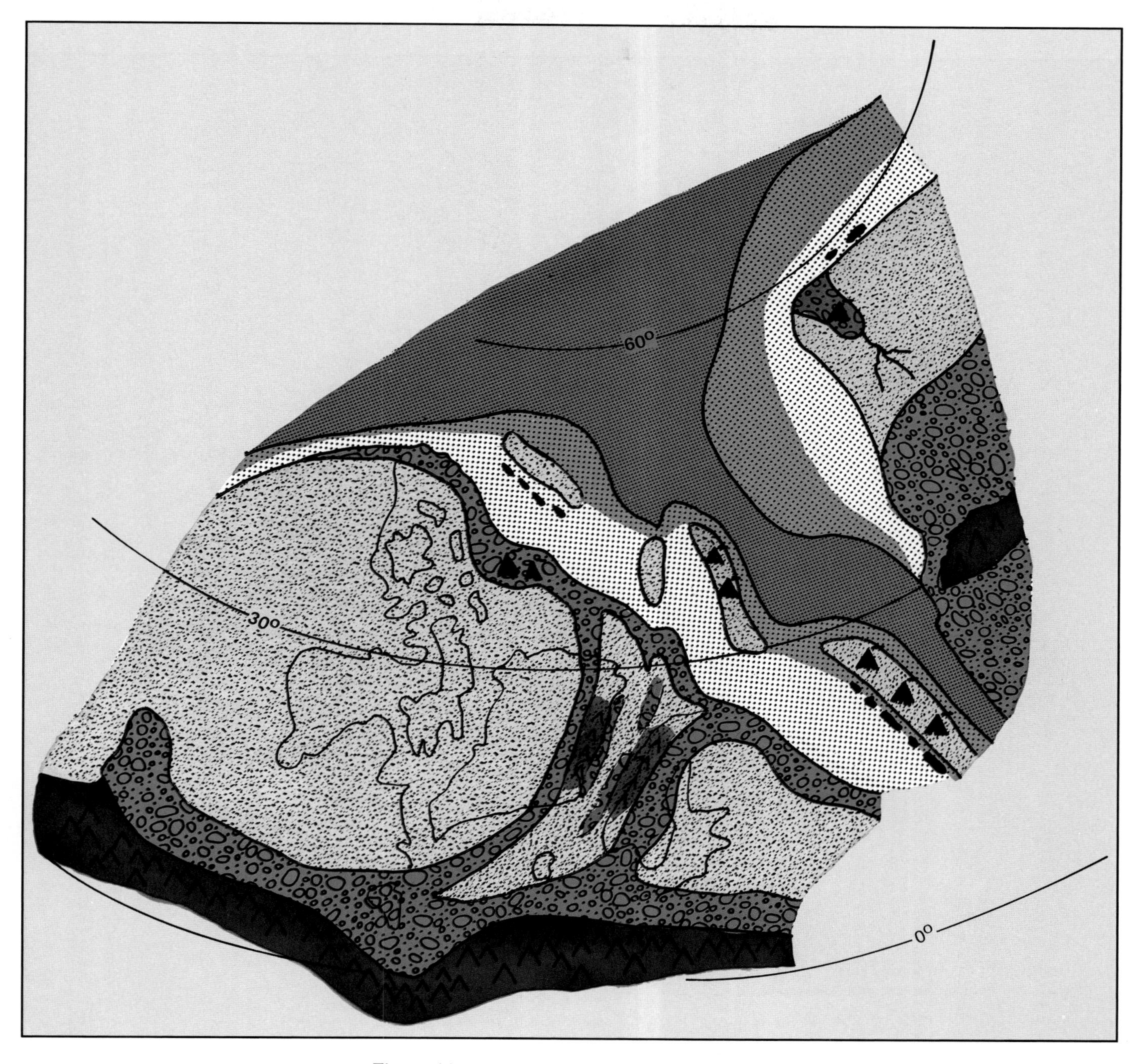


Figure 11. *Late Carboniferous paleogeography.*

of 25 km (15 mi) and pierce more than 10 km (6 mi) of sediments. This part of Norway was transgressed by a shallow sea that withdrew during Early Triassic, giving way to non-marine deposition in the Middle Triassic (Gloppen and Westre, 1982). Organic-rich Triassic source rocks have been found on Svalbard and in wells in the Barents Sea (Bjorøy and Hall, 1983).

The Sverdrup basin and the North Slope–Chukotka–New Siberian Islands block underwent rifting, differential uplift, and subsidence during Permian–Triassic time. As a consequence of these movements, the carbonate deposits that accumulated in these areas during the Carboniferous were replaced by thick, clastic sequences such as the Sadlerochit of the North Slope, the Trold and Bjorne formations of the Canadian Arctic Islands, and similar units in Chukotka.

This fundamental change from carbonate to clastic deposition occurred at the end of the Permian in the Sverdrup basin. For the remainder of its history, the basin was filled with clastic sediments (Meneley, Henao, and Merritt, 1975). After the interval of erosion in the latest Permian, subsidence rates in the Sverdrup basin increased sharply to values as high as 11 cm/1,000 yrs (4 in) in the earliest Triassic (Sweeney,1977). Large, sand-rich deltas prograded into the rapidly subsiding basin from the southeast and the southwest. These deltas deposited

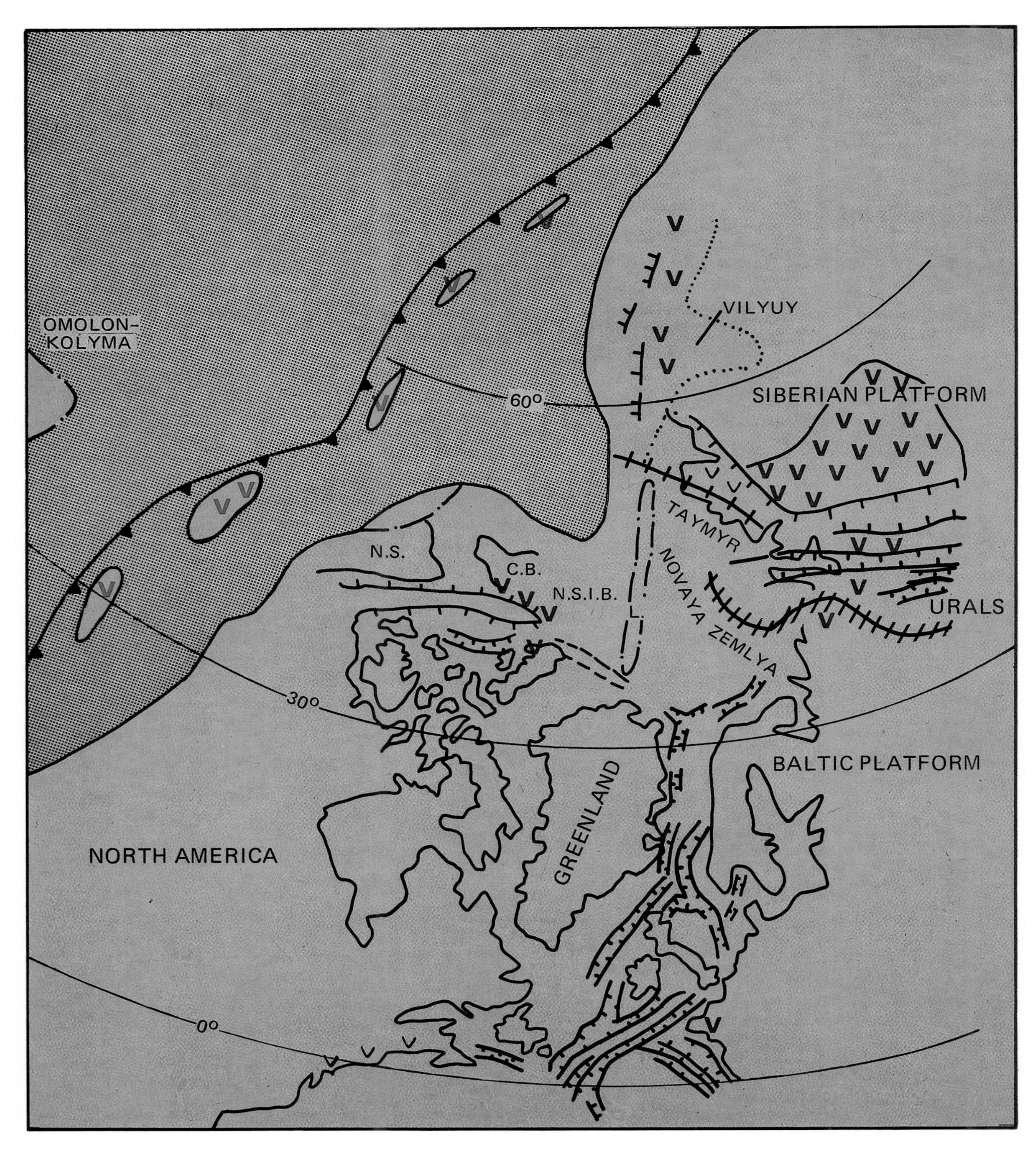


Figure 12. *Permian–Triassic reconstruction.*

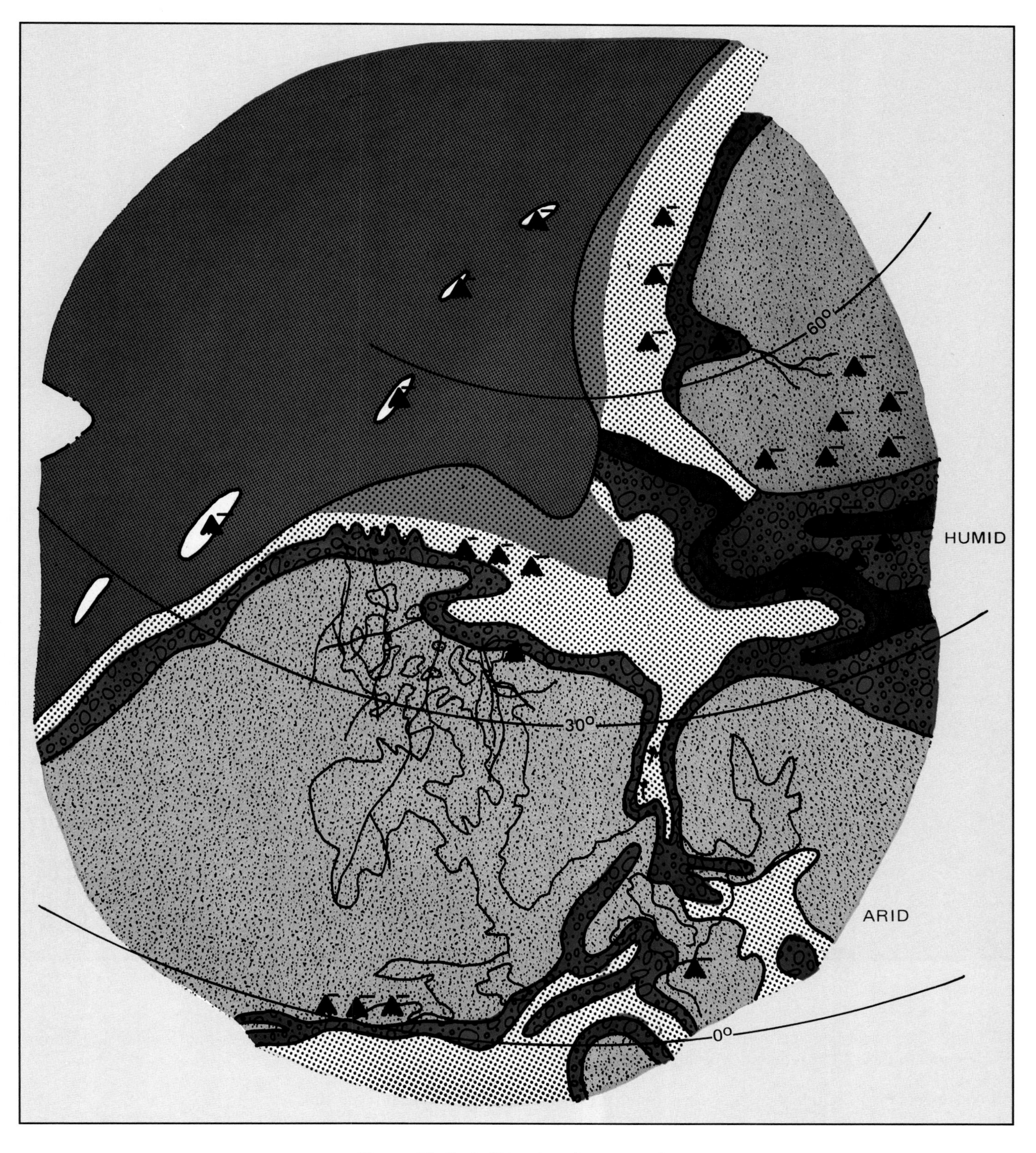


Figure 13. *Early Triassic paleogeography.*

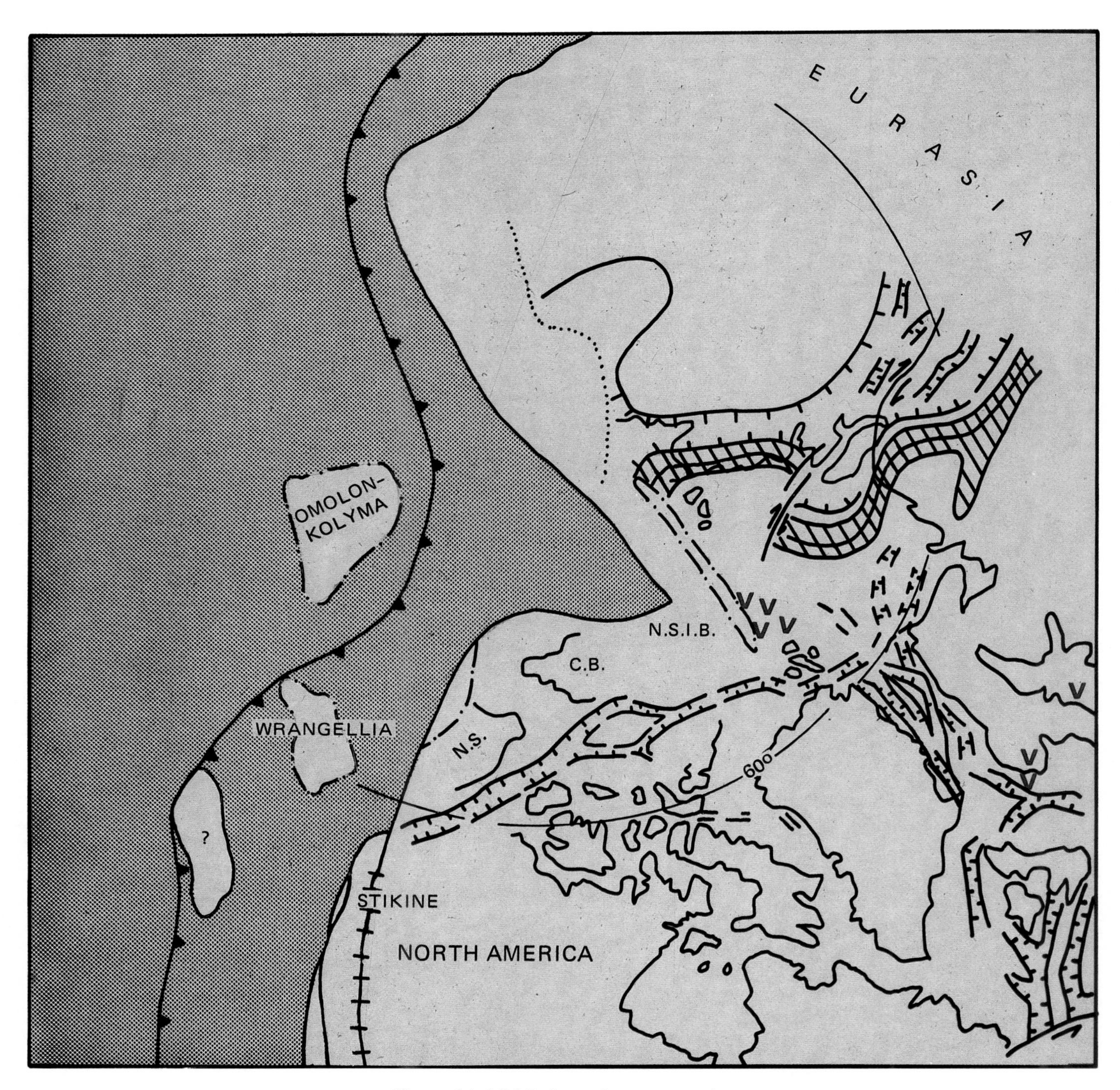


Figure 14. *Middle Jurassic reconstruction.*

4–5 km (2–3 mi) of clastic rocks during the Triassic, a thickness nearly equal to that of the remainder of the Mesozoic section (Balkwill, 1978).

A convergent plate boundary and a series of island arcs separated the Pacific Ocean from the proto-Arctic Ocean and the passive margins of Siberia and North America (Savostin and Shirshov, 1984). Several of the exotic terranes now exposed in Alaska and eastern Siberia formed at this time in the Pacific. The position of Omolon–Kolyma is schematic but based in part on latitudinal constraints derived from faunal evidence, sedimentary facies, and paleomagnetism (Gorodnitsky, Zonenshayn, and Mirlin, 1978).

## Middle Jurassic (165 Ma)

From the Middle Triassic to the Middle Jurassic, the continental blocks drifted to a more polar position (Figures 14 and 15).

The Urals remained a sediment source for the foreland trough on the Baltic Platform, but the sediment source now extended along strike to include Novaya Zemlya. The Urals also shed sediment into the West Siberian basin, which continued to

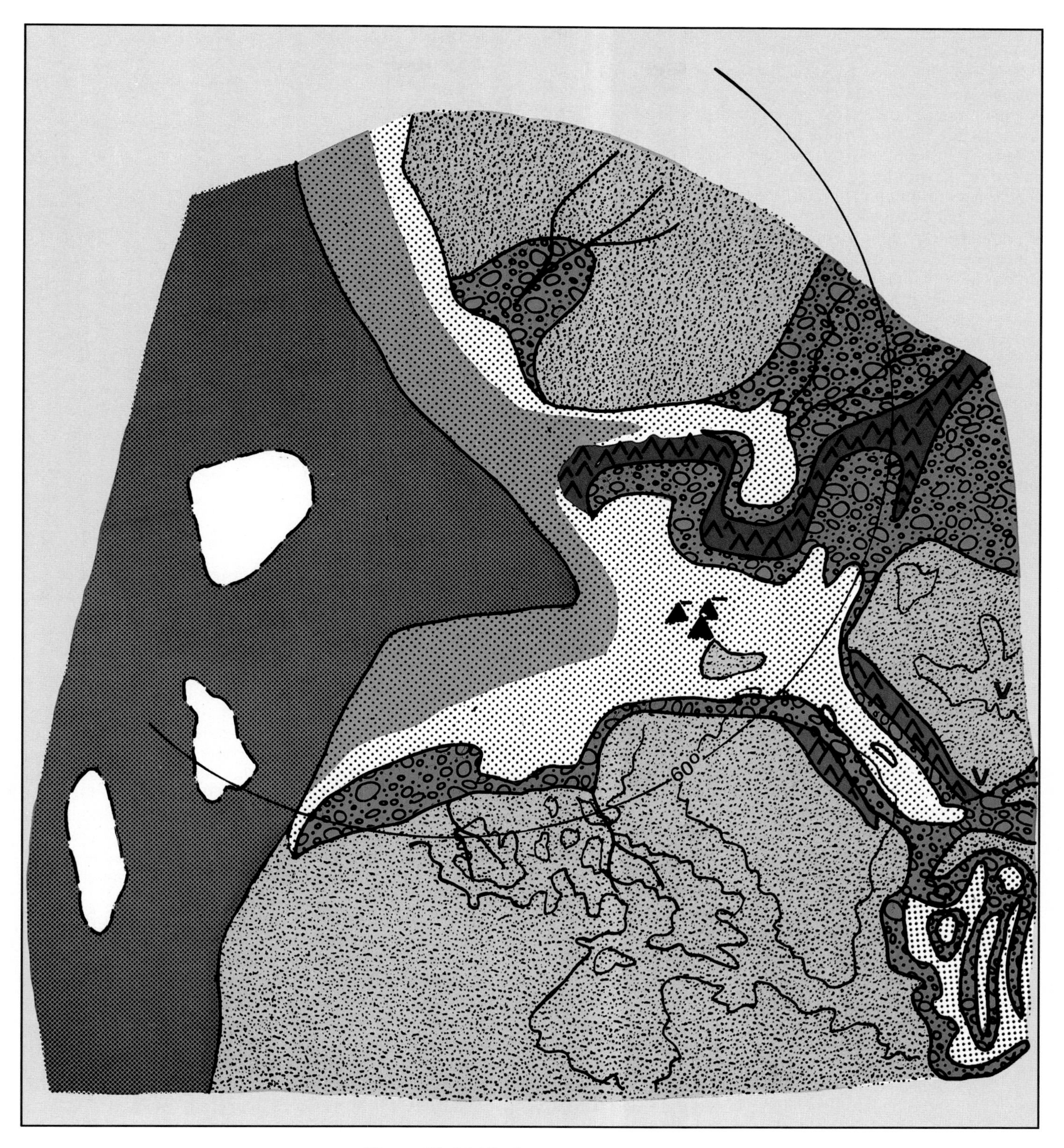


Figure 15. *Middle Jurassic paleogeography.*

subside. These sediments are primarily non-marine, lacustrine, coal-bearing, clastic rocks of deltaic facies that prograded onto a complex horst-and-graben topography.

During Early and Middle Jurassic, the rifting and subsidence in northern Europe combined with a gradual, eustatic rise of sea level resulting in joining of the Tethys and Arctic–Pacific seas (Ziegler, 1981). Along the west-central coast of Norway, these same events caused a regional transgression and deposition of estuarine, clastic rocks that interfinger with shallow marine shales (Hollander, 1982). A small outcrop along the northwestern Norwegian coast (Andøya Island) consists of shallow marine sands and shales of this age (Dalland, 1975). Blocks of Middle Jurassic, near-shore to non-marine sediment have been found locally, north of Trondheim, Norway. These contain coal, siderite, ironstone, and calcareous sandstone and indicate a nearshore environment (Oftedahl, 1971). Offshore, the Helgeland basin subsided and filled during the Late Triassic and Early Jurassic, but sedimentation slowed by the Middle Jurassic (Price and Rattey, 1984). Lower to Middle Jurassic sands, deposited in coastal settings such as bays and tidal flats, serve as reservoirs in the Trømso area.

The Middle Jurassic deformation in the Barents Sea was mainly extensional. Large displacement and normal faults that subparallel the Caledonian and Svalbardian grain formed at this time. Rift-related doming occurred in the Hammerfest basin at the end of the Middle Jurassic (Gloppen and Westre, 1982) as did the Hammerfest basin fault system (Gabrielsen, 1984).

In the North Sea, a major episode of rifting (mid-Kimmerian phase) reactivated Permian–Triassic grabens. A large volcanic center developed at the triple junction of the Central, Viking, and Moray Firth grabens (Ziegler, 1981). A broad region flanking the graben triple junction uplifted, while shallow marine to non-marine sands and shales were deposited within the graben.

In Milne Land, east central Greenland, Middle Jurassic sediments with boreal affinities directly overlie the Caledonian metamorphic basement. Root-horizons in the lower part of the sedimentary section are overlain by beds bearing ammonites. This contact records the first marine transgression in the area. Giant foreset beds indicate transport of large volumes of sediment from the (present-day) west and northwest. Such depositional features require steep topographic gradients and coarse sediments (Callomon and Birkelund, 1980). The sand-rich Middle Jurassic deposits along the Scandinavian and Greenland margins were derived from high topographic relief caused by active rifting, which was occurring from east Greenland and Scotland north to Spitsbergen.

Clastic deposition in Canada expanded from the Sverdrup basin to an elongated area across a broad shelf, which extended from Ellesmere Island on the east to the Yukon on the west. In the Yukon, a northwest trending (present-day) trough received a thick section of sands and shales. This trough indicates a renewed period of rifting, which culminated in the separation of the North Slope–Chukotka block from the Canadian Islands.

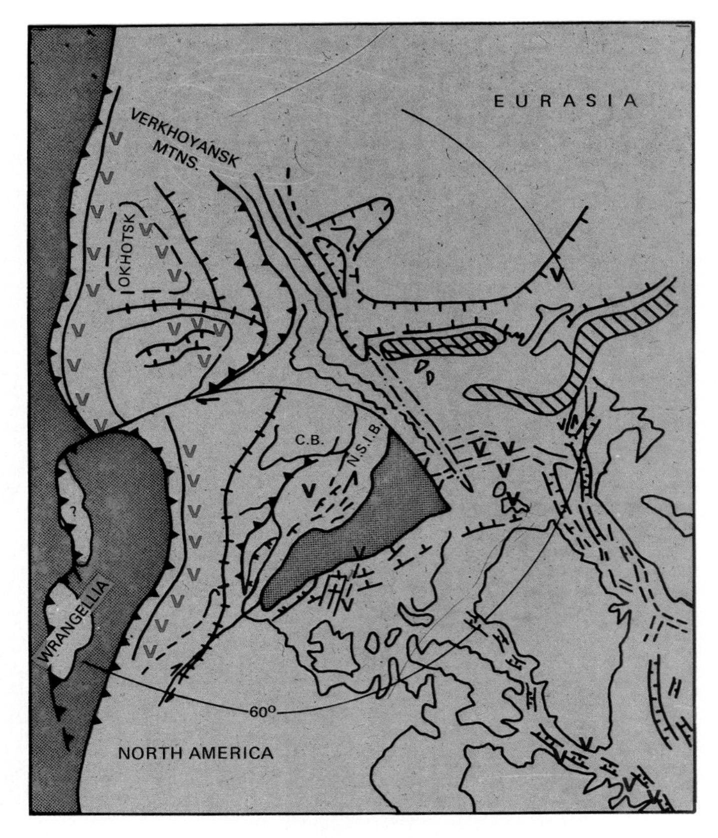


Figure 16. *Late Jurassic–Early Cretaceous reconstruction.*

### Late Jurassic–Early Cretaceous (120–150 Ma)

Major tectonic events had profound effects on basin formation and sedimentation patterns in the Arctic during this interval (Figures 16 and 17).

A series of allochthonous terranes collided with the North American and Siberian plates and isolated the Pacific Ocean from the new Arctic Ocean (Churkin and Trexler, 1980, 1981). The Stikine terrane may have collided with western North America by Middle Jurassic. In post-Early Cretaceous time, Wrangellia collided in the area of the Yukon (Coney, Jones, and Monger, 1980).

By Late Jurassic, the Okhotsk block had collided with the Siberian Platform (Fujita, 1978). The docking of Okhotsk against Siberia was followed to the north by that of the Omolon–Kolyma block collision that formed the Cherskiy suture. A belt of large, granitic intrusions, dated as 120–140 Ma (Putintsev and Zagruzina, 1980), parallels the Cherskiy suture and borders both Kolyma and Okhotsk. The collision of these allochthonous terranes formed the Verkhoyansk Mountains, an orogenic belt that folded and thrust the massive sedimentary wedge that had accumulated along the passive Verkhoyansk margin of the Siberian Platform since the Devonian. The Verkhoyansk fold and thrust belt is arc-shaped and widest where the embayment of the Vilyuy Rift extends into the continent.

While collision events occurred on the Pacific continental margin of Eurasia, the North Slope–Chukotka block and the New Siberian Islands block, which had rifted away from the Arctic Islands of Canada, rotated counter-clockwise to open the Canada Deep (Carey, 1958; Tailleur, 1973; Sweeney, Irving, and Geuer,1978; Sweeney, 1981). During the opening of the new Arctic Ocean, the major transcontinental, transform fault systems, such as the Tintina fault and Kaltag fault, were active (Poulton, 1982; Jones, 1982). Thus, the opening of the Canada Deep may have been a combination of strike-slip and rotational translation of the North Slope–Chukotka block. This strike-slip, related rotation comes by comparing paleomagnetic data from Devonian and Carboniferous rocks on the North Slope with data from rocks of the same age in nuclear North America (Newman, Mull, and Watkins, 1979). The North Slope–Chukotka block collided with the Kolyma block in Siberia and other allochthonous terranes as it drifted westward. During Early Cretaceous, the Brooks Range–Anyuy suture formed and the

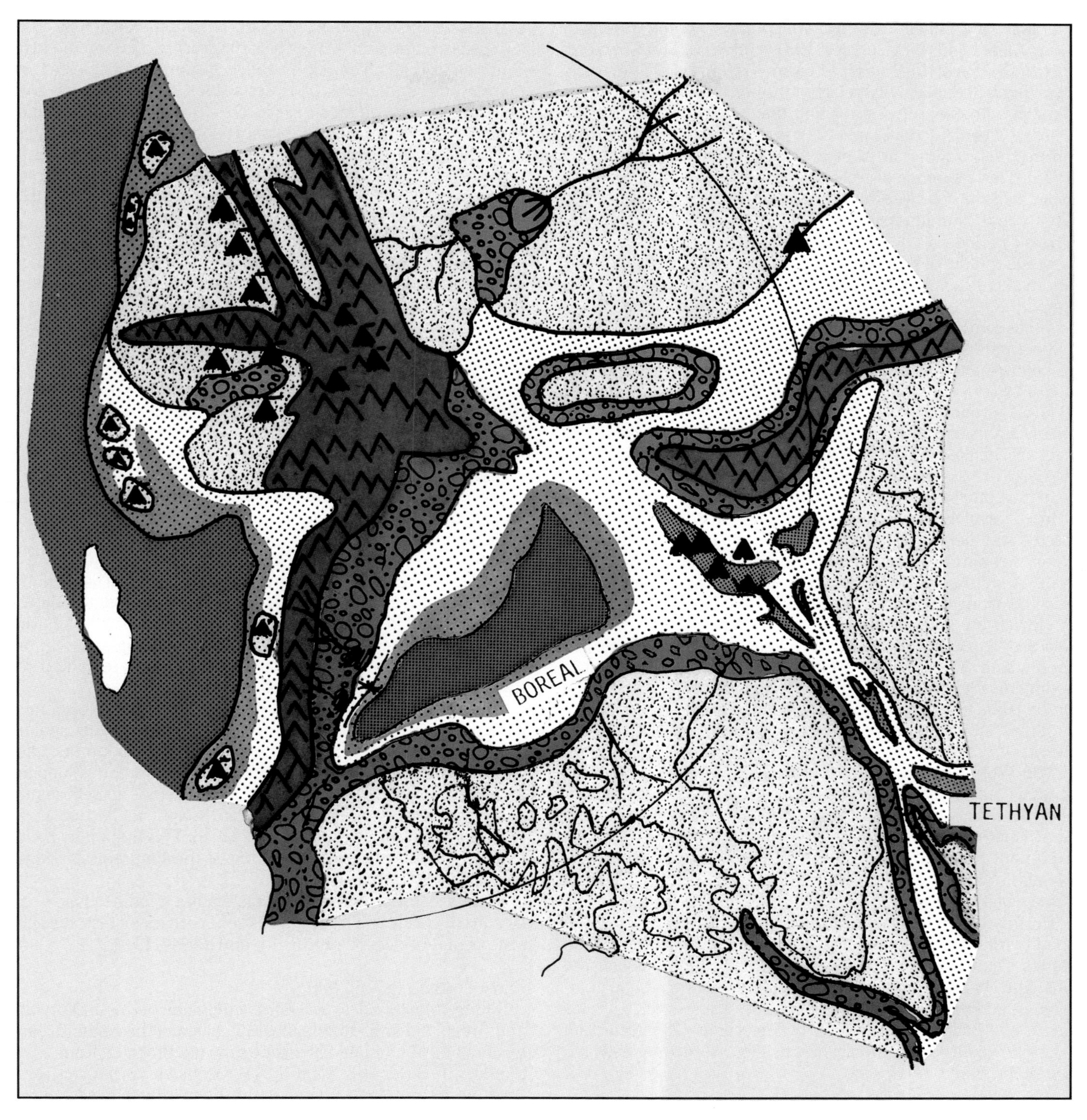


Figure 17. *Early Cretaceous paleogeography.*

Brooks Range thrusting began. The sediments derived from the uplift prograded toward the new ocean, opposite to the direction of previous sediment transport in northern Alaska.

While widespread, convergent, volcanic-plutonic activity occurred along the Pacific margin of Siberia and southern Alaska, rift-related alkaline dikes and flows were active around the periphery of the Canada Deep. Balkwill and Fox (1982) suggest that the faulting and massive, northeast-striking, gabbro dikes, which penetrate the western Sverdrup basin, are remnants of a Jurassic–Cretaceous, incipient rift zone. The isotopic ages range from 152 Ma (Oxfordian) to 123 Ma (Hauterivian). Intrusions in east central Spitsbergen (Parker, 1966, 1967) relate to active faulting at the Jurassic–Cretaceous boundary (Harland et al, 1974). Lower Cretaceous basalt flows and dolerite dikes, of a tholeiitic nature, are common on the islands of Kong Karls Land (Birkenmajer, 1981) and Franz Josef Land on the north edge of the Barents Platform. Harland (1973) suggests that this igneous activity relates to a hot dike injection zone, parallel with the initial spreading that originated the Canada Deep. Volcanic activity was also present at this time in the northern New Siberian Islands (Vol'nov et al, 1970; Vol'nov, 1975). This sequence of tectonic events repeats the pattern of plate collision followed by rifting and the formation of a new ocean that characterizes the Devonian and the Permian–Triassic periods.

In the middle of the southern Canada Deep, there are linear anomalies resembling spreading lineations in both shape and amplitudes (Vogt and Kovacs, 1984). A prominent magnetic low that coincides with a gravity high may be the location of an extinct, spreading center. The anomalies have been interpreted as belonging to the "M" reversal sequence, indicating a crustal age of 110–155 million years (Vogt and Kovacs, 1984). Taylor et al (1981) propose that sea-floor spreading occurred in the Canada Deep from 153 Ma to 127 Ma (Late Jurassic–Early Cretaceous). The basement depth and heat flow are consistent with an Early Cretaceous age (Baggeroer and Falconer, 1982).

In the Canada Deep, approximately 200 km (120 mi) from the north coast of Alaska, a thick section (2–3 km or 1–2 mi) of compacted sediment (2.7–3.3 km/sec or 8,800–10,800 ft/sec) lies upon oceanic crust (Mair and Lyons, 1981). This sediment may be Upper Jurassic and thick, Lower Cretaceous, distal shale, derived from the south, as the Canada Deep opened. On Andøya Island, off northern Norway, measured P-wave velocities are about 2.5 km/sec (8,200 ft/sec) in Cretaceous sediments and 3.1 km/sec (10,200 ft/sec) in Upper Jurassic sediments (Sellevol, 1975). The growth of the Mackenzie delta in the embayment, formed by the opening of the Canada Deep, provides an additional constraint on the timing of tectonic events.

The West Siberian basin subsided slowly during the Late Jurassic and Early Cretaceous, following the intense Triassic–Early Jurassic rifting. Thick, bituminous clays were deposited in an epicontinental sea under warm, humid climatic conditions. Luxuriant organic life flourished in quiet, restricted waters (Krylov and Korzh, 1984). Subsidence rates increased in the Early Cretaceous.

During the latest Jurassic and earliest Cretaceous, late Kimmerian rifting reactivated the graben systems of the Grand Banks, Celtic Sea, Western Approaches, North Sea, Rockall and Shetland Troughs, east Greenland, and the Norwegian margins. In addition, new rifts formed in the Labrador Sea, Davis Strait, and Baffin Bay during Early Cretaceous. This tectonism was followed by Early Cretaceous sea-floor spreading between the Grand Banks and Iberia.

### Middle Cretaceous (90–115 Ma)

By middle Cretaceous time, the North Slope–Chukotka block had rotated and moved along transform faults, nearly 70 degrees, to its present position relative to North America. Also, the New Siberian Islands block had collided with the Siberian block (Figures 18 and 19).

We postulate that a thinned and intruded continental fragment (Alpha Ridge) rifted away from the Barents Platform as the Fletcher basin opened. Taylor et al (1981) tentatively identify sea-floor spreading magnetic anomalies 34 through 21 (80–53 Ma) in the Fletcher (Makarov) basin. The Alpha Ridge may contain pieces of the proto-Arctic oceanic crust in addition to older continental crust. The Alpha Ridge began to subside into a marine environment as Late Cretaceous silicoflagellates and radiolarians accumulated near its crest (Ling, McPherson, and Clark, 1973; Jackson and Johnson, 1984). The spreading center in the Fletcher basin was connected by a transform fault through the Nares Strait to the rift basin in Baffin Bay, and from there to the Labrador Sea where the sea floor is as old as 70–80 m.y. (Srivastava, 1978).

Precise dating of the collision along the southern Anyuy suture and its westward extension, south of the New Siberian Islands block, is difficult. In the eastern part of the suture, collision probably occurred in Barremian to Aptian time, although intrusive, acidic, and intermediate volcanic rocks are slightly younger (110 m.y.) (Putintsev and Zagruzina, 1980). Convergence, following the Cherskiy collision between Omolon–Kolyma and Eurasia, continued through much of the Cretaceous. Total shortening, during the Cretaceous in the Verkhoyansk fold and thrust belt, may have approached 600 km (370 mi). As the collision of the New Siberian Islands block with Siberia continued, folding followed the right-angled bend in the former passive margin in the area of the Lena River delta. Substantial deposition persisted in the Vilyuy, Khatanga, and West Siberian basins.

At the same time, along the Pacific sides of both the Omolon–Kolyma and Okhotsk blocks, there was a convergent, accretionary margin developing. Calc-alkaline intrusives that are dated at roughly 100 m.y. are common in this region. In Alaska, the right-lateral, Porcupine-Kaltag fault system became active, or reactivated, in the middle of the Cretaceous and remained so through the Late Cretaceous and early Tertiary.

Offshore of northern Norway, a rapid basement subsidence with thick sections of Lower and middle Cretaceous shales accumulated. During the Late Cretaceous, deposition became discontinuous, and hiatuses of this age are common (Gloppen and Westre, 1982). The absence of Upper Cretaceous sediments along the northern Barents shelf is thought to be due to a regional uplift of the area (Harland, 1973). This uplift may have been the initial phase of the rifting of the Lomonosov Ridge away from the Barents Platform.

The Møre/Inner Voring basin system is a middle to Late Cretaceous depocenter. Fault patterns confirm a major tectonic episode of mid-Cretaceous age (Price and Rattey, 1984).

### Early Tertiary (55–65 Ma)

Rifting progressed to sea-floor spreading between Norway and Greenland and into the Eurasian basin of the Arctic Ocean by about 58 Ma, or late Paleocene time (La Brecque, Kent, and Cande, 1977) (Figures 20 and 21). Extensive volcanism from Scotland to east central Greenland (Evans, Fitch, and Miller, 1973; Brooks and Dawson, 1980) and the Norwegian shelf (Bugge, Prestvik, and Rokoengenk, 1978) accompanied the start of spreading.

The Upper Cretaceous to lower Tertiary Kap Washington volcanics erupted at the northern tip of Greenland (Batten et al, 1981). The volcanic rocks were preceded by dolerite dikes, which cut Upper Cretaceous shales but not the younger volcanic flows. These alkaline volcanics are the result of crustal thinning and doming prior to the rifting of the Eurasian basin (Larsen, 1982).

The sea-floor spreading along the Nansen–Gakkel Ridge, in the Arctic Ocean, separated the Lomonosov Ridge from the Eurasian continental margin. At the same time, spreading in the Fletcher (Makarov) basin, which began in the Late Cretaceous, (anomaly 34, 80 Ma), continued and did not end until early Eocene (anomaly 21, 53 Ma) (Taylor et al, 1981).

As the Labrador Sea spread during the early Tertiary, the southern end of Greenland rotated northeastward, causing compression in northwestern Greenland and Ellesmere Island. Balkwill and Bustin (1980) called this localized transpressional event the Eurekan orogeny. From the latest Cretaceous, through the Eocene, to the earliest Miocene, the compression of the Eurekan orogeny was accompanied by rifting in the Arctic Islands (Kerr, 1982; Miall, 1984; Riediger, Bustin, and Rouse, 1984). Thus, a tectonic linkage was active from Baffin Bay, through the Arctic Islands, to the Kaltag–Porcupine fault system, into the Bering Sea and the Norton basin.

At its eastern end, the Nansen–Gakkel Ridge intersects Siberia, and it appears to continue into thinned, continental crust. Seismicity extends under the Lena delta, and the position of the delta itself indicates that rifting and subsidence occurred at least that far into the Siberian continent.

Much of the land bordering the Arctic Ocean was uplifted at the beginning of the Cenozoic, perhaps in response to the opening of the new Eurasian Ocean. As a result, deposition of sediments outside the ocean basin was localized in the West Siberian, Vilyuy, Anadyr, Hope–Chukotka, and Norton basins, and in Baffin Bay. For example, erosion was widespread during the early Tertiary in the southern Barents Sea; no lower Paleocene deposits exist there (Gloppen and Westre, 1982). In contrast, Paleocene deposition continued in the West Siberian basin and consisted mainly of diatomaceous and radiolarian shales.

Late Cretaceous and Paleogene deep-sea sediment cores from the central Arctic Ocean indicate that open-ocean, polar upwelling had developed in this land-locked sea now separated from the Pacific Ocean (Kitchell and Clark, 1982).

### Late Eocene (40–50 Ma)

In the late Eocene, sea-floor spreading in the Labrador Sea ceased, but a wide, left-lateral ductile, megashear zone continued to be active on both sides of the Nares Strait (Hugon, 1983) (Figures 22 and 23). Transpression also continued between the northeast corner of Greenland and Svalbard, where the North American plate sheared past the southwestern margin of the Barents Sea. Convergence along this plate boundary caused folding and uplift.

On the other side of the Arctic, the Aleutian Island arc formed. A portion of the oceanic Kula plate was trapped when subduction shifted from the present-day Bering Sea continental margin to the Aleutian Trench (Ben-Avraham and Cooper, 1981). Relaxation of compression and increased divergent, strike-slip faulting along the old Pacific convergent margin developed extensional, successor basins like Anadyr, Khatyrka, St. George, and Navarin.

The wide shelves of the Arctic were exposed at various times during the middle and upper Tertiary. Sedimentation around the margins of the Arctic Ocean was not extensive except in localized deltas and fans. For example, as sea level fell during mid-Tertiary, sediment flooded into the northern end of Baffin Bay. Also, across the continental margin south of Bear Island, thick middle to upper Tertiary sediments accumulated upon new oceanic crust.

## PETROLEUM OCCURRENCE IN ARCTIC BASINS

The sedimentary basins of the Arctic contain large reserves of petroleum. Giant oil and gas fields have been found in the West Siberian basin, the Pechora basin, the Vilyuy basin, the North Slope of Alaska, the Beaufort Sea and Mackenzie delta, and the Sverdrup basin (Figure 3). The unexplored sedimentary basins bordering the Arctic Ocean are among the most promising candidates for large, future oil and gas discoveries (Nehring, 1978).

In the Arctic, as in any geologically complex region, certain elements of the geological history were conducive to hydrocarbon generation, migration, and entrapment, and others were not. Based on knowledge from previous exploration and considering the proposed geologic evolution of the Arctic described above, we can make the following general observations about the petroleum geology of the sedimentary basins of the Arctic.

### Tectonostratigraphic Setting

Most of the productive, sedimentary basins in the Arctic formed after the Middle Devonian. Many lie upon continental crust that has thinned by rifting and igneous intrusion. The collision of continental blocks in the Arctic caused multiple periods of deformation within many basins. Such a structurally complex history can be beneficial to hydrocarbon accumulation. For example, in the Prudhoe Bay field of Alaska, the Permian–Triassic clastics derived from the north (present-day) in an environment conducive to the deposition of excellent reservoir sediments but poor for source rock accumulation. In the latest Jurassic and Early Cretaceous, the suturing from the south uplifted the Brooks Range and reversed the polarity of sediment transport from north to south. Thus, the rich, distal, Lower Cretaceous marine shales were deposited in a restricted sea and juxtaposed with the much older, underlying reservoir sands (Bushnell, 1980; Bird, 1983).

Many of the large interior rift basins of the Arctic developed during the Jurassic and Cretaceous, which were periods of time when excellent source and reservoir rocks were deposited worldwide (Grunau, 1983) (Figure 24).

### Reservoir Sediments

Productive and potential reservoirs in the Arctic range in age from Precambrian to Tertiary. Thus far, sandstone reservoirs predominate, and some have excellent characteristics of thickness, uniformity, high porosity, and permeability. In the Pechora basin, Devonian sands, which contain more than 70% of the hydrocarbon reserves, were deposited during the early rift phase of basin formation. Additional reservoirs in this basin range from Ordovician to Triassic in age.

Almost 95% of the reserves in the West Siberian basin are in Cretaceous sands. All the giant and large oil fields in the southern half of the basin have reservoirs of Valanginian to Hauterivian age. All the supergiant and giant gas fields in the northern part of the basin produce from Cenomanian sandstones. Good reservoir sands are also found in Middle and Upper Jurassic rocks. A Soviet study placed the Kara Sea gas resources at 388 tcf (Oil and Gas Journal, 1982).

In the Vilyuy basin, the main reservoirs are Upper Permian

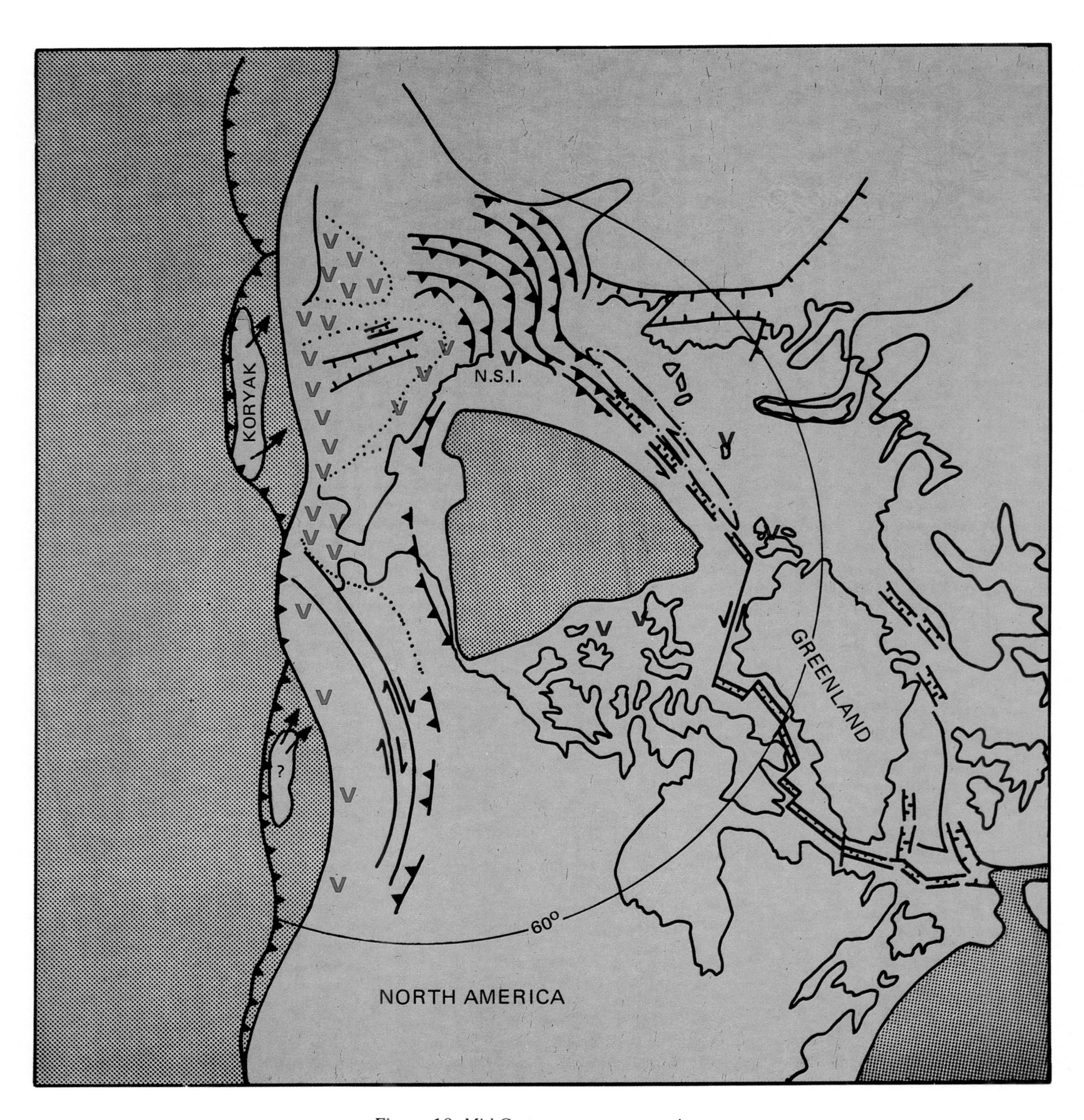


Figure 18. *Mid-Cretaceous reconstruction.*

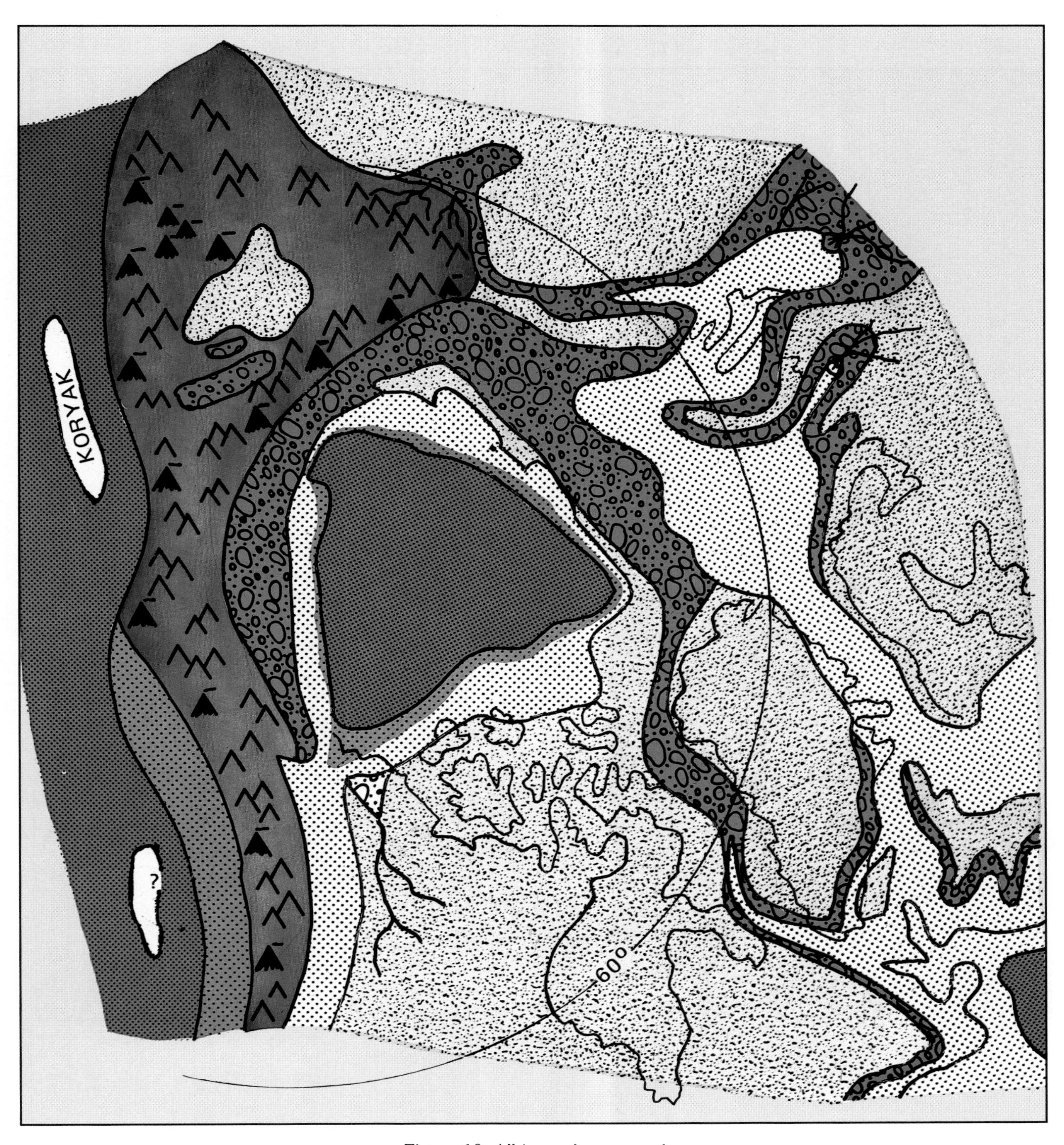


Figure 19. *Albian paleogeography.*

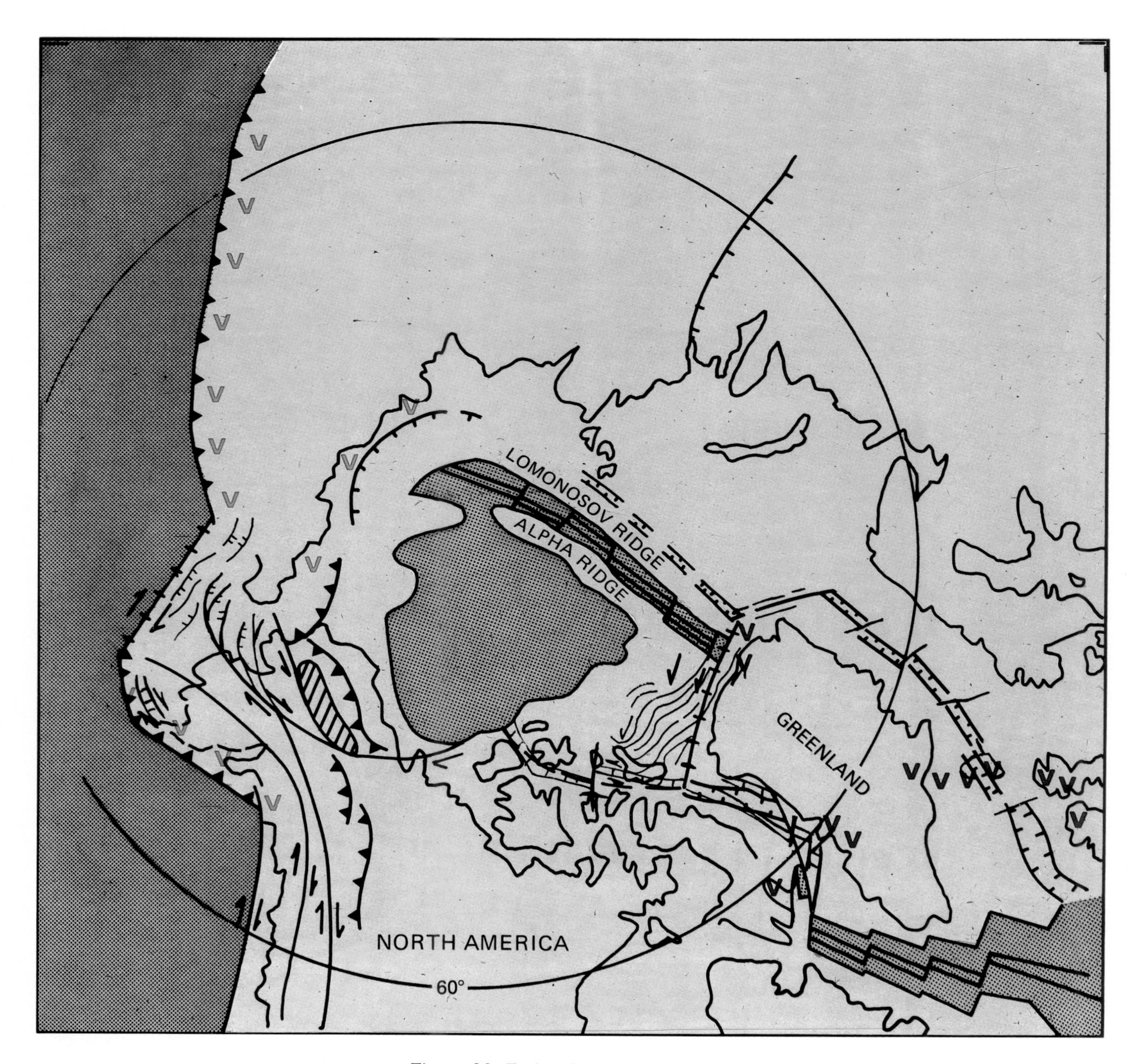


Figure 20. *Earliest Tertiary reconstruction.*

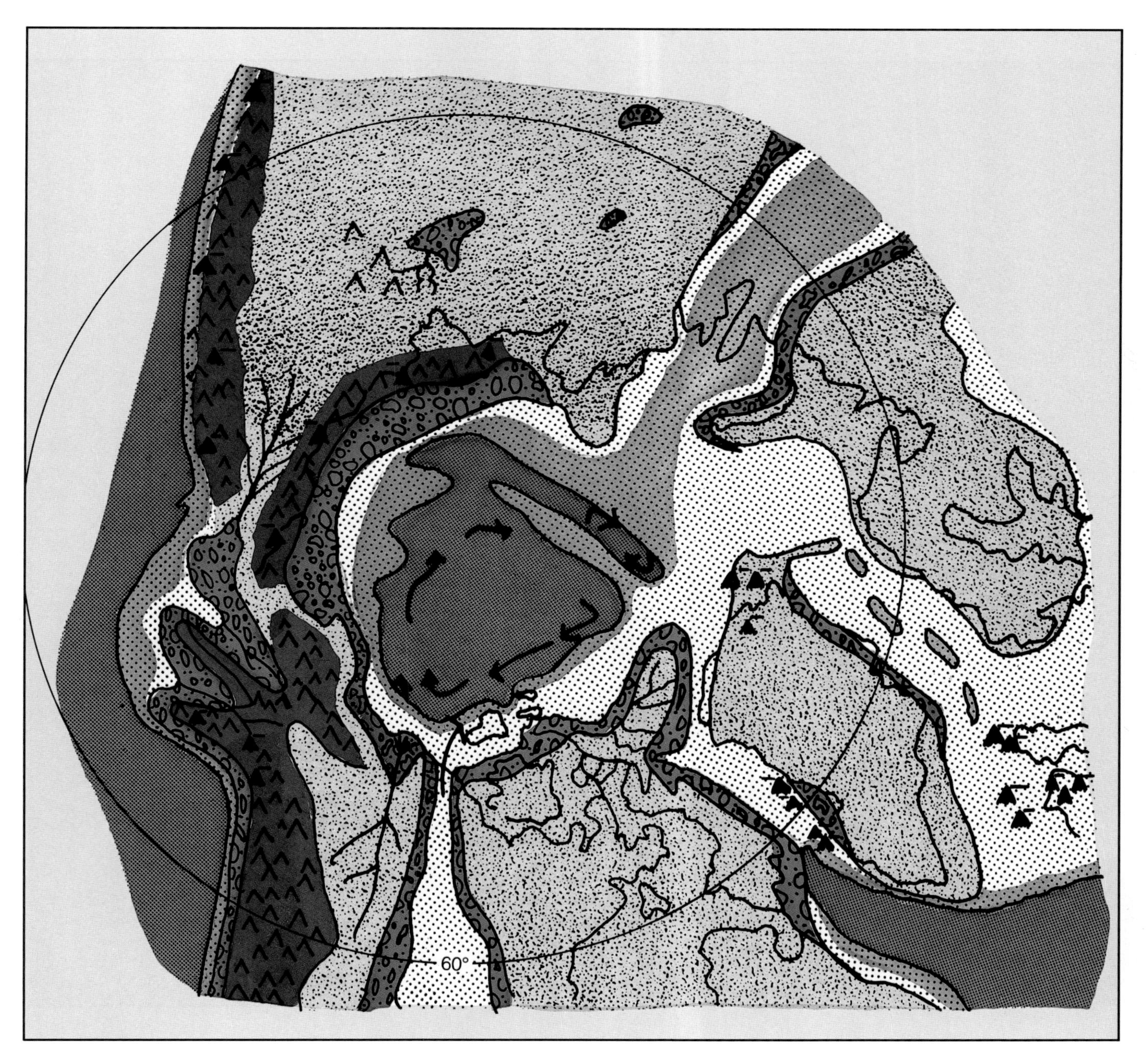


Figure 21. *Earliest Tertiary paleogeography.*

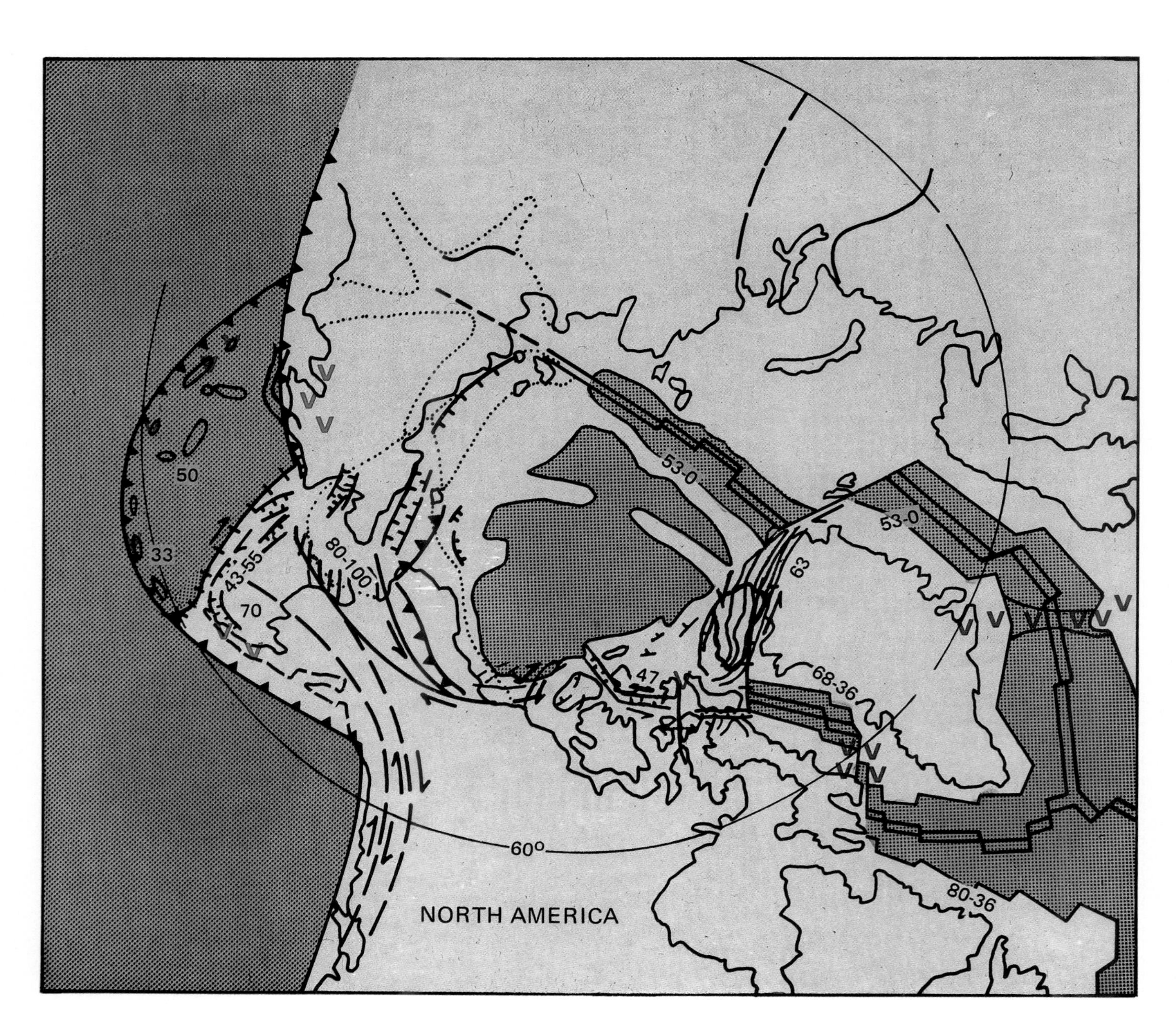


Figure 22. *Eocene reconstruction.*

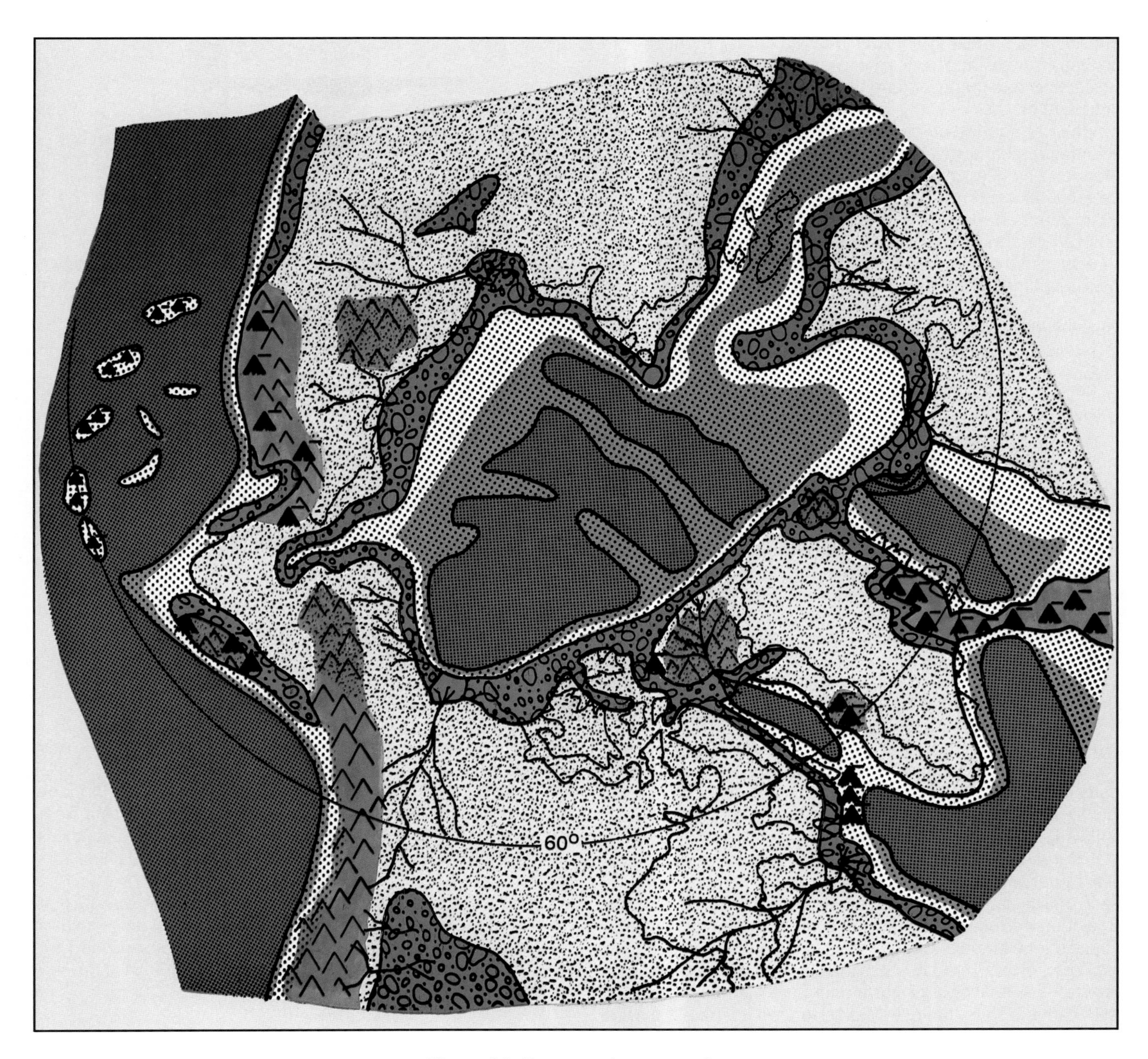


Figure 23. *Eocene paleogeography.*

and Lower Triassic, braided stream and deltaic sandstones. Similar facies of Permian–Triassic age serve as the major reservoirs on the North Slope of Alaska (Sadlerochit Group). Additional reservoirs on the North Slope occur in Carboniferous carbonates (Lisburne Group) and in Jurassic and Cretaceous sands. Upper Triassic and Lower Jurassic sands produce gas in the Sverdrup basin. These sands were deposited during rapid subsidence and penecontemporaneous movement of basement fault blocks.

In the Hammerfest basin off northern Norway, Lower to Middle Jurassic shallow-water sands are hydrocarbon reservoirs (Gloppen and Westre, 1982). Sands of this age are important reservoirs in the North Sea.

The youngest reservoirs in the Arctic that contain commercial hydrocarbon reserves are the Tertiary sands in the Mackenzie delta. Tertiary reservoirs are present in offshore Norway, east Greenland, Baffin Bay, and the Bering Sea.

### Source Sediments

Known source rocks in the Arctic range in age from Upper Devonian to Tertiary; however, the dominant source intervals are in Upper Devonian, Permian, Triassic, Upper Jurassic, and Lower and middle Cretaceous sediments. Thus far, all known source rocks are shales that were deposited in restricted marine environments. Such environments existed frequently in the aborted rift basins and isolated seaways so common throughout the geological history of the Arctic.

In the Pechora basin, Upper Devonian, organic-rich, flammable, bituminous shales are the obvious source rocks. Excellent Devonian source rocks have been found on Severnaya Zemlya and thus may be present in the northeastern Kara Sea. In the West Siberian basin, vast reserves originated in Upper Jurassic to Lower Cretaceous bituminous shales. Total organic matter in these rocks reaches 20% or more over large areas (Kontorovich et al, 1975). Lower Cretaceous and uppermost Jurassic bituminous shales are probably the source rocks for the North Slope of Alaska (Magoon and Claypool, 1983).

In the Vilyuy basin, source rocks are, presumably, the coals that are abundant in Permian, Triassic, Jurassic, and Cretaceous deposits. Triassic shales in the Sverdrup basin are the likely source rock for the discoveries in Triassic sandstones (Balkwill, 1978).

Off the west central coast of Norway, source rocks are thought to be Upper Jurassic black shales (Bjorøy and Virgran, 1979). In the Trømso–Hammerfest basins off northern Norway, marine circulation during the Triassic and again during the Late Jurassic became restricted, and excellent source rocks were deposited. The Lower Cretaceous marine shales in this area contain less than 2% total organic carbon and are fair source rocks at best. There are organic-rich Middle to Upper Triassic marine shales on Svalbard and one well in the Barents Sea encountered marine Triassic shales that had moderate source potential (Gloppen and Westre, 1982).

### Hydrocarbon Traps

The complex tectonic history of the Arctic has created a wealth of potential structural and stratigraphic traps. To date, the dominant trap types include rotated, basement-involved fault blocks and related drape closures; regional stratigraphic pinchouts upon basement highs; large fan deposits close to clastic sources; detached, down-to the basin faults; and related diapiric closures. Thrust belt closures are common in the Arctic but have not yet been extensively explored.

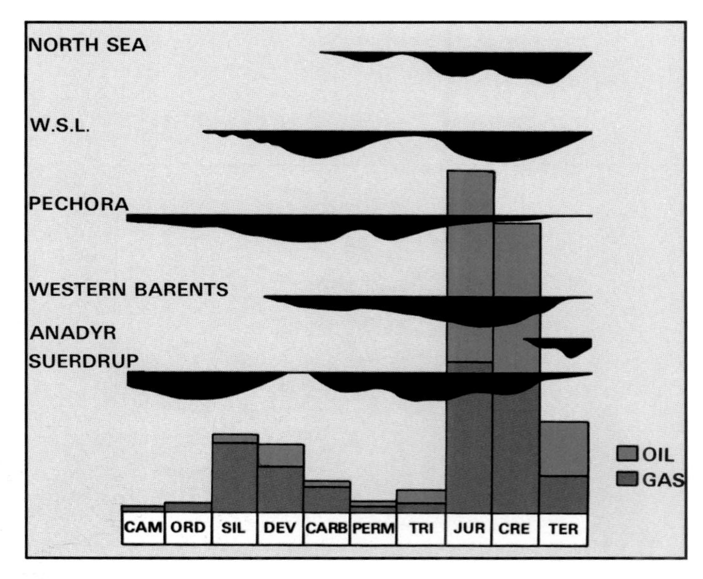


Figure 24. *Phanerozoic oil and gas occurrence—basin formation.*

### Maturation

Many rift basins and passive continental margins in the Arctic have been heated to moderate maturity. Temperature, of course, depends on local burial history, conductivity, and heat flow. In many basins of the Arctic these variables appear to have combined in a favorable manner.

## SUMMARY

Wallace Pratt published a map of the Arctic in 1944 that depicts areas in which oil has been seen and areas that are geologically favorable for prospecting (Pratt, 1944) (Figure 25). The production of oil and gas in many of the Arctic basins from sources and reservoirs of numerous geologic ages confirms Wallace Pratt's optimism for the petroleum potential of the Arctic.

Pratt (1944, p. 112) concluded his paper by reminding us that the Great Ice Age is behind us and urged exploration geologists and geophysicists to "follow up its retreat and to reclaim for mankind the vast empire released to us by the return of the sun to the friendly Arctic."

## REFERENCES

Baggeroer, A. B., and R. Falconer, 1982, Array refraction profiles and crustal models of the Canada basin: Journal of Geophysical Research, v. 87, n. B7, p. 5461–5476.

Balkwill, H. R., 1978, Evolution of Sverdrup basin, Arctic Canada: AAPG Bulletin, v. 62, n. 6, p. 1104–1028.

——, and R. M. Bustin, 1980, Late Phanerozoic structures, Canadian Arctic Archipelago: Paleogeography, Palaeoclimatology, Palaeoecology, v. 30, n. 3–4, p. 219–227.

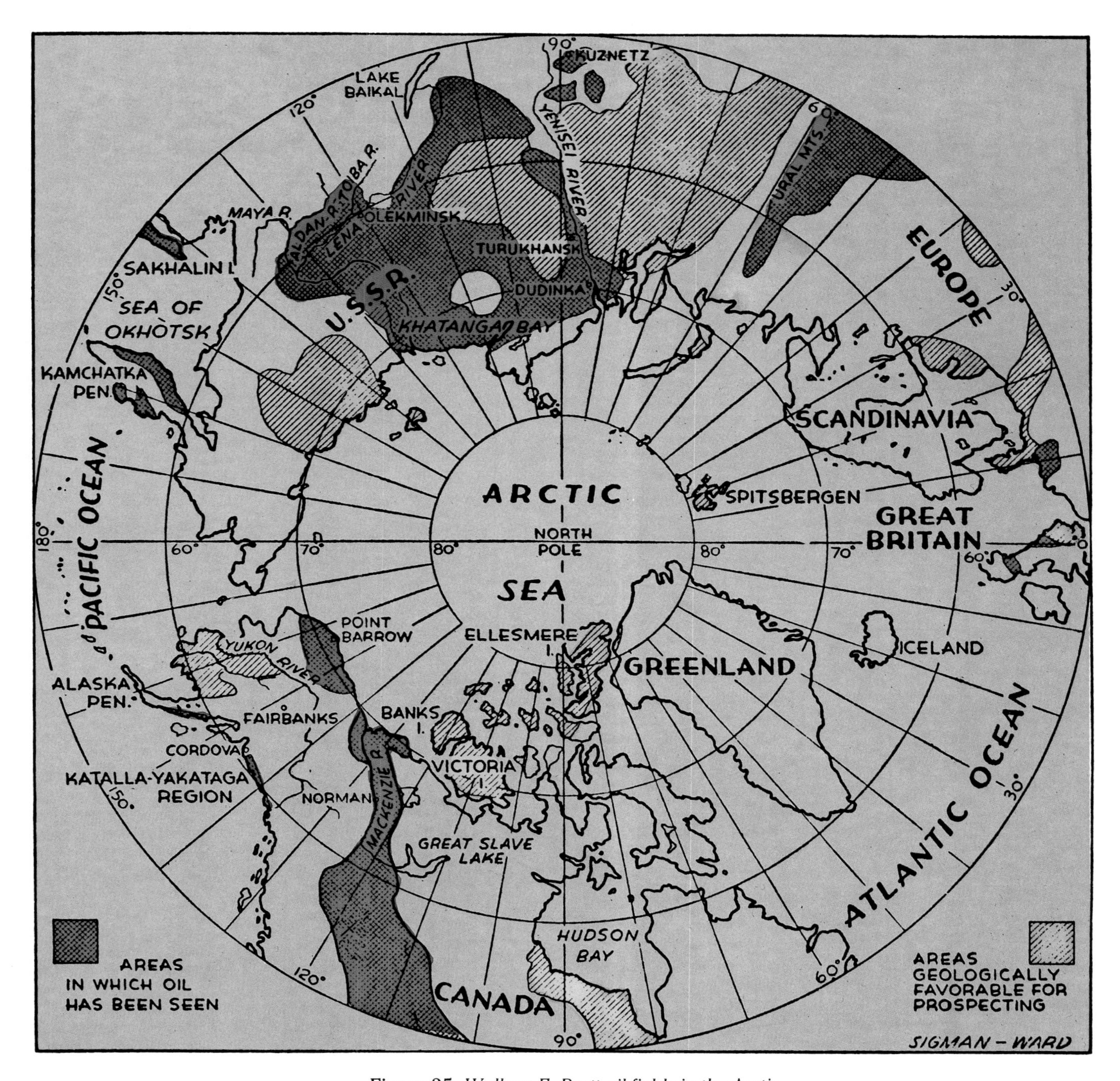


**Figure 25.** *Wallace E. Pratt oil fields in the Arctic.*

——, and N. E. Haimila, 1978, K/Ar ages and significance of mafic rocks, Sabine Peninsula, Melville Island, District of Franklin: *in* Current Research Part C, Paper—Geological Survey Canada (CAN) (CGSPAC), n. 78–1C, p. 35–38.

——, and F. G. Fox, 1982, Incipient rift zone, western Sverdrup basin, Arctic Canada, *in* A. F. Embry and H. R. Balkwill, eds., Arctic geology and geophysics: proceedings of Third International Symposium on Arctic Geology: Canadian Society of Petroleum Geologists, p. 171–187.

Batten, D. J., P. E. Brown, P. R. Davies, A. K. Higgins, B. E. Koch, I. Parsons, and N. J. Soper, 1981, Peralkaline volcanicity on the Eurasia basin margin: Nature, v. 294, p. 150–152.

Ben-Avraham, Z., and A. K. Cooper, 1981, Early evolution of the Bering Sea by collision of oceanic rises and North Pacific subduction zones: Geological Society of America Bulletin, v. 92, n. 7, p. 485–495.

Bird, K. J., 1983, Oil and gas resources of the North Slope, Alaska, *in* U. S. Geological Survey Polar Research Symposium: USGS Circular 911, p. 27–30.

Birkenmajer, K., 1981, The geology of Svalbard, the western part of the Barents Sea, and the continental margin of Scandanavia, *in* A. E. M. Nairn, M. Churkin, and F. G. Stehli, eds.,

The ocean basins and margins, Vol. 5, the Arctic Ocean: Plenum Press, p. 265–330.

Bjorøy, M., and J. Virgran, 1979, Source rock studies on outcrop samples from the Norwegian continental shelf: Trondheim, Continental Shelf Institute, 63 p.

——, and P. B. Hall, 1983, A rich Middle Triassic source rock in the Barents Sea area: Offshore Technology Conference, OTC 4623, p. 379–394.

Brooks, C. K., and M. R. Dawson, 1980, Episodic volcanism, epeirogenesis and the formation of the North Atlantic Ocean: Palaeogeography, Palaeoclimatology, Palaeoecology, v. 30, n. 3–4, p. 229–242.

Bugge, T., T. Prestvik, and K. Rokoengenk, 1978, Lower Tertiary volcanic rocks off Kristiansund-Mid Norway: Trondheim, Continental Shelf Institute, Contribution n. P-155/11/78, 19 p.

Bushnell, H. P., 1980, Unconformities: key to major oil accumulations: North Slope, Alaska: AAPG Bulletin, v. 64, n. 5, p. 684.

Callomon, J. H., and T. Birkelund, 1980, The Jurassic transgression and the Mid-Late Jurassic succession in Milne Land, central east Greenland: Geology, v. 117, n. 3, p. 211–226.

Carey, S. W., 1958, A tectonic approach to continental drift, *in* Continental drift, A symposium: Hobart, Tasmania University, p. 177–355.

Christie, R. L., 1979, The Franklinian geosyncline in the Canadian Arctic and its relationship to Svalbard: Norsk Polarinstitutt Skrifter, n. 167, p. 263–314.

Churkin, M., Jr., W. J. Nokleberg, and C. Huie, 1979, Collision-deformed Paleozoic continental margin, western Brooks Range, Alaska: Geology, v. 7, n. 8, p. 379–383.

Churkin, M., Jr., and J. H. Trexler, Jr., 1980, Circum-Arctic plate accretion: isolating part of a Pacific plate to form the nucleus of the Arctic basin: Earth and Planetary Science Letters, v. 48, n. 2, p. 356–362.

——, and ——, 1981, Continental plates and accreted ocean terranes in the Arctic, *in* A. E. M. Nairn, M. Churkin, and F. G. Stehli, eds., The ocean basins and margins, Vol. 5, the Arctic Ocean: Plenum Press, p. 1–20.

Coney, P. J., D. L. Jones, and J. W. H. Monger, 1980, Cordilleran suspect terranes: Nature, v. 288, p. 329–333.

Dalland, A., 1975, The Mesozoic rocks of Andoy northern Norway: Norges Geologiske Undersoekelse (Publikasjoner), v. 316, p. 271–287.

Demenitskaya, R. M., and K. L. Hunkins, 1970, Shape and structure of the Arctic Ocean, *in* A. E. Maxwell, ed., The sea: New York, Wiley-Interscience, v. 4, p. 223–249.

Dillon, J. T., G. H. Pessel, J. H. Chen, and N. C. Veach, 1980, Middle Paleozoic magmatism and orogensis in the Brooks Range, Alaska: Geology, v. 8, p. 338–343.

DOBEX, 1980, The petroleum potential of the Arctic (unpublished report).

Duckworth, G. L., A. B. Baggeroer, and H. R. Jackson, 1982, Crustal structure measurements near Fram II in the Pole Abyssal Plain: Tectonophysics, v. 89, n. 1–3, p. 173–215.

Eisbacher, G. H., 1983, Devonian-Mississippian sinistral transcurrent faulting along the cratonic margin of western North America: a hypothesis: Geology, v. 11, p. 7–10.

Evans, A. L., F. J. Fitch, and J. A. Miller, 1973, K-Ar age determination on some British Tertiary igneous rocks: Journal of Geological Society of London, v. 129, p. 419–443.

Faleide, J. I., S. T. Gudlaugsson, and G. Jacquart, 1984, Evolution of the western Barents Sea: Marine and Petroleum Geology, v. 1, p. 123–150

Forsyth, D. A., 1978, Review of Arctic crustal studies, *in* J. F. Sweeney, ed., Arctic geophysical review: Canadian Earth Physics Branch Publication, v. 45, n. 4, p. 75–85.

Fujita, K., 1978, Pre-Cenozoic tectonic evolution of Northeast Siberia: Journal of Geology, v. 86, n. 2, p. 159–172.

Gabrielsen, R. H., 1984, Long-lived fault zones and their influence on the tectonic development of the southwestern Barents Sea: Journal of Geological Society of London, v. 141, p. 651–662.

Gjelberg, J., and R. Steel, 1983, Middle Carboniferous marine transgression, Bjørnøya, Svalbard: facies sequences from an interplay of sea level changes and tectonics: Geological Journal, v. 18, p. 1–19.

Gloppen, T. G., and S. Westre, 1982, Petroleum potential big off northern Norway: Oil and Gas Journal, p. 114–135, June 1982.

Goodfellow, W. D., and I. R. Jonasson, 1984, Ocean stagnation and ventilation defined by $^{34}S$ secular trends in pyrite and barite, Selwyn basin, Yukon: Geology, v. 12, p. 583–586.

Gorodnitsky, A. M., L. P. Zonenshayn, and E. G. Mirlin, 1978, Rekonstruktsii polozheniya materikov v Fanerozoe (Reconstructions of continent position during the Phanerozoic): Moscow, Nauka Press, 121 p. (Russian).

Gramberg, I. S., et al, 1984, Geological structure of the Arctic continental margin of the USSR, *in* Arctic Geology, 27th International Geological Congress, Colloquium 04, Reports v. 3, Moscow, p. 3–12.

Green, A. R., 1985, Integrated sedimentary basin analysis for petroleum exploration and production: Offshore Technology Conference, v. 1, p. 9–20.

Grunau, H. R., 1983, Abundance of source rocks for oil and gas worldwide: Journal of Petroleum Geology, v. 6, n. 1, p. 39–54.

Harland, W. B., 1969, Contribution of Spitsbergen to understanding of tectonic evolution of North Atlantic region, *in* Marshall Kay, ed., North Atlantic Geology and Continental Drift: AAPG Memoir 12, p. 817–851.

——, 1973, Tectonic evolution of the Barents Shelf, *in* M. G. Pitcher, ed., Arctic Geology, Proceedings of Second International Symposium on Arctic Geology, AAPG Memoir 19, p. 599–608.

——, J. L. Cutbill, P. F. Friend, D. J. Gobbett, D. W. Holliday, P. I. Maton, J. R. Parker, and J. H. Wallis, 1974, The Billefjorden fault zone, Spitsbergen, *in* Norsk Polarinstitutt Skrifter, v. 161, 72 p.

Henriksen, N., and A. K. Higgins, 1976, East Greenland Caledonian fold belt, *in* Geology of Greenland: Copenhagen, The Geological Survey of Greenland, p. 183–246.

Heselton, L. R., Jr., 1968, The continental shelf: CNA Research Contribution No. 106, Institute of Naval Studies, Center for Naval Analyses.

Hollander, N. B., 1982, Evaluation of the hydrocarbon potential offshore mid-Norway: Oil and Gas Journal, p. 168–172. April, 1982.

Hugon, H., 1983, Ellesmere-Greenland fold belt: structural evidence for left-lateral shearing: Tectonophysics, v. 100, p. 215–225.

Jackson, M. R., and G. L. Johnson, 1984, Structure and history of the Amerasian basin, *in* Arctic Geology, 27th International Geological Congress, Colloquium 94, Reports v. 4, p. 143–151.

Johnson, G. L., P. T. Taylor, P. R. Vogt, and J. F. Sweeney, 1978, Arctic basin morphology: Polarforschung, v. 48, n. 1–2, p. 20–30.

Jones, P. B., 1982, Mesozoic rifting in the western Arctic Ocean basin and its relationship to Pacific seafloor spreading, *in* Arctic Geology and Geophysics: Proceedings of the Third International Symposium on Arctic Geology, Canadian Society of Petroleum Geologists, p. 83–99.

Kerr, J. W., 1977, Cornwallis fold belt and the mechanism of basement uplift: Canadian Journal of Earth Science, v. 14, p. 1374–1401.

——, 1982, Evolution of sedimentary basins in the Canadian Arctic: Philosophical Transactions of the Royal Society of London, A305, p. 193–205.

Kitchell, J. A., and D. L. Clark, 1982, Late Cretaceous–Paleogene paleogeography and paleocirculation: evidence of North Polar upwelling: Palaeogeography, Palaeoclimatology, Palaeoecology, v. 40, p. 135–165.

Kontorovich, A. E., I. I. Nesterov, F. K. Salmanov, V. S. Surkov, A. A. Trofimuk, and Yu. G. Ervje, 1975, Geologiya nefti i gaza Zapadnoi Sibiri (Geology of oil and gas of West Siberia): Moscow, Nedra Press, 680 p. (Russian).

Krems, A. Ya., B. Ya. Vasserman, and N. D. Matvievskaya, 1974, Usloviya formirovaniya i zakonomernosti razmechsheniya zalezhei nefti i gaza (The environment of formation and the regularity of distribution of oil and gas pools (Pechora basin)): Moscow, Nedra Press, 335 p. (Russian).

Kristoffersen, Y., E. S. Husebye, H. Bungum, S. Gregersen, 1982, Seismic investigations of the Nansen Ridge during the FRAM I experiment: Tectonophysics, v. 82, n. 1–2, p. 57–68.

Krylov N., and M. Korzh, 1984, Upper Jurassic black bituminous shales in Western Siberia: Sedimentary Geology, v. 40, p. 211–215.

La Brecque, J. L., D. V. Kent, and S. C. Cande, 1977, Revised magnetic polarity time scale for Late Cretaceous and Cenozoic time: Geology, v. 5, n. 5, p. 330–335.

Larsen, O., 1980, The age of the Kap Washington Group volcanics, north Greenland: Bulletin of the Geological Society of Denmark, v. 31, p. 49–55.

Levashov, K. K., 1979, Paleorift structure in the eastern environs of the Siberian Platform: International Geology Review, v. 21, n. 2, p. 188–200.

Ling, H. Y., L. M. McPherson, and D. L. Clark, 1973, Late Cretaceous silicoflagellates from the Alpha Cordillera of the Arctic Ocean: Science, v. 180, p. 1360–1361.

Magoon, L. B., and G. E. Claypool, 1983, Petroleum source rock richness, type and maturity for four rock units on the Alaskan North Slope—are they sources for the two oil types? *in* U.S. Geological Survey Polar Research Symposium, USGS Circular 911, p. 30–32.

Mair, J. A., and J. A. Lyons, 1981, Crustal structure and velocity anisotropy beneath the Beaufort Sea: Canadian Journal of Earth Science, v. 18, n. 4, p. 724–741.

Mel'nikova, G. B., 1983, Petrology of andesites of the main volcanogenic zone of the Urals: International Geology Review, v. 25, n. 1, p. 21–38.

Meneley, R. A., D. Henao, and R. K. Merritt, 1975, The northwest margin of the Sverdrup basin: Canadian Society of Petroleum Geology, Memoir 4, p. 531–544.

Miall, A. D., 1976, Devonian geology of Banks Island, Arctic Canada, and its bearing on the tectonic development of the circum-Arctic region: Geological Society of America Bulletin, v. 87, p. 1599–1608.

——, 1984, Sedimentation and tectonics of a diffuse plate boundary: the Canadian Arctic Islands from 80 Ma B. P. to the present: Tectonophysics, v. 107, p. 261–277.

Milanovskiy, E. E., 1978, Faults of rift zones: International Geology Review, v. 20, n. 7, p. 757–569.

——, 1981, Aulacogens of ancient platforms: problems of their origin and tectonic development: Tectonophysics, v. 73, p. 213–248.

Nehring, R., 1978, Giant oil fields and world oil resources: Rand Corporation Report R-2284-CIA, 162 p.

Newman, G. W., C. G. Mull, and N. D. Watkins, 1979, Northern Alaska paleomagnetism, plate rotation, and tectonics, *in* A. Sisson, ed., The relationship of plate tectonics to Alaskan geology and resources: Proceedings of the 6th Alaska Geological Society Symposium, n. 6, p. C1–C7.

Oftedahl, C., 1971, A sideritic ironstone of Jurassic age in Beitstadtfjorden, Trondelag: Norsk Geologisk Tidsskrift, v. 52, p. 123–134.

Oil and Gas Journal, Feb. 22, 1982, p. 39–42.

Parker, J. R., 1966, Folding, faulting, and dolerite intrusions in the Mesozoic rocks of the fault zone of central Spitsbergen, *in* Norsk Polarinstitutt Arbok 1964, p. 47–55.

——, 1967, The Jurassic and Cretaceous sequence in Spitsbergen: Geology v. 104, n. 5, p. 487–505.

Pogrebitsky, Yu. E., 1971, Paleotektonicheskii analiz Taymyrskoi skladchatoi sistemy (Paleotectonic analysis of Taymyrian folded system): Leningrad, Transcript of the Arctic Institute of Geology, Issue 166, 248 p. (Russian).

Poulton, T. P., 1982, Paleogeographic and tectonic implications of the Lower and Middle Jurassic facies patterns in northern Yukon Territory and adjacent Northwest Territories, *in* A. F. Embry and H. R. Balkwill, eds., Arctic geology and geophysics: Proceedings of Third International Symposium on Arctic Geology, Canadian Society of Petroleum Geologists, p. 83–99.

Pratt, W. E., 1944, Oil fields in the Arctic: Harper's, v. 188, p. 107–112.

Price, I., and R. P. Rattey, 1984, Cretaceous tectonics off mid-Norway: implications for the Rockall and Faeroe-Shetland Troughs: Journal of the Geological Society, v. 141, pt. 6, p. 985–992.

Putintsev, V. K., and I. A. Zagruzina, 1980, Magmatism of the Mesozoides of the Soviet East: International Geology Review, v. 24, n. 4, p. 431–438.

Riediger, C. L., R. M. Bustin, and G. E. Rouse, 1984, New evidence for the chronology of the Eurekan Orogeny from south-central Ellesmere Island: Canadian Journal of Earth Science, v. 21, p. 1286–1295.

Savostin, L. A., and P. P. Shirshov, 1984, Mesozoic paleogeodynamics and paleogeography of the Arctic region, *in* Palaeoceanography, 27th International Geological Congress, Colloquium 03, Reports volume 3, Moscow, p. 217–238.

Schwerdtner, W. M., and K. Osadetz, 1983, Evaporite diapirism in the Sverdrup basin: new insights and unsolved problems: Bulletin of Canadian Petroleum Geology, v. 31, n. 1, p. 27–36.

Sellevol, M. A., 1975, Seismic refraction measurements and continuous seismic profiling on the continental margin of Norway between 60°N and 69°N: Norges geologiske undersokelse, offprint NGU 316, p. 219–235.

Srivastava, S. P., 1978, Evolution of the Labrador Sea and its bearing on the early evolution of the North Atlantic: Geophysical Journal of the Royal Astronomical Society, p. 313–357.

Steel, R. J., 1976, Devonian basins of western Norway—sedimentary response of tectonism and to varying tectonic context: Tectonophysics, v. 36, p. 207–227.

Stuart, B. A., I. B. Pringle, and D. Roberts, 1975, Caledonian Nappe sequence of Finnmark, Northern Norway, and the timing of orogenic deformation and metamorphism: Geological Society of America Bulletin, v. 86, p. 710–718.

Surlyk, F., 1983, Source rock sampling, stratigraphical and sedimentological studies in the upper Paleozoic of the Jameson Land basin, East Greenland: Greenland Geological Survey, Report 115, p. 88–93.

——, and J. M. Hurst, 1984, The evolution of the early Paleozoic deep-water basin: Geological Society of America Bulletin, v. 95, p. 131–154.

Sweeney, J. F., 1977, Subsidence of the Sverdrup basin, Canadian Arctic Islands: Geological Society of America Bulletin, v. 88, p. 41–48.

——, J. F., E. Irving, and J. W. Geuer, 1978, Evolution of the Arctic basin, *in* J. F. Sweeney, ed., Arctic geophysical review: Canadian Earth Physics Branch, Publication v. 45, n. 4, p. 91–100.

——, 1981, Arctic seafloor structure and tectonic evolution, *in* M. W. McElhinny and D. A. Valensio, eds., Paleoreconstruction of the continents: Geodynamics Series, v. 2, p. 55–64.

Tailleur, I. L., 1973, Probable rift origin of Canada basin, Arctic Ocean, *in* M. G. Pitcher, ed., Arctic Geology: AAPG Memoir 19, p. 526–535.

Taylor, P. T., L. C. Kovacs, P. R. Vogt, and G. L. Johnson, 1981, Detailed aeromagnetic investigation of the Arctic basin, 2: Journal of Geophysical Research, v. 86, n. B7, p. 6323–7333.

Til'man, S. M., A. N. Afitskiy, and A. D. Chekhov, 1977, Comparative tectonics of the Alazey and Oloy zones (Northeast USSR) and the problem of the Kolyma Massif: Geotectonics, v. 11, n. 4, p. 245–251.

Trettin, H. P., H. R. Balkwill, and R. A. Price, coordinators, 1979, Contributions to the tectonic history of the Innuitian Province, Arctic Canada: Canadian Journal of Earth Science, v. 16, n. 3, part 2, p. 748–769.

Ustritsky, V. I., 1977, Permskiy etap razvitiya Novoi Zemli (The Permian stage of development of Novaya Zemlya), *in* Tektonika Arktiki, skladchatyy fundament shel'fovykh sedimentatsionnykh basseinov (Tectonics of the Arctic, a folded basement of the shelf sedimentary basins): Leningrad, Institute of the Geology of the Arctic Press, p. 44–54. (Russian).

Van der Voo, R., and C. Scotese, 1981, Paleomagnetic evidence for a large (approximately 2,000 km) sinistral offset along the Great Glen fault during Carboniferous time: Geology, v. 9, n. 12, p. 583–589.

Vogt, P. R., and L. C. Kovacs, 1984, Amerasian basin, Arctic Ocean: magnetic anomalies and their decipherment, *in* Arctic Geology, 27th International Geological Congress, Colloqiuum 04, Reports v. 4, Moscow, p. 152–161.

Vogt, T., 1936, Orogensis in the region of Paleozoic folding of Scandinavia and Spitsbergen, 16th International Geological Congress (Washington, 1933), Report, v. 2, p. 953–955.

Volk, V. E., et al, 1984, Crustal structure of the Arctic inferred from geophysical data, *in* Arctic Geology, 28th International Geological Congress, Colloqiuum 04, Reports v. 4, Moscow, p. 30–41.

Vol'nov, D. A., 1975, Istoriya geologicheskogo razvitiya raiona Novosibirskikh ostravov (The history of the geological development of the New Siberian Islands), *in* Geology and mineral resources of the new Siberian Islands and Wrangel Island: Leningrad Institute of the geology of the Arctic, p. 61–71. (Russian).

——, V. N. Vojtsekhovskiy, O. A. Ivanov, D. S. Sorokov, and D. S. Yashin, 1970, Geologiya Novosibirskikh ostrovov (New Siberian Islands), *in* Geology of the USSR, v. 26, Islands of the Soviet Arctic, Geological descriptions, Moscow, Nedra Press, p. 324–374. (Russian).

Weber, J. R., 1979, The Lomonosov Ridge experiment, "Lorex 79": EOS, American Geophysical Union, Transactions, v. 60, n. 42, p. 715–721.

Ziegler, P. A., 1981, Evolution of sedimentary basins in northwest Europe, *in* L. V. Illing and G. D. Hobson, eds., Petroleum Geology of the continental shelf of north-west Europe: London, Heyden and Son Ltd., p. 3–39.

# Alaska: Potential for Giant Fields

J. J. Hohler
W. E. Bischoff
*Sohio Petroleum Company*
*Houston, Texas*

The demise of Mukluk is forcing government and the oil industry into the most remote and hostile areas of Alaska at a more rapid rate than originally envisioned. Even giant fields may, in some circumstances, prove marginal or non-commercial. The unexplored offshore waters in the Arctic and the Bering Sea hold the greatest promise.

## INTRODUCTION

Alaskan offshore waters comprise some of the most hostile areas for oil and gas exploration in the world. The Gulf of Alaska and the southern Bering Sea provide the starting point for most of the Pacific storms that affect North America. Storms in the northern Bering Sea and off north Alaska are given even sharper teeth by the pack ice that accompanies them. In the Beaufort and Chukchi Seas, ice movement is further affected by ocean currents (Figure 1) that, even without storm activity, maintain a fairly consistent east to west movement of the pack ice.

In the Gulf of Alaska and the Bering Sea, the technology for drilling exploratory wells is relatively conventional and normally makes use of large, semi-submersible drilling vessels. These are, however, subject to seasonal restrictions because of storm intensity or ice development or both. In the Beaufort and Chukchi Seas, ice is the major problem. In shallow waters artificial gravel islands have solved the problem; however, the islands have been limited to a 60 ft (18.3 m) water depth. In greater depths, moveable structures of sufficient mass to resist the ice forces appear to provide the only viable solution and may ultimately prove to be more economical in shallower waters as well.

These methods all have one thing in common: high costs. In 60 ft (18.3 m) or more of water along the Arctic Coast, the initial wildcat wells will cost on the order of $100–200 million and the existing technology for drilling in more than 200 ft (61 m) of water in these areas is, at best, questionable. Field sizes on the order of 150–450 million bbl of oil may be economic at water depths of 100–200 m (331–662 ft) (U.S. Geological Survey, 1981). Our experience in much shallower areas of the Beaufort would suggest that these numbers are extremely optimistic. It is, of course, very difficult to give a firm, lower economic limit because the estimate will depend on other accumulations found in the area, and on their size and geographic distribution.

As an example, the Kuparuk Field on the North Slope would not be economical without the pipeline and infrastructure built for Prudhoe Bay, despite the fact that it has several billion barrels of oil in place. Similarly, other "small" accumulations near Prudhoe Bay, such as Milne Point, the Endicott Pool, and others, will eventually be developed, at varied levels of profitability, only because Prudhoe Bay and the Trans-Alaska Pipeline provide the necessary infrastructure.

In the Bering Sea, minimum field sizes can be smaller because the enormous ice forces of the northern seas are absent. The U.S. Geological Survey (1981) estimates that the economic limits are 75–200 million bbl of oil in the 100–200 m (330–655 ft) depth range and 50–125 million bbl of oil in the 30–100 m (100–330 ft) depth range. No infrastructure whatever exists in these areas, and except for the Norton and Bristol Bay areas, it will be extremely costly to bring oil onshore by pipeline for processing and tanker loading. Offshore loading, in terms of technology, is a difficult but not insurmountable problem, but fishing and other interest groups may object.

If the discovered hydrocarbons turn out to be gas, then the medium-term (10 years ±) commercial outlook is poor. Before producers develop any gas in the northern waters, a pipeline must be built from the known gas fields of the North Slope to the lower 48 states. Given the existing U.S. gas bubble and the financial and construction-permit restrictions of such a pipeline, plus the time required for actual construction, producers probably will not develop new gas discoveries onshore or offshore in this century. In the Norton and Bristol Bay (North Aleutian) areas, gas could be brought ashore for liquefaction and tanker shipping, but it is difficult to predict how competitive this procedure will be compared to alternative sources in Indonesia, Malaysia, or Australia.

Studies carried out within Sohio prior to the lease sales suggest that in the Navarin and St. George basins, on-site liquefaction and tanker loading offer the only reasonable method of developing gas accumulations, and are probably economically unattractive for the foreseeable future.

In summary, exploration drilling is possible with present technology, to perhaps 150 ft (45 m) in the northern waters, and improvements in such technology promise to extend this capability to water depths of 200–300 ft (60–90 m) during the next decade. Costs will, however, be extremely high and development will be practical only for very large oil fields. In the Bering Sea basins the engineering problems are less severe, but their remote location and lack of infrastructure mean that relatively large oil reserves will still be required to justify development.

## ESTIMATES OF ALASKAN OFFSHORE POTENTIAL

Many oil and gas companies, individual geologists, and government groups have stated at one time or another that Alaska, and especially offshore Alaska, is the only remaining area in the U.S. with the promise of significant numbers of giant or supergiant fields. By implication, because such fields also comprise a significant proportion of major reserves, the Alaskan offshore should offer the potential for discovering a large percentage of the as-yet-undiscovered recoverable resources remaining in the U.S.

U.S. Geological Survey Circular 860 (1981) estimates that the Alaskan offshore potential ranges from 4,600 to 24,200 million bbl of oil and 33.3 to 109.6 trillion cu ft (tcf) of gas, with the respective means at 12,200 million bbl of oil and 64.6 tcf of gas. As a general statement, the U.S.G.S. estimates that the Alaskan offshore will yield between one-third and one-half of all future U.S. offshore discoveries, and between one-tenth and one-fourth of the total U.S. future discoveries. This estimate does, however, include undiscovered resources allocated to areas where the Arctic pack ice presently precludes development.

The Rand Corporation report (Nehring, 1975), published prior to Circular 860, is somewhat less optimistic. It does not separate onshore and offshore Alaska, but we believe that, with a few exceptions, the potential for future large hydrocarbon discoveries lies in the offshore areas. The Rand report estimates undiscovered Alaskan resources to comprise 6,500–11,200–23,400 million bbl of oil and 31.3–49.0–83.8 tcf of gas for the low, mean and high cases, respectively. When comparing these numbers with those of Circular 860, we find that they are surprisingly similar, although the Rand report differs by allocating a much higher proportion of future discoveries to Alaska. This difference, 45% versus 70% and 20% versus 40% of total undiscovered U.S. oil and gas respectively, is due to Rand's

→

**Figure 1.** *Location of Alaskan basins.*

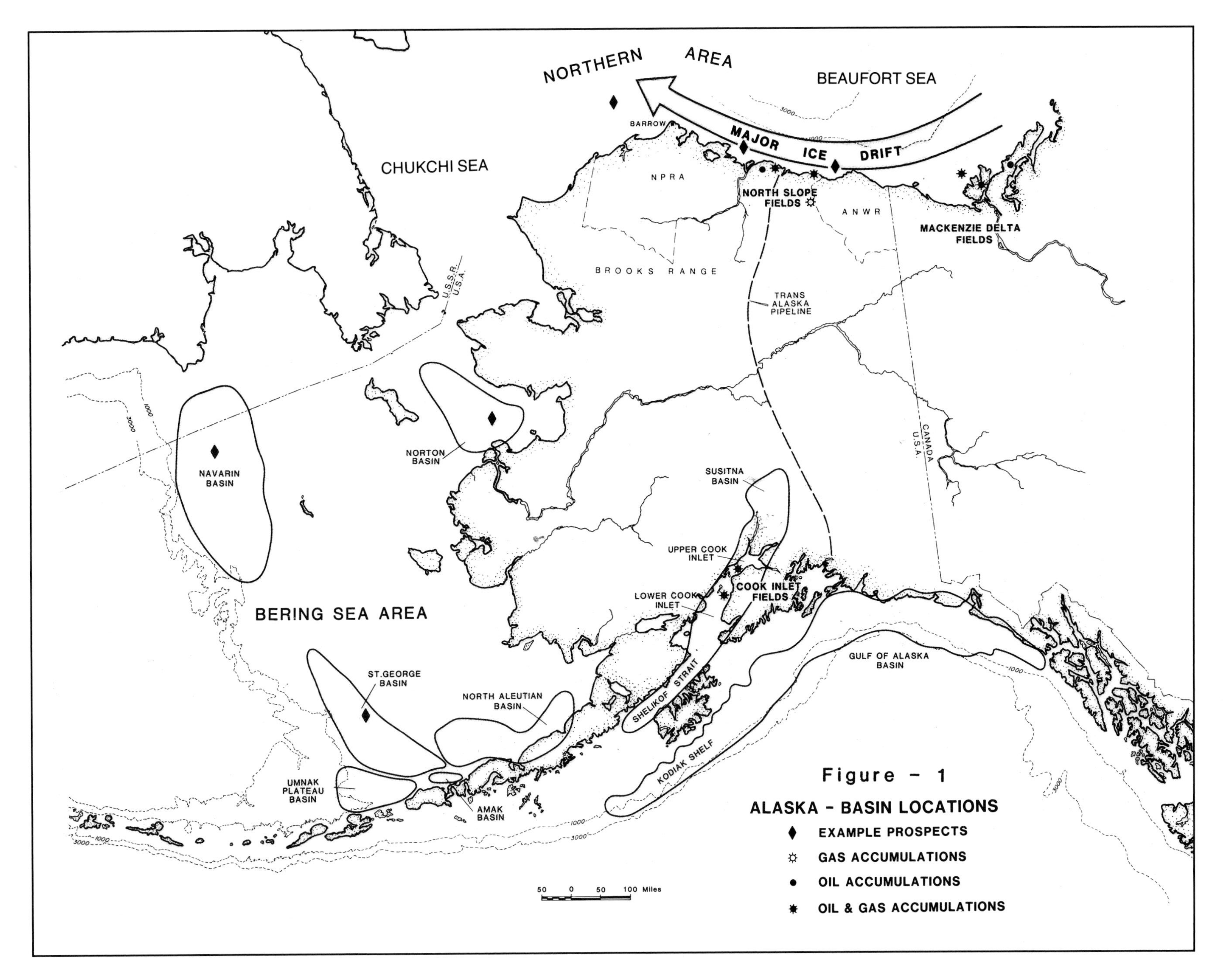


Figure – 1
ALASKA - BASIN LOCATIONS

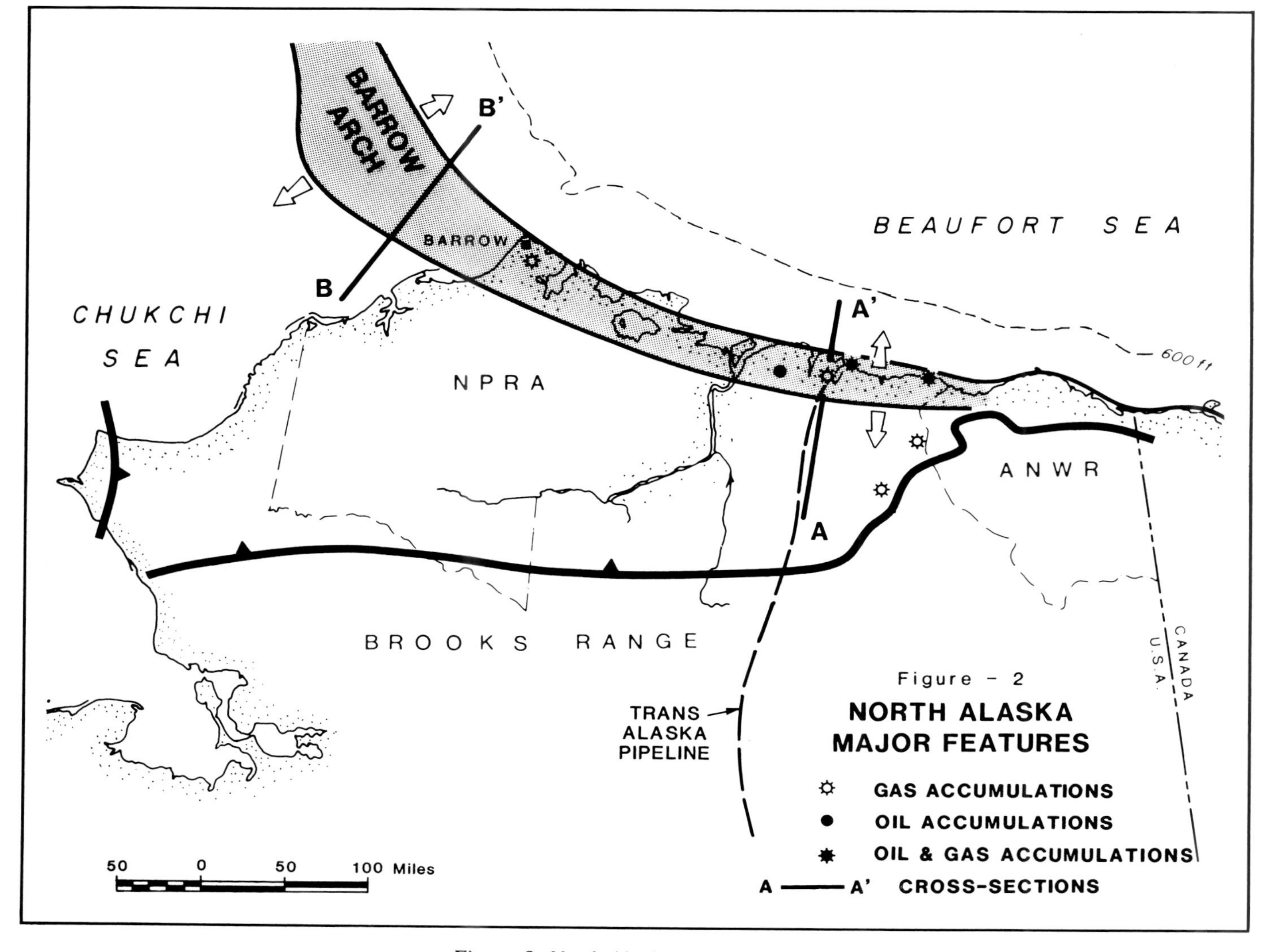


Figure 2. *North Alaska major features.*

much lower estimate of undiscovered resources in the lower 48-state onshore and offshore areas.

In both studies, the most likely outcome could be filled by a single field of Prudhoe Bay dimensions (9,600 million bbl of oil), while in the maximum case, each study would provide for nearly three Prudhoe Bay type fields. The lower end of the range could be accomplished by discovering a reasonable number of marginally economic (depending on location) fields in the 100–300 million bbl of oil category, but this amount would provide little incentive for exploration, and development will be protracted. Only super-giant discoveries in the 1–10 billion bbl range, depending on location, would provide the incentive and generate the enthusiasm, on the part of industry and government, required for rapid development.

Several of the basins we will discuss will probably fail to contain significant or commercial quantities of oil, in which case the anticipated future discoveries will all have to come from one or two basins where geological conditions have been favorable. In fact, the current and scheduled drilling will, by the end of 1985, provide industry with a good estimate of the Bering Sea's potential.

## Prospective Area

To help discussion, we have grouped the major prospective basins into the Northern and Bering Sea Areas (Figure 1). The Northern Area comprises the Beaufort and Chukchi Seas from the Canadian border to the U.S./U.S.S.R. demarcation line. The Bering Sea area consists of the Norton, St. George, North Aleutian, and Navarin basins, which are all located on the Bering Sea Shelf (Ehm, 1983).

The prospective parts of the Northern Area are subject to severe ice problems and are geologically related to the prolific oil and gas accumulations of the Central Arctic Slope, for example Prudhoe Bay and Kuparuk. The Bering Sea basins resemble each other in geologic style, but do not, as yet, contain any producing fields. However, they are not subject to the extreme ice conditions of the north.

In the interest of space, and in deference to proprietary information, we will give a brief overview of a portion of Alaska's prospective areas.

## Northern Area

Considered the most prospective offshore area in the U.S.,

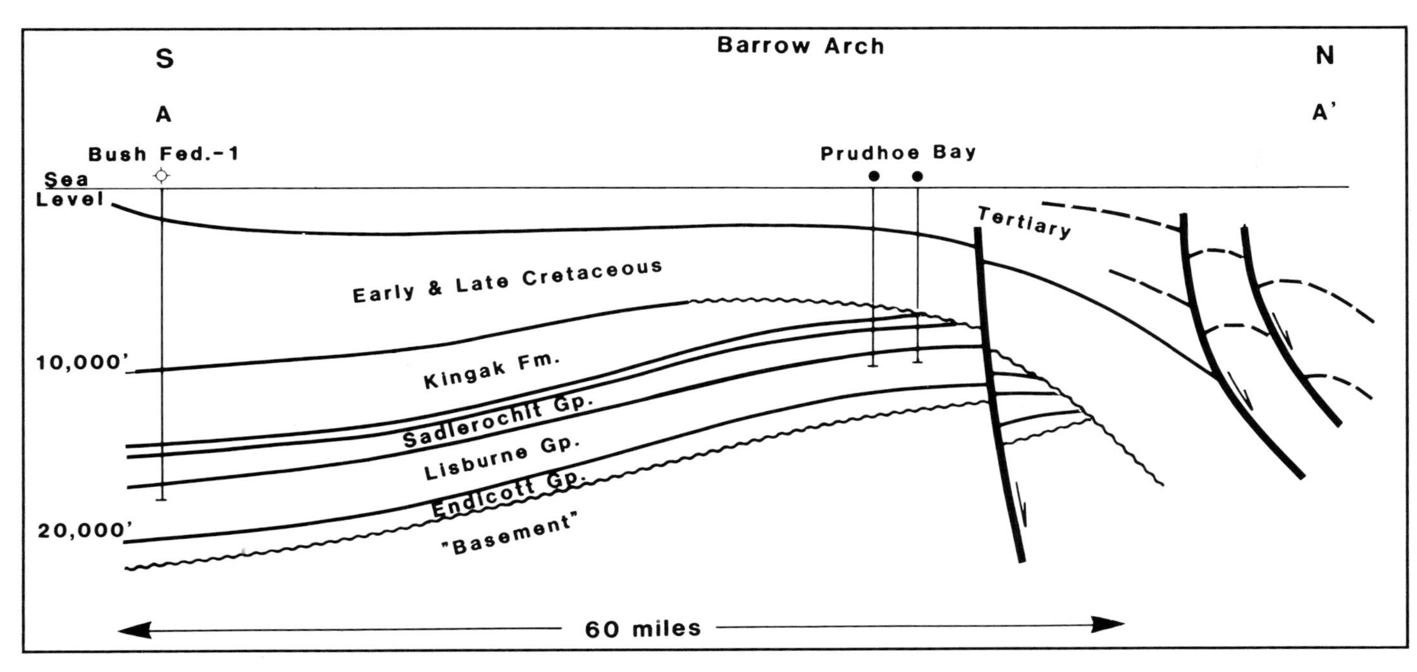


Figure 3. *Diagrammatic cross-section of Eastern Arctic Slope.*

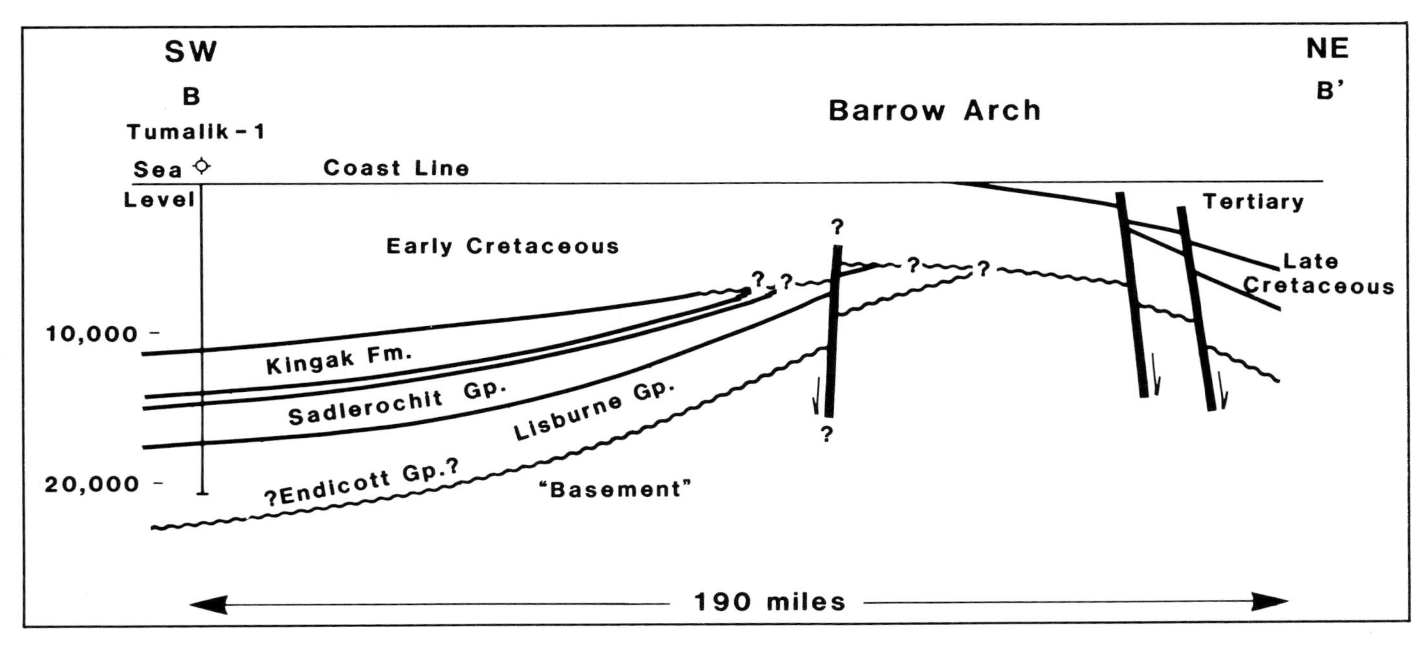


Figure 4. *Diagrammatic cross-section of the Chukchi Sea area.*

the Northern Area will be the first Alaskan offshore area, outside the Cook Inlet, to be developed as accumulations such as the Endicott Field are brought into production. This relatively "small" (350 million bbl of oil) field, however, benefits from its proximity to the Prudhoe Bay facilities and its extremely shallow (5–10 ft or 1.5–3.0 m) water depths.

Geologically, the Barrow Arch (Figure 2) dominates the Northern Area, stretching from the National Wildlife Range westward into the Chukchi Sea. Not a true arch, this remnant feature was formed by normal faulting on the northern side and regional dip to the south (Figures 3 and 4).

The Mississippian through Jurassic sediments came from a northern source prior to the rifting that formed the Barrow Arch. These sediments include clastic reservoirs (Endicott Group, Sadlerochit Group, Sag River Sandstone), carbonate reservoirs (Lisburne Group), and potential source rocks in the Kingak and Shublik Formations. Lower Cretaceous sediments comprise both south- and north-derived sediments. The latter includes such actual and potential reservoirs as the Kuparuk Formation, the Point Thomson Sandstone, and the Kemik

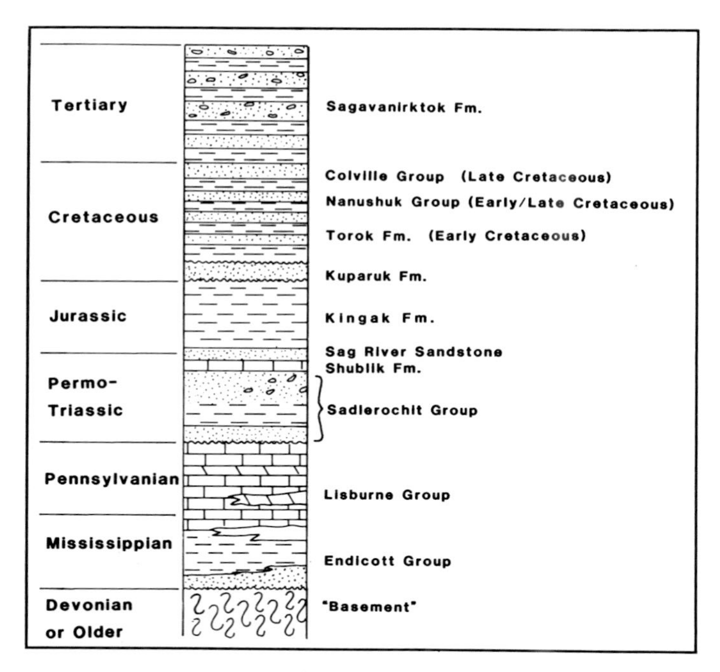


Figure 5. *Generalized North Slope stratigraphic column.*

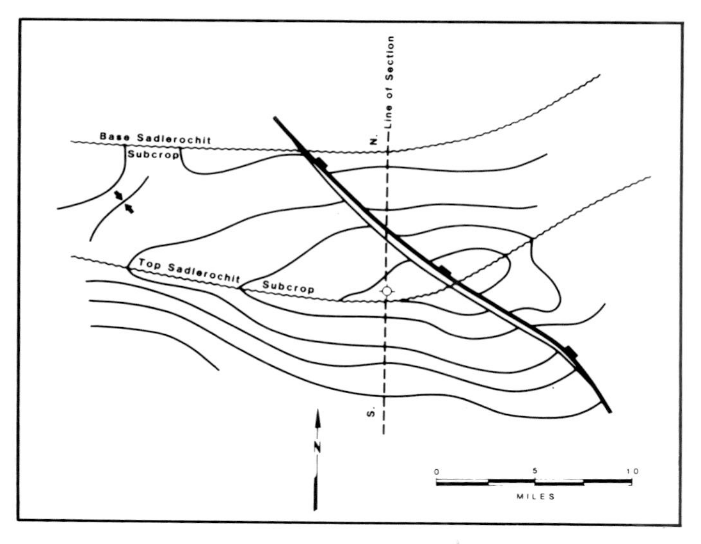


Figure 6. *Form-line map of Mukluk prospect; contour interval: 100 ft.*

Sandstone, which are all associated with the Prudhoe Bay Unconformity. Overlying these rocks (Figure 5) is a characteristic shale with a marked gamma-ray signature, which is also a potential source rock. The south-derived sediments include the extremely thick Lower Cretaceous fill in the Colville Trough (Figure 5).

All Upper Cretaceous and Tertiary sediments are of southern provenance and include a variety of marine and non-marine sandstones, coals, siltstones, and shales. These include a variety of potential and actual reservoirs and source rocks.

## Barrow Arch Province

Structural–stratigraphic traps, such as Prudhoe Bay, predominate along the Barrow Arch. These traps are formed by gentle structures, with or without faulting, truncated during Early Cretaceous time. Potential reservoirs are primarily of Permian–Triassic or Mississippian age. The Mukluk structure is an excellent example of this type of feature and exemplifies the fact that even with excellent quality seismic data and reasonable geologic knowledge and well control, there are many factors that cannot be predicted in the search for major oil fields.

The Mukluk structure is approximately 20 mi (32.2 km) long and 9 mi (14.5 km) wide. It is an extremely gentle feature, but appears to have several hundred feet of present-day closure (Figure 6). A north to south cross-section (Figure 7) emphasizes the gentle nature of the structure and the truncation that forms its northern edge. Had it been oil bearing, this combination of structure and reservoir could have held on the order of 1–1.5 billion bbl of recoverable oil. There are other features along the Barrow Arch capable of forming similar combination traps of variable size. Exxon has drilled another such feature near Cape Halkett during 1984–1985. The lack of any positive announcement suggests that no hydrocarbon accumulation was discovered.

Geologists will continue to speculate as to why Mukluk proved water-bearing despite indications that, at one time, there either had been oil in the structure or oil had migrated through it. The two most probable possibilities are the lack of an effective seal, or the absence of structural closure during some crucial stage in the prospect's geological development. In the first instance, a thin, Lower Cretaceous sandstone directly overlying the truncating unconformity, and therefore also the Ivishak (Sadlerochit Group) reservoir, could have provided a conduit that either prevented an accumulation from forming or allowed slow leakage over time. In the second case, difficulties in carrying mapping intervals from the east side of Mukluk to the west side may have prevented recognition of a temporary lack of paleo-closure.

No present-day structural closure may exist, and the seismically mapped closure to the west is caused by unpredicted velocity changes. The relative precision with which velocities, and hence depths, were predicted at the well location makes this last alternative unlikely.

Additional work may or may not allow a definite statement to be made because our data may not be sufficiently precise to identify the stratigraphic or structural nuances responsible. Mukluk, interestingly, was identified in 1969–1970, put up for bid in 1982, and drilled in 1983–1984, using techniques not significantly improved over those available in the early 1970s.

East of Point Barrow, the Barrow Arch swings northwestward into the Chukchi Sea (Figure 2) and may offer opportunity for major accumulations. The seismic data shot over this area indicate that the pre-Cretaceous sediments have been faulted and probably locally truncated. Normal faulting and tilting of fault blocks continued during the Cretaceous, resulting in some very large structures that could provide hydrocarbon traps. Figure 8 illustrates such a feature west of Point Barrow. It comprises a tilted fault block with a very large area of possible structural closure. The example seismic line (Figure 9) indicates that considerable structural growth occurred through time with major expansion faulting on the northwest side and onlap to the southeast. From our knowledge of the sedimentary section, as seen in the Point Barrow Area 100 mi (161 km) to the east, we

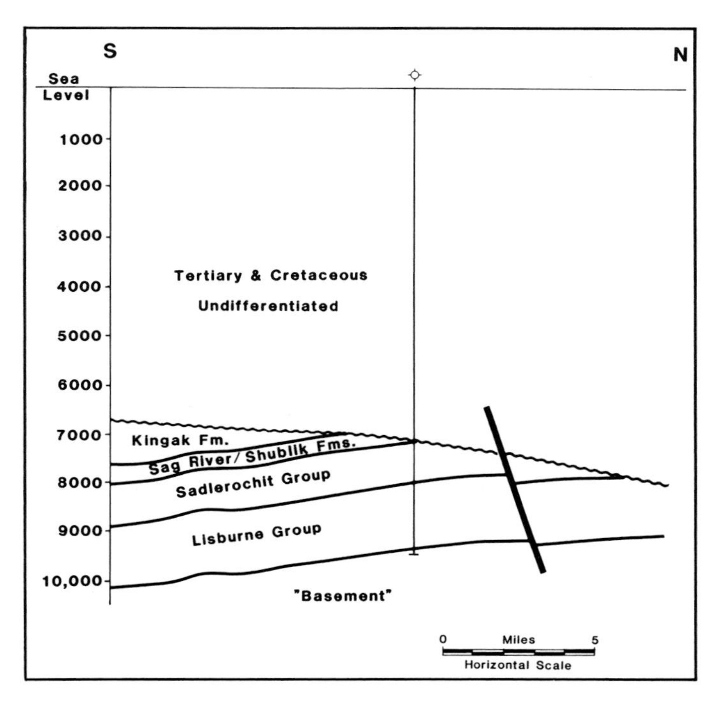


Figure 7. *Cross-section of Mukluk prospect; vertical exaggeration X8.3.*

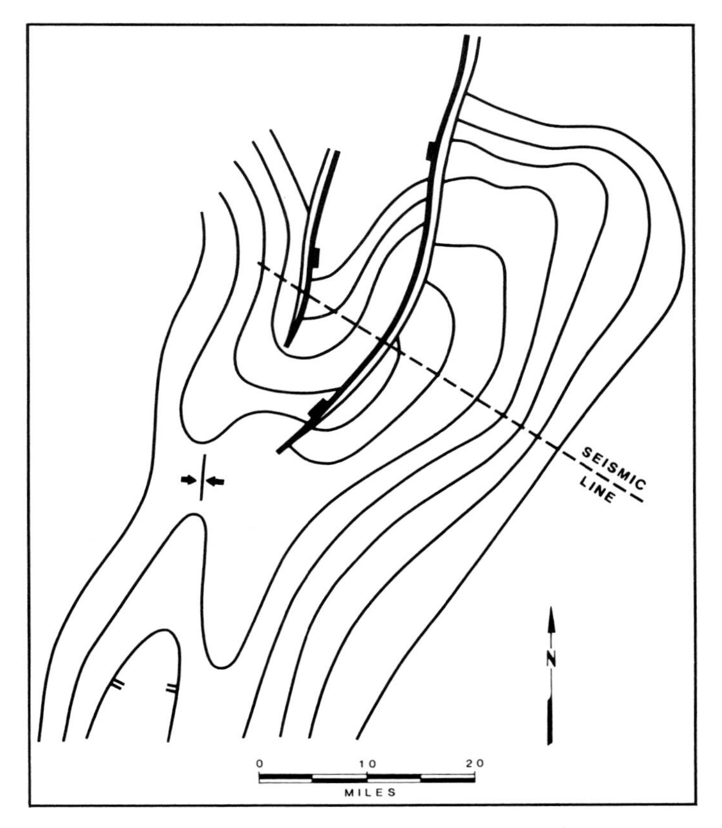


Figure 8. *Form-line map of Barrow Arch: Chukchi example; contour interval: 500 ft.*

could expect the Lisburne and Sadlerochit groups and the overlying Kingak Formation, as well as some of the thinner Mesozoic intervals, such as the Shublik Formation and the Barrow Sand, to be present on the flanks.

Many potential prospects appear at a variety of stratigraphic levels and in a variety of traps in the area. Lower Cretaceous sands that are of northeastern provenance and are equivalent in stratigraphic position to those productive in the Kuparuk and Point Thomson fields could result in extensive reservoirs over the example structure. They could also provide an enormous trap in terms of area. The thickness of such sands, however, is open to speculation. The seismic section (Figure 9) illustrates the possibility for growth fault or clastic wedge development modified by truncation on the northwest side and onlap closures to the southeast. Such differing traps are not mutually exclusive, and in the Prudhoe Bay Area, accumulations in Mississippian sandstones (Endicott), Mississippian–Pennsylvanian carbonates (Lisburne), Permian–Triassic sandstone (Ivishak), Lower Cretaceous sandstones (Kuparuk), and Upper Cretaceous–Tertiary sandstones are all clustered within a relatively small area.

While the general stratigraphic sequence is reasonably predictable, the facies's development, this far from any well control, is open to speculation. Reservoir and source rock qualities are, therefore, also speculative, but given the basin's history in terms of generating oil to the east, and assuming similar depositional and tectonic histories along the entire Barrow Arch, hydrocarbons may have been generated and trapped in the Chukchi Sea. Earlier,we spoke of the most likely Alaskan offshore potential in the range of 10–12 billion bbl of oil and 50–60 tcf of gas. The Chukchi Sea has the potential for fields of this magnitude, that is, Prudhoe Bay size in water depths of only 120–150 ft (36.5–45.7 m), but at distances of 100 mi (161 km), or more, from shore and in an area of extremely severe ice conditions. An accumulation of more than 5 billion bbl of recoverable oil will probably be required in this remote and difficult area to initiate development. There can be little doubt that such potential exists.

## Beaufort Sea Province

The Beaufort Sea Province comprises the thick wedge of clastic sediments, mainly of Late Cretaceous and Tertiary age, north of the Barrow Arch (Figure 2).The section has been penetrated by wells along the coast between the Colville and Canning Rivers, where it thickens eastward from a few thousand to over 15,000 ft (4,570 m). Onshore, it is a mixed marine and nonmarine interval comprising shales, siltstones, sandstones, and coals. The sandstones range from fine-grained to conglomeratic and potential source rocks occur within the section. In addition, at least the southern part of the province will be underlain by Lower Cretaceous shales that, while thin, have excellent source rock characteristics.

Late Cretaceous and Tertiary deposition into the rapidly subsiding Camden Bay subbasin and onto the subsiding shelf resulted in a thick prograding section that, typical of such sedimentation, developed growth faults of major proportions. The sediments were, of course, derived from the Brooks Range. The subbasin and shelf were subsequently modified by uplift and tilting related to the post-Cretaceous thrusting of the Eastern Brooks Range exemplified by the Sadlerochit Mountains. As a result, numerous potential hydrocarbon traps formed. An

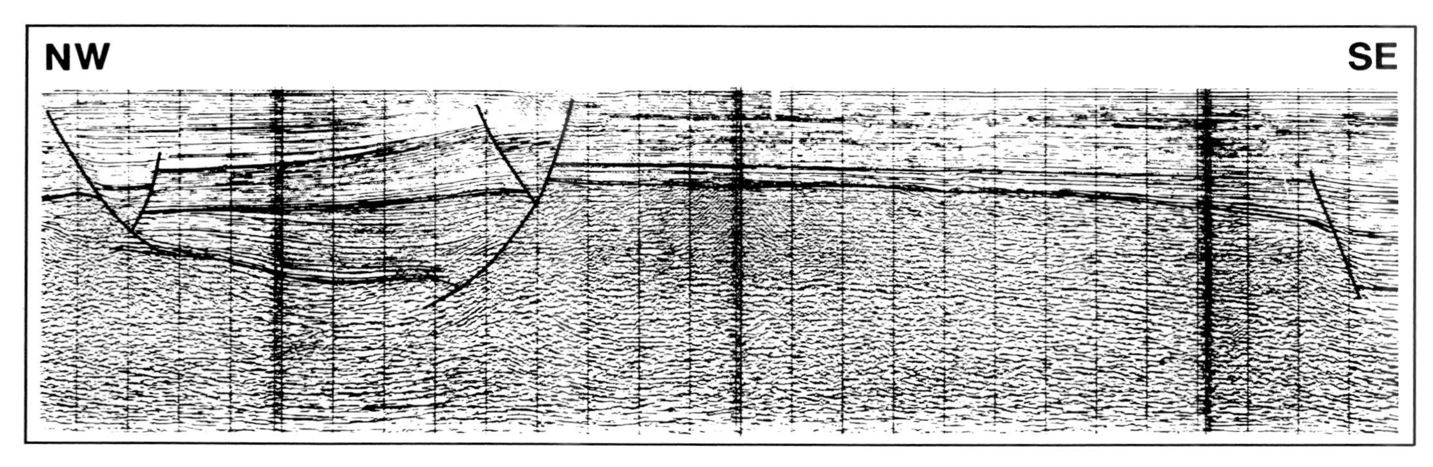


Figure 9. *Northwest to southeast seismic line, Barrow Arch (Chukchi) example.*

example of one of these is a large, relatively gentle anticlinorium cut by several normal faults. The dimensions are again impressive because the anticlinorium approaches 30 mi (48.3 km) in length (northwest to southeast) and 20 mi (32.2 km) in width (Figure 10).

The large, down-to-the-north faults (Figure 11) provide local closure but also have the effect of complicating the exploratory drilling program. In the event of successful exploratory drilling, they will also tend to complicate developmental drilling. The overall structure is sufficiently large to hold, at a maximum, 5–10 billion bbl of recoverable oil, or recoverable gas in the order of 20–40 tcf, or a wide range of combinations of oil and gas between these end points. High-amplitude anomalies visible on the seismic data (Figure 11) may indicate a relatively high probability of finding at least some gas.

The economic viability of this and other similar prospects in the area is, of course, greatly dependent on finding large volumes of oil that can be recovered with relatively few production platforms or structures and hence a limited number of wells. Water depths are 100–125 ft (30–38 m) in an area of extreme ice activity. The Tertiary sands seen in wells 40–50 mi (65–80 km) southwest of the prospect appear to offer excellent reservoir characteristics, but are typical of deltaic sands in that their average thickness is 100 ft (30 m) or less. They can be expected to thin offshore. In order to achieve adequate production rates to warrant development, the industry will have to find stacked reservoirs. Given the lack of a North Slope gas pipeline, discovery of large gas volumes, a not unreasonable outcome in these Tertiary rocks, will not prove commercial.

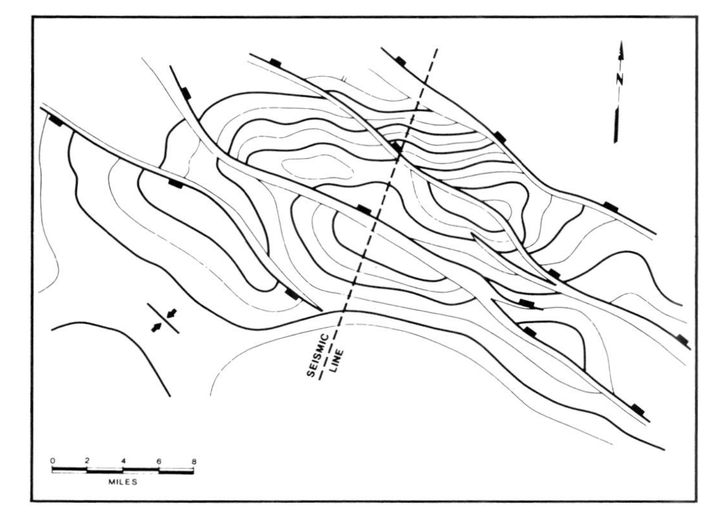


Figure 10. *Form-line map, Beaufort Sea example; contour interval: 500 ft.*

### Bering Sea Area

The Bering Sea Shelf covers an enormous area (Figure 1) of which, unfortunately, only the isolated Norton, Navarin, St. George, and North Aleutian (Bristol Bay) basins contain a sufficient thickness of sediment to be considered prospective for hydrocarbons. First-round sales have been held in all these basins except for the North Aleutian. Drilling is presently underway in the Norton and St. George basins, with the Navarin to follow in the summer of 1985. This activity will greatly increase our knowledge of these basins and, we hope, encounter significant quantities of hydrocarbons along with better-than-predicted reservoir rocks.

Because the North Aleutian basin has a first-round sale scheduled for November 1985, we will confine this discussion to the other three basins for proprietary reasons.

### General

The Bering Sea basins are believed to be largely of Tertiary age. The Navarin and St. George basins (Figure 1) are wrench fault related, the North Aleutian basin is essentially a back-arc feature, and the Norton appears to be primarily extensional. Most of the remaining shelf seems to have been remarkably stable throughout Tertiary time and has only a thin veneer of sediment over a variety of older rocks. Water depths are reasonable over the entire shelf and over the basins.

### Norton Basin

The Norton basin has seen considerable drilling activity during 1984, and, from released information, with disappointing initial results. It is, however, probably too early to consider the basin totally non-prospective. The seismic line in Figure 12 illustrates the general structural style of prospects in the basin. It covers a large prospect 12 mi (19.3 km) long and 4 mi (6.4 km) wide. The line indicates the presence of a large basement struc-

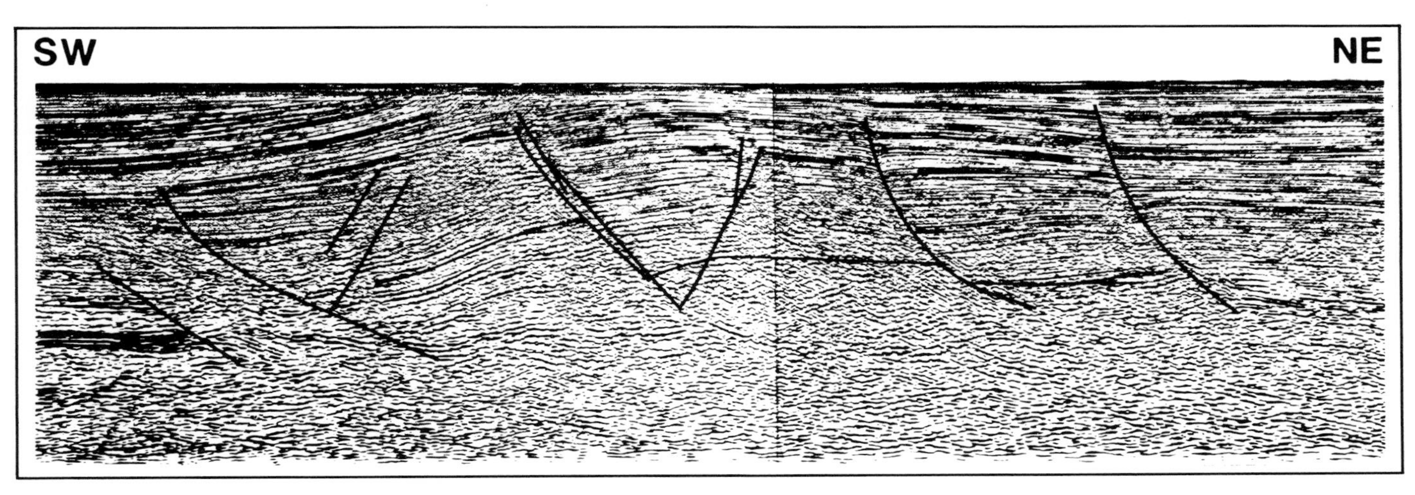


Figure 11. *Northeast to southwest seismic line, Beaufort Sea example.*

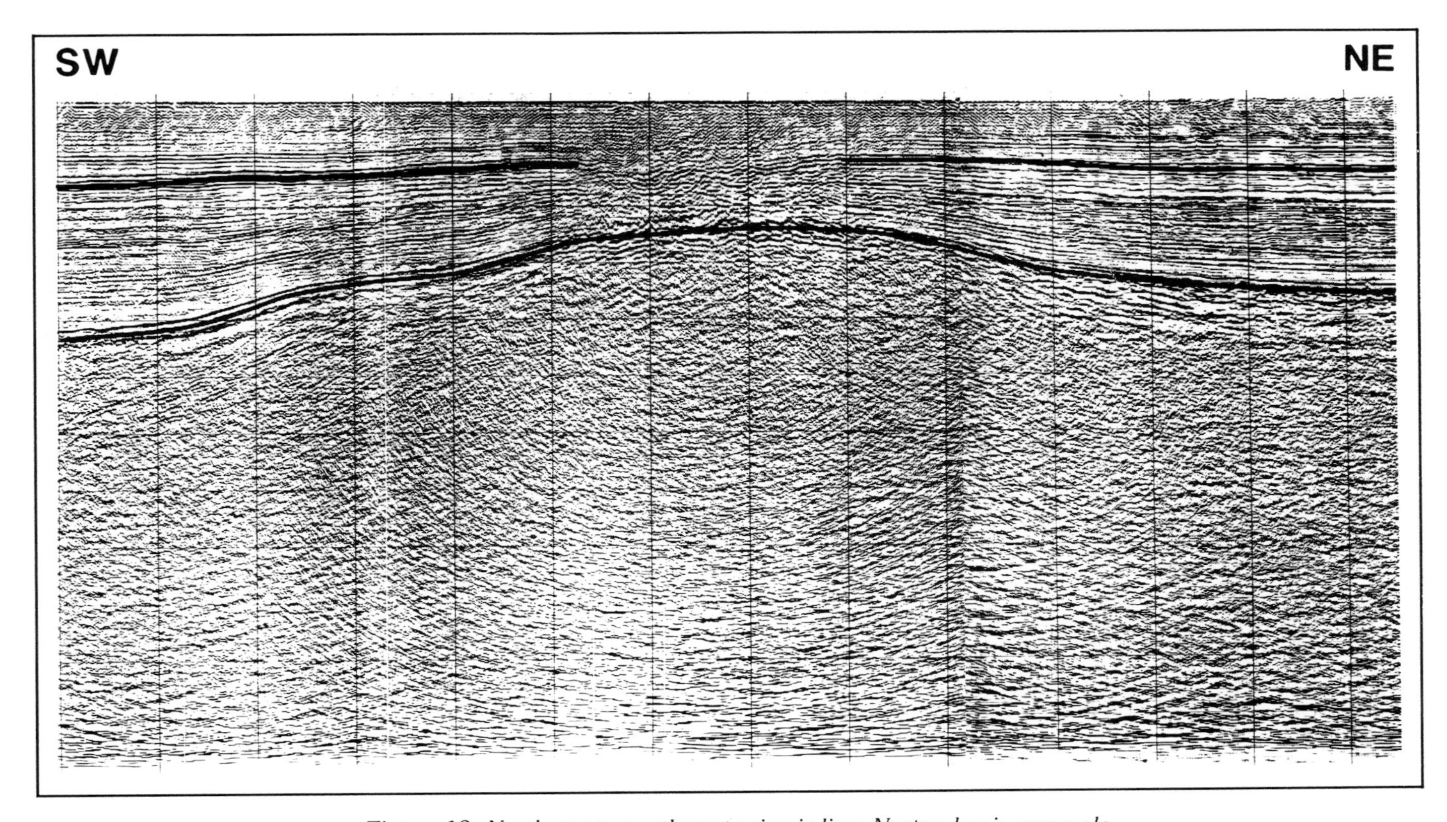


Figure 12. *Northeast to southwest seismic line, Norton basin example.*

ture, with sediments draping over it, resulting in a progressively less-pronounced structure upward. It appears that the crest is, at less than 2,500 ft (760 m), extremely shallow.

Stratigraphically, the Norton basin section comprises a series of marine and non-marine sandstones, siltstones and shales, and several coal-bearing sequences, plus volcanics in the lower Tertiary rocks. The sandstones are generally fine-grained and commonly are micaceous. No massive sandstones occur that could readily provide high-capacity reservoirs of regional extent. Preliminary indications are that reservoirs in the basin will be relatively poor, and we tend to believe that conventional, oil-prone source rocks are absent. Numerous coals and carbonaceous shales, however, are present that could certainly provide a source for gas and, given the development of the right type of coal facies, might also serve as an oil source.

The structure shown in Figure 13 has dimensions of 10 mi by 5 mi (16.1 km by 8.1 km) and is certainly of sufficient size to hold very large hydrocarbon volumes if the requisite reservoirs exist. A pronounced, low-velocity anomaly exists over the crest of the structure (Figure 12), indicating that the sediments may be gas charged and meaning that the presence of producible gas is debatable. Low gas saturation in unconsolidated sediments

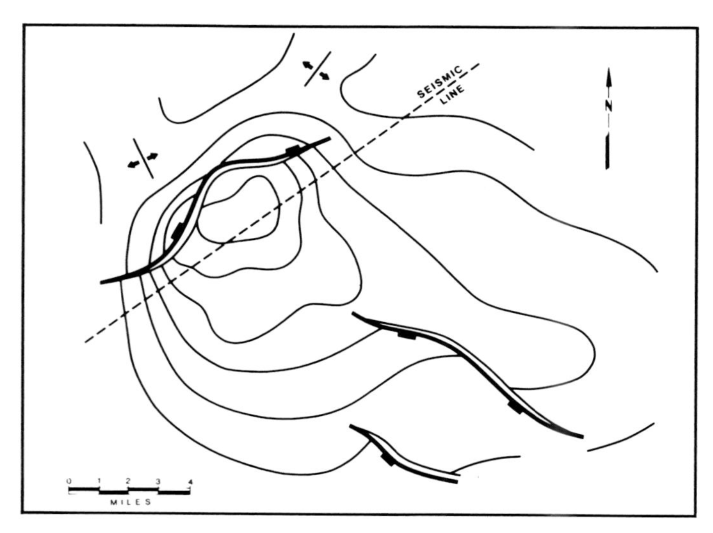


Figure 13. *Form-line map, Norton basin example; contour interval: 200 ft.*

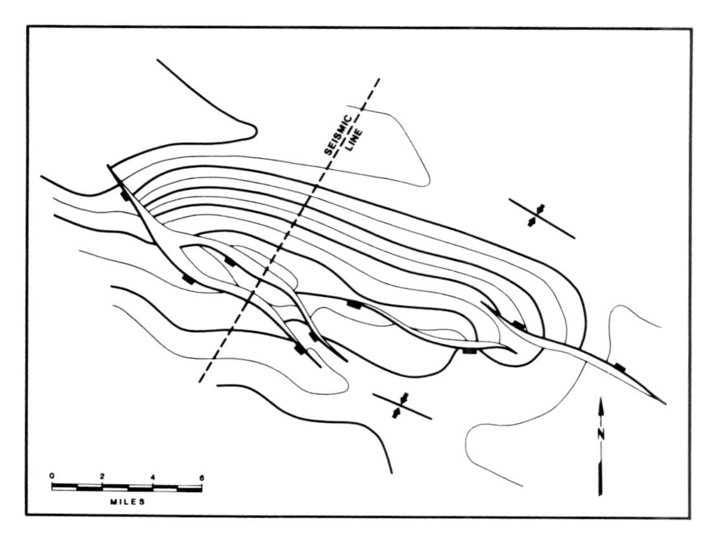


Figure 14. *Form-line map, St. George basin example; contour interval: 500 ft.*

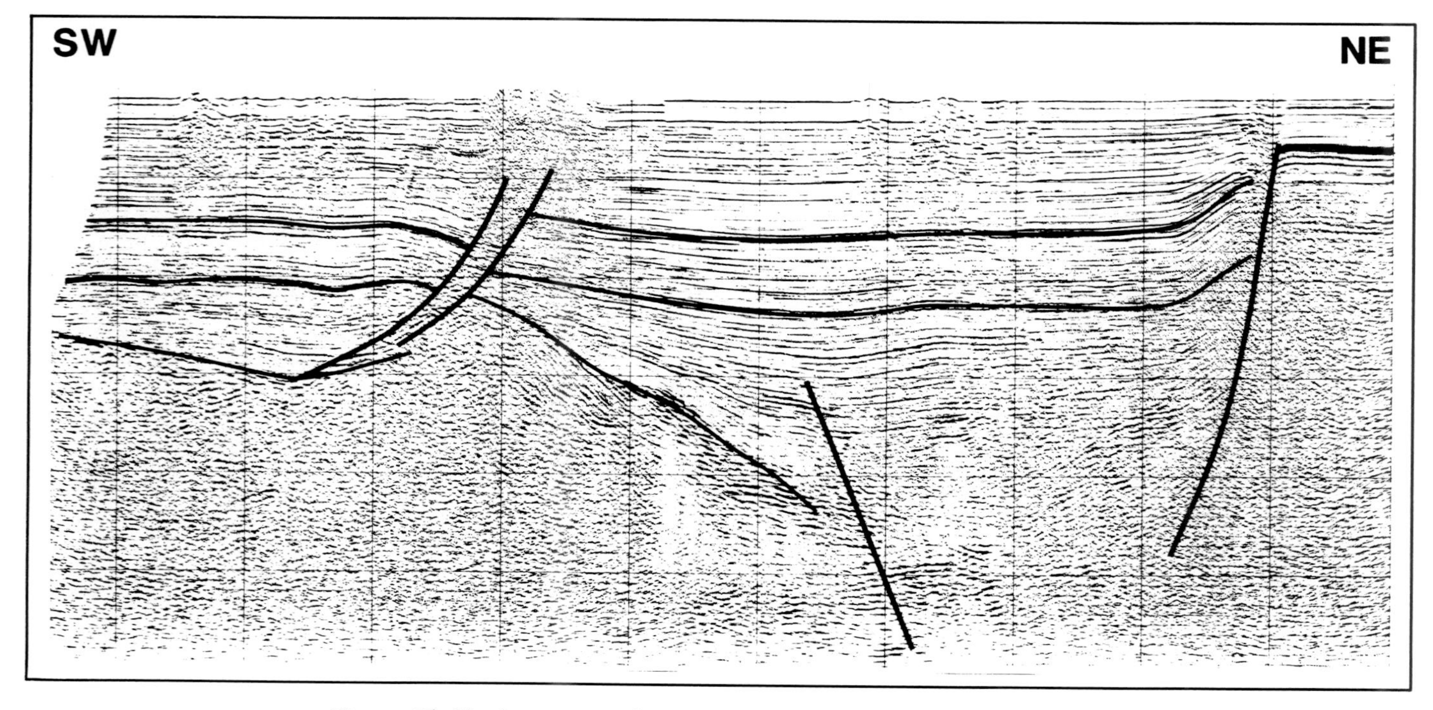


Figure 15. *Northeast to southwest seismic line, St. George basin example.*

could give similar effects. A well, however, has recently been drilled on the crest of this structure and abandoned without a discovery announcement.

## St. George Basin

Located inside the Aleutian Island chain and near the shelf edge, the St. George basin is an elongate feature (Figure 1) with a northwest to southeast, long-axis dimension of perhaps 200 mi (320 km) compared to a width of about 75 mi (120 km). OCS Sale #70 saw the leasing of nearly 100 blocks to industry, and explorers have begun drilling. To date they have announced no successes, but it is too early to make a definitive statement as to what bearing this lack of success has on the basin's prospectiveness.

The pronounced tensional expression indicated by the basin dimensions is reflected in the prevailing northwest to southeast trend of fault systems and structures. Structures are basement controlled and range from broad, gentle features to much more pronounced, albeit generally smaller, fault-related structures. Geologists identified many prospective structures in the basin prior to OCS Sale #70. Sediments are draped over the basement features, and structural development decreases upward.

The example used lies in the northwest part of the basin and in style is one of the more pronounced types of structures. It is

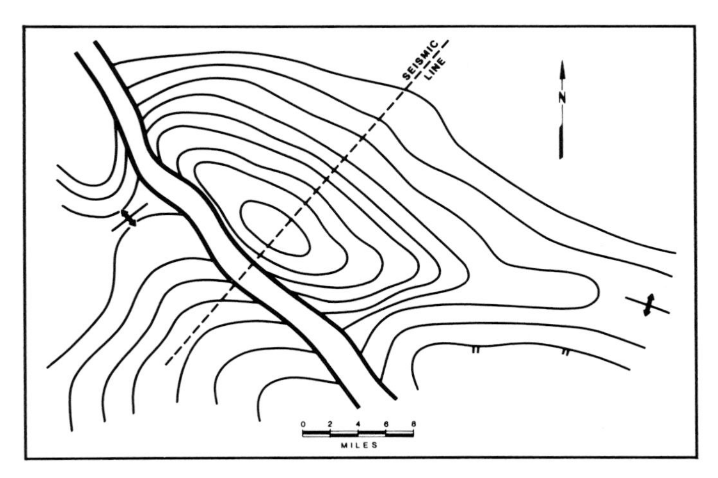


Figure 16. *Form-line map, Navarin basin example; contour interval: 1,000 ft.*

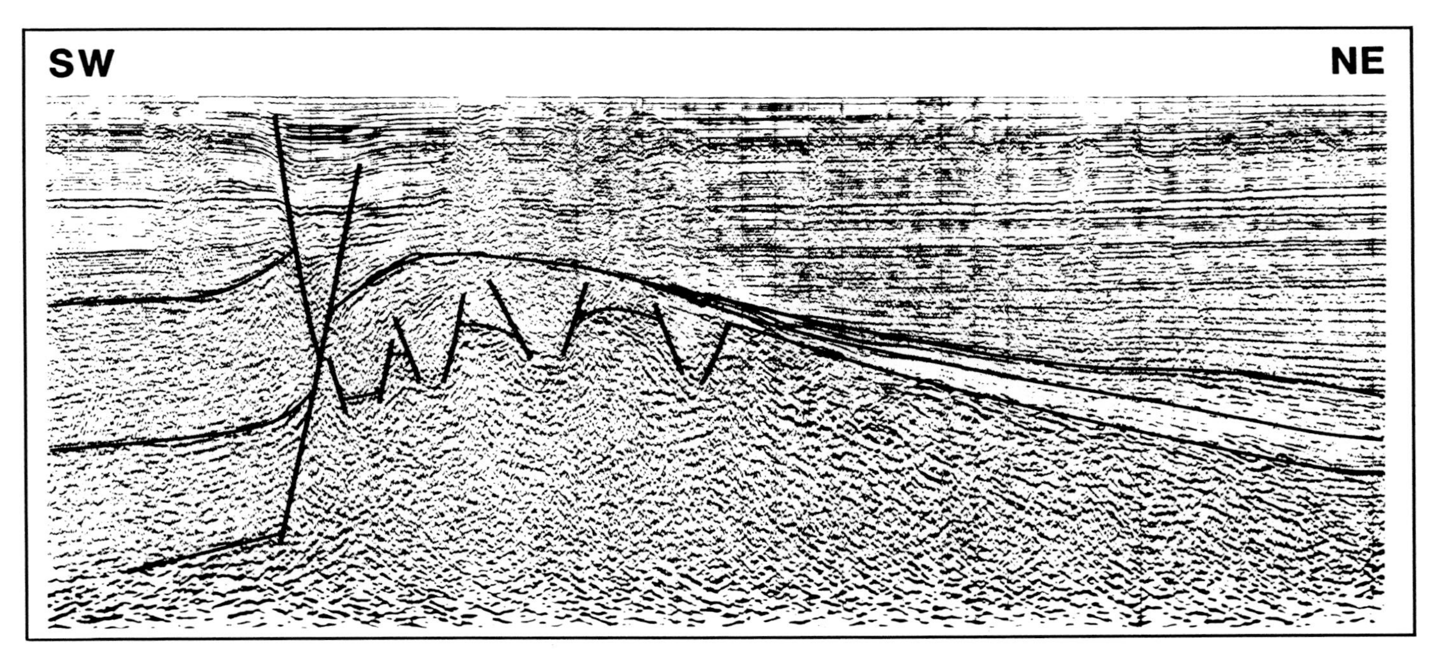


Figure 17. *Northeast to southwest seismic line, Navarin basin example.*

approximately 20 mi (32.2 km) long and 5 mi (8.1 km) wide, with over a thousand feet of closure at the mapped level (Figure 14). The seismic section (Figure 15) shows the position of the example feature in relation to the deep flanking graben to the northeast. The pre-Tertiary section gives indications of bedding on the northeast side of the structure and it is possible that these are Mesozoic sediments. If such a section is present and contains oil-prone, mature source rock, it would significantly improve the oil potential of the area. Porous and permeable Mesozoic sediments, if sealed by overlying Tertiary rocks, could in this and similar positions also form subcrop traps.

The Tertiary basin fill consists largely of siltstones, mudstones, and occasional, predominantly lithic, sandstones. The lower Tertiary part of the section contains, at least locally, a high proportion of volcaniclastic material. No coals have been reported, but moderately rich, gas-prone source rocks appear to be present. The reservoir characteristics of the Tertiary sandstones are not outstanding at shallow depths and, because of the lithic fragment content, will tend to deteriorate markedly with depth.

On the positive side, however, the structures in the basin are large, the sampling extremely limited (two C.O.S.T. wells), and there is more than ample space for unforeseen changes in lithology.

### Navarin Basin

The Navarin basin is the most westerly of the Bering Sea basins (Figure 1). The western boundary of OCS Sale #83 was formed not by the basin limits but by an informal demarcation line between U.S. and Soviet waters. As stated earlier, the Navarin is the largest of the western basins. One can obtain an idea of its size from the sale notice: it comprised over 5,000 tracts or 28 million acres (11,331,600 ha.). At the sale, 186 blocks were leased.

Like the St. George basin, the Navarin is a tensional basin formed in response to Late Cretaceous–early Tertiary wrench faulting superimposed on what presumably was a Mesozoic fore-arc basin. The structures are large, fault-related types. The example used is about 25 mi (40.2 km) long and 12 mi (19.3 km) wide (Figure 16) with an area of closure in excess of 10,000 acres (4,050 ha.). Seismic data (Figure 17) suggest that a variety of structural and structural–stratigraphic traps are possible. These include simple, drape-type closures over the crest onlap on the flanks, sands in the crestal grabens, and clastic wedges along the large fault on the southwest side.

Stratigraphically, the basin fill comprises Eocene and younger Tertiary rocks underlain by at least several thousand feet of Upper Cretaceous marine and non-marine sediments. The former consist of siltstones, mudstones, and graywackes, and the latter of sandstones, siltstones, and coal. Older Mesozoic rocks, probably volcaniclastics and siltstones with some local sandstones, could be present. Numerous dikes and sills have locally intruded the Upper Cretaceous sediments.

Reservoirs appear to be limited and largely confined to the upper Oligocene and lower Miocene sediments. These sandstones have a high lithic content that could, with deeper burial, adversely affect porosity and permeability. At shallower depths reservoir characteristics are reasonable, but considerable improvement would be required for the high capacity wells needed for truly profitable development. In contrast to the Norton and St. George basins, however, the Navarin contains mature marine source rocks, at least theoretically capable of generating oil, and the Anadyr basin, which is the Soviet equivalent, has produced minor quantities of oil.

There are structures in the basin that could contain between 1 and 10 billion bbl of oil. Several productive structures would probably be required in order to make the Navarin a truly profitable venture but, since offshore loading is required in any event, even a single large field might prove attractive.

## CONCLUSIONS

The Alaskan offshore waters could provide the only remaining U.S. potential for discovering significant numbers of giant or super-giant oil and gas fields. Each basin will, however, probably require a super-giant oil field to reach the threshold level needed for development. Present gas economics suggest poor development possibilities, prior to 2000 A.D., for any gas accumulations discovered.

Given the fact that the Bering Sea may prove largely gas prone, and that development in the pack ice areas of the Arctic will be difficult and expensive, the reasonable estimate of ultimately recoverable oil probably should not exceed 15–20 billion bbl. There may, however, be many "small" fields discovered in the offshore Arctic that will prove noncommercial.

## ACKNOWLEDGMENTS

This brief overview is based on data, interpretations, and concepts generated by Sohio and BP Alaska staff over more than 15 years. The authors wish to thank all those whose work has been used in arriving at the conclusions. We also extend appreciation to Sohio Petroleum Company for permitting publication.

## REFERENCES CITED

Ehm, A., (compiler), 1983, Oil and gas basins map of Alaska: Alaska Department of Natural Resources.

Nehring, R. G., 1975, The discovery of significant oil and gas fields in the United States: Washington, D.C., Department of Energy, U.S. Government Printing Office, 162 p.

U.S. Geological Survey, 1981, Estimates of undiscovered recoverable conventional resources of oil and gas in the U.S., Circular 806: Washington, D.C.: U.S. Government Printing Office.

# Oil and Gas Fields in the East Coast and Arctic Basins of Canada

Robert A. Meneley
*Meneley Enterprises Ltd.*
*Calgary, Alberta, Canada*

The East Coast and Arctic basins of Canada have been under serious hydrocarbon exploration for over 20 years. While the density of drilling is low, extensive seismic control has outlined a high proportion of the structures in these basins, and the stratigraphic framework of the basins is known from the 600 wells that have been drilled. The five most thoroughly explored basins include the Beaufort, Sverdrup, Labrador, East Newfoundland, and Scotian basins. Examples of discoveries from each of these basins illustrate the factors that control the accumulation of hydrocarbons.

The physical environment in these Canadian basins is severe, and while great strides have been made in coping with the environment, the cost of doing so is becoming increasingly onerous. There is a growing sense of appreciation regarding the cost, risk, and time that will be involved in developing these resources. The vast reserves of oil in the tar sands of western Canada provide a competitive ceiling that will limit the future development in the East Coast and Arctic basins of Canada to those projects in which production costs are not significantly higher than those of the tar sands.

## INTRODUCTION

Because oil finders draw heavily on analogies in developing new exploration play concepts, the geological study of major oil and gas fields is an essential tool for directing ongoing exploration. The fields must be viewed in their basin setting, and a critical evaluation of the parameters that control the entrapment and preservation of hydrocarbons is essential.

The role that the basins in the East Coast and Arctic of Canada will play as future petroleum provinces will be determined by the combination of their geological potential and the economics of finding and developing the discovered fields. We must compare the economics to the alternatives for energy development elsewhere in Canada, and most specifically to the alternative of tar sands development, which represents a massive available resource base (Proctor, Taylor, and Wade, 1983).

This paper will review the five most thoroughly explored frontier basins in Canada: the Beaufort, Sverdrup, Labrador, East Newfoundland, and Scotian basins (Figure 1). The basic structure and stratigraphy of these basins are known from extensive geophysical surveys that have led to drilling almost 600 exploratory and delineation wells over more than 20 years of serious and well-directed exploration.

I will describe one or two examples of typical, large discoveries in each basin, emphasizing those attributes that are most critical to entrapping and preserving the hydrocarbons. In all the examples, reservoir quality and fluid deliverability are excellent and I will not comment on that. In addition, I will not discuss source rock potential because others have already established the presence of source rocks in each of these basins. Finally, maturation levels and type and age of source rocks, as controls on the size and content of oil and gas fields, should be the object of further study.

Discovered reserves in the East Coast and Arctic basins of Canada include 568 million cu m (3.6 billion bbl) of oil, 127 trillion cu m (4.5 tcf) of associated gas, 998 trillion cu m (35.4 tcf), and 98 million cu m (614 million bbl) of condensate.

Discovered reserves are the best technical estimate of the proved, probable, and possible recoverable reserves without regard to the economics or feasibility of producing those reserves. I have considered only those fields in which at least one well has tested hydrocarbons, and have used available geological and geophysical data to describe the trap geometry and capacity.

For the purposes of this study, gas has been converted to oil-equivalent bbl on the basis of 10,000 cu ft/bbl (1,772 cu m of gas/cu m of oil). The reserves in each basin have been displayed on a single, pool-size array that ranks the discoveries by size and shows the relative contribution of oil and gas to the total reserves in the basin.

## BEAUFORT BASIN

### Regional Setting

The Beaufort basin (Figure 2) lies on the northern mainland coast of Canada and includes the modern Mackenzie River delta, the Tuktoyaktuk Peninsula, and the Beaufort Sea. The Beaufort basin is a Jurassic to Recent sedimentary basin, developed over a down-to-the-north truncated terrane of Precambrian to Devonian age (Figure 3). The sediment fill came from the uplifted Cordillera to the southwest and the continental area to the south, and was deposited by an ancestral Mackenzie River drainage system.

Wrench and normal faults involving the basement ran parallel to the Tuktoyaktuk Peninsula and affected the basin's sedimentation up to the Lower Cretaceous. Tertiary (Laramide) compression has modified the growth-faults, mud diapirs, and shale-cored anticlines that occur in the basin center.

Hydrocarbons have been found in three different facies-cycle wedges (White, 1980). Wedge-base pools (Type I) are found in Jurassic and Lower Cretaceous sand reservoirs or sub-unconformity reservoirs or both. These pools are found in a northeasterly trend along the Tuktoyaktuk Peninsula. Wedge-top pools (Type II) of Tertiary age occur in the area of the Mackenzie River delta and extend seaward to just beyond the 20 m (65.6 ft) bathymetric contour. Tertiary turbidite sands (Type III) form reservoirs that were deposited further seaward from the delta complex. All three exploration plays pool hydrocarbons in structural traps; no regional stratigraphic traps have been discovered to date.

The pool size array (Figure 4) ranks the 14 largest discoveries in the basin on an oil-equivalent basis. The ranking number for each pool has been used to identify its geographic position on the map (Figure 2), while the Roman numeral identifies the play type.

The discovered pools contain 213 million cu m (9.7 tcf) of gas.

### Parsons Lake Gas Field

The Parsons Lake gas field (Cote, Lerand, and Rector, 1975) is an example of a Type I accumulation (Figures 2 and 5). The Parsons structure developed along the Tuk Flexure Zone that marks the southeastern margin of a rift basin which actively subsided in Jurassic to Early Cretaceous time. This flexure zone is a narrow belt comprising horst and graben structures in an area dominated by northeasterly trending faults. The Tuk Flexure Zone is bounded by major, high-angle fault systems that show right lateral strike-slip movement, and subsidiary faults splay off these major systems. Folding commonly strikes northerly, sub-parallel to the major faults, and often shows an en-echelon pattern.

The Upper Jurassic to Lower Cretaceous wedge-base sequence was accommodated by substantial differential subsidence along the southeastern-bounding fault (Figures 6 and 7) that marked the edge of the original rift basin. The wedge-base cycle started with deposition of organic-rich Jurassic shale (Langhus, 1980) and culminated with the deposition of the excellent reservoir sands in the upper part of the Parsons Group, which forms the main reservoir for the field. The sands are overlain by a thick, Lower Cretaceous mudstone top seal.

The gas column in the field is over 320 m (1,050 ft), and the average net gas pay is 46 m (150 ft). The gas is dry and only carries 0.04 cu m of gas/1,000 cu m (7–8 bbl/million cu ft or mmcf)

**Figure 1.** *Canadian sedimentary basins, showing the five most thoroughly explored frontier basins and the oil and gas fields and pipelines in Canada.*

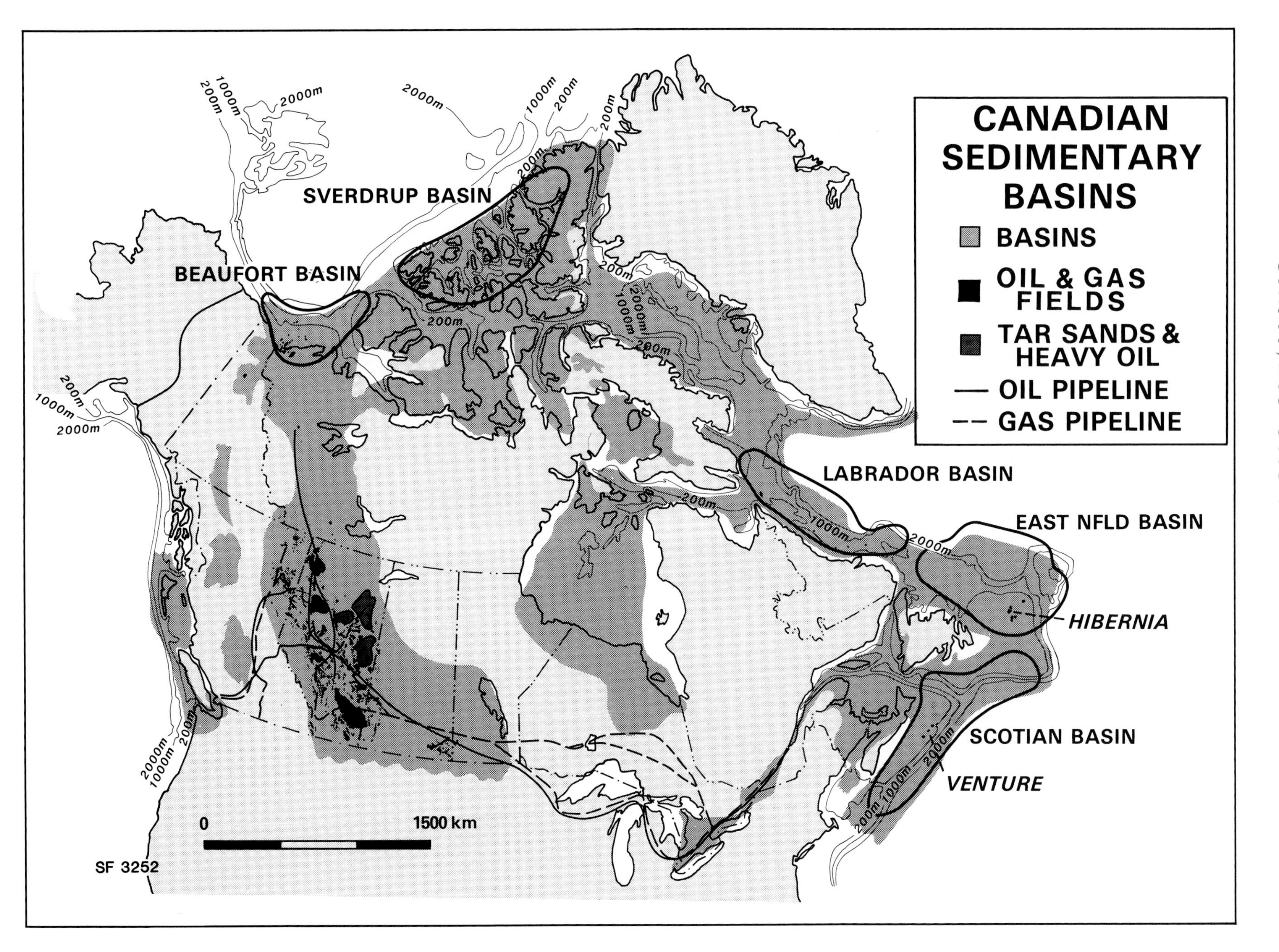
CANADIAN SEDIMENTARY BASINS
BASINS
OIL & GAS FIELDS
TAR SANDS & HEAVY OIL
OIL PIPELINE
GAS PIPELINE
SVERDRUP BASIN
BEAUFORT BASIN
LABRADOR BASIN
EAST NFLD BASIN
HIBERNIA
SCOTIAN BASIN
VENTURE
0
1500 km
SF 3252
200m
1000m
2000m

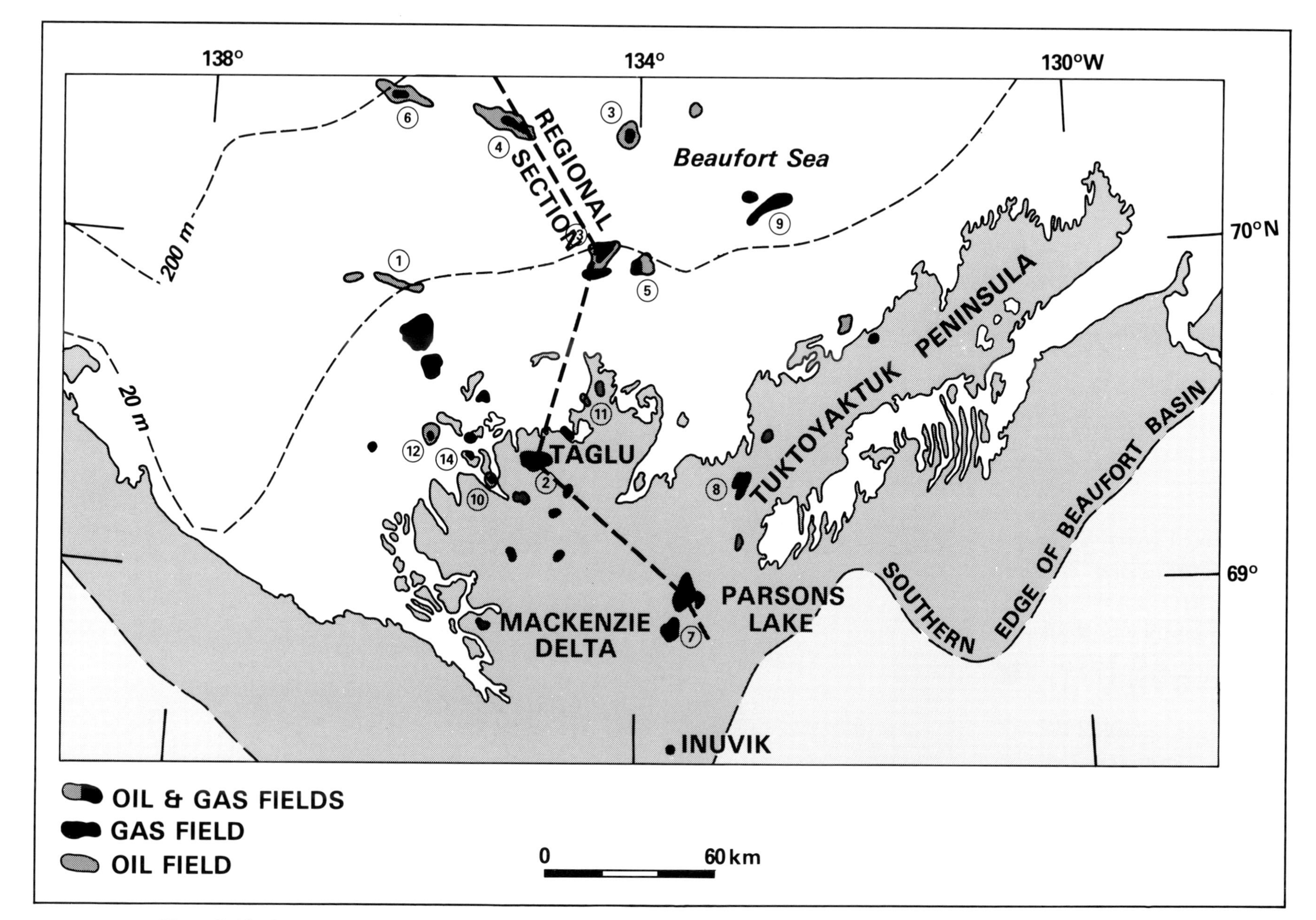


Figure 2. *The Beaufort basin, showing the oil and gas discoveries. The numbers on the fields refer to the ranking numbers on Figure 4.*

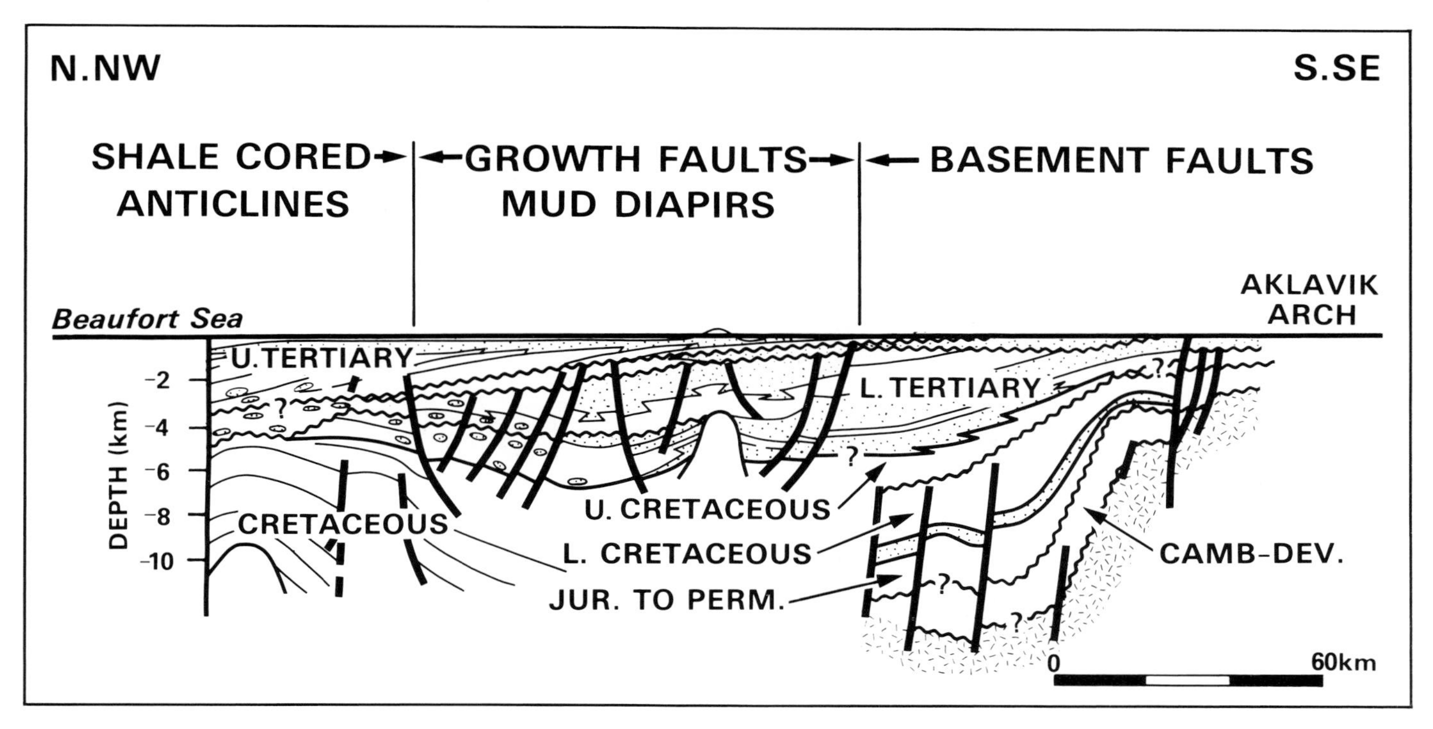


Figure 3. *Beaufort basin regional cross-section. See Figure 2 for the location of the section.*

of condensate. Two separate gas pools and one small oil pool have been defined with seismic and drilling. The combined field area covers 151 sq km (58.3 sq mi).

The factors that appear to have been critical for accumulating hydrocarbons at Parsons Lake and other Type I fields are the combination of differential subsidence along the rift margin faults; deposition of high-quality, reservoir sands; and localizing of structural culminations by wrench tectonics shortly after the reservoir and overlying shales were deposited. There has been no late structural uplift.

The Type I play trend on the Tuktoyaktuk Peninsula varies from larger, dry gas fields in the southwest, to more condensate-rich gas, and to small oil accumulations, in the northeast.

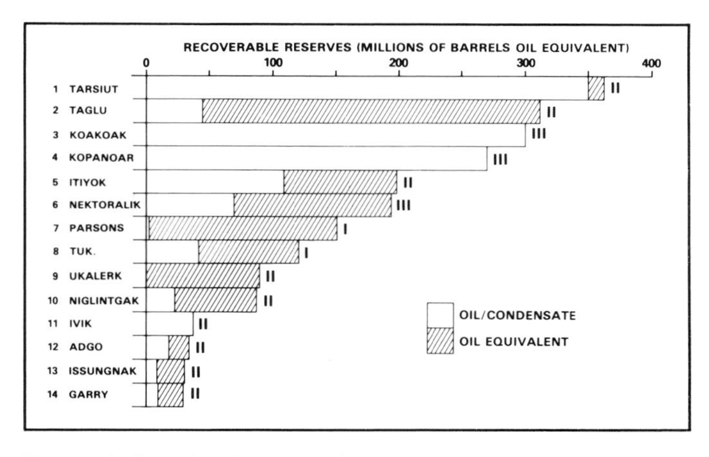


Figure 4. *Beaufort basin pool size array. The Roman numeral refers to the play type, and the ranking numbers have been used to locate the fields geographically on Figure 2.*

## Taglu Gas Field

The Taglu gas field (Hawkins and Hatlelid, 1975; Bowerman and Coffman, 1975) is an example of a Type II accumulation (Figures 2 and 8). The northern part of the modern Mackenzie delta was deformed initially by a series of synsedimentary, down-to-basin growth faults induced by rapid sediment loading over an undercompacted, pro-delta shale fill. Wrench faulting and Late Cretaceous to mid-Tertiary compression have also been operative in structural development. The Taglu structure is a high-side closure at the intersection of a north-trending anticline and an east-to-west-trending normal fault (Figures 9 and 10).

The Taglu gas reservoirs are found in the upper part of the Reindeer Formation, which is a sand-prone formation at the top of an early Tertiary, regressive depositional cycle. The overlying shales mark a significant shoreward transgression of the Beaufort Sea.

The sand reservoirs were deposited as part of a river-dominated delta sequence during a time of rapid, coarse-clastic influx (Dixon, 1980). The overlying shales are overpressured, which contributes to the efficacy of the top seal.

Both the Reindeer Formation and the overlying shale show thinning onto the structure, which confirms that the structure developed very early and was, therefore, available to pool hydrocarbons immediately after deposition of the reservoir.

The gas pays lie in two zones that are separated by a sealing shale. The gas column is up to 518 m (1,700 ft), which almost fills the 640 m (2,100 ft) of structural relief. The field covers 45 sq km (17 sq mi) and the net pay averages 84 m (276 feet). The

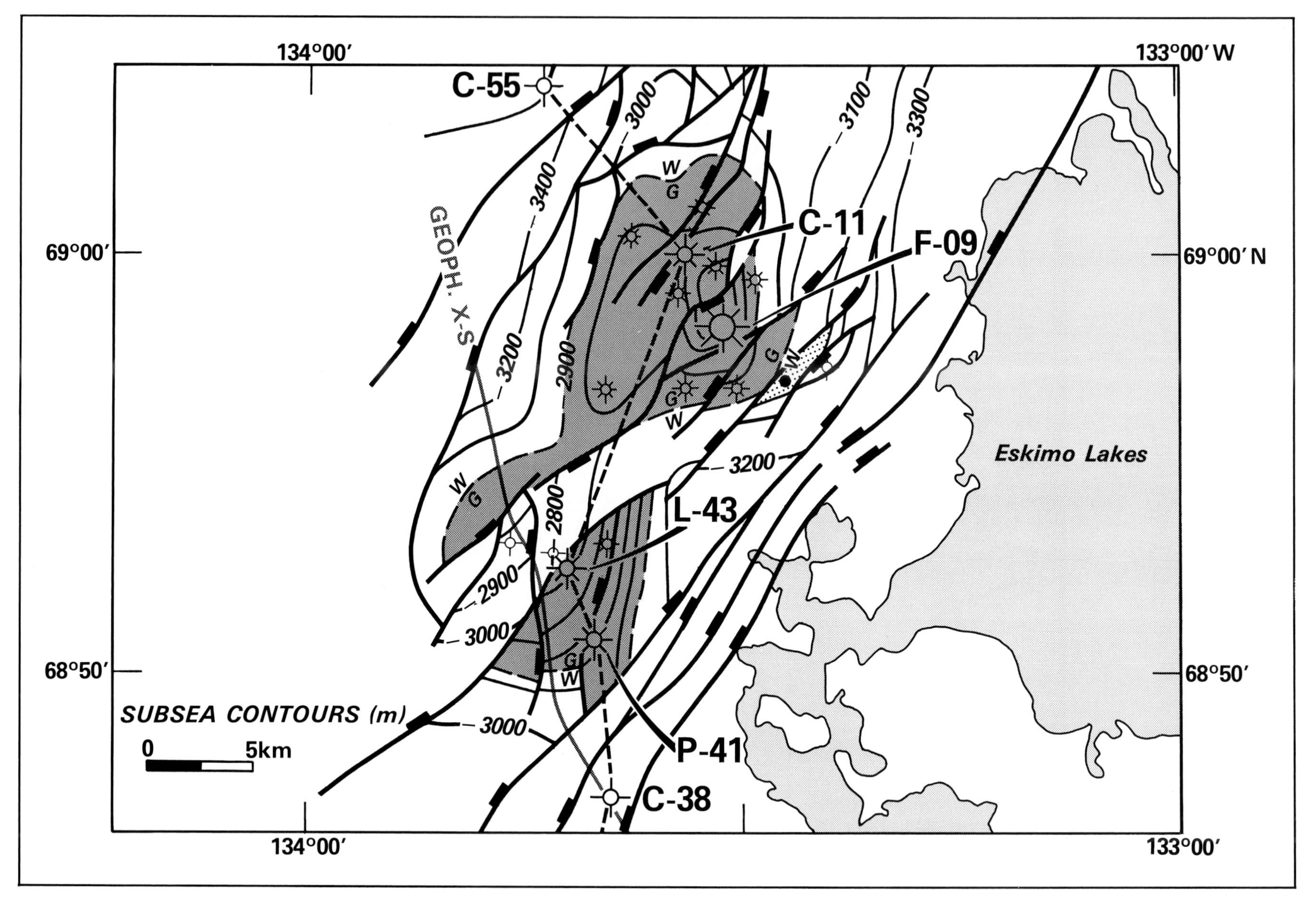


Figure 5. *Structure map of the Parsons Lake structure drawn on the top of the Parsons Group.*

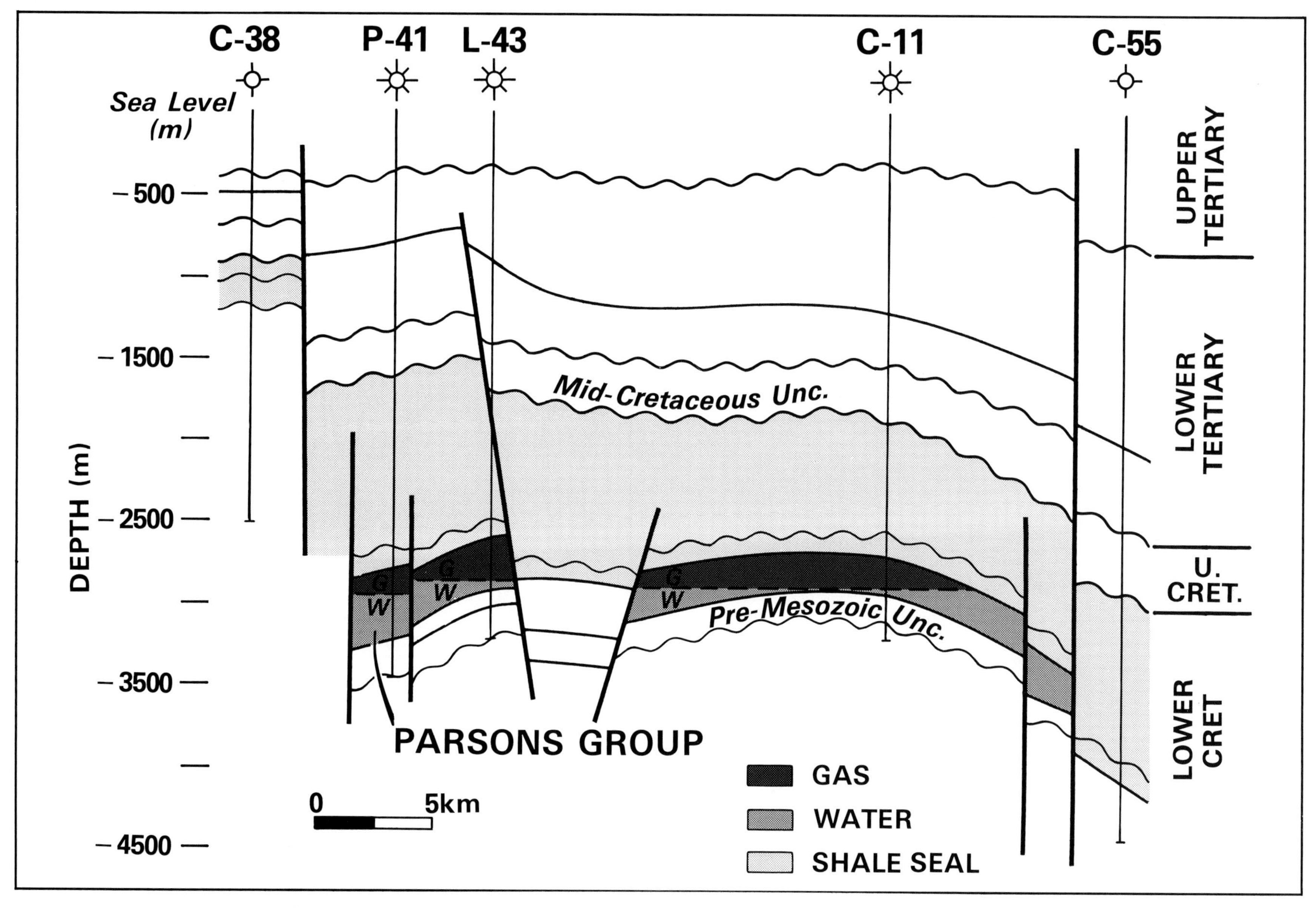


Figure 6. *Structural cross-section across the Parsons Lake field. See Figure 5 for the location of the cross section.*

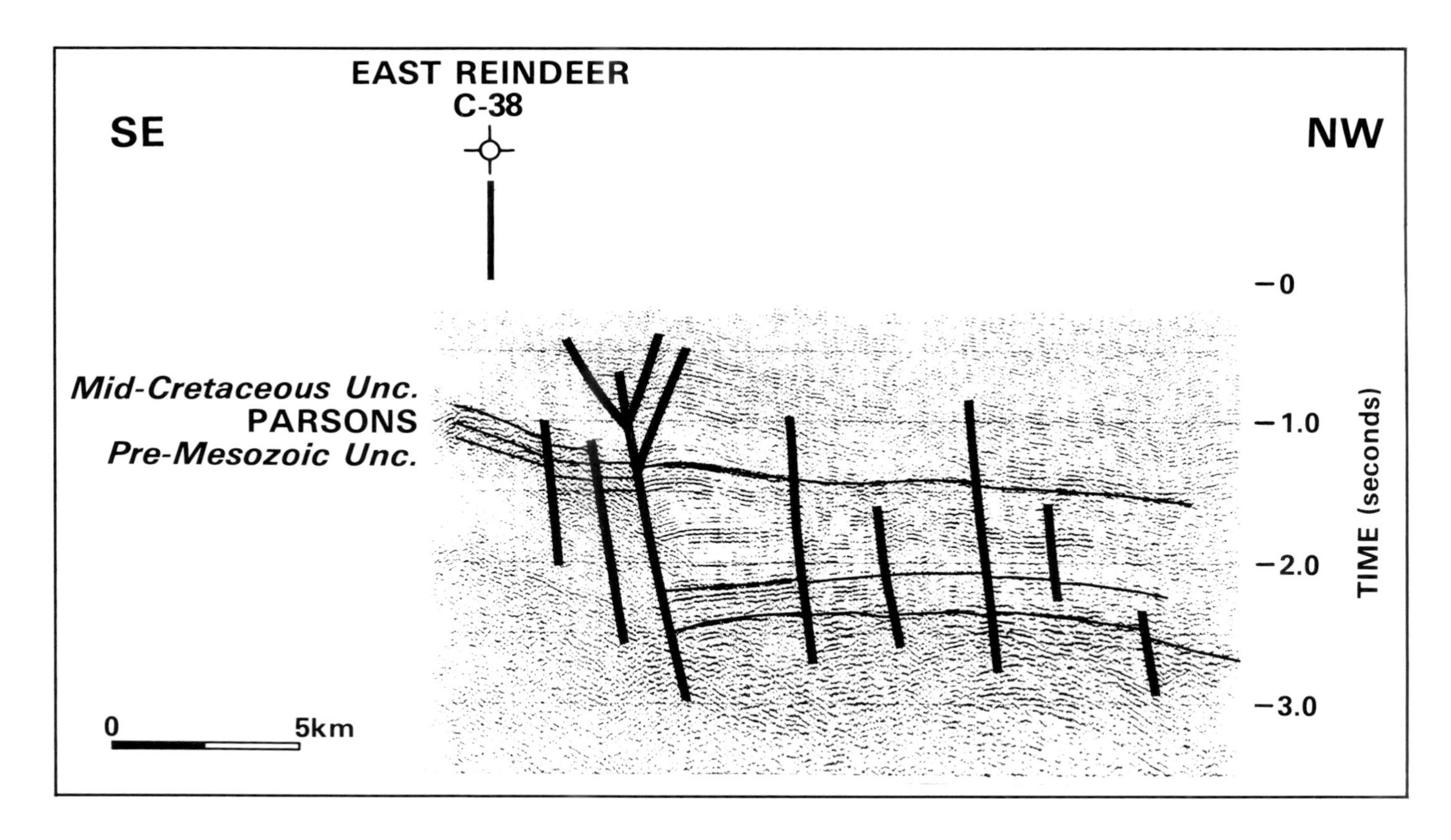
EAST REINDEER
C-38
SE
NW
Mid-Cretaceous Unc.
PARSONS
Pre-Mesozoic Unc.
-0
-1.0
-2.0
-3.0
TIME (seconds)
0
5km

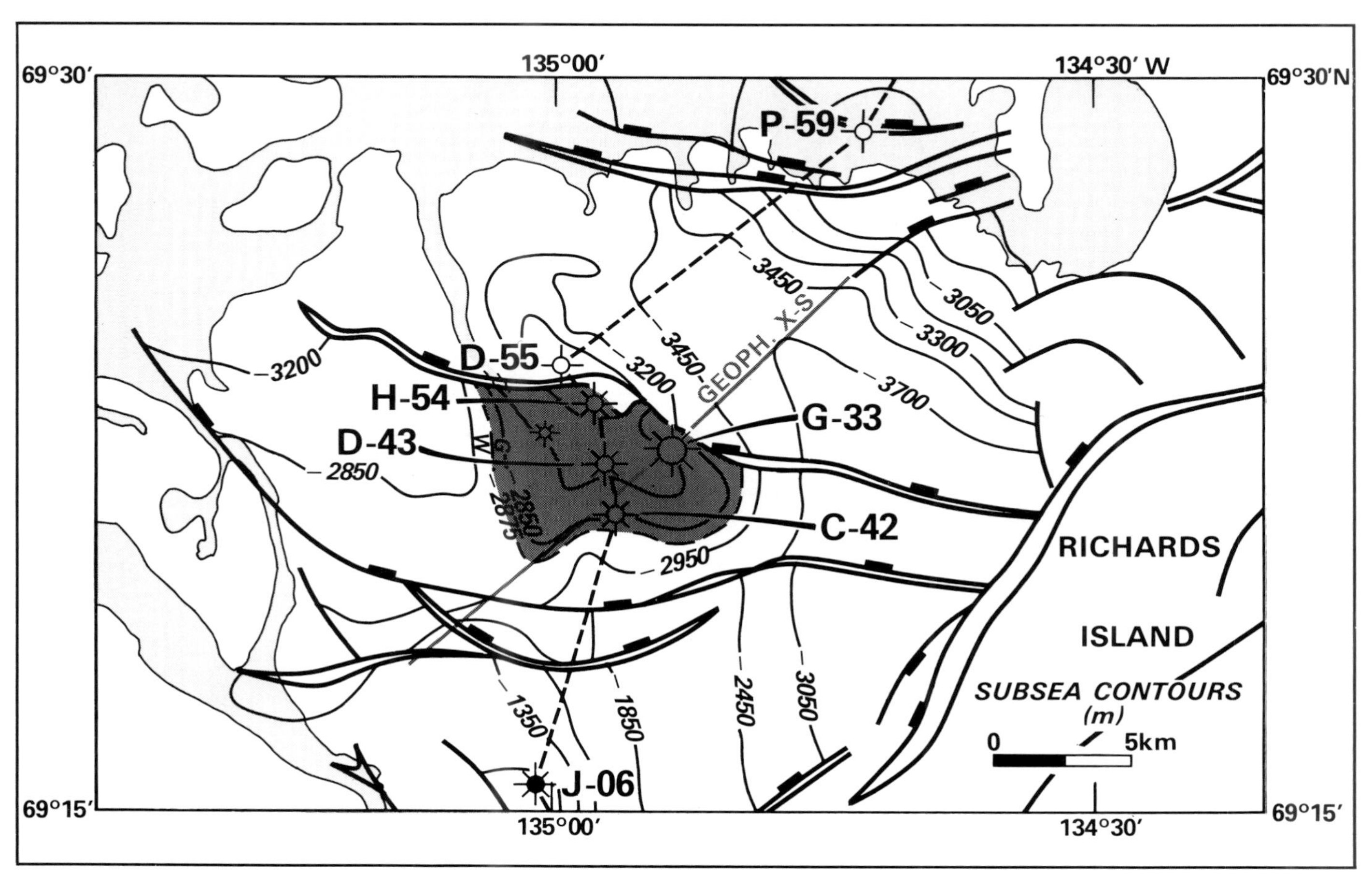
69°30′
135°00′
134°30′ W
69°30′N
P-59
D-55
H-54
D-43
G-33
C-42
J-06
GEOPH. X-S
3450
3050
3300
3700
3200
2850
2950
2875
1350
1850
2450
RICHARDS
ISLAND
SUBSEA CONTOURS
(m)
0
5km
69°15′
135°00′
134°30′
69°15′

Figure 7. *Geophysical cross-section across the Parsons Lake South structure. See Figure 5 for the location of the section.*

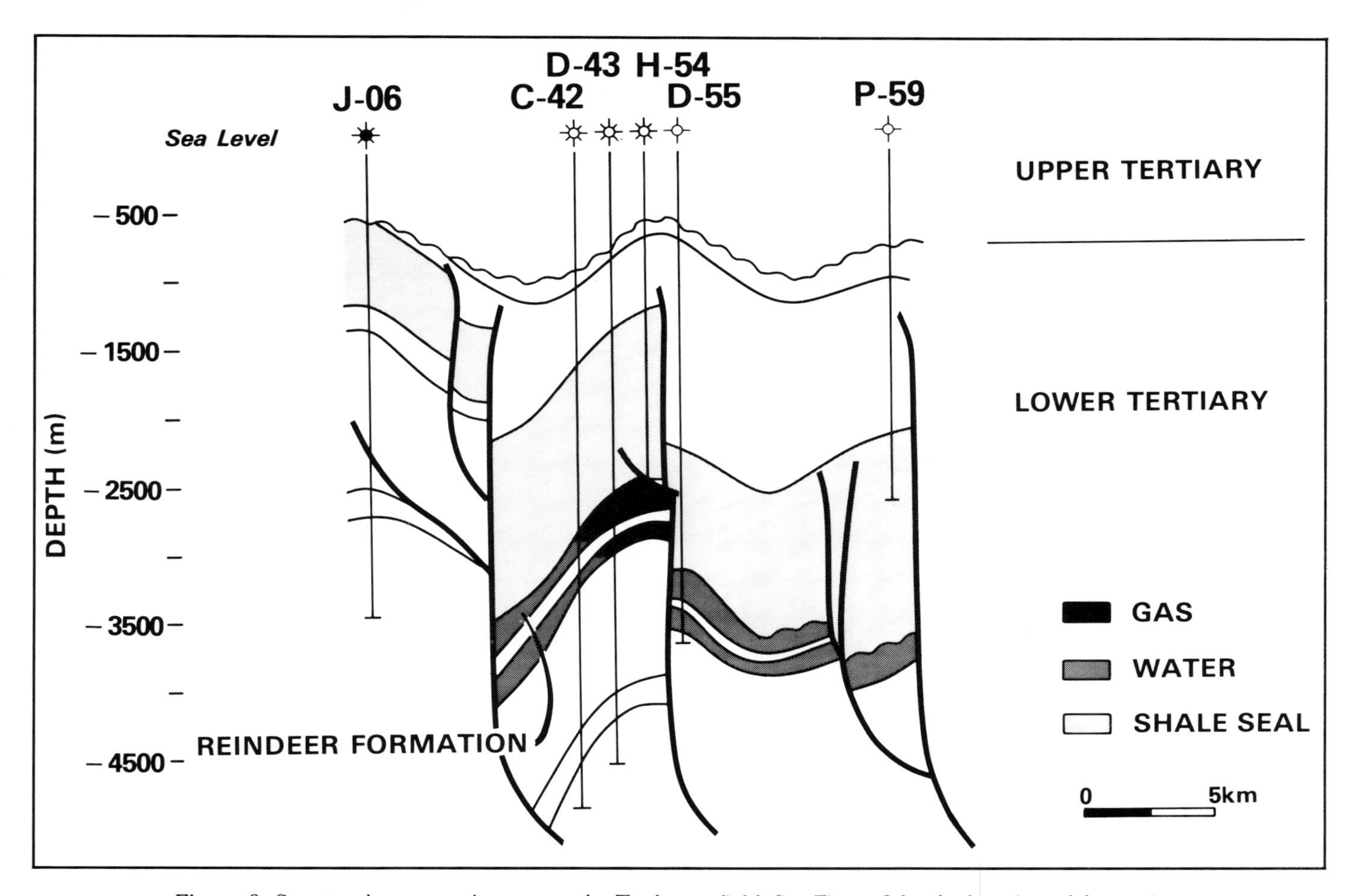


Figure 9. *Structural cross-section across the Taglu gas field. See Figure 8 for the location of the section.*

Figure 8. *Structure map on the top of the Reindeer Formation at the Taglu gas field.*

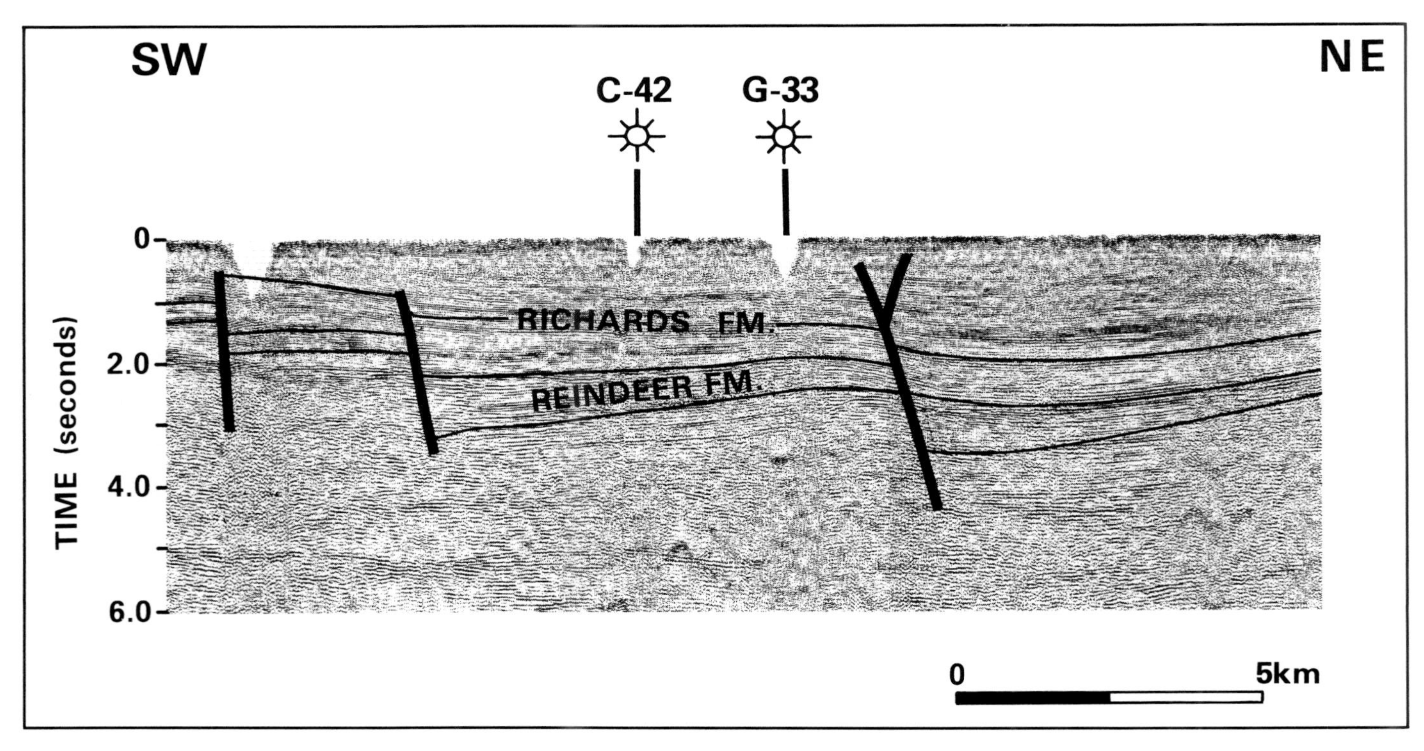


Figure 10. *Geophysical section across the Taglu gas field. See Figure 8 for the location of the section.*

gas contains up to 0.085 cu m/1,000 cu m (15 bbls/mmcf) of condensate.

The factors critical for accumulation are the structure's early development ; the efficient shale top seal which permitted retention of a thick gas column; and absence of late structural uplift and diapirism disrupting the trap.

Type II discoveries are found beneath the modern Mackenzie River delta and in the area out to about the 20 m (66 ft) bathymetric contour. Both oil and gas have been found in this play. There appears to be a general trend toward an increasing pool size and a greater propensity for oil accumulation toward the north.

### Development

Development of oil and gas fields in the Beaufort basin faces severe environmental challenges because of the Arctic climate. The most accessible area is the onshore and offshore portion of the Mackenzie River delta out to the vicinity of the 20 m (66 ft) bathymetric contour. In this area, pads constructed of coarse, granular material can be built to defend against the effects of permafrost and the shore-fast offshore ice. In the deeper-water offshore areas, the moving polar ice pack produces a much more severe and dynamic ice regime that will be significantly more difficult and expensive to defend against.

## SVERDRUP BASIN

### Regional Setting

The Sverdrup basin (Figure 11) underlies the archipelago of the Queen Elizabeth Islands in the high Arctic of Canada. The low-relief western islands are separated by broad ice-scoured channels that deepen rapidly to maximum depths of just over 500 m (1,640 ft).

The Sverdrup basin is an Early Carboniferous to Tertiary basin (Balkwill, 1978) that developed over a strongly uplifted and eroded lower Paleozoic, Proterozoic, and possibly continental basement surface (Figure 12). Upper Paleozoic basin-center evaporites were deposited in a carbonate-rimmed basin, followed by a series of southerly sourced sandstone units interbedded with shale. Sedimentation was almost continuous in the basin center, while on the basin margins minor unconformities truncate portions of the sedimentary section.

Extensional faulting affected the early history of the basin, while compressional tectonics in Late Cretaceous to Oligocene time impressed a westerly diminishing structural grain on the area. Diapiric and non-diapiric salt tectonics were significant throughout the area of the Carboniferous salt basin. Gabbroic sills and dikes are prominent, intruding the sedimentary section to varying levels. A characteristic of the western Sverdrup basin's productive area is that the intrusives lie below the producing horizons. Where the intrusives cut the potential reservoirs, no discoveries have been made. In the western Sverdrup basin, there is a prominent suite of northeast-striking, normal faults that cut Cretaceous and older rocks (Balkwill and Fox, 1982), and that are significant in hydrocarbon pooling in the basin.

The discovered fields in the Sverdrup basin contain 15 million cu m (532 million bbl) of oil and 452 trillion cu m (16 tcf) of gas. Hydrocarbons have been found in two structural settings. The largest pools, Drake Point and Hecla (Figure 13), are found on the basin margin and are combined structural/stratigraphic traps (Type I). The remaining pools are usually found in the cen-

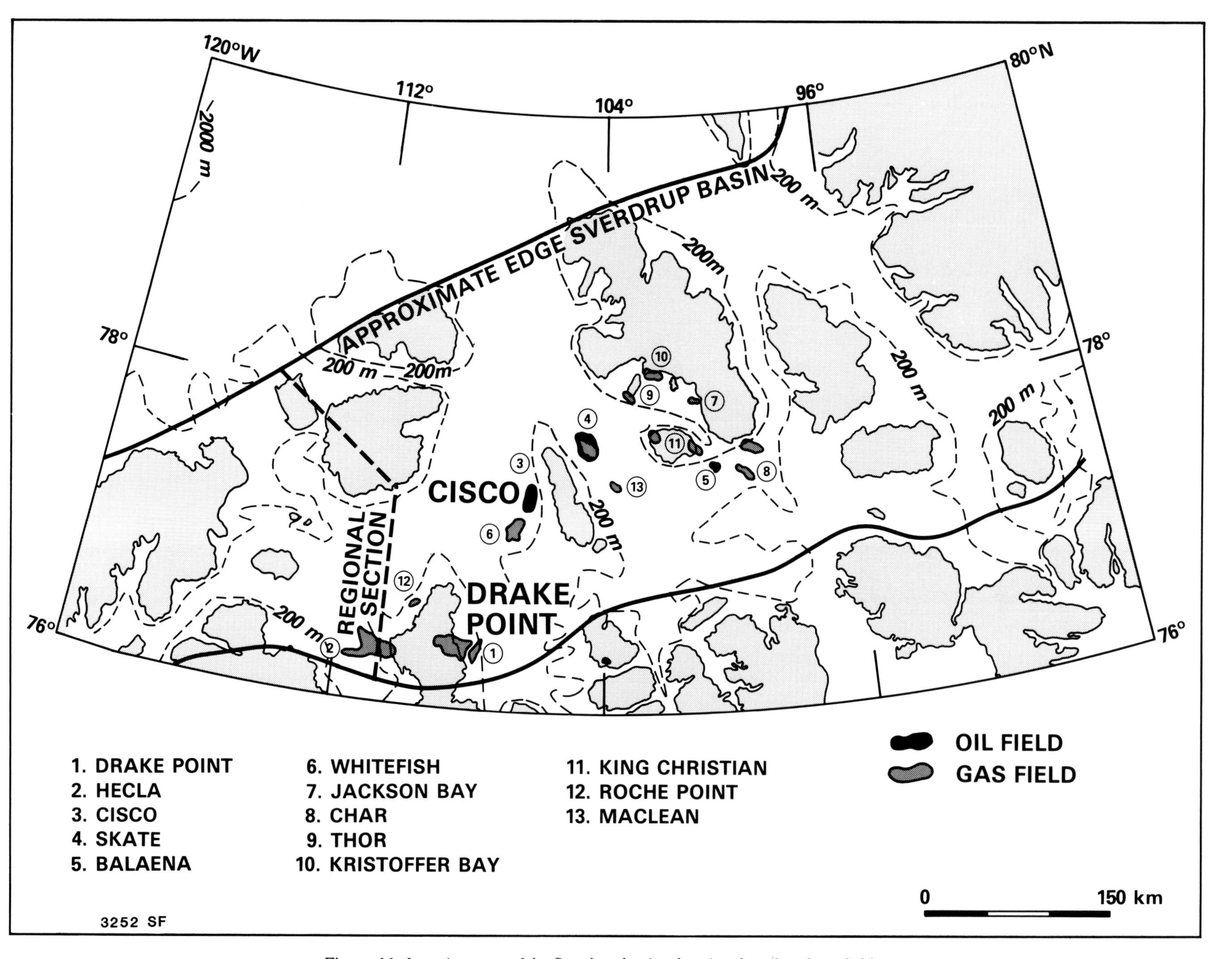


*Figure 11. Location map of the Sverdrup basin, showing the oil and gas fields.*

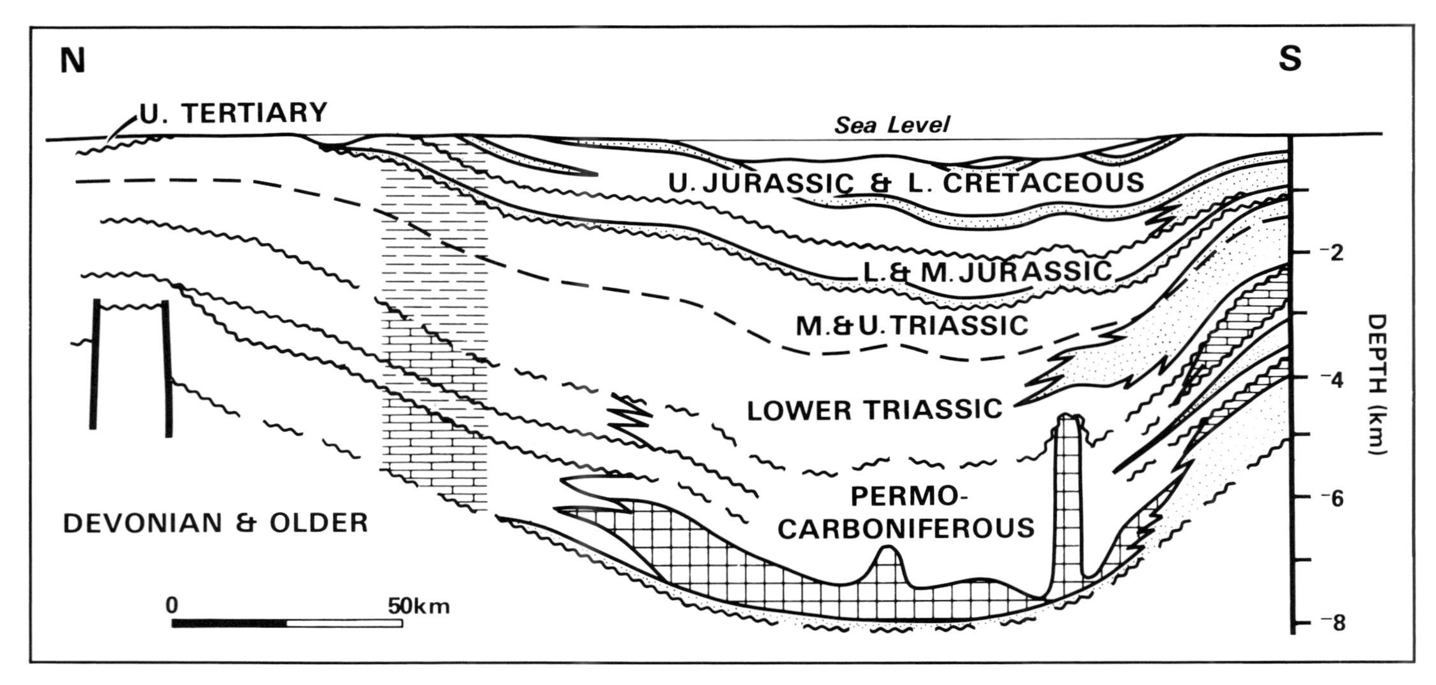


Figure 12. *Regional cross-section across the Sverdrup basin. See Figure 11 for the location of the section.*

tral part of the basin or in high-relief halokinetic structures (Type II).

## Drake Point Gas Field

The Drake Point gas field straddles the shoreline of the Sabine Peninsula on Melville Island (Figure 14) and is an example of a Type I accumulation. The Drake Point anticline is located close to the southern margin of the Sverdrup basin. It is a large, low-relief structure that plunges to the northwest. The northeast-striking normal faults, usually in the form of keystone grabens, segment the field into different pools with gas–water contacts that rise toward the southeast, up the plunge of the structure.

The Lower Jurassic, King Christian Formation is the main reservoir in the field (Figure 15). The reservoir is a beach-bar sandstone that has its depositional strike parallel to the basin margin, which accounts for the remarkably consistent presence and quality of the reservoir throughout the field area and further to the west in the Hecla field, as well.

The reservoir sands are truncated by a minor basin-margin unconformity that lies at the base of the upper unit of the King Christian Formation. The mudstones that underlie the reservoir sands and the silty, glauconitic beds in the upper King Christian combine as the top seal–base seal set to form the stratigraphic component of the Drake Point trap.

The main period of folding that produced the Drake Point anticline was no earlier than Late Cretaceous, as the structure can be mapped at surface (Figure 16). There may have been some earlier subtle drape over a deep-seated structure that localized the early accumulation within the stratigraphic pinch-out; however, we have no concrete evidence for the presence of a paleostructure at either Drake Point or Hecla.

The Drake Point anticline has about 244 m (800 ft) of structural relief and is filled close to spill-point by the 183 m (600 ft) gas column. The total field covers 436 sq km (168 sq mi). The

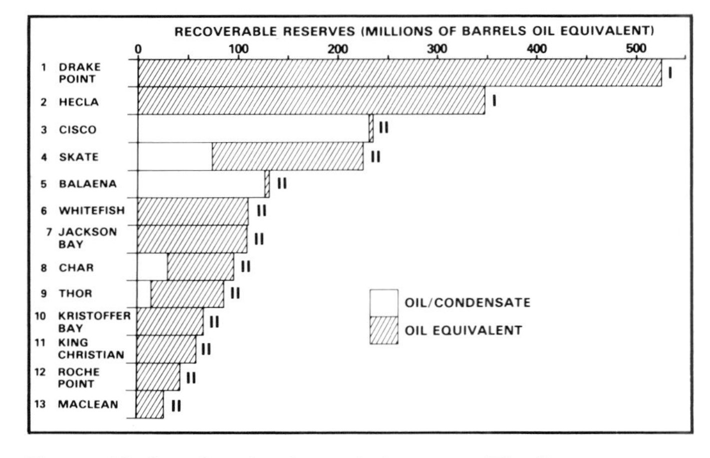


Figure 13. *Sverdrup basin pool size array. The Roman numerals refer to the play type. The ranking numbers have been used to locate the fields geographically on Figure 11.*

net pay averages 20 m (65 feet). The gas is sweet and dry, containing over 97% methane and no hydrogen sulphide.

The northeast-trending faults that provide the plunge closure on the structure are not effective in sealing the thick sand reservoir in the Triassic Bjorne Formation, and only a minor amount of gas is trapped on a local closure in the vicinity of the Drake D-68 well (Figures 14 and 15).

The critical factors in accumulating gas at Drake Point are the combination of the regional stratigraphic trap, the development of the Drake Point anticline, and the presence of the set of northeast-trending faults. Unless the Drake Point structure had a subtle, early expression, the accumulation at Drake Point either took place later than Late Cretaceous or the gas remigrated from the early regional pinch-out trap to the late structure.

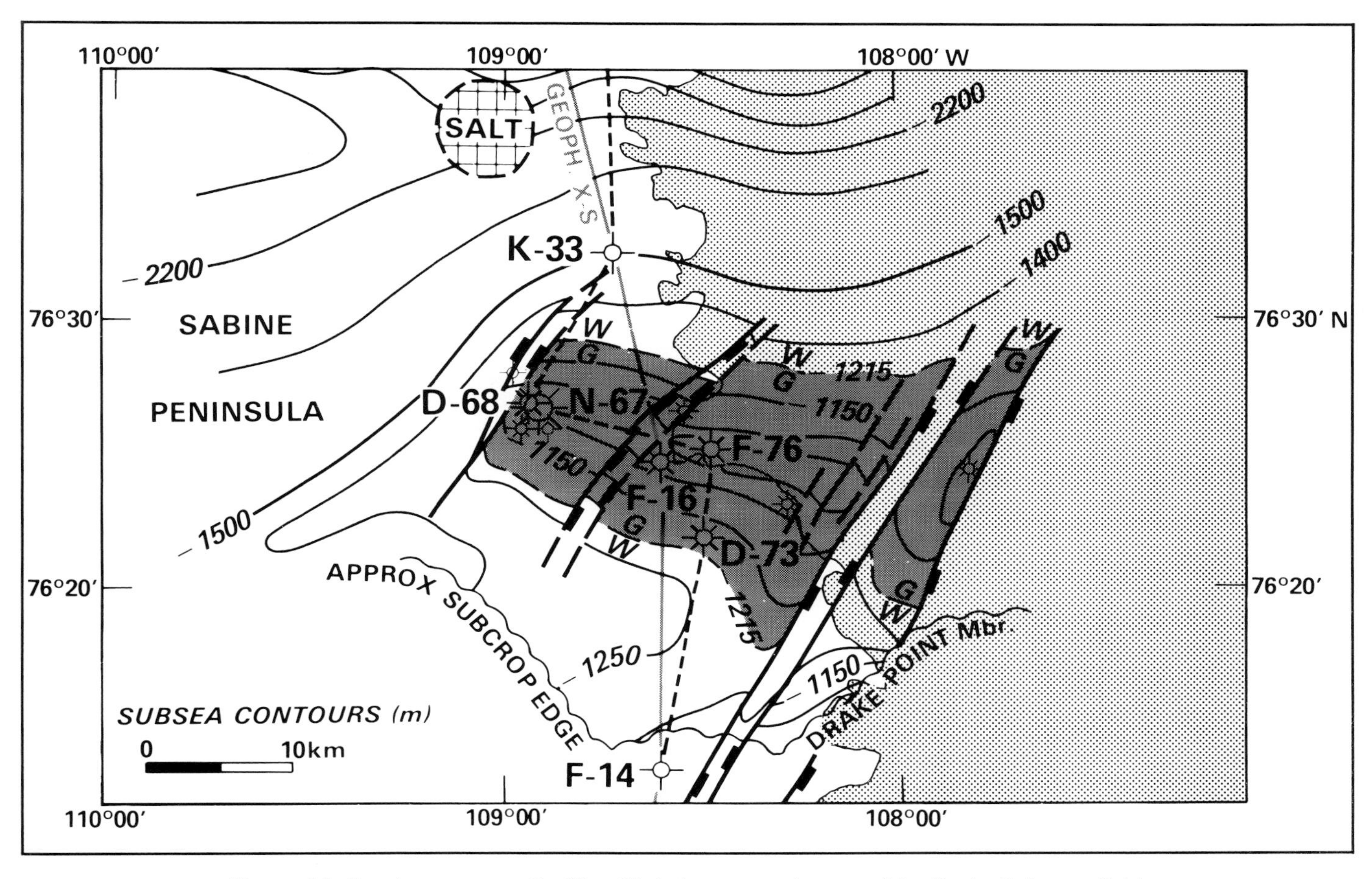


Figure 14. *Structure map on the King Christian reservoir zone of the Drake Point gas field.*

The two Type I fields at Drake Point and Hecla (Figure 13) are the largest hydrocarbon accumulations in the Sverdrup basin. While other similar traps may be present along the southern basin margin, it is unlikely that more fields of the size already found remain to be discovered.

### Cisco Oil Field

The Cisco oil field is located in the western Sverdrup basin offshore from Lougheed Island (Figures 11 and 17). The Cisco structure is a low-relief, northwest-trending anticline. Because the structure is located in an area of thin Carboniferous salt (Balkwill and Fox, 1982, figure 2), it has had less halokinetic structuring than the more typical Type II structures in areas of thicker salt. The fold trend is parallel to the Eurekan structures that are so prominent in the eastern Sverdrup basin. The northeast-trending normal faults are part of the set already discussed. They break the Cisco oil field into pools with different oil–water contacts.

The Upper Jurassic Awingak Formation is the primary reservoir (Figure 18) in the Cisco oil field. The Awingak is one of the Mesozoic sandstone units that prograded a significant distance across the Sverdrup basin. The reservoir is in delta-front sands that lie close to the northern limit of sand deposition in this stratigraphic unit. The grain size and permeability in the sands decreases rapidly toward the north within the field area, while the reservoir is underlain and overlain by shale.

High-quality seismic is difficult to obtain when shooting on the ocean ice in the Arctic. The seismic section shown on Figure 19 illustrates the subtle expression of the structure. It is difficult to map the faults accurately and, therefore, difficult to delineate the individual fault blocks within the field.

At the Awingak zone, the Cisco structure has 110 m (360 ft) of structural relief, but only 55 m (180 ft) of oil column is present. The productive area covers 124 sq km (48 sq mi) and the net oil pay averages 10 m (33 ft). The oil is light gravity, 0.816 SG (42° API), but it has a high wax content and a pour point of 6°C (43°F). The gas–oil ratio is 89 cu m/cu m (500 scf/bbl). The deeper King Christian reservoir contains only a small quantity of oil and gas (Figure 18).

The critical factors for trapping at Cisco include the presence of the Awingak reservoir close to the structure's northern depositional limit; development of a low-relief, halokinetic structure; and the absence of late structural uplift or diapirism. The underfilled nature of this trap and almost all the Type II discoveries suggest that there is a limit to the amount of a hydrocarbon column that can be contained by the overlying shale top seal. While the fractures associated with the northeast-trending faults may have served as conduits for hydrocarbon migration, as suggested by Balkwill and Fox (1982), they may be most significant as escape routes for oil and gas.

Oil in the Sverdrup basin is found mainly in Upper Jurassic and Cretaceous reservoirs (Figure 13) in the axial portion of the

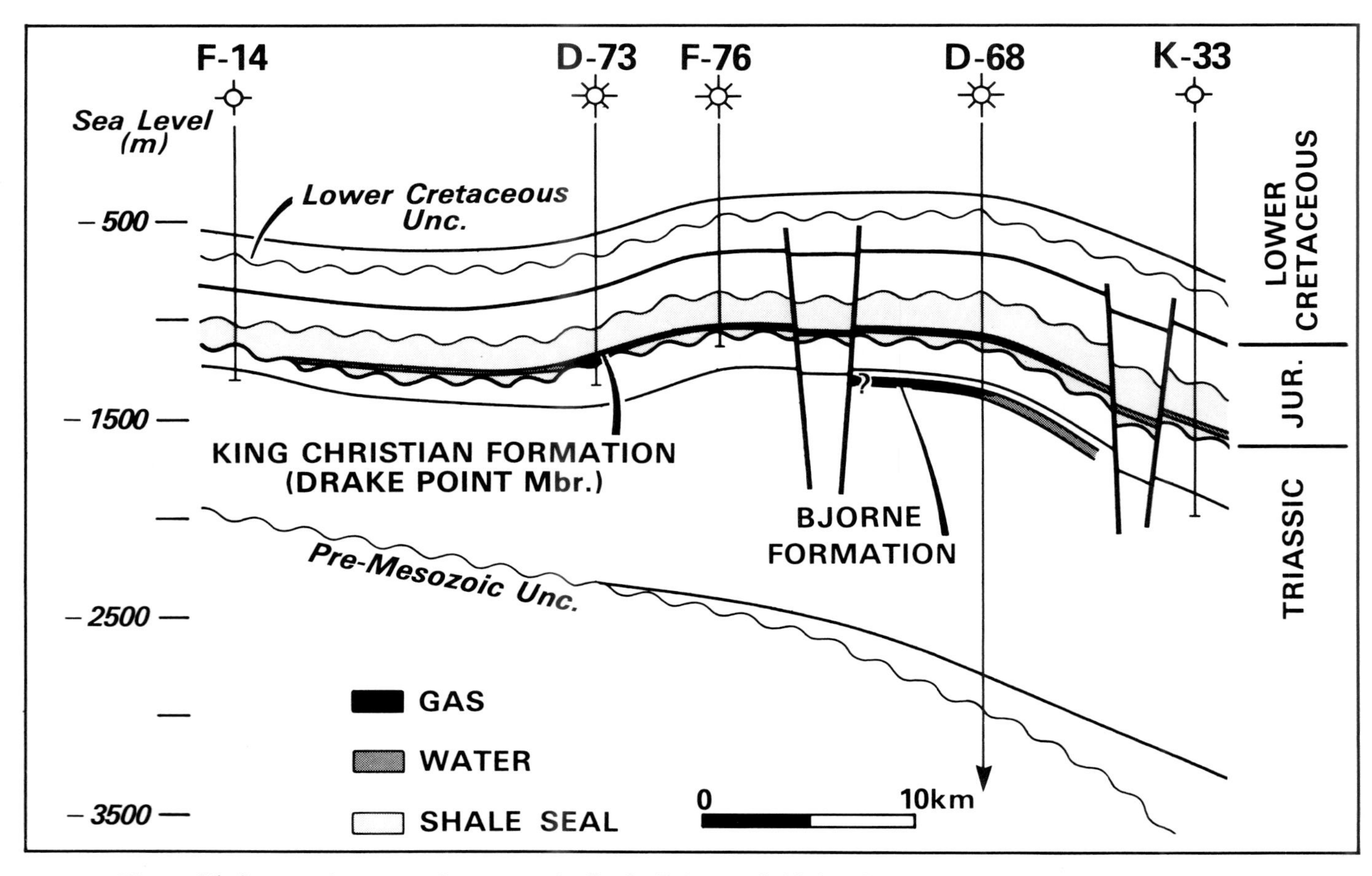


Figure 15. *Structural cross-section across the Drake Point gas field. See Figure 14 for the location of the section.*

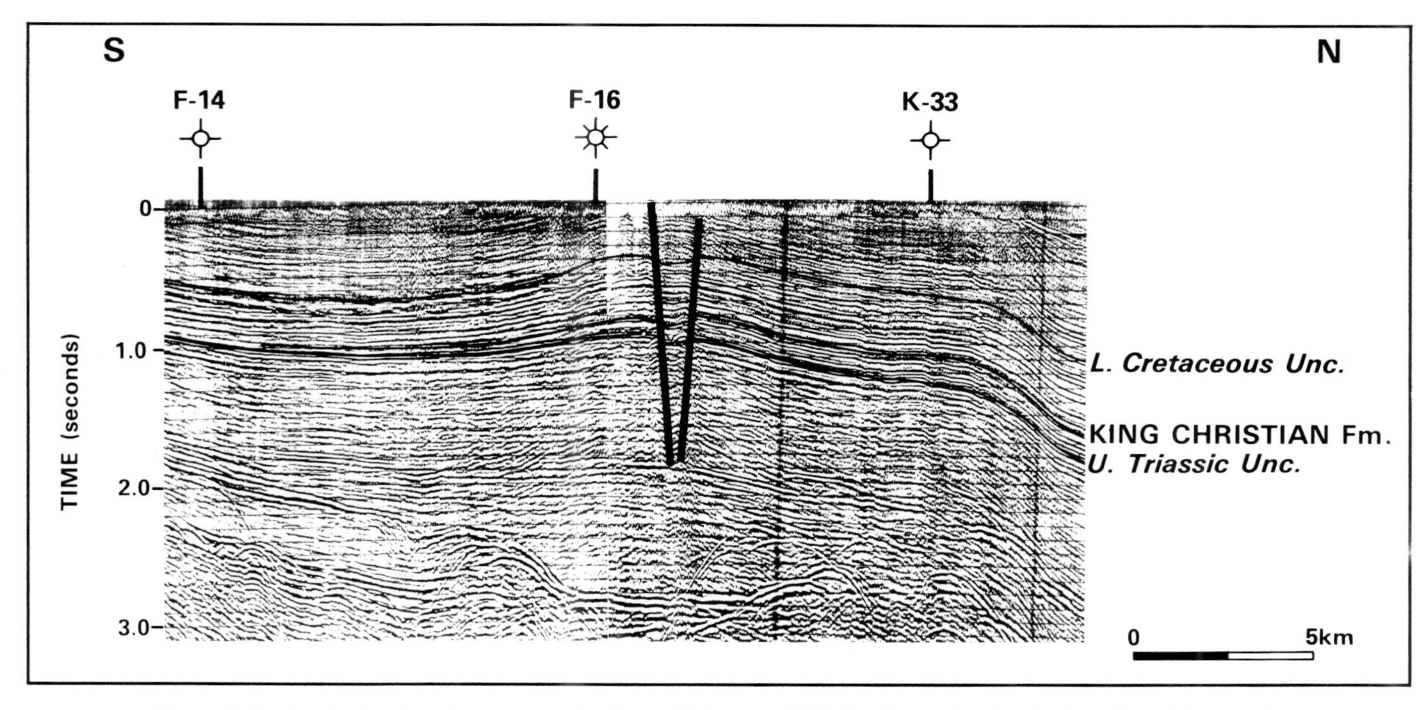


Figure 16. *Geophysical section across the Drake Point gas field. See Figure 14 for the location of the section.*

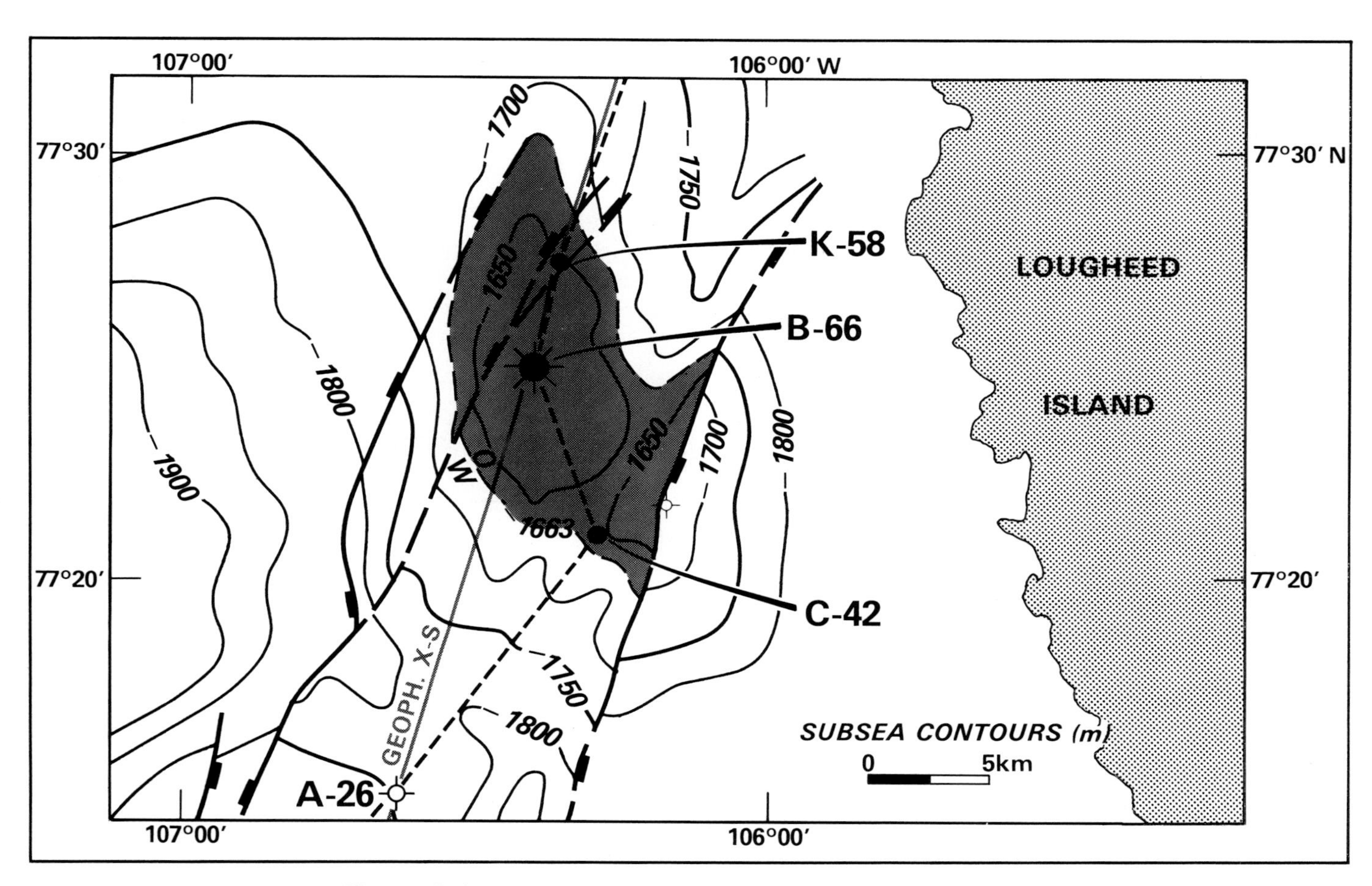


Figure 17. *Structure map on the Awingak sand at the Cisco structure.*

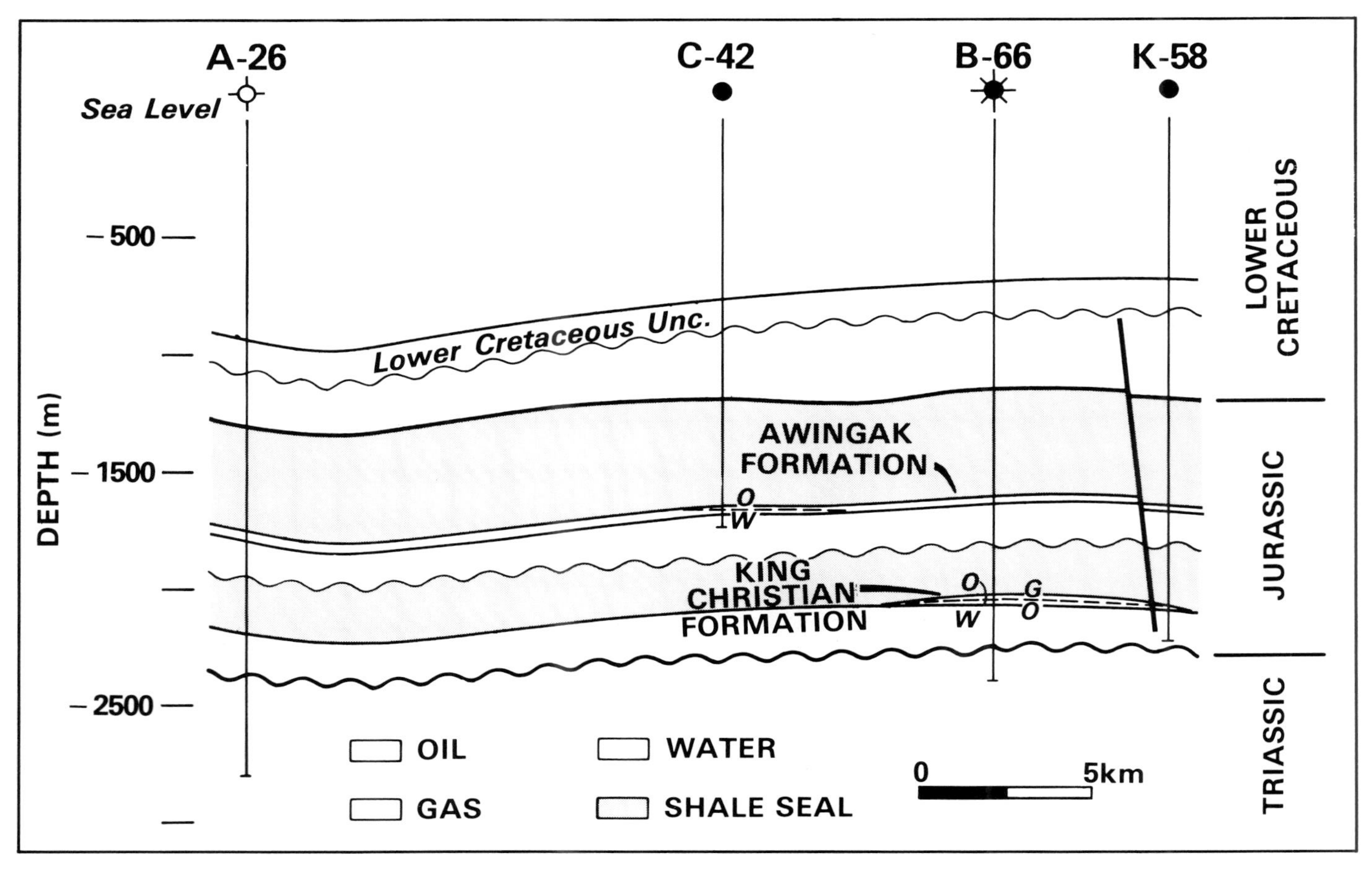


Figure 18. *Structural cross-section across the Cisco structure. See Figure 17 for the location of the section.*

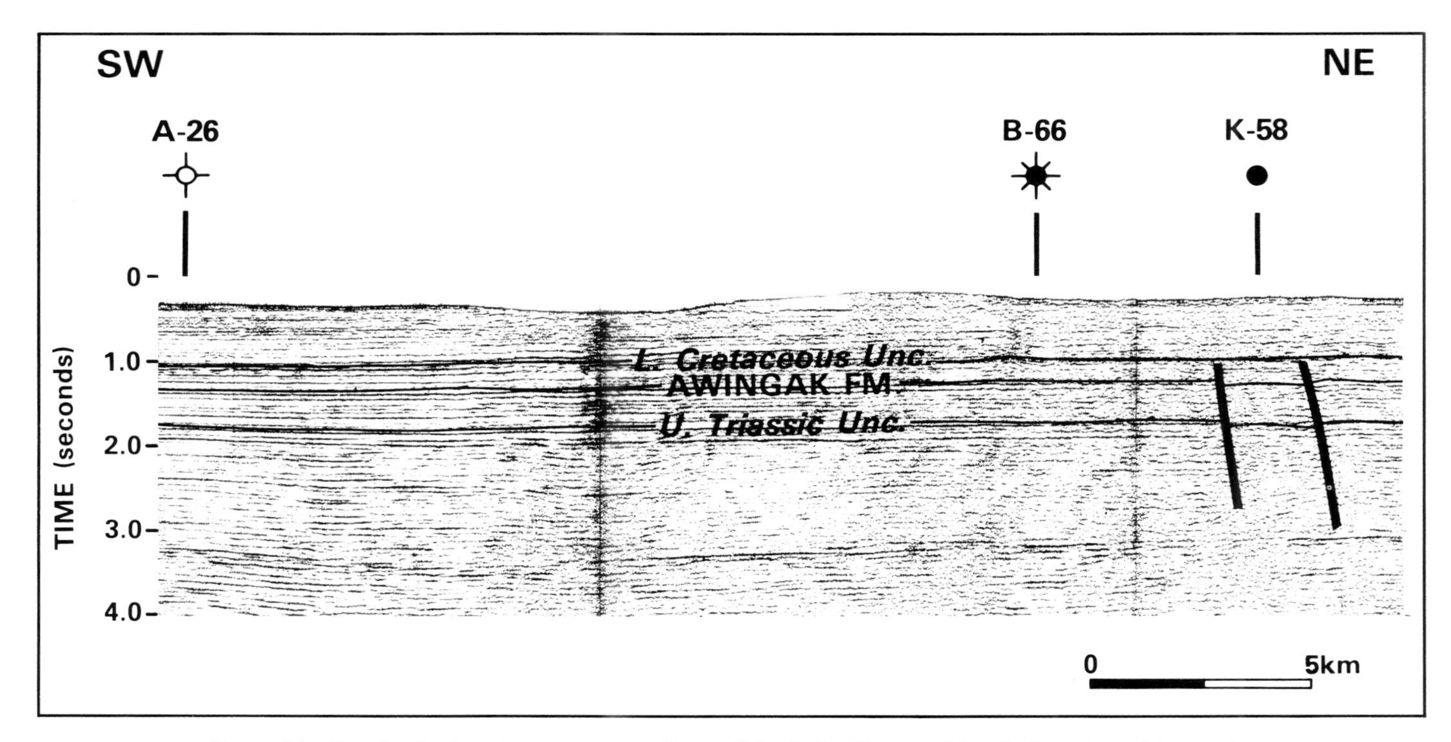


Figure 19. *Geophysical section across the Cisco oil field. See Figure 17 for the location of the section.*

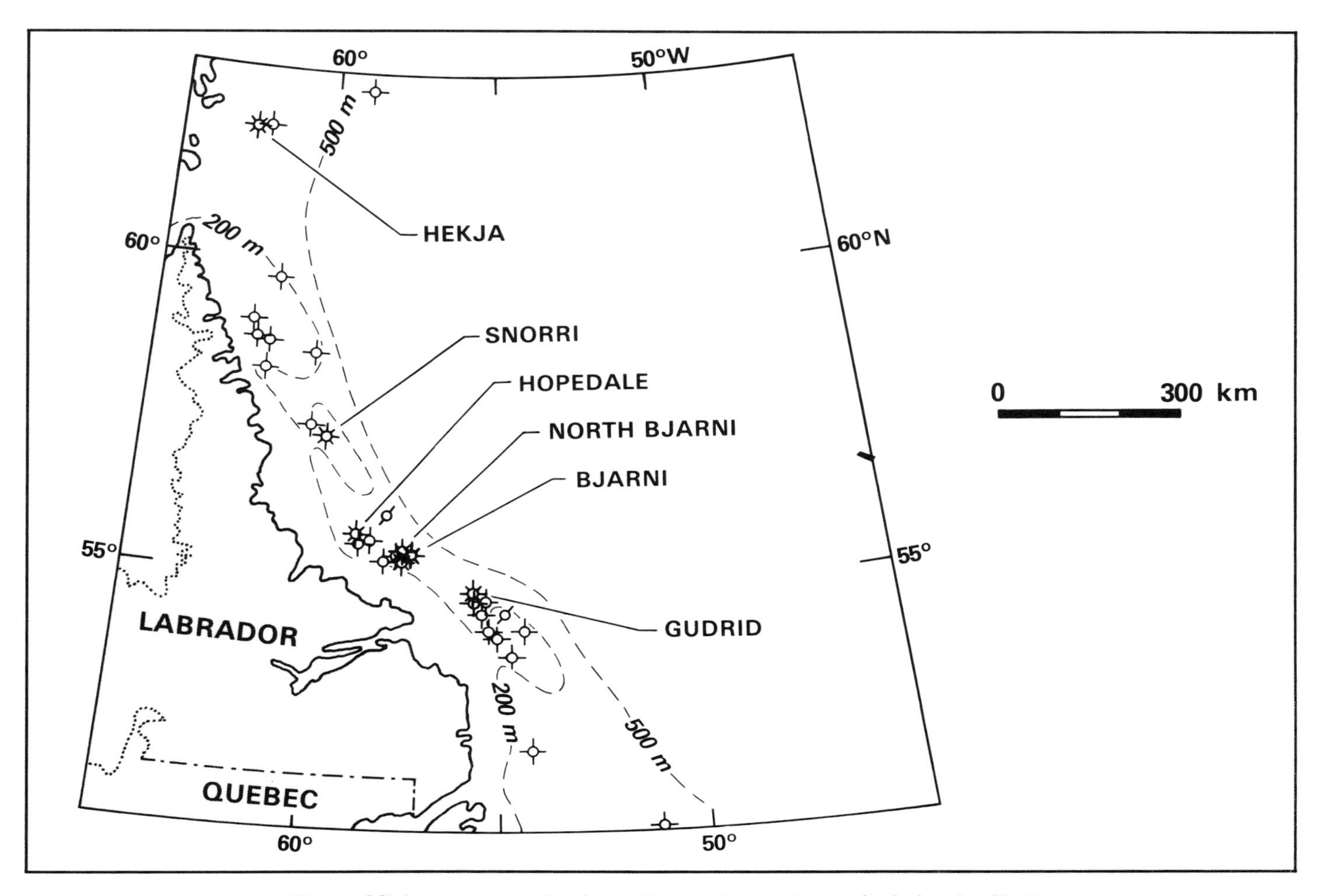


Figure 20. *Location map for the wells and discoveries on the Labrador Shelf.*

basin. The areal extent of these productive stratigraphic units is controlled by erosion at the surface or the sea floor, which limits the prospective area for future discoveries in these zones.

### Development

Developing and marketing oil and gas from the Arctic Islands faces the presence of single to multi-year pack ice throughout the Sverdrup basin and along the shipping route as far as Baffin Bay. Use of ice-breaking tankers carrying either oil or liquified natural gas appears technically feasible, providing that the projects are economically viable. Development of the offshore areas that face the dual problem of deep water and an almost continuous but mobile ice cover present technological challenges that have yet to be resolved.

## LABRADOR BASIN

### Regional Setting

The Labrador basin (Figures 1 and 20) lies beneath the deep continental shelf offshore from the rugged coast of Labrador. The Labrador basin is the second youngest of the four geological provinces that are found offshore of Canada's east coast. These four provinces reflect the sequential evolution of the North Atlantic as first the African plate, then the Iberian plate, and finally the European–Greenland plate separated from its margins.

Geophysical work in the 1960s demonstrated the existence of a continental terrace wedge (MacMillan, 1973) that extended for 1,400 km (870 mi), along the Labrador margin. Subsequent drilling and seismic have confirmed that the middle and outer parts of the shelf are underlain by a seaward-thickening wedge of Cretaceous and Tertiary clastics, which are locally as thick as 10 km (6.2 mi).

The Labrador terrace wedge lies partly on a Precambrian, sialic crust that underwent large-scale extension in the Early and mid-Cretaceous, and partly on a thick pile of Cretaceous and Paleocene basalt flows that may constitute transitional cratonic–oceanic crust, or oceanic crust (Figure 21). The basal sediments of the terrace wedge are river-deposited Neocomian to mid-Upper Cretaceous, immature, felspathic quartz sandstones. These strata were deposited syntectonically in grabens during regional crustal extension. Succeeding, mid-Upper Cretaceous to Pliocene strata consist of terrigenous clastics that accumulated in shallow to deep marine settings during seaward tilting and massive thermal subsidence of the extended continental margin.

Discoveries in the Labrador basin contain 109 billion cu m (3.9 tcf) of gas and 27 million cu m (171 million bbl) of condensate (Figure 22). Two exploration plays have been recognized,

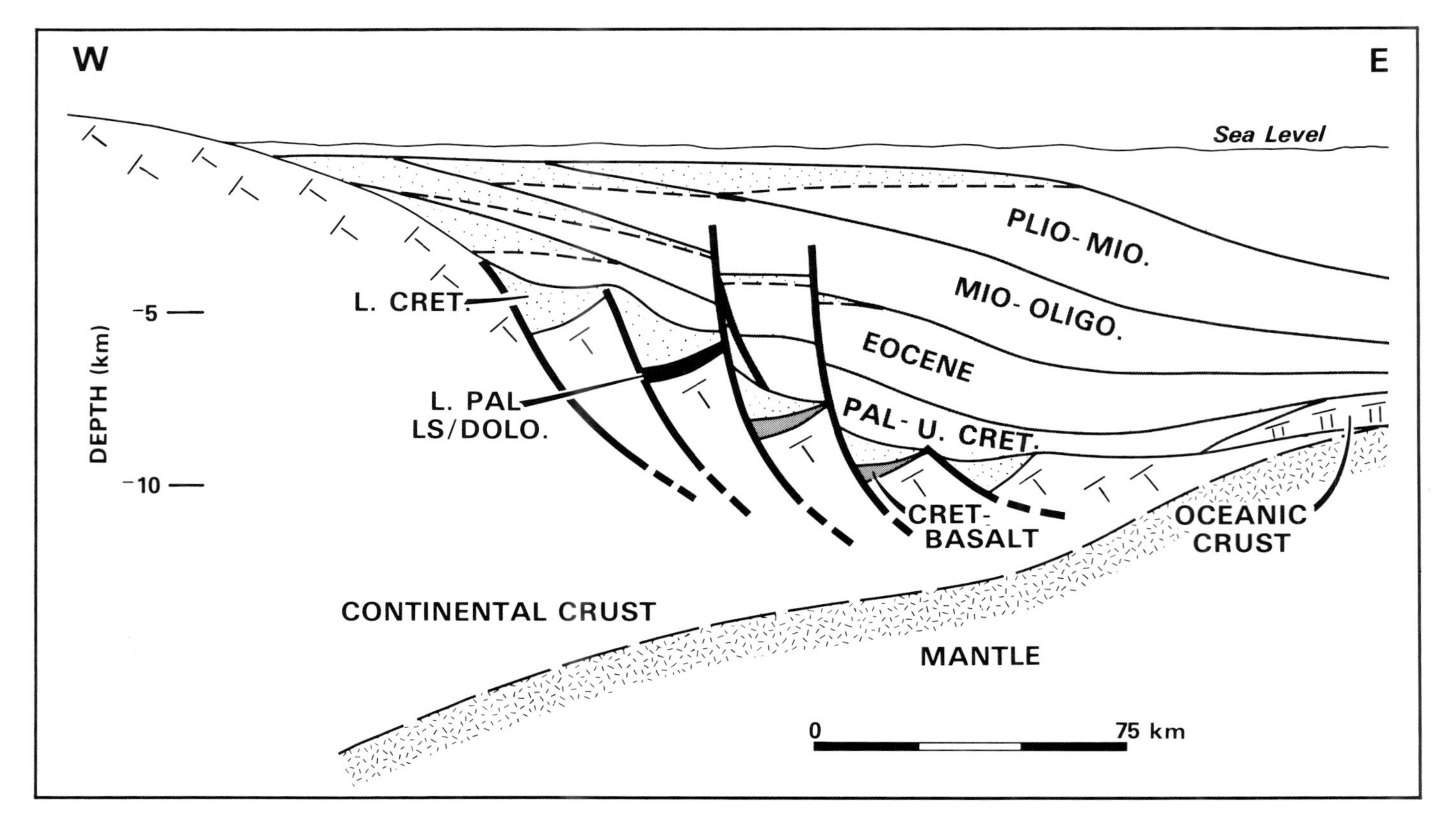


Figure 21. *Schematic cross-section across the Labrador at approximately lat. 55°N.*

and the bulk of the exploration effort has been directed toward the wedge-base play, which has yielded only gas and condensate and a very minor show of oil. The reservoirs involved in the wedge-base play are mainly Lower Cretaceous sands, but include Paleozoic carbonates where they are preserved and porous beneath the pre-Cretaceous unconformity. A wedge-top play involving the regressive Tertiary clastics on growth-fault traps has received only two tests and the potential of this play is not yet known.

## Bjarni and North Bjarni Gas Fields

The Bjarni and North Bjarni gas fields are located just north of lat. 55°N on the Labrador Shelf (Figures 20 and 23). They are closely related in a structural and stratigraphic sense and are typical of the wedge-base type of prospects in the Labrador basin.

Acoustic basement under the southern Labrador Shelf is mainly granite and high-grade metamorphic rocks (Figure 21). Paleo-erosional outliers of Ordovician (and possibly Carboniferous) carbonate rocks lie on the Precambrian surface. The carbonate outliers and the irregularly distributed Neocomian Alexis Basalts are seismically difficult to distinguish from basement.

Bjarni Formation syn-rift clastics are as thick as 1,200 m (3,900 ft) in some of the northwestward-striking grabens and half-grabens under the southern part of the Labrador Shelf (Umpleby, 1979; McWae et al, 1980). The formation consists of two regionally mappable seismic–stratigraphic sequences, separated by an unconformity that is locally angular.

The lower Bjarni sequence ranges in age from Neocomian to

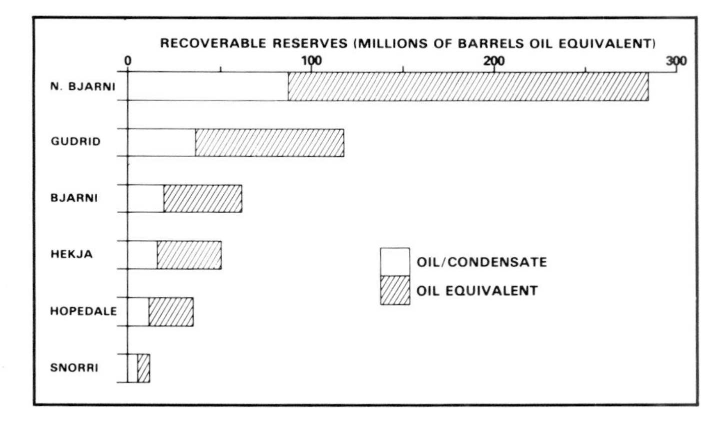


Figure 22. *Pool size array for the discoveries on the Labrador Shelf.*

mid-Aptian and abuts the faulted edges of some horst blocks, onlaps the flanks of syndepositionally rotated blocks, and oversteps the crests of a few horsts. The lower Bjarni strata consist mainly of fine to coarse felsphathic quartz sandstone that was deposited in the grabens by high-gradient, braided rivers.

The upper Bjarni commonly oversteps the lower Bjarni sequence, but is mainly confined to the grabens (Figure 23). The upper Bjarni is finer-grained and contains late Aptian and Albian marginal and shallow marine fossils. In some places fault blocks carrying lower Bjarni sands were rotated into, or remained in, structurally high positions as rifting progressed and remained topographically high, possibly as emergent

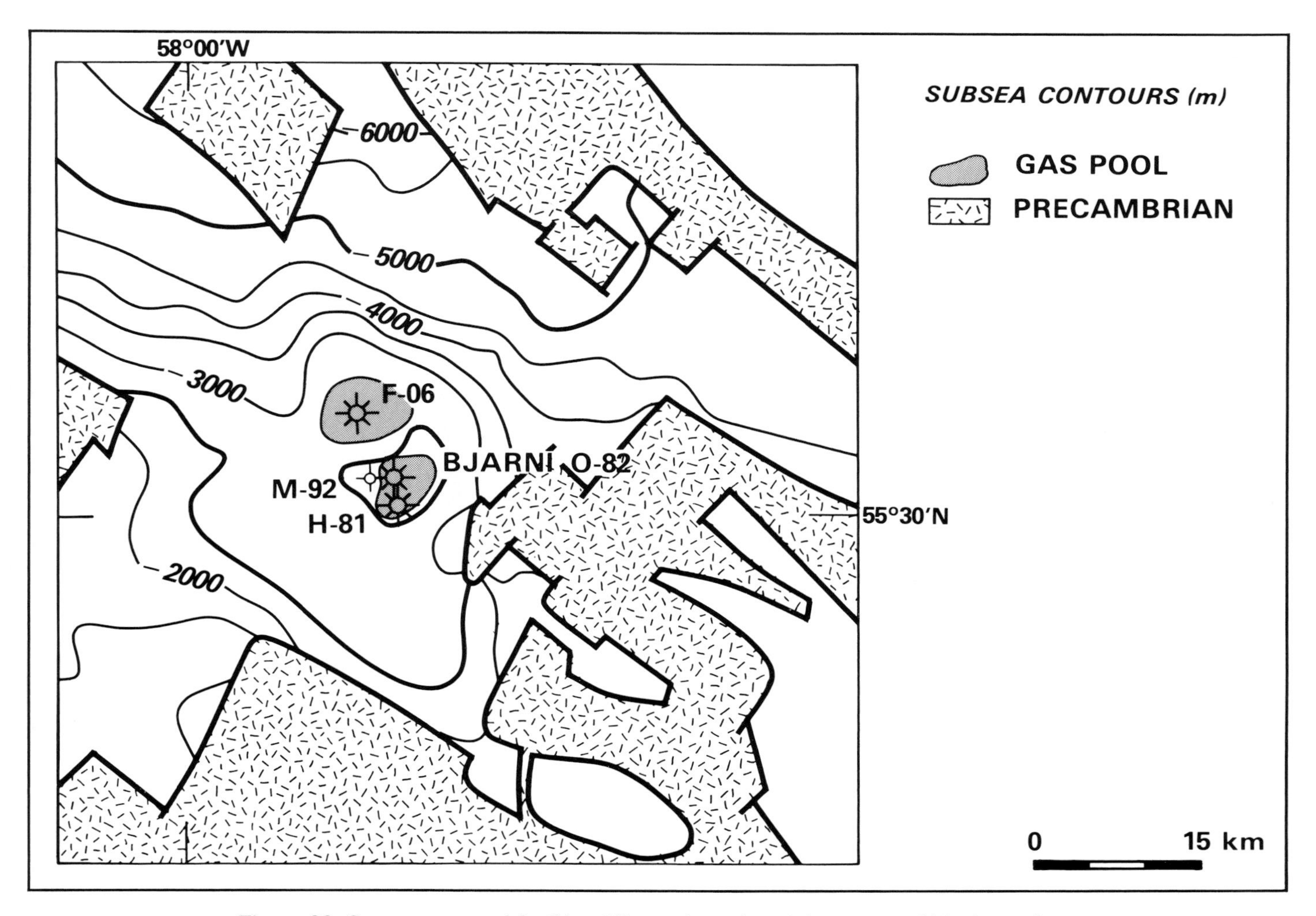


Figure 23. *Structure map of the Bjarni Formation where it is present within the grabens.*

ridges, during the late Bjarni marine incursion.

The seismically derived structural map on the top of the Bjarni Formation (Figure 23) illustrates how even the most widespread, upper Bjarni unit is confined to the rectangularly faulted graben system. The pronounced regional dip to the northeast, produced by seaward rotation during Tertiary subsidence, is also evident. Although the obvious stratigraphic traps on the updip graben flanks have not been tested within this map area, tests of such prospects have proven unsuccessful elsewhere on the Labrador Shelf. The Bjarni and North Bjarni fields (Figure 24) are located on local, structural highs within the graben system.

A thick, widespread succession of Upper Cretaceous marine shales, the Markland Formation, deposited during the early stages of regional subsidence, onlaps and is draped over the Bjarni strata. Most of the structural relief on productive structures was attained by the end of the Cretaceous.

Four wells, three on the Bjarni structure and one at North Bjarni, have been drilled. The lower Bjarni reservoirs in the successful wells consist of felspathic quartz sand partly cemented by calcite and kaolinite. Porosity is derived from calcite decementation. The original calcite cement was emplaced before substantial compaction had taken place and provided a rigid framework that inhibited further compacting. Subsequent dissolution of calcite, probably when the lower Bjarni beds were sub-areally exposed in Late Cretaceous, created the excellent reservoir that characterizes these fields.

The seismic section (Figure 25) across the Bjarni structure illustrates the pronounced differential thinning of both the Bjarni and the Markland, which characterizes the productive structures in this play. There has been no late structural movement.

The North Bjarni F-06 field covers 31 sq km (12 sq mi) and the net gas pay is 79.5 m (260 ft) out of a gross gas column of 177 m (580 ft). The critical spillpoint closure is to the southwest, and the basin rotation during the Tertiary has tended to reduce the capacity of the trap.

The critical factors for accumulating hydrocarbons in this play appear to be, first deposition of the reservoir facies and the subsequent enhancement of the porosity by subareal exposure in Late Cretaceous. A second factor is the development of early structural closure on horst blocks, which was augmented by further differential subsidence as the overlying Upper Cretaceous marine shales were deposited.

The Labrador basin covers a huge area and the limited exploration so far has found only gas and some small shows of oil. Individual graben basins may well have their unique propensity to generate either oil or gas.

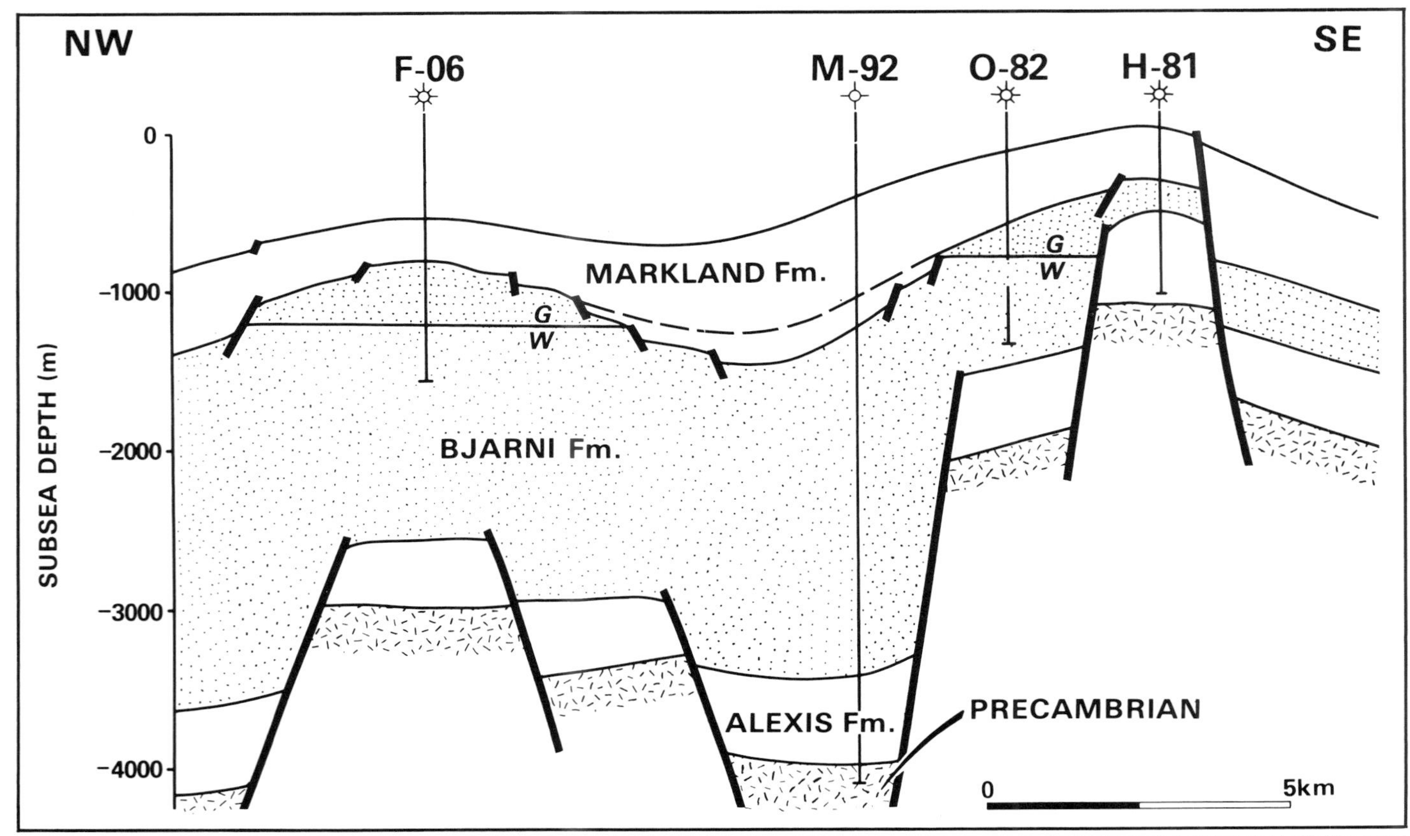
NW
SE
F-06
M-92
O-82
H-81
0
-1000
-2000
-3000
-4000
SUBSEA DEPTH (m)
MARKLAND Fm.
G
W
BJARNI Fm.
ALEXIS Fm.
PRECAMBRIAN
0
5km

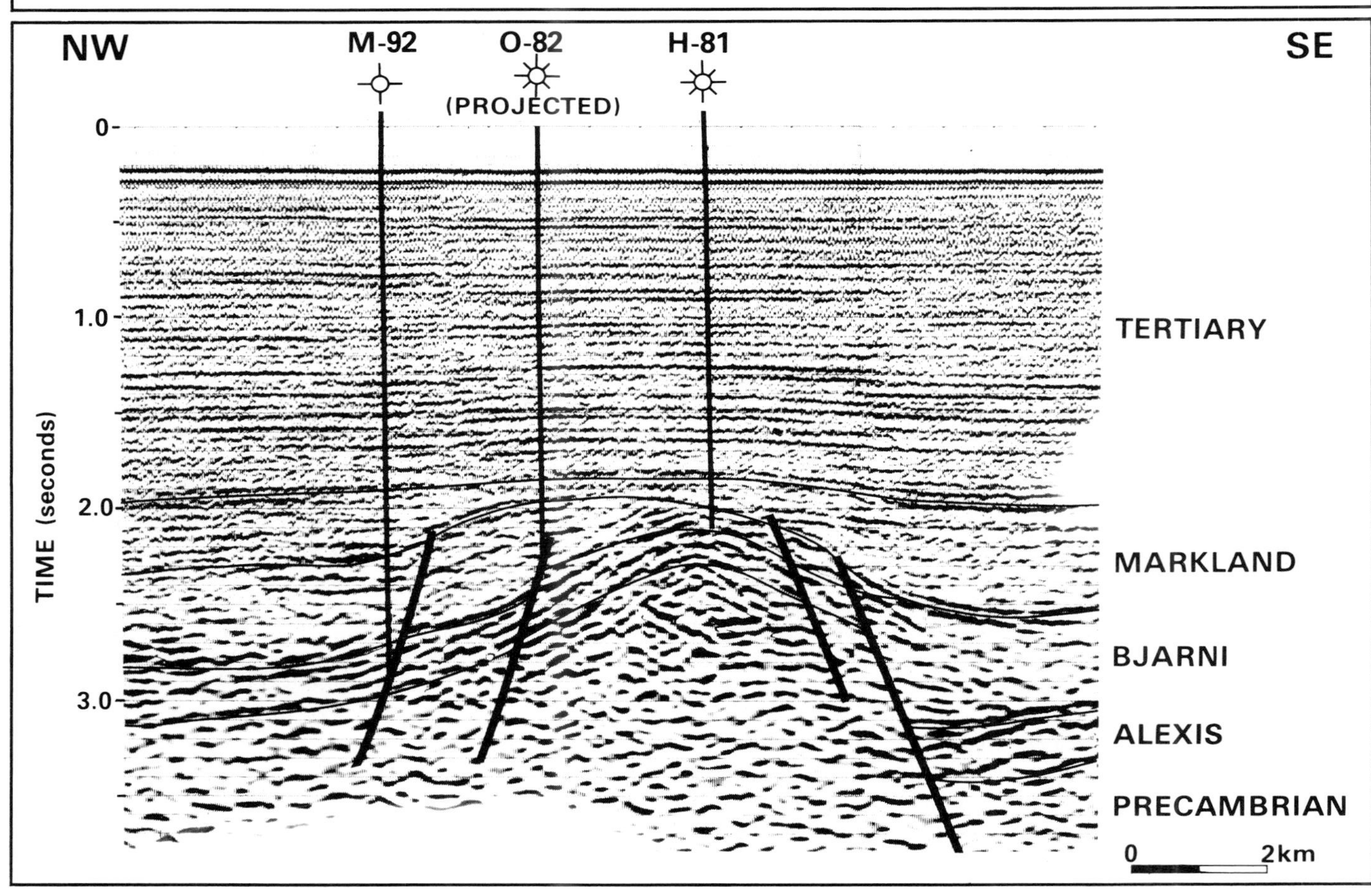
NW
SE
M-92
O-82
(PROJECTED)
H-81
0
1.0
2.0
3.0
TIME (seconds)
TERTIARY
MARKLAND
BJARNI
ALEXIS
PRECAMBRIAN
0
2km

Figure 24. *Structural cross-section across the Bjarni–North Bjarni gas fields. The location of the section can be determined from the well locations on Figure 23.*

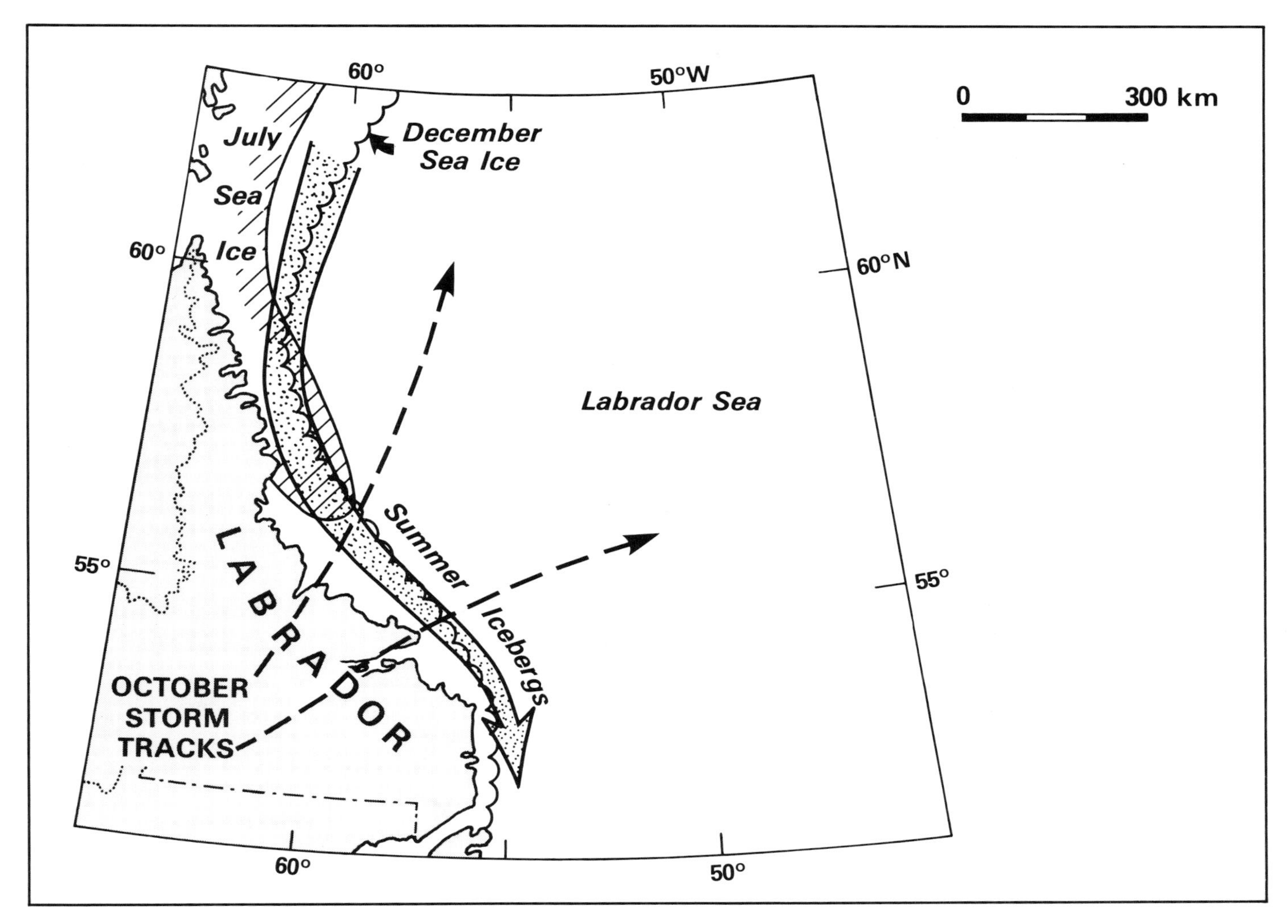


Figure 26. *Operating constraints for exploratory and potential production operations on the Labrador Shelf.*

Development

The Labrador basin is faced with a limited operating season of about 100 days per year. The start date for the drilling season is set by the exodus of pack ice from the vicinity of any specific location (Figure 26), while the end of the season is dictated by winter storms of increasing severity that force the dynamically positioned drillships to leave the area. During the summer drilling season, dynamically positioned drilling rigs are required to avoid the icebergs that are carried southward each year by the Labrador current. Diverting the icebergs by towing has proven to be highly successful in minimizing the number of wellhead disconnects caused by the proximity of icebergs.

## GRAND BANKS BASIN

Regional Setting

The Grand Banks, offshore from the province of Newfoundland, form a broad, shallow-water shelf outlined by the 200 m bathymetric contour (Figures 1 and 27). The Grand Banks are underlain by parts of two major Mesozoic basins, the East Newfoundland basin and the Scotian basin, which are separated by the Avalon Uplift (Sherwin, 1973).

The Grand Banks formed the hub of Laurasia during the Mesozoic and witnessed several episodes of rifting in the course of the sequential separation of the North American and

Figure 25. *Geophysical cross-section across the Bjarni gas field showing the differential subsidence during Bjarni and Markland deposition. See Figure 23 for the location of the section.*

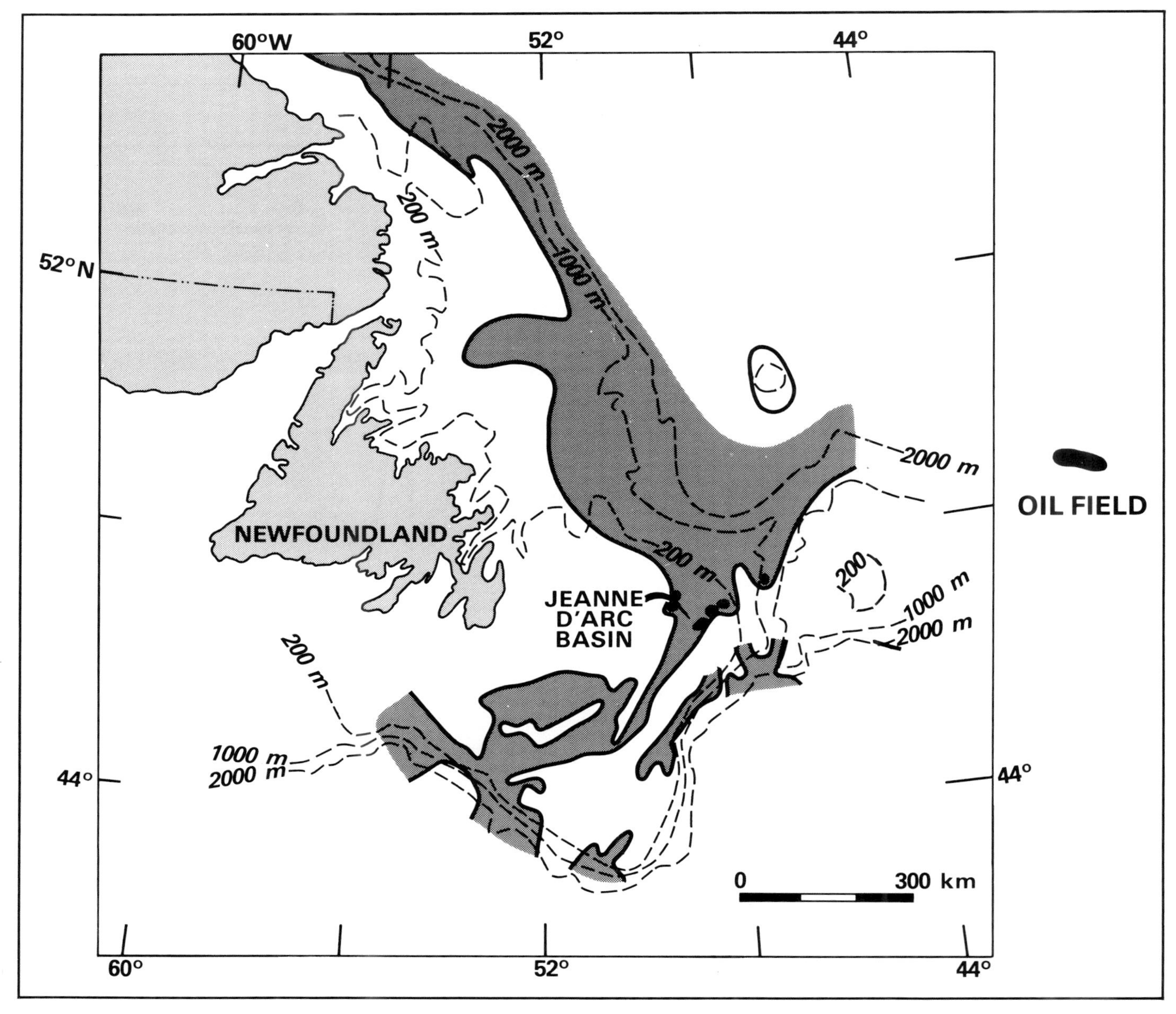


**Figure 27.** *The Grand Banks area, showing the outline of the pre-mid-Cretaceous Mesozoic basin. Note the location of the oil discoveries within the Jeanne D'Arc basin.*

European plates.

The pre-Mesozoic basement comprises a suite of late Precambrian and Cambrian volcanic and sedimentary rocks of the Avalon Terrane (Williams and Hatcher, 1983). Some of the younger Paleozoic sequence that is preserved on the southern Grand Banks contains porous sandstone. Basement lineaments inherited from Proterozoic and Paleozoic tectonism appear to have exerted a significant control on the development of Mesozoic fault patterns.

Early exploration defined a series of Mesozoic basins (South Whale, Whale, Horshoe, Jeanne D'Arc, and Carson) that contain significant thicknesses of pre-mid-Cretaceous, Mesozoic sediments (Amoco Canada and Imperial Oil, 1973, figure 7). All the discoveries that have been made are confined to the Jeanne D'Arc basin, the southwestern arm of the much larger East Newfoundland basin that extends far to the north, into deeper water beyond the Grand Banks.

The Mesozoic basins on the Grand Banks are overlain by a wedge of post-mid-Cretaceous sediments. This young wedge thickens markedly toward the margins of the banks. Where Tertiary sedimentation could not keep pace with the post-rift thermal subsidence, the water depths increase dramatically: for example, the north end of the Jeanne D'Arc basin.

## Jeanne D'Arc Basin

The Jeanne D'Arc basin is bounded by prominent faults with a major, listric normal fault offset by wrench faults forming the western boundary (Figure 28). Active subsidence along the

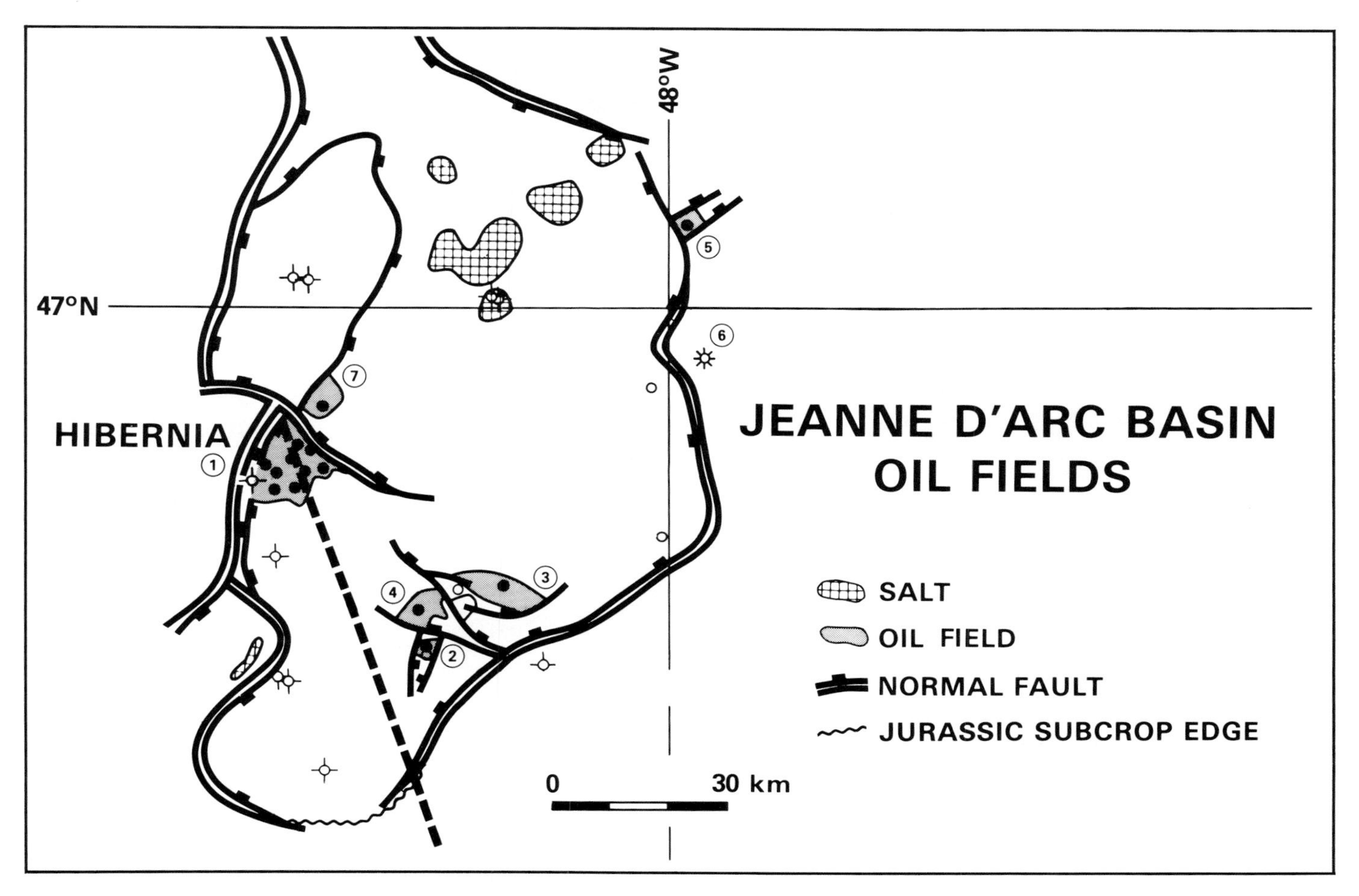


Figure 28. *Jeanne D'Arc basin location map. The numbers on the discoveries are the ranking numbers from the pool size array in Figure 31.*

western boundary fault (the Murre fault) has resulted in major reservoir facies stacking, particularly in the Hibernia oil field. The eastern ramp of the half-graben is structured with a suite of faults rather than a single fault system. Deposition on this ramp resulted in onlapping of thin, but laterally extensive, reservoir zones such as those found in the Terra Nova, Hebron, Ben Nevis, and South Tempest discoveries. The uptilted eastern edge of the half-graben forms the Central Ridge Complex (also referred to as the Outer Ridge Complex), which is dissected by cross-cutting faults. Non-deposition and erosional stripping have led to thinning, which has removed much of the Lower Cretaceous and Upper Jurassic section containing the oil-bearing reservoirs in the Jeanne D'Arc basin from the Central ridge Complex.

The geometry and stratigraphy of the Jeanne D'Arc basin are illustrated in the two cross-sections (Figures 29 and 30) that extend from northwest to southeast across the basin. Rotation along the Murre fault is reflected in the very thick Jurassic section. The thickest accumulations of Mesozoic sediments exceed 14 km (8.7 mi). The basin plunges to the north and, as a result, progressively more of the stratigraphic section is preserved at the various unconformities northward along the plunge axis of the basin. A cluster of shallow, piercement salt diapirs are found in the northern depocenter of the basin; to the southwest a smaller area of diapiric salt is associated with the Egret Structure. Deeper-seated halokinetic and halotectonic salt structures are also evident on seismic (Figure 30).

Figure 29 summarizes the stratigraphy of the Jeanne D'Arc basin. Deposition of Triassic red beds and Triassic–Jurassic evaporites was followed by a Lower to Middle Jurassic shale-carbonate sequence. Renewed rifting led to deposition of the interbedded sandstone-shale units that contain the reservoir facies for the oil accumulations in the basin. The rift phase of sedimentation was terminated with the mid-Cretaceous (Aptian) breakup–unconformity event.

Discovered reserves in the Jeanne D'Arc basin include 297 million cu m (1.9 billion bbl) of oil and 46 billion cu m (1.6 tcf) of associated gas, found in seven discoveries (Figure 31).

### Hibernia Oil Field

Significant hydrocarbon discoveries in the Jeanne D'Arc basin are confined to uppermost Jurassic and Lower Cretaceous beds. There are four principal reservoir zones (Figure 29): the Jurassic Jeanne D'Arc sandstones, and the Lower Cretaceous Hibernia, Catalina, and Avalon sandstones. Because each of these stratigraphic intervals was tectonically controlled, they have equivalents about the margin of the basin (Arthur et al, 1982).

A fifth zone not recognized at Hibernia was developed during the Aptian transgression over the mid-Cretaceous unconform-

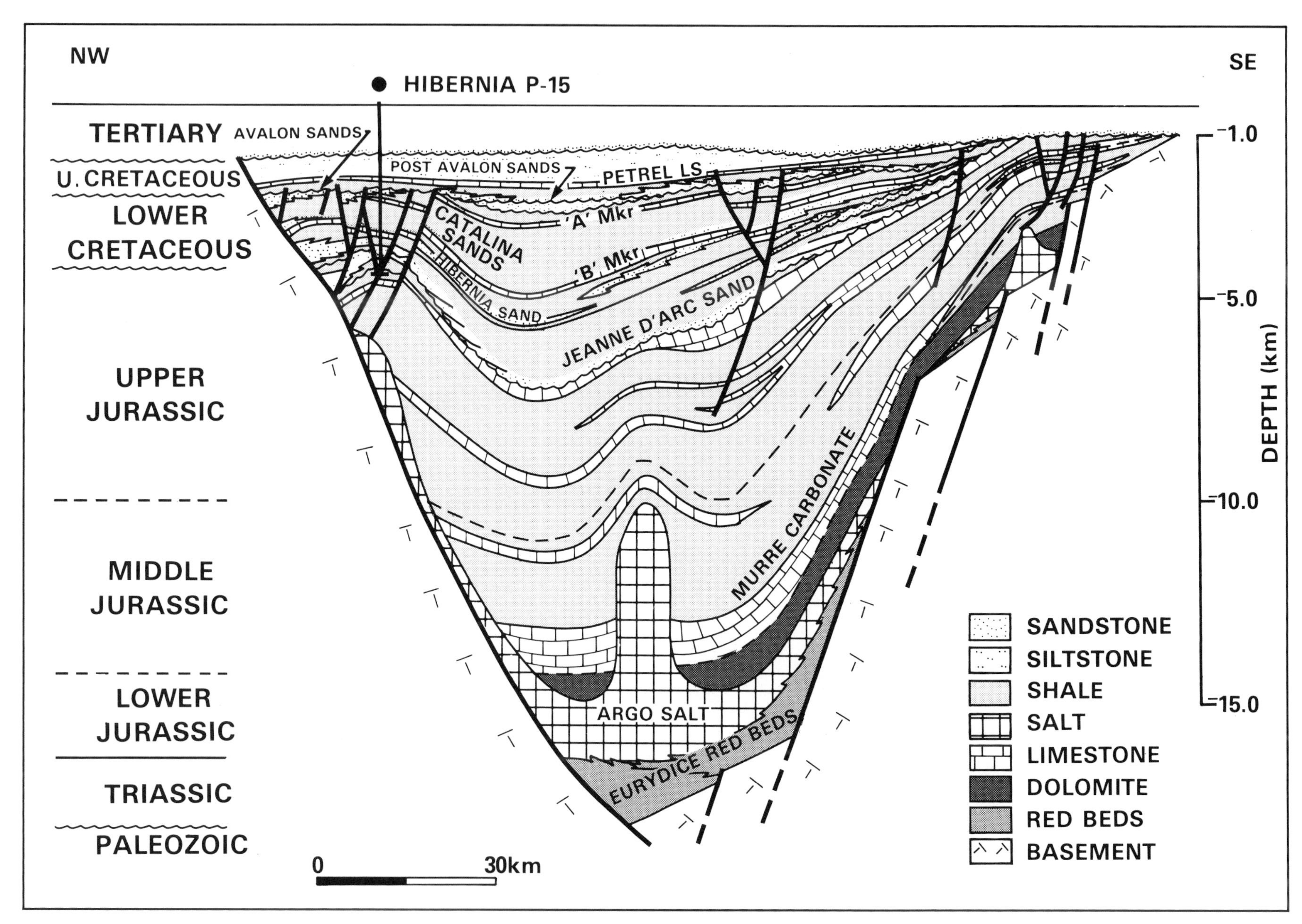


Figure 29. *A geological cross-section across the Jeanne D'Arc basin using both seismic and well control. See Figure 28 for the location of the section.*

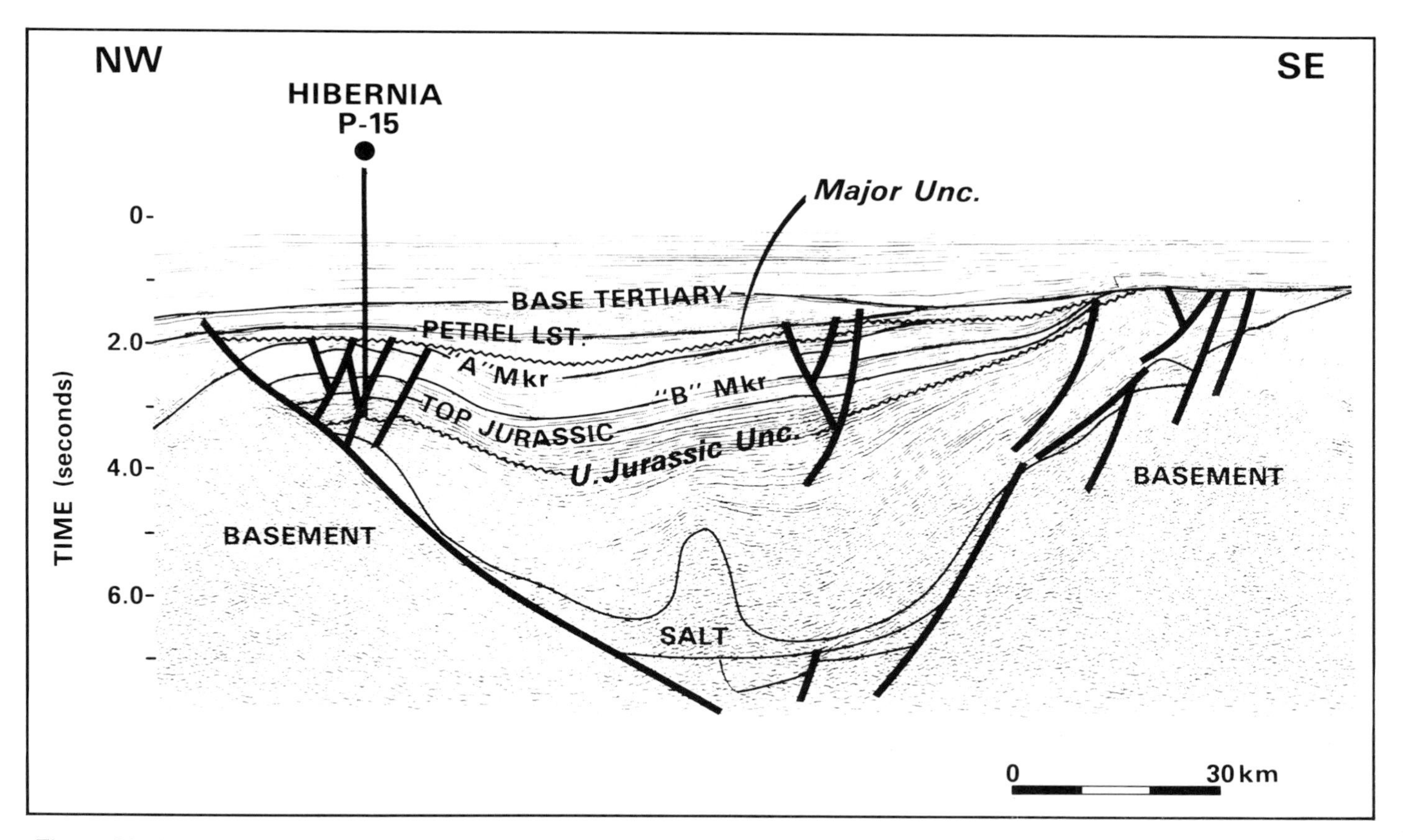


Figure 30. *A geophysical section across the Jeanne D'Arc basin, approximately along the same line of section used for Figure 29.*

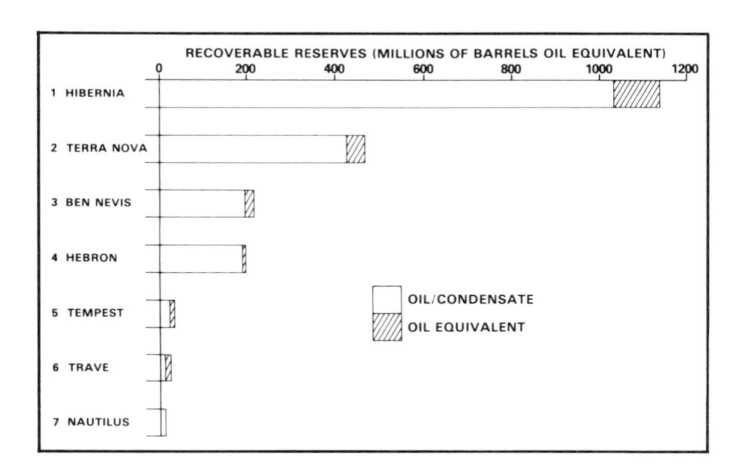


Figure 31. *Pool size array for the discoveries in the Jeanne D'Arc basin. The ranking numbers have been used to locate the fields geographically on Figure 28.*

ity. These post-Avalon sands were deposited as an onlapping sequence of sediments.

The Late Jurassic and Early Cretaceous was punctuated by several movements of sea level, resulting in carbonate deposition or cutting unconformities. These bounding limestone and unconformity surfaces form characteristic seismic reflectors that are the basis for seismic–stratigraphic studies.

The Hibernia oil field (Figure 32) is an example of an oil field developed along the major bounding fault side of an asymmetric basin where boundary-fault-controlled sedimentation is important. Synsedimentary subsidence along the Murre fault produced a typical rollover anticline (Figure 32). Hibernia I-46 penetrated Lower Jurassic shale and evaporites, showing that the anticlinal form was complicated by early salt tectonics. The combination of extensional tectonism and diapirism have resulted in stratigraphic and structural complexity. The quartzose sandstone reservoirs are characterized by vertical stacking and thickening toward the boundary fault. Both primary and secondary porosity are developed, the latter by dissolution of carbonate cements.

The oldest reservoir interval comprises interbedded mudstones, sandstone, and conglomerates of the Jeanne D'Arc interval, which were deposited in braided fluvial environments. Hibernia P-15 penetrated a 100 m (328 ft) thick interval of overpressured section that provided the initial show of oil in this discovery well. Test results have already been reported (Arthur et al, 1982, table 1). Subsequent drilling has shown that the Jeanne D'Arc zone does not contain significant reserves in the Hibernia Field.

The Hibernia zone is one of the two major reservoir intervals at Hibernia. Hibernia sandstone deposition reflects processes in low-sinuosity, coastal-plain rivers that formed fan deltas along the basin's margin. Sediments were probably derived from the rift-margin highlands to the west. Sandstone packages comprise erosively-based, quartzose sands for which the predicted connectedness for each lenticular sandstone package is good (> 1). The Hibernia interval is extensively dissected by minor faults that will be significant in this reservoir's drainage. The

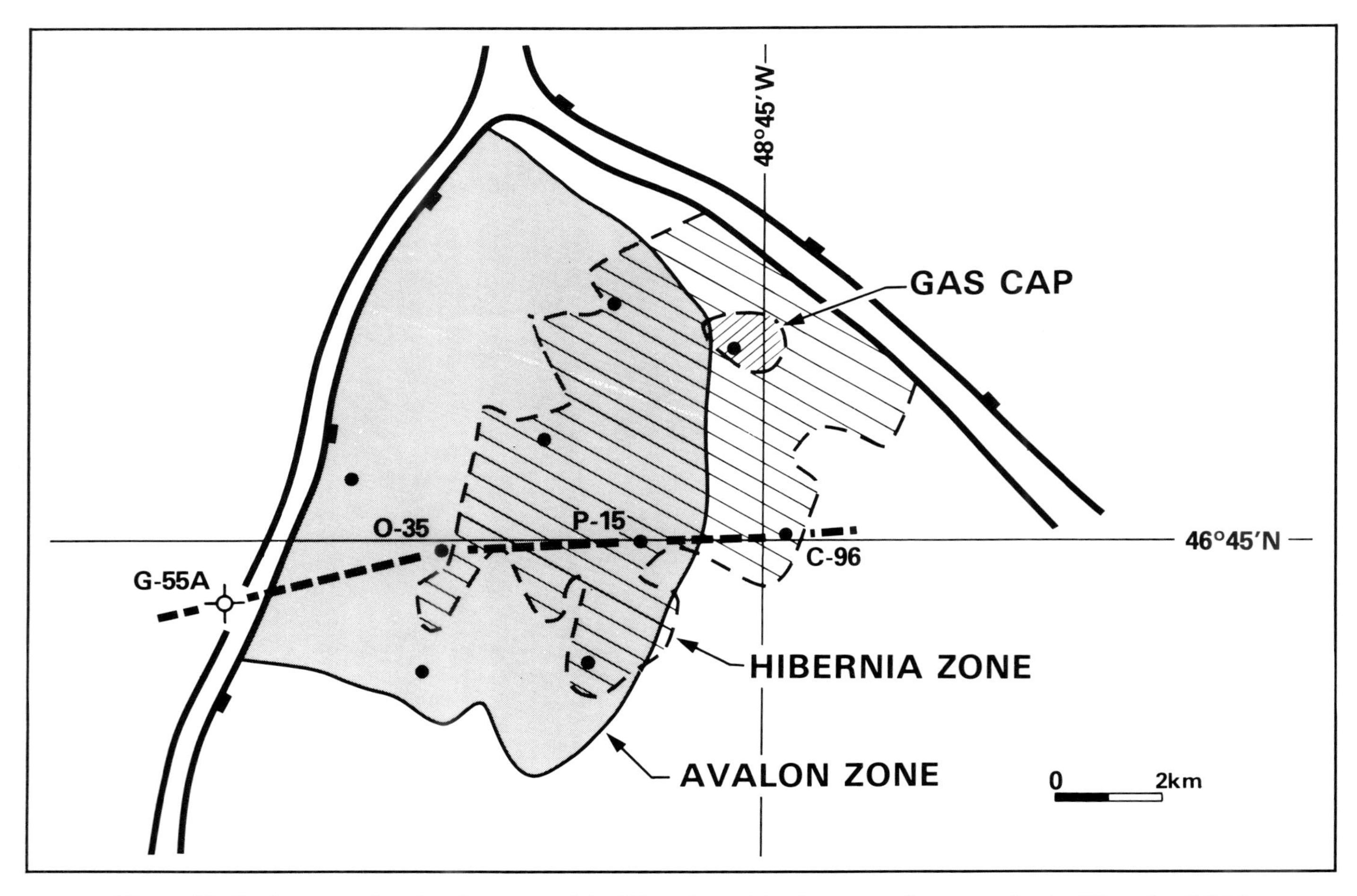


Figure 32. *Outline map showing the extent of the Hibernia and Avalon zone oil reserves for the Hibernia oil field.*

trapping mechanism is the combination of the rollover anticline and the overlying shale top-seal; the wells adjacent to the bounding fault lie below the oil–water contact for this pool.

The total hydrocarbon column in the Hibernia zone is about 700 m (2,300 ft) from the crest of the structure in the gas cap at Hibernia B-08 (Figure 32) to the lowest oil–water contact. The hydrocarbon column exhibits a significant vertical gradient in both oil gravity and gas–oil ratio, and the oil is undersaturated in the lower part of the reservoir.

The next regressive phase deposited the Catalina (B or C sandstone of some partners) sandstone in marginal marine environments. This zone contributes little to the overall field's reserves.

The Avalon is the second major producing zone in the Hibernia oil field, and it is offset to the west from the deeper Hibernia zone pool. In contrast to the Hibernia architecture, the Avalon zone contains at least three major offlapping sand bodies separated by prominent shale beds. Deposition occurred mainly in subaqueous, micro-tidal environments that include an inner shelf and shallow delta front, subordinating shore zone and lagoonal facies. The shelf–delta front facies comprises interbedded mudstones and fine- to very fine-grained sandstone. Each lobate reservoir body terminates in a zone of shore reworked sandstones possessing optimal reservoir properties. The overall trend of the reservoir quality increases upward in each package.

The total hydrocarbon column in the Avalon zone is 576 m (1,890 feet) from the structurally highest well (Hibernia O-35) to the lowest oil–water contact. Although the oil here tends to have a lower gravity and gas–oil ratio than that of the Hibernia zone, there appears to be a similar vertical gradient to both of these attributes, as is found in the Hibernia zone.

The trapping mechanism for the Avalon zone is much less obvious than for the Hibernia. The Avalon trap formed by the combination of the overcropping, Lower Cretaceous shale unit above the mid-Cretaceous unconformity and a very subtle fault trap that lies between the water-laden sands of the G-55A well and the fully oil-saturated section at a comparable structural elevation in the closest oil well east of the fault (Figures 32 and 33). Had the G-55A well been drilled first, it would have taken a great deal of courage to predict that a billion-barrel oil field lay immediately downdip.

The critical factors for trapping and preserving the Hibernia oil field are the development of the Hibernia structure contemporaneously with the deposition of the reservoir zones; the presence of an overcropping, Lower Cretaceous top-seal; and the absence of late, structural uplift or diapirism.

### Eastern Flank of the Jeanne D'Arc Basin

On the eastern ramp of the Jeanne D'Arc basin, exploration and delineation drilling are actively in progress. A number of single well discoveries have occurred, and some general com-

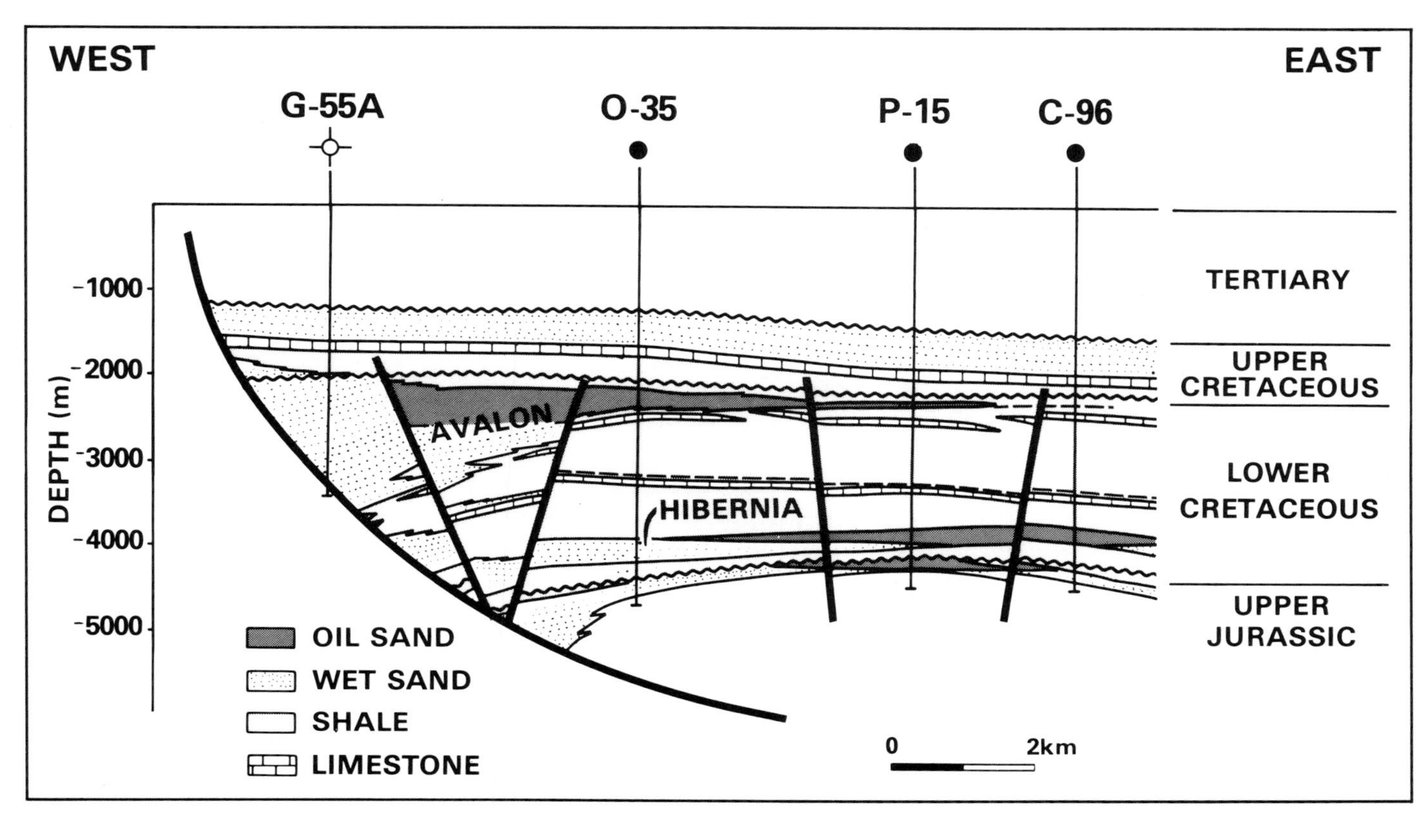


Figure 33. *Geological cross-section across the Hibernia oil field. See Figure 32 for the location of the section.*

ments can be made about the similarities and differences between this area and the better-defined Hibernia oil field.

The same basin controls are overprinted on the stratigraphy of the entire basin, leading to regional stratigraphic continuity of major units. Relative uplift of the eastern ramp is reflected in a series of unconformities. Onlap onto unconformities and thinner but laterally more continuous reservoirs are characteristic along the eastern margin. Significant discoveries have been made at Terra Nova, Hebron, Ben Nevis, and South Tempest.

Productive intervals include the Jeanne D'Arc sands, which tend to be the most consistent; Hibernia; Avalon; and the post-Avalon sand, which onlaps the mid-Cretaceous unconformity. Pay sections tend to be thinner and less well-stacked than at Hibernia. Oil gravities show considerable well-to-well and depth variation. Biodegradation has reduced the oil gravity and increased its viscosity in the post-Avalon sands. At South Tempest, the mid-Cretaceous unconformity has stripped all but the Upper Jurassic reservoirs.

The slower rate of subsidence on the eastern flank of the basin and the convergence of unconformities result in complex stratigraphy and complex subcrop and overcrop patterns that affect the traps in this area.

### Development

Development of oil pools in the Jeanne D'Arc basin will face a severe operating environment (Figure 34). The combination of heavy sea-state, incursion of the pack ice, and the risk of iceberg impact will require new and innovative engineering and management technology during the development stage and throughout the subsequent productive life of the field. Development planning is underway, but the mode of development has not yet been decided.

## SCOTIAN BASIN

### Regional Setting

The Scotian basin (Figures 1 and 35) occupies the continental shelf and slope, mainly offshore from the province of Nova Scotia. It is a large, passive-margin basin that overlies a pre-rift basement complex of Hercynian metasediments. The rift stage sediments include Triassic and Lower Jurassic clastics and evaporites, while a very thick, post-rift sequence of Jurassic to Tertiary age forms a seaward-thickening wedge of sediments. The Scotian basin is a regional downwarp overlying a complex of rift-stage half grabens. The geology and hydrocarbon potential of the Scotian basin have been extensively discussed in the literature (Sherwin, 1973; Jansa and Wade, 1975; Given, 1977; Eliuk, 1978; Purcell et al, 1980).

A major depocenter in the vicinity of Sable Island, the Sable subbasin, was the focus of Upper Jurassic and Lower Cretaceous clastic deposition (Figure 36). The depocenter marks a significant reentry into the regional trend of the Jurassic carbonate bank-edge that elsewhere is approximately coincident with the present continental shelf edge.

Unlike the strongly structurally controlled sedimentation in the Jeanne D'Arc basin, regional downwarping caused by post-rift thermal decay has resulted in deposition of regionally continuous sandstones beneath the Scotian Shelf. Sedimentary loading caused the sequential development of down-to-basin

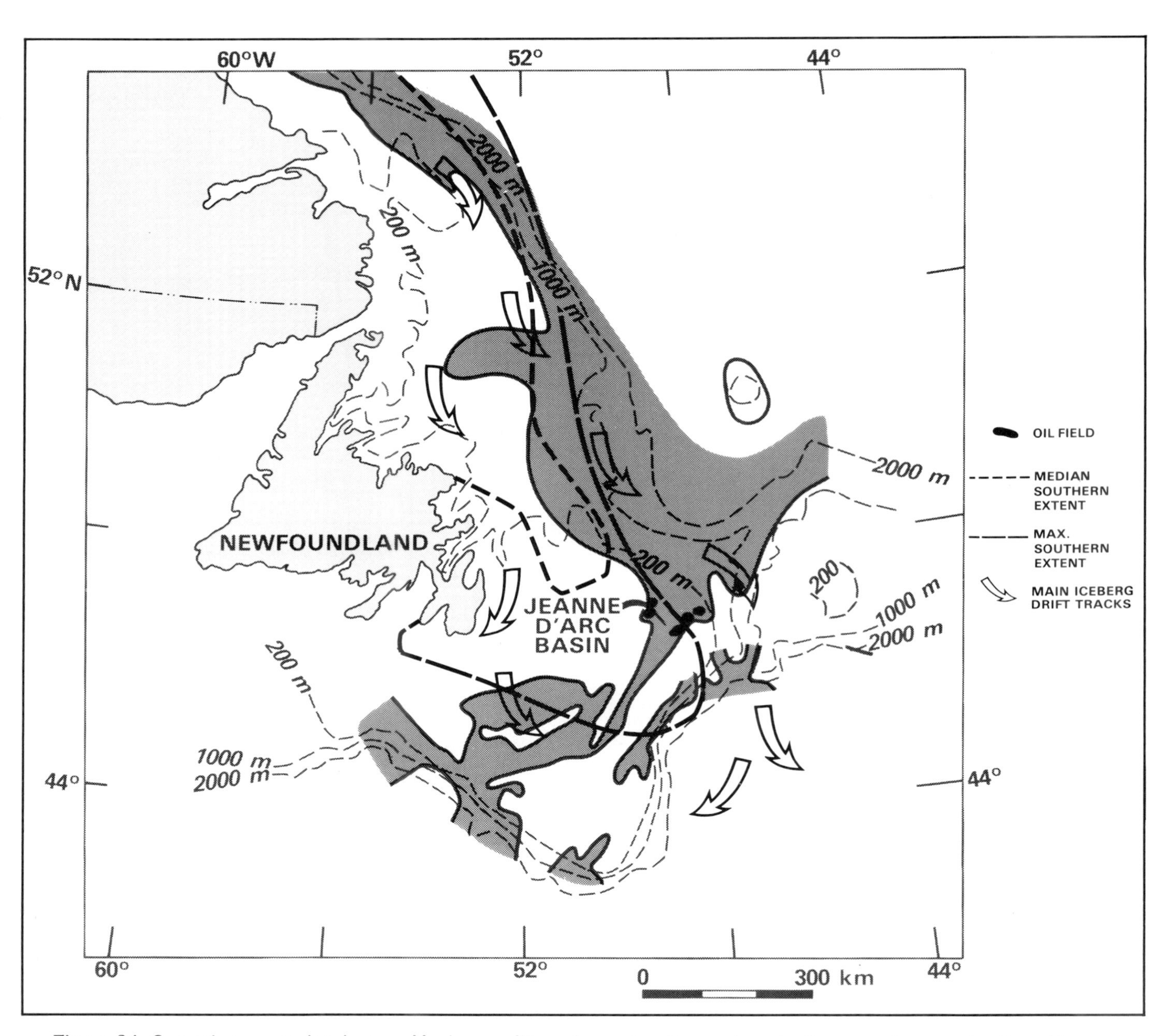


Figure 34. *Operating constraints imposed by ice conditions for exploration and production operations on the Grand Banks.*

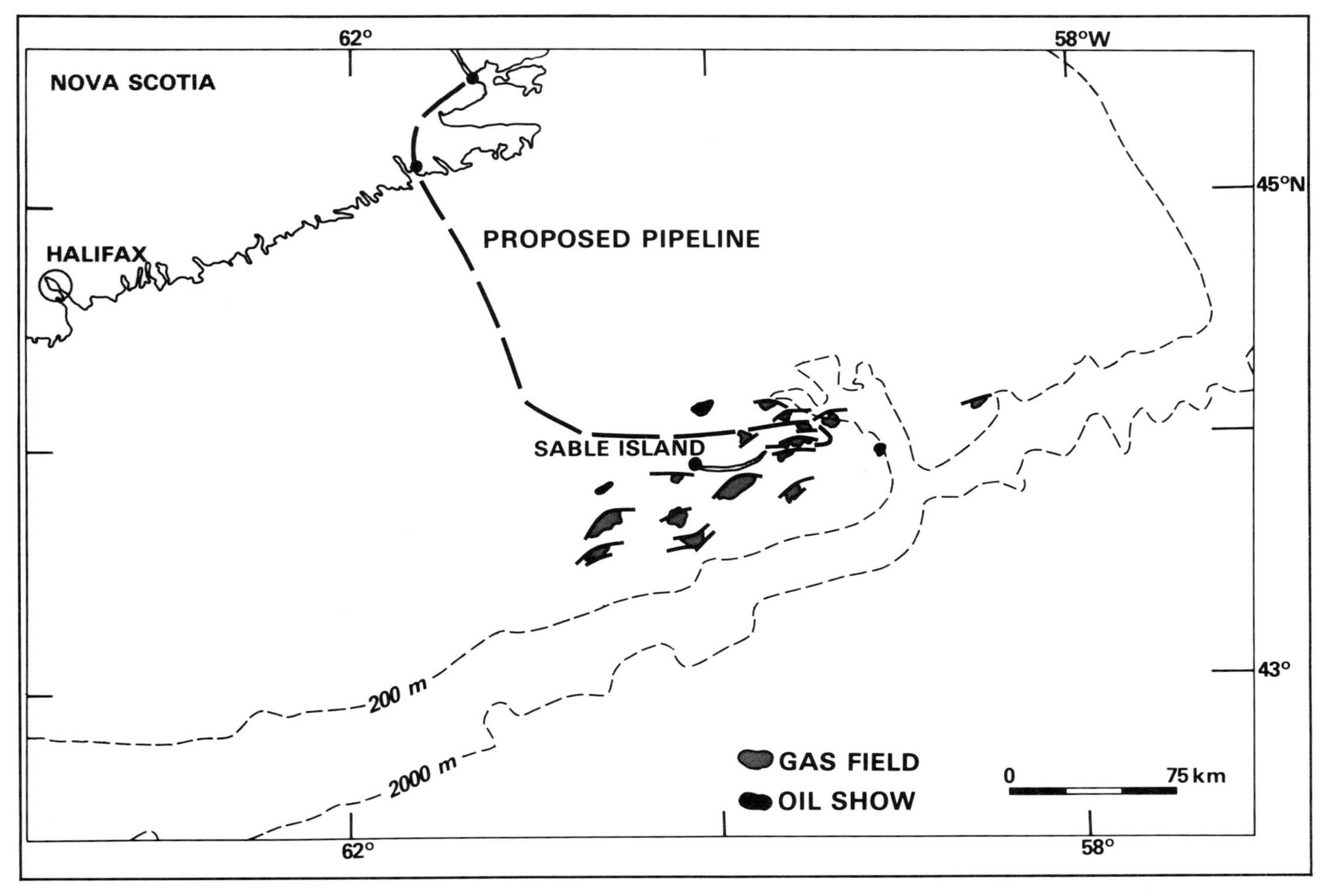


Figure 35. *Location map for the Scotian Shelf, showing the location of the gas discoveries, oil shows and the proposed pipeline route from the Venture gas field.*

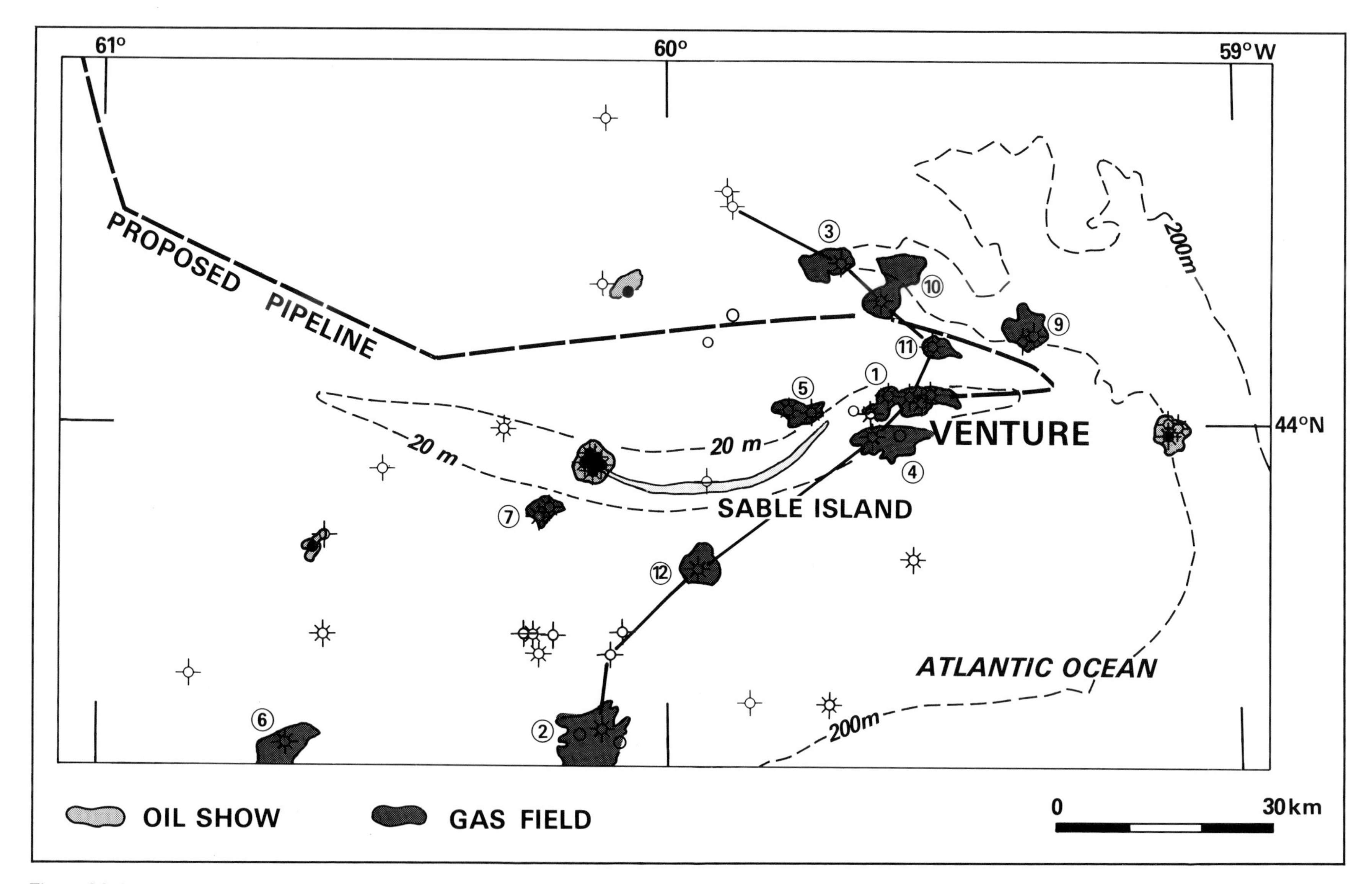


Figure 36. *Location map for the Sable Island area showing gas discoveries and oil shows. The numbers on the discoveries are the ranking numbers on the pool size array in Figure 38.*

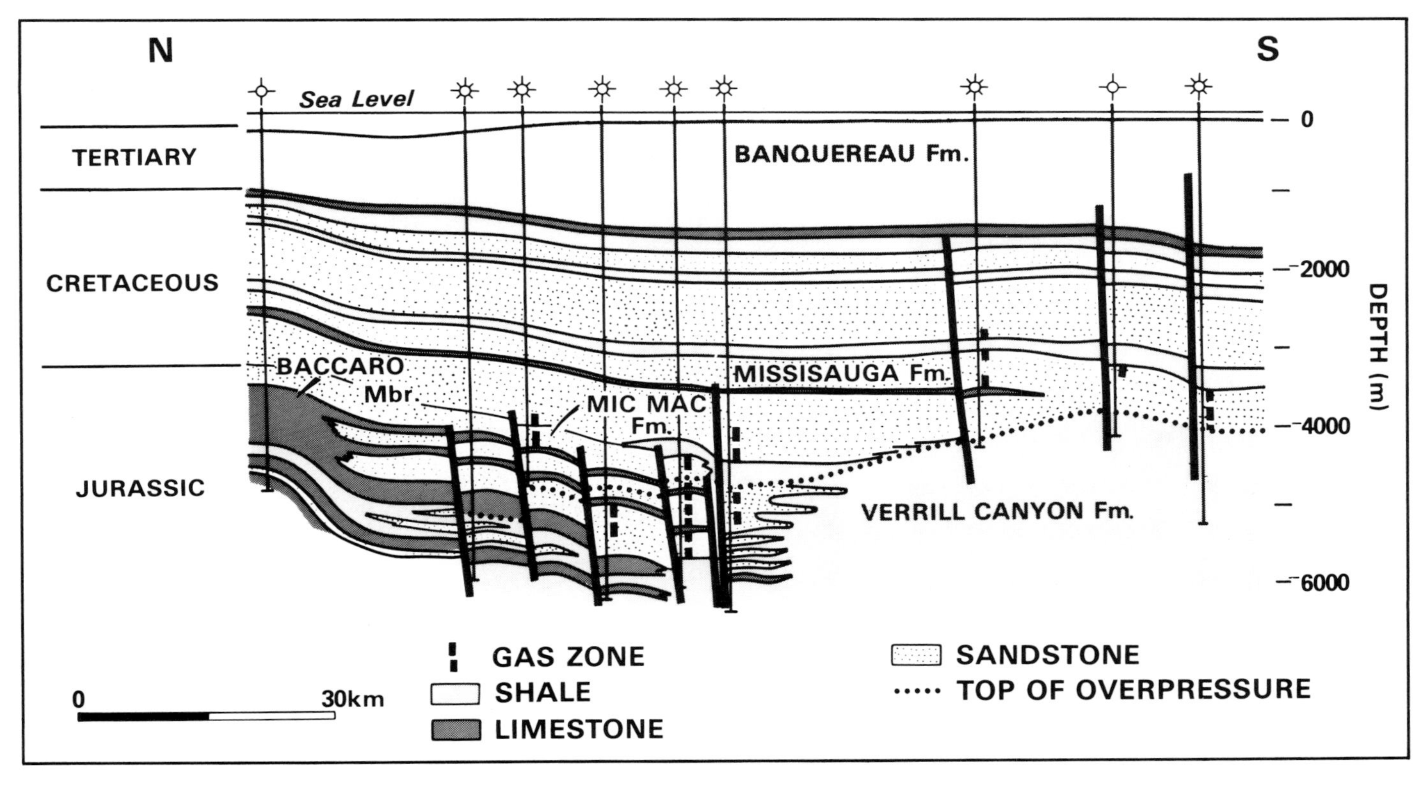


Figure 37. *Geological cross-section across the Sable subbasin. See Figure 36 for the location of the section.*

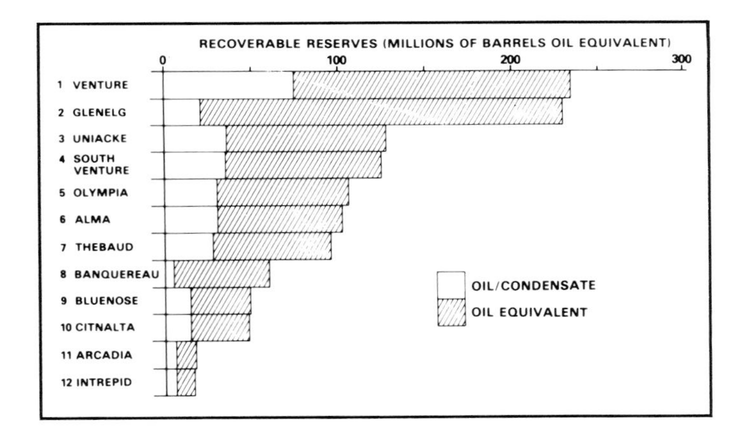


Figure 38. *Pool size array for the discoveries on the Scotian Shelf. The ranking numbers have been used to locate the fields geographically on Figure 36.*

listric growth faults that are the prime trapping mechanism in the basin. Salt diapirs, which pierce the sedimentary section to varying stratigraphic levels, also developed under the sediment load, and some salt movement may be involved in the listric fault structures as well. The shallow piercement salt structures have been more destructive than constructive in a hydrocarbon pooling sense.

The regional cross section across the Sable subbasin (Figure 37) illustrates the stratigraphy, the timing of structural movements, and the distribution of the gas zones. In general, the age of structural development and hydrocarbon accumulation decreases from northeast to southwest across the basin. The older and deeper reservoirs are overpressured and richer in condensate, while normally pressured, dryer gas is found in a seaward direction in younger reservoirs.

Twelve gas discoveries in the Scotian basin contain 257 billion cu m (9.1 tcf) of gas and 48 million cu m (303 million bbl) of condensate (Figure 38). Only shows of oil have been found. All the significant indications of oil and gas have been confined to the vicinity of the Sable subbasin.

### Venture Gas Field

The Venture gas field (Figure 39) is the largest and best defined discovery on the Scotian Shelf. The field is bounded on the north and south by east-to-west-trending growth faults of Late Jurassic to Early Cretaceous age (Figures 40 and 41). A rollover anticline parallel to the faults combines to form the structural trap. The structure can be mapped on a number of different levels; however, Figure 39 is drawn on the #6 sand, which lies just above the "Purple" seismic marker (Figure 41). This reservoir is the best in the field. The productive area covers 2,640 ha (6,500 acres) with 120 m (394 ft) of structural closure.

The productive section at Venture lies primarily in the Upper Jurassic, with a minor amount found in the Lower Cretaceous. The reservoirs are contained in the gradational facies of the Missisauga and Mic Mac formations, which form the southward, prograding, coarse clastic facies shelfward and which overlie the prodelta Verril Canyon facies (Figure 37). The gross reservoir interval is about 1,500 m (4,920 ft) and comprises interbedded sandstone, siltstone, and shale, with minor oolitic limestone beds. The sands appear to have been sourced as sheets from the north and northwest. They are predominantly

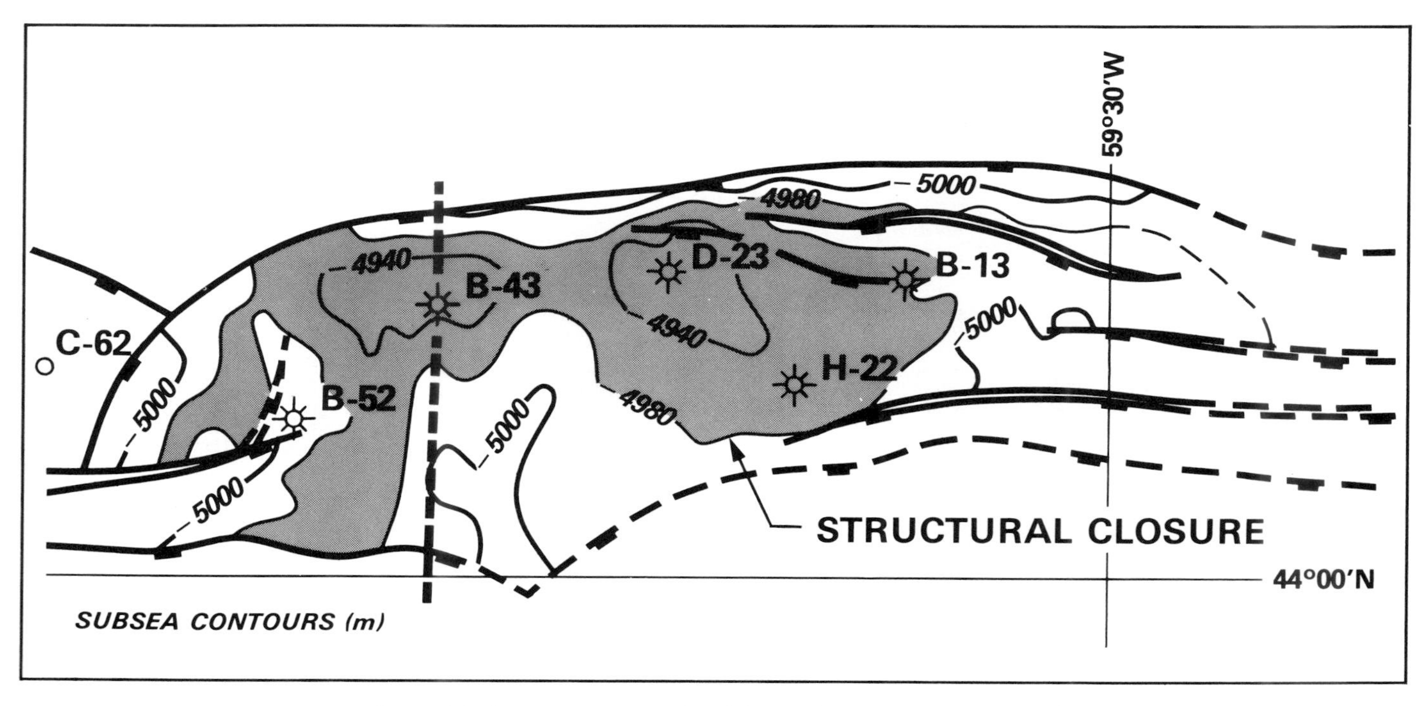


Figure 39. *Structure map on the # 6 sand in the Venture gas field.*

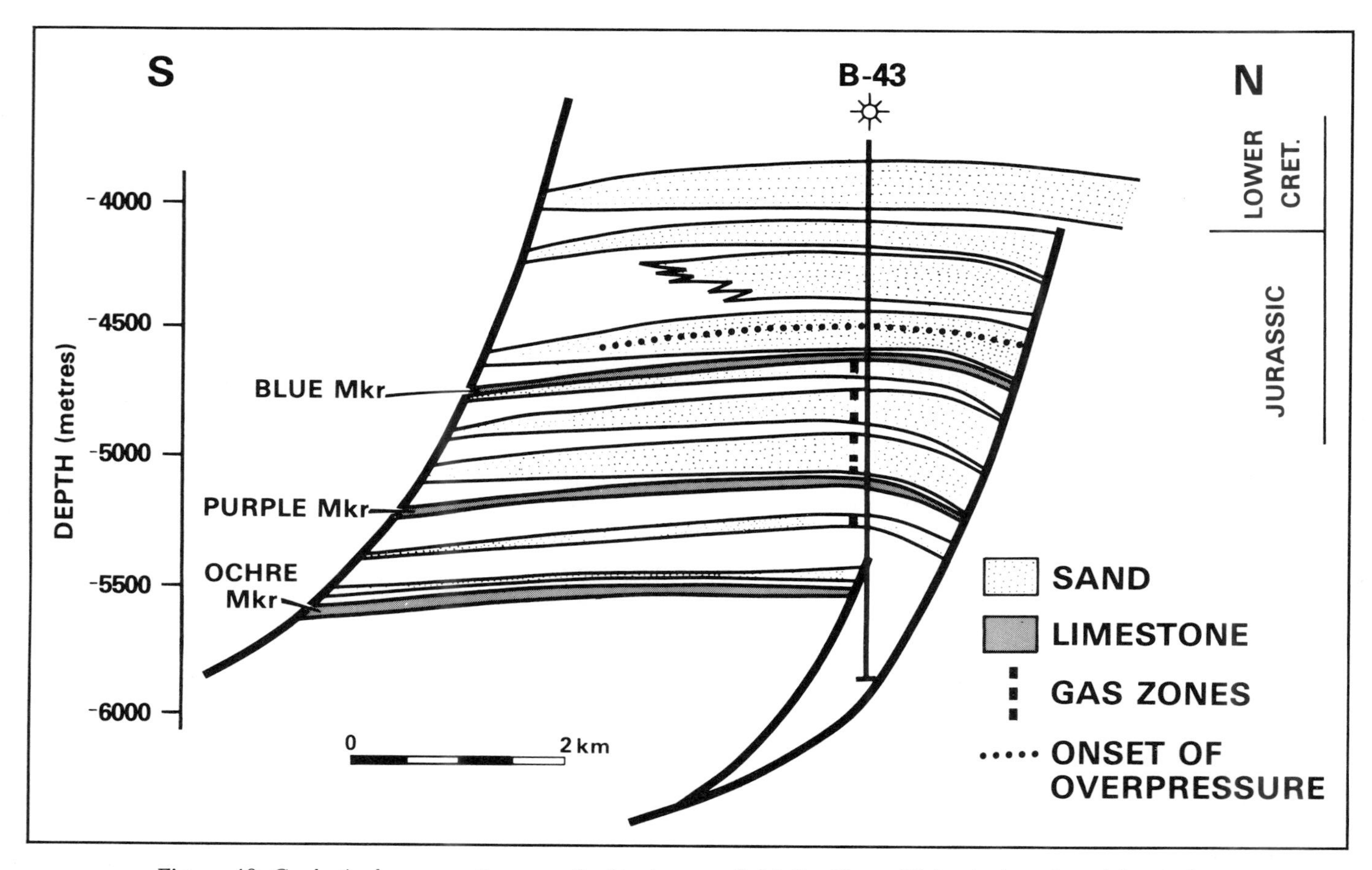


Figure 40. *Geological cross-section over the Venture gas field. See Figure 39 for the location of the section.*

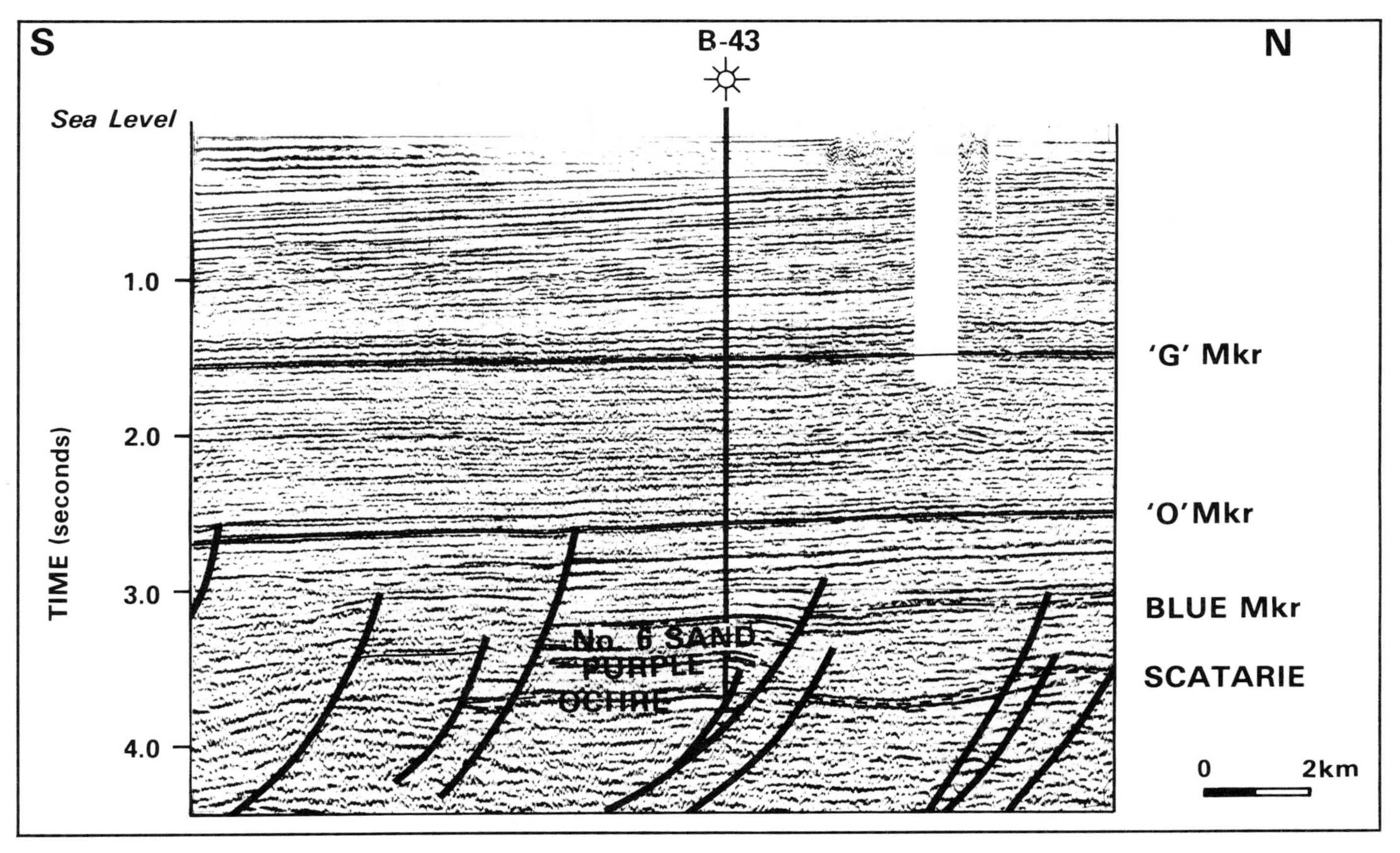


Figure 41. *Geophysical section across the Venture gas field. See Figure 39 for the location of the section.*

quartz arenites to litharenites, very fine- to coarse-grained, and poorly to moderately sorted. The best reservoir sands are in the upper shoreface facies of a deltaic system and average 10 to 25 m (33 to 82 ft) in thickness. The thickness and reservoir quality of these sands appears to deteriorate toward the south and southwest.

The bulk of the reserves are contained in the reservoirs that lie at the top of the overpressured section (Figure 40). Diagenetic alteration and the effects of overpressuring have combined to preserve excellent reservoir characteristics in these sands in spite of their age and burial depth. The gas pays exhibit continuity along the strike of the structure, but the pay sections are spread among a number of individual sand reservoirs, each with its own gas–water contact. The greatest individual gas column in any sand is about 65 m (213 ft).

The factor that has been most critical in gas entrapment at Venture is the timely, early development of the rollover structural trap shortly after deposition of the interbedded sands and shale during the Late Jurassic and Early Cretaceous. The reservoir quality has been enhanced by diagenesis and the absence of late structural uplift or diapirism.

### Development

The Scotian Shelf presents no unique physical environmental problems that will complicate development. Conventional, bottom-founded, steel jacket platforms are proposed, and the development planning process is proceeding.

## SUMMARY AND CONCLUSION

In the East Coast and Arctic basins of Canada, more than 20 years of exploration have involved the drilling of nearly 600 exploratory and delineation wells, resulting in the discovery of 666 million cu m (4.2 billion bbl) of oil and condensate and 1,125 billion cu m (39.9 tcf) of gas.

The discoveries are almost entirely contained in clastic reservoirs, in structural traps that show coeval deposition of the reservoir facies and development of the structural trap. Late stage uplift is conspicuously absent in the discovery structures.

On an oil-equivalent basis, almost equal quantities of oil and gas have been found. The reserves are nearly equally divided among reservoir rocks of Tertiary, Early Cretaceous, Late Jurassic, and Early Jurassic age. The Tertiary and Lower Cretaceous reservoirs contain proportionately more oil than the Jurassic reservoirs. Practically all the oil reserves have been found in the Jeanne D'Arc and Beaufort basins.

The best oil discoveries, those in the Jeanne D'Arc basin, have been made since 1979—relatively late in the exploration history of these basins. The discoveries happened because geologists recognized an area where good geology results in the concentration of high quality reserves in a small area.

An understanding of the critical factors that control the accumulation of oil and gas in the East Coast and Arctic basins of Canada is useful in guiding ongoing exploration in those basins and in other petroleum provinces in the world.

The future role that these Canadian basins will play in supplying hydrocarbons will depend on the basins' ability to compete economically with the alternative sources of hydrocarbon supplies, particularly the tar-sands resources of Western Canada.

## ACKNOWLEDGMENTS

This paper has been published with the permission of Petro-Canada. The material has been assembled by the Exploration Division of Petro–Canada Resources using in-house and public domain information. Gulf Canada assisted with information on Parsons Lake, Esso Resources on Taglu, and Panarctic Oils Ltd. on Drake Point and Cisco. I acknowledge the specification assistance of Derek Lee, High Balkwill, Tony Tankard, Larry Grieve, and Eugene Olynyk, who were supported by the staff of Petro-Canada Resources.

## REFERENCES

Amoco Canada Petroleum Company Ltd., and Imperial Oil Limited, 1973, Regional geology of the Grand Banks: Bulletin of Canadian Petroleum Geology, v. 21, p. 475–503 and AAPG Bulletin, v. 58, p. 1109–1123.

Arthur, K. R., D. R. Cole, G. G. L. Henderson, and D. W. Kushnir, 1982, Geology of the Hibernia discovery, *in* M.T. Halbouty, ed., The deliberate search for the subtle trap: AAPG Memoir 32, p. 181–195.

Balkwill, H. R., 1978, Evolution of the Sverdrup basin, Arctic Canada: AAPG Bulletin, v. 62, p. 1004–1028.

——, and F. G. Fox, 1982, Incipient rift zone, Western Sverdrup basin, Arctic Canada, *in* A. F. Embry, and H. R. Balkwill, eds., Arctic geology and geophysics: Canadian Society of Petroleum Geologists, Memoir 8, p. 171–187.

Bowerman, J. N., and R. C. Coffman, 1975, The geology of the Taglu gas field in the Beaufort basin, N. W. T., *in* C. J. Yorath, E. R. Parker, and D. J. Glass, eds., Canada's continental margins and offshore exploration: Canadian Society of Petroleum Geologists, Memoir 4, p. 649–662.

Cote, R. P., M. M. Lerand, and R. J. Rector, 1975, Geology of Lower Cretaceous Parsons Lake gas field, Mackenzie delta, Northwest Territories, *in* C. J. Yorath, E. R. Parker, and D. J. Glass, eds., Canada's continental margins and offshore exploration: Canadian Society of Petroleum Geologists, Memoir 4, p. 613–632.

Dixon, J., 1980, Sedimentology of the Eocene Taglu delta, Beaufort–Mackenzie basin, example of a river dominant delta: Geological Survey of Canada Paper 80–11.

Eliuk, L. S., 1978, The Abenaki Formation, Nova Scotia Shelf, Canada--a depositional and diagenetic model for a Mesozoic carbonate platform: Bulletin of Canadian Petroleum Geology, v. 26, p. 424–514.

Given, M. M., 1977, Mesozoic and early Cenozoic geology of offshore Nova Scotia: Bulletin of Canadian Petroleum Geology, v. 25, p. 63–91.

Hawkings, T. J., and W. G. Hatlelid, 1975, The regional setting of the Taglu field, *in* C. J. Yorath, E. R. Parker, and D. J. Glass, eds., Canada's continental margins and offshore exploration: Canadian Society of Petroleum Geologists, Memoir 4, p. 633–648.

Jansa, L. F., and J. A. Wade, 1975, Geology of the continental margin off Nova Scotia and Newfoundland, *in* Offshore Geology of Eastern Canada: Geological Survey of Canada Paper 74–30, v. 2, p. 51–105.

Langhus, P. G., 1980, Generation and migration of hydrocarbons in the Parsons Lake area, N. W. T., Canada, *in* A. D. Mial, ed., Facts and principals of world petroleum occurrence: Canadian Society of Petroleum Geologists, Memoir 6, p. 523–534.

McMillan, W. J., 1973, Shelves of the Labrador Sea and Baffin Bay, *in* R. G. McCrossan, ed., The future petroleum provinces of Canada: Canadian Society of Petroleum Geologists, Memoir 1, p. 473–517.

McWhae, J. R. W., R. Elie, K. C. Laughton, and P. R. Gunther, 1980, Stratigraphy and petroleum prospects of the Labrador Shelf: Bulletin of Canadian Petroleum Geology, v. 28, p. 460–488.

Proctor, R. M., G. C. Taylor, and J. A. Wade, 1984, Oil and natural gas resources of Canada: Geological Survey of Canada Paper 83–31.

Sherwin, D. F. 1973, Scotian Shelf and Grand Banks, *in* R. G. McCrossan, ed., The future petroleum provinces of Canada: Canadian Society of Petroleum Geologists, Memoir 1, p. 519–559.

Umpleby, D. C., 1979, Geology of the Labrador Shelf: Geological Survey of Canada Paper 79–13.

White, D. A., 1980, Assessing oil and gas plays in facies–cycle wedges: AAPG Bulletin, v. 64, p. 1,158–1,178.

Williams, H., and R. D. Hatcher, 1983, Appalachian suspect terranes: Geological Society of America Memoir 158, p. 33–53.

# The Continental Margin of Eastern Canada: Geological Framework and Petroleum Potential

A. C. Grant
K. D. McAlpine
J. A. Wade
*Atlantic Geoscience Centre*
*Geological Survey of Canada*
*Bedford Institute of Oceanography*
*Dartmouth, Nova Scotia*

The continental margin of Eastern Canada is underlain by a series of Mesozoic sedimentary basins that have been targets of hydrocarbon exploration for over two decades. Discoveries have been made in three major geological/geochemical regions. On the Scotian Shelf, there have been 15 gas and condensate discoveries out of 75 structures drilled. Five of these are in overpressured reservoirs. On the eastern Grand Banks, 28 wildcat wells have yielded 9 light oil discoveries, including the giant Hibernia oil field. Exploration of the Labrador–Southeast Baffin Shelf has yielded 6 gas/condensate discoveries out of 28 wildcats.

The Geological Survey of Canada has developed hydrocarbon generation concepts to explain the regional variation in oil and gas occurrence, and to assess the future potential of the margin in terms of the thermal maturity of the source rocks, type of organic material, and time of trap formation.

## INTRODUCTION

The continental margin of eastern Canada is regarded as a classic "Atlantic-type" margin, in which a prism of sedimentary deposits accumulated in Mesozoic and Cenozoic time as the adjacent marine areas widened and deepened. Geophysical evidence, indicating a thick section of sedimentary rocks beneath the continental shelves, was provided by Canadian and American research institutes and government surveys beginning in 1951. The interpretations of these early, refraction seismic and potential field studies stimulated petroleum industry interest in the area, which appeared to contain all the prerequisites of a major petroleum province. Recognition of the hydrocarbon potential of the eastern Canadian continental margin was widespread in the oil industry by the early 1960s. By 1971, exploratory drilling off eastern Canada had advanced northward from the Scotian Shelf and the Grand Banks to the Labrador Shelf, and exploration permits had been filed to the northern reaches of Baffin Bay (Figure 1). Geophysical exploration defines the principal basins along the eastern Canadian continental margin (Figure 2), including Baffin Bay. However, drilling has not been carried out north of the Labrador Sea. Thus, in terms of assessing the future of this petroleum province, the northern one-third of the more than 5,000 km (3,100 mi) of prospective margin has not yet been tested by a single exploratory well. Apart from local concentrations of wells on the Scotian Shelf and the northeastern Grand Banks, exploratory drilling to date provides only sparse coverage of these offshore areas, and well control is almost totally lacking in the slope-rise zone.

The well data used in this paper are those released to the public by the Canada Oil and Gas Lands Administration (COGLA) at the end of a two year period of confidentiality for wildcat wells, or a 90 day period for delineation wells. The geophysical data are mainly multichannel, reflection seismic lines shot by industry and held confidential for five years by COGLA. By the end of October 1984, data were released for 162 wells of a total of 206 drilled or drilling within the study area, and available multichannel coverage amounted to about 600,000 km (372,830 mi).

This paper is necessarily an interim report on offshore petroleum potential, as exploration programs on the Scotian Shelf and Grand Banks are currently in full flight. Our knowledge is perhaps most mature for the Scotian Shelf; on the Grand Banks, the exploration activity triggered by the Hibernia discovery is still gaining momentum, and reporting at this time is provisional. The Labrador Shelf has not been an area of high priority for the past year because no drilling rigs were operating there. However, this temporary lull in drilling will provide an opportunity for critical reappraisal of exploration results.

## REGIONAL GEOLOGIC SETTING

The Mesozoic-Cenozoic sediments of Canada's east-coast petroleum province lie almost entirely offshore (Figure 2). On the Scotian Shelf, Grand Banks, and east of Newfoundland, these sediments onlap the Paleozoic and Precambrian rocks of the Appalachian orogen; to the north, in the Labrador Sea and Baffin Bay, they onlap Paleozoic sedimentary strata and crystalline rocks of the Precambrian Shield. The stratigraphic range of the Mesozoic sediments on the Scotian Shelf and Grand Banks contrasts with that of the Labrador–Southeast Baffin Shelf (Figure 3), presumably because of differences in the sea-floor spreading histories of these regions. Whereas the Atlantic is believed to have commenced opening in the Early Jurassic, seafloor spreading apparently did not begin in the Labrador Sea until the Late Cretaceous, and eventually progressed northward to Baffin Bay. We interpret the red bed–evaporite sequence that is a prominent element of the early Mesozoic stratigraphy of the Scotian Shelf and the Grand Banks as the product of a narrow, shallow, restricted seaway and generally arid climatic conditions. The overlying carbonates reflect development of more open marine conditions, as the Atlantic Ocean widened and deepened. By the end of the Jurassic, deposition of clastic sediments predominated in these regions. On the Labrador Shelf the oldest Mesozoic beds encountered to date are Lower Cretaceous volcanics and continental clastics, which give way to Upper Cretaceous marine clastic beds.

While the plate tectonic scenario outlined above provides a first-order model for describing some of the gross aspects of the geology, such models may not be relevant at more local scales. For example, the fault-bounded basins that characterize the Grand Banks are the product of tensional conditions presumed to be related to the pre-drift phase of the North Atlantic development. However, the structural complexities of these basins can be related to plate tectonics only in terms of speculations that they somehow express transition between the seafloor spreading regimes of the Atlantic and the Labrador Sea. Also, in contrast to assumptions of plate tectonic reconstructions that the Labrador Sea did not exist before the Late Cretaceous (e.g., Srivastava, 1978), drilling confirms that the northeasterly trending Appalachian orogen turns northward off southern Labrador, and forms at least part of the "economic basement" for some distance to the north. Seismic data suggest that Paleozoic rocks may be a component of the basement all along the Labrador Shelf, and connect with lower Paleozoic rocks that have been mapped and sampled on the shelf off southern Baffin Island. These occurrences of marine Paleozoic rocks indicate the probable existence of a Paleozoic seaway off Labrador, extending at least as far north as Davis Strait. Because Paleozoic rocks are the reservoir for two of the gas discoveries off Labrador, the occurrence of these units is of economic as well as academic interest. Moreover, plate tectonic models for Baffin Bay have been challenged on the basis of geological evidence from the adjacent land masses (e.g., Dawes and Kerr, 1982). Obviously, such uncertainties about the origin of the basement rocks underlying the offshore sedimentary basins is an important consideration in assessing their hydrocarbon potential.

## THE SCOTIAN BASIN

### Exploration History

Since exploration began in the Scotian basin in 1959, 89 wildcat and 20 delineation wells have resulted in 15 significant discoveries (all in the Sable Island area) as well as numerous gas and oil shows (Figure 4). The region has experienced two major cycles of exploration. The initial drilling, which resulted in discoveries near Sable Island, was followed by a period of disappointing results that prompted large reductions in land holdings. The discovery of gas at Venture in 1979 led to the current round of intensive exploration focused on overpressured reservoirs.

Comprehensive descriptions of the subsurface stratigraphy of this area have been published by McIver (1972), Amoco and

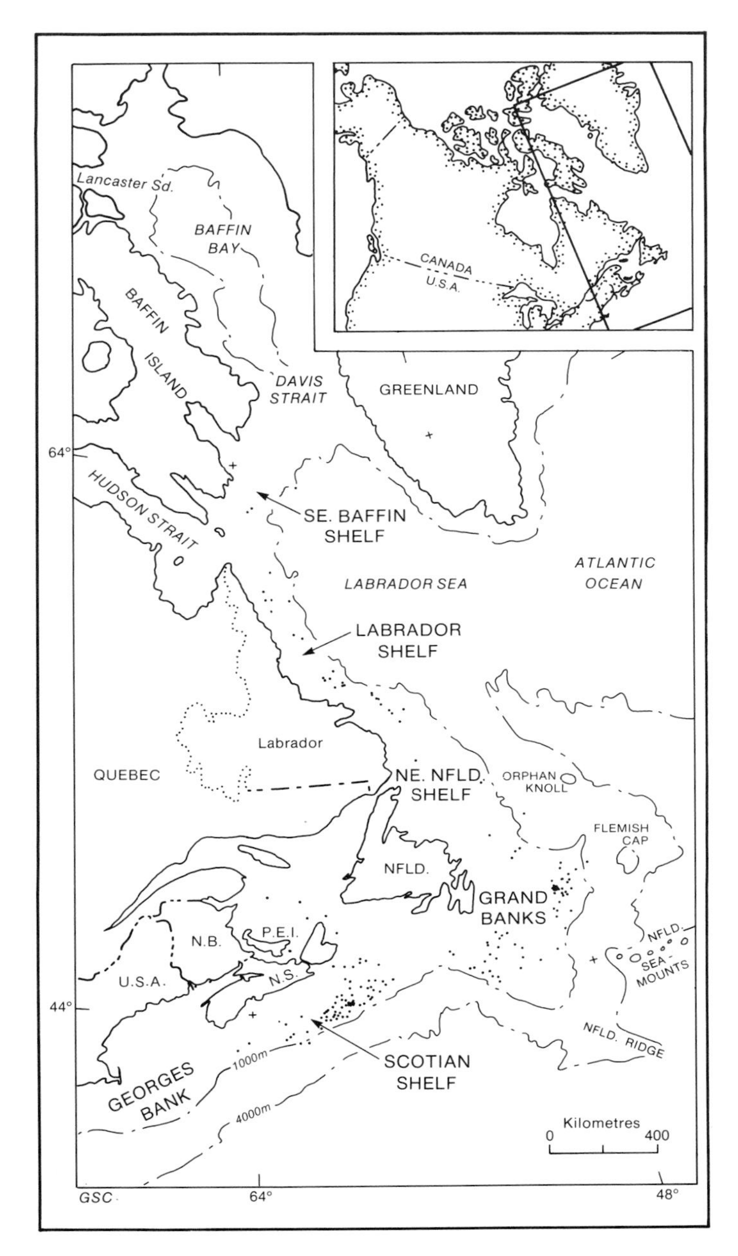


Figure 1. *Bathymetric map showing principal physiographic features of eastern Canada's continental margin and locations of exploratory wells (small dots).*

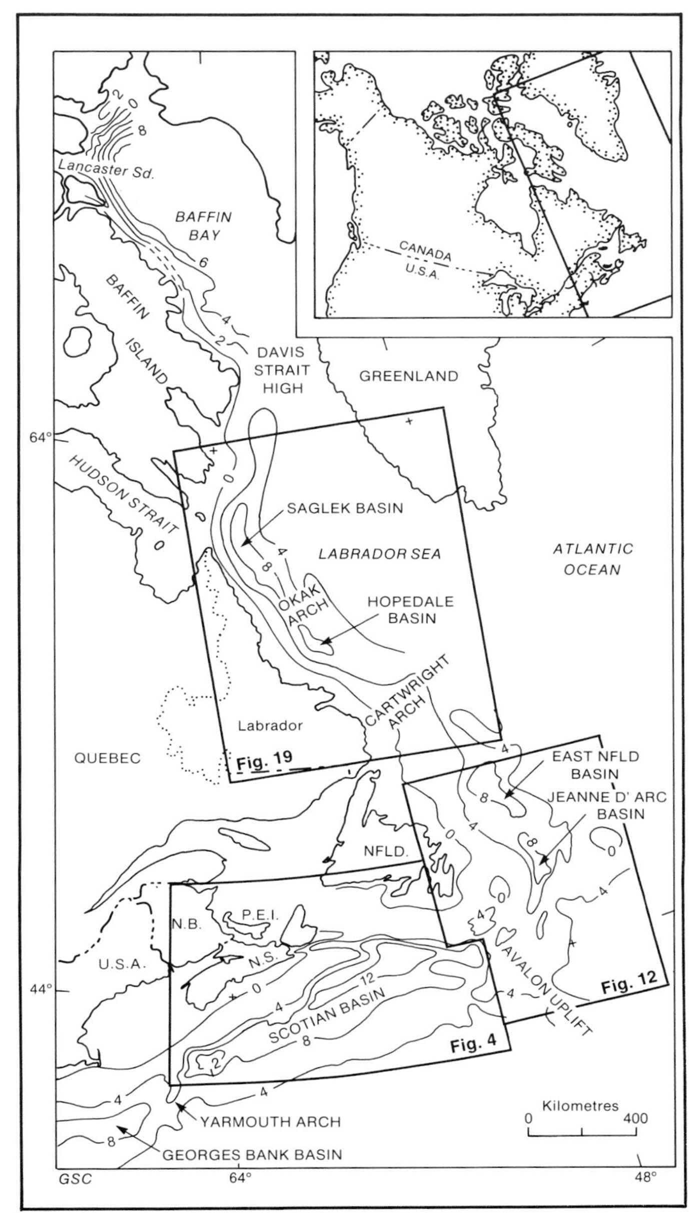


Figure 2. *Index map of offshore sedimentary basins showing total Mesozoic-Cenozoic sediment fill in kilometers (after Jackson, 1985). Areas outlined are detailed in Figures 4, 12, and 19.*

Imperial Oil (1973), Jansa and Wade (1975), Williams (1975), Given (1977), Eliuk (1978), Wade (1978, 1981), and McWhae (1981), and are briefly summarized below.

## Geology

The Scotian basin, a major site of Mesozoic–Cenozoic deposition, extends from the Yarmouth Arch in the southwest to the Avalon Uplift in the northeast, a distance of nearly 1,200 km (746 mi) (Figure 2); and from the edge of sediments southeastward into a deep oceanic sequence, a distance of over 250 km (155 mi) (Figure 4).

The Scotian basin onlaps the southeastern flank of the Appalachian orogen and distally overlies oceanic crust. It contains a number of structural elements (Figure 4). The Yarmouth Arch, LaHave Platform, and Canso Ridge are stable elements that formed the western and northern margins of the basin during its early development. The Avalon Uplift is a Late Jurassic–Early Cretaceous element that forms its eastern flank. The depth to basement on these more positive areas is generally less than 4,000 m (13,123 ft). A major basement hinge zone separates the stable elements from a series of interconnected depocenters that occur beneath the outer continental shelf and slope. In these areas, prolonged subsidence has resulted in the accumulation of 12,000 m (39,370 ft) or more of sedimentary strata (Figure 4).

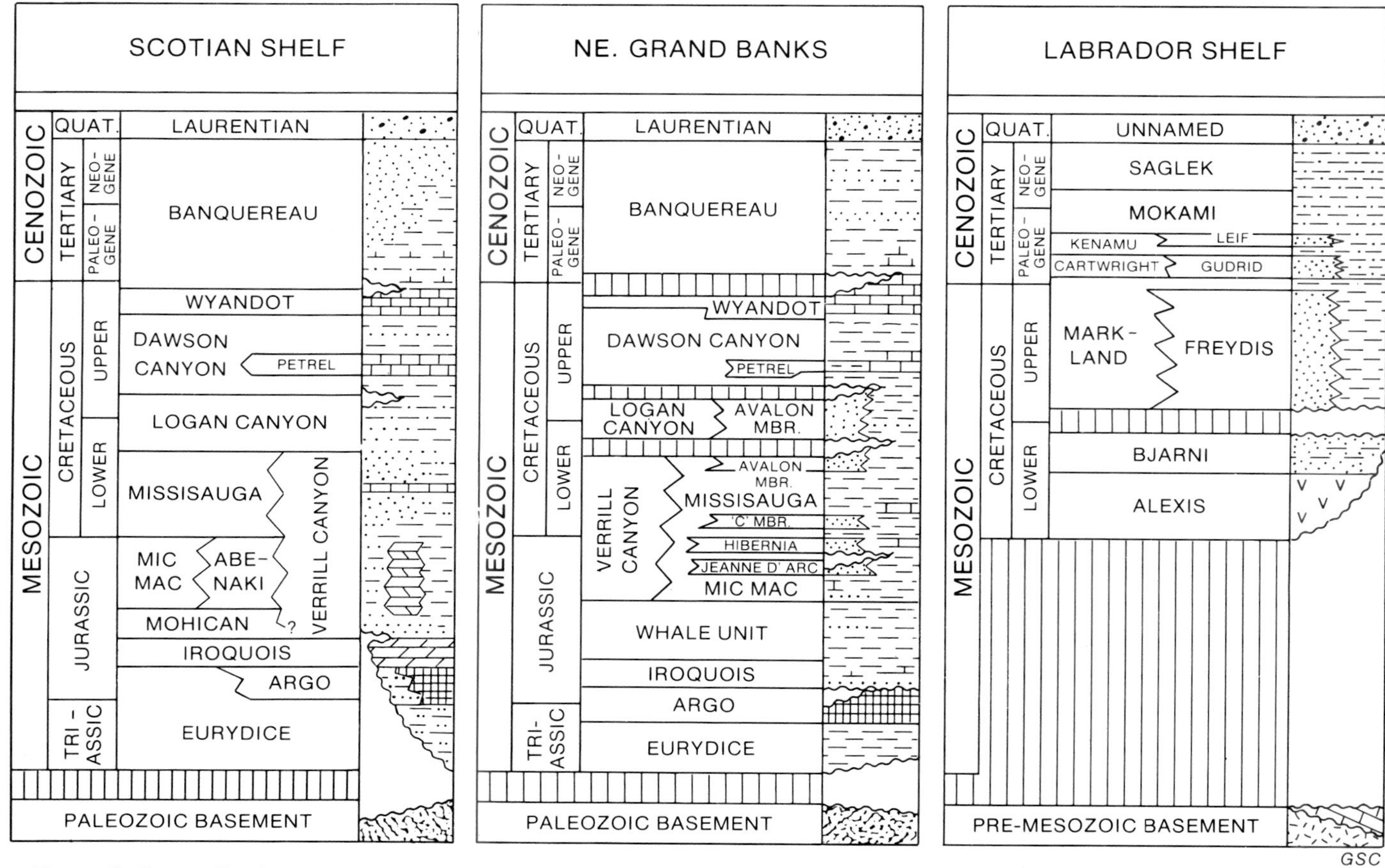


**Figure 3.** *Generalized stratigraphic columns for the three principal areas of hydrocarbon exploration offshore eastern Canada.*

The rifting stage of continental breakup produced a northeast-trending rift valley system flanked with ancillary grabens, some of which formed the early sites of deposition in the Scotian basin. The basal sediments are terrestrial, Upper Triassic–Lower Jurassic red beds of the Eurydice Formation (Figures 5 and 6), which were deposited in fluvial and lacustrine environments.

Evaporite deposition, mainly in the form of salt, was widespread northeast of the Yarmouth Arch during the latest Triassic and Early Jurassic, with more than 3,000 m (9,843 ft) of Argo Formation accumulating in some of the early grabens. Salt deposition ceased abruptly in the Early Jurassic, followed by deposition of evaporitic dolostones of the Iroquois Formation and continental clastic facies of the Mohican Formation (Figure 5). Across the LaHave Platform and in parts of Nova Scotia, a major unconformity developed in conjunction with the latter sequence. It was accompanied, locally, by basaltic dykes and sills, including the North Mountain Basalt of Nova Scotia, which has yielded an average age of 191 ± 2 m.y. (Hayatsu, 1979). This break-up unconformity (Falvey, 1974) marks the initiation of sea-floor spreading in this part of the proto-Atlantic (Jansa and Wade, 1975).

This sea-floor spreading initiated a normal marine regime, and during the Middle and Late Jurassic, a complex set of facies was deposited along the margin of the Scotian basin. Most prominent is the Abenaki Formation carbonate bank complex that developed along the outer part of the LaHave Platform. This development is flanked landward by the Mic Mac and Mohawk formations, which consist of fine- and coarse-grained clastic facies. Seaward of the carbonate bank are marine shales of the Verrill Canyon Formation (Figure 6). East of the LaHave Platform, across a northerly trending hinge zone, the Abenaki Formation is less well developed and interfingers with clastics of the Mic Mac Formation. This change coincides with the transition to the Abenaki and Sable subbasins (Figure 4), and a corresponding increase in the thickness of these formations toward the east.

In response to local downwarping in the Late Jurassic, small, deltaic systems developed within the Mic Mac Formation in the Sable Island area. By the end of the Jurassic, a widespread regression was underway in the Scotian basin. This regression was due to a sediment oversupply caused by the uplifting of source areas and the elevation of the Avalon Uplift, which has been attributed to the separation of the North American and European plates (Jansa and Wade, 1975). These adjustments resulted in a much larger fluvial-deltaic system, the Missisauga Formation, which, in the Early Cretaceous, covered most of the eastern Scotian Shelf and southwestern Grand Banks. This Missisauga delta was fed by an extensive, ancestral St. Lawrence River drainage system. Across the stable platform and into the Abenaki subbasin, up to 1,000 m (3,280 ft) of delta-plain sediments were deposited. These deposits consist mainly of thick, fine- to coarse-grained sand bodies separated by relatively thin shale interbeds. A lower Missisauga delta-front to prodelta

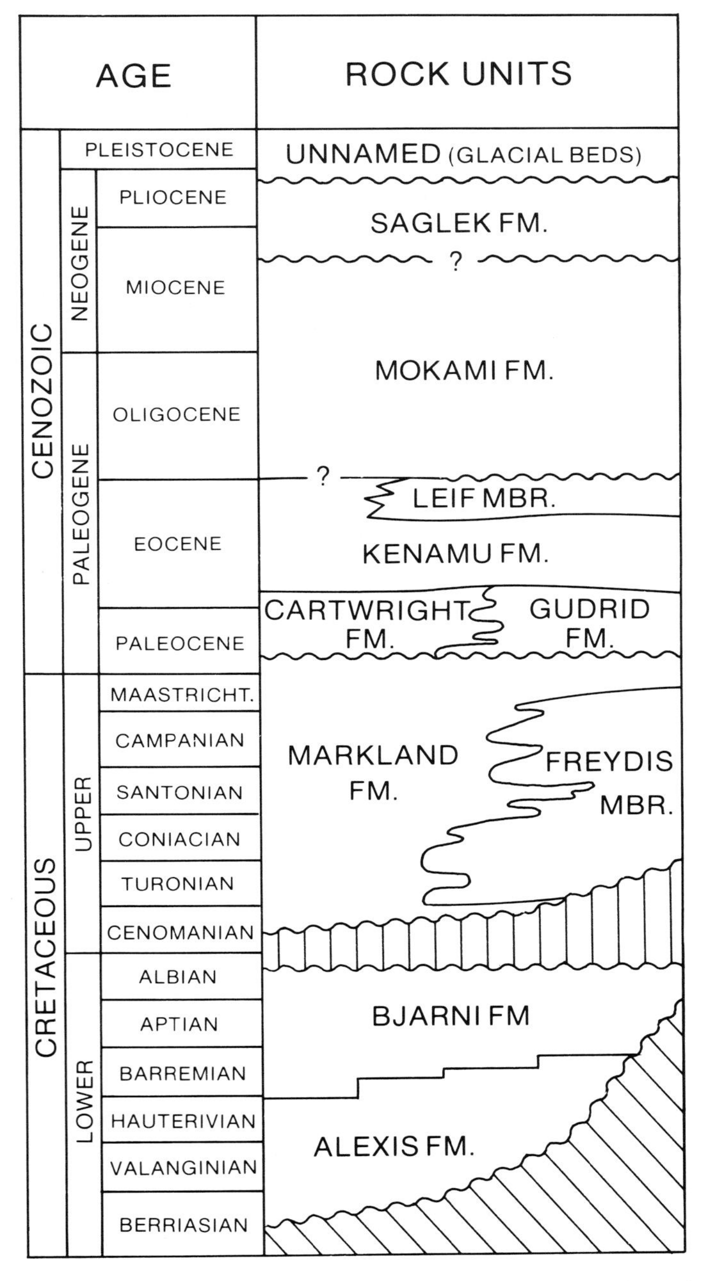


Figure 20. *Generalized stratigraphy of the Labrador Shelf (modified from McWhae et al, 1980).*

Both the Gudrid and Cartwright formations are overlain unconformably by brown-grey marine mudstone of the Kenamu Formation. Thin beds of microcrystalline dolomitic limestone are present throughout the formation. McWhae et al (1980) suggest an early to late Eocene age for this unit, but they note evidence that it may extend into earliest Oligocene. The Leif Member of this formation is a relatively clean, fine-grained sandstone with generally good porosity, regarded as a neritic (possibly tidal) deposit.

The Kenamu is overlain disconformably by the Mokami Formation, which consists of partly consolidated neritic claystone and soft shale, with thin interbeds of siltstone, sandstone, calcareous sandstone, and limestone. The age of this unit is estimated to range from late Eocene or early Oligocene to middle Miocene. The period of its deposition is marked by shallowing, hiatuses, and increased climatic cooling (Gradstein and Srivastava, 1980). The upper surface of the Mokami is a widely developed unconformity overlain by the Saglek Formation.

The Saglek Formation is a very porous, shelly sandstone and conglomerate containing abundant lignite and glauconite and subordinate siltstone and claystone interbeds. It is overlain unconformably by Quaternary boulder beds. The deposits of the Saglek Formation reflect an environment ranging from shallow neritic to continental and lagoonal conditions.

The structural style of the Labrador margin differs considerably from that of the two previously discussed areas where salt deformation plays a key role. Basement block faulting and the associated drape of overlying reservoir beds constitute the main hydrocarbon plays on the Labrador Shelf. Other plays include reservoirs in porous Paleozoic facies associated with basement highs, stratigraphic closures at the depositional edges of the Bjarni and Gudrid formations, and growth faulting in a thick, unstable (possibly due to overpressure) sediment pile in the vicinity of the Corte Real well (Figure 19).

The lack of reservoir beds is a problem in the Labrador Shelf region. The Bjarni Formation, the thick, arkosic, continental sandstone that is widespread in the southern end of the Erik graben, has an average porosity of only 12%. The Gudrid Formation has porosities up to 25%, but this unit is relatively thin, and its distribution, from seismic and well data, is limited to the western flank of the basin. Similar conditions apply to the Leif Member sand. A fourth reservoir unit occurs locally in the form of porous Paleozoic carbonates. These carbonates are part of the basement complex, and are charged with hydrocarbons from overlying Mesozoic source rocks.

The northern part of the area, along the southern Baffin Island Shelf, overlies a complexly faulted zone. The area is further complicated, from a petroleum standpoint, by the presence of Tertiary volcanics. Plays include fault-block structures, Lower Cretaceous sandstones on and around horsts, and reservoirs in Paleozoic outliers (Klose et al, 1982).

### Geochemistry

The types of kerogen present in the sediments, their geochemical characteristics, and their degree of thermal maturation indicate that most of the sedimentary sequences on the Labrador Shelf will be gas-prone (Bujak, Barss, and Williams, 1977a, 1977b; Powell, 1979; Umpleby, 1979; Rashid, Purcell, and Hardy, 1980; Purcell, Umpleby, and Wade, 1980; Powell and Snowdon, 1983). This interpretation is supported by the hydrocarbons discovered to date. The predominant organic matter is liptinite-rich, resinite-poor Type III, which has matured, in this relatively warm basin, to yield a gas/condensate product. Total organic carbon levels from Labrador wells are above average and indicate a good potential hydrocarbon source.

One possible exception to the gas-prone nature of the sediments occurs in the northern part of the Labrador Shelf. Here marine, lower Eocene beds in the Kenamu Formation, containing an average organic carbon content of 3.4%, occur at depths below 2,500 m (8,200 ft) where ambient temperatures are greater than 70°C (158°F). In this area, light gas analyses from canned cuttings range from 50–60,000 ppm, of which 60–80%

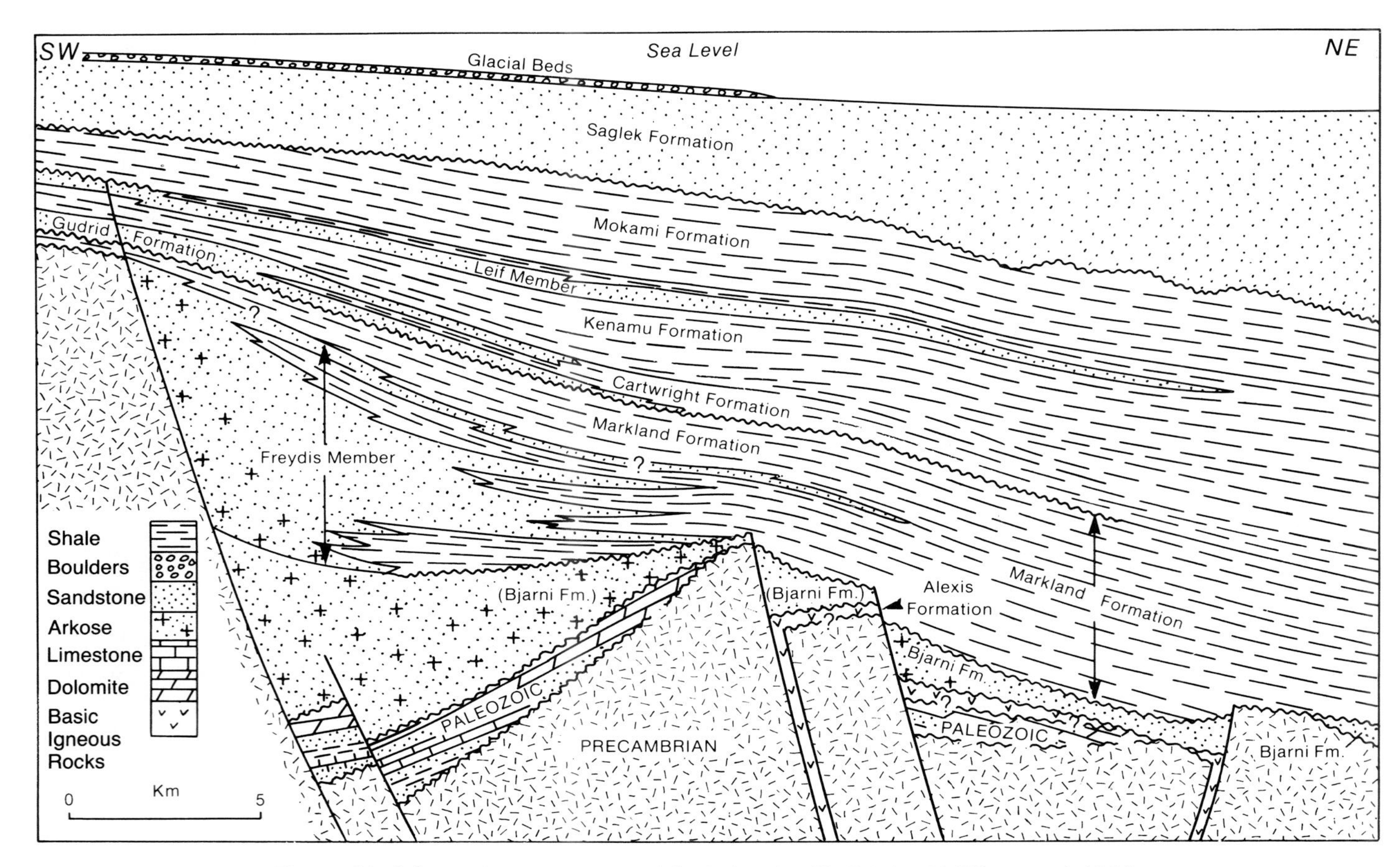


**Figure 21.** *Schematic cross section of the Labrador Shelf (after McWhae et al, 1980).*

are $C_2+$, suggesting that the source rocks can generate liquids if given the proper temperature regime. However, the oil potential of this sequence may not be high. On average, 7% of the organic matter is bitumen (that is, it is soluble in organic solvents) and 20% of this bitumen consists of saturated and aromatic hydrocarbons. Thus, the hydrocarbon generation potential is only 16 mg/g organic carbon. Plotting this quantity and the percentage of hydrocarbons in the bitumen on polar coordinates (Powell, 1978) indicates a probable dry gas or wet gas product with only a minor amount of oil.

Vitrinite reflectance data from six Labrador Shelf wells indicate that onset of full maturity (0.7% $R_o$) occurs between 2,600 and 3,400 m (8,530 and 11,155 ft), depending on basin position and age of sediments, and the base of the mature zone (1.3% $R_o$) will occur between 3,900 and 4,700 m (12,795 and 15,420 ft). When combined with the hydrocarbon occurrence data, these analyses support Powell and Snowdon's 1983 interpretation that the gas condensate was generated off structure in the mature to overmature deep basin, while the waxy oil in North Leif was probably generated locally. On the other hand, some geologists (e.g., McWhae et al, 1980) interpret the gas and condensate recoveries as a product of biogenic degradation or early maturation of Type III organic matter.

**Significant Discoveries**

There are five significant discoveries on the Labrador Shelf and one on the Southeast Baffin Shelf (Figure 19). All are gas and condensate (Table 3). The Gudrid well encountered gas in a 142 m (466 ft) thick Carboniferous dolomite in a faulted basement block in the center of the Erik graben (Figure 21). At this location, the Markland Formation shale immediately overlies the reservoir. The thermal regime, structural relationship, and geochemical characteristics of the Markland Formation, including the gas-prone nature of the kerogen, indicate that it is the source of the gas.

The Bjarni structure is an upfaulted basement block over which interbedded sandstones and shales of the Bjarni Formation drape. This structure also occurs in the central part of the Erik graben. Of the two delineation wells drilled, the first failed to find any pay, while the second encountered 59 m (194 ft) of pay. The Markland Formation again overlies the reservoir and is considered the source of the gas. Petro-Canada drilled a Bjarni look-alike structure, North Bjarni, in 1981, and also discovered gas in the Bjarni sand. The Chevron et al Hopedale E-33 well discovered gas and condensate both in Paleozoic limestone and in Bjarni sandstone reservoirs.

The fifth Labrador Shelf discovery is Eastcan et al Snorri J-90. This well, drilled on an upfaulted basement block, discovered gas in the Gudrid sand.

In 1980, a significant wet gas discovery was made at the Aquitaine et al Hekja O-71 location on the southeast Baffin Shelf. The well, drilled to test a deep basement structure, encountered gas in an upper Paleocene, shallow-marine to deltaic sand (Klose et al, 1982) at a depth of 3,212 m (10,538 ft) (Table 3).

Although not considered a discovery, the Petro-Canada et al North Leif I-05 well is significant in that it recovered the only

Table 3. *Gas discoveries, Labrador—Southeast Baffin Shelf.*

| Well (date) | Interval m | Oil Rate bbl/day | Gas Rate MMCFGD | Gravity °API | Remarks |
|---|---|---|---|---|---|
| Bjarni H-81 (1974) | 2151–2256 | 100 (cond.) | 12.9 | 55 | Bjarni sand |
| Gudrid H-55 | 2663–2723 | 114 (cond.) | 20 | 50 | Carboniferous dolomite |
| (1974) | 2756–2772 | 60 (cond.) | 8.1 | 50 | " " |
| Snorri J-90 (1976) | 2493–2502 | 235 (cond.) | 9.8 | | Gudrid sand |
| Hopedale E-33 | 1948–1959 | 310 (cond.) | 14.2 | 60.1 | Bjarni sand |
| (1978) | 1983–1997 | 500 (cond.) | 19.5 | 58.6 | Paleozoic carbonate |
| North Bjarni F-06 (1981) | 2585–2604 | | 7–8 (est.) | | Bjarni sand (inconclusive test) |
| Hekja A-72 (1980) | 3212–3251 | 106 (cond.) | 9.4 | 54 | Paleocene "Hekja" sand |

Table 4. *Discovered and undiscovered oil and gas resources (recoverable).*

| | Oil Potential ($10^9$ bbl) | | | Gas Potential (tcf) | | |
|---|---|---|---|---|---|---|
| | High Confidence | Average Expectation | Speculative Estimate | High Confidence | Average Expectation | Speculative Estimate |
| Scotian basin | 0.5 | 2.2* | 3.5 | 8.2 | 25.4 | 46.0 |
| Grand Banks | 3.5 | 8.6 | 15.1 | 8.7 | 13.4 | 23.0 |
| Labrador–Southeast Baffin Shelf | 0.1 | 0.9 | 1.7 | 8.4 | 29.0 | 58.5 |

These numbers vary slightly from Procter et al (1984) because of revisions and different geographic combinations.
* includes condensate

free oil on the Labrador margin. Drilled in 1981 to test the pinchout edge of the Bjarni Formation on the flank of a plunging anticline, this well encountered heavy oil staining throughout the lower part of the formation, and 22.8 bbl of mature, waxy oil (33.1° API) were recovered from the pipe in a drillstem test.

The best, current estimate of resources discovered to date in this area is 2.7 tcf of gas and 25.8 million bbl of liquid hydrocarbons (Procter, Taylor, and Wade, 1984).

### Hydrocarbon Potential

Most significant to the estimates of oil and gas potential is the consensus in geochemical, geological, and hydrocarbon occurrence data that the Labrador Shelf is a gas-prone region, and that major accumulations of liquid hydrocarbons are not likely to occur (Table 4). At the same time, however, there remains a sizable liquid hydrocarbon resource at the speculative level that is mainly attributable to an oil play in the Saglek basin.

A detailed, prospect-by-prospect assessment of the Labrador Shelf, together with a play assessment of the less-well-known Southeast Baffin Shelf, indicates an average expectation for gas and oil of more than 29 tcf and 0.9 billion bbl respectively, including discovered resources. The majority of this potential is in Bjarni sand and Paleozoic carbonate reservoirs in structural prospects.

### Summary

Analysis of the Labrador pool size distribution data indicates a relatively low probability for discovery of the several major gas accumulations required to establish an economic threshold in this relatively remote and climatically severe area. However, at the speculative estimate level of Procter, Taylor, and Wade (1984), which is more than twice the average expectation, we can anticipate several major accumulations. These accumulations, however, would have to be considerably larger than any of the discoveries made thus far if they are to be viable economically.

## OVERVIEW

Geochemical data, together with the discoveries and shows recorded to date, indicate that the sedimentary basins off Nova Scotia and Labrador are largely gas-prone, whereas the Jeanne d'Arc basin is a proven oil province with potential for significant gas reserves as well. At the present time, the Hibernia oil discovery off Newfoundland and the Venture gas discovery off Nova Scotia are believed to be economically viable, and likely will be brought into production within the next 5 to 10 years. In terms of present economic conditions, the gas discoveries off Labrador are too small, lie in too harsh an environment, and are too far from markets to be considered producible. Although situated in a more favorable environment, nothing is known of the size of recent gas discoveries in the Jeanne d'Arc basin.

Undoubtedly the most important point to bear in mind at the present time regarding the petroleum potential of a region as vast as the continental margin of Eastern Canada is that the exploration of this region is still at a very early stage. Numerous structures remain to be tested, particularly at greater depths and in deeper water, not to mention a variety of more subtle stratigraphic traps. In addition, well control is increasingly sparse to the north in the Labrador Sea, and the Canadian margin of Baffin Bay and its adjacent waterways are completely untested. If we take a long-term view of the resource potential of eastern Canada's continental margin beyond the 1980s and 1990s and into the 21st century, the developments that we have witnessed to date most assuredly represent an auspicious beginning.

## ACKNOWLEDGMENTS

We are pleased to acknowledge the extensive contribution to this paper that derives from a long association with colleagues in the Geological Survey of Canada in offices at Dartmouth and Calgary. We thank J.S. Bell, Atlantic Geoscience Centre, and G.R. Campbell and D.F. Sherwin, Canada Oil and Gas Lands Administration, Ottawa, for their critical review of the manuscript. Special thanks to G.M. Grant and G.L. Cook, who drafted the illustrations.

## REFERENCES

Amoco Canada Petroleum Company Ltd., and Imperial Oil Ltd., 1973, Regional Geology of the Grand Banks: Bulletin of Canadian Petroleum Geology, v. 21, p. 479–503.

Arthur, K. R., D. R. Cole, G. G. L. Henderson, and D. W. Kushnir, 1982, Geology of the Hibernia Discovery, *in* M. T. Halbouty, ed., The Deliberate search for the subtle trap: AAPG Memoir 32, p. 181–196.

Auzende, J. M., J. L. Olivet, and J. Bonnin, 1970, La marge du Grand Banc et la fracture de Terre-Neuve: Comptes Rendus de l'Academie des Sciences, Paris, Series D, v. 271, p. 1063–1066.

Barss, M. S., J. P. Bujak, and G. L. Williams, 1979, Palynological zonation and correlation of sixty-seven wells, eastern Canada: Geological Survey of Canada Paper 78-24, 118 p.

——, J. P. Bujak, J. A. Wade, and G. L. Williams, 1980, Age, stratigraphy, organic matter type and color, and hydrocarbon occurrences in forty-seven wells offshore eastern Canada: Geological Survey of Canada Open File Report 714; 6 p., 50 figs.

Benteau, R. I., and M. G. Sheppard, 1982, Hibernia—a petrophysical and geological review: Journal of Canadian Petroleum Technology, v.21, p. 59–72.

Bujak, J. P., M. S. Barss, and G. L. Williams, 1977a, Offshore eastern Canada—Part I, organic type and color and hydrocarbon potential: Oil and Gas Journal, v. 75, n. 14, p. 198–202.

——, M. S. Barss, and G. L. Williams, 1977b, Offshore eastern Canada—Part II, organic type and color and hydrocarbon potential: Oil and Gas Journal, v. 75, n. 15, p. 96–100.

Dawes, P. R., and J. W. Kerr, eds., 1982, Nares Strait and the drift of Greenland; a conflict in plate tectonics: Meddelelser om Gronland, Geosciences 8, 392 p.

Eliuk, L. S., 1978, The Abenaki Formation, Nova Scotia Shelf, Canada—a depositional and diagenetic model for a Mesozoic carbonate platform: Bulletin of Canadian Petroleum Geology, v. 26, p. 424–514.

Falvey, D. A., 1974, The development of continental margins in plate tectonic theory: APEA Journal, v. 14, p. 95–106.

Given, M. M., 1977, Mesozoic and early Cenozoic geology of offshore Nova Scotia: Bulletin of Canadian Petroleum Geology, v. 25, p. 63–91.

Gradstein, F. M., G. L. Williams, W. A. M. Jenkins, and P. Ascoli, 1975, Mesozoic and Cenozoic stratigraphy of the Atlantic continental margin, eastern Canada, *in* C. J. Yorath, E. R. Parker, and D. J. Glass, eds., Canada's Continental Margins and offshore petroleum exploration: Canadian Society of Petroleum Geologists Memoir 4, p. 103–131.

——, and S. P. Srivastava, 1980, Aspects of Cenozoic stratigraphy and paleoceanography of the Labrador Sea and Baffin Bay: Paleogeography, Paleoclimatology, Paleoecology, v. 30, p. 261–295.

——, and G. L. Williams, 1981, Stratigraphic charts of the Labrador and Newfoundland shelves: Geological Survey of Canada Open File Report 826, 11 p., 4 figs.

Grant, A. C., 1972, The continental margin off Labrador and eastern Newfoundland—morphology and geology: Canadian Journal of Earth Sciences, v. 9, p. 1394–1430.

——, 1975a, Structural modes of the western margin of the Labrador Sea: Geological Survey of Canada Paper 74-30, p. 217–231.

——, 1975b, Geophysical results from the continental margin off southern Baffin Island, *in* C. J. Yorath, E. R. Parker, and D. J. Glass, eds., Canada's Continental Margins and Offshore Petroleum Exploration: Canadian Society of Petroleum Geologists Memoir 4, p. 411–431.

——, 1980, Problems with plate tectonics: the Labrador Sea: Bulletin of Canadian Petroleum Geology, v. 28, p. 252–278.

Hallam, A., 1978, Eustatic cycles in the Jurassic: Paleogeography, Paleoclimatology, Paleoecology, v. 23, p. 1–32.

Handyside, D. D., and W. I. Chipman, 1983, A preliminary study of the Hibernia field: Journal of Canadian Petroleum Technology, v. 22, p. 67–78.

Hardy, I. A., and A. E. Jackson, 1980, A compilation of geochemical data; eastcoast exploratory wells: Geological Survey of Canada Open File Report 694, 10 p., 54 figs.

Jackson, A. E., 1985, Sediment Thickness Map, offshore eastern Canada: Geological Survey of Canada Open File Report.

Jansa, L. F., and J. A. Wade, 1975, Geology of the continental margin off Nova Scotia and Newfoundland: Geological Survey of Canada Paper 74-30, p. 51–105.

——, F. M. Gradstein, I. M. Harris, W. A. M. Jenkins, and G. L. Williams, 1976, Stratigraphy of the Amoco-IOE Murre G-67 well, Grand Banks of Newfoundland: Geological Survey of Canada Paper 75-30, 14 p.

——, J. P. Bujak, and G. L. Williams, 1980, Upper Triassic deposits of the western North Atlantic: Canadian Journal of Earth Sciences, v. 17, p. 547–559.

Keen, C. E., D. L. Barrett, K. S. Manchester, and D. I. Ross, 1972, Geophysical studies in Baffin Bay and some tectonic implications: Canadian Journal of Earth Sciences, v. 9, p. 239–256.

Kerr, J. W., 1967, A submerged continental remnant beneath the Labrador Sea: Earth and Planetary Science Letters, v. 2, p. 283–289.

Klose, G. W., E. Malterre, N. J. McMillan, and C. G. Zinkan, 1982, Petroleum exploration offshore southern Baffin Island, northern Labrador Sea, Canada, *in* A. F. Embry and H. R. Balkwill, eds., Arctic geology and geophysics: Canadian Society of Petroleum Geologists Memoir 8, p. 233–244.

Kristoffersen, Y., 1978, Seafloor spreading and the early opening of the north Atlantic: Earth and Planetary Science Letters, v. 38, p. 273–290.

LePichon, X., R. D. Hyndman, and G. Pautot, 1971, Geophysical study of the opening of the Labrador Sea: Journal of Geophysical Research, v. 76, p. 4724–4734.

McIver, N. L., 1972, Cenozoic-Mesozoic stratigraphy of the Nova Scotia Shelf: Canadian Journal of Earth Sciences, v. 9, p. 54–70.

McKenzie, R. M., 1980, The Hibernia structure, prepared by the Frontier Group of Mobil Oil of Canada Ltd. and presented to the C.S.E.G. luncheon meeting on December 9, 1980, Calgary: Reported in Oilweek, December 15, 1980, p. 59.

McMillan, N. J., 1973, Shelves of Labrador Sea and Baffin Bay, Canada, *in* R. G. McCrossan, ed., The future petroleum provinces of Canada—their geology and potential: Canadian Society of Petroleum Geologists Memoir 1, p. 473–517.

——, 1980, Geology of the Labrador Sea and its petroleum potential, *in* Tenth World Petroleum Congress Proceedings (Bucharest), v. 2, Exploration—Supply and Demand, Heyden, London, p. 165–175.

——, 1982, Canada's East Coast; the new super petroleum province: Journal of Canadian Petroleum Technology, v. 21, p. 1–15.

McWhae, J. R. H., 1981, Structure and spreading history of the northwestern Atlantic region from the Scotian Shelf to Baffin Bay, *in* J. W. Kerr and A. J. Ferguson, eds., Geology of the North Atlantic Borderlands: Canadian Society of Petroleum Geologists Memoir 7, p. 299–332.

——, R. Elie, K. C. Laughton, and P. R. Gunther, 1980, Stratigraphy and petroleum prospects of the Labrador Shelf: Bulletin of Canadian Petroleum Geology, v. 28, p. 460–488.

Nantais, P. T., 1983, A reappraisal of regional hydrocarbon potential of the Scotian Shelf: Geological Survey of Canada: Open File Report 1175, 179 p.

Powell, T. G., 1978, An assessment of the hydrocarbon source rock potential of the Canadian Arctic Islands: Geological Survey of Canada Paper 78–12, 82 p.

——, 1979, Geochemistry of Snorri and Gudrid condensates, Labrador Shelf; implications for future exploration, *in* Current Research, Part C: Geological Survey of Canada Paper 79-1C, p. 91–95.

——, 1982, Petroleum geochemistry of the Verrill Canyon Formation; a source for Scotian Shelf hydrocarbons: Bulletin of Canadian Petroleum Geology, v. 30, p. 167–179.

——, 1984, Hydrocarbon-source relationships, Jeanne d'Arc and Avalon basins, offshore Newfoundland: Geological Survey of Canada Open File Report 1094, 12 p.

——, and L. R. Snowdon, 1983, A composite hydrocarbon generation model: Erdöl und Kohle-Erdgas-Petrochemie, v. 36, p. 163–170.

Procter, R. M., G. C. Taylor, and J. A. Wade, 1984, Oil and natural gas resources of Canada—1983: Geological Survey of Canada Paper 83–31, 59 p.

Purcell, L. P., D. C. Umpleby, and J. A. Wade, 1980, Regional geology and hydrocarbon occurrences off the east coast of Canada, *in* A. D. Miall, ed., Facts and principles of world petroleum occurrence: Canadian Society of Petroleum Geologists Memoir 6, p. 551–566.

Rashid, M. A., L. P. Purcell, and I. A. Hardy, 1980, Source rock potential for oil and gas of the East Newfoundland and Labrador Shelf areas, *in* A. D. Miall, ed., Facts and principles of world petroleum occurrence: Canadian Society of Petroleum Geologists, Memoir 6, p. 589–608.

Sherwin, D. F., 1973, Scotian Shelf and Grand Banks, *in* R. G. McCrossan, ed., The future petroleum provinces of Canada: Canadian Society of Petroleum Geologists Memoir 1, p. 519–559.

Srivastava, S. P., 1978, Evolution of the Labrador Sea and its bearing on the early evolution of the north Atlantic: Geophysical Journal of the Royal Astronomical Society, v. 52, p. 313–357.

Swift, J. H., and J. A. Williams, 1980, Petroleum source rocks, Grand Banks area, *in* A. D. Miall, ed., Facts and principles of world petroleum occurrence: Canadian Society of Petroleum Geologists Memoir 6, p. 567–588.

——, R. W. Switzer, and W. F. Turnbull, 1975, The Cretaceous Petrel limestone of the Grand Banks, Newfoundland, *in* C. J. Yorath, E. R. Parker, and D. J. Glass, eds., Canada's Continental Margins and offshore petroleum exploration: Canadian Society of Petroleum Geologists Memoir 4, p. 181–194.

Umpleby, D. C., 1979, Geology of the Labrador Shelf: Geological Survey of Canada Paper 79–13, 34 p.

Upshaw, C. F., W. E. Armstrong, W. B. Creath, E. J. Kidson, and G. A. Sanderson,1974, Biostratigraphic framework of the Grand Banks: AAPG Bulletin, 58, p. 1124–1132.

Van der Linden, W. J. M., 1975, Crustal attenuation and seafloor spreading in the Labrador Sea: Earth and Planetary Science Letters, v. 27, p. 409–423.

Wade, J. A., 1978, The Mesozoic-Cenozoic history of the northwestern margin of North America: Offshore Technological Conference Proceedings, v. 3, p. 1849–1859.

——, 1981, Geology of the Canadian Atlantic margin from Georges Bank to the Grand Banks, *in* J. W. Kerr and A. J. Ferguson, eds., Geology of the North Atlantic Borderlands: Canadian Society of Petroleum Geologists Memoir 7, p. 447–460.

——, G. R. Campbell, R. M. Procter, and G. C. Taylor, in press, Petroleum resources of the Scotian Shelf: Geological Survey of Canada Paper.

Williams, G. L., 1975, Dinoflagellate and spore stratigraphy of the Mesozoic-Cenozoic, offshore eastern Canada: Geological Survey of Canada Paper 74–30, p. 107–161.

——, and W. W. Brideaux, 1975, Palynological analysis of late Mesozoic-Cenozoic rocks of the Grand Banks off Newfoundland: Geological Survey of Canada Bulletin 236, 162 p.

# Oil and Gas Potential of the Amazon Paleozoic Basins

Raul Mosmann
Frank U. H. Falkenhein
*Petrobrás*
*Rio de Janeiro, Brazil*

Alfredo Gonçalves
*Petrobrás*
*Manaus, Brazil*

Francisco Nepomuceno Filho
*Petrobrás*
*Belém, Brazil*

The Upper, Middle and Lower Amazon basins are large elliptical, elongated syneclises located between the Guyana and the Brazilian Precambrian shields, covering a total area of about 850,000 sq km (327,000 sq mi). These basins, or sags, contain multiple cycles of Phanerozoic sediments interrupted by unconformities. Sediments are predominantly Paleozoic with minor Cretaceous covering. Regional, positive elements include upwarps, swells, platforms, structural terraces, and major interbasinal arches, entirely caused by basement tectonics. Crustal stretching also caused widespread rifting and associated basic volcanism. Tectonic modifiers comprise mostly wrenching, within or at basin margins, especially in the Upper Amazon basin, where folds and reverse faults trap large gas accumulations.

Exploration in the Amazon basins began in 1950, with seismic and gravimetric surveys along major rivers. Based on these surveys, 130 wells were drilled prior to 1971. Because of constraints imposed by poor seismic resolution and also because of massive diabase sills, only regional, positive features and minor, subdued structures were identified and the exploratory drilling results were disappointing. From 1976 to 1984, however, an impressive enhancement on seismic reflection methods combined with airborne magnetometry and gravity maps allowed the detailing of important structural features, and exploration success began.

The ultimate natural gas resources of the Upper Amazon basin are presently rated at 120 billion cu m (4.2 tcf). The Middle and Lower Amazon basins still remain as frontier areas for hydrocarbon exploration.

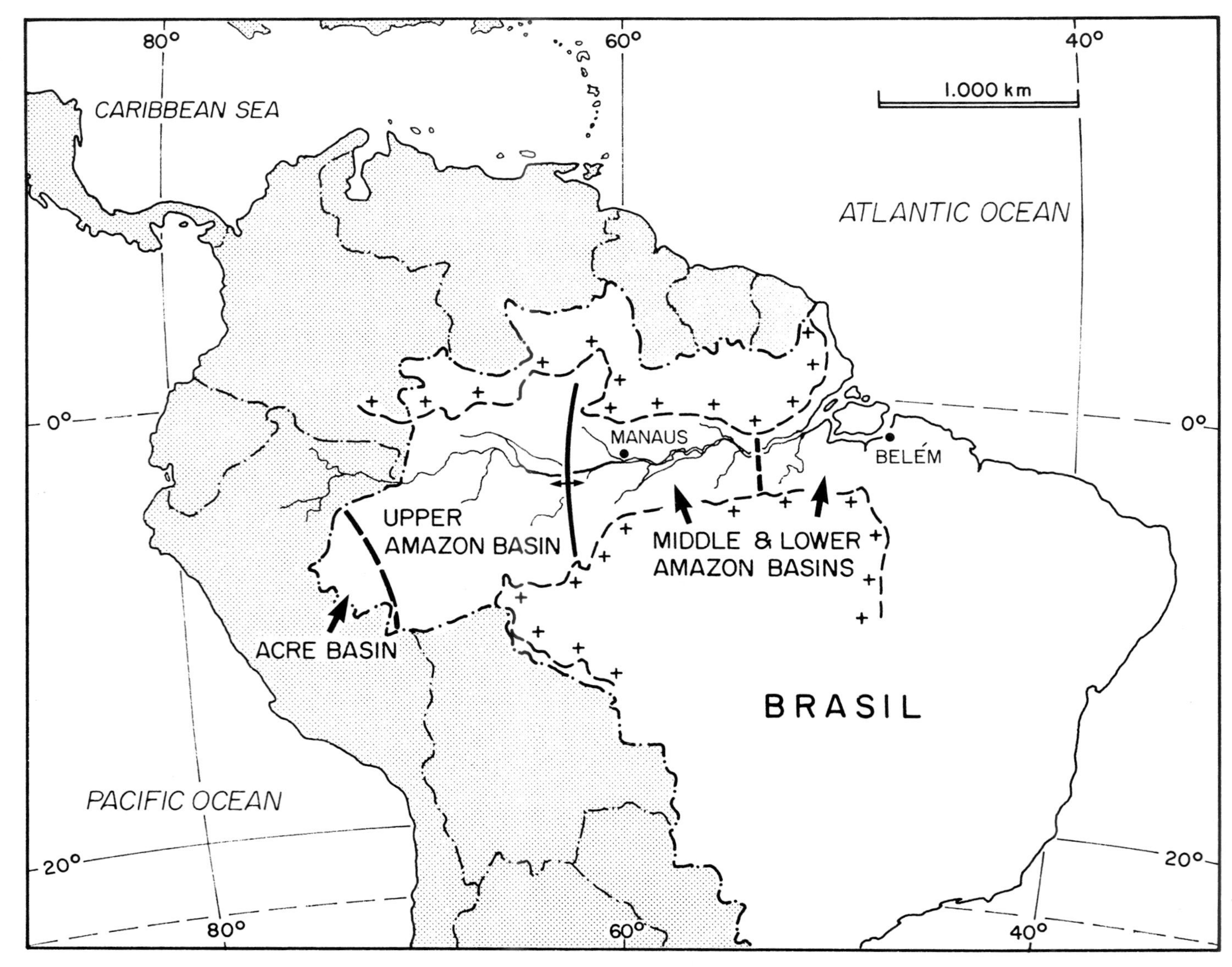


Figure 1. *Location map of Amazon Paleozoic basins.*

## INTRODUCTION

The Paleozoic Amazon basins, covering an area of 850,000 sq km (327,000 sq mi) are elongated, symmetrical, intracratonic syneclises filled with up to 6,000 m (19,700 ft) of sediments, separated by basement arches, and located in continental interior areas (Figure 1).

The Upper, Middle, and Lower Amazon basins, as they have been named since the beginning of exploration in the area, extend in an east to west direction for about 2,500 km (1,500 mi), which is equivalent to the distance from Phoenix, Arizona to Atlanta, Georgia. The preserved width of their Paleozoic sedimentary filling reaches a maximum of 535 km (320 mi). The main course of the Amazon River coincides with or parallels the longer axis of the basins, except in the Upper Amazon. The Amazon sedimentary basins are entirely covered by tropical jungle.

The first subdivision of the basins was based on their geographic position with respect to the river course. Now, it seems that only the Purus Arch, which separates the Upper and Middle Amazon basins, remained as a positive feature during most of Phanerozoic time. Thus, the distinctions between the two basins are evident in their sedimentary thicknesses and structural style. The differences between the Middle and Lower Amazon are less remarkable, although these basins show changes in tectonic style.

Systematic exploration for hydrocarbons in this region started in 1950, and by the end of that decade, a good knowledge of the Paleozoic stratigraphy of the basins was already available. Workers obtained this information by geological surface mapping along the narrow outcrops of Paleozoic rocks in the Middle and Lower Amazon basins' margins, and by geophysical surveys and drilling of stratigraphic and exploratory wells, mainly along the rivers. The Upper Amazon was the lesser-known basin, because the Cretaceous and Tertiary rocks cover entirely the underlying Paleozoic depositional sequences. For this reason and also because of a complete lack of hydrocarbon shows in the drilled wells, geologists rated the

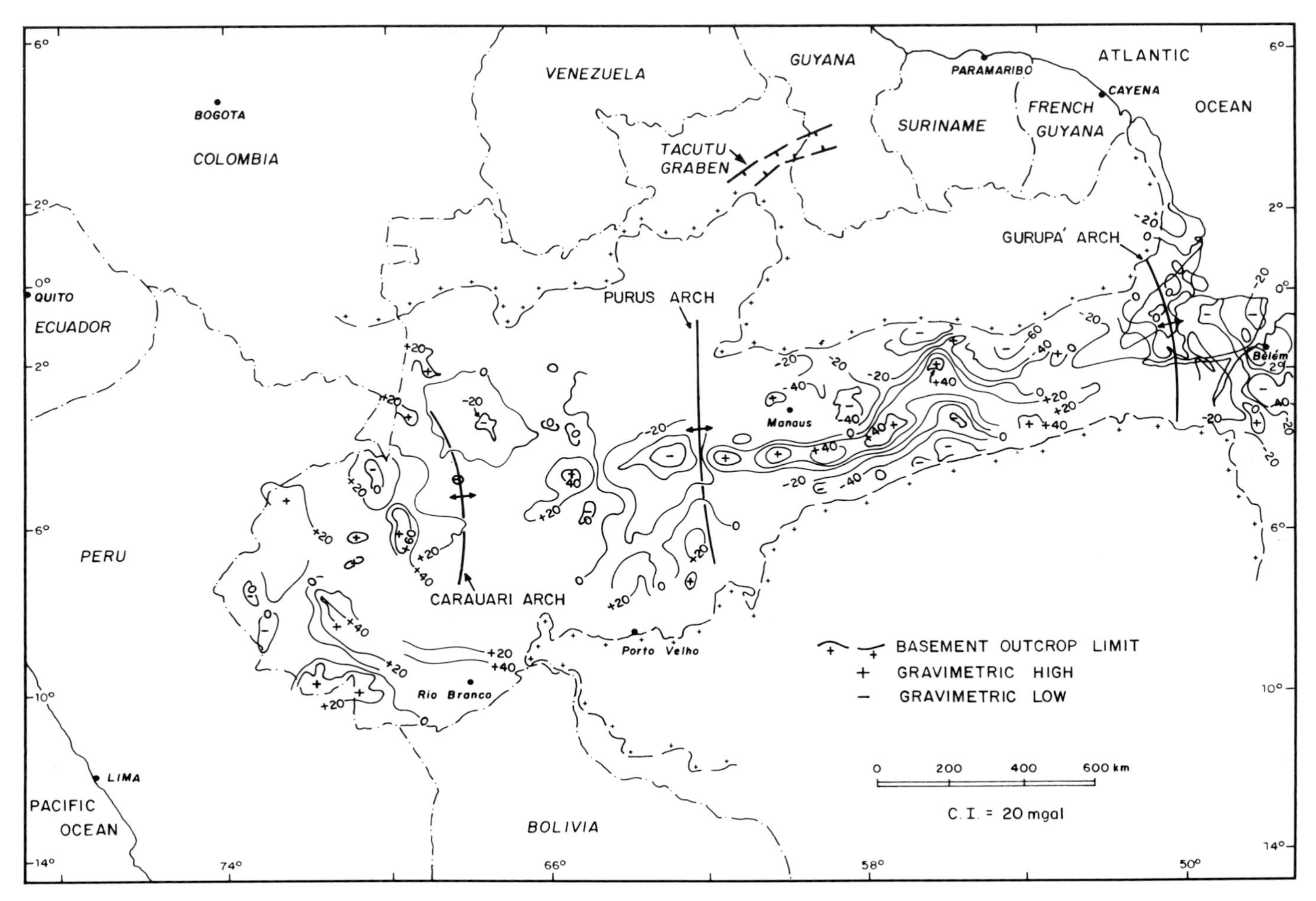


Figure 2. *Bouguer gravity map of the Amazon basins, showing two distinct patterns of anomalies separated by the Purus Arch. The axial high gravity anomalies flanked by gravity lows in Middle Amazon, near Manaus, contrast with the random-like anomaly pattern of Upper Amazon.*

Upper Amazon very low in terms of hydrocarbon potential.

Until the middle of the 1970s, the lithologic characteristics of the Amazon basins hindered the seismic identification of structural features, and exploratory drilling was unsuccessful. These basins contain thick layers of high-velocity diabase sills interbedded with evaporites and carbonates overlying the Upper Carboniferous reservoir, which is the main exploratory target. Further problems were related with difficulties in operating in the tropical jungle, especially away from the main rivers.

By this time the Amazon basins were considered to be Paleozoic intracratonic basins with very little tectonic deformation. Geologists attributed most structures to either faulting associated with the basic volcanism or to compaction over gentle basement irregularities or over uplifted arches.

The available tectonic models were unable to explain the evolution of the basins and to predict areas of more intense structural deformation. Nevertheless, some singular facts were observed in early exploration of the area, like the strong, linear, positive Bouguer anomalies coinciding with the axis of the syneclises, mainly in the Middle Amazon (Figure 2). These anomalies were first observed by Linsser (1958), who related them to high-density intrabasement masses at great depths (40 km or 25 mi). According to his modelling, these denser, intrabasement masses reflect crustal thinning by rifting and consequent mantle rising. The rift basins thus formed were filled with pre-Silurian sediments.

Later, with the establishment of the theory of plate tectonics, the idea of the South American continent drifting away from Africa and the recognition of the large, east to west fracture zones in the mid-ocean ridges of the South Atlantic made some geologists relate the effects of such movements within the interior of the continent, like in northern Brazil, with the present configuration of the long Amazon Trough. Loczy (1970) postulated, for the first time, the presence of a left-lateral transcurrent fault, going across South America from the Marajó Island, on the Atlantic coast of Brazil, to Huancabamba in Peru. Later, several authors also referred to such an idea. Using structural evidences based mainly in the Middle and Lower Amazon, Rezende and Brito (1973) proposed a right-lateral east to west shear couple causing strike-slip faults with northwest and northeast directions, cutting the Paleozoic rocks.

These ideas evolved until the technological advances achieved in acquiring seismic data allowed the recording of fairly good seismic reflection lines in the Paleozoic areas, mainly in the Upper Amazon basin. This fact, coupled with the advent of heli-transported seismic crews and heli-rigs, allowed

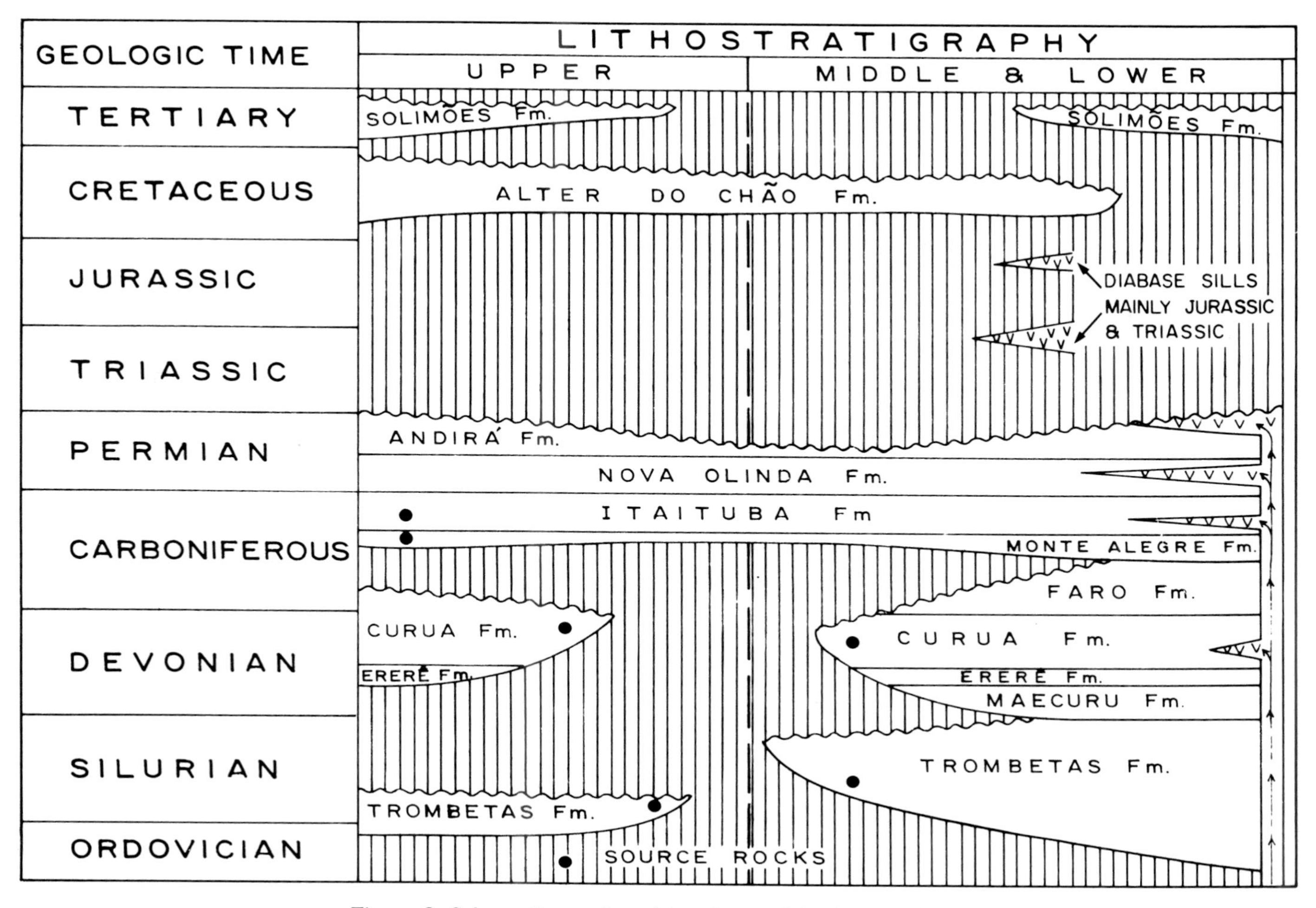


Figure 3. *Schematic stratigraphic column of the Amazon basins.*

a better investigation and resulted in the discovery of a large gas province in the Juruá River area. In this area, several structures are associated with a highly deformed transpressional zone, extending for about 650 km (400 mi) in a northeast-to-southwest direction. Wrench tectonics, forming reverse faults and associated folds, constitute the already mapped structural elements. These structures were considered improbable in such a basin until 1975.

## STRATIGRAPHY

As simple intracratonic basins—Klemme's type 1—or interior sags (Kingston, Dishroon, and Williams, 1983), the Amazon basins comprise multiple Paleozoic deposit cycles, from Ordovician to Permian (Figure 3).

During the Ordovician and Silurian, about 700 m (2,300 ft) of fluvio-deltaic to marine clastic sediments were deposited, corresponding to the Trombetas Formation (Figure 4), with source rocks of marine black shales of the Pitinga Member (Figure 5). In the Upper Amazon basin, this entire sequence reaches locally only 100 m (330 ft) thickness.

The Devonian sedimentation began after a regional unconformity and corresponded to a transgressive-regressive marine, siliciclastic sequence. The lower part of this sequence is represented by deltaic to tidal-flat sandstones, corresponding to the Maecuru and Ereré formations. The upper part of the Devonian is a shallowing-upward marine sequence, beginning with prodelta and offshore shales and topped by a major deltaic sedimentation. Short episodes of glacial and periglacial sedimentation occurred also during Late Devonian. The radioactive Barreirinha shales from the Curuá Formation (Figure 6) constitute the best source rock of the Amazon basins. All Devonian sandstones are fairly good reservoir rocks. Fluvio-deltaic sandstones of the Faro Formation were deposited during the Early Carboniferous, mostly in the Lower Amazon basin.

The Devonian and Lower Carboniferous sedimentary sequences comprise up to 1,600 m (5,200 ft) in the Middle and Lower Amazon elongated sags, but barely reach 200 m (660 ft) in the Upper Amazon basin (Figure 7).

During the Late Carboniferous, a continental sedimentation resumed after a regional unconformity that especially eroded the Carauari and Purus Arches (Figure 8 and 9). The Monte Alegre fluvial to nearshore basal sandstones, about 50 m (165 ft) thick, were then deposited and constitute the principal reservoir rocks of the basins.

Sedimentation continued in Late Carboniferous and Permian, resulting in an alternating sequence of open and

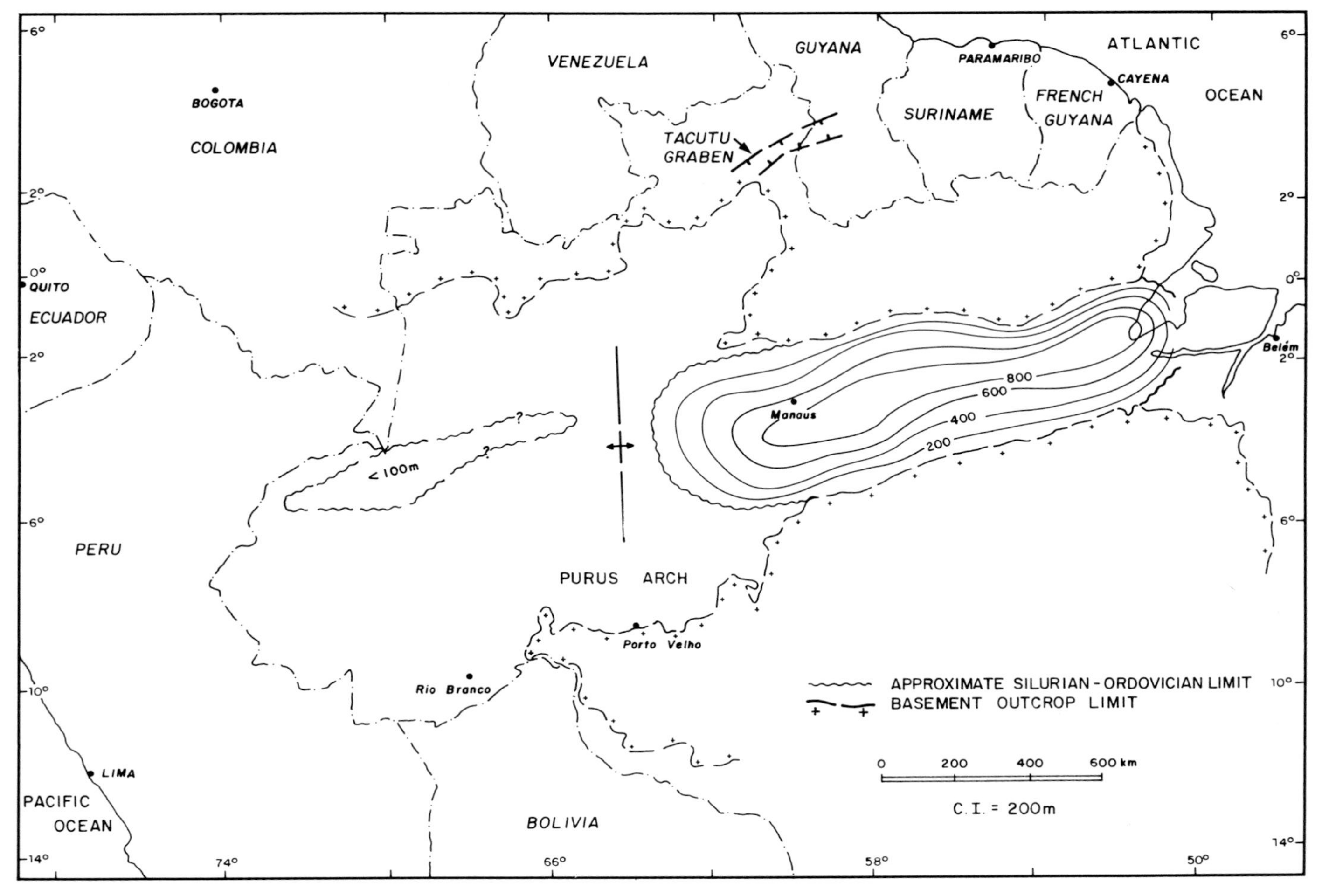


**Figure 4.** *Isopach map of the Silurian–Ordovician sequence. The thickness of this unit is less than 100 m, west of the Purus Arch in the Upper Amazon basin.*

restricted marine carbonates, evaporites, and fine-grained siliciclastics. This sequence, corresponding to the Itaituba, Nova Olinda, and Andirá formations, overlies the Monte Alegre Formation.

Evaporites on the top of the Monte Alegre sandstone are the seal rocks for this reservoir in the Upper, Middle, and part of the Lower Amazon basin. The Permian–Carboniferous sediments were deposited in three principal sags, oriented along the west-to-east direction and reaching thicknesses of up to 2,200 m (7,200 ft), as shown in Figure 10.

The entire Paleozoic section is very thick in the Middle and Lower Amazon, where up to 4,400 m (14,400 ft) of sediments were identified (Figure 11). In the Upper Amazon, the Paleozoic sedimentation reached 2,400 m (7,800 ft) on the eastern part and only up to 1,200 m (3,900 ft) on the western part of the Carauari Arch. Therefore, we can identify three, well-defined Paleozoic sags.These sags, active also during the Permian–Carboniferous, controlled the alternating restricted to open-marine sedimentation in these basins.

Basic volcanism affected the Amazon basins during Triassic and Jurassic. This volcanism is represented by diabase sills and dikes that intruded the Paleozoic section. Diabase sills, up to 500 m (1,600 ft) in thickness, were preferentially intruded in the evaporite rocks. Radiometric datations of the diabases show ages ranging from 220 m.y. to 140 m.y. The total thickness of diabase sills identified in the Upper Amazon is up to 800 m (2,600 ft), while in the Middle and Lower Amazon these volcanics reach 500 m (1,600 ft). A volume of 340,000 cu km (81,500 cu mi) of basic magma was injected as sills in the Paleozoic rocks of the Amazon basins.

The Cretaceous sedimentation started after a regional unconformity and was characterized by fluvial coarse-clastic deposits reaching thicknesses of 400 m (1,300 ft) in the Upper, Middle, and western part of the Lower Amazon (Figure 12). This Cretaceous sedimentation overlies an angular unconformity in the western part of the Upper Amazon basin, reflecting continuing sagging in other parts of the basins and consisting of syntectonic deposits eastward of the Gurupá Arch (Marajó Trough). The differentiation into three basins—Upper, Middle, and Lower Amazon—resulted from these stratigraphic characteristics because of interbasinal arches, as well as separate tectonic evolutions.

The final Tertiary sedimentary infill affected both the Upper and Lower Amazon, where respectively 100 m (330 ft) and 1,000 m (3,300 ft) of fluvio-lacustrine sandstones and shales were deposited (Figure 13).

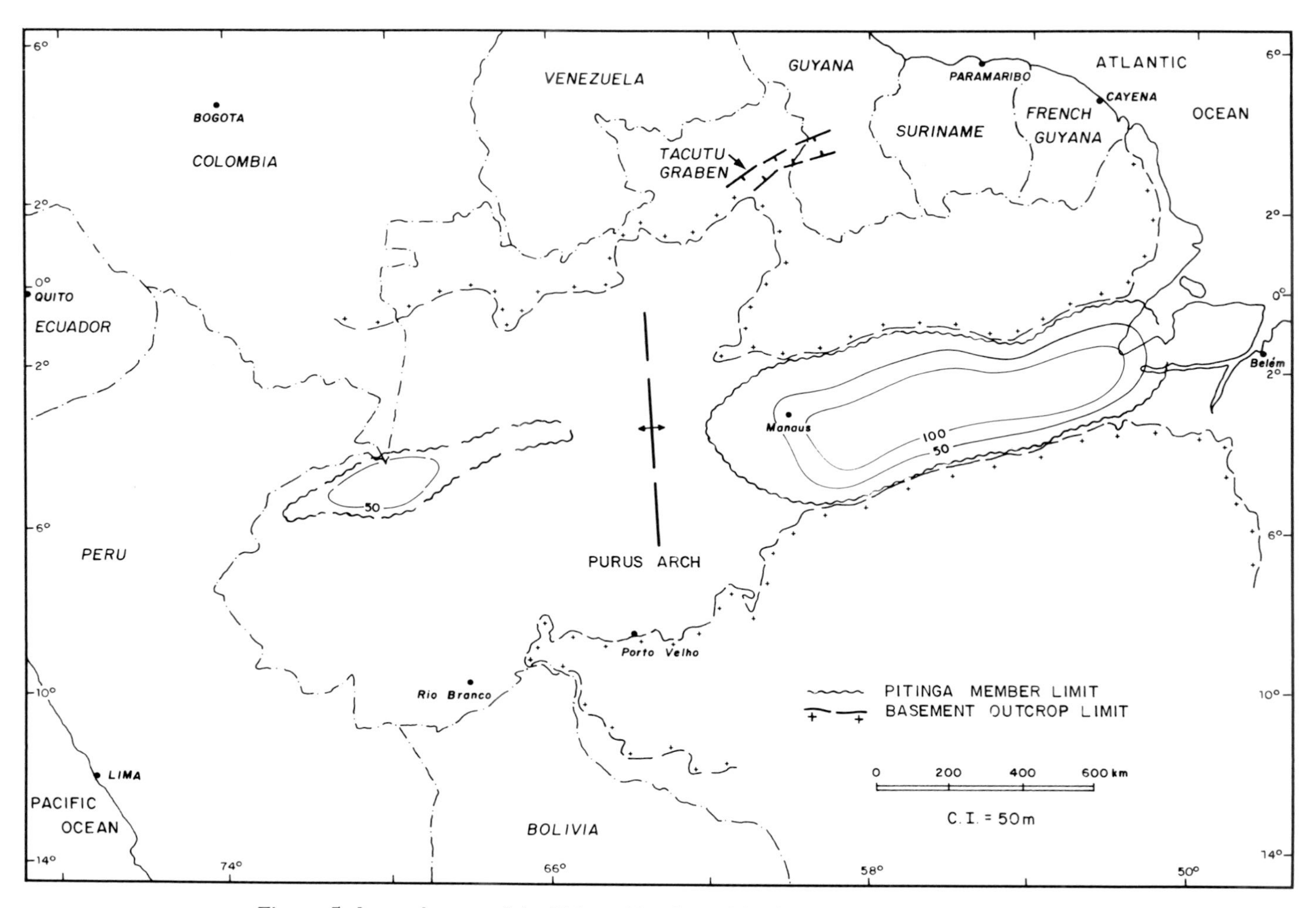


Figure 5. *Isopach map of the Pitinga Member of the Silurian Trombetas Formation.*

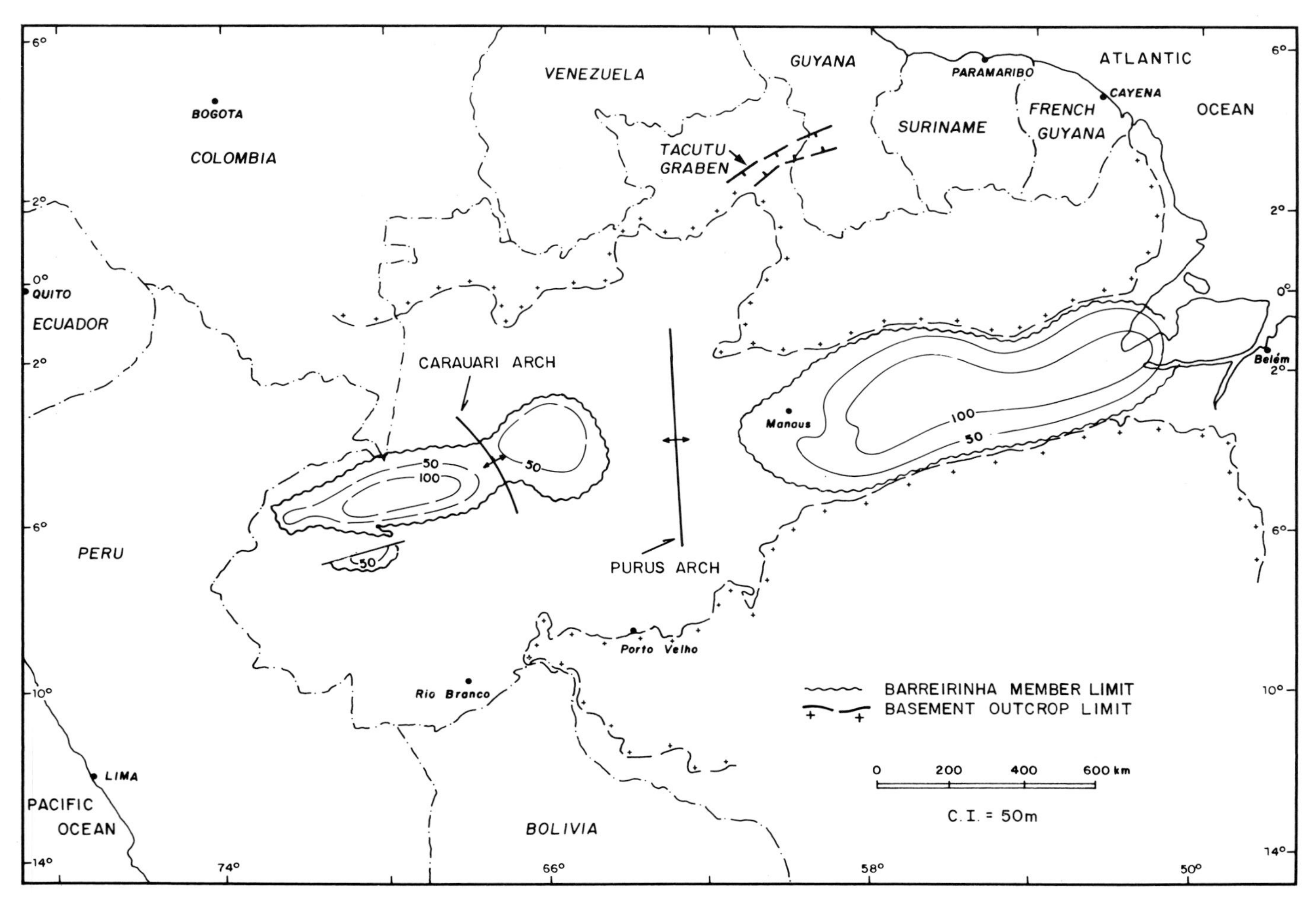


Figure 6. *Isopach map of the Barreirinha Member of the Devonian Curuá Formation.*

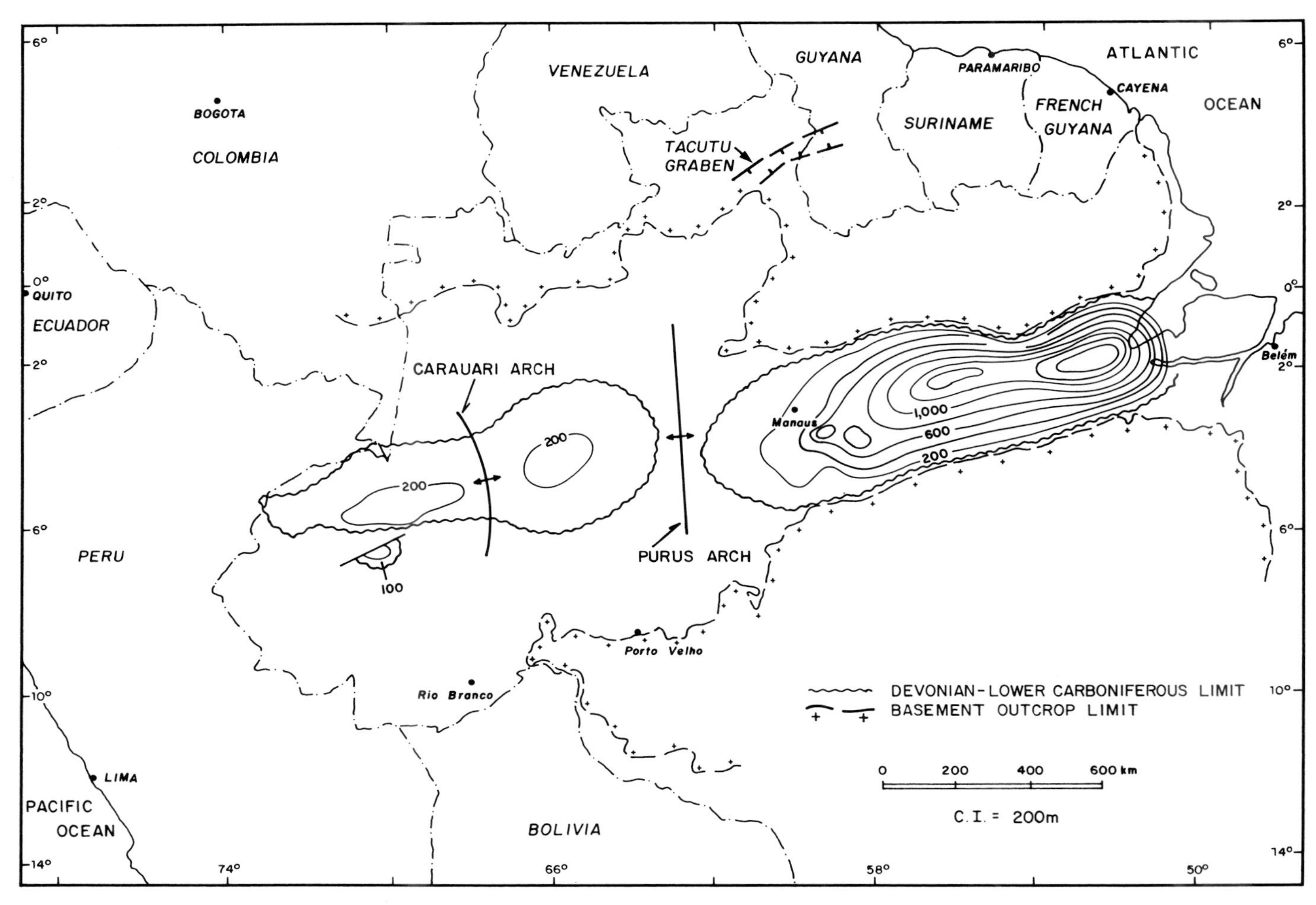


Figure 7. *Isopach map of the Devonian–Lower Carboniferous sequence. The Lower Carboniferous Faro Formation is only identified in the Middle and Lower Amazon.*

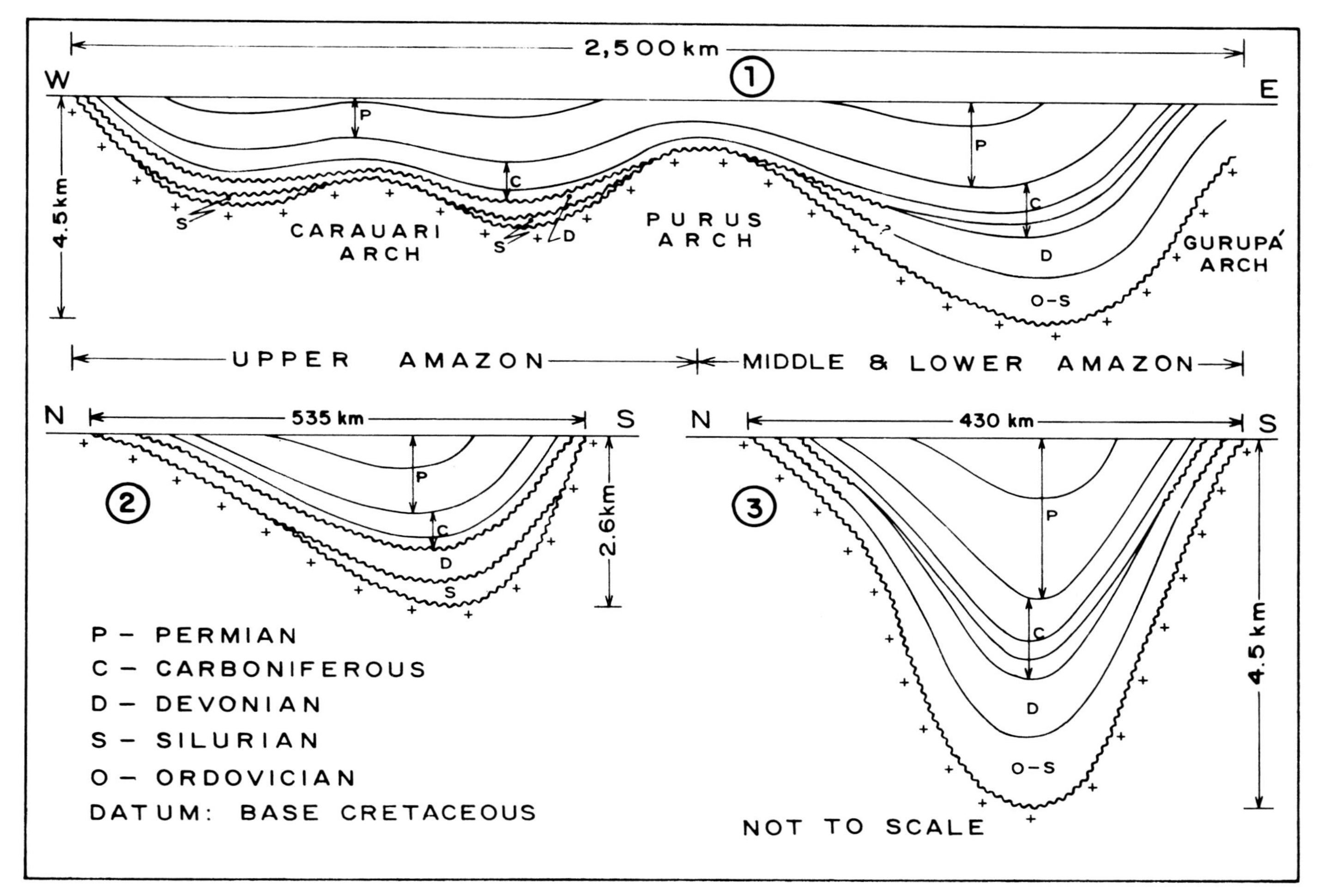


Figure 8. *Schematic regional stratigraphic cross-sections (1, 2 and 3) through the Amazon basins, showing regional unconformities and the main arches oriented in an approximate north to south direction (lines 1, 2 and 3 in Figure 9).*

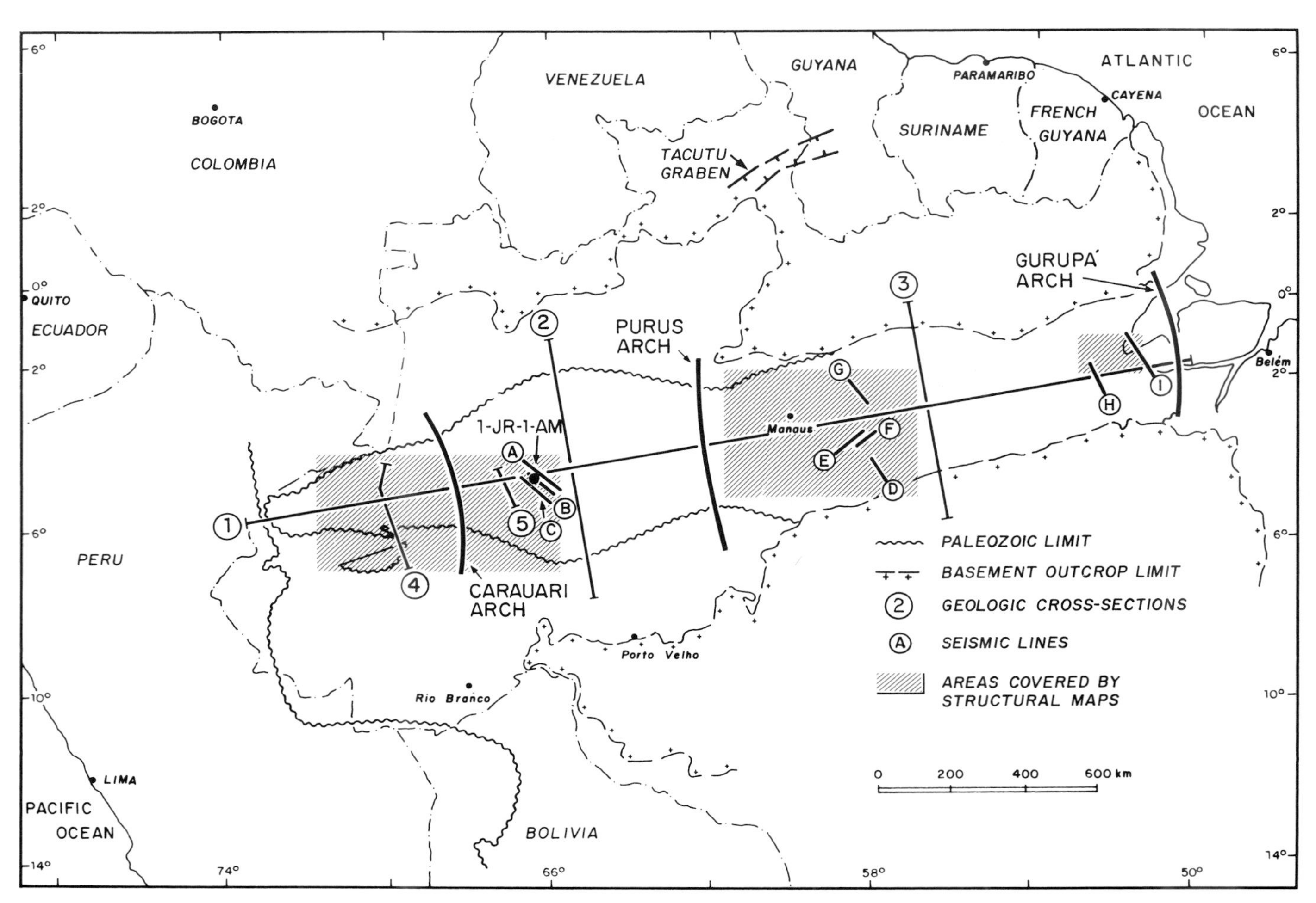


Figure 9. *Index map showing the limits of the Paleozoic basins in the Amazon, and the areas covered by simplified structural maps (Figures 14, 15, and 16), geologic cross-sections (Figures 8 and 35) and seismic lines (Figures 25 to 33) used in this paper.*

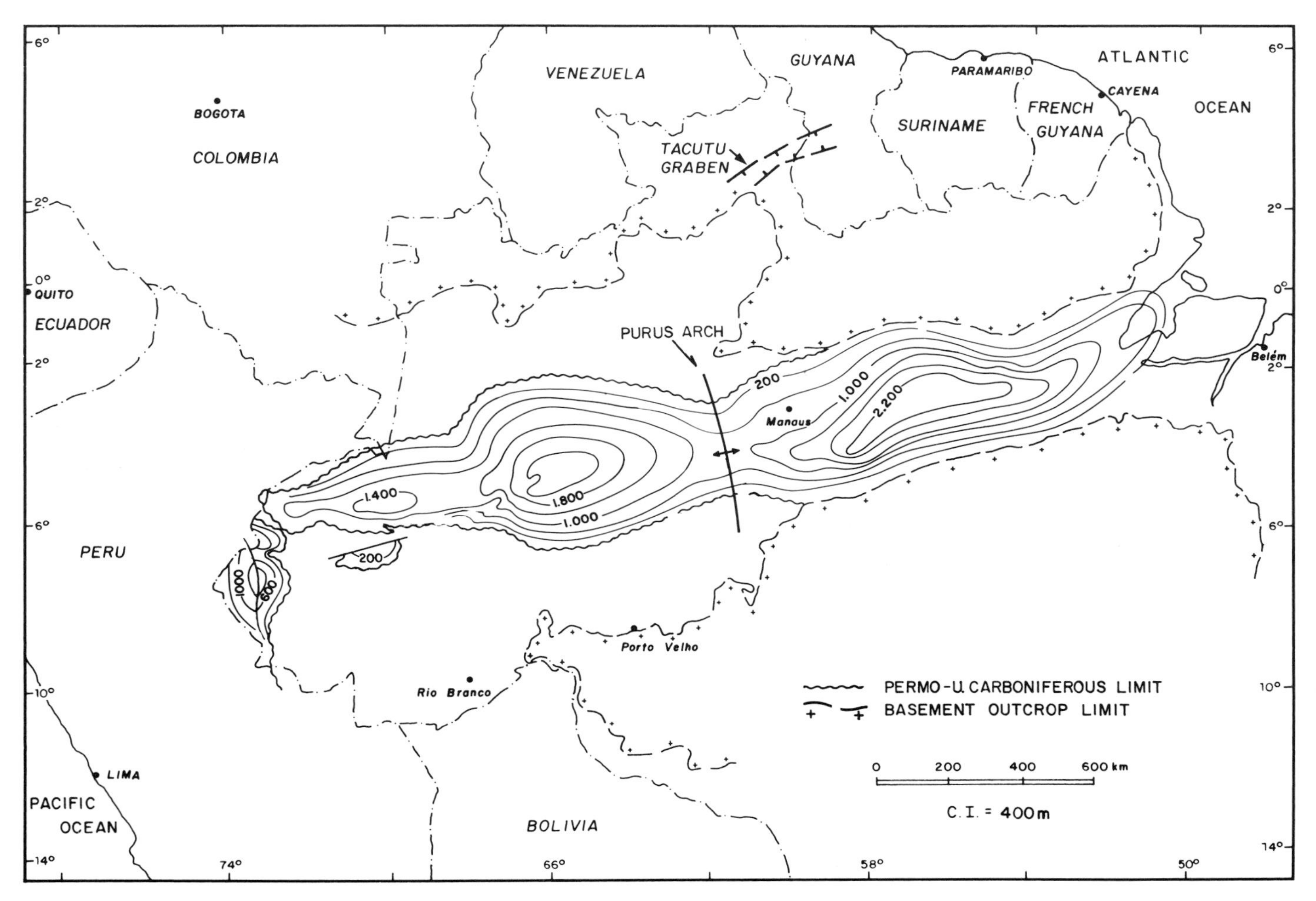


Figure 10. *Isopach map of the Permian–Upper Carboniferous sequence, which includes the Monte Alegre, Itaituba, Nova Olinda and Andirá Formations.*

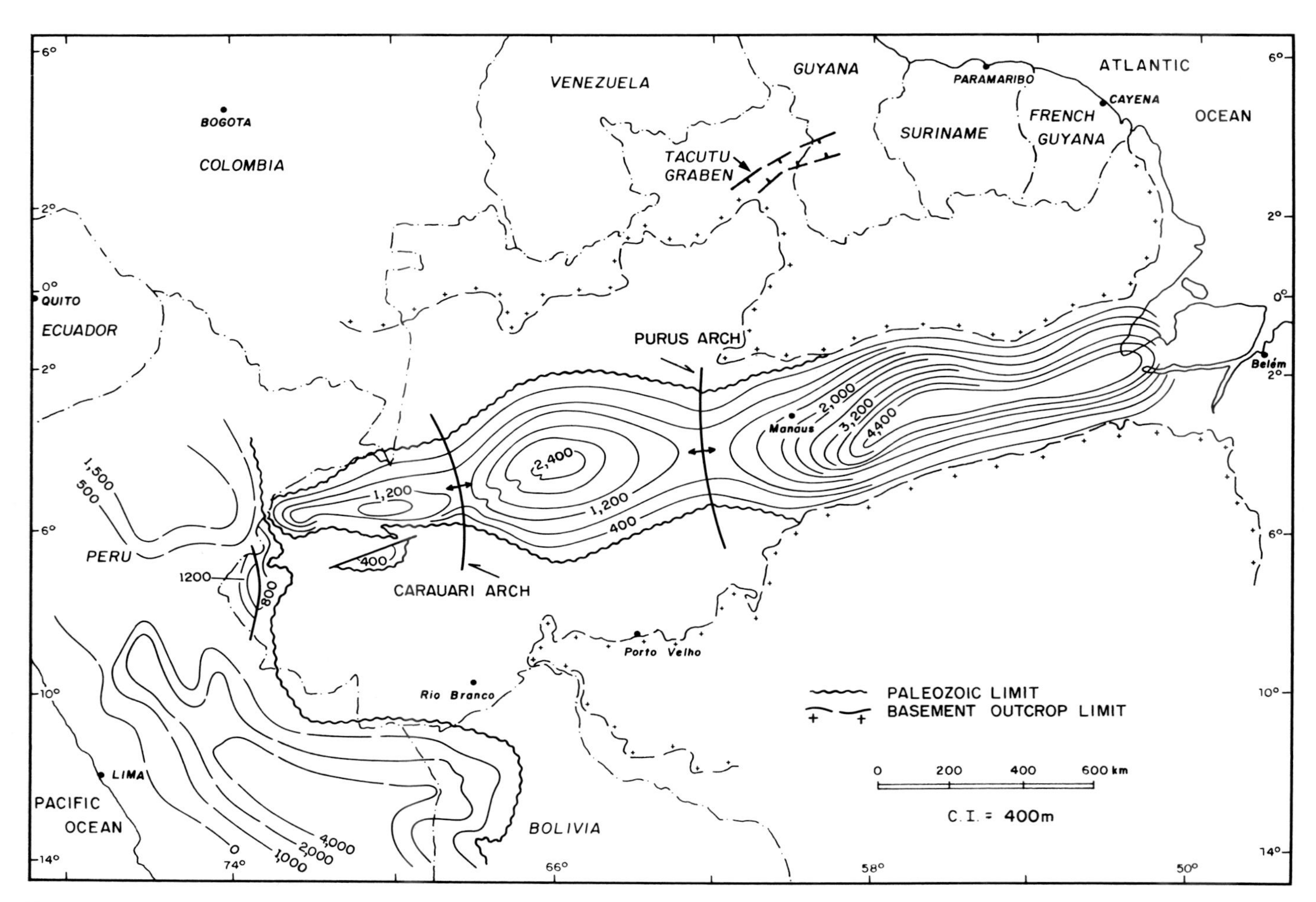


Figure 11. *Isopach map of the preserved Paleozoic section (Upper Ordovician to Permian). The Paleozoic sequence crops out only in narrow bands along the northern and southern margins of the Middle and Lower Amazon basins.*

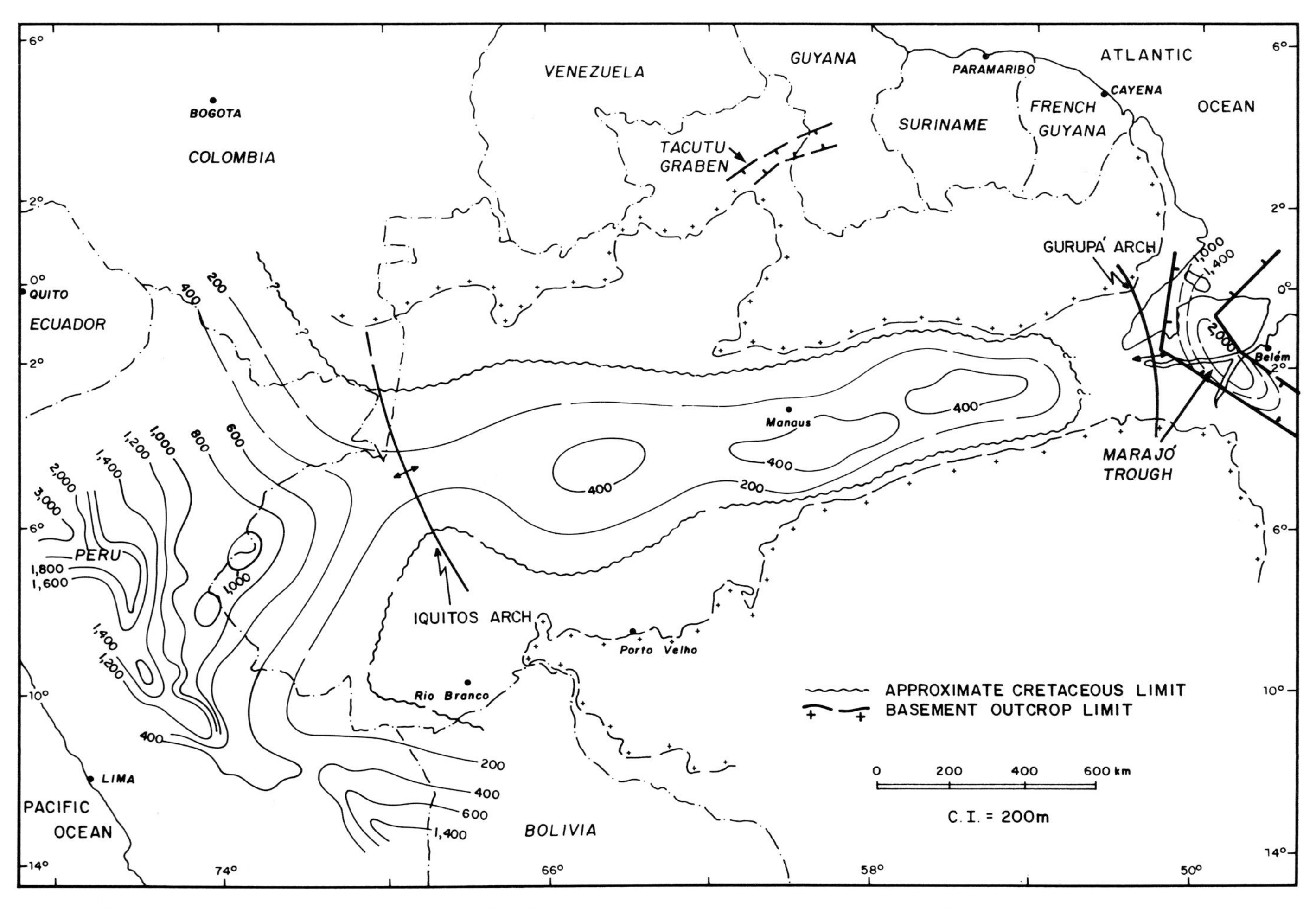


Figure 12. *Isopach map of the Cretaceous Alter do Chão Formation in the Amazon basins. The Iquitos Arch is the Cretaceous boundary between the Amazon basins, the Brazilian Acre basin, and the Peruvian Cretaceous basin.*

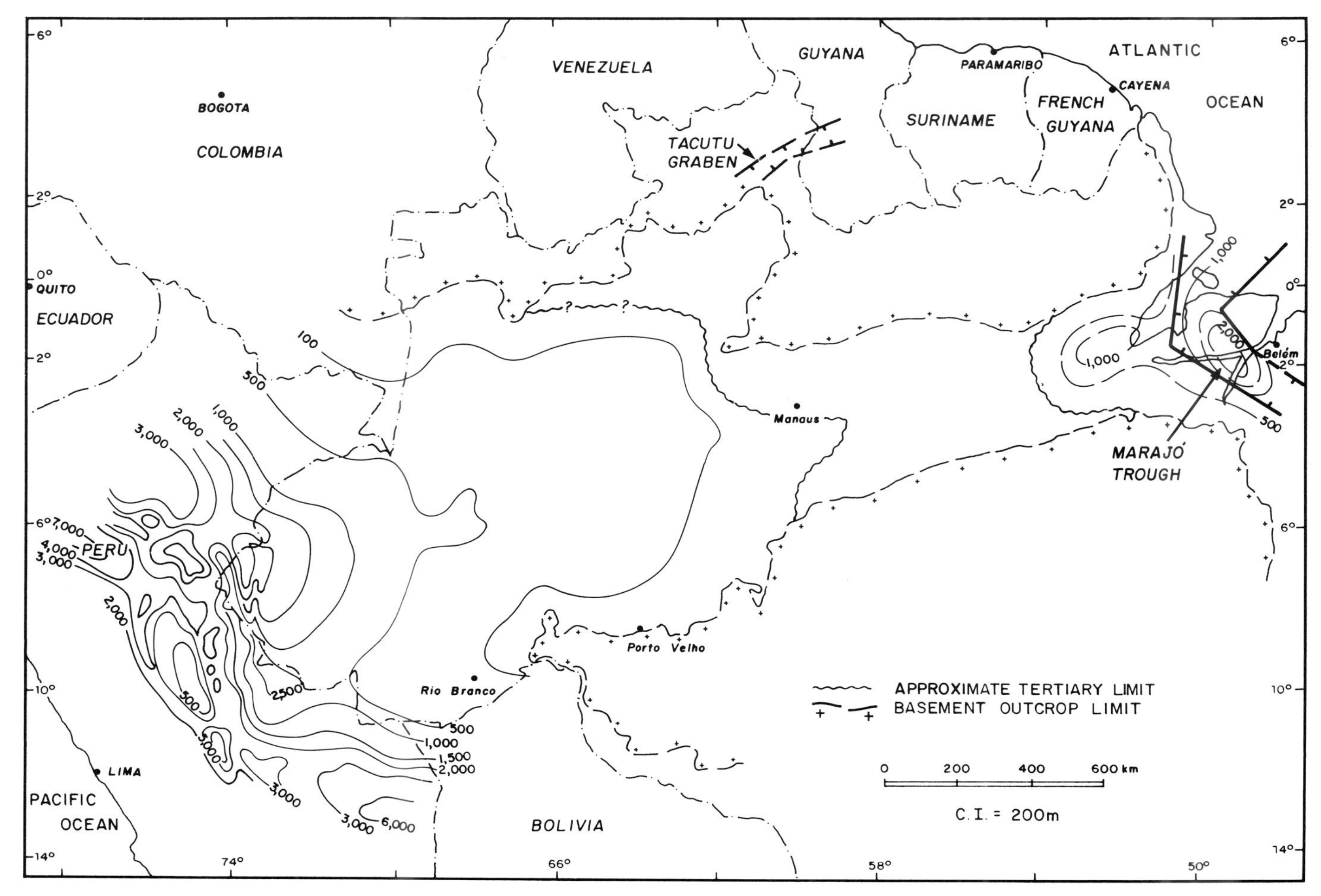


Figure 13. *Isopach map of the Tertiary Solimões Formation in the Amazon basins. This formation is practically absent in the Middle Amazon basin and presents great thicknesses west of the Brazilian/Peruvian border, in the Lower Amazon basin, and in the Marajó Trough.*

## TECTONISM AND STRUCTURAL FRAMEWORK

The literature identifies three principal driving mechanisms to explain vertical motions of the lithosphere in intracratonic sag basins located in areas of divergence but not at plate margins. The mechanisms are related to thermal events, crustal thickness changes, and addition or removal of load (Falvey, 1977; McKenzie, 1978; Royden, Sclater, and Von Herzen, 1980).

Cooling and, therefore, thickening of the lithosphere, thinning of the crust, and loading cause subsidence. Heating, which causes thinning of the lithosphere, thickening of the crust, and load removal by erosion, causes uplift of interbasinal warps or arches. However, the structural genesis of basement sags is not well understood because they have multiple origins. Sags lack diagnostic features, such as plate-boundary associations, that could suggest a particular type of deformation. Differential regional subsidences or uplifts have intuitive appeal as a cause of many intraplate sags and arches. Differential vertical motion could have occurred either because of irregularities in subcrustal processes or because of large, inhomogeneities within the crust.

Green (1977), while reviewing crustal processes, stated that the most important mechanism of such sites of continental lithospheric thinning and subsidence results from differential lithospheric cooling, phase changes, subcrustal ductile flowage, injection and stopping of dense material and related magma-chamber collapse, and surface erosion following thermal uplift. Deep-seated processes can also modify structural responses to crustal inhomogeneities.

The origin of domal features, such as the Carauari, Purus, and Gurupá Arches, contrasting with linear or blocklike structures, is much more speculative because simple plate tectonics may not be the exclusive cause of this deformation.

Many basement warps in intraplate settings suggest, however, first-order vertical movement in their development, although they lack consistent orientation, and apparently lack dependency on faulting for their development.

One of the most accepted thermo-mechanical models for explaining the tectono-sedimentary evolution of intracratonic basins is the "Steers Head" model of Dewey (Dewey and Pitman, 1984). This model, developed from several British basins and from the North Sea (Harding, 1983), includes polyphase stretching, from pre-rift sagging through rifting, and a major ter-

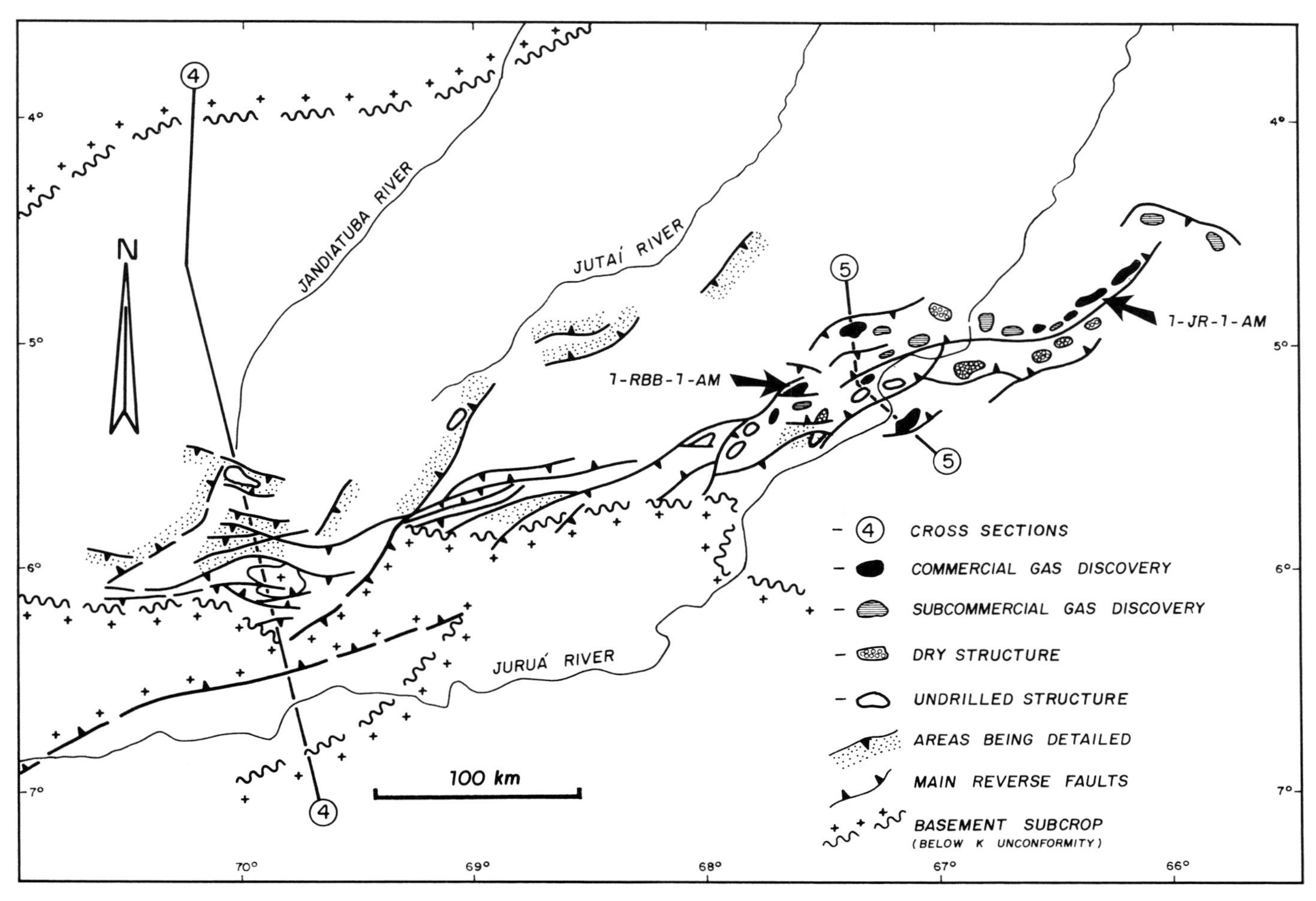


Figure 14. *The Juruá structural gas-bearing trend of the Upper Amazon basin (See Figure 35 for cross-sections indicated by lines 4 and 5).*

minal sagging because of cooling.

Nunn and Aires (1984), studying the subsidence history of the Middle Amazon basin from backstripped well data combined with gravimetric analyses, noticed that thermal subsidence from a single rifting event in the early Paleozoic is not compatible with the evolution of these basins. A single heating event cannot simultaneously explain slow subsidence in the Permian, with more than 200 m.y. duration of basin subsidence (Figure 8). Nunn and Aires (1984) concluded that after a small rifting event in the Early Ordovician, a uniform erosion was followed by a major rifting stage in the Early Devonian. Then a short erosional episode at the basin's margins was followed by the largest rifting event in the Carboniferous, and finally a 200 m (660 ft) uniform erosion occurred during the pre-Cretaceous Mesozoic. The Upper, Middle, and Lower Amazon basins had, however, different tectonic modifiers.

Vertical movements in the Upper Amazon basin occur within a large scope of horizontal plate translations. Szatmari (1984) interpreted strike-slip faults caused by wrench tectonics to be related to counterclockwise rotation of the Northwest South American microplate, during the separation of North and South America.

Although lithospheric segments currently away from the intraplate boundaries are not common habitat for wrench tectonics, strike-slip faults in a midplate setting are known worldwide. In North America, especially in the Illinois basin, which had a similar tectonic evolution to that of the Amazon basins, well-identified wrench faults with small displacements are the Cottage Groove and the Rough Creek fault zones located in the southern part of this basin.

Wrench tectonics, active mostly during the Jurassic, affected the Upper Amazon basin in the western and southwestern parts, generating reverse faults and associated folds, and deformed the Paleozoic sedimentary sequence and the Jurassic–Triassic volcanics (Figure 14). The stress-field analysis of this area by Szatmari (1983) suggests a shear-couple that produced a strike-slip fault zone in the Juruá area as part of a major midplate, left-lateral transcurrent fault called Pisco-Juruá. According to Szatmari (1983), this northeast-to-southwest fault zone would extend from Pisco in Peru to the border between Brazil and Guyana, in the Tacutu graben. The Cretaceous and Tertiary sequences, deposited after a regional erosional unconformity, were not affected by this tectonism.

In the structural framework of the Middle Amazon basin, geologists identified several major features, such as structural terraces called Manaus, Canumã, and Mamuru Platforms, and a major, elongated depocenter coinciding with the axis of the basin. In the central part of this basin, we interpret that, as in

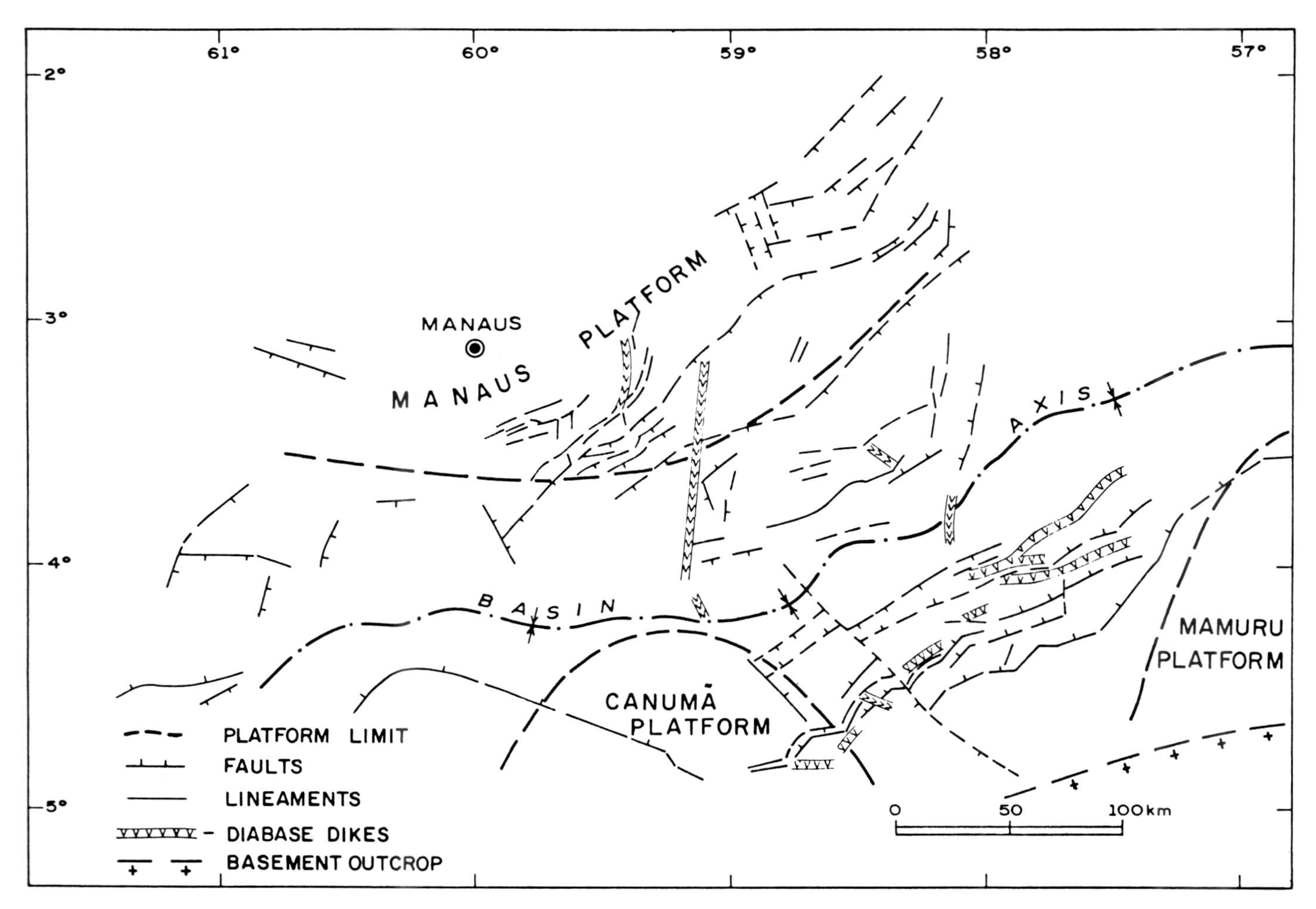


Figure 15. *Main structural features identified in the Middle Amazon basin. Shallow platforms at the basin margins and northeast-to-southwest-trending faults are the most prominent structures. For location of this area, refer to Figure 9.*

the Williston–Blood Creek basin (Thomas, 1984), lineament-block tectonics generated northwest and northeast normal faults with minor intrablock shearing (Figure 15). This shearing developed local, mild to subdued folds. On the margins of this basin, some recent seismic data with better resolution suggest the presence of structural features looking like "flower structures," which could be related to incipient wrench tectonics.

In the Lower Amazon basin, a recent seismic survey identified several normal faults representing Carboniferous to mostly Cretaceous and Tertiary rifting events. We interpret that in this basin minor structures are related to shoulders of half-grabens or drapes over fault planes. The presence of two sets of fault systems, one northwest-to-southeast and another northeast-to-southwest, suggests continuation of the lineament-block tectonics identified in the Middle Amazon (Figure 16). Local apparent "flower structures" may also represent minor wrench events.

## SOURCE ROCKS

Intracratonic basins worldwide have a low source-rock potential, and many of them are predominantly gas-prone, contributing 79% of the world's gas supply (Klemme, 1980).

Two marine black shale layers are the most important source rocks for hydrocarbon generation in the Amazon basins: the Silurian Pitinga Member (Figure 17) and the Devonian Barreirinha Member (Figure 18). Of smaller significance are thin, Carboniferous Itaituba shales.

The organic geochemistry data and thermal maturation analyses in the Upper Amazon indicate that the Barreirinha shales have a good organic matter content and reached an overmature, diagenetic stage (Figure 19). The thermal history of the Upper Amazon basin and the application of the Lopatin method indicate that gas generation occurred during the Permian and that overmaturity was reached during the Jurassic. The Itaituba shales are only locally moderate source rocks.

The overall geochemical picture in the Middle Amazon basin is a little different because the Pitinga shales are also gas-prone and have a fairly good organic-matter content. The most important source rock according to the vitrinite reflectance and thermal alteration index is the Barreirinha shale, which presents all three maturation stages in this basin (Figure 20).

The Lower Amazon shows geochemical conditions and thermal history very close to those of the Upper Amazon, but all over the basin source rocks reached the overmature diagenetic stage early; therefore, the basin is gas-prone (Figure 21). The

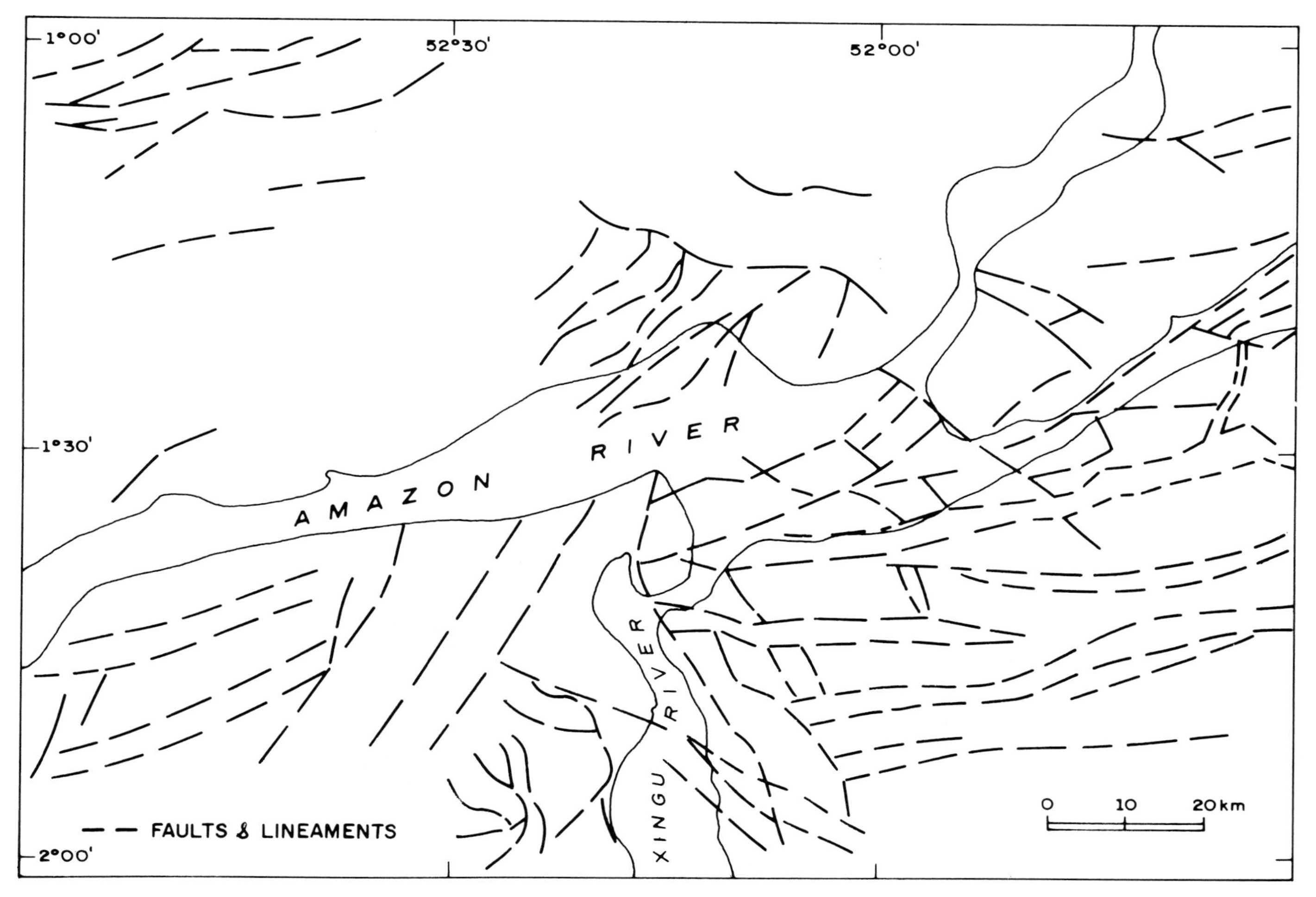


Figure 16. *Map of the Lower Amazon basin showing two main sets of predominantly normal faults. Along some fault zones, shearing may have occurred causing wrench-type structures. For location of the represented area, refer to Figure 9.*

main exploratory problem, however, is the lack of evaporite seals for the Monte Alegre reservoirs in the eastern part of this basin.

## EXPLORATION HISTORY

Besides minor, noncommercial oil accumulations discovered during the 1950s and found in stratigraphic traps in the Middle Amazon basin, geologists located nothing important in the three basins until 1978 because of the poor quality of seismic data.

Geophysical surveys in the Amazon basins began in 1950. Up to 1970, 82,000 km (50,800 mi) of gravimetric surveys and 7,200 km (4,400 mi) of seismic refraction data, as well as 18,000 km (11,000 mi) of analogic seismic reflection data were recorded along major rivers. Based on these data, 130 wells were drilled. The results of the drilling of structural prospects, however, were disappointing.

The major problems for exploring these basins until 1975 were low seismic resolution at the target level and diabase intrusions, and high drilling costs because of the difficult access to the Amazon jungle.

Since 1978, the use of Common Depth Point (CDP) and digital seismic reflection methods has played an increasingly important role in exploring for oil and gas in the Amazon basins, and about 36,000 km (22,300 mi) of new seismic reflection data were acquired (Figure 22).

The first commercial success was achieved in 1978, with the discovery of the Juruá gas field. This discovery followed the mapping of a regional, seismic reflector corresponding to an Upper Devonian sandstone. Since then Petrobrás and other companies have drilled 42 additional exploratory wells in the Amazon basins (Figure 23).

In 1980, about 113,000 km (70,200 mi) of aeromagnetic surveys were conducted and interpreted (Figure 24). The resulting magnetic maps were used extensively as reconnaissance tools and supported, along with gravity Bouguer maps, the outline of local, detailed seismic programs.

The importance in defining the evolution of the crust underneath these basins led to an additional seismic survey with recording time of up to 20 seconds and using techniques similar to those in COCORP's surveys. In late 1984, two lines located in the Upper and Lower Amazon basins for searching deep reflectors were being shot.

With over six years of hard work by geophysicists, the quality of seismic data was enhanced substantially and relates to

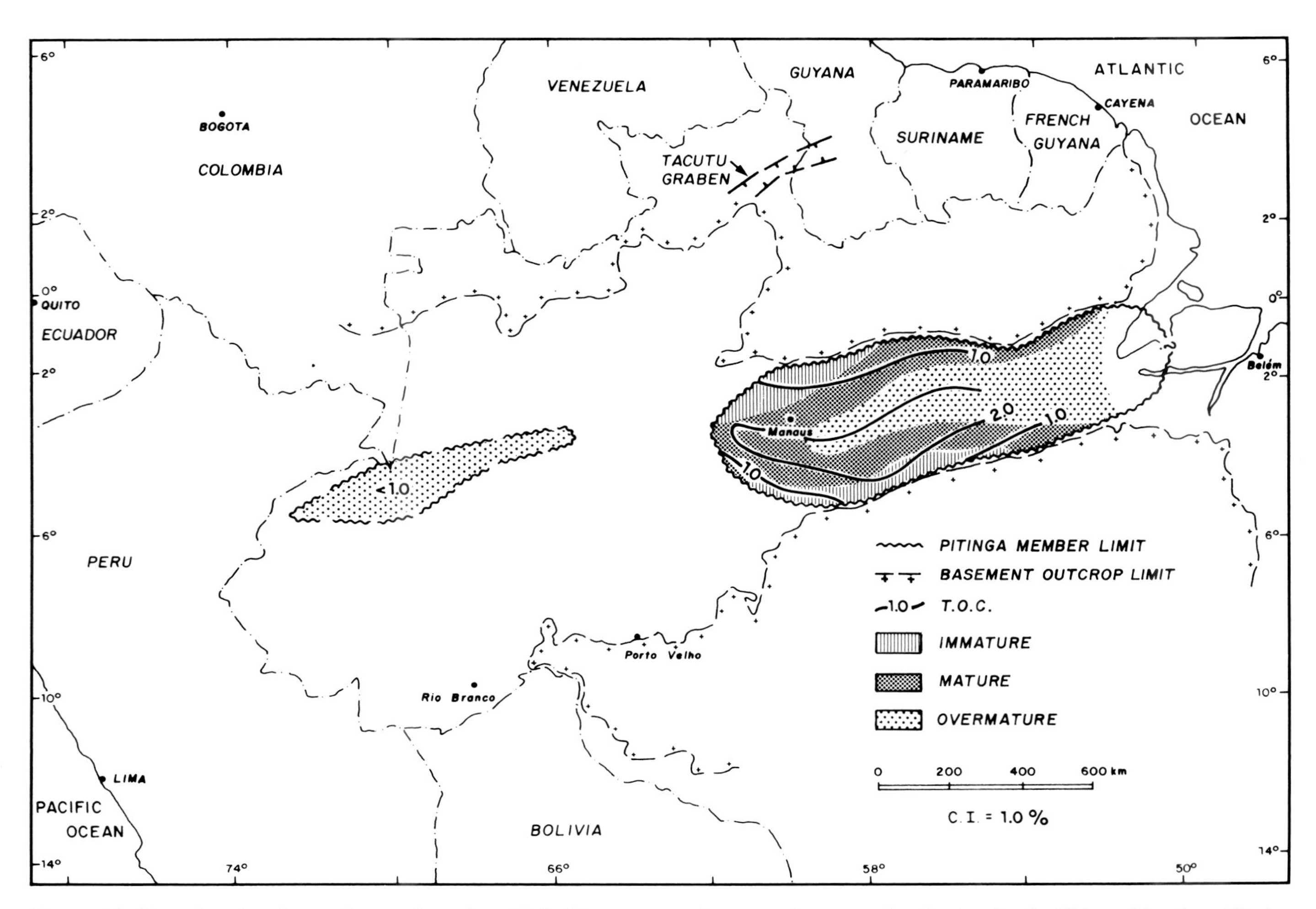


Figure 17. *Map showing the total organic carbon (T.O.C.) content and maturation zone distribution for the Pitinga Member, Silurian Trombetas Formation. In the Upper Amazon basin as well as in most Lower Amazon, this unit is overmature.*

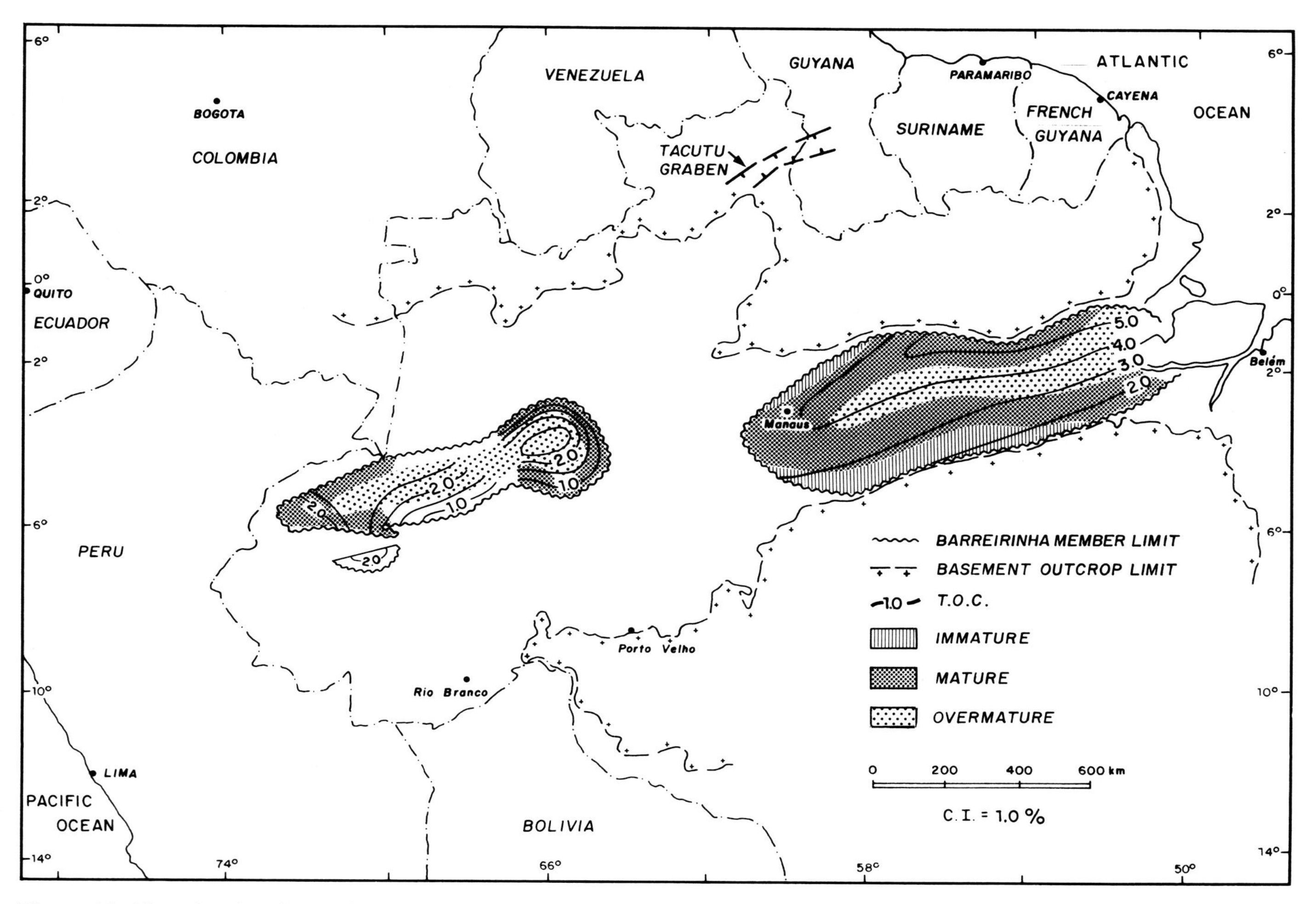


Figure 18. *Map showing the total organic carbon (T.O.C.) content and maturation zone distribution for the Barreirinha Member, Devonian Curuá Formation.*

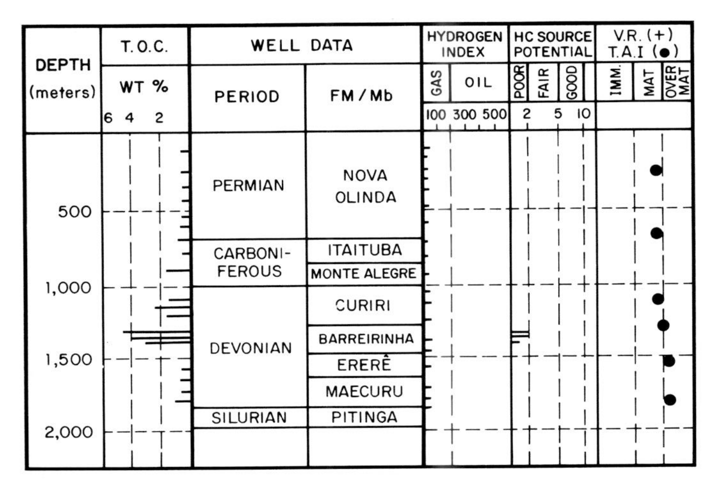


**Figure 19.** *Upper Amazon geochemical summary. Barreirinha shale is by far the main source rock for the gas found in this basin. Minor quantities of hydrocarbons could be generated by the Devonian Curiri Member and by the Upper Carboniferous section. (HC source potential in kg of HC/ton of rock; V.R. = Vitrinite reflectance; T.A.I. = Thermal alteration index).*

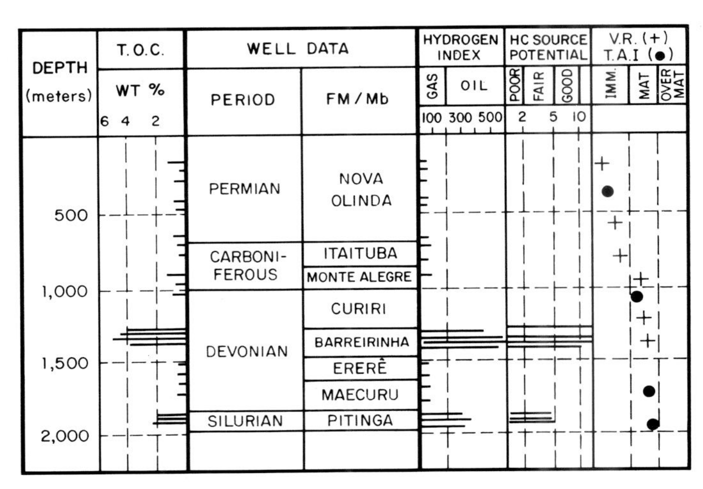


**Figure 20.** *Middle Amazon geochemical summary. Barreirinha Member presents good geochemical characteristics for oil generation. In this basin this unit presents all three stages of organic matter maturation. Overmaturation was reached in the central part of the basin. (HC source potential in kg of HC/ton of rock; V.R. = Vitrinite reflectance; T.A.I. = Thermal alteration index).*

re-examining of data acquisition and improving processing routines.

Re-examining of how we acquire data led to the following techniques:

1. Short field arrays with a large range of source-receiver offsets.
2. Anti-aliasing geophone arrays.
3. CDP fold coverage enlargement from 1,200% to 4,800% by increasing the number of channels from 24 to 96.
4. Statics control improvement by using shallow refraction surveys.
5. Recording equipment change from DFS-IV to DFS-V and later to Sercel-348 with a telemetric system.
6. Helicopter support to get into difficult jungle areas.

These procedures permitted a performance increase from 60 to 200 km (36 to 120 mi) per crew-month, and consequently a more cost-effective operation.

Improving data processing routines included using the first break static, using a velocity filter, applying deconvolution processes, thus increasing the statical signal analysis; automatically correcting residual static; synthesizing the source-receiver array on the computer; controlling better the velocity and data quality; and directing processing procedures toward both structural and stratigraphic analyses.

## THE FUTURE EXPLORATION

The role of seismic reflection in exploring the Amazon basins can be viewed with optimism because the data quality can be further improved in future surveys.

Recent seismic surveys allow more precise tectonic interpretation and, consequently, a better understanding of these basins. Therefore, geologists can identify various structural features, of different degrees of complexity, from monotonous plat-

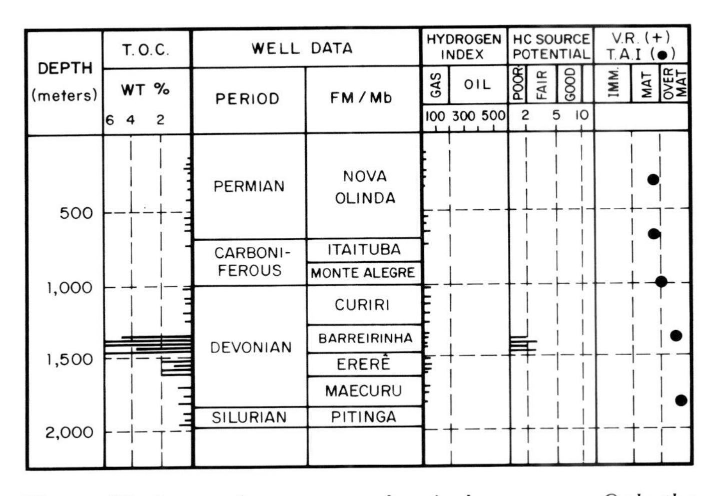


**Figure 21.** *Lower Amazon geochemical summary. Only the Barreirinha Member has conditions to generate fair quantities of gas in this basin. (Hydrocarbon source potential in kg HC/ton of rock; V.R. = Vitrinite reflectance; T.A.I. = Thermal alteration index).*

forms to highly deformed areas.

Figures 25 to 27 illustrate the evolution of seismic data in the Upper Amazon basin from 1976 to 1982, showing reverse faults and associated folds affecting Paleozoic sediments and Triassic volcanics. They also show the undeformed Cretaceous and Tertiary sediments overlying an unconformity in the structural trend of Juruá gas field. Figures 28 to 31 show the quality of seismic records in the Middle Amazon basin and illustrate its tectonic styles, with several half-grabens and possibly minor wrench structures. Figures 32 and 33 show rifting and possible wrench structures in the Lower Amazon basin, which might be related to a younger strike-slip fault event.

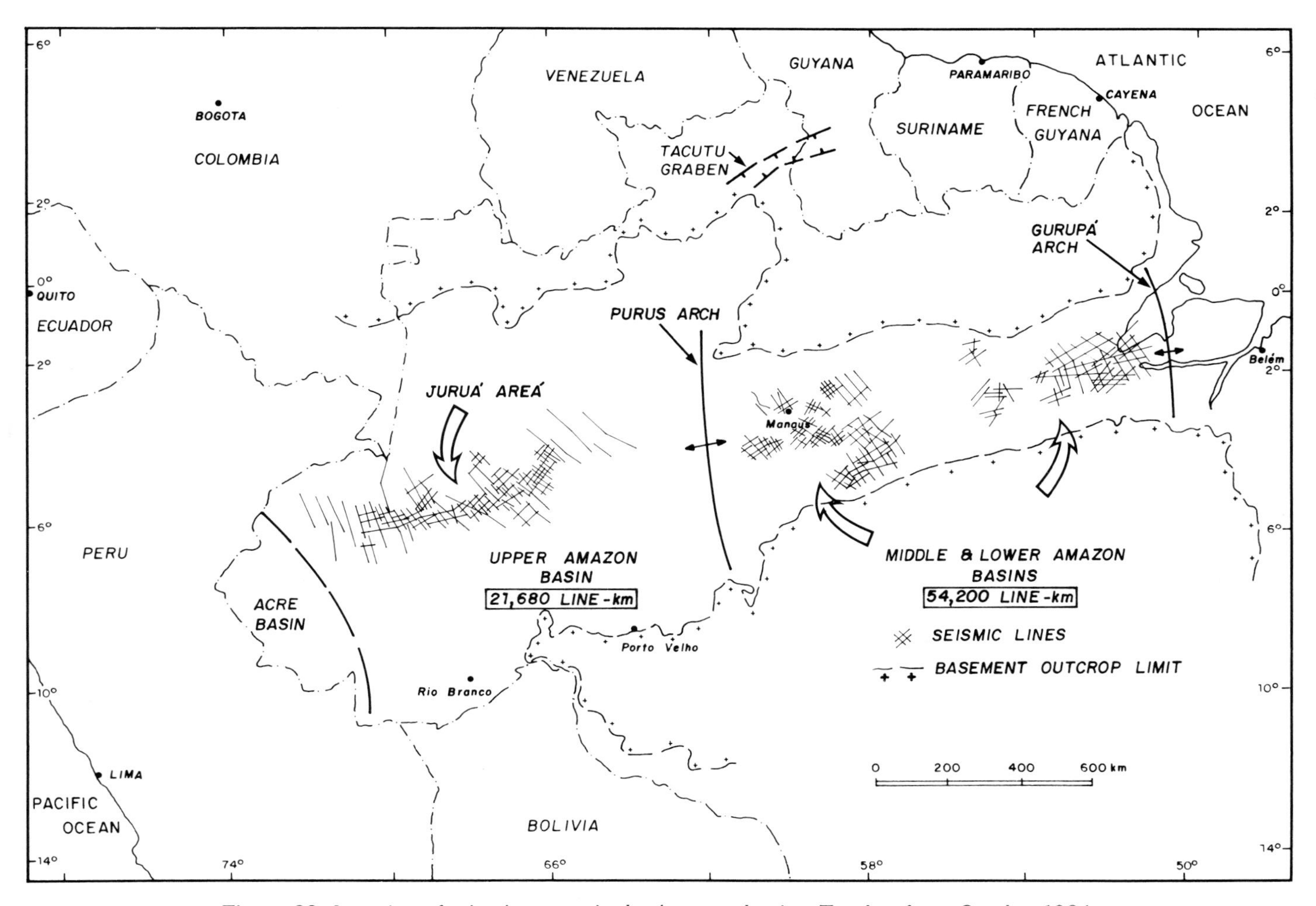


Figure 22. *Location of seismic survey in the Amazon basins. Totals refer to October 1984.*

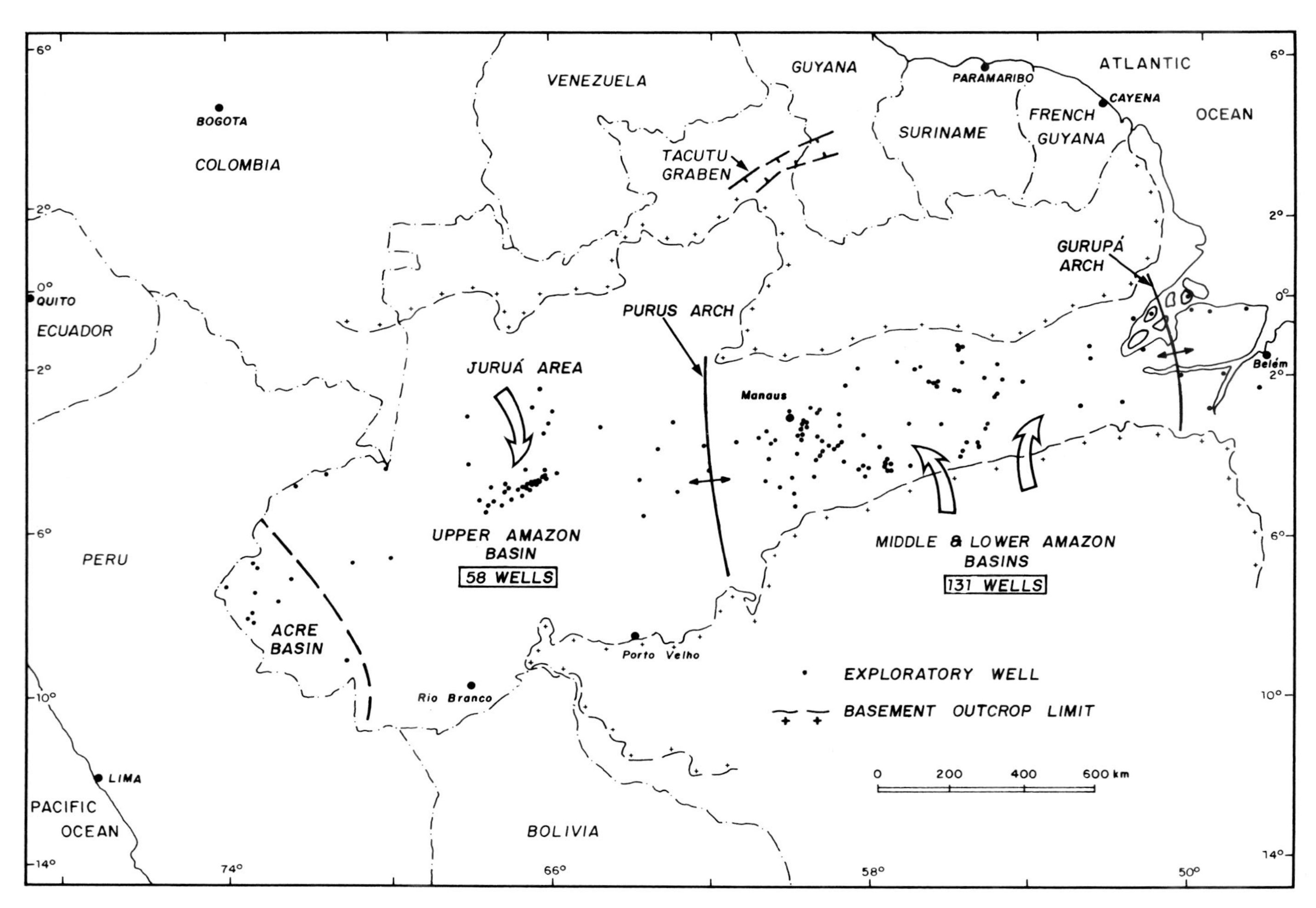


Figure 23. *Location of exploratory wells drilled in the Amazon basins until October 1984.*

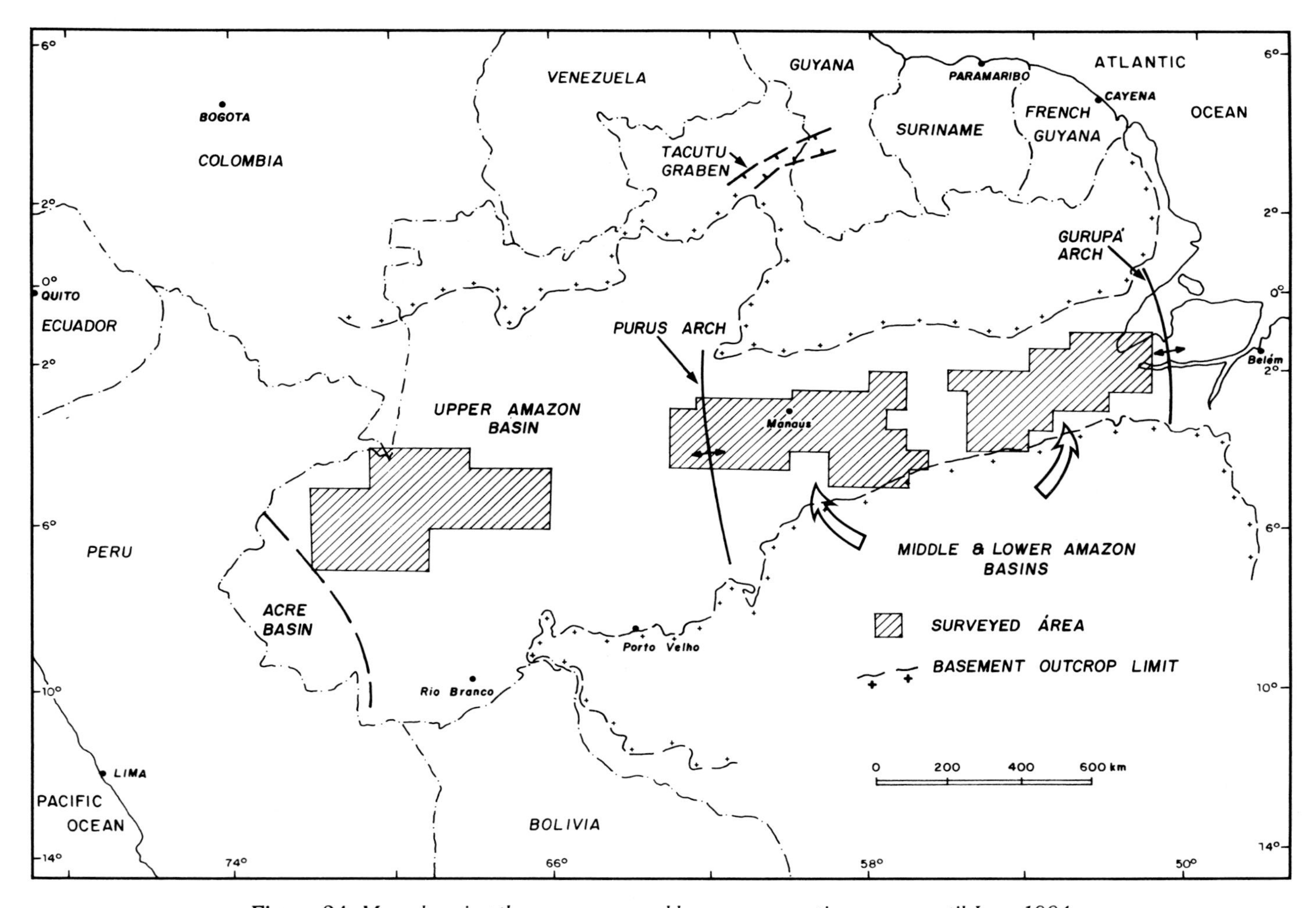


Figure 24. *Map showing the areas covered by aeromagnetic surveys until June 1984.*

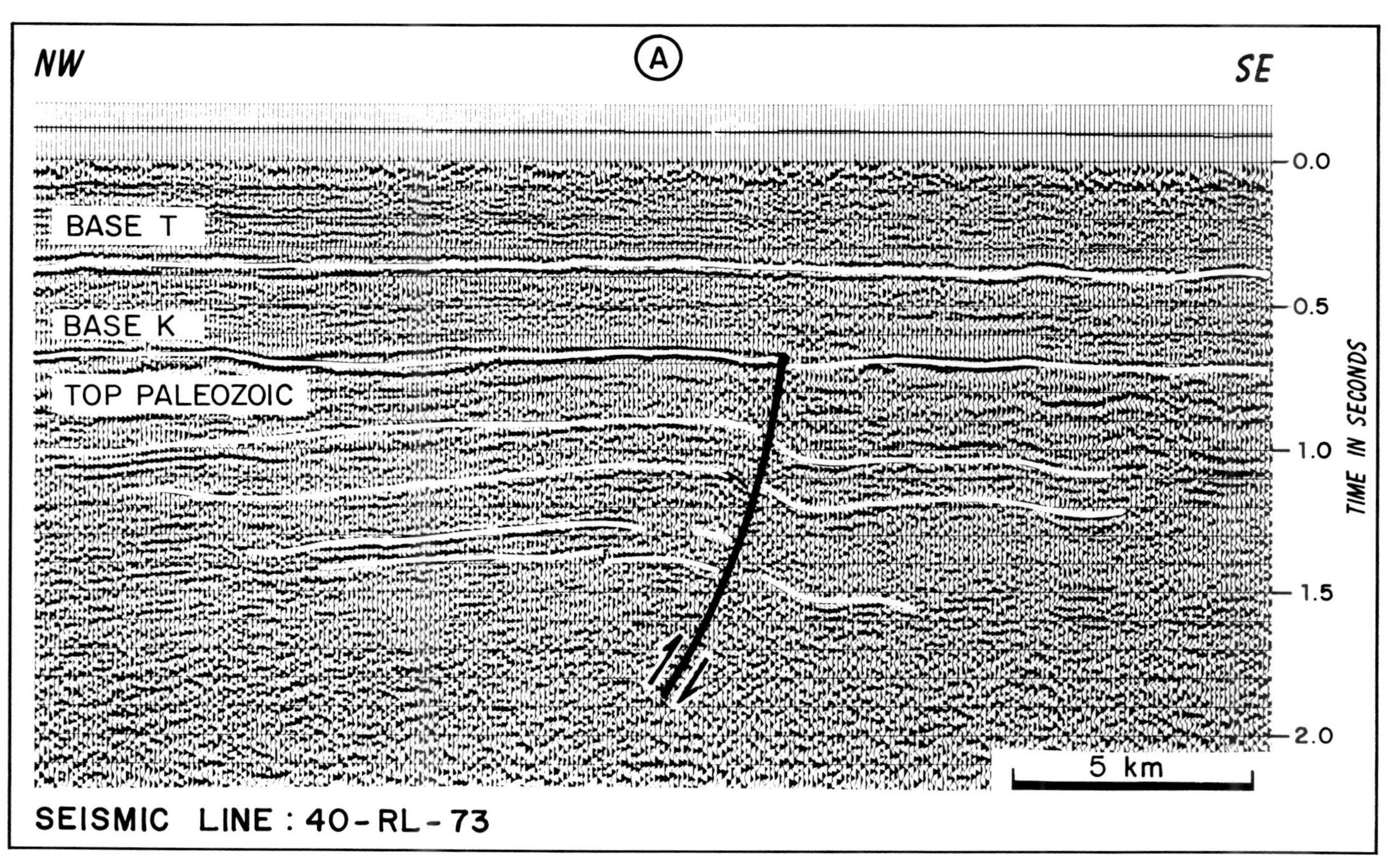


Figure 25. *Upper Amazon basin: seismic line shot in 1976 in the Juruá area, which showed, for the first time in a Brazilian Paleozoic basin, faint indications of structural deformations due to compressive stress (Line A in Figure 9).*

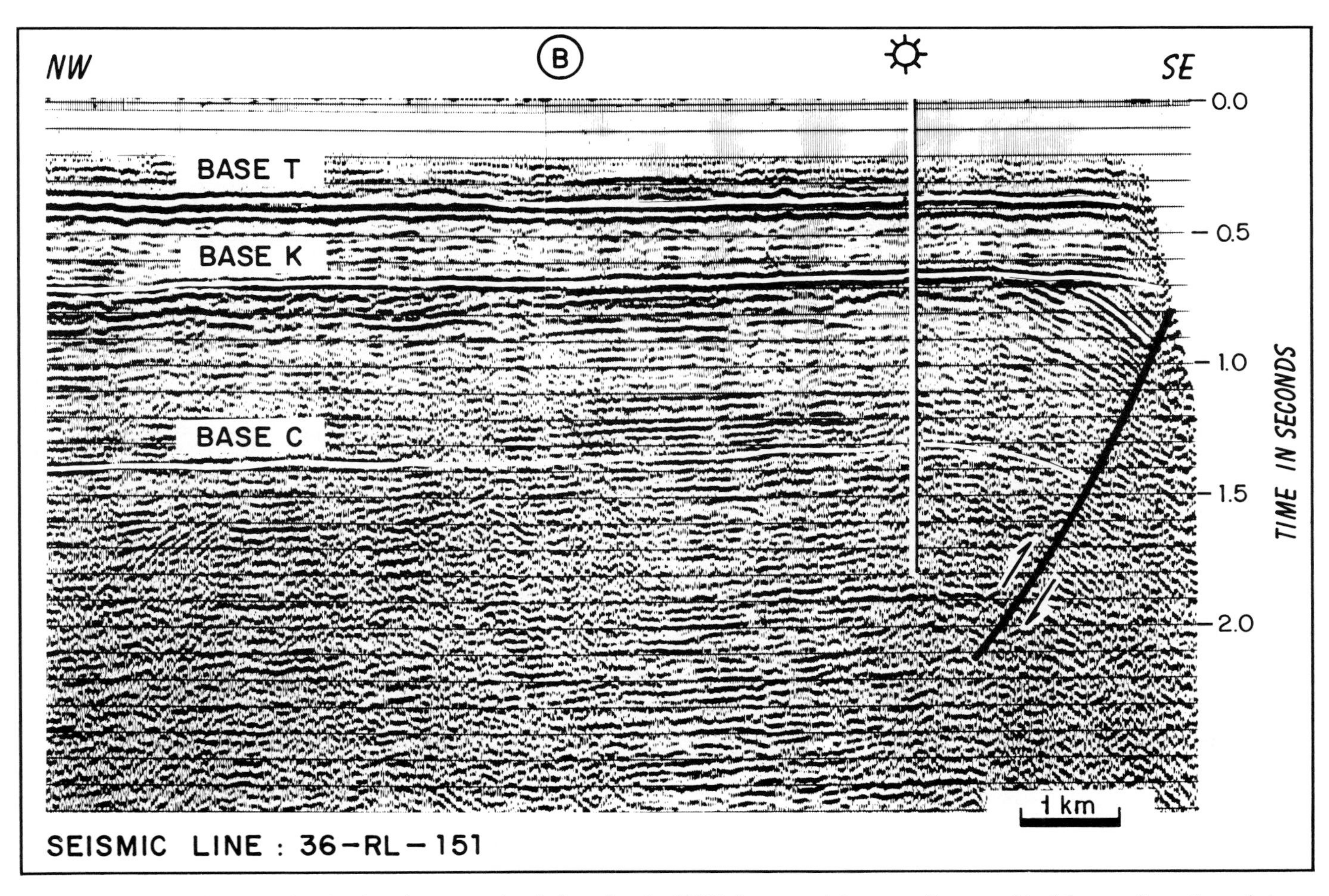


Figure 26. *Upper Amazon basin, Juruá area: seismic line shot in 1977. Improved data quality permitted the confirmation of structures caused by compressive stress. This line shows clearly fold and reverse fault affecting Paleozoic rocks and Mesozoic volcanics, truncated by a pre-Cretaceous unconformity (Line B in Figure 9).*

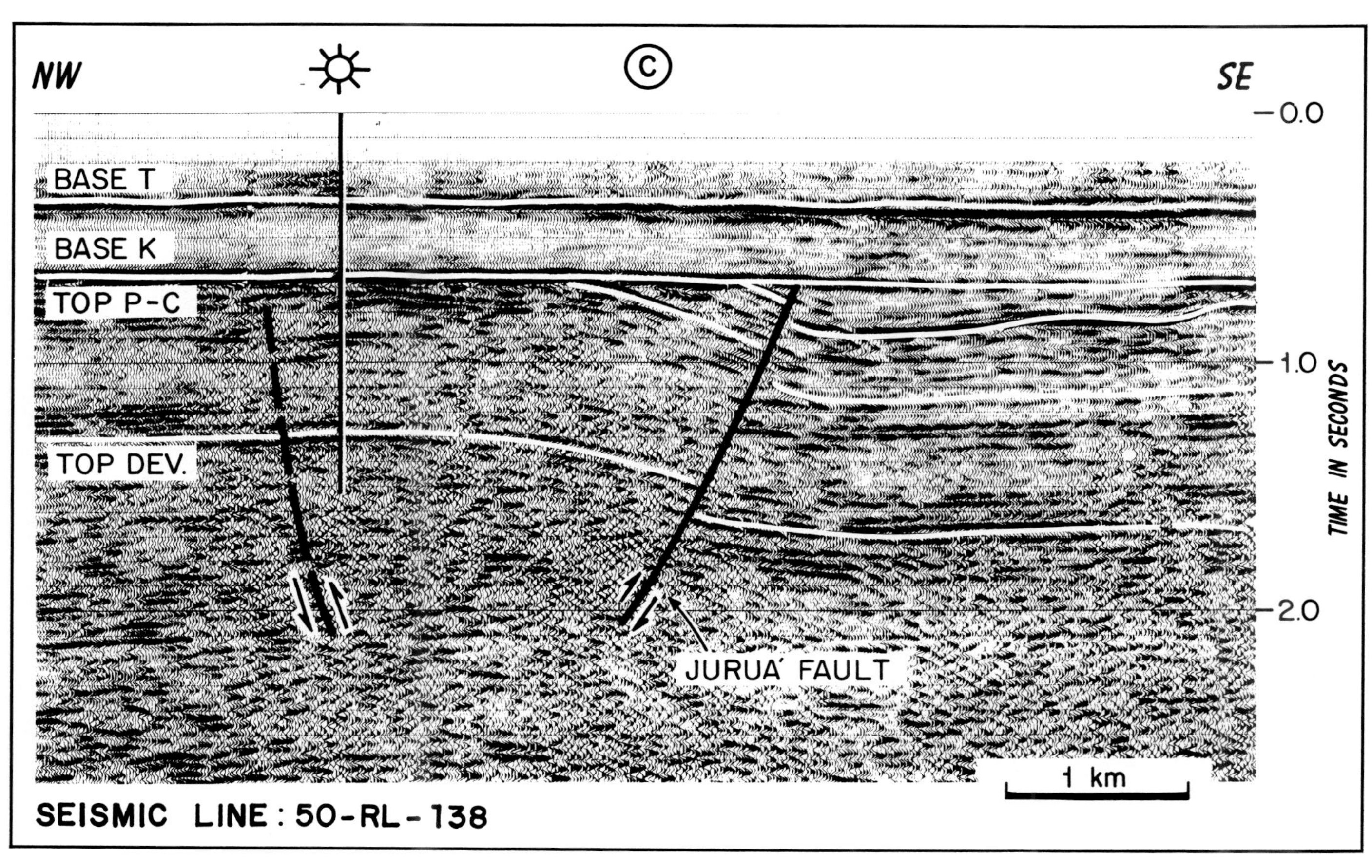


Figure 27. *Upper Amazon basin, Juruá area: seismic line shot in 1982. One can clearly observe the great improvement in data quality since 1976, when comparing this line with line A in Figure 25 (Line C in Figure 9).*

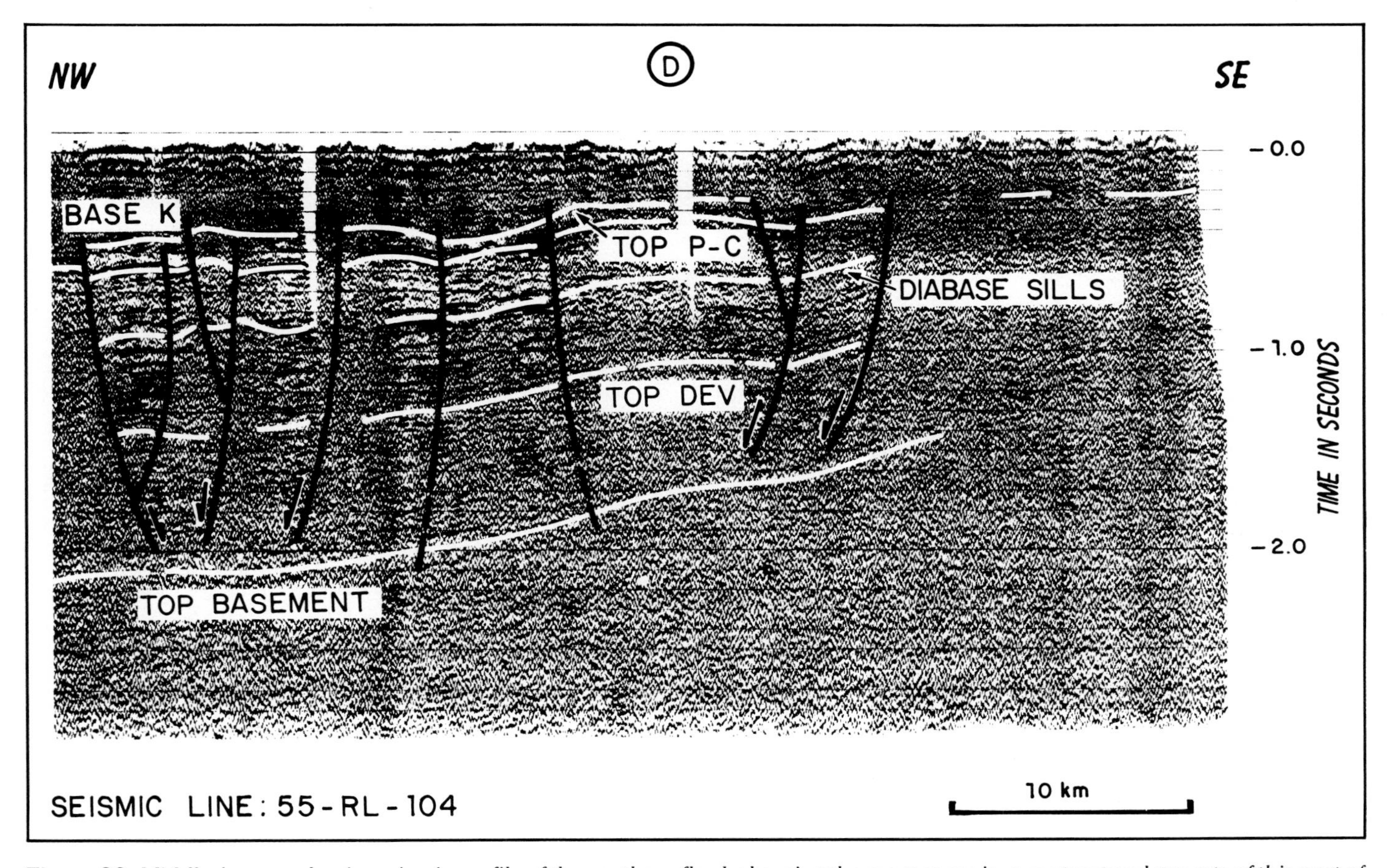


Figure 28. *Middle Amazon basin: seismic profile of the southern flank showing the most conspicuous structural aspects of this part of the basin. Note data quality deterioration in the right side of the section. Some faults could be due to wrenching, while some local deformations and lack of reflections could be caused by diabase dikes and sills intrusions (Line D in Figure 9).*

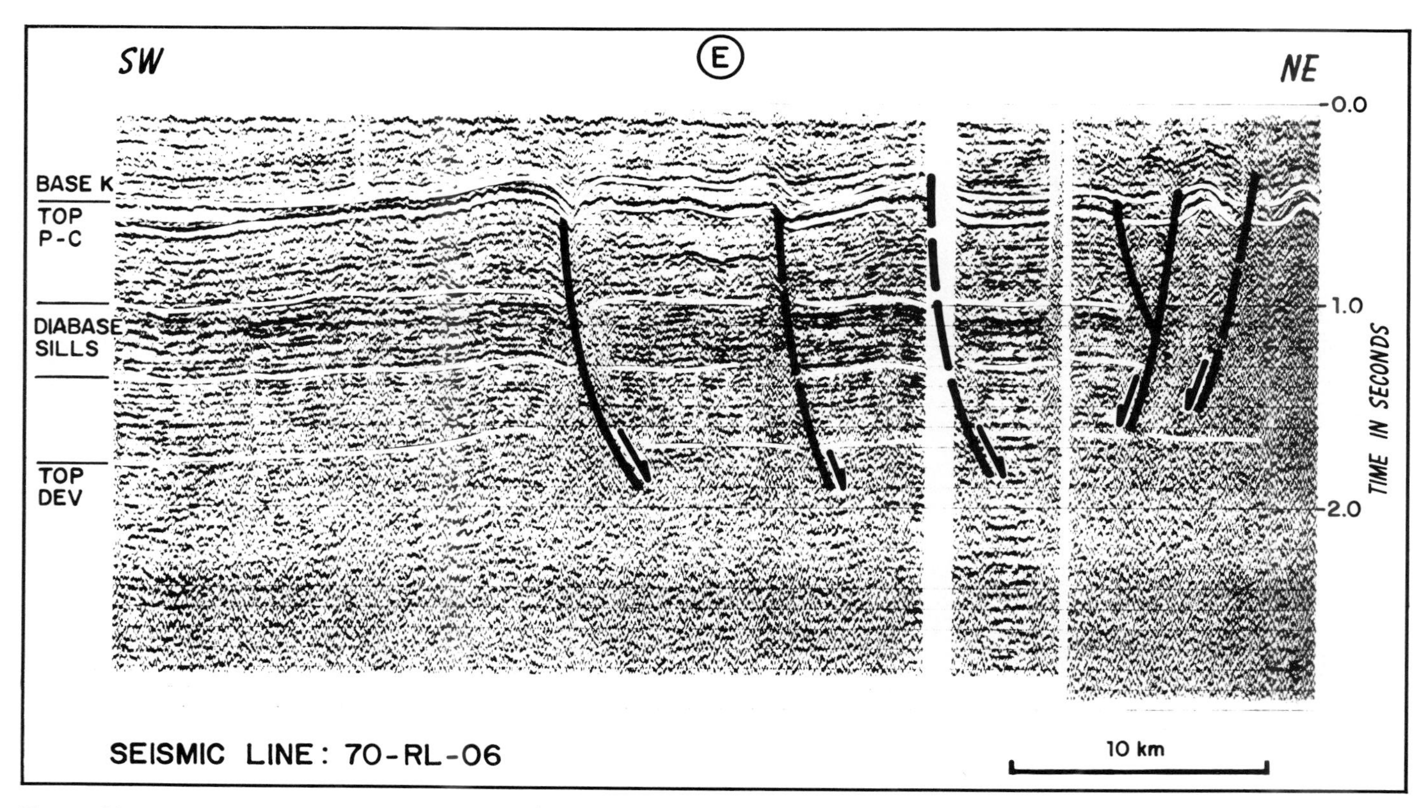


Figure 29. *Middle Amazon basin: seismic line showing structural characteristics of the central part of the basin. Note deformation affecting the upper part of the Permian–Carboniferous section and also Cretaceous strata. In this part of the basin, deformations due to salt movement are expected (Line E in Figure 9).*

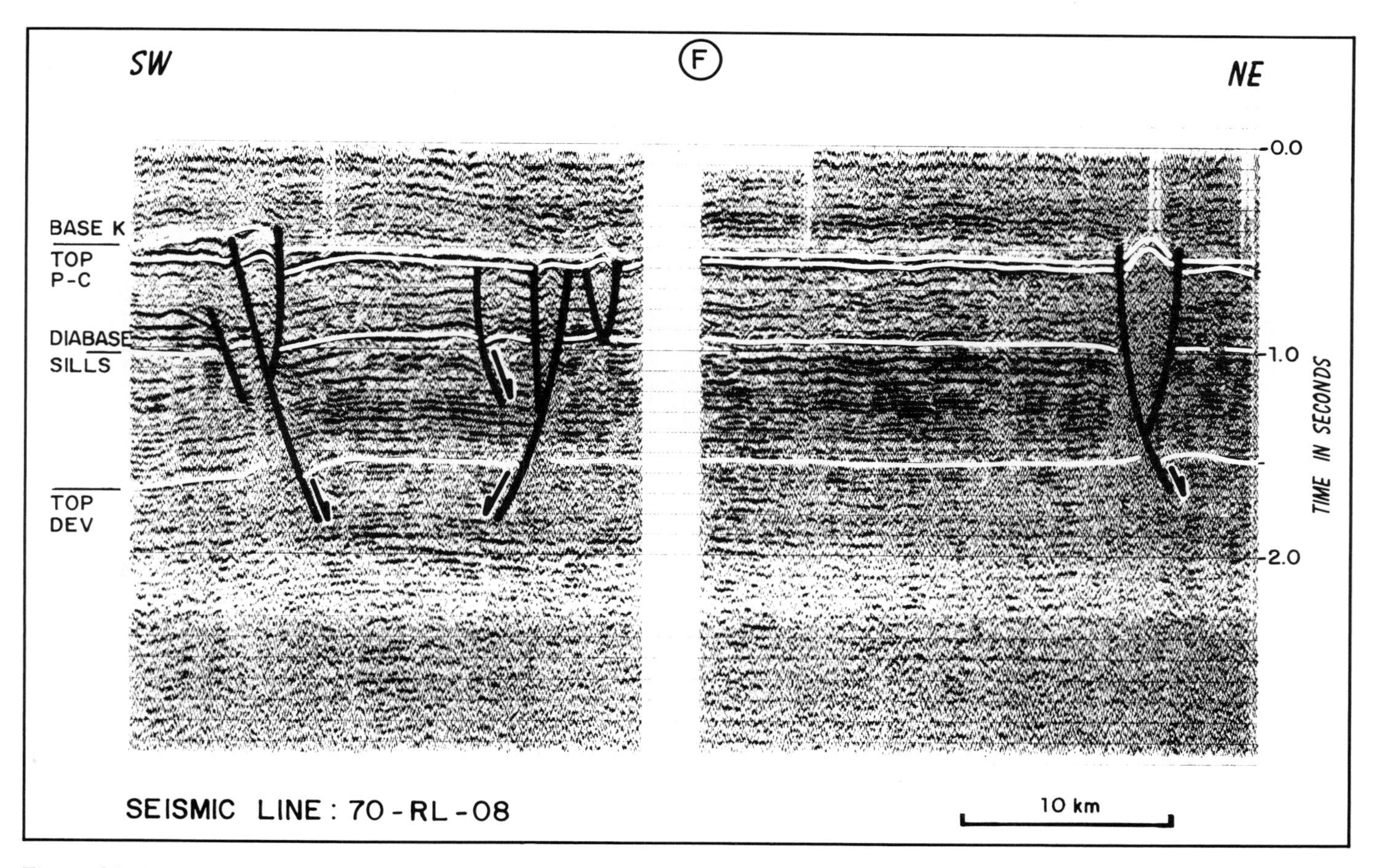


Figure 30. *Middle Amazon basin: seismic line showing structural characteristics of the central part of the basin. Note absence of reflections in places where upper horizons are deformed. Many faults are highly speculative, and salt movement-related deformations are possible. Anomalous reflections, like those around 1.0 second (left end of line), are probably related to diabase intrusions (Line F in Figure 9).*

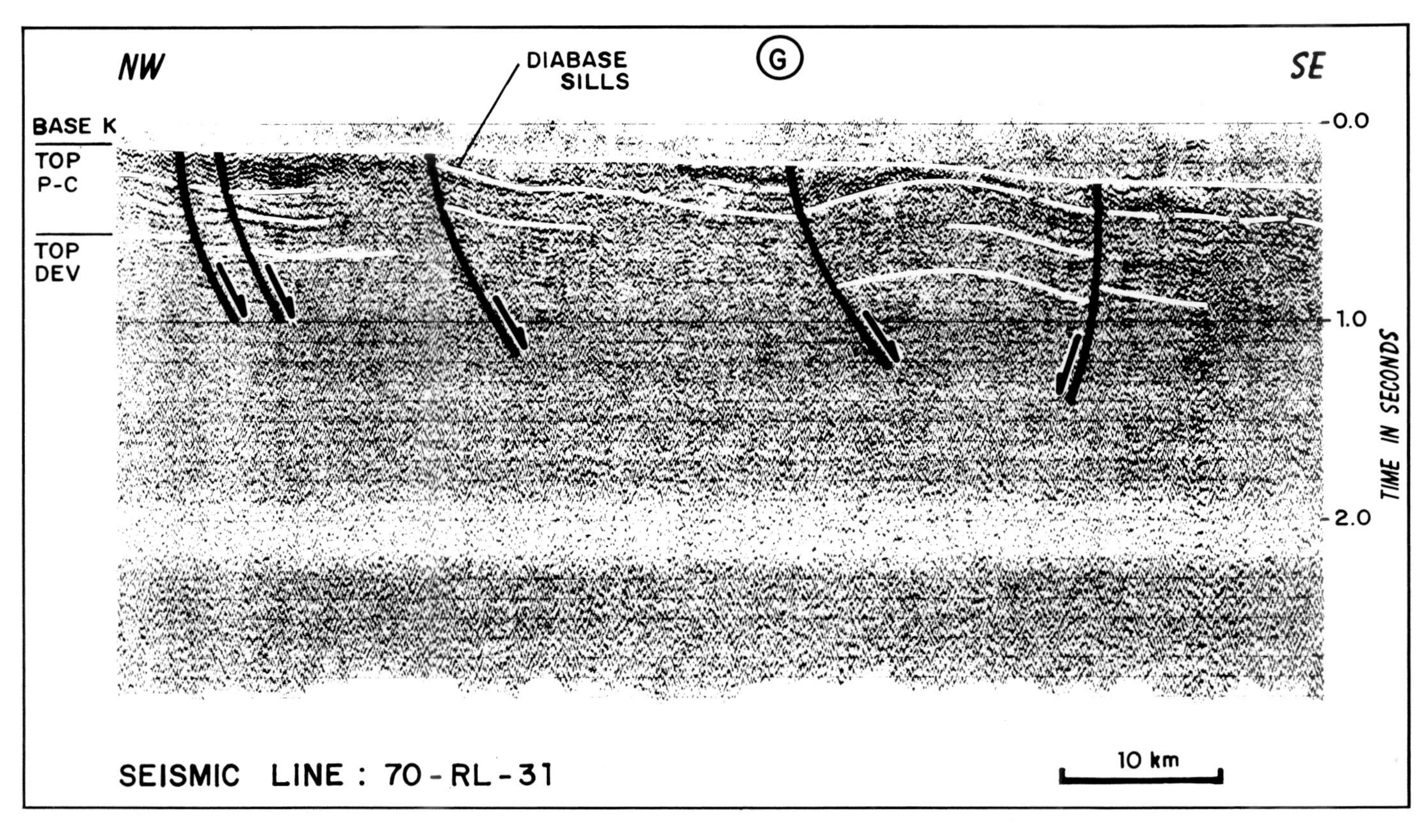


Figure 31. *Middle Amazon basin: seismic line showing structures detected in the northern flank of the basin. Note the poor quality of seismic data in this part of the basin (Line G in Figure 9).*

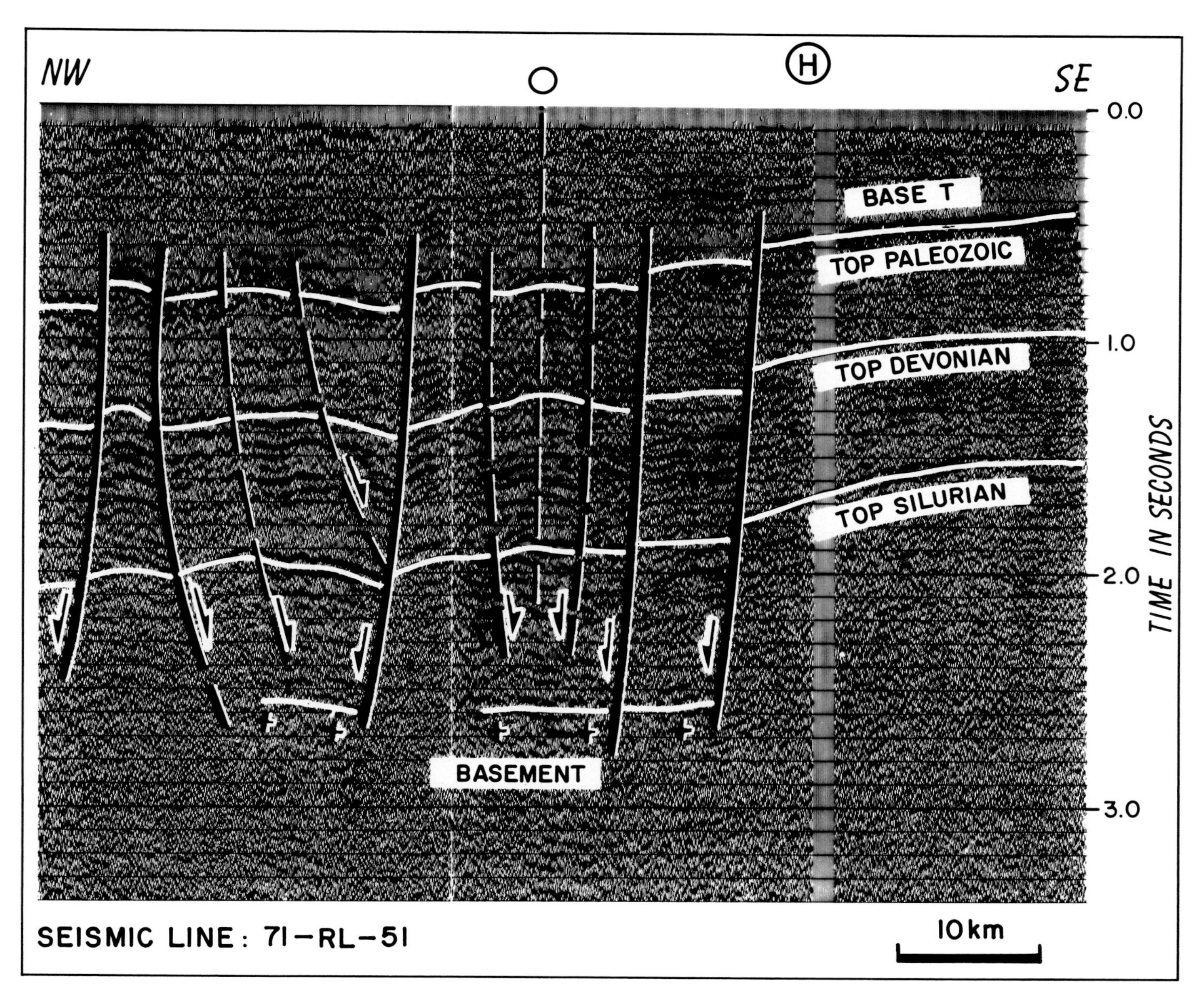


Figure 32. *Lower Amazon basin: seismic line showing intense faulting in the central portion of the basin. Faults cut through the entire sedimentary section. Note the poor quality data at right end of the line. The dashed vertical line represents proposed wildcat (Line H in Figure 9).*

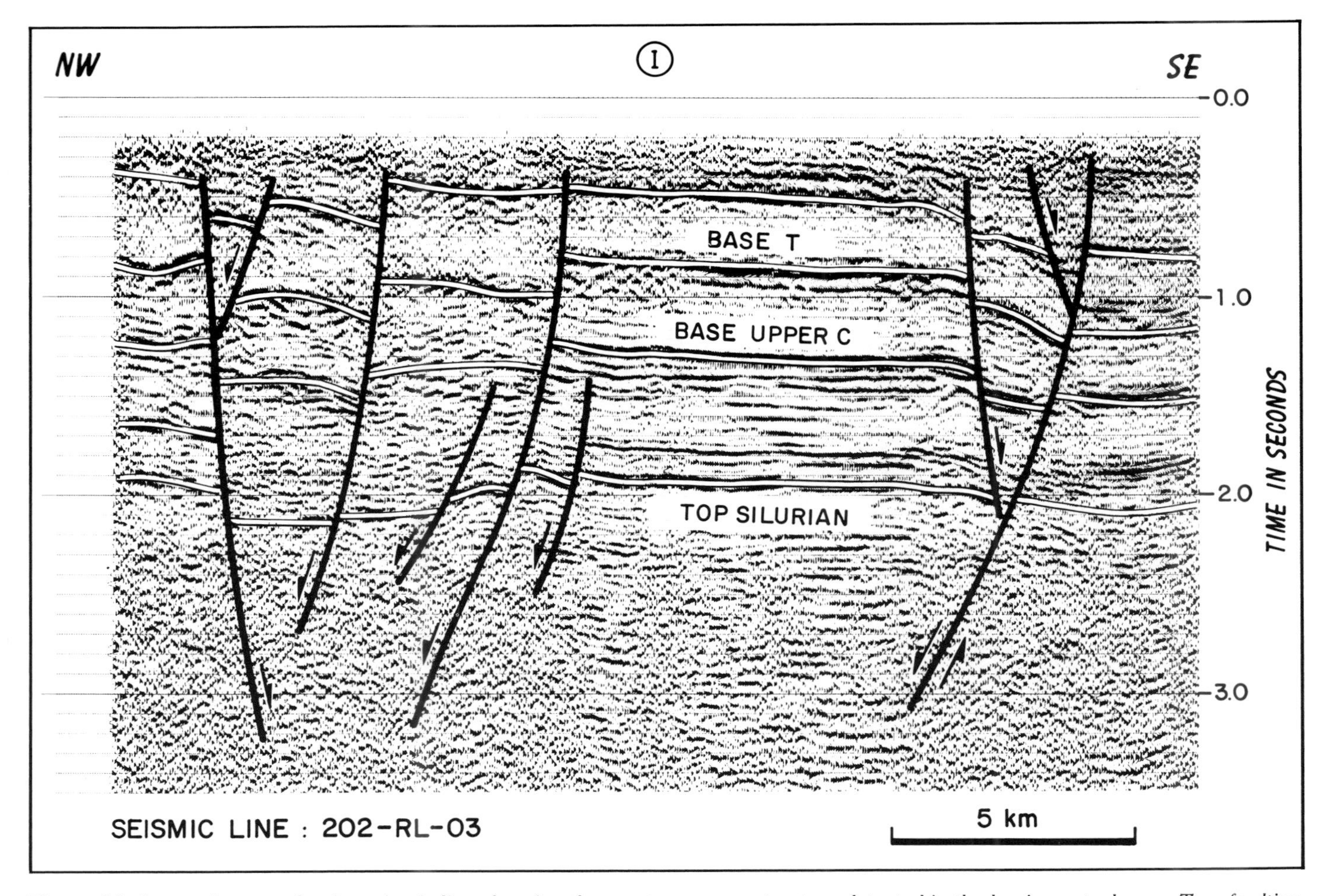


**Figure 33.** *Lower Amazon basin: seismic line showing the most common structure detected in the basin central area. Two faulting episodes are observed: one affecting only the lowermost section and the other affecting the entire sedimentary sequence, including Tertiary rocks (Line I in Figure 9).*

Geophysicists can acquire useful seismic reflection data in the Amazon basins, but must learn about the signal-noise ratio and how to improve it.

The future exploration in the Upper Amazon basin relies on further drilling in already well-defined seismic structures on the Monte Alegre and Itaituba Formations. The Monte Alegre reservoir rocks are quartzarenites with average porosity of 18%, average permeability of 50 mD, and average net-pay zones of 15 m (50 ft) (Figure 34). Porosity is usually of secondary origin in Juruá. Itaituba sandstones, locally with fairly good reservoir conditions and interlayered with evaporites, constitute minor gas accumulations. The Monte Alegre play in the Upper Amazon relates to Permian maturation and Jurassic migration of gas, generated in the Devonian Barreirinha shale. The paths of migration are reverse fault planes that acted as conduits. Because of very high heat flow in this basin, we can only expect gas accumulations. Traps are essentially related to several domal structures, elongated along the upthrown side of reverse faults (Figure 14 and 35). So far, twelve of these folds hold commercial gas accumulations. The present main objective is to detect the southwest extension of the hydrocarbon trend. Until now, the exploring strategy has been to drill only one wildcat well per structure. Therefore, although only 6 billion cu m (210 bcf) of gas are proven reserves, the ultimate natural gas resources are presently estimated as 120 billion cu m (4.2 tcf). More than 10 wildcat locations have been approved for further drilling.

Future exploration in the Middle Amazon relies on the results from the Manaus Platform seismic survey. Another key area is the southern part of the basin, especially between the Canumá and Mamuru Platforms (Figure 15). There, small folds from lateral shear generated by lineament-block tectonics are expected to bear oil accumulations.

Exploratory drilling in the Lower Amazon basin can now become reliable because recent enhancement of seismic resolution showed several drapes and other minor folds located on shoulders of the upthrown sides of normal faults.

The Carboniferous faulting and the younger rifting stage identified in the Middle and Lower Amazon had a favorable tec-

→

**Figure 34.** *Typical log of part of the Upper Carboniferous–Devonian sequence. Well 1-RBB-1-AM.*

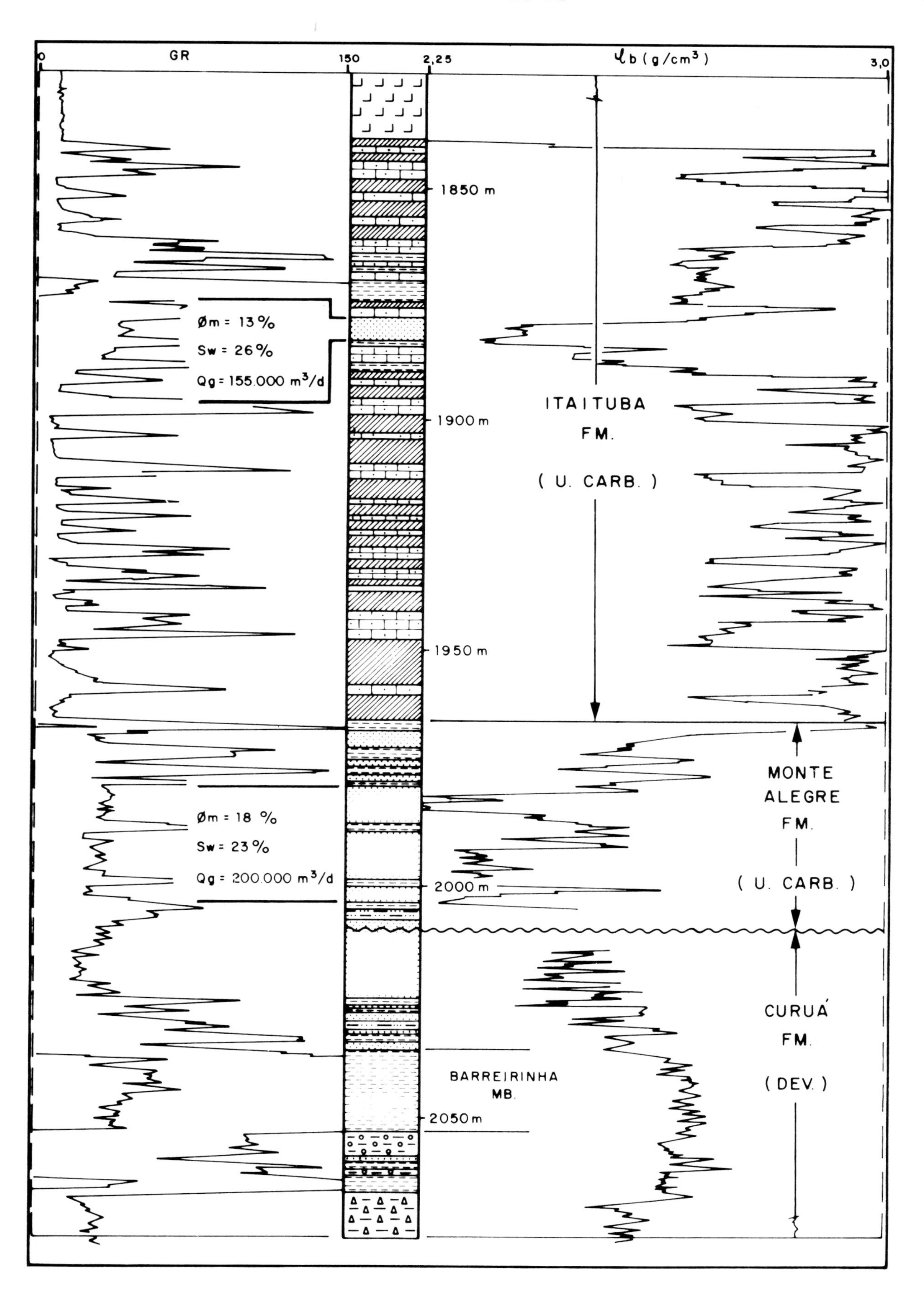

0
GR
150
2,25
ρb (g/cm³)
3,0
1850 m
Øm = 13%
Sw = 26%
Qg = 155.000 m³/d
1900 m
ITAITUBA
FM.
( U. CARB. )
1950 m
MONTE
ALEGRE
FM.
Øm = 18 %
Sw = 23%
Qg = 200.000 m³/d
2000 m
( U. CARB. )
CURUÁ
FM.
BARREIRINHA
MB.
(DEV.)
2050 m

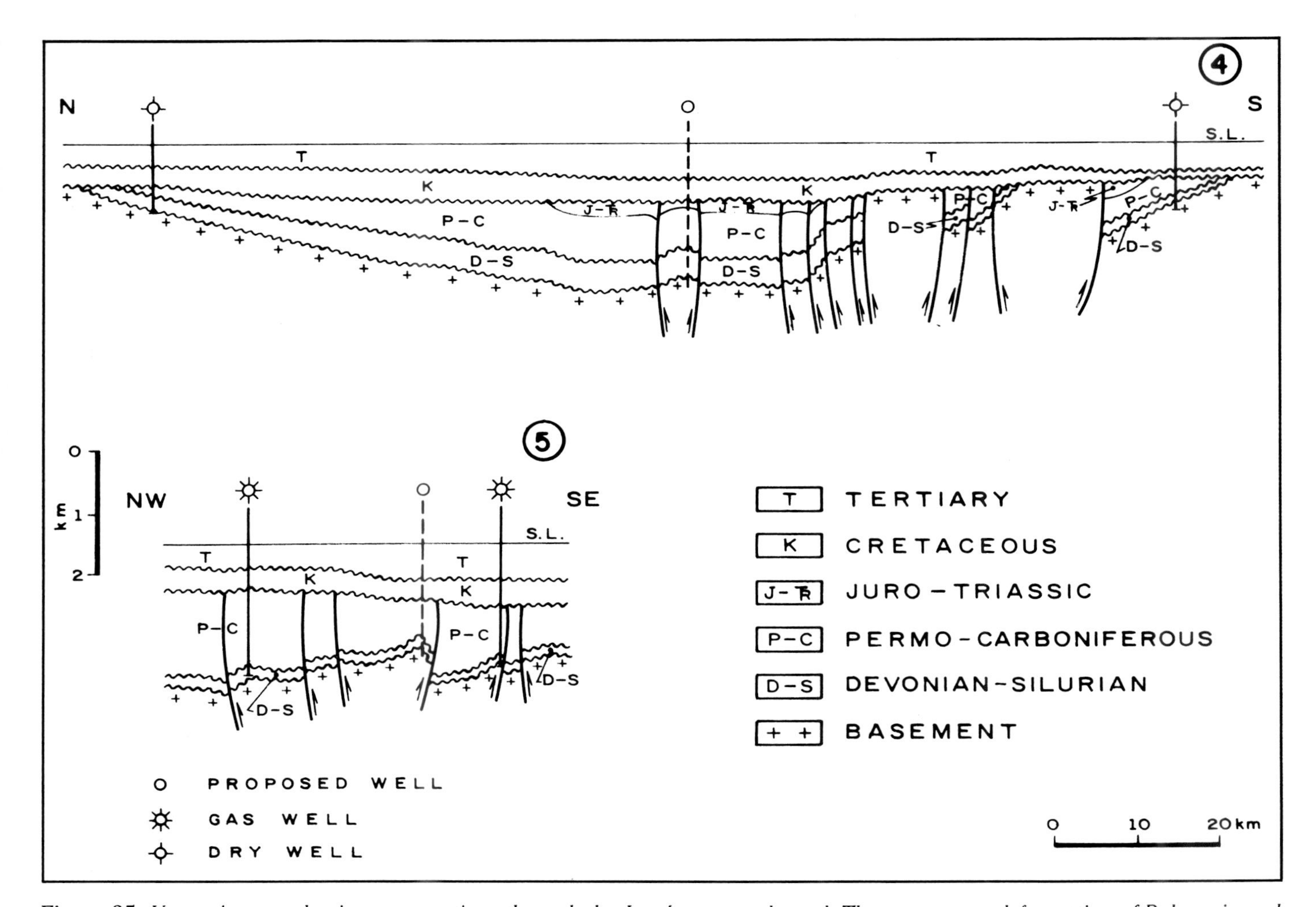


Figure 35. *Upper Amazon basin: cross-sections through the Juruá structural trend. The very severe deformation of Paleozoic and Juro-Triassic strata due to compressive stress affected the southern half of the basin (Lines 4 and 5 in Figures 9 and 14).*

tonic overprint for thermal maturation and migration of hydrocarbons, because of a late and large thermal event in the basin's evolution and because of intense faulting. This faulting might have acted as pathways for migration and caused the formation of traps.

## CONCLUSIONS

The Upper, Middle, and Lower Amazon are intracratonic Phanerozoic basins, evolved from narrow grabens during the rifting stage through syneclises during widespread Paleozoic sagging.

Wrench tectonics differentially affected these basins. The Upper Amazon basin was more intensively deformed by wrenching during Jurassic, when the entire Paleozoic sedimentary sequence and Mesozoic volcanics were folded and faulted. In the Middle and Lower Amazon, the lineament-block tectonics formed several northwest and northeast faults, which are the predominant structural features. Structures related to wrench tectonics located at basin margins are thus far local, subdued features and are possibly younger than those detected in the Upper Amazon. The Lower Amazon is affected also by these lineament-block tectonics, but the overprint of Carboniferous to Cretaceous rifting developed well-defined grabens, which are presently being identified on seismic lines.

Geologists considered the Upper Amazon basin a frontier area until 1978. Since then, the geological and geophysical data obtained helped establish a tectonic model that explained fairly well the structural trends initially mapped. Further exploratory programs in this basin will consider the proposed model in order to extend the already known trends and identify new ones.

In the Middle and Lower Amazon basins, according to previous exploratory results, the factors controlling hydrocarbon occurrence do exist. These basins are considered of high potential for hydrocarbon exploration because of abundant source rocks, the presence of good reservoirs, and favorable geochemical conditions. These basins, however, are still considered frontier areas because identifying structural features and integrating them to establish the regional tectonic framework depend on more and better-quality seismic data and new ideas.

Reinterpretation of data based on new ideas is certainly one of the most, if not the most, important factors for finding oil, as stated by Wallace E. Pratt: "Oil must be sought first of all in our minds. Where oil really is, then, in the final analysis, is in our own heads."

## ACKNOWLEDGMENTS

We express our gratitude to Hilton Muhlmann of Petrobrás for his review of the manuscript and his valuable effort in helping to prepare the illustrations. We thank Dr. Carlos Walter Marinho Campos, Director of Petrobrás, for permission to publish this paper. We acknowledge also the contributions of many geologists and geophysicists of the Exploration Department and Research Center of Petrobrás who recorded their interpretations in several internal reports of the company.

## REFERENCES

Dewey, J. F., and W. C. Pitman III, 1984, The origin and evolution of sedimentary basins: Short Course Notes, Tectan Inc., 398 p.

Falvey, D. A., 1977, Isostatic constraints in 2-D gravity modeling: Bulletin of Australian Society of Exploration Geophysics, v. 8, p. 106–110.

Green, A. R., 1977, The evolution of the earth's crust, *in* The earth's crust—its nature and physical properties: American Geophysical Union, Geophysical Monthly, n. 20, p. 1–17.

Harding, T. P., 1983, Graben hydrocarbon plays and structural styles: Geologie en Mijnbouw, v. 62, p. 3–23.

Kingston, D. R., C. P. Dishroon, and P. A. Williams, 1983, Global basin classification system: AAPG Bulletin, v. 67, p. 2175–2193.

Klemme, H. D., 1980, Petroleum basins—classification and characteristics: Journal of Petroleum Geology, v. 3, p. 187–207.

Linsser, H., 1958, Interpretation of the regional gravity anomalies in the Amazonas area: Petróleo Brasileiro S. A.-PETROBRÁS, DEPEX, Internal Report no. 3250, 12 p.

Loczy, L. de, 1970, Role of transcurrent faulting in South American tectonic framework: AAPG Bulletin, v. 54, p. 2111–2119.

McKenzie, D., 1978, Some remarks on the development of sedimentary basins: Earth Planetary Science Letters, v. 40, p. 25–32.

Nunn, J. A., and J. R. Aires, 1984, Subsidence history and tectonic evolution of the Middle Amazon basin, Brazil: PETROBRÁS-Internal Report, 40 p.

Rezende, W. M., and C. G. de Brito, 1973, Avaliaçâo geológica da bacia paleozóica do Amazonas: Anais do XXVII Congresso da Sociedade Brasileira de Geologia, v. 3, p. 27–45.

Royden, L., J. G. Sclater, and R. P. Von Herzen, 1980, Continental margin subsidence and heat flow: AAPG Bulletin, v. 64, p. 173–187.

Szatmari, P., 1983, The Amazon rift and Pisco-Juruá fault: their relation to the separation of North America from Gondwana: Geology, v. 11, p. 300–304.

———, 1984, Tectonismo da faixa de dobramentos do Juruá: Anais do II Symposium Amazônico (Departamento Nacional da Produçâo Mineral-DNPM), Manaus, Brasil, v. 2, p. 117–128.

Thomas, G. E., 1974, Lineament-block tectonics: Williston-Blood Creek basin: AAPG Bulletin, v. 58, p. 1305–1322.

# Deep Water (200–1,800 Meters) Hydrocarbon Potential of the U.S. Gulf of Mexico

P. Dolan
*JEBCO Seismic Limited*
*London, England*

The paper reports on the upper continental slope within the U.S. portion of the Gulf of Mexico, which comprises several submarine geomorphological units.

The updip limit of the area is the 650 ft (200 m) isobath that, until recently, has been an outer limit to hydrocarbon exploration. The downdip limit of 6,000 ft (approximately 1,800 m) is governed in part by geology, in part by the feasibility of resource exploitation in the medium term, but most importantly by the availability of exploration data.

The size of the area is about 51,560 sq mi (133,550 sq km) or 33 million acres (13.4 million ha). For comparison, this area represents approximately 23% of the total Western, Central, and Eastern Planning Areas.

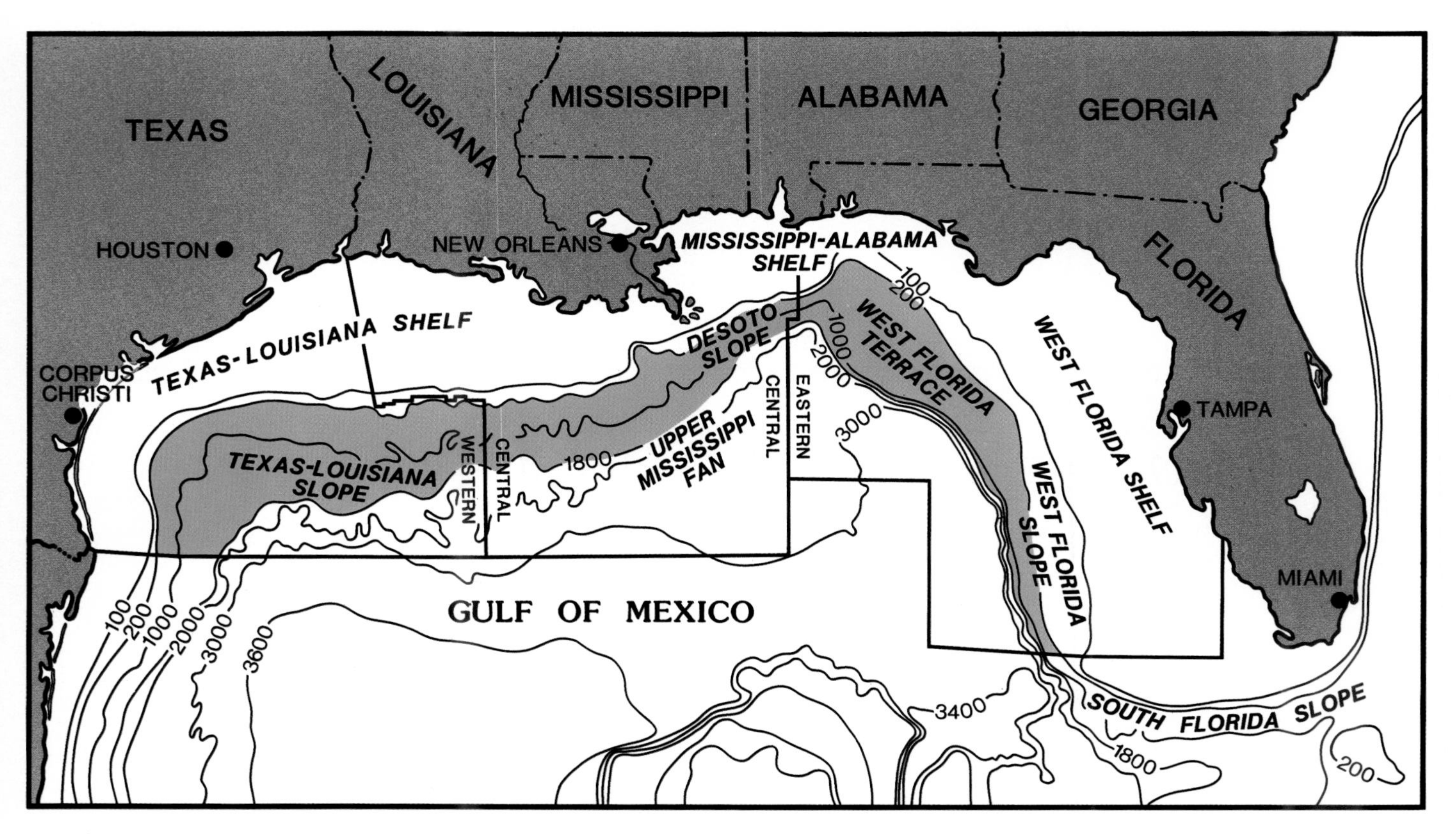


Figure 1. *Location Map.*

## THE ECONOMIC IMPORTANCE OF THE AREA

The U.S. portion of the North American upper continental slope in the Gulf of Mexico (Figure 1) comprises several geomorphological units. The U.S. Minerals Management Service (MMS) has named these units the Texas-Louisiana slope, the upper Mississippi fan, the Desoto slope, the west Florida terrace and the west Florida slope.

Within the context of national oil and gas production in the U.S., offshore production has been a valuable component for the last 20 years, according to Weise et al (1983) (Figure 2). The Gulf of Mexico itself has contributed well in excess of 90% of this offshore oil and gas production. Geologists must take into account nontechnical considerations when they assess the continuing contribution that the Gulf of Mexico will make to the national total.

Of particular significance to geologists is the withdrawal of industry access to vast areas of onshore federal lands, mainly designated as wilderness areas. A summary of recent industry estimates (Figure 3) indicates that 80 million acres (32.4 million ha) of onshore land have been withdrawn over the last 20 years, with another 30 million acres (12.1 million ha) likely to follow in the next 5 years. This loss of 80 million acres (32.4 million ha) represents the same area as the entire Western and Central Planning Areas of the Gulf of Mexico.

The future focus of U.S. activity is likely, therefore, to be on offshore areas.

Considering the U.S. outer continental shelf alone, the Gulf of Mexico has been of paramount importance, because it contributed 100% of offshore production through 1967 and still accounted for an overwhelming 90.6% in 1982 (Figure 4). This dominance will almost certainly persist partly because the oil industry and environmental interests have largely avoided conflicts, and, barring a major catastrophe, withdrawal of significant amounts of acreage in the Gulf of Mexico is unlikely. In contrast, various moratoria over the last 3 years have caused an increase in the rate of withdrawal of offshore federal areas outside the Gulf of Mexico, which now total over 52 million acres (0.4 to 21 million ha), Crowe (1984).

These 'non-technical' constraints have resulted in a recent surge of deep water leasing activity during 1983 and 1984. Prior to mid-1983, the oil industry had acquired some 600,000 acres (0.24 million ha) of leases in water depths over 650 ft (200 m). With the introduction of area-wide leasing and a series of significant discoveries, the industry has acquired an additional 2.5 million acres (1 million ha) in little over a year (Hagar, 1984).

## GEOLOGICAL SETTING

### Gulf of Mexico Structural Evolution

The hydrocarbon potential of a basin can best be assessed if there is a reliable understanding of the basin framework. Unfortunately, the tectonic evolution of the Gulf of Mexico is still a contentious matter, despite its relatively long history of petroleum exploration.

A series of four maps has recently been published (Hall et al, 1983) illustrating various contemporary ideas that prevail

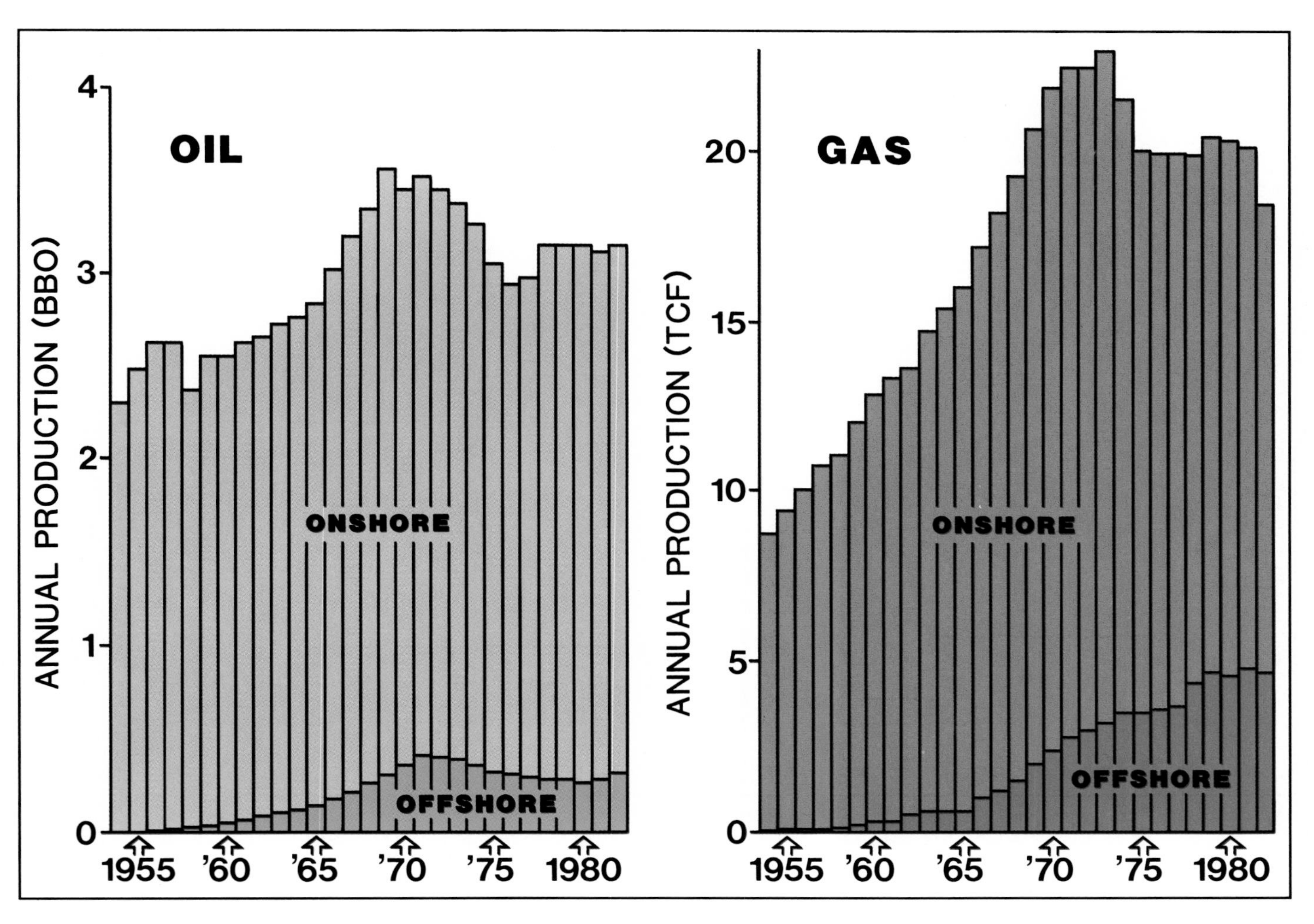


Figure 2. *U.S. oil and gas production: offshore and onshore. Annual production is given in billion bbl of oil (BBO) and trillion cu ft of gas (tcf).*

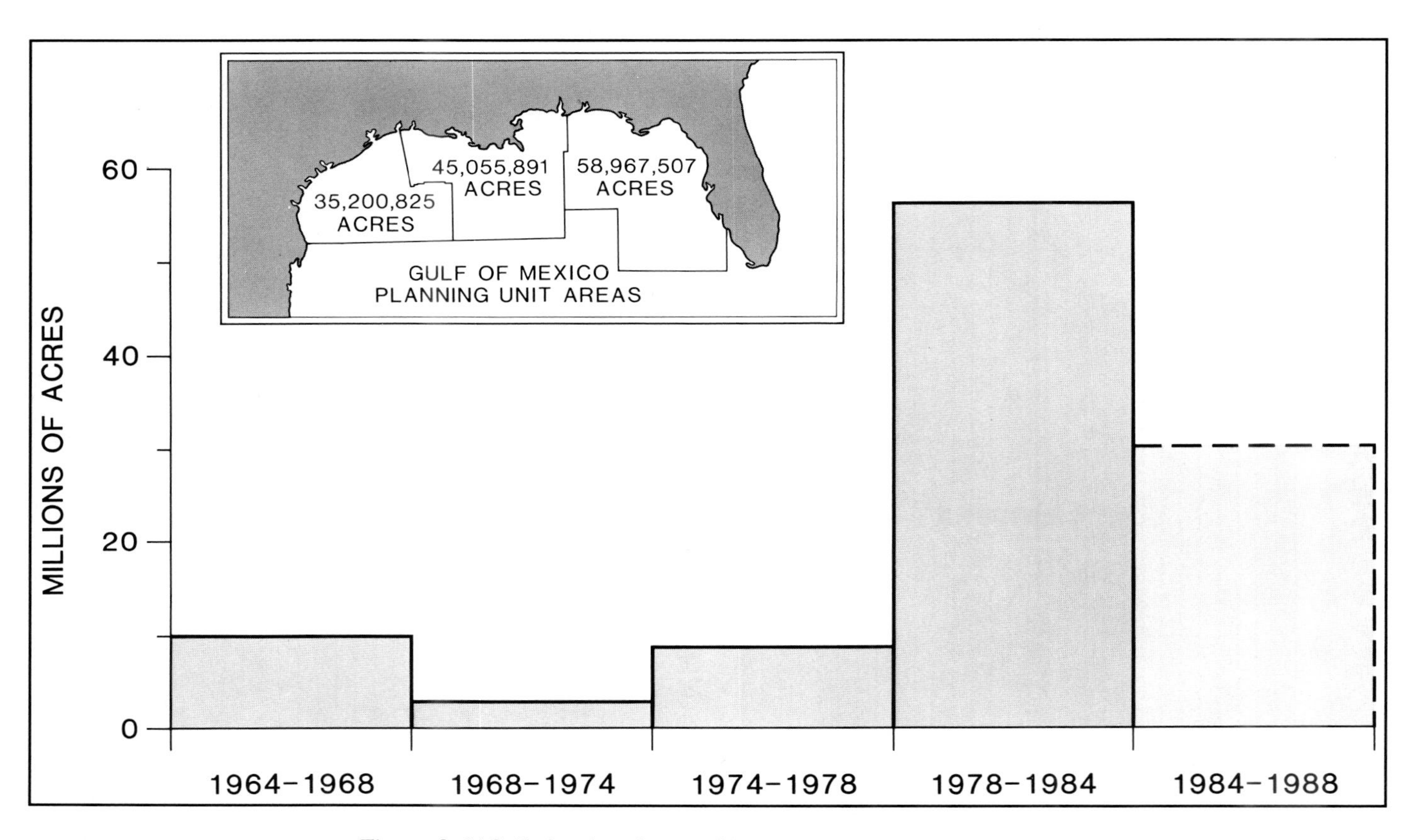


Figure 3. *U.S. Federal onshore wilderness designations: 1964–1988.*

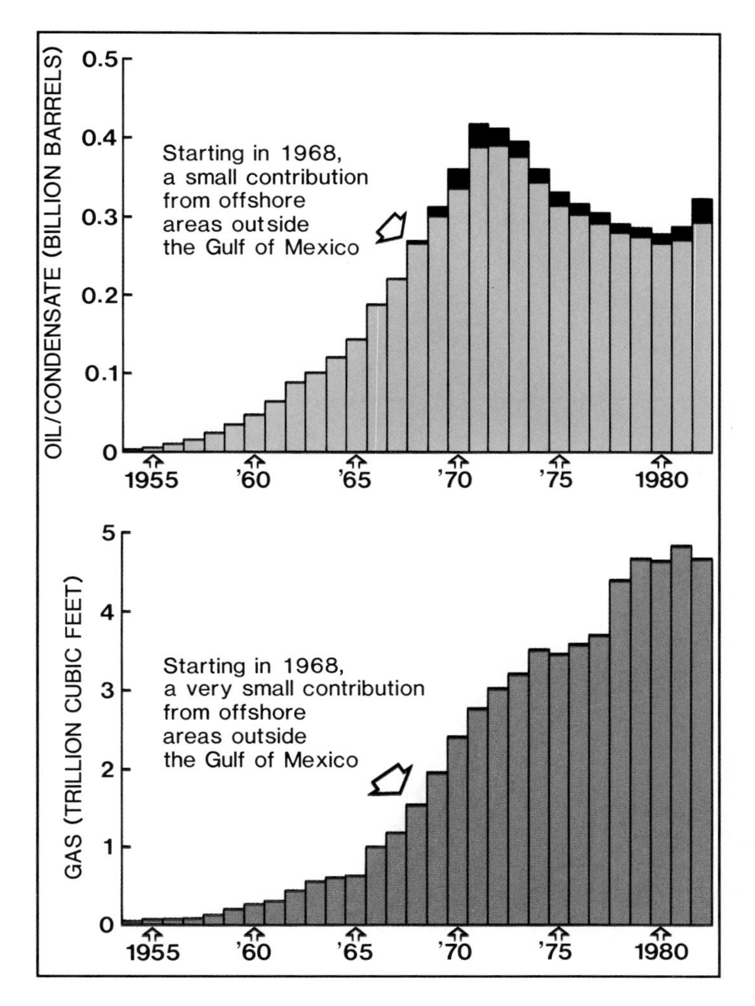


Figure 4. *U.S. outer continental shelf (OCS) and Gulf of Mexico oil and gas production.*

regarding the evolution of the basin configuration (Figure 5). In this paper, the favored model is that of Hall et al (1983), based on two considerations.

First, narrow, linear, positive gravity anomalies are developing on opposing sides of the Gulf of Mexico (Figure 6). Geologists know of and have documented such anomalies along the outer edges of many rifted continental margins. This disposition of the gravity anomalies suggests that the Yucatan block rotated clockwise during the Mesozoic, away from Mexico and Texas, resulting in a 'saw-tooth' series of alternating rifted and transform margins along the northern arc of the Gulf of Mexico.

Second, micro-plates in the area possibly fit and support the rotational model. A spatial evolution is proposed (Figure 7) which may partially explain the development of the Eastern Planning Area, a matter not specifically resolved by Hall et al (1983). This model invokes the Yucatan block rotation as proposed by Hall et al (1983), as well as the Nicaraguan block occupying the eastern Gulf of Mexico.

The initial break up of the region involved the parting of North America from the combined South America and Africa plates. By Middle Jurassic times, the Yucatan micro-plate began rotating out of the incipient Gulf of Mexico and a proto-Greater Antilles island arc may have formed along a subduction zone. During the Cretaceous, this island arc may have thrust the Bahama Block northeastward. A contemporary event could have been that the Nicaraguan Block rotated across a spreading center, which extended eastward from the Pacific, ahead of the main South America plate from which it split. As Nicaragua continued to rotate counterclockwise, the Pacific spreading center was over-ridden (perhaps the Cocos Ridge is a remnant); the Pacific plate salients became separate within the orbit of the Caribbean; and, as a consequence, the Greater Antilles subduction zone became quiescent. The structural history of Cuba, Hispaniola, and the Lesser Antilles is, however, still enigmatic and such a pattern of development is simplistic, requiring rigorous corroboration.

If essentially correct, however, this model implies that a better appreciation of the hydrocarbon potential of the Eastern Planning Area could result by studying the geology of the now northern margin of the Nicaraguan block. It is, perhaps, more than coincidence that wells drilled to date along the northern margin of the Nicaraguan block have found success as elusive as those drilled along the west Florida shelf.

### History of Sedimentation

Geophysical evidence suggests that during the initial development of the Gulf of Mexico, a basal sequence of sediments was deposited, perhaps mainly clastics of Early to Middle Jurassic age. A sequence of salt and evaporites then began to develop. Contrary to many other examples of rift-margin salt deposition, where the salt is restricted behind the outer basement ridge (for example, opposing sides of the South Atlantic and Red Sea), the salt of offshore Texas and Louisiana is developed in both this restricted position and also over a thin pre-salt section that sits on early oceanic crust basinward of the outer basement ridge (Figure 8).

Previous authors have noted that under the outer continental shelf, the zone of very thick mobile Louann salt is restricted principally to the area west of the Mississippi delta. Perhaps, the structural model of Hall et al (1983) can explain this (Figure 9). If the model is restored such that the gravity highs off Louisiana and the Yucatan Peninsula abut one another, then a westerly restricted salt basin is created, while to the east is a relatively unrestricted basin, open to the south.

Basin fill developed as a Cretaceous and Tertiary wedge, dominated by clastics in the Western and Central Gulf and a carbonate platform in the Eastern Gulf. The principal reason for this dichotomy was the presence of many rivers that, because of the prevailing paleogeography of the North American continent, drained mainly into the Western and Central Gulf areas.

This enormous clastic load is critical to the petroleum geology of the Gulf of Mexico in several ways. While the sediments provide both the source rocks and reservoir rocks for hydrocarbon occurrences, they also provide the load that has triggered listric faulting and mobilization of both shales and the underlying salt, thereby setting up many of the structural traps.

## PETROLEUM GEOLOGY

### Data Base

The Gulf of Mexico will continue to be one of the most intensively explored and developed outer continental shelf regions

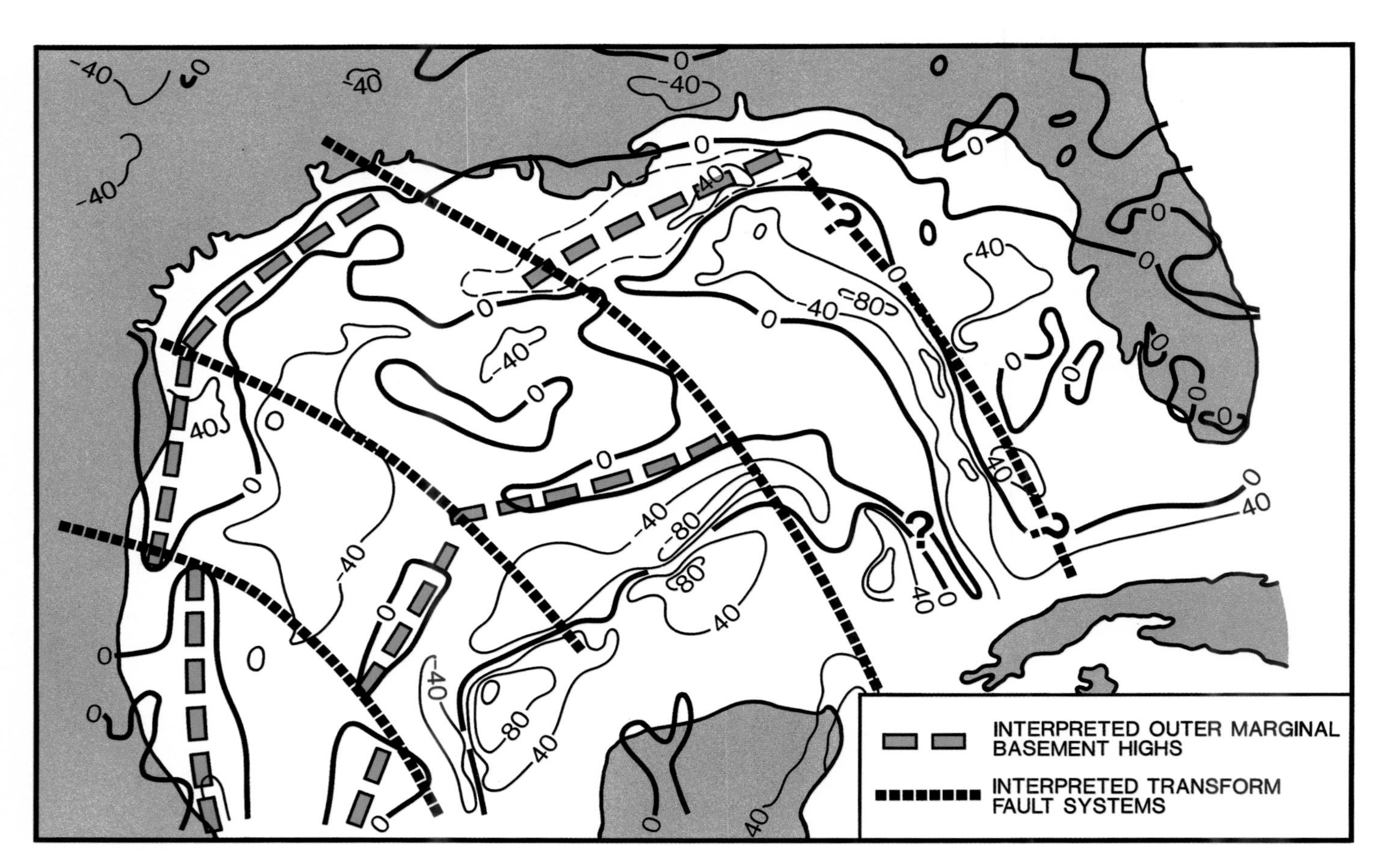


Figure 5. *Gulf of Mexico: various recent interpretations of crustal type.*

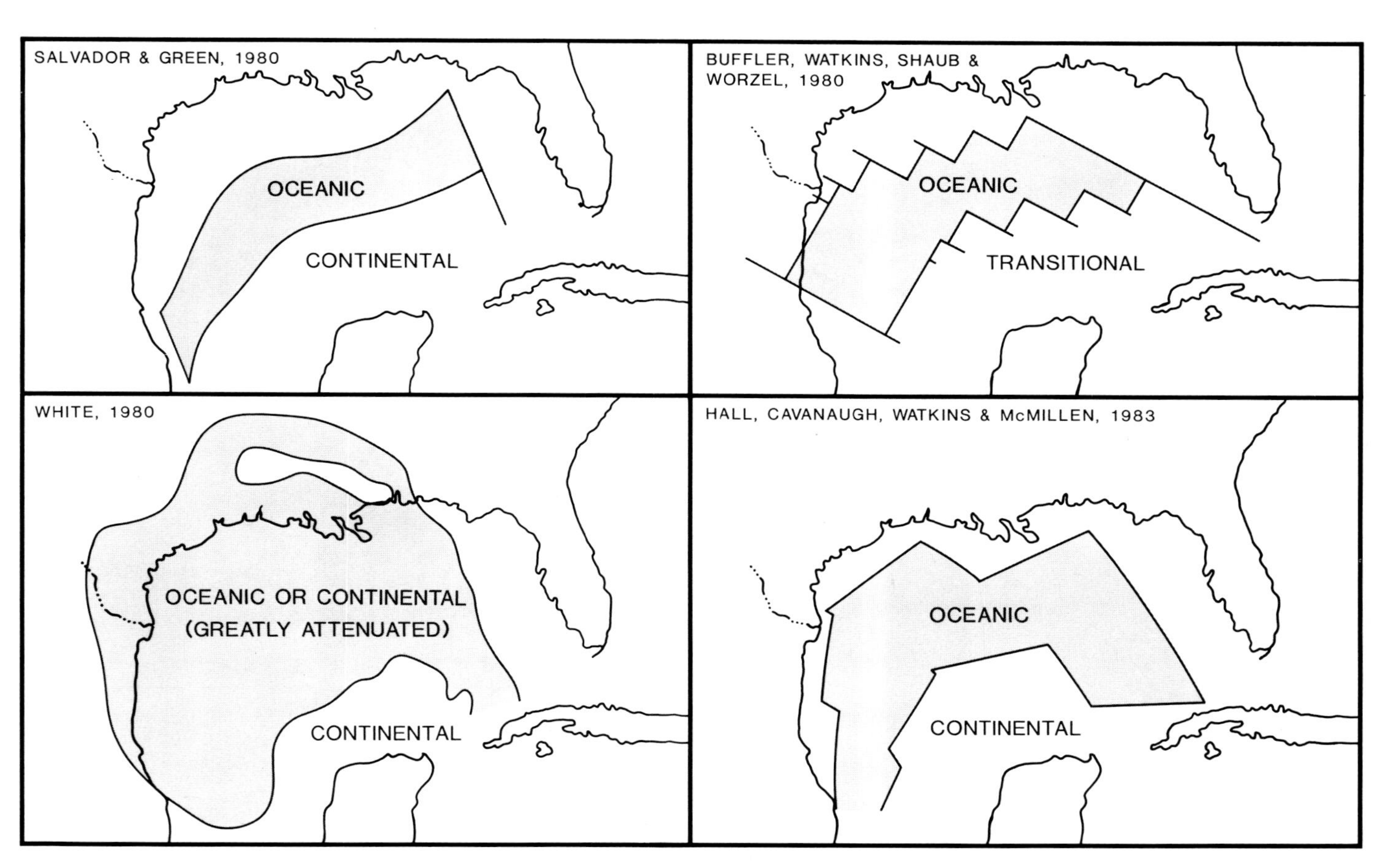


Figure 6. *Regional gravity data in the Gulf of Mexico.*

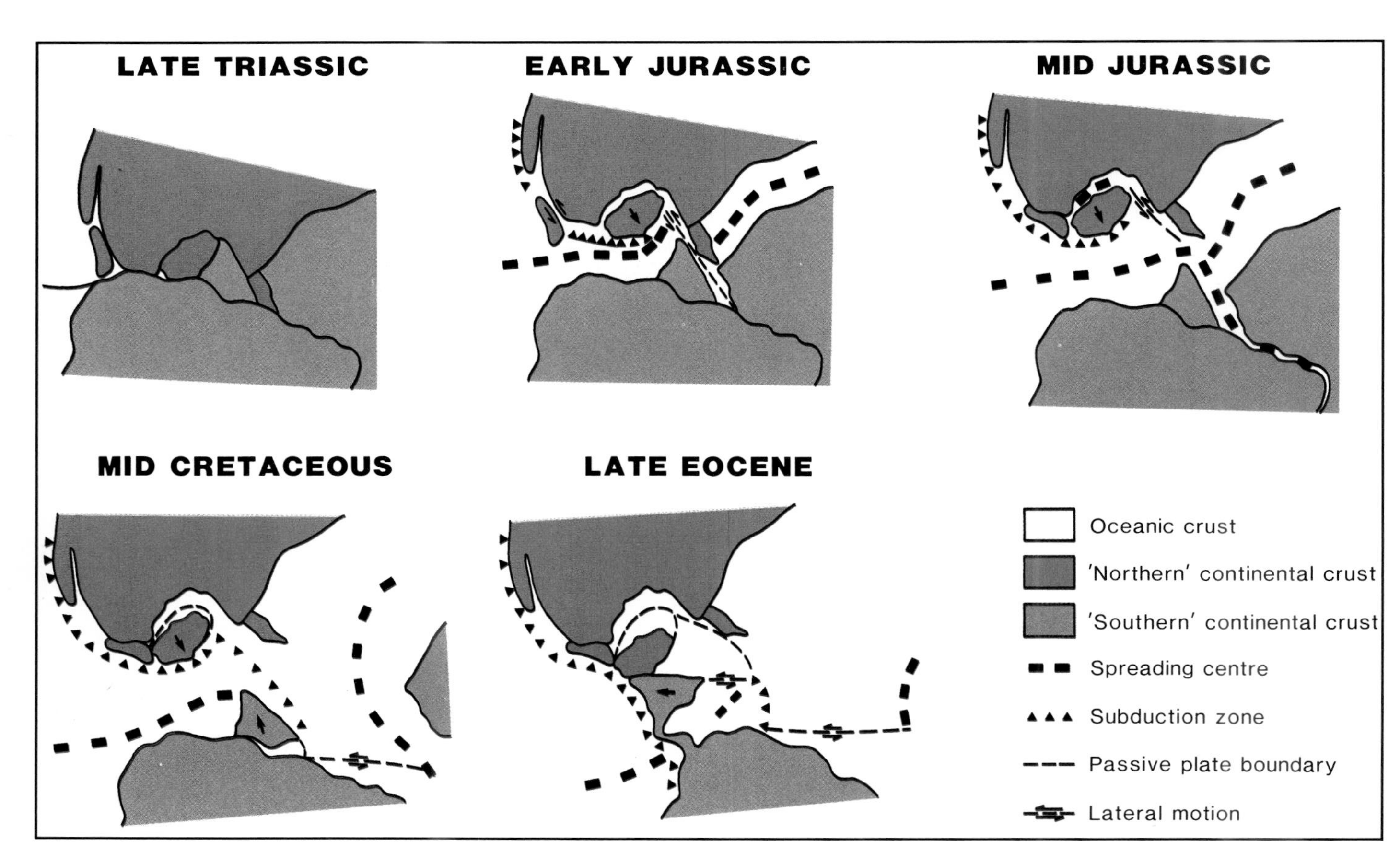


Figure 7. *Simplified micro-plate motions proposed for the Gulf of Mexico/Caribbean region.*

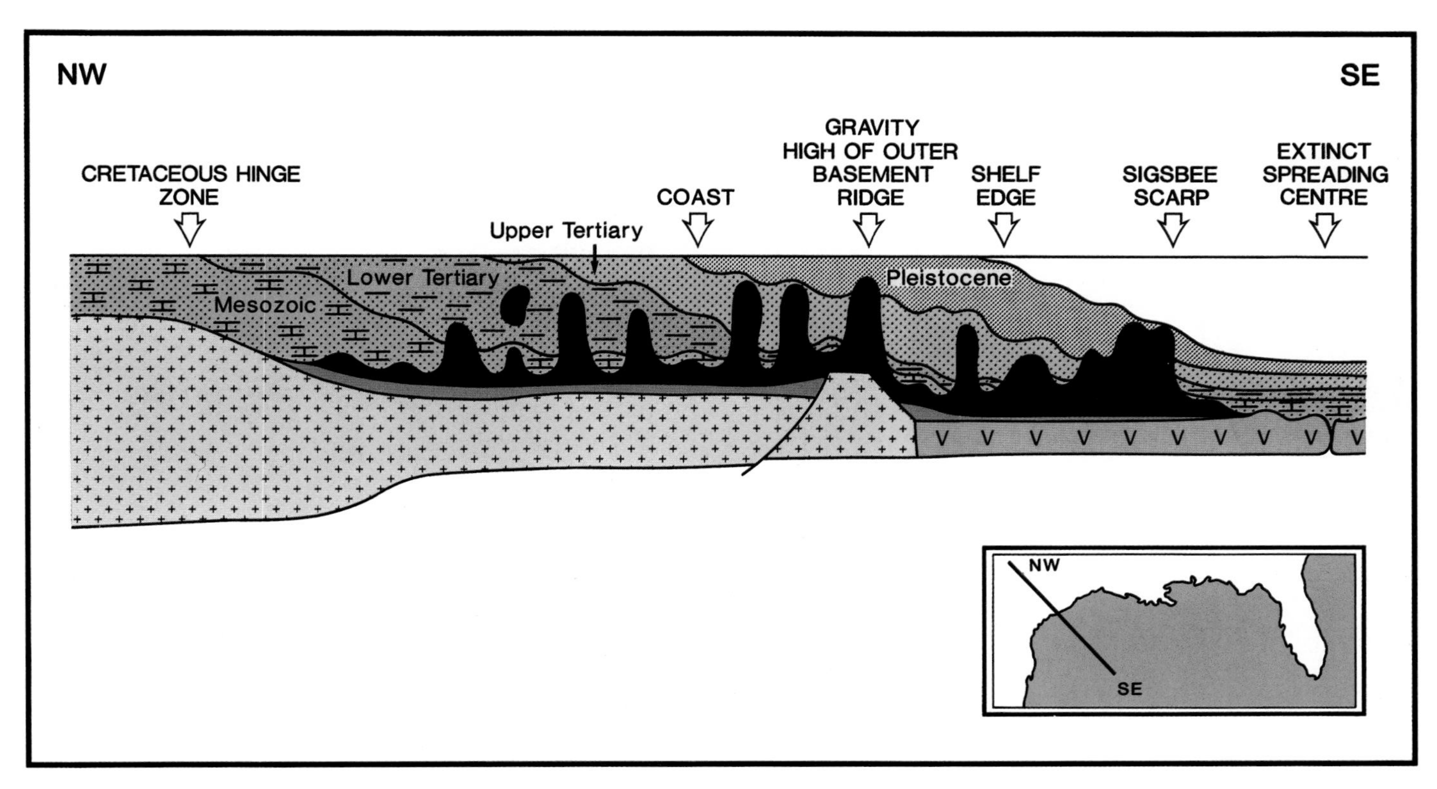


Figure 8. *Development of basin fill in the Gulf of Mexico.*

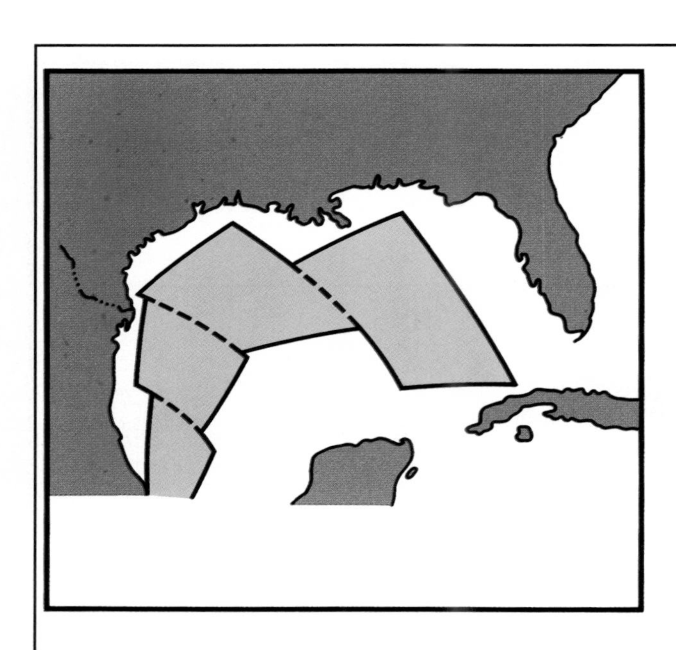


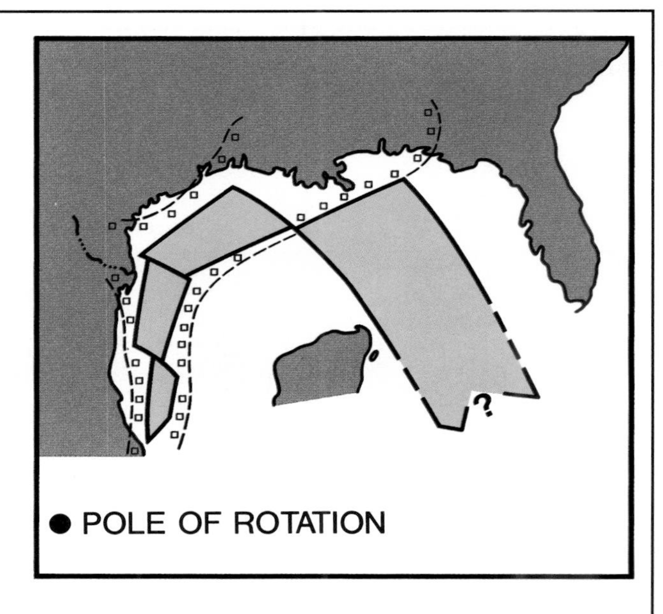


Figure 9. *Structural control of Louann Salt basin.*

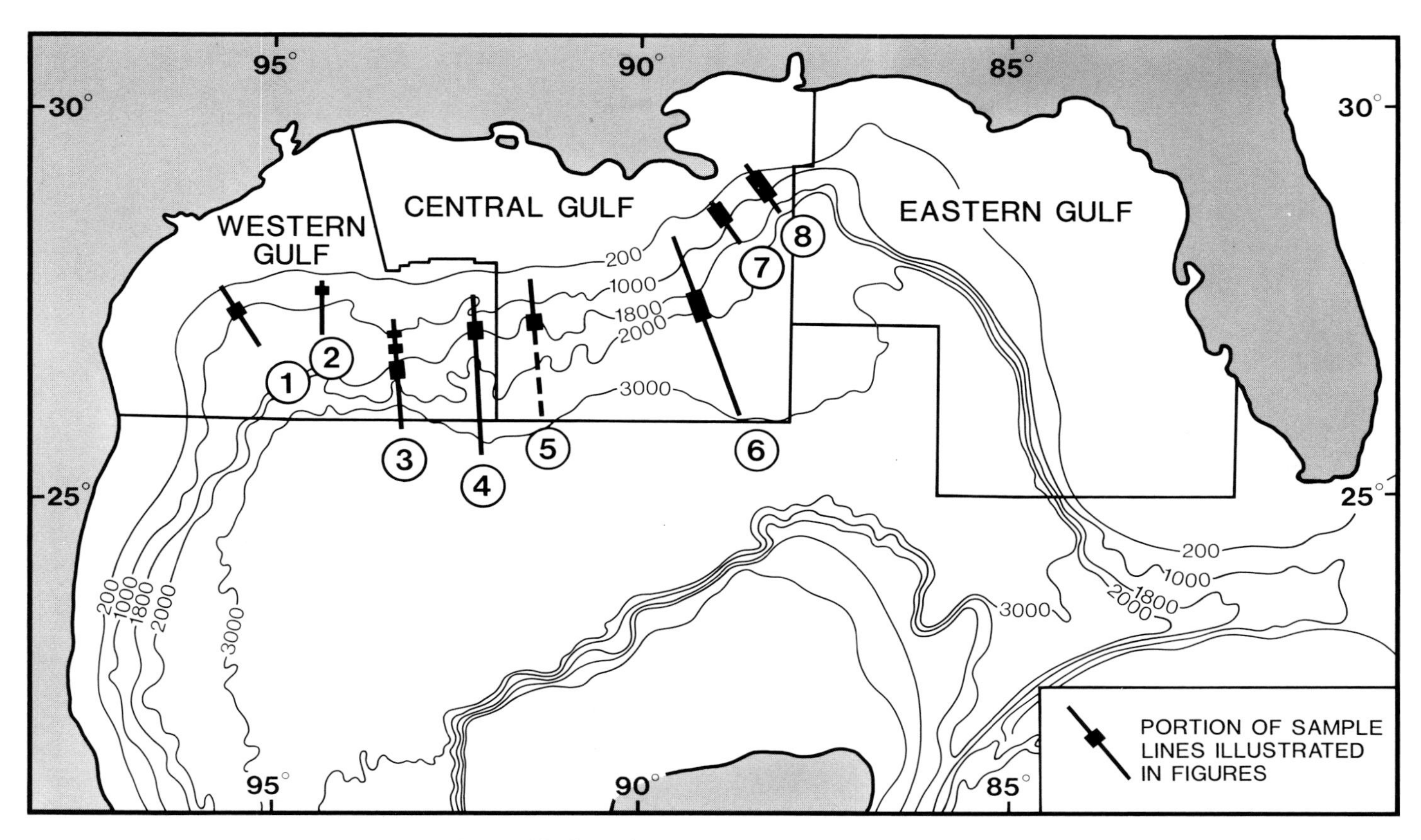


Figure 10. *Sample seismic sections: location map.*

in the world, despite both seismic and drilling activity having already been in progress for over 50 years in waters less than 650 ft (200 m) deep.

Until recently, evaluating an area with greater than 650 ft (200 m) of water cover has had to rely mainly on an *ad hoc* collection of shallow core holes; widely spread Deep Sea Drilling Project (DSDP) boreholes; several thousand line miles of shallow penetration, single channel sparker profiles; and a limited amount of reconnaissance, multichannel seismic data. However, the oil industry is expanding its effort to establish a more comprehensive and coherent seismic and drilling data base across the upper continental slope. While the oil companies gathered the well information as proprietary data, the geophysical contracting industry has acquired much of the seismic data on a speculative basis.

### Structural Styles and Traps

Figure 10 shows the approximate location of the dip-oriented portions of eight seismic lines. These lines (Figures 11–18) illustrate various structural styles and confirm that the proven productive seismic sequences of the shelf area (albeit diachronous in part) extend well down the continental slope across the area of interest.

#### Line 1

Figure 11, from the western part of the East Breaks area, provides a good example of a salt feature with structural overhang. The explorationist should consider such aspects of the seismic section as the water bottom and top salt multiples and the possibility of a semi-coherent set of deep data.

#### Line 2

The second line (Figure 12) further east, but still in the East Breaks area, is again dominated by mobile salt. However, the shallow section is sufficiently undisturbed to reveal an extensive area of counter-regional dip, down to the north. If such a trend prevails for considerable distances along the strike, with the obvious implications for hydrocarbon migration, then explorationists may have to step boldly from the 650 ft (200 m) water depth out to the 1,300 to 1,600 ft (400 to 500 m) zone where broad dip reversal occurs.

#### Line 3

Situated in the western part of the Garden Banks area, the third line (Figure 13) is also pervaded by mobile salt. Three portions of the line illustrate the variety of salt movement: (from north to south) a salt spine, a massive salt ridge, and an area of chaotic salt piercement.

#### Line 4

From the evidence of line 4 (Figure 14), which is dominated by massive salt near the sea floor, the zone of mobile salt extends throughout the upper continental slope of the Garden Banks area.

#### Line 5

Line 5 (Figure 15) indicates that the western part of the Green

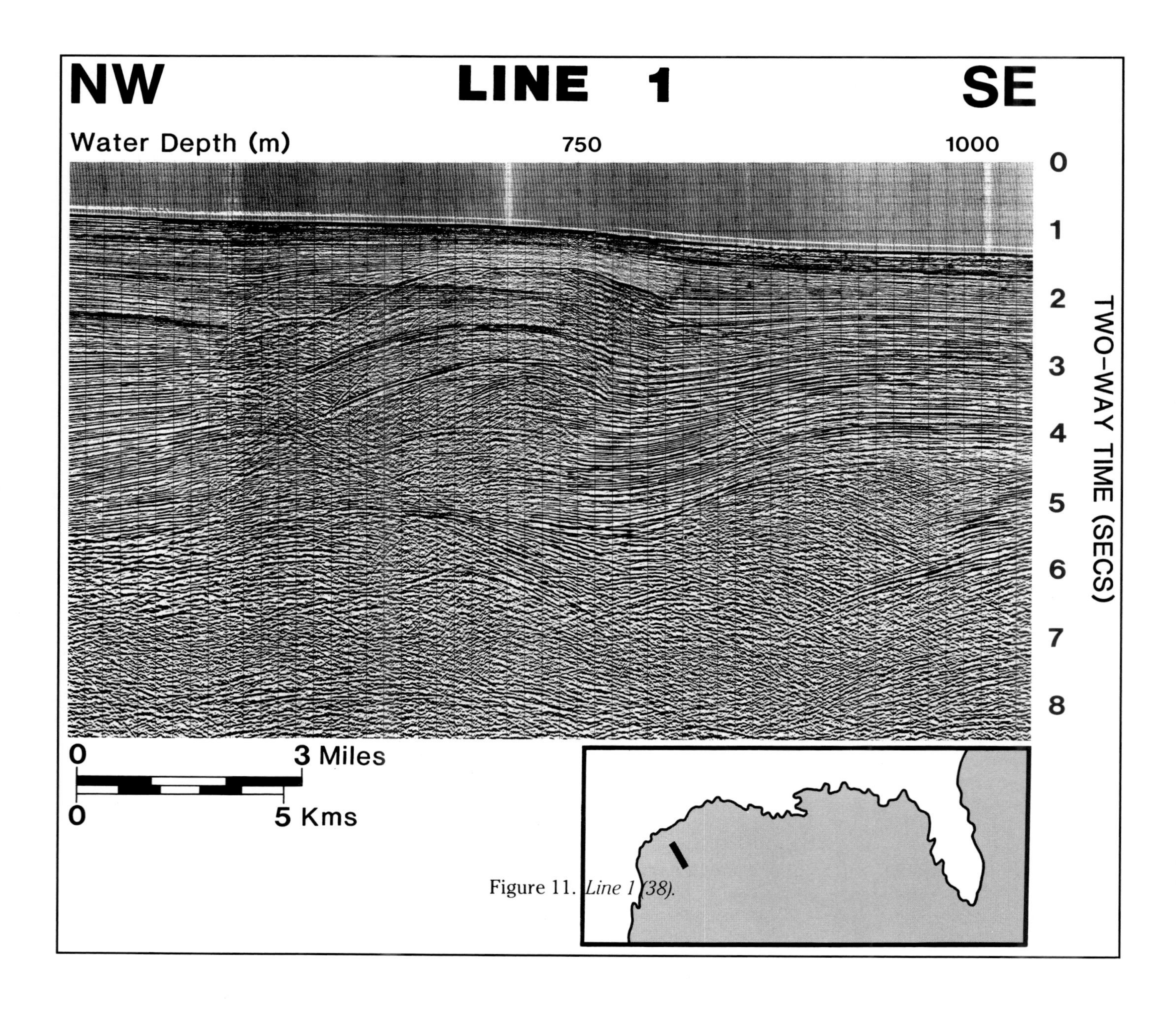


Figure 11. *Line 1 (38).*

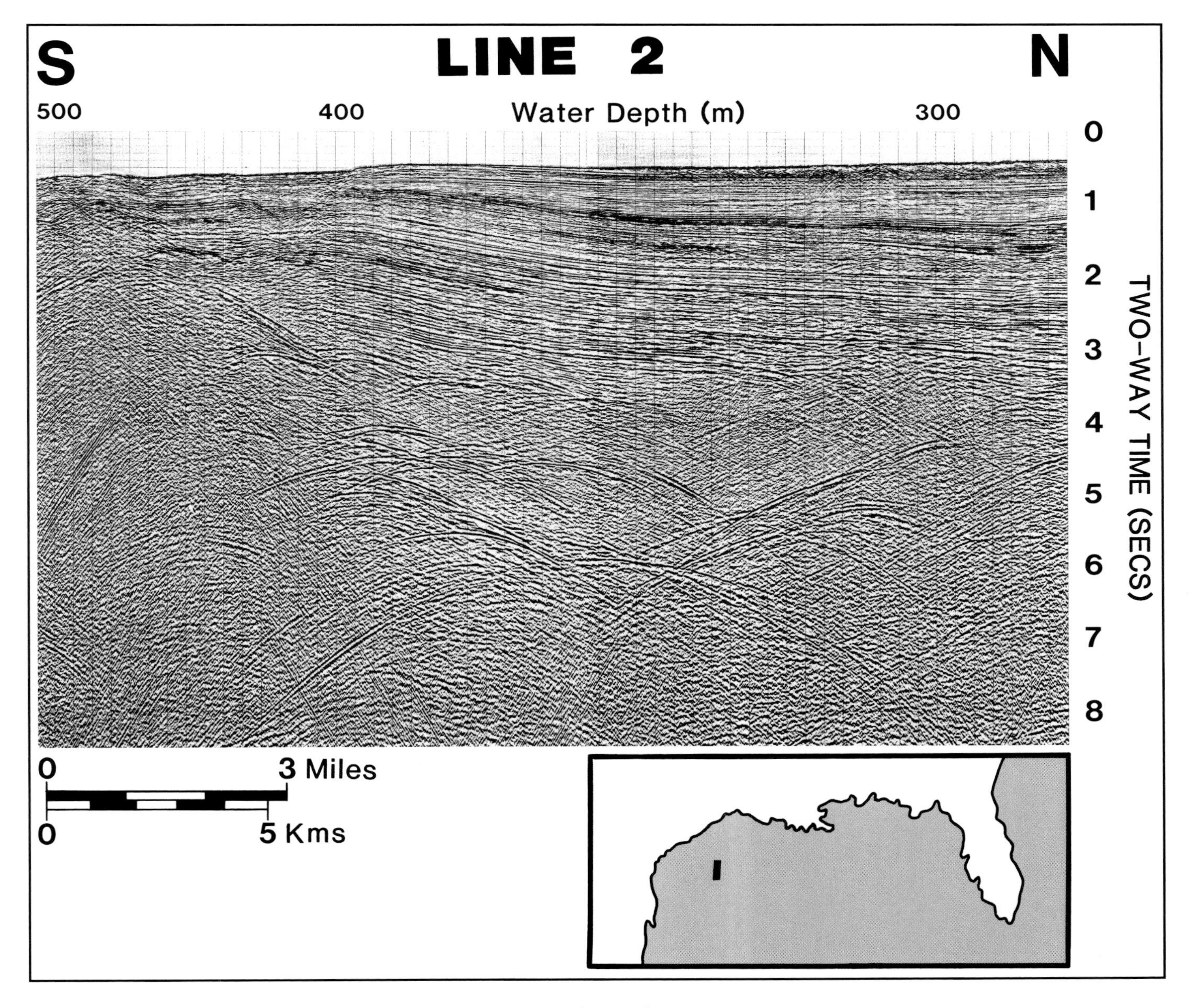


Figure 12. *Line 2 (140).*

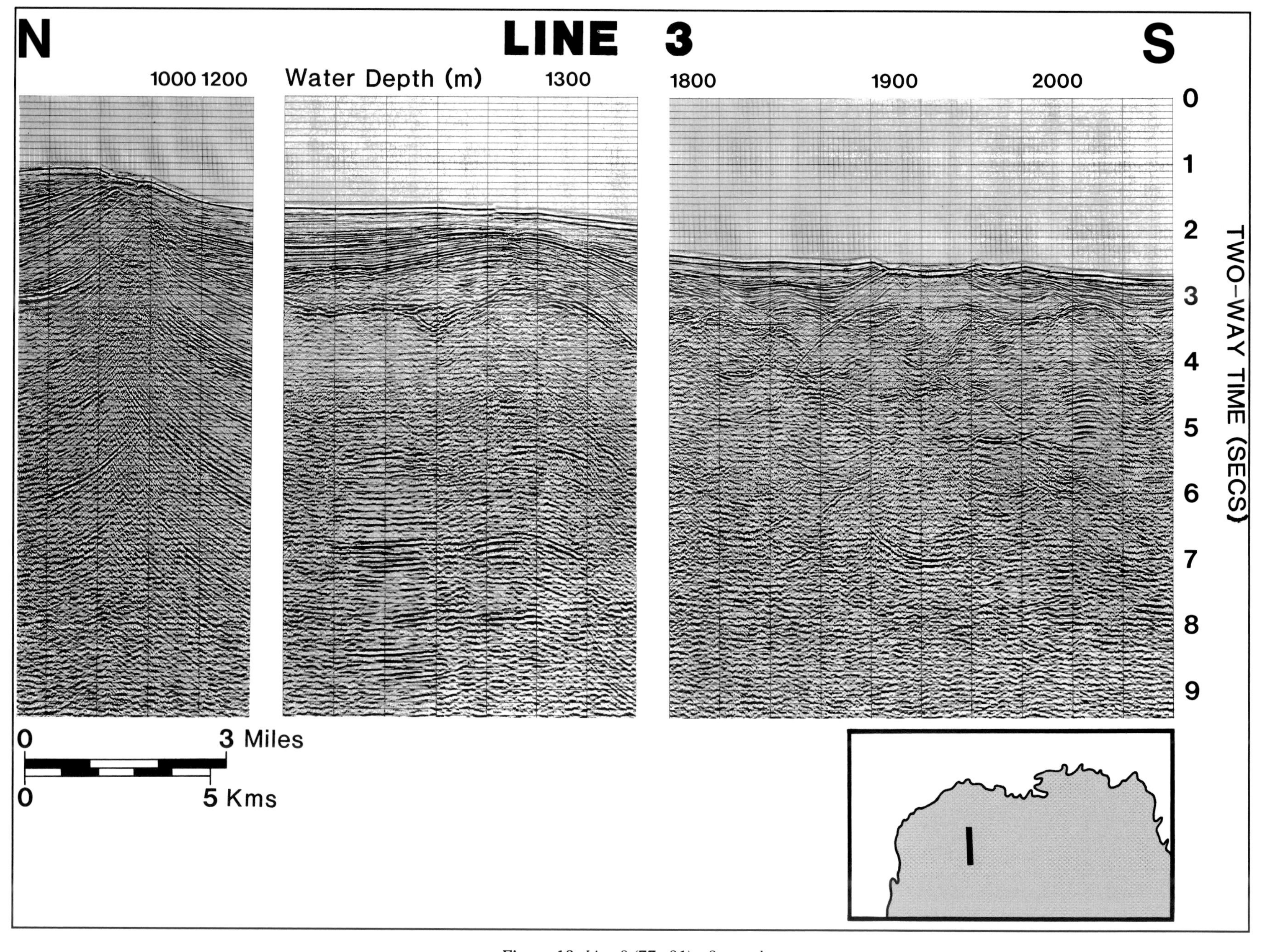


Figure 13. *Line 3 (77 - 31)—3 panels.*

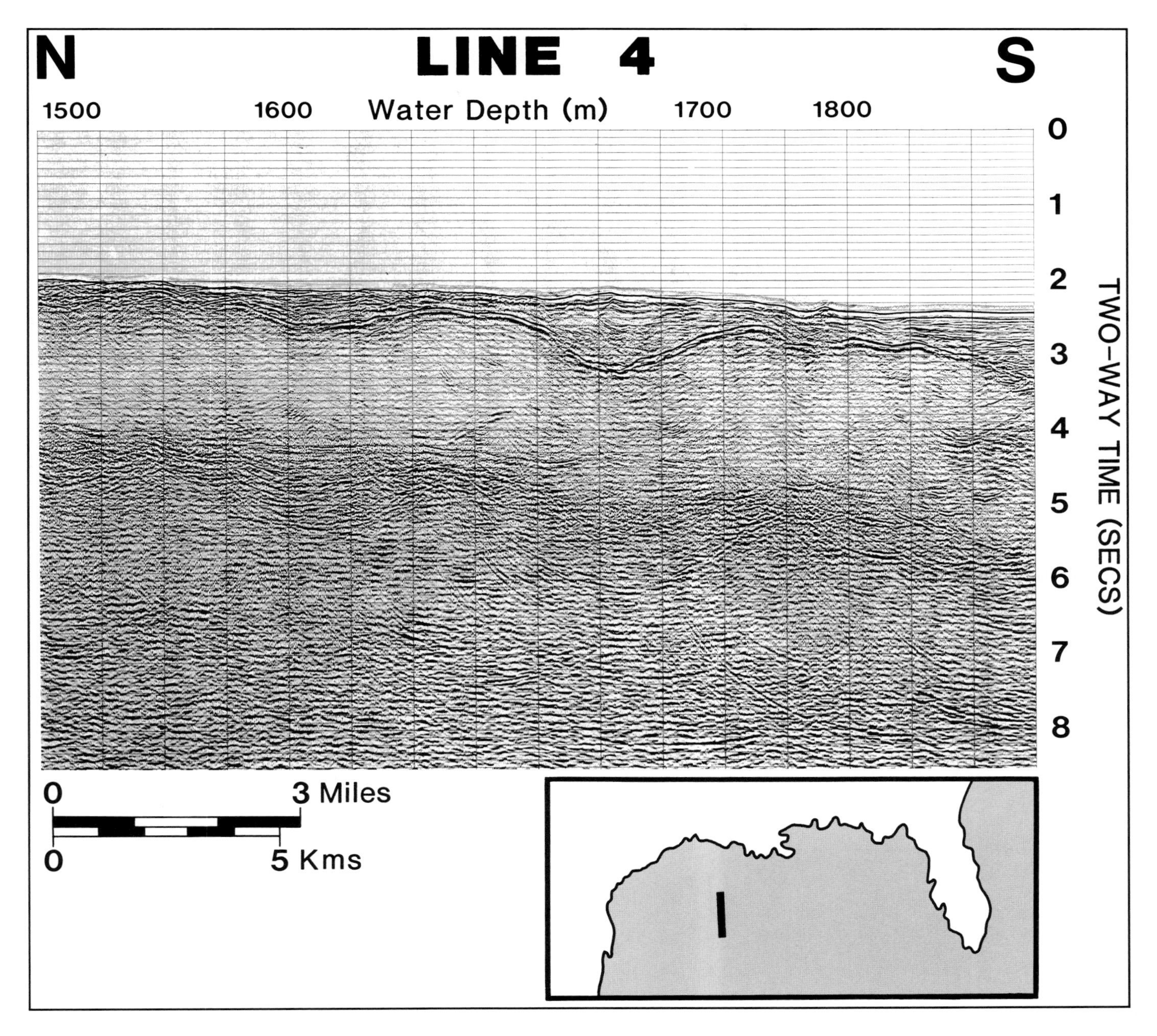


Figure 14. *Line 4 (77 - 23).*

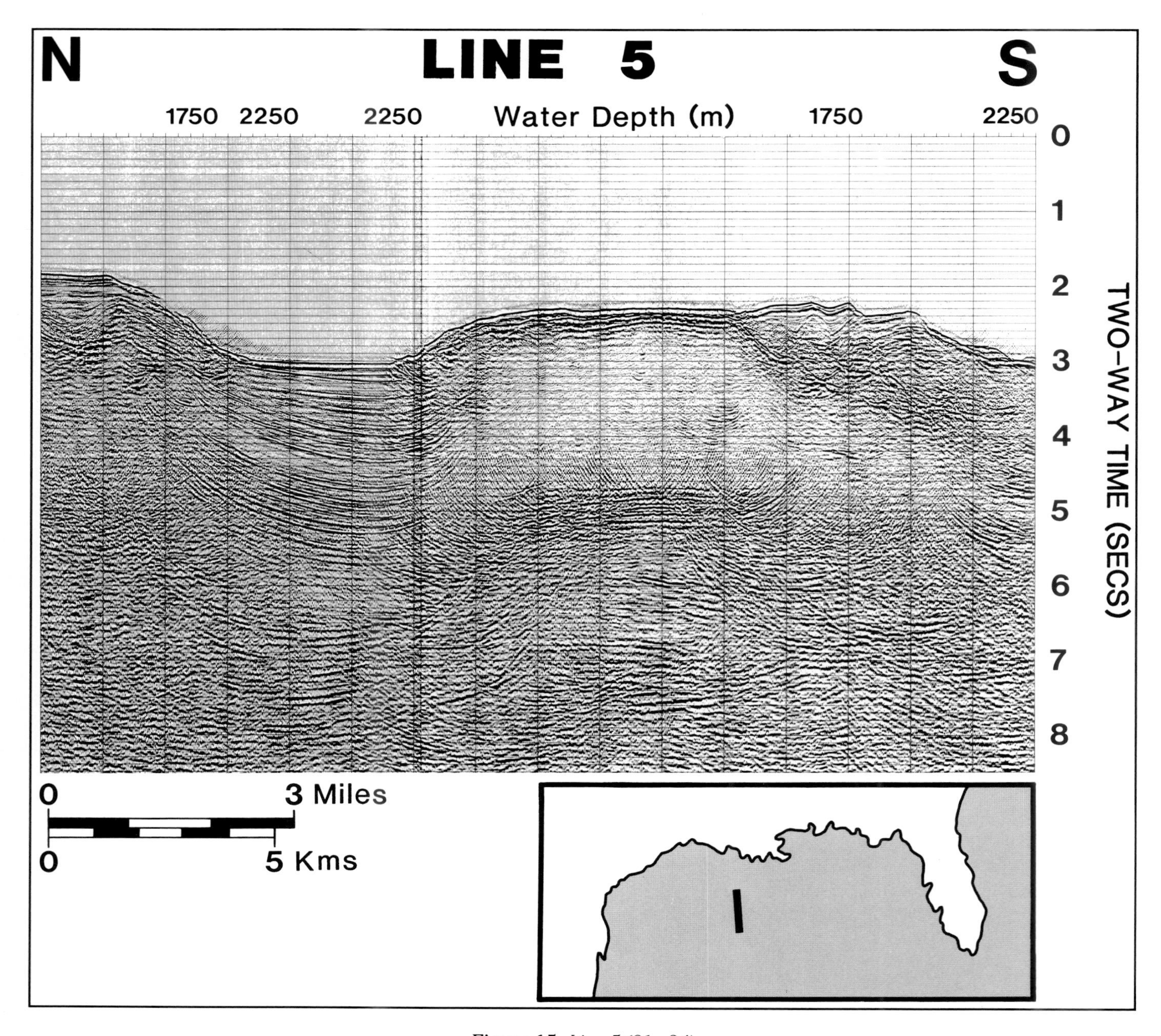


Figure 15. *Line 5 (81 - 34).*

Canyon area is still within the region where mobile salt prevails.

### Line 6

The sixth line (Figure 16) appears to be near the eastern margin of the Louann Salt basin and extends into an area of an undisturbed, generally flat-lying section. At first sight, this section appears to be relatively devoid of trapping mechanisms. However, this may not be the case because the line indicates both deep counter-regional dip and shallow stratigraphic anomalies, in particular a turbiditic sequence.

### Line 7

Further east, in the Mississippi Canyon area, is a deltaic sequence typical of that found regionally updip on the continental shelf. Line 7 (Figure 17) clearly illustrates the possibility of finding large, prospective, deep-seated structural reversals.

### Line 8

Finally, line 8 (Figure 18) has all the hallmarks of a deltaic sequence: mobile shales, regional and counter-regional faults, slump blocks resting on listric fault planes, expanded sections, disconformities, etc. The deep faulting may be induced by ultra-deep mobile salt or shale (the traditional view) or, in part, by tectonic block faulting (as preferred in this paper).

→

Figure 16. *Line 6 (81 - 19).*

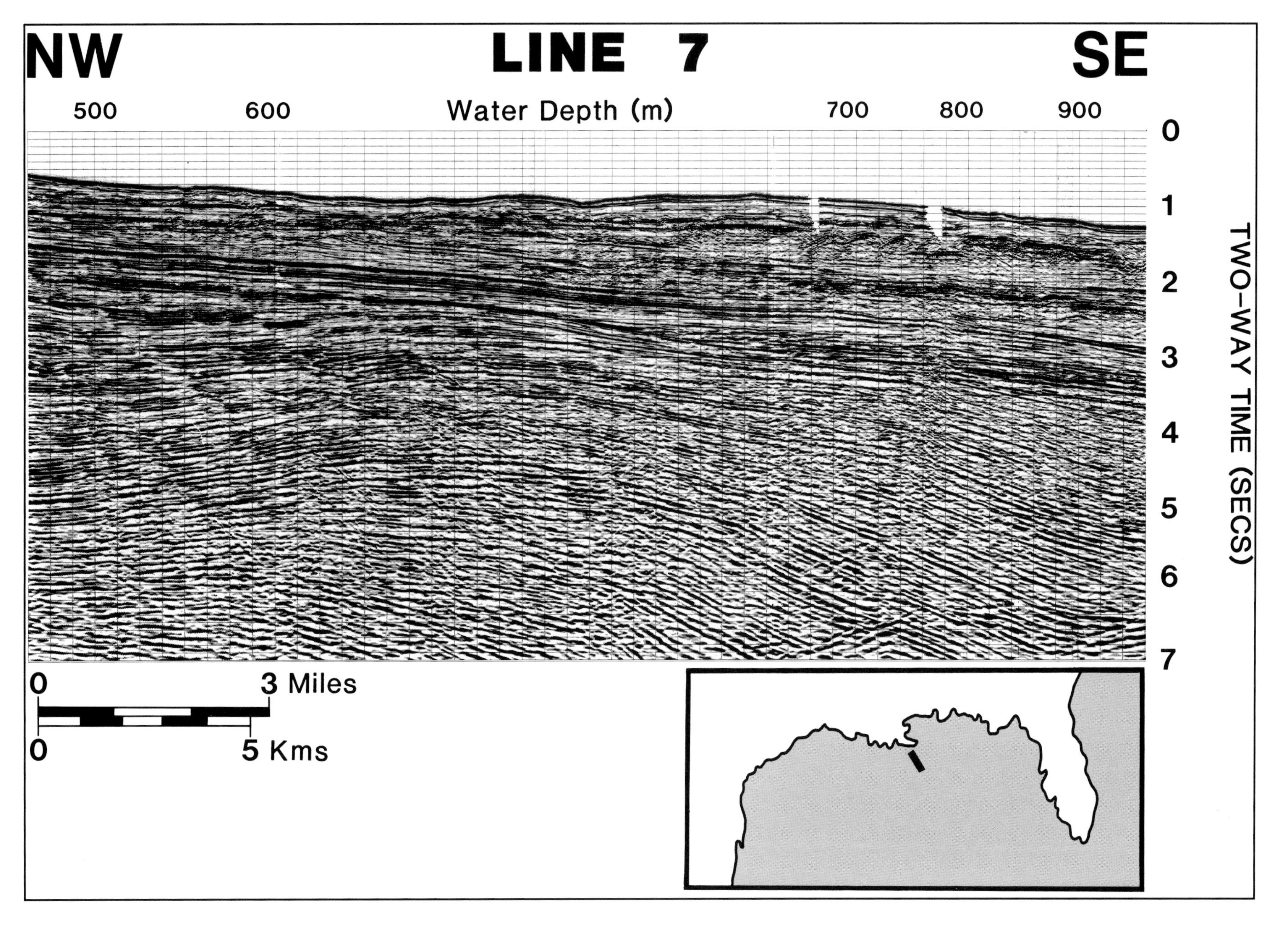


Figure 17. *Line 7 (MC 256).*

Table 1. *Demonstrated original and remaining recoverable reserves in 521 fields on the continental shelf.*

| | Original Recoverable | | Cumulative Production | | Remaining Reserves | |
|---|---|---|---|---|---|---|
| | Oil (billion bbl) | Gas (tcf) | Oil (billion bbl) | Gas (tcf) | Oil (billion bbl) | Gas (tcf) |
| Western Gulf | 0.288 | 1.256 | 0.085 | 5.240 | 0.198 | 7.320 |
| Central Gulf | 8.710 | 103.004 | 5.755 | 57.100 | 2.955 | 34.600 |
| Eastern Gulf | 0 | 0 | 0 | 0 | 0 | 0 |
| | 8.998 | 104.260 | 5.840 | 62.340 | 3.153 | 41.920 |

Source: Hewitt, Brooke and Knipmeyer, 1984

### Reservoirs

Historically, geologists have been concerned about the reservoir quality of the sandstones along the upper continental slope of Texas and Louisiana. The seismic sections of Figures 11 to 18 contain sufficient data to indicate, under normal circumstances, the presence of extensive interbedded shales and sandstones. However, various industry sources suggest that the abundance of reflection data on the continental slope is not as clearly related to lithological boundaries as it is on the continental shelf. Apparently, changes in the physical characteristics of otherwise relatively homogenous lithologies can generate reflections. The possibility of this phenomenon being widely applicable should be recognized when evaluating all deep water areas.

Despite this problem, exploration has already confirmed the presence of reservoir quality sands that high-energy turbidity currents deposited on the Texas and Louisiana continental slope.

### Source Rocks

Similarly, prior to the recent spate of exploration, geologists were concerned about source rock potential. The consensus was that the sedimentary section was very young (Pleistocene in age), possibly immature and gas prone. Well results have, however, confounded this view, and substantial discoveries of oil with gravities of between 26° and 44° API, along with condensate and gas, have been reported (Hagar, 1984). Although still a matter for debate, the view taken in this paper is that most of the hydrocarbons in deep water Pleistocene reservoirs come from contemporaneous, very fine-grained clastics, and that no long-distance migration paths need to be invoked.

## HYDROCARBON POTENTIAL

### Continental Shelf

Before attempting to estimate the hydrocarbon potential of the deep water area (600 ft or 200 m to 6,000 ft or 1,800 m) of the Gulf of Mexico, it is instructive to review the much better-known potential of the area shallower than 600 ft.

Hewitt et al (1984) (Table 1) estimated original recoverable reserves at 8.998 billion bbl of oil and 104.26 trillion cu ft (tcf) of gas, with remaining, recoverable reserves of 3.153 billion bbl of oil and 41.92 tcf of gas. These numbers relate to 521 fields (16 now abandoned) and do not include an additional 51 fields discovered in the last 2 years for which, as yet, there are insufficient data.

The distribution of the recoverable reserves is concentrated stratigraphically in the post-lower Miocene section. Of particular significance to the deeper water areas is the trend toward increasing reserves in the southerly Pleistocene sediments of the shelf area (Figure 19).

As noted, the figures of Hewitt et al (1984) relate to recoverable reserves and do not include undiscovered, recoverable resource estimates for the shelf area. Havran (1981) of the U.S. Geological Survey has estimated that these resources are in the order of 4.0 billion bbl of oil and 45.3 tcf of gas, suggesting, very approximately, a doubling of known, remaining, recoverable reserves.

Although speculative in nature, published data therefore suggest ultimate recoverable reserve figures for the shelf area in the order of 13 billion bbl of oil and 150 tcf of gas.

### Continental Slope

Havron (1981) estimated undiscovered recoverable resources of 2.6 billion bbl of oil and 26.5 tcf of gas (for 600 to 8,200 ft or 200 to 2,500 m water depth)—in other words, about 20% and 18% respectively of those to be found on the shelf. These estimates are, obviously, the product of considerable expertise and knowledge. However, they comprise an important element in the estimate of future U.S. hydrocarbon resources that will be available for exploration in environmentally undisturbed areas during the medium and long term, and they warrant a degree of critique before we accept them.

Apart from the previously described seismic evidence of a suitable sedimentary section, with a variety of trap styles, there are two reasons to be optimistic about exploitable resources.

First, the initial exploration results have yielded early and high success ratios as drilling moves into deeper waters (Figure 20).

Secondly, the U.S. Geological Survey believes that oil and natural gas resources exist in the area designated by that body as the Maritime Boundary Region of the Central Gulf of Mexico (Figure 20), Foote et al (1983). This area encompasses that part of the Gulf of Mexico where jurisdiction over natural resources has not yet been established. Studies suggest that the sediments in this area have in-place resources of 2.24 to 21.99 (mean = 9.11) billion bbl of oil and 5.48 to 44.4 (mean = 18.77) tcf of gas. The natural corollary is, of course, that the area between the proven potential of the shelf and this untested Maritime Boundary Region, that is, the continental slope, has considerable speculative potential.

Figure 18. *Line 8 (MC 230).*

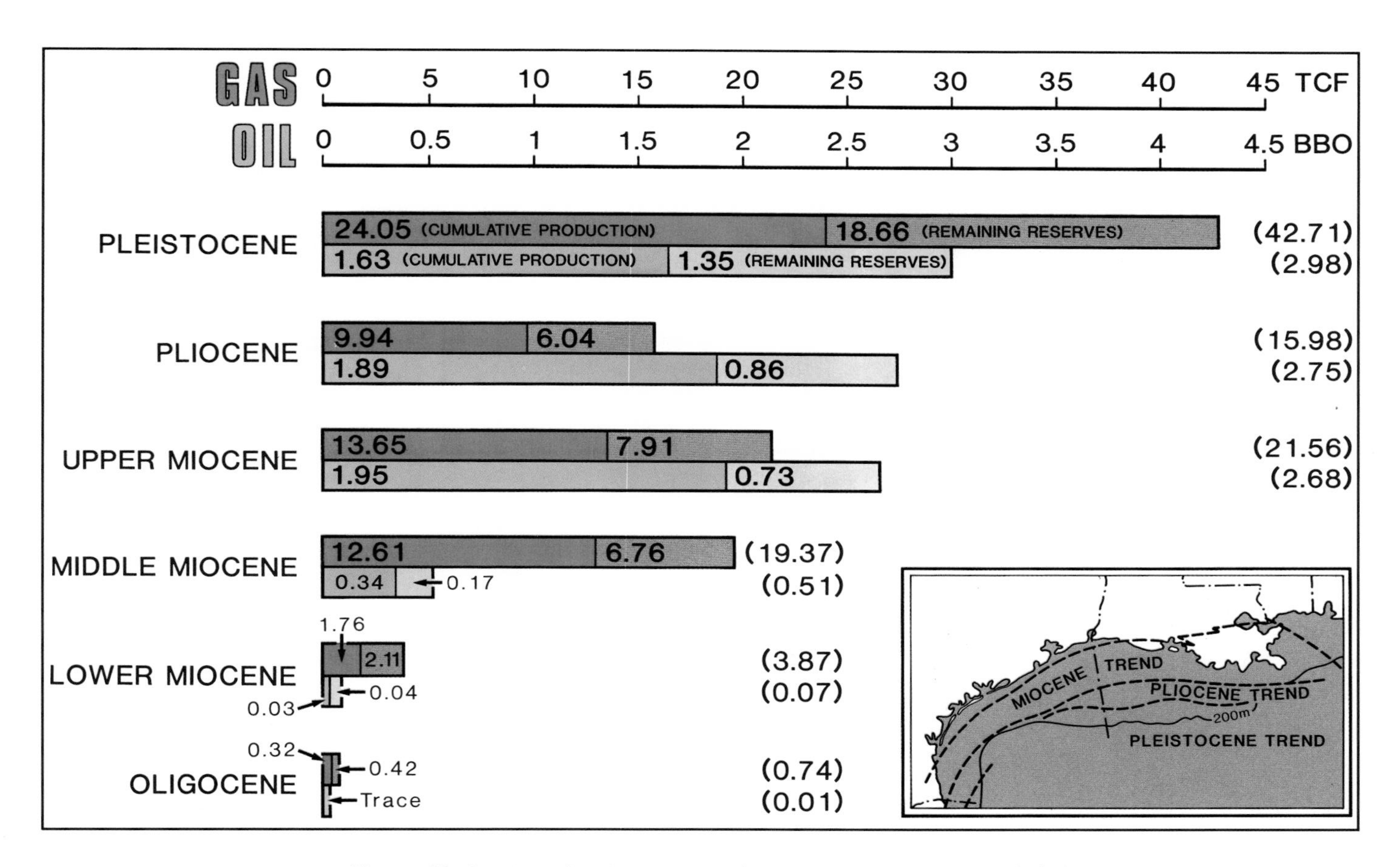


Figure 19. *Stratigraphic distribution of reserves on the continental shelf.*

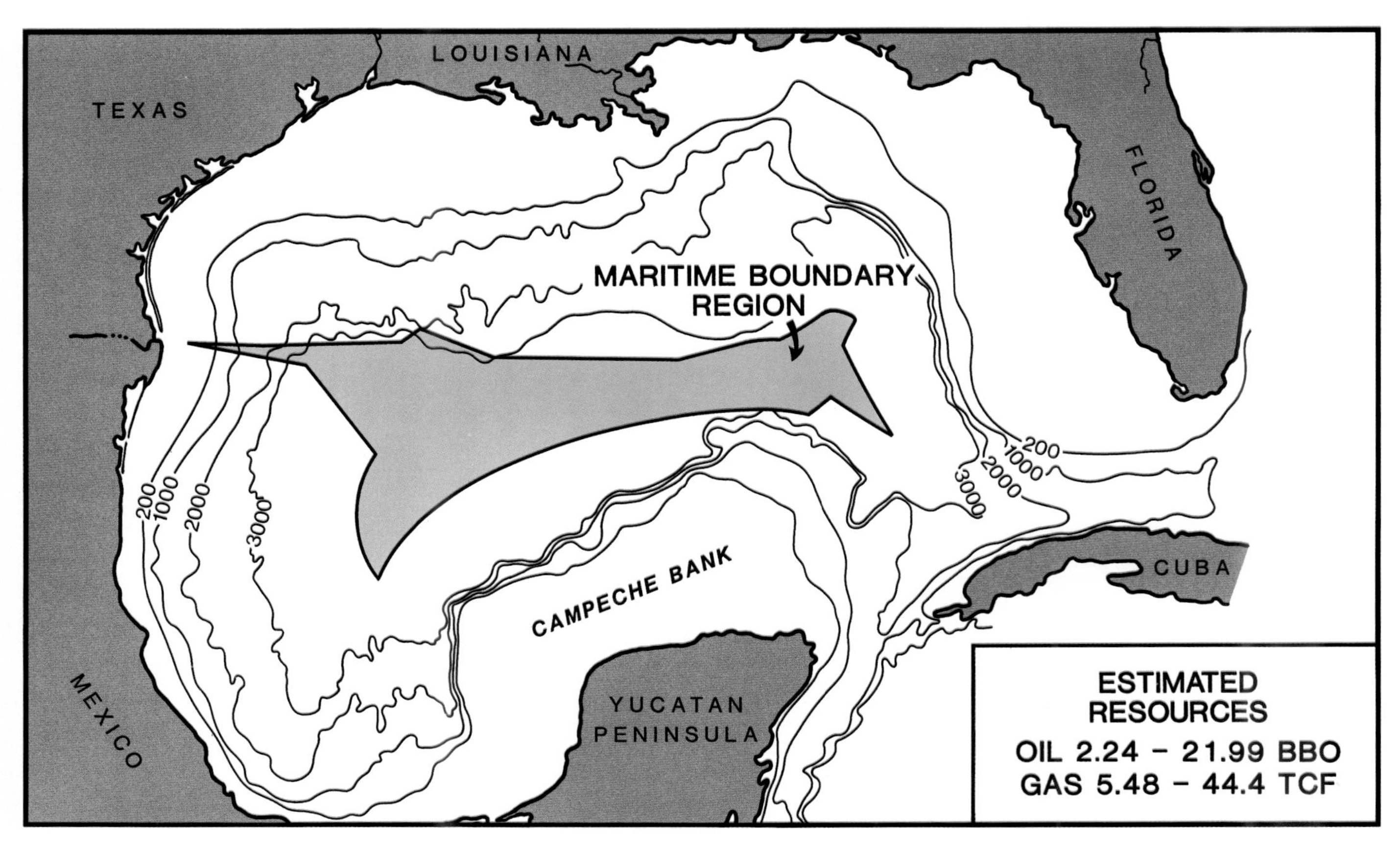


Figure 20. *Maritime boundary region: hydrocarbon potential.*

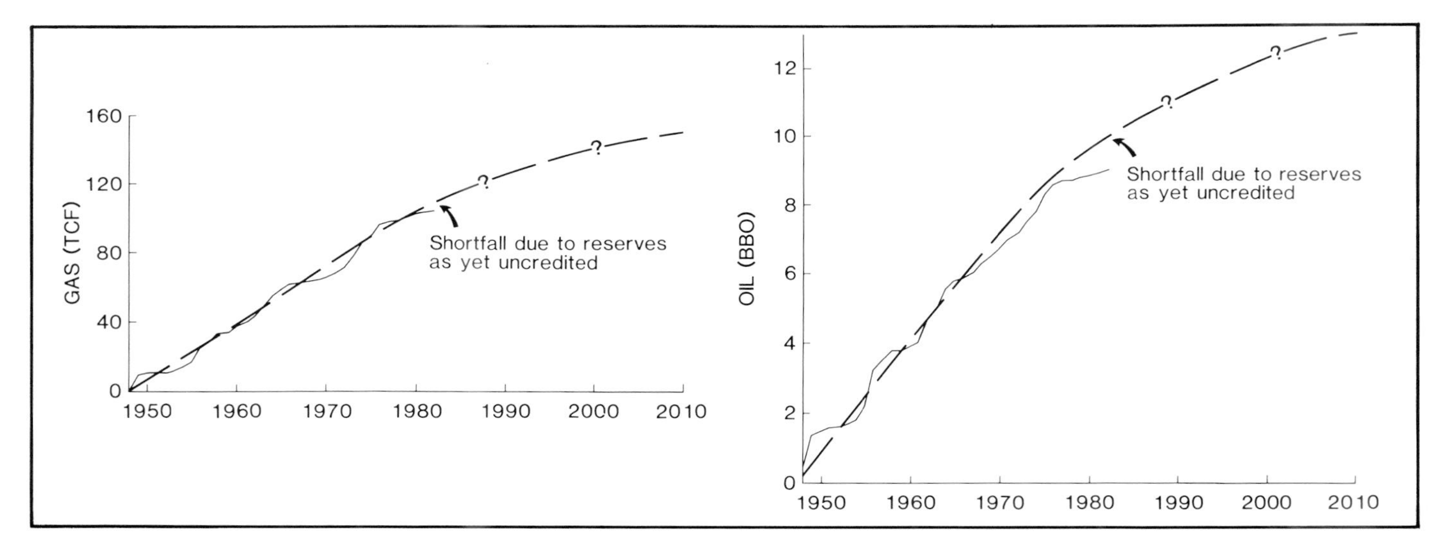


Figure 21. *Continental shelf reserves discovered on an annual basis. Oil reserves are shown in billions of bbl (BBO), gas reserves in trillion cu ft (tcf).*

## CONCLUSIONS

To quantify the potential of the upper continental slope is a hazardous exercise for even the most perspicacious explorationist. It would, therefore, be presumptious to offer a new set of resource estimates based, as it were, on little more than a hunch. However, the currently published U.S. Minerals Management Service figures of 2.6 billion bbl of oil and 26.5 tcf of gas may be very conservative, for the following reasons:

1. An historical analysis of reserves discovered by year for the shelf area (Figure 21), indicates a remarkably linear

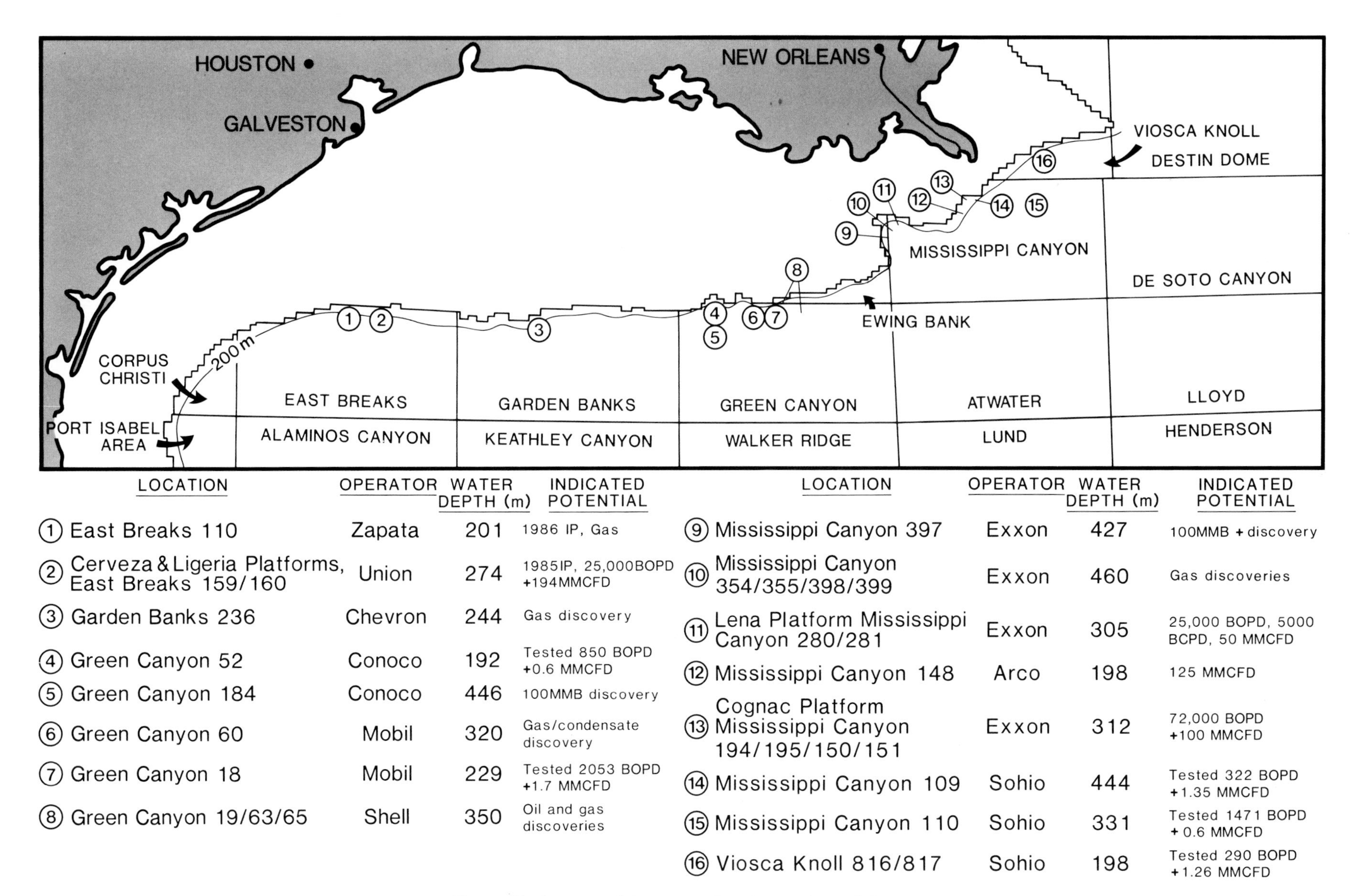


| | LOCATION | OPERATOR | WATER DEPTH (m) | INDICATED POTENTIAL |
|---|---|---|---|---|
| 1 | East Breaks 110 | Zapata | 201 | 1986 IP, Gas |
| 2 | Cerveza & Ligeria Platforms, East Breaks 159/160 | Union | 274 | 1985IP, 25,000BOPD +194MMCFD |
| 3 | Garden Banks 236 | Chevron | 244 | Gas discovery |
| 4 | Green Canyon 52 | Conoco | 192 | Tested 850 BOPD +0.6 MMCFD |
| 5 | Green Canyon 184 | Conoco | 446 | 100MMB discovery |
| 6 | Green Canyon 60 | Mobil | 320 | Gas/condensate discovery |
| 7 | Green Canyon 18 | Mobil | 229 | Tested 2053 BOPD +1.7 MMCFD |
| 8 | Green Canyon 19/63/65 | Shell | 350 | Oil and gas discoveries |
| 9 | Mississippi Canyon 397 | Exxon | 427 | 100MMB + discovery |
| 10 | Mississippi Canyon 354/355/398/399 | Exxon | 460 | Gas discoveries |
| 11 | Lena Platform Mississippi Canyon 280/281 | Exxon | 305 | 25,000 BOPD, 5000 BCPD, 50 MMCFD |
| 12 | Mississippi Canyon 148 | Arco | 198 | 125 MMCFD |
| 13 | Cognac Platform Mississippi Canyon 194/195/150/151 | Exxon | 312 | 72,000 BOPD +100 MMCFD |
| 14 | Mississippi Canyon 109 | Sohio | 444 | Tested 322 BOPD +1.35 MMCFD |
| 15 | Mississippi Canyon 110 | Sohio | 331 | Tested 1471 BOPD + 0.6 MMCFD |
| 16 | Viosca Knoll 816/817 | Sohio | 198 | Tested 290 BOPD +1.26 MMCFD |

Figure 22. *Recent activity reported on the continental slope.*

relationship for both oil and gas. Because the geological conditions on the upper continental slope are similar to those on the shelf, we can forecast a generally similar progression. If the ultimate reserves are indeed only 2.6 billion bbl of oil and 26.5 tcf of gas, then the initial, significant deep-water successes reported by Hagar (1984) exhibit a very large element of serendipity!

2. The volume of Pleistocene sediments involved in the upper continental slope play is broadly similar to those of the same age on the continental shelf that have already been estimated to have ultimate recoverable reserves of 2.98 billion bbl of oil and 42.71 tcf of gas.
3. As exploration moves into deeper waters (Figure 22), closer to the distal end of the Mississippi delta wedge, the older Tertiary and Mesozoic sediments become more accessible to the drill. These sequences, already proven to be prospective onshore, may provide a considerable resource potential. Evaluating these resources, however, will require further work to establish a more comprehensive understanding of the Gulf of Mexico's early architecture and structural evolution.
4. The continuing advances in drilling, completing, and development techniques will result in a significant increase in recovery factors for the deeper water fields of the future. Justification for this optimism exists because, historically, the U.S. Gulf of Mexico outer continental shelf has been the test-bed for many new offshore techniques, and this innovativeness is likely to continue in the forseeable future as it becomes the first intensively explored, deep water area of the world.

## ACKNOWLEDGMENTS

I wish to thank the managements of Geco Geophysical Company, Inc., Petty-Ray Geophysical and JEBCO for the release of considerable volumes of seismic data from which I selected sample lines. I also wish to acknowledge invaluable comments from J. E. Bobbitt, G. E. Landry, and G. B. Matthews, but take full responsibility for any remaining errors or omissions. Finally, Claygate Drawing Services Limited skillfully compiled the illustrative material.

## REFERENCES

Crow, P., 1984, Critical issues in OCS activity could be resolved this summer: Oil and Gas Journal, v. 82, n. 27, 2 July, p. 19–24.

Foote, R. Q., R. G. Martin, and R. B. Powers, 1983, Oil and gas potential of the Maritime Boundary Region in the Central Gulf of Mexico: AAPG Bulletin, v. 67, n. 7, p. 1047–1064.

Hagar, R., 1984, Deepwater drilling hits record pace in Gulf of Mexico: Oil and Gas Journal, v. 82, n. 41, 8 October, p. 25–28.

Hall, D. J., T. D. Cavanaugh, J. S. Watkins, and K. J. McMillen, 1983, The rotational origin of the Gulf of Mexico based on regional gravity data, *in*, J. S. Watkins and C. L. Drake, eds., Studies in continental margin geology: AAPG Memoir 34, p. 115–126.

Havran, K. J., 1981, Gulf of Mexico Summary Report 2, August 1981. A revision of Outer Continental Shelf Oil and Gas Activities in the Gulf of Mexico and their Onshore Impacts: A Summary Report, September 1980: U.S. Geological Survey Open-File Report 81-620, p. 1–79.

Hewitt, J. E., J. P. Brooke, and J. H. Knipmeyer, 1984, Estimated oil and gas reserve, Gulf of Mexico outer continental shelf and continental slope, December 31, 1983: Department of the Interior/Minerals Management Service, OCS Report MMS 84-0020, p. 1–21.

Weise, D. J., D. L. Slitor, and C. A. McCord, 1983. Gulf of Mexico summary report, September 1983. Outer continental shelf oil and gas activities in the Gulf of Mexico and their onshore impacts: Department of the Interior, Minerals Management Service, Outer Continental Shelf Oil and Gas Information Program; Contract No. 14-08-0001-19719, p. 1–106.

# Exploration History and Future Prospects of the U.S. Atlantic Margin

Joel S. Watkins[1]
Anthony M. Pytte[2]
Robert E. Houtz[3]
*Gulf Research and Development Co.*
*Houston, Texas*

The U.S. Atlantic continental margin contains four basins: The Georges Bank basin, the Baltimore Canyon Trough, the Carolina Trough, and the Blake Plateau basin. For the purposes of this paper, the Southeast Georgia Embayment and Bahamas Platform are included with the Blake Plateau basin, to which they are stratigraphically related. The shelves of all but the Carolina Trough have been tested to varying extents; the slopes have been tested only minimally. The tests have included 51 wells at a cost of over two billion dollars. No commercial discoveries have been reported.

Reservoir rock is ubiquitous and of good quality throughout the predominantly clastic Cretaceous sections of the Georges Bank basin, Baltimore Canyon Trough, Southeast Georgia Embayment, and, by inference, in the Carolina Trough. Source rock is of fair-to-good quality, although gas-prone. Source quality is poor in older rocks drilled on the shelf and encountered in Deep Sea Drilling Project (DSDP) holes in deep water.

The predominantly carbonate Blake Plateau rocks have not been drilled, but drilling of similar, coeval facies in the Bahamas Platform and Cuba has yielded only a little oil in Cuba. Source quality was apparently not reported in these wells drilled in the 1950s and 1960s.

In spite of the disappointing results, many plays remain untested. These are generally in deep water or at greater depths than previously drilled, and many are likely to produce only deep gas. Renewal of exploration interest in the Atlantic outer continental shelf (OCS) probably awaits significantly higher oil and gas prices, especially gas.

[1]Present address: Department of Oceanography, Texas A&M University, College Station, Texas
[2]Present address: Chevron Overseas Petroleum, Inc., San Ramon, California
[3]Present address: Chevron Oil Field Research Co., La Habra, California

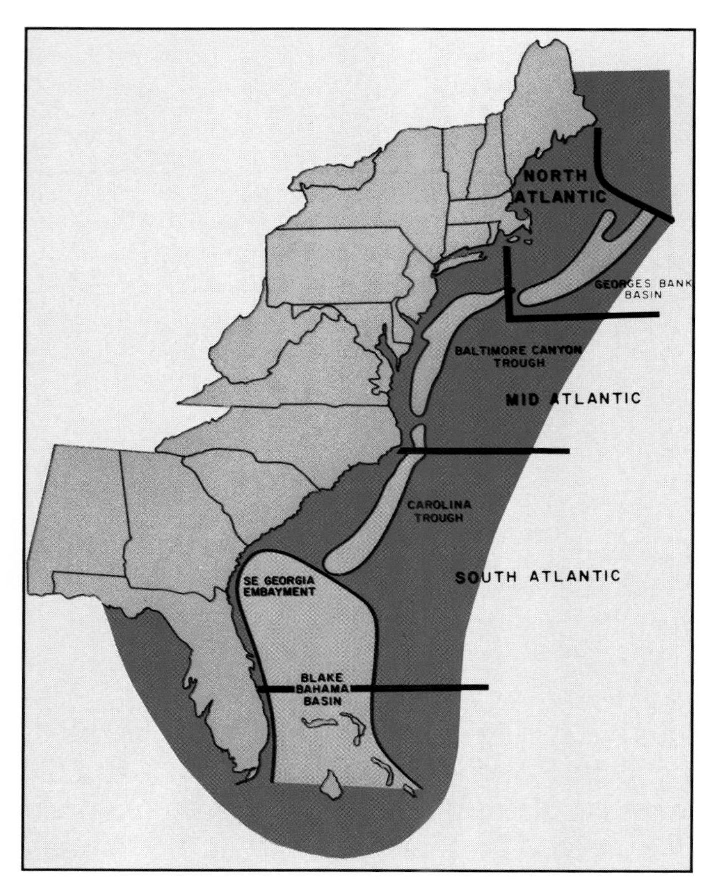


**Figure 1.** *The East Coast of the United States showing major basins and outer continental shelf (OCS) bid areas. The Blake Plateau, Southeast Georgia Embayment, and Bahamas Platform are shown as a single basin because of stratigraphic continuity.*

## INTRODUCTION

The U.S. Atlantic basins lie along the continental margin between the oil and gas provinces of eastern Canada on the north and the immense oil accumulations of southern Mexico to the southwest. In the 1960s and 1970s, many thought that hydrocarbons would be found in great quantities in this gap. Thus, the petroleum industry lavished time, effort, and money on the area, drilling 51 wells (including continental offshore stratigraphic test, referred to as COST, wells), spending two billion dollars on leases and drilling, and testing in three of the four basins. The drillers were frustrated in attempts to find large hydrocarbon accumulations. Chastened, industry shifted its attention elsewhere and is now largely ignoring the Atlantic margin. Current and planned drilling is at a low level. Shell Oil Company recently completed several tests of the Cretaceous-Jurassic carbonate bank edge and other structures in deep water without finding commercial accumulations. Plans for the 1985 drilling season were modest.

Many plays, however, remain untested. Large structures are known to exist in deep water beneath the slopes; the Blake Plateau is entirely untested and only skimpily surveyed, and the few tests of the Bahama Platform have been poorly situated.

The most important question about the hydrocarbon potential of the U.S. Atlantic margin may be the availability of mature source rock. Middle Cretaceous to latest Jurassic source rocks in the Scotian Shelf fields have a maturity threshold averaging about 3 km (2 mi) depth. In U.S. margin basins, these rocks lie above the maturity threshold, which averages about 4 km (2.5 mi) in depth, and are consequently immature. Hibernia source rocks were deposited mainly during the rifting of Greenland and Canada, which occurred later than rifting in the U.S. outer continental shelf. Consequently, open ocean conditions prevailed in the U.S. Atlantic outer continental shelf, and no comparable source rocks appear to have been deposited on the U.S. Atlantic outer continental shelf. It is possible that source rocks analogous to those in Hibernia were deposited during the rift stage of the U.S. Atlantic margin, but they would be very deep and overmature.

The untested plays remaining tend to be concentrated either at greater depths beneath the shelf and slope, in deeper water of the continental slope and rise, or in the carbonates of the Carolina Trough, Blake Plateau, and Bahamas Platform. For the most part, exploration of these areas awaits improved and cheaper deep-water technology or higher profit margins for oil and gas or both. One or both of these conditions should prevail before the end of the century, and exploration will again proceed on the Atlantic outer continental shelf.

In this paper, we will briefly review the geology of the major U.S. Atlantic basin areas: the Georges Banks, the Baltimore Canyon Trough, the Carolina Trough, and the Southeast Georgia–Blake Plateau–Bahamas. Each is different to some extent in both geologic history and in lithology, and these differences have influenced the extent to which they have been explored. We will then review the exploration history and the status of current leasing and drilling. Finally, we will summarize and speculate about future trends.

## GEOLOGY

The east coast of the United States contains four major basins: the Georges Bank basin, Baltimore Canyon Trough, Carolina Trough, and the Blake Plateau basin (Figure 1). Structurally, a transform fault separates the deeper Blake Plateau basin from the adjacent Bahamas Platform; but stratigraphically, the basins are largely coeval. Hence, they and the Southeast Georgia Embayment, an arm of the Blake Plateau basin largely filled with nearshore clastic facies, are treated as a single basin. The Scotian basin, while outside U.S. borders stratigraphically, resembles the Georges Bank basin and Baltimore Canyon Trough to the extent that the stratigraphic sequence developed for the Scotian Shelf has been extended southward into the other two basins (Wade, 1977; Poag, 1982; Libby-French, 1984). Hence, we will frequently refer to it. (The Hibernia basin developed in a later stage of rifting that ultimately led to eastern Canada separating from Greenland. Hence, its value as an analog is of less interest.)

The geology of the Baltimore Canyon Trough shelf is fairly well-known, but data from the slope are sparse. The same is true of the Georges Bank geology. The geologies of the Carolina Trough and the Blake Plateau are poorly known, being based almost exclusively on seismic data. Information available from drilling in the southeast Georgia Embayment and from wells drilled on the Bahamas Platform is extrapolated into the Blake Plateau basin.

The four basins straddling the continental shelf slope boundary are filled with 10–20 km (6.2–12.4 mi) of sedimentary rock. A basement hinge zone separating thick, little-rifted crust on the landward side from thinned, more intensely faulted and rifted crust on the seaward side marks the tectonic landward edge of the basins. Thick sediments of the basins extend seaward over the rifted crust and onto oceanic crust, with the East Coast Magnetic Anomaly marking the crustal boundary. A mainly carbonate paleoshelf edge is prominent within the sedimentary section of the Baltimore Canyon Trough and Georges Bank basin. A sea-level drop, followed by a sharp sea-level rise and attendant shelf retrogradation, dramatically terminated the prograding carbonate bank seen in seismic dip sections in the area. The paleoshelf margin is buried in the Georges Bank basin, Baltimore Canyon Trough, and parts of the Carolina Trough, but is exposed subaqueously along the outer edge of the Blake Plateau and locally in submarine canyons off New England. It also underpins parts of the present-day Bahamas Shelf edge. Making up much of the paleoshelf is a carbonate bank of Early Cretaceous age and older rocks in Georges Bank and the Scotian Shelf, of middle Cretaceous and older rocks in the Baltimore Canyon Trough, and of earliest Tertiary and older rocks beneath the edge of the Blake Plateau (Jamsa, 1981). All exploration drilling to date has been on this paleoshelf or landward of it.

Rifting of the U.S. Atlantic margin appears to have begun in Triassic time and continued into the Jurassic. The onset of drifting is not precisely established, but Leg 76 of the Deep Sea Drilling Project (DSDP) drilled oceanic basement of Callovian (Late Jurassic) age or about 155 m.y. (Sheridan et al, 1982). Magnetic anomalies indicate that the sea-floor opening proceeded from south to north (Sheridan et al, 1982). Evaporites, indicative of sea-water invasion during the rift or early drift phases or both, are found in Upper Triassic rocks of the Grand Banks, Upper Triassic–Lower Jurassic rocks of the Scotian basin wells, Lower Jurassic and possibly older rocks of the Georges Bank (Jansa, 1981), and in Upper Jurassic rocks of the Bahamas (Tator and Hatfield, 1975). The age of the Bahamas evaporite relative to the age of oceanic crust immediately north of the Blake Plateau suggests that these evaporites may have been deposited in a lagoon environment after the initiation of drifting rather than during a rifting. Linear chains of salt diapirs are located just seaward of Georges Bank and the Carolina Trough paleoshelf edges (Figure 2). These were probably deposited during late-rift or early-drift stages, but their ages have not been established.

A carbonate bank extended discontinuously over a distance of 6,000 km (3,730 mi) from the Grand Banks to the Bahamas and thence possibly to the Yucatan during the Jurassic and part of the Early Cretaceous (Jansa, 1981). As the North American continent drifted northward, carbonate deposits progressively gave way to clastic deposits. In the Scotian basin the carbonate deposits ceased in the Early Cretaceous (Jansa and Wiedmann, 1982; Wade, 1977). In the Georges Banks COST wells, carbonate deposition ceased in mid-Cretaceous (Jansa, 1981). In the Blake Plateau basin, the bank edge was drowned by earliest Tertiary. Only off the Bahamas does it remain active today.

In the deeper parts of the Blake Plateau, current deposition is dominantly calcareous silt and clay, while shallow-water carbonate deposition continues on the Bahama Platform.

**Georges Bank Basin**

The Georges Bank basin (Figure 3) is structurally more complex than other basins along the eastern U.S. margin. It consists of several subbasins (Schlee and Frisch, 1982) and contains numerous basement normal faults that delineate grabens and a few horsts (Grow, Bowin, and Hutchinson, 1979, Figure 4). These faults generally displace only the pre-Mesozoic basement. The structural complexity of the Georges Bank basin probably results from a large, strike-slip or oblique slip component of transform separation of this part of the North American plate from West Africa. Normal rather than strike-slip faulting characterized the other basin openings. Sediments of presumed Triassic age fill the Georges Bank basin's grabens (Poag, 1982), and salt structures are observed in several isolated basins (Figure 4). An unconformity bevelled the relatively shallow, pre-Triassic basement rocks. This unconformity, the "break-up unconformity" (Grow, 1981), also truncated the sediments within the grabens and is thought to lie near the Triassic-Jurassic boundary. Underlying the synrift, sequences vary in thickness from 0 to more than 10 km (0–6.2 mi) (Klitgord and Behrendt, 1979), and fill in the grabens by onlap against relatively steep basement slopes. The depth to the top of the Triassic increases seaward from the hinge line to depths in the range of 6–7 km (3.7–4.3 mi) (Schlee and Fritsch, 1982). Triassic rocks may also occur in the deeper parts of the nearshore Atlantic basin on oceanic crust where a few diapirs are observed (Uchupi, Ballard, and Ellis, 1977).

Jurassic and younger sediments in the Georges Bank basin area also are variable, both in lithology and in thickness, because of complex basement structure. Although basement rocks do not protrude significantly above the break-up unconformity, basement highs caused differential compaction in younger sedimentation. A mainly dolomitic and evaporitic sequence was deposited early in the Jurassic in a shallow basin located in the central part of the present basin (Poag, 1982). By the Middle Jurassic, terrigenous sediments were deposited closer to the continent, and a calcareous sequence developed in an embayment along the southwestern part of the present basin. A major reef complex developed at the Georges Bank basin shelf edge during the Late Jurassic, as it did along most of the U.S. margin (Figure 2).

Based on seismic stratigraphy, Schlee and Fritsch (1982) divide Georges Bank basin sediments into six units that range from synrift Triassic (?) red beds to relatively thin, Tertiary marine and glacial marine sediments. Poag (1982) constructs a type section for Georges Bank basin using Canadian formation names, and confirms earlier seismic correlations by Wade (1977). These and other investigators infer the following history for the Georges Bank basin.

Following deposits of Mohican clastics, the shelf edge began prograding in mid-Jurassic times. Similar sequences in the Scotian basin are described by Eliuk (1978) as carbonates and clastics, but this relation has not been established in the Georges Bank area. Schlee and Fritsch (1982) infer that the prograding sequence terminates abruptly just northeast of Oceanographer Canyon. This termination coincides with the shoreward projection of the New England Seamount Chain. They also show that an Upper Jurassic carbonate platform lies just seaward of the shelf break. Submarine canyons cut the Jurassic shelf edge and expose it north of the New England Seamounts where Ryan et al (1978) found reef-like facies in dredge hauls. Postrift sedi-

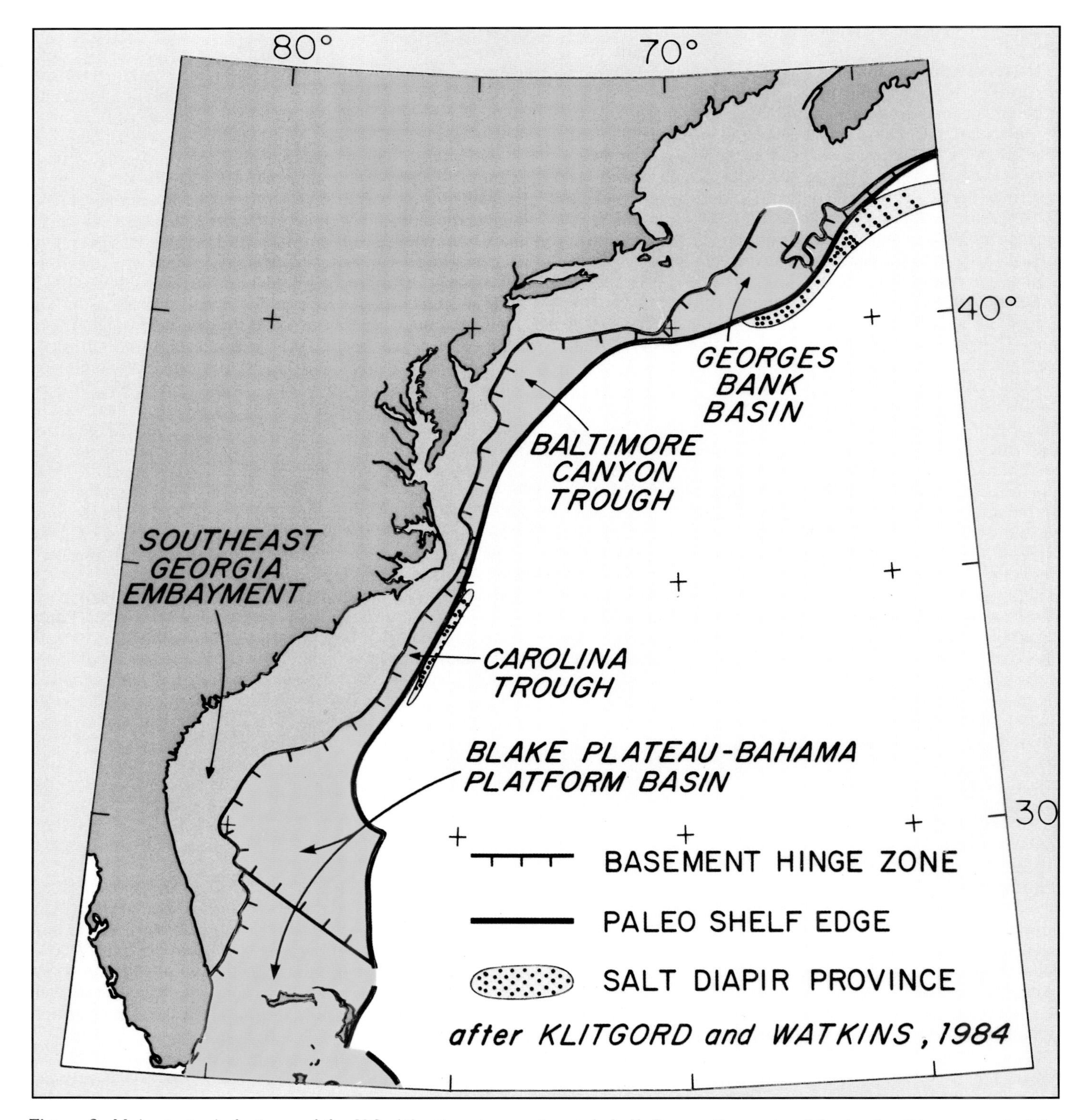


Figure 2. *Major tectonic features of the U.S. Atlantic outer continental shelf. Prospective parts of the basins thin seaward of the paleoshelf edge.*

ments in COST wells are, however, largely terrigenous (Scholle and Wenkham, 1982).

Scholle and Wenkham (1982) report Argo salt in the COST, G-2 well and tentatively identify several salt swells in their sections. Uchupi, Ballard, and Ellis (1977) show salt structures extending from the Scotian basin west of Northeast Channel (see Figure 2).

## Baltimore Canyon Trough

The Baltimore Canyon Trough is approximately 300 km (186

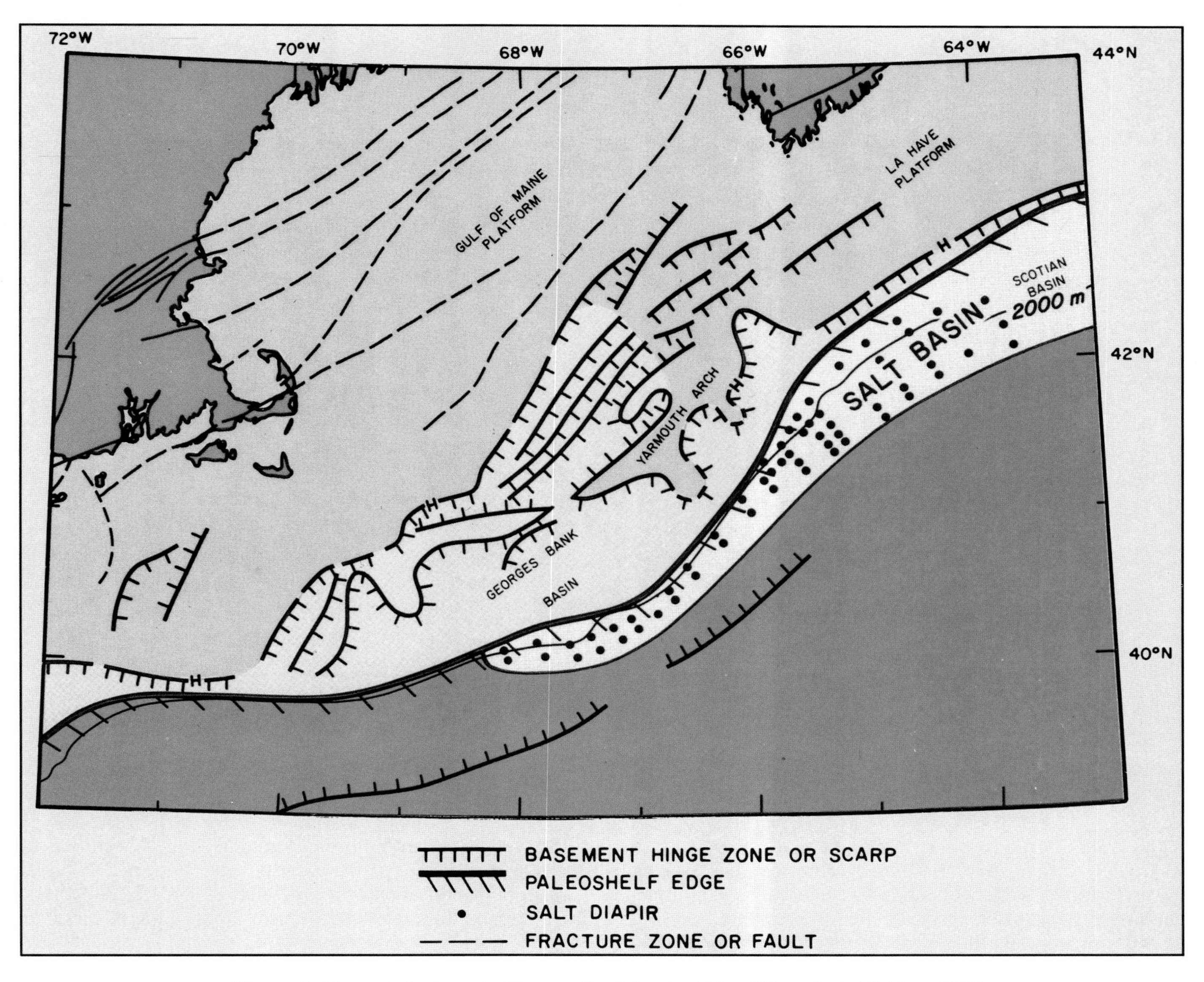


**Figure 3.** *Tectonic features in Georges Bank basin. (After Klitgord and Schlee, 1982).*

mi) long, lies offshore New Jersey, Delaware, and Maryland and has a maximum width of about 150 km (93 mi) (Figure 1). Sediment thickness is the greatest near the outer shelf, where it ranges perhaps to 20 km (12.4 mi) (Scholle, 1977). At the base of the upper slope, at a water depth of about 2 km (1.2 mi), sediment thickness exceeds 9 km (5.6 mi) (Figure 5).

An interpretive section of the Baltimore Canyon Trough in Figure 7 highlights the major structural elements. Rifting of the continent commenced in Late Triassic or Early Jurassic times and led to the separation of Africa from North America. Most subsequent sediment accumulation is believed to be composed of Mesozoic age sediments. Tertiary sediments do not exceed about 1.5 km (1 mi) in thickness (Mattick, Schlee, and Bayer, 1981). In the deeper part of the basin, the oldest sediments reached by the drill are of Late Jurassic age (Libby-French, 1984). Major unconformities existing at the top of the Jurassic between the Hauterivian and the Barremian, and in the Oligocene, are readily mapped in seismic reflection sections. Rocks of Paleocene age are thin or absent in the holes drilled to date.

The crustal structure consists of the following features (from west to east):

1. A continental crust platform under the coastal plain that thins toward the zone of rifting. This platform was not affected significantly by faulting, and sediments are generally thin.
2. A basement hinge zone that separates the shallow platform from the subsided rift zone. Sediments thicken markedly on the seaward side of the zone, thus representing the structural boundary of the basin.
3. The rift zone is stretched and downfaulted continental crust. Close to the hinge zone, the rifting is clearly developed in continental crust. Seaward from the hinge zone, a thick cover of normally faulted sediments masks the basement. This zone probably represents the change from continental crust to transitional crust.
4. Oceanic crust of the ocean basin floor.

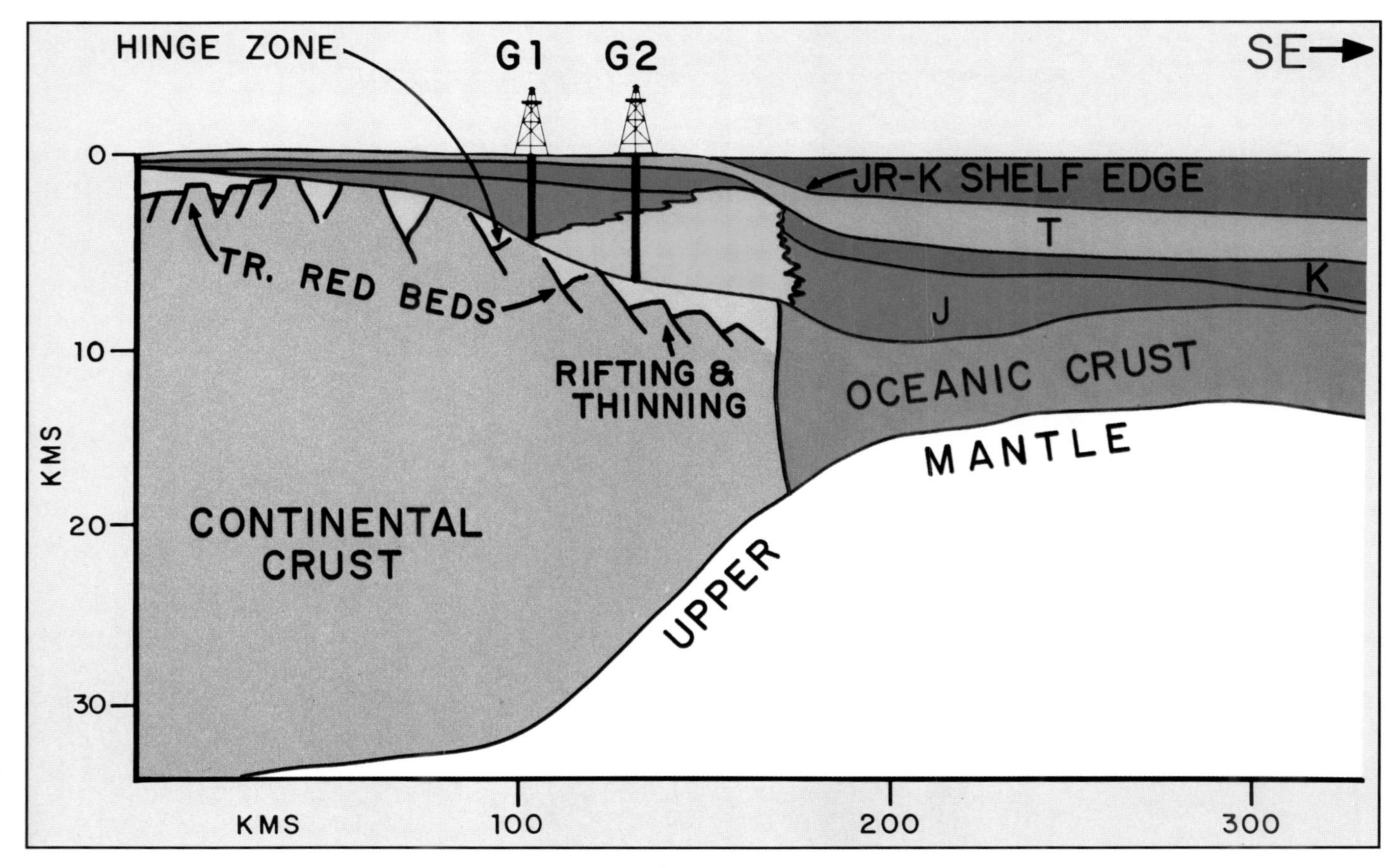


Figure 4. *Cross section across Georges Bank basin showing COST wells G-1 and G-2 superimposed on generalized structure. (After Grow, Bowin, and Hutchinson, 1979.)*

The Baltimore Canyon Trough contains a well-developed, prograding, carbonate paleoshelf-edge that extends seaward from the vicinity of the boundary between oceanic and thinned continental crust. A similar sequence of elements occurs in other basins, but is perhaps most clearly expressed in the Baltimore Canyon Trough.

The presumed crustal boundary is roughly parallel to the shoreline and is buried by more than 12 km (7.5 mi) of sediments. It is marked by the East Coast Magnetic Anomaly, which may result from the intrusions of volcanic rocks along the principal rift (the East Coast Boundary fault) associated with the opening of the Atlantic (Alsop and Talwani, in press). The East Coast Magnetic Anomaly can be used to trace the position of the principal rift along the U.S. East Coast margin, northward to the New England Seamount Chain. North of the Seamount Chain, the anomaly extends seaward in deeper water, and its association with the principal rift is not clear.

The principal rift probably had a strong influence on the location of the initial shelf edge of the young Atlantic Ocean in Early Jurassic or Middle Jurassic time. The earliest sedimentation, the Mohican clastics, flooded the rapidly subsiding margin just after continental separation. In the Baltimore Canyon Trough, the shelf edge subsequently prograded seaward for a distance of more than 50 km (31 mi), while the basin subsided more than 4 km (2.5 mi) during the remainder of the Jurassic (Grow, 1981; Figure 6). The paleoshelf edge appears to have had two stationary positions for long intervals of time (Gamboa, Truchan, and Stoffa, 1985); one in the present middle shelf of mid-Jurassic age, and the other below the present, upper slope of Late Jurassic age. At both locations, the shelf edge built directly up rather than out, and then retreated after the Upper Jurassic reef stillstand.

The prograding shelf edge appears to consist of carbonate rocks, as shown by the results from the Texaco 598-2 and Exxon 599 wells released by the National Geophysical Data Center. Expanding spread profiles (ESP's) reported by Gamboa, Truchan, and Stoffa (1985) show anomalously high velocities at or slightly above the shelf edge. A velocity inversion below the shelf edge suggests that clastic rocks may underlie the carbonates. This condition would seem to be a normal development in a prograding shelf environment where platform detritus and possibly terrigenous sediments were deposited as slope facies. Upper Jurassic sediments behind the shelf edge contain substantial amounts of sandstone and shale with some lignite. There are indications of changes seaward as well as downward to limestone and dolomite (Libby-French, 1984), while an unconformity occurs near the top of the Jurassic.

The Lower Cretaceous consists of interbedded sandstone and shale with shale increasing seaward. A major unconformity

Figure 5. *Sediment thickness in the Baltimore Canyon Trough (after Scholle et al, 1977).*

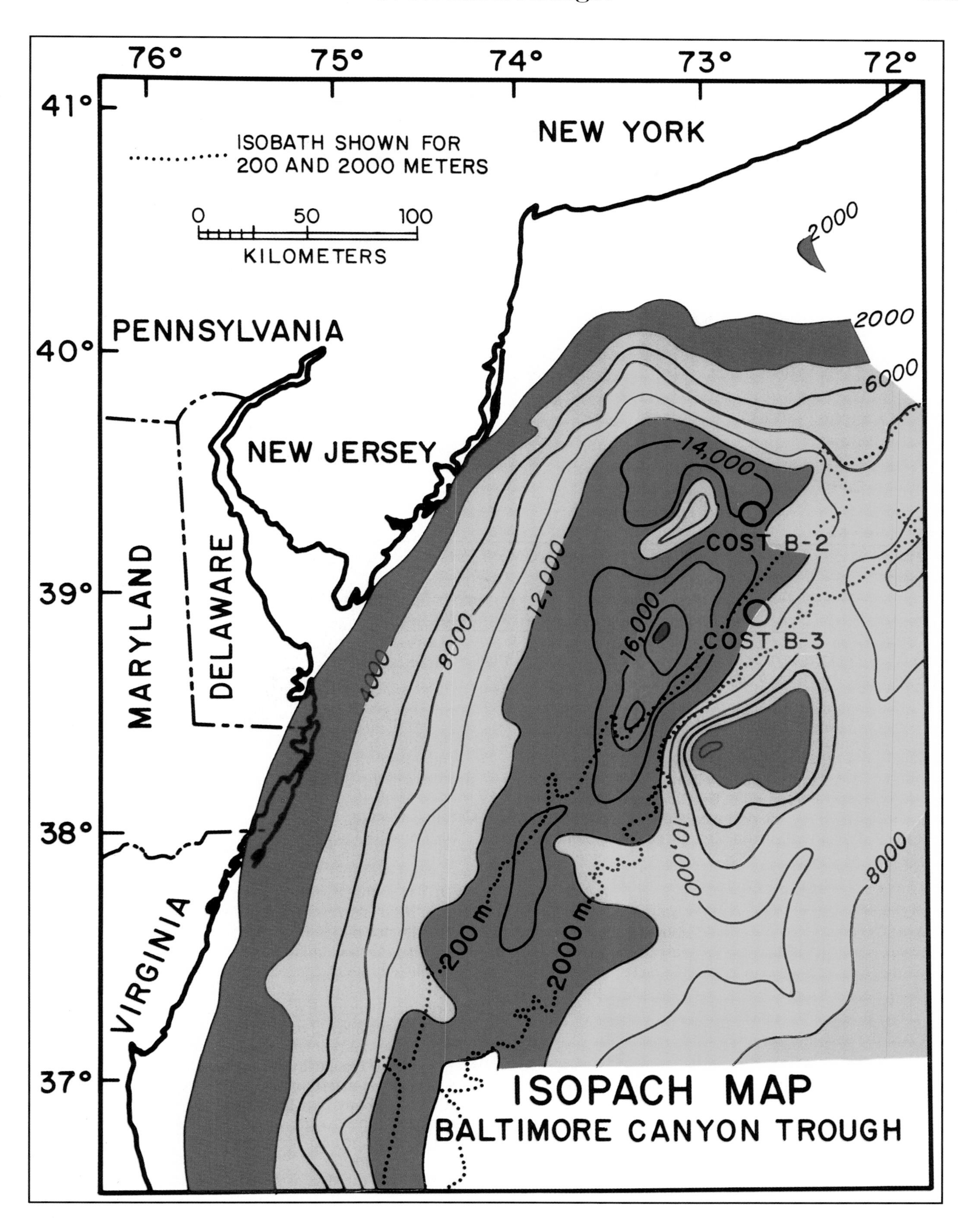
76°
75°
74°
73°
72°
41°
40°
39°
38°
37°
NEW YORK
ISOBATH SHOWN FOR
200 AND 2000 METERS
0
50
100
KILOMETERS
PENNSYLVANIA
NEW JERSEY
MARYLAND
DELAWARE
VIRGINIA
2000
2000
6000
14,000
12,000
8000
4000
16,000
COST B-2
COST B-3
10,000
8000
200 m
2000 m
ISOPACH MAP
BALTIMORE CANYON TROUGH

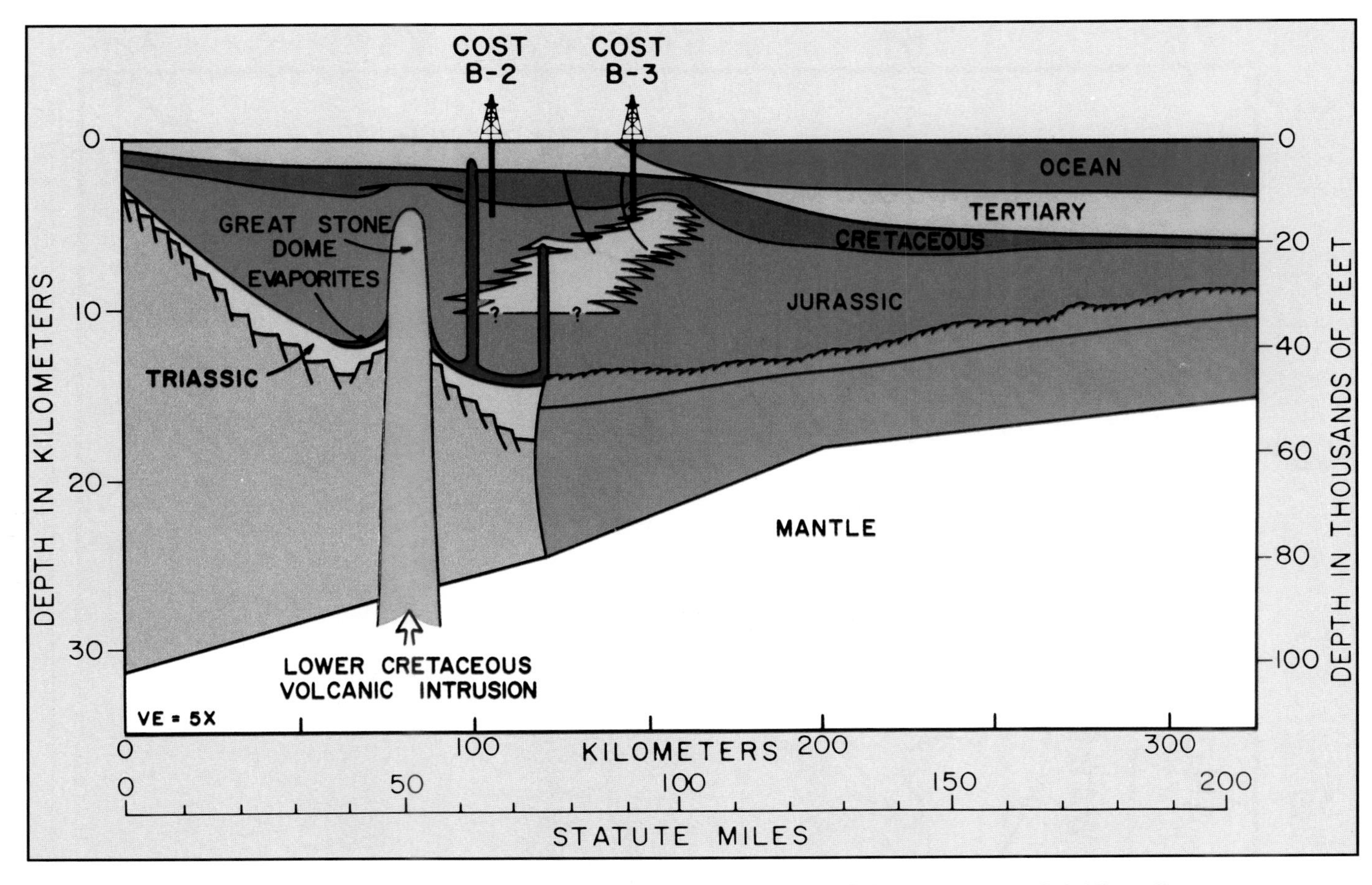


**Figure 6.** *Generalized cross section of the Baltimore Canyon Trough showing COST wells B-1 and B-2. The yellow area represents the prograding Jurassic shelf edge (after Grow, 1981).*

marks the boundary between the Hauterivian and Barremian.

The lower Tertiary and Upper Cretaceous comprise a transgressional sequence—chiefly shale, calcareous claystone, and limestone deposited under outer shelf and slope environments. Major unconformities occur in the Oligocene and near the Tertiary–Cretaceous boundaries; the Paleocene is often thin or absent. The Baltimore Canyon Trough generally contains a thin, Quaternary section; a relatively thick, Neogene (chiefly Miocene) section; and a thin, Paleogene sedimentary section. Neogene sediments consist of deltaic sand of a shelf progradational sequence in the mid-shelf area and shale in the outer shelf and upper slope environments (Libby-French, 1984).

Salt diapirs, growth faults, and a large volcanic plug occur in the Baltimore Canyon Trough (Figure 6). Salt structures localized in the northeast part of the basin are small and isolated. Grow (1980) identifies deeper and less pronounced salt swells in the same area, but they were not mapped elsewhere.

Growth faults can be seen in USGS Line 25, about 15 miles (24 km) south of the zone where salt structures developed (Figures 7 and 8), but the largest growth faults occur in association with salt domes. The largest fault extends down into mid-Jurassic carbonates.

Lippert (1984) states that the large volcanic plug, "The Great Stone Dome" (Figure 6), formed a subaerial high that was about a mile above sea level before being eroded over a period of 10 m.y. The basic intrusion was implanted during the Aptian and was covered unconformably by Albian sandstones. The overlying section is now 2.3 km (7.6 mi) thick.

Aside from these localized features, there are relatively few structures in the Baltimore Canyon Trough.

## Carolina Trough

The Carolina Trough is a narrow, linear basin (60–80 km or 37–50 mi wide, and 250 km or 155 mi long), located between the Baltimore Canyon Trough and the Blake Plateau basin (Figure 9). The continental margin off the Carolinas is fairly broad to the south but narrows toward Cape Hatteras. The trough is mainly situated under the present slope and upper rise. It has a thick, Mesozoic section (in places more than 10 km or 6.2 mi) and a much thinner Cenozoic section at the shelf and slope. Relative to the other basins along the margin, the Cenozoic section is very thick beneath the continental rise, where its thickness locally exceeds 3 km (2 mi).

The northern edge of the trough is separated from the Baltimore Canyon Trough by an arch off Cape Hatteras. To the south, it is separated from the Blake Plateau basin by the Blake Spur Fracture Zone. The basement hinge line marks the landward edge of the trough, and a salt ridge or zone of diapirs marks the seaward edge (Figures 1 and 9). The Carolina Trough salt zone coincides for the most part with the East Coast Magnetic Anomaly (ECMA), as does the diapir zone of the Georges Bank basin (Figure 2). Ridges or arches at either end of the Caro-

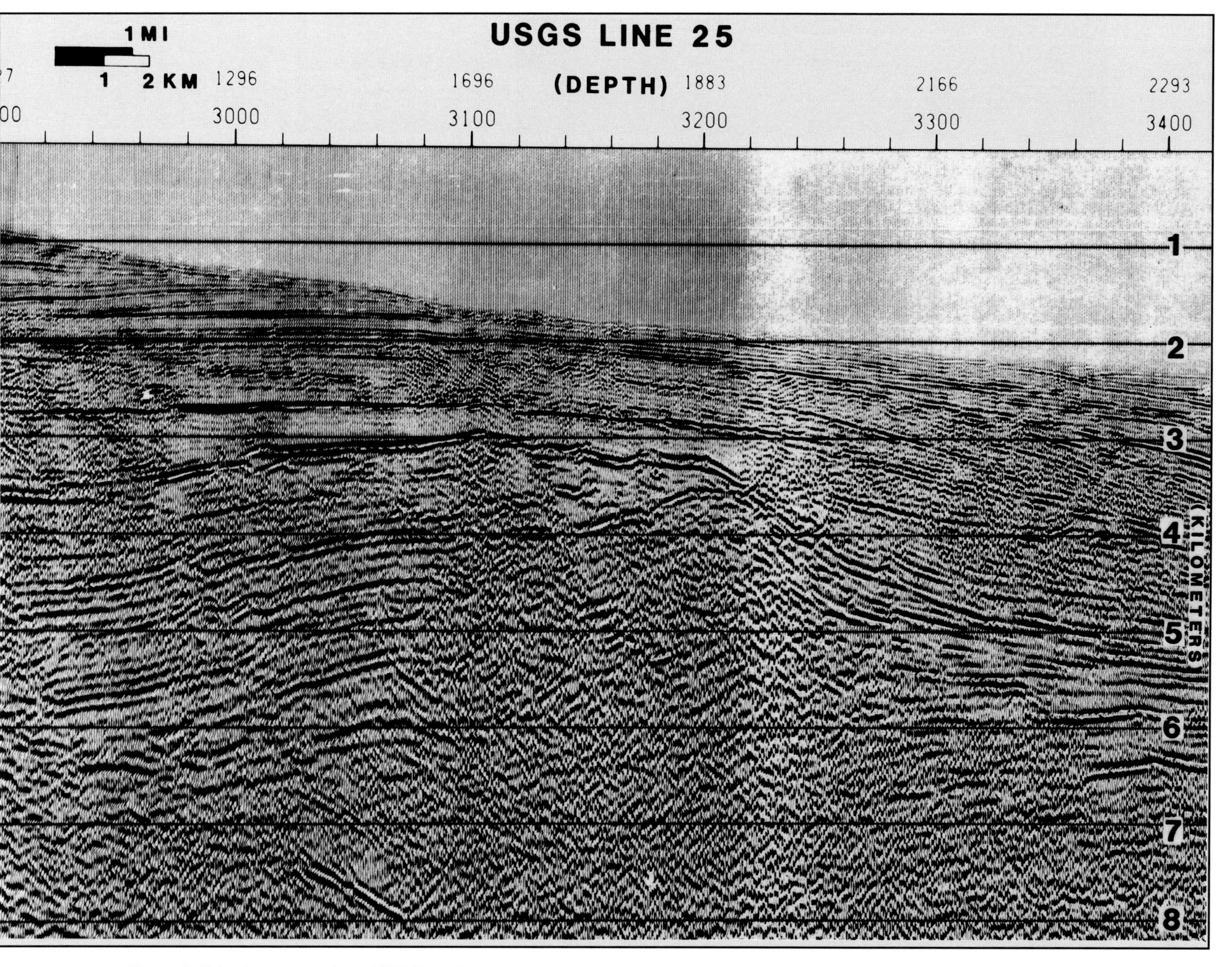


Figure 7. *Seismic depth section of USGS line 25, showing growth faults and prograding shelf edge in the Baltimore Canyon Trough.*

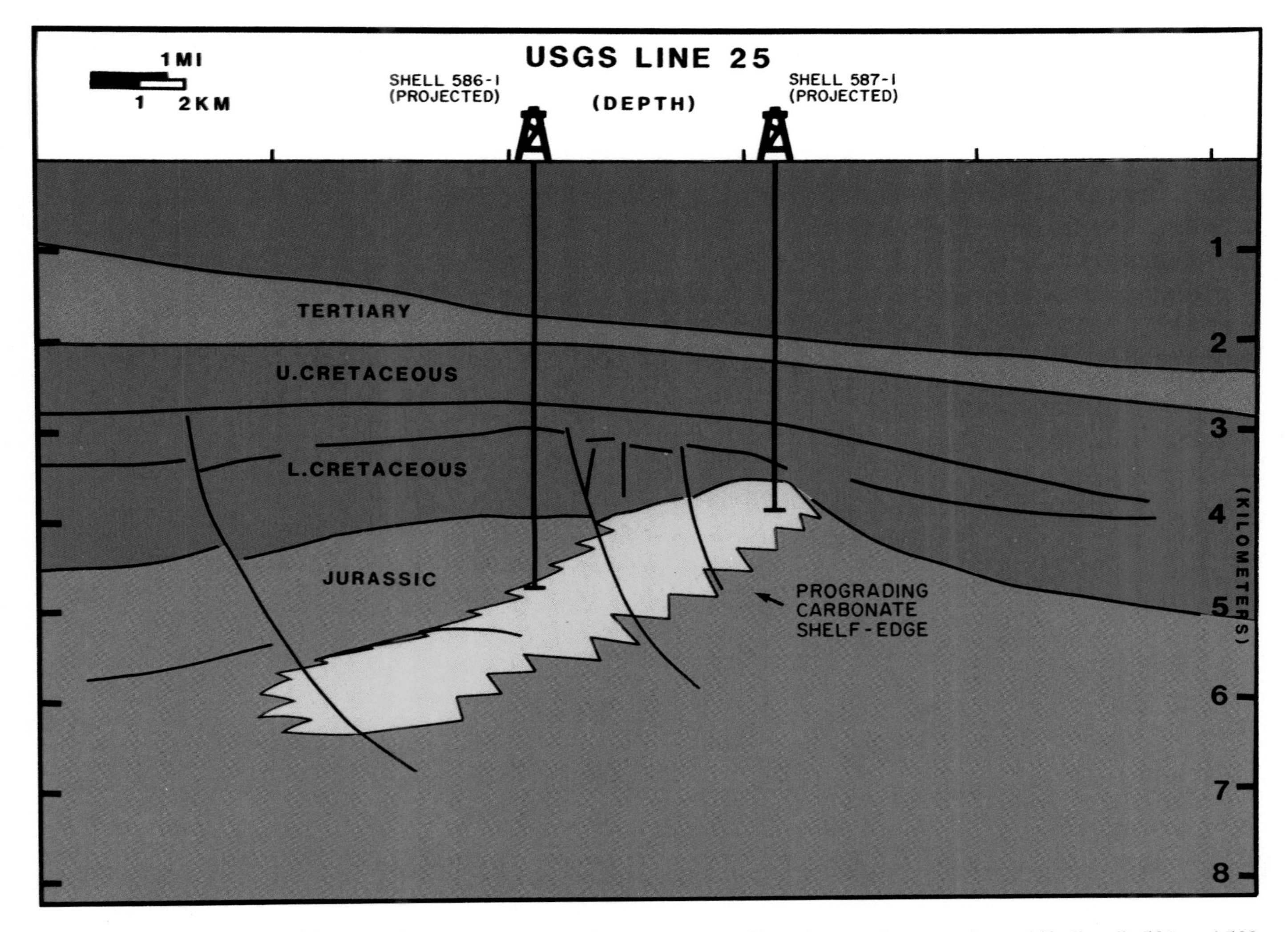


**Figure 8.** *Interpretation of USGS line 25 shown in Figure 7 showing projected locations and penetrations of Shell wells 581 and 582.*

lina Trough restricted circulation during early rifting and probably caused the salt accumulation. These boundaries effectively sealed off the Carolina Trough from the adjoining basins from the time of its formation to Middle and Late Jurassic times. As a consequence, about half the shelf sediments were deposited at a time when they were separated from the adjoining basins. This situation could have resulted in significant source rock deposits.

Although older rocks of the Carolina Trough have not been penetrated, inferences as to their character can be drawn from coeval rocks on the African margin. Plate reconstructions (K. D. Klitgord, 1984, personal communication) suggest that the Carolina Trough and the northern Senegal basin were adjacent formerly. The earliest Triassic–Liassic rift sediments on both sides of the Atlantic are coeval, red-bed, evaporitic sequences, capped by relatively massive salt (VanHouten, 1977). Seismic sections from the Senegal basin (Wissman, 1982) show that its basic features are similar to those of the Carolina Trough, with a hinge line, a rift basin filled with Jurassic sediments that onlap continental rocks, and salt diapirs near the boundary between transitional rift and oceanic crust. A diapiric lineament between Cape Blanc and the Senegal River (Seibold, 1982) matches the lineated diapir field of the Carolina Trough, as shown by Dillon et al (1984).

Between Cape Fear and Cape Hatteras, Brown, Miller, and Swain (1972) show 11 wells that penetrate to basement. The Hatteras Light # 1 is the northernmost of these coastal wells, and was used by the U.S. Geological Survey to identify, tentatively, reflectors in the Carolina Trough. These 11 wells indicate a thickening of the basin between Cape Fear and Cape Hatteras and define the Cape Fear Arch. Fine- to medium-grained basal sands on the Cape Fear Arch are Early Cretaceous in age. Basement onshore at Cape Fear is about 2.5 km (1.5 mi) higher than at Hatteras, whereas there is no expression of the arch in the Carolina Trough. There may be some relict arch structure in the Carolina Platform, but it has not been described in the literature. Offshore and south of the Cape Fear Arch, a large, deltaic deposit, probably Late Cretaceous in age, is identified on USGS seismic line BT1 and may be derived from the degradation of the arch.

Dillon et al (1982) and Hutchinson et al (1982) examine the seismic data from the Carolina Trough. They find that the basement is shallow near shore below the present shelf; the basement defines a steep hinge zone under the outer shelf and

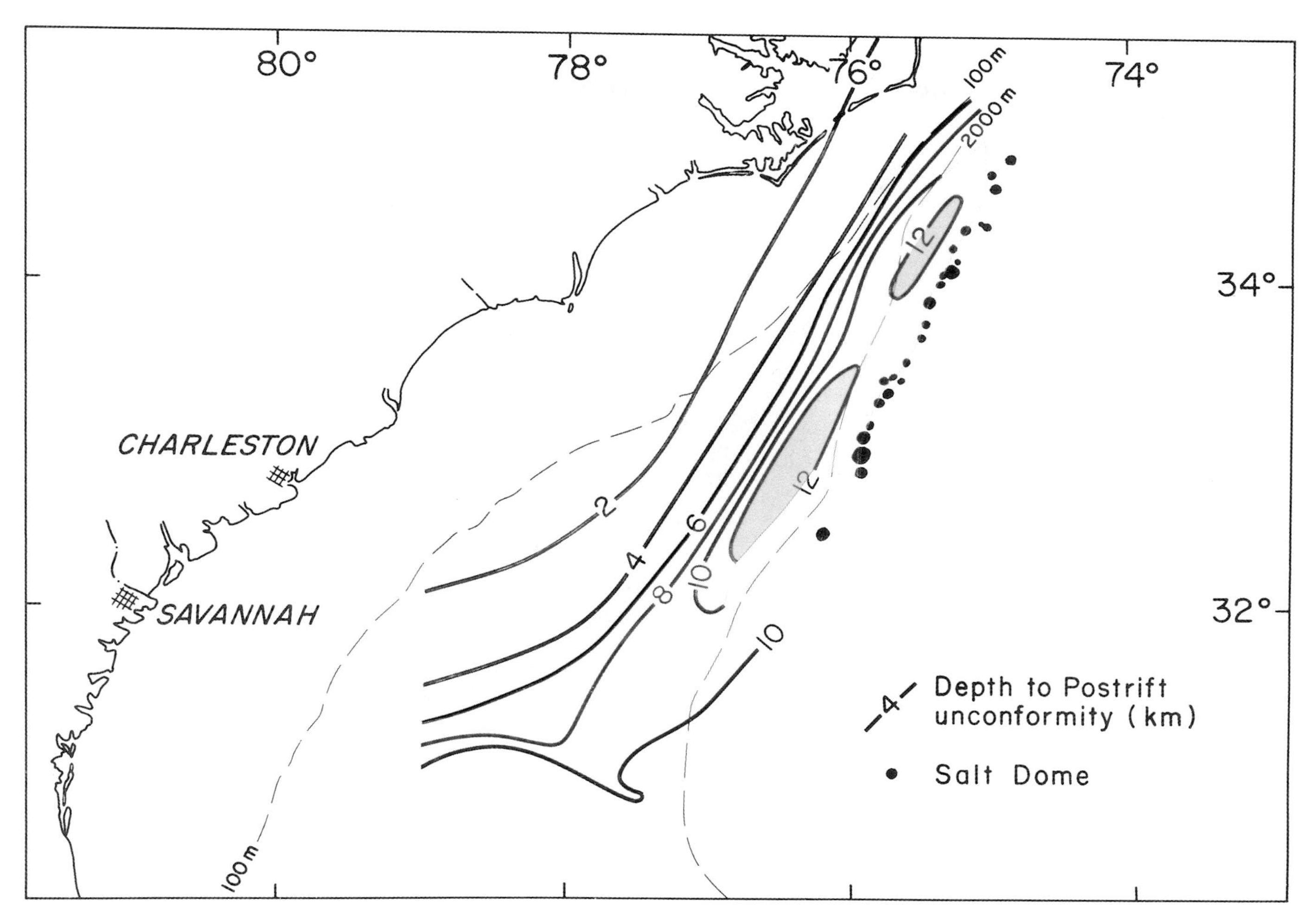


Figure 9. *Structure contours on break-up unconformity in the Carolina Trough (after Dillon et al, 1982).*

slope, and is deeper than the recorded section below the slope and upper rise. In the deeper part of the basin, a conspicuous, high-amplitude reflector is considered by Hutchinson et al (1982) to be the top of an evaporitic sequence. This reflector onlaps the basement.

In the central part of the basin, the Upper Jurassic section seismically resembles the carbonate section in the Baltimore Canyon Trough; however, in much of the Carolina Trough, this sequence was deeply eroded during the Oligocene. Dillon et al (1982) point out that they can observe no carbonate build-up in their Carolina Trough sections.

### Blake Plateau–Bahamas Platform Basin

The Blake Plateau basin proper encompasses the Blake Plateau, parts of the eastern margin of Georgia and northeastern Florida, and the northern portion of the Bahama Platform (Jansa, 1981). In this paper we also include the remainder of the Bahamas Platform. Thus, for present purposes, the Blake Plateau–Bahamas Platform fronts the Atlantic Ocean to the east and abuts Cuba to the south. The basin ranges from lat. 32°N to below lat. 20°N off Hispaniola, and from roughly long. 71°W at Navidad Bank to about long. 81°W in southern Florida. It is the largest of the basins underlying extensive Mesozoic carbonate banks found from the Yucatan around the Gulf of Mexico and into the Canadian Atlantic margin (Figures 1 and 2).

Data are not extensive but appear adequate to define the overall structure and setting of the area. Three deep wells, the Great Isaac-1 (total depth-5441 m) on Great Bahama Bank, Andros-1 (total depth-4447 m) on Andros Island, and Long Island-1 (total depth-5353 m), were drilled between Florida and Cuba. Other deep wells have been drilled off Cuba and both on and offshore Florida. These wells are described by Meyerhoff and Hatten (1974) and Tator and Hatfield (1975) (Figure 13).

Six, shallow-penetration ($\geq$ 300 m or 984 ft) holes were drilled in 1965 by the C/V Caldrill, while in 1975, the D/V Glomar Challenger drilled two holes on the Blake Nose. Finally, in 1976, the D/V Glomar Conception drilled holes on the Florida Shelf and margin. In addition, the Glomar Challenger has drilled a number of holes in deep water east of the Blake Plateau and Bahamas Platform. All these wells have been discussed in a number of reports. (See, for example, Hollister et al, 1972; Sheridan et al, 1982; and Sheridan and Enos, 1979.)

Although not a good record area, the Blake–Bahama basin has recently yielded interpretable multifold seismic reflection data. Shipley, Buffler, and Watkins (1978) report results of a line extending northeastward from the Florida coast at long. 30°N across the Blake Plateau; Dillon et al (1978) report on the Southeast Georgia Embayment and Northern Blake Plateau; Sher-

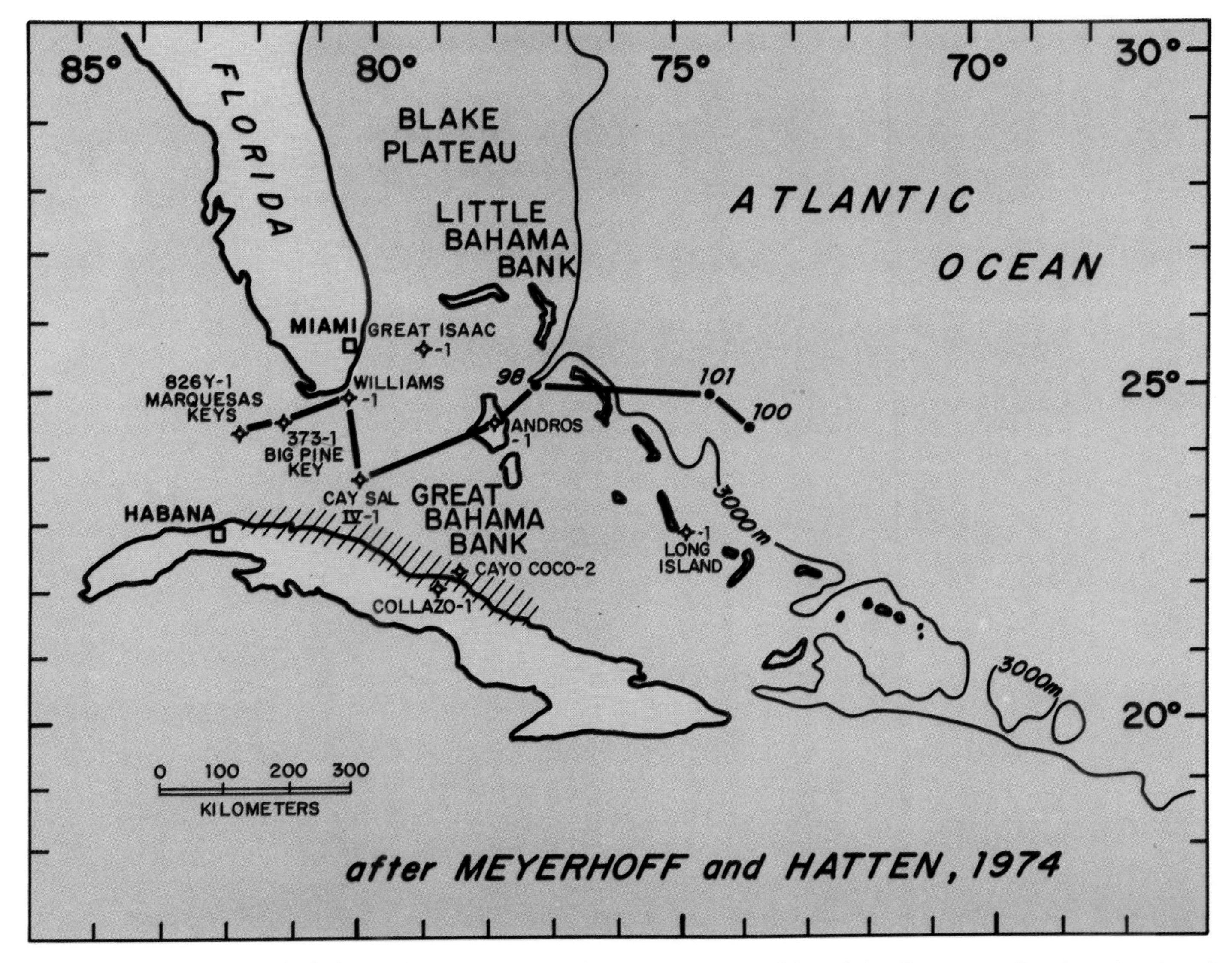


**Figure 10.** *Selected wells in the Bahamas Platform, offshore Florida and on- and offshore Cuba. East-to-west line shows location of cross section in Figure 13 (after Meyerhoff and Hatten, 1974).*

idan et al (1978) report on the Blake Escarpment; and Sheridan et al (1981) report on the southern Blake Plateau and Northern Bahamas Platform. We do not know much of the Bahamas Platform, where shallow water inhibits acquiring seismic data.

Blake Plateau basement depths are greater than those beneath the Bahamas Platform, ranging down to more than 12 km (7.5 mi), whereas Bahamas basement depths generally range from 5–10 km (3.1–6.2 mi) with an average of 7–8 km (4.4–5.0 mi). The only well to penetrate the basement offshore in either area is the Great Isaac well on Great Bahama Bank (Figure 10), which encountered basement at a depth of 5.3 km (3.3 mi). The Great Isaac well spudded on a magnetic anomaly (Meyerhoff and Hatten, 1974); hence, basement depth and lithology may be anomalous. Magnetic data suggest that the well lies on an extension of the Florida Platform that spreads southeastward down the spine of the Bahamas Platform. Meyerhoff and Hatten (1974) infer a trough with basement depths of 8–9 km (5.0–5.6 mi) separating the spinal high from a subparallel basement high extending along the seaward edge of the Bahamas Platform. To the south, the basement dips toward the extinct Cuban subduction zone, reaching depths of 10–12 km (6.2–7.5 mi). Basement depths beneath the Little Bahama Bank appear to be intermediate between those of the Blake Plateau and the Great Bahama Bank.

Offshore South Carolina and Georgia, seismic reflection data (Dillon et al, 1978) show an acoustic basement dipping gently southeastward toward the hinge line, slightly landward of the Brunswick Magnetic Anomaly. A right-lateral offset in the hinge line at its intersection with the landward projection of the Blake Spur Fracture Zone suggests transform motion. Southeast of the hinge line, the basement lies deeper than 12 km (7.5 mi), then rises to 8–9 km (5.0–5.6 mi) beneath the outer edge of the Plateau (Figure 11), according to Shipley, Buffler, and Watkins (1978), and Dillon et al (1978). Sheridan et al (1981), on the other hand, interpret the Dillon et al (1978) basement to be a westward-dipping, Triassic–Jurassic volcaniclastic sequence

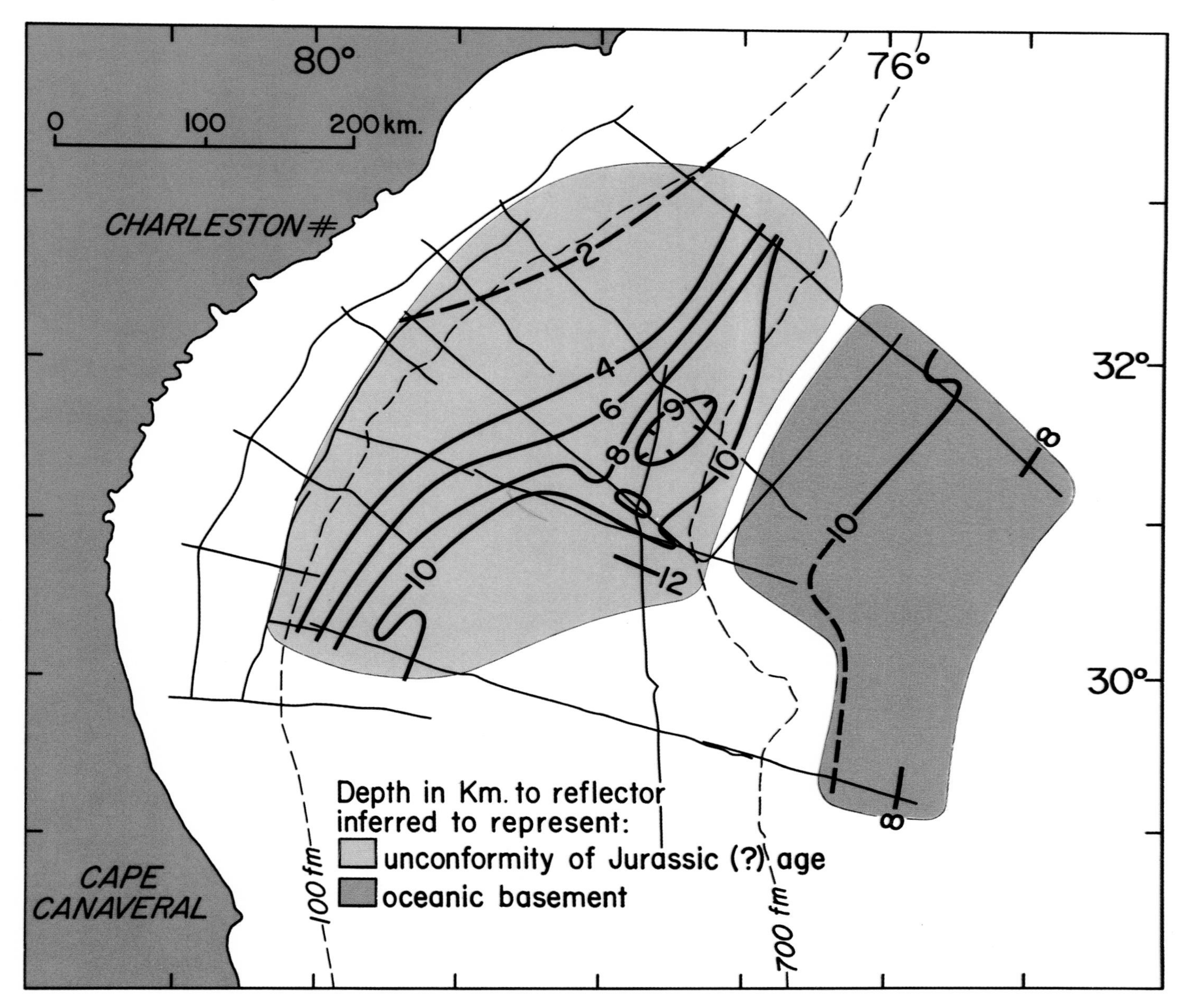


**Figure 11.** *Structure contours on deep Jurassic unconformity and inferred oceanic basement in the Southeast Georgia Embayment and northern Blake Plateau (after Dillon et al, 1978).*

8–9 km (5.0–5.6 mi) deep, which overlies metamorphic basement at 11–14 km (6.8–8.7 mi). Sheridan et al (1981) believe that the inferred volcaniclastic sequence originally formed on the African margin, and was detached by a spreading center jump in the Middle Jurassic. Figure 12 shows a cross section from Florida to the Atlantic basin across the Blake Plateau.

Basement lithologies are highly inferential except for the volcaniclastics encountered in the Great Isaac well. Magnetic anomalies beneath the Blake Plateau, Little Bahama Bank, and Great Bahama Bank (Klitgord and Behrendt, 1978) differ mainly in amplitude and in gradient, with deeper basements having lower magnitudes and gentler gradients, and shallower basements having higher amplitudes and steeper gradients. Except along structural boundaries, all exhibit poorly organized magnetic anomalies without evident linear trends. Assuming isostatic compensation at 40 km (25 mi) and standard crustal columns, we calculate the thickness of metamorphic and igneous crust beneath the Blake Plateau to be in the range of 8–12 km (5.0–7.5 mi). It could represent either thickened oceanic crust, rifted and severely attenuated continental crust, or transitional crust. The calculated crustal thickness of the Bahamas Platform of approximately 20 km (12.4 mi) indicates that it is rifted and thinned continental crust. Figure 15 shows a generalized cross-section through the Blake Plateau.

Drilling on the Blake Plateau indicates that Lower Cretaceous rocks were deposited in shallow water. Water depth at the edge of the plateau (DSDP site 392) increased from marginal marine to bathyal during the Aptian (Benson et al,1978). Sheridan and

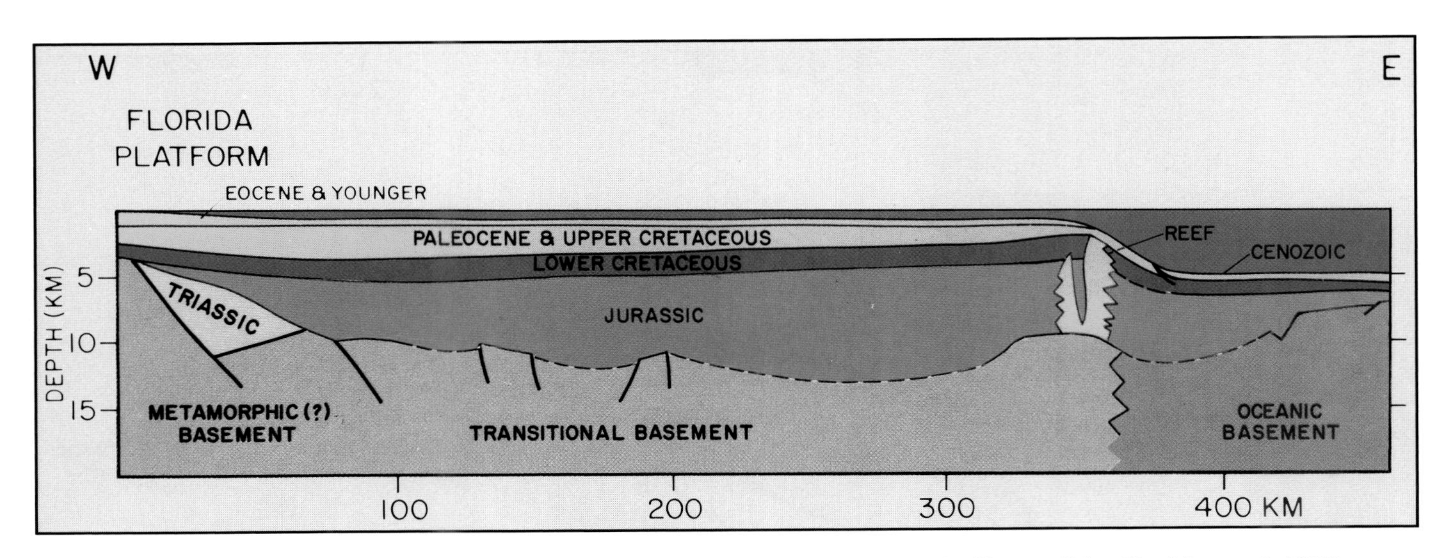


Figure 12. *Cross section from Florida to Atlantic abyssal plain across the Blake Plateau (After Sheridan et al, 1978).*

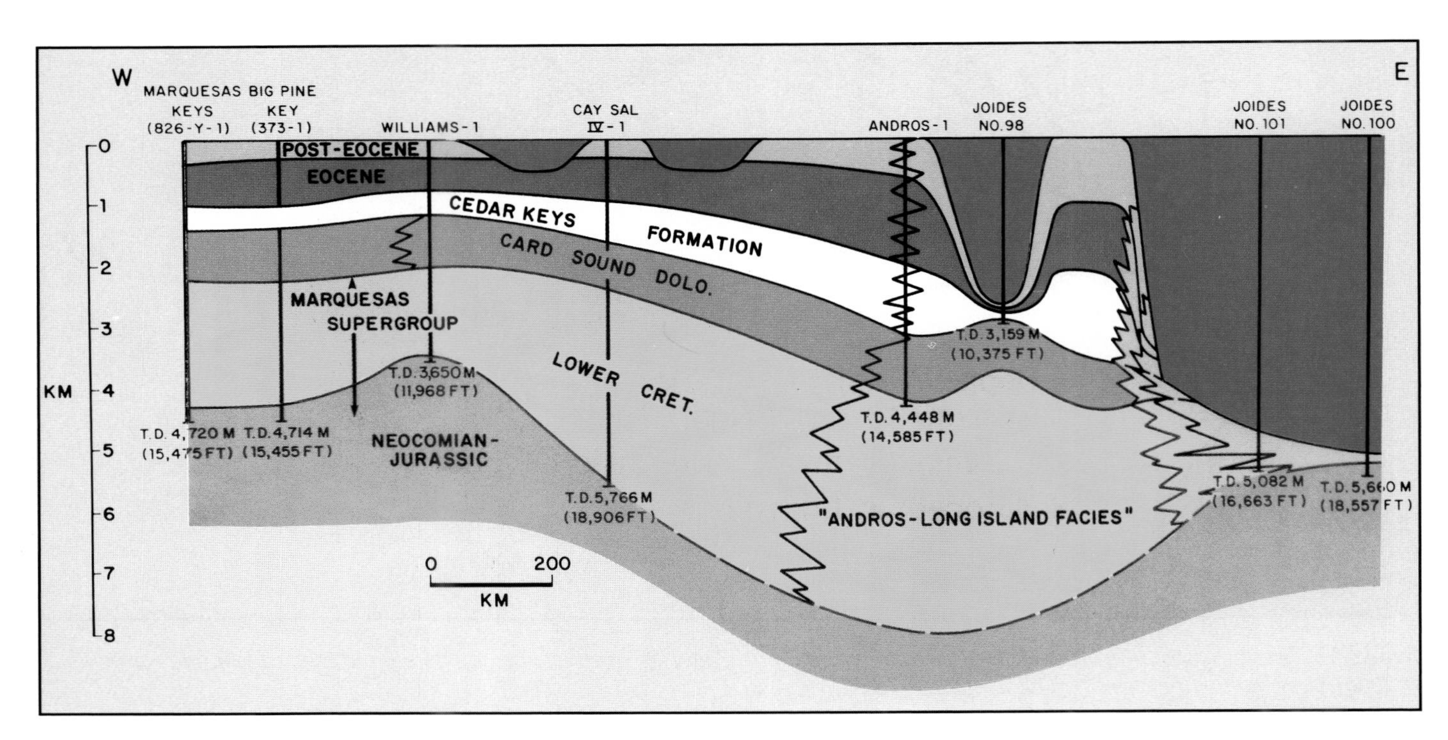


Figure 13. *Cross section from Florida Keys, across Bahamas Platform to Atlantic abyssal plain, showing major stratigraphic units (after Meyerhoff and Hatten, 1974).*

Enos (1979) interpret their seismic data as indicating that only the outer edge foundered at that time, with a reef, immediately landward of sites 390 and 392, remaining near sea level through Santonian–Coniacian. Drilling on the central part of the plateau indicates a bathyal environment through the interval penetrated: the Holocene–Paleocene. The present-day Blake Plateau shelf edge was established at least by Paleocene and possibly by Maastrichtian–Campanian time (Sheridan and Enos, 1979).

As previously mentioned, one offshore well, the Great Isaac (total depth-5441 m), penetrated the basement. Immediately above basement rhyolitic volcaniclastics are found arkosic sandstones and conglomerates with interbedded red shales (Tator and Hatfield, 1975), probably of Jurassic or Triassic age. The top of the Jurassic is poorly marked, but is roughly 500–600 m (1,640–1,970 ft) above the arkose. Most of the Jurassic sediments belong to the Marquesas Supergroup (Meyerhoff and Hatten, 1974), a thick, shallow-water lagoonal, carbonate-evaporite sequence of Late Jurassic and Early Cretaceous age found in wells in eastern Florida, northern Cuba, and on nearby islands (Figure 13). Descriptions by Tator and Hatfield (1975) suggest that the Marquesas Supergroup is approximately 3 km

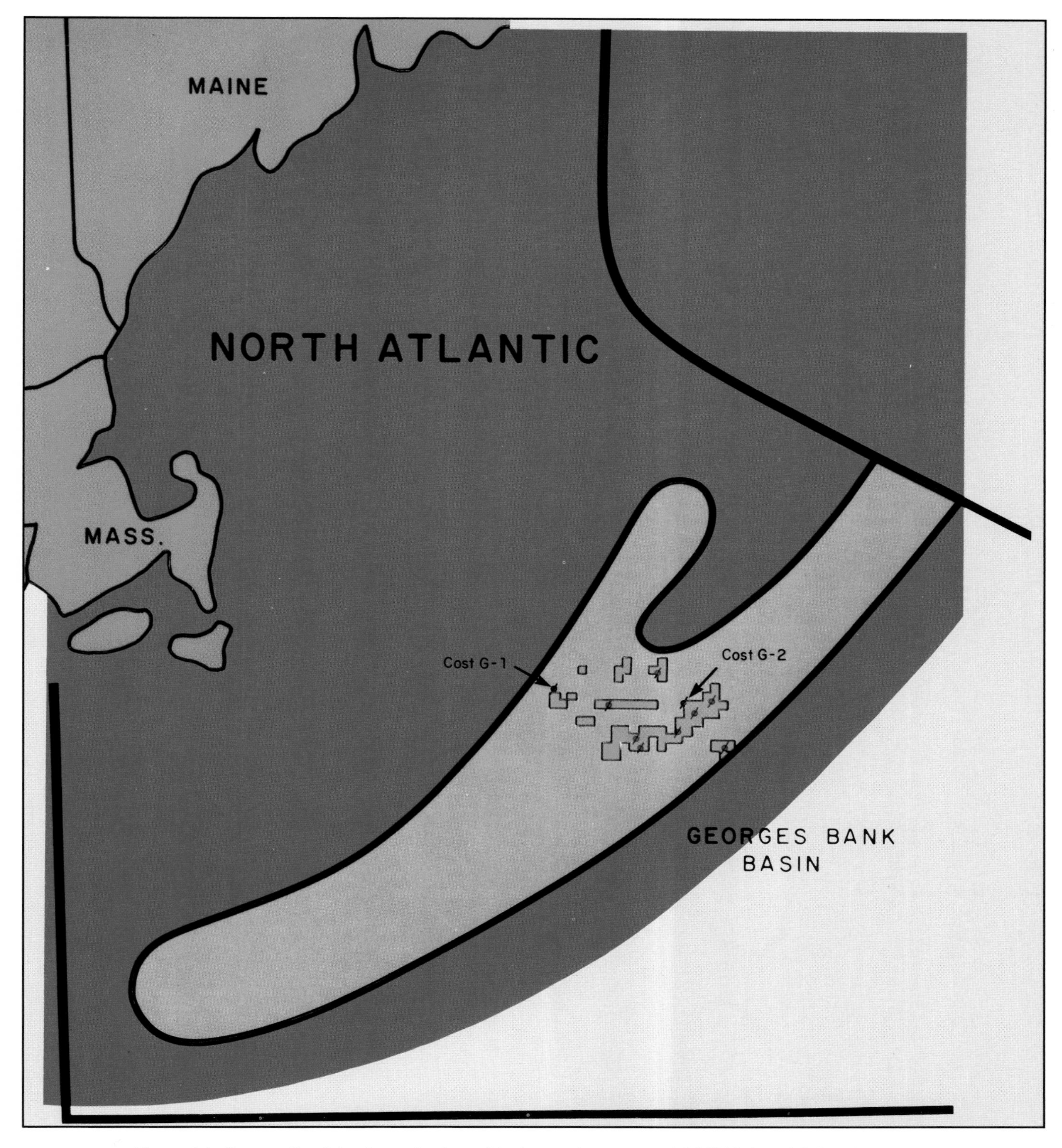


**Figure 14.** *Georges Bank basin, active lease blocks, exploratory and COST G-1 and G-2 well locations.*

(1.9 mi) thick in the Great Isaac well.

The Cay Sal well (total depth-5764 m) bottomed while still in the Marquesas Supergroup after drilling through it for 2.9 km (1.8 mi) (Meyerhoff and Hatten, 1974). The Cay Sal well is anomalous in that it encountered evaporites near the bottom of the well. Onshore, the Amerada 2 Cowles Magazine well (total depth-3887 m) found 2 km (1.2 mi) of Marquesas lying on metamorphic basement rocks (Tator and Hatfield, 1975), while the

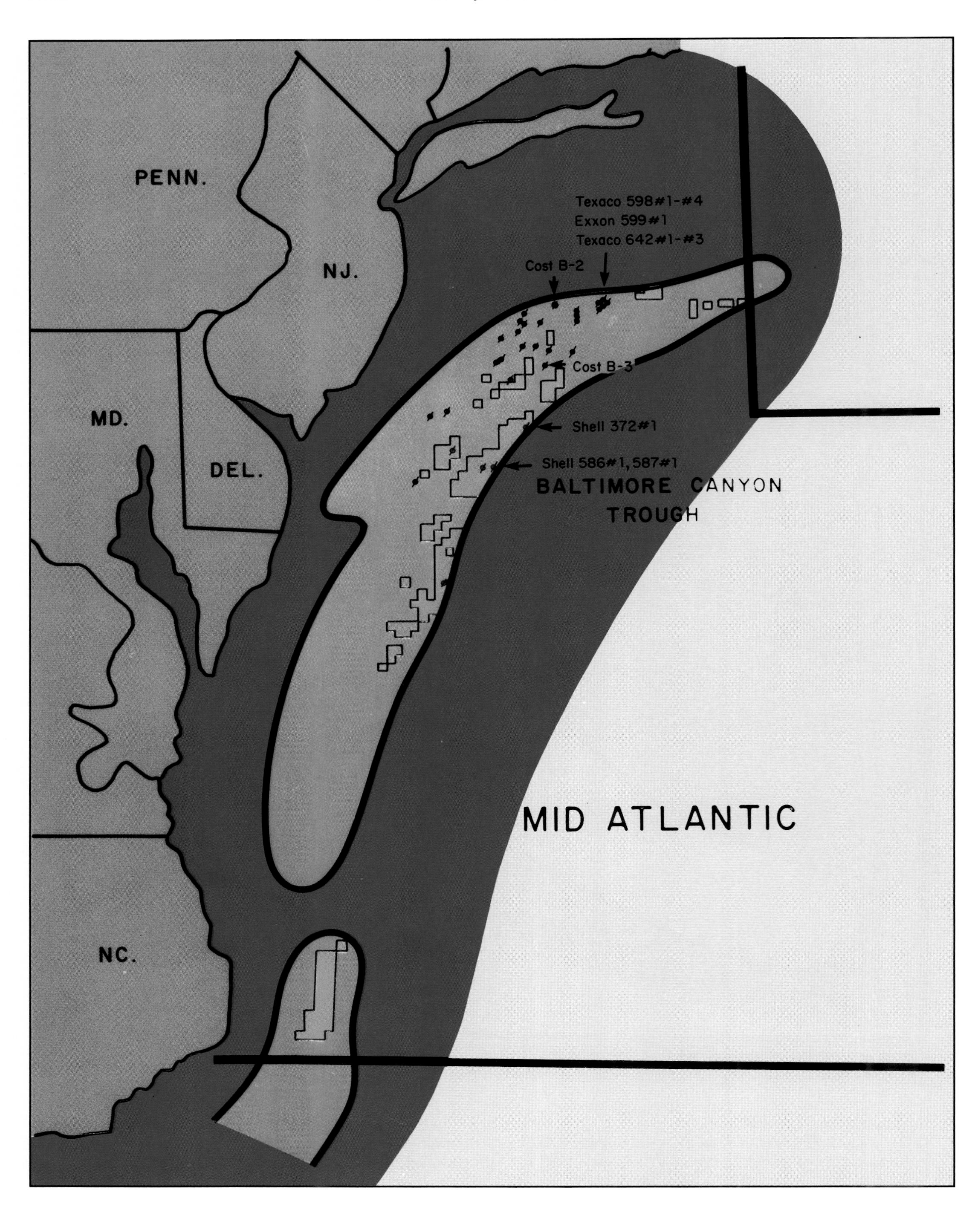
PENN.
NJ.
MD.
DEL.
NC.
Texaco 598#1-#4
Exxon 599#1
Texaco 642#1-#3
Cost B-2
Cost B-3
Shell 372#1
Shell 586#1, 587#1
BALTIMORE CANYON
TROUGH
MID ATLANTIC

Cayo Coco-2 (total depth-3227 m) bottomed in the Marquesas (Meyerhoff and Hatten, 1974). Tectonically disturbed clastic and carbonate facies of equivalent age found in Cuba suggest that the Marquesas Supergroup extended over much of the western Bahamas Platform and adjacent Florida (Figure 13). Depositional environments were relatively stable during the Marquesas deposition. Marquesas rocks may have formed a megabank extending over the Blake Plateau, Bahamas Platform, and south toward Cuba during much of Late Jurassic and Early Cretaceous times.

Upper Cretaceous rocks found in the Great Isaac and Cay Sal wells belong to Cedar Keys, Pine Key, and Card Sount Formations, formations recognized onshore Florida (Meyerhoff and Hatten, 1974).

Meyerhoff and Hatten (1974) report that the Long Island (total depth-5353 m) and Andros Island (total depth-4447 m) wells drilled through reef-backreef facies, but Corso (1983) finds that the Long Island cuttings and cores consist mainly of restricted, platform interior facies. The upper 1,676 m (5,499 ft) were deposited in more open marine conditions than were lower carbonates. Corso (1983) interprets descriptions from the Andros Island well (Goodell and Garman, 1969) as indicating similar depositional environments. He believes that the reef facies is largely missing along much of the Bahama Escarpment because of submarine erosion.

The Florida Straits formed in the Cenomanian, and appear to have scoured and eroded the seafloor repeatedly during times of lowered sea level (Sheridan et al, 1981).

Tertiary carbonates tend to be thin throughout the Blake-Bahama area, with thicknesses generally amounting to less than a kilometer but ranging to more than 2 km (1.2 mi) in the Andros Island well (Meyerhoff and Hatten, 1974).

## EXPLORATION HISTORY

### Georges Bank Basin

On December 18, 1979, OCS Sale No. 42, the only lease sale in the Georges Bank basin, was held after almost two years of delay. As with previous lease sales further south, it concentrated in a principal Mesozoic basin. Most of the bidding was for lease tracts over easily discernible structures.

Prior to the December 18 sale, lithologic data from two COST wells (G-1 and G-2) were available to the industry and the government. The principal basin was known to be filled with Jurassic carbonates and evaporites with an overlying, thin clastic section. Several major prospect types were recognized. One was determined by a large, bright spot located on non-proprietary U.S. Geological Survey seismic data. The location of the bright spot play at the shoreward edge of the basin suggested a greater potential for thick, coarse, clastic reservoirs. The acreage overlying this prospect was leased by Exxon, and was drilled during the winter of 1981–1982. The well was plugged and abandoned as a dry hole. The bright spot had been caused by an evaporite deposit.

A second prospect type was associated with a series of domal and anticlinal structures formed by salt movement and basement highs. These features are located in the center and southeastern end of the main basin. The lithologies below 1,830 m (6,000 ft) were known to consist primarily of carbonates, and the plays made in the region were for carbonate-stratigraphic traps along the crest or flanks of these structures. Localized carbonate patch reefs or shoals may also have been targeted. Seven dry holes were drilled over these prospects (Figure 14).

Predictions of source rock quality have been based on trends observed in the COST G-1 and COST G-2 wells, drilled landward of and in the center of the main basin, respectively. Shallow, thermally immature, gas-prone sediments penetrated by the COST G-1 well were expected to be more deeply buried and more thermally mature toward the basin depocenter. Industry hoped that deeper, older sediments, devoid of significant organic material in the landward direction, would contain more abundant, more marine, and more oil-prone organic matter further seaward. To date, none has been reported.

Certain other predictions of lithology and source-rock quality have been borne out. Seaward of the COST wells, the sediments reflected increasingly open marine conditions, while Jurassic carbonates become more prevalent to the southeast. Organic material becomes more marine and oil-prone toward the shelf edge. The quantity of organic matter preserved decreases dramatically toward the shelf edge, indicating an open marine environment with good circulation of Jurassic, oxygen-rich water over the shelf. These conditions are good for the production of organic material, but bad for its preservation.

No major carbonate reservoir rocks have been reported in the Georges Bank area. Poorly understood and locally controlled variables affect generating and destroying carbonate porosity and prevent accurate prediction of the presence of carbonate reservoir rocks in the Georges Bank basin prior to drilling.

Drilling in the Georges Bank basin has been restricted to a small area, and facies, reservoir and source characteristics may differ in other parts of the basin.

### Baltimore Canyon Trough

The Baltimore Canyon Trough was the first U.S. eastern outer continental shelf basin to be drilled, and is currently the only one to have entered a second phase of exploration farther offshore.

Exploring began in the central portion of the trough with a great deal of enthusiasm. Sediment thickness was estimated to be over 12 km (7.5 mi) in the depocenter (Scholle, 1977). The COST B-2 well was drilled in this region prior to the first lease sale, OCS Sale 40 held on August 17, 1976, and penetrated a predominantly clastic section into the Upper Jurassic at 4,891 m (16,047 ft) total depth (Figure 15). Numerous potential sandstone reservoirs were encountered, while potential hydrocarbon source rocks units were found in the Tertiary and Lower Cretaceous sediments. Tertiary organic matter is oil- and gas-prone, whereas Lower Cretaceous organic matter is terrestrially-derived and gas-prone. Both are thermally immature. The low geothermal gradient in the Baltimore Canyon Trough requires burial at a depth of 3–4 km (1.9–2.9 mi) for significant liquid hydrocarbon generation (Scholle, 1977). None-

**Figure 15.** *Baltimore Canyon Trough showing active lease blocks, exploratory well locations, selected well numbers and COST B-2 and B-3 well locations.*

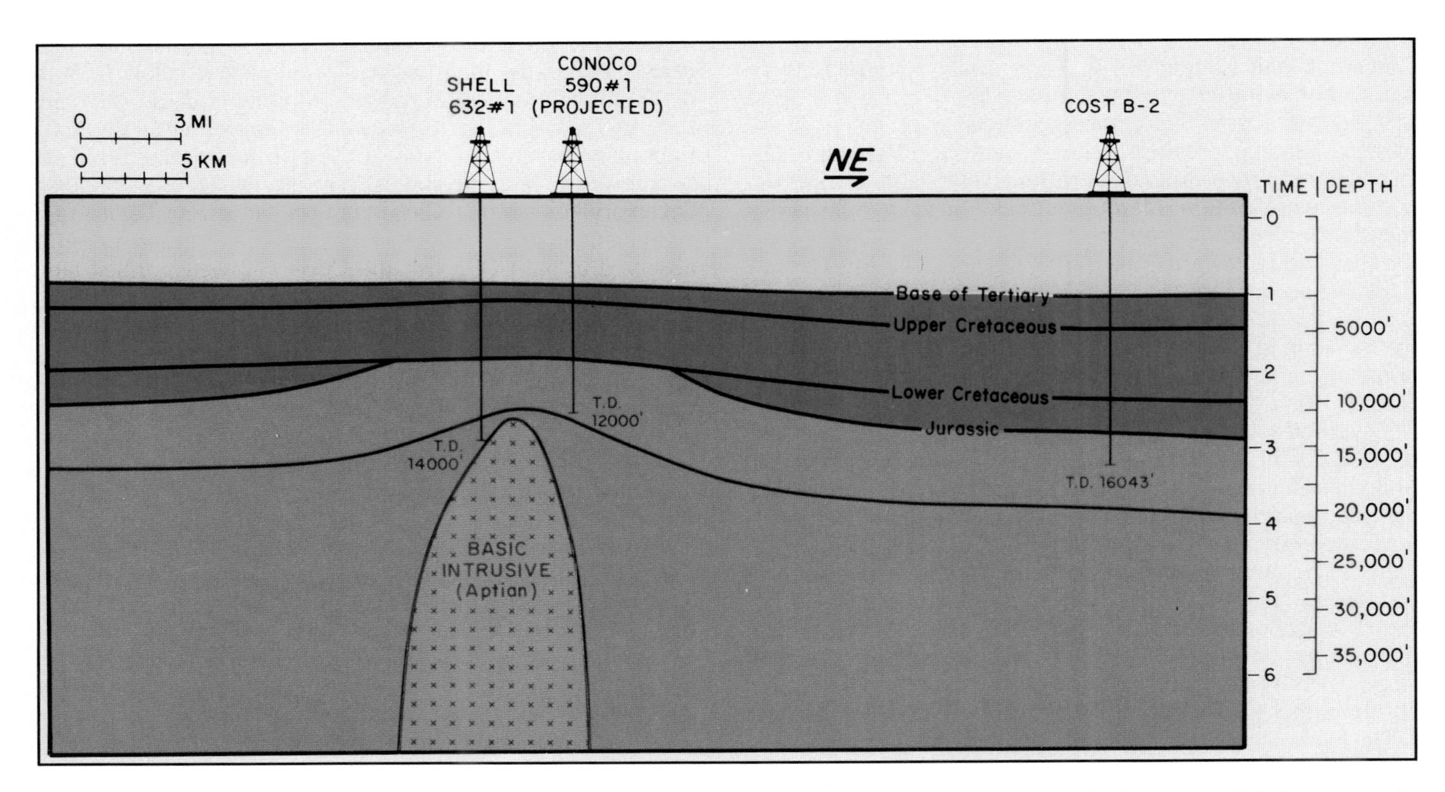


Figure 16. *Schematic cross section of the "Great Stone Dome," an Aptian age intrusion. Shell well 632, Conoco well 590 (projected), and COST well B-2 are shown.*

theless, good potential source and reservoir rocks inspired the industry to bid heavily in the first lease sale on the U.S. Atlantic margin.

Bidding in OCS Lease Sale 40 focused on three types of structural plays. The structure around the Great Stone Dome, a 108 million-year-old, igneous intrusion, drew the highest bids (Figures 15 and 16). The Mobil group bid over $100 million for one block over the play. An enormous domal structure overlies this intrusion, and had a seal been established before hydrocarbon migration, drillers might have found enormous reserves. Instead, hydrocarbons generated by the thermal plume associated with the intrusion migrated to the surface and were lost. Seven dry holes were drilled over the crest and flanks of the structure (Lippert, 1983). One well, drilled by Houston Oil and Minerals, penetrated salt, suggesting that salt migration accompanied the igneous intrusion. The intrusion corrected one problem; it raised the local geothermal gradient tremendously for a short time, creating an aureole of thermally mature and overmature Lower Cretaceous source beds.

Salt-induced structures may also have raised the geothermal gradient locally, though to a much lesser extent. These structures are analogous to Argo salt structures of the Scotian Shelf basin of eastern Canada (Given, 1977; Powell, 1982). The only prospect on the U.S. Atlantic margin to test as significant for hydrocarbons overlies a deep-seated salt feature. Four lease blocks, Tenneco 642, Texaco 598, and Exxon 599 and 643 of the Hudson Canyon Protraction, overlie this diapir that uplifts, but does not pierce, Jurassic prograding carbonate units. Wildcats and step-out wells drilled into the overlying structure penetrate the uppermost portion of the carbonates at depths of 5–6 km (3.1–3.7 mi). Shallower Cretaceous and Jurassic producing sands (from 2.5–4.0 km or 1.6–3.7 mi) may have been sourced by more thermally mature, more deeply buried carbonates (Mattick et al, 1981). Numerous faults connect the shallow sands and deep carbonates and may have served as migration pathways for the generated gas and condensate; the hydrocarbons are trapped by fault closures. Other salt-related prospects have been drilled in the basin, but all have been dry. The deep, prograding carbonates have been penetrated by few wells.

A third type of play in this region is that of fault closure within the shallow clastic section, which has no direct relationship to a salt dome. Many of these growth fault features occur here, but they do not appear to contain hydrocarbons.

Bidding in OCS Lease Sale 49, held on February 29, 1979, after the drilling of nine dry holes in the area, was subdued, and relatively few companies bid on the offered acreage. Although the plays were similar to the smaller prospects pursued during the first sale (OCS Sale 40), bids were much lower.

After drilling nearly 30 wells, including the two COST wells, industry retreated from exploring the shallow shelf in the U.S. Mid-Atlantic area. Their disappointment was hardly surprising considering the initial U.S. Geological Survey predictions for undiscovered recoverable reserves on the shallow Atlantic continental margin (less than 200 m or 656 ft water depth) of 10 billion to 20 billion bbl of liquid petroleum and 55 trillion to 110 trillion cubic feet (tcf) of natural gas (Foote, Mattick, and Behrendt, 1974). The fundamental problem has been a lack of thermally mature source rock.

A second phase of exploring began in the Mid-Atlantic area on December 8, 1981, when industry enthusiasm revived for OCS Lease Sale 59. Active bidding took place over a new type of prospect believed to have tremendous potential for petroleum

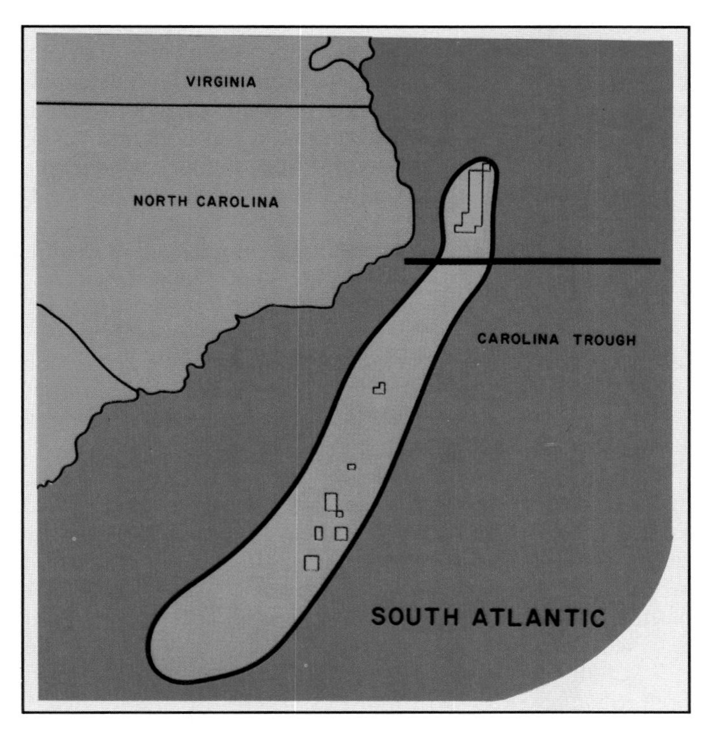


Figure 17. *Carolina Trough basin showing active lease blocks.*

reserves. This acreage is in deeper water over Jurassic shelf-edge carbonates and Lower Cretaceous deltaic clastics, at the eastern edge of the main basin. The acreage leased during Sale 59 and the follow-up OCS Sale 76 (April 26, 1983) are in the early stages of testing.

### Southeast Georgia Embayment

The Southeast Georgia Embayment, a landward extension of the Blake Plateau basin, is filled by Cretaceous and Tertiary coastal plain sediments. Prior to OCS Lease Sale No. 43, the COST GE-1 well was drilled in the area of maximum Cretaceous and Tertiary subsidence within the embayment (Mattick, Schlee, and Bayer, 1981). The well bottomed in Devonian metasediments at 3,974 m (13,038 ft) depth. Overlying the Paleozoic basement were 3,366 m (11,043 ft) of predominantly clastic Cretaceous and Tertiary sediments. Lower Cretaceous coastal sands show good reservoir potential, though no evidence for potential hydrocarbon source rock was found (Scholle et al, 1979).

The lease sale was held on March 28, 1978, and had low participation, due to the scarcity of good structures and lack of encouraging source-rock information from the COST GE-1 well. In the two years following the sale, six dry holes were drilled into Paleozoic and Lower Cretaceous structures. There has been no further interest in the region.

### Carolina Trough

The Carolina Trough is as yet untested, although a number of leases have been taken (Figure 17).

### Blake Plateau–Bahamas Platform Basin

No recent tests have been made in the Blake Plateau-Bahamas Platform basin.

### Reservoir Potential

The presence of reservoir rock has not been a major problem in the majority of the wells drilled on the U.S. Atlantic outer continental shelf. The most promising reservoirs are Mesozoic clastic sediments penetrated in the Southeast Georgia Embayment and the Baltimore Canyon Trough. Paleozoic metasediments penetrated by two wells off the east coast (COST No. GE-1 well, and COST No. G-1 well) have little significant porosity or permeability.

Upper Triassic and Lower to Middle Jurassic sediments have been penetrated in the Georges Bank basin wells, COST G-1 and COST G-2 (Figures 4 and 14). These sediments are non-marine to shallow-marine clastics, carbonates, and evaporites with minimal reservoir potential (Scholle, 1980).

The Upper Jurassic to lowermost Cretaceous sediments generally grade from a non-marine deltaic and shallow-marine clastic facies on the western side of the basins to shallow-marine, carbonate-platform and possibly reefal facies to the east. The reservoir characteristics of the clastic facies are variable, with some cleaner sands having excellent potential. Significant quantities of gas flowed from Jurassic sands in wells on blocks 642, 598, and 599 of the Hudson Canyon area (Figure 15).

Diagenesis could have created excellent reservoir potential in the Upper Jurassic and Lower Cretaceous carbonate facies. Little data, however, are presently available for these sediments, although where penetrated behind the shelf edge in COST B-3, Tenneco 642 #2, Texaco 598 #2, (see Figure 15) and in the Georges Bank basin, they have poor reservoir characteristics. A shelf-edge wildcat drilled on the Scotian Shelf, the Shell Demascota G-32 well, penetrated 168 m (550 ft) of porous dolomite and limestones with some reef facies. Porosities in the limestone range as high as 14%, and fracture porosities in the dolomites are of unknown magnitude (Given, 1977; Eliuk, 1978; Mattick, 1982). However, data from test wells being drilled at the eastern edge of these carbonates offshore of New Jersey are not yet available. Porosities of shelf-edge carbonates in Bahama wells are also excellent (Meyerhoff and Hatten, 1974). Fore-reef slope carbonates, as yet undrilled, may also have reservoir potential. A major portion of Cretaceous–Jurassic production in Mexico comes from such rocks.

The reservoir potential of Scotian Shelf, Baltimore Canyon Trough, and Georges Bank basin carbonates may have been greatly enhanced by subaerial exposure that created secondary porosity (Epstein and Friedman, 1983). Sea level curves and regional unconformities suggest this event may have happened frequently during the Jurassic and Early Cretaceous. Preserving secondary porosity depends on locally controlled, poorly understood variables that are not predictable prior to drilling.

In the Baltimore Canyon Trough, Lower Cretaceous deltaic sandstones, possibly related to the Missisauga and Logan Canyon sandstones of the Scotian Shelf, may have the greatest reservoir potential (Libby-French, 1984). These sandstones are associated with a prograding delta that covered the Jurassic carbonate shelf, and can be traced by seismic data out to the paleoshelf edge. The porosity and permeability of these sands are variable, ranging from 15 to 30%, and 0.1 to 500 md, respectively (Libby-French, 1984).

Lower Cretaceous coastal sands penetrated by the COST GE-1 well may also have good reservoir characteristics, but their lateral extent is unknown. The major deltaic facies present in the Baltimore Canyon area is not present in the Georges Bank basin. Instead, a 3.5 km (2.2 mi) thick, marginal-to-shallow marine carbonate-clastic section was deposited. Porosity and permeability of these sediments range from poor to moderate (Scholle, 1982).

Although not drilled, the Carolina Trough is thought to be stratigraphically similar to the Baltimore Canyon Trough. The Tertiary section in the Carolina Trough is considerably thicker, however.

The Upper Cretaceous and Cenozoic sediments are more marine in nature, with primarily fine-grained clastics and carbonates in the Georges Bank basin and the Baltimore Canyon Trough. Corresponding deep-water carbonates are present in the Blake Plateau area, yet the reservoir potential of these sediments is minimal.

### Source Rock

The most important discovery in exploring the U.S. Atlantic margin to date is the apparent absence of thermally mature, oil-prone source rock. Although large, untested structures remain, and good, potential reservoirs abound, industry is reluctant to test the area further without evidence of oil generation.

Initial hopes of abundant mature source rock analogs found in the Scotian Shelf basin and the Grand Banks basin have been dashed. The source rock for the gas fields of the Scotian Shelf is the Verrill Canyon Formation (Powell, 1982). Verrill Canyon samples derived predominantly from prodelta shales and were buried beneath 3 to 4 km (1.9 to 2.5 mi) of overburden. Equivalent units in the Georges Bank basin and the Baltimore Canyon Trough are thermally immature because of a lack of overburden and a low geothermal gradient, respectively.

In the Hibernia oil field of the Grand Banks east of Newfoundland, Kimmeridgian sediments are considered the most likely source rock (Arthur et al, 1982). These sediments were deposited in a failed rift, formed during the Late Jurassic rifting of Greenland from Canada. In contrast with the restricted circulation in the failed rift, open marine conditions were predominant at this time along the U.S. Atlantic margin and did not preserve organic material.

In the Baltimore Canyon Trough, most of the plays were targeted at structures in Cretaceous and Upper Jurassic strata. Good reservoir potential exists in Lower Cretaceous, deltaic sandstones and Upper Jurassic, marine sands, but this section is almost devoid of preserved organic material, except for material that is gas-prone. Upper Cretaceous and Tertiary shales have excellent potential as oil-prone source rocks but are immature. The geothermal gradient is very low in this basin (approx. 1.2 to 1.3°F/100 ft or 22–24°C/km), and the main phase of the oil generation window lies below 3.5–4.0 km (2.2–2.5 mi) (Scholle, 1977, 1980).

The lack of any known hydrocarbon shows in the eight wildcats drilled in the Georges Bank basin suggests that the source potential of lithologies encountered may not have differed much from those encountered in the COST G-1 and COST G-2 wells. Although the COST well data suggest that slightly less overburden is required to reach thermally mature sediments (about 3.0 km or 1.9 mi), the sediments below this depth contain little organic carbon.

## SUMMARY AND DISCUSSION

Wells drilled in the Atlantic outer continental shelf and Scotian Shelf found no lack of reservoir rock. Cretaceous sandstones were widespread, porous, and permeable. Jurassic sandstones were less porous and permeable, but locally exhibited good reservoir quality. Backreef carbonates, where drilled, were tight, but bank-edge carbonates in the Bahamas, Baltimore Canyon Trough, and Georges Bank exhibited good porosity. Thus, oil, if present, should have no difficulty finding a reservoir.

Traps were large and visible although less abundant than in the Gulf of Mexico and California outer continental shelf. A few traps beneath the shelf were tested; many beneath the slope and rise remain untested. A number of stratigraphic traps will probably remain unrecognized in these basins. We know little of the deeper structure of the Blake Plateau-Bahamas Platform because we lack modern seismic data, and only the near-surface structure is evident in the existing data.

An adequate amount of mature source rock is the major uncertainty. Cretaceous and Tertiary shales are good source rock, but they appear to lie above the anomalously deep hydrocarbon maturation threshold (3.5–4.0 km or 2.2–2.5 mi), and are consequently immature. Jurassic–Lower Cretaceous basinal shales are the source for the Scotian Shelf fields, but they occur in a re-entrant that has no known analog below the U.S. border. Jurassic deep-water shales encountered in DSDP Hole 105 are poor in quality and gas-prone. Clearly, the future of the Atlantic outer continental shelf must depend on other sources.

At least three possibilities exist for such sources. Jurassic basinal shales were probably deposited; indeed, seismic velocities in the zone beneath the prograding carbonate shelf edge are consistent with basinal shale velocities at comparable depths. Examining the Atlantic outer continental shelf seismic data suggests, however, that potentially mature, Jurassic basinal shales lie either at considerable depths beneath the shelf, with any hydrocarbon present probably having been converted to gas; or beneath the slope and rise. In either case, exploring and producing are probably uneconomic, given current oil and gas prices.

Rift-stage shales constitute a second potential source bed depositional event. The Kimmeridge Clay in the North Sea is an example of this type of source rock deposited in anoxic conditions during rifting, and similar conditions appear to have contributed to the hydrocarbons found in the Hibernia region. If such an event occurred in the Atlantic outer continental shelf, it is probably deeply buried and overmature, so that any accumulations would not be economic, given the high cost of deep gas.

Self-sourcing carbonates are the third possibility. Murris (1980) and Harris and Frost (1984) indicate that many large fields in Middle East carbonates derive from sediments deposited in anoxic intrashelf basins. Similarly, the Sunniland field of southwestern Florida is thought to be sourced from within the Sunniland Formation. Our limited knowledge of Blake Plateau–Bahamas Platform carbonates precludes evaluation of this possibility. No deep wells have been drilled on the Blake Plateau, and all but one of the wells in the Bahamas, the Great Isaac, have been drilled along the margins of the present-day platform. Interestingly, some oil has been found associated with coeval Cuban carbonates (Meyerhoff and Hatten, 1974).

It thus seems premature to write an obituary for the Atlantic outer continental shelf. It consists of a large area with few wells, most of which are concentrated on even fewer plays, with many plays untested. It has produced little to date, but the possibility must be considered that it is a sleeping giant awaiting its call.

Furthermore, the shelf lies in U.S. waters, and thus is not subject to the political problems of many overseas areas; it is close to U.S. markets, which could make large gas accumulations economic; and both deep drilling and deep-water drilling are amenable to technological improvement. Greater experience in these areas will bring prices down in the next decade, decreasing the minimum size of economic accumulations and pushing the frontiers to greater drilling depths and into deeper waters.

## ACKNOWLEDGMENTS

During the last decade, the Branch of Atlantic Marine Geology, U.S. Geological Survey, Woods Hole, Massachusetts has been a leader in the studies of the Atlantic outer continental shelf. The senior author participated with Dr. Kim Klitgord of this organization in a Department of Interior symposium on the U.S. Exclusive Economic Zone during the early stages of preparation of this paper, and some of the ideas presented here were derived during this time. Dr. Klitgord and his colleagues graciously made available data and illustrations in areas where the authors were unable to obtain release of proprietary data. We are greatly indebted to them for this.

We are also appreciative of the permission of Gulf Research and Development Company and Gulf Oil Exploration and Production Company to present this work.

## REFERENCES CITED

Alsop, L., and M. Talwani, in press, The East Coast Magnetic anomaly: Science.

Arthur, K., D. Cole, G. Henderson and D. Kushnir, 1982, Geology of the Hibernia discovery, *in* M. Halbouty, ed., The deliberate search for the subtle trap: AAPG Memoir 32, p. 181–196.

Benson, W. E., et al, 1978, Site 392: south rim of Blake Nose, *in* W. E. Benson and R. E. Sheridan, et al, Initial reports of Deep Sea Drilling Project, V.LXIV: Washington, U.S. Government Printing Office, p. 337–393.

Brown, P., J. Miller, and F. Swain, 1972, Structural and stratigraphic framework, and spatial distribution of permeability of the Atlantic Coastal Plain, North Carolina to New York: U.S. Geological Survey Professional Paper 796, 79 p.

Bryan, G., and J. Heirtzler, 1984, Ocean margin drilling program regional data synthesis series: Atlas 5 Mar. Sci. International, Woods Hole Oceanographic Institution.

Buffler, R., J. Watkins, and W. Dillon, 1979, Geology of the offshore Southeast Georgia Embayment, U.S. Atlantic margin, based on multichannel seismic reflection profiles, *in* J. Watkins and C. L. Drake, eds., Studies in continental margin geology: AAPG Memoir 34, p. 21–48.

Corso, W., 1983, Sedimentology of rocks dredged from Bahamian Platform Slopes: University of Miami, unpublished MS thesis, 74p.

Dillon, W. P., C. K. Paull, R. T. Buffler and J. -P. Fail, 1982, Structure and development of the Southeast Georgia Embayment and northern Blake Plateau; preliminary analysis, *in* J. S. Watkins et al, eds., Geological and geophysical investigations of continental margins: AAPG Memoir 29, p. 27–41.

———, P. Popenoe, J. A. Grow, K. D. Klitgord, B. A. Swift, C. K. Paull and K. V. Cashman, 1984, Growth faulting and salt diapirism: their relationship and control in the Carolina Trough, Eastern North America, *in* J. Watkins and C. L. Drake, eds., Studies in continental margin geology: AAPG Memoir 34, p. 21–48.

Eliuk, L., 1978, The Abenaki Formation, Nova Scotia Shelf, Canada—a depositional and diagenetic model for a Mesozoic carbonate platform: Bulletin of Canadian Petroleum Geologists, v. 26, p. 424–514.

Epstein, S. A., and G. M. Friedman, 1983, Depositional and diagenetic relationships between Gulf of Elat (Aqaba) and Mesozoic of United States East Coast Offshore: AAPG Bulletin, v. 67, p. 953–962.

Foote, R. Q., R. E. Mattick, and J. C. Behrendt, 1974, Atlantic outer continental shelf resource and leasing potential: U.S. Geological Survey Open File Report 74-34B, 33 p.

Folger, D. W., W. P. Dillon, J. A. Grow, K. D. Klitgord and J. S. Schlee, 1979, Evolution of the Atlantic continental margin of the United States, *in* M. Talwani et al, eds., Deep drilling results in the Atlantic Ocean: continental margins and paleoenvironment: American Geophysical Union, Maurice Ewing Series 3, p. 87–108.

Gamboa, L., M. Truchan, and P. Stoffa, 1985, Middle and Upper Jurassic depositional environments at the outer shelf and slope of the Baltimore Canyon Trough: AAPG Bulletin, v. 69, p. 610–621.

Given, M., 1977, Mesozoic and early Cenozoic geology of offshore Nova Scotia; Bulletin of Canadian Petroleum Geologists, v. 25, p. 63–91.

Goodell, H. G., and R. K. Garman, 1969, Carbonate geochemistry of Superior deeptest well, Andros Island, Bahamas, AAPG Bulletin, v. 53, p. 513–536.

Grow, J., 1980, Deep structure and evolution of the Baltimore Canyon Trough in vicinity of the COST B-3 well, *in* P. Scholle, ed., Geologic studies of the COST B-3 Well: U.S. Geological Survey Circular 833, p. 117–125.

———, 1981, The Atlantic margin of the United States: AAPG Course Note Series 19, 41 p.

———, C. Bowin, and D. Hutchinson, 1979, The gravity field of the U.S. Atlantic Continental Margin: Tectonophysics, v. 59, p. 27–52.

Harris, P. M. and S. H. Frost, 1984, Middle Cretaceous carbonate reservoirs, Fahud Field and Northwest Oman: AAPG Bulletin, v. 68, p. 649–658.

Heroux, Y., A. Chagnon, and R. Bertrand, 1978, Compilation and correlation of major thermal maturation indicators: AAPG Bulletin, v. 62, p.128–144.

Hollister, C. D., et al, 1972, Initial reports of the Deep Seal Drilling Project, V. XI: Washington, U.S. Government Printing Office.

Hutchinson, D., et al, 1984, Deep structure and evolution of the Carolina Trough, *in* J. Watkins, and C. L. Drake, eds., Studies in continental margin geology: AAPG Memoir 34, p. 129–154.

Jansa, L. F., 1981, Mesozoic carbonate platforms and banks of the eastern North American margin: Mar. Geol., v. 44, p. 97–117.

———, and J. Wiedmann, 1982, Mesozoic–Cenozoic development of the eastern North American and northwest African continental margins; a comparison, *in* U. vonRad et al, eds., Geology of the northwest African continental margin: Berlin, Springer-Verlag, p. 215–272.

Klitgord, K. D., and J. C. Behrendt, 1979, Basin structure of the U.S. Atlantic margin, *in* J. S. Watkins et al, eds., Geological and geophysical investigations of continental margins: AAPG Memoir 29, p. 85–112.

Libby-French, J., 1984, Stratigraphic framework and petroleum potential of northeastern Baltimore Canyon Trough, Mid-Atlantic Outer Continental Shelf: AAPG Bulletin, v. 68, p. 50–73.

Lippert, R., 1984, The 'Great Stone Dome'—a compaction feature, *in* A. W. Bally, ed., Seismic expressions of structural styles: AAPG Study Series, v. 1.

Mattick, R., J. Schlee, and K. Bayer, 1981, The geology and hydrocarbon potential of the Georges Bank-Baltimore Canyon area, *in* J. W. Kerr and A. J. Fergusson, eds., Geology of the North Atlantic borderlands: Canadian Society of Petroleum Geologists Memoir 7, p. 461–486.

McWhae, J., 1981, Structure and spreading history of the northwestern Atlantic region from the Scotian basin to Baffin Bay, *in* J. Kerr, and A. Fergusson, eds., Geology of the North Atlantic borderlands: Canadian Society of Petroleum Geologists Memoir 7, p. 299–332.

Menard, H. W., 1978, Frontier oil regions: Address at the Presidential Energy Technology Seminar, Atlanta, Ga., 5p.

———, 1979, Oversight hearing—oil off East Coast: Testimony to U.S. House of Representatives, September 17, 1979, [Washington, GPO] 12 p.

Meyerhoff, A. A., and C. W. Hatten, 1974, Bahamas Salient of North America: tectonic framework, stratigraphy, and petroleum potential: AAPG Bulletin, v. 58, p. 1201–1239.

Murris, R. J., 1980, Middle East—Stratigraphic evolution and oil habitat: AAPG Bulletin, v. 64, p. 597–618.

Poag, C., 1982, Stratigraphic reference section for Georges basin—depositional model for passive New England margin: AAPG Bulletin, v. 66, p. 1021–1041.

Powell, T., Petroleum geochemistry of the Verril Canyon Formation: a source for Scotian Shelf hydrocarbons: Bulletin of Canadian Petroleum Geologists, v. 30, p. 167–179.

Purcell, L., M. Rashid, and J. Hardy, 1979, Geochemical characteristics of sedimentary rocks in the Scotian basin: AAPG Bulletin, v. 63, p. 87–105.

———, D. Umpleby, and J. Wade, 1980, Regional geology and hydrocarbon occurrence off the east coast of Canada, *in* A. D. Miall, ed., Facts and principles of world petroleum occurrence: Canadian Society of Petroleum Geologists, Memoir 6, p. 551–556.

Rashid, M., and J. McAlary, 1977, Early maturation of organic matter and genesis of hydrocarbons as a result of heat from a shallow piercement salt dome: Journal Geochemical Exploration, v. 8, p. 549–569.

Ryan, W. B. F., et al., 1978, Bedrock geology in New Engand submarine canyons: Oceanologica Acta, v. 1, p. 233–254.

Schlee, J. S., and J. Fritsch, 1982, Seismic stratigraphy of the Georges Bank basin complex, offshore New England, *in* J. S. Watkins, and C. L. Drake, eds., Studies in continental margin geology: AAPG Memoir 34, p. 223–251.

Scholle, P. A., ed., 1977, Geological studies of the COST No. B-2 well, U.S. Mid-Atlantic outer continental shelf area: U.S. Geological Survey Circular 750, 71 p.

———, 1980, Geological studies of the COST No. B-3 well, United States Mid-Atlantic continental slope area: U.S. Geological Survey Circular 833, 132 p.

Scholle, P. A. and C. R. Wenkham, 1982, Geological studies of the COST Nos. G-1 and G-2 wells, United States North Atlantic outer continental shelf: U.S. Geological Survey Circular 861, 193p.

Seibold, E., 1982, The Northwest African margin—an introduction, *in* U. vonRad et al, eds. Geology of the northwest African margin: Berlin Springer-Verlag, p. 3–22.

Sheridan, R. E., and P. Enos, 1979, Stratigraphic evolution of the Blake Plateau after a decade of scientific drilling, *in* M. Talwani et al, eds., Deep drilling results in the Atlantic Ocean continental margins and paleoenvironment: American Geophysical Union, Maurice Ewing Series 3, p. 109–122.

———, et al, 1982, Early history of the Atlantic Ocean and gas hydrates of the Blake Outer Ridge: results of the Deep Sea Drilling Project Leg 76: Bulletin of the Geological Society of America, v. 93, p. 876–885.

Sheridan, R. E., J. T. Crosby, G. M. Bryan and P. L. Stoffa, 1981, Stratigraphy and structure of southern Blake Plateau, northern Florida Straits, and northern Bahama Platform from multichannel seismic data: AAPG Bulletin, v. 65, p. 2571–2593.

Shipley, T. H., R. T. Buffler, and J. S. Watkins, 1978, Seismic stratigraphy and geologic history of Blake Plateau and adjacent western Atlantic continental margin: AAPG Bulletin, v. 62, p. 792–812.

Sumpter, R., 1979, Prospects darken for Baltimore Canyon: Oil and Gas Journal, v. 77, n. 10, p. 72–72.

Tator, B. A., and L. E. Hatfield, 1975, Bahamas present complex geology: Oil and Gas Journal, v. 73, n. 43, p. 172–176; and n. 44, p. 120–122.

Uchupi, E., R. Ballard, and J. Ellis, 1977, Continental slope and upper rise off western Nova Scotia and Georges Bank: AAPG Bulletin, v. 61, p. 1483–1492.

Valentine, P., 1982, Upper Cretaceous subsurface stratigraphy and structure of coastal Georgia and South Carolina: U.S. Geological Survey Professional Paper 1222, 23 p.

VanHouten, F., 1977, Trassic–Liassic deposits of Morocco and eastern North America: Comparison: Bulletin of the Geological Society of America, v. 61 p. 79–99.

Wade, J., 1977, Stratigraphy of the Georges Bank basin—Interpretation from seismic correlation to the western Scotian Shelf: Canadian Journal of Earth Science, v. 14, p. 2274–2283.

Weigel, W., G. Wissman, and P. Goldflam, 1982, Deep seismic structure (Mauritania and Central Morocco), *in* U. vonRad et al, eds., Geology of the northwest African continental margin: Berlin, Springer-Verlag, p. 132–159.

Wissman, G., 1982, Stratigraphy and structural features of the continental margin basin of Senegal and Mauritania, *in* U. vonRad et al, eds. Geology of the northwest African continental margin: Berlin, Springer-Verlag, p. 160–181.

# California and Saudi Arabia Geologic Contrasts

Roger G. Alexander, Jr.
*Chevron Overseas Petroleum Inc.*
*San Francisco, California*

Analysts can quantitatively and systematically compare the geology of relatively unexplored basins with that of maturely explored, "analog" basins. This can be a useful tool for assessing the unexplored basins' potential for future hydrocarbon discoveries. This paper shows how two important basin areas, California and Saudi Arabia, can be compared. Geologists and geophysicists can use the techniques discussed as an exploration checklist to help recognize and isolate potential hydrocarbon-producing areas.

## INTRODUCTION

In assessing potential future petroleum provinces of the world, one must refer to analogous basins or provinces already having significant histories of hydrocarbon exploration and discovery. Rarely does a data base exist that is complete enough to form the sole basis for predicting hydrocarbon occurrences from the new basin being analyzed. Referral to geologically analogous basins can suggest elements of geology that may occur in the new basin, and therefore limit the predictions.

The new basin and the analog should be viewed simultaneously, using comparable scales for spatial and linear dimensions, geologic elements and history, as well as known and forecast field sizes. This method allows explorationists to evaluate quantitatively the similarities or differences between basins. Basin ranking involves making judgments as to which basin may contain more, or less, undiscovered hydrocarbons. Comparing these basins with one or more maturely explored analogs can influence the process constructively.

California and Saudi Arabia are two contrasting, maturely explored geologic provinces, with different numbers and sizes of oil and gas fields. Both have very different geologic histories. I have chosen them as statistical "end members," one representing the results of a very active, dynamic geologic life, and the other, a peaceful and quiet one. Their comparison will demonstrate analytical techniques that can be applied to other basins.

California is an important oil and gas province in the United States, with significant new discoveries being made offshore, and large numbers of projects that are increasing existing productive capacity. Saudi Arabia is, of course, renowned as the single country in which the most oil has been discovered to date; it also contains the world's largest oil field, Ghawar. Reported discoveries indicate probable future expansion of gas and oil reserves there in Paleozoic rocks.

Table 1 shows that the amount of reported oil discovered in Saudi Arabia is roughly ten times that discovered in California, and the reported gas four times as much as California's. In this paper, I summarize the most important geologic elements responsible for these differences.

In considering Saudi Arabian geology, I will only treat those elements bearing on oil in the Jurassic and Cretaceous rocks. [See Al Laboun, this volume, for a discussion of Saudi Arabian Paleozoic exploration.]

Table 1. *California and Saudi Arabia oil and gas discovered to date (from U.S. Geological Survey, Oil and Gas Journal, American Petroleum Institute, California Division of Mines).*

| | California | Saudi Arabia |
|---|---|---|
| **Oil (billion bbl)** | | |
| Produced to 12/31/83 | 20.0 | 48.2 |
| Reserves 12/31/83 | 5.3 | 166.0 |
| Discovered to date | 25.3 | 214.2 |
| **Gas (trillion cu ft)** | | |
| Produced to 12/31/83 | 28.4 | 14.9 |
| Reserves 12/31/83 | 5.8 | 121.0 |
| Discovered to date | 34.2 | 135.9 |

## PLATE TECTONICS

Examining a geologic province in terms of plate tectonic evolution can shed light on the cause of geologic styles associated with its oil and gas accumulations. Basins having analogous plate tectonic histories may not have equally potent combinations of source beds, conduits, reservoirs, seals, and detailed geologic history to produce identical volumes of hydrocarbon reserves. However, certain types of stratigraphic suites, structural style and stability, trap characteristics, and burial history are associated with regions that have undergone a complex genesis, whereas other types would tend to occur in regions having a simple and quiet history.

Figure 1 interprets an Early Jurassic worldwide distribution of land masses and areas of shallow and deeper ocean water. (Note: The generalized world maps illustrate the marked contrast in the probable histories of the areas I will discuss. They do not reflect new or detailed interpretations in many other parts of the world).

California was situated on the western edge of North America where subduction is believed to have been taking place along a northwest-trending zone, immediately southwest of a linear and parallel magnetic arc (Dickinson, 1981). This subduction was near the end of a period during which a "Japanese type" continental margin existed involving trenches, island arcs, and marginal seas. The evidence for this identification lies in the Jurassic sedimentary, metamorphic, ultrabasic, and acidic igneous rocks found on the northeastern side of the 600 mi (970 km) long, northwest-trending, San Andreas regional strike-slip fault.

Saudi Arabia at this time was the site of a broad, stable platform at least 850 mi (1,350 km) wide in a northeast to southwest direction. Shallow-water carbonate deposition started, and was the dominant, regional sedimentational type until latest Paleocene time. Only limited pulses of clastic and evaporite deposition interrupted this pattern. The relative stability of the broad Jurassic platform is shown by the extensive, near-parallelism of bedding planes. This shallow-water platform was on the northeast corner of the African plate, the Saudi Arabian Peninsula still being an integral part of it. The seaway was an arm of Tethys, near the equator, which was apparently far from the active arc in Central Iran and, therefore, unaffected by it. Passivity in a warm climate was the rule.

By Cretaceous time, as shown in Figure 2, subduction was still occurring along the west coast of North America as it moved to the northwest away from Africa. During the Jurassic and the Cretaceous periods, California was the site of "Andean Type" tectonics, with an area of northeastern plutonic intrusions, a medial belt of forearc basin filling, and farthest to the west a developing subduction complex (Dickinson, 1981). The complicated origin of most of the resulting mixed Mesozoic sediments, metamorphics, melanges, and intrusives is outlined by Graham and Ingersoll (1981) and Kleist (1981). The least complicated rock sequence developed at this time was the forearc basin fill, the Great Valley Sequence, in the upper part of which occurs commercial dry gas in the northern part of the Great Valley (Graham, 1981a). Significant hydrocarbons occur in California only in Upper Cretaceous and Cenozoic rocks.

One can see a setting for Saudi Arabia in the Late Cretaceous (Figure 2) that was very similar to that shown for it in Early Jurassic time (Figure 1). This setting reflects continued stability

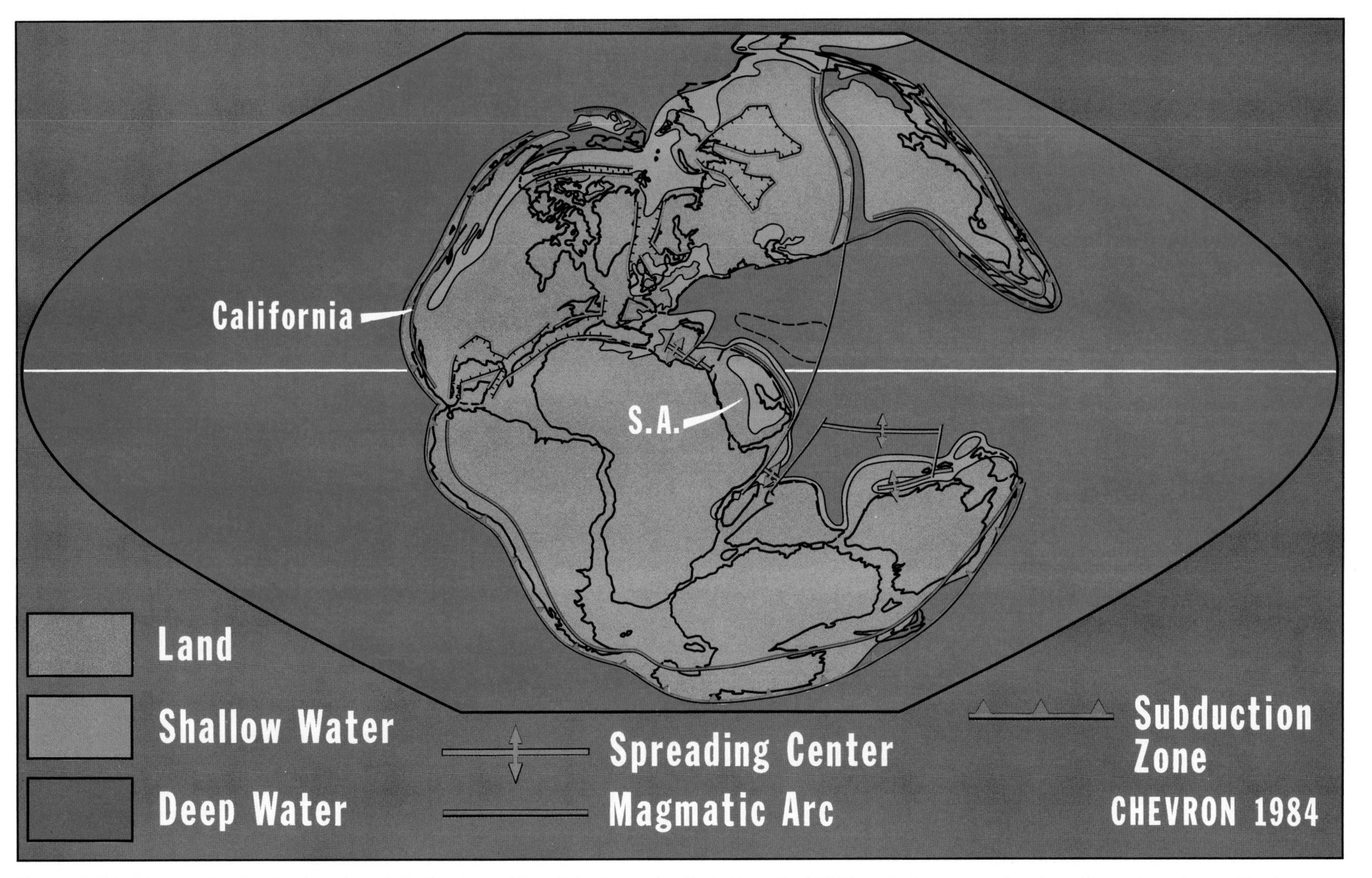


Figure 1. *World map showing the location of California and Saudi Arabia in the Early Jurassic, 190 Ma, relative to postulated positions of continental land masses, oceans, and lines of tectonic activity at that time.*

for the northeastern corner of the greater African plate throughout the 120 million years separating these two times. Though Iran was the site of orogenic and magmatic activity, Arabia was protected by its crustal rigidity, suffering only gentle regional tilting and growth (a few degrees or less) over basement-controlled structures, or others believed to have formed over mobile basal Paleozoic salt. This gentle growth took place in pulses, with a major pulse occurring in the Upper Cretaceous at the base of the Aruma formation. By then, a gentle tilting toward the northeast was also taking place. A vast area in eastern Saudi Arabia underwent only minor flexing with regional and local dips being only a few degrees or less, and rarely as much as ten degrees.

From the Mesozoic to present, California has undergone radical changes, while Saudi Arabia has changed only superficially.

Figure 3 shows that North America continued its relative westward migration from the mid-Atlantic spreading center. However, the geologic environment along its western border in the sector of California had changed—note the lack of any symbol indicating subduction here. During the earliest part of the Cenozoic, convergence, with subduction, was still taking place as the oceanic Farallon Plate descended northeasterly under adjacent continental rocks (Dickenson, 1981). However, by Miocene time, the San Andreas transform fault developed, along which right-lateral motion occurred between the Pacific plate and the North American plate. Total northwestward movement of the Pacific plate is some 350 mi (560 km) along this 600 mi (970 km) transform zone, with major displacement being about 180 mi (290 km) along the San Andreas fault. In a complicated fashion still not fully understood, individual sedimentary basins formed both before and after the advent of right-lateral movement along the transform zone (Ernst, 1981). As individual structural blocks were depressed, uplifted, or tilted, low areas received chiefly clastic sediments—both marine and terrestrial. Movement of the blocks was due to compressional and extensional stress regimes, caused by the change in direction and angular relationships of the multiple faults making up the San Andreas transform zone (Crowell, 1974). Volcanism took place locally. In summary then, coastal California was fragmented during the Cenozoic Era.

In direct contrast, Saudi Arabia was modified only gently during the Cenozoic, chiefly in two ways: First, the west-central part of the country was broadly and gently uplifted in a domal sense, while the eastern part of the country was concurrently depressed. Thus, a 600 mi (970 km) wide, nearly planar segment of a local craton tilted eastward and flexed slightly in a north to south direction. Resulting regional dips are a degree or less (Brown, 1972). Second, existing structures continued to grow, increasing their flank dips by a few degrees and, in some cases, less than a degree.

The area of the large Saudi Arabian oil fields was insulated from any significant structural modifications resulting from dynamic crustal action along the borders of the Arabian plate. It was during the Tertiary that the opening of the Red Sea took place. The Arabian plate thus moved north-northeast, away from Africa. Its convergence and subduction under the Iranian terrain caused the Arabian plate to tilt slightly to the northeast and to form the spectacular family of anticlines and thrusts of the Iranian Zagros Mountains on the northeast side of the Persian Gulf. The central part of the Arabian craton was rigid and remained essentially unchanged from the start of the Cenozoic.

## REGIONAL GEOLOGY AND BASINS

### California

The Generalized Geologic Map (Figure 4) shows California's key geologic characteristics. Low-elevation terrain occurs in the areas shown as "Alluvial/Cenozoic Non-Marine," while hilly to mountainous terrain occurs in the other catagories.

Paleozoic rocks occur as minor isolated masses in the northeastern part of the state, often associated with Mesozoic intrusives. Paleozoics also occur in separate desert ranges in the southeastern part of the state. Mesozoic intrusives dominate a broad medial band extending for many hundreds of miles in a north-northwesterly direction. The emplacement of these rocks was associated with the subduction that took place during the Jurassic and Cretaceous. In the northwestern part of the state, a broad terrain, particularly on the western side of the northern end of the Great Valley, occurs of highly structured, chiefly Mesozoic (minor Paleogene) slope and basinal sediments, melange, ultrabasics, and metamorphics. Taken all together, they comprise the "subduction complex" formed from Jurassic to Paleogene time (Dickenson and Seely, 1979; Kleist, 1981). These rocks play only a minor role as oil and gas traps.

In the southern, near-coastal part of California, widespread, numerous, fault-separated outcrops of Cenozoic marine rocks occur adjacent to two fault zones trending generally northwest, separated by a central zone trending east to west. These faults make up the transform system along which the oldest rocks southwest of the San Andreas fault are believed to have been moving, perhaps several hundred miles northwesterly relative to those to the northeast, since at least early Miocene. The impressive San Andreas fault can be seen for over 500 mi (800 km), from a point northwest of the Salton Sea (SS), to the coast northwest of San Francisco (SF), where it extends offshore.

One can sense how Tertiary marine transgressions took place over an everchanging group of structural blocks, at times extending into the southern part of the Great Valley. The first development of an ancestral "Great Valley" (Figure 4) took place in its present general position and form as a forearc-basin that filled during the Mesozoic (Dickinson, 1981). Since then, it has been receiving sediment continuously, though not uniformly. It was never regionally dissected by any individual, strike-slip fault, as happened in the transform area to the southwest.

The oil and gas of California occur in those depressed basinal areas generally seaward of the larger Mesozoic intrusives, where significant amounts of Upper Cretaceous or Cenozoic rocks or both were retained. The distributions of California's larger oil and gas fields are shown by means of colored field outlines on Figure 5.

Central inland California is dominated by the combined Sacramento and San Joaquin basins, which together underlie the Great Valley immediately west of the Sierra Nevada. This depression is asymmetric, being deeper along the western margin (Morrison et al, 1971; Zieglar and Spotts, 1978; Callaway, 1971).

To the southwest, on the seaward side of the San Andreas fault, are a host of basins occurring in part onshore, and in part offshore. In southern California the Los Angeles basin, one of the world's most prolific basins for its size, is situated mostly onshore. The city of Los Angeles is located in its northwest end (Gardett, 1971). To the southwest of the coastline there is a

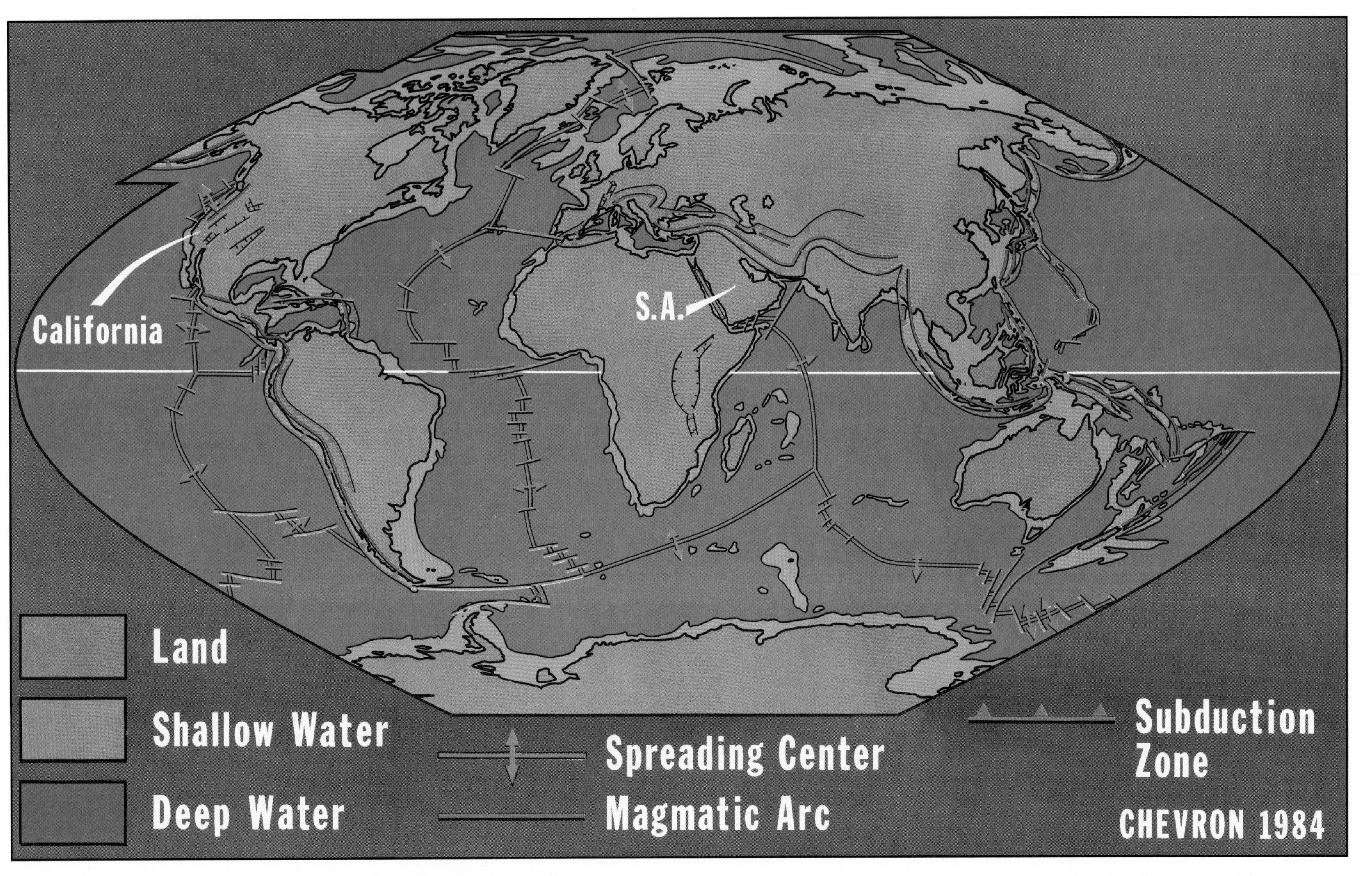


Figure 2. *World map showing the location of California and Saudi Arabia in the Late Cretaceous, 70 Ma, relative to postulated positions of continental land masses, oceans, and lines of tectonic activity at that time.*

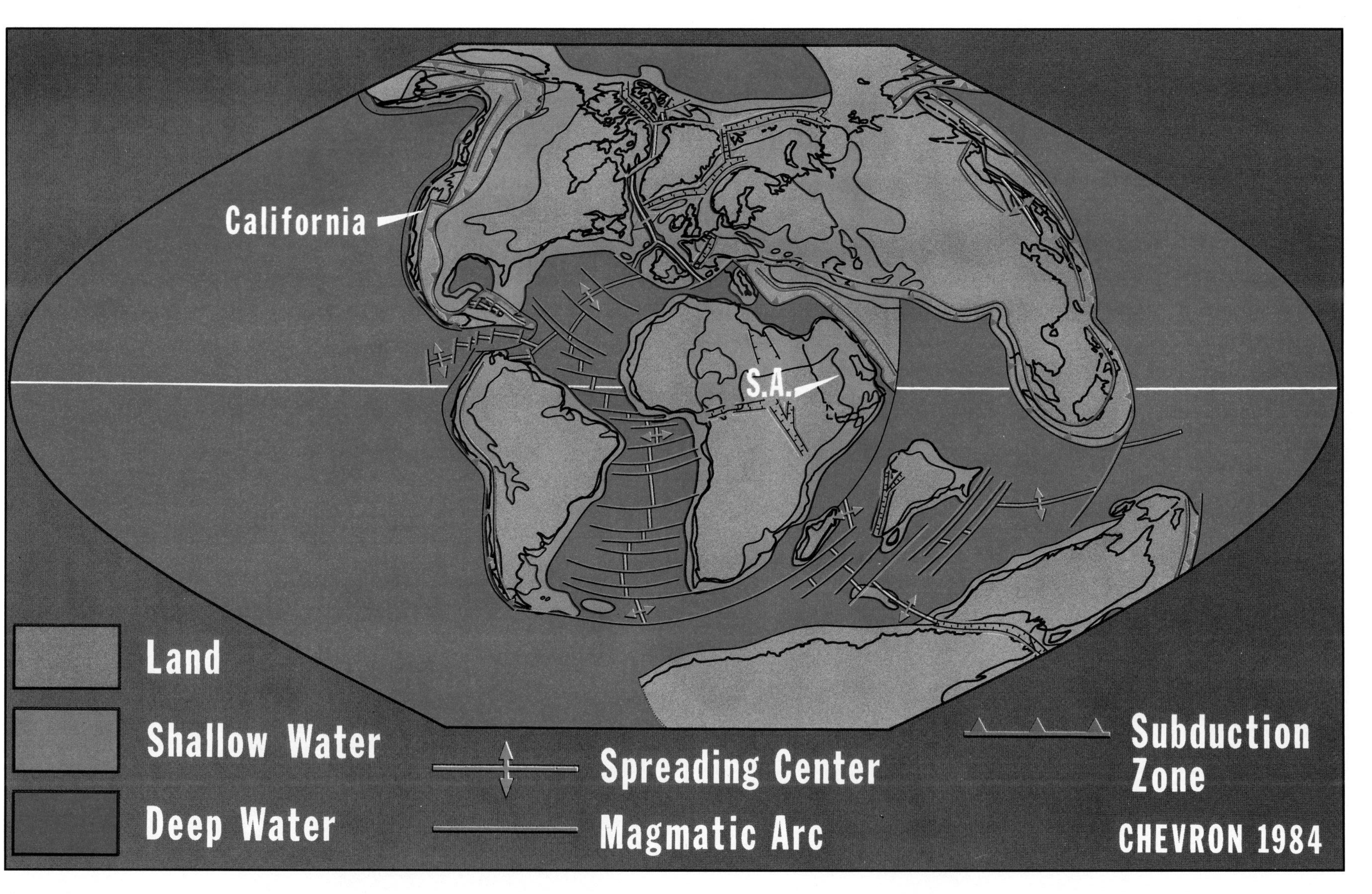


Figure 3. *World map showing the present-day location of California and Saudi Arabia relative to the current positions of continental land masses, oceans, and lines of tectonic activity.*

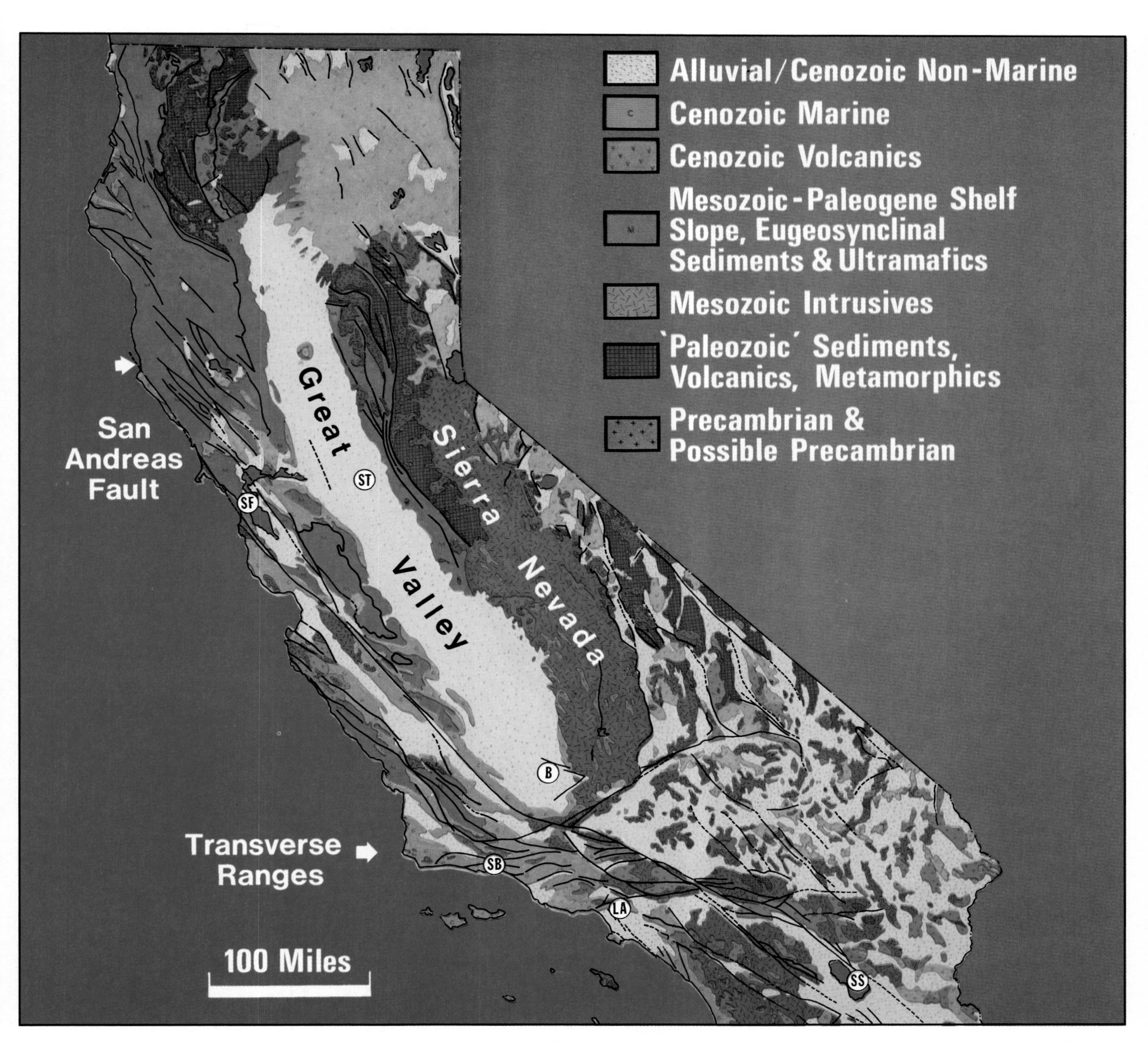


Figure 4. *Generalized geologic map of California (after United States Geological Survey and California Division of Mines and Geology, 1966). SF = San Francisco, ST = Stockton, B = Bakersfield, SB = Santa Barbara, LA = Los Angeles, SS = Salton Sea.*

fragmented area, approximately 120 mi (300 km) square, with significant depocenters between various faults (up to 13,000 ft or 3,900 m of sediments [Parker, 1971]) but in which no significant hydrocarbons have been discovered to date.

Between Los Angeles and Santa Barbara, to the northwest, lies a province of east-to-west-oriented geology, the uplifts of which form the Transverse Ranges, and between which are significant depressions. One of these, 100 mi (160 km) long, consists of a chiefly onshore portion on the east, called the Ventura basin, and a largely offshore portion to the west, called the Santa Barbara Channel basin. Note, in Figure 5, that this is the site of the greatest estimated thickness of the combined Upper

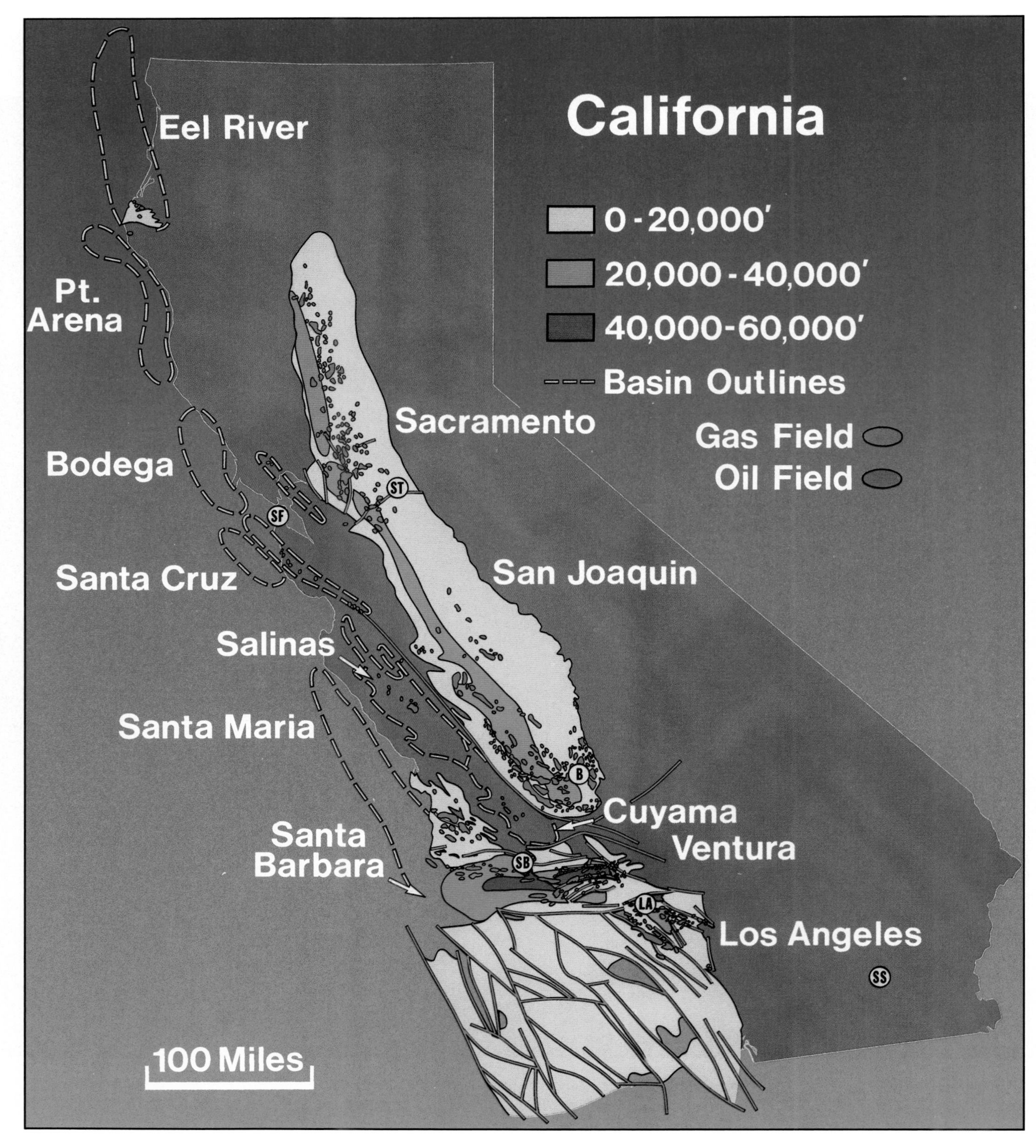


Figure 5. *Generalized map of California showing the location of oil and gas fields and of the chief sedimentary basins containing discovered hydrocarbons, or potential for such. Estimated total sediment thicknesses shown are generalized. (Onshore after Ogle, 1968; Rudkin, 1968; Morrison et al, 1971; Callaway, 1971; Nagle and Parker, 1971; Gardett, 1971; Crawford, 1971. Offshore after Currant et al, 1971; Parker 1971) Offshore dashed basin outlines are from Zieglar and Cassell, 1978, where no thickness data were available. See Figure 4 for place name initials.*

Table 2. *California: hydrocarbons discovered to 1/1/79—by basin (from California Division of Oil and Gas, 1978).*

| Basin | Recoverable Reserves Oil (million bbl) | Recoverable Reserves Gas (trillion cu ft) | Approximate Basin Area (sq mi) |
|---|---|---|---|
| Sacramento | — | 7.8 | 9,200 |
| San Joaquin | 10.0 | 11.3 | 11,350 |
| Coast Ranges | 0.9 | 0.5 | 2,450 |
| Santa Maria | 0.8 | 0.8 | 1,000 |
| Ventura | 2.1 | 4.2 | 3,600 |
| Los Angeles | 8.1 | 7.3 | 1,450 |
| | 21.9 | 31.9 | 29,050 |

Cretaceous and Cenozoic strata (Nagle and Parker, 1971; Curran, Hall, and Heron, 1971). The existence of this local segment of east-to-west structural elements is not fully understood, but may be related to the presence of the more easterly oriented segment of the San Andreas fault.

At the west end of the east to west oriented Santa Barbara basin (Figure 5) geologists are documenting how this basin merges with the northwest-to-southeast trending offshore Santa Maria basin, as successful offshore operations continue here.

The dashed outlines for the offshore basins to the northwest (Figure 5) are those shown in Zieglar and Cassell's 1978 discussion of offshore well information; Hoskins and Griffiths also refer to these basins in their 1971 discussion of California's hydrocarbon potential. Total sediment thickness data are not available for these basins, nor for the Cuyama, Salinas, and two, more northerly, unnamed onshore basins. The other prospective offshore areas, north of the southern end of the offshore Santa Maria basin, are only lightly explored.

At this point we can see that California's hydrocarbons have been discovered in a segmented, continental borderland. The recoverable oil and gas, reported discovered to January 1, 1979, is distributed in the various basinal areas as listed in Table 2.

### Saudi Arabia

The surface geology of the Arabian Peninsula is shown in Figure 6. Four characteristics of this area contrast it markedly with California. The first is the presence of an extensive and broad western shield area of Precambrian rocks, over 700 mi (1,100 km) long, and up to 400 mi (640 km) wide, with some Cenozoic extrusive volcanics present in its western part.

The second characteristic is the long, simple, and arcuate flanking outcrops to the east of the shield, involving successive bands of Paleozoic, Mesozoic, and then Cenozoic rocks. The relative simplicity of this outcrop pattern is the result of the erosion of the gently warped sedimentary rocks of these ages from the higher western portion of a slightly domed cratonic plate. The central belt of north-trending and east-dipping Paleozoic and Mesozoic rocks extends continuously for nearly 900 mi (1,400 km), and the basal Tertiary strata do so for over 500 mi (800 km). These outcrops are generally at elevations of 1,500 to 3,000 ft (450 to 900 m) above sea level. The majority of the Precambrian terrain is between 3,000 and 6,000 ft (900 and 1,800 m) above sea level—the regional topography is very subdued.

The third characteristic in contrast to California is the broad extent of Cenozoic sediments and Quaternary surficial deposits (including much drifting sand) that exists for hundreds of miles to the east of the Mesozoic outcrops. The vast majority of this young terrain is less than 2,000 ft (600 m) above sea level. The long, northeast-trending Rub Al Khali, or Empty Quarter, floored with Quaternary, is the site of sand dunes many hundreds of feet high separating level and flat sabkha interdune surfaces.

The fourth characteristic of the area is the rarity of faults cutting the exposed sedimentary rocks. The U.S.G.S. (1963) 1:2,000,000 "Geologic Map of the Arabian Peninsula," from which the patterns on Figure 6 were derived, documents this lack of faulting. Though linear tonal features are suggested on some modern satellite images of the Arabian Peninsula, I am not aware that the results of modern exploration have documented any significant surface faults there. Faulting within the Precambrian affects the exposed sedimentary rocks only rarely. (I have excluded the extensive southwesterly thrusting known to have occurred at the extreme southeast corner of the Arabian Peninsula in Oman. This thrusting is far enough away not to significantly alter the generalities being made about the areas of the Saudi Arabian oil fields).

In total then, the geologic map documents the presence of the simplest surface geology imaginable, with very broad and unbroken extents of low-dipping, unfaulted Cenozoic rocks, underlain in the west by continuous Mesozoic only rarely faulted. This pattern overlies nearly parallel but varying thicknesses of Paleozoic.

Figure 7 shows the area of exposed Precambrian (in red), structure contours on the base of Mesozoic, the oil fields of Saudi Arabia, and the concession areas still held by The Arabian American Oil Company (ARAMCO). Oil and gas fields occurring in countries other than Saudi Arabia have been omitted from this map. (The lines of section A–A′ and B–B′ are the indices for the two cross sections appearing in Figure 10).

It is noteworthy that the bulk of the Saudi Arabian oil fields are concentrated in a relatively limited area. Here, the Mesozoic base does not exhibit only east-homoclinal dip off the exposed shield to the west, but, is characterized by some generally north-plunging noses. Similar northeast-plunging interruptions to uniform regional dip occur near the southeast end of Section B–B′ where there are oil and gas fields.

The contours depicted here were transposed from those published by Brown (1972). At the contour interval chosen for Figure 7, individual anticlinal closures cannot be shown, as have been published for Ghawar and other Arabian oil fields (ARAMCO, 1959; El Aquar, 1960; Tleel, 1973). One would suppose that there have been more structural details documented outside and inside the present Retained Areas during ARAMCO's long exploration and relinquishment history. Yet, Figure 7 does show that Arabia's vast oil reserves exist in a locally structured segment of a flexed, but relatively unbroken plate of strata. The maximum dip rate to the northeast along Cross Section A–A′ is 100 ft (30 m) to the mile, or 1°.

For a direct comparison, the basins of California are shown in Figure 7 at the same scale as those of Saudi Arabia. A more detailed comparison will be made below.

## STRUCTURE

### California

Figure 8 shows four, true-scale cross sections illustrating the

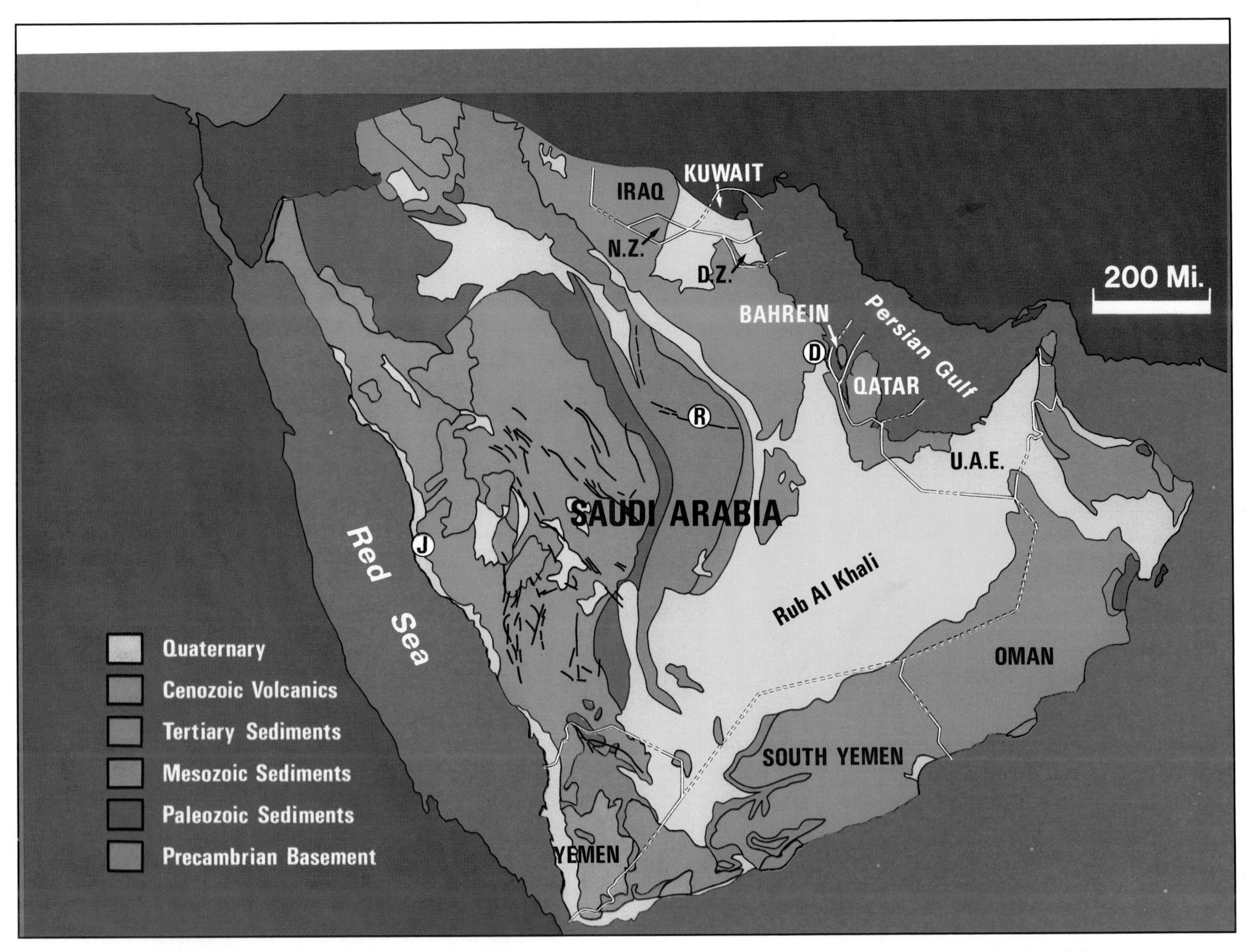


Figure 6. *Generalized Geologic Map of the Arabian Peninsula. (Modified from Steineke et al, 1958; U.S.G.S., 1963; Powers, 1968).*

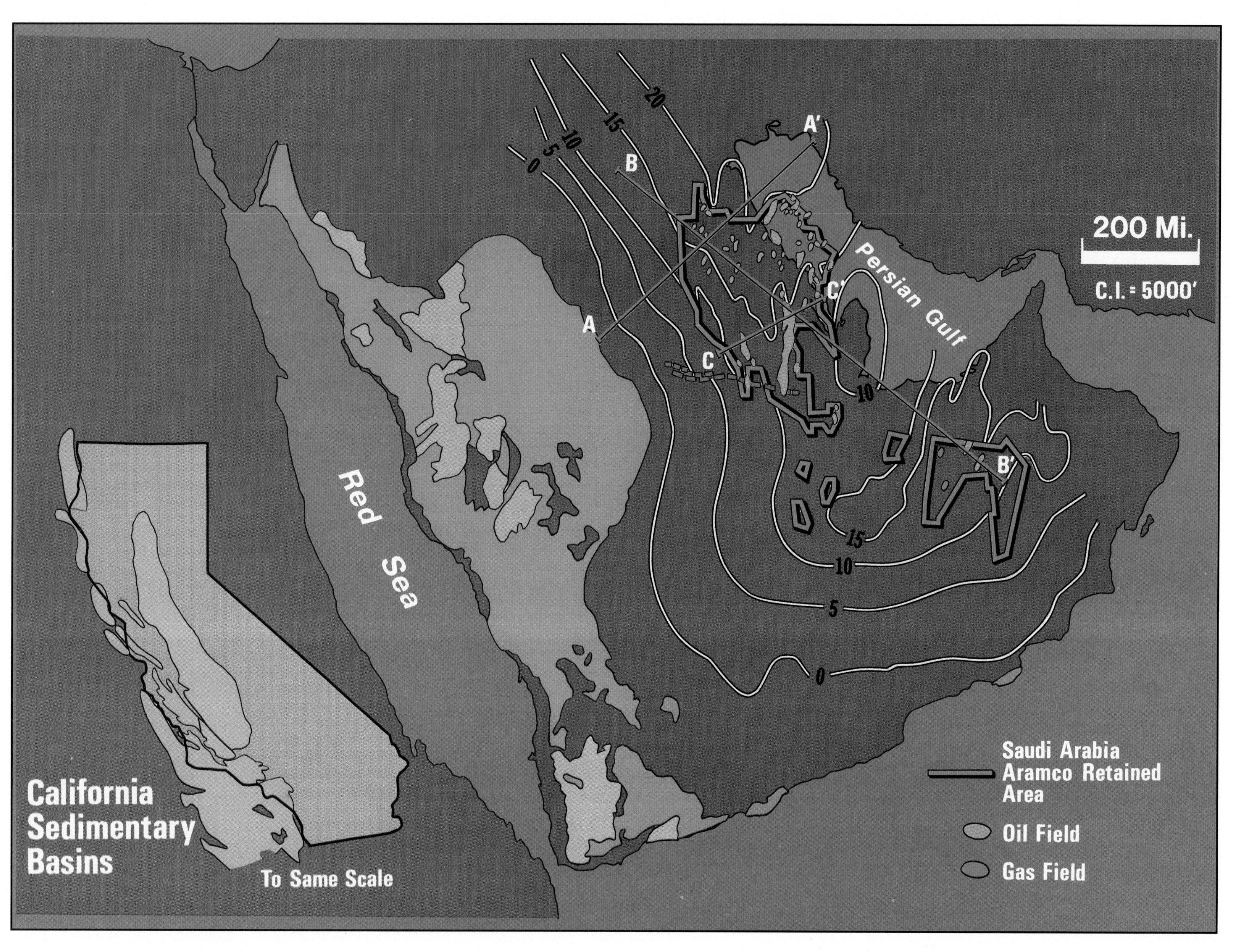


Figure 7. *Generalized structure contour map of the eastern Arabian Peninsula, on the approximate base of Mesozoic (after Brown, 1972), showing the general location of oil and gas fields and the areas retained by ARAMCO (ARAMCO, 1984).*

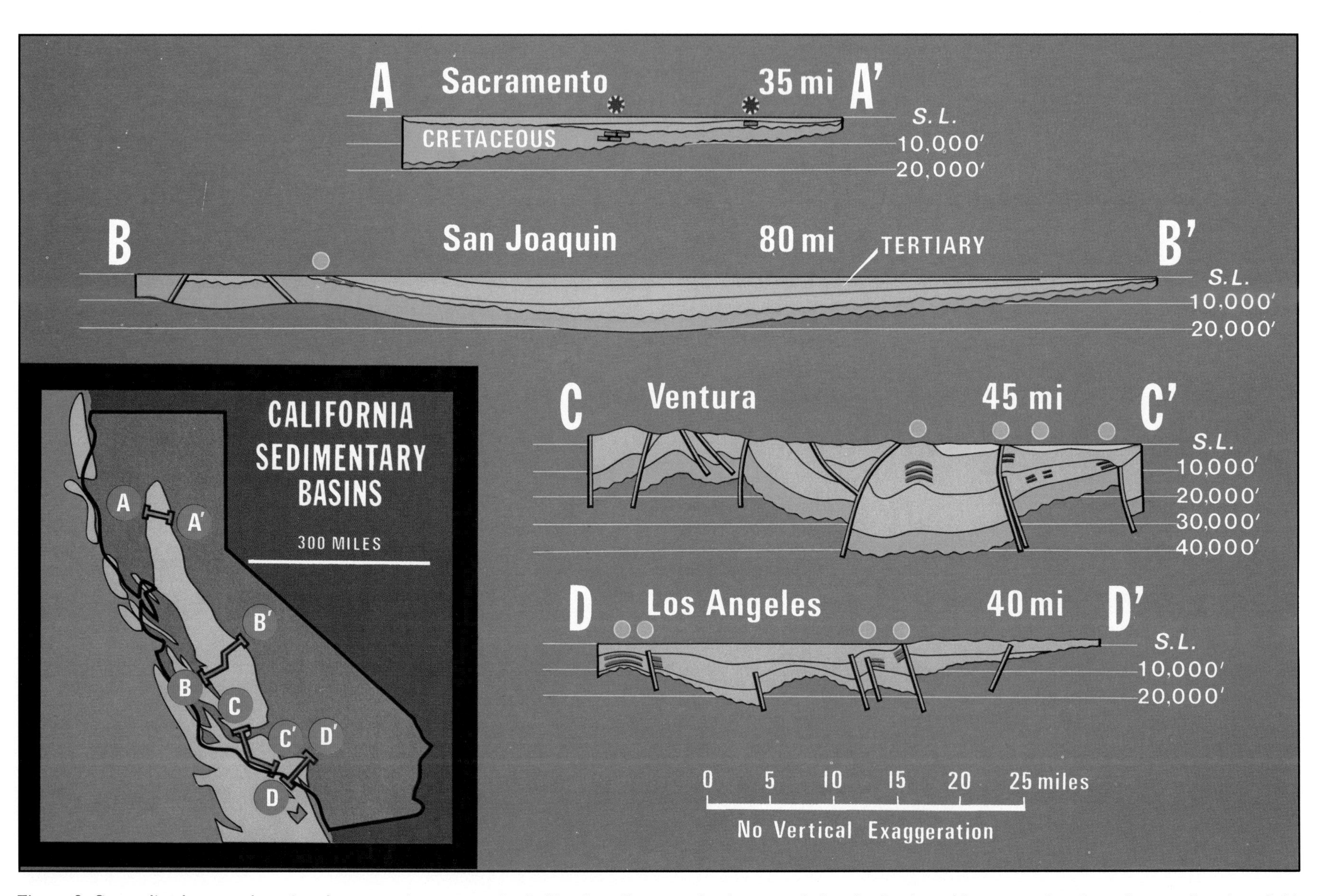


**Figure 8.** *Generalized true-scale regional cross sections across the California sedimentary basins named, showing basin architecture and settings of some oil and gas fields (After Morrison et al, 1971; Callaway, 1971; Nagle and Parker, 1971; and Gardett, 1971).*

kinds of structure that are characteristically involved with California's oil and gas deposits. Section A–A′ crosses the Sacramento portion of the Great Valley and B–B′ the southern San Joaquin portion. They show relatively simple geometry on their eastern ends, the result of the uplift of the Sierra Nevada to the east with a synchronous depression of the Great Valley basins to the west. One can sense that a great amount of the western-thickening sedimentary section was derived from the east. Were section A–A′ to have been continued farther to the west, one would see Cretaceous beds upturned and exhibiting steep east dips. The gas accumulations shown here are typical of the fact that the Sacramento basin has yielded significant gas and not oil. Traps in the Sacramento area include anticlines and faulted folds, but there are also many stratigraphic traps (Graham, 1981a).

Section B–B′, located south of A–A′ (Figure 8), approaches a disturbed belt adjacent at its western end to the San Andreas fault. Note how the Tertiary has been uplifted and eroded away in places, nearer the coastline, as the result of increased structural activity during Tertiary time. The oil accumulations shown in the west on Section B–B′ are Miocene and Paleogene stratigraphic accumulations involving both stratigraphic pinchouts and tar seals at the outcrop (Coalinga and Guijarral Hills fields). In the San Joaquin basin south of Section B–B′, many anticlines formed, either by wrench tectonics associated with the San Andreas and other major strike-slip or wrench faults, or both. Large numbers of oil fields occur here where traps are largely anticlines, often involving faulting and stratigraphic trapping elements (Nehring, 1981).

Sections C–C′ and D–D′ of Figure 8 illustrate the spectacular structural style developed within the coastal area southwest of the San Andreas fault. Tremendous structural relief is common here over very short lateral distances, the result of extreme local block uplift, rotation, and depression. High angle faulting—normal and reverse—often bounds structural blocks with oblique-slip movement or significant strike-slip movement on many of the faults.

Section C–C′ crosses the east-to-west-oriented Ventura basin where it merges with the offshore Santa Barbara Channel basin. At the northern end, we can see a structurally high area, comprised of uplifts associated with long, generally east-to-west faults that characterize the Transverse Ranges. In the center of the section, we can see the deep Tertiary depocenter, the Santa Clara Trough. This center contains the giant Ventura Avenue field, comprised of multiple Pliocene reservoirs in a highly-faulted, tightly-folded anticline that is in the footwall of a large, north-dipping thrust. The most southern and thinner Tertiary sedimentary prism contains the Montalvo field pools (Figure 8, under number 45), and the Oxnard field, updip to the south, involving truncation entrapment. This Tertiary prism dips to the north and thins to the south. The south end of this section crosses the western plunge of the east-to-west-trending Santa Monica Mountains, a site of local Miocene volcanics.

Section D–D′ of Figure 8, crossing the Los Angeles basin in a northeast to southwest direction, demonstrates the northwest-trending Tertiary depocenter, flanked on both sides by subparallel lines of fault-associated, anticlinal uplifts that converge at the northwest end of the basin. At the west end of this section, near the coastline, are the offshore Belmont and onshore Seal Beach fields. Farther inland, on the northeast flank of the depocenter, are the East Coyote and Olinda fields.

**Saudi Arabia**

The true-scale regional structure of eastern Saudi Arabia is difficult to present graphically because of the extent, flatness, and relative thinness of the sedimentary rock prisms. Figure 9 shows a "dip" cross section, A–A′, and a "near-strike" one, B–B′ (see Figure 7 for their location). These cross sections were constructed from contours appearing on the Tectonic Map of the Arabian Peninsula (Brown, (1972), using comprehensive surface, subsurface, and geophysical data.

The true dimensions of the sedimentary packet are shown by the two sections. These sections best document the regional stability and rigidity of the continental crust underlying this area in Cenozoic and Mesozoic times, and, to a more questionable degree, in Paleozoic time. ARAMCO geologists were among the first to emphasize this stability (Powers et al, 1966). This depiction of the Paleozoic reflects no new variations from this simplified pattern documented by exploratory work carried out in the last thirteen years.

On the dip section A–A′ in Figure 9, the regional dip on the base of Tertiary, from the outcrop northeastward to the letter 'a' in the word *Tertiary*, amounts to 15 ft/mi (3 m/km), or one-sixth of a degree. The regional dip on the base of Mesozoic, from its outcrop to this same letter 'a', amounts to about 60 ft/mi (11m/km), or two-thirds of a degree.

Local flank structure at Ghawar (Figure 9), reaches a maximum dip of approximately 2°. Maximum local flank dips near the top of Jurassic are 5–10° for Ghawar (ARAMCO, 1959); 4° for Abqaiq, immediately to the north (El-Aguar, 1960); and $2^1/_2$° for Harmaliya, immediately to the east (Ibrahim, Kahn, and Khatib, 1981). Mesozoic rocks in the area have been flexed only slightly since their time of deposition.

Section B–B′ of Figure 9, southeast of Ghawar, received lesser thicknesses of Mesozoic sediments and greater thicknesses of Paleozoic sediments than has the area to the northwest of Ghawar. As the Arabian craton slowly subsided during the Mesozoic, it was very subtlely segmented into blocks involving lesser or greater subsidence. The recognition of local thickening or thinning of stratigraphic units usually requires very careful data gathering and analysis because bounding bedding planes are so nearly parallel.

Figure 10 shows a slightly more detailed cross section of the Ghawar and Khurais oil fields. Though the structure is subtle, regional data indicate that both anticlines have longer, near-planar northeast and east flanks, in contrast to shorter west flanks. A detailed, published map of Ghawar (ARAMCO, 1959) shows variously steepened local flanks (up to 10°) and documents a close similarity in form between a Bouguer gravity anomaly over the Ghawar structure and the form of known subsurface structure near the top of the Jurassic. The ARAMCO staff (1959) suggest that this anticline is basement controlled.

This explanation concurs with the concept that Khurais is also basement controlled. Structure here represents a passive reaction of at least Mesozoic and Cenozoic sediments to the slight eastward tilting of a supporting basement block that is fault-bounded on the west flank of Khurais and extends northeastward to the syncline just west of Ghawar.

The productive structures of eastern Saudi Arabia are primarily anticlines of depression. Total downward movement of strata through time, from an original site of deposition near sea level, far exceeded any short-term uplifts and truncation. During Mesozoic time when the entire area was sinking, sedimen-

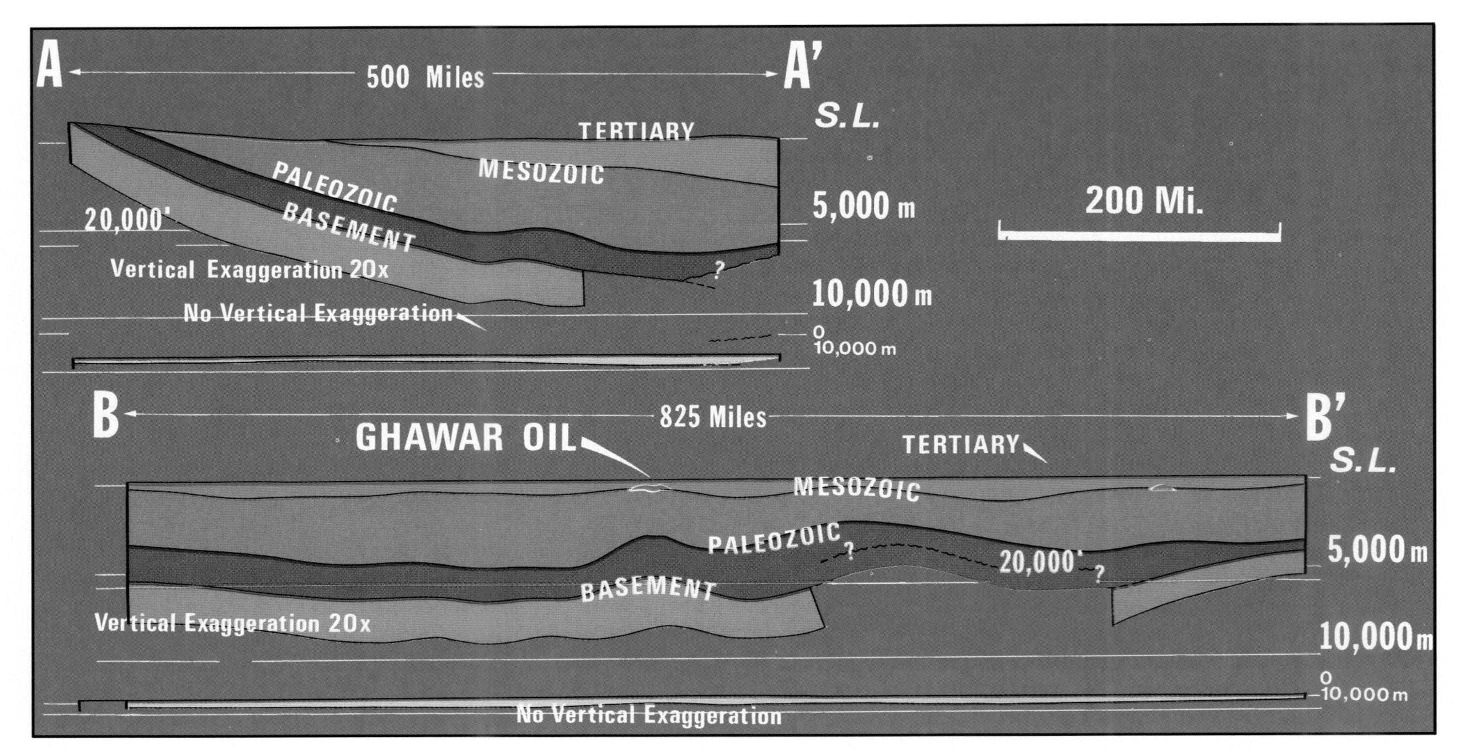


Figure 9. *Generalized, vertically exaggerated, and equivalent true scale, regional cross sections crossing the eastern Arabian Peninsula (See Figure 7 for their location. Sections constructed from map data of Brown, 1972). The general positions of two oil accumulations are shown on Section B–B′.*

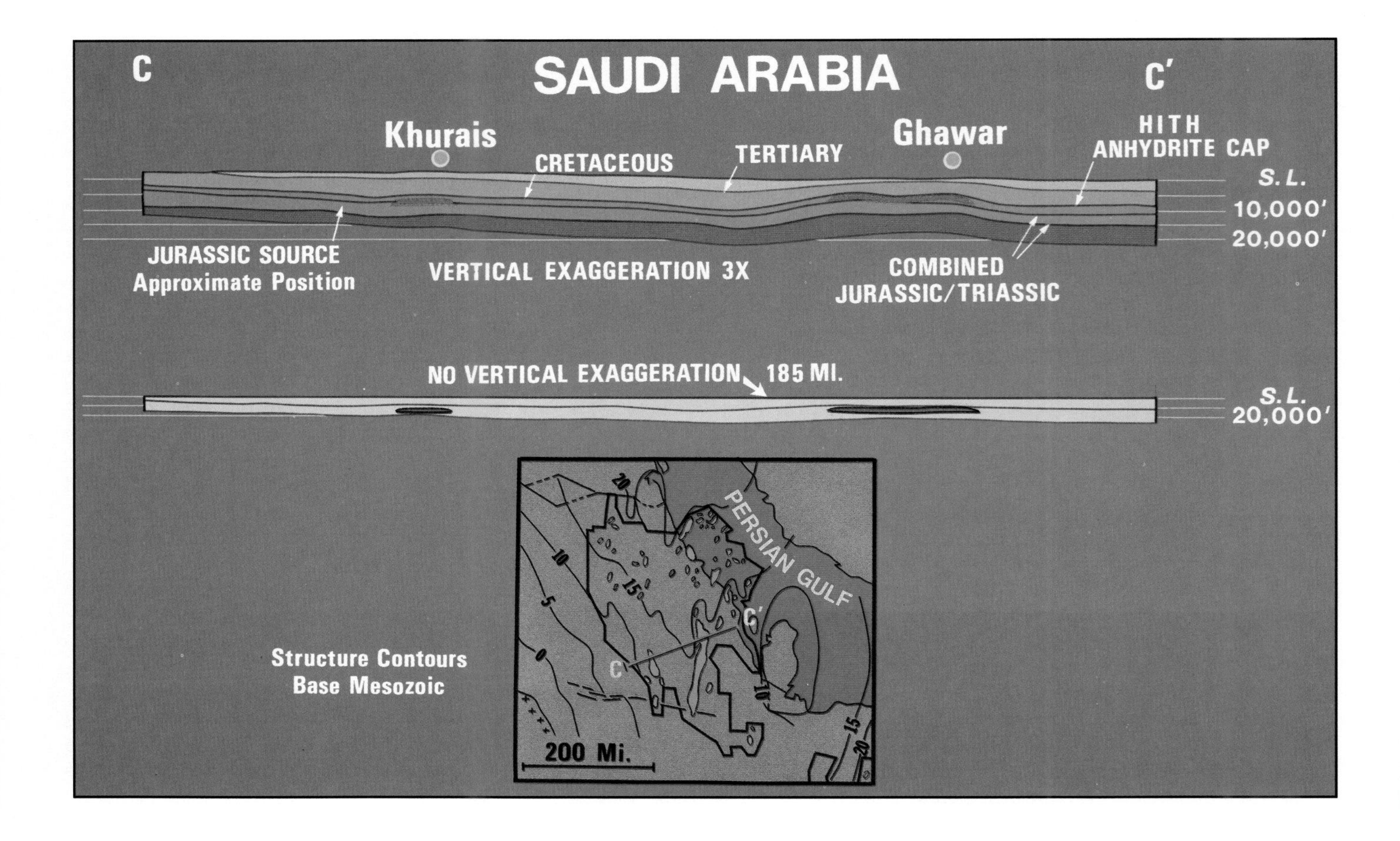

tation equaled subsidence. However, anticlinal areas did not sink as far as synclinal areas. Highs developed over basement in southwesterly fields and over probable mobile salt (Precambrian to Cambrian) in the more northeasterly fields.

Minor faulting does occur in Arabian oil fields. In one part of Ghawar there are possible "normal faults of small displacement which do not alter the fundamental characteristics of the structure" (ARAMCO, 1959). A published map at Harmaliyah, east of Ghawar, shows faults affecting the Jurassic, with throws up to 180 ft (55 m). Although minor bending occurred, relatively small crestal and flank faults of small displacement would be expected in the carbonates and sandstones dominating the Mesozoic column. Coincidentally, such faulting probably exists in the Hout field, in the Divided Zone offshore (D.Z. on Figure 6). This field is immediately north of the Saudi Arabian border and involves crestal normal faulting in middle Cretaceous beds, where faults with up to 600 ft (180 m) of throw die out downward within 3,000 ft (900 m) (Hosoi and Murakami, 1971). Flank dips here are 8.5 and 5.5 degrees.

Figure 11 depicts the cross-sectional contrast between California's basin architecture and Saudi Arabia's architecture. Three features of the cross sections stand out:

1. The extent of continuous, unbroken rocks in this part of Saudi Arabia as opposed to the much narrower, or shorter segments, of unfaulted rocks in California.
2. The comparative size of the source areas from which oil could migrate laterally into the crest of structures, and
3. The widespread gentleness of Arabian dips, as opposed to those found throughout many of the California basins.

Figure 12 emphasizes the contrast between California's and Saudi Arabia's basin geometries, oil field sizes and distributions. This view directly compares two areas of similar length in a northwest-to-southeast direction, but very different widths in a northeast-to-southwest direction. It highlights the continuous, widespread prism of Arabian sedimentary rocks, as opposed to the separated and segmented, smaller sedimentary prisms in California.

Published areal data on California oil fields indicate that the surface areas (Nehring, 1981) of the eleven fields having the largest ultimate reserves (Williams, 1984a), listed on Table 3, range from nearly 29,000 acres (Midway-Sunset) to approxi-

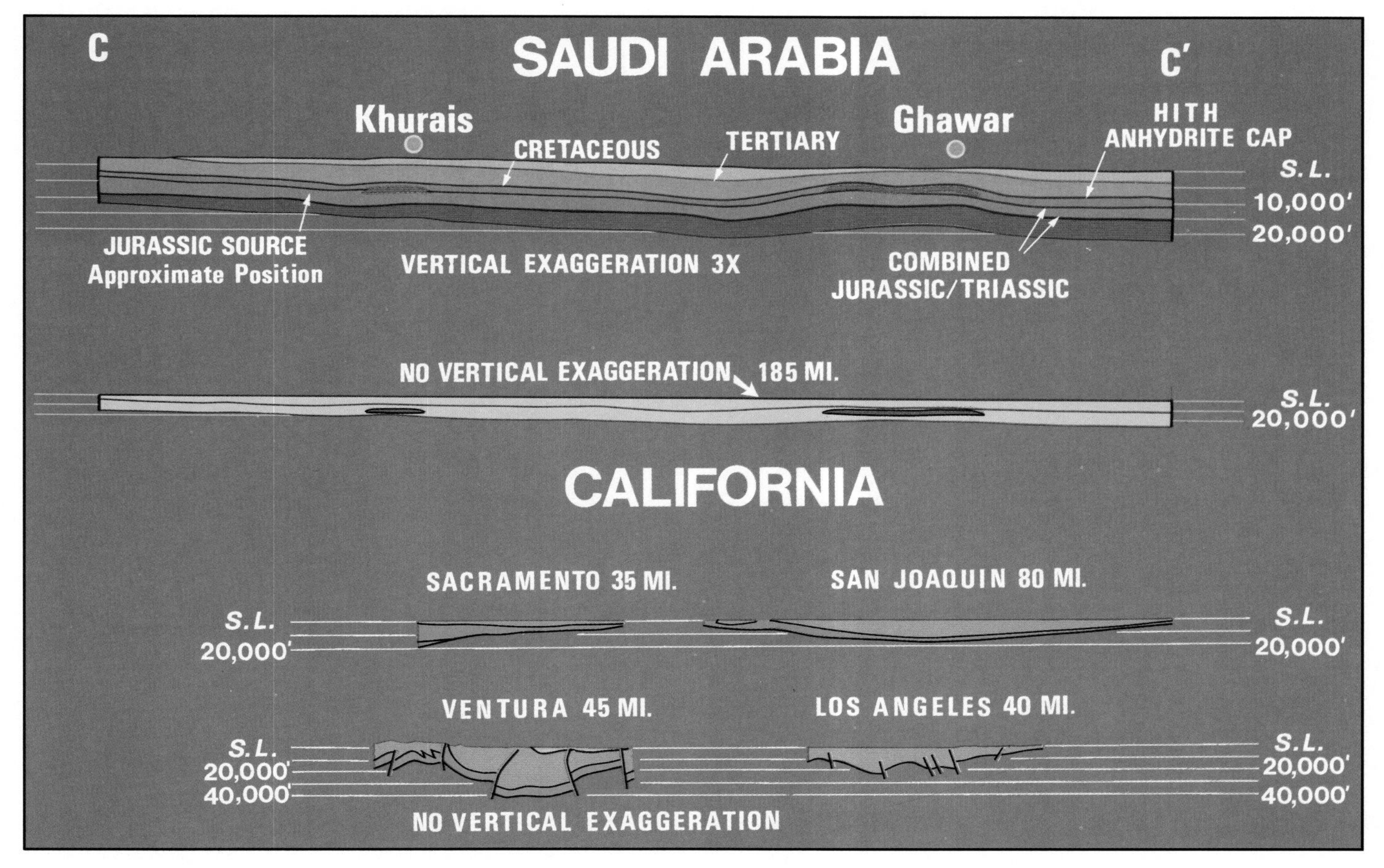


Figure 11. *Generalized regional cross sections of California and Saudi Arabia, for direct comparison at the same true scale. (See Figures 8 and 10 for locations and source of data.)*

Figure 10. *Regional cross sections across the Khurais and Ghawar anticlines, exaggerated and true scale. (Based on map of Brown 1972; and Powers et al, 1966; ARAMCO, 1959; Ministry of Petroleum and Mineral Resources, 1981; and Ayers et al, 1982.)*

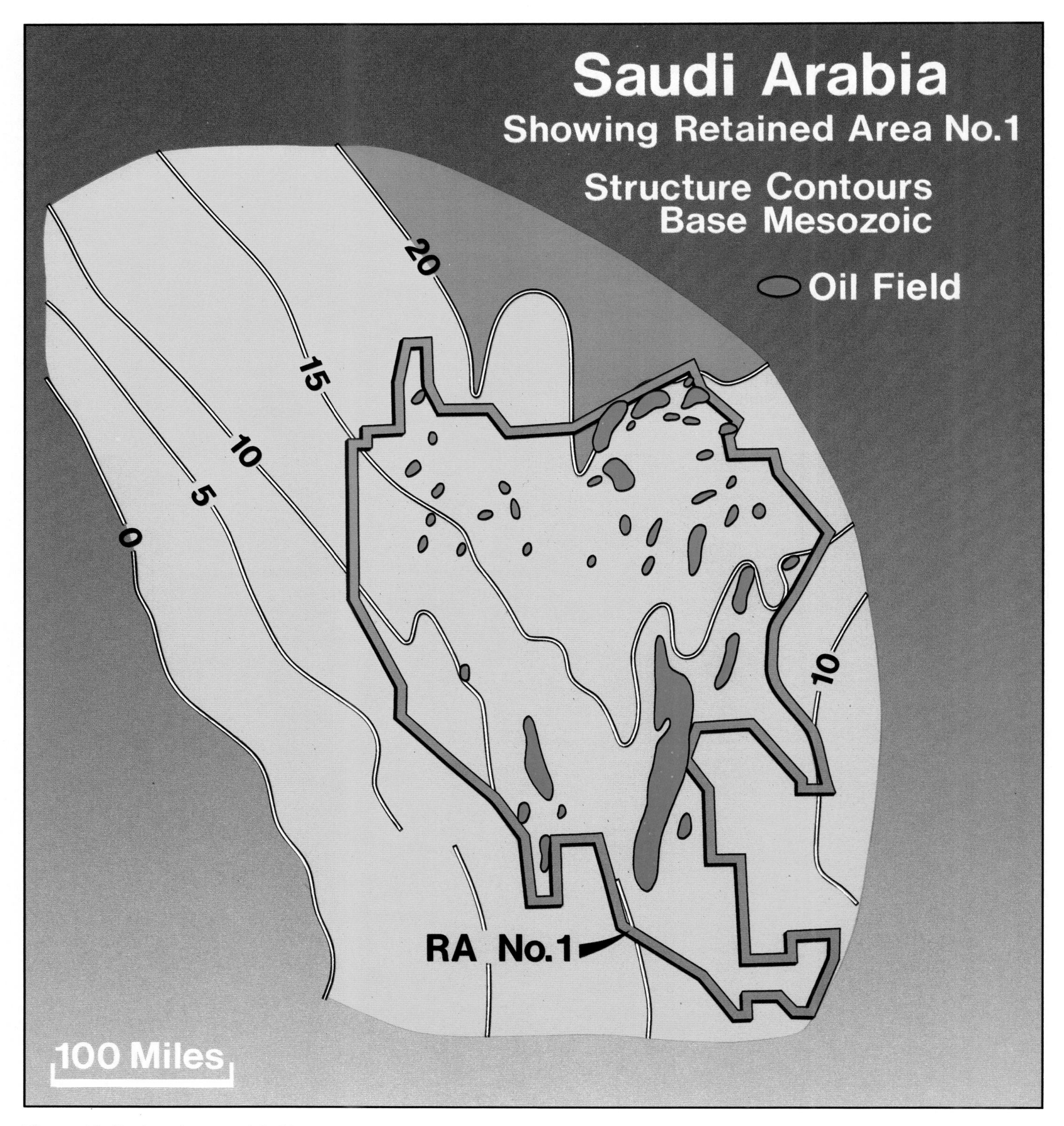


Figure 12. *Regional maps of California and that part of the eastern portion of the Arabia Peninsula containing ARAMCO Retained Area Number 1, both at the same scale.*

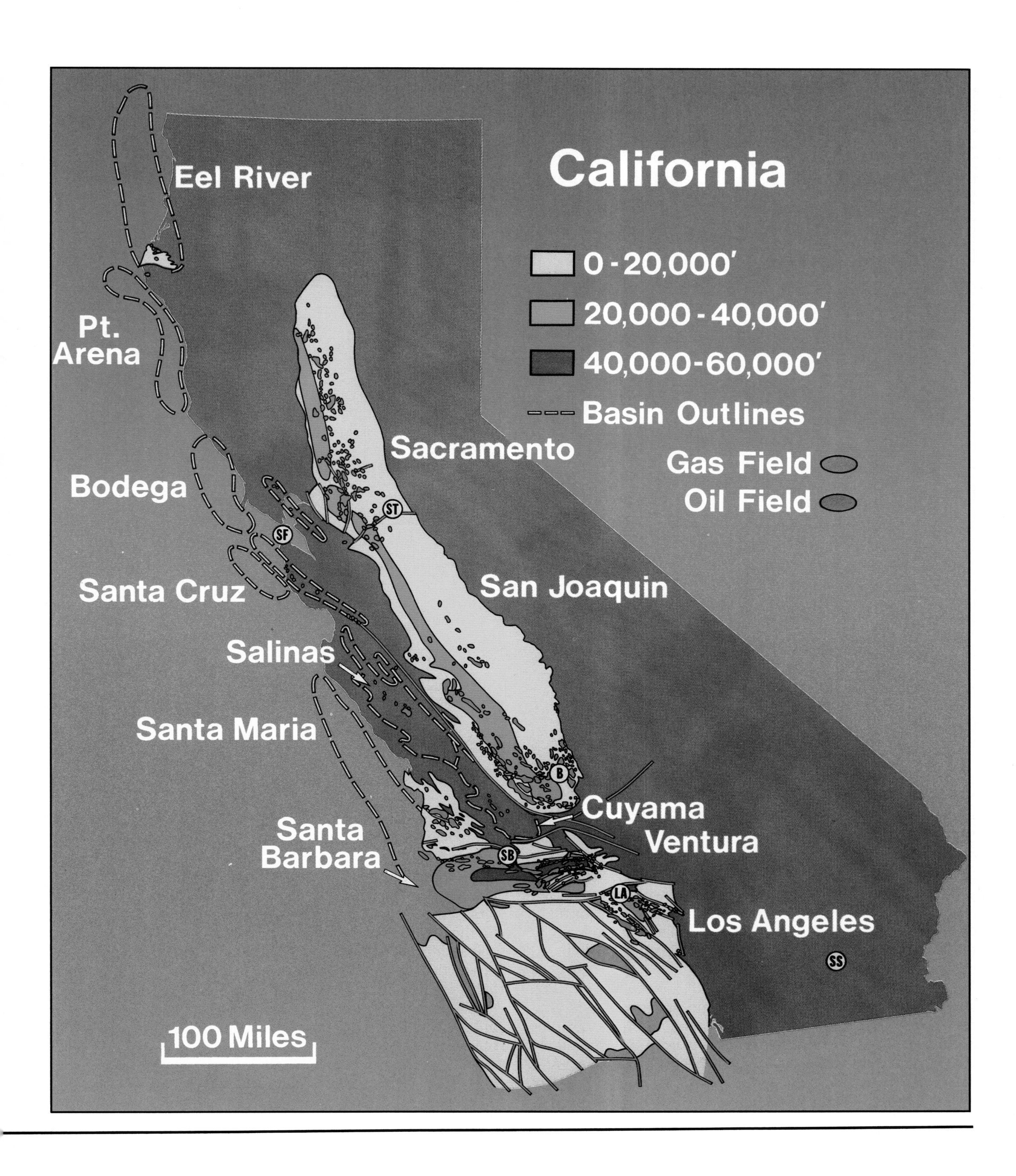
California
0 - 20,000′
20,000 - 40,000′
40,000-60,000′
Basin Outlines
Gas Field
Oil Field
Eel River
Pt. Arena
Bodega
Santa Cruz
Salinas
Santa Maria
Santa Barbara
Sacramento
San Joaquin
Cuyama
Ventura
Los Angeles
ST
SF
B
SB
LA
SS
100 Miles

Table 3. *California: field size distribution—by basin (from California Division of Oil and gas, 1978; and Oil and Gas Journal, 1/30/84).*

| Billion Barrels | San Joaquin | Coast Ranges Santa Maria | Ventura Santa Barbara | Los Angeles | (Total Fields) |
|---|---|---|---|---|---|
| 3— | | | | | |
| | Midway Sunset—2.1 | | | Wilmington—2.5 | (2) |
| 2— | Kern River—1.9<br>Elk Hills—1.5 | | Ventura—1.0 | Huntington Beach—1.1 | (4) |
| 1— | Coalinga—.8<br>Belridge—.8<br>Buena Vista—.7 | | | Long Beach—.9<br>Santa Fe Springs—.6 | (5) |
| 0.6— | (12) | (5) | (6)* | (11) | (34) |
| 0.1—<br>0— | (100) | (16) | (70)* | (55) | (241) |
| Total | (118) | (21) | (77) | (70) | (286) |

Note: numbers in parentheses = Number of fields (e.g. 118)
* Does not include several offshore fields

After Oil & Gas Journal—Jan., 1984
Calif. Div. of Oil & Gas, 1978

mately 1,500 acres (Santa Fe Springs), with an average of 12,250 acres per field (ultimate reserves for these fields equal the sums of cumulative production to January 1984, plus the estimated remaining reserves as of that date). These fields have a total areal extent of about 135,000 acres, with total reported reserves of 13.9 billion bbl. The average productivity for these fields thus equates to 103,000 bbl/acre. Individual productivity ranges from a calculated 40,000 bbl/acre (Coalinga), to 540,000 bbl/acre (Long Beach).

Measurements made from a published map (ARAMCO, 1982) of the areal dimensions of the eleven largest fields in Saudi Arabia (Table 4) having the largest quoted ultimate reserves (Nehring, 1978) suggest sizes ranging from approximately 1,000,000 acres (Ghawar) to approximately 30,000 acres (Abu Safah) (ultimate reserves for these fields equal the sums of cumulative production to December 31, 1975, plus reserves listed for that date). The average size, including Ghawar, approximates 195,000 acres, and for the ten fields other than Ghawar, 110,000 acres. Hence, the areal sizes of California's larger oil fields are smaller than the sizes of the equivalent number of Saudi Arabia's large fields, by an order of magnitude.

Calculated areal productivities in Saudi Arabia appear to vary from approximately 60,000 bbl/acre in Khurais, to 210,000 bbl/acre in Abu Safah. The calculated average for the eleven largest fields, including Ghawar, is about 90,000 bbl/acre, and for the ten fields other than Ghawar, about 105,000 bbl/acre.

In California 40 fields of over 100 million bbl have an average productivity of 86,000 bbl/acre (Nehring, 1981). Nineteen such fields in the San Joaquin basin average 66,000 bbl per acre, nine equivalent fields listed for Coastal areas average 76,000 bbl per acre, and 12 in the Los Angeles basin average 177,000 bbl per acre.

Mature source beds are obviously present in the deeper parts of each California basin as proven by the presence of the large oil and gas fields. The size of these fields also indicates that the source units have abundant hydrocarbons, because drainage areas are limited to adjacent relatively short, local flanks ending in nearby synclines.

Table 4. *California and Saudi Arabia: comparative oil field sizes (from AAPG, Oil and Gas Journal, California Division of Oil and Gas, and Nehring, 1978).*

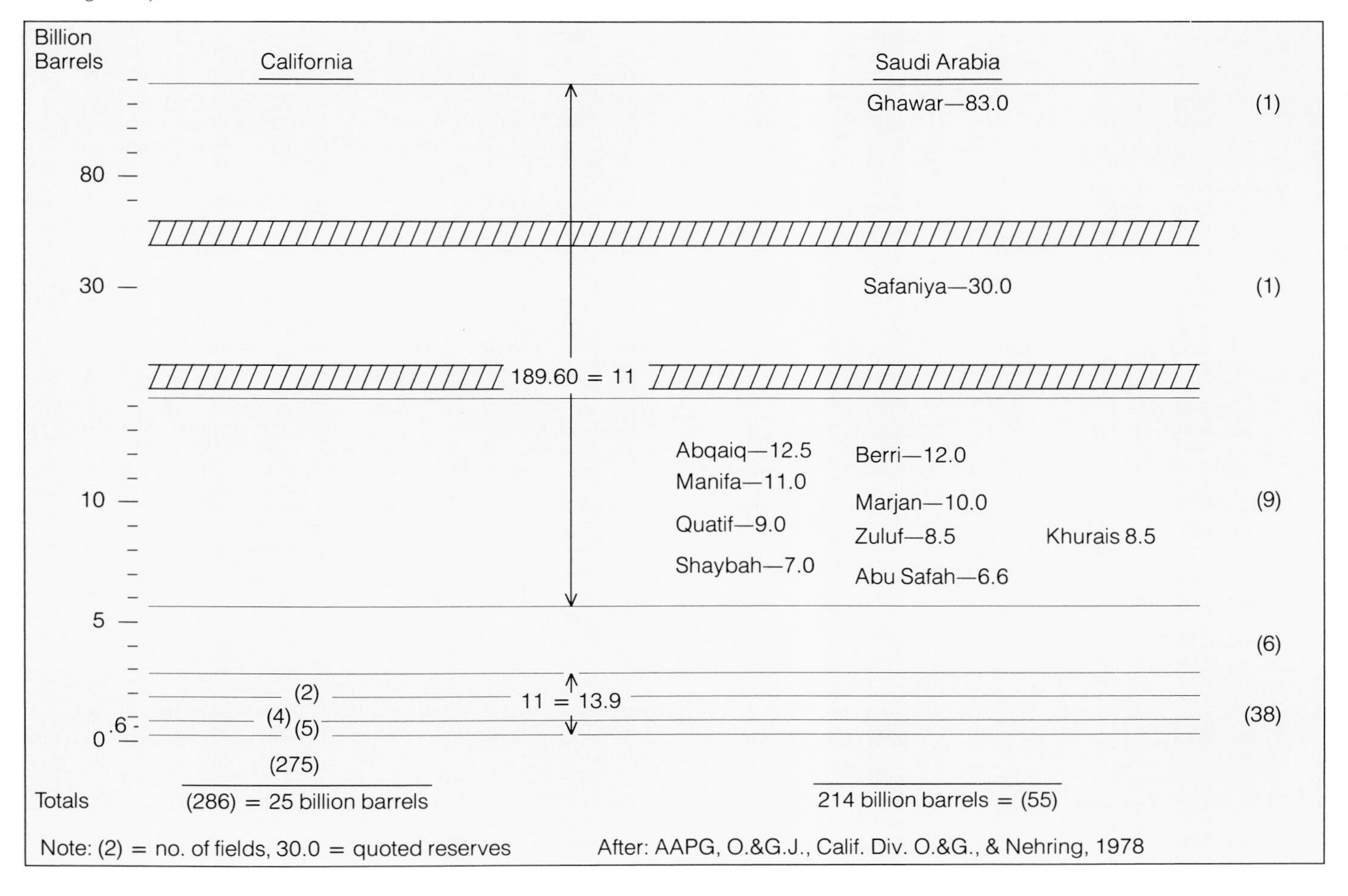


## STRATIGRAPHY

### California

The stratigraphy of California is characterized by several elements directly related to the complicated Mesozoic and Cenozoic geologic history along the western edge of North America. The rock sequence is dominantly clastic, with ubiquitous sandstone, siltstone, and claystone. Local basins were often characterized by large-scale relief. Deep-water depocenters were frequently close to shorelines with relatively high source areas on the landward side. Facies changes can be very rapid, resulting from the short distances between flood plain, deltaic, littoral and bathyal sediments. Fluvial, paralic, shoreline, fan, and turbiditic clastics all occur. Two major instances of quiet and deep water, fine-grained, siliceous-carbonate-phosphatic sedimentation took place, once in the Eocene and once in the Miocene. Some short-term volcanism occurred, with local deposition of lava flows and ash falls.

Mesozoic and Cenozoic deposition in California resulted in filling local depressions in this tectonically active area, convergent in the early stages shifting to transform movement in the later stages (Graham, 1981b).

Figure 13 shows two simple sections that briefly suggest the dominant stratigraphic style of California hydrocarbon-bearing basins. On the left is a section near the geographic center of the southern portion of the San Joaquin basin. At the northwest end of this section, the strata are thinner and older than is the case to the southeast. The general presence of sandstone is

Table 5. *California and Saudi Arabia: geologic contrasts (Mesozoic and Cenozoic).*

| | California | Saudi Arabia |
|---|---|---|
| Plate tectonics: | Active | Passive |
| Regional structure: | Complex<br>Depression & uplift | Simple<br>Depression |
| Basins: | Localized | Broad |
| Source rocks: | Local, though potent | Widespread & potent |
| Stratigraphy: | Clastics (minor other)<br>Variable & local | Uniform & broad<br>Carbonates & evaporites<br>Lesser clastics |
| Seals: | Local | Widespread |
| Structure: | Dips highly variable<br>Unprotected | Dips very low<br>Protected |

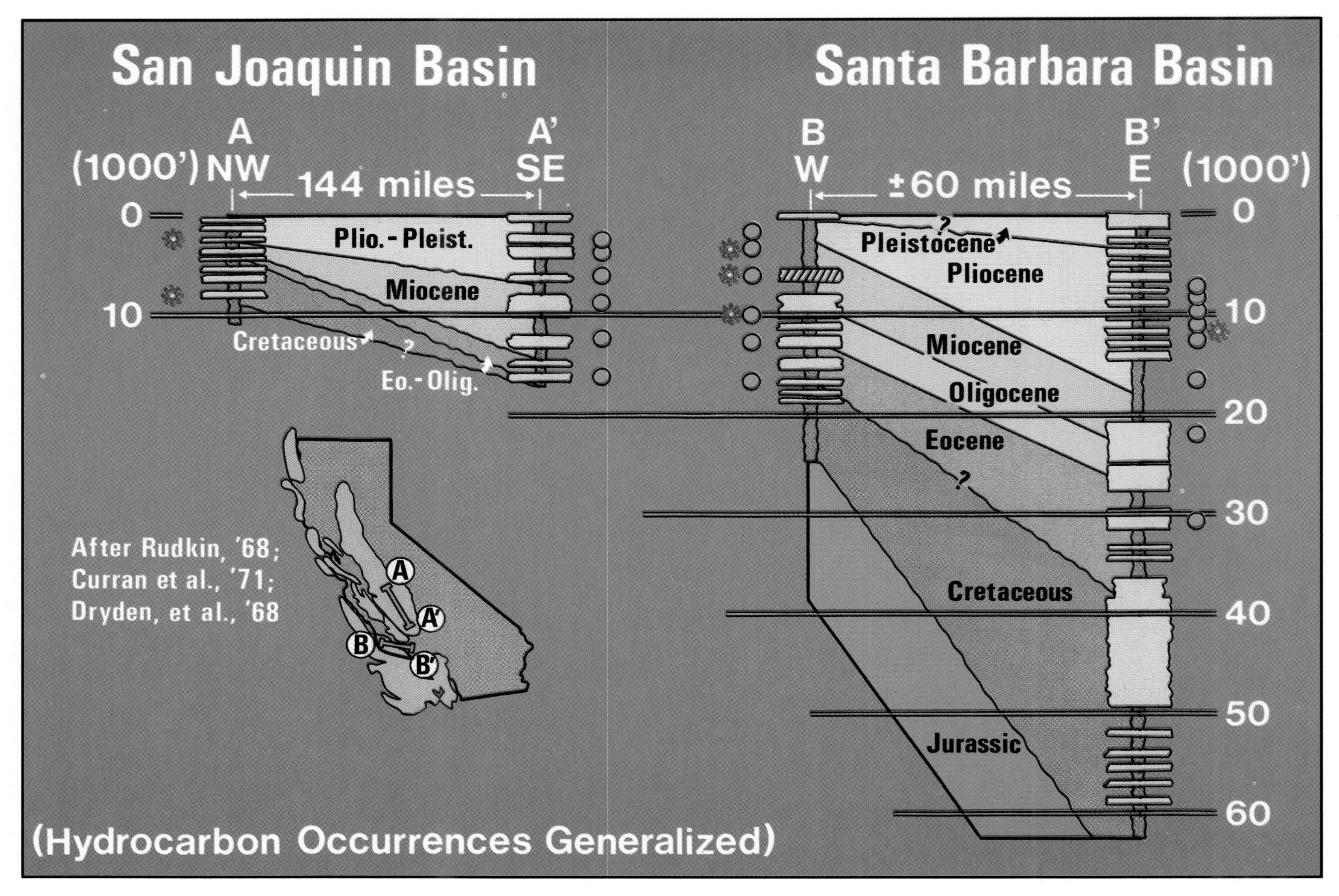


Figure 13. *Generalized representative stratigraphic columns and cross sections from the San Joaquin and Santa Barbara basins, California.*

shown as yellow on the stylized log patterns. This section is northeast of the deeper part of the local basin (Figure 5).

On Figure 5 we can see that the northern end of the San Joaquin basin and the more northerly Sacramento basin are the sites of gas fields. Insignificant oil fields have been omitted. Cretaceous, gas-prone source rocks are within the generative window, whereas any Tertiary oil-prone source beds are too shallow (Zieglar and Spotts, 1978). In the southern part of the San Joaquin Valley, in contrast, two depocenters show much greater thicknesses and much deeper burial of Tertiary oil-prone source material. Reservoirs in these areas show a complicated structural history. This structure provided proper conditions for numerous oil fields. Throughout the combined Sacramento and San Joaquin basins, sandstones are the dominant reservoir rock (Zieglar and Spotts, 1978; Nehring, 1981), and upper Miocene diatomites contain large volumes of oil. Enhanced recovery programs are increasing the percentage of recovery in these fields (Williams, 1984a, b and c).

In the more southerly basins of California, southwest of the San Andreas fault, the basinal dimensions are smaller. The thickness of sediments, however, can be large, and the hydrocarbon production of the basinal system can be locally impressive.

The triangular-shaped Los Angeles basin is the most southeastern of the major basins in California with most of its fields located onshore (Figure 5). Nehring (1981) summarizes facts regarding 30 of the largest fields in this basin, and lists their reservoirs in sandstone, with one exception producing from schist. Maximum sediment thickness in this basin is greater than 30,000 ft or 9,000 m (Gardett, 1971).

The combined east-to-west-oriented Ventura (onshore) and Santa Barbara Channel (offshore) basins, adjacent to the Transverse Ranges, contain the thickest section of significant Upper Cretaceous and Cenozoic strata of any basin in California (Figure 5). A generalized east to west onshore section along the northern border of the Santa Barbara Channel appears in Figure 13 (cross section B–B′). Using the combined implications of Figures 5, 8, and 13, we can sense what a dynamic depocenter this was in both Cretaceous and Cenozoic time.

As shown in Figure 13, sandstones are prominent, separated by fine-grained rocks, largely siltstones and claystones. The majority of oil in the easterly, onshore Ventura basin oil fields occurs in sandstone. In the more westerly fields, both offshore and onshore, including the onshore Santa Maria basin, the middle and upper Miocene Monterey formation is an important reservoir. It receives ongoing notoriety as the host for large, newer discoveries made in the Santa Barbara Channel basin and near Pt. Arguello, where this basin joins the offshore Santa Maria basin. Many predict that new Monterey discoveries will be made offshore farther to the northwest. This is alluded to in annual review articles on offshore California (B. Williams, 1984a and 1985a) and indicated by continued industry interest in having leases made available in appropriate areas to facilitate new discoveries here.

The Monterey warrants special note. First, as much as 2 billion bbl of offshore reserves have been discovered in this formation since 1981 in this 100 mi (160 km) long coastal area. This formation is made up of "biogenous and diagenetic silica, detrital minerals, carbonate rocks, apatite, and organic matter" (Isaacs, 1980). Outcrop evidence along the northern onshore exposures bordering the Santa Barbara Channel suggest increasing diagenesis to the west, because of greater original burial (Isaacs, 1979). These Monterey strata, both offshore and farther onshore, also record an episode of phosphorite and sandstone formation in which unusual coastal conditions caused organic-rich deep-water rocks to be deposited "within or near fluctuating boundaries of a well developed oxygen minimum zone" (Garrison, 1981). It is important that such a rich, combined "source-fractured reservoir" rock developed locally along an active continental margin.

Monterey Formation exploitation is currently limited to a portion of California. Neither its full offshore exploratory potential nor its onshore enhanced-recovery potential has yet been attained.

### Saudi Arabia

The area of the Saudi Arabian oil fields near the Persian Gulf has a thin sheet of Cenozoic sedimentary rocks; a significant, widespread, thick prism of Mesozoic rocks; and what must be a Paleozoic sequence showing similar sheet-like geometry, though probably thinner and more locally variable.

As Figure 6 shows, the Paleozoic outcrops continue for 600 mi (960 km) in a curving, generally north to south direction, a short distance west of the city of Riyadh (*R* on Figure 6). From its most eastward position, southwest of Riyadh, for 200 mi (300 km) to the southwest, it consists of only the Permian Khuff carbonate formation (U.S. Geological Survey, 1963), between 750 and 1,000 ft (230 and 300 m) thick, resting on basement. Farther northwest, this unit continues displaying very uniform thicknesses along the strike. The Paleozoic outcrops become wider, as lower and middle Paleozoic formations appear subcropping below the base-Khuff unconformity. At the northern end of the northwest-trending straight outcrops, the combined Khuff and Cambrian–Devonian clastic formations attain thicknesses approaching 6,500 ft or 2,000 m (Brown, 1972).

Brown (1972) estimated that along the east coast of Saudi Arabia, the Paleozoic may also be about 6,000–7,000 ft (1,800–2,100 m) thick. Even though significant gas is reported to have been discovered in the Khuff Formation in eastern Saudi Arabia and geologists are interested in Paleozoic oil, I will not discuss rocks of this age.

We can appreciate the regional dimensions of the Mesozoic sequence by considering the true-scale aspects of the combination of Figures 6, 7, 9, and 10. The compositional makeup of this sequence is shown in Figure 14.

The column (Figure 14) graphically describes the sequence found in the area of the Arabian oil fields. The Mesozoic section varies from zero at its western outcrop edge to an estimated thickness of 16,000 ft (4,900 m) or more near the coastline, 325 mi (520 km) to the east (Brown, 1972). The degree to which the total sequence may vary in thickness is shown in the cross sections on Figure 9.

As shown by the lithologic symbols in Figure 14, the Mesozoic is dominated by carbonates, but it also has a significant sequence of evaporites and some prominent and important sandstones. In black are shown the level at which some 26 reservoir zones occur in the various oil fields (after Ibrahim, Kahn, and Khatib, 1981). Dunnington (1967) graphically showed that oil in the northeasterly offshore corner of the large, Retained Area No. 1 occurs in Cretaceous reservoirs, whereas oil farther south and southwest tends to occur in Jurassic rocks. (This speculation was later amplified by Halbouty et al, 1970; and Ayers et al, 1982).

A remarkable aspect of Arabian Mesozoic stratigraphy is the existence of many geographically widespread porous zones overlain by effective cap rocks. The petrology of these rocks—calcarenites, calcarenitic limestones, dolomites, micrites, and evaporites, developed over an expansive shelf that existed over thousands of square miles—has been amply discussed by others (Steineke, Bramkamp, and Sanders, 1958; Bramkamp and Powers, 1958; Powers, 1962; Murris, 1984). Capping can occur by micritic limestone, shaly beds, dolomitic beds, and evaporites. The famous Upper Jurassic Hith anhydrites, in a section 500 ft (150 m) thick at Ghawar, extend over most of eastern Saudi Arabia. The lowermost Hith anhydrite is the effective cap for many tens of billions of bbl of oil in the Arab D reservoir throughout the 160 mi (260 km) long Ghawar anticline, as well as in other fields.

The sandstone shown in the Cretaceous in Figure 14 is the main rock body, having a lesser lateral extent than most others. This unit pinches out easterly and undergoes a facies change to marine shales and limestones along a generally north to south line 100 mi (160 km) east of Ghawar (Powers, 1968). It is still sheet-like and widespread enough to form excellent reservoirs in offshore Saudi Arabia.

Eastern Arabian Mesozoic rocks extend over many thousands of square miles, with individual stratigraphic units often being correlated for hundreds of miles. Murris (1984) has comprehensively described and illustrated this relationship. Of particular value are the two cross sections he published, extending east from the Ghawar oil field for a distance of nearly 450 mi (720 km). In these cross sections, he shows the amazing continuity of rock units in the interval Middle-Jurassic to Albian. Over this distance, the largest, average regional thickness change, from a maximum total thickness of 5,850 ft (1,780 m) at the west, is less than 2 ft/mi (0.4 m/km). The maximum stratigraphic dip shown is less than one-half of one degree. He documents alternating porous and non-porous rocks, as well as two prominent source rocks.

On the right-hand side of Figure 14 is the stratigraphic position of the two source rocks identified by Murris (1984). They are in the area of Qatar for the lower one, just east of Saudi Arabia, and farther to the east by 50–200 mi (80–320 km) for the upper one.

On the left-hand side of Figure 14 are source rock zones identified in Saudi Arabia, with the one in the Tuwaiq Mountain formation the most important (Ayers et al, 1982). In this unit, total organic content is reported to average three to five percent by weight; bitumen content is high, often exceeding kerogen; and visual analysis shows the kerogen to be dominantly amor-

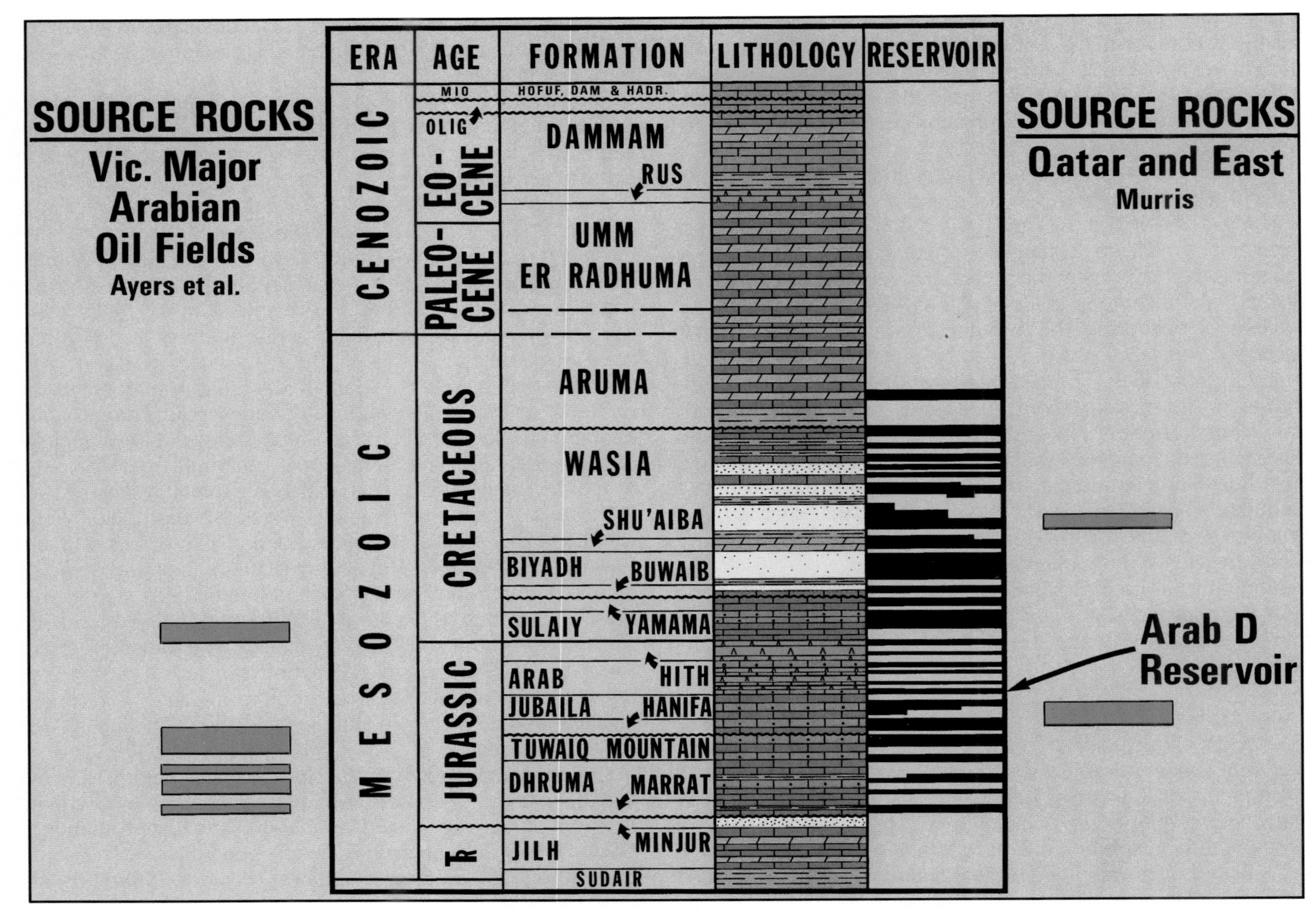


Figure 14. *Generalized post-Paleozoic stratigraphic column, eastern Saudi Arabia, Vicinity of ARAMCO Retained Area 1. The stratigraphic positions of local oil reservoirs are shown in black. The reported stratigraphic positions of local source rocks are shown in green on the left, and those for more easterly source rocks in green on the right.*

phous. Electric log responses were used in determining the extent of this source rock, its distribution and volume in Figure 15 showing this.

The major Jurassic accumulations, interestingly, occur in the area of this source, near and above it stratigraphically. It appears that oil was released from the Tuwaiq Mountain source in latest Mesozoic and Tertiary time and migrated vertically through fractures until reaching appropriate cap rocks over anticlines.

The production of offshore oil in Cretaceous reservoirs from sources farther east and north also seems a reasonable possibility, along with some local vertical migration.

The Mesozoic stratigraphy of Saudi Arabia varies drastically from that of California. Relatively shallow-water platform carbonates, evaporites, fine clastics, and limited coarser clastics are the dominant rock types, as opposed to the predominant clastics in California. Strata were deposited at or very near sea level on a low-relief shelf extending for hundreds of miles. Sedimentation kept up with platform subsidence, and facies changes normally took place over much greater distances, with rare exceptions. Potential reservoir and caprock existed over great areas. Their continuity was usually not broken by erosion during the formation of large-scale anticlines in the Mesozoic. Unconformities are rare in the rock record. Minor structural activity in the area reduced erosion to the degree that reservoir-caprock pairs were rarely breached. Traps were protected from any significant rupture or erosion through geologic time, and remained available to migrating oil.

## CONCLUSIONS

In his opening talk at the Pratt Memorial Conference (1984), Michel T. Halbouty reminded us that explorationists must not only appraise frontier areas, but must also reevaluate mature regions. He also stressed the need for new and better use of geology, and other allied scientific disciplines.

Comparing basinal areas objectively in a systematic manner comprises one way of "using geology better." Doing this for California and Saudi Arabia facilitates drawing the following conclusions:

1. Assembling facts regarding the geology of California and

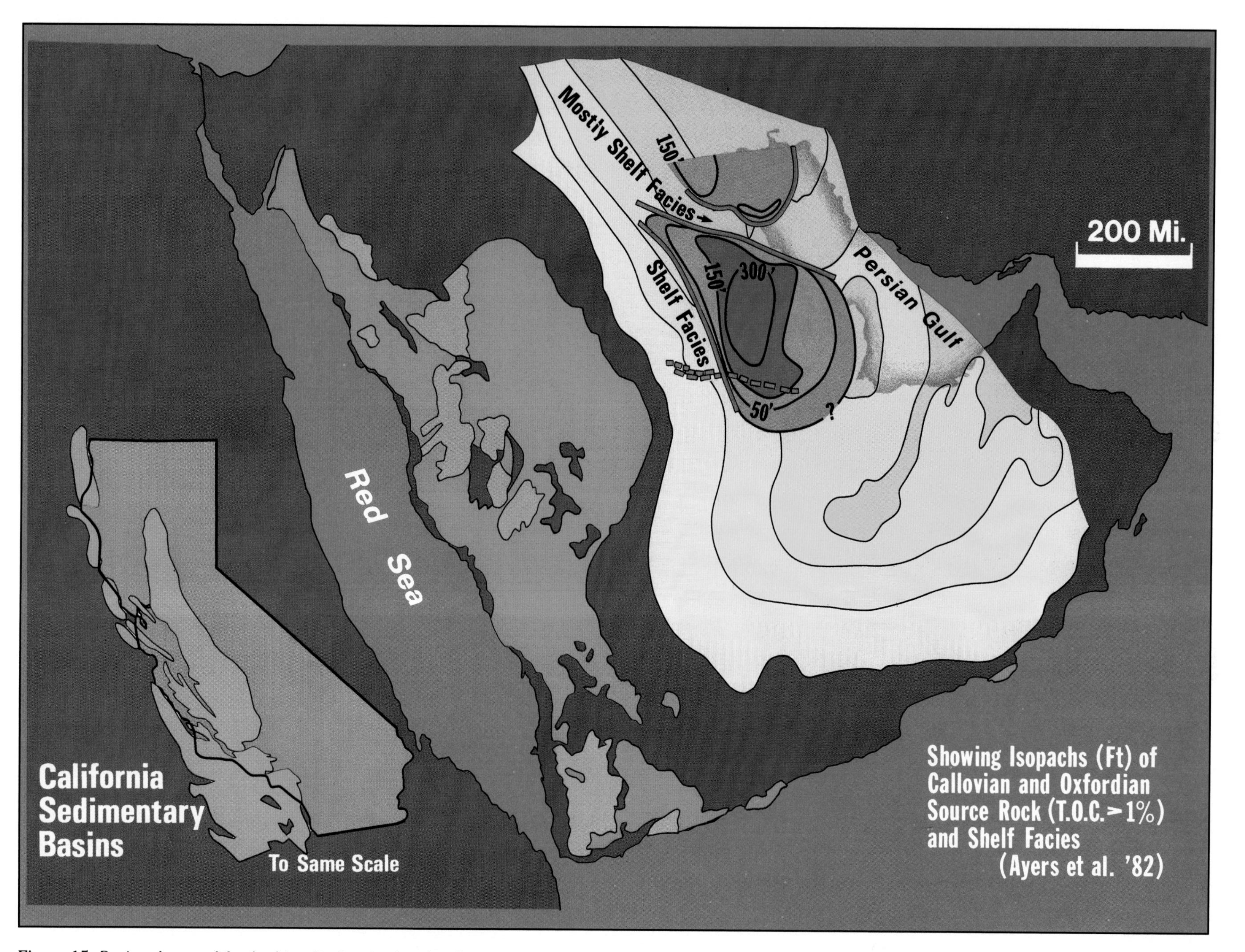


Figure 15. *Regional map of the Arabian Peninsula showing the extent reported Callovian and Oxfordian source rock (Ayers et al, 1982). Regional brown and tan bands represent the form of structure contours on the approximate base of Mesozoic from Figure 7. California sedimentary basins are plotted at the same scale.*

Saudi Arabia in a similarly-ordered fashion for both areas proved valuable. Even though most of the facts were well known, analyzing them this way highlighted the specific elements causing these provinces to contrast so markedly.

2. These elements resulted from the differing geologic histories of the areas in terms of plate tectonics. California had an active history, characterized by complicated tectonism; Saudi Arabia had a passive history, characterized by the most gentle tectonism. The differences are summarized in Table 5.
3. The areal size of the eleven largest fields in Saudi Arabia exceeds that of the eleven largest fields in California by an order of magnitude.
4. For the same largest fields in both areas, the average reserves per acre in California equals that for Saudi Arabian fields. Two California fields have reserves per acre exceeding any in Saudi Arabia.
5. That such a large amount of oil accumulated in Saudi Arabia, as compared to California, is attributable to the comparative simplicity and immense areal size of traps, source areas, and source-rock systems. It is not attributable to any greater sourcing or trapping power of a local nature.
6. Continued basin modelling in both areas will be valuable for learning more about the individual contributions of sourcing power, migration paths, available reservoir volumes, capacity of seals, and geologic timing.
7. The completion of this paper reemphasized the importance of two questions facing world explorationists: First, is there another Saudi Arabia somewhere? And second, where are those immaturely explored basins that will ultimately prove to be analogs of those containing California's oil?

## ACKNOWLEDGMENTS

I was stimulated by students and faculties at several west coast universities where I was invited to discuss this subject. I wish to thank the management of the Chevron Corporation and Chevron Overseas Petroleum, Inc. for sponsoring this paper. I also wish to thank the following Chevron personnel for their role in helping with this study: Messrs. John P. McLaughlin and J. W. Fusso for assembling data; Mr. F. D. Pruett for early suggestions as to content; Messrs. Gerard J. DeMaison, Peter J. Pickford, and Donald L. Ziegler for their critical review of the final manuscript; Mrs. Barbara Lind for superb drafting; Mrs. Jeannie Bloxham for effective stenographic help; and Ms. Michele Morris for patient typing.

## REFERENCES CITED

Arabian American Oil Company Staff (ARAMCO), 1959, Ghawar oilfield, Saudi Arabia: AAPG Bulletin, v. 43, p. 434–454.

ARAMCO, 1982, ARAMCO 1981: Public Relations Department, Aramco, Dhahran, Saudi Arabia, p. 1–33 (unnumbered).

——, 1984, ARAMCO 1983: Public Relations Department, Aramco, Dhahran, Saudi Arabia, p. 1–29.

Ayers, M.G., 1982, Hydrocarbon habitat in main producing areas, Saudi Arabia: AAPG Bulletin, v. 66, p. 1–9.

Barbat, W. F., 1958, The Los Angeles basin area: AAPG, A Symposium–Habitat of Oil, p. 62–77.

Bramkamp, R. A., and R. W. Powers, 1958, Classification of Arabian carbonate rocks: Geological Society of America Bulletin, v. 69, p. 1305–1318.

Brown, G.F., 1972, Tectonic map of the Arabian Peninsula: Ministry of Petroleum and Mineral Resources—Directorate General of Mineral Resources, Kingdom of Saudi Arabia, Arabian Peninsula Series Map AP-2.

California Division of Mines and Geology, 1966, Geology Map of California 1:2,500,000.

California Division of Oil and Gas, 1979, 64th Annual Report of the State Oil and Gas Supervisor—1978: Sacramento, p. 1–164.

Callaway, D.C., 1971, Petroleum potential of San Joaquin basin, California: *in* Future Petroleum Provinces of the United States: Their geology and potential, Ira H. Cram, ed., v. 1, AAPG Memoir 15, p. 239–253.

Crawford, F. D., 1971, Petroleum potential of Santa Maria province, California, *in* Future Petroleum Provinces of the United States—Their geology and potential, Ira M. Cram, ed., v. 1, AAPG Memoir 15, p. 316–328.

Crowell, J.C., 1974, Origin of late Cenozoic basins in southern California: SEPM Special Paper 22, p. 190–214.

Curran, J. F., K. B. Hall, and R. F. Herron, 1971, Geology, oil fields, and future potential of Santa Barbara Channel area, California: *in* Future Petroleum Provinces of the United States: Their geology and potential, Ira H. Cram, ed., v. 1, AAPG Memoir 15, p. 192–211.

Dickinson, W. R., 1981, Plate tectonics and the continental margin of California: Geotectonic Margin of California–Rubey Volume 1: Englewood Cliffs, N.J., Prentice Hall, p. 1–28.

——, and D. R. Seely, 1979, Structure and stratigraphy of forearc regions: AAPG Bulletin, v. 63, p. 2–31.

Dryden, J. E., R. C. Erickson, and T. Off, 1968, Gas in Cenozoic rocks in Ventura and Santa Maria basins, California: *in* Natural gases of North America, a symposium, B. Warren Beeke, ed., v. 1, AAPG Memoir 9, p. 135–148.

Dunnington, H. V., 1967, Stratigraphic distribution of oil fields in the Iraq– Iran–Arabian basin: Journal of the Institute of Petroleum, v. 53, no. 520, p. 129–153.

El-Aquar, M. A., 1960, Reservoir performance under gas and water injection, Abqaiq field, Saudi Arabia: Second Arab Petroleum Congress, The League of Arab States, p. 5–25.

Ernst, W. G., 1981, Summary of the geotectonic development of California: Geotectonic Development of California, v. 1: Englewood Cliffs, N.J., Prentice-Hall, p. 601–613.

Gardett, P. H., 1971, Petroleum potential of Los Angeles basin, California: *in* Future Petroleum Provinces of the United States: Their geology and potential, Ira H. Cram, ed., v. 1, AAPG Memoir 15, p. 298–308.

Garrison, R. E., 1981, Lithofacies and depositional environments of Monterey shale, California (abs.): AAPG Bulletin, v. 65, p. 929.

Graham, S. A. , 1981a, Stratigraphy and depositional patterns and hydrocarbon occurrence, Sacramento Valley, California: Pacific Section AAPG v. 50,

——, 1981b, Tectonics, sedimentation, and petroleum geology of transform margin of central California (abs.), AAPG Bulletin, v. 65, p. 932.

——, and R. V. Ingersoll, 1981, Field guide to the Mesozoic-

Cenozoic convergent margin of northern California—objectives and geologic overview: Pacific Section AAPG, v. 50, p. 3–12.

Halbouty, M. T. et. al. 1970, World's giant oil and gas fields, geologic factors affecting their formation, and basin classification, *in* Geology of giant petroleum fields, a symposium, Michel T. Halbouty, ed., AAPG Memoir 14, p. 502–555.

Hemer, D. 0., J. F. Mason, and G. C. Hatch, 1981, Middle East-1980: AAPG Bulletin, v. 65, p. 2134–2153.

Hoskins, E. G., and J. R. Griffiths, 1971, Hydrocarbon potential of northern and central California offshore: *in* Future Petroleum Provinces of the United States: Their geology and potential, Ira H. Cram, ed., v. 1, AAPG Memoir 15, p. 212–228.

Hosoi, H., and R. Murakami, 1971, Radiated rupture faults on the Arabian type oil field: Journal of the Japanese Petroleum Technologists, v. 36, p. 10–17.

Ibrahim, M. W., M. S. Kahn, and H. Khatib, 1981, Structural evolution of Harmaliyah oil field, Eastern Saudi Arabia: AAPG Bulletin, v. 65, p. 2403–2416.

Isaacs, C. M., 1979, Diagenesis in Monterey Shale, Santa Barbara Coast, California (abs.): AAPG Bulletin, v. 63, p. 473.

———, 1980, Monterey rocks along Santa Barbara Coast, California (abs.): AAPG Bulletin, v. 74, p. 444.

Jennings, C. W., R. G. Strand, and T. M. Rogers, 1977, Geologic Map of California 1:750,000: California Division of Mines and Geology.

Jones, V., 1968, Occurrence of natural gas in Mesozoic rocks of northern California: AAPG Memoir 9, v. 1, p. 597–602.

Kleist, J. R., 1981, The Franciscan complex of northern California: Pacific Section AAPG, v. 50, p. 91–98.

Ministry of Petroleum and Mineral Resources, 1981, Saudi Arabia Oil and Gas World: Kingdom of Saudi Arabia, p. 1–59.

Morrison, R. R., W. R. Brown, W. F. Edmonson, J. N. Thompson, and R. J. Young, 1971, Potential of Sacramento Valley gas province: *in* Future Petroleum Provinces of the United States: Their geology and potential, Ira H. Cram, ed., v. 1, AAPG Memoir 15, p. 329–338.

Murris, R. J., 1984, Middle East: stratigraphic evolution and oil habitat: *in* Petroleum geochemistry and basin evaluation, G. Demaison and R. J. Murris, eds., AAPG Memoir 35, p. 353–372.

Nagle, H. E., and E. S. Parker, 1971, Future oil and gas potential of onshore Ventura basin, California: *in* Future Petroleum Provinces of the United States: Their geology and potential, Ira H. Cram, ed., v. 1, AAPG Memoir 15, p. 254–297.

Nehring, R., 1978, Giant oil fields and world oil resources: Santa Monica, Rand Corporation.

———, 1981, The discovery of significant oil and gas fields in the United States: Santa Monica, Rand Corporation.

Ogle, B. A., 1968, Natural gas in Eel River basin, Humboldt County, California, *in* Natural gases of North America, a symposium, B. Warren Beeke, ed., AAPG Memoir 9, v. 1, p. 68–75.

Parker, F. S., 1971, Petroleum potential of southern California offshore: *in* Future Petroleum Provinces of the United States: Their geology and potential, Ira H. Cram, ed., v. 1, AAPG Memoir 15, p. 178–191.

Powers, R. .W., 1962, Arabian Upper Jurassic carbonate reservoir rocks, *in* Classification of carbonate rocks, as symposium, W. H. Ham, ed., AAPG Memoir 1, p. 122–192.

———, 1968, Fascicule 10 b 1, Arabie Saoudite: Lexique Stratigraphique International, v. III, Union Internationale des Sciences Geologiques.

———, L. F. Ramirez, C. D. Redmond, and E. L. Elberg, Jr., 1966, Sedimentary geology of Saudi Arabia: U.S. Geological Survey, Professional Paper 560-D.

Rudkin, G. H., 1968, Natural gas in San Joaquin Valley: *in* Natural gases of North America, a symposium, B. Warren Beeke, ed., AAPG Memoir 9, p. 113–134.

Safonov, A., 1968, Stratigraphy and tectonics of Sacramento Valley: *in* Natural gases of North America, a symposium, B. Warren Beeke, ed., AAPG Memoir 9, v. 1, p. 611–635.

Steineke, M., R. A. Bramkamp, and N. J. Sanders, 1958, Stratigraphic relations of Arabian Jurassic oil, AAPG Symposium, Habitat of Oil, p. 1294–1329.

Tleel, J. W., 1973, Surface geology of Dammam dome, Eastern Province, Saudi Arabia, AAPG Bulletin, v. 57, p. 558–576.

U.S. Geological Survey, 1963, Geologic map of the Arabian Peninsula, 1:2,000,000: U.S. Geological Survey Miscellaneous Geologic Investigations, Map I 270 A.

U.S. Geological Survey and California Division of Mines and Geology, 1966, Geologic map of California, 1:2,500,000.

Williams, B., 1984a, Hot Santa Maria basin play may cool soon: Oil and Gas Journal, 1/9/84.

———, 1984b, New California field added to list of United States Oil Giants: Oil and Gas Journal, 1/30/84.

———, 1984c, Oil steamflood projects thriving in Kern County, California: Oil and Gas Journal, 12/10/84, p. 45–50.

———, 1985, Big finds spur push off California: Oil and Gas Journal, 1/14/85, p. 55–64.

Zieglar, D. L., and J. K. Cassell, 1978, The synthesis of OCS well information, offshore central and northern California, Oregon, and Washington: Pacific Section, AAPG, 53rd Annual Meeting, 1978.

———, and J. H. Spotts, 1978, Reservoir and source-bed history of Great Valley, California: AAPG Bulletin, v. 62, p. 813–826.

# Rifting: Lithospheric Versus Crustal Extension as Applied to the Ridge Basin of Southern California

G. D. Karner
J. F. Dewey
*Department of Geological Sciences*
*University of Durham, England*

Isostatic readjustment of rifted crust introduces a lateral thermal perturbation into the lithosphere, dissipation of which results in the familiar negative exponential subsidence characteristic of intracratonic sedimentary basins. Many small, pull-apart basins, however, exhibit a rapid but short-lived subsidence with little or no thermal recovery of the lithosphere. Such behavior indicates either (1) inhomogeneous, essentially upper crustal stretching without major sub-crustal lithospheric involvement, or (2) the rapid lateral removal of heat from the lithosphere and therefore enhancing basin subsidence, or both. The degree of sub-crustal lithospheric involvement depends on the mechanical properties of the lithosphere during rifting. For small basins (< 20 km or 12 mi), the finite flexural strength of the lithosphere precludes any significant participation of the sub-crustal lithosphere in controlling basin subsidence. An example of such a basin is the Ridge basin of southern California, a small basin that formed in association with major strike-slip motion across the San Gabriel fault during the late Miocene. During this period, basin infilling was controlled by the rate of sediment input rather than the rate of basement subsidence. We propose that the Ridge basin owes its existence to extension within the upper crust by block formation and rotation (or flakes), decoupled along a major, intra-crustal decollement. Independent rotation of the blocks, evidenced by paleomagnetic and seismicity data, results in a surface gap. Mechanical compatibility is maintained laterally by an equivalent amount of coeval overthrusting of the rotated blocks. Rift basin formation, therefore, is balanced by the creation of a compressional (flexural) basin. The tectonic setting of the Ridge basin was significantly modified by both the late Miocene establishment of the San Andreas fault system onshore and the continuing relative westward migration of the central Californian segment of the North American plate boundary (associated with Basin and Range lithospheric extension). During the early Pliocene, compression accompanying the progressive bending of the San Andreas fault (the Big Bend) caused basin tilting and truncation and mountain uplift/rejuvenation. Transpression eventually forced continental lithospheric subduction of the Pacific plate under the North American plate.

## INTRODUCTION

An approach to help define future petroleum provinces of the world lies in understanding how and why sedimentary basins form; that is, to determine a theoretical framework in which to view basin development. Necessarily, however, our approach addresses basin evolution on a regional scale.

Quantitative modelling of sedimentary basins has proved that it is possible to predict the general stratigraphy of a sedimentary basin by monitoring the form and time behavior of the tectonic driving force responsible for initiating basin subsidence (e.g. Haxby, Turcotte, and Bird, 1976; Beaumont, 1981; Watts, Karner, and Steckler, 1982; Quinlan and Beaumont, 1984). All sedimentary basins are the result of the isostatic adjustment of the lithosphere to a tectonic driving force, enhanced by subsequent sediment loading. In the case of foreland basin development, it is primarily the emplacement and migration of thrust sheets that constitutes the driving subsidence (Price, 1973; Beaumont, 1981), while with intracratonic basins, such as the Michigan, Williston, and Eromanga basins, the cooling and contraction of the continental lithosphere follow rifting (Sleep, 1971; Watts, Karner, and Steckler, 1982). Fundamental in understanding basin development, therefore, is the need for a detailed understanding of the driving subsidence.

While this subsidence is the cause of basin formation (but the result of some geodynamic process), the surface deformation into which sediments will accumulate is ultimately controlled by the strength of the lithosphere. If the lithosphere is highly fractured, basin development will tend to be compensated locally. For a rigid lithosphere, however, the deformation associated with even a local driving subsidence will tend to produce a regionally compensated basin. Therefore, in addition to knowing the form of the driving subsidence, we must also know how the lithosphere will redistribute the driving subsidence.

Small basins, such as pull-apart basins, exhibit some of the same characteristics as their larger counterparts, while showing some major differences. For example, small basins tend to subside extremely quickly, even faster than oceanic lithosphere, suggesting that their subsidence must be controlled by more than just the thermal recovery of the lithosphere following rifting. But should we even expect that the lithosphere is involved in the same way, if at all, for small basins? After all, the degree of lithospheric involvement in any geologic process should relate to its spatial wavelength. If sub-lithospheric heat is not involved, why do some of these basins show excellent hydrocarbon potentials? Likewise, if the strength of the lithosphere is important in basin development, and given the lack of a major sub-lithospheric control in small basins, then the lithosphere should be relatively cold and hence rigid. During sedimentation, therefore, these basins should become areally extensive, which of course is not the case.

In an effort to understand small basin development, many workers (e.g. Turcotte and McAdoo, 1981; Chadwick, 1985; Dickinson et al, in press) have borrowed the concepts and modelling procedures used to study intracratonic and passive margin basins, such as lithospheric stretching (McKenzie, 1978) and back-stripping (Steckler and Watts, 1978). While a useful first step, the basin mechanism models need extensive modification to address secondary characteristics of small basins. However, we believe that it is these secondary problems that negate the general use of the same models that describe large basins.

The purpose of this paper is, therefore, twofold: first, to review the thermal and mechanical properties of the lithosphere and their importance to intracratonic basin and passive margin development, and second, to test the general applicability of these concepts to small basin development using the southern Californian pull-apart basins—in particular, the Ridge and Soledad basins—as examples.

## LITHOSPHERIC AND CRUSTAL ISOSTASY

Isostasy is a fundamental physical principle that determines the equilibrium of the lithosphere to an applied stress or load. A density anomaly represents such a load, and as topography is a major density contrast, isostasy is usually described in terms of the compensation of mountain topography. However, any density distribution on, within, or beneath the lithosphere represents a load. Compensation is the process whereby an equal but opposite density contrast opposes the load. Figure 1 summarizes various isostatic schemes generally discussed in the literature. For example, Airy isostasy (Airy, 1855) refers to the production of a local compensating "root" to mountain topography (i.e., crustal thickening), whereas Pratt isostasy (Pratt, 1855) relates the density of the topography to its elevation.

These isostatic schemes describe only the static arrangement of load and compensating density distributions, without making any attempt to understand the mechanism by which the density distributions were created, particularly the compensating density distribution. We must address isostasy through a mechanistic approach in which the density anomalies are determined by such geodynamic processes as basin formation mechanisms or plate tectonics.

The flexural model, the third isostatic scheme shown in Figure 1, predicts a regional in contrast to a local compensation of the surface load. Recent studies of loading within the oceans and continents establish that flexure is the predominant mechanism by which the lithosphere supports emplaced loads (e.g., Watts, 1978; Bodine and Watts, 1979; Caldwell and Turcotte, 1979). The exact form of regional compensation is determined by its flexural strength—for high, flexural rigidities, loading creates a shallow but broad deformation. Alternatively, a low flexural rigidity results in a relatively larger amplitude deformation that is less regionally developed. The total cross-sectional area of the deformation, however, remains constant. In terms of mountain topography, a high rigidity would imply the lack of discrete, compensating roots, whereas a low rigidity would produce a deformation that would approximate a root. Figure 1 shows that Airy isostasy is an end-member of the regional model for which the lithosphere has zero flexural strength. The crustal root is in this case the deformation of the lithosphere by the topographic load. A zero strength lithosphere, seemingly an unrealistic representation of lithospheric behavior, may be approximated during extensive heating or faulting of the crust or both; for example, during rifting.

Interpretation of oceanic and continental loading suggests that the lithosphere responds to geological loads in a manner analogous to a thin, elastic plate overlying a weak fluid (Walcott, 1970; Watts and Cochran, 1974). The flexural strength of the lithosphere is marked by its effective, elastic thickness,

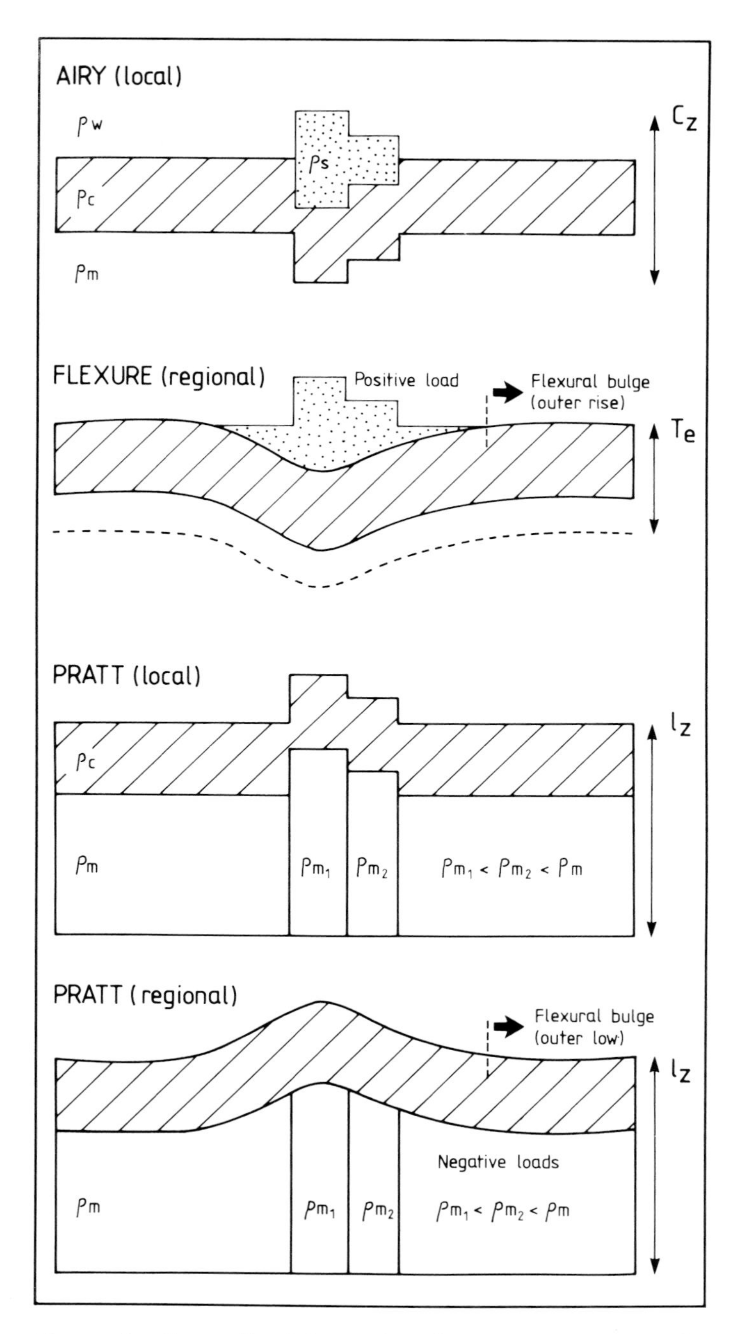


**Figure 1.** *Generalized isostatic schemes representing the response of the lithosphere to applied loads (density contrasts). Surface loading, when the lithospheric rigidity is low, is similar to the classical Airy isostatic model. Compensation predominantly occurs across the Moho Rigidity is parameterized by Te, the effective elastic thickness of the lithosphere. For increasing flexural strength of the lithosphere, the crustal compensation becomes distributed with respect to the load. Alternatively, subsurface loading is similar to the classical Pratt isostatic model, and the compensating interface is now the surface topography. Again, the form of the compensating interface is controlled by the flexural strength of the lithosphere. In sedimentary basin development, both isostatic schemes independently co-exist, with surface loading being associated with sediment deposition and sub-surface loading with the driving subsidence.*

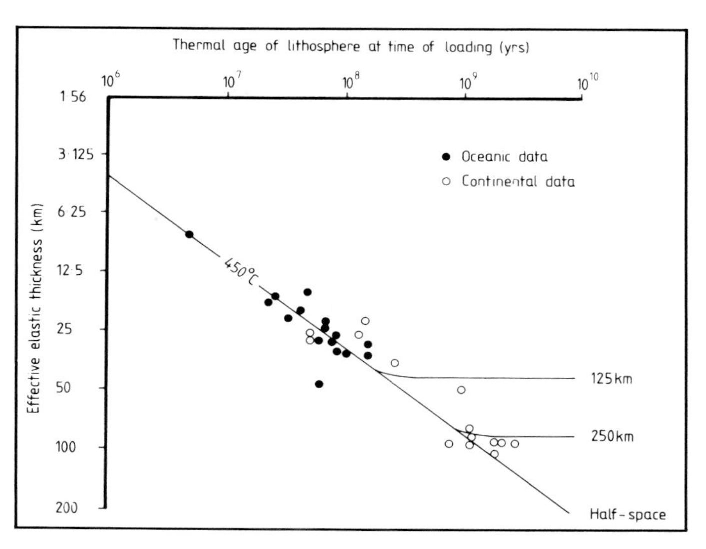


**Figure 2.** *Log-log plot of the thermal age of continental and oceanic lithospheres at the time of loading versus the corresponding effective elastic thickness. The solid circles represent oceanic loading data and the open circles, continental data. Superimposed is the 450° C (842° F) isotherm from Parsons and Sclater (1977). The first major conclusion from this plot is that the flexural strength of the lithosphere increases as it cools, with the depth to the 450° C (842° F) isotherm approximately tracking the effective elastic thickness of the lithosphere; and the second, that the old continental lithosphere appears to have an asymptotic thickness of 200–250 km (124–155 mi). Because of its age, the thermal structure is essentially indistinguishable from a cooling half-space.*

$T_e$. Typically, $T_e$ ranges from 5–40 km (3–25 mi) in the oceans and 5–100 km (3–62 mi) in the continents (Figure 2). Though useful in describing the deformation of the lithosphere, $T_e$ does not necessarily imply the existence of a real geological boundary at that depth; it implies that the deformational characteristics of an elastic plate of thickness $T_e$ are analogous to the integrated, rheological properties of the lithosphere.

A contemporary view of Airy isostasy considers loads on or within the crust compensated by deformation of the Moho. In contrast, Pratt isostasy considers loads (e.g., temperature-induced density anomalies) acting beneath or within the lithosphere. In this latter case, topography represents the response or compensating interface to the subsurface load (Figure 1) rather than being the load. The form of the compensating interface, be it topography or deformation of the Moho, is ultimately determined by the flexural strength of the lithosphere.

Isostasy, therefore, predicts the deformation of an interface, the lithosphere, by an applied force. The interface may act passively, as in the case of local compensation, or actively, as in the case of regional compensation, in that it redistributes the load-induced deformation (the compensation) according to the flexural strength of the lithosphere. This mechanistic approach suggests that isostasy is a natural consequence of rock strength and therefore lithospheric flexure (where the flexural deformation is determined by the rheological and mechanical properties of the lithosphere, not necessarily the elastic model, Beaumont, 1978). In addition, schemes such as those of Airy and Pratt represent, in principle, differences in the representation (and generation) of the load. With this approach, concepts such as uncompensated load or density distributions become meaningless.

## THERMAL AND MECHANICAL PROPERTIES OF OCEANIC AND CONTINENTAL LITHOSPHERE

Studies of oceanic and continental loading (e.g., Watts, 1978; Beaumont, 1981; Bodine, Steckler, and Watts, 1982; Karner, Steckler, and Thorne, 1983; McNutt, 1983; Karner and Watts, 1983; Lyon-Caen and Molnar, 1983; Quinlan and Beaumont, 1984) demonstrate a simple relationship between the flexural strength of the lithosphere and its temperature structure. For the oceanic lithosphere, temperature structure and age are simply related: older lithosphere is characterized by relatively higher rigidities when compared with younger lithosphere (Figure 2). In contrast, the continental lithosphere relates not only to its absolute age, but also to any modification or resetting of its temperature structure by thermal events (such as rifting). Analagous to the increase of rigidity with age away from a mid-oceanic ridge, continental rigidities appear to increase with time following a major thermal event (Karner, Steckler, and Thorne, 1983). Conversely, reheating the lithosphere reduces its flexural rigidity. Figure 2 shows the observed relationship between $T_e$ and lithospheric thermal age (as opposed to absolute age) and approximates the depth to the 450°C (842°F) isotherm (Watts, 1978) as predicted from the cooling plate model of Parsons and Sclater (1977).

## ISOSTASY AND BASIN FORMATION MECHANISMS

Sedimentary basins are dominated during their evolution by vertical movements of the lithosphere. Although an individual basin may change its tectonic setting during its evolution, we can classify most basins as occurring in either rifted (extensional) or collisional (compressional) settings. All sedimentary basins are the result of a driving force or subsidence that creates a depression in which sediments accumulate. Sediment infill further amplifies the driving subsidence by flexural loading of the lithosphere. There is now general agreement that the main cause of the tectonic subsidence of basins is thermal (Sleep, 1971; Sleep and Snell, 1976). The subsidence form of passive margin basins is strikingly similar to the isostatic subsidence of the oceanic lithosphere, following its creation and subsequent cooling as it moves away from mid-ocean ridges. The similarity in tectonic setting and subsidence for passive margin basins implies a common origin: thermal contraction of the lithosphere following heating during continental rifting. Intracratonic basins are characterized by essentially the same negative exponential subsidence form.

Researchers have proposed numerous mechanisms to account for basin subsidence: phase changes or intrusion of dense material at depth (Falvey, 1974; Haxby, Turcotte, and Bird, 1976); crustal doming, rifting and erosion (Sleep, 1971), crustal and lithospheric necking and stretching (Artemjev and Artyushkov, 1971; McKenzie, 1978); and sub-crustal/sub-lithospheric tectonic erosion (Bott, 1971, 1981; Spohn and Neugebauer, 1978). A combination of these mechanisms may be responsible for basin development, but because of the mathematical versatility, predictability, and testability of the McKenzie lithospheric stretching model, we will concentrate on aspects of this model to investigate the development of intracratonic sedimentary basins.

Stretching (and hence thinning) of the crust and lithosphere by a factor of $\beta$ results in the passive rise of asthenosphere to infill the thinned lithosphere. As the lithosphere/asthenosphere boundary represents an isotherm (ca. 1200°–1400°C or 2,192°–2,552°F, Parsons and Sclater, 1977), thermal re-equilibration by vertical and lateral heat conduction produces the familiar negative, exponential, driving subsidence characteristic of passive margin and intracratonic basins. Isostatic considerations help predict the interaction between changing crustal thickness and sub-crustal heating. The crustal thinning accompanying lithospheric stretching results in a surface depression that, in an isostatic sense, represents negative topography. Compensation of this topography results in a flexural uplift of the Moho. If the lithospheric rigidity is low, as might be expected during rifting and hence fracturing of the crust, the crustal configuration will approximate the familiar Airy scheme for basin compensation. Sub-crustal thinning leads to the introduction of a thermal anomaly within the lithosphere, compensation of which follows a Pratt-type isostatic scheme. Flexural compensation of the thermal load produces a compensating uplift whose wave-length and amplitude will again be controlled by the strength of the lithosphere, competing with the surface depression associated with crustal thinning. The resultant basement response, uplift or subsidence, depends upon the ratio of the crustal to lithospheric thickness, but is generally negative and is called the initial, or fault-controlled, subsidence. Note that basin subsidence is at a maximum in the absence of this competing Pratt or thermal effect.

As the thermal anomaly dissipates, the Pratt-type compensation progressively diminishes so that isostatic compensation of the sedimentary basin eventually is achieved entirely within the crust. During the cooling process, lithospheric rigidities will increase (as shown in Figure 2) so that the driving subsidence is flexurally redistributed according to an ever-increasing rigidity, as is the basement deformation associated with sediment loading. The result is a complex basement response that represents the interaction of the driving subsidence (or perhaps more appropriately, the driving load) and sediment loads with the spatial and temporal variations in lithospheric flexural rigidity.

Figure 3 summarizes the basic and possible variations of the McKenzie stretching model. Figure 3a represents the pre-stretching lithosphere configuration and equilibrium temperature structure. During depth-independent stretching, a thermal anomaly proportional to the degree of stretching is introduced into the lithosphere (Figure 3b). The Central Bass basin in Australia is an example of this (Etheridge, Branson, and Stuart-Smith, 1985). However, studies of east Canadian passive margin subsidence (Keen, 1979; Keen and Barrett, 1981) suggest that additional heat may be involved during the rifting process than could be accounted for by the crustal thickness of the margin. If stretching of the crest is denoted by $\delta(x)$ and sub-crustal heating by $\beta(x)$, then Figures 3c-3e represent the Royden and Keen (1980) model of depth-dependent stretching across a detachment surface. In particular, for $\delta = 1$ and $\beta > 1$, the Pratt thermal load is compensated by a surface uplift (Figure 3c), but being thermal in origin, it eventually subsides. An

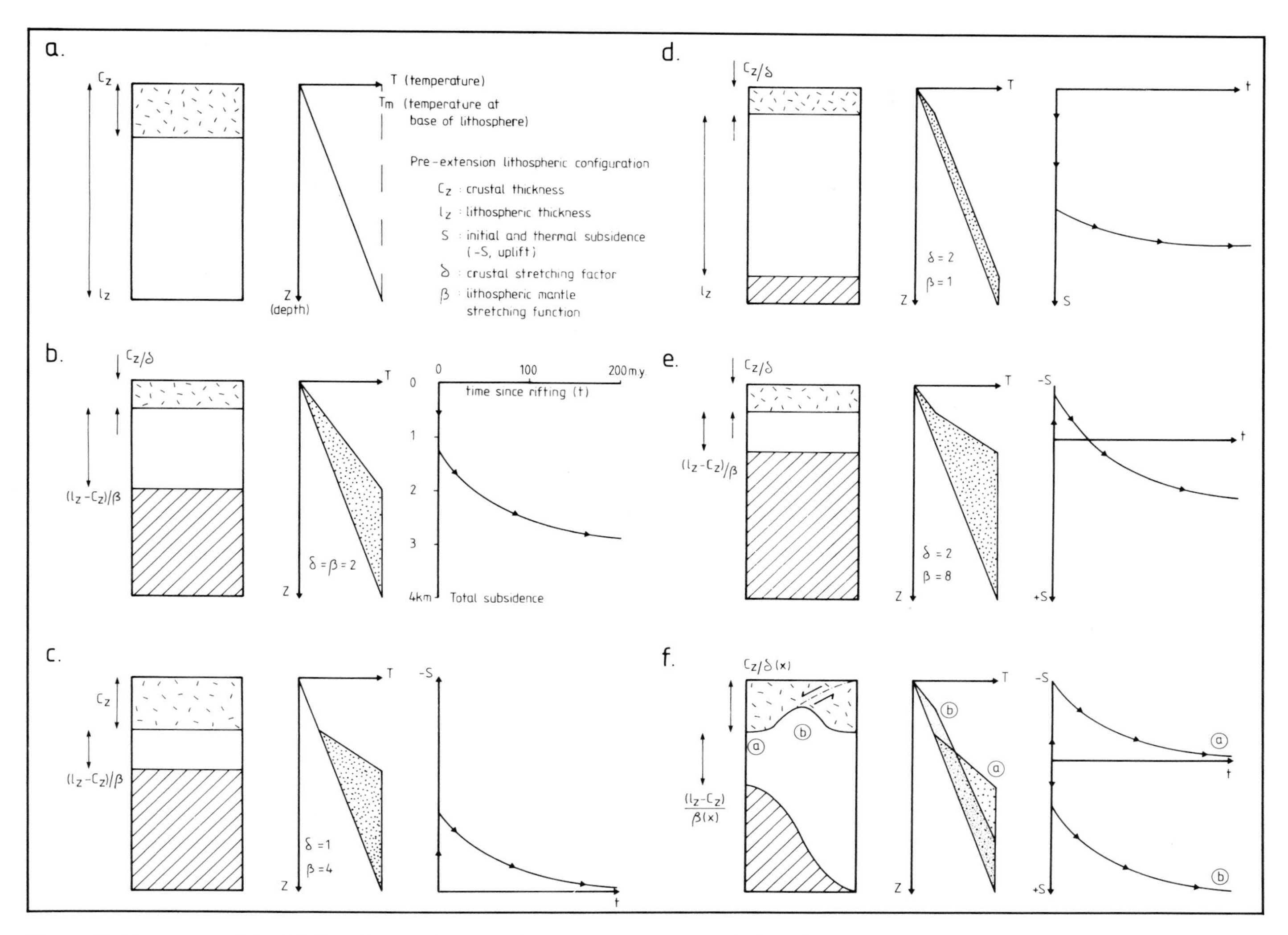


Figure 3. *Summary of the McKenzie stretching model, and its variations. Lithospheric stretching initially modifies the thickness of both the crust and sub-crustal lithosphere: β represents stretching above the decollement (usually considered to be the Moho) and β represents either stretching below the decollement or the degree of sub-crustal heating. The simplest model of basin formation is depth-independent stretching (δ = β) in which the pre-event lithospheric configuration and temperature structure (a) is modified to give (b). Dissipation of the introduced thermal anomaly results in the familiar negative exponential subsidence of passive margins. If the crust is not thinned during sub-crustal extension, major uplift occurs in response to the isostatic compensation of the thermal anomaly (c). Alternatively, crustal thinning in the absence of sub-crustal extension significantly modifies the temperature structure of the lithosphere (d). The general model, therefore, is for depth-dependent stretching across a decollement (e). If the decollement is itself depth-dependent, stretching is portioned across the decollement such that there is a physical separation of the initial and thermal subsidences (f).*

example is the Murray basin forming peripherally to the uplift of the eastern Australian highlands (Karner and Weissel, 1984). For $\delta > 1$ and $\beta = 1$, rapid crustal subsidence is produced with only minor thermal recovery of the lithosphere, as shown in Figure 3d and exemplified by the Wessex basin in southern England (Lake, Karner and Dewey, in press). Figure 3e shows the general depth-dependent stretching for the case, $\delta > 1$, $\beta > 1$; the Pannonian basin in Hungary is an example (Sclater et al, 1981). When the detachment surface is itself depth-dependent, then the stretching functions and resulting interaction become more complex (Figure 3f). This stretching scheme predicts a physical separation between the initial uplift/subsidence rift subbasin and the thermal or flexural subsidence. An example of this scheme is the Basin and Range physiographic province in the southwestern U.S. (Wernicke, 1981).

Lithospheric stretching philosophies using detachment or decollement surfaces are only viable provided mass can be conserved, or strain compatibility maintained, across the detachment surface. As an example, Figure 4 considers the forward modelling of an intracratonic basin and underlines the importance of flexure and strain compatibility in basin development. Consider the possible stretching of the lithosphere by the stretching functions $\delta(x)$ and $\beta(x)$ of Figure 4a. For a lithosphere with zero flexural strength, the rift (the dotted regions of Figures 4b-4d) and thermal subbasins are spatially conformable, with discontinuities in $\delta(x)$ translating directly into basement highs that are then maintained for the entire development of the basin. Further, there is a one-to-one correlation between basin development and lithospheric stretching

Even though a low flexural strength may be applicable to the lithosphere early in basin development, thermal cooling following rifting leads to its mechanical recovery so that the basin-driving subsidence and the sediment loading will undergo a progressively increasing flexural rigidity with time (Figure 4c). Basement highs remain as such only during the formation of the rift subbasin, and affect only the rift basin stratigraphy. The flexural redistribution of the thermal subbasin is compensated by an increased Moho topography. The low rigidity assumed during rifting predicts the existence of local and unreasonable crustal roots compensating the basement highs. The response to stretching, however, should be dominated by the depth-dependent rheological properties of the lithosphere. As tensional failure of the upper crust is undoubtedly governed by a brittle process, the crustal stretching function $\delta(x)$ is necessarily discontinuous. On increasing confining pressure and temperature, brittle failure is replaced by a ductile mechanism such that lower crustal/sub-crustal stretching is a continuous function, i.e., $\beta(x)$. Strain compatibility demands that the integrated brittle and ductile strains be equal, as is the case in Figure 4a, and results in a detachment surface only within the stretched region (Karner, 1984). If the lower crust deforms ductilely, crustal roots will not be produced; and in order to maintain a mass balance, the average Moho deformation is reduced. The resultant basin, Figure 4d, shows all the characteristics of observed intracratonic basins, namely, two phases of basin formation: the rift and thermal subbasins, a general increase of basin width with time, a smooth Moho, and rifting complexities such as fault related basement highs contained within the rift subbasin.

Further, by monitoring the spatial and temporal variations of lithospheric rigidity following rifting, it is also possible to predict the generalized stratigraphy of intracratonic and passive margin basins (Watts, Karner, and Steckler, 1982; Beaumont, Keen, and Boutilier, 1982). Note that philosophically, these basins are identical because an intracratonic basin represents a symmetric passive margin, but without the intervening ocean basin. Figure 5 allows us to compare and contrast theoretical extensional basin structure and stratigraphy with that observed from intracratonic (Watts, Karner, and Steckler, 1982) and passive continental margin basins (Steckler, 1981; Watts, 1981). The main stratigraphic difference is that the basement onlap pattern that characterizes the early development of the thermal basin is eventually replaced by offlap (e.g., North Carolina coastal plain after the Upper Cretaceous). Given that basement onlap is caused by the increasing flexural strength of the lithosphere following rifting, this observed offlap must be due to a factor not included in the modelling. One such factor is the temporal variation in sea level, which theoretically controls the level of sediment infilling or the paleo-water depth (Watts, 1982; Watts and Thorne, 1984). Figure 6 predicts the stratigraphy of an intracratonic basin, but now includes the effects of both a eustatic sea-level rise and fall during basin development. We assume that there is always sufficient sediment to fill the basin to sea-level. With a sea-level rise (Figure 6a), the basin stratigraphy is expanded by the increase in water level and, when the sea-level rise is greater than the topography of the flexural bulge, sediment spills out over the peripheral regions onto the continental

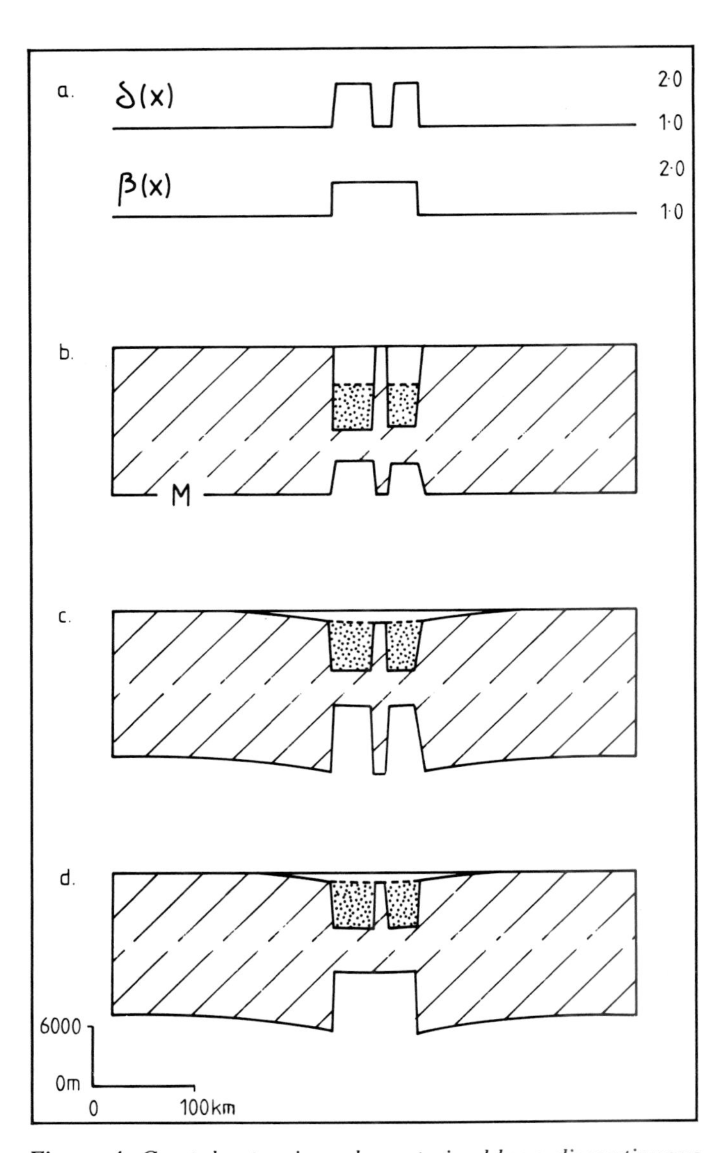


**Figure 4.** *Crustal extension, characterized by a discontinuous stretching function. Such a situation highlights the importance of lithospheric flexural strength during basin development. In (a), brittle upper crustal stretching is represented by $\delta(x)$ and ductile lower crustal/subcrustal stretching by $\beta(x)$. In (b), we see that if the lithosphere has no strength (i.e., Airy compensation), the rift and thermal subbasins have the same distribution and directly relate to the form of the stretching functions. Non-stretched regions become intrabasinal highs for the entire history of the basin. (c) A finite flexural strength during the thermal subsidence phase (lithospheric cooling) produces a broad basin regionally developed over the rift basin. Basin subsidence outside the rift zone is not directly related to lithospheric stretching but to flexural redistribution of the driving subsidence. Therefore, the subsidence form within the flank region retains its negative exponential (or more precisely, the form of the driving subsidence) even though the underlying lithosphere is unstretched. (d) The formation of short-wavelength crustal roots remains unacceptable. However, Moho topography can be evenly redistributed within the rift zone (such as that associated with ductile extension) providing that the integrated $\delta(x)$ and $\beta(x)$ remain equal.*

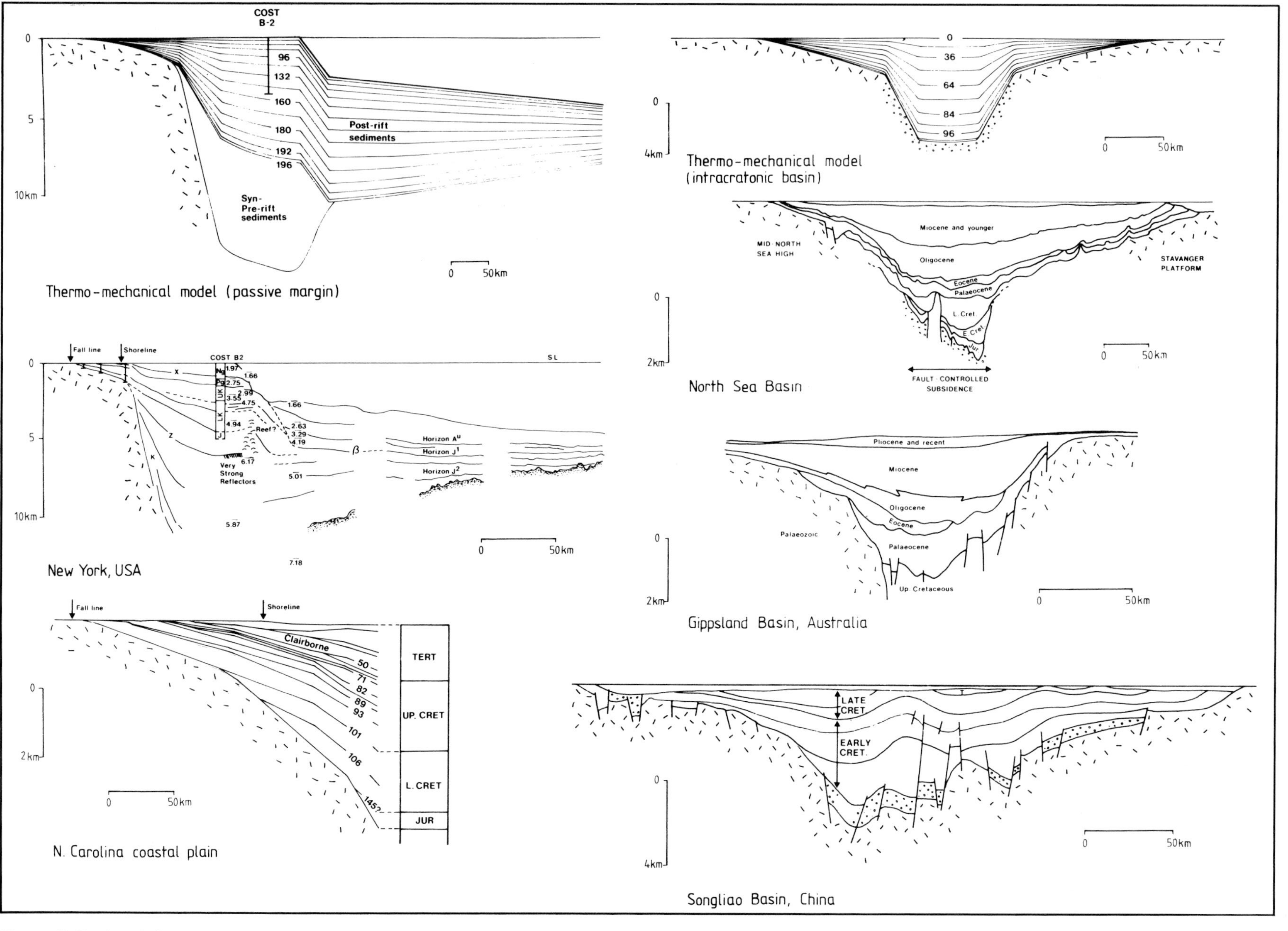


**Figure 5.** *Regional characteristics of observed intracratonic and passive margin basins, compared with theoretical basins produced by integrating stretching models with the thermal and mechanical properties of the lithosphere. The modelled passive margin and intracratonic basins are from Watts (1981), and Watts et al (1982) respectively. The models generally reproduce the main features of the observed basins.*

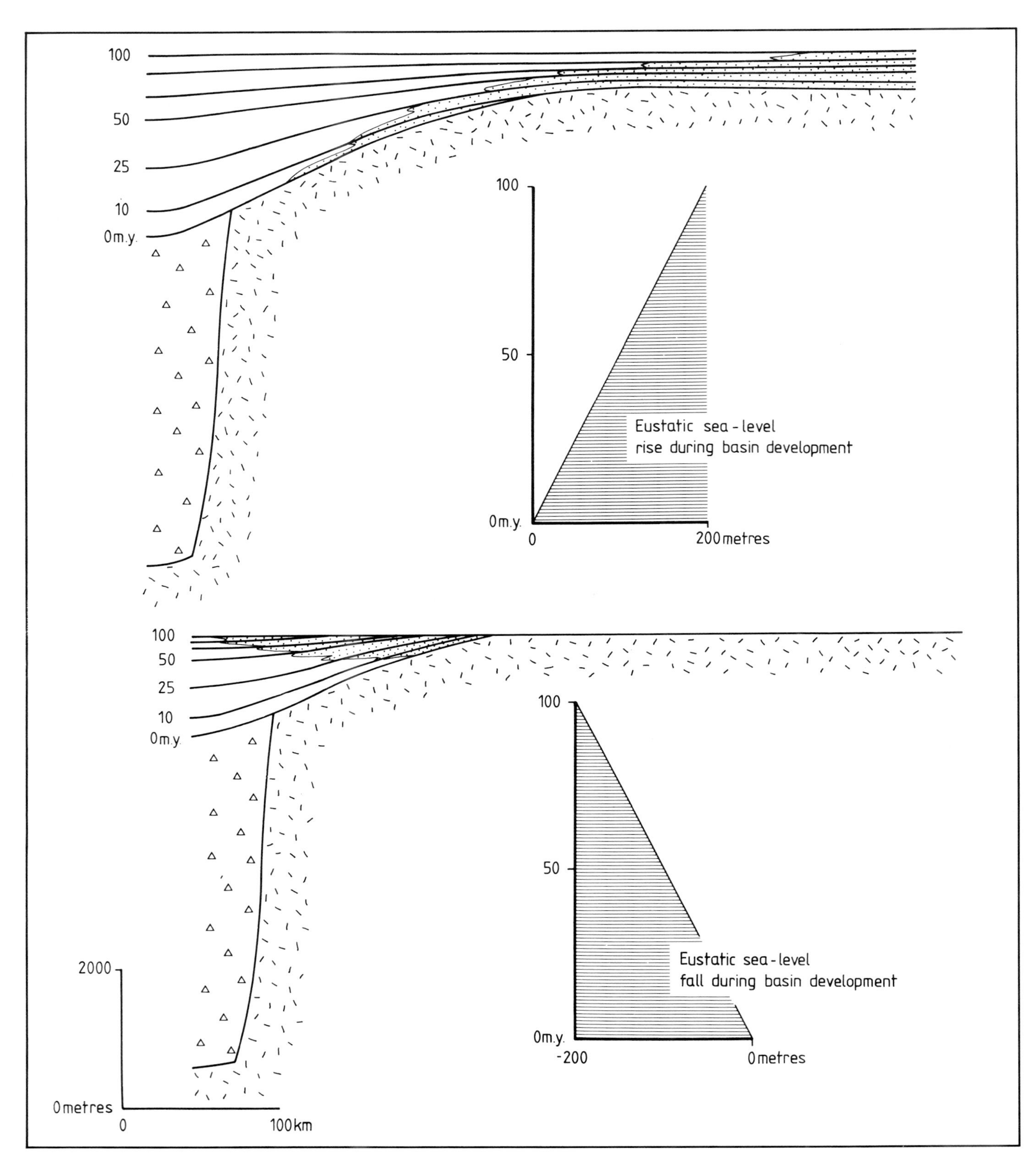


**Figure 6.** *The stages of basin development, showing later stage regional offlap in contrast to the predicted onlap. Such onlap may be related to either sea-level variations or erosion or both. During a sea-level rise, the areal extent of sediment deposition rapidly exceeds the basin proper, especially when the sea-level rise breaches the flexural outer-rise. During a sea-level fall, however, basin development becomes progressively restricted, thereby constraining the area available for sediment deposition. The result is an offlap sequence, which becomes even further enhanced by erosional downcutting.*

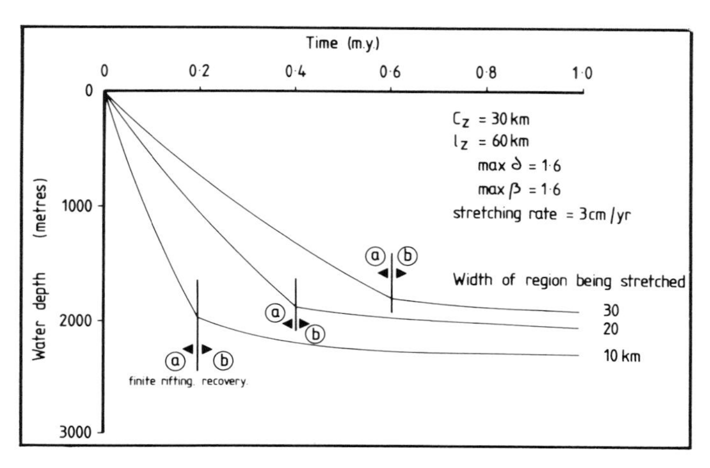


**Figure 7.** *Enhanced basin subsidence rates by finite rifting of the lithosphere. Subsidence rates are primarily determined by the rate at which heat is removed from the rift zone. Decreasing the strain rate and the area being stretched tends to increase the subsidence rate, the limit being when no heat is involved in the extension process at all. Such a situation is approximated by upper crustal extension.*

platform (Watts and Thorne, 1984). In contrast, a sea-level fall generally condenses the stratigraphy, reduces the surface area for sedimentation thereby producing an offlap pattern (Figure 6b), an effect accentuated by surface erosion (Watts, Karner, and Steckler, 1982; Watts and Thorne, 1984). Therefore, by the simple augmentation of sea level variations into the thermo-mechanical model, both the basement onlap and surface offlap generally observed within intracratonic and passive margin basins are explained.

## SMALL, PULL-APART SEDIMENTARY BASINS

Many small basins, particularly pull-apart basins along major strike-slip fault zones, are characterized by extreme driving subsidence rates. For example, the Cenozoic subsidence rates in southern California were 700 m/m.y. in Eocene-Miocene time, and greater than 1,000 m/m.y. post-Miocene (Yeats, 1978). For comparison, the isostatic subsidence of the oceanic lithosphere within the first 10 m.y. is less that 400 m/m.y. (Parsons and Sclater, 1977). Following this rapid, initial phase, however, subsequent driving subsidence is either absent or insignificant (Dewey, 1982). A major implication of lithospheric extension and crustal rifting was the introduction of a thermal anomaly that, on dissipation, produced the temporal driving subsidence for intracratonic basin development. Given that these small basins are the result of crustal fracturing, the lack of a lithospheric thermal recovery phase is problematic, at least in terms of the McKenzie stretching model (1978).

Assumptions in the McKenzie stretching model preclude the effects of lithospheric flexure, finite rifting time (compared with instantaneous rifting), and lateral heat flow. Figure 4 details the importance of lithospheric flexure in controlling thermal subbasin development, a phase generally missing from small basins. The problem becomes, therefore, one of explaining the absence of this thermal subbasin. First, McKenzie (1978) assumed that the process of stretching occurred essentially instantaneously, thereby maximizing the heating of the lithosphere with no heat loss from the rift zone during the stretching process ('instantaneous' was redefined by Jarvis and McKenzie in 1980 to be less than 20 m.y.). Second, the effect of lateral heat flow is accentuated in small basins because the peripheral regions may act as major heat sinks. In fact, lateral heat flow becomes so efficient that it replaces vertical heat conduction as the dominant basin-forming mechanism (Steckler, 1981; Cochran, 1983). Together, finite rifting and lateral heat flow maximize the rate of thermal recovery of the lithosphere during the rift phase, thereby significantly reducing the potential subsidence of the thermal subbasin. Figure 7 models the effectiveness of finite rifting and lateral heat flow, using the methodologies of Cochran (1983) and Karner, Steckler, and Haxby, (*in press*), for a range of basin widths, 10, 20, and 30 km, (6, 12, and 19 mi), with crustal and lithospheric parameters representing southern California (Zandt and Furlong, 1982). The stretching rate of 3 cm (1.2 in)/yr, typical of plate velocities, defines a strain rate of $10^{-12}$–$10^{-13}$ $sec^{-1}$ and rifting ceases when extension has reached a $\delta$ and $\beta$ of 1.6. For very small basins (10 km or 6 mi), subsidence rates can approach 10 km/m.y. (6 mi/m.y.), suggesting that the McKenzie model, modified to include finite rifting rates and lateral heat flow, may apply equally well to small and pull-apart basins.

Though elegant, the McKenzie model poses a credibility problem when the size of the basin is considered. A 10 km (6 mi) wide rift basin, producing an asthenospheric diapir or plume of 10 km (10 mi) width, and amplitude 50 km (31 mi), is considered unreasonable because a lithosphere/asthenosphere involvement in any process will be necessarily regional. Further, crustal faulting, while significantly reducing the strength of the crust, need not destroy the integrated, flexural strength of the lithosphere (Karner, 1984). For example, volcanoes within the rift zone of the North Atlantic mid-ocean ridge system are isostatically maintained by an elastic plate with a $T_e$ of 5–10 km (3–6 mi) (Tapponnier and Francheteau, 1978). Flexural redistribution of the above diapir with $T_e$ as low as 10 km (6 mi) causes an 80% reduction in amplitude, thereby significantly reducing the vertical extent of lithospheric heating. Redistribution also produces uplift within the peripheral regions. Decreasing the amplitude of this thermal plume proportionally increases the basement subsidence, and suggests that the integrated flexural strength of the lithosphere during crustal stretching may play a major role in determining the degree of sub-crustal involvement in basin subsidence and hence development. The most efficient basement subsidence mechanism for small basins would appear to be crustal rifting without sub-crustal interference (the rifting mechanism of Figure 3d), and, consequently, we need to investigate the implications and feasibility of the sub-crustal lithosphere's non-involvement in the formation of pull-apart basins such as the Ridge and Soledad basins of southern California.

## REGIONAL TECTONIC SETTING

In the past 20 m.y., the tectonics of central and northern California have been dominated by the northward migration of the Mendocino triple junction along the Pacific-North American plate boundary (Atwater, 1970). The rates of transform motion

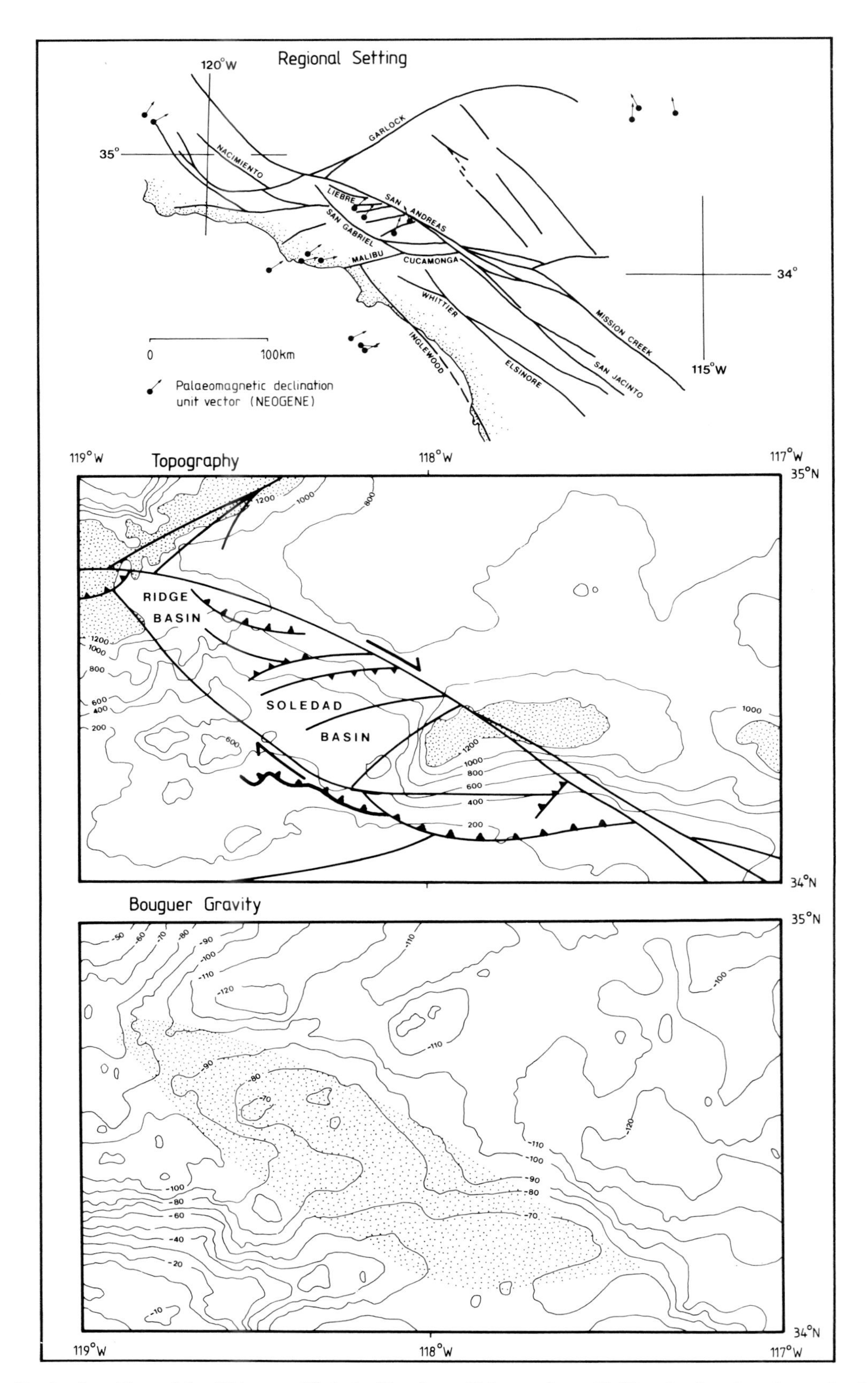


**Figure 8.** *Regional tectonic setting of the Ridge and Soledad basins within southern California showing the major faults, strike-slip, and thrust. The San Gabriel and San Bernardino Mountains are outlined by the 1,000 m (3,281 ft) contour line. The Bouguer gravity is also shown (contour interval, 10 mgal). Note the lack of correlation between the observed gravity and basin distribution, and the lack of compensating roots to the topography.*

were variable: ca. 4 cm (1.6 in)/yr between 29 and 21 Ma, 1.3 cm (0.5 in)/yr between 21 and 10 Ma, 4 cm (1.6 in)/yr from 10 to 5 Ma, and 5.5 cm (2.2 in)/yr to the present (Atwater and Molnar, 1973). The boundary between the Pacific and North American plates is a broad, continental deformation zone (Figure 8), and during the late Cenozoic the crust failed via simple shear, allowing pull-apart basin formation, fault block compression, uplift, and rotation. The main plate boundary, the San Andreas transform fault system and its main constraining bend, "the Big Bend," are only relatively recent features, formed during the Pliocene (Crowell, 1974a, 1981). From 12–5 Ma, the San Gabriel fault zone was the principal strike-slip fault of the plate boundary, though it probably joined with the San Andreas at depth. Certainly during the Pliocene, the San Gabriel fault connected directly with the San Andreas (Crowell, 1975) and extended northwestward through to the Coast Ranges. At about 5 Ma, the San Andreas transform fault itself played a major role in the opening of the Gulf of California, thereby abandoning the San Gabriel fault, creating the Big Bend, and rapidly compressing and uplifting the Transverse Ranges (Crowell, 1975, 1979). Since the Miocene, the total right-lateral movement on the fault has been approximately 320 km (199 mi) (Crowell, 1979).

The Transverse Ranges, consisting of the San Gabriel in the west and the San Bernardino Mountains in the east, are a west-trending interruption of the predominantly north-west structural grain of California. The San Gabriel (<3500 m or 11,484 ft) and San Bernardino (<3700 m or 12,140 ft) Mountains are, in general, higher than the Coast Ranges (<1800 m or 5,906 ft) to the northwest and the Peninsula Ranges (<1500 m or 4,992 ft) to the south. Though ancestral Transverse Ranges existed during the Miocene, the major uplift phase occurred in the Pleistocene with a severe north-south shortening (Woodburne, 1975), as evidenced by the folding and faulting of Pliocene-Pleistocene sediments. Gravity studies (Grannell, 1971; Oliver, 1980; Sheffels and McNutt, in press) (Figure 8) and seismic studies (Hadley and Kanamori, 1977; Raikes and Hadley, 1979) indicate the lack of compensating crustal roots beneath the Transverse Range—i.e., the Bouguer gravity anomaly is characterized by a broad gradient rather than a discrete anomaly as expected for a topographic root (Figure 8). This feature is common to high lithospheric, flexural rigidity support of mountains (Karner and Watts, 1983), in this case, by the North American lithosphere (Sheffels and McNutt, in press), or possibly, in-plane compression generated across the San Andreas fault (Yeats, 1981). Lithospheric thicknesses within the Big Bend and the Transverse Range region are 60–70 km (37–43 mi) with crustal thicknesses of 25–35 km (15–22 mi) (Zandt and Furlong, 1982), implying that at the time of basin formation in southern California, a higher regional temperature gradient must have existed.

Major transpression along the San Andreas in the Miocene and Pliocene saw the crustal fragmentation of the Pacific plate (Crowell, 1975). Paleomagnetic data, mainly from Miocene igneous rocks in southern California west of the San Andreas fault, suggest that large regions have undergone syn/post-Miocene clockwise rotations (70–80°, Figure 8), implying the existence of independent crustal blocks (Greenhaus and Cox, 1979; Luyendyk, Kamerling, and Terres, 1980). Such crustal blocks include the eastern Transverse Ranges and parts of the offshore borderlands. In contrast, the Mojave Desert area may have undergone a counterclockwise rotation of about 14° (Garfunkel, 1974; Calderone and Butler, 1984). Block fragmentation and rotation probably began in the late Oligocene and ceased in the late Miocene, coinciding with the opening of the Gulf of California.

Focal mechanisms of small earthquakes at mid-crustal depths west of the San Gabriel Mountains suggest that north-dipping thrust faults near the surface may pass downward into flat thrust faults (Hadley and Kanamori, 1977), suggesting that the west-central Transverse Ranges may rest on a detachment thrust fault or decollement and are, therefore, allochthonous (Yeats, 1981). Refraction data also suggest a deep velocity discontinuity, possibly a decollement (Malin and Leary, 1979), at about 12 km (7 mi) in this same region. Similarly, the San Bernardino Mountains, even though cut by branches of the San Andreas fault system on the south, appear to have been uplifted in the north along thrust faults that dip southward into and beneath the range (Richmond, 1960). Many of the surface traces of these thrust/reverse faults show late Quaternary reactivation. Geologists strongly suspect the existence of a regional detachment surface within the upper crust because of the lack of correspondence between surface topography and the inferred mantle structure.

Seismic tomographic imaging of the mantle beneath southern California has revealed a prominent, high-velocity (2–3% faster than surrounding mantle) anomaly directly beneath the Transverse Ranges, appearing as a slab-like feature that dips slightly to the north (Humphreys, Clayton, and Hager, 1984). Its maximum depth is 250 km (155 mi) beneath the San Bernardino Mountains (eastern Transverse Ranges) and about 100 km (62 mi) beneath the San Gabriel Mountains (western Transverse Ranges). Given the transpressive nature of the San Andreas fault, we concur with Bird and Rosenstock (1984); Humphrey, Claytin, and Hager, (1984); and Sheffels and McNutt (in press) that this feature represents the subducted lower mantle/subcrustal lithosphere of the Pacific plate. The failure of the upper crust to be subducted results in its obduction over the North American plate (cf., Oxburgh, 1972), thereby helping to produce the San Bernardino Mountains. Compression, crustal shortening and consequent thrust fault generation to the south of the San Gabriel Mountains represent the initial stages of continental subduction/obduction.

## CHARACTERISTICS OF THE RIDGE AND SOLEDAD BASINS

The Ridge basin (Figures 8 and 9) developed in the late Miocene–early Pleistocene between the right-lateral San Gabriel and San Andreas faults (Crowell, 1974a). The basin is 30–40 km (19–25 mi) long and 6–15 km (4–9 mi) wide, with a maximum driving subsidence of 4–5 km (2–3 mi). Sediments accumulated at a rate of about 1–2 km (0.6–1.2 m)/m.y. (Crowell, 1974a; Yeats, 1978; Hempton and Dunne, 1984) and, because of the pull-apart tectonic setting of the basin, allowed a total stratigraphic thickness of 12 km (7 mi) to accumulate. Such rapid subsidence is also characteristic of Cenozoic sediments within the Ventura, Los Angeles, and Soledad basins. Yeats (1978) showed that those basins exhibiting accelerated subsidence rates all cluster around the Big Bend of the San Andreas fault.

Figure 10 schematically represents the general stratigraphic succession of the Ridge basin. Basin initiation began in the

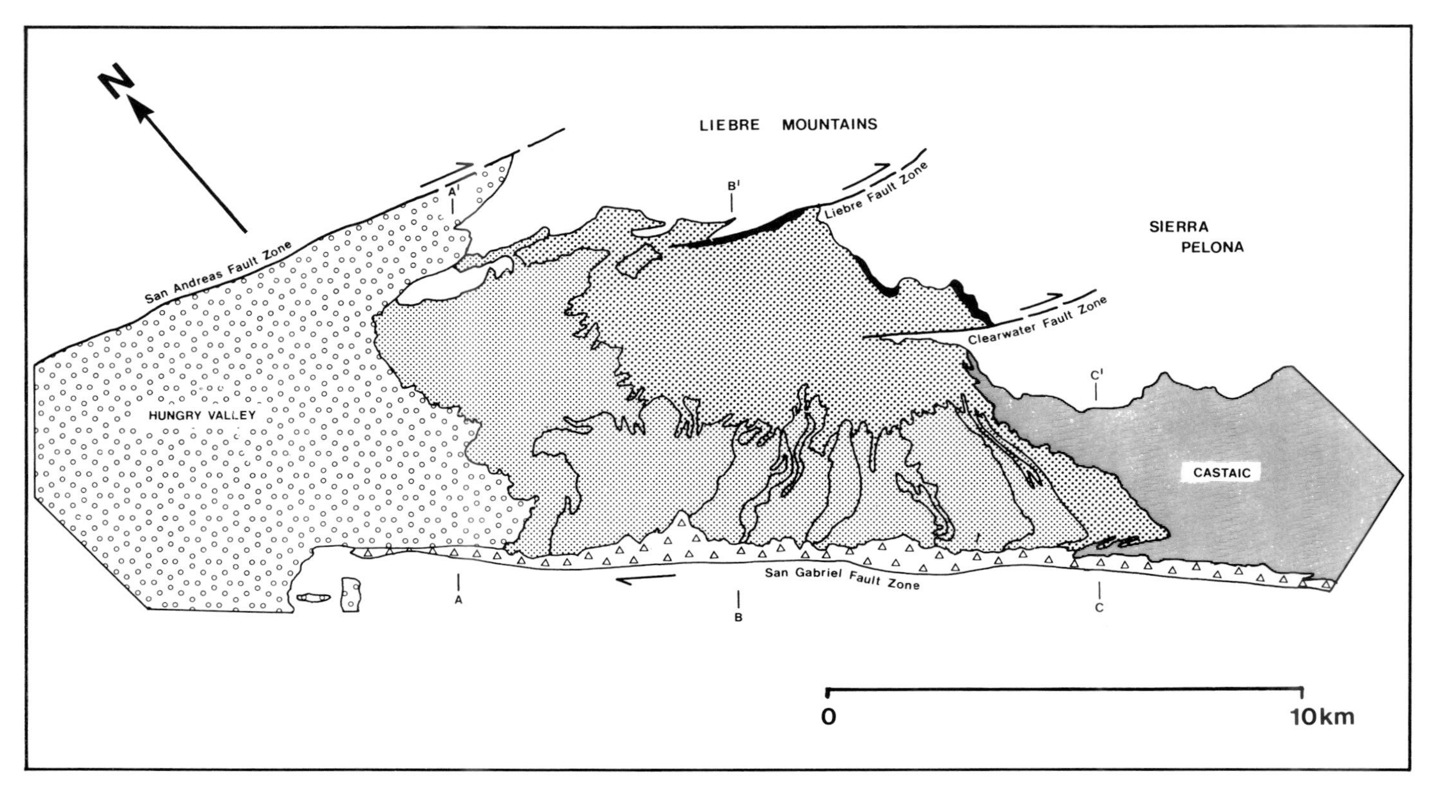


Figure 9. *General surface geology of the Ridge basin. The oldest rocks, those of the Castaic Formation, crop out in the southwest of the basin while the youngest, the Hungry Valley Formation, occur in the northwest. Cross-sections A, B, and C are shown in Figure 10.*

upper Miocene with the deep-water, restricted environment facies of the Castaic Formation, which consists of thinly bedded, diatomaceous and organic-rich shales and minor sandstones and siliceous, shaly and fossiliferous limestones. The upper Miocene-lower Pliocene Ridge Basin Group unconformably overlies the Castaic Formation and consists of dark grey and brown shales and siltstones with interbedded sandstones. The facies and distribution patterns (Figures 9 and 10) of this Group show it to be a major coalescence of deltas/alluvial fans, originating from the east. The succession is conformably capped by the upper Pliocene-lower Pleistocene Hungry Valley Formation, a continental facies of coarse sandstones and conglomerates. In agreement with their deltaic origin, sediments grade sharply from conglomerates and sandstones in the northeast, to predominantly shales southwest, along the axis of the basin. Along the San Gabriel fault zone exists a course sedimentary breccia, the Violin Breccia, which interfingers with the Castaic and Hungry Valley Formations and the Ridge Basin Group. This interfingering dates the initiation of activity on the San Gabrial fault at about 12 m.y. ago and hence the development of the Ridge basin (Woodburne, 1975). Clasts within the Violin Breccia show that they have been derived from a limited source area now offset by 28 km (17 mi) (Crowell, 1974a).

In addition to the longitudinal migration of deposits parallel to the San Gabriel fault, tectonic activity also migrated laterally with time (Ensley and Verosub, 1982) (Figures 9 and 10) and was probably related to inherited weaknesses within the detached crustal block between the San Gabriel and its various splay faults. Initially, deposition was constrained between the San Gabriel and Clearwater faults (12.0–8.0 m.y.), then the San Gabriel and Liebre faults (8.0–5.0 m.y.), and finally the San Andreas fault (ca. 5 m.y.).

Continued strike-slip fault motion on the San Gabriel fault allowed the cumulative (but lateral) deposition of up to 12 km (7 mi) of sediments within the basin with successively younger units to be deposited directly onto basement (Crowell, 1974a). The current, northwesterly sediment basement onlap (Crowell, 1974a), particularly within the Castaic Formation, requires a southeasterly dip of the basement during deposition. The basement block, given its current northwesterly dip, has therefore undergone substantial tectonic tilting since the formation of the basin (post-Pliocene), which presumably explains the uplift and truncation of the deep-water Castaic Formation in the northwest of the basin. Miocene-Pliocene basin development was generally accompanied by local thrusting or thrust reactivation (Woodburne, 1975; Ehlig, 1981).

Relatively little is known of the Soledad basin, and the following summary is based on Muehlberger (1958), Nagle and Parker (1971), and Woodburne (1975). The Soledad basin is situated north of the San Gabriel Mountains and south of the Sierra Pelona. In general, the sedimentary units at the surface are progressively younger from east to west. A regional marine unit, the Martinez Formation of Paleocene age, represents basement for both the Ridge and Soledad basins. The oldest rocks of the Soledad basin proper belong to the Oligocene Valquez Formation, which generally rests unconformably on basement (though occasionally it has a fault contact). Unconformably overlying the Vasquez Formation is the non-marine, middle

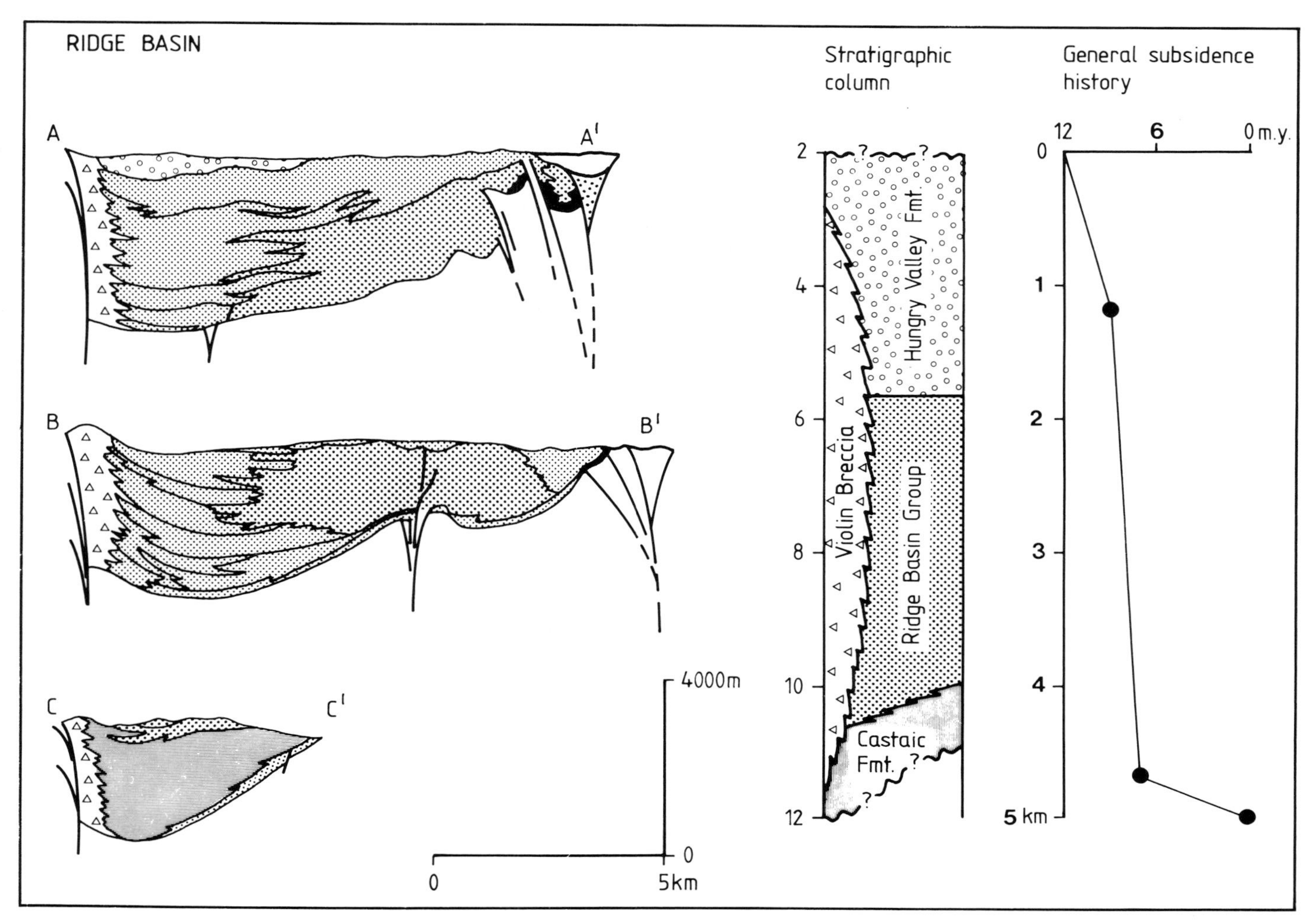


**Figure 10.** *Representative geological sections across the Ridge basin showing the complex interfingering of deltaic units within the Ridge basin Group. Extensive regional, northeasterly tilting during the formation of the Big Bend in the late Miocene resulted in sediment truncation and the exposure of the deep water Castaic Formation in the southwest. A generalized stratigraphic column and subsidence history are also shown.*

Miocene Tick Canyon Formation, which in turn is unconformably overlain by the non-marine Mint Canyon Formation. These non-marine formations invariably consist of fluvial and lacustrine sediments and are, in turn, unconformably overlain by the marine, upper Miocene Castaic Formation. The Castaic Formation represents a sedimentary bridge between the Ridge and Soledad basins. Since the Castaic Formation unconformably overlies the Mint Canyon Formation, the Soledad basin must predate the Ridge basin. The Pliocene marine sequence, which overlies both the Castaic and Mint Canyon Formations in the Ridge-Soledad area, is a potential source rock for hydrocarbons. However, this sequence more appropriately relates to the eastern margin of the hydrocarbon-rich Ventura basin, rather than the Ridge-Soledad basins (Nagle and Parker, 1971).

Workers have recognized three main faulting episodes in the Soledad basin: (1) normal faulting during the Oligocene-early Miocene; (2) in the late Miocene, east to west extension and sympathetic but local north to south compression (?thrusting) coeval with Ridge basin initiation; and (3) Pleistocene regional north to south thrusting and tilting.

## MODELS FOR PULL-APART BASIN FORMATION

The Ridge basin appears to have developed within the restraining bend of a major, right-lateral, strike-slip fault and as such, is not consistent with the configuration usually ascribed (Figure 11a) to pull-apart basins. These basins (e.g., Niksar, Susehri, and Erzincan basins along the North Anatolian fault; Hempton and Dunne, 1984) form along a plate boundary whose local bends do not impede the strike-slip motion of the plates, but instead allow local divergence of the plate boundary. Nevertheless, the basins of southern California are referred to as pull-apart basins in the more general definition of Mann et al (1983). The present fault configuration within southern California, however, generally post-dates Ridge basin formation.

Crowell (1974a, 1974b) developed a number of kinematic models to explain the formation of sedimentary basins within a restraining bend of major strike-slip faults. His basic contention is that inside such restraining bends, the lithosphere is compressed, shortened, and hence elevated, whereas outside the bend, the lithosphere becomes stretched, thereby forming a

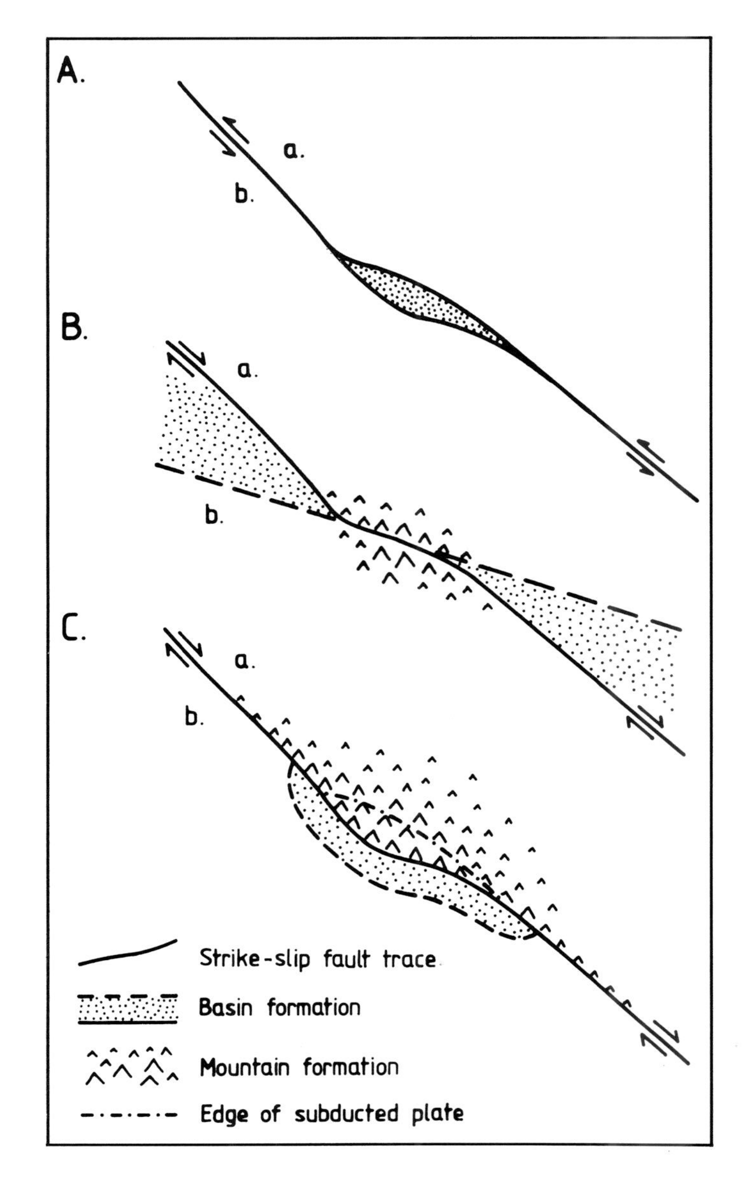


Figure 11. *Models for pull-apart basin formation with respect to the fault configuration of southern California. (a) The standard model in which basin formation occurs across a restraining bend within a left-lateral strike-slip fault between two plates, "a" and "b". (b) As movement across the San Andreas fault is right-lateral, strike-slip motion causes compression across the restraining bend leading to mountain formation. Possible basin formation develops on both sides of the strike-slip fault, but outside the bend region. However, the Ridge basin has formed within this bend. (c) If plate "a" overrides plate "b," a flexural basin forms within plate "b" around the bend. Isostatic compensation of the obducted plate "a" results in mountain formation. Mountain development along the fault is determined by the rigidity of plates "a" and "b" while interior mountain development is related to the degree of underriding of plate "b". In southern California, young, thrust-related mountain ranges exist on the same plate that is undergoing basin development and form contemporaneously with the basins. The association of mountains and basins suggests that the present configuration of the San Andreas fault trace has played a minor role, if any, in the formation of the Ridge basin.*

basin (Figure 11b). By symmetry, basins should form on both sides of the fault bend, but, as shown in Figure 8, the Ridge basin has actually formed within the bend.

Depending on the magnitude of the compression across the bend, stress release may occur by the obduction of one plate (plate 'a') over the other (plate 'b'). The sense of obduction will be controlled by the relative strengths or rigidities of the opposing plates, with the softer plate being overthrust by the stronger plate (Figure 11c) (Molnar and Atwater, 1978; Cloetingh, Wortel, and Vlaar, 1984; Sheffels and McNutt, in press). If their flexural strengths are similar, obduction does not occur and the configuration shown in Figure 11b develops. Isostatic compensation to obduction results in significant topography along the edge of the flexurally stronger plate 'a', whereas a regional foreland-type basin is produced along the fault zone by flexure of the softer plate 'b'. The width of the basin is determined by the effective elastic thickness of plate 'b'. The distribution of topography is now determined not only by the rigidity of plate 'a', but also by the amount of obduction, shown as a dashed-dot line in Figure 11c. Generally, therefore, mountain topography will be distributed regionally relative to the basins, with both mountains and basins parallel to the fault zone. Continuing convergence results in increased mountain and basin development along the fault. The most complicated geological zone will be along the bend where basin sediment and crustal blocks are reworked into nappes and thrust sheets (cf., the Sub-molasse zone of the Alps). For the Ridge basin, this model fails to explain the fractured crust and hence basin development to the north of the mountain topography, and the topography actually belonging to plate 'b' rather than plate 'a'. Obduction of plate 'b' over plate 'a' is ruled out from the seismic tomography results of Humphreys, Clayton, and Hayes, (1984).

Crowell (1974a) also recognizes that the braiding and splaying of major strike-slip faults will result in crustal block generation that, depending on their position relative to the bend, may be forced to either subside or uplift. For example, tilting of the block between the transpressive San Gabriel and San Andreas faults resulted in both the formation of the Ridge basin and the San Gabriel Mountains. It is this concept, and a modification to the model of Figure 11b, that we wish to extend.

## APPLICATION TO RIDGE BASIN DEVELOPMENT

Our model for the origin of the Ridge basin attempts to integrate both the geological and geophysical data for the Ridge basin in particular and southern California in general. The rapid subsidence of the basin suggests that sub-crustal involvement must have been minor, therefore negating the need to consider the quantitative models of lithospheric stretching with finite rifting and horizontal heat flow. This realization leaves the most efficient means of creating a basin to extension within the crust, in this case, by upper crustal block rotation above a detachment zone. Isostasy plays a major role in maintaining and compensating the resulting extensional and compressional basins (Figure 12a and 12b) through the flexural deformation of the lithosphere during block rotation/movement.

Consider a major strike-slip fault zone in compression. Such

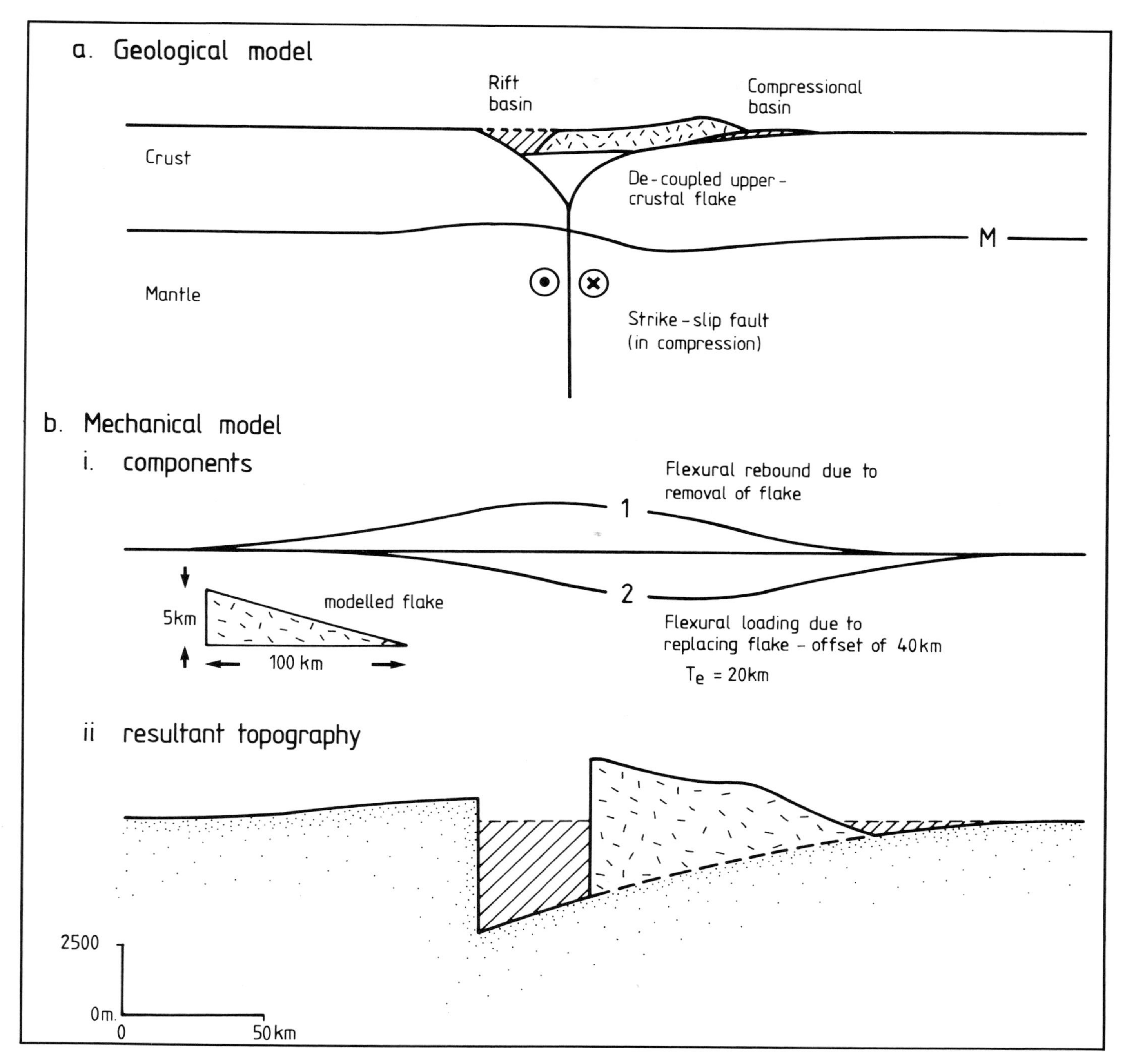


Figure 12. *(a) Mass redistribution during block rotation, caused by strike-slip motion (pure or transpressive) acting on the base of the block. This redistribution is responsible for both basin and mountain formation. The decoupled crustal flake is produced by the converging of strike-slip faults at depth. Basins and mountains form as compatibility structures at the margins of the rotating blocks. (b) Flexural adjustment to block movement. This adjustment results in uplift of the footwall block in addition to the rotated block. Even with relatively low rigidities, the small basin size results in only minor, flexural deformation of the Moho.*

tectonic settings are invariably characterized by "flower structures" (e.g., Harding, 1974) involving sediments or basement or both and the numerous splay faults, such as the San Gabriel-San Andreas fault system. These fault systems represent the wings of major, intracrustal, transpressional flower structures (Figure 12a). Simple strike-slip motion at depth translates to a torque acting within the upper crust. If the crust is fractured and splayed, as is the case along the Pacific and North American plate boundary, individual crustal blocks will rotate clockwise along the detachment surface. This rotation may be controlled by the brittle/ductile transition or the strain rate, thereby creating gaps between the blocks (cf., Crowell, 1974a; Figure 12a). The rate of production of these intercrustal gaps will depend on the relative slip rates along the faults.

Extending only the upper crust results in a strain incompatibility across the detachment surface. However, strain, and isostatic and mechanical equilibria continue if the detached crustal block is forced to override, or obduct, onto adjacent crust (Figure 12a). Mechanical and isostatic equilibria persist because the mass redistribution associated with rotation of the crustal block flexurally unloads the lithosphere within the pull-apart basin region while flexurally loading the adjacent lithosphere, resulting in either the compression of pre-existing basins or the formation of a foreland-type basin (Figure 12b). The interference between loading and unloading the lithosphere produces regional uplift beneath the extensional basin and subsidence to form the foreland basin. Subsequent sediment loading flexurally amplifies the driving subsidence within both basins.

Figure 12b also indicates why small basins do not show large flexural effects during block rotation and sediment deposition, even though the lithospheric rigidities may be high. The interference between flexural unloading and flexural loading is critically determined by the degree of block rotation, and, because of the relatively small lateral offsets of the blocks, the effects almost cancel each other. Additional subsidence by sediment loading will likewise be approximately counterbalanced by the flexural rebound associated with local topographic erosion. These results underline the dependency between load wavelength and the flexural response of the lithosphere (Beaumont, 1978; Lambeck and Nakiboglu, 1981).

Fundamental to this model is the contemporaneous formation of a compressional basin (or destruction of previous basins) during the formation of the extensional basin. Basin destruction is via either the overthrusting of the rotated crustal block over the pre-existing basin, or by the inversion of normal faults, which created earlier pull-apart basins (and therefore the development of local reverse/thrust faults). Both are observed characteristics of the late-stage development of many southern California basins, for example, the Ventura (Nagle and Parker, 1971; Yeats, 1981, 1983), Los Angeles (Gardett, 1971), Cuyama (Baldwin, 1971), and Soledad (Meuhlberger, 1958) basins. In addition to modifying pre-existing basins, block obduction will in itself result in significant local topography (because of the isostatic compensation of the obducted block) and further, may reactivate thrusts associated with pre-existing allochthonous blocks, such as the Sierra Pelona block.

The extreme rates of basement subsidence help to explain the general stratigraphic facies of these small pull-apart basins. Plotted in Figure 13 is a theoretical, paleo-water depth calculation using a basin subsidence rate typical for these basins and an opposing (but extreme) sedimentation rate (applicable to the Himalayas). Initially, the sedimentation rate is incapable of pacing basement subsidence, thereby leading to the formation of a restricted deep-water facies, probably marked by the presence of black, organic-rich shales (cf., Castaic Formation). When the driving subsidence ceases abruptly, sediment infilling has a chance to fill the basin, especially because erosion has been reactivated by the formation of thrust/obducted related topography. Delta and alluvial fan formation, encroachment, and facies interfingering eventually fill the basin (cf., Ridge Basin Group), and in so doing, isostatically load the basement (Figure 13). In marked contrast to the ordered subsidence and time-line distribution of intracratonic basin subsidence (Figure 5), the stratigraphy and time-line distribution of these pull-apart basins will be controlled by the rate of sediment input, details of the

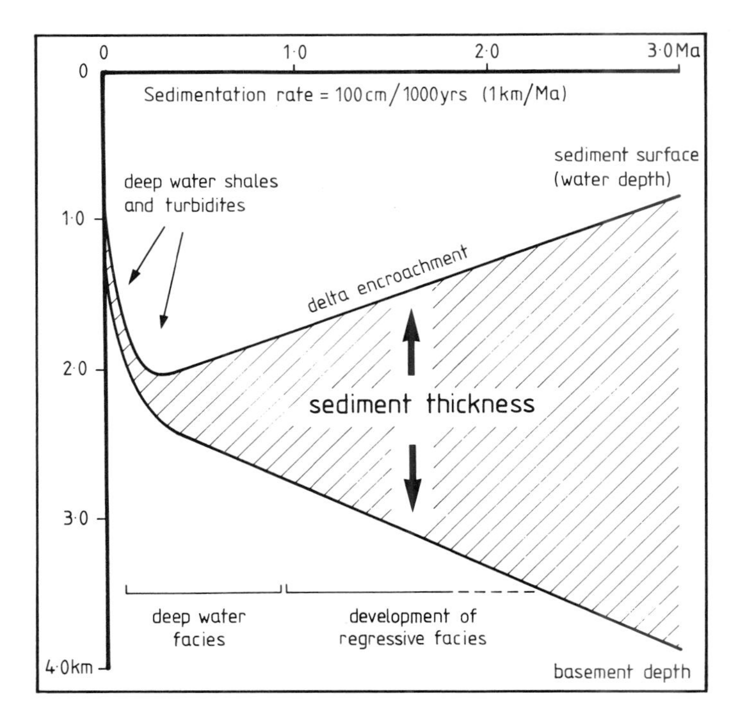


**Figure 13.** *Block rotation in initiating a basin maximizes basin subsidence rates. Initially the inability of sedimentation to pace these subsidence rates produces a sediment-starved basin. As sedimentation begins to fill the basin, generally by the converging of deltas/alluvial fans, the facies changes from deep-water, through distal turbidites, to the proximal facies of the delta foreset beds. Such a regressive facies is potentially favorable for hydrocarbon generation, with an organic rich source bed being buried by the interbedded clean sands of the turbidites and deltas.*

depositional process (e.g., delta versus braided stream versus alluvial fan, etc.), and the geometry of the basement. Necessarily, therefore, sediment distribution and accumulation within the basin will be spatially and temporally discontinuous, thereby leading to the production of many (?erosional) unconformities.

Hydrocarbon generation and migration will depend on the availability of suitable source and reservoir rocks. In the Ridge basin, the Castaic Formation and the proximal deltaic facies of the Ridge Basin Group (and possibly units of the Hungry Valley Formation) represent source and reservoir units, respectively. Even the lack of an anomalous heat source for maturation, as usually supplied by lithospheric stretching in intracratonic basin formation (Figure 3), is not a serious limitation in the case of southern California because the relatively small lithospheric thickness of 60–70 km (37–44 mi) (Zandt and Furlong, 1982) ensures a high regional geothermal gradient. Hydrocarbon maturation, therefore, can easily be obtained at depths of 3–5 km (2–3 mi) within the Ridge basin via the temperature re-equilibration of the sediments by vertical, but principally lateral, heat flow from adjacent, thermally unperturbed regions. No doubt a major, complicating factor for both temperature gradients and hydrocarbon migration will be water circulation within the fractured upper crust (Andrews and Oxburgh, 1984).

We think that the model summarized in Figure 12a and 12b

satisfactorily explains not only the crustal (or mechanical) origin of the Ridge basin but also, crustal block interactions along any (pure or transpressional) anastomosing strike-slip fault system. However, it is still necessary to address the actual mechanism responsible for the Ridge basin within the context of the interacting Pacific and North American plates. The post-Miocene (< 5 m.y.) period within southern California was a critical time that witnessed the major re-organization of strike-slip and thrust fault geometries and the Pacific–North American plate boundary itself, in particular (1) the Ridge basin ceased developing and its northern boundary became disrupted by thrusts; (2) the strike-slip San Gabriel fault was abandoned as motion was transferred to the San Andreas fault; (3) regional compression began with the uplift of the San Gabriel and San Bernadino Mountains along north and south dipping thrust faults, respectively; (4) the Ridge basin tilted northward and truncated; (5) the Big Bend began within the San Andreas fault system; (6) the increasing curvature of the San Andreas fault with time has transferred motion to newly created, presumably more favorably oriented, faults as evidenced by the creation of the San Jacinto fault in the Pleistocene and the Elsinore fault in the Holocene; (7) Pliocene-Quaternary folding along the Western Margin of the Great Valley; (8) the continental subduction/obduction started beneath the present-day Transverse Ranges; (9) the Gulf of California opened; and (10) the Big Bend region migrated westward by some 40-50 km (25-31 mi).

Prior to 5 Ma, the principal transform displacement zone between the Pacific and North American plates lay along the Californian Borderland as a series of long transform/short pull-apart segments. We believe that these Miocene pull-apart basins are an important setting for future important oil discoveries. Onshore basins, in contrast, such as the Ridge basin, lay in a zone of distributed shear and anastomosing fault systems (e.g., San Gabriel). Basins in this regime formed as either compatibility structures at the margins of rotating upper-crustal blocks (flakes) or as pull-aparts at transpressional relays between strike-slip faults. At 5 Ma, the principal offshore transform zone jumped into the continent to form the San Andreas fault, detaching the Salinian Block and forcing it to slide north-northwesterly with respect to North America, thereby opening the Gulf of California. This transform reconfiguration probably resulted from the cooling and hence strengthening of the borderland lithosphere following the impingement of the Pacific/Farallon Ridge with the Farallon/North American Trench.

The easterly transfer of the offshore right-lateral shear system fundamentally altered the tectonic regime along the Californian margin of North America. Whatever controlled the formation and possible margration of the Big Bend apparently also controlled the geological development of the region. The major factor connecting these observations is the late Miocene to Holocene lithospheric extension within the Basin and Range province (cf., Bohannon and Howell, 1982). Differing extension rates between the northern and southern Basin and Range provinces are accommodated by left-lateral slip along the Garlock fault (Davis and Burchfiel, 1973). The fundamental consequence of this extension is the westward migration of the Sierra Nevada Mountains and the Great Valley, and, therefore, the western boundary of the North American plate. Conservative estimates (e.g., Garfunkel, 1974) suggest that basin and range extension may have displaced the Sierra Nevada Mountains by 150–200 km (93–124 mi) to the west. Restoring this stretched crust results in the present restraining bend of the San Andreas system diminishing to either a straight or gently transpressive fault, or even becoming a releasing bend. Westward migration of the Sierra Nevada Mountains may have progressively straightened the releasing bend of the San Gabriel fault system until about 5 Ma, when the bend became reversed, forming a transpressive restraining bend. Pull-apart basin formation, therefore, would have abruptly ceased. Within the restraining bend, transpression resulted in the formation of mountains probably antecedent to the San Gabriel Mountains. This same southerly uplift introduced a northwesterly tilt to the Ridge basin basement and to the exposure of the Castaic Formation. As the westward bend continued to grow, the newly activated strike-slip faults progressively became locked, forcing the creation of even newer faults in order to relieve the compressive strains and remove the curvature across the Big Bend. For example, during the Quaternary and Holocene, strike-slip motion was transferred from the San Andreas to the San Jacinto and Elsinore faults, respectively. Transpression associated with strike-slip motion across the restraining bend was further enhanced by the opening of the Gulf of California, which eventually led to crustal shortening and continental lithospheric subduction. Subduction was initiated in the Pleistocene as evidenced by the rapid uplift rates of both the San Gabriel and San Bernardino Mountains (ca., 5 km/m.y. or 3 mi/m.y.).

The possibility of continental subduction, though strongly supported by seismic tomographic imaging of a slab beneath the Transverse Mountains, is problematic. Generally, subduction is identified by its Benioff zone or characteristic volcanicity, neither of which are present. The youth of the subduction zone (<2 m.y.) may help explain the lack of slab melts. The asymmetric depth of the slab, 250 km (155 mi) in the east to 100 km (62 mi) in the west, is explained in terms of the progressive, westerly migration of the restraining bend as the Pacific plate continues to migrate north. However, continental subduction zones are invariably associated with major foreland basins such as the Ganges basin of the Himalayas and the Molasse basin of the Alps. Obduction of upper-crustal blocks during continental subduction from the Pacific to the North American plate could feasibly destroy any foreland basin, thus explaining the absence of such a feature to the south of the Transverse Ranges. The rapid uplift rates of the San Bernardino Mountains, their high elevation, and their regional isostatic compensation suggest that the mountains represent obducted upper-crust from the Pacific plate (cf., Oxburgh, 1972). The decollement beneath the San Bernardino Mountains is, therefore, the surface of the North American plate.

In contrast, obduction has yet to occur within the San Gabriel Mountains as it would destroy the Ridge basin, but given the decreasing westward subduction of the Pacific lithosphere, subduction is probably only in its infancy. The decollement beneath the Ridge basin probably represents the original detachment surface that allowed upper-crustal block rotation during basin development.

## CONCLUSIONS

From this study we conclude that, regarding the mechanisms of basin formation:

(1) The formation of sedimentary basins is a direct conse-

quence of the principle of isostasy. Isostasy describes the deformation of the lithosphere by applied loads, in particular, the basin-driving force or subsidence and subsequent sediment loads. The spatial and temporal form of the deformation defines the geometry and regional stratigraphy of the developing sedimentary basin.

(2) Lithospheric rifting, as quantified by McKenzie (1978) and later modified for the thermal and mechanical properties of the lithosphere, is an adequate description of the regional, tectonostratigraphic development of intracratonic and passive margin basins. The apparent regressive facies that characterizes the late development of many basins is most easily explained in terms of either an eustatic sea-level fall during the latter stages of basin formation or surface erosion or both.

(3) The simplest initiating mechanism for basins exhibiting a rapid but short-lived subsidence is for crustal stretching, in contrast to intracratonic and passive margins, without significant sub-crustal (i.e., the lithosphere/asthenosphere boundary), and hence heat, involvement. Strain compatibility is maintained by balancing crustal extension (normal faulting) with adjacent but contemporaneous crustal obduction (thrust faulting); that is, both normal and reverse faulting are responsible for basin development.

(4) Such a basin mechanism is a simple consequence of crustal block rotations within a tectonic regime dominated by intracrustal detachment zones or crustal fracturing along major continental transform zones or both. In the case of southern California, detached crustal blocks are produced by the convergence of splay faults at depth (e.g., San Gabriel, Clearwater, Liebre, and San Andreas faults) into a ductile shear zone. Strike-slip motion supplies the driving force for block rotation and hence basin subsidence. Given the right-lateral motion along the San Andreas fault system, detached blocks will rotate clockwise.

(5) The upper Miocene Ridge basin of southern California resulted from the rotation of detached, upper crustal blocks (effectively, upper crustal stretching) and represents a classic pull-apart basin. Contemporaneous thrusting within the older Soledad basin and local topographic uplift maintain the strain compatibility within the upper crust. The apparent restraining bend in which the Ridge basin now resides is a consequence of the westward migration of the Sierra Nevada Mountains (and to a lesser extent the Mohave block) because of upper Miocene-Holocene Basin and Range province crustal extension.

(6) Because sedimentation rates are unable to pace the early basin subsidence, we expect a sedimentary facies succession from deep-water turbidites, distal and proximal deltaic/alluvial fan deposits, and, finally, coarse sands and conglomerates. This succession is represented in the Ridge basin by the Castaic, Ridge Basin Group, and Hungry Valley Formation respectively.

(7) Because basin initiation by block rotation essentially represents cold rifting, equilibrium temperatures are maintained within adjacent areas. Lateral and vertical heat flow into the basin with time re-establishes a steady-state temperature structure. Given the modest temperatures required for hydrocarbon generation (70°–100°C or 158°–212°F) and the present lithospheric thickness of southern California, we predict a liquid window at 3–5 km (2–3 mi) within the Ridge basin.

(8) Southern California underwent a major transformation of the North American–Pacific plate boundary in the late Tertiary: (a) pull-apart basin formation because of strike-slip motion along the San Gabriel fault from 12–5 Ma, and (b) the abandonment of the San Gabriel fault in preference for the San Andreas fault while creating a restraining bend as the Sierra Nevada Mountains, and hence the western boundary of the North American plate migrated westward from 0–5 Ma. Plate boundary migration was in response to lithospheric extension within the Basin and Range province. Increasing transpression across the restraining bend, with time, resulted in the subduction of the Pacific plate under the North American plate. Recent obduction within the eastern Transverse Range is responsible for the San Bernardino Mountains (cf. the flake tectonic concept of Oxburgh, 1972). Subduction beneath the western Transverse Ranges is only in its infancy.

(9) Mountains are of two types: (a) regional mountains (e.g., the Transverse Ranges) produced by either crustal obduction or crustal shortening, and (b) local mountains (e.g., Liebre Mountains) representing the underthrust edge of upper-crustal block, rotated during pull-apart basin formation.

## ACKNOWLEDGMENTS

A special thank you to Walter Pitman, III for encouragement and support during the preparation of this manuscript and access to unpublished material, in particular Figures 7 and 13. Our sincere thanks also to Mike Steckler, Stuart Lake, Kevin Brown, and Robin Bell for discussions and to Marcia McNutt for supplying a pre-print of her gravity paper. We thank Andy Reid for his fine drafting efforts. G.D.K. gladly acknowledges support from the Durham University Research Foundation and the Society of Fellows.

## REFERENCES

Airy, G. B., 1855, On the computation of the effect of the attraction of mountain-masses as disturbing the apparent astronomical latitude of stations of geodetic surveys: Philosophical Transactions of the Royal Society, v. 145, p. 101–104.

Artemjev, M. E., and E. V. Artyushkov, 1971, Structure and isostasy of the Baikal rift and the mechanism of rifting: Journal of Geophysical Research, v. 76, p. 1179–1211.

Atwater, T., 1970, Implications of plate tectonics for the Cenozoic tectonic evolution of western North America: Geological Society of American Bulletin, v. 81, p. 3513–3536.

——— , and P. Molnar, 1973, Relative motion of the Pacific and North American plates deduced from sea-floor spreading anomalies in the Atlantic, Indian, and south Pacific Oceans, *in* R. L. Kovach, and A. Nur, eds., Proceedings, Conference on tectonic problems of the San Andreas fault system: Stanford University Publications, Geological Science, v. XIII, p. 136–148.

Baldwin, T. A., 1971, Petroleum potential of California Central

Coast Ranges, *in* I. H. Cram, ed., Future petroleum provinces of the United States — their geology and potential: AAPG Memoir 15, v. 1, 803 p.
Beaumont, C., 1978, The evolution of sedimentary basins on a viscoelastic lithosphere: theory and examples: Geophysical Journal of the Royal Astronomical Society London, v. 55, p. 471–498.
———, 1981, Foreland basins: Geophysical Journal of the Royal Astronomical Society London, v. 65, p. 291–329.
———, C. E. Keen, and R. Boutilier, 1982, A comparison of foreland and rift margin sedimentary basins: Philosophical Transactions of the Royal Society London, v. A 305, p. 295–317.
Bird, P., and R. W. Rosenstock, 1984, Kinematics of present crust and mantle flow in southern California: Geological Society of America Bulletin, v. 95, p. 946–957.
Bohannon, R. G., and D. G. Howell, 1982, Kinematic evolution of the junction of the San Andreas, Garlock, and Big Pine faults, California: Geology, v. 10, p. 358–363.
Bodine, J. H., and A. B. Watts, 1979, On lithospheric flexure seaward of the Bonin and Mariana trenches: Earth and Planetary Science Letters, v. 43, p. 132–148.
———, M. S. Steckler, and A. B. Watts, 1981. Observations of flexure and the rheology of the oceanic lithosphere: Journal of Geophysical Research, v. 86, B5, p. 3695–3707.
Bott, M. H. P., 1971, Evolution of young continental margins and formation of shelf basins: Tectonophysics, v. 11, p. 319–327.
———, 1981, Crustal doming and the mechanism of continental rifting: Tectonophysics, v. 73, p. 1–8.
Calderone, G., and R. F. Butler, 1984, Paleomagnetism of Miocene volcanic rocks from southwestern Arizona: tectonic implications: Geology, v. 12, p. 627–630.
Caldwell, J. G., and D. L. Turcotte, 1979, Dependence of the elastic thickness of the elastic oceanic lithosphere on age: Journal of Geophysical Research, v. 84, p. 7572–7576.
Chadwick, R. A., 1985, Seismic reflection investigations into the stratigraphy and structural evolution of the Worcester graben: Journal of the Geological Society London, v. 142, p. 187–202.
Cloetingh, S. A. P. L., M. J. R. Wortel, and N. J. Vlaar, 1984, Passive margin evolution, initiation of subduction and the Wilson cycle: Tectonophysics, v. 109, p. 147–163.
Cochran, J. R., 1983, Effects of finite rifting times on the development of sedimentary basins. Earth and Planetary Science Letters, v. 66, 289–302.
Crowell, J. A., 1974a, Sedimentation along the San Andreas fault, California, *in* R. Dott, and R. Shaver, eds., Modern and ancient geosynclinal sedimentation: SEPM Special Publications v. 19, p. 292–303.
———, 1974b, Origin of late Cenozoic basins in southern California, *in* W. R. Dickinson, ed., Tectonics and Sedimentation, SEPM Special Publications v. 22, p. 190–104.
———, 1975, The San Andreas fault in Southern California: Calif. Division of Mines and Geology Special Report, 118, 272 p.
———, 1979, The San Andreas fault system through time: Journal of the Geological Society London, v. 136, p. 293–302.
———, 1981, An outline of the tectonic history of southeastern California, *in* W. G. Ernst, ed., The geotectonic development of California (Rubey Vol. 1): Englewood Cliffs, N. J., Prentice Hall, 706 p.
Davis, G. A., and B. C. Burchfiel, 1973, Garlock fault: an intracontinental transform structure, southern California: Geological Society of America Bulletin, v. 84, p. 1407–1422.
Dewey, J. F., 1982, Plate tectonics and the evolution of the British Isles: Journal of the Geological Society, v. 139, p. 371–412.
Dickinson, W. R., R. A. Armin, N. Beckvar, T. C. Goodline, S. U. Janecke, R. A. Mark, R. D. Norris, G. Radel, and A. A. Wortman, in press, Geohistory analysis of rates of sediment accumulation and subsidence for selected California basins: University of California, Los Angeles, Rubey Colloquiem volume (expected 1986).
Ehlig, P. L., 1981, Origin and tectonic history of the basement terrane of the San Gabriel Mountains, Central Transverse Ranges, *in* W. G. Ernst, ed., The geotectonic development of California (Rubey v. 1) Englewood Cliffs, N. J., Prentice Hall, 706 p.
Ensley, R. A., and K. L. Verosub, 1982, A magnetostratigraphic study of the sediments of the Ridge basin, southern California and its tectonic and sedimentologic implications: Earth and Planetary Science Letter, v. 59, p. 192–207.
Etheridge, M. A., J. C. Branson, and P. G. Stuart-Smith, 1985, Extensional basin-forming structures in Bass Strait and their importance for hydrocarbon exploration. Journal of the Australian Petroleum Exploration Association, in press.
Falvey, D. A., 1974, The development of continental margins in plate tectonic theory. Australian Petroleum Exploration Association Journal, v. 14, p. 95–106.
Gardett, P. H., 1971, Petroleum potential of Los Angeles basin, California, *in* I. H. Cram, ed., Future petroleum provinces of the United States — their geology and potential: AAPG Memoir 15, v. 1, 803 p.
Garfunkel, Z., 1974, Model for the late Cenozoic tectonic history of the Mojave Desert, California, and for its relation to adjacent regions: Geological Society of America Bulletin, v. 85, p. 1941–1944.
Grannel, R. B., 1971, A regional gravity survey of the San Gabriel Mountains, California (abs.): Geological Society America Abstracts with Programs, v. 3, p. 127.
Greenhaus, M. R., and A. Cox, 1979, Paleomagnetism of the Morro Rock-Islay Hill complex as evidence for crustal block rotations in Central Coastal California: Journal of Geophysical Research, v. 84, p. 2392–2400.
Hadley, D. M., and H. Kanamori, 1977, Seismic structure of the Transverse Ranges, California: Geological Society of America Bulletin, v. 88, p. 1469–1478.
Harding, T. P., 1974, Petroleum traps associated with wrench faults: AAPG Bulletin, v. 58, p. 1290–1304.
Haxby, W. F., D. L. Turcotte, and J. M. Bird, 1976, Thermal and mechanical evolution of the Michigan basin: Tectonophysics, v. 36, p. 57–75.
Hempton, M. R., and L. A. Dunne, 1984, Sedimentation in pull-apart basins: active examples in eastern Turkey: Journal of Geology, v. 92, p. 513–530.
Humphreys, E., R. W. Clayton, and B. H. Hager, 1984, A tomographic image of mantle structure beneath southern California: Geophysical Research Letter, v. 11, p. 625–627.
Jarvis, G. T., and D. P. McKenzie, 1980, Sedimentary basin formation with finite extension rates: Earth and Planetary Science Letters, v. 48, p. 42–52.

Karner, G. D., 1984, Continental tectonics — a quantitative view of the thermal and mechanical properties of the continental lithosphere in compressional and extensional stress regimes: Toulouse, France, Centre National d'Etudes Spatiales, Summer School of Space Physics.

———, and A. B. Watts, 1983, Gravity anomalies and flexure of the lithosphere at Mountain ranges: Journal of Geophysical Research, v. 88, B12, p. 10449–10477.

———, and J. K. Weissel, 1984, Thermally induced uplift and lithospheric flexural readjustment of the eastern Australian highlands (abs.): Geological Society of Australia Abstracts, v. 12, p. 293–294.

———, M. S. Steckler, and J. A. Thorne, 1983, Long-term thermo-mechanical properties of the continental lithosphere: Nature, v. 304, n. 5923, p. 250–253.

———, ———, and W. F. Haxby, in press, Thermal transients associated with sedimentary basin evolution: Journal of Geophysical Research, (expected 1986).

Keen, C. E., 1979, Thermal history and subsidence of rifted continental margins. Evidence from wells on the Nova Scotia and Labrador Shelves: Canadian Journal of Earth Science, v. 16, p. 505–522.

———, and D. L. Barrett, 1981, Thinned and subsided continental crust on the rifted margin of eastern Canada: crustal structure, thermal evolution and subsidence history: Geophysical Journal of the Royal Astronomical Society London, v. 65, p. 443–465.

Lake, S. D., G. D. Karner, and J. F. Dewey, in press, The flexural development of the onshore Wessex basin in southern England: Journal of the Geological Society of London, (expected 1986).

Lambeck, K., and S. M. Nakiboglu, 1981, Seamount loading and stress in the ocean lithosphere, 2: viscoelastic and elastic-viscoelastic models: Journal of Geophysical Research, v. 86, p. 6961–6984.

Luyendyk, B. P., M. J. Kamerling, and R. Terres, 1980, Geometric model for Neogene crustal rotations in southern California: Geological Society of America Bulletin, v. 91, p. 211–217.

Lyon-Caen, H., and P. Molnar, 1983, Constraints on the structure of the Himlaya from an analysis of the gravity anomalies and a flexural model of the lithosphere: Journal of Geophysical Research, v. 88, p. 8171–8192.

McKenzie, D. P., 1978, Some remarks on the development of sedimentary basins: Earth and Planetary Science Letters, v. 40, p. 25–32.

McNutt, M. K., 1983, Influence of plate subduction on isostatic compensation in northern California: Tectonics, v. 2, p. 399–415.

Malin, P. E., and P. C. Leary, 1979, Seismic velocity structure near Palmdale, California (abs.): EOS, Transactions of the American Geophysical Union, 60, p. 314.

Mann, P., M. Hempton, D. Bradley, and K. Burke, 1983, Development of pull-apart basins: Journal of Geology, v. 91, p. 529–554.

Molnar, P., and T. Atwater, 1978, Inter-arc spreading and cordilleran tectonics as alternatives related to the age of subducted oceanic lithosphere: Earth and Planetary Science Letters, v. 41, p. 330–340.

Muehlberger, W. R., 1958, Geology of northern Soledad basin, Los Angeles County, California: AAPG Bulletin, v. 42, p. 1812–1844.

Nagle, H. E., and E. S. Parker, 1971, Future oil and gas potential of onshore Ventura basin, California. *in* I. H. Cram, ed., Future petroleum provinces of the United States — their geology and potential: AAPG Memoir 15, p. 254–297.

Oliver, H. W., 1980, Transverse Ranges. *in* H. W. Oliver, ed., Interpretation of the gravity map of California and its continental margin: Sacramento, California, Bulletin 205, California Divisions of Mines and Geology.

Oxburgh, E. R., 1972, Flake tectonics and continental collision: Nature, v. 239, p. 202–204.

Parsons, B., and J. G. Sclater, 1977, An analysis of the variation of ocean floor bathymetry and heat flow with age: Journal of Geophysical Research, v. 82, p. 803–827.

Pratt, J. H., 1855, On the attraction of the Himalaya Mountains, and of the elevated regions beyond them, upon the plumb line in India: Philosophical Transactions of the Royal Society, v. 145, p. 53–100.

Price, R. A., 1973, Large-scale gravitational flow of supracrustal rocks, southern Canadian Rockies: *in* K. A. deJong and R. Scholten, eds., Gravity and tectonics: New York, J. Wiley and Sons, p. 491–502.

Quinlan, G. M., and C. Beaumont, 1984, Appalachian thrusting, lithospheric flexure, and the Paleozoic stratigraphy of the Eastern Interior of Northern America: Canadian Journal of Earth Science, v. 21, p. 973–996.

Raikes, S. A., and D. M. Hadley, 1979, The azimuthal variation of teleseismic P-residuals in southern California: implications for upper mantle structure: Tectonophysics, v. 56, p. 89–96.

Richmond, J. F., 1960, Geology of the San Bernardino Mountains north of Big Bear Lake, California: California Division of Mines and Geology, Special Report 65, p. 68.

Royden, L., and C. E. Keen, 1980, Rifting processes and thermal evolution of the continental margin of eastern Canada determined from subsidence curves: Earth and Planetary Science Letters, v. 51, p. 343–361.

Sclater, J. G., L. Royden, F. Horvath, B. C. Burchfiel, S. Semken, and L. Stegena, 1981, The formation of the intra-Carpathian basins as determined from subsidence data: Earth and Planetary Science Letters, v. 51, p. 139–162.

Sheffels, B., and M. McNutt, in press, The role of subsurface loads and regional compensation in the isostatic balance of the Transverse Ranges, California: evidence for intracontinental subduction: Journal of Geophysical Research.

Sleep, N. H., 1971, Thermal effects of the formation of Atlantic continental margins by continental break-up: Geophysical Journal of the Royal Astronomical Society London, v. 24, p. 325–350.

———, and N. S. Snell, 1976, Thermal contraction and flexure of midcontinent and Atlantic marginal basins: Geophysical Journal of the Royal Astronomical Society London, v. 45, p. 125–154.

Spohn, T., and H. J. Neugebauer, 1978, Metastable phase transition models and their bearing on the development of Atlantic-type geosynclines: Tectonophysics, v. 50, p. 387–412.

Steckler, M. S., 1981, The thermal and mechanical evolution of Atlantic-type continental margins: New York, Ph.D. Dissertation, Columbia University, 261 p.

———, and A. B. Watts, 1978, Subsidence of the Atlantic-type

continental margin off New York: Earth and Planetary Science Letters, v. 41, p. 1–13.

Tapponier, P., and J. Francheteau, 1978, Necking of the lithosphere and the mechanics of slowly accreting plate boundaries: Journal of Geophysical Research, v. 83, p. 3955–3970.

Turcotte, D. L., and D. McAdoo, 1979, Thermal subsidence and petroleum generation in the southwest block of the Los Angeles basin, California: Journal of Geophysical Research, v. 84, p. 3460–3464.

Walcott, R. I., 1970, Flexural rigidity, thickness, and viscosity of the lithosphere: Journal of Geophysical Research, v. 75, p. 3941–3954.

Watts, A. B., 1978, An analysis of isostasy in the world's oceans: 1. Hawaiian-Emperor seamount chain: Journal of Geophysical Research, v. 83, p. 5989–6004.

——— , 1981, The U.S. Atlantic continental margin: subsidence history, crustal structure and thermal evolution, *in* The geology of passive continental margins: history, structure and sedimentologic record: AAPG Education Course Note Series, 19.

——— , 1982, Tectonic subsidence, flexure, and global changes of sea level: Nature, v. 297, p. 469–474.

——— , and J. R. Cochran, 1974, Gravity anomalies and flexure of the lithosphere along the Hawaiian-Emperor seamount chain: Geophysical Journal of the Royal Astronomical Society London, V. 38, p. 119–141.

——— , and J. Thorne, 1984, Tectonics, global changes in sealevel and their relationship to stratigraphic sequences at the U.S. Atlantic continental margin: Marine and Petroleum Geology, v. 1, p. 319–339.

——— , G. D. Karner, and M. S. Steckler, 1982, Lithospheric flexure and the evolution of sedimentary basin formation, *in* The evolution of sedimentary basins: Philosophical Transactions of the Royal Society, Series A, v. 305, 338 p.

Wernicke, B., 1981, Low-angle normal faults in the basin and range province: nappe tectonics in an extending orogen: Nature, v. 291, p. 645–648.

Woodburne, M. O., 1975, Cenozoic stratigraphy of the Transverse Ranges and adjacent areas, Southern California: Geological Society of America Special Paper 162, 91 p.

Yeats, R. S., 1978, Neogene acceleration of subsidence rates in southern California: Geology, v. 6, p. 456–460.

———,1981, Quaternary flake tectonics of the California Transverse Ranges: Geology, v. 9, p. 16–20.

———,1983, Large-scale Quaternary detachments in Ventura basin, southern California: Journal of Geophysical Research, v. 88, p. 569–583.

Zandt, G., and K. P. Furlong, 1982, Evolution and thickness of the lithosphere beneath coastal California: Geology, v. 10, p. 376–381.

# African Oil — Past, Present, and Future

Andy C. Clifford
*BHP Petroleum, (U.K.) Ltd.*
*London, England*

Eighty-three sedimentary basins cover 50% of Africa's 11.7 million square miles. Hydrocarbons are produced from 20 of these basins and a further 13 have non-produced hydrocarbons. Yet of all the basins, only 18% have been explored through a semi-mature stage. This paper presents a simplified classification of basins in Africa.

Africa's recoverable hydrocarbon reserves of 100 billion barrels of oil equivalent account for 7% of the world's total, proven reserves. This paper documents that reserve base's development through time together with the significant discoveries. Oil and gas occurrences and the play potential of each basin type are systematically described, with examples demonstrating exploration problems regarding particular basins.

Africa contains an abundance of unexplored sedimentary basins that should demonstrate relatively high success ratios but which must be evaluated with some caution because of political, fiscal, and logistical concerns.

## INTRODUCTION

To write a fully comprehensive paper on such a huge topic could easily occupy this entire volume if it were not for numerous, excellent publications by many authors. As such, this paper can be no more than a synopsis of ideas developed over the past few decades by these and other writers. I cannot describe each basin and every hydrocarbon occurrence in detail because that would be too lengthy and the result too confusing. Rather, I have grouped together sedimentary basins having common genesis and similar geologic characteristics and have highlighted their important factors relating to hydrocarbon occurrences. The paper should give an overview of African hydrocarbons, explaining their occurrence related to a basin classification system, and demonstrating the use of some common basin assessment procedures. The paper should also show the vast potential for future hydrocarbons in Africa within certain basin types rather than others, stressing the importance of exploration models, both structural and stratigraphic.

Because a paper on African oil embraces an overwhelming volume of subject matter, it has been necessary to include only examples of hydrocarbon accumulations pertaining to the basins' special characteristics. Rather than include references to other papers concerning African petroleum geology within the body of the paper, a list of useful references is included at the end.

The paper explains the statistical methods used in basin assessment, and gas-chromatographic analysis used in geochemistry, along with several illustrative examples of each. A basin classification for Africa leads into a systematic basin-type by basin-type description of hydrocarbon occurrences and plays. The paper includes some statistical results concerning maturity of exploration and success ratios, and compares the commercial attractiveness between African countries. This comparison demonstrates the use of economic evaluation and analysis.

## BASIN ASSESSMENT PROCEDURES[1]

An assessment of a sedimentary basin gives a range of hydrocarbon reserves generated within that basin, of which some already may have been produced or proven. A summation of individual play assessments gives the best result, but, if data are lacking, "lookalike" criteria can be used. A lookalike is a geologic similarity, or analogy, between a known productive basin (or play) and the basin (or play) being assessed. Three common procedures in assessment that this paper uses are field size distributions, success ratios, and exploration maturity.

Field size distributions are plots of field size (in oil-equivalent barrels, referred to as BOE) against the probability of a field size that large or larger occurring. They are derived by listing the sampled fields in ascending order of size, then assigning a series of probabilities to those field sizes. When plotted on log probability paper, a best-fit straight line can be drawn. A better fit is derived if the fields sampled are large in number and related to each other by discrete plays and trap-types. A most useful parameter obtained from the graph is the $P_{50}$ value (approximately mean field size). Together with other percentile values, geologists can compare basins of similar or dissimilar genetic type. Such mean field sizes and ranges can be applied to virgin basins to estimate likely field sizes, especially when used in conjunction with risking factors and prospect assessments.

Success ratios are strictly defined as the number of discoveries greater than a specific size, divided by the number of new-field wildcats drilled on prospects capable of holding discoveries greater than that specified size. However, such detailed information is usually lacking, so success ratios are more commonly the number of commercial fields, divided by the number of new-field wildcats. A further refinement is the use of historical success ratios, that is, the progression of success ratios through time. These can show the development of a particular hydrocarbon play or the effects of political or fiscal changes. If sufficient data are available, then success ratios can be applied to individual trap types, such as glide-plane rollovers, tilted fault blocks, and diapir flank traps.

The level of exploration maturity within a basin can be measured in numerous ways. This paper counts the area tested per new-field wildcat (NFW) as determined by measuring the total basin area and dividing by the total number of new-field wildcats drilled within that basin. Stratigraphic tests, core holes, water wells, and appraisal wells are not included in the well count. Thus, arbitrary ranges of exploration maturity can be assigned that help to compare basins. The ranges and categories used throughout this paper are as follows:

| | |
|---|---|
| No NFW's in basin | = Untested |
| $>1,000$ mi$^2$ per NFW | = Immature |
| 500–1,000 mi$^2$ per NFW | = Semi-mature |
| 100–499 mi$^2$ per NFW | = Mature |
| $<100$ mi$^2$ per NFW | = Very mature |

## GEOCHEMICAL ANALYSIS

The paper illustrates examples of gas chromatograms that determine the presence or absence of "geochemical fossils" such as naphthenes, isoprenoids, and normal alkanes. Their presence indicates a degree of immaturity and aids identification of source material, while their absence indicates maturity. The spikes on these gas-chromatogram profiles represent normal alkanes and the area beneath these spikes represents naphthenes. Pristane and phytane are examples of isoprenoids. Apart from being important marker points on the profile, their ratio is important in assessing the type of organic source material. A commonly used source-type and maturation indicator is the Carbon Preference Index (CPI), a ratio of odd-numbered, normal alkanes to even-numbered, normal alkanes, most commonly in the $C_{25}+$ region.

## AFRICAN BASIN CLASSIFICATION

Africa's total area of 11.7 million sq mi (30.3 million sq km) can easily accomodate the United States, Europe (excluding U.S.S.R.), India, Southeast Asia, Japan, and New Zealand within its coastlines (Figure 1). Sedimentary basins cover 5.85 million

[1]Note: All reserves quoted within the paper, for fields, basins or countries, are recoverable unless stated otherwise.

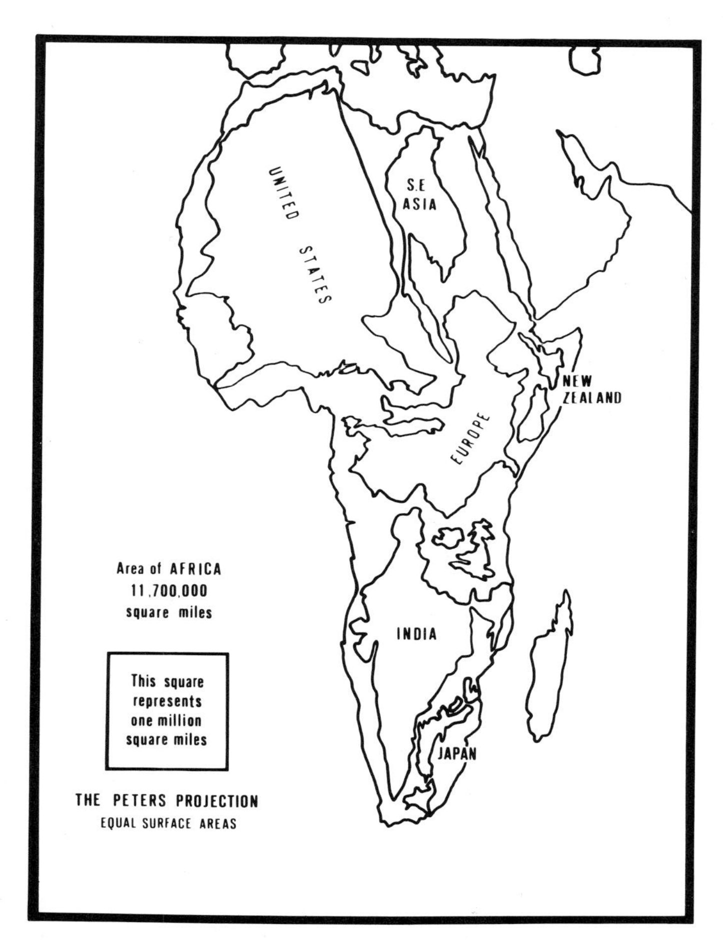


**Figure 1.** *Comparison of the size of Africa versus other continents. The Peters Projection is used because it displays the continents on an equal-area basis.*

sq mi (15.15 million sq km), exactly 50% of its area. Out of 83 identified sedimentary basins, 20 are productive of hydrocarbons while a further 13 have hydrocarbon indications with no production as yet. The Sirte basin of Libya has 13 giant fields, and another 9 African basins contain at least one giant. A giant hydrocarbon accumulation is classed as one capable of producing either 500 million bbl of oil or 3 trillion cubic feet (tcf) of gas.

The paper follows Kingston et al (1983) and Klemme (1975) in classifying the 83 basins (Figure 2) according to their genesis, but only considers the dominant cycle type, although all basins have had a polycyclic history. Four primary basin types are recognized.

There are 14 interior sag basins, 4 foreland or marginal sag/interior sag basins, 34 interior fracture basins, and 19 marginal sag basins. In addition, 3 secondary basin types are recognized, including 2 deltaic sags, 10 wrench-modified basins and 2 fold belts. The fold belts are not included in the basin count and are not covered by this paper.

### African Reserves

Significant highlights in the discovery of African oil and gas are shown in Figure 3. Not shown are the early discoveries flanking the Gulf of Suez in Egypt, such as Gemsa (discovered in 1886), Hurghada (1913), and Ras Gharib (1938). The latter was significant with estimated, ultimate recoverable reserves of 260 million bbl of oil. Other small oil finds were made earlier in North Africa, indicating the presence of hydrocarbons. Note especially the discovery of the first Egyptian giant in 1955, the first Algerian giant at Hassi Messaoud in 1956, the first Libyan giant in 1958, and the first Nigerian giant in the same year, following the first oil discovery four years earlier.

Africa accounts for about 7% of the world's petroleum reserves, Libya having the largest reserves within a country, followed by Algeria, Nigeria, Egypt, and Tunisia. Figure 4 shows a comparison of reserve estimates for African countries and those for the rest of the world. West Africa plots as one area with 5 billion BOE. Table 1 shows the individual reserve estimates for the African countries with the total proven hydrocarbon reserves varying, depending on source, between 86.0 and 100.5 billion BOE.

## CHARACTERISTIC HYDROCARBON OCCURRENCES AND PLAYS BY BASIN TYPE

The following describes the characteristics of oil occurrences and hydrocarbon plays within specific basin types.

### Interior Sag Basins

The interior sag basins are large, nearly circular, intracratonic downwarps within a continent, and may overlie earlier aborted interior fracture basins. They are formed by simple thermal sagging of the continental crust and are characterized by low geothermal gradients. This basin-type contains both marine and non-marine deposits, often Paleozoic in age, and may not be very deep. Examples of these interior sag basins outside of Africa with significant hydrocarbon production are the Illinois, Michigan, and Williston basins of North America. The model for interior sag basins in Africa is the Illizi basin of Algeria.

The Illizi basin is separated from the Ghadames basin to the north by an east-to-west-trending arch. The Paleozoic succession is preserved as a series of saucers subcropping the Mesozoic at the Hercynian Unconformity. The whole succession consists of sandstones and shales, except for marls and limestones of the Upper Carboniferous. In the south, Cambrian sandstones rest unconformably on the Precambrian basement, while on the northern edge, erosion removed the oldest Paleozoic formations during the Hercynian orogeny.

The basin has a series of north to south axes caused by reactivation of old lines of weakness. Most of the anticlines in the basin are related to faults, but some are due to intrusion of dolerite laccoliths and sills.

The chief source rocks are Silurian shales, whose thickness varies from 800–1,000 ft (244–305 m). There are sandstone reservoirs throughout the Paleozoic, but reservoir quality decreases down the succession; for example Devonian, Carboniferous, and Triassic sands have porosities of up to 25%, with permeabilities ranging from 100 md to several darcies, but Cambrian–Ordovician sandstones have primary porosities of only 5–8% and rely on secondary fracture porosity to produce. Interbedded shales provide cap rocks.

The first discovery in the basin was made at Edjeleh in 1956. This field has estimated reserves of 220 million bbl of oil. The largest fields are Zarzaitine (estimated ultimate recoverable

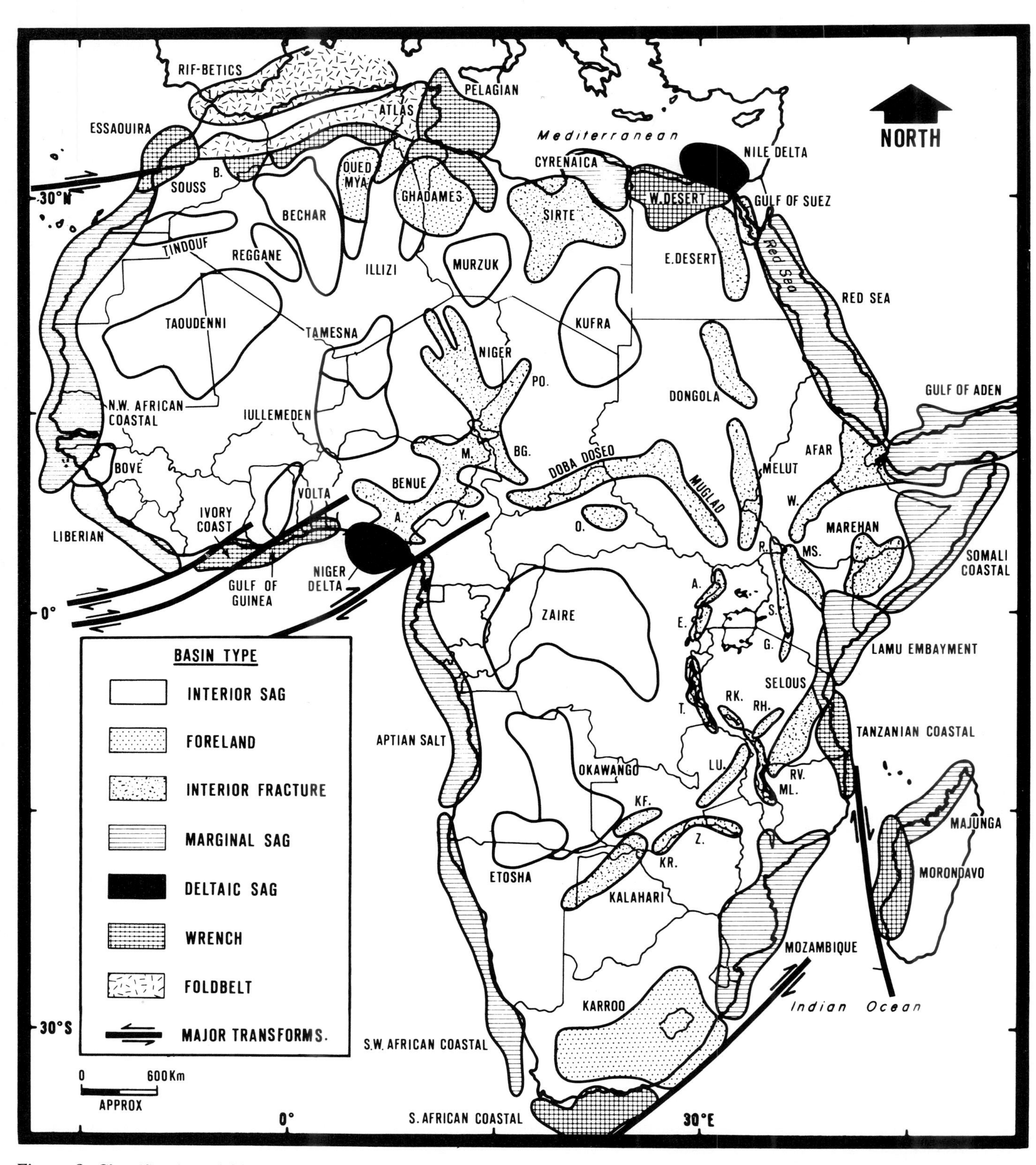


Figure 2. *Classification of African basins. Seven basin types are recognized: four primary and three secondary or modified types. Although most basins have had a polycyclic history, only dominant cycles are used in classification.*

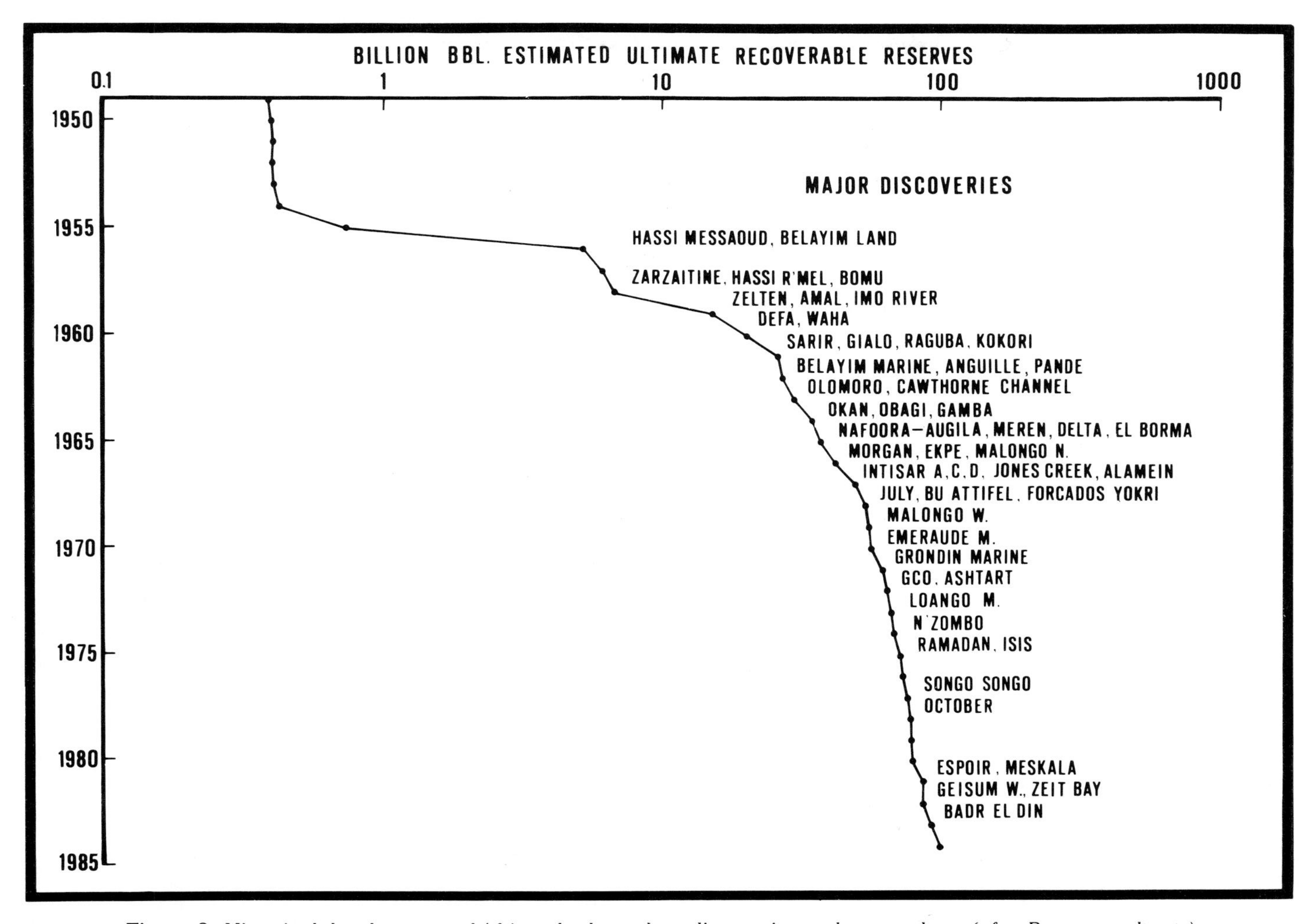


Figure 3. *Historical development of African hydrocarbon discoveries and reserve base (after Petroconsultants).*

reserves, referred to as EURR, of 900 million bbl of oil plus 2.8 tcf of gas), discovered in 1957, and Alrar (EURR of 4.7 tcf of gas), discovered in 1960.

The oil-rich Silurian Tannezzuft Formation shale is part of a coarsening-upward deltaic sequence that is Early Silurian to Early Devonian in age. The distribution of marine Silurian rocks, of which oil-prone black graptolitic shales constitute a major part, is shown by the Silurian paleogeographic map of Figure 5. Most fields sourced from the Silurian in Africa have currently been found in the Algerian Sahara, chiefly in the basins of Oued Mya, Illizi, and Ghadames (also in Libya). However, as can be seen from the map, there is a very extensive distribution of similar, high-quality source rock across Africa that offers potential for future Silurian-sourced oil discoveries. Areas having the highest potential are the Reggane and Ahnet basins of Algeria, the Bové basin of Guinea, the Taoudenni basins of Mauritania and Mali, the Tamesna basin of Niger, the Keta basin of Ghana, and the Murzuk and Kufra basins of Libya.

Two interior sag basins with little past exploration are the Taoudenni and Zaire basins. The vast size of both of these is evident when compared to Texas, as in Figure 6. With only three wildcats in the Taoudenni (area of 320,000 sq mi or 828,740 sq km) and two in Zaire (area of 350,000 sq mi or 906,430 sq km), they have well densities of only 1/106,667 sq mi (1/276,246 sq km) and 1/175,000 sq mi (1/453,215 sq km) respectively. The Taoudenni basin and Texas both are flat and sandy, and the former may, in time, prove to be as prolific as the latter for oil.

The presence or absence of Silurian shale in the Taoudenni basin is the key to its future potential. A Lower Devonian unconformity is known locally to remove the Silurian, which may also have a sandy facies. Seismic velocity analysis is not adequate to resolve the problem. Potential plays are Middle Devonian reefs, Lower Devonian and Ordovician sandstones. As illustrated in Figure 7, basic intrusions can cause problems in some areas, but there is an excellent match between magnetic and seismic expression of these, such that detailed modelling can help separate igneous from reefal anomalies.

The Zaire basin, illustrated in Figure 8, contains some very rich source rocks, although locally they may not have reached geothermal maturity.

The Upper Jurassic Stanleyville shales, with total organic carbon (TOC) of up to 18.5% in outcrop and sapropelic organic matter, form the best regional seal to potential reservoirs in Karroo (Carboniferous to Triassic) equivalents. However, they may or may not have reached maturity. Likewise, the Lower Cretaceous Loia shales, with TOCs of 13–25% in outcrop, tend to be

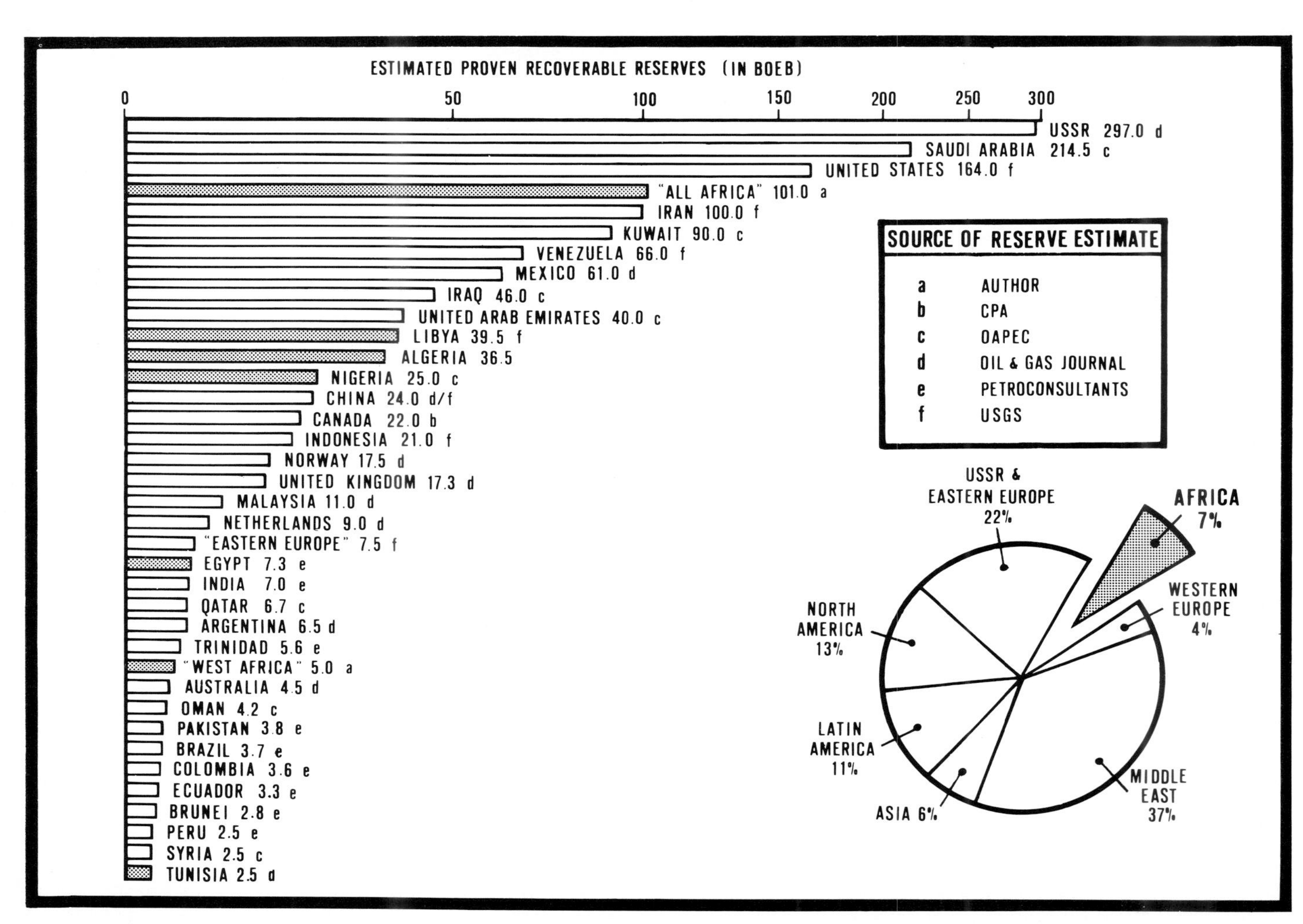


Figure 4. *African hydrocarbon reserves compared with the rest of the world (given in billion BOE, or BOEB). Because many reserve estimates for various producing countries are available, a best estimate has been used for the countries displayed. The source of the various reserve estimates is indicated.*

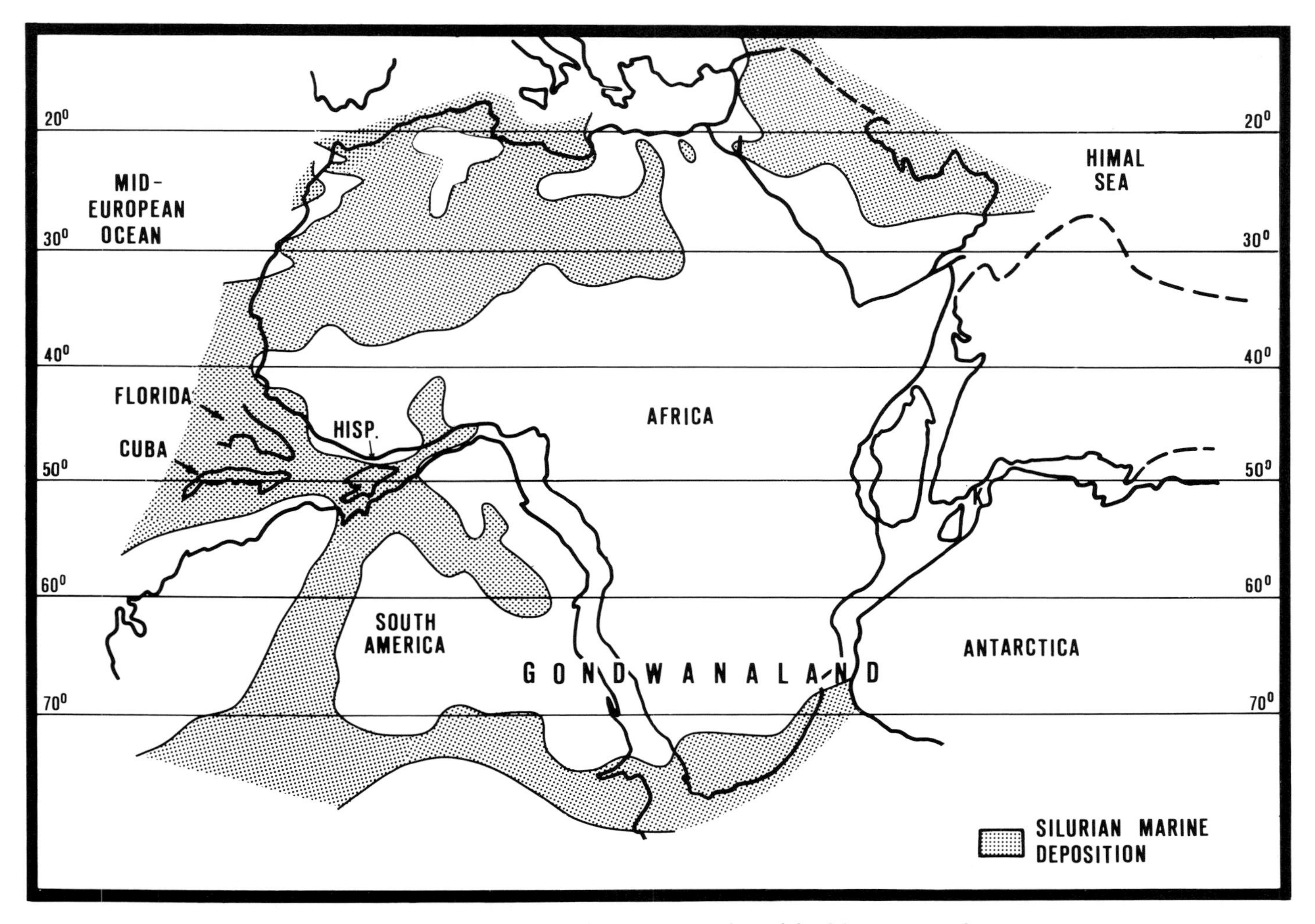


Figure 5. *Paleogeographic reconstruction of the Silurian period.*

Table 1. *Estimates of recoverable proven hydrocarbon reserves in Africa.*

| Country | Source of Reserve Estimate | | | | |
|---|---|---|---|---|---|
| | Author | OGJ | PC* | USGS* | OAPEC* |
| Algeria | 36,475 | 27,620 | 15,400 | 19,160 | 14,800 |
| Angola | 1,770 | 1,960 | 1,800 | | |
| Benin | 20 | 100 | 22 | | |
| Cameroon | 1,320 | 1,220 | 810 | | |
| Chad | 150 | 0 | 150 | | |
| Congo | 1,250 | 760 | 600 | | |
| Egypt | 6,800 | 4,635 | 7,300 | 6,710 | 5,500 |
| Gabon | 1,450 | 575 | 1,450 | | |
| Ghana | 20 | 5 | 25 | | |
| Ivory Coast | 430 | 610 | 120 | | |
| Libya | 27,790 | 24,835 | 36,000 | 39,510 | 35,800 |
| Morocco | 450 | 220 | 17 | | |
| Mozambique | 500 | 0 | 0 | | |
| Namibia | 50 | 0 | 0 | | |
| Niger | 100 | 0 | 0 | | |
| Nigeria | 17,605 | 22,360 | 20,000 | 17,370 | 25,100 |
| Senegambia | 0 | 750 | 0 | | |
| South Africa | 85 | 180 | 0 | | |
| Sudan | 1,650 | 300 | 230 | | |
| Tanzania | 335 | 35 | 0 | | |
| Tunisia | 1,915 | 2,515 | 1,700 | 1,190 | 2,200 |
| Zaire | 355 | 115 | 200 | | |
| Undifferentiated | | | | 5,740 | 6,200 |
| Total | 100,520 | 88,580 | 86,000 | 89,680 | 89,600 |

All figures are million BOE assuming a conversion factor of 6,000 cubic feet of gas to 1 barrel of oil.
OGJ = Oil and Gas Journal, PC = Petroconsultants.
*These estimates exclude recoverable gas reserves.

immature. Permian–Carboniferous varved lake shales with TOCs up to 1.8% and woody-coaly organic matter could also constitute a possible source.

Figure 9 shows a gas chromatogram of $C_{15}+$ saturates for a sample from the Stanleyville shales occurring at outcrop in the eastern part of the Zaire basin. There is a predominance of component A ($nC_{25-33}$), which represents humic, nonmarine souce material, over component B ($nC_{15-23}$), which represents marine source material. These shales were deposited in a lacustrine environment. Throughout this paper, lacustrine source rocks will continue to appear as prolific hydrocarbon producers throughout Africa.

Large structures, possibly caused by infra-Cambrian salt movements, are known (Figure 8), and high-angle normal faulting is present (possible evidence of an earlier interior fracture cycle). A huge northwest-to-southeast-trending basement arch extends 250 mi (400 km) across the basin. However, there is little or no structure above the Triassic.

All the components of a productive hydrocarbon province have been realized to date, though not yet in circumstances favorable for production. With only two wildcats and a number of stratigraphic holes, this basin cannot be written off as nonprospective.

The Etosha basin in Southern Africa, like the Zaire basin, contains a sequence of Precambrian to Karroo sediments. The basin is very sand-prone but with seismic indications of infra-Cambrian reefs, as in the Zaire and Taoudenni basins. Karroo black shales and others interbedded with the carbonates are potential source beds to the fractured reefal carbonates or Paleozoic sands.

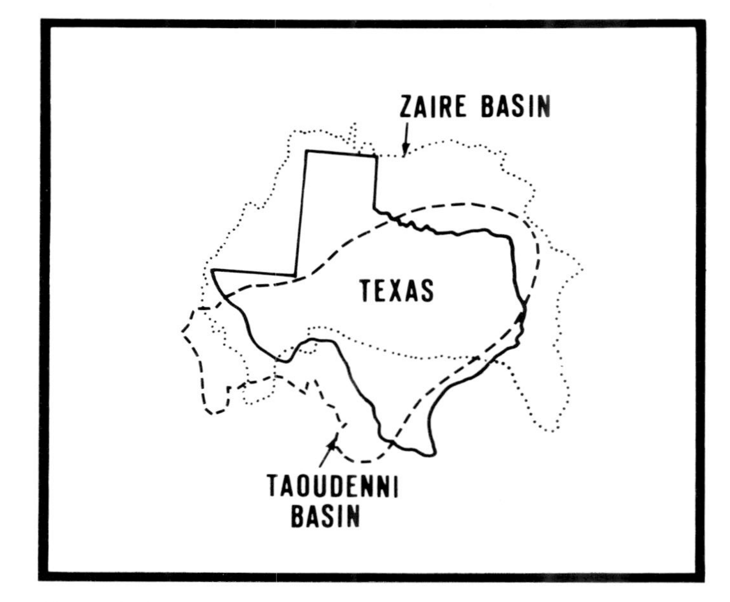


Figure 6. *Comparison of size of the Taoudenni and Zaire basins with the state of Texas.*

The Okawango basin, which lies north of the Etosha in Angola, has never been explored despite an area of 180,000 sq mi (466,000 sq km). It should share characteristics of the Zaire and Etosha basins.

In summary, the interior sag basins occupy 29% of the

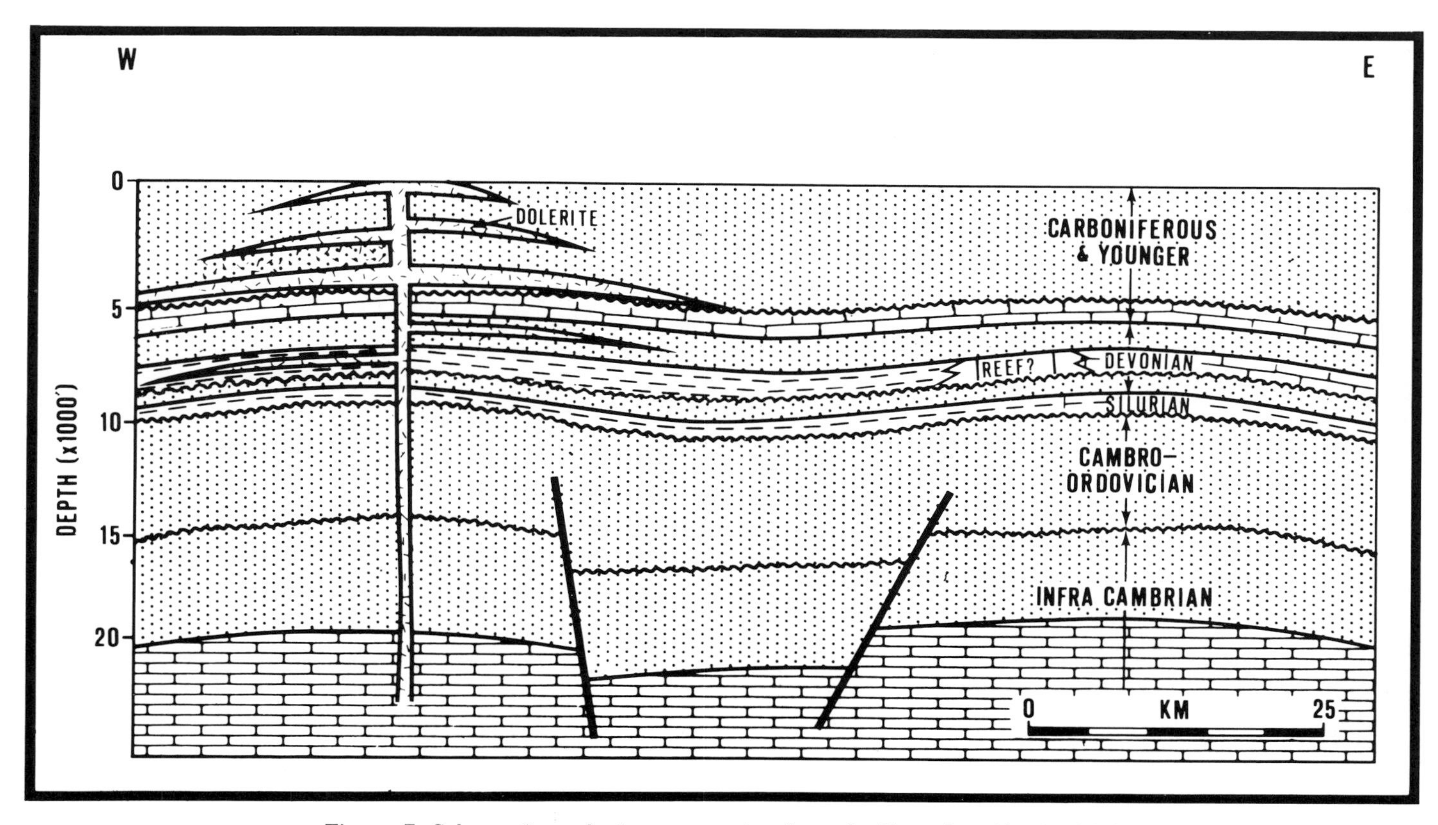


Figure 7. *Schematic geologic cross-section from the Taoudenni basin, Mali.*

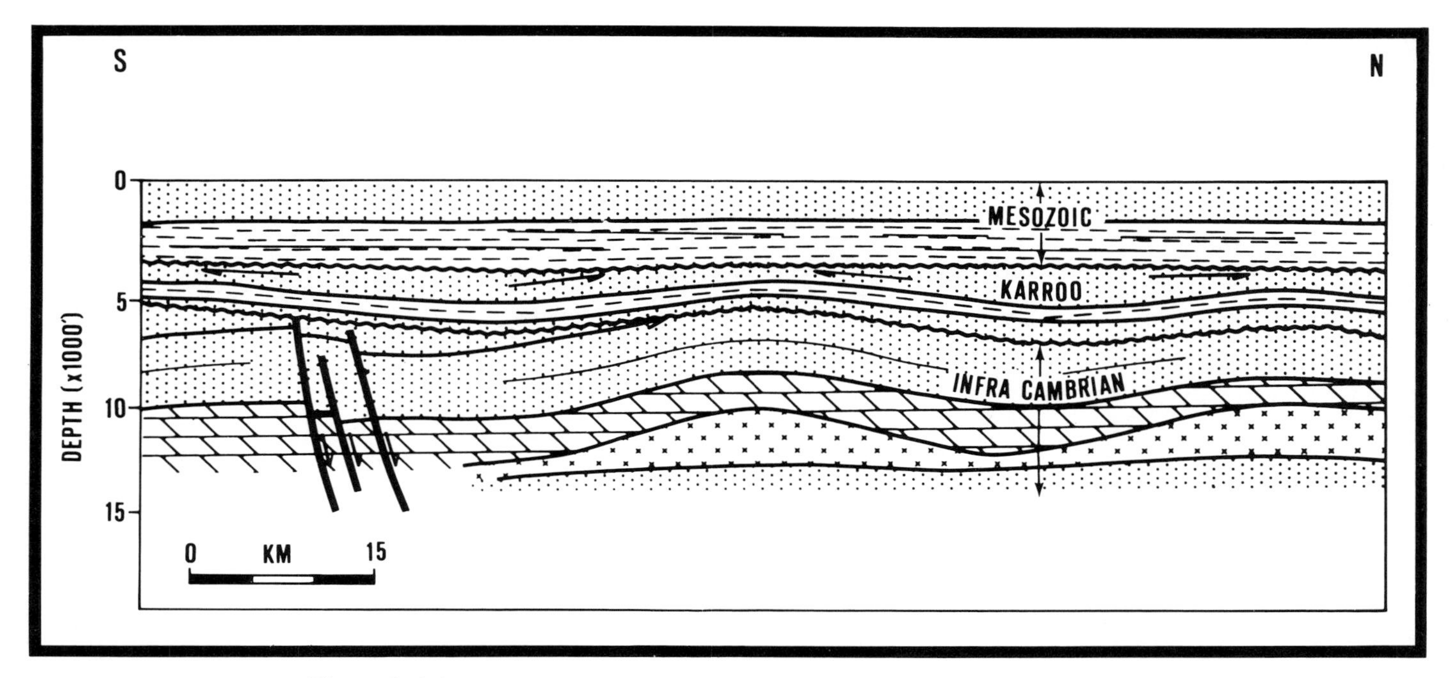


Figure 8. *Schematic geologic cross-section across the Zaire basin, interior Zaire.*

basinal areas of Africa and represent 17% of all basins in number. Only 7% of these basins have been explored beyond the immature stage, so they represent vast, unexplored regions with enough encouraging signs to stimulate further interest. The logistical problems are great, so these basins would fit into an exploration portfolio of ventures at the high-risk/high-reward end of the spectrum.

### Foreland Basins

Whereas the interior sag basins represent a single cycle of tec-

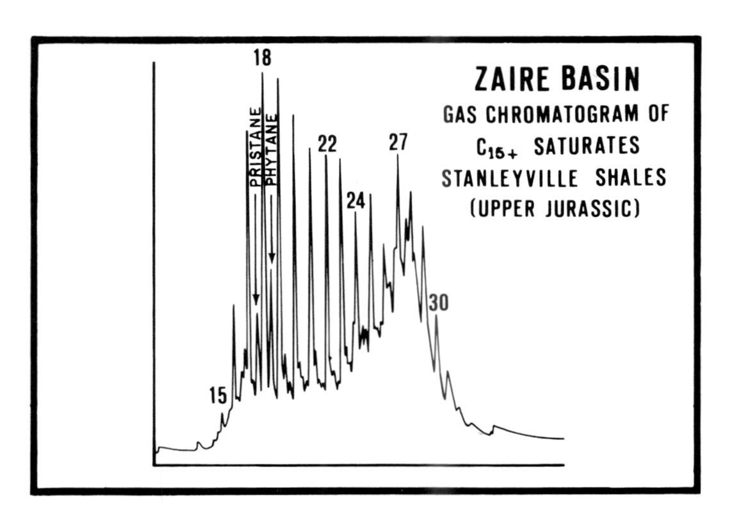


Figure 9. *Gas chromatogram of $C_{15}+$ saturates for Upper Jurassic Stanleyville Formation shale sample from outcrop. The dominance of phytane over pristane suggests a rich, marine source material in a strongly-reducing environment; however, the strong $C_{23}+$ component and the predominance of odd-numbered n-alkanes over even alkanes support a non-marine lacustrine source material. The high naphthene envelope and wide $C_{23-30}$ naphthene peak indicate a surface sample with immaturity.*

tonic development, the marginal sag/interior sag or foreland basins are polycyclic.That is, they have more than one cycle of development. They constitute a major, primary basin type with many examples worldwide producing significant hydrocarbons.

These marginal sag/interior sag basins comprise marginal sag basins overlain by interior sag basins, created by subsequent formation of a foldbelt on the seaward side. Such foldbelts formed along either convergent or conservative plate boundaries. The upper parts of these basins are characterized by sheet sands and carbonates with coarser sediments in the former, landward direction. Temperatures are higher than in either interior sag or marginal sag basins alone because of the interplay with the accreted foldbelt, but, they are not as high as in interior fracture basins. Evaporites are often present and large fields are often found on regional arches. Over 75% of the world's gas has been found in these basins.

Examples of marginal sag/interior sag basins from around the world are the West Siberian basin of the U.S.S.R., and the Permian, Powder River, Big Horn, Anadarko, and Alberta basins of North America. In Africa, such basins exist in the Algerian Sahara and in South Africa.

In Algeria, the Oued Mya and Ghadames basins, which lie south and southeast of the Atlas foldbelt, are marginal sag/ interior sag basins that overlie earlier interior sag basins. In each, a Paleozoic interior sag basin is truncated by the Hercyn-

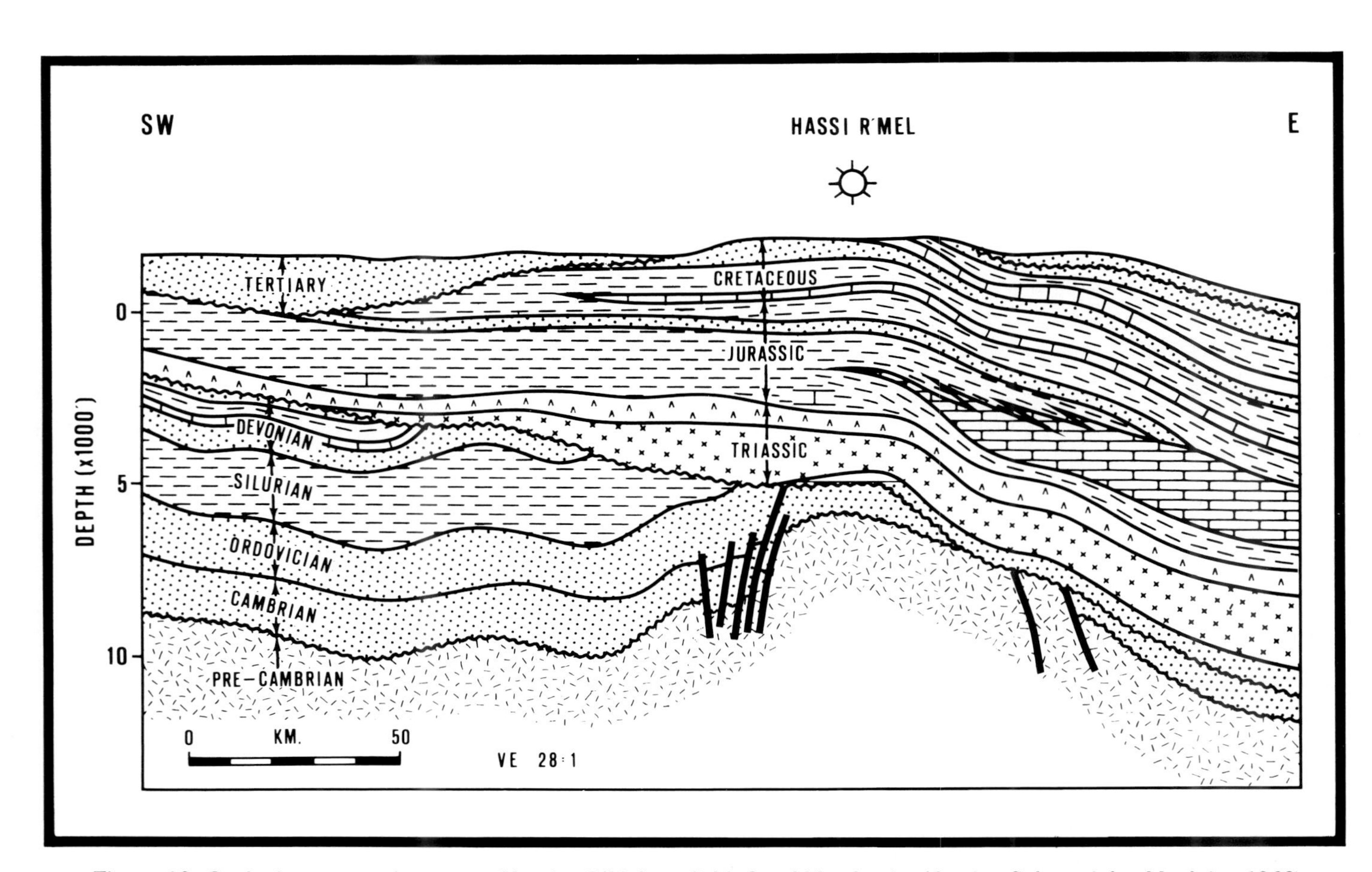


Figure 10. *Geologic cross-section across Hassi er R'Mel gas field, Oued Mya basin, Algerian Sahara (after Magloire, 1968).*

ian Unconformity and overlain by a Mesozoic marginal sag basin. Wrench movements in the Late Cretaceous to early Tertiary formed the Atlas foldbelt that isolated these marginal sag basins from the Tethyan Ocean to the north. Thereafter, interior sag basins developed and filled with coarse clastic debris shed from the rising mountain chain.

The largest Algerian fields, Hassi Messaoud (9 billion bbl of oil) and Hassi R'Mel (70 tcf of gas plus 2.6 billion bbl of crude), are situated in these two basins. The former, which straddles a north-to-south-trending regional arch on the western side of the Ghadames basin, produces oil from fractured Cambrian quartzitic sandstones capped by impervious Ordovician sandstones and sourced from Silurian shales. The latter field, on the northwest side of the Oued Mya basin, has gas and condensate reservoirs in Triassic sands that unconformably overlie subcropping Paleozoics on a regional arch similar to Hassi Messaoud (Figure 10). Within these two basins, Triassic salt forms an overlying regional seal, although Triassic silts and shales also seal locally. The subcropping Silurian shale is the source of the oil found in both basins.

Many Triassic discoveries have been made in both the Ghadames and Oued Mya basins. However, recent finds tend to be small and stratigraphic traps are common. Field-size distributions for these Algerian Sahara basins can be calculated separately for fields occurring on arches (high blocks) and in troughs (low blocks). The large difference in $P_{50}$ values between the former ($P_{50}$ = 370 million BOE) and the latter ($P_{50}$ = 34 million BOE) can be seen on Figure 11. All 26 fields sampled share a combined $P_{50}$ of 68 million BOE.

The Karroo basin of South Africa is a marginal sag/interior sag basin of Paleozoic age which has yet to see successful exploration for hydrocarbons. Minor, shallow occurrences of oil or gas in the basin are related either to local coal bed sources, matured by juxtaposition with dolerite intrusions, or are residual pools, developed because of a lack of regional sealing strata for hydrocarbons, at the time of generation.

To the basin's south, paleotemperatures were raised by deep burial prior to and during formation of the Cape foldbelt in Late Permian time. Numerous dolerite intrusions have also had a detrimental effect on hydrocarbon preservation. The Permian Ecca Series contains some rich (TOC of 2–5%) prodelta shales that constitute a good source rock if not overmature. These represent the closing stages of the marginal sag cycle prior to the Cape folding, and subsequent nonmarine deposition in an interior sag basin.

In the north of the basin, the Permian shales are not overmature (Vitrinite reflectances range from 1.00–1.35), but contemporaneous sandstones have only fair porosities and permeabilities at shallow depths. Sealing shales for potential traps are not common, however.

In general, hydrocarbon potential within the Karroo basin increases northward from the Cape foldbelt. If discoveries are made, however, they will be small, shallow, and probably stratigraphic.

The marginal sag/interior sag basins represent only 5% of all African basins, cover 7% of the total sedimentary area, but have production from 75% of them, all from Algeria. There, the exploration effort has concentrated around the giant fields discovered in the mid 1950s. The Karroo basin, with an area of 220,000 sq mi (570,000 sq km), is classified as being in an immature stage of exploration chiefly because of its vast size.

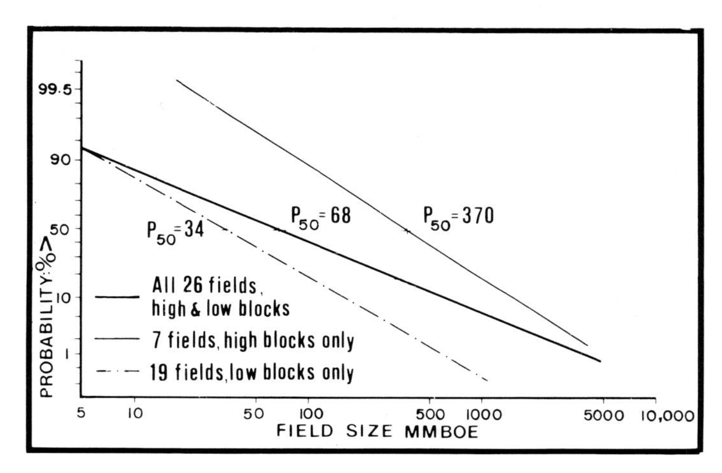


Figure 11. *Field-size distributions (given in million BOE or MMBOE) for the Triassic sand play in the Algerian Saharan basins. The mean field-size ($P_{50}$) varies immensely between low blocks and high blocks. Hassi er R'Mel is an example of a high-block field, but most fields are small with some stratigraphic element. It is important, therefore, to break down field-size distributions into discrete plays such as this.*

Large areas, however, are unprospective.

Therefore, these basins are not considered to have potential as prolific hydrocarbon producers. The areas in the Algerian Sahara have been well explored, and geologists believe that they will find only smaller fields there.The Karroo basin suffers from poor source and reservoir preservation, because of burial and diagenetic effects, except in the north where some small, shallow accumulations might be found.

### Interior Fracture Basins

Interior fracture basins are rather small, elongated, extensional basins located entirely on continental crust, either inland on present plates or near the plate margin. If located near the margin, their axes are usually at an angle to that margin, they may be in isolation, or they may form part of an extensional graben system. They are caused by extensional shear, and plate separation can either have been aborted, in which case an interior sag basin usually overlies it, or have occurred, in which case a marginal sag basin develops. In the latter case, therefore, interior fracture basins are preserved under many present-day marginal sag basins, with one-half on each side of the newly created ocean.

These basins commonly show horst-and-graben development, often with better reservoirs associated with the highs, and better source rocks associated with the lows, throughout basin development. There is usually an initial, restricted stage during which coarse, nonmarine clastics and lacustrine shales accumulate. A mature, semi-restricted stage follows, with widespread occurrences of rich source rocks, good reservoirs, both clastic and carbonate, together with evaporites. Structural rejuvenation occurs during this mature stage. Higher-than-average temperatures and heat flow are normally associated with these basins.

Africa contains two of the best-known hydrocarbon producing examples of interior fracture basins: Egypt's Gulf of Suez, and Libya's Sirte basin.

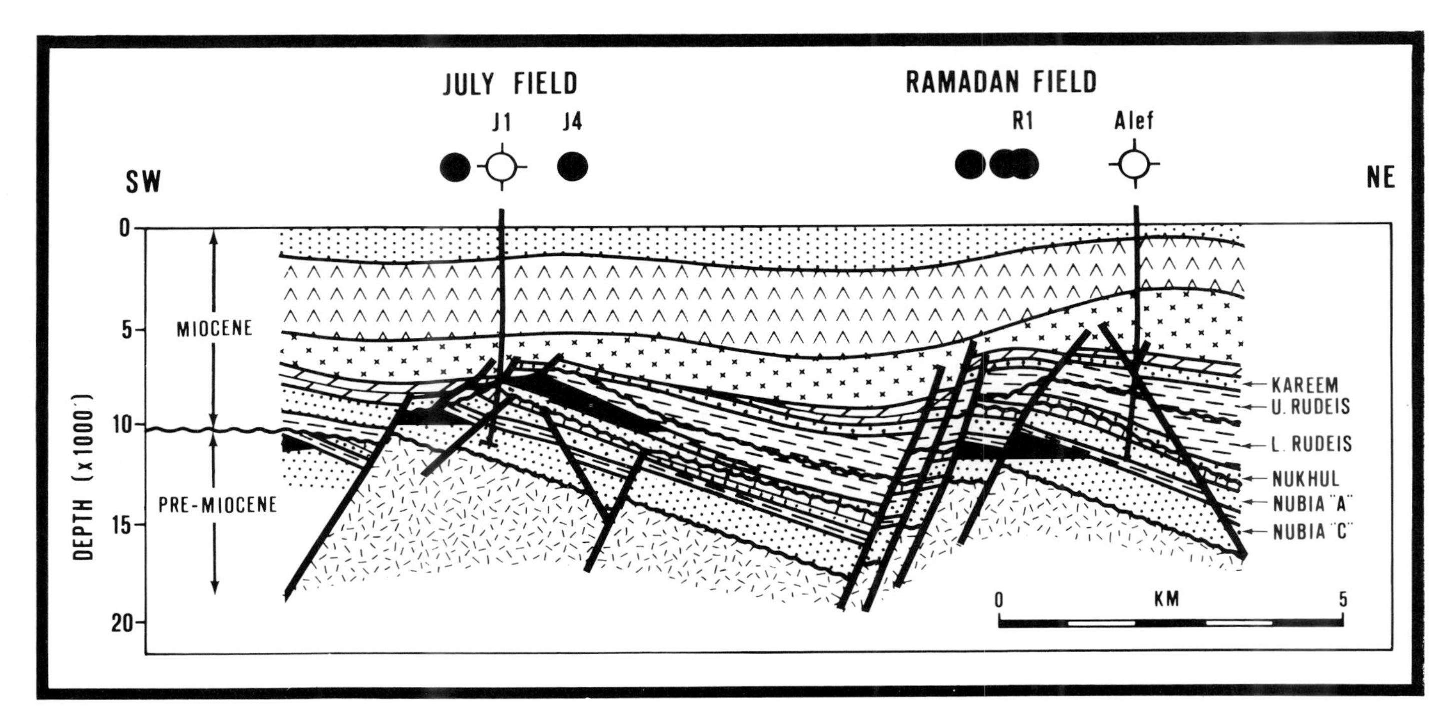


Figure 12. *Geologic cross-section across July and Ramadan oil fields, Gulf of Suez, Egypt. The July field contains estimated reserves of 750 million bbl oil in Rudeis (API 34° oil) and Nubian sands (API 31.5° oil). It took three wells to find this huge field. The Ramadan field, with estimated reserves of 350 million bbl oil in the Nubian sands (API 31.7° oil), was found with the second well. In both fields, and in other parts of the basin, the Miocene salt, draping the pre-Miocene fault blocks, distorts the seismic character of deep structure (after Brown, 1978).*

Figure 12, a geologic cross-section across the July and Ramadan oil fields in the Gulf of Suez, illustrates a few particular problems in the basin while also demonstrating two huge fields having different pay zones. Note the tilted fault blocks of the pre-Miocene sequence and the different thicknesses and attitude of the South Gharib salt over the two fields. Tilted fault blocks face different ways in different parts of the basin, these zones being separated by northeast-to-southwest-trending shears. Other cross faults, smaller in magnitude but related to the same overall extensional stress regime, are important in forming the trap-door structures that are so prospective. The structural picture is severely distorted on seismic data by the thick, Miocene evaporites, and the cross faults are difficult to recognize.

The first well on the July structure, in late 1966, was located at the top and base salt structural crest as then mapped on seismic. It was to test the Miocene Rudeis, Nukhul, and pre-Miocene Nubian sands. The well was abandoned as a dry hole after encountering water-wet Nubia "C" sands and oil shows in Cretaceous and Nukhul fractured carbonates. No Miocene sands were encountered. A second well to the north, drilled in 1967, was also dry and abandoned. Six years later, well J-4 was planned to test the massive Nubia "C" sands, 1,000 ft (305 m) higher than in J-1, on the upthrown side of a fault block. However, although the Nubian was found to be structurally lower than assumed, the Miocene clastics were very thick, with over 1,000 ft (305 m) of lower Rudeis and 800 ft (244 m) of net pay. While drilling a delineation well to this lower Rudeis discovery, a Carboniferous Nubia "C" discovery was made a little to the west. Later, an upper Rudeis sand pay was discovered while drilling a lower Rudeis water-injection well. The northern and southern parts of the field are cut by a cross fault with sinistral shear. This cross fault has been highly significant in developing the field's sand bodies and in causing local scalping of tilted fault blocks.

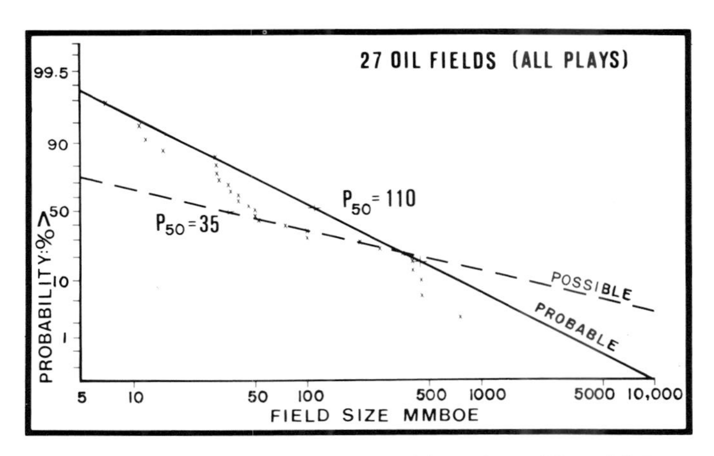


Figure 13. *Field-size distribution (given in million BOE, or MMBOE) for the Gulf of Suez, Egypt. Ambiguities in choice of best-fit straight lines, as illustrated here, can be avoided if field-size distributions are made for discrete plays.*

The Ramadan structure was first tested with the Alef-1 well in late 1965. It, too, was spudded on the top and base salt structural crest. It terminated in Cretaceous Nubia "A" sands without encountering shows. There was insignificant sand in the Mio-

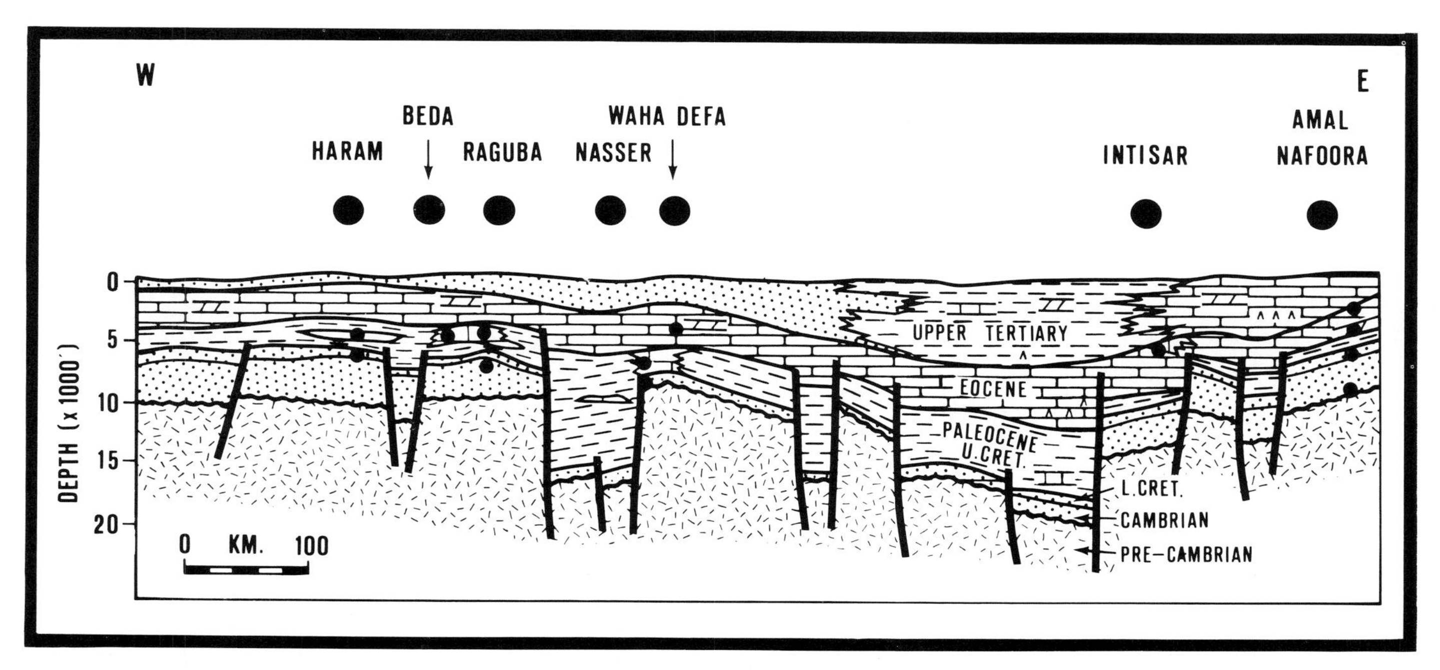


Figure 14. *Highly schematic, geologic cross-section across the Sirte basin, Libya. The fields illustrated above have the following estimated reserves: Beda (90 million bbl oil), Raguba (750 million bbl oil), Nasser (2,200 million bbl oil), Waha (1,400 million bbl oil), Defa (1,800 million bbl oil), Intisar A–E (2,400 million bbl oil), Amal (4,250 million bbl oil), and Nafoora (1,800 million bbl oil). Libya's largest field, Sarir (6,500 million bbl oil), is not shown.*

cene Kareem and Rudeis formations and all Cretaceous sands were water-wet. The dipmeter, together with later, improved seismic data, showed that the well was drilled downdip and to the east of the pre-Miocene structure. Nine years later, the discovery well found oil in Cretaceous sand intervals. Total net pay in this well was 1,362 ft (415 m), the Nubia "C" having 1,110 ft (440 m) net pay.

The best location for the first exploratory well on such structures appears to be at the structural crest of the deepest mappable Miocene horizon, with more than one well probably required. State-of-the-art seismic processing techniques are essential and dipmeter logs should determine, as accurately as possible, the subsurface structural configuration. Outcropping, tilted fault blocks onshore serve as excellent models for the study of subsurface, pre-Miocene structures. Although the Gulf of Suez is classed as a very mature basin, with an area of only 11,000 sq mi (28,500 sq km), it still holds a considerable amount of promise, and the exploration history of the July and Ramadan fields illustrates how elusive such large fields can be.

In the Gulf of Suez, all plays are structurally related. However, the various carbonate and sandstone reservoirs are quite different in their development, and, hence, there is difficulty in generating one field-size distribution for all plays within the basin (Figure 13). A maximum $P_{50}$ of 110 million BOE and a minimum of 36 million BOE can be derived. Field-size distributions should be made for discrete plays, leading to better-fitting straight lines.

The Miocene Globigerina Marl has traditionally been considered to be the best source rock in the Gulf of Suez, but other rich, potential source rocks exist in the Eocene Thebes Formation and the Upper Cretaceous Brown Limestone Formation, with average TOC s of 2% and greater than 3%, respectively.

In the Sirte basin, a system of horsts and grabens has had a major influence on depositional control, particularly of the reefal buildups that constitute the main reservoirs in the basin (Figure 14). In addition, fracture porosity has been generated in sands on the edges of horsts and tilted fault blocks, while sourcing shales were deposited in the grabens, especially during the Late Cretaceous. The thinning of the pre-Cretaceous clastics into the basin center suggests initial arching and deep erosion of the crest prior to extension and basin subsidence. A north-northwest-to-south-southeast-trending structural framework has persisted to the present since initiation in the Early Cretaceous. Thick, non-marine, Lower Cretaceous sand overlies Cambrian quartzitic sand similar to that in the Ghadames and Illizi basins of Algeria. Above the Lower Cretaceous sands are thick, Upper Cretaceous shales with tight, micritic carbonate marking the top of the Mesozoic. Such shales thicken markedly in the troughs and are widely thought to provide the best source rocks, with TOC's averaging 2%. Lower Tertiary shales also have source potential. Oil in Lower Cretaceous and Cambrian sands reservoirs, being more waxy and paraffinic, may have a non-marine source similar to that in other Lower Cretaceous failed rift systems, such as the Central African Rift System discussed later.

The Paleocene contains major carbonate buildups within a carbonate-shale succession. Lower Eocene shale in the basin center passes upward into a sequence of evaporites and carbonates and, finally, into marl and shale. Oil is trapped primarily in Paleocene reefs that developed on the crests of deeper horst blocks. The field-size distribution diagram of the Sirte basin (Figure 15) illustrates the difficulty, as in the Gulf of Suez, of obtaining a "best fit" for all 21 Paleocene carbonate fields sampled in the basin.

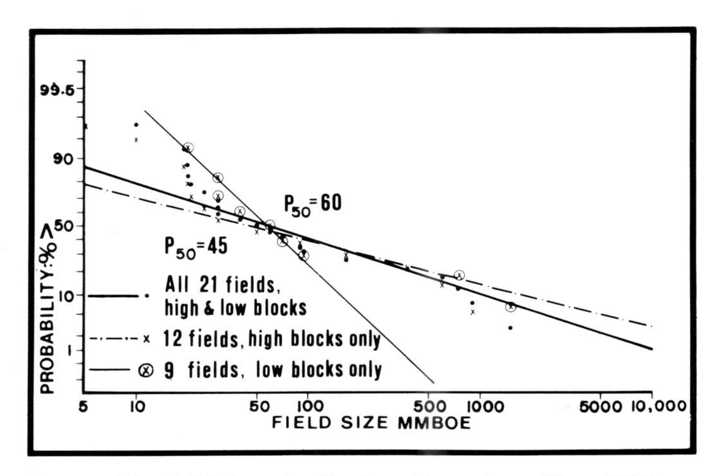


Figure 15. *Field-size distribution (given in million BOE, or MMBOE) of Paleocene carbonate play, Sirte basin, Libya. There is difficulty in applying a best-fit line to the data points, as in the Gulf of Suez (Figure 13), but separation of the 21 fields into high and low blocks gives a better result.*

However, separation into horsts and grabens, as with Algeria, allows a better delineation of the trends. Common to all 21 fields and grabens only, are $P_{50}$s of 60 million BOE; horsts show a $P_{50}$ of only 45 million BOE. Other reservoirs of importance are Cambrian and Lower Cretaceous sands, the latter being the principle reservoir of one of Africa's largest oil fields, Sarir, with 6.5 billion bbl oil recoverable. The presence of effectively sealing shales is generally the most important parameter in determining trap integrity in the Sirte basin.

The Niger basin, which straddles the Upper Cretaceous seaway, shown on the Upper Cretaceous paleogeographic map (Figure 16), has a Lower Cretaceous, non-marine rift fill overlain by an Upper Cretaceous to Paleocene marine wedge marking the transition from an interior fracture to an interior sag basin. Non-marine deposition was re-established in the Tertiary. Plays are present in all three depositional cycles at the basin edges because of juxtaposing marine shale source beds and sandstone reservoirs at optimum burial depths. In the basin center, discontinuous turbidite sands of the uppermost cycle are also prospective. Over 40,000 ft (12,200 m) of sediment has accumulated in this rapidly subsiding basin.

The Central African Rift System comprises numerous Lower Cretaceous interior fracture basins extending from Nigeria in

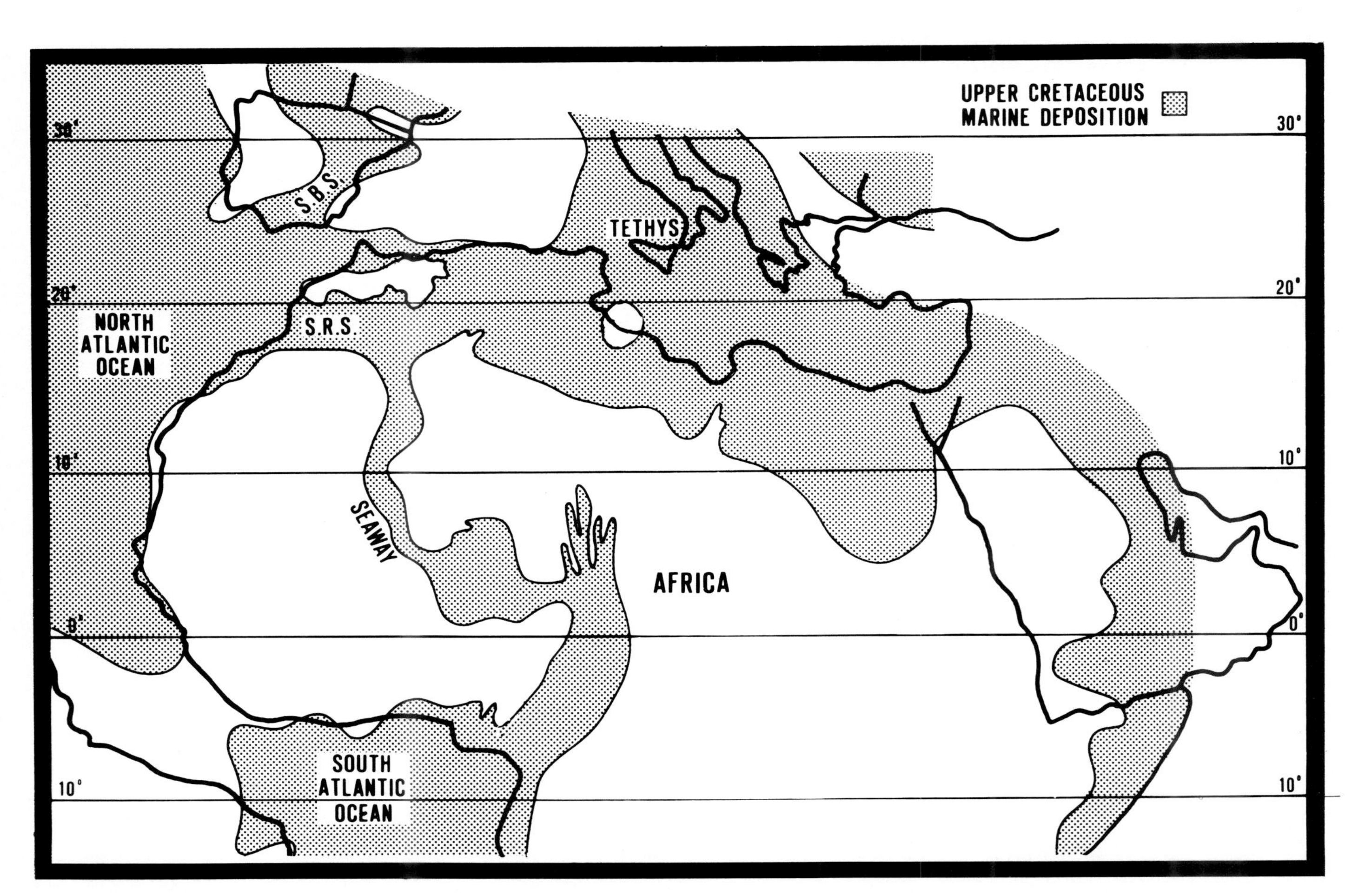


Figure 16. *Paleogeographic reconstruction for the Upper Cretaceous showing the Trans-Saharan seaway. The Upper Cretaceous marine shales, which source the Libyan oil accumulations, were widely distributed across northern Africa. Source potential in such shales is witnessed in West Africa and might yet be found in East Africa.*

the west through Niger, Chad, C.A.R., and Sudan to Kenya in the east. It is considered as one system because of the remarkable similarities of the individual basins, as illustrated by the burial history curves in Figure 17. These curves show rapid early subsidence of 500–1,000 ft (150–300 m)/million years during the rifting phase, which lasted about 25 million years. Although a Cenomanian unconformity truncates most of the early faulting, reactivation of some early faults is common. Subsidence rates decreased to 100–200 ft (30–60 m)/million years during the transition phase when some local uplift and erosion was related to wrenching and associated basin inversion. At the end of the Cretaceous, thermal subsidence with rates of only 10–50 ft (3–15 m)/million years was established and resulted in development of an interior sag basin at the same time that the Lower Cretaceous, lacustrine source rocks entered and passed through the oil-generating window.

Within interior fracture basins, clastic sediments are carried in longitudinally and laterally. However, the reservoir quality of sands can be highly variable, depending upon their provenance and subsequent burial history with concomitant diagenesis. Generally, a fairway of interfingering fluvial sands and lacustrine shales exists along the graben flanks, with only poor reservoirs, if they are present, characterizing the basin center (Figure 18). Lacustrine turbidite sands may be encountered in axial portions of the basin and may offer a secondary play. If the rate of sediment supply into the basin was constant while basin subsidence rates gradually decreased (as demonstrated in Figure 17), a progradation of fluviatile sediments across the basin would have occurred. This situation gives rise to less reservoir risk in the basin center and has led to the discovery of several fields with oil sourced by underlying lake shales and reservoired in fluvial sands. Individual depositional systems can often be recognized by seismic facies analysis.

Prior to the separation of southern Africa and South America, black lacustrine shales and fluviatile and turbidite sands accumulated in a series of grabens and half-grabens, the distribution of which is shown on the Lower Cretaceous paleogeographic map (Figure 19). These now underlie the marginal sag basins of West Africa and Brazil and provide their principal oil source. Figure 19 also indicates the Central African Rift System, the full extent of which has still to be established. The limits of the Central African Rift System have been extended toward the Sirte basin, the Eastern Desert of Egypt, and the Lamu Embayment of Kenya. Oil has already been established in Niger, Chad, and Sudan, while commercial quantities of crude have been proven through a highly successful effort in Sudan, allowing support for an export pipeline project.

In the Central African Rifts, oil has been found both in shallow and in deep plays. Migration from deeply-buried lacustrine source rocks into younger, fluviatile sand reservoirs leads to the former, and short migration into small, prograding delta-front sands or turbidites leads to the latter (Figure 20). More extensive fluvial and braided stream sands higher up may give rise to larger fields if sufficient regional seal is developed.

Oil tends to be waxy and paraffinic from these lacustrine source rocks. Generally, lighter and more mature oils are found at greater depths while heavier, less-mature, and sometimes water-washed oils are shallower (Figure 20). There is also an important porosity cut-off of 10–15% that often coincides with the top of the oil window, at burial depths of about 9,000 ft (2,740 m). The basin's geothermal history relates the destruction of reservoir quality in the sands, by pressure solution and the growth of authigenic minerals, to the onset of hydrocarbon generation from the source beds. Present-day temperature gradients, calculated from wells, vary between 1.2°F/100 ft and 1.8°F/100 ft depending on the well's position within the basin.

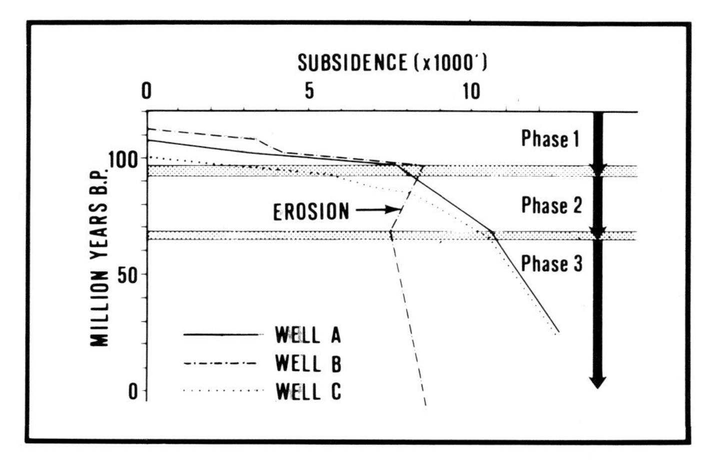


Figure 17. *Burial history curves from the Lower Cretaceous, Central African Rift System. Phase 1 represents the initial rift phase lasting about 25 million years with rapid subsidence rates of 500–1,000 ft/million years. During this phase, the basins were largely starved because sediment input could not keep up with subsidence, thus large lakes formed in which dark source shales accumulated. Phase 2 saw a decrease of subsidence to 100–200 ft/million years and some basin-inversion accompanying wrenching occurred in certain areas. As subsidence rates decreased, clastics prograded across the basins. These fluviatile sands constitute the best reservoir targets in the basins. At the end of the Mesozoic, Phase 2 brought the onset of passive thermal subsidence with rates of only 10–15 ft/million years. The Phase 1 shales were thus matured sufficiently to generate oil.*

Figure 21 shows an example of a typical deep-play oil field, fictitiously named the Bongo Field. Oil is reservoired in thin, non-extensive sands within thick, organic-rich lacustrine shales (average TOC of 2%). Not illustrated, but having excellent exploratory potential, are large inter- or intra-basin uplifts sourced by long-distance hydrocarbon migration.

The fields found to date, which range in size from 10–500 million bbl oil and average 60–70 million bbl oil, tend to be structurally and sedimentologically complex, causing unpredictable appraisal drilling. The largest fields are complexes of several, structurally related pools.

The non-marine source characteristics of this oil are remarkably similar to those of the Aptian Salt basin of west Africa, suggesting similar, non-marine sources in both cases. A gas chromatogram of $C_{15}+$ saturates for an oil sample from the typical oil field of Figure 21 is shown in Figure 22. The strong component of $C_{23}+$ normal alkanes, the slight suggestion of a preference of odd-numbered C-molecules over even (above $C_{19}$), and the pristane/phytane ratio, all indicate non-marine source material.

Success ratios, as previously defined, are useful in play assessment but are not as enlightening as historical success ratios. These historical ratios show the progression of a play through time, as in the case of onshore Sudan (Figure 23). A

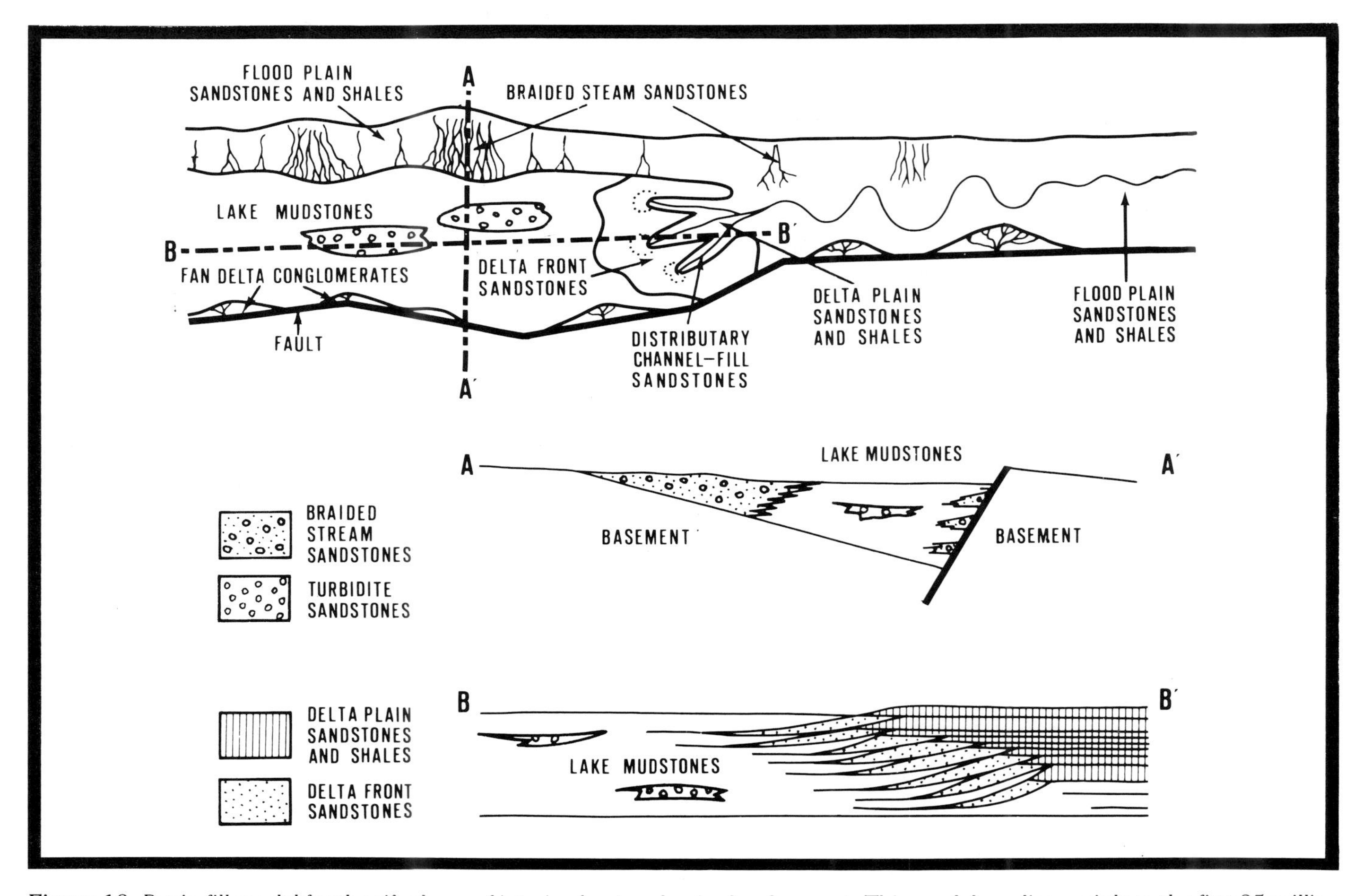


Figure 18. *Basin fill model for the rift phase of interior fracture basin development. This model applies mainly to the first 25 million years of Figure 17. It can also be applied to interior fracture basins underlying present-day marginal sag basins, for example, the Aptian Salt basin. Laterally prograding braided stream sandstones, sourced at or near the basin margins, contrast with longitudinally prograding deltas that are chiefly responsible for introducing finer-grained deposits into the lake basin. Turbidite sands, deposited in the lake during storm conditions, constitute an important reservoir during this phase of basin evolution. Lateral progradation of sediment becomes more important as subsidence decreases and sediments become coarser with lakes and deltas absent.*

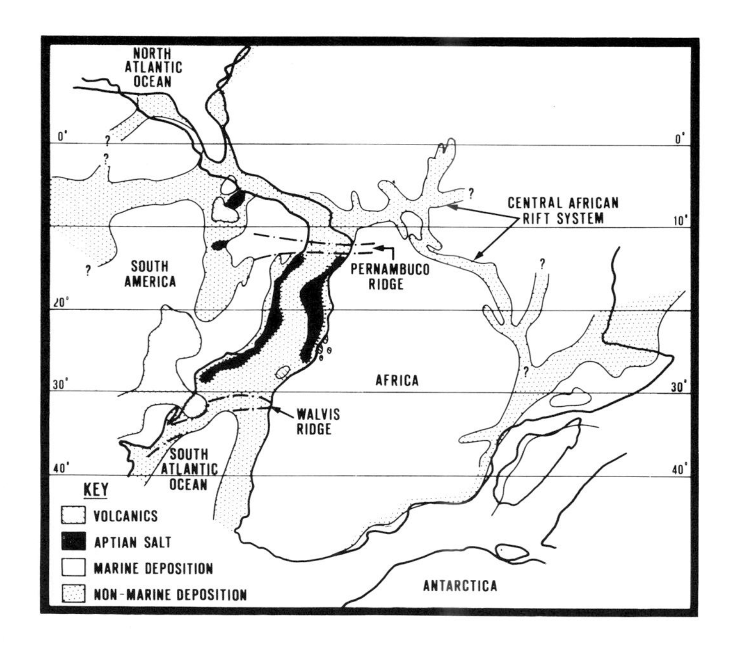


Figure 19. *Paleogeographic reconstruction of the Lower Cretaceous showing a system of interior fracture basins. Lower Cretaceous lacustrine source beds have been identified in Central and West Africa and in Brazil. They source oil found in either nonmarine pre-drift or marine post-drift reservoirs. The limit of Aptian salt deposition in West Africa was controlled by the two structural ridges shown. The trend of Central African rifts follows ancient tectonic lineaments.*

frontier play having early erratic success, after less than two years, matures after only one and one-half years, when success ratios even out at about 33%. This example from Sudan suggests that there is a better understanding of the play. Other examples, illustrated later in the paper, show the effects of political and fiscal changes in west Africa.

Failed rifts, neighboring the Red Sea and Gulf of Aden, have recently attracted attention following the North Yemen oil discovery by Hunt Oil. The geologic cross-section of Figure 24 illustrates the stratigraphic sequence and main structural fea-

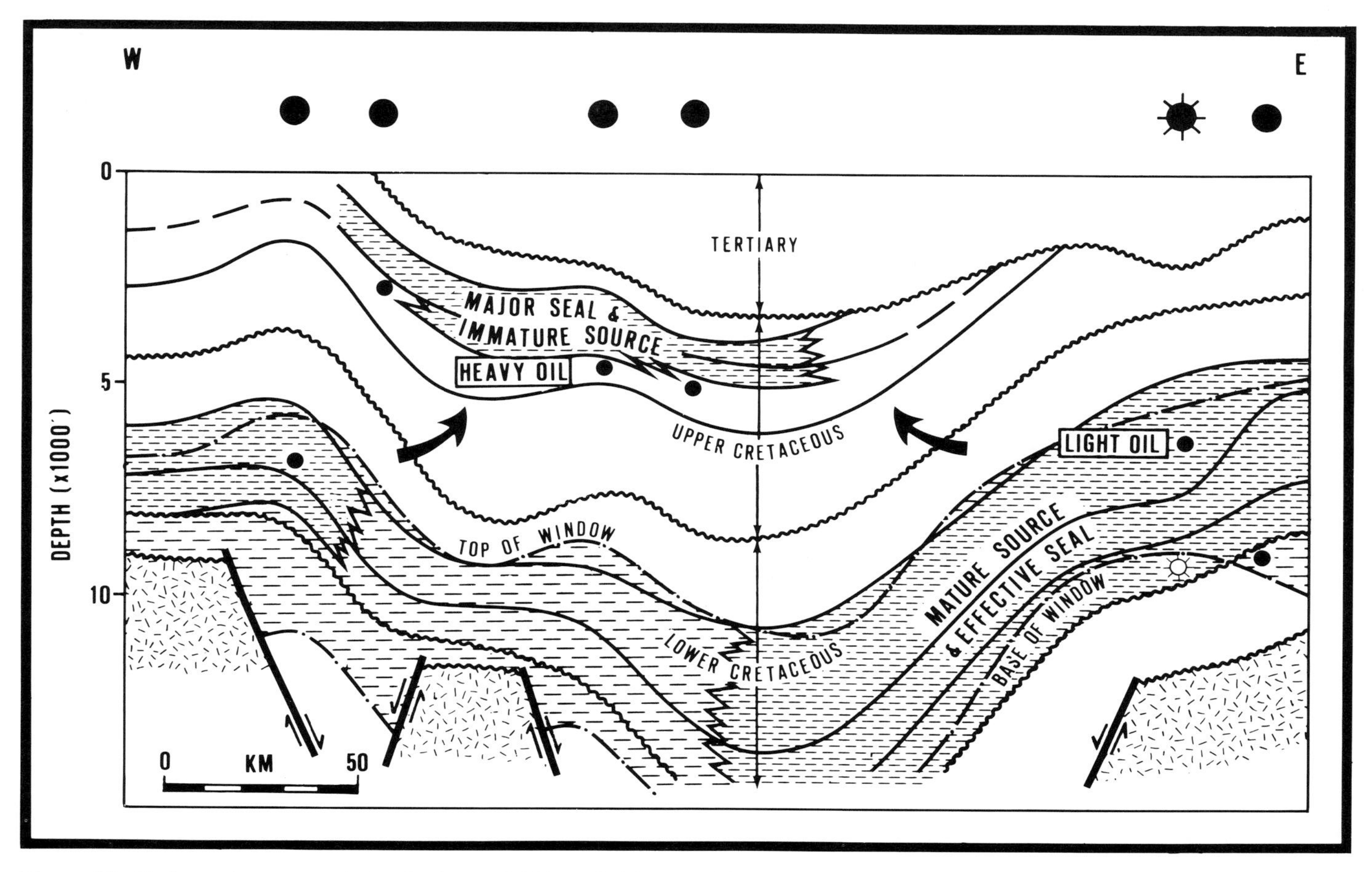


Figure 20. *Hydrocarbon scenario for the Central African Rift System. Two distinct sets of hydrocarbon occurrences are found in these basins. The deeper play is characterized by delta-front and turbidite sand reservoirs, interbedded with and sourced from rich oil-prone lacustrine shales. The shallow play is determined by the presence of sealing shales, which would be source rocks if mature. This oil, which often shows paraffin and naphthene depletion from bacterial decay, is hosted in more extensive, coarser, fluviatile sands.*

tures in the Guban area of Somalia. Basal Adigrat sands of Early Jurassic age, which are an excellent reservoir target in east Africa, are overlain by Jurassic carbonates and shales, followed by a Cretaceous Nubian clastic succession similar to the Gulf of Suez. Mixed, lower Tertiary sediments, including anhydrite, complete the sequence, and rifting begun in the Tertiary is still active. Red Sea and Gulf of Aden transform faults project onshore and have had a controlling effect on structural and stratigraphic development within these interior fracture basins. The strike of the basins is perpendicular to the $\sigma_3$ extensional stress axis and not parallel to either the Red Sea or the Gulf of Aden, which form two arms of a triple-junction.

Other interior fracture basins in Africa that have had little, if any, past exploration are the Karroo grabens of southern Africa: for example the Kalahari, Ruaha, Luangwa, and Zambezi grabens. The Zambezi graben extends from onshore Mozambique into the offshore beneath upper Mesozoic and Tertiary marginal-sag-basin sediments. A Karroo play could exist beneath Lower Jurassic salt with oil potential where burial depth has not been sufficient to cause overmaturation of any oil generated. One can envision structuring caused by drape over basement fault blocks. Shallower Mesozoic plays in Mozambique are gas-prone because of the type of organic material in the shales of the Cretaceous Domo and Grudja formations. These, and their equivalents, have sourced the gas deposits discovered to date in Mozambique and Tanzania.

In summary, the interior fracture basins represent 41% of all African basins while covering only 18% of the total basin area. Although 6% have hydrocarbon production, only 7% have been explored through a semi-mature stage. Many interior fracture basins remain unexplored with 68% untested and 26% with immature exploration.

These basins have been, and will continue to be, the most attractive of all African basin types with higher-than-average success rates (25–35%) and large, often giant, field sizes. They offer low–moderate risk for moderate–high reward ventures, the most seductive combination.

### Marginal Sag Basins

Marginal sag basins are located on the outer portions of the continental crust, always parallel to the continental/oceanic crust boundary. They are extensional basins usually initiated by, and hence underlain by, interior fracture basins prior to continental break-up. Often the presence of such interior fracture basins makes the marginal sag basins prospective.

Africa contains several of the world's classic marginal sag

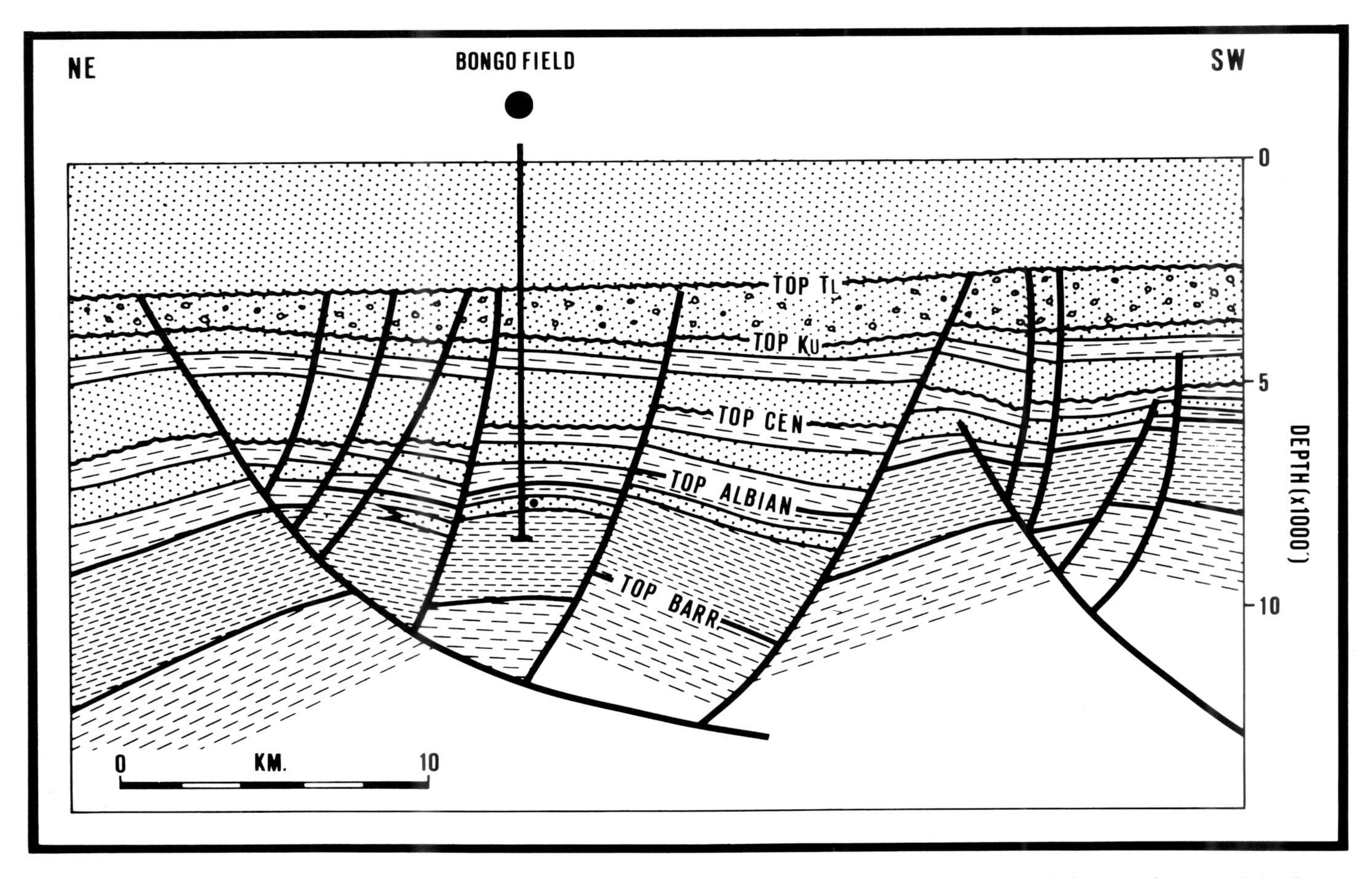


Figure 21. *A typical oil field in the Central African Rift System. This example of a deeper-play oil field, fictitiously named the Bongo Field, contains estimated recoverable reserves of 35 million bbl of 38° API oil. The reservoirs are 15–30 ft (5–9 m) turbidite sand lenses with porosities of about 15% and permeabilities of 100–300 md. Oil saturation can reach 84% in such fields. The sands are surrounded by over 3,000 ft (914 m) of rich, oil-prone, lacustrine shales. Present-day geothermal gradients of about 1.6°F/100 ft are common.*

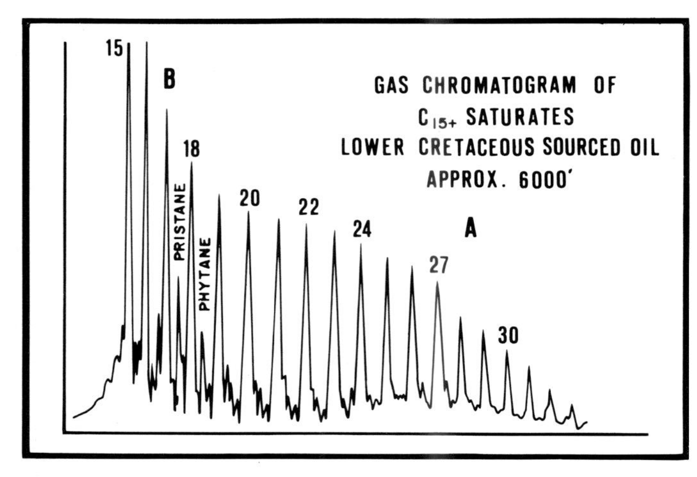


Figure 22. *Gas chromatogram of $C_{15}+$ saturates for Lower Cretaceous lacustrine-sourced oil from the Central African Rift System. Some immaturity is suggested by the $C_{22-30}$ naphthene bulge, but the strong component of $C_{23}+$ n-alkanes, a slight suggestion of odd-over-even preference and the pristane/phytane ratio indicate a non-marine source material.*

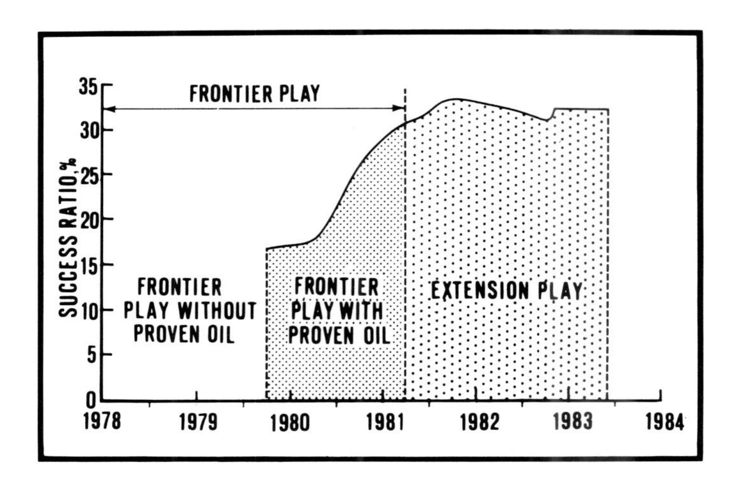


Figure 23. *Historical success rates, Sudan. The first wildcat in the interior of Sudan was drilled by Chevron in December 1977, and the first discovery was made at Abu Gabra in September 1979. Erratic success followed until, in early 1981, success rates exceeded 30% and then stabilized. At this point, an extension play rather than a frontier play was entered into.*

basins, flanking all present-day coastlines of the continent. The Aptian Salt basin of west Africa mirrors the Campos basin of Brazil, with equivalent success, while the Northwest African Coastal basin shares its misfortune with the East Coast of the United States.

Both the Red Sea and Gulf of Aden themselves, and not the neighboring aborted rift basins, are examples of the transition from interior fracture to marginal sag basin types with trailing-edges developed at either flank of new oceanic crust. High heat flows accompanying early uplift, intrusion, and spreading are partly responsible for the Red Sea being gas-prone. Geothermal gradients range from 3°F/100 ft in the northern part to over 12°F/100 ft in the southern parts of the Red Sea.

Figure 25 illustrates one such gas accumulation. The Suakin gas/condensate field, offshore Sudan, demonstrates how structuring is related to salt tectonics. Growth-faulting is listric into Miocene evaporites, with the hydrocarbons held in Miocene sand reservoirs sealed by an upper salt layer. The source beds of the Miocene Globigerina Marl, so rich in the Gulf of Suez further north, are here overmature for oil due to the high temperature gradients. As in the Gulf of Suez, intense salt tectonics complicate seismic interpretation of deep pre-Miocene structures. Three gas discoveries have been made to date in the Red Sea, which, with an area of 350,000 sq mi (906,430 sq km), has had little exploration.

The Red Sea holds great promise as a future gas province, but oil potential should not be totally precluded. A subcommercial oil discovery was made recently offshore South Yemen in a somewhat analogous situation to that described above. Shallow oil accumulations might yet be found in the Red Sea.

In the offshore Mauritania portion of the Northwest African Coastal basin, a carbonate platform developed during Late Jurassic to Early Cretaceous times (Figure 26). A high-energy bank front was developed at the shelf margin, with patch reefs landwards of it on the carbonate platform itself. Thick, Upper Cretaceous and Paleocene basinal shales became diapiric at the toe of listric growth-faults, while a series of deltas prograded out over the shelf through Late Cretaceous and Tertiary times. Turbidites were deposited within these basinal shales. This part of the basin contains potential reservoirs in the shelf-edge carbonates, deltaic and turbidite Cretaceous sands, and in Miocene turbidites. Oil-prone Cretaceous rocks, having fair-to-good source potential, are present; and structuring caused by diapiric shales and deep Triassic salt, and growth faulting, have resulted in the formation of numerous potential traps. However, oil exploration to date has been unsuccessful for reasons not entirely clear. Shallower plays in the east lack a seal. Perhaps many deeper tests were drilled not on valid structural closures, but on seismic anomalies instead. The Lower Cretaceous reefal carbonates, which were never tested, hold some potential, as do deeper, lower Mesozoic or Paleozoic clastic plays, perhaps in grabens not yet discovered.

Further south in Guinea-Bissau, early wells were aimed at Gulf-of-Mexico-type traps such as salt domes, flank traps, and turtles. The most attractive potential is in less mature salt swells and in sub-paleoslope traps. The geology is similar to that existing offshore Mauritania (Figure 27). However, much of the best potential is in deep waters. A potentially prolific play exists where Albo-Aptian delta-front sands are truncated by the Senonian paleoslope, and sealed by Maestrichtian basinal shales. Light-gravity oil (32–36° API) was recovered on the drill-stem test from such sands in well PGO-3, which was drilled on a glide-plane rollover structure. Other similar structures in the area hold the potential for smaller-sized fields, but are situated in shallower water depths.

Two heavy-oil occurrences have been discovered in northwest Africa, at Cap Juby, offshore Southern Morocco, and at Dome Flore, offshore Casamance, Senegambia. In the former, oil occurs in Upper Jurassic, shelf-edge carbonates, but suffers from either biodegradation or water-washing, resulting in an API gravity of 11° with 4% sulphur. In the latter case, oil generated in deeply-buried, Lower Cretaceous prodelta shales migrated up the flanks of a salt piercement structure into Oligo-

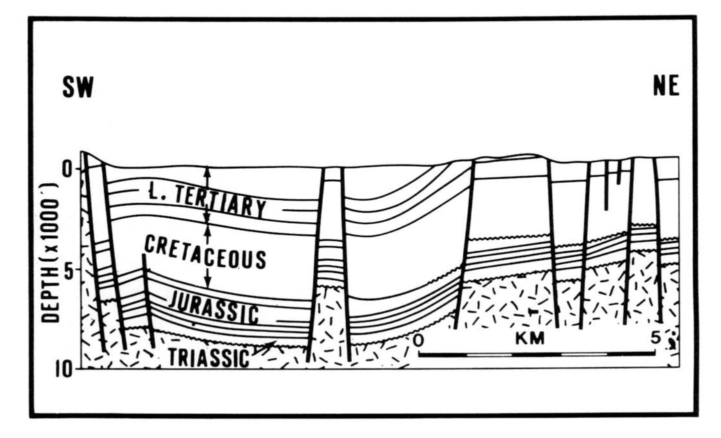


Figure 24. *Geologic cross-section across a typical Gulf of Aden aulacogen basin. Although rifting only began in the Tertiary and is still active, a thick, prospective succession of Triassic to Tertiary sediments is present.*

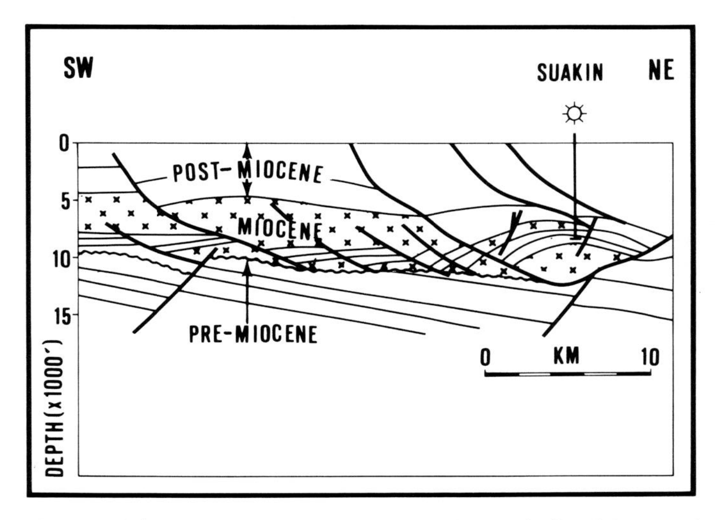


Figure 25. *Geologic cross-section through Suakin gas/condensate field in the Red Sea basin, offshore Sudan. Suakin field contains estimated reserves of 880 billion cubic feet of gas plus 22 million bbl of condensate. The reservoir is a 28 ft (8.5 m) sand interbedded with shales and marls below upper Miocene salt. Porosity is 20%, and water saturation 35%. The temperature gradient is greater than 2° F/100 ft. Numerous other prospects in the area hold potential for wet gas and condensate.*

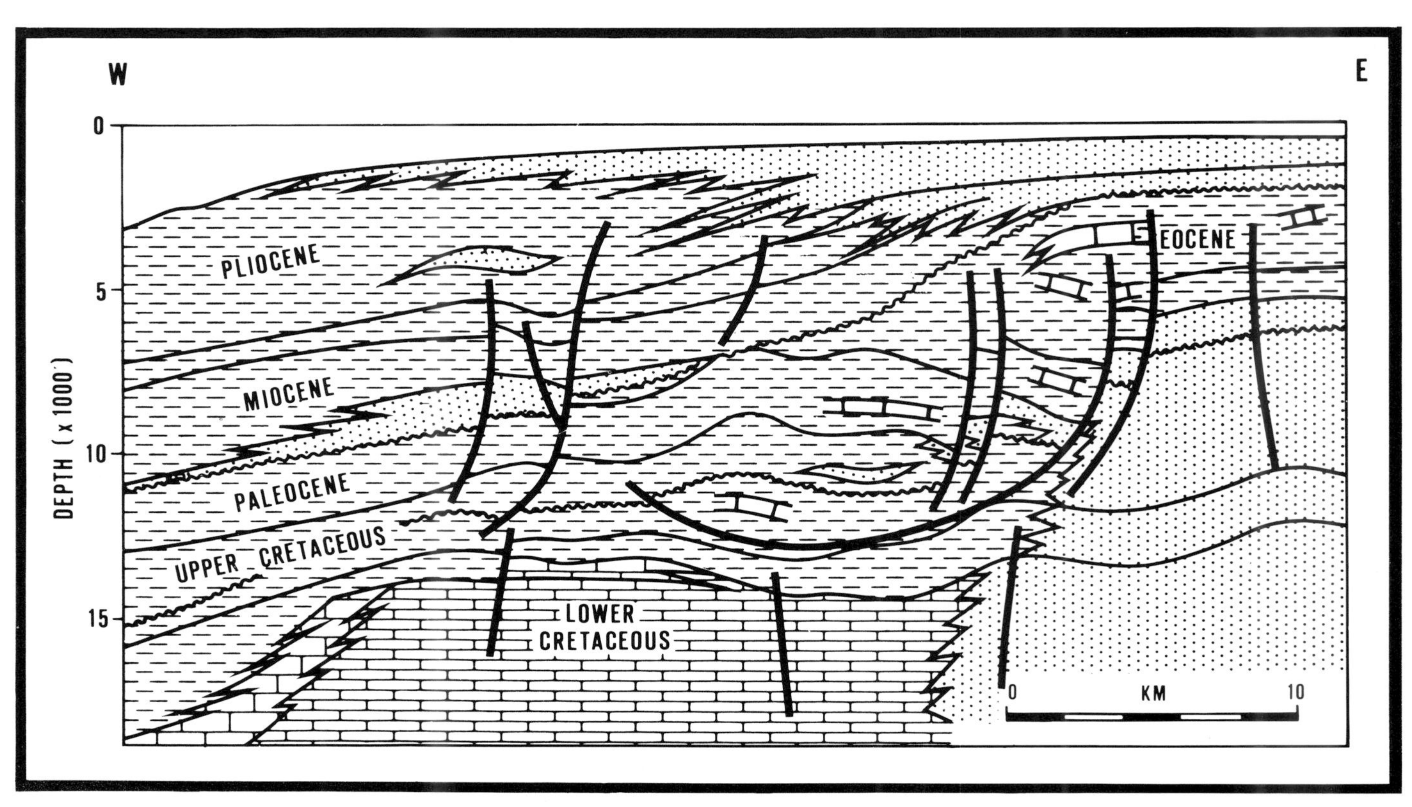


Figure 26. *Geologic cross-section across the Northwest African Coastal basin, offshore Mauritania. Several plays exist in this basin, but commercial hydrocarbon accumulations have not been found to date. Oil-prone source rocks (TOC up to 5%) within the Albian–Cenomanian and reservoir-quality Upper Cretaceous deltaic sands with porosities of 17–25% have been identified. The Lower Cretaceous carbonate play is untested in this part of the basin. Miocene turbidite fans are also prospective.*

cene foraminiferite at very shallow depths. This oil has an API of 10° with 1.6% sulphur. Workers have reported shows of lighter oil from well CM-7 in the Albian sands flanking the salt structure at depth.

These and other minor oil occurrences within the basin demonstrate that liquid hydrocarbons were generated in the Northwest African Coastal basin, but no commercial accumulation has yet been found. The problem is possibly one of timing of hydrocarbon migration, and the need for an effective sealing shale. However, many of the earlier exploration wells were drilled on structures based on interpretations of poor seismic data. These seismic data suffered from single- and low-fold, common-depth-point stacking and severe multiple problems; low seismic resolution and no migrated data were available. Many of these wells are part of a hurried drilling campaign with some based on unusual geologic concepts. Hence, a large number of the wells in this basin might not be valid tests.

The Aptian Salt basin is a marginal sag basin with established oil production. This name applies to the subbasins of Campo, Rio Muni, Gabon, Congo, Cuanza, and Benguela, extending north to south from Cameroon to Angola along the Atlantic coast of Africa. Each of these subbasins shares common structural and stratigraphic characteristics, and hence they are grouped together. A Jurassic to Lower Cretaceous, non-marine rift sequence is followed by a transgressive sheet sand before deposition of the Aptian salt. Although numerous evaporite cycles occur in the south above this salt, generally the Albian is characterized by a marine carbonate platform. The overlying, Upper Cretaceous to Tertiary sequence is mostly clastic, except for a few minor carbonate cycles. Source rocks are chiefly the pre-salt, non-marine lacustrine shales with minor sources in basinal shales coeval with the marine sandstones.

The chief reservoirs are pre-salt fluviatile and turbidite (lacustrine) sands, pre-salt freshwater carbonate reefs, post-salt marine carbonates, and post-salt intertidal channel or estuarine sands. The transgressive sheet sand (Gamba in Gabon, Chela in Congo, and Upper Cuvo in Angola) is an excellent oil-field reservoir where so structured. In addition, it serves as a migratory pathway and as an indicator in the interpretation, from seismic data, of the deeper structure over which it drapes. The cross-section of the central portion of the Aptian Salt basin in Figure 28 shows the tilted fault blocks, and transgressive sand draping them, beneath the salt layer. Structural development has strongly influenced sedimentation: for example, the growth of Toca freshwater reefs on the edges of basement highs.

Tilted fault blocks in the pre-salt interior fracture basin, and salt movements, stimulated by either fault reactivation or differential basin subsidence, are the chief structuring mechanisms in trap formation. In the southern parts of the basin, post-salt oil deposits are dependent upon a migratory pathway from the pre-salt source beds. One such mechanism is via listric faulting, accompanying salt tectonics, with faults soling out at the base salt. Also, localized salt solution may offer a migratory pathway. The post-salt play is restricted in a landward direction by a lack

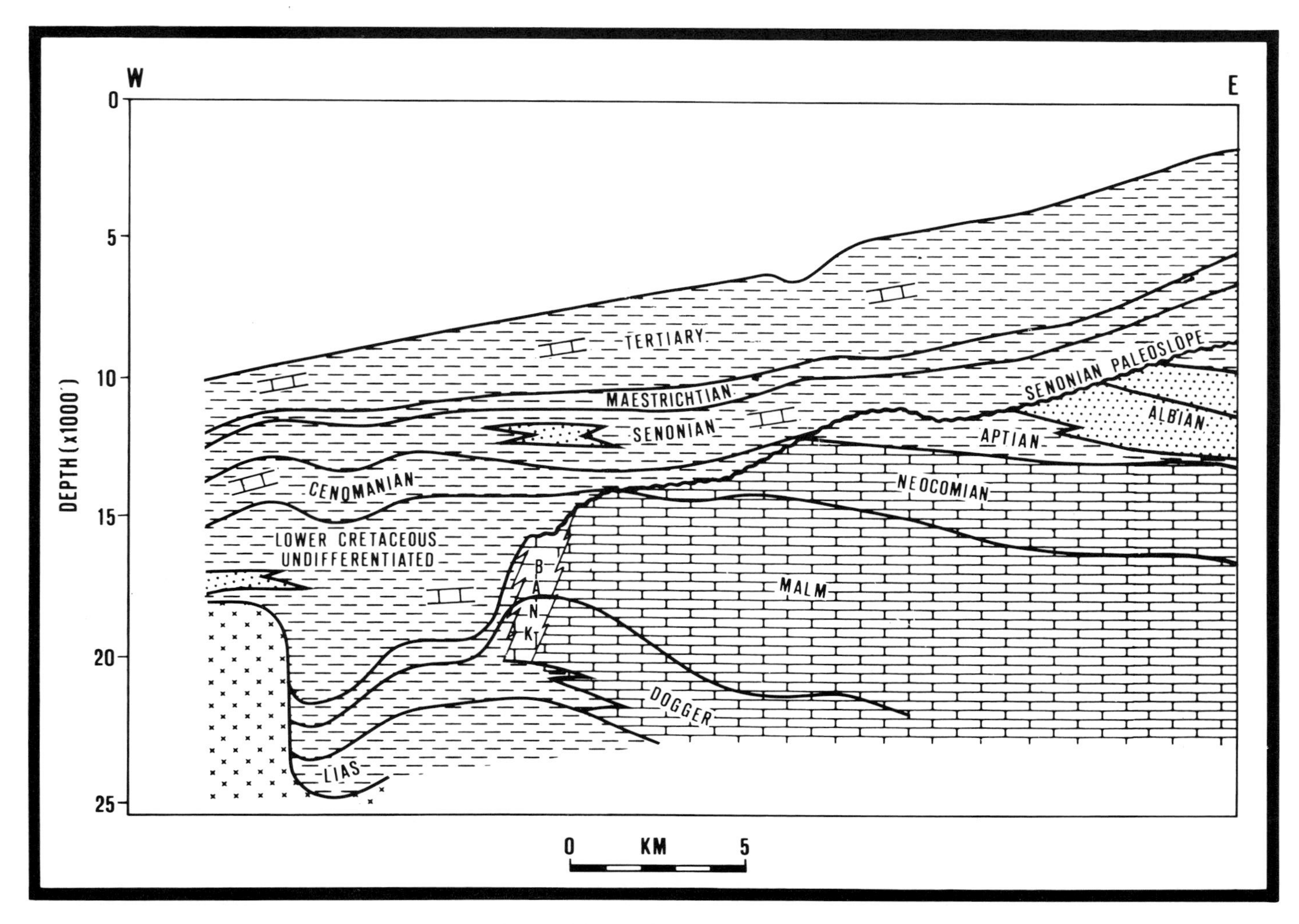


Figure 27. *Geologic cross-section from the Northwest African Coastal basin, offshore Guinea-Bissau. The Albian–Cenomanian Geba delta prograded out across the carbonate platform providing reservoir-potential sands and source-potential shales. Both the carbonates and the deltaics subcrop the Senonian paleoslope, offering interesting stratigraphic trap potential.*

of an effective top seal.

Within the Aptian Salt basin, geochemical fingerprint analyses of both post- and pre-salt oil are almost identical. Gas chromatograms of oil hosted in post-salt reservoirs show a non-marine origin (see Figure 29; compare with Figure 22). The rise in the base envelope shows the presence of naphthenes, and this particular one suggests maturity. The spikes above the base envelope, representing individual normal alkanes, have an odd-over-even C-molecule preference (above $C_{17}$), which is typical of Type III kerogen. The pristane/phytane ratio also shows non-marine terrigenous plant material, and there is a strong component of $C_{23}+$ normal alkanes. Although non-marine source rocks rarely give rise to giant accumulations of oil, they have been and are continuing to prove to be significant contributors to hydrocarbon reserves.

The Upper Cretaceous Batanga sandstone is a significant, producing formation in the Gabon subbasin. This sand is considered to be an intertidal channel sand deposit, with its own coeval basinal shale source. Recent discoveries at Oguendjo and Obando are from these sands, which have been structured by deep-seated salt features. Figure 30 illustrates the field-size distribution of eight Gabonese oil fields producing from the Batanga sandstone. A best-fit line gives a $P_{50}$ of 40 million BOE. However, the largest field in such sands is Grondin, with 190 million bbl oil recoverable, which, like other large fields in the area, is located further west where salt structures are less mature, broader, and flatter. Smaller accumulations are related to piercement structures.

For the pre-salt sequence, see the two field-size distributions in Figure 31, one for the whole of the Aptian Salt basin and the other for Gabon alone. Respectively, they yield a $P_{50}$ of 74 and 46 million BOE, both of which are larger than the post-salt $P_{50}$ of 40 million BOE.

Fields with larger-size reserves occur at Gamba and Malongo, where pre-salt-generated hydrocarbons have been trapped over large basement arches separating the Atlantic margin from the interior grabens. The Malongo pools of Cabinda include Malongo, which has solely post-salt production, and Malongo West and North, which are both pre-salt (Figure 32). The transgressive Chela sand, immediately pre-salt, is draping basement horst blocks, and fault reactivation after salt deposition has led to some oil migrating into post-salt traps.

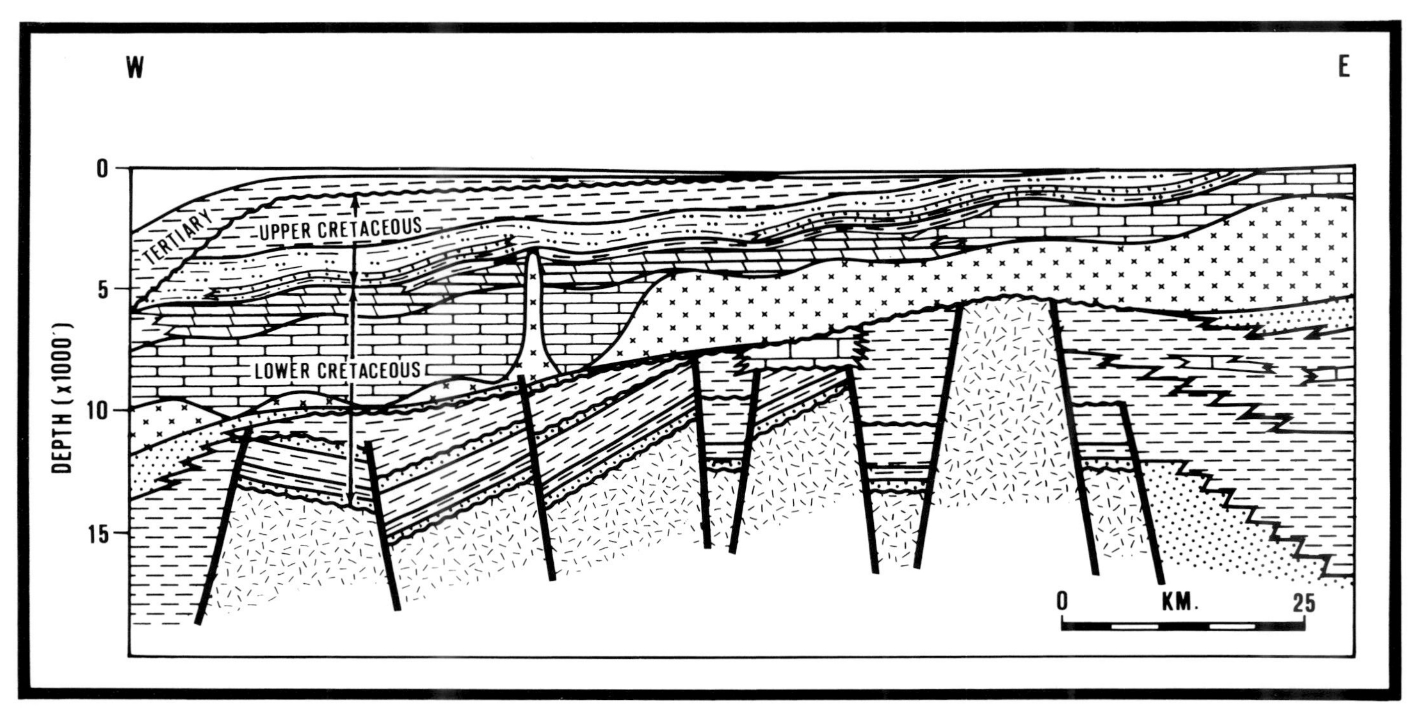


Figure 28. *Schematic, geologic cross section across the Aptian Salt basin, offshore Congo. The deep pre-salt structure is masked by Aptian salt, which forms post-salt structural traps. Pre-salt lacustrine source beds feed sands in tilted fault block traps and freshwater carbonate banks that grew around horst blocks. Post-salt reservoirs are sourced via reactivated faults or salt solution hollows.*

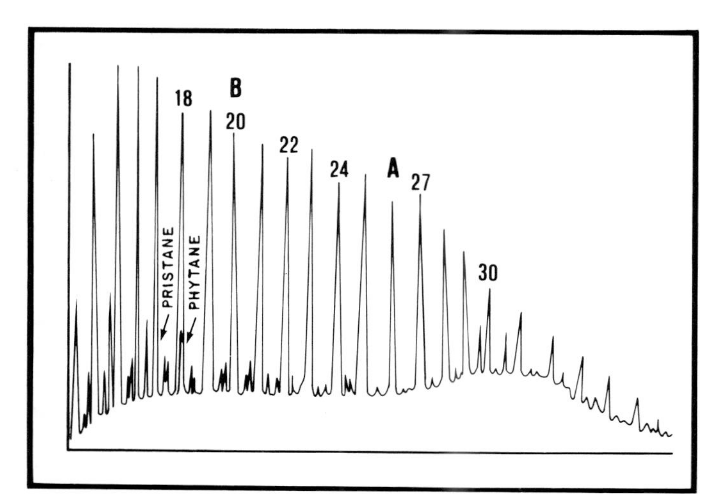


Figure 29. *Gas chromatogram of $C_{15}+$ saturates for pre-salt lacustrine sourced oil from the Aptian Salt basin of west Africa. This oil sample, from a post-salt reservoir, shows evidence of a pre-salt origin. There is a strong, odd-over-even preference, typical of Type III (woody-coaly) kerogen, and the pristane/phytane ratio shows non-marine terrestrial plant-derived material. There is a strong component of $C_{23}+$ n-alkanes. Some degree of immaturity is suggested by the naphthene hump between $C_{30}$ and $C_{35}$.*

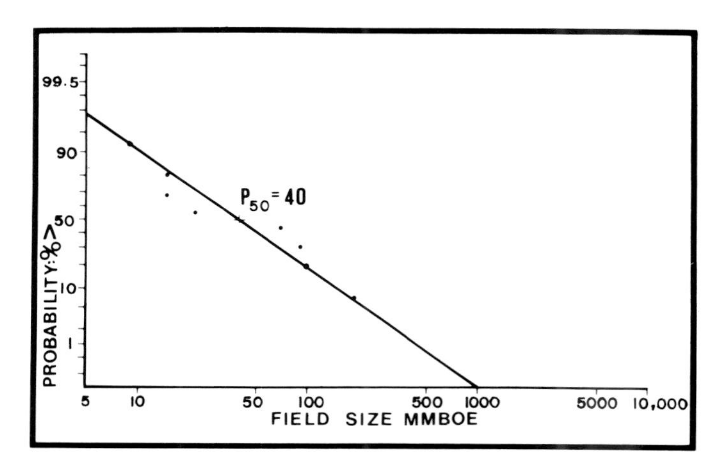


Figure 30. *Field-size distribution (given in million BOE, or MMBOE) of post-salt Batanga sandstone play, Aptian Salt basin, Gabon.*

Further success with big finds can be expected in the poorly explored pre-salt of this basin. The chief problem in this basin for the explorationist is recognizing the true structural events at depth, and predicting stratigraphy in the pre-salt because of the effects of not only salt tectonics, but also inter-bed multiples. Several seismic techniques are now available to help combat these problems, including beam-steering and depth migration. Another problem is interpretation of structural timing as it relates to hydrocarbon migration.

The onshore succession in the Cuanza subbasin of Angola shows non-marine, pre-Aptian, lower Cuvo sands with some volcanics, representing a rift sequence, succeeded by a transgressive, upper Cuvo sand, then a series of evaporite–carbonate cycles, related to subsidence and sea-level changes. This classic

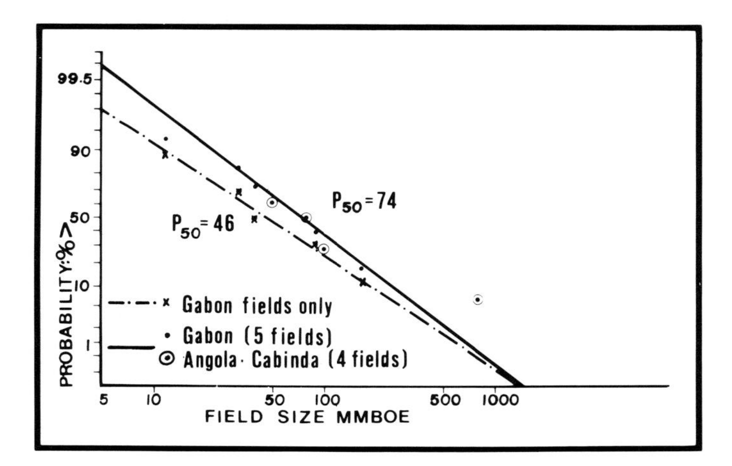


Figure 31. *Field-size distribution (given in million BOE, or MMBOE) of pre-salt non-marine sandstone play, Aptian Salt basin.*

onshore post-salt sequence of evaporite cycles becomes a more basinal succession offshore, with turbidite sands dominating over more sheet-like sands (Figure 33). Islands, formed by uplifted basement blocks, were the sites of platform carbonates overlying salt and clean, pre-salt sands at depth. Onshore, the extent of post-salt deltaic sand wedges was guided by salt movements and overall basin subsidence, the axis of which shifted through time.

Salt-evacuation grabens, clearly recognized on seismic data, may occasionally host secondary salt diapirs. Turtle structures and all other salt features are common to the Cuanza subbasin, providing numerous potential traps. However, the quality of reservoirs can vary with post-salt carbonates needing fracture porosity to produce and pre-salt sands being unpredictable in distribution, facies development, and reservoir attributes.

Studies of the historical success ratios in both Angola (Figure 34) and coastal Zaire (Figure 35) demonstrate a few interesting points. In Angola, the decline of onshore discoveries and the increasing importance of the offshore plays is clear, with only a minor change accompanying Angolan independence from Portugal in 1974. In Zaire, an unsuccessful onshore exploration phase, lasting seven years (contrasted with only two years in Angola), was followed by two years of rapid offshore success prior to nationalization. As a result both of political and economic uncertainties, it took a further four years before activities resumed previous success rates.

The coastal basins of East Africa have suffered from a lack of aggressive exploration, in part because of prejudices based

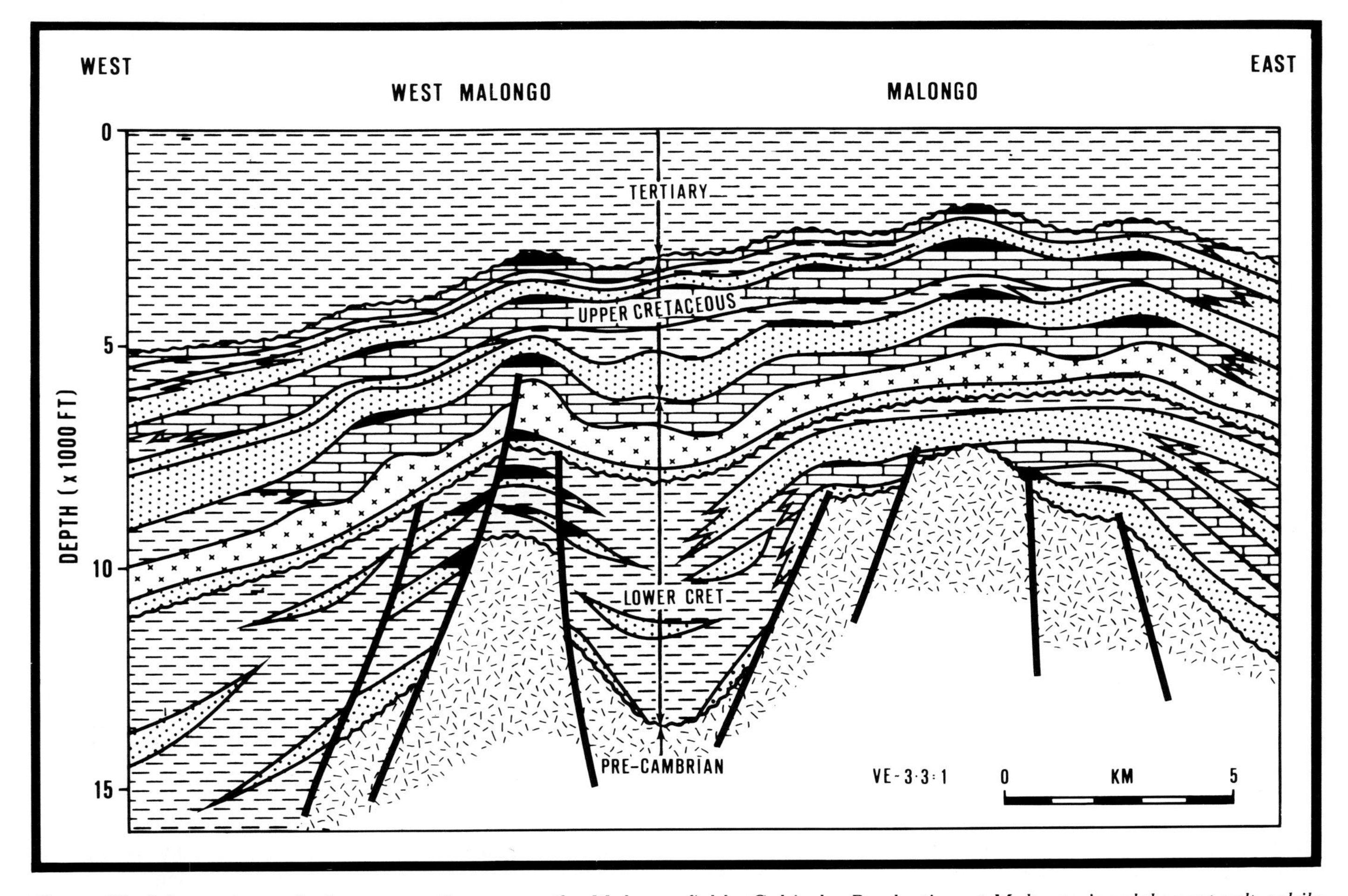


Figure 32. *Schematic, geologic cross section across the Malongo fields, Cabinda. Production at Malongo is solely post-salt, while Malongo West is pre-salt. Estimated reserves for all Malongo fields are over 800 million bbl oil.*

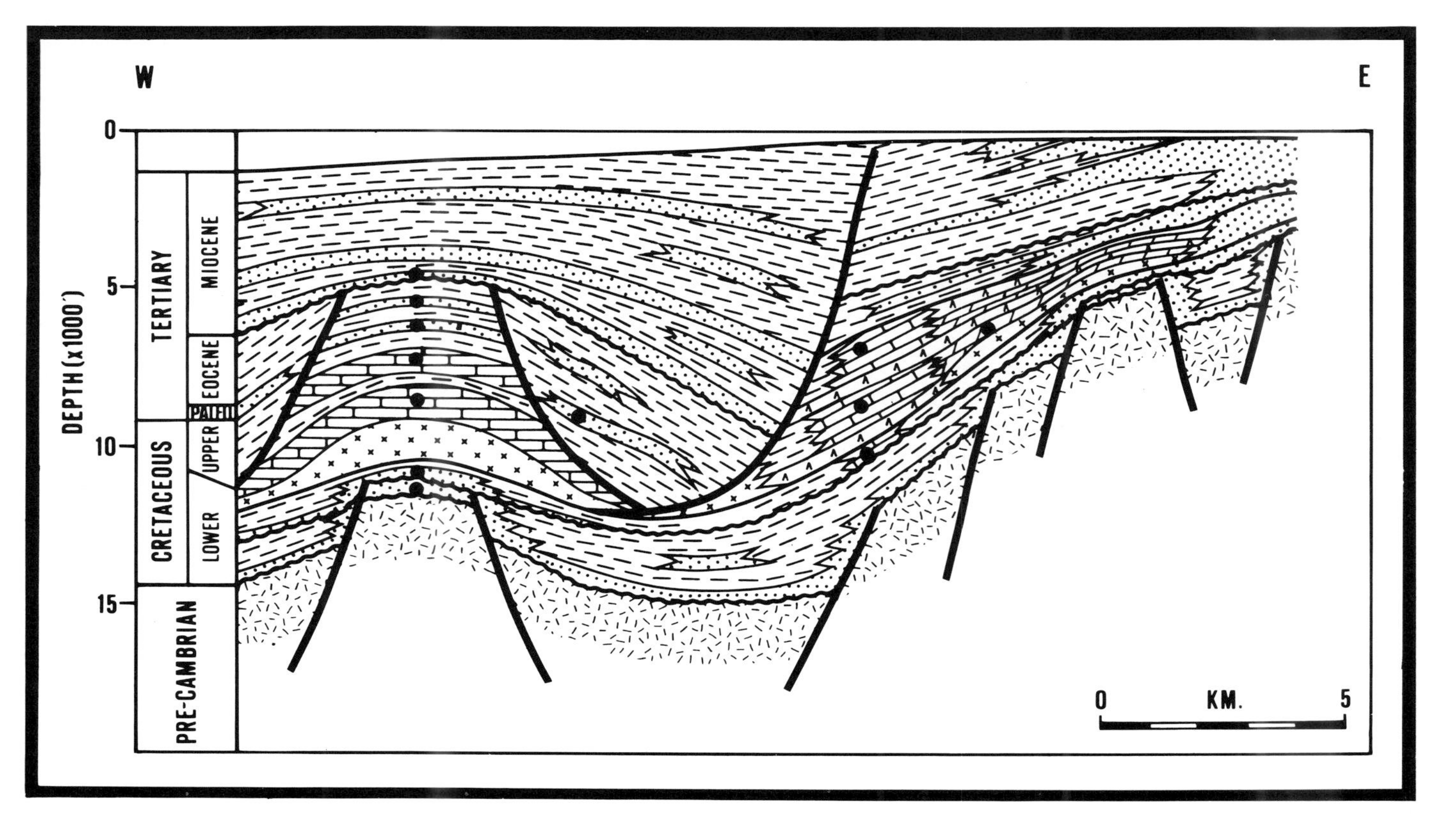


Figure 33. *Schematic geologic cross section showing plays, offshore Angola. Salt evacuation grabens, as illustrated, are common in the Cuanza and Benguela subbasins of the Aptian Salt basin.*

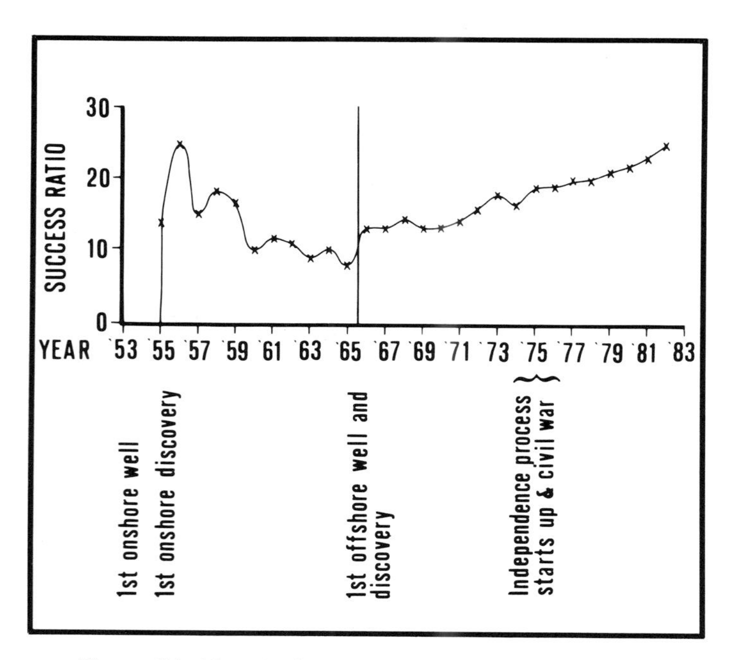


Figure 34. *Historical success ratios, Angola/Cabinda.*

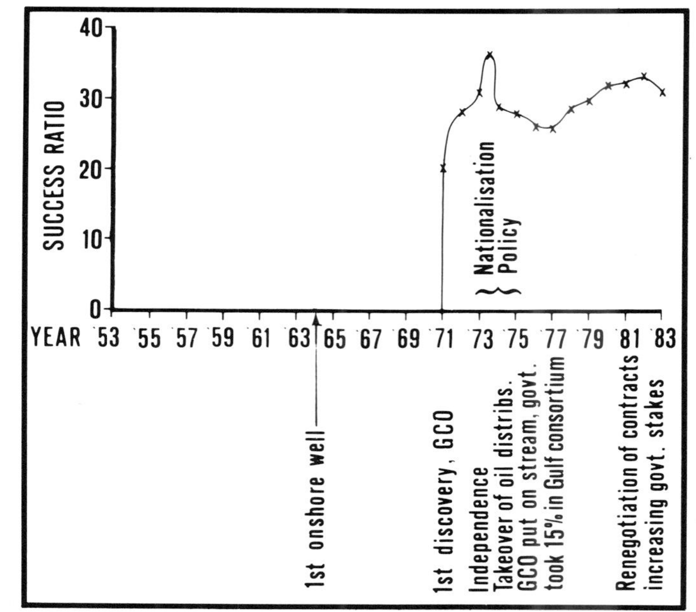


Figure 35. *Historical success ratios, coastal Zaire.*

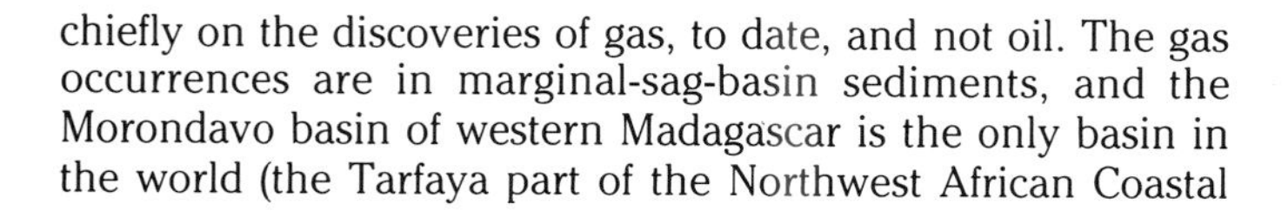

chiefly on the discoveries of gas, to date, and not oil. The gas occurrences are in marginal-sag-basin sediments, and the Morondavo basin of western Madagascar is the only basin in the world (the Tarfaya part of the Northwest African Coastal basin may qualify as another) with major heavy oil/tar sand deposits, but lacking major hydrocarbon production.

The tar sands at Bemolanga in Madagascar are probably earliest Jurassic in age. The sands, covering 260 sq mi (67 sq km),

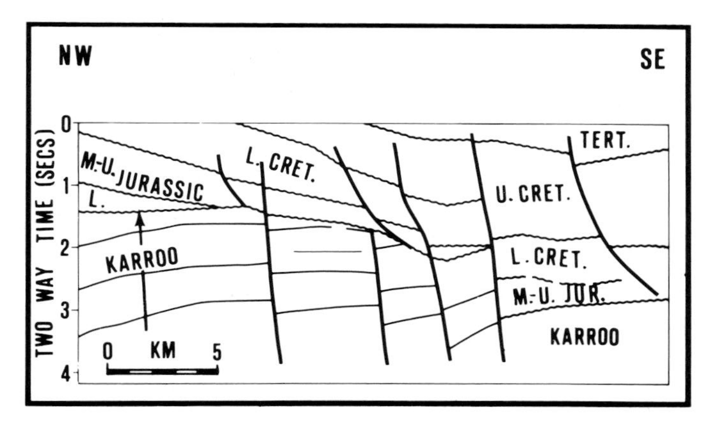


Figure 36. *Geologic cross section across the Selous basin, onshore Tanzania. Lower Jurassic and Upper Triassic (Karroo) sands sourced and sealed by Jurassic marine shales constitute the invoked hydrocarbon play.*

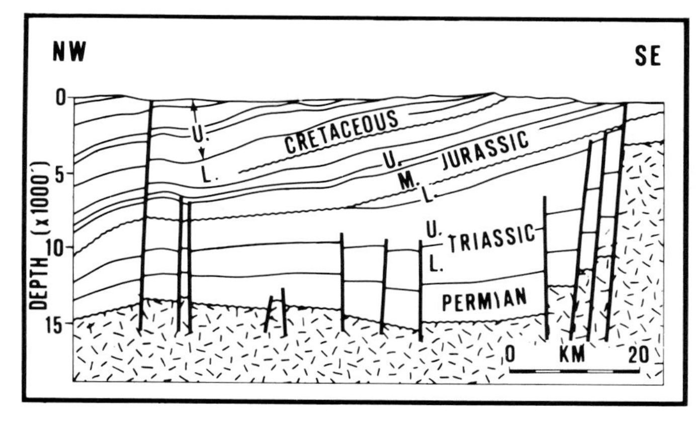


Figure 37. *Geologic cross section across the Majunga basin, northern Madagascar. This is a lookalike basin to the Selous basin of Tanzania with the same play potential.*

were deposited in deltaic distributary channels, and reserves are estimated at 2 billion bbl of oil in place with 25% recoverable by surface mining.

The Tsimiroro heavy-oil pool is structurally lower, with the same Lower Jurassic strata overstepping Triassic onto basement. Over an approximately similar area, reserves are estimated at 4 billion bbl of oil in place. The oil, which has an API gravity of 12–14°, has flowed at rates of only 1/2 bbl/day on test.

The Selous basin of Tanzania is at the seaward end of the Ruvu Valley, itself a Karroo-age interior fracture basin, formed as the Gondwana continent broke up and Madagascar moved away east and south-eastward. The Selous basin consists of a Mesozoic marginal sag basin overlying a pre-drift Karroo interior fracture basin (Figure 36). The other half of this interior fracture basin is thought to be the Majunga basin of Northern Madagascar (Figure 37). The transform fault system that separated these two basins is shown in Figure 2. In both Figures 36 and 37, the Middle Jurassic break-up unconformity separates non-marine clastics and lacustrine shales below, from marine carbonates, marls, shales, and marginal-to-marine clastics above. The picture is similar to West Africa except in the weaker development of evaporites, which are latest Triassic or earliest Jurassic in age in East Africa, similar to Northwest Africa, but older than the Aptian evaporites of Gabon and Angola. The salt deposits mark the transition from non-marine to marine conditions as the break-up of Gondwana progressed.

The play invoked in East Africa is one of Karroo, (Permian–Triassic) or Lower Jurassic, fluvio-deltaic sands, on the high-side of tilted fault blocks, sourced and sealed either by Jurassic marine marls and shales or by lacustrine shales within the Karroo itself. This scenario is encouraged by demonstrated source potential for oil in Middle Jurassic marine shales and some Karroo shales in southern Africa. Also, the situation is analogous to the pre-salt play in West Africa. The top-loading source and seal situation is akin to Prudhoe Bay in Alaska, where shales both above and below the break-up unconformity have source potential. The previously mentioned heavy-oil/tar-sand pools of Madagascar and small, shallow oil pools in the north of the Karroo basin of South Africa probably result from the lack of an effective seal when the Karroo source beds matured. The presence of an effective seal is the key to this exciting, new East African play.

The marginal sag basins have the largest areal spread of any basin type in Africa, covering 35% by area, and representing 23% of basins. Only 16% have production and, West Africa apart, exploration in these basins has been disappointing. However, a new look at these basins, accompanied by an aggressive exploration effort, might result in success.

### Wrench-Modified Basins

Several basins in Africa show varying degrees of wrench-modifying effects. These basins are classified separately in this paper as wrench-modified basins, having had a dominating wrench effect superimposed on their primary basin type. The primary types in Africa recognized as being modified by such wrenching are marginal sag and marginal sag/interior sag basins. The transform fault systems, responsible for basin modification, are shown in Figure 2. These fault systems splay out on continental crust into a swath of anastomosing and en echelon faults. Such a splay of faults, with accompanying folds and fractures, takes up surplus lateral motion so that the transform fault does not penetrate far onshore. The exception to this rule is the Atlas Flexure of North Africa, which is actually a conservative plate margin.

The prospectivity of wrench basins varies, depending on the stage of maturity of the wrenching. If wrenching has not progressed too far, oil could be trapped in en echelon folds within sandstone reservoirs and sourced from shales deposited between the folds. Wrenching gives rise to an abundance of structure types, but field sizes tend to be small.

Wrenching has modified the Morondavo basin of Madagascar by the dextral movement of Madagascar away from Kenya and Tanzania. The Karroo subbasin, which lies east of the Ilovo fault, shows a series of en echelon folds oriented northwest-to-southeast, suggesting dextral shear and supporting the idea that Madagascar came from the north rather than from Mozambique to the west.

The Gulf of Guinea contains several basins showing such wrench effects. One such basin is the Ivory Coast basin (Figure 38).

A strong break-up unconformity in the middle Cenomanian

overlies and truncates the underlying, essentially non-marine, sediments. The overlying sequence is that of a marginal sag basin, with turbidites proximal and distal to prograding deltaics. The cross section suggests the basin's sand-poor nature. The oil and gas discoveries to date occur in turbidites, as at Belier, and in locally-derived shallow marine sands, as at Espoir. The upper edges of tilted fault blocks formed a string of islands around which such sands were deposited. These sand ridges are related to a series of anastomosing dextral shears controlling deposition through Albian–Aptian times that should continue out into deep waters. The turbidite sands of the Upper Cretaceous, however, are poorly connected and not extensive. Source rocks in the Upper Cretaceous are both oil- and gas-prone, but Albian–Aptian, prodelta shale source beds may be present.

To the east, the Tano subbasin becomes more sand-prone, but the few discoveries that have been made there are of non-commercial size because of very intense wrench structuring. Seismic data are also poor, making interpretation difficult. Further east, the Central Ridge separates the latter from the Keta basin; this whole area has been poorly explored. The area displays complex wrench structuring, with flower structures and hinge zones showing occasional scissor movements.

The Saltpond field, offshore Ghana, lies adjacent to one such hinge zone with oil trapped in Devonian fluviatile sands, interbedded with shales that are the source rocks. The reservoirs are at depths of 7,500–8,500 ft (2,280–2,590 m). On the south side of the flanking strike-slip fault, over 15,000 ft (4,570 m) of Cretaceous is preserved. Such a situation is typical in the Gulf of Guinea.

In South Africa, several subbasins trending northwest-to-southeast, and separated by arches, are collectively called the South African Coastal basin. These grabens are related to the right-lateral motion of South Africa relative to the Falklands (Malvinas) Plateau of South America. Commercial gas discoveries have been made offshore, and potential remains here in stratigraphic traps for Lower Cretaceous turbidites, or in deeper, fluvial–estuarine sands subcropping the Lower Cretaceous, deep-water shales.

In Morocco, the Meskala gas/condensate field was thought to be a significant find in 1980. It opened a new, more rewarding play because previous Moroccan production in the area was

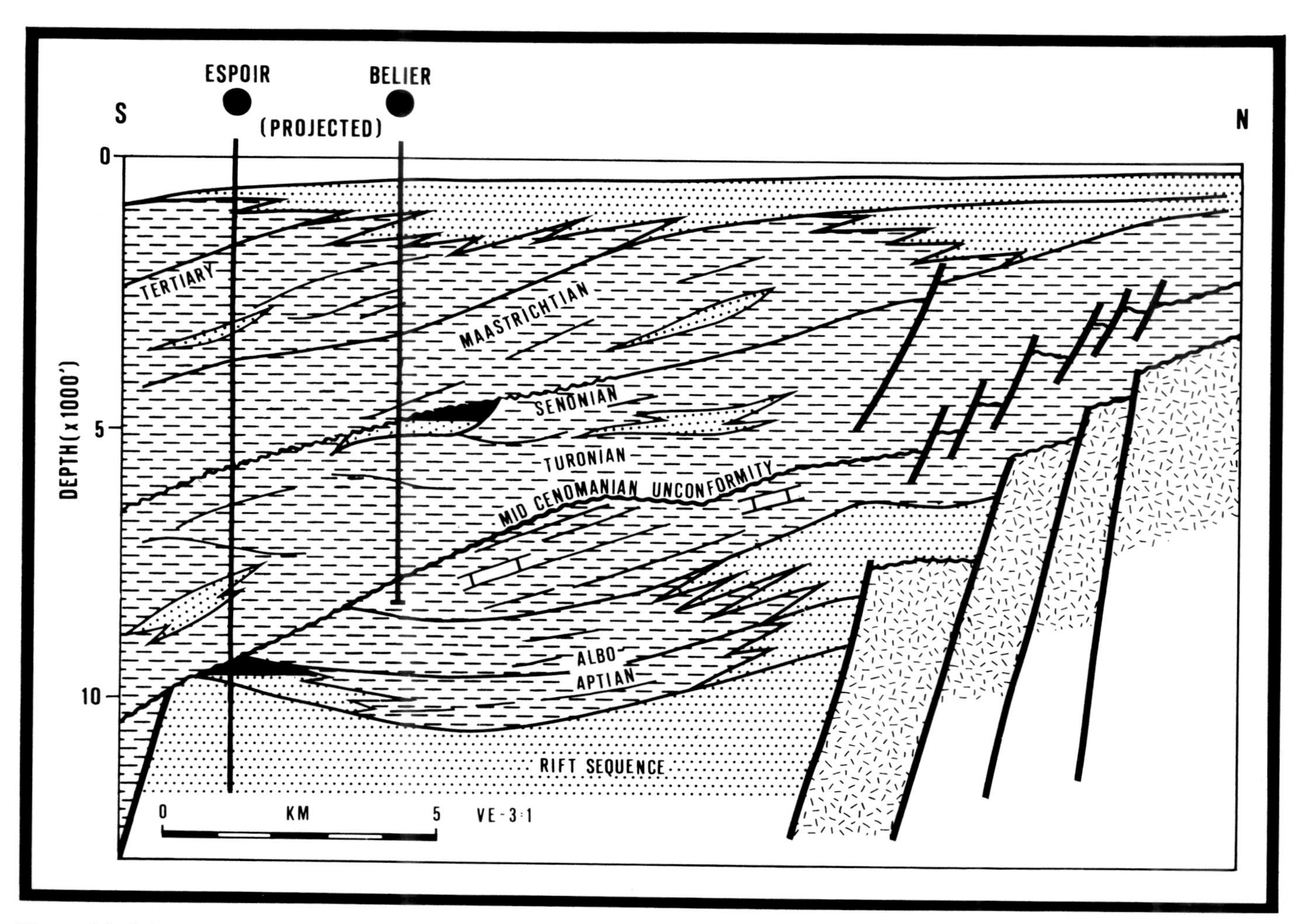


Figure 38. *Schematic geologic cross section across the Ivory Coast basin. The Espoir field, with estimated reserves of 150–200 million bbl oil plus some gas, has an Albian–Cenomanian shallow-water sand reservoir whereas the Belier field, with estimated reserves of 20 million bbl oil, has a Senonian turbidite sand reservoir.*

very small and came from post-salt Jurassic carbonates. Meskala was the last post-salt structure drilled because it was small; however, it revealed a surprise in the pre-salt Triassic and Paleozoic, opening up a new play in Northwest Africa as well as encouraging deeper drilling of other shallow post-salt structures. Just as in the Algerian Saharan Oued Mya basin, the Triassic and Paleozoic sands are sourced from subcropping Silurian black shales (Figure 39).

In the Essaouira basin, in which this field occurs, geologists have observed inversion of both Triassic saliferous and Jurassic anhydritic sequences. Diapiric salt walls follow sigmoidal trends, suggesting dextral shear associated with the South Atlas Flexure. Initial transform movements occurred as the North Atlantic began to open in early Mesozoic, but much evidence suggests earlier Paleozoic movements and reactivation of old lines of weakness.

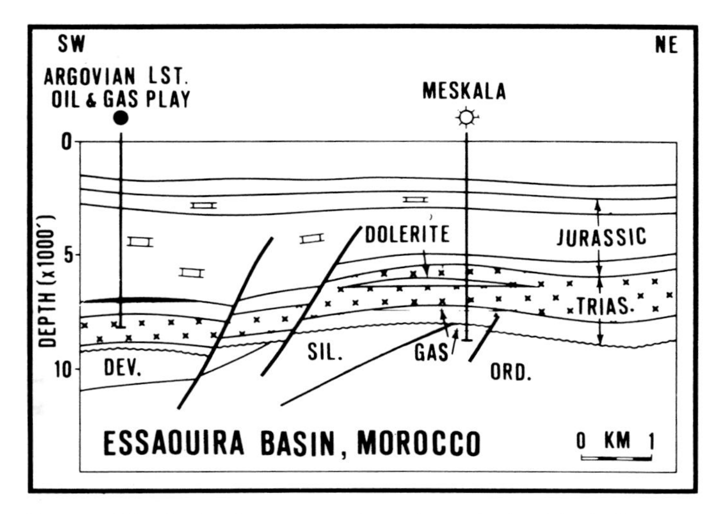


Figure 39. *Geologic cross section through the Meskala gas/condensate field, Essaouira basin, Morocco. Estimated reserves in Meskala are 1.1 tcf of gas with 100 million bbl of condensate.*

Similarly, over much of Africa, a regmatic shear pattern of northwest-to-southeast and northeast-to-southwest trends within Precambrian shield areas is repeated in younger strata in every area. For example, the pattern occurs in the Red Sea and Gulf of Aden; the East Coast of Africa and Madagascar; the Central African Rift System (especially the Niger basin parallel to the Muglad basin of Sudan, with pure extension; the Doba/Doseo basin of Chad, with superimposed dextral shear from the Borogop fault movements); and the Gulf of Guinea.

Additional basin inversion caused by wrenching and with associated salt tectonics is seen further east along the Atlas Flexure. The onshore portion of the Pelagian basin in Tunisia is such an example, with Upper Cretaceous inversion of a Lower Cretaceous basin over Triassic salt intrusions. Eocene (Metlaoui) reefal buildups developed on the flanks of paleohighs (Figure 40). Wrenching has had a significant influence on basin inversion, reactivation of old faults—particularly reversing throws, and triggering salt movements. The area is very complex structurally, and many fields may be found, though small in size because of advanced wrenching.

In the offshore portion of the Pelagian basin, inversion of an earlier marginal sag basin occurred in the Late Cretaceous, with further inversion during the Miocene. These tectonic events reflect plate interactions along conservative margins in North Africa and the western Mediterranean.

Many recent finds in Tunisia and northern Libya occurred in the Metlaoui Nummulite Trend. Figure 41 shows the theoretical development of these nummulite banks on a lower Eocene shelf margin. The trend appears not to be a sharply-defined linear belt, as previously envisaged, but a wider zone with numerous embayments. The relationship of different facies across the

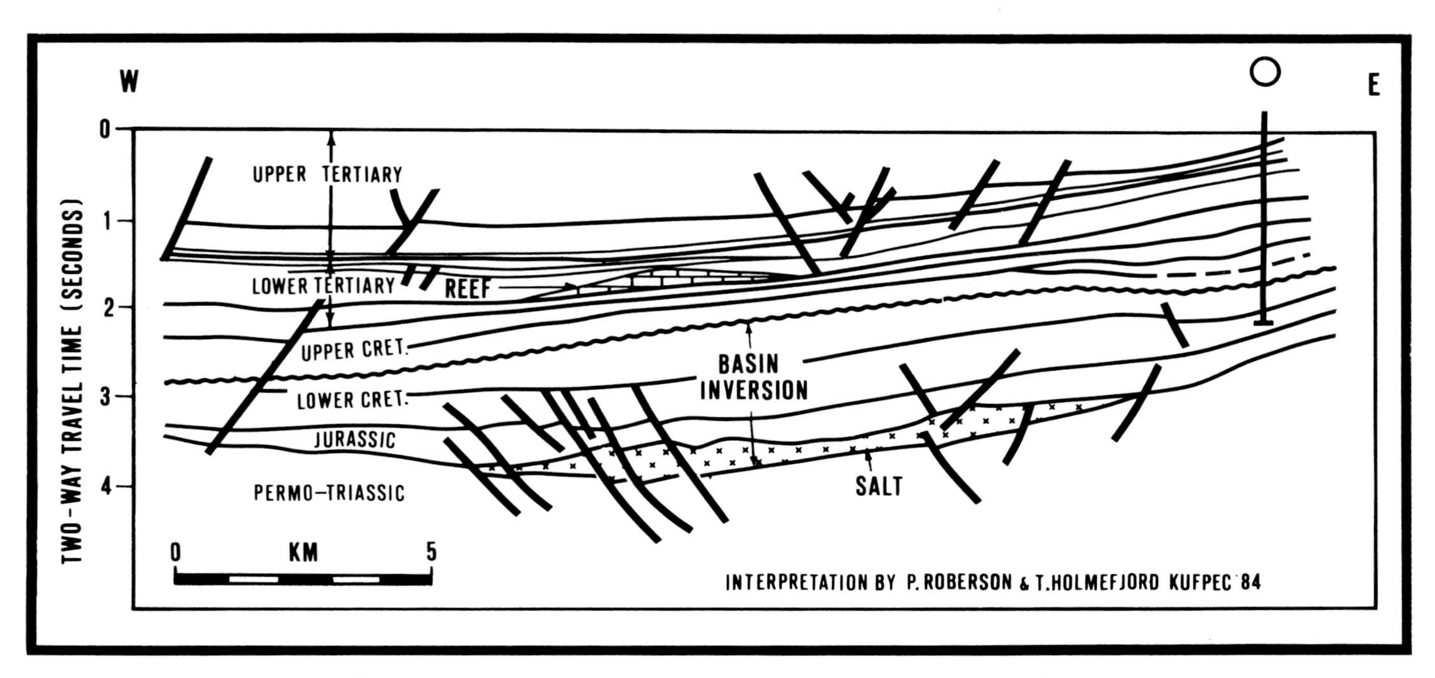


Figure 40. *Geologic cross section from the Kairouan area, onshore Tunisia.*

shelf has been ascertained by field traverse along the north–south axis of onshore Tunisia (Figure 42). The globigerinid facies of the basinal Bou Dabbous Formation has the source potential within the Metlaoui play. Reservoirs occur not only in the nummulite banks, which have major potential, but also in lagoonal Nara-Kralif gastropod facies and in El Garia nummulithoclastic facies. Moldic solution of skeletal tests is the most common porosity type.

Further east, the Western Desert of Egypt previously held promise as an extension of the Sirte basin. The stratigraphic succession is very similar, with a clastic sequence from Paleozoic to Lower Cretaceous with intercalations of limestones, dolomites, and some evaporites. This succession was followed in the mid-Cretaceous by carbonates with interbedded shales; clastics and evaporites are minor constituents. A number of small-to-medium-size oil finds have been made in Cretaceous reservoirs.

The most important reservoir is the Aptian Alamein Dolo-

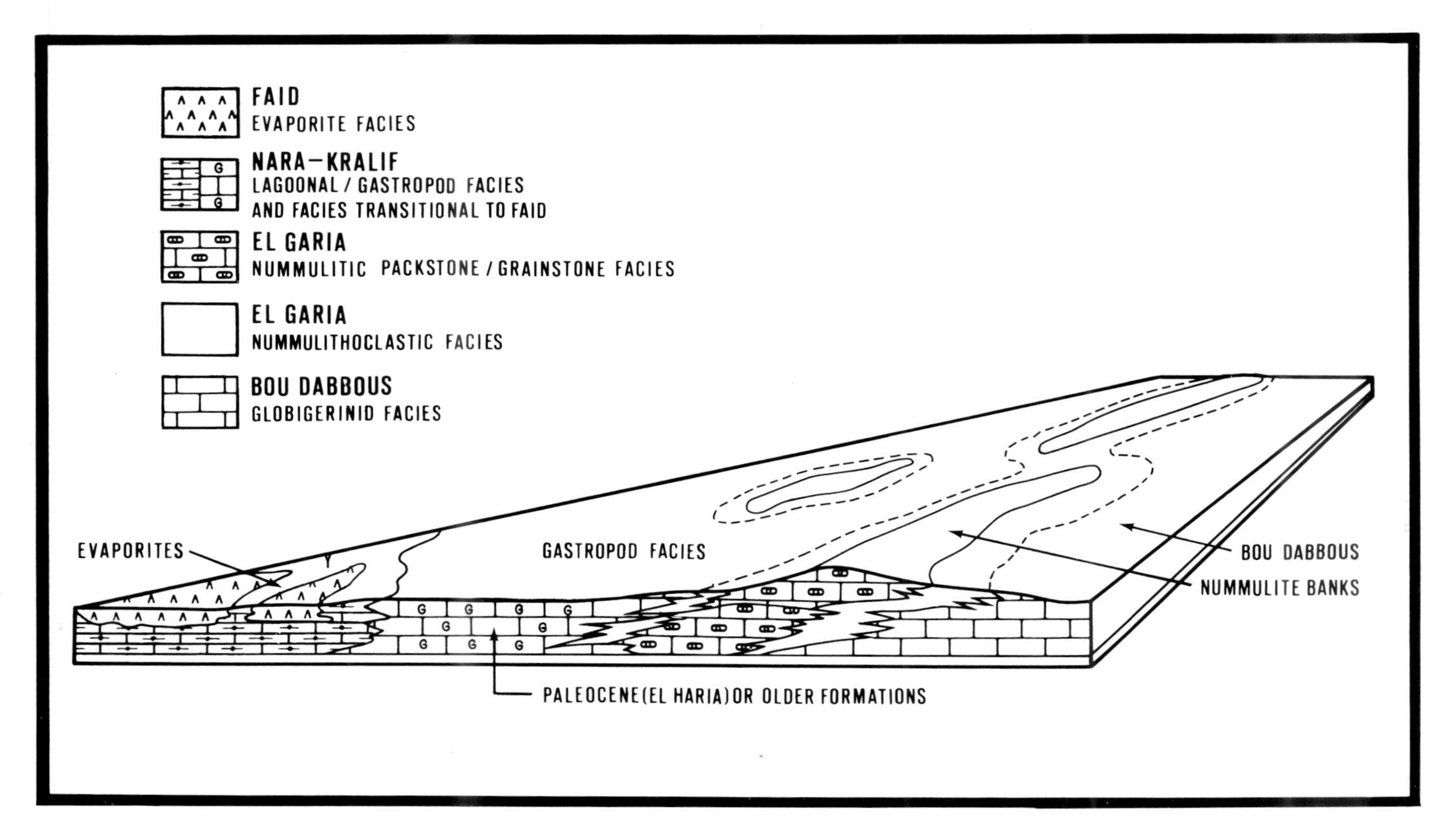


Figure 41. *Idealized diagram of nummulite banks and associated facies in Tunisia (after Arni, 1963; Moody, 1984).*

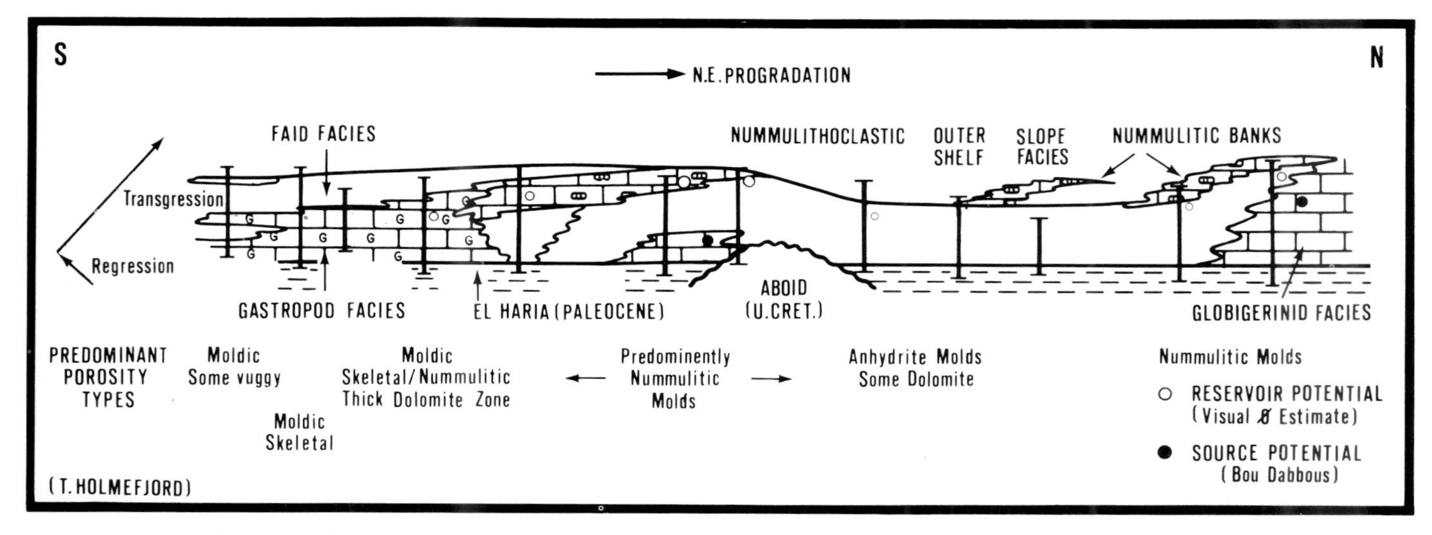


Figure 42. *Field traverse along north–south axis showing Eocene shelfal facies, onshore Tunisia.*

mite with sealing Dahab shale. Turonian Abu Roash sands and carbonates are also important reservoirs. Thick Khatatba shales of the Middle Jurassic, with TOC up to 4%, have source potential, as well as middle Cretaceous, Abu Roash shales.

Several subbasins in the Western Desert have been explored but the Abu Gharadig, in which the recent Badr el Din oil and Abu Gharadig gas/condensate fields are located, has resulted in the most success. Figure 43 illustrates the latter field, discovered in 1969; it is typical of the area. Oil reservoirs occur in Turonian Abu Roash sands in several zones. Some 40 million bbl oil (33–38° API) are estimated as recoverable.

Gas and condensate are present in Cenomanian sands deeper in the horst block (over 1 tcf recoverable). Other subbasins include the Alamein, which contains the Qatar-Alamein ridge on which the Alamein, Yidma, Razzak, and East Razzak fields were discovered, and the Matruh subbasin, in the northwest, which contains the Umbarka and Meleiha fields. Several shelf-edge plays hold potential, especially in these latter two subbasins.

A field-size distribution for the Western Desert indicates a $P_{50}$ of 23 million BOE for all fields, which, although small, are commercially attractive in this area. The reasons for the discovery of only small fields to date, and low success ratios, include poor reservoir continuity, poor source-rock development, and later wrench faulting that decreased the sizes of structures and affected structural timing. The sealing integrity of faults appears to be significant, and it is apparent that different oil–water and gas–water contacts in different fault blocks occur. In the Abu Gharadig field, fault blocks have different hydrocarbon-water contacts because faults and shales in adjacent blocks seal, suggesting sand lensing.

The Abu Gharadig structure is a complex of horst blocks, the main part of which trends northeast-to-southwest. All the structuring is dated from Santonian times, coincident with the inversion of the Atlas Gulf of Morocco and Algeria to create the present-day Atlas Mountains. The inversion of the latter interior fracture basin was caused by right-lateral wrench movements along the Atlas Transform Fault System. There is evidence, in fault patterns, reversal of throws on faults, and in fold trends, of the same conservative plate boundary existing beneath the Western Desert.

Wrench-modified basins represent 12% of African basins by number and 9% by area. However, 60% of them have production, although field sizes are rather small. Success rates in these basins are lower than average (10–23%), reflecting, to some degree, the complexity of such basins. They would probably be represented by high risk/low reward ventures in an exploration portfolio.

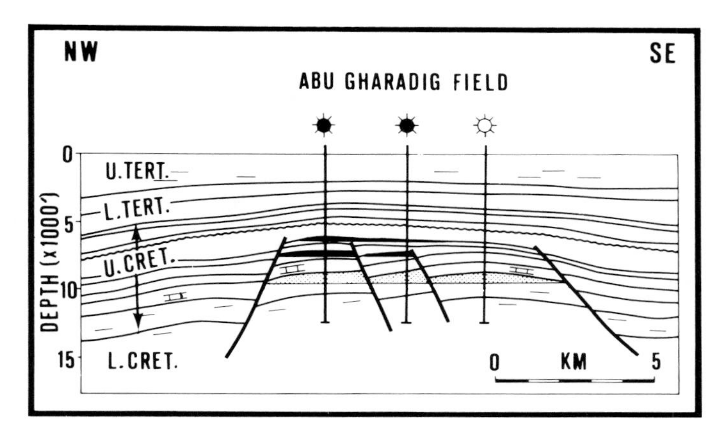


Figure 43. *Geologic cross section across the Abu Gharadig gas/condensate field in the Egyptian Western Desert.*

## Deltaic Sag Basins

The last category of basin type this paper describes is the Tertiary delta, or marginal sag/deltaic sag basin. The term "marginal sag/deltaic sag" comes from the fact that such deltas are always located on marginal sag basins. They contain topset, bottomset, and foreset beds prograding out and downwarping into ocean areas. Distinct facies of delta-top, delta-front, and prodelta can be recognized with both non-marine and marine clastics present. Delta-front sands, interfingering with prodelta shale source beds, are the most significant hydrocarbon-producing reservoirs, followed in importance by turbidite sands within the source itself. Field sizes tend to be small because traps relate to down-to-the-basin glide-plane faulting, or to salt or shale diapirism or both.

Examples of producing deltas outside of Africa are the Mississippi of the United States and the Mahakam of Indonesia. The two recognized marginal sag/deltaic sag basins in Africa are the Niger delta and the Nile delta basins, both having hydrocarbon production (the former oil and the latter gas).

The Niger delta is an excellent example of a producing delta with distinct delta-top (Benin Formation), delta-front (Agbada Formation), and prodelta (Akata Formation) facies development. Nearly all production is from delta-front sands interfingered with prodelta shales, the best source rock in the basin. As the proto-Niger delta prograded out into the Atlantic Ocean from the failed arm of a triple-junction, prodelta shales of the Akata Formation became unstable and began to flow. In distal parts of the delta, fields produce from the tops and flanks of shale diapirs with their associated growth faults. Turbidite sands produce in the downdip delta. Both Cameroon and Equatorial Guinea have portions of the Niger delta that contain hydrocarbons.

Growth faulting is very common in the Niger delta, with primary and secondary synthetic and antithetic faults developed (Figure 44). Typical migration paths of Akata Formation sourced hydrocarbons are shown, together with the complexity of traps with or without gas caps. The rollover anticlines are known as collapse crest structures.

Gas chromatograms of $C_{15}+$ saturates from two oil fields in the delta indicate the admixture of two hydrocarbon types. An example of indigenous hydrocarbons from woody-coaly, gas-prone organic matter with lesser amounts of herbaceous plant debris is shown in Figure 45. A very waxy, paraffin-base type of crude occurs, in distinct contrast to oils from sapropelic, oil-prone types of organic matter, largely restricted to burial depths of over 10,000 ft (3,050 m). There is a greater predominance, at depth, of the latter type hydrocarbons. Two components stand out on both figures: Component A ($nC_{25-33}$) represents humic, non-marine source material in the extract, while component B ($nC_{18-23}$) represents a marine source material. Therefore, indigenous, non-marine oil appears at Gbokoda in Figure 45 and migrated marine oil appears at Meren in Figure 46.

A field size distribution of 138 Niger delta fields in Figure 47 shows an excellent straight-line best-fit. All production is from the Agbada delta-front sands, in traps similar to that shown in Figure 44. A $P_{50}$ of 72 million BOE is derived, with fields rarely achieving giant status. This designation is probably more a function of the complexity of structuring than any other factor.

Several gas discoveries have been made in the Nile delta of Egypt since the Abu Madi discovery in 1967. This field, together with Abu Qir discovered two years later and NAF discovered in 1978, all have a basal Pliocene sand reservoir called the Abu Madi Formation. Field sizes are larger than 1 tcf of gas, which is largely methane, plus some condensate. This deltaic basin

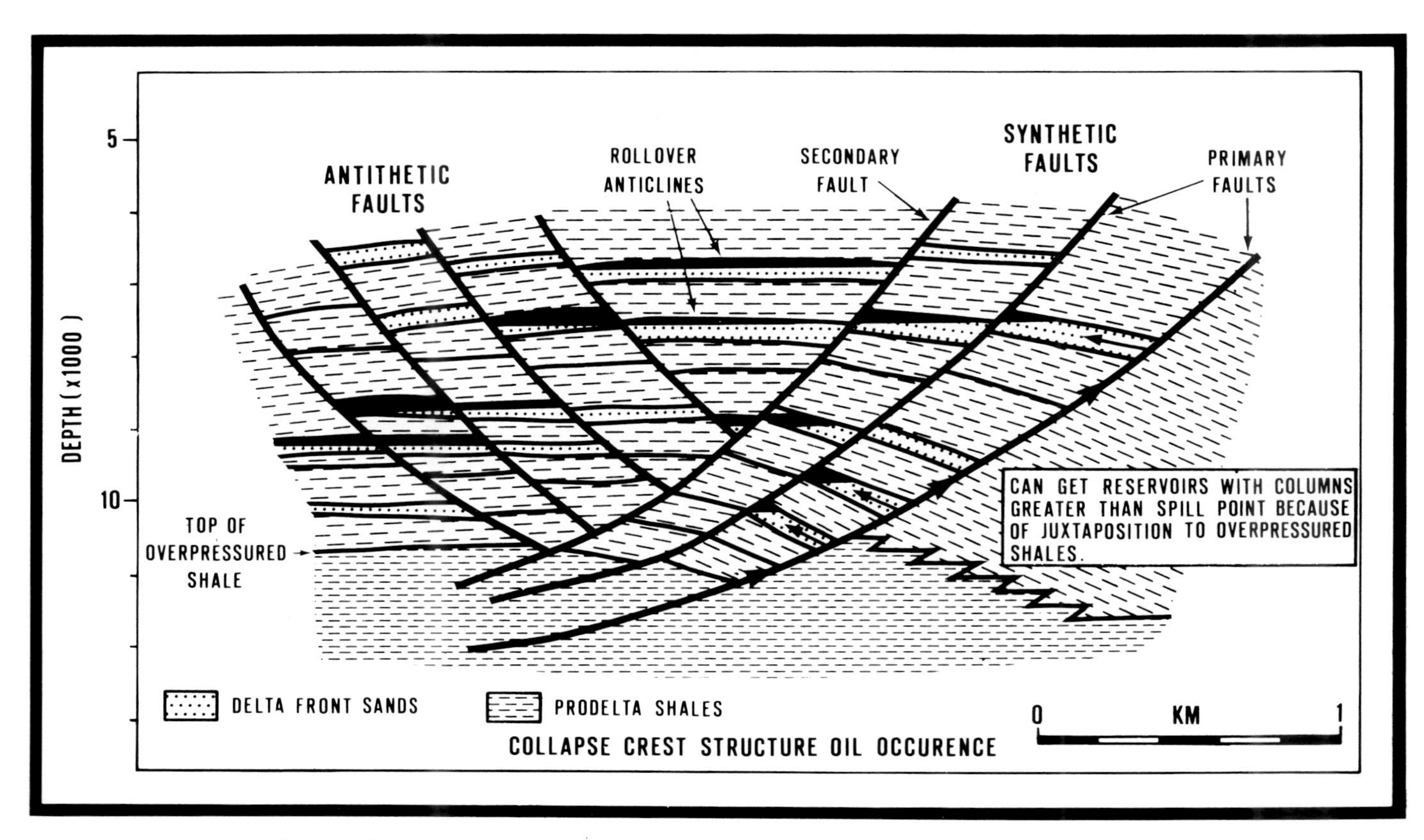


Figure 44. *A typical oil occurrence associated with growth-faulting in the Niger delta.*

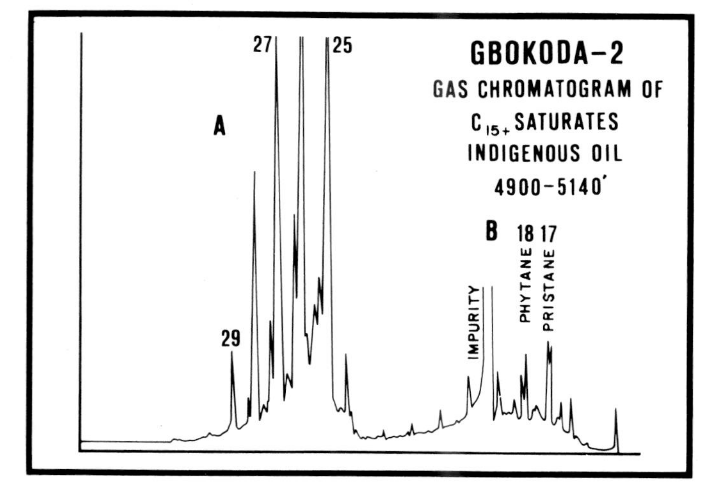


Figure 45. *Gas chromatogram of $C_{15}+$ saturates, Gbokoda-2 well, Niger delta, Nigeria. The very large spikes between $C_{24}$ and $C_{29}$ together with the pristane/phytane ratio indicate non-marine source material. Immaturity is suggested by a $C_{22-30}$ sterane and the triterpane bulge in the naphthene envelope (after Fisher, 1979).*

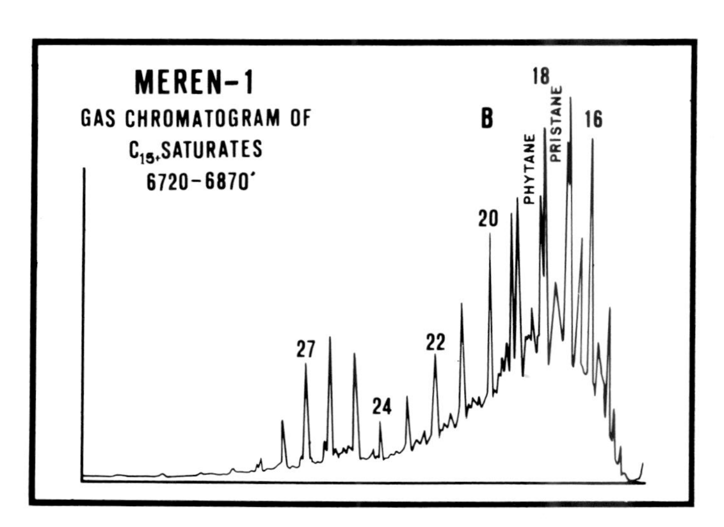


Figure 46. *Gas chromatogram of $C_{15}+$ saturates, Meren-1 well, Niger delta, Nigeria. The large naphthene envelope suggests more maturity than at Gbokoda (see Figure 45). The $C_{18-23}$ n-alkane predominance and pristane/phytane ratio indicate marine source material (after Fisher, 1979).*

holds considerable potential for future gas discoveries.

These delta basins, although covering only 2% of Africa's sedimentary area, are very important for hydrocarbon production. They have proved to be, worldwide, one of the most enticing of all basin types because of the interplay of deltaic sands, prodelta shales, and very rapid subsidence, accompanied by listric faulting and shale diapirism. However, no new major deltas remain to be explored in Africa that would be sites of major oil accumulations.

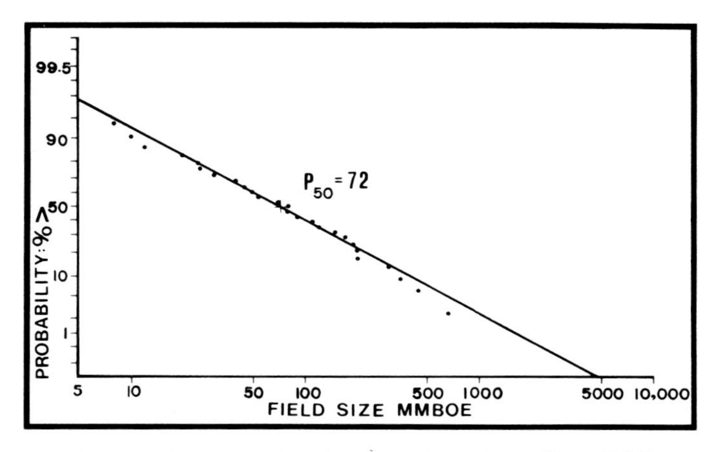


Figure 47. *Field-size distribution (given in million BOE, or MMBOE) of Agbada delta-front sand play, Niger delta, Nigeria.*

## OVERVIEW

Generally, in Africa as in the rest of the world, commercial attractiveness is directly proportional to field-size. However, Figure 48 illustrates that there are a few exceptions to this rule, reflecting more onerous state-take with increasing field size in some countries. More attractive fiscal terms are obviously available from non-producing countries, while several countries are notable exceptions and have attractive opportunities as well as significant exploration potential, despite having established production.

Twenty-one basins from around the world are plotted on Figure 49 for comparison of success ratios. Among these are 12 African basins, of which four are interior fracture, three are

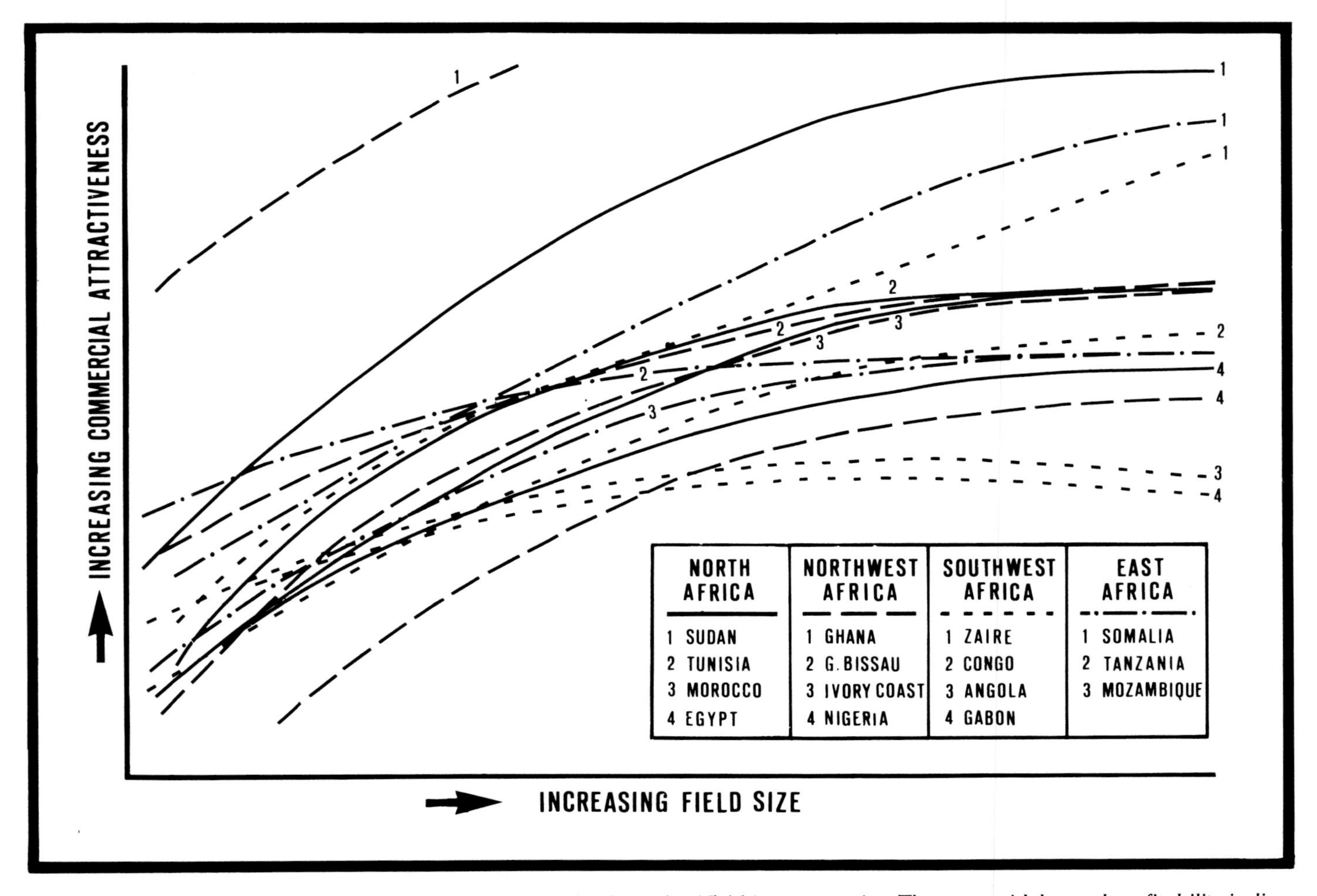


Figure 48. *Measure of commercial attractiveness versus field-size for 15 African countries. The most widely-used profitability indicators for petroleum venture assessment are discounted cash flow rate of return, net present value, profit/investment ratio, and profit/risk ratio.*

marginal sags, three are wrench-modified, and one each is marginal sag/deltaic sag and interior sag. As mentioned earlier, this method of using success ratios has pitfalls and, in fact, a few anomalies are included. For example, the Niger delta has a low threshold of commercial value; that is, fields with recoverable reserves of only 1 million bbl oil can be put on stream because of Nigeria's infrastructure and because the deposits are onshore and close to markets. The Tanzanian Coastal basin also has had rather high success rates for gas discoveries after only little exploration.

The interior fracture basins appear as consistently the most successful on this chart, with the wrench-modified basins much less successful. The interior sag basins of Africa are barely explored so conclusions about their chances of success cannot be made.

Finally, Figure 50 details the exploration maturity of individual African sedimentary basins as defined earlier in the paper. Significantly, 37% of all the basins are untested and 45% are immature as far as oil exploration is concerned. Together, these add up to 82%, leaving 9% of all basins as semi-mature, 7% as mature, and only 2% as very mature. The mature basins are in the Algerian Sahara, the Illizi basin (also in Algeria), the Western Desert, and the Gabon and Congo subbasins. The very mature basins are the Niger delta and the Gulf of Suez. Surprisingly, perhaps, the Sirte and the Ghadames basins of northern Africa are only semi-mature.

## SUMMARY

The author hopes that this paper will serve as a current report on the African oil situation and that some of the conclusions and facts presented will serve as a stimulus for future exploration of this huge, largely-unexplored continent.

Political uncertainties, together with logistical problems, have discouraged the higher levels of exploration characteristic of other, more attractive, parts of the world. In some African countries, nationalization or increased levels of state participation have supported such prejudices and, together with heavy state-takes, have deterred many companies. But some countries are learning that compromise by both parties, the country and the oil company, is necessary for concessions to be taken and wells drilled. The increasing involvment of developmental agencies, such as the World Bank, has occurred over the past decade, and that helps to bring the two sides together with mutual understanding.

African basins such as the interior sag basins of Algeria, Mali, Niger, and Zaire, have had little exploration, yet hold great promise for larger fields, while the rift basins of central and southern Africa and the coastal marginal sag basins contain fields of more modest size. How commercially attractive such small or moderate-size fields are depends upon the fiscal terms offered by, or negotiated with, the host government. Also to be considered, naturally, are the locations of any hydrocarbons discovered and their closeness or access to markets. Within one particular basin, one country may offer more attractive terms than another, but the prospectiveness of that part of the basin may be less. This situation leads to the value of applying basin, play, and prospect assessment procedures, linked with risked geologic and economic evaluations, in order to achieve a fully-balanced exploration portfolio containing ventures ranging from high risk/high reward to low risk/low reward.

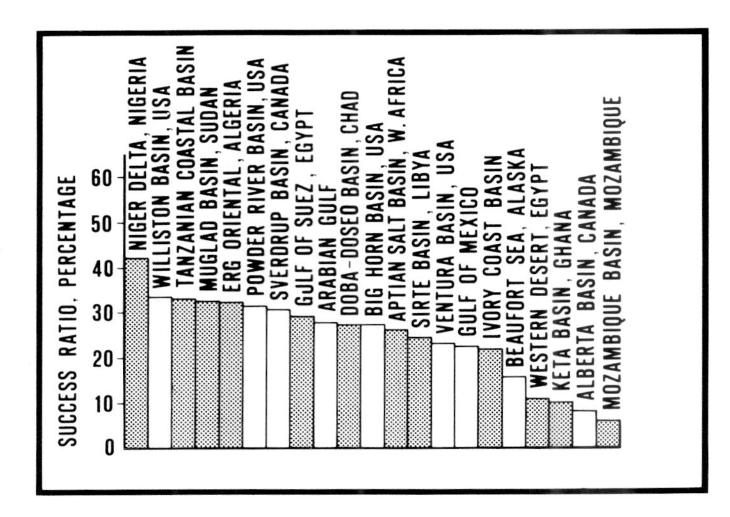


Figure 49. *Comparison of African success ratios with success ratios for the rest of the world (for definition of success ratio see text).*

## ACKNOWLEDGMENTS

I wish to express my thanks to the management of both Kuwait Foreign Petroleum and BHP for permission to publish this paper, to colleagues for their ideas and suggestions, and particularly to Andy Pearson for editing the manuscript. However, the views expressed in this article are not necessarily those of Kuwait Foreign Petroleum or BHP and are largely my own.

This collation of data and ideas would have been impossible without the wealth of published and unpublished information that I have drawn upon, only some of which I have been able to reference.

I am particularly grateful to Christine Srnecz for her drafting and to Dawn Longhurst for her typing.

## SELECTED REFERENCES

Aadland, A. J., and A. A. Hassan, 1972, Hydrocarbon potential of the Abu Gharadig basin in the Western Desert, Arab Republic of Egypt: Proceedings of the 8th Arab Petroleum Congress, Algiers, 19 p.

Brice, S. E., and G. Pardo, 1980, Hydrocarbon occurrences in non-marine, pre-salt sequence of Cabinda, Angola: Presented at AAPG Convention, Denver, 1980.

Brink, A. H., 1974, Petroleum geology of Gabon basin: AAPG Bulletin, v. 58, p. 216–235.

Brognon, G. P., and G. R. Verrier, 1966, Oil and geology in Cuanza basin of Angola: AAPG Bulletin, v. 50, p. 108–158.

Brown, R. N., 1978, History of exploration and discovery of Morgan, Ramadan, and July oilfields, Gulf of Suez, Egypt: Presented at OAPEC Conference, Kuwait (Egypt Paper), 1978.

Conant, L. C., and G. H. Goudarzi, 1967, Stratigraphic and tectonic framework of Libya: AAPG Bulletin, v. 51, p. 719–730.

Decrouez, D., and E. Lanterno, 1979, Les "Bancs a Nummulites" de l'Eocene Mesogeen et leurs implications: Arch. Sc. Geneve, v. 32, p. 67–94.

Egbogah, E. O., and D. O. Lambert-Aikhionbare, 1980, Possible new oil potentials of the Niger Delta: Oil and Gas Journal,

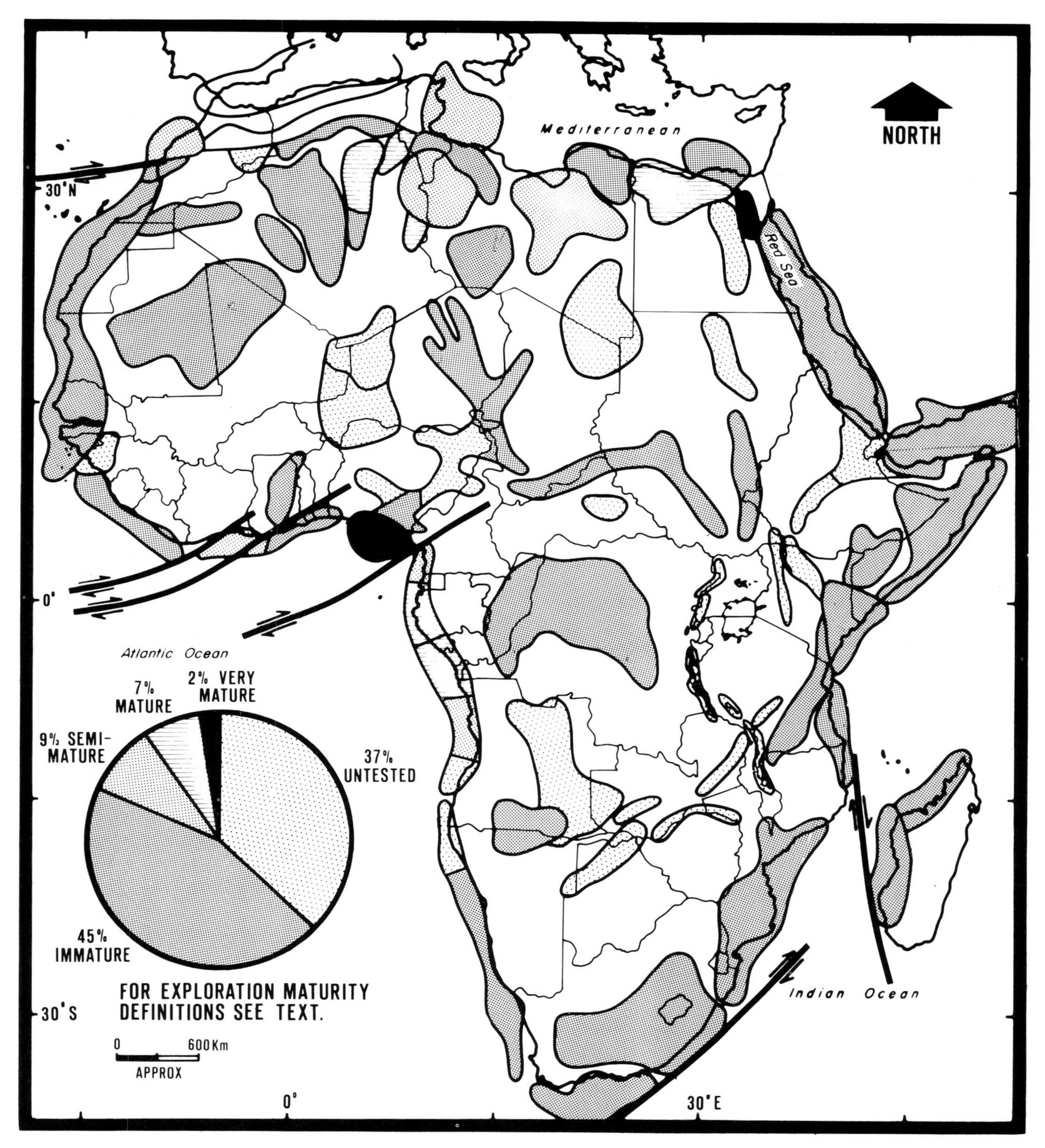


Figure 50. *Exploration maturity of African basins (for definitions of maturity levels see text).*

April 14, 1980.

Evamy, D. D., et al, 1978, Hydrocarbon habitat of Tertiary Niger Delta: AAPG Bulletin, v. 62, p. 1–39.

Fisher, A., 1979, Petroleum source-rock studies of selected wells, Niger Delta Complex, West Nigeria: PhD Thesis, Imperial College, London. 357 p.

Flores, G., 1973, The Cretaceous and Tertiary sedimentary basins of Mozambique and Zululand: Association of African

Geological Surveys, Paris, p. 81–111.
Kamen-Kaye, M., 1982, Mozambique-Madagascar geosyncline, 1: Deposition and architecture: Journal of Petroleum Geology, v. 5, p. 3–30.
Kamen-Kaye, M., and S. U. Barnes, 1979, Exploration geology of northeastern Africa-Seychelles basin: Journal of Petroleum Geology, v. 2, p. 23–45.
Kingston, D. R., C. P. Dishroon, and P. A. Williams, 1983, Global basin classification system: AAPG Bulletin, v. 67, p. 2175–2193.
Klemme, H. D., 1975, Geothermal gradients, heat flow and hydrocarbon recovery, *in* A. G. Fischer and S. Judson, eds, Petroleum and Global Tectonics: Princeton, New Jersey, Princeton Univ. Press.
Magloire, P. R., 1968, Triassic gas field of Hassi er R'Mel, Algeria: AAPG Memoir 14 (1970), p. 489–501.
Momper, J. A., 1982, The Etosha basin reexamined: Oil and Gas Journal, April 5, 1982.
Moody, R. T. J., 1984, Report on the fieldwork programme carried out on behalf of the Kuwait Foreign Petroleum Company in Central and Southern Tunisia: Kingston Earth Science Resources Centre, unpubl. rpt.
Parker, J. R., 1982, Hydrocarbon habitat of the Western Desert, Egypt: Presented at EGPC Sixth Exploration Seminar, Cairo, 1982.
Parsons, M. G., A. M. Zagaar, and J. J. Curry, 1979, Hydrocarbon occurrences in the Sirte basin, Libya, *in* A. Miall, ed., Facts and principles of world petroleum occurrence: Can. Soc. Petr. Geol. Memoir 6, p. 723–732.
Radelli, L., 1975, Geology and oil of Sakamena basin, Malagasy Republic (Madagascar): AAPG Bulletin, v. 59, p. 97–114.
Rizzini, A., et al, 1976, Stratigraphy and sedimentation of Neogene-Quaternary section in the Nile delta area: Presented at EGPC Fifth Exploration Seminar, Cairo, 1976.
Rowsell, D. M., and A. M. J. de Swardt, 1976, Diagenesis in Cape and Karroo sediments, South Africa, and its bearing on their hydrocarbon potential: Trans. Geol. Soc. S. Afr., v. 79. p. 81–145.
Sonatrach, 1979, Geology of Algeria—The hydrocarbon bearing provinces, *in* Schlumberger Well Evaluation Conference 1979, Algeria.
Villemur, J. R., 1967, Reconnaissance geologique et structurale du nord du bassin de Taoudeni: Mem. Bur. Rech. Geol. Min., No. 51, 151 p.

# Stratigraphy and Hydrocarbon Potential of the Paleozoic Succession in Both Tabuk and Widyan Basins, Arabia

Abdulaziz Abdullah al-Laboun
*ARAMCO*
*Dhahran, Saudi Arabia*

The Tabuk and Widyan basins of the greater Arabian basin contain more than 5,000 m (16,400 ft) of oil- and gas-prospective clastic and carbonate Paleozoic sediments. The Ha'il-Rutbah Arch divides the region into two main basins: the Tabuk basin to the west and the Widyan basin to the east. The Tabuk basin is filled with Paleozoic sediments of Cambrian to Devonian age, whereas the Widyan basin contains sediments ranging in age from Cambrian through Late Permian. The north to south trending basement rock of the Summan Platform separates the Widyan basin from the eastern sector of the Arabian basin.

The significant factor relating to oil and gas prospects in both the Tabuk and Widyan basins is the presence of thick potential source rocks, alternating with good reservoir rocks. The sediments show lateral facies variations, with sandstone grading into shale basinward. The facies changes, combined with several unconformities, folding and faulting, enhance the hydrocarbon potential of the two basins. The pre-Unayzah unconformity is an important regional, angular unconformity where the clastics transgress subcrops of various possible source rocks, particularly the shale and carbonate members of the Tabuk and Jauf Formations. Also, known shows and recoveries of oil and gas from the Paleozoic section in the eastern part of the Arabian basin further support the possibility of the presence of commercial quantities of hydrocarbons in these basins.

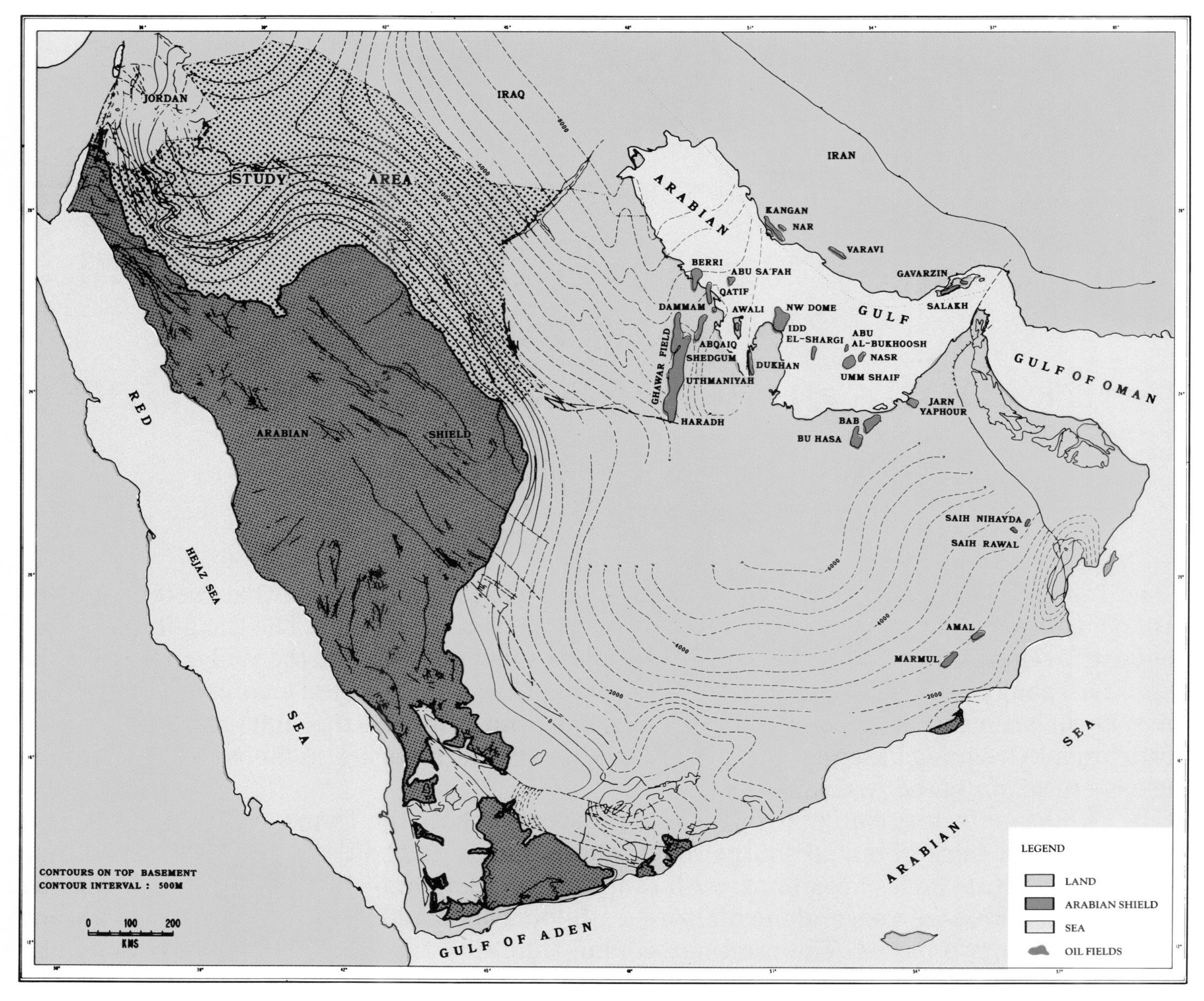


Figure 1. *The location map of study area and some of the Paleozoic oil and gas fields of the Arabian basin. Contours and top basement after Brown (1972).*

## INTRODUCTION

The current hydrocarbon discoveries in the Arabian basin are predominantly from north, northeast, and northwest trending anticlines and dome-shaped structures of the foreland shelf and the orogenic belt. These structures are mainly controlled by earlier structures of the Precambrian and Paleozoic orogenies. The sedimentation in the basin was controlled by slow, progressive subsidence of the Arabian foreland, which was interrupted intermittently by epeirogenic movements, movements that are reflected in thickness variations, facies changes, and sedimentational breaks.

Hydrocarbon production, (both oil and gas) from deep, Paleozoic reservoirs in the eastern sector of the Arabian basins (Arabia, Iran, Oman, Qatar, the United Arab Emirates, and Bahrain), has stimulated interest in both the Tabuk and the Widyan basins (Figure 1). Most of the Paleozoic hydrocarbon production to date is from the Khuff Carbonates of Late Permian age, although lesser amounts have been discovered in older Paleozoic reservoirs.

Little data have been released on the Paleozoic in the subsurface of the Arabian basin. This paper synthesizes available data from wells drilled in both the Tabuk and Widyan basins, for a regional understanding of this part of the basin's geology. Two regional, structural cross sections, one stratigraphic cross section, and a series of isopach maps compile both the surface and subsurface data; but these are somewhat generalized due to the lack of control. A paleogeology map shows that the relationships of the pre-Unayzah Paleozoic rocks in both basins has also been constructed.

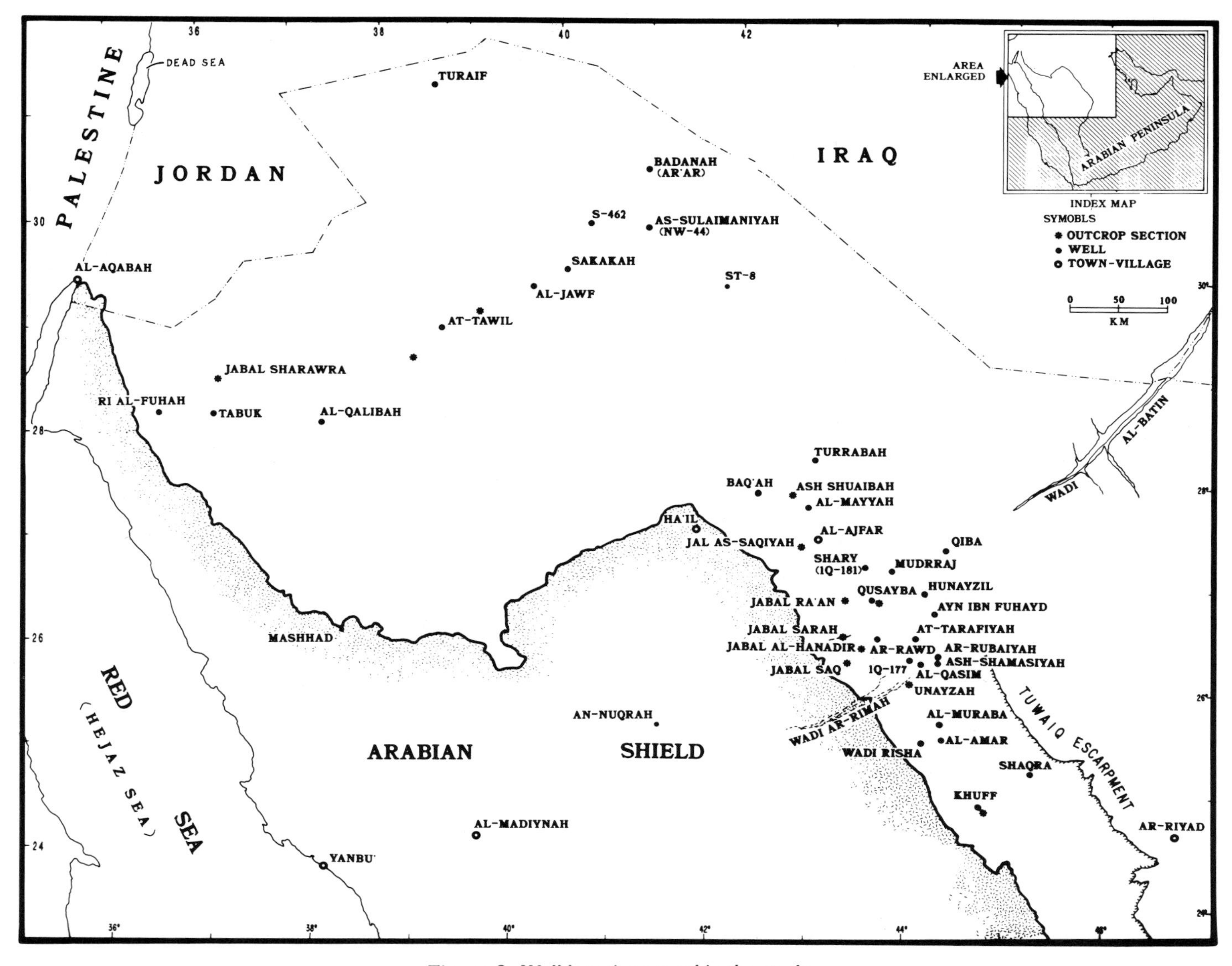


Figure 2. *Well locations used in the study.*

Within the past twenty years, scores of shallow and deep wells have been drilled into Paleozoic strata in the study area, to obtain water for farming, household and industrial use (Figure 2). These wells provide abundant information including cores, cuttings samples, mechanical logs, and palynological as well as paleontological control. Interpretation of these borehole data demonstrate the lateral continuity of the Paleozoic rocks in the subsurface.

The Paleozoic sedimentary rocks are exposed as a great curved belt around the Arabian shield in northwestern Arabia (Figure 3). These rocks generally dip to the east, north, and northeast, to disappear in the subsurface. The successive marine cycles intertongue with the wedge-shaped continental sands surrounding the shield. The generalized, stratigraphic column (Figure 4) represents a compilation of the best penetrated and exposed Paleozoic sedimentary rocks in central and northwestern Arabia.

## Location

The greater Arabian basin is bounded by the Arabian shield and the Dead Sea Rift on the west, the southeastern Turkey overthrust on the north, the Zagros Fold Belt and the Oman Mountains to the east and the Hadramut and the Huqf Archs on the south.

The area under study covers al-Qasim, al-Jawf, Turaif, and Tabuk regions (Figures 1 and 2). In general the area is a northwest-to-southeast-trending belt extending from the Tuwaiq Escarpment west of ar-Riyad to the border with Iraq on the northeast and both the Arabian shield and Jordan on the west and northwest. The study area is bounded by lat. 24° and 32° N and long. 46° and 35° E.

## Scope of Present Study

The present study identifies and maps the Paleozoic sedimentary rocks in the subsurface of the Tabuk and Widyan basins. Further, the study establishes a type-section for the Carboniferous–Permian clastics exposed subjacent to the Khuff escarpment in al-Qasim. In addition, the study defines the clastic section bracketed between the basal shale unit of the Unayzah Formation and the carbonates of the Hammamiyat

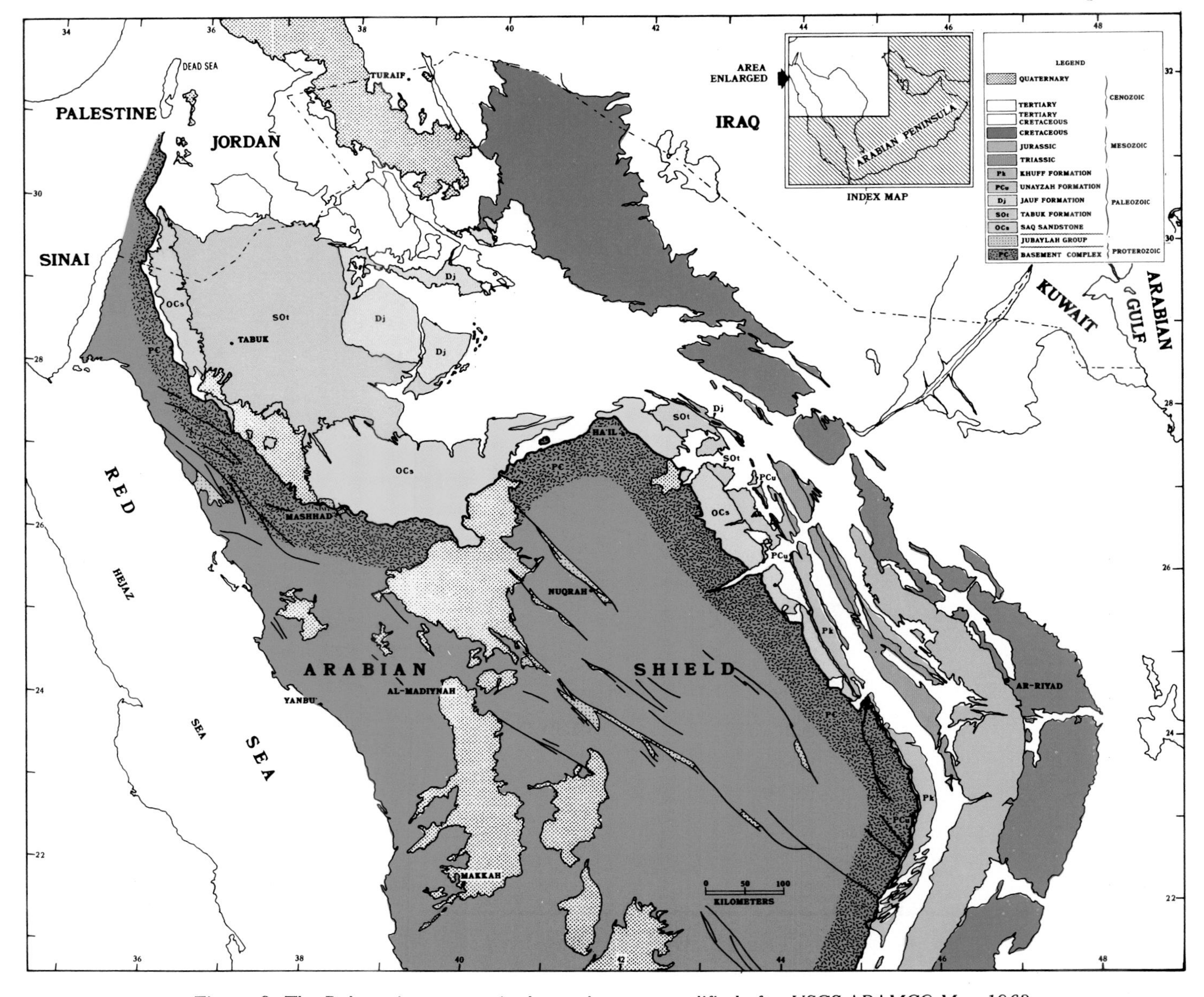


Figure 3. *The Paleozoic outcrops in the study area, modified after USGS-ARAMCO Map 1963.*

Limestone Member of the Jauf Formation in the northern part of the Widyan basin. Finally, the study investigates the hydrocarbon potential of these Paleozoic sediments.

## Geological Setting

The study area is bounded on the south, southwest, and west by the Arabian shield (Figure 5). On the east, it is bordered by both Wadi al-Batin and the Jurassic Tuwaiq Escarpment. Immediately behind this escarpment is the Central Arabian graben and trough system. On the north and northeast, it is bounded by the plains of Jordan and Iraq. The Ha'il-Rutbah Arch separates the study area into the Tabuk basin on the west and the Widyan basin on the east. In the northern and northeastern parts of the Widyan basin, Mesozoic rocks overlap the Paleozoic succession and much of the area is covered by Quaternary eolian sands commonly known as an-Nafud.

The Tabuk basin is outlined by the outcrop of the Saq Sandstone to the west and south and is bordered on the east by the Ha'il-Rutbah Arch. The basin axis trends north-northwest. Within the Tabuk basin, rocks of Cambrian to Devonian age crop out away from the westward and southward positioned basement, with the youngest being in the central portion of the basin.

The axis of the Widyan basin trends in a north-northwest to south-southeast direction. The basin is bounded on the west by the low-lying shield area and the Ha'il-Rutbah Arch, and on the east by postulated basement rocks of the Summan Platform (beneath the northern Tuwaiq segment of the Interior Homocline) as well as by Wadi al-Batin. Paleozoic rocks ranging in age from Cambrian to Permian crop out and subcrop in an offlap manner away from the Arabian shield.

## Previous Work

First reports of Paleozoic rocks in Arabia were documented

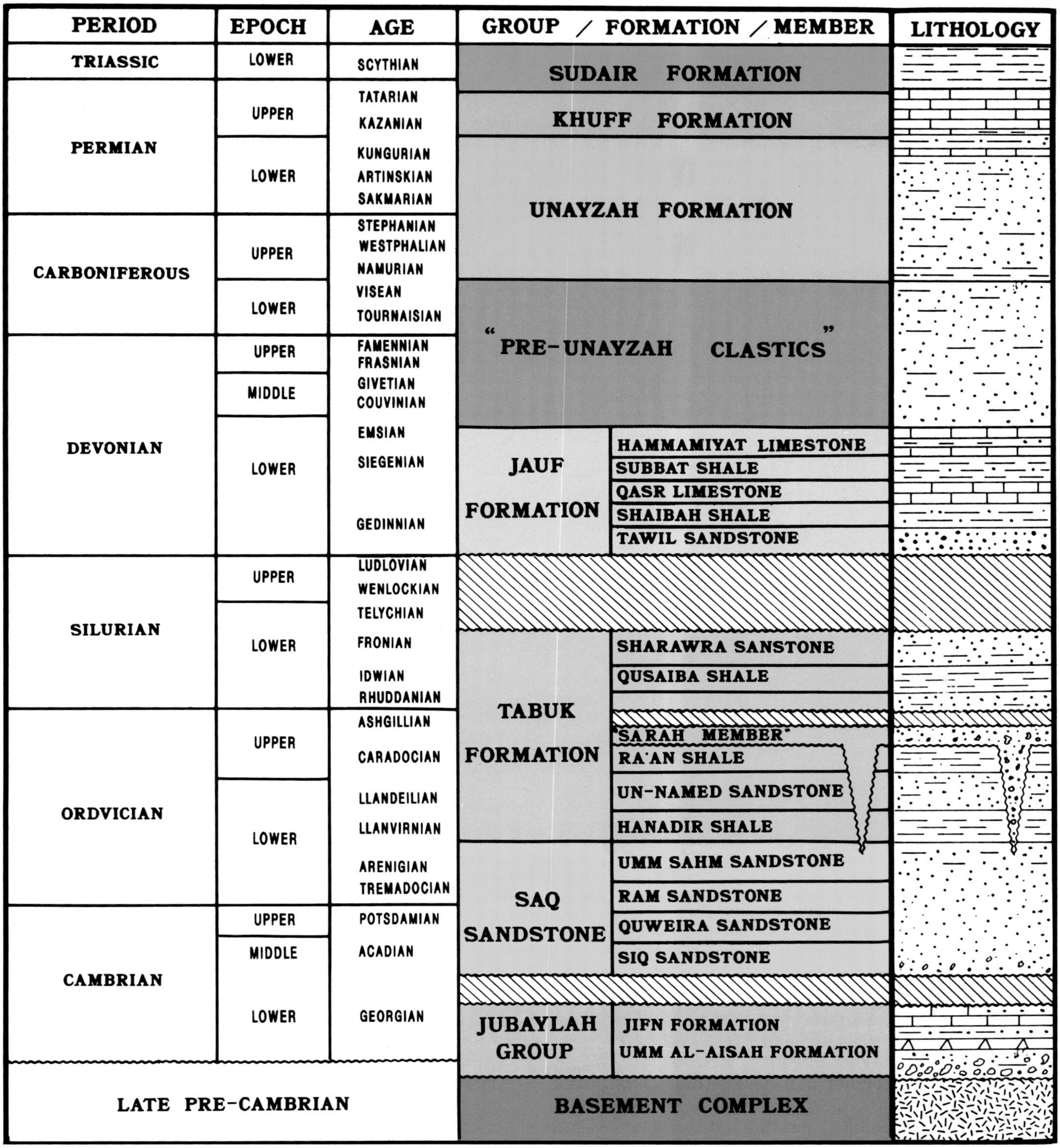

| PERIOD | EPOCH | AGE | GROUP / FORMATION | MEMBER |
|---|---|---|---|---|
| TRIASSIC | LOWER | SCYTHIAN | SUDAIR FORMATION | |
| PERMIAN | UPPER | TATARIAN<br>KAZANIAN | KHUFF FORMATION | |
| PERMIAN | LOWER | KUNGURIAN<br>ARTINSKIAN<br>SAKMARIAN | UNAYZAH FORMATION | |
| CARBONIFEROUS | UPPER | STEPHANIAN<br>WESTPHALIAN<br>NAMURIAN | UNAYZAH FORMATION | |
| CARBONIFEROUS | LOWER | VISEAN<br>TOURNAISIAN | "PRE-UNAYZAH CLASTICS" | |
| DEVONIAN | UPPER | FAMENNIAN<br>FRASNIAN | "PRE-UNAYZAH CLASTICS" | |
| DEVONIAN | MIDDLE | GIVETIAN<br>COUVINIAN | "PRE-UNAYZAH CLASTICS" | |
| DEVONIAN | LOWER | EMSIAN<br>SIEGENIAN<br>GEDINNIAN | JAUF FORMATION | HAMMAMIYAT LIMESTONE<br>SUBBAT SHALE<br>QASR LIMESTONE<br>SHAIBAH SHALE<br>TAWIL SANDSTONE |
| SILURIAN | UPPER | LUDLOVIAN<br>WENLOCKIAN | | |
| SILURIAN | LOWER | TELYCHIAN<br>FRONIAN<br>IDWIAN<br>RHUDDANIAN | TABUK FORMATION | SHARAWRA SANSTONE<br>QUSAIBA SHALE |
| ORDVICIAN | UPPER | ASHGILLIAN<br>CARADOCIAN | TABUK FORMATION | "SARAH MEMBER"<br>RA'AN SHALE |
| ORDVICIAN | LOWER | LLANDEILIAN<br>LLANVIRNIAN | TABUK FORMATION | UN-NAMED SANDSTONE<br>HANADIR SHALE |
| ORDVICIAN | LOWER | ARENIGIAN<br>TREMADOCIAN | SAQ SANDSTONE | UMM SAHM SANDSTONE<br>RAM SANDSTONE |
| CAMBRIAN | UPPER | POTSDAMIAN | SAQ SANDSTONE | QUWEIRA SANDSTONE |
| CAMBRIAN | MIDDLE | ACADIAN | SAQ SANDSTONE | SIQ SANDSTONE |
| CAMBRIAN | LOWER | GEORGIAN | JUBAYLAH GROUP | JIFN FORMATION<br>UMM AL-AISAH FORMATION |
| LATE PRE-CAMBRIAN | | | BASEMENT COMPLEX | |

Figure 4. *Generalized Paleozoic stratigraphic column of the Tabuk and Widyan basins.*

by Pascha (1908), Blanckenhorn (1914), and Kober (1919). They conducted reconnaissance field studies while Arabia was part of the Othman Caliphate, neither preparing detailed maps nor establishing formation names. Later on, both Abel (1935) and Picard (1937, 1941) described Paleozoic fossils from Arabia.

Intensified interest in the geology of Arabia followed granting of an oil concession (May 29, 1933) to the California Arabian

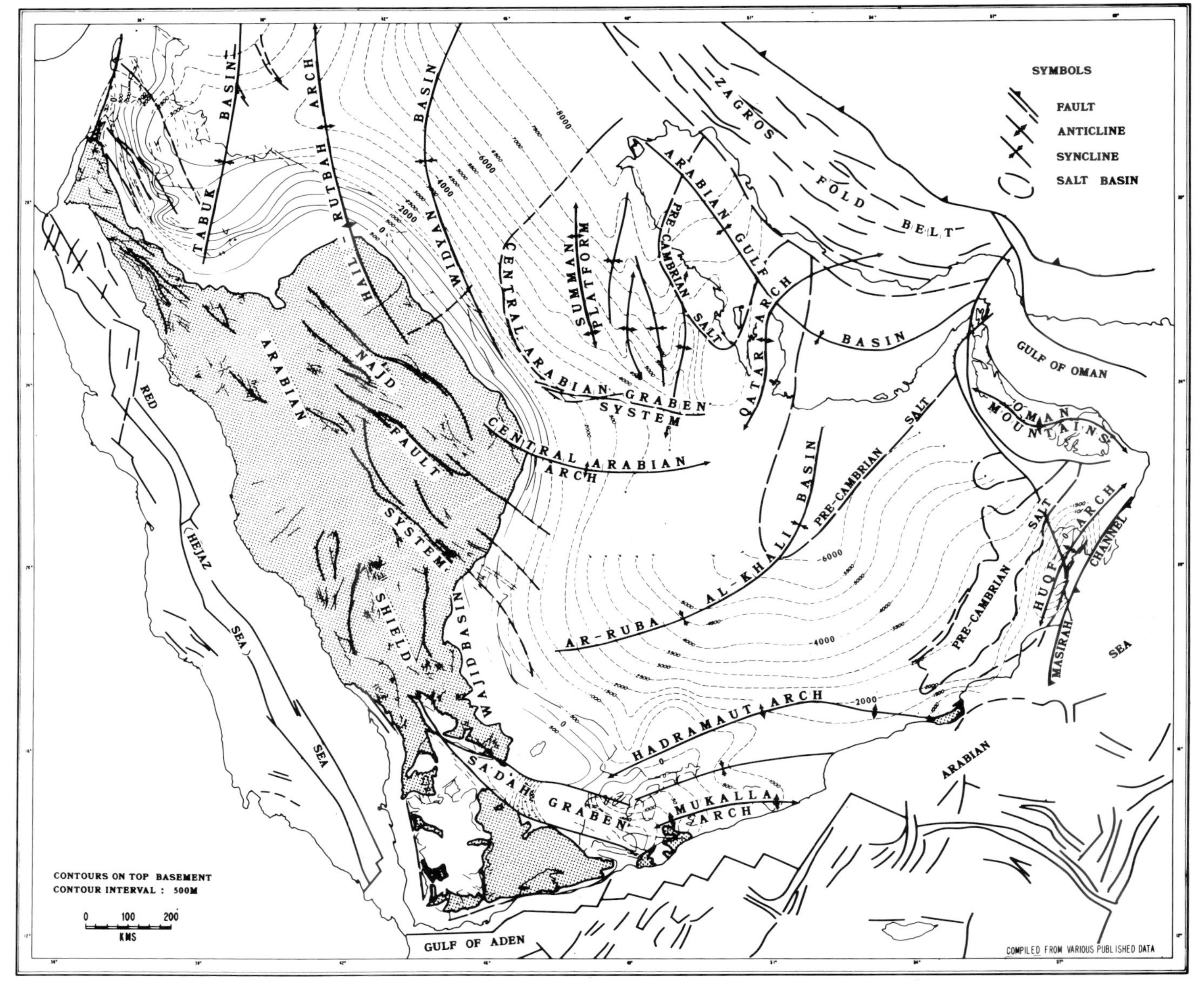


Figure 5. *Regional structural elements of the Arabian Peninsula, compiled from various published data.*

Standard Oil Company (CASCO), which later became the Arabian American Oil Company (ARAMCO). In 1934 the first field party penetrated the interior of Arabia. Certain findings of these field parties were published by Thralls and Hasson (1956), Steineke, Bramkamp, and Sanders (1958), and Bramkamp (1964). Powers et al (1966) and Powers (1968) summarized ARAMCO's work.

Helal (1964a, 1964b, 1965, and 1968), Hemer (1965) Hemer and Nygreen (1967a, 1967b), Delfour (1967 and 1970), McClure (1968), Bigot and Lafoy (1970), Bigot (1970), Brown (1972), Hadley (1974), Thomas (1977), Cloud, Awramik and Hadley (1979), Clark-Lowes (1980), El-Khayal, Chaloner, and Hill (1980), Bahafzallah, Jux and Omara (1981a and 1981b), Sharief (1982) Delfour et al (1982), Vaslet, Brosse, and LeHindre (1982), and Basahel et al (1984) contributed valuable information on the Paleozoic in central and northwestern Arabia.

Figure 6 summarizes the stratigraphic nomenclature of the Paleozoic rocks in the outcrops and subsurface of Arabia.

## Stratigraphy

The generalized stratigraphic column (Figure 4) summarizes both the best outcrop and subsurface Paleozoic sections in the two studied basins (Tabuk and Widyan basins). The time-stratigraphic assignment is based on both paleontologic evidence and stratigraphic position. Rock units are picked on biofacies and lithofacies changes, which usually fit wireline log configuration. In this regard, Powers (1968) states: "One interesting aspect of Arabian stratigraphy has been the fact that most Arabian rock units, originally defined solely on the basis of the criteria for rock-stratigraphic units, have proved over the years to approximate time-stratigraphic units as well. This is the case both along the length of outcrop and in the subsurface."

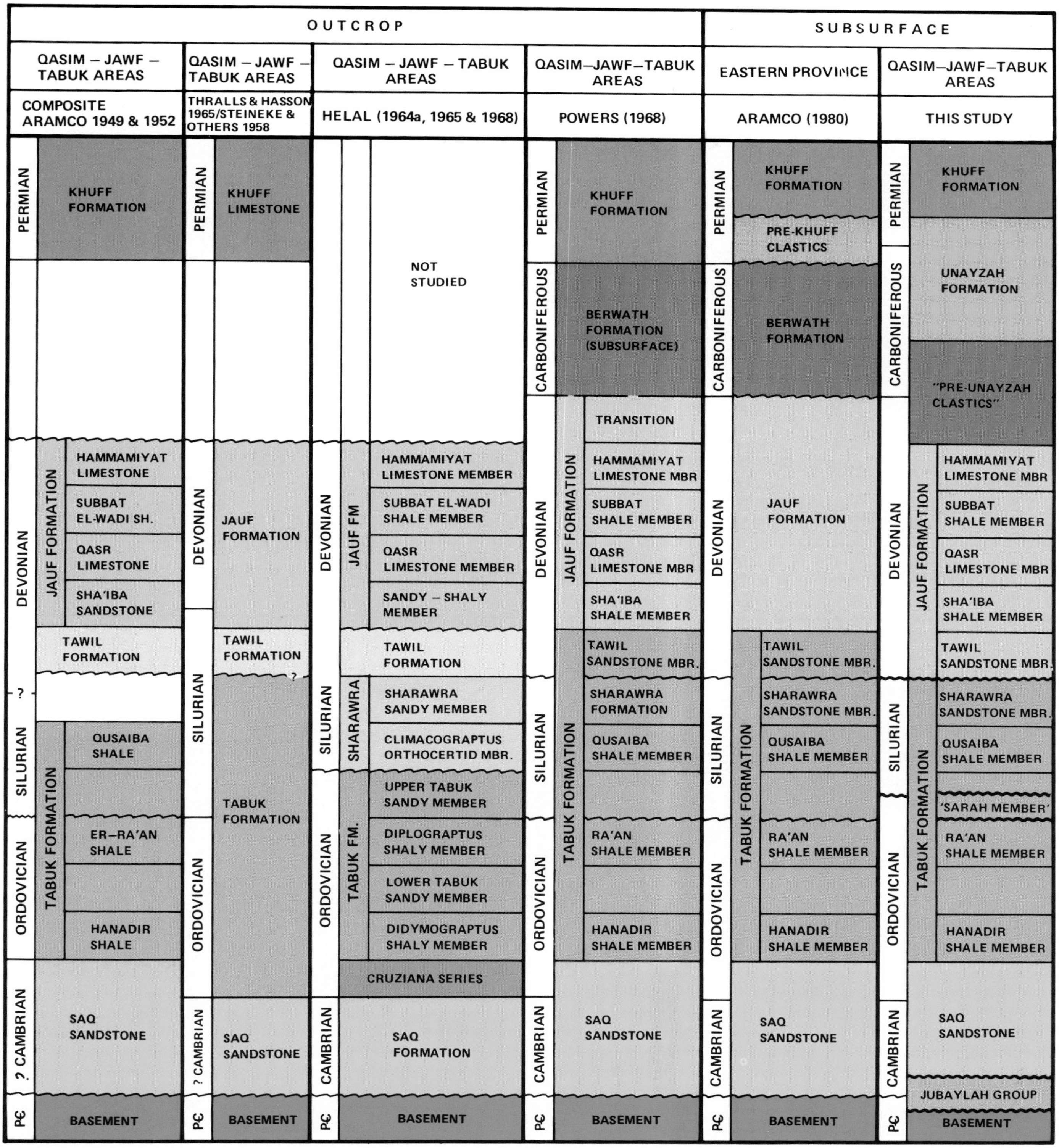

| OUTCROP | | | | SUBSURFACE | |
|---|---|---|---|---|---|
| QASIM – JAWF – TABUK AREAS | QASIM – JAWF – TABUK AREAS | QASIM – JAWF – TABUK AREAS | QASIM–JAWF–TABUK AREAS | EASTERN PROVINCE | QASIM–JAWF–TABUK AREAS |
| COMPOSITE ARAMCO 1949 & 1952 | THRALLS & HASSON 1965/STEINEKE & OTHERS 1958 | HELAL (1964a, 1965 & 1968) | POWERS (1968) | ARAMCO (1980) | THIS STUDY |
| PERMIAN: KHUFF FORMATION | PERMIAN: KHUFF LIMESTONE | NOT STUDIED | PERMIAN: KHUFF FORMATION | PERMIAN: KHUFF FORMATION | PERMIAN: KHUFF FORMATION |
| | | | | PRE-KHUFF CLASTICS | |
| | | | CARBONIFEROUS: BERWATH FORMATION (SUBSURFACE) | CARBONIFEROUS: BERWATH FORMATION | CARBONIFEROUS: UNAYZAH FORMATION |
| | | | | | "PRE-UNAYZAH CLASTICS" |
| | | | DEVONIAN: JAUF FORMATION: TRANSITION | | |
| DEVONIAN: JAUF FORMATION: HAMMAMIYAT LIMESTONE | DEVONIAN: JAUF FORMATION | DEVONIAN: JAUF FM: HAMMAMIYAT LIMESTONE MEMBER | HAMMAMIYAT LIMESTONE MBR | DEVONIAN: JAUF FORMATION | DEVONIAN: JAUF FORMATION: HAMMAMIYAT LIMESTONE MBR |
| SUBBAT EL-WADI SH. | | SUBBAT EL-WADI SHALE MEMBER | SUBBAT SHALE MEMBER | | SUBBAT SHALE MEMBER |
| QASR LIMESTONE | | QASR LIMESTONE MEMBER | QASR LIMESTONE MBR | | QASR LIMESTONE MBR |
| SHA'IBA SANDSTONE | | SANDY – SHALY MEMBER | SHA'IBA SHALE MEMBER | | SHA'IBA SHALE MEMBER |
| TAWIL FORMATION | SILURIAN: TAWIL FORMATION | SILURIAN: TAWIL FORMATION | TABUK FORMATION: TAWIL SANDSTONE MBR. | TABUK FORMATION: TAWIL SANDSTONE MBR. | TAWIL SANDSTONE MBR. |
| ? | TABUK FORMATION | SHARAWRA: SHARAWRA SANDY MEMBER | SILURIAN: SHARAWRA FORMATION | SILURIAN: SHARAWRA SANDSTONE MBR. | SILURIAN: TABUK FORMATION: SHARAWRA SANDSTONE MBR. |
| SILURIAN: TABUK FORMATION: QUSAIBA SHALE | | CLIMACOGRAPTUS ORTHOCERTID MBR. | QUSAIBA SHALE MEMBER | QUSAIBA SHALE MEMBER | QUSAIBA SHALE MEMBER |
| | | ORDOVICIAN: TABUK FM.: UPPER TABUK SANDY MEMBER | | | 'SARAH MEMBER' |
| ORDOVICIAN: ER–RA'AN SHALE | ORDOVICIAN | DIPLOGRAPTUS SHALY MEMBER | ORDOVICIAN: RA'AN SHALE MEMBER | ORDOVICIAN: RA'AN SHALE MEMBER | ORDOVICIAN: RA'AN SHALE MEMBER |
| | | LOWER TABUK SANDY MEMBER | | | |
| HANADIR SHALE | | DIDYMOGRAPTUS SHALY MEMBER | HANADIR SHALE MEMBER | HANADIR SHALE MEMBER | HANADIR SHALE MEMBER |
| | | CRUZIANA SERIES | | | |
| ? CAMBRIAN: SAQ SANDSTONE | ? CAMBRIAN: SAQ SANDSTONE | CAMBRIAN: SAQ FORMATION | CAMBRIAN: SAQ SANDSTONE | CAMBRIAN: SAQ SANDSTONE | CAMBRIAN: SAQ SANDSTONE |
| | | | | | JUBAYLAH GROUP |
| P€: BASEMENT | P€: BASEMENT | P€: BASEMENT | P€: BASEMENT | P€: BASEMENT | P€: BASEMENT |

Figure 6. *Comparison of Paleozoic stratigraphic nomenclature of Arabia.*

We can identify several stratigraphic breaks in the Paleozoic section. These breaks can be correlated with the Assyntic, Caledonian, and Hercynian global tectonic cycles events.

A more complete section of the Paleozoic rocks was encountered basinward. There, in addition to the Cambrian, Ordovician, Silurian, and Early Devonian age rocks found in the outcrops, Middle-Late Devonian, Carboniferous, and Early-Middle Permian age rocks were recognized in the subsurface.

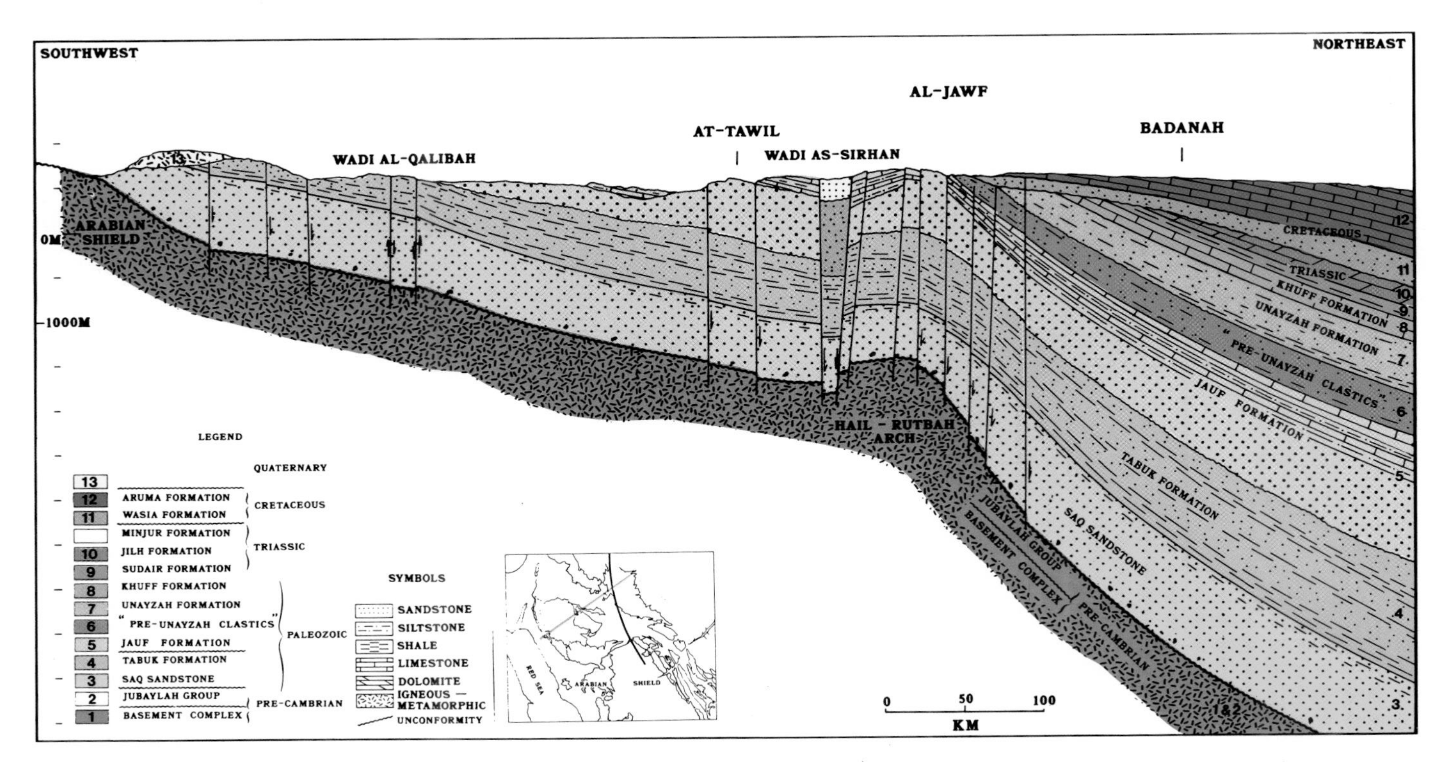


Figure 7. *Structural cross section of Paleozoic relationships in the Tabuk-Widyan basins, in part schematic.*

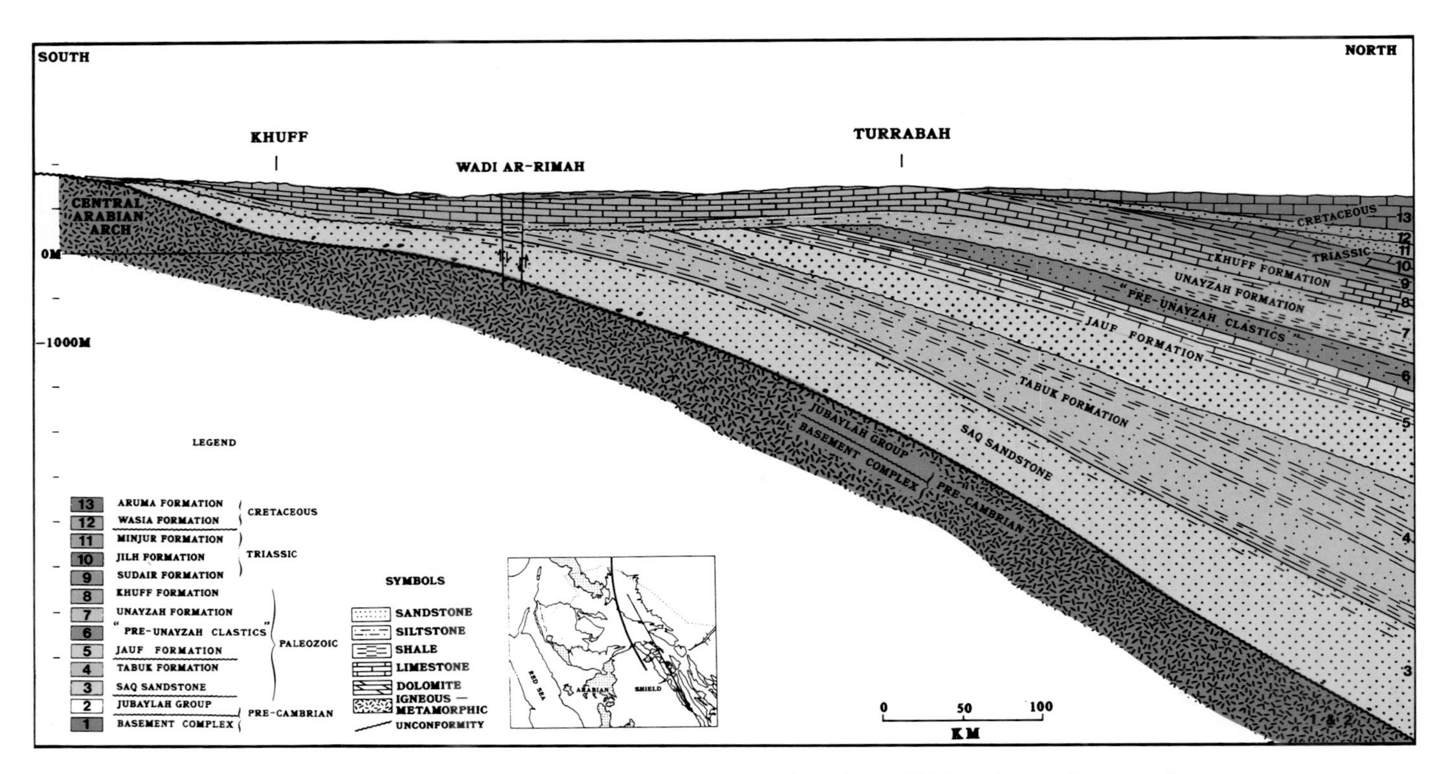


Figure 8. *Structural cross section of Paleozoic relationships in the Widyan basin, in part schematic.*

Following stabilization of the Arabian shield, the maturely peneplaned surface of the igneous and metamorphic rocks supplied the foreland shelf with an influx of clastics that reflect continental to shallow-marine depositional environments. A primary provenance of these sedimentary rocks is believed to have been the igneous and metamorphic terrain of the Arabian shield. These rocks, which are almost lacking or very poor in non-quartz minerals, represent long-distance transportation or

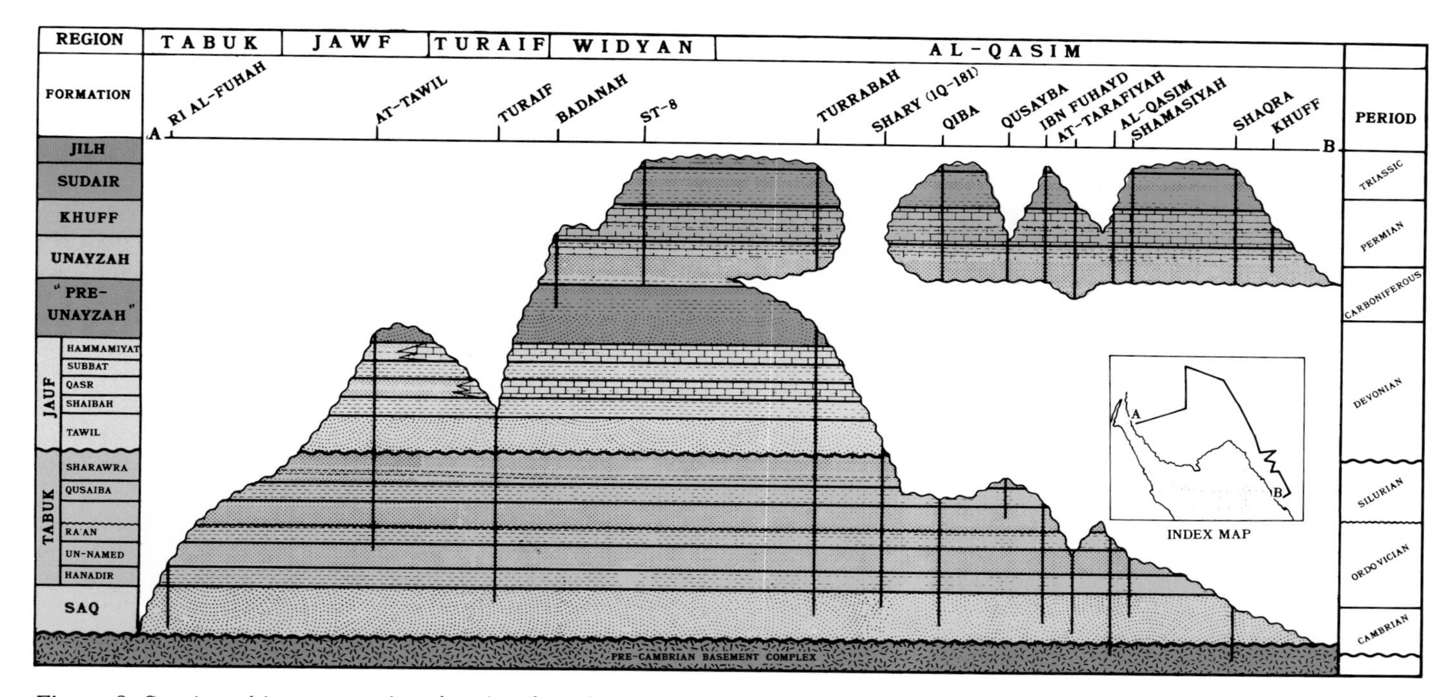


Figure 9. *Stratigraphic cross section showing the subsurface relationships of the Paleozoic rocks in the Tabuk and Widyan basins.*

reworking of older sandstones. The limit to which the Paleozoic rocks extended over the shield cannot be extrapolated because much of the sediment must have been eroded. However, with the low dip off the basement rocks and the presence of the Jubaylah Group in several areas on the shield, the Paleozoic seas might have extended far westward of the present-day outcrop area. Composite sections of Paleozoic rocks contain over 5,000 (16,400 ft) of sediments. Epicontinental seas moved back and forth across the craton, depositing marine sediments between non-marine sandstones. Depositional cycles are usually terminated by regional unconformities related to global, tectonic movements or worldwide eustatic changes. Figures 7, 8, and 9 show the subsurface stratigraphic relationships of the Paleozoic sediments in both the Tabuk and the Widyan basins.

## BASEMENT COMPLEX

Where the basement and Paleozoic sedimentary rocks come into contact, as seen in al-Qasim, the Saq Sandstone rests on either metamorphic rocks (sericite and chlorite schists) or plutonic rocks (mainly granites) that have intruded the metamorphic rocks. The basement rocks were reached by few wells in both the Tabuk and al-Qasim regions. Stoeser and Elliott (1980) dated many of the granites of the shield area as 525–650 million years old. These dates are comparable to those assigned to the Pan-African thermotectonic episode by Kennedy (1964). Thus, many of the granites in the basement complex may in fact be early Cambrian in age.

## PALEOZOIC SEDIMENTARY ROCKS

### Late Precambrian–Cambrian: Jubaylah (Jibala) Group

The pre-Saq Paleozoic section is represented by the strongly folded and unmetamorphosed conglomerates, limestones, sandstones and andesitic and basaltic lavas and pyroclastics of the Jubaylah Group (Delfour, 1966, 1970). These rocks are believed to be taphrogeosynclinal sediments, as they are found only in the northwest-trending grabens of the Najd Fault System in the Arabian shield (Figure 3).

In the type-locality, north of Jabal Jubaylah, Delfour (1966) subdivided the Jubaylah Group into three units that consist of the following, from base to top: (1) conglomerates and sandstones, (2) andesite and basalt flows with tuffs, and (3) clayey or gritty limestone bed intercalated with cherts. The Jubaylah Group rocks rest unconformably on older metasediments and metaigneous rocks of the Precambrian basement complex.

Equivalent units of the Jubaylah Group were recognized in other localities in the Arabian shield. In the Mashhad area, near al-Ula, Hadley (1973) correlated with the Jubaylah Group a section composed of about 875 m (2,870 ft) of the clastic lower Rubatyn Formation, the andesitic basalt middle Batayi Formation, and the carbonates and clastics of the upper Muraykhah Formation. The Jubaylah Group is more uniform in the Musayna'ah-Nuqrah area where two lithologically distinct formations can be recognized (Delfour, 1975) (Figure 4). The Umm al-Aisah Formation is composed of about 500 m (1,640 ft) of basal polygenetic conglomerates overlain by about 300 m (920 ft) of interbedded limestone, chert, dolomite, marl, and argillite. The Umm al-Aisah Formation is conformably overlain by the Jifn Formation, which is composed of about 2,500 m (8,200 ft) of interbedded sandstones, siltstones, argillite, and limestone. The medial andesitic basalts of the Batayi Formation of the Mashhad area are missing in this area, but basaltic-andesitic flows and pyroclastic rocks interrupted the sediments of the two formations at various levels (Delfour, 1982).

In Wadi Fatima Basahel et al (1984) correlate the Jubaylah Group with the Fatima Formation, which is composed of almost unaltered and highly folded and faulted carbonates, and clastics intercalated with volcanics.

Radiometric dating of the Jubaylah andesite gives Cambrian age (Hadley, 1975), Cambrian with some reservation (Brown and Jackson, 1960), and an age between 520 and 500 Ma (Delfour, 1982). Cloud, Awramik and Hadley (1979) report the first biologically definable fossils from both the cherty limestone of the Muraykhah and the stromatolitic limestone of the Umm al-Aisah formations. These fossils include the coiled tabular filaments of the oscillatorialean, blue-green algae, *Obruchevella para* and the conical stromatolites, *Conophyton*. The age of the Jubaylah Group, from these fossils, appears to be close to the Proterozoic–Phanerozoic transition (Cloud, Awramik, and Hadley, 1979). Similar fossils were reported from the Fatima Formation, Wadi Fatima by Basahel et al (1984). They (op. cit.) also reported the presence of archaeocyathids in the Fatima Formation, thus establishing a Cambrian age for at least a portion of that formation.

### Cambrian–Ordovician: Saq Sandstone

Except for the taphrogeosynclinal sediments of the Jubaylah Group in the Arabian shield, sedimentation, following the stabilization of the shield, started with deposition of thick, continental sandstones of the Saq Sandstone (Figure 3). The outcrops of the Saq Sandstone as well as the subsurface information indicate that the formation can be divided into two distinct facies: a lower fluvial and an upper littoral to shallow marine. The lower facies is a thick, extensive sandstone sequence most likely deposited by braided river systems that drained the nearly peneplaned source area. The sandstones consist of well to poorly sorted red, pink, and white quartz, and medium and coarse-grained quartz. Thin micaceous siltstone and varicolored shale partings locally occur in the section. Borings, identified as *Scolithus* spp., are common to the Saq Sandstone. The Saq Sandstone rests nonconformably on the maturely peneplaned igneous and metamorphic basement rocks. The contact of the Saq with the underlying basement is marked by pebble to boulder conglomerates derived from the basement complex (Powers et al, 1966). However, in places where the contact was observed (near both ar-Ras and al-Idwah in al-Qasim), the cross-bedded, medium- to coarse-grained sandstones of the Saq rest on weathered plutonic and metamorphic rocks. Locally, thin lenses of quartz pebble conglomerates are found at the base of the formation. Absence of boulder conglomerates suggests that the Saq Sandstone was in part deposited far from the sediment source. The Saq–basement contact marks one of the major unconformities in the region and is probably related to the Assyntic orogeny (Figure 4). In the northern part of the shield, the Siq Sandstone, the lowermost unit of the Saq Sandstone, rests unconformably in the Mashhad area on the Jubaylah Group (Hadley, 1973). In the subsurface, the basement–Saq contact is picked by wireline log character at the distinct lithology change from sandstone to granite or metamorphic rocks. The upper part of the Saq Sandstone consists of littoral to shallow marine-deltaic, argillaceous sandstones, micaceous siltstones, and shales. This facies transitionally gave way in Early Ordovician time to the first regional marine invasion in Arabia. The gradation marks the contact of the Saq Formation with the overlying Tabuk Formation.

The Saq Sandstone equivalent units in northwestern Arabia are the Siq, Quweira, Umm Sahm, and Ram Sandstones (Figure 4). The upper units of these sandstones contain siltstones and shales with arthropod tracks identified as species of the genus *Cruziana*, establishing a probable Early Ordovician (Tremadocian-Arenigian) age for the upper portion of the Saq. The lower units, the Siq and Quweira sandstones, are assigned to the Cambrian based on the presence of Middle Cambrian trilobites in interbedded limestone and shale units near the Dead Sea. Thus, the Saq Sandstone may record Middle Cambrian to Early Ordovician sedimentation (Powers et al, 1966).

The sandy facies in the Saq formation, in general, increases in thickness toward the craton from the al-Qasim well. Simultaneously, the continential lower part of the Saq decreases in thickness toward the basin, while the littoral upper part increases, indicating more marine influence. This thickness variation suggests a wedge-like deposition in a continental-marine, littoral environment for the Saq Sandstone. The shales in the upper part of the Saq at both Qiba and al-Qasim wells suggest more marine influences basinward. The lowermost Saq sediments encountered in wells located in the more central part of the basin may be older than those resting directly on the basement in the outcrop, or penetrated in wells positioned nearer to the craton. The most complete section of the Saq Sandstone, 482 m (1,580 ft) was encountered in the al-Qasim well, while the thickest section, 928 m (3,040 ft), was drilled in the Ri al-Fuhah well in the Tabuk region. Basement rocks were not reached in that well. Figure 10 shows a generalized, isopach map for the Saq Sandstone.

### Ordovician–Silurian: Tabuk Formation

The Tabuk Formation consists of more than 1,700 m (5,570 ft) of rhythmically alternating marine shales and continental to marginal-marine sandstones of Ordovician and Silurian age. Six units, representing periods of transgressions and regressions in the formation, could easily be observed (Figure 4). The shale members are the Hanadir, the Ra'an, and the Qusaiba. Because of a lack of diagnostic fossils, non-distinct character and poor outcrop patterns, partly due to a widespread calcrete blanket, the intervening sandstones between the shales are difficult to recognize in outcrops. Recognizing each sandstone unit in the subsurface depends on shale identification and stratigraphic position. The post-Qusaiba Silurian sandstones, siltstones and claystones are named the Sharawra Sandstone Member, while both the post-Ra'an Ordovician and Silurian sandstones and the post-Hanadir sandstones have not been named. The present study demonstrates that these members have large areal distribution in both outcrop and subsurface. Thus, we should keep these member names as used here. The Hanadir, Ra'an, and Qusaiba Shale Members represent periods of maximum transgressions. These are widespread and are probably coeval in part with the underlying and overlying sandstones that are continental to marginal marine sediments. The sandstone units are readily recognizable along the depositional strike, but basinward, the sandstones interfinger with the shales. Figure 11 shows a generalized isopach map for the total Tabuk Formation, indicating a pronounced increase in thickness basinward.

#### *Hanadir Shale Member*

The maximum phase of the first transgression in the area resulted in the deposition of the Hanadir Shale Member on the Saq Sandstone. The Member is composed of gray to dark gray, brown, red, and greenish claystones, with thin, silty sandstone beds. Because this shale is deposited over a large area, it serves

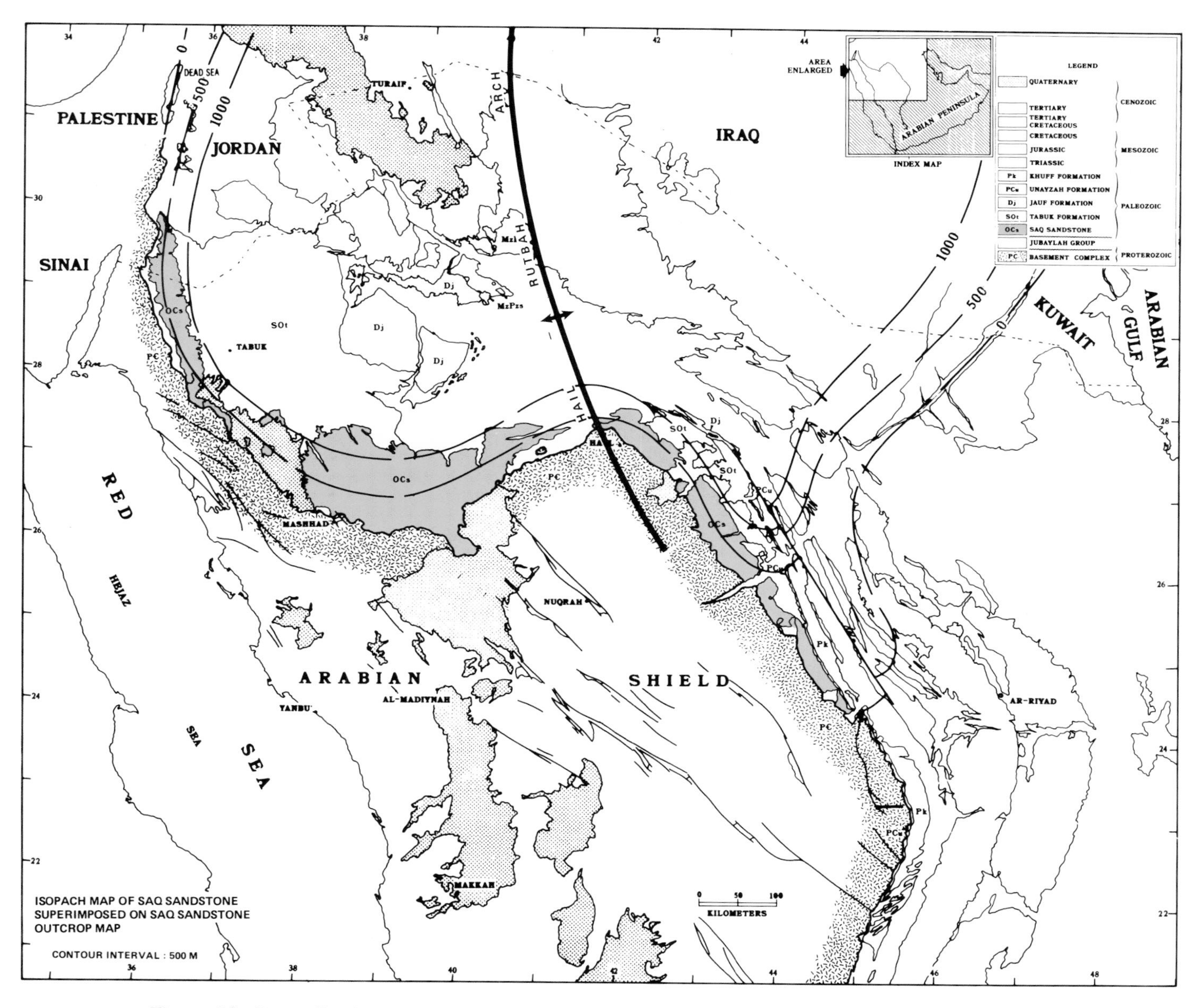


Figure 10. *Generalized isopach map of Saq Sandstone superimposed on Saq Sandstone outcrop map.*

as an excellent marker in the Paleozoic. In the subsurface the Hanadir Shale Member is peneterated in all wells that reached the Saq Sandstone, except in the Shaqra well where the Hanadir Shale Member and the upper part of the Saq Sandstone are missing, probably due to erosion. Based on the presence of *Didymograptus protobifidus* Elles, the Hanadir was considered to be Early Ordovician (Arenig) in age (Powers et al, 1966). Thomas (1977) identified trilobites that were collected from the Jabal al-Hanadir as *Paesiacomia vacuvertis* sp.nov., while Forety and Morris (1982) identified *Neseuretus tristani* (Desmarest, 1817) from the same locality. These trilobitic remains confirm the Early Ordovician, Llandoverian age of the Hanadir Shale Member. Palynomorphs recovered from the Hanadir shales, in the subsurface, also confirm this age assignment.

The Hanadir increases in thickness toward the center of the basin, where it attains more than 100 m (328 ft).

*Ordovician Sandstone Member*

Sea level fluctuation resulted in deposition of thick, fine-to-medium-grained, micaceous sandstones with thin siltstone and shale intercalations, conformably overlying the Hanadir Shale Member. In both Qiba and al-Qasim wells, near-shore conditions prevailed, and thick, massive sandstones, siltstones, and thin shales were deposited. Basinward, at Turrabah, these sandstones lose their identity and become more marine, with a pronounced increase in thickness and interfingering with shale beds. The stratigraphic position of these beds, between the Hanadir and Ra'an Shales, suggests a Llandeilian age. This sandstone varies in thickness from 40 m (130 ft) to more than 400 m (1,310 ft), basinward.

*Ra'an Shale Member*

During the Late Ordovician, short-lived transgressive and

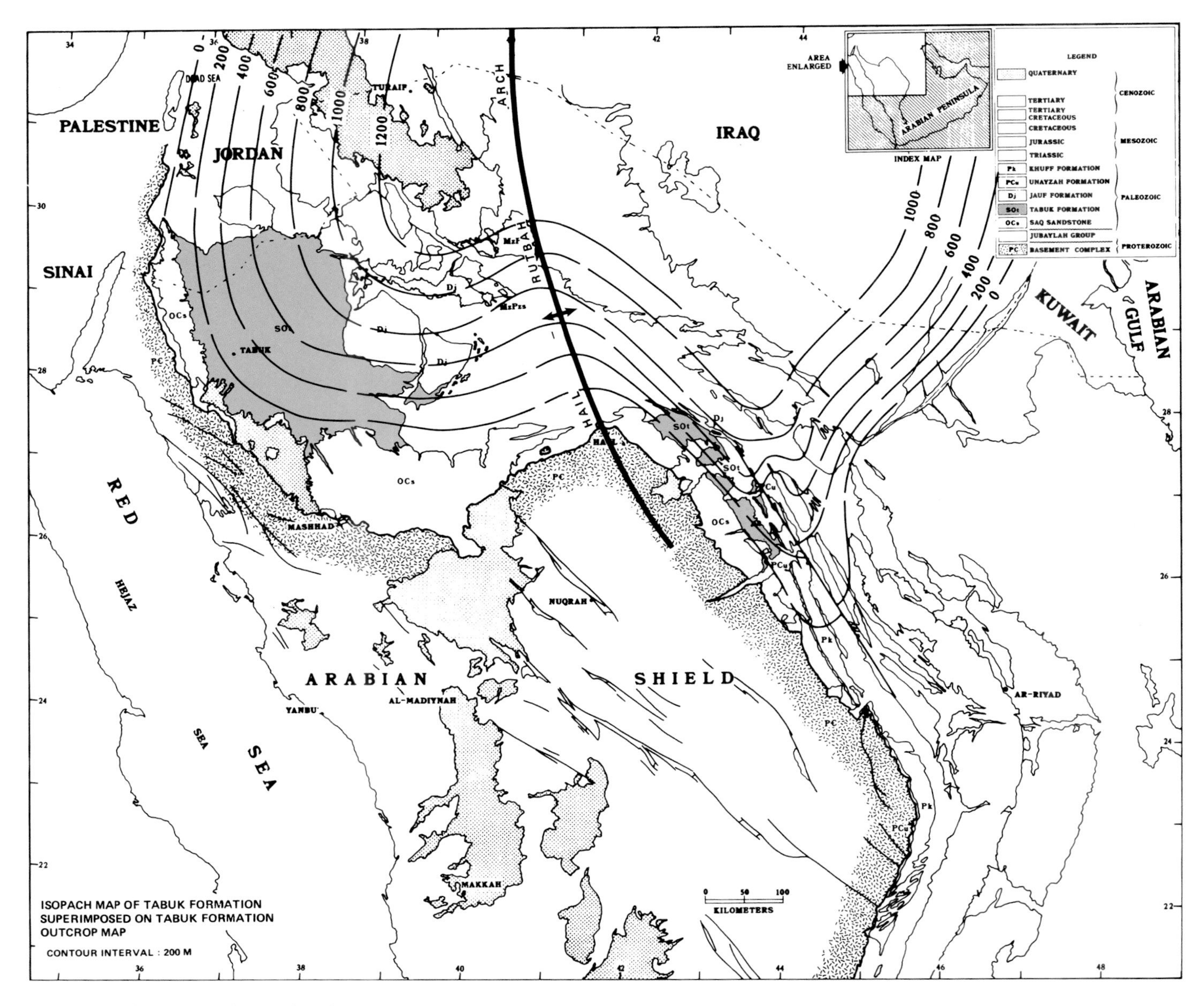


Figure 11. *Generalized isopach map of Tabuk Formation superimposed on Tabuk Formation outcrop map.*

regressive cycles resulted in the deposition of alternating shales, siltstones, and sandstones mainly under deltaic conditions. The Ra'an Shale Member represents the uppermost, well-developed shale unit of the Ordovician part of the Tabuk Formation. It is composed of gray-purple, green, micaceous, graptolite-bearing claystones and thin-bedded sandstones and siltstones. *Diplograptus?* sp., *Climacograptus* sp. cf. *C. brevis* Elles and Wood, of Late Ordovician age, were identified from the Ra'an shales (Powers et al, 1966).

In the subsurface, Late Ordovician palynomorphs have been obtained from these shales. The Ra'an Shale Member represents the last transgression of the Ordovician sea in the area. Under certain conditions, the Late Ordovician Ra'an Shale Member may conformably overlie the Early Ordovician Hanadir Shale Member as the intervening sandstone pinches out basinward in the two basins. In both Khashim ar-Ra'an and Jal as-Saqiyah areas (Figure 2) large sandstone boulders are found near the top of the Ra'an Shale Member in a few localities. These boulders are elongated with their large axis exceeding 1 m (3.3 ft) and up to 40 cm (16 in) thick. Some of the concretions have shallow, narrow, parallel striations on their upper surface, but pluck marks were not observed. These concretions are at approximately the same stratigraphic level as channels filled with pebbles, cobbles, and boulders. Striations indicate general west to east movement and may be explained by the movement of boulders of sedimentary rocks that cut channels in the Ra'an Shale Member.

The Ra'an Shale Member varies in thickness from 14 m (46 ft) to about 100 m (330 ft) in the Tabuk basin. The thickness variation probably is due to local structures. The absence of the Ra'an Shale Member in most of the wells drilled in the southern part of the Widyan basin may be attributed to the severity of later erosion represented by the post-Ordovician unconformity. The absence of the Ra'an Shale Member in wells immediately to the east of the al-Qasim well suggests a down-to-the-west normal fault. An offset along this fault is also inferred in younger members of the Tabuk Formation. The name "Tarafiyah fault" is proposed here for this fault (Figure 11).

*Ordovician–Silurian Sandstone Member*

A major sea level drop in Late Ordovician resulted in severe erosion and truncation of older units in the two basins.

In the Tabuk area, Bigot and Lafoy (1970) report a 1 m (3.3 ft), heterogeneous horizon containing igneous and metamorphic pebbles and boulders overlying the Ra'an Shale Member at both Ri al-Fuhah and Bir Hirmas outcrops. This conglomeratic section is overlain in outcrop by silty, micaceous, cross-laminated sandstone and silty shale.

In the Jabal Sarah area, al-Qasim, the outcrops of the Ordovician sediments, including the upper part of the Saq Sandstone, were deeply truncated by paleovalleys. The Jabal Sarah outlier (Figure 2) represents a paleovalley sedimentary fill, lying unconformably on the Saq Sandstone. This ridge stands up because of weathering away of the enclosing sediments. The term "Sarah Member" was proposed by Clark-Lowes (1980) for the widespread and lithologically distinct conglomerates, conglomeritic sandstones, sandstones, and minor shales exposed in Jabal Sarah in al-Qasim. The "Sarah Member" represents the lower part of the intervening sandstone section between the Ra'an and Qusaiba shales (Figure 4).

At both Khashim ar-Ra'an and Jal as-Saqiyah, al-Qasim, heterogeneous conglomerates up to 1 m (3.3 ft) in thickness fill narrow channels cutting in the Ra'an Shale Member. This section is overlain by fine- to medium-grained quartz sandstones, with thin beds of green, silty claystone and siltstone. These sandstones contain granite and metamorphic exotics, ranging from pebble size up to boulders exceeding 1 m (3.3 ft) in diameter. Many of the small boulders have one or two flat surfaces with striations. When seen in place, the striated or smooth surface is on top of the boulder. Some of these exotics and the sandstone concretions are striated, polished, grooved, and faceted. Pluck marks have not been observed on any of these boulders, and many larger boulders do not have striations or flattened surfaces. Few of these boulders were found in situ, while many are lag boulders, having been let down several meters as the enveloping sandstones were eroded. These exotics extend upward for at least 12 m (40 ft) above the Ra'an Shale Member.

Such boulders are enigmatic because of their distance from an apparent source, their widespread distribution both laterally and vertically, and their many different types. Clark-Lowes (1980) interpreted the conglomerates, conglomeritic sandstones, and dismictites of the Sarah Member as fluvial and mud flow deposits, although McClure (1978), had proposed glaciation or ice rafting that brought the boulder at Jal as-Saqiyah from the basement area. However, the apparent local distribution of these boulder beds, the isolated occurrence of the boulders, and their vertical stratigraphic distribution (which ranges to over 12 m (40 ft) above the base of the sandstone and possibly into the overlying Qusaiba Shale Member), are arguments against widespread glaciation. I suggest that the exotic boulders, in the Jal as-Saqiyah, could be transported from an upthrown block of basement rocks east of the at-Tarifiyah fault rather than the westward shield area. This suggestion would explain their linear distribution and apparent decrease in size westward toward the shield area.

If these conglomeritic sandstones are of glacial origin, then the hiatus between the Ordovician and Silurian may represent a period of continental glaciation in Arabia. Except in two wells in the Qasim region, these sandstones are barren of both chitinozoa and acritarchs. However, being conformably overlain by the well-dated Qusaiba Shale Member, these sandstones are considered to be, at least in part, of earliest Silurian (Rhuddanian age of the Llandoverian).

*Qusaiba Shale Member*

Cyclic sedimentation continued during the Silurian, beginning with sandstones, followed by the transgressing sea that laid down the Qusaiba Shale Member.

During the Idwian age (of the Llandoverian), the marine Qusaiba Shale Member was deposited over the region. It represents the last major shale unit of the Tabuk Formation. The Qusaiba Shale Member contains *Monograptus* spp., other graptolites and some pelecypods of Silurian age (Powers et al, 1966). Thomas (1977) identified trilobites collected from the Qusaiba shales at *Platycorphe dyaulasc* sp. nov. of the Convolutus zone, Idwian age. The Qusaiba shales contain abudant Early Silurian palynomorphs in the subsurface.

The Qusaiba Shale Member consists of green, gray, mustard-yellow graptolitic shales and hematitic siltstones exposed in a vertical, east-facing cliff at Qusayba. Data collected from boreholes in both the Tabuk and Widyan basins demonstrate the widespread continuity of the Qusaiba Shale Member.

The Qusaiba Shale Member in the subsurface of the Widyan basin varies in thickness from 34 m (110 ft) to 210 m (690 ft). At Qiba well, the Qusaiba is unconformably overlain by the Unayzah Formation. The Qusaiba Shale Member is neither exposed nor penetrated in any well south of the Qusayba area. In the Tabuk basin, the Qusaiba increases in thickness basinward from 124 m (400 ft) to 260 m (850 ft).

The Qusaiba Shale Member is gradationally followed by and interfingering with a sandstone, siltstone, and a silty shale section of the Sharawra Sandstone Member. The gradational contact is arbitrarily placed at the base of the lowermost, fine-grained micaceous sandstones and siltstones of the Sharawra that Powers et al (1966) included in the Qusaiba, at the type-locality. This interfingering relationship may explain the increase of thickness of the Qusaiba toward the central part of the Widyan and Tabuk basins.

Outcrops of the Qusaiba Shale Member in the vicinity of al-Ajfar are in part covered with cobbles and boulders of igneous, metamorphic, and sedimentary rocks. Most of the cobbles and smaller boulders are well worn and appear to be of fluviatile origin. However, some of the quartzite boulders exceed 1 m (3.3 ft) in greatest dimension, are sharp edged, and show no apparent rounding or striations. If these larger boulders are fluviatile in origin, having moved over 50 km (30 mi), then the large, exotic boulders in the underlying sandstone at Jal as-Saqiyah may be lag from similar deposits.

*Sharawra Sandstone Member*

Following deposition of the Qusaiba Shale Member, a fluctuating, regressive sea deposited the Sharawra Sandstone Member. The Sharawra is composed of alternating, fine-grained micaceous sandstones, siltstones, and claystones. It is exposed at Jabal Sharawra in the Tabuk area. Fine-grained sandstones

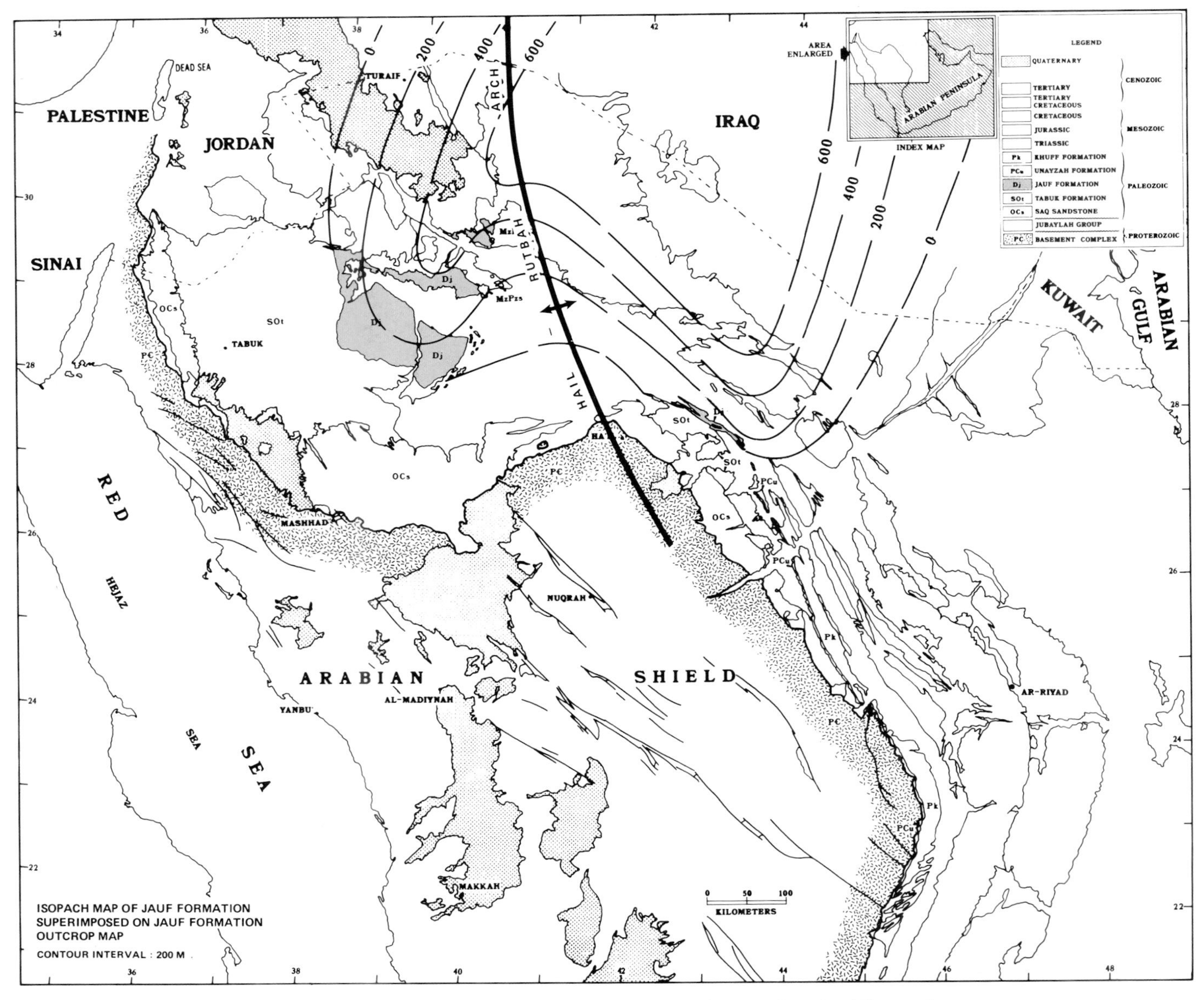


Figure 12. *Generalized isopach map of Jauf Formation superimposed on Jauf Formation outcrop map.*

overlying the Qusaiba Shale Member at Qusayba correlate with the Sharawra Sandstone Member.

The basal part of the Sharawra Sandstone Member contains shales similar to those of the Qusaiba Shale Member. The Sharawra contains typical Silurian chitinozoans and acritarchs which are identical to those found in the Qusaiba Shale Member. The Sharawra is considered to be Llandoverian in age. Thus, the term "Sharawra Sandstone Member" should be applied to all post-Qusaiba Silurian sandstones. The Sharawra Sandstone Member is well developed in at-Tawil well, where 765 m (2,510 ft) of shales and sandstones were encountered. At this locality, the lower part of the Sharawra is almost entirely shale, whereas the upper part is mostly sandstone.

The deposition of the Sharawra was followed by a period of regional uplift and erosion. The Late Silurian is believed to be not represented in Arabia, probably because of major erosion or non-deposition.

### Devonian: Jauf Formation

The Jauf Formation marks a distinct change in the depositional pattern of the entire basin, where carbonates appeared in the section for the first time in the area. This depositional cycle began with the continental Tawil Sandstone Member, which is followed by an alternating clastic and carbonate section. These sediments crop out in the central part of both the Tabuk and the Widyan basins (Figure 3). In the latter basin, two narrow, alternating carbonate and clastic bands are exposed at Jal ash-Shuaybah and Jal al-Hayla, southeast of an-Nafud (Figure 3). The most complete section of the Jauf Formation was penetrated in the Turrabah well (Figure 13), where all Jauf members are present. The Jauf Formation, as defined in this paper, is subdivided into five members from bottom to top: the Tawil, Shaibah, Qasr, Subbat, and Hammamiyat.

The Middle–Late Devonian age clastics overlying the Hammamiyat Limestone Member, informally known as the "Transi-

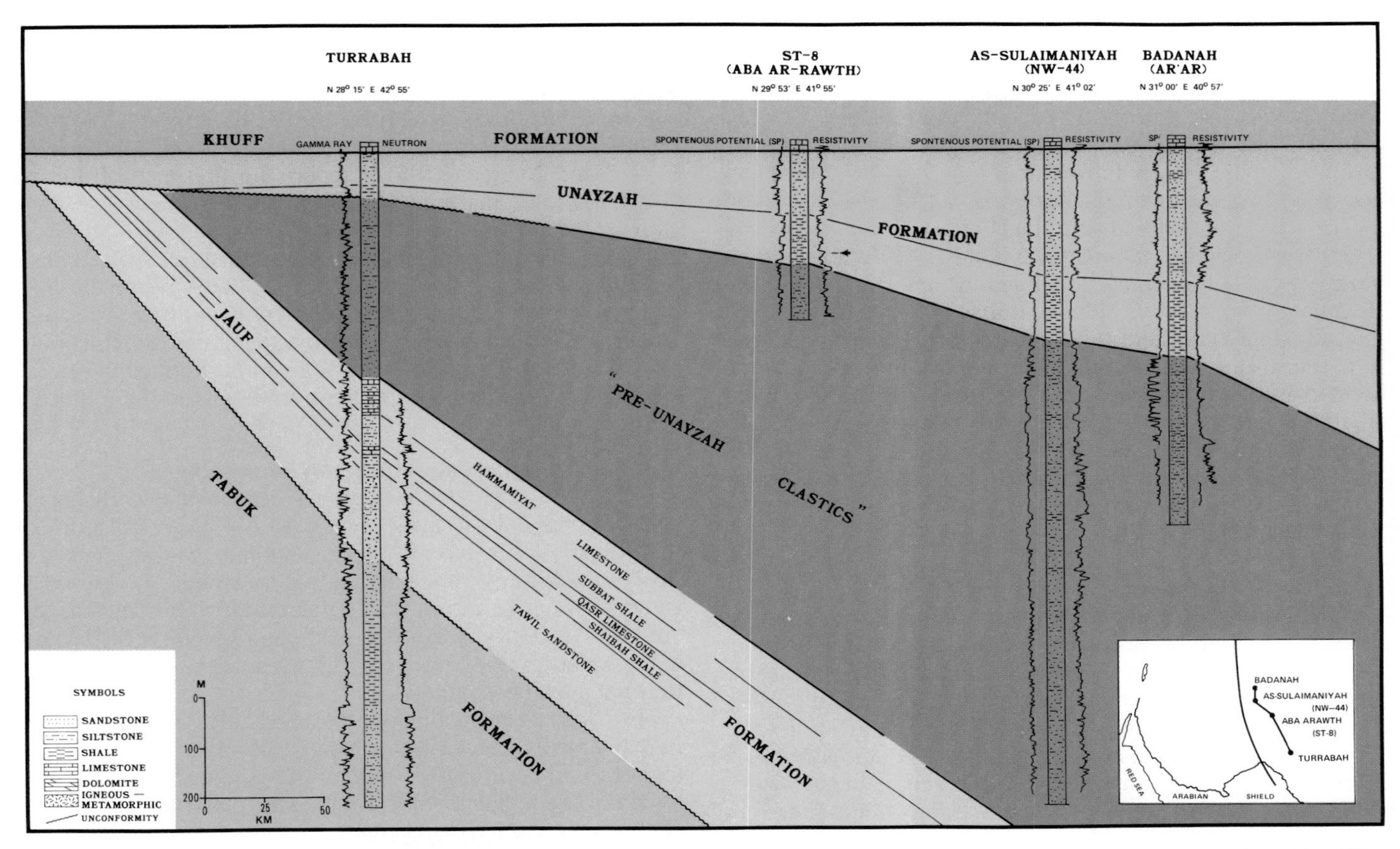


Figure 13. *Stratigraphic cross-section showing the stratigraphic relationships of the Unayzah "pre-Unayzah Clastics" and Jauf Formation in the northern part of the Widyan basin. Arrow points to top Berwath Formation, at type locality, well ST-8.*

tional Zone," are removed from the Jauf Formation and considered as the basal part of the "pre-Unayzah clastics" (informal name) (Figures 4 and 13). In the at-Tawil well, the limestone members are not developed (Figure 9) and the Jauf Formation is composed of 236 m (775 ft) of undifferentiated sandstones and shales, which may be, in part, coeval with the limestones in the outcrops of al-Jawf area. The Jauf Formation was considered to be of probable Early and Middle Devonian age (Powers, 1968). The total Jauf Formation thickness, as shown in Figure 12, is increasing north-northeastward.

### *Tawil Sandstone Member*

The Tawil Sandstone Member consists of medium- to coarse-grained, gravelly and pebbly, thickly cross-bedded quartz sandstones with occasional shales. The Tawil Sandstone Member outcrops are believed to be restricted to at-Tawil and al-Jawf area of the Tabuk basin.

The continental Tawil Sandstone Member is barren of diagnostic fossils, with the exception of *Cruziana* found in certain outcrops. It is given an Early Devonian, Gedinnian age because of its stratigraphic position. The lithologic nature of the Tawil Sandstone Member indicates rapid erosion and continental deposition (Figure 4). The contact between the Tawil Sandstone Member of the Jauf Formation and the Sharawra Sandstone Member of the Tabuk Formation marks a major hiatus. The Upper Silurian is missing (Figure 4) from Arabia, and the Lower Devonian Tawil rests on Lower Silurian, Llandoverian sediments. This regional hiatus represents a period of severe Paleozoic uplift and erosion in Arabia, probably related to the Caledonian movement.

In the subsurface of the Widyan basin the Tawil Sandstone Member extends as far south as the ash-Shary well, and drillers have not encountered it in any of the wells drilled to the south. The Tawil Sandstone Member ranges in thickness from 50 m (160 ft) to 189 m (620 ft).

### *Shaibah Shale, Qasr Limestone, Subbat Shale and Hammamiyat Limestone Members*

The influx of terrigenous material of the Tawil Sandstone Member was gradually replaced by near-shore to shallow-marine clastics and carbonates of the Shaibah Shale, Qasr Limestone, Subbat Shale, and the Hammamiyat Limestone Members (Figure 13). These four members crop out in the center of the Tabuk basin and in a narrow band in the Widyan basin. These members in the Turrabah well contain palynomorphs diagnostic of Siegenian and Emsian ages of the Early Devonian.

The transitional Shaibah Shale was followed by marine encroachment that continued and culminated in depositing the Qasr Limestone Member. The Shaibah Shale Member is composed of sandstones and shales. The Qasr Limestone Member consists of limestone, shale, and sandstone. In the al-Jawf well, the lower limestone member, the Qasr Limestone Member, is probably not developed (Figure 9).

This short-lived cycle was terminated by a general withdrawal of the sea. The Subbat Shale was deposited in shallow-marine to continental environments during this regression. This member is composed of sandstones and shales situated between the Qasr and the Hammamiyat Members. Another marine invasion occurred during the Siegenian, depositing the carbonates of the Hammamiyat Limestone Member. The member consists of fossiliferous, locally coralline, thinly bedded limestone alternating with thin beds of gray, and green, silty shale. The top of the Hammamiyat Limestone Member is arbitrarily placed at the top of the uppermost limestone beds. The Hammamiyat Limestone Member transitionally grades upward into silty shales and sandstones of Middle–Late Devonian age.

In the al-Huj area, the Jauf Formation appears to grade upward into silty shales, siltstones, and sandstones. Bigot (1970) suggests that the Sakakah Sandstone may be Devonian in age. If Bigot's age assignment is valid, the continental Middle–Late Devonian age sandstones encountered in the subsurface can correlate with, at least, the lower portion of the Sakakah Sandstone in outcrop.

### Devonian–Carboniferous: "Pre-Unayzah Clastics"

In the northern part of the Widyan basin, a thick section of Permian, Carboniferous, and Devonian clastics was penetrated subjacent to the Khuff Formation (Figure 13). The term "Berwath Formation" was introduced by Hemer and Powers (Powers, 1968) for the Lower and Upper Carboniferous (Tournaisian–?Westphalian) sandstones, siltstones, and shales based on their microflora content. Hemer and Powers (Powers, 1968) established the type locality of the Berwath at the Wadi aba-Arawth well ST-8, in the Widyan basin, between drilled depths of 1328.3 and 1518.5 m (4358 and 4982 ft).

The Berwath Formation is overlain by 200 m (660 ft) of clastics. The Unayzah Formation of al-Qasim region is correlated with this clastic section, including a well-developed basal shale unit that serves as a distinct, lithologic break separating the Unayzah Formation from the underlying clastics. The lower part of this shale interval, 52 m (170 ft) in thickness, was included as the upper part of the Berwath, at the type locality, because it contains Carboniferous microspores.

The Berwath Formation, as introduced by Hemer and Powers (Powers, 1968) is very difficult to recognize because recognition of the formation is based on its Carboniferous microfloral content, rather than on its lithology. The base of the formation is not defined, because it was not penetrated in the type locality well. The contact with superadjacent formation, also clastics, is based only on geological age assignment, at the change from sandstone containing Carboniferous microfloral assemblage to superadjacent varicolored sandy shale containing an Upper Permian microfloral assemblage. The Berwath is restricted to the subsurface. Based on previous evidence, the Berwath is too poorly defined to deserve formation status and, thus, should be discarded.

The clastic section bracketed between the Hammamiyat Limestone Member of the Jauf Formation and the basal, well-developed shale unit of the Unayzah Formation (Figure 13) is informally introduced here as the "pre-Unayzah Clastics." The age of these clastics ranges from Middle–Late Devonian to Carboniferous. The main reason for considering this clastic section, which includes the Transition Zone of the upper part of the Jauf Formation and the overlying clastics (part of the Berwath Formation), as an informal unit is the lithic homogeneity of the facies and the difficulty in separating them into different formations. In the Turrabah well, (Figure 13) 413 m (1,350 ft) of continental sandstones with minor shales of Middle–Late Devonian age were penetrated overlying the Hammamiyat Limestone Member. A similar section was penetrated in well S-462 where Hemer and Nygreen (1967b) report the presence of 408 m (1,340 ft) of non-marine Late and Middle Devonian age sandstones.

These sandstones are the lower part of the pre-Unayzah Clastics and have not yet been found to crop out because of erosion, or because they are overlapped by younger strata. The pre-Unayzah Clastics show a pronounced increase in thickness toward the northern part of the Widyan basin (Figure 14). No Carboniferous sediments are present west of the Ha'il Rutbah Arch.

### Carboniferous–Permian: Unayzah Formation

The Khuff Formation, as originally described by early geologists, includes a basal clastic section. ARAMCO geologists informally restricted the term Khuff Formation to the carbonate and the interbedded claystone and anhydrite sequence. The subjacent shales and sandstones were referred to, informally, as the "pre-Khuff Clastics" (Figure 6). Although lithologic distinction of these clastics from the overlying carbonate sequence is clear, they have not been defined.

The Unayzah Formation is introduced here for the clastic section bracketed between the Khuff carbonates and the pre-Unayzah unconformity, with various older units in the al-Qasim region (Figure 8) and at the base of the well-developed shale unit, penetrated in the northern part of the Widyan basin (Figure 13).

#### *Type Section*

The type section (Figure 15) of the proposed Unayzah Formation is located alongside the Unayzah–Buraydah road within the Unayzah Town.

#### *Reference Section*

The reference section (Figure 16) for the Unayzah Formation is established at the escarpments of the Qusayba area.

#### *Occurrence and Thickness*

The Unayzah Formation at type locality exceeds 33 m (100 ft) as the base of the formation is not exposed (Figure 15). At the Qusayba reference section, the Unayzah is 35 m (115 ft) (Figure 16). The Unayzah thickens to the north, basinward, where it is more than 430 m (1,400 ft) thick in the Badanah well (Figures 13 and 17). In the southern part of the Widyan basin, in al-Qasim, the Unayzah ranges from 30 m (98 ft) to 180 m (590 ft). These varying thicknesses indicate that the Unayzah is filling a narrow trough, the axis of which passes near Mudarraj southeastward of ash-Shamasiyah. Another northwest-to-southeast-trending thick section of Unayzah is mapped from al-Amar to Khuff (Figure 17). The shallow arch that separates the trends of the two, thick sections of the Unayzah Formation may be related to faulting.

The Unayzah Formation marks the beginning of a major change in the sedimentation and the evolution of the Arabian basin. The Unayzah clastics and impure carbonates blanketed the entire basin, overlapping truncated and block- faulted, pre-existing rocks (Figure 8).

Delfour et al (1982) describe a similar sequence of sandstones

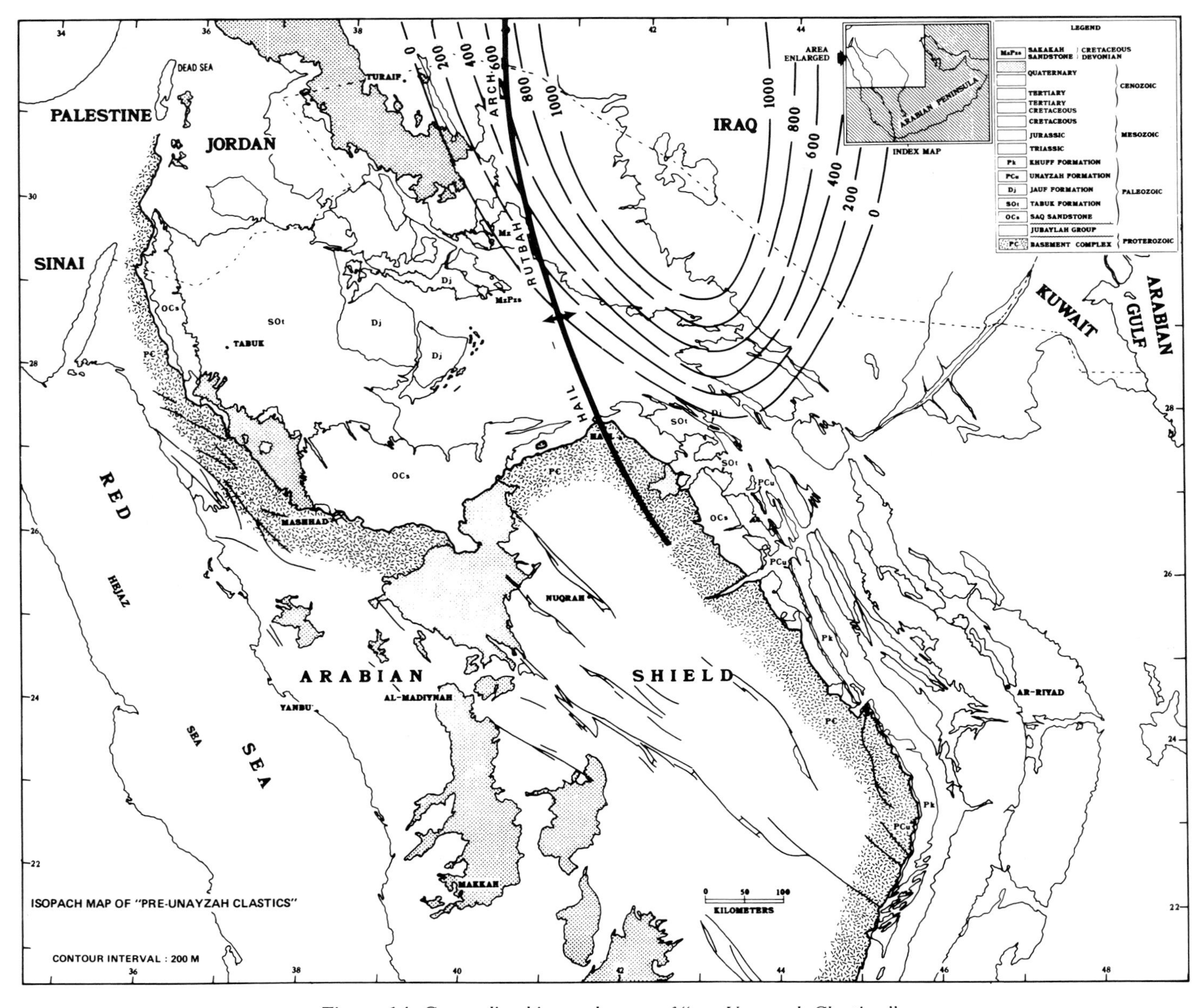


Figure 14. *Generalized isopach map of "pre-Unayzah Clastics."*

and shales of Permo-Carboniferous age beneath the Khuff in the ad-Dawadimi quadrangle. The Unayzah Formation is exposed as a thin band parallel to the Khuff Formation outcrop (Figure 3). A similar sequence of shales, sandstones and limestones of the same age and stratigraphic position are mapped in other areas. The Faraghan Formation (Szabo and Kheradpir, 1978) in the Zagros basin of Iran, and the Lower Haushi Formation of Oman, are the time-stratigraphic equivalents of the Unayzah Formation. Therefore, Unayzah may be present in the subsurface of much of the greater Arabian basin.

*Lithology*

The Unayzah Formation, in general, is composed of cycles of cross-bedded, fine- to coarse-grained sandstones; siltstones; green, grey, and purple claystones; and thin beds of impure limestone (Figures 15 and 16). In the northern part of the Widyan basin, a thick shale section developed in the lower part of the Unayzah Formation (Figure 13). These shales increase in thickness northward from 49 m (160 ft) at well ST-8 to 165 m (540 ft) at Badanah.

*Fossils and Age*

The lowermost sandstone, exposed type locality, begins with a plant-rich (about 1/2 to 2 in. or 1.3 to 5 cm) thick limonite bed. Certain of the impure limestones and sandstones contain casts of unidentifiable clams. At Qusayba, the shales and siltstones contain large fragments of silicified trunks of trees.

Plant remains from Unayzah outcrops permitted El-Khayyal, Chaloner, and Hill (1980) to assign a Permian–Carboniferous age to these clastic beds. In most of the wells penetrating the Unayzah Formation in the Widyan basin, palynomorphs of Westphalian, Stephanian, Sakmarian, Aertinskian, and

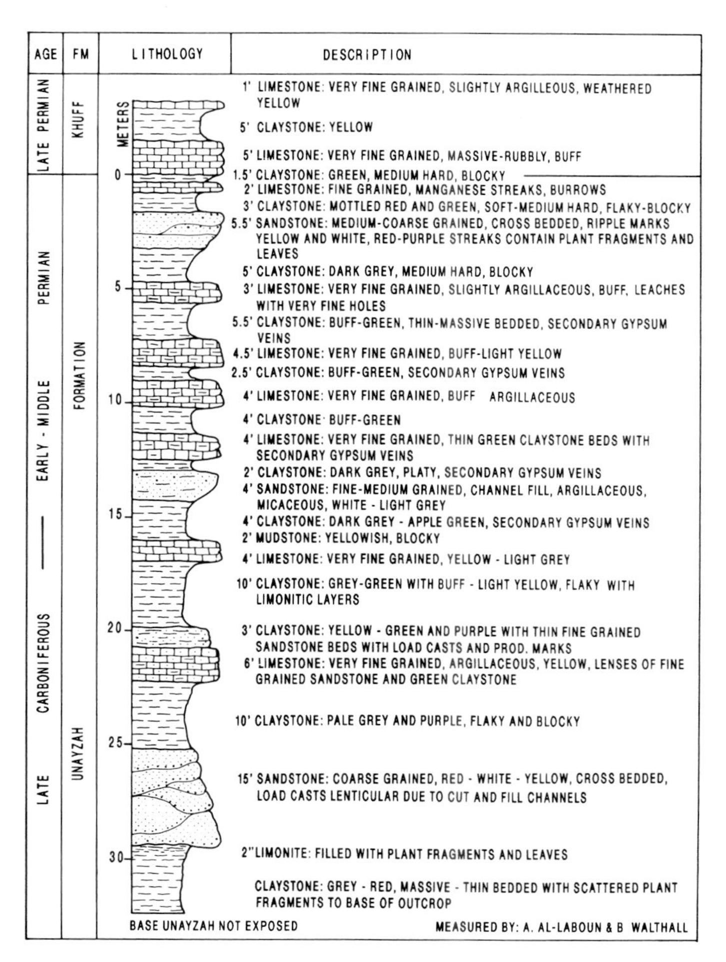


Figure 15. *Composite section of Unayzah Formation in type locality, Unayzah—al-Qasim.*

Kungurian were extracted and identified. (Unpublished ARAMCO reports).

### *Underlying Formations*

The base of the Unayzah Formation is not exposed at the type locality but in a nearby well, IQ-177, the Unayzah Formation unconformably overlies the Silurian part of the Ordovician–Silurian Sandstone Member of the Tabuk Formation. In the Qusayba area, the Unayzah Formation unconformably overlies the Sharawra Sandstone Member of the Tabuk Formation. The contact is well-marked by a fine- to coarse-grained, friable, limonitic, weathered, well-cemented, rubbly, ferruginous, burrowed, pisolitic, lateritic duricrust paleosurface of the Unayzah Formation on the whitish shaly siltstones of the Sharawra Sandstone Member. In the Badanah well, the Unayzah Formation conformably overlies the pre-Unayzah Clastics (Figures 7, 8 and 13). Based on subsurface data, the Unayzah Formation unconformably overlies and overlaps various older units, from Carboniferous to Cambrian, before directly overlapping the Precambrian basement complex (Figure 7). This regional, angular, and pronounced unconformity is hereby referred to as the "pre-Unayzah unconformity." The unconformity probably represents the maximum phase of the Hercynian orogeny in the area.

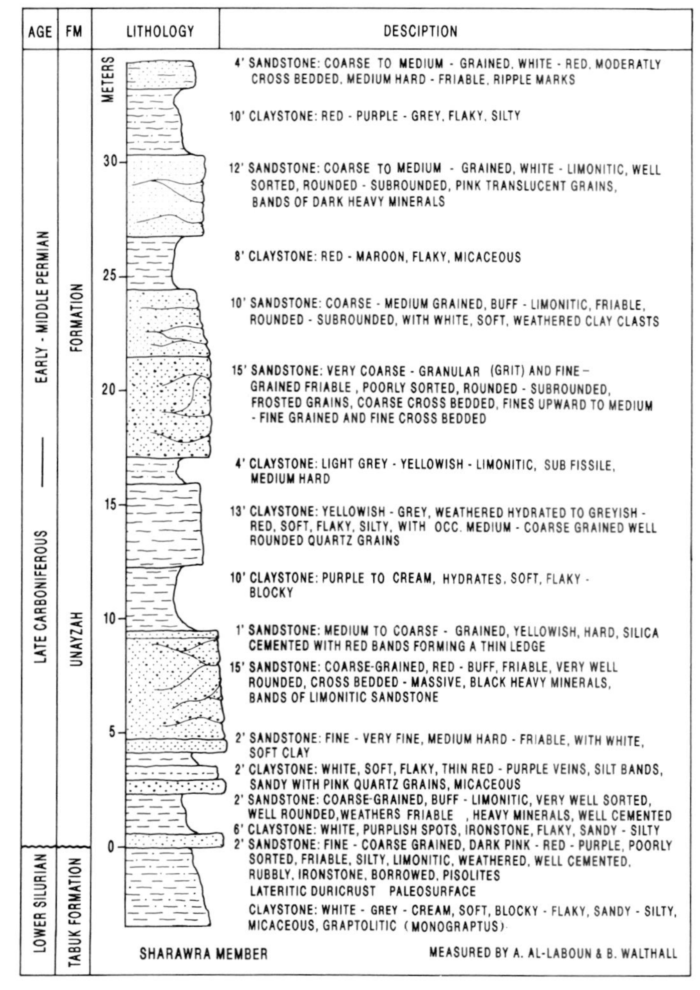


Figure 16. *Composite section of Unayzah Formation, reference section, Qusayba village—al-Qasim.*

### *Overlying Formation*

In outcrop as well as in the subsurface, the Unayzah–Khuff contact is picked at the base of the lowest, well-developed limestone bed of the Khuff Formation overlying the varicolored shales and sandstones and thin, impure limestone beds of the Unayzah Formation (Figure 15). The relationship between the Khuff and Unayzah is conformable if not transitional, and palynomorphs of all ages from Late Carboniferous to Late Permian have been extracted from cuttings and cores.

### *Depositional Environments*

The cyclic sandstones, shales, and impure limestones in the Unayzah suggest a fluctuating sea level. The ironstone may have been siderite at the time of deposition and, along with the abundance of fern fronds and other plant remains, suggests swampy conditions. The fluctuating Permian sea occasionally invaded the al-Qasim region, depositing impure limestones. Some of the limestone contains unidentified molluscs as do some of the thin sandstones. Channel sandstones in shales suggest stream erosion; but the green shales were probably deposited in a reducing environment. Foreset beds and ripple marks are associated with the lenticular sandstones in the upper part

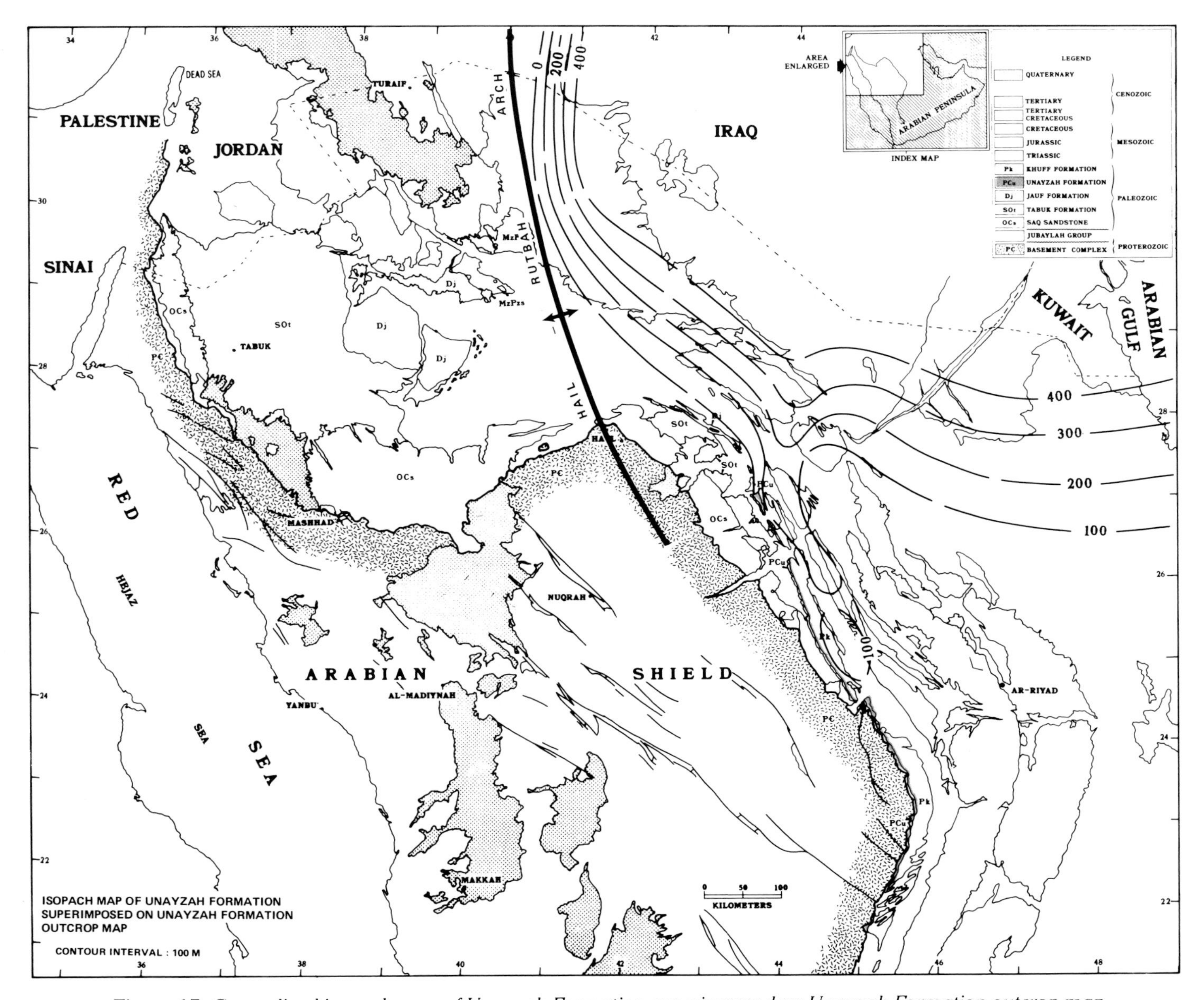


Figure 17. *Generalized isopach map of Unayzah Formation superimposed on Unayzah Formation outcrop map.*

of the Unayzah. Some of the sandstones have symmetrical ripples oriented N35°E, suggesting reworking of sands by seas transgressing from the northwest. These data conflict with the concept that the sediments came directly from the shield area to the west. Although the Unayzah outcrop is essentially parallel with the basement-sediment contact to the west, cross-bedding and ripple marks indicate current direction from the north-northwest. The southward transgressing Permian seas, therefore, reworked unconsolidated continental sediments, filling channels and depositing lenses of sandstones, leaving fingerprints of north to south currents.

### Late Permian: Khuff Formation

During Permian time, the Arabian peninsula was invaded by the first, major carbonate sediment that blanketed the entire region. The Khuff Formation marks the continuation of subsidence and the beginning of what is called the "layer cake" sediment of the Arabian basin. That evolved as a recognizable entity about this time. The Khuff Formation thins gradually toward central Arabia from the depocenter to the east and northeast. In the study area, the subsurface of the Khuff Formation can be traced as far as Badanah in the northern part of the Widyan basin (Figure 18). The Khuff Formation consists of a cyclic sequence of limestones, dolomites, dolomitic limestones, and shales, with thin anhydrite beds. The Khuff Formation conformably overlies the clastics of the Unayzah Formation and is conformably overlain by Sudair Shales (Figure 4).

## STRUCTURE AND TECTONIC SETTING

Figure 5 shows the main structural features of the Arabian Peninsula. The major structural elements in the study area are the Ha'il-Rutbah Arch, the Tabuk basin, the Widyan basin, and

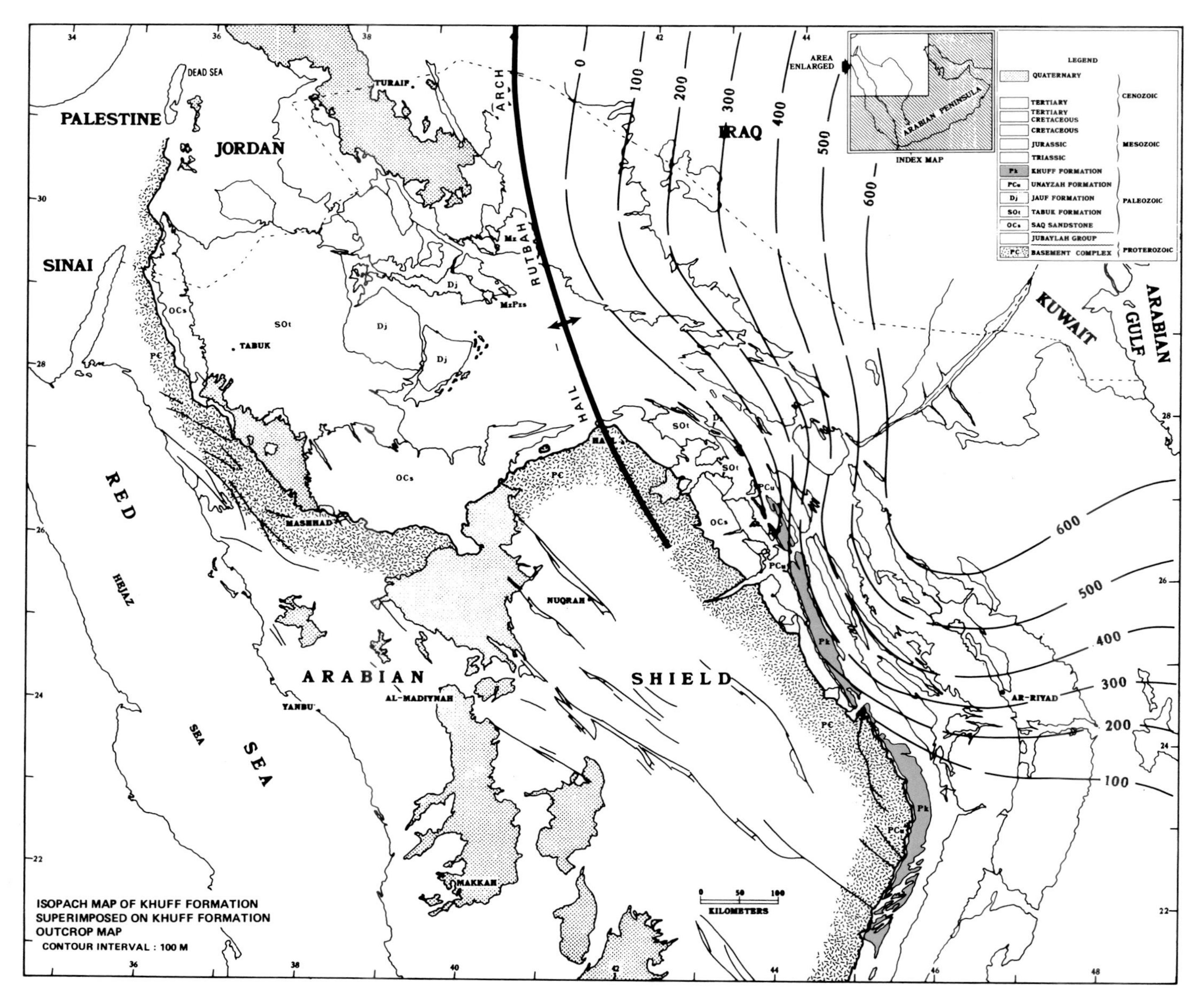


Figure 18. *Generalized isopach map of Khuff Formation superimposed on Khuff Formation outcrop map.*

the Summan Platform to the east of the Widyan basin (Figure 5). Other structural elements include the Central Arabian Graben and Trough System and its postulated northeastward extension beneath Wadi al-Batin; the poorly defined at-Tarafiyah fault; several faults mapped within the outcrop of the Saq, Tabuk, and Jauf Formations by previous investigators; and the Central Arabian Arch to the south.

### Ha'il-Rutbah Arch

Powers et al (1966) considered the Ha'il Arch as a northerly plunging anticline separating the Tabuk segment of the Interior Homocline from the Widyan basin margin. A general shift in the strike of the lower Paleozoic sedimentary rocks around Ha'il defines the arch. Subcrop patterns of Pre-Unayzah rocks (Figure 19) suggest that the Ha'il-Rutbah Arch is at least Late Devonian–Early Carboniferous age. Throughout the Phanerozoic, the arch must have been of low relief, possibly not higher than now, and may not have been a major source of sediments into either the Tabuk basin to the west or the Widyan basin to the east. The absence of Permian and Carboniferous sediments west of Badanah and Wadi as-Sulaimaniyah (NW-44) suggests that the Tabuk basin was filled by the end of the Devonian, and the Ha'il-Rutbah Arch may have been a barrier to the transgressing Carboniferous and Permian seas.

The Jawf region is complexly block-faulted while the carbonates and clastics of the Jauf Formation are highly deformed. This area is an uplifted block forming a horst where younger sediments were eroded.

### Tabuk Basin

West of the Ha'il-Rutbah Arch is the Tabuk basin. This basin is outlined by both outcrops and subcrops of Paleozoic units (found in boreholes) and the axis of the basin trends in a north-northeasterly direction (Figure 5). The western and southwest-

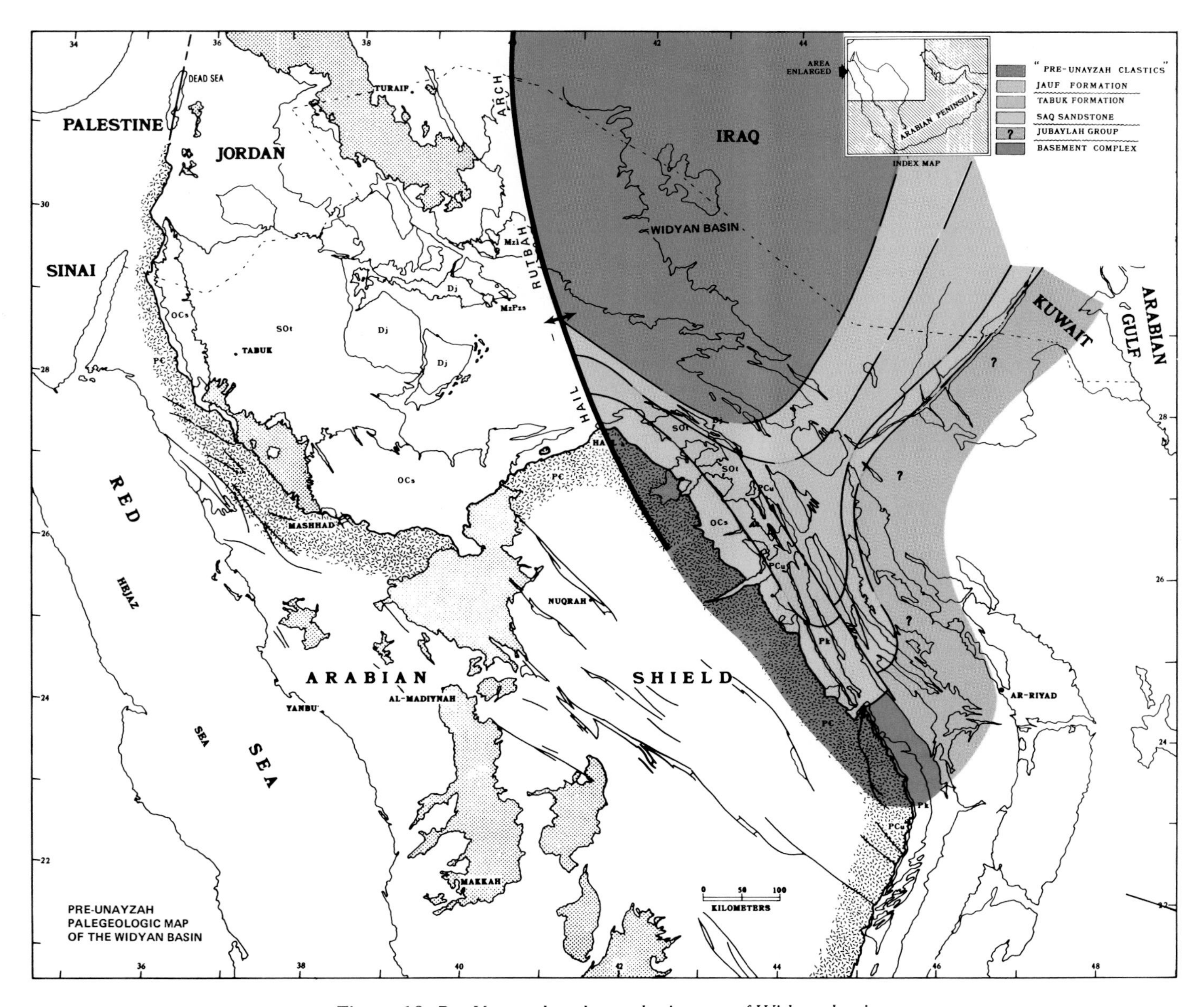


Figure 19. *Pre-Unayzah paleogeologic map of Widyan basin.*

ern flanks of the basin are outlined by Cambrian–Ordovician sandstones, while to the east, it is bounded by the Ha'il-Rutbah Arch. The northward extent of the basin is unknown, and such an investigation is beyond the scope of this study.

In the Tabuk basin, Cambrian–Devonian rocks are slightly dipping toward the east and northeast with several northwest striking faults dissecting the Saq and Tabuk Formations. The al-Qalibah and Wadi Dabl faults are the two main regional faults in the Tabuk region. The Wadi as-Sirhan and Khawr Umm Wu'al grabens represent the main, northwest striking faults in the region. The fault system in the Tabuk region may represent a surface reflection of deep seated basement faults. In general, these faults are high angle, normal faults with regional northwest trends.

## Widyan Basin

Powers et al (1966) named the eastern flank of the Ha'il-Rutbah Arch the Widyan Basin Margin. To the east of the arch, the Upper Cretaceous and Eocene rocks show a gentle northeast dip. Based on both outcrop and subcrop patterns, this study shows that the Widyan Basin Margin is the western flank of a north-northwest trending basin situated between the Ha'il-Rutbah Arch and the Summan Platform (Figure 5). Outcropping along the east flank of this subsurface basin is a segment of the Central Arabian Graben and Trough System with vertical displacement of several hundred feet. The axis of the Widyan basin is clearly outlined by subcrops of both the Unayzah Formation and older units (Figure 19). The western flank of the basin is outlined by subcrops and outcrops paralleling the edge of the basement-sediment contact to the southeast of Ha'il. The central part of the basin is filled with thick clastics of Middle–Late Devonian to Carboniferous age (Figures 13 and 19). All the wells studied in the Widyan basin (except those drilled on outcrops of Devonian and older rocks) encountered the

Carboniferous–Permian Unayzah Formation.

The geometry of the Widyan basin and the apparent northwest-to-southeast current direction may explain some of the stratigraphic problems seen in outcrops. These in particular are (1) the mature sandstones found adjacent to, or near, the shield; (2) the lateral distribution of the boulders considered to be ice rafted; and (3) the N35°E-trending ripple marks found in the sandstones of the Unayzah Formation in outcrop.

As mentioned previously, the Paleozoic sandstones are almost lacking in non-quartz minerals, even though many of these sediments crop out adjacent to, or only a few tens of kilometers from, the Arabian shield. The igneous and metamorphic rocks of the adjacent low-lying shield may have been important but were not the sole provenance of these sediments. The postulated eastward highlands and distant low-lying basement to the south also may have been important sources of these feldspar-free clastics.

In the Widyan basin, the Cambrian to Devonian rocks form a northeasterly dipping Qasim homocline. The Carboniferous–Permian Unayzah Formation, with a well-pronounced regional angular unconformity, overlies successively, from north to south, the lower part of the pre-Unayzah clastics, then the Jauf, Tabuk, and Saq Formations before resting on the Central Arabian Arch segment of the Arabian shield (Figures 8 and 13).

The Paleozoic rocks to the east and north of the basin are buried under a thick succession of younger sediments. The Middle Cretaceous Wasia Formation, in the Sakakah region to the north, is believed to rest with another major angular unconformity on the pre-Unayzah clastics as the pre-Wasia unconformity truncated the Lower Cretaceous, Jurassic, Triassic, Permian, Carboniferous, and Devonian rocks.

The Wadi ar-Rimah and Wadi ar-Risha represent the main, northeast striking faults that cut the basement complex and the overlying Paleozoic rocks.

### At-Tarafiyah Fault

The term "at-Tarafiyah fault" is proposed for a poorly defined, subsurface, normal fault with a down-to-the-west throw. The postulated fault, named after the at-Tarafiyah well, can be visualized from the offset in the subcrop pattern of the members of the Tabuk Formation (Figure 11) in the southern part of the Widyan basin. This fault is assumed to offset the underlying Saq (Figure 10) and perhaps the overlying Silurian members of the Tabuk. The amount of fault movement is unmappable on the basis of available data. Possible strike-slip movement along this fault, however, cannot be over emphasized.

### Central Arabian Graben and Trough System

Hancock, Al-Khatieb, and Al-Kadhi (1981) studied computer enhanced Landsat images and confirmed that the Wadi al-Batin, in part, occupies a trough that may be an extension of the Central Arabian Graben and Trough System. They proposed that the 1,000 km (620 mi) arcuate graben and trough system marks the boundary of a major tectonic subunit, and is a reflection of a weakness zone in the basement complex. The Central Arabian Graben and Trough System may be an antithetic fault system in Phanerozoic rocks along a zone of weakness in the basement, most probably a down-to-the-west normal or a left-lateral strike-slip fault. This fault may mark the contact between two types of basement rocks with low-grade meta-sediments to the east, and plutonics and highly metamorphosed rocks to the west of the fault (Figure 5). These low-grade metamorphics may in fact be a fold belt consisting of metasedimentary rocks overthrust westward onto crystalline rocks. Such being the case, the narrow Dibdibah basin mentioned by Powers et al (1966) may represent a taphrogenic breakdown behind the fold belt.

If the Central Arabian Graben and Trough System (Figure 5) marks the eastern margin of the Widyan basin, the graben beneath Wadi al-Batin may also reflect the northeastern edge of the plate as well as the margin of older Paleozoic rocks on the eastern flank of the Widyan basin. A thin sedimentary section of the Paleozoic rocks may be expected on and adjacent to the western edge of the overthrust plate. Furthermore, pre-Unayzah rocks of Paleozoic age are probably absent from the westernmost part of this plate.

### Summan Platform

The central and northwestern region of the Arabian basin is separated from the eastern sector by an arch or block trending north to south beneath the northern Tuwaiq segment of the Interior Homocline. This structurally high block is named the Summan Platform (Figure 5) and is composed of highly deformed and steeply dipping low-grade metamorphic sedimentary rocks.

### Hydrocarbon Potential

The thickness of Paleozoic sediments in the Tabuk and Widyan basins (Figure 7 and 8), as well as their lateral facies variations combined with tectonic activity, are fundamental for the evaluation of these sediments as potential hydrocarbon reservoirs. Continued subsidence and marine sedimentation were interrupted by cyclic epeirogenic uplift movements allowing the marine organic-rich shales to alternate with the non-marine potential reservoir rocks. Wedge-out in these clastic sediments plays a significant role as a stratigraphic trap.

The thick cyclic nature of the clastic sedimentation of the Ordovician and Silurian Tabuk Formation provides a favorable condition for hydrocarbon generation and entrapment. The graptolitic shale members and the fine-grained silty marine sediments may be regarded as good, organic-rich, source rocks. The thin, coralline limestone members of the Devonian Jauf Formation could serve as a source as well as reservoir beds as these limestones thicken and are interbedded with shales and siltstones.

The Carboniferous–Permian Unayzah clastics show important stratigraphic trap potential as they transgress over subcrops of older units (Figure 8). These clastics are significant targets where they overlie the Qusaiba Shale Member and other potential source rocks.

The carbonates of the Khuff Formation have all characteristics of favorable source and reservoir rocks. As these carbonates become thicker to the east and northeast, they have a significant hydrocarbon potential.

The Tabuk and Widyan basins have experienced considerable tectonic activity. This, combined with the stratigraphic aspects of the basins, makes them attractive future oil and gas provinces possessing both structural and stratigraphic traps (Figure 7).

The Ha'il-Rutbah Arch (Figure 5) is the major antcline recognized in the region by surface mapping and borehole data. This arch is thought to separate the Tabuk homocline to the west and the Qasim or Widyan homocline to the east (Figure 7). Complex fault systems cutting the Paleozoic outcrops mapped in the two basins may represent a surface reflection of deep-seated basement faults. The faults are mainly high-angle, normal block faults with a regional, northwest-to-southeast strike. The wadi as-Sirhan and Khawr Umm Wu'al grabens and al-Qalibah and Wadi Dabl faults are the major faults in the study area.

Large reserves of hydrocarbons were discovered from similar Paleozoic sediments in the eastern part of the Arabian basin (Figure 1). More than half of the nonassociated gas reserves of Arabia are in the Permian Khuff Formation in the Haradh, Ain Dar, Shyedgum, Uthmaniyah, and Hawiyah areas of the Ghawar Field. Gas is also found in Abqaiq, Berri, Damman, Qatif, and Abu Sa'fah (Oil and Gas Journal 1983). Pre-Khuff Paleozoic gas was discovered at the Haradh and Uthmaniyah areas of the Ghawar Field and at the Abu Sa'fah Field.

In the Gulf States, gas was discovered in the deep reservoirs of the Khuff and pre-Khuff Paleozoic section. The biggest known gas accumulation in the Khuff reservoir is the North Dome Structure, offshore Qatar. In Oman, the Permian–Carboniferous and Ordovician reservoirs contain oil.

## CONCLUSIONS

The Tabuk and Widyan basins, separated by the Ha'il-Rutbah Arch, received more than 5,000 m (16,400 ft) of Paleozoic sediments.

Paleozoic Rocks are recognized both in outcrop and in the subsurface where a more complete section of Cambrian–Permian is encountered.

The clastic section, subjacent to the Khuff Formation, has been named the Unayzah Formation with its type locality at the city of Unayzah and a reference section at the Qusayba area, al-Qasim. The contact of the Unayzah with the overlying Khuff Formation is conformable, while the lower contact represents a major unconformity with Late Carboniferous age rocks overlie Early Carboniferous or truncated older units that locally have been subjected to block faulting. The Unayzah Formation, in the study area, is limited to the Widyan basin.

The Unayzah marks a major change in the geologic history of the region as it is thickest in low areas and thinnest over old structures, resulting in a near peneplain whereon the Khuff Carbonates were deposited.

The clastic section, penetrated in the northern part of the Widyan basin, subjacent to the Unayzah Formation and superadjacent to the Hammamiyat Limestone Member of the Jauf Formation, is here informally introduced as the "pre-Unayzah Clastics".

The deposition patterns around the Ha'il-Rutbah Arch suggest that it is Precambrian in age, but the absence of both the Carboniferous and Permian sediments to the west of the arch indicate a rejuvenation of the arch's movement in Late Devonian–Early Carboniferous time.

The depocenter east of the Ha'il-Rutbah Arch is outlined, mapped, and named the Widyan basin. The subcrop patterns suggest that the eastern margin of the basin may be coincident with the central Arabian Graben and Trough System and Wadi al-Batin.

The Arabian shield may not, therefore, have been the primary source of clastics within the Widyan basin. The metasediments of the Summan Platform to the east and southeast may, therefore, have been the primary source of sediments.

The igneous and metamorphic exotic cobbles and boulders found on the outcrop of the intervening sandstone between both the Ra'an and Qusaiba Shale Members, and on top of the Qusaiba Shale Member in al-Ajfar area, are thought to be fluvial and mud-flow deposits transported from an upthrown block of basement rocks east of the at-Tarafiyah fault rather than glacially transported from the westward shield area.

The continuous, thick, alternating marine and non-marine sediments consist of organic rich shales overlain by potential reservoir rocks with considerable tectonic activity. This activity makes the two basins, especially the Widyan basin, attractive targets as future oil provinces possessing both structural and stratigraphic traps.

The Unayzah clastics, in particular, possess stratigraphic trap potential as they transgress over older units with angular unconformity. The Unayzah clastics are important targets as they overlie possible, mature source rocks of the Silurian and Ordovician shales.

The Upper Permian Khuff carbonates of the Widyan basin have all the characteristics of potential source and favorable reservoir rocks. They offer the greatest hydrocarbon potential toward the northeast of the Widyan basin, where they are thicker and more deeply buried. The Khuff Formation is a major gas reservoir in the Gulf region.

## ACKNOWLEDGMENTS

This paper is a extracted with some modification, from my Ph.D. dissertation which was carried out at King Abdulaziz Univerisity under the direction of both Professor S. Omara of that University and Dr. Bennie H. Walthall of ARAMCO. The manuscript was typed by A. S. Haq.

I would like to thank the Ministry of Agriculture and Water for supplying samples and wireline logs, and for granting permission to publish this study. I am also thankful to the Ministry of Petroleum and Mineral Resources for their approval to publish it. My thanks are extended to the Geological Department of the Arabian American Oil Company who permitted me to undertake this study as a special project.

## REFERENCES

Abel, O., 1935, Vorzeitige Lebensspuren—Jena.

ARAMCO, 1980, Generalized Saudi Arabian stratigraphic section—formational and zonal nomenclature with generalized lithology and drilling notes, Dhahran.

Ayres, M. G., et al, 1982, Hydrocarbon habital in main producing areas, Saudi Arabia: AAPG Bulletin, v. 66, n. 1, p. 1–9.

Bahafzallah, A., A. Jux, and S. Omara, 1981a, Stratigraphy and facies of the Devonian Jauf Formation Saudi Arabia: Neues Jahrbuch für Geologie und Paläontologie Monatshefte, 1, 1–18, Stuttgart.

Basahel, A. N., et al, 1984, Early Cambrian platform of the Arabian shield: Neues Jahrbuch für Geologie und Paläontologie Monatshefte H.2, 113–128.

Bigot, M., 1970, Geology of the Tabuk and Jauf Formations in the Wadi Al-Fajr area: Bureau de Recherches Geologiques et Minieres, 70, Jed. n. 28; Jeddah.

——, and C. Lafoy, 1970, The Paleozoic series in the Tabuk basin: Bureau de Recherches Geologiques et Minieres, 70, Jed. n. 11, Jeddah.

Blackenhorn, M., 1914, Syrien Arabian and Mesopotamien, Handbüuch der regionalen geologic: Heidelberg 5, Abt. 4, H. 17.

Bramkamp, R. A., and L. F. Ramirez, 1958, Geologic map of Northern Tuwayq Quadrangle, Kingdom of Saudi Arabia: United States Geological Survey Miscellaneous Geologic Investigations Map 1-207A.

——, et al, 1963, Geologic map of the Wadi as Sirhan quadrangle: United States Geological Survey Miscellaneous Geologic Investigations Map I-200 A.

——, G. F. Brown, and A. E. Pocock, 1963, Geology of the Wadi Ar Rimah Quadrangel, Kingdom of Saudi Arabia: United States Geological Survey Miscellaneous Geologic Investigations Map I-206 A.

Brown, G. F., 1972, Tectonic map of the Arabian peninsula: Saudi Arabia, Dir. Gen. Min. Res. Geol. Map AP-2.

——, and R. O. Jackson, 1960, The Arabian shield: Copenhagen, 21 International Geological Congress, 9: 69–77.

Clark-Lowes, D. D., 1980, Sedimentology and mineralization potential of the Saq and Tabuk Formations: Qassim district: Imperial College Open-File Report CRC/1C-7.

Cloud, P., S. M. Awramik, and D. G. Hadley, 1979, Earliest Phanerozoic or latest Proterozoic fossils from the Arabian shield: Directorate General of Mineral Resources, Saudi Arabia, Proj. Rep. 260: 1–39; Jeddah.

Delfour, J., 1967, Report on the mineral resources and geology of the Hulayfah-Musayna'ah region: Bureau de Recherches Geologiques et Minieres Open-file report 66-A-8, 139 p.

——, 1970, The J'balah group, a new unit of the Arabian shield: Bureau de Recherches Geologiques et Minieres report 70-JED-4.

——,1975, Geology and mineral exploration of the Nuqrah quadrangle, 25/41 A: Bureau de Recherches Geologiques et Minieres report 75-JED-28.

——, 1977, Geologic map of the Nugrah quadrangel 25E, Directorate General of Mineral Resources. Map-28 scale 1:250,000.

——, 1982, Geology and mineral resources of the northern Arabian shield, a synopsis of Bureau de Recherches Geologiques et Minieres investigation 1965–1975: Open-file report Bureau de Recherches Geologiques et Minieres-of-02-30.

——, et al, 1982, Explanatory notes to the geologic map of the ad-Dawadimi quadrangle: Sheet 24G.

El-Khayal, A. A., W. G. Chaloner, and C. R. Hill, 1980, Paleozoic plants from Saudi Arabia: Nature, v. 285, 33–34.

Gorin, G. E., L. G. Racz, and M. R. Walter, 1982, Late pre-Cambrian–Cambrian sediments of Huqf Group, Sultanate of Oman: AAPG Bulletin v. 66, n. 12, p. 2609–2627.

Hadley, D. G., 1973, Geology of the Sahl al-Matran quadrangle, northwestern Hijaz: Saudi Arabia, Dir. Gen. Min. Res. Geol. Map GM-6 scale 1:100,000.

——, 1974, The taphrogeosynclinal Jubaylah group in the Mashhad area, northwestern Hijaz: Saudi Arabia Dir. Gen. Min. Res. Bull. v. 10, p. 189.

——, 1975, Geology of the Qal'at as-Sawarah quadrangle: U.S. Geological Survey map, GM-24.

Hancock, P. L., and A. Kadhi, 1978, Analysis of mesosopoc fractures in the Dhruma—Nisah segment of the Central Arabian Graben System: Journal Geological Society of London, v. 135, p. 339–347.

——, S. O. Al-Khatieb, and A. Al-Kadhi, 1981, Structural and photogeological evidence for the boundaries to an East Arabian block: Geological Magazine, v. 118, n. 5, p. 533–538.

Helal, A. H., 1964a, On the occurrence of lower Paleozoic rocks in Tabuk area: Saudi Arabia—Neues Jahrbuch für Geologie und Paläontologie Monatshefte, 7, 391–414, Stuttgart 1964.

——, 1964b, On the occurrence and stratigraphic position of Permo-Carboniferous tillites in Saudi Arabia: Geologisches Rundschau, 54, 193–207, Stuttgart.

——, 1965, General geology and litho-stratigraphic subdivision of the Devonian rocks of the Jawf area, Saudi Arabia: Neues Jahrbuch für Geologie und Paläontologie Monatshefte, 9, 527–551, Stuttgart.

——, 1968, Stratigraphy of outcropping Paleozoic rocks around the northern edge of the Arabian shield: Z. Dt. Geol. Ges., 117 (jq. 1965): 506–543; Hannover.

Hemer, D. O., 1965, Application of palynology in Saudi Arabia: Fifth Arab Petroleum Congress, Cairo.

——, 1968, Diagnostic Palynological fossils from Arabian Formations: 2nd Regional Technical Symposium, Society of Petroleum Engineers of American Institute Mining Engineers, Saudi Arabia Section, Dhahran.

——, and P. W. Nygreen, 1967a, Algae, acritarchs and other microfossils incertae sedis from the Lower Carboniferous of Saudi Arabia: Micropaleontology v. 13, n. 2, p. 183–194.

——, and ——, 1967b, Devonian palynology of Saudi Arabia: Review of Paleaeobotany and Palynology, v. 5, p. 51–61; Amsterdam.

Kennedy, W. Q., 1964, The structural differentiation of Africa in the pan-African (±500 ma) tectonic episode: Res. Int. Afr. Geology Leeds Univ., 8th Ann. Rep., p. 48–49.

Kober, L., 1919, Geologische Forschungen in Vorderasien. 2. Teil: Das Nordliche Hegaz—Denkchr. Akad. Wiss. Wien, mat.—nat. Kl., 96, 779–820, Wien.

al-Laboun, A. A., 1982, The subsurface stratigraphy of pre-Khuff Formations in central and northwestern Arabia: Jeddah, Saudi Arabia, King Abdulaziz University, unpublished Ph.D. thesis, 102 p., 32 fig., 21 pl.

McClure, H. A., 1978, Early Paleozoic glaciation in Arabia: Palaeogeography Palaeclimatology, Palaeoecology, v. 25, n. 4, p. 315–326.

Oil and Gas Journal, 1983, Middle East Report; October 10.

Pascha, A., 1908, Die Hedschazbahn, 2. Teil: Ma'an bis El-Ula-Peterm. geogr. Mitt., Erg.—H., 161, Gotha.

Picard, L., 1937, On the structure of the Arabian Peninsula: Bulletin of the Geology Department, Hebrew University, 3. Series 1, 1–12, Jerusalem 1937.

——, 1941, The Precambrian of the North Arabian-Nubian Massif: Bulletin of the Geology Department, Hebrew University, v. 3. n. 3–4, p. 1–30, Jerusalem.

——, 1942, New Cambrian fossils and Paleozoic Problematica from the dead Sea and Arabia: Bulletin of the Geology Department, Hebrew University, v. 4, n. 1, Jerusalem.

Powers, R. W., 1968, Lexique stratigraphique international: Asie, v. 3, Fasc. 10, b. 1, Saudi Arabia. Centre National de la Recherche Scientifique, Paries.

——, et al, 1966, Geology of the Arabian Peninsula—

Sedimentary Geology of Saudi Arabia: U.S. Geological Survey Professional Paper, 5460-D: Washington.

Sharief, F. A., 1982, Lithofacies distribution of the Permian-Triassic rocks in the Middle East: Journal of Petroleum Geology, v. n. 4, p. 3.

Steineke, M., R. A. Bramkamp, and N. J. Sander, 1958, Stratigraphic Relations of Arabian Jurassic Oil *in* L. G. Weeks, ed., Habital of Oil, AAPG symposium: AAPG, Tulsa, Oklahoma, 1384 p.

Stoeser, D. B., and J. E. Elliott, 1980, Post-orogenic peralkaline and calc-alkaline granites and associated mineralization of the Arabian shield, Kingdom of Saudi Arabia, *in* I.A.G. Bulletin n. 3, Evolution and mineralization of the Arabian Nubian shield, v. 4, published for Faculty of Earth Sciences, King Abdulaziz University, Jeddah: Oxford, Pergamon Press.

Szabo, F., and A. Kheradpir, 1978, Permian and Triassic stratigraphy, Zagros basin, south-west Iran: Journal of Petroleum Geology, 1, 2, pp. 57-82.

Thomas, A. T., 1977, Classification and phylogeny of homalonotid trilobites: Palaeontology, v. 20, pt. 1., pp. 159-178, pls. 23-24.

Thralls, W. H., and R. C. Hasson, 1956, Geology and oil Resources of Eastern Saudi Arabia: 20 Congrso Geologico International, 2, Asia & Oceania; Mexico City.

Vaslet, D., J. M. Brosse, and Y. M. Le Nindre, 1982, Geology of the phanerozoic of the ad-Dawadimi Quadrangle: Sheet 24G, Bureau de Recherches Geologiques et Minieres -OF-01-29.

# The Southern Mozambique Basin: The Most Promising Hydrocarbon Province Offshore East Africa

M. De Buyl
*Western Research, Division of Western Geophysical Co.*
*Houston, Texas*

G. Flores
*Consultant*
*Impruneta, Italy*

Recent offshore acquisition of approximately 12,800 km (7,950 mi) of seismic reflection data, with some gravity and magnetic profiles encompassing the southern half of the Mozambique basin, reveals new facets of the subsurface geology. Integrated interpretation of these new geophysical data with old well information allows the development of depositional and tectonic models that enhance the basin's hydrocarbon potential. Previous drilling was sparse, however, predating modern seismic technology and exploration philosophy, and left the area classified as a frontier province. Despite some encouraging hydrocarbon shows, previous operators considered the entire basin gas-prone and immature. Recent interpretation suggests that the basin's geothermal history must have been more favorable than generally was inferred from present-day gradients, and oil-prone source rocks may be buried below the sequences drilled during wildcatting. Also, marine oil-prone source rocks may be present. Furthermore, the liquid hydrocarbon sourcing potential of the Karroo coal measures should not be underestimated.

Wildcatters have not, in retrospect, optimally positioned exploratory wells in the vicinity of ample structural closures; nor have they drilled to sufficient depth, owing to poorly resolved, outdated seismic data and to erroneous age assignments to lava flows. Therefore, the viability of prospects mapped along a major offshore extension of the East African Rift System, delineated by this new survey, has not been adequately assessed. Finally, a well remains to be drilled within the rift to assess the presence of reservoirs and the maturity, richness, and volume of source rock sequences not necessarily represented over the eroded structures drilled to date.

## INTRODUCTION

The probability of finding liquid hydrocarbons along a continental margin increases with the size of the offshore expression of coastal basins, the width of the continental shelf, and the volume of sediments deposited. Geologic factors modify this simplistic statement in controlling the distribution of source and reservoir rocks and their respective qualities. However, and more importantly, preserving the source intervals available for subsequent maturation requires syndepositional euxinic environments.

On passive, continental margins, this condition chiefly depends either upon the development of paralic troughs or upon the existence of a ridge in the distal part of the basin limiting the extent of the sedimentary wedge. As Hedberg (1964) points out, "We know of few major oil fields formed in a simple homoclinal apron of sediments extending out from the continental margin into the ocean depths." Subsequently, the basin's geothermal and burial history controls the maturation, while structural and stratigraphic traps have to develop to allow the resulting hydrocarbons to accumulate in pools of economic interest. This paper presents favorable evidence that highlights the hydrocarbon potential of the Southern Mozambique basin in comparison with adjacent basins along the east coast of the African continent.

Along some 7,000 km (4,350 mi) of the East African coast between Cape Guardafui and Port Elizabeth, five sedimentary basins have developed since late Karroo (Liassic) times (Figure 1). Their respective sizes, including their offshore portions, vary considerably, as does the width of the continental shelf toward their eastern boundaries where exploratory and production drilling would be economical.

The geological history of these coastal basins acquires a common character from the Early Cretaceous marine transgression onward, whereas considerable differences may exist in their Middle and Late Jurassic development. At present, this earlier part of their growth is better understood in those basins located north of lat. 10° S, where Middle and Upper Jurassic marine sequences are known both in outcrop and in the subsurface.

The Somali Coastal basin in the extreme north of East Africa stretches for about 1,900 km (1,180 mi) along the shoreline southward to Mombasa (Kenya) where the shelf width barely reaches 25 km (15 mi) (Figure 1). This basin is the largest in East Africa and encompasses the 300 km (185 mi) Lamu embayment. The sedimentary continental margin widens to the south of Lamu and continues astride the Rovuma river through the Tanzanian basin into northern Mozambique. The Rovuma basin is about 150 km (93 mi) long onshore and less than 20 km (12 mi) long offshore.

A narrow sedimentary strip 10 to 20 km (6 to 12 mi) wide borders the large bulge of the Precambrian metamorphic Mozambique Belt. Further south, the onshore part of the Mozambique basin from lat. 17° to 26°40′ S covers a surface of approximately 265,000 sq km (102,325 sq mi) encompassing 1,825 km (1,135 mi) of coastline. Along the Zambezi River re-entrant, the Mozambique basin is 300 km (185 mi) wide and reaches a maximum width of 440 km (273 mi) around lat. 22°30′ S between Pafuri and the coast.

In comparison with other coastal basins of East Africa, the Mozambique basin is second in size only to the Somali Coastal basin, whereas its offshore extension on the continental shelf is for the most part much wider. The Tanzanian basin, including its southern appendix of the Rovuma basin in northern Mozambique, is only 85,000 sq km (32,820 sq mi), while the Lamu embayment covers 115,000 sq km (44,405 sq mi).

The continental shelf portion of the Southern Mozambique basin between Beira and the South African border is the widest of offshore East Africa. From south to north along the East African coast, the shelf is only about 50 km (30 mi) wide offshore Durban, widening to 130 km (80 mi) between lat. 26° and 25° S and to 140 km (85 mi) offshore in the Zambezi River delta. In Tanzania and Kenya, the shelf is seldom over 30 km (19 mi) wide, becoming even narrower between Lamu and Cape Guardafui.

Deep wells in Mozambique on- and offshore are sparse and far between. Only 12 exploratory tests offshore on the inner continental shelf and 44 wells onshore were drilled by former operators (SECH, n.d.). These two groups of wells represent densities of one well per 8,750 sq km (3,380 sq mi) and one well per 4,065 sq km (1,570 sq mi), respectively. In addition, because the thickness of sediments is greater on the Southern Mozambican shelf than elsewhere, the statistics of well density per unit of sediment volume further indicate the sparse sampling of the hydrocarbon potential, thus leaving very large areas untested.

Onshore at Pande, Buzi, and Temane (Figure 2), the discovery of gas with condensates, in addition to the several gas and oil shows in many of the wells drilled offshore (SECH, n.d.), indicates that thermogenic conditions do exist in this large basin. Because of favorable maturation and sedimentary conditions and of well density considerations, we believe that this area should rank as the most promising offshore in East Africa.

Historically, the presence of gas seeps in the Inhaminga area (Figure 2) led to the first investigations, with the first deep well drilled in 1936. Thereafter, between 1948 and 1967, Gulf Oil alone (1948–1958), and later in association with Pan Am (1959–1967), explored a large portion of the Southern Mozambique basin both onshore and offshore out to water depths of 30 m (98 ft). They discovered three gas fields onshore at Pande, Buzi, and Temane, with daily production tests from Upper Cretaceous reservoirs of 25.7, 8.7, and 6.7 thousand cu ft (mcf), respectively. While Pande's reserve estimates are set at 1.3 trillion cu ft (tcf), the other two fields were considered uneconomical.

After giving up a large portion of the onshore acreage in 1967, geologists investigated the offshore area (Figure 3), leading to the 1970 drilling of two wells: Sofala-1 and Nemo-1. Each encountered gas-bearing zones.

Because of its considerable spatial extent, the prospective structural closures along the Nemo–Sofala axis may trap as much as 5.1 tcf of dry gas in the Domo Sand reservoirs. Geolo-

Figure 1. *Regional geologic settings of East Africa. The size of the South Mozambique basin relative to the North Sea emphasizes the area's significance as a future petroleum province. Note the variation in the shelf width [shoreline to the 200 m (656 ft) isobath] along the East African shore.*

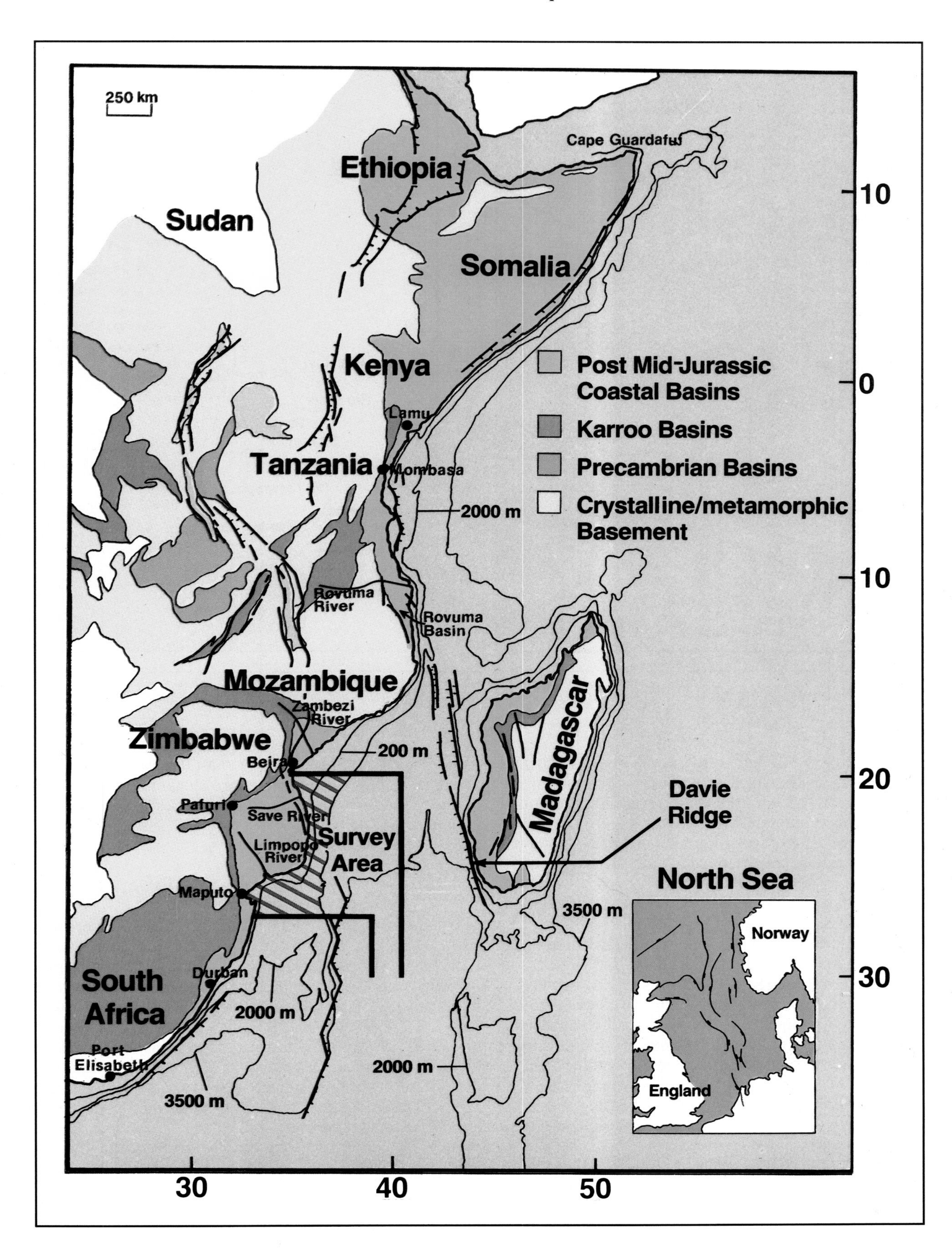
250 km
Cape Guardafui
Ethiopia
Sudan
Somalia
Kenya
Lamu
Tanzania
Mombasa
2000 m
Post Mid-Jurassic Coastal Basins
Karroo Basins
Precambrian Basins
Crystalline/metamorphic Basement
Rovuma River
Rovuma Basin
Mozambique
Zambezi River
Zimbabwe
200 m
Beira
Madagascar
Davie Ridge
Pafuri
Save River
Limpopo River
Survey Area
Maputo
North Sea
3500 m
Norway
England
South Africa
Durban
2000 m
Port Elisabeth
3500 m
2000 m
10
0
10
20
30
30
40
50

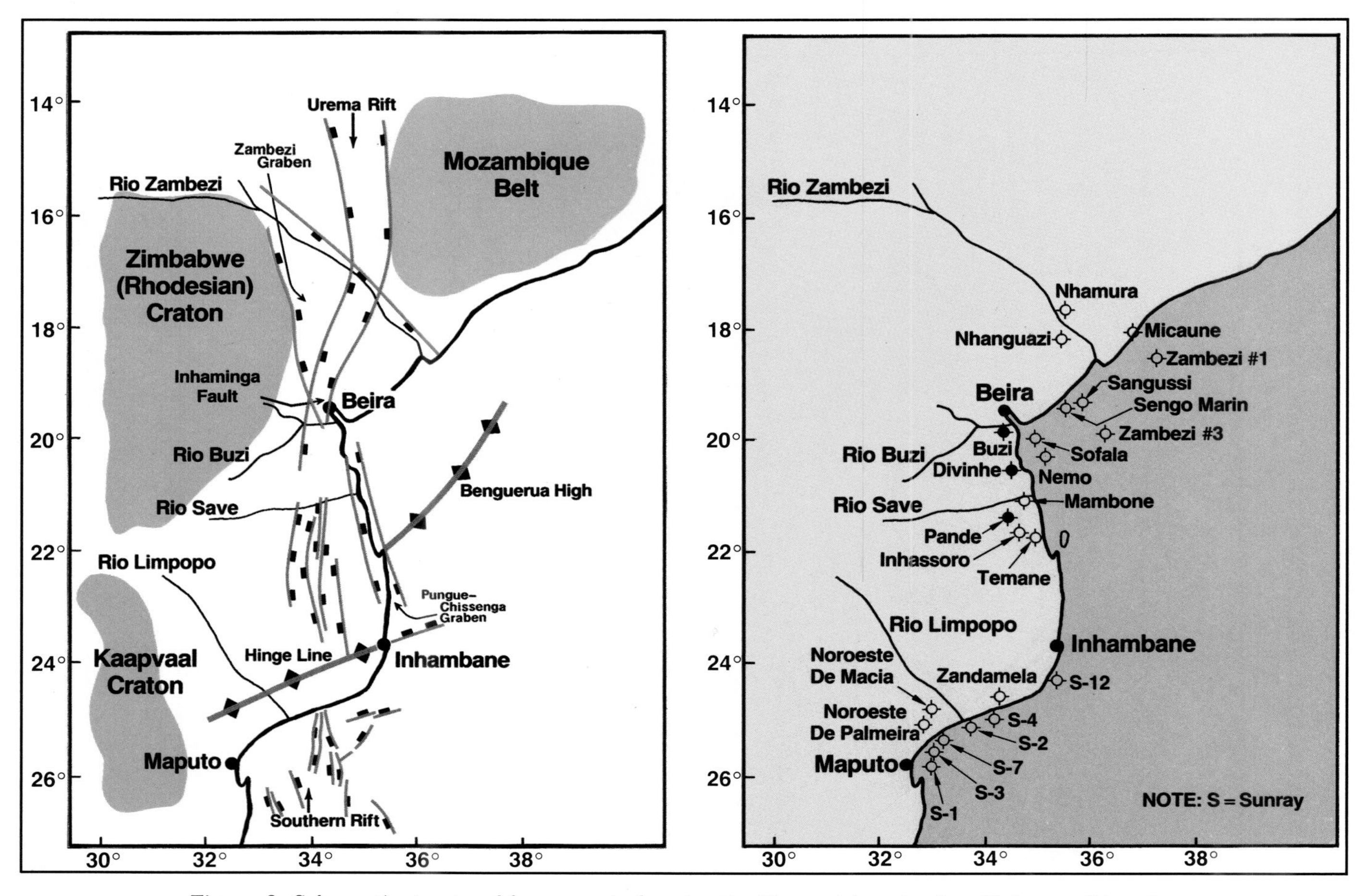


Figure 2. *Schematic structural framework showing the Mozambique basin with key well locations.*

gists encountered 30° API oil shows in the Grudja sands and heavy oil staining in the Oligocene deltaic sediments in the Temane and Zambezi-3 well. Elsewhere, drilling continued until 1972 without commercial oil discoveries.

Pessimistic views regarding the liquid hydrocarbon potential in the Mozambique basin stem from common observations:

(1) The only exploitable hydrocarbon discoveries to date have been gas.
(2) The only known source rocks are of mid-Cretaceous age, contain predominantly land-derived organic carbon, and were, hence, considered to be gas prone.
(3) Current geothermal gradient is such that the oil window is below the basement interpreted from the old seismic data, suggesting that existing hydrocarbons result from biodegradational processes not generally conducive to oil generation.
(4) The old seismic survey did not reveal sizable structural traps.

Modern seismic data, characterized by both higher resolution and deeper penetration, cover large portions of the shelf previously unexplored (Figure 3), thus allowing prior interpretations to be challenged. In comparison with the number of dry wells drilled in the North Sea prior to the first economic oil discovery, well density four times lower than that of this similarly sized province characterizes the entire Mozambique basin (onshore and offshore), hence, leaving the area classified as a frontier area.

## GEOLOGIC SETTINGS

The entire Mozambique basin encompasses over 300,000 sq km (115,839 sq mi) onshore and on the continental shelf and is the largest African sedimentary basin below the equator. Figure 1 shows that this basin is comparable in size and tectonic style to the North Sea. The basin is bounded:

(1) To the north by the Mozambican fold belt affected by the Pan–African orogeny (500 Ma), overprinting a dominant northwest to southeast grain on the chiefly metamorphic formations that have undergone multiple, Precambrian tectonic phases;
(2) To the northwest by the Precambrian shield of the Zimbabwe (Rhodesian) craton, whose structural grain orientated in a northeast to southwest direction; and
(3) To the southwest by the Kaapvaal craton delineated on its eastern flank by the volcanic chain of the Lebombo, of Liassic age.

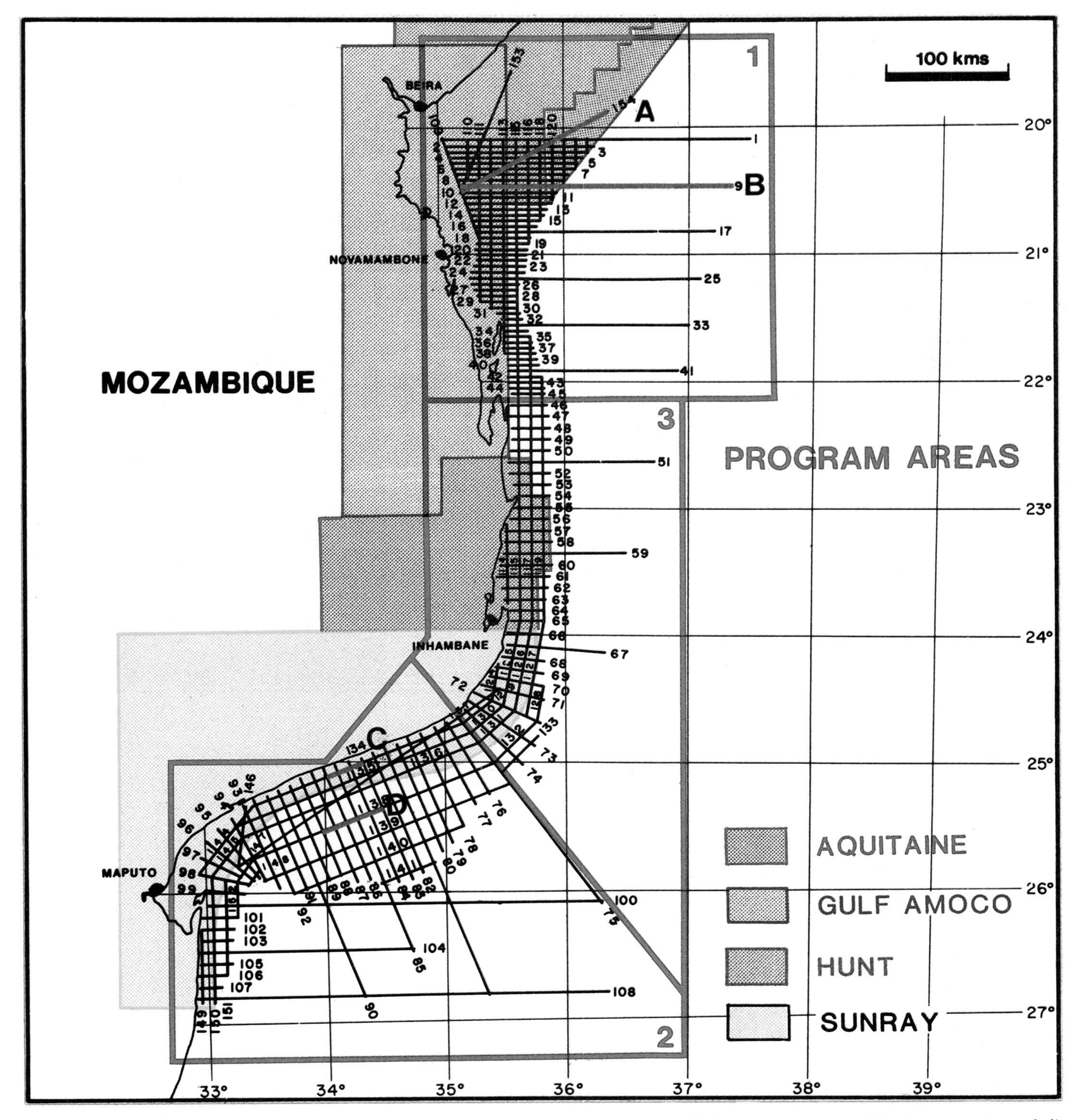


Figure 3. *Seismic survey and sample line location map shows previously held (pre-1984) concession areas. The program was subdivided into three areas of different line spacings (1, 2, and 3); the survey straddles the continental shelf and slope between the 15 and 500 m (50 and 1,640 ft) isobaths, while regional lines extend to the 2,000 m (6,560 ft) isobath.*

We can divide the geologic evolution of the southern part of the Mozambique basin into two distinct phases, separated by a major unconformity:

(1) An intra-cratonic rifting phase, and
(2) A passive-margin/coastal-basin phase.

Both phases are characterized by contrasting depositional regimes represented by the following sedimentary units (Figure 4):

(1) A thick, Karroo sequence of a postulated continental nature that culminates in volcanic episodes represented by pyroclastics and lava flows, increasing in frequency toward the top of the sequence, in response to rifting; and
(2) A transgressive sequence represented by a gradual encroachment of marine clastic sediments regulated by global sea level changes (Figure 5).

## Stratigraphy

The Karroo Supergroup comprises continental sequences ranging from Permian–Carboniferous to Liassic in age, and is only represented in Mozambique by two units, from top to bottom:

(1) The Stormberg Series: represented by volcaniclastics, lava flows, and sandstones, and correlating with the Isalo Formation in Madagascar. (The Beaufort Series, characterized by sandstone and shales in South Africa, is missing in both Mozambique and Madagascar.)
(2) The Ecca Series: associated with the deposition of coal measures, shale, siltstone, and sandstone, and correlating with the Sakamena Formation in Madagascar.

Based upon outcrop data and limited subsurface evidence onshore, geologists have shown that the marine transgression began at least in the Early Cretaceous. Indeed, marine sediments, assigned to the Maputo Formation of Albian–Aptian age (Flores, 1973), have been found interbedded between the Lebombo basalts and rhyolites south and west of the capital city, Maputo. Based on seismic evidence and palynology at the Nhamura well, we believe that an Upper-Jurassic transgression took place in Mozambique, as observed in Tanzania, Kenya, and Madagascar. Furthermore, the eastward increase in marine character of the Ecca Formation and of its time equivalent, the Sakamena Formation in Madagascar, westward (Flores, 1970), may testify to the onset of a marine incursion as early as in the Middle Permian, possibly associated with crustal stretching preceding rifting. Toward the east, the Maputo clastic sequences grade upward into a lower Cenomanian sandstone sequence characterized by a local turbiditic facies and a rich ammonitic fauna. The Maputo Formation is overlain in the subsurface by the Domo Formation consisting of carbonaceous shales with sandy intercalations. Onshore, however, the Lower Cretaceous marine Maputo Formation grades northwestward into the thick, continental Sena Formation.

Below the Maputo Formation, the Sunray well Noreste de Palmeira (Figure 2) contains a section of "Red Beds" about 100 m (330 ft) thick resting on basalt (K/AR 285.6 $\pm$14.3 Ma). These red beds are probably of Jurassic age and in a continental facies likely to become marine eastward. Similar red beds outcrop in the lower Zambezi in the same stratigraphic position (Flores, 1964, 1973) and are called the Belo Formation.

The lower Domo Shales, which may have been deposited in a pelagic environment and under euxinic conditions, contain up to 1% organic carbon of humic origin and were at one time postulated to be the source rock for the wet gas discovered in Grudja sands onshore. However, measurements of vitrinite reflectance performed on samples from the onshore wells suggest that these black shales are immature. The maximum values as reported in SECH's archives (SECH, n.d.) did not exceed 0.6%.

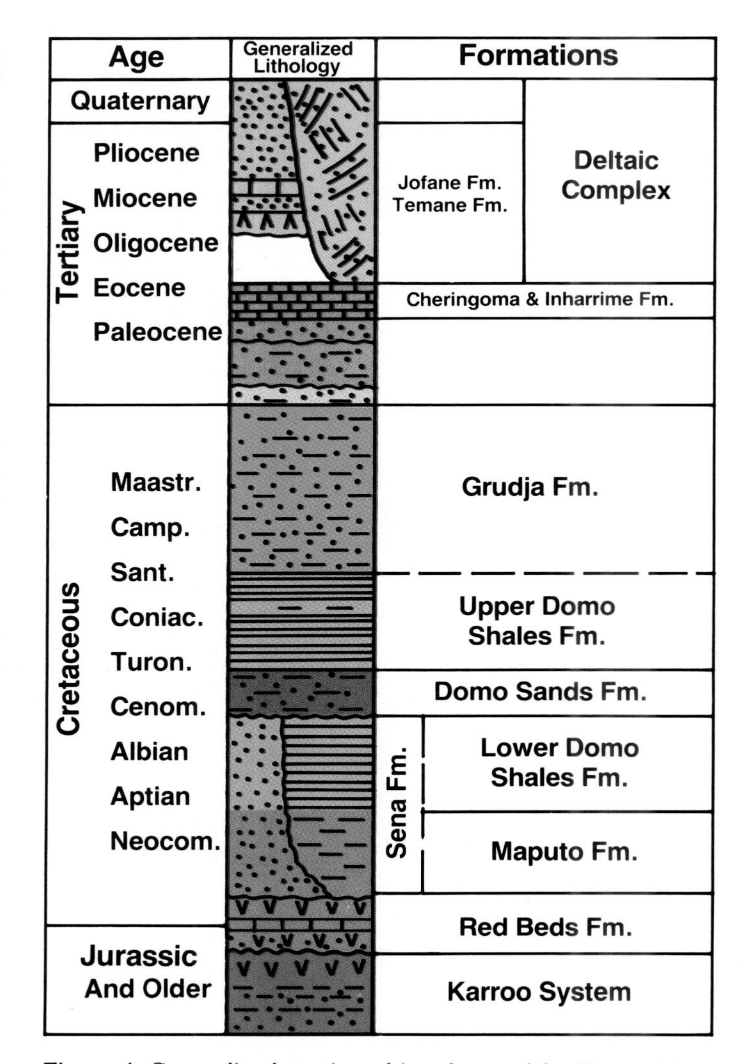


Figure 4. *Generalized stratigraphic column of the Mozambique basin. Note: Color assignments to formation or age intervals is consistent (where possible) in the subsequent figures.*

The upper and lower Domo Shales are separated by a sandy episode constituting the reservoir rock for the methane gas discovered offshore at the Nemo-1 well location.

The less carbonaceous upper Domo Shales progressively grade upward into more detritic lithologies represented by the Grudja Formation, which contains the gas encountered in Pande Buzi, Temane, Divinhe, and, to a lesser extent, in most of the wells drilled in Mozambique. This formation is truncated offshore by a major erosional unconformity that is in turn overlain by clay, silt, and shales. Seismic stratigraphic interpretations suggest that these latter deposits mark the base of a major prograded sequence of Oligocene age believed to represent the paleo-Zambezi delta (Figure 6).

In the northern portion of the survey area, a turbiditic sequence of Maestrichtian, Paleocene, and early Eocene age

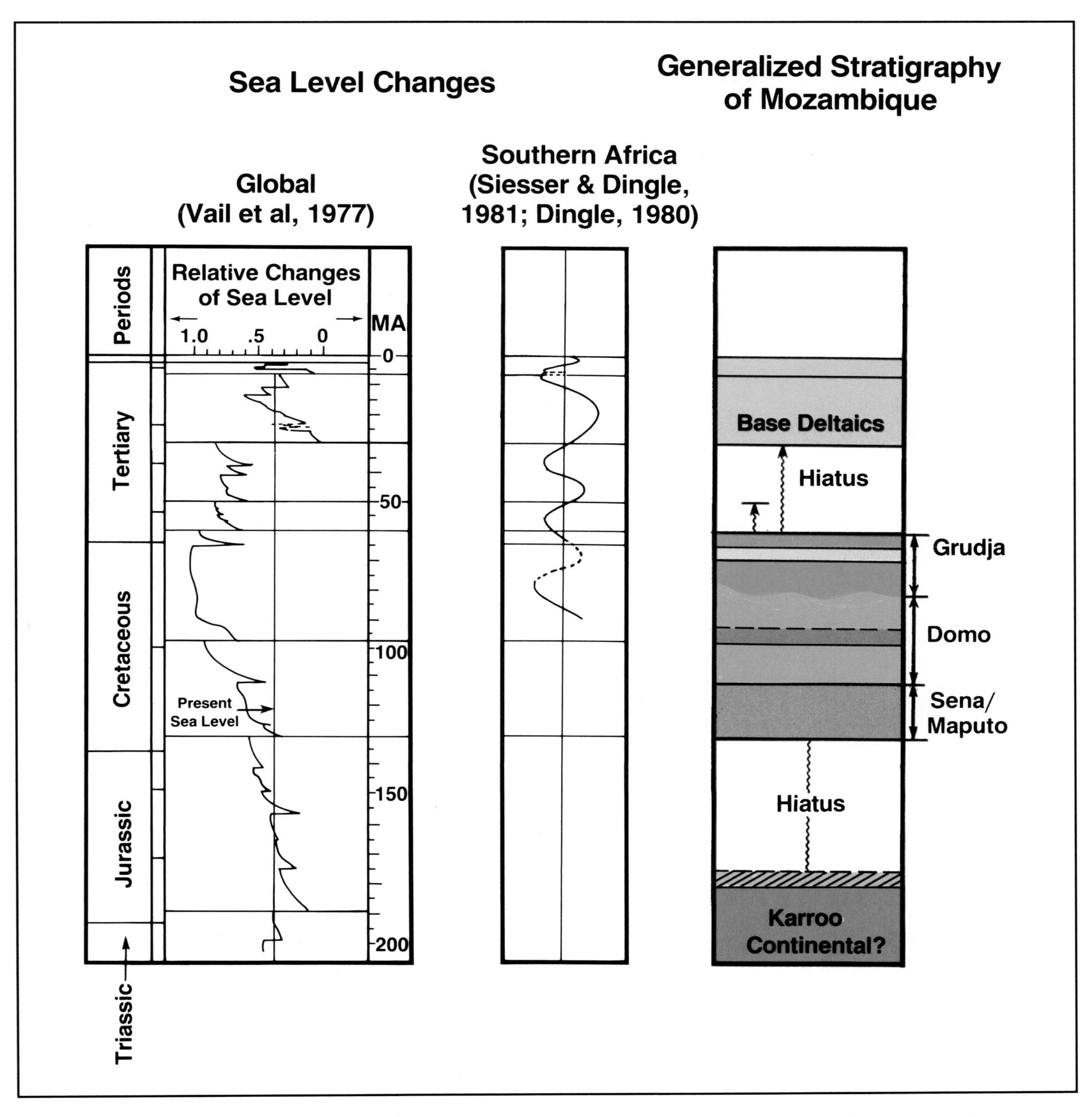


Figure 5. *Correlation of South African and global sea level changes with geology of Mozambique. Correlation analyzed from seismic and well data from offshore Mozambique.*

fills in a depocenter hypothetically located below the paleo-Zambezi delta. Carbonate lithologies of calcarenites, calcilutites, and marls encountered in the Zambezi-3 well represent the upper portion of this sequence and exhibit graded bedding (Figure 7). The youngest member of this sedimentary and erosional wedge offshore may be correlated onshore with the lower Eocene Cheringoma Formation described in outcrops as limestones rich in nummulitic fauna (Flores and Barbieri, 1959; Flores, 1973).

The base of the Oligocene clastics is characterized in Zambezi-3 by the presence of lubricated shales, bentonite, and tar that is perhaps a telltale indicator of earlier oil migration from the deeper part of the basin into the postulated entrapment updip.

Farther south, away from the Zambezi depocenter, the rise of the depositional basement affords interesting opportunities for stratigraphic traps as we observe seismic marker onlap on several seismostratigraphic sequence boundaries. The effect of the sea-level changes, noted in Figure 6, is also more pronounced to the south, with a number of distinctive unconformities apparent on the seismic data.

Lower Cretaceous marine sediments gradually engulfed the pre-Cretaceous paleotopography. Locally, anoxic conditions may have been isolated between the low reliefs, while coarse-grained deposits accumulated on these highs or on their flanks as turbidite flows or as detrital screes. Therefore, the accumulation would provide a favorable juxtaposition of Lower Cretaceous source with reservoir rocks. Reliefs of this kind become more important southward as the intensity and magnitude of faulting increases.

### Tectonic and Depositional History

Before the breakup of Gondwana, the area of the Mozambique sedimentary basin formed part of a stable plate with the deposition of dominantly continental facies. Some authors (Förster, 1975; Kamen-Kaye, 1982, 1983; and Flores, 1970, 1984) suggest that before the Late Jurassic, the Mozambique Channel originated either as a gently subsiding trough (Förster and Kamen-Kaye) or as a consequence of rifting beginning in the Permian (Flores). In both cases, sedimentation would tend to become more marine eastward and also may have been, at least initially, associated with anoxic conditions before the opening of the Mozambique Channel. The question of plate tectonic reconstructions in this area still is debated and has some relevance to the hydrocarbon potential of offshore Mozambique. Nevertheless, prolific source rocks can be found in euxinic seaways or lacustrine conditions resulting from initial intracratonic rifting, regardless of the drift model adopted.

Views by Flores (1970, 1984) and Green (1972) place Madagascar against Mozambique and South Africa and invoke easterly or northeasterly movement. However, the following data support a northerly position for Madagascar against Kenya and Tanzania:

(1) Paleomagnetic data in the Mozambique basin (Embleton and McElhinny, 1975; Segoufin, 1978), in the Somali basin (Segoufin and Patriat, 1980), in the Natal Valley (Martin, Goodlad, and Salmon, 1982), and in the southwest Indian Ocean (Fisher and Sclater, 1983).
(2) Evidence from deep, crustal seismics (Segoufin and Recq, 1980; Sinha, Louden, and Parsons, 1981; Rabinowitz, Coffin, and Falvey, 1981; Coffin and Rabinowitz, 1985).
(3) Early geographical correlation (Du Toit, 1937). In this latter model, Antarctica/Madagascar/India would have separated from South America/Africa at approximately 150 Ma in Upper Jurassic, while Madagascar would have come to rest in the Late Cretaceous (Norton and Sclater, 1979; Lawver, Sclater, and Meinke, 1985). The north-to-south-trending coast of northern Mozambique would have been a sheared continental margin extending southward into the Davie Ridge with its subparallel counterpart of the southern Mozambique Ridge offset in a right lateral sense. The southwest-to-northeast-trending segments of the coast of Mozambique would correspond to rifted continental margins (Kinsman, 1975).

This model, however, does not seem to be corroborated by the new seismic survey. On the contrary, the data suggest the opposite model, where north to south directions coincide with rifting and northeast to southwest trends correspond to transform faults, hence, supporting a northeasterly tensional regime. On the other hand, early tectonic trends of Jurassic age may have been rejuvenated under a different stress regime during the severance of South America and Antarctica from Africa in the mid-Cretaceous, rifted margins being reactivated as transform margins, and vice versa. In addition, the present-day crustal thicknesses of the Mozambique Channel and the Northern Natal Valley modeled from gravity data exceed those normally expected for oceanic crust (Darracott, 1974). Considerable stretching of the continental crust is believed to have taken place regardless of the identity of the landmass once present in the channel. This hypothesis, based on gravity data, does not involve accretion of an oceanic plate, so that the interpretation of paleomagnetic anomalies in the Southern Mozambique Channel should be regarded with caution.

In contrast, there is no evidence of tectonic activity offshore of the northern portion of the survey since Early Cretaceous times (Figures 6 and 7); faulting appears to be related solely to compaction effects. However, intense tensional faulting was experienced onshore where the northeast to southwest Inhaminga fault system cuts across the older northwest to southeast Zambezi graben (Flores, 1973). This fault was sporadically reactivated during the Tertiary (Figure 8), deflecting the sediment load of the Zambezi River toward Beira to the south (Flores, 1965). The disparity between the size of the Pungue River and the geomorphology of its valley supports this hypothesis further. Moreover, the volume of sediments deposited in the lower Tertiary offshore Beira and evident on the seismic data is apparently unaccountable by the modern flow of the Pungue River.

In addition to activity of the Inhaminga fault system, tilting regional and eustatic sea level changes also controlled the influx of Tertiary sediments offshore Beira as interpreted from the seismic stratigraphy.

In the southern portion of the survey, the occurrence and amplitude of faulting generally increases offshore to reach a maximum in a zone of normal faulting, extending north to south away from the Mozambique coast, 150 km (93 mi) offshore from the capital city of Maputo (Figure 9). In the transition

Figure 6. *Line A. This 30-fold seismic reflection (migrated) profile was shot between the Nemo-1 and Zambezi-3 wells (see Figure 3 for location). This interpretation illustrates, from top to bottom, a classic example of prograding deltaic sequences, the updip erosional truncation of a calcareous turbidite wedge, and the basinward pinchout of the Domo sand reservoir drilled at the Nemo-1 well. Refer to Figures 4 and 5 for color to formation equivalence. Note the positions (A, B, C, and D) where detailed velocity analyses were conducted (Figure 20).*

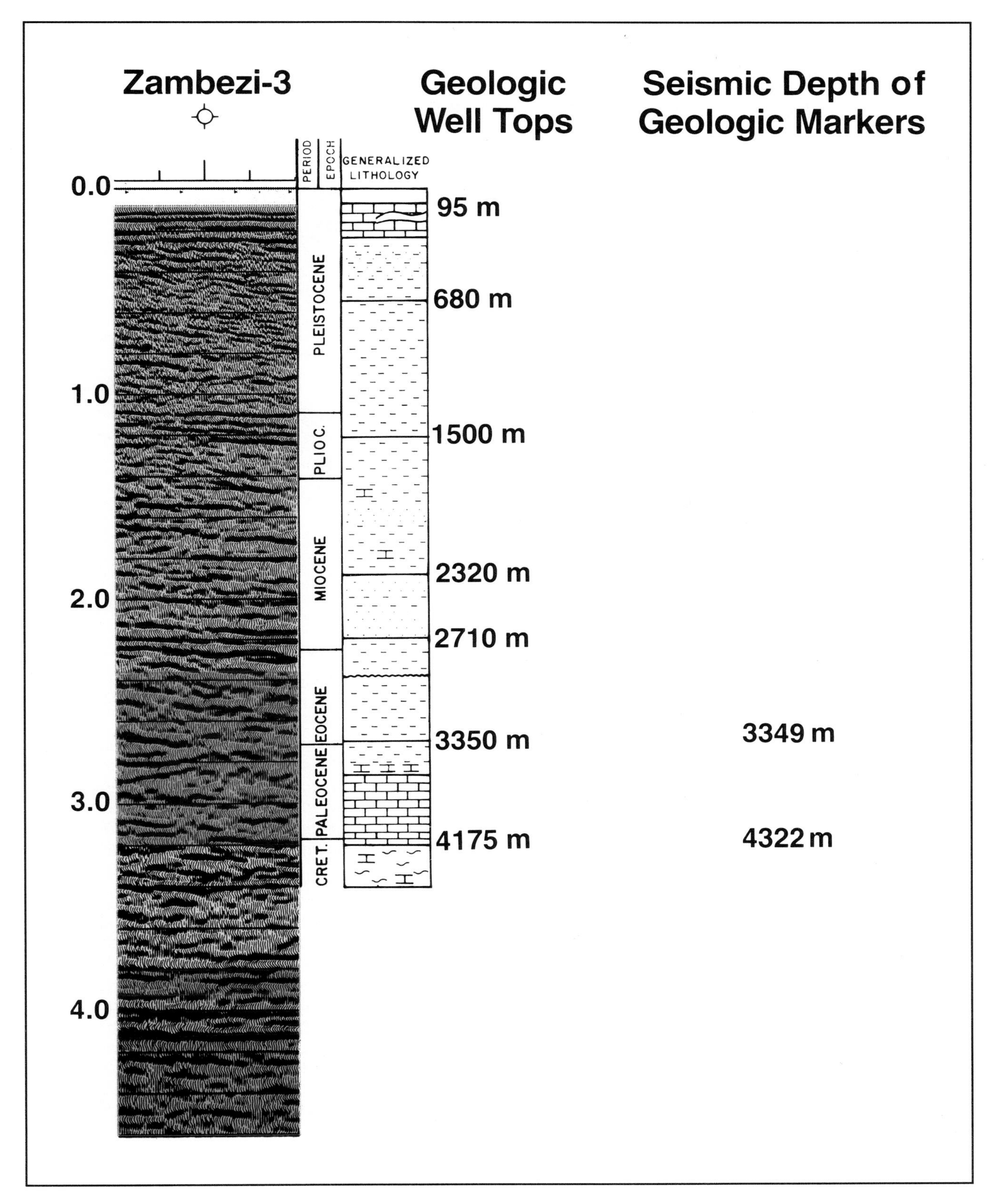


Figure 7. *Correlation between seismic and well data. The chronostratigraphy of the Zambezi-3 well location comes from the former operator's interpretation, available in SECH's archives in Maputo, Mozambique. Note: Hunt's report refers to the presence of graded bedding in the calcarenites and calcilutites comprised in the Upper Cretaceous/Paleocene calcareous interval at the bottom of the well. These observations substantiate the turbiditic nature of the carbonate wedge interpreted from the seismic data.*

zone, south of the Save River, Tertiary faulting was less intense onshore, allowing a period of stability associated with the localized deposition of the thin Temane evaporite formation during the Oligocene.

From a hydrocarbon exploration standpoint, extensive erosional truncation of Upper Cretaceous and Paleocene sequences provides favorable trapping conditions for oil accumulation offshore Beira, where we have mapped over 200 sq km (77 sq mi) of areal closure straddling the present-day shelf edge. The hydrocarbon prospectivity of resulting wedges is enhanced further by regional tilting of the basin to the northeast during the Tertiary, which may have induced updip migration and entrapment of oil drained from a large portion of the northern part of the basin.

We can observe faulting of the Domo Sands related to both differential compacting and flowage of the lower Domo Shale, providing occasional prospective closures. These minor faults, though not obviously related to basement faulting, form a pattern generally oriented in a north-northwest to south-southeast direction.

This orientation corresponds to the trend of the Pungue–Chissenga mid-Cretaceous graben not covered by this survey. The eastern edge of this graben is represented by a normal fault that offsets the Domo Sand against the upper Domo Shale, providing the updip closure for the gas discovered at the Nemo and Sofala wells. Those wells are apparently located somewhat off-structure, as suggested by the recent interpretation of data.

Below the continental slope, toward water depths exceeding 2,000 m (6,560 ft), the overall thickness of sediments gradually decreases as the depositional basement rises along the Benguerua High (Figure 10). This feature would have provided favorable circumstances for the deposition and preservation of potential source rocks. Figure 11 provides an approximate depth map of the Karroo surface that illustrates the persistence of this high axis until the present. After compensation for travel-time delays through a variable water column, the resulting depth map exhibits a differential relief of 1,500 m (4,920 ft) between the apex of the ridge and the lower elevation of that surface westward in the vicinity of the Nemo-1 well location.

Southward from the Zambezi depocenter, the sedimentary sequence also thins progressively. This thinning is notably more pronounced for the Tertiary sediments, testifying to the increased subsidence of the depocenter and tilting of the basin to the northeast (Figure 12) during this period.

Faulting becomes more widespread south of 23°, with the dominant orientation being east-to-west to northeast-to-southwest. A major offshore fault along this trend appears to coincide with the gravimetric Hinge Line described onshore (Flores, 1973).

In the southernmost area, two typical sections parallel to the coast (Figures 13 and 14) illustrate the tectonic and depositional features characterizing the offshore extension of the East African rift system. We have interpreted fault throws in excess of 2,000 m (6,560 ft) as the top Karroo marker in the main rift. In the absence of well control within the rift, the presence of thick Jurassic and Lower Cretaceous sedimentary sequences in the graben can only be inferred involving fault displacements significantly greater and older than previously inferred. The former operators' contention regarding the onset of rifting and the age and type of early sedimentary infill may well be invalidated when stratigraphic data within the rift become available.

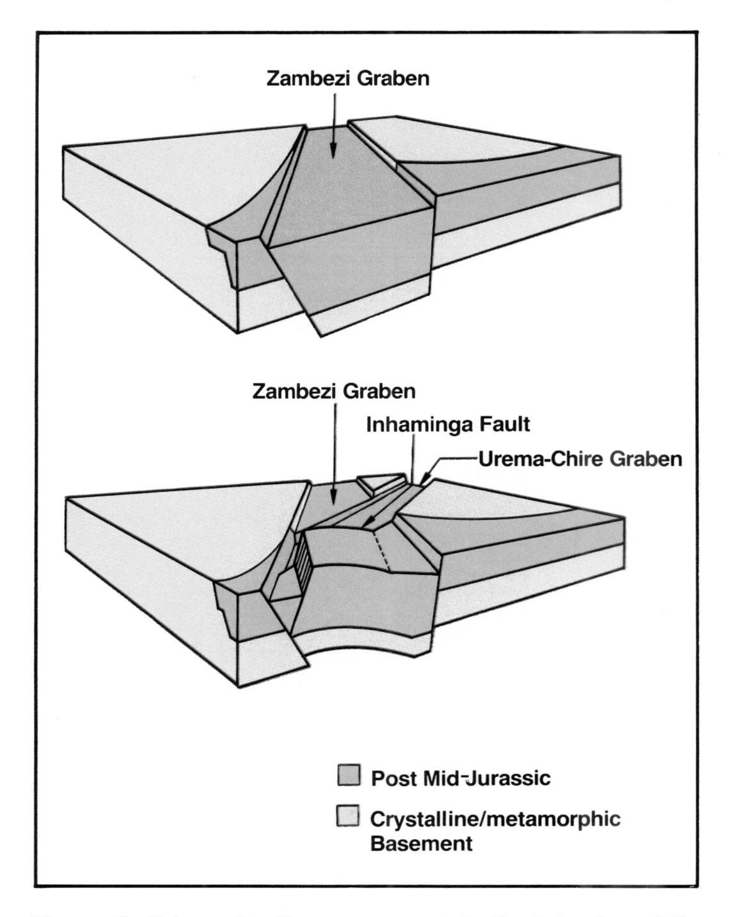


Figure 8. *Schematic diagram represents the interplay of the Inhaminga fault, the Urema Rift, and the Zambezi graben (modified after Flores, 1973). The Zambezi graben trending northwest to southeast intersects in the Upper Cretaceous with the Urema Rift. Its eastern boundary, the Inhaminga Fault, remained active until late Tertiary time, thereby affecting the lower course of the paleo-Zambezi River, the direction and amount of the sediment influx in the basin, and, hence, the paleoposition of the Zambezi delta during Tertiary times.*

In contrast to brittle-fracture faulting typifying the base of the rift (Karroo), listric growth faults affect the Middle and Upper Cretaceous and, in places, lower Tertiary sequences. The younger faulting is apparently decoupled from the deeper tectonics by the presence of late- or post-Karroo ductile sediments interpreted from the seismic data. Those sediments would also be responsible for large-scale, differential compaction and flowage structures within the graben. Ongoing differential compaction of the rift infill has deformed successive Tertiary unconformities.

The main faulted zone of the southern rift is from 20 to 60 km (12 to 37 mi) wide and at least 100 km (62 mi) long under and beyond the present-day continental shelf. East of this zone,

Figure 9. *Regional two-way reflection time map of the top Karroo surface, or its time equivalent, marking the base of the Cretaceous or of locally earlier marine transgression.*

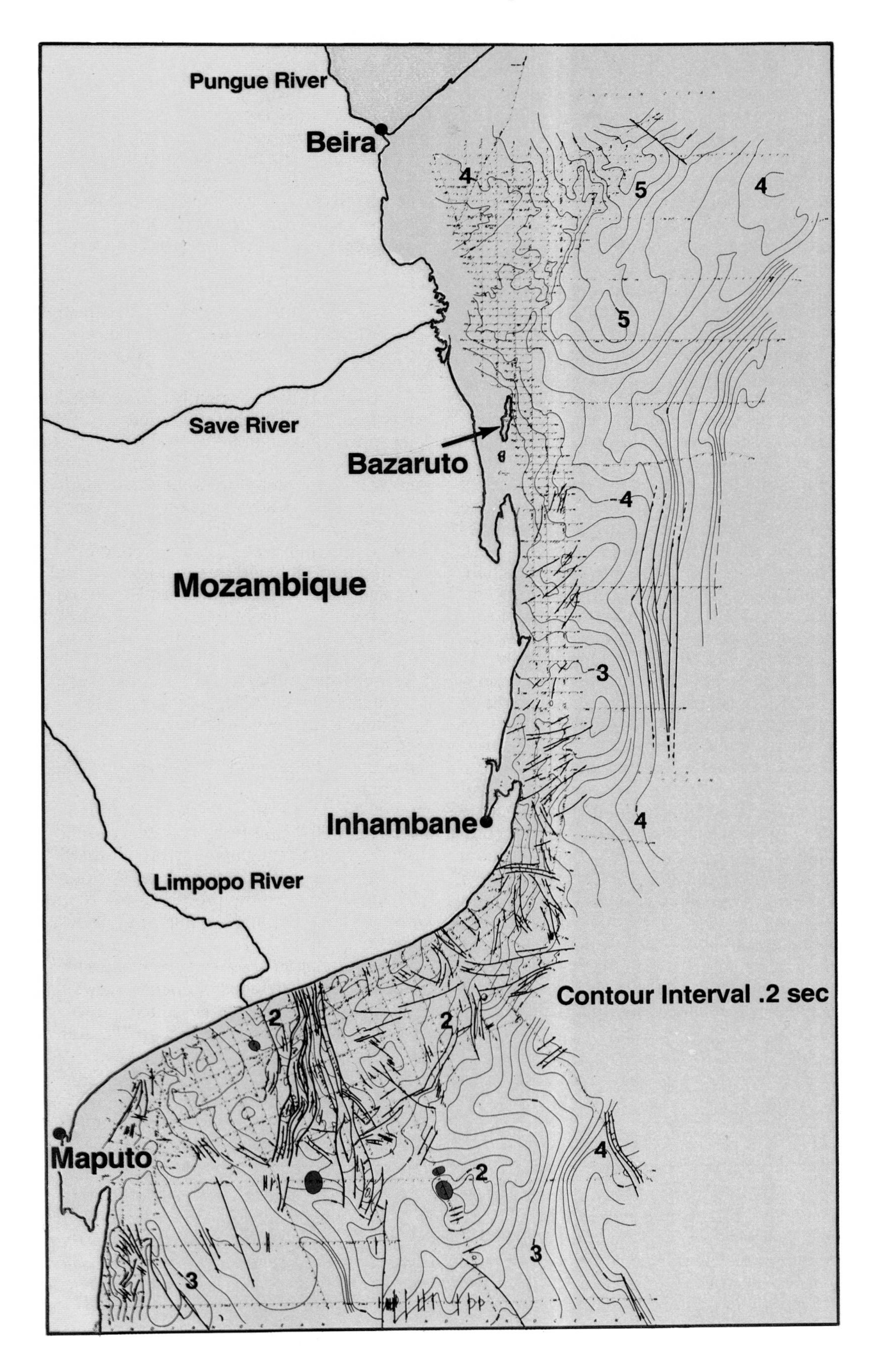
Pungue River
Beira
4
5
4
5
Save River
Bazaruto
4
Mozambique
3
Inhambane
4
Limpopo River
Contour Interval .2 sec
2
2
Maputo
4
2
3
3

faulting exhibits a general northeast to southwest orientation. In comparison, the morphology of the rift zone and the isochron thinning of pre-rift seismic markers is compatible with the model proposed by Holmes (1978, p. 652) for the development of the Rhine graben over a regional upwarp.

With regard to plate tectonic reconstructions, the seismic evidence in the southern part of the survey suggests the presence of a dominant east to west tensional regime until Upper Jurassic time. The onset of rifting, however, is speculative because little or no chronostratigraphic data exist on the age of the infill sequences. The rift orientation is perpendicular to, hence incompatible with, the most likely regional tensional regime associated with both the drift of Antarctica from Africa in Kimmeridgian time and separation of the Falkland Plateau in South America from South Africa, in Early Cretaceous time. Thus, major faulting of the rift area may either be associated with a localized uplift predating the opening of the Mozambique channel, or alternate plate tectonic reconstructions should be invoked, such as a northeasterly drift of Madagascar.

The Southern Mozambique rift exhibits some particular characteristics that may indicate the presence of doming, possibly centered around the intersection of the rift trend and the Hinge Line. We have noted that the width of the rift zone increases southward from 20 to 60 km (12 to 37 mi) beyond the shelf edge, whereas the fault throws decrease in the same direction, reaching marginal dimensions toward the South African border. Hence, the amplitude of faulting and the thickness of subsequent sedimentary infill increase northward, and so too, probably, does the age of the onset of faulting. Accordingly, the South Mozambique rift system may only be the tectonic expression of a localized bulge, not necessarily to be interpreted in a regional plate tectonics context.

Nonetheless, the rift aligns with an ancient zone of weakness characterized by rifting of different ages along East Africa, stretching from the Afar Triangle through the great lakes region down to the Malawi and Urema rifts and plunging under the Cenozoic sediments of the Mozambican coastal plain. Accordingly, the southern Mozambique rift appears to be the offshore evidence of a much more regional stress field, and, hence, it may indeed relate to the early breakup of Gondwana under an east-to-west tensional regime. If so, the initial, relative plate motion between Madagascar and Africa would have been directed west to east, during, say, Carboniferous–Permian times, before the Late Jurassic and Cretaceous drift that produced the paleomagnetic anomalies that support most plate reconstructions.

The evolution of the rift can be illustrated by four stages (Figure 15):

(1) The earlier development of faulting, producing initially the westernmost graben, was probably induced by localized doming in Liassic time.
(2) Thereafter, this narrow rift was offset by right lateral transform faults trending east-northeast to west-southwest (Hinge Line Trend).
(3) A series of rotated fault blocks tilted to the east and were downthrown to the west in response to continuing uplift in the Early Cretaceous.
(4) Finally, an asymmetric readjustment occurred with the development of a series of north to south listric faults affecting the Middle and Upper Cretaceous sequences.

These faults were evidently induced by differential compacting within the rift fill that continued until Tertiary time. We cannot verify speculation on the shaley mature and sourcing potential of the early rift fill. We can, however, justify it by analogy with other rift basins around the world.

The seismic evidence of pre-Cretaceous tectonic activity certainly upgrades the hydrocarbon potential of the basin. Favorable source and reservoir rocks heretofore not encountered by drilling may be found in a mature state and effective trapping configuration, offshore southern Mozambique. We will discuss later the influence of Cretaceous volcanism in the graben area on the geothermal history of the basin.

### Seismic Interpretation

We interpreted and mapped seven seismic horizons corresponding to geologic well tops, as well as important seismic stratigraphic markers, such as the Top Karroo basalts, the Top and Base Domo Sand, and several Cretaceous and Tertiary unconformities.

We based the horizon identification on the comparison between well depths (to chronostratigraphic markers) with seismic depth values, (computed from seismic times and evaluation of interval velocities derived from stacking velocities). Figures 16 and 7 illustrate the process at the Nemo-1 and Zambezi-3 wells offshore Beira, while Figure 17 illustrates the well to seismic tie for Sunray 7-1.

The correlation of seismic events with most geologic markers picked by the previous operators shows good consistency of identification. Seismic evidence, however, indicates that basalts, which were ascribed to the Karroo Supergroup in

Figure 10. *Line B. This 30-fold seismic reflection (migrated) profile was shot in an east to west direction between the vicinity of the Nemo-1 well to the left of the section and eastward to the 2,000 m (6,560 ft) isobath (see Figure 3 for location). Note the Benguerua High relief on the right side of the section. This distal part of the basin exhibits signs of higher vertical mobility than the shelf area. In particular, the oldest seismostratigraphic interval of marine character is base lapping westward on a depositional basement that could consist of stretched continental crust, oceanic crust, the Karroo Stormberg volcanics, or time equivalent deposits. An Upper Cretaceous carbonate reef-like buildup coincides with the Benguerua High axis that would imply considerable subsidence during the Tertiary. This feature may have played an important role in controlling the distribution of Cretaceous sediments and preserving the organic content of source rock intervals by isolating restricted circulation and euxinic conditions from the open marine environment. Also note the classic examples of turbidites and contourites in the distal part of the upper Tertiary section.*

certain wells, are in fact interbedded in a seismic stratigraphic sequence of late Mesozoic and even Cenozoic age. Age revisions ensured correlation consistency at other well locations. The resulting ties fit with the eustatic sea level chart (Vail, Mitchum, and Thompson, 1977), and the K/AR age dating results (Flores, 1984).

Furthermore, the Karroo lava flows lie over older sequences locally exhibiting a pronounced marine seismic character. Accordingly, post-Jurassic basalts may have been erroneously interpreted as Stormberg basalts of the Karroo Supergroup. Alternatively, the Karroo Supergroup contains marine sequences. The former must have had great impact on the success of past exploratory drilling, since wells that encountered basalts were abandoned under the assumption that the Karroo had been reached. For the remaining horizons, consistency in well ties demonstrates that various operators have made compatible geologic correlations and that they have successfully correlated and mapped seismic horizons across the area.

### Hydrocarbon Potential

Mozambique's hydrocarbon potential hinges on three main geologic aspects:

(1) The sediment and fresh-water influx in the basin, and the amount and type of organic carbon in potential source rocks.
(2) The problems of maturation in the basin's geothermal context.
(3) The vertical and spatial distribution of reservoir and source rocks, and the possible migration and trapping mechanisms for hydrocarbon accumulations.

### Sedimentologic Aspects

The Karroo source and reservoir facies in Mozambique were not considered prospective since the dominantly continental nature of these sediments implies gas-prone humic and vegetal carbon concentrations and highly variable reservoir quantities. However, new seismic data indicate the local existence of possible marine or lacustrine sequences within as well as above the upper Karroo. These sequences may be widespread if they were deposited during subsidence preceding the final breakup of Gondwana. Hence, favorable geologic circumstances for the sedimentation of oil-prone source rocks are likely to have existed before as well as after the Liassic.

In the southern third of the survey area, the existence of major rifting of Cretaceous and of earlier age increases the likelihood of restricted-marine Karroo sediments. In addition, the hydrocarbon accumulations in Madagascar of probable Karroo origin have prompted the idea that geologists should now consider the Karroo as a "viable exploration objective" (Kamen-Kaye, 1983).

Most of the Mesozoic and Cenozoic sediments, deposited during the large-scale marine transgression, are clastic in nature. From the Grudja Formation upward, considerable amounts of prograding clastic sediments were deposited in the Mozambique basin, developing locally into deltaic fans. Figure 6 shows a typical example of these deltaic sequences in the northern half of the survey (the paleo-Zambezi delta featured on Line A). This section also illustrates the strong, erosional unconformity truncating the Upper Cretaceous wedges.

Figure 18 illustrates the depositional model proposed for these wedges. This model implies that, toward the end of the Mesozoic Era, the reduced erosional activity onshore modified the depositional regime in the central part of the basin. A simultaneous decrease in sediment and a fresh-water influx allowed a carbonate environment to develop on both the shelf and the continental slope. In a model suggested by Figure 19, a reef buildup can thrive on the shelf edge, thereby isolating a near-shore, paralic, back-reef environment from an open marine, hemipelagic depositional area. The shaded area on Figure 19 represents the sequence eroded during the base Oligocene hiatus, as interpreted from the survey data between the Nemo-1 and Zambezi-3 wells (Figure 6).

Core descriptions in the bottom part of the Zambezi-3 well support the notion that these wedges consist of carbonate rocks; that is further corroborated by detailed seismic velocity analyses. Figure 20 features a block model generated by independently picking the optimum time-velocity pairs on constant velocity stacks and applying the Dix equation (1955) for calculating the interval velocity. Subsequently, the previous time interpretation was overlaid onto the model. The match between the model and horizon-time selection testifies to the consistency of the results, to their lateral stability, and, in turn, to the reliability of the velocity model.

This model indicates a sharp increase in velocity at the upper boundary of the wedge interval. The hummocky seismic character of sections A and B (Figures 6 and 10) suggests that this interval is turbiditic. Because high porosity and often lower velocity than that of clastic, undisturbed sediments typifies turbidites, the increase in velocity indicates the calcareous nature of the whole interval.

Figure 10 also highlights the possible presence at the apex of the Benguerua High of a carbonate buildup of comparable age that could have surrounded the Zambezi depocenter in a crescent-shaped atoll. The eastern half of this seismic profile features seismostratigraphic evidence of the basin's distal segment, which is expected in the presence of thinned continental crust.

This implied break in the clastic depositional regime, possibly responding to a change in climatic conditions toward the end of the Cretaceous Era, is important in considering the northern area's liquid hydrocarbon prospects. Significant concentrations of planktonic life forms into oil-prone source rocks are not likely for as long as massive terrigenous and fresh-water influx dominate the marine sediment deposition.

In addition, the burial of most marine Cretaceous sediments is too rapid in the proximal part of the basin to allow winnowing and sorting of the clastics by currents and wave action, which would result in better reservoir properties. Consequently, long and complex migration paths would be required if any oil derived from a marine source rock were to be trapped in reservoir rocks chiefly distributed in the near-shore area. The proposed model would explain why near-shore wildcatting offshore Beira discovered only biogenic gas in the Domo and Grudja sand bodies. These sandy reservoirs are usually intercalated with carbonaceous shales whose organic content released biogenic methane gas.

In contrast, the possibility of a drastic, sedimentary environment change upgrades the liquid hydrocarbon potential of marine sources offshore Mozambique, as the marine carbon productivity can be expected to increase when the fresh-water influx is reduced.

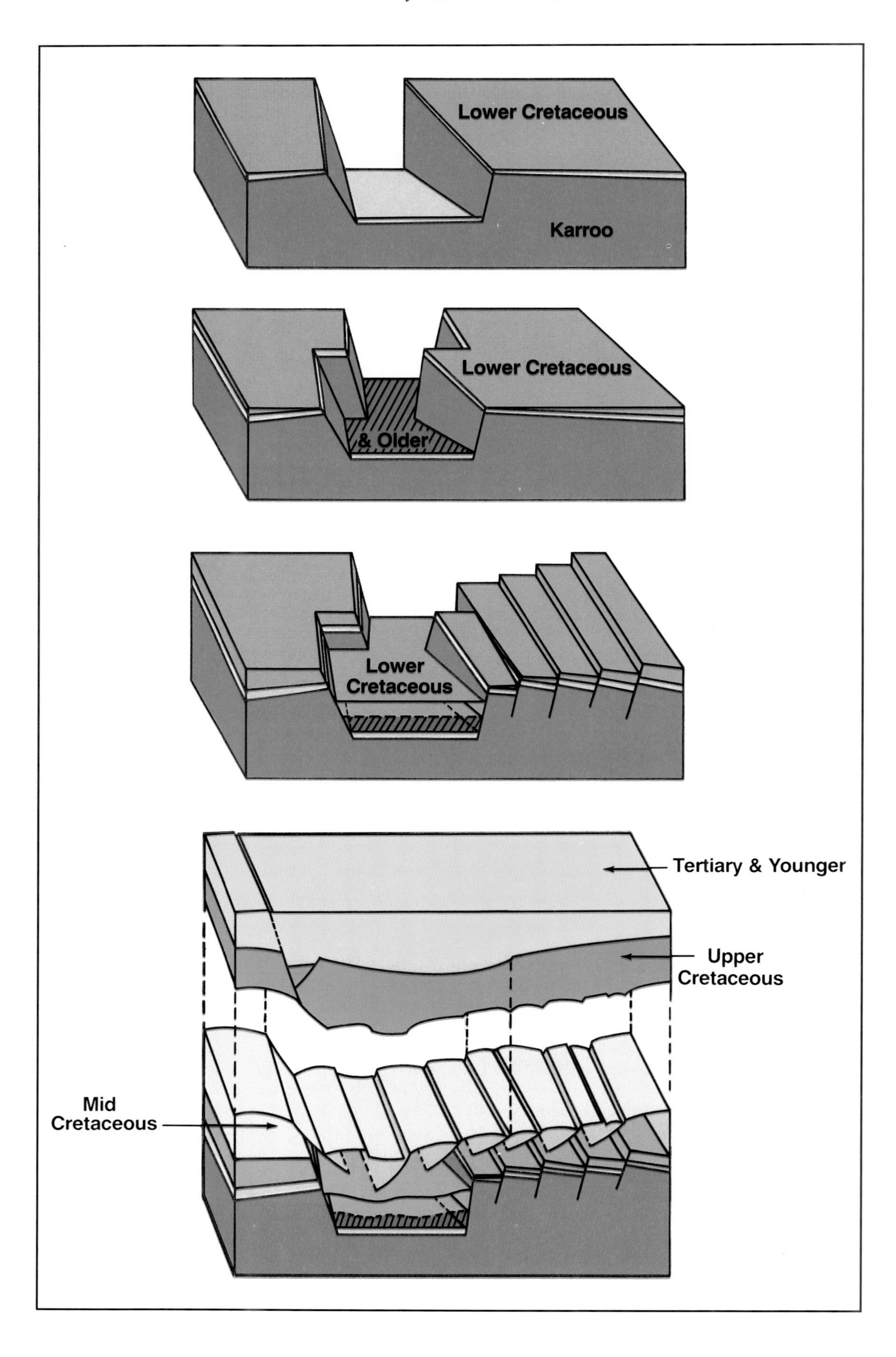
Lower Cretaceous
Karroo
Lower Cretaceous
& Older
Lower
Cretaceous
Tertiary & Younger
Upper
Cretaceous
Mid
Cretaceous

Figure 15. *Palinspastic model of the South Mozambique rift. A. The early tectonic development led to the formation of a graben deepening and narrowing northward. B. Subsequent right lateral transform fault offset the graben, its early fill, and a thin, Lower Cretaceous wedge on its flanks. C. Later readjustment resulted in the development of tilted fault blocks dipping to the east, downthrowing to the west, and affecting the Lower Cretaceous wedge. D. The middle Cretaceous infill of the above paleotopography was subjected to differential compaction that later induced growth faulting, affecting the Upper Cretaceous section as well as the Tertiary to a lesser extent.*

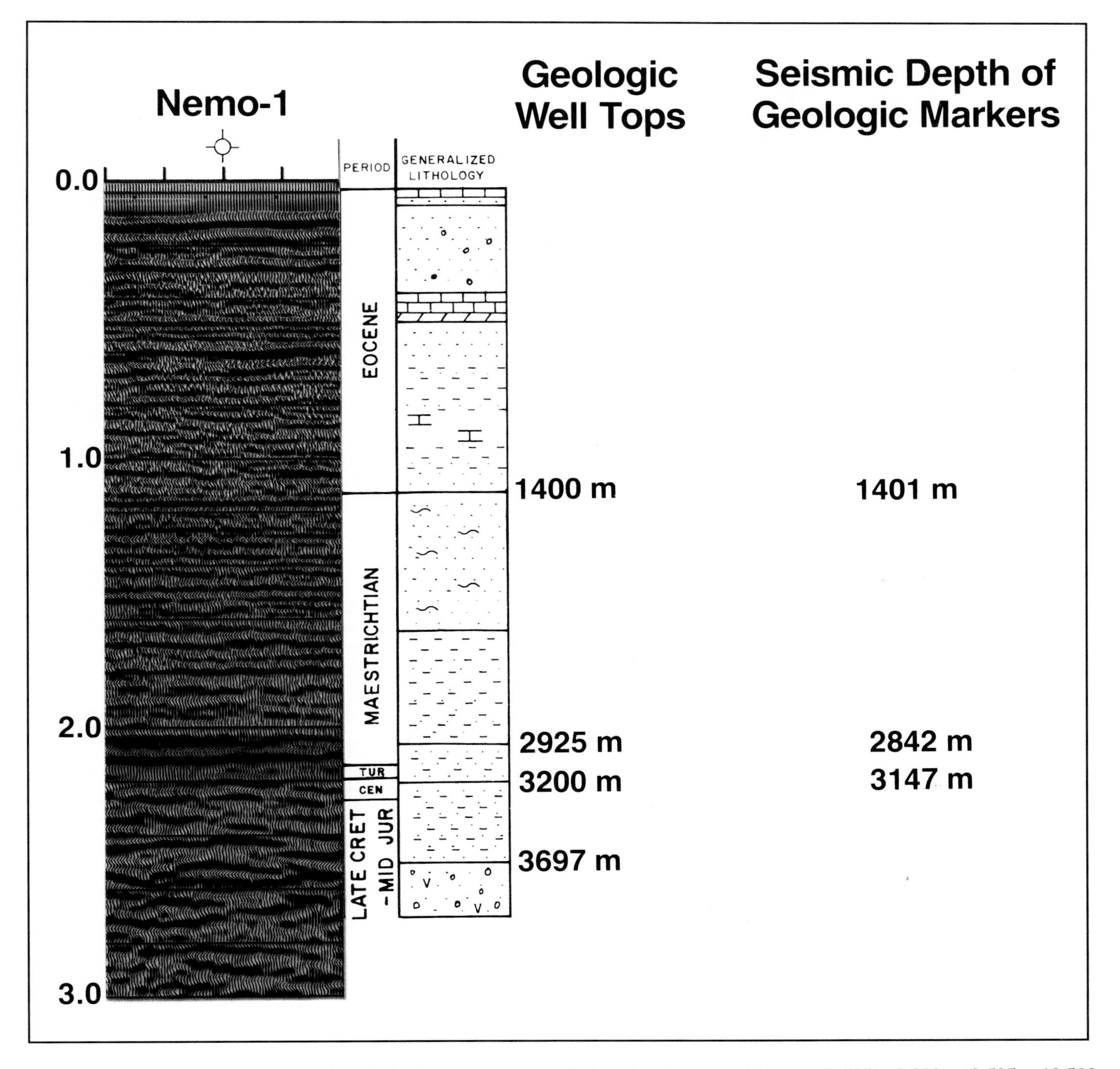


Figure 16. *Correlation of seismic and geologic data at Nemo-1 well. Note the Domo sand interval [2,925 to 3,200 m (9,595 to 10,500 ft)] where a significant amount of dry gas was encountered.*

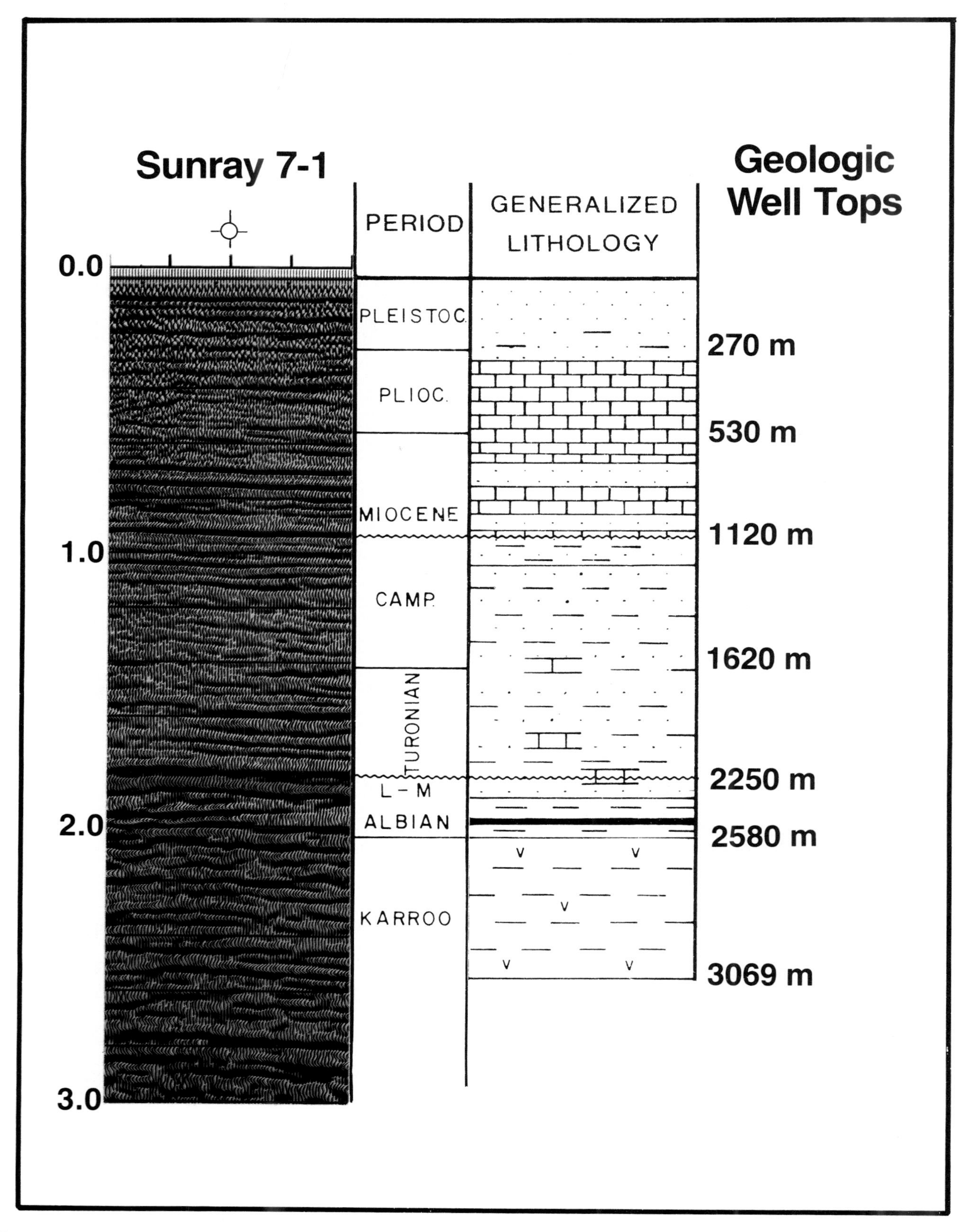


Figure 17. *Correlation of seismic and geologic data done at Sunray 7-1 well. Note the presence of two distinct unconformities.*

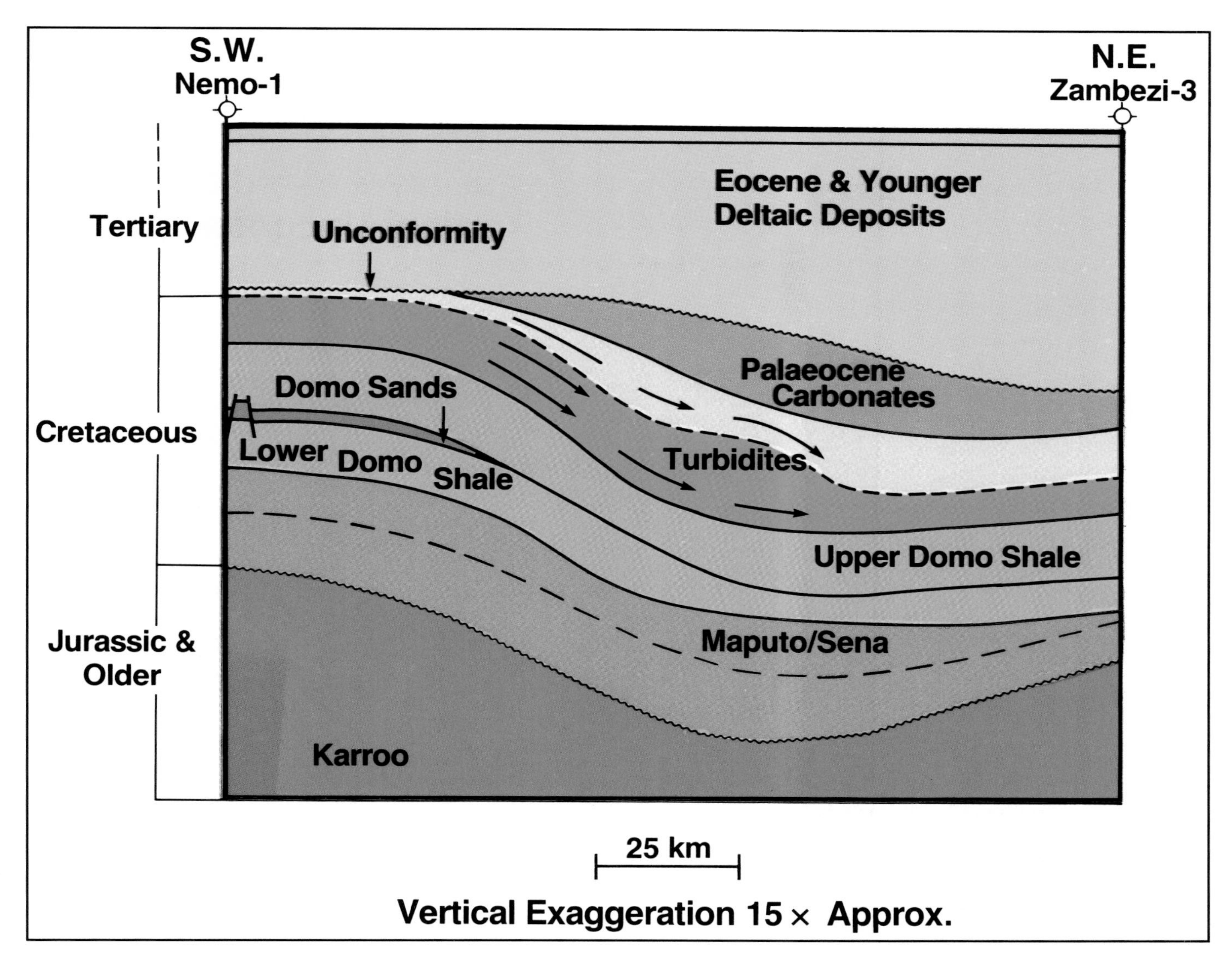


Figure 18. *Schematic structural cross section — offshore Beira. The cross section was done along Line A.*

On the other hand, the importance of the thick Karroo coal deposits in the evaluation of the hydrocarbon potential of Mozambique should not be overlooked. Indeed, significant oil discoveries in Cabinda, Angola, in the McKenzie, Niger, and Mahakam deltas, and in the Gippsland basin originate from land-derived carbon sources and coal measures (Durand, Parrate, and Bertrand, 1983). Durand points out that "Rock-Eval pyrolysis and the chromatography of pyrolysis effluents from coals of all geological ages show that such coals produce up to 20% of their weight in hydrocarbons. These hydrocarbons are mainly liquid hydrocarbons ($C_{+5}$) up to a rank corresponding to about 1.5% of the reflecting power of vitrinite.

"This is why appreciable quantities of oil should be formed during the coalification of most coals, no matter what their age of deposition may be. The quasi-general absence of oil accumulations in paleozoic coal series thus should be due to geological factors and not to specific geochemical features. The oil formed in such ancient series was probably dissipated by migration during successive sedimentary cycles.

"For equal rank, the variability of the amounts and distributions of hydrocarbons formed by pyrolysis is just as great for coals as for other types of organic matter."

The liquid hydrocarbon potential of Karroo coal measures and other carbon-bearing rocks should, therefore, be reconsidered.

In the northern portion of the survey offshore Beira, we have observed extensive closure of the interpreted calcareous turbidite wedges of Maestrichtian to Eocene age along the outer continental shelf. The lower Eocene sequence corresponds to pro-delta clays, grading upward in Oligocene time into the arenaceous sediments of the paleo-Zambezi delta, represented by sigmoid foresets in the upper part of the seismic section (Figure 6). The presence of these clayey deposits is critical in providing an effective updip seal to the underlying carbonate wedges. A larger closure of this stratigraphic trap is provided by the Tertiary tilt to the northeast and the ongoing subsidence of the southwest flank of the depocenter, bounded to the east by the Benguera High and to the west by the edge of the paleoshelf.

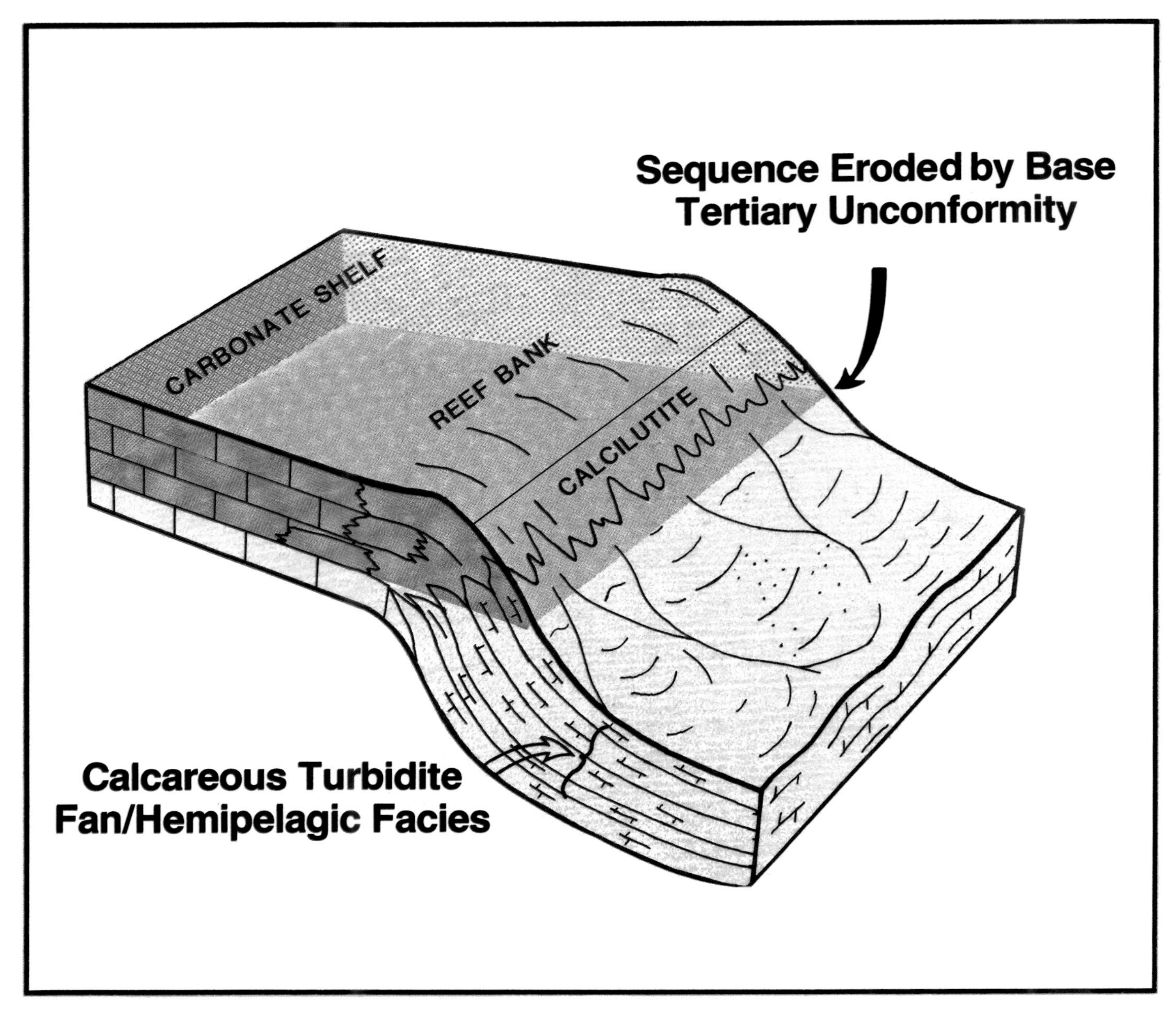


Figure 19. *Carbonate buildup in post-rift basin. This model (modified from Brown and Fischer, 1977) explains the absence of seismic evidence of reef development in the vicinity of the present-day shelf edge by the erosional hiatus represented by the Oligocene unconformity. Such reef development, however, has been identified on the Benguerua High as well as farther north along the Mozambique shoreline and onshore (see Figure 10).*

In the southern areas, early rifting may have been associated with marine incursions and deposition of anoxic shales penecontemporaneous with the formation of the Stormberg or later volcanic effusives on the subaerial reliefs surrounding the rift.

### Geothermal Gradient and Structural Considerations

In the northern area, we need to explain the paradox of gas occurring with condensate and of 30° API oil shows in Temane and the presence of immature gas-prone rocks. South of Beira, onshore and in the near offshore area, the current geothermal gradient does not exceed 2°C per 100 m (1.1°F per 100 ft), which implies that the oil window is not reached before 5,000 m (16,404 ft), a depth below which the carbonaceous Domo Shales, allegedly the source rocks of the Pande gas, have never been buried in this area. To explain the presence of the hydrocarbons, we postulate that older source rocks have matured either under sporadic geothermal activity related to the reactivation of the East African rift system and associated magmatic intrusion, or perhaps, under more stable geothermal conditions during longer periods. In our first hypothesis, direct contact, or lower temperature regional metamorphism of the Karroo coal measures and of post-Karroo organic-rich sediments, may have taken place during periods of increased volcanic activity. Since both Karroo and post-Karroo sediments may be thicker within

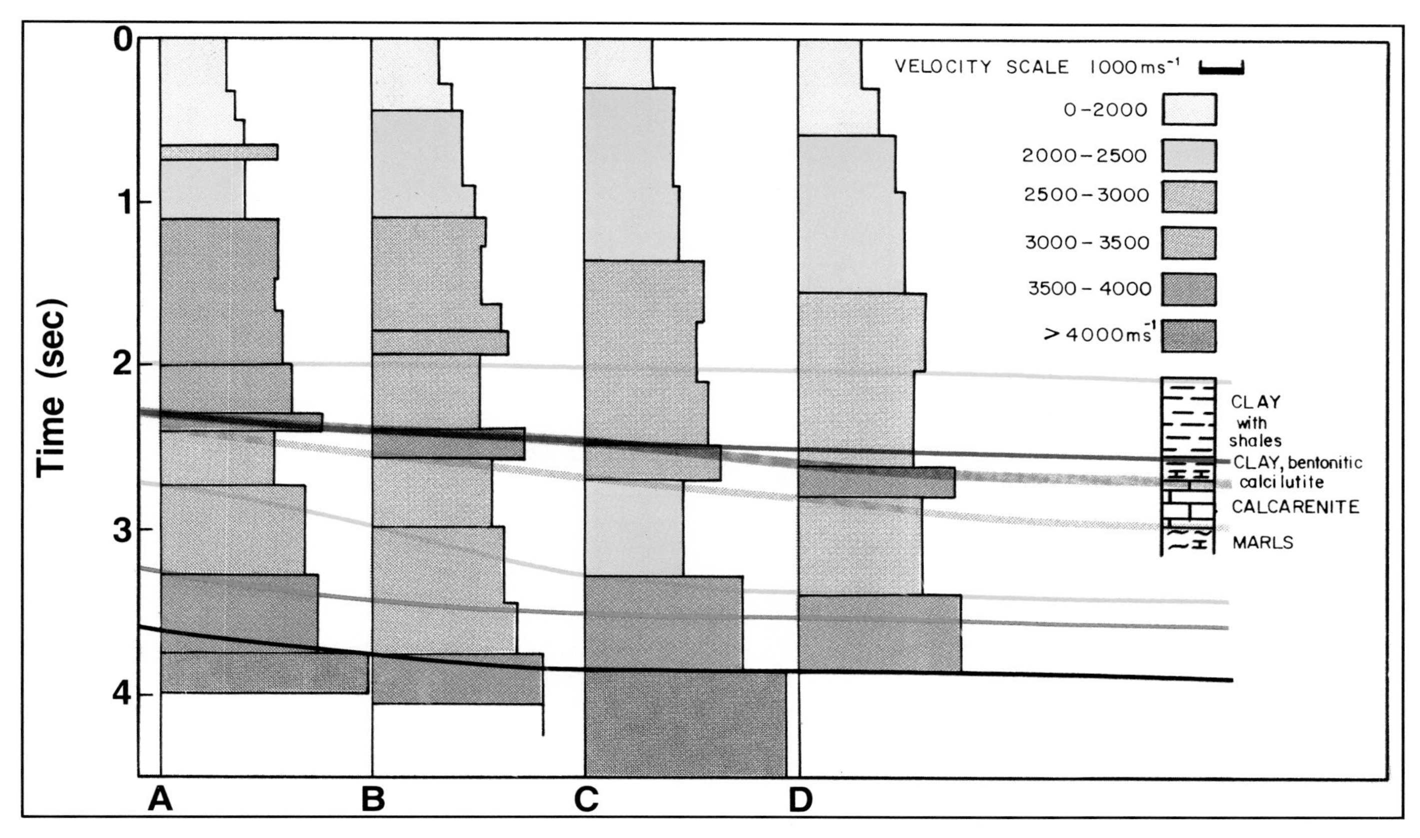


Figure 20. *Dix interval velocity model of the carbonate wedge. The locations of each block model offshore Beira are shown on Figure 6 (Dix, 1955). Note the sharp increase in velocity across the base Tertiary erosional unconformity between the pro-delta shale and the calcareous wedge, represented by calcilutite and calcarenite at the Zambezi-3 well. The remaining part of the wedge, interpreted as turbidites because of the hummocky seismic character of the interval, also is hypothesized to be of a calcareous nature based on its high interval velocity.*

the rift system, considerable amounts of oil and thermogenic gas may have been generated. Thus, this model would explain the Pande and Buzi gas discoveries as the result of thermogenic processes. In the nearshore area, the presence of methane gas at Nemo and Sofala has already been discussed and postulated as biogenic in nature. This portion of the offshore basin is relatively unaffected by tectonism and, hence, may be characterized by a lower geothermal regime. However, in our second hypothesis, a milder geothermal regime also may have been sufficient to mature source rocks deeper in the post-Karroo section, hence hinting at the possibility of migrated oil in an older reservoir below the discovered gas pools.

As demonstrated by the regional survey lines (Figure 11), the depositional basement rises sharply toward the east in the northern area, while the pelagic sediments gradually become thinner. Accordingly, the apparent monoclinal structure of the basin on the shelf area corresponds to the western limb of an asymmetric depression straddling the current continental slope and is limited to the east by the Benguerua High axis. This ridge may have isolated euxinic conditions favorable to the preservation of the organic content of potential source rocks, conditions that seem to have prevailed offshore since post-Karroo times.

In the southern area, the question of a favorable geothermal history for hydrocarbon maturation poses less of a problem offshore, particularly when considering the likelihood of a higher heat flow during periods of rifting and associated volcano activity. Furthermore, in this area potential Jurassic and Cretaceous source rocks have been faulted down to depths adequate for hydrocarbon maturation, even under lower geothermal gradients.

## Trapping Mechanisms

In the northern area, the proven reservoir rocks are confined to the Domo and Grudja Formations, both onshore and offshore. At Pande and Buzi, the main producing level consists of a sand wedge elongated in a southwest to northeast direction. A closure on the order of 20 m (66 ft) may provide considerable reserves in a reservoir whose thickness does not exceed 25 m (82 ft). The discontinuous nature of the producing sand interval and sedimentological considerations suggest that the depositional environment corresponds to coastal barriers or longshore bars.

The Nemo–Sofala structure is closed updip by fault truncation of a gentle anticline plunging to the northeast. The north to northwest direction of this faulting corresponds to the so-called Pungue–Chissenga graben (Figure 2).

East of this fault system, the fault pattern remains aligned with the structural grain, although faulting chiefly relates to differential compaction and gravity movements. Another gentle anticlinal closure on the Domo Sand is almost exclusively associated in the northern area with shale flowage and drape-over Karroo erosional remnants.

Of considerably more interest to explorationists is the erosional truncation and sealing by Oligocene pro-delta shale of the top Maestrichtian, Paleocene, and Eocene carbonates. Figure 18 shows the sedimentary/truncational Cretaceous and Paleocene wedges as well as the geometry of the basinward pinchout of the Domo Sand.

Faulting related to shale flowage and basement structure becomes more prevalent to the south. Our seismic interpretation of the new data has revealed the existence of combined trapping mechanisms associated with formation truncation of the Domo Sand time-equivalent, although the lithology of this interval may not necessarily be consistent over such a large area. Figure 12, representing the schematic structural cross section of the area, illustrates this consideration and also shows the pre-Cretaceous beds rising to the south. As Mesozoic and Cenozoic sediments become thinner in that direction, better quality seismic data from shallower pre-Top Karroo structures, unconformities, and wedges allow a clear definition of the economical prospect.

Although Karroo rocks are not renowned for their reservoir quality in South Africa, we should not overlook the Karroo hydrocarbon potential because continental lithofacies cannot be safely extrapolated over large distances. Furthermore, the Karroo deposits in South Africa are generally found in interior basins whose evolution and sedimentary infill is likely to differ from that in coastal basins. Indeed, the presence of Karroo age salt formations and lacustrine deposits in Kenya and Tanzania corroborates the view that marine sediments may have been deposited in a proto–East African margin and proto-Mozambique Channel. While the importance of marine sediments of Karroo age has already been mentioned in connection with sourcing, we also should not overlook the argument for the presence of marine reservoir rocks of Karroo age. Immature, coarse sediments derived from the subaerial erosion of the edges of the early graben may have been deposited as fanglomeratic scree against the buttress of the normal faults bounding the rift. This process would result in the development of highly porous and permeable reservoir intervals in a combined stratigraphic and structural trapping configuration in the close vicinity of potentially mature source rocks. In the current survey, the southernmost area probably represents the greatest concentration of hydrocarbon traps that necessitate the shortest migration paths to be charged by hydrocarbons.

## CONCLUSIONS

If the new survey shows so many prospective structures, why is it that none of the 12 wells drilled offshore to date has been successful in finding significant quantities of oil or economic quantities of gas?

Part of the explanation is that 12 wells in such a large area of continental shelf is a very low drilling density, so all the wildcats drilled were on the most obvious features determinable from the then-current seismic data. The new survey has not only revealed more prospective structures, such as the wedge in offshore Beira and the rift offshore the capital city, but also has highlighted some possible reasons for the failure of certain wells to pay.

In the south, many wells were presumably drilled on the structure known at the time, but that was not proven in three dimensions (Sunray 2-1), had less closure than previously thought (Sunray 1-1A), or was situated on the flanks of a major structure (Sunray 4-1B, 7-1). Our discussion for the first time publicly details the existence of this rift system offshore southern Mozambique and adds new information to both the regional tectonic history and the local hydrocarbon potential. Furthermore, the existence of a classic rift system is unique offshore East Africa and therefore upgrades the prospects of the Mozambique coastal basin relative to its African neighbors on the Indian Ocean. In particular, potential for both hydrocarbon sourcing and trapping is much greater in the rift than around the previously explored offshore areas of south Mozambique.

The past exploration philosophy was clearly to regard Karroo volcanics as the economic basement. As stated above, we contend that this philosophy is outdated, particularly where seismic evidence suggests the presence of both a non-continental series underneath the volcanics, and an earlier structural development. Therefore, in hindsight, these wells should have penetrated apparently thin Karroo volcanics to test possible reservoirs below the Stormberg basalt.

Moreover, the seismic and K/AR age data demonstrate that the "Karroo" rocks in which Sunray 2-1 and 4-1B bottomed are of Late Cretaceous age. Geologists previously have confused a basalt layer extruding locally in a region of intense faulting with Karroo Stormberg basalts. Therefore, in the case of these two wells, the Lower Cretaceous and Karroo sediments were not tested, nor were the wells optimally located with respect to structure.

Finally, there is no information regarding the maturity of the organic carbon contained in both pre-rift conventional source rock, in Karroo coals, and in the sediments deposited during and shortly after rifting took place. These sediments may be carbonaceous and may have been deposited under anoxic conditions within the rift. Drilling evidence for these hypotheses does not exist in such an unexplored province, since few wells drilled were aimed at structural highs and often were too shallow to reach mature source rocks.

## ACKNOWLEDGMENTS

We are indebted to Mr. Mario Marques, the Director General of the Mozambican Secretariat for Coals and Hydrocarbons (SECH) for his permission to publish this article and for allowing us to review much invaluable, confidential information in SECH's archives in Maputo.

Our appreciation also goes to Western Geophysical Company seismologists John A. Austin and Ian Webster, who have skillfully picked some 13,000 km (8,078 mi) of seismic data while fully aware of the pitfalls in interpreting it.

Finally, our special thanks to Rosa Maria Davis, who painstakingly deciphered our scribblings and typed innumerable versions of the manuscript, without losing her diplomacy.

## REFERENCES CITED

Brown, L. F., and W. L. Fischer, 1977, Seismic stratigraphic interpretation of depositional systems: examples from Brazilian rift and pull-apart basins, *in* C. E. Payton, ed., Seismic stratigraphy—applications to hydrocarbon exploration: AAPG Memoir 26, p. 213–248.

Coffin, M., and P. Rabinowitz, in press, The fit of Madagascar in Gondwana: Tectonics. A.G.U. (expected 1985).

Darracott, B. W., 1974, On the crustal structure and evolution of Southeastern Africa and the adjacent Indian Ocean: Earth Planetary Science Letter, 24, p. 282–290.

Dingle, R. V., 1980, Marine Santonian and Campanian Ostracods from a borehole at Richards Bay, Zululand: Annales South African Museum, v. 82, p. 1–70.

Dix, C. H., 1955, Seismic velocities from surface measurements: Geophysics, v. 20, p. 68–86.

Durand, B. M. Parrate, and P. H. Bertrand, 1983, Le potentiel en huile des charbons: une approche géochimique: Revue De L'Institut Francais Du Pétrole, v. 38, n. 6 Editions Technip.

Du Toit, A. L., 1937, Our wandering continents: Edinburg, Oliver and Boyd, 366 p.

Embleton, B. J. J., and M. W. McElhinny, 1975, The palaeoposition of Madagascar: palaeomagnetic evidence from the Isalo group: Earth Planetary Science Letter, v. 27, p. 329–341.

Fisher, R. L., and J. G. Sclater, 1983, Tectonic evolution of the Southwest Indian Ocean since the mid-Cretaceous: plate motions and stability of the pole of Antarctica/Africa for at least 80 Myr: Geophysical Journal, Royal Astronomical Society, v. 73, p. 553–576.

Flores, G., 1964, On the age of the Lupata rocks, lower Zambezi River, Mozambique: Transactions of the Geological Society of South Africa, v. 67, p. 111–118.

——, 1965, Observacoes sob o curso inferior do Zambeze: Primeiras Jornadas De Engenharia en Mocambique. Lorenzo Marques.

——, 1970, Suggested origin of the Mozambique Channel: Transactions of the Geological Society of South Africa, v. 73, p. 1–16.

——, 1973, The Cretaceous and Tertiary sedimentary basins of Mozambique and Zululand, *in* G. Blant, ed., Sedimentary basins of the African coasts, pt. 2, south and east coasts: Association of African Geological Survey, Paris, p. 81–111.

——, 1984, The SE Africa triple junction and the drift of Madagascar: Journal of Petroleum Geology, v. 7, n. 4, p. 403–418.

——, and F. Barbieri, 1959, Outline of the stratigraphy of Mozambique between the Maputo and Zambezi rivers: unpublished information available in Secretariat for Coals and Hydrocarbons archives, Maputo.

Förster, R., 1975, The geological history of the sedimentary basin of southern Mozambique Channel. Palaeogeography, Palaeoclimatolo, Palaeoecology, v. 17, p. 267–287.

Green, A. G., 1972, Seafloor spreading in the Mozambique Channel: Nature, v. 236, p. 19–21.

Hedberg, H. D., 1964, Geologic aspects of the origin of petroleum: AAPG Bulletin, v. 48, n. 11, p. 1755–1803.

Holmes, A., 1978, Principles of physical geology (3rd ed.): London, Nelson, 730 p.

Kamen-Kaye, M., 1982, Mozambique–Madagascar geosyncline, I: deposition and architecture: Journal of Petroleum Geology, v. 5, p. 3–30.

——, 1983, Mozambique–Madagascar geosyncline, II: petroleum geology: Journal of Petroleum Geology, v. 5, n. 3, p. 287–308.

Kinsman, D. J. J., 1975, Rift Valley basins and sedimentary history of trailing continental margins, *in* A. G. Fischer and S. Judson, eds., Petroleum and global tectonics: Princeton, N.J., Princetown University Press, p. 83–126.

Lawver, L. A., J. G. Sclater, and L. Meinke, 1985, Mesozoic and Cenozoic reconstructions of the South Atlantic: Tectonophysics (in press) Proceedings of the IUGG Symposium on Geophysics of Polar Regions XVIII Assembly - Hamburg, Germany, 1983.

Martin, A. K., S. W. Goodlad, and D. A. Salmon, 1982, Sedimentary basin infill in the northernmost Natal Valley, hiatus development and Agulhas current palaeo-oceanography. Journal of the Geological Society of London, v. 139, p. 183–201.

——, et al, 1982, Cretaceous palaeopositions of the Falkland Plateau relative to southern Africa using Mesozoic seafloor spreading anomalies: Geophysical Journal, Royal Astronomical Society, v. 71, p. 567–579.

Norton, I. O., and J. G. Sclater, 1979, A model for the evolution of the Indian Ocean and the breakup of Gondwanaland: Journal of Geophysical Research, v. 84, p. 6803–6830.

Rabinowitz, P. D., M. F. Coffin, and D. Falvey, 1983, The separation of Madagascar and Africa: Science, v. 220, p. 67–69.

SECH, n.d., Secretariat for Coal and Hydrocarbons, unpublished operators' reports archived in Maputo, Mozambique.

Segoufin, J., 1978, Anomalies magnétiques mésozoiques dans le bassin de Mozambique: Comptes Rendus de l'Academie des Sciences Paris, 287, D, p. 109–112.

——, and P. Patriat, 1980, Reconstructions de l'ocean Indien occidental pour les époques des anomalies M21 et 34, Paléoposition de Madagascar. Bulletin de la Societé Géologique de France, v. 22, p. 469–479.

——, and M. Recq, 1980, La transition entre le canal de Mozambique et le bassin de Mozambique: Bulletin de la Societé Géologique de France, v. 22, n. 3, p. 469–479.

Siesser, W. G., and R. V. Dingle, 1981, Tertiary sea-level movements around Southern Africa: Journal of Geology, v. 89, p. 83–96.

Sinha, M. C., K. E. Louden, and B. Parsons, 1981, The crustal structure of the Madagascar Ridge: Geophysical Journal, Royal Astronomical Society, v. 66, p. 351–377.

Vail, P. R., R. M. Mitchum, and S. Thompson, 1977, Seismic stratigraphy and global changes of sea level, part 4: Global cycles of relative changes of sea level, *in* C. E. Payton, ed., Seismic stratigraphy—applications to hydrocarbon exploration: AAPG Memoir 26, p. 83–98.

# Oil and Gas Possibilities On-and Offshore Ghana

G. O. Kesse
*Geological Survey Of Ghana*
*Accra, Ghana*

Ghana, located on the west coast of Africa, has been a focal point of hydrocarbon exploration since 1896. To date, 57 wells have been drilled: 36 offshore and 21 onshore. Onshore exploration efforts are concentrated in three sedimentary basins: the Tano, Keta and Voltaian basins; but their oil and gas potentials have not yet been fully evaluated. Offshore exploration is progressing on the Ghanaian continental shelf. The most promising offshore area is at the Ghanaian end of the Tano basin, adjoining the Ivory Coast.

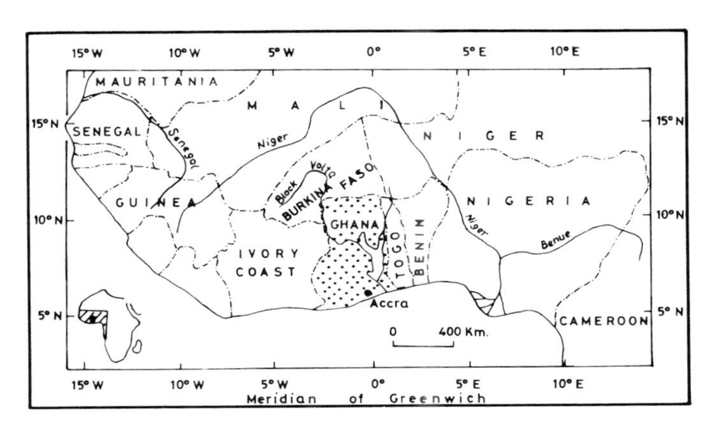


Figure 1. *Ghana's position in West Africa.*

## INTRODUCTION

The Republic of Ghana (population 12,205,574 in 1984) is one of the West African countries that lie along the Gulf of Guinea. It ranges between latitude 4°44′N above the Equator at Cape Three Points in the south and latitude 11°11′N in the north. The country's western and eastern extremities lie at longitudes 3°15′W and 1°12′E. It is roughly rectangular in shape with an average width of about 357 km (222 miles) and length an average of 670 km (414 miles), making an area of 239,460 sq km (91,800 sq mi). Its coastline, 555 km (345 mi) long, is cut by the Greenwich Meridian at Tema 24 km (15 miles) to the east of the capital, Accra.

To the east of Ghana lies the Republic of Togo, beyond which are the Republics of Benin and Nigeria. On the west is the Ivory Coast and on the north and northwest the Republic of Burkina Faso (formerly, Upper Volta) (Figure 1).

## GEOLOGY

Ghana falls mostly within the Precambrian Guinea Shield of West Africa. Ghana's main Precambrian rock units are the metamorphosed and folded Birimian, Tarkwaian, and Dahomeyan Systems, and the Togo Series and Buem Formation (Figure 2).

The Birimian and the Tarkwaian rocks occur mainly in the Eastern, Central, and Western Regions as well as in Ashanti, Brong Ahafo, and the western part of the Northern and Upper Regions of the country. The Dahomeyan, Togo, and Buem occur in the Volta and Eastern Regions of Ghana.

Late Proterozoic to Paleozoic rocks of the Voltaian System overlie the Precambrian rocks. This Voltaian System consists of sandstones, shales, mudstones, conglomerates, limestones, and tillites.

Younger rock units than the Voltaian System occur at several places along the coast and include the Early or Middle Devonian Accraian Series, Middle Devonian-Lower Cretaceous Sekondian Series, Upper Jurassic to Lower Cretaceous Amisian Formation, Upper Cretaceous Apollonian Formation, Tertiary to Recent unconsolidated marine, coastal, lagoonal, fluviatile sediments and deposits (Bates, 1962).

Intruded into the Birimian rocks are large masses of granitoids known as Cape Coast and Winneba (G.1) rock types, smaller masses of granitoids called Dixcove type (G.2), and rare Bongo potassic granitoids found mainly in the northern part of the country. Table 1 outlines the stratigraphic succession.

Table 1. *The stratigraphic succession of Ghana.*

| Era | Period/Epoch | System/Series/Formation |
|---|---|---|
| Cenozoic | Quaternary-Recent | Unconsolidated clays and sands of lagoon, delta, and littoral areas. |
| | Tertiary-Eocene | Partly consolidated red continental deposits of sandy clay and gravel. |
| Mesozoic | Cretaceous | Apollonian Formation (Upper Cretaceous Cenomanian-Campanian) |
| | Jurassic | Amisian Formation (Upper Jurassic-Lower Cretaceous) |
| Paleozoic | Devonian | Sekondian Series (Middle Devonian-Lower Cretaceous); Accraian Series (Early or Middle Devonian) |
| | Cambrian | Voltaian System (late Proterozoic to early Paleozoic 300–1,000 m.y.) |
| Proterozoic | Upper Precambrian | Buem Formation<br>Togo Series<br>Dehomeyan System. Age uncertain (middle to late Precambrian, probably reactivated at about 550 m.y. before present)<br>Tarkwaian System (Age unknown possibly 1,650–1,850 m.y.) |
| | Middle Precambrian | Birimian System (Age uncertain approximately 1,800–2,100 m.y.) |

## MAIN AREAS OF OIL/GAS EXPLORATION IN GHANA

The source rocks for petroleum and natural gas are mainly sedimentary rocks of marine origin. We must know exactly where these sedimentary rocks occur within Ghana. They will receive first consideration in evaluating oil and gas presence within any previously unproductive regions, because the chances of eventually finding commercial oil/gas are proportional to the volume of sediments present in any given area.

Fortunately, sedimentary rocks cover nearly half of Ghana's total area, about 135,000 sq km (42,100 sq mi). Table 2 and Figure 3 show these rocks occurring mainly in four different areas of the country, i.e., (1) Tano basin, (2) Keta basin, (3) Volta basin, and (4) the continental shelf.

Table 2. *Major sedimentary deposits in Ghana.*

| Area | Sq Mi | Sq Km | On/Offshore |
|---|---|---|---|
| 1. Tano basin | ... 450 | 1,165 | Onshore |
| 2. Keta basin | ... 850 | 2,200 | Onshore |
| 3. Voltaian basin | ...40,000 | 103,600 | Onshore |
| 4. Continental Shelf | ...10,642 | 27,562 | Offshore |
| | 51,942 | 134,537 | |

Ghana, therefore, possesses as much sedimentary cover as the Niger delta that, at present, is producing nearly two million barrels of oil per day. Exploration has concentrated in these

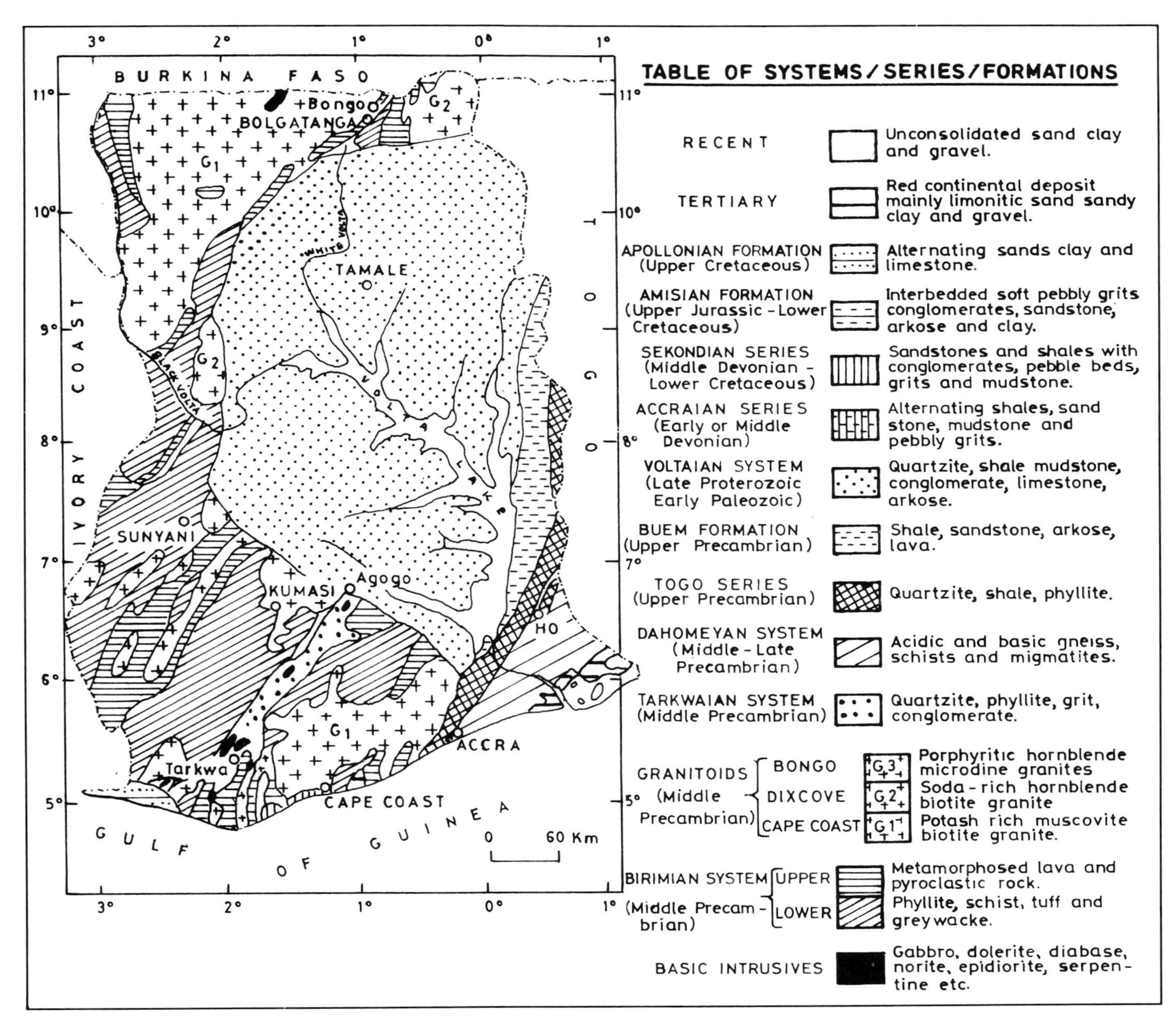


Figure 2. *Geological map of Ghana.*

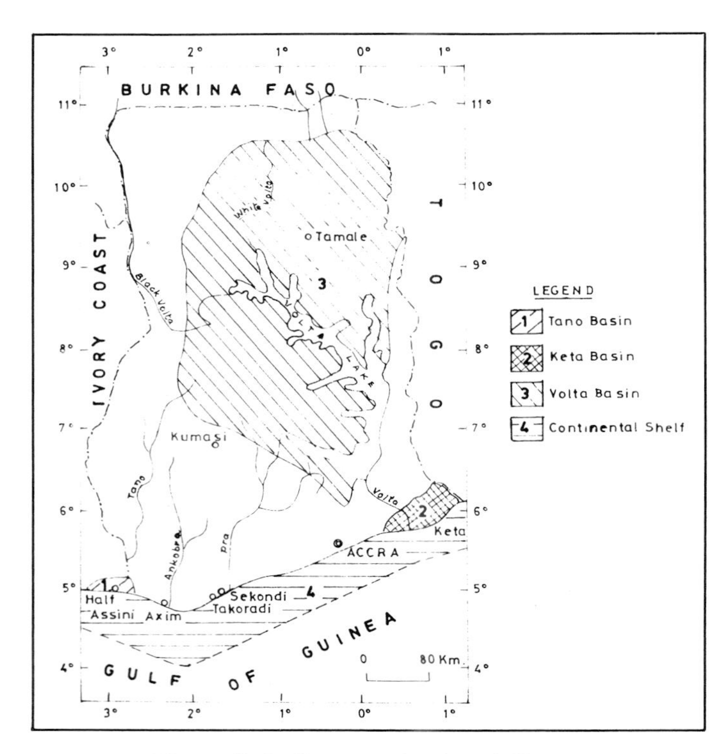


Figure 3. *Sedimentary basins of Ghana.*

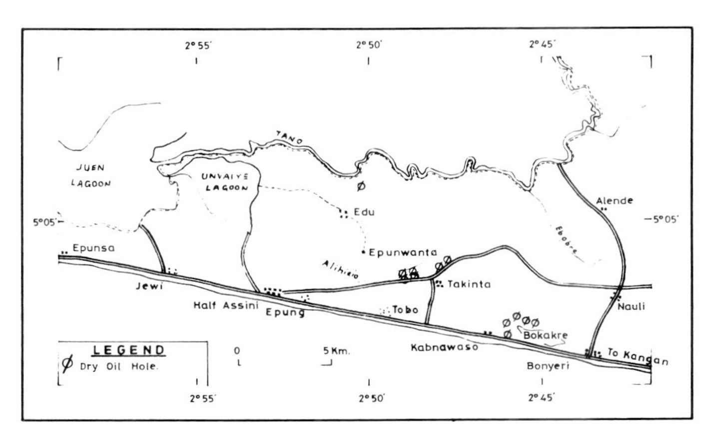


Figure 4. *Locations of wells drilled in the Tano basin from 1896–1925.*

four areas, that is, three onshore and one offshore. Let us now briefly examine these basins and the efforts that have gone into finding oil/gas in them.

## THE TANO BASIN, WESTERN REGION

Despite the uncertainty of finding oil in commercial quantities in Ghana, this country was among the first African countries to attract the attention of oil companies. They concentrated their early efforts in the Tano basin because, during the 19th century, explorationists noted oil seeps and saturated superficial sands here and in the north Ahanta-Nzima area.

### Geology

The Tano basin area, underlain by Upper Cretaceous Apollonian sediments, is low-lying and largely covered by swamps and lagoons. The geology is obscured by the swamps and lagoons and by a capping of Pleistocene to Recent sands and gravels. The structure and stratigraphical relationship of the sediments cannot, therefore, be accurately determined except by digging pits, drilling, or using geophysical methods.

The Apollonian consists of rapidly alternating sands and clays with occasional thin beds of gravel and fossiliferous limestone. The sands and clays are more compacted at depth and pass into sandstones and shales. Nodules of pyrite or marcasite are common in the clays and shales, and muscovite occurs in the sandy bars.

The rocks strike at west-northwest and, for the most part, dip at very low angles. On the southern side of the ridge, which extends from Kangan through Nauli to Edu, the regional dip of the beds is to the south or south-southwest at low angles of 2° to 6°. On the immediate north side of the ridge between Takinta and Nauli or Nawuli, geologists have noted a few dips to the north at angles up to 10° to 15° but do not know whether these are persistent or only local. Figure 4 shows rare, low, westerly and easterly dips in the vicinity of Takinta camp.

We can trace the line of of fossiliferous limestone outcrops from the beach at Kangan in a west-northwesterly direction through a point 800 m (2,600 ft) north of Nauli to the Tano River, 24 km (15 mi) north-northwest of Edu. Fossiliferous clays containing pyrite or marcasite and gypsum occur with the limestone (Cudjoe, 1972).

### Prospecting

Between 1896 and 1925, explorationists drilled at least 12 boreholes through the underlying Apollonian rocks in the Tano basin near Bonyere and Takinta to locate oil/gas pools in oil seepage areas. Two deep holes were sunk between 1923 and 1925, where explorationists had noted oil seepages. All the known seepages occurred at the southern side of the Takinta-Nauli ridge except for some in the valley of the Ekwogum stream 2.8 km (2 mi) northwest of Tikwabo (Junner, 1939). Geologists associated the seepages between the ridge and the sea with an oil sand that extends past the slopes of the Kangan-Edu ridge and that occurs at a shallow depth in the area between Tobo, Epunwanta, Nauli, and Bonyere (Figure 4).

### The Four Drilling Periods in the Tano Basin

According to available records, the four drilling periods in this area were 1896–1897, 1909–1913, 1923–1925, and 1956–1957 (Junner, 1939).

#### *1896–1897: Drilling by the West African Oil and Fuel Company*

In the first period, the West African Oil and Fuel Company sank four wells. We do not know the final depths of the holes, but the first well passed through 183 m (600 ft), cutting a bed of sand containing paraffin at 166 m (545 ft). According to the records, the second well on the Takinta concession produced five barrels of oil per day. The third well reached a depth of 236.5 m (776 ft). The fourth showed oil at 174 m (570 ft) (Junner, 1939).

In 1903, this company sank another well through alternating sands and clays to a depth of 433 m (1,420 ft). They encountered

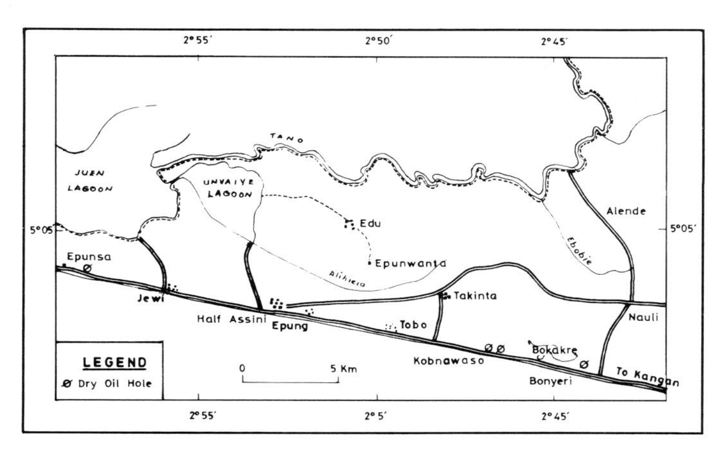


Figure 5. *Locations of wells drilled by the Gulf Oil Company during 1956–1957.*

a limestone bed at 56 m (184 ft), water with only a small show of oil at 219 m (718 ft) and shale with shows of oil 238 m (780 ft).

### *1909–1913: Drilling by the Société Française de Petrole (Junner, 1939)*

From 1909 to 1913, the Société Française de Petrole (SFP) sank six wells close to oil and gas seepages near Bokakre on Domini Lagoon (Figure 4). The results were as follows:

No. 1 well struck oil at 10 to 17 m (33 to 56 ft), and yielded seven barrels of oil per day.
No. 2 borehole was abandoned in 1912.
No. 3 well has no available records.
No. 4 well was sunk 46 m (150 ft) from the second well and reached a depth of 340 m (1,115 ft). Very good traces of light oil were encountered at the three horizons of the hole.
No. 5 well was sunk to 668 m (2,190 ft). It showed gas with traces of oil. Gas with some paraffin accumulating on the casing at 549 m (1,800 ft).
No. 6 was drilled to 829 m (2,720 ft) on Tano River bank. Strong traces of gas and oil were encountered at 152 m (498 ft). Shale saturated with oil was also encountered at 549 m (1,800 ft). There were traces of gas and oil at 792 m (2,598 ft).

### *1923–1925: Drilling by the African and Eastern Trade Corporation (Junner, 1939)*

Two holes were sunk in 1923 and 1925 on the Takinti-Nauli ridge by the African and Eastern Trade Corporation (Figure 4). They reached depths of 1,221 m and 916 m (4,005 and 3,005 ft), respectively, and bottomed in flat-bedded shales and sandstones of the Apollonia sediments. The drillers encountered heavy oil in an oil sand at 17 to 25 m (56 to 82 ft) and found traces of lighter oil at 144 to 150 m (472 to 492 ft), 339 to 374 m (1,112 to 1,226 ft) and 427, 549, 732, 808 and 1,015 m (1,400, 1,800, 2,400, and 3,330 ft; respectively). In a second borehole near Takinta, they found traces of oil and gas at 36 to 38 m (118 to 125 ft) and from 310 to 311 m (1,017 to 1,020 ft).

### *1956–1957: Drilling by the Gulf Oil Company*

The Gulf Oil Company of Ghana sank four boreholes along the coast between June, 1956, and November, 1957 (Figure 5). The drilling site was 11.2 km (7 mi) east of Half Assini. In their first well, Kobnawaso 1, they discovered small, noncommercial shows of oil near the surface and also at 1,232 m (4,042 ft). They plugged and abandoned the hole at 3,252 m (10,670 ft) in sediments believed to be of Middle Cretaceous age.

During the first quarter of 1957, they drilled the second well, Epunsa 1, as a stratigraphic test-hole about 11 km (7 mi) west of Half Assini and about 22.5 km (25 mi) west of Kobnawaso 1. The drilling site was bordered on one side by the sea and on the other by Epunsa Lagoon.

The drillers completed the well at 2,071 m (6,795 ft) in black, thin-bedded silt containing numerous pebbles and cobbles of igneous and metamorphic rocks. The only oil shows were small lumps of tar in ditch samples from approximately 73 m (240 ft). After completing the drilling program, the drillers had not proven the structural nose predicted by seismic survey. The Gulf Oil Company geologists suggested a periodically active fault between Kobnaswaso 1 and Epunsa 1.

Gulf's third well, Bonyere 1, was located about 0.75 km (0.5 mi) west of Bonyere and 4.8 km (3 mi) east of Kobnaswaso 1. The hole was completed at 2,497 m (8,192 ft) in grey and grey-green silt and sandstone, probably of Middle Cretaceous age. The geologists observed traces of oil, detectable in most cases only with the fluoroscope, in partings of very fine sand at irregular intervals of 1.6 to 24.4 mm (0.063 to 1.0 in.) in black shale cores between 1,096 m (3,596 ft) and the top of a conglomerate at 1,160 m (3,806 ft).

Their last well, Kobnaswaso 2, was drilled at a point approximately 610 m (2,000 ft) west of Kobnaswaso 1 and as near to the beach as was practicable. They drilled the well without difficulty to a depth of 1,376 m (4,514 ft) and completed in grey-green sands and shales 120 m (394 ft) below the base of the black shale. The geologists observed traces of oil in a near-surface sand at the approximate level of a small showing in Kobnaswaso 1. Cores from the top of a sandstone immediately beneath the black shale showed streaks of saturation. When the drillstem tested the zone, it showed streaks of salt water but no oil or gas. They found no faulting above the top of the Grey Series.

## The Romanian Oil Experts and the Geological Survey Department

In 1962, the Ghana government invited a team of Romanian oil experts to examine the oil potentialities of Ghana's three sedimentary basins: Tano, Keta, and Voltaian. They recommended gravity surveys in these basins so experts could estimate basin thickness and map the topography. The Ghana Geological Survey Department, therefore, conducted detailed gravity surveys in the Tano and Keta basins, finding major faults, the Kangan and Ahonjuri faults, suitable for trapping oil in the Tano basin. The fault, trending northwest to southeast, divided the basin into two parts. On the up-thrown side, or northern edge, of the basin, the sedimentary complex was not as thick as that on the down-thrown side, where the thickness increased toward the Republic of Ivory Coast.

Geologists drew the following conclusions concerning gas and oil potential in the Tano basin:

1. The maximum thickness of the sedimentary rocks in this basin was more than 3,000 m (9,800 ft) along the coast, increasing toward the Ivory Coast border.

2. The oldest rocks encountered in the survey were of Middle Cretaceous age.
3. Rocks of marine origin were found at 1,770 m (5,800 ft), and were separated from non-marine rocks by an angular discordance.
4. Two horizons indicated oil: one, near the surface (i.e., the Nauli limestone horizon) and the other at greater depth (that is, the black shale horizon).
5. The most promising area for oil accumulation was immediately south of the major fault.

## KETA BASIN, VOLTA REGION

The Keta basin covers an area of 3,755.50 sq km (1,450 sq mi), of which 2,201.50 sq km (850 sq mi) are onshore and the remainder offshore. Regionally, the whole basin is a part of the great coastal sedimentary basin which extends from Nigeria through Benin, Togo, Ghana and the Ivory Coast. Pliocene deposits cover the basin to the north and Recent deposits cover it to the south.

In the past, geologists obtained all geological information about this area from logs of shallow boreholes that, in rare cases, attained a maximum depth of about 304 m (995 ft). These were drilled by the Water Department, now known as the Ghana Water and Sewerage Corporation (Khan, 1970).

### Geology

The rocks of the Keta basin are mainly sand, gravel, siltstones, shale, and clays with layers of fossiliferous limestone. The rocks near the surface have a gentle dip of about 2° toward the southeast. The rocks of the basement may be the Dahomeyan, similar to that cropping out to the north of the basin, which might also include some basic intrusions (gabbros, dolerites) as found to the north and northeast.

In 1962–63, a Romanian team conducted a regional gravity survey over the whole area, followed, in 1966–67, by a more detailed survey by the Geological Survey (Khan, 1968).

According to Kirton (1967), the Keta basin has on the whole a tectonic block structure bounded by a fault or fault systems on its northern flank. This fault system essentially trends northeast to southwest, but near Galotse in the western half of the area, the fault system becomes bifurcated with the northern part becoming north-northeast to south-southwest oriented and the southern part east to west.

The gravity survey indicated that the basin's deepest part is in its center, forming a longitudinal trough. The gravity values decrease both from the northern margin (35 mgal) and southern coast line (0 mgal) toward the center (–15 mgal) (Kirton et al, 1974).

### Prospecting

Beginning in 1962, explorationists used the gravity survey method to look for oil in this basin and the Volta delta. In 1966 and 1967, a Romanian team drilled two test wells at Atiavi and Anloga (Figure 6). These wells showed (1) that the sedimentary rocks in this area are at least 2,133 m (7,000 ft) thick; (2) that the rocks here are marine and non-marine, rich in organic matter, and are, therefore, a potential source of hydrocarbons; (3) that the sand, sandstones, and limestones present in the sediments are potential reservoir rocks; and (4) that marine sediments thicken in a down-dip direction toward the sea, and this trend may continue offshore.

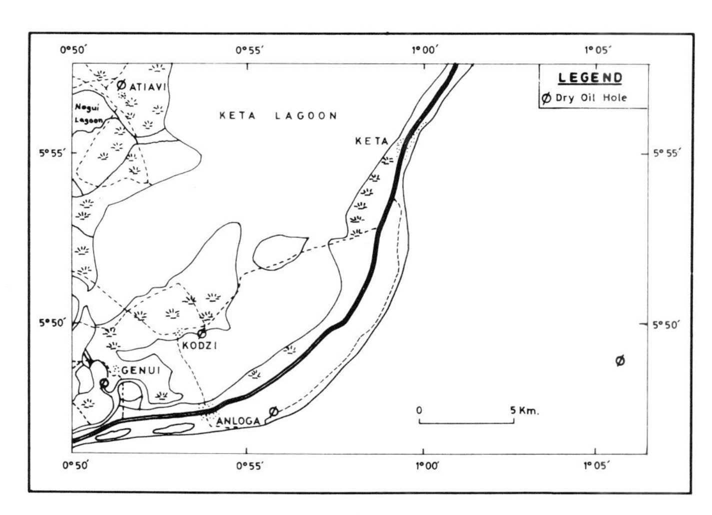


Figure 6. *Locations of wells in the Keta basin.*

### *Atiavi 1 Well*

The first Atiavi well was drilled through Quaternary, Eocene, Cretaceous, and Devonian sediments and bottomed at 1,568.8 m (5,147 ft) in Precambrian gneiss. This well provided valuable stratigraphic information about the presence of a 583.6 m (1,915 ft) thickness of Devonian shales and sands.

The drilling of this well was abandoned because of the presence of a fractured dolerite between the interval of 866–936 m (2,840–3,070 ft). Large pieces of this dolerite sloughed into the borehole almost to the final depth. This seriously affected the performance of the drilling bit and the drilling collars. As a result, drilling had to be suspended on a number of occasions; it finally reached the crystalline basement at a depth of 1,568.80 m (5,147 ft).

### *Anloga 2 Well*

Anloga 2, drilled by the Ghana government in 1967, was located on the coast and 16 km (10 mi) south of the Atiavi 1 well. The well bottomed at 2,133 m (6,998 ft) after penetrating Quaternary, Miocene, and a thick, marine Cretaceous section at 2,133 m (6,998 ft). Electric logs indicated traces of gas at the following intervals: 797 to 799 m (2,615 to 2,621 ft), 804 to 806 m (2,638 to 2,644 ft), and 808 to 811 m (2,651 to 2,660). The well was abandoned because of serious loss of mud during drilling.

### Other Drilling Activities in the Basin

After abandoning these two wells, Volta Petroleum Company Limited and others, in 1970, drilled a dry hole at approximately 17 km (11 mi) southeast of Keta. In 1973, Mesa of Ghana Limited also drilled a dry well approximately 11 km (7 mi) northwest of Anloga. Some oil shows, however, were encountered at certain depths (Kesse, 1975).

Lately, Texas Pacific Ghana, Incorporated has acquired concessions in the Keta basin and have completed the seismic surveys of the area prior to drilling.

### Stratigraphy of the Keta Basin

Based on the two wells the Romanians drilled at Atiavi and

Table 3. *Stratigraphy of the Keta basin (after Khan, 1970).*

| Unit No. | Lithology and Thickness | Age |
|---|---|---|
| I | Beach deposits of loose sands and gravels<br>Thickness: 30–60 m (98–197 ft) | Recent |
| | Unconformity | |
| II | Glauconitic, fossiliferous clays<br>Thickness: 180 m (591 ft) | Miocene |
| | Unconformity | |
| III | Calcareous clays interbedded with fossiliferous limestones<br>Thickness: 250–700 m (820–2,297 ft) | Eocene |
| IV | Bentonitic clays, fossiliferous<br>(120–240 m or 394–787 ft) | Paleocene |
| V | Bluish-grey clays, fossiliferous, interbedded with limestones | Maestrichtian |
| | Unconformity | |
| VI | Brown, reddish-brown, grey, fine to medium, grained sandstones with subordinate shales<br>Thickness: 400–550 m (1,312–1,804 ft) | Campanian |
| VII | Grey, greyish-white, coarse to medium-grained sandstones; gravels interbedded with mudstones and shales<br>Thickness: 370 m (1,214 ft) | Albian |
| VIII | Greenish-grey, grey, poorly sorted sandstones; siltstones and shales<br>Thickness: 579 m (1,900 ft) | Aptian |
| | Dolerite | |
| IX | Dark grey, Micaceous, often varve-like shales, and siltstones; fossiliferous<br>Thickness: 610 m (2,001 ft) | Devonian |

Anloga, Khan (1970) proposed the basin's stratigraphic sequence as listed in Table 3. This sequence shows rocks of Devonian, Cretaceous (Aptian, Albian, Campanian, and Maestrichtian), Paleocene, Eocene, and Miocene to Recent ages.

## THE VOLTAIAN BASIN

The Voltaian basin is an expansive sedimentary basin that covers an area of about 103,600 sq km (40,000 sq mi). It is drained by the Volta River and its tributaries.

### Geology

Rocks of the Precambrian to lower Paleozoic epicontinental Voltaian System underlie the basin. They are a thick sequence of marine and continental sediments that are exposed in impressive cliffs at the edges of the basin. The younger rocks cover most of the older beds, so that the older members are exposed where they overlie the Precambrian basement of igneous and metamorphic rocks only along the northern and western edges of the basin. The younger members are generally flat-lying or have very low dips of 3° to 5°, generally toward the center of the basin. Because of this fact and the predominantly continental nature of the exposed sediments, the basin was, until recently, regarded as shallow, structureless, and unlikely to have any hydrocarbon accumulations. Now, geologists realize its thickness (Anan-Yorke, 1971).

To the east of the basin occur a series of sandstones, quartzites, phyllites, and schists intruded by the Togo Series and the Buem Formation, with basic and ultrabasic igneous rocks exposed (Bates, 1962). These formations are strongly folded and faulted in places and strike south-southwest to northeast. Recent geological mapping suggests that these rocks are not contemporaneous with the Voltaian System and, together with the Dahomeyan System, collectively represent a Mobil belt at the eastern edge of the Voltaian basin.

### Prospecting

Under the terms of a contract between the Ghana government and the U.S.S.R., a Soviet geological survey team in 1961–1966 conducted a hydrogeological survey to study groundwater conditions in the Voltaian basin. The team planned to drill 11 key boreholes, each to a depth of over 700 m (2,300 ft). Unfortunately, they could complete only four boreholes: Tamale—249.50 m (819 ft); Tibagona—604.60 m (1,984 ft); Yendi—699 m (2,293 ft); and Nasia—760.70 m (2,496 ft) (Soviet Geological Team, 1965). The deepest borehole intercepted only the upper and middle parts of the middle Voltaian rocks and encountered no basement rocks (Figure 7).

The results showed that the lower members of the Voltaian have been folded, with the team recording dips of up to 40°. The Voltaian contains limestones and sandstones that could serve as cap rocks for the accumulation of hydrocarbons, and the team intercepted chlorocalcic waters, usually associated with hydrocarbon occurrences. Their preservation suggests favorable conditions for hydrocarbons in this basin. The limestone cores recovered from boreholes at Nasia, Prang, Buipe, and Tamale showed traces of heavy oil. The team reported a gas show from the Tamale borehole, also revealing that the basin deepened in the center and that stratigraphic traps were likely to occur.

In 1962, the Ghana government requested that a Romanian oil prospecting team evaluate the petroleum prospect of the basin. After studying the available data, the Romanian team

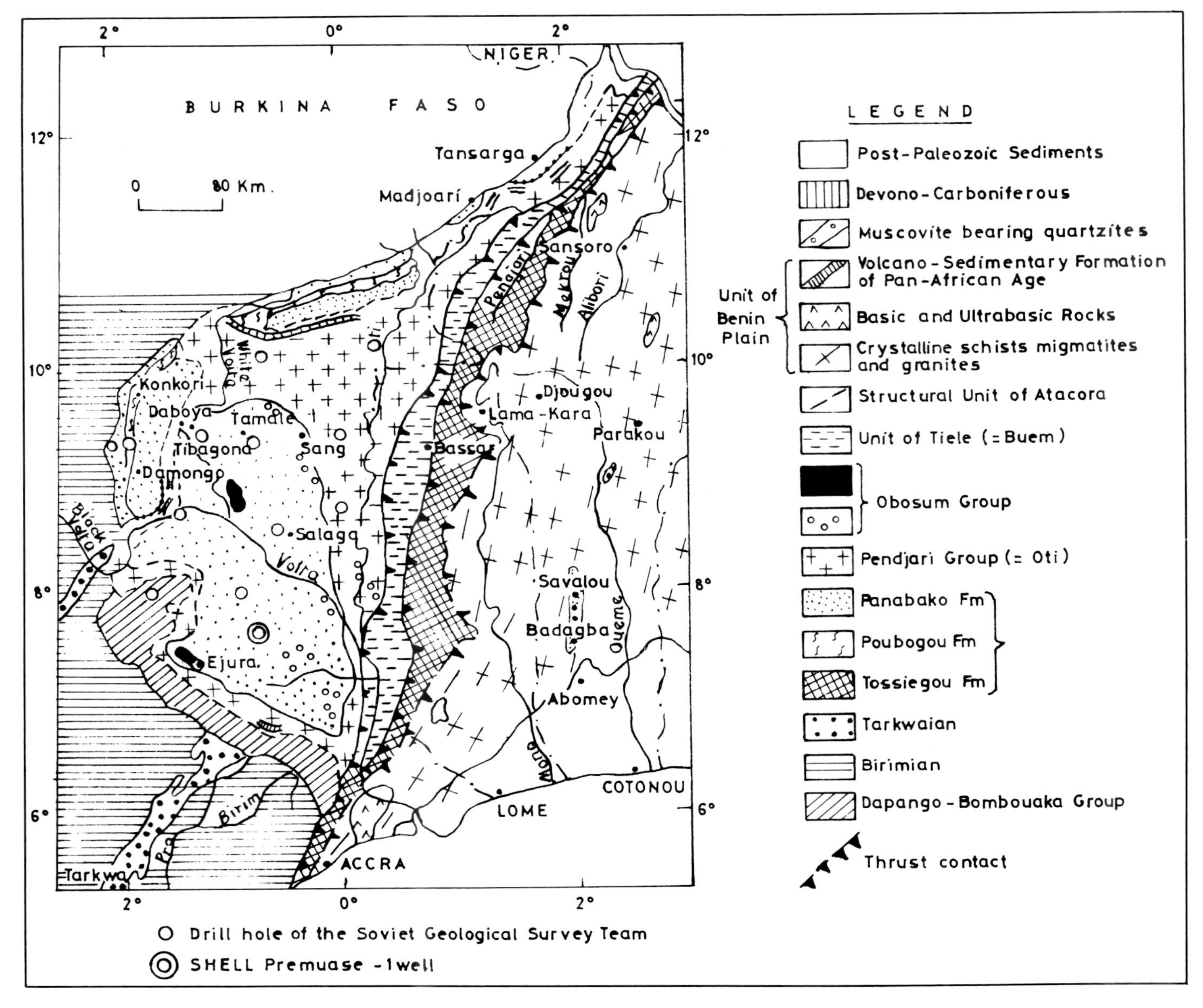


Figure 7. *Drill holes in the Voltaian basin.*

recommended that rapid gravimetric and magnetic surveys could reveal the approximate depths of the basement and the configuration of the buried structures in the basin.

Unfortunately, the Geological Survey could only undertake the gravimetric survey in 1965 in Tamale. The first profiles ran to the edge of the basin in the north. In 1974, Proca, geophysicist of the Geological Survey Department, interpreted the results (Proca, 1974).

Another geophysical survey occurred in January, 1971, under a contract with the Compagnie Generale de Geophysique of France that compiled and interpreted an airborne magnetometric survey over Ghana's mainland (Compagnie Generale de Geophysique, 1971). This contract was on behalf of Shell Oil Company (Ghana) Limited. The survey showed that 4 to 5 km (2 to 3 mi) of non-magmatic rock underlies the eastern part of the Volta depositional area near Lake Volta east of Kwadjokrom. Selecting Shell's concession blocks around this aeromagnetic deep assumed that a substantial part of the basin fill might be prospective sedimentary rocks.

After these surveys, many oil companies showed interest in obtaining concessions in the Voltaian basin. The government divided the basin into 39 blocks for oil companies to prospect (Kesse, 1975) (Figure 8).

On January 10, 1975, Shell Exploration and Production Company (Ghana) Limited received a petroleum prospecting license for five years, covering 10 blocks totalling 28,490 sq km (11,000 sq mi) in the southern part of the basin between latitudes 6°30′N and 8°30′N and longitudes 0°30′W and 0°30′E.

## Shell Exploration and Production Company (Ghana) Limited's Program

After Shell received the petroleum prospecting license, the

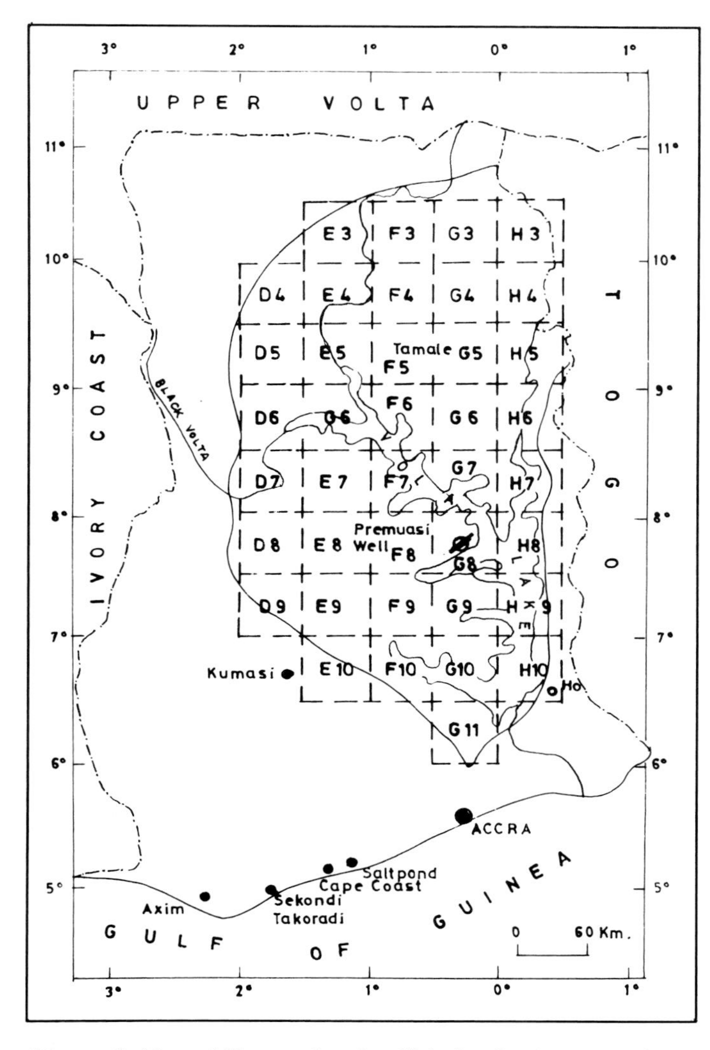


Figure 8. *Map of Ghana showing Voltaian basin concessions.*

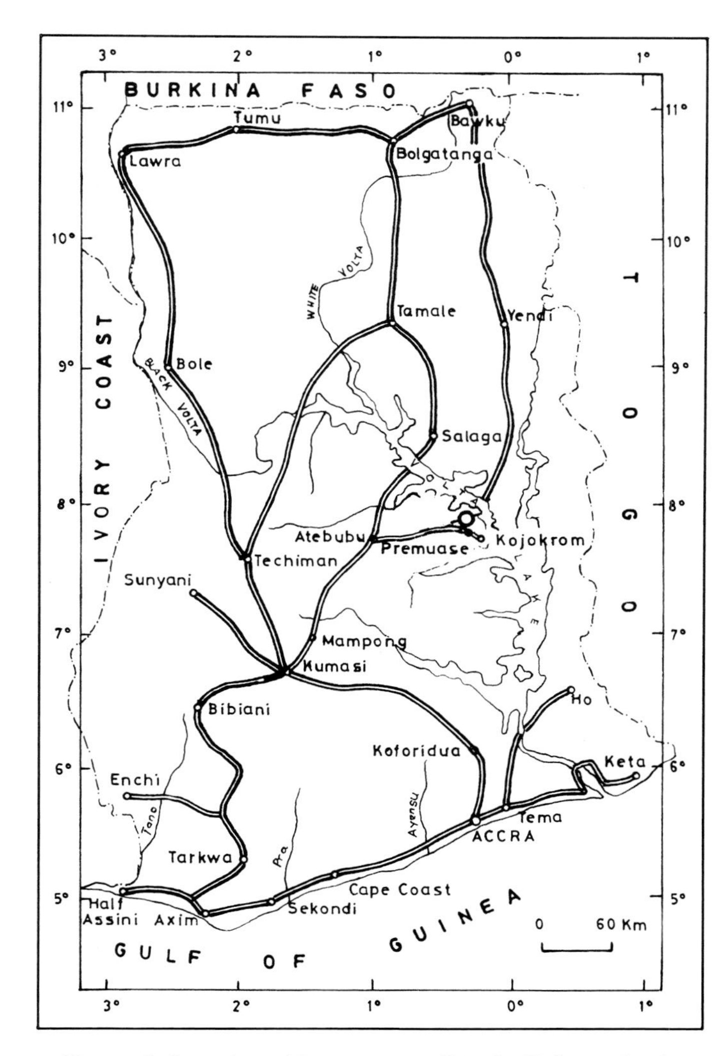


Figure 9. *Location of Premuase well in the Voltaian basin.*

Company first photogeologically interpretated the aerial photographs covering the southern part of the basin.

Between December, 1975, and March, 1976, Shell Company contracted with Seismograph Services of the United Kingdom to carry out a reconnaissance seismic survey in the area between Atebubu and Lake Volta where geologists expected the thickest representative sedimentary section.

The objectives of the seismic survey were to

1. check the feasibility of the seismic method in the Volta basin;
2. check and calibrate aeromagnetic results; and
3. investigate the structural picture of the basin fill.

Four seismic lines were shot, comprising a total of 208 km (130 mi) using a semi-portable mode. Seismograph Services processed and interpreted the recorded data during May, 1976. The quality of the records varied but generally provided the reconnaissance information Shell required (Seismograph Service Limited, 1976).

The results of the seismic survey showed that the middle Voltaian barely reached a thickness in excess of 1,000 m (3,280 ft) with a maximum thickness of 1,400 m (4,600 ft) for the Voltaian basin north of latitude 7°30′N.

The results of the seismic survey also showed flat to gentle dips in the middle Voltaian without any significant dip-closed structures. The underlying pre-middle Voltaian reached 6,000 m (19,700 ft) and were intruded at depth by igneous rocks.

At Premuase, a village located about 8 km (5 mi) east of Atebubu and 14 km (9 mi) west of Kwadjokrom on Lake Volta in Brong Ahafo Region (Figure 9), a gentle, southwesterly plunging faulted nose may be the most promising structural feature in the area.

### Premuase Well, Brong Ahafo Region

Premuase Well was drilled on a structural nose at Premuase village (7°47′45″N and 0°17′55″E) adjacent to the north side of the Atebubu/Kwadjokrom road. The well was to obtain lithostratigraphic information on the middle Voltaian and the upper part of the lower Voltaian.

The well was spudded on 2nd July, 1977. Drilling was completed on 8th August, 1977, to 1167.5 m (3,830 ft)—in Precambrian sandstone. The geologists encountered no trace of hydrocarbons in any part of the section. The well was plugged

Table 4. *Sequence of Premuase well penetration.*

| Period/Epoc | Series | | Depth (m) | Depth (ft) |
|---|---|---|---|---|
| Middle Voltaian: | Tamale Red Series | ... | 0–54 m | 0–177 ft |
| (Cambrian?) | Upper Greenish-Grey Series | ... | 54–602 m | 177–1,975 ft |
| | Variegated Series | ... | 602–1,132 m | 1,975–3,714 ft |
| Lower Voltaian: | Lower Greenish-Grey Series | ... | 1,132–1,267.5 m | 3,714–3,830 ft |
| (Precambrian) | | | | |

and abandoned as dry on 10th August, 1977 (Watt, 1977).

The nature of the formations encountered during drilling meant slow progress, particularly near the surface. In the last 200 m (650 ft) above basement, the chocolate-brown and green shales, siltstones, claystones, and limestone of the Variegated Series presented a layered sequence that became more and more resistant to drilling.

### Geological Section

The Premuase Well is the deepest hole drilled to date in the Volta basin. The section consisted of very hard, compact, mainly fine-grained Precambrian clastics, ranging from siliceous mudstone and hard shale through hard, greenish-grey siltstones to fine sandstones with occasional stringers of hard, grey limestone.

Table 4 shows the sequence penetrated:

The middle Voltaian mainly consists of fine, hard sandstone, siltstones, claystones, shales, and thin limestone bands. The Precambrian was a uniform, greenish-grey, fine, hard, argillaceous sandstone.

### Oil/Gas Possibilities Onshore Ghana

Before we move on to oil/gas possibilities offshore Ghana, we should evaluate the oil/gas possibilities onshore Ghana, in the three onshore basins:

1. Tano basin
2. Keta basin
3. Inland Voltaian basin

#### *Oil/Gas Possibilities in the Tano Basin*

Geologists have noted several places in the onshore Tano basin with bitumen seepages, tar sands, or oil/gas shows. Oil shows have appeared in the Bonyere, Takinta, Nauli, Kobnawaso, Tikobo, and Ebuaso areas for more than 100 years and have repeatedly enticed wildcat drillers to the area since 1896.

The best near surface oil shows occur in the Domini Lagoon area. In particular, SFP borehole No. 1, drilled at the northeastern end of the lagoon in 1909, found a 73 m (240 ft) oil sand at a depth of 10 m (33 ft) that flowed at 7 barrels of oil per day (BOPD). In 1912 SFP drilled borehole No. 5, 0.8 km (0.5 mi) to the southeast. They penetrated a 6 m (20 ft) oil sand at a depth of 15.24 m (50 ft) that yielded a strong oil flow. Studies report several other shows from 1.6 km (1 mi) to 4.8 km (3 mi) north of the lagoon.

These Tano basin shows are mainly light gravity oil that will flow when pressure is released, for example, by swabbing. A laboratory analysis of an oil sample shows that it has a gravity of 36° API, a viscosity of 3.7%, stokes at 40° C (104° F), is of middle maturation and is from an algal source (Robertson Research International Ltd., 1981).

The sample of tar sand the laboratory examined, showed sand impregnated with soluble oil of light fractions from land-derived organic matter. Technically, therefore, these are not "tar sands" in the accepted sense, but appear to be seeps of light gravity oil moving from a down dip hydrocarbon source to the surface where the oils lose their light ends to the atmosphere and form an asphalt residue or incipient "tar-mat."

Technically, "tar sands" are sands with a highly viscous, crude hydrocarbon content that is not recoverable in its natural state by ordinary production methods (Alberta Oil and Gas Conservation Act). The oils have gravities of less than 10° API at 60° F (15.6° C) and may be immature. They may not have been buried to sufficient depths nor subjected to sufficiently high temperatures to permit their cracking into higher gravity fractures.

Because of their high viscosity, tar sands must be stimulated with steam injection or fire flood to permit recovery unless they are close to the surface, as in the case of Athabasca deposits in Canada. The numerous gas seeps reported from the lagoon also appear to support this theory. If a sufficiently wide tar-mat forms a surface seal, a near-the-surface accumulation of light oil could be present.

Table 5 summarizes the oil and gas shows in the Tano basin.

Table 6 summarizes the analysis of oil and tar sand from the Tano basin.

#### *Proposed Exploration of the "Tar Sands"*

Robertson Research International Limited (1981) proposed a preliminary evaluation of the "tar sands" in the Tano area. The Company suggested that this exploration program would require a minimum of 72 km (45 mi) of seismic line and 5 coreholes (Figure 10). This program should be part of the work program commitment for any oil exploration company wishing to explore this area.

They suggested the following seismic lines:

1. Line AB. 40 km (25 mi) along the coast, east from Half Assini and far enough to intersect the trace of the gravity fault.
2. Line CD. 19.2 km (12 mi), to follow the road arching inland from Ahobre to Bonyeri.
3. Line EF. 12.8 km (8 mi) northeast from the coast road at Domini lagoon to the Prah River.

They suggest drilling Corehole Nos. 1–4 to sufficient depth so that they penetrate the Nauli limestone marker. The boreholes should intersect, unless interrupted by faulting, at the following depths:

### Oil/Gas Possibilities in Onshore Keta Basin

Three onshore wells have so far been drilled in the Keta basin. Although geologists have recorded hydrocarbon shows

Table 5. *Oil and gas shows in the Tano basin (Source: Robertson Research International Ltd, 1981).*

| No. | Borehole/Seep | Location | Area | Comment |
|---|---|---|---|---|
| 1. | Seep? | 5°1.8′N 2°47.3′W | N. Ebwazo | Asphalt on path |
| 2. | Gulf Oil Company of Ghana, Kobnawaso No. 1 | 5°1.8′N 2°47.3′W | E. Ebwazo | Small oil shows near surface sand. Carbonaceous black shale 1,173–1,254 m (3,848–4,114 ft) Core at 1,254 m (4,114 ft) has streaks of oil. DST at 1,254 m (4,114 ft) and has no shows. |
| 3. | Gulf Oil Company of Ghana, Kobnawaso No. 2. | 5°1.4′N 2°46.9′W | E. Ebwazo | Traces of oil in near surface sand. Carbonaceous black shale 1,173–1,254 m (3,848–4,114 ft) Core at 1,254 m (4,114 ft) has streaks of oil DST at 1,254 m (4,114 ft) and has no shows. |
| 4. | Societe Francaise de Petrole (SFP) Borehole No. 3 | 5°1.5′N 2°46.2′W | E. Ebwazo | Drilled beside surface oil seep. No records |
| 5. | Seepage | 5°1.7′N 2°46.4′W | E. Ebwazo | No description (mapped only) |
| 6. | Seepage | 5°1.8′N 2°46.6′W | E. Ebwazo | No description (mapped only) |
| 7. | Borehole No. 5 de Petrole (SFP) | 5°1.5′N 2°45.4′W | Domini lagoon | Strong flow of oil bearing sands at 15–21 m (49–69 ft). Traces of gas at 357 m (117 ft). Strong gas show at 549 m (1,801 ft) |
| 8. | SFP Borehole No. 1 | 150′ from No. 4 | Domini lagoon | Oil sand 10–17 m (16–56 ft) 7 BOPD |
| 9. | SFP Borehole No. 4 | 5°1.7′N 2°45.9′W | Domini lagoon | No records. Drilled beside seep |
| 10. | SFP Borehole No. 2 | 5°1.7′N 2°45.6′W | Domini lagoon | No information |
| 11. | Gulf Bonyeri No. 1 | 5°0.9′N 2°44.1′W | N.W. Bonyeri | Trace of oil in sandstones formation. Stringers in black shale at 1,068–1,157 m (3,504–3,796 ft). |
| 12. | Borehole and seep | 5°2.3′N 2°49.7′W | Tobo | No description. Mapped only. |
| 13. | WAO & F | 5°2.9′N 2°49.7′W | N. Tabo | Yielded 5 BOPD. |
| 14. | Seepage | 5°2.1′N 2°47.4′W | E. Ebwazo | No description. Mapped only. |
| 15. | Oil Sand | 5°2.4′N 2°45.6′W | N. Domini lagoon | No description. Mapped only. |
| 16. | Gulf Epunsa No. 1 | 5°3.8′N 2°58.7′W | E. Epunsa | Lumps of tar in ditch samples at 73 m (239 ft). |
| 17. | Seepage | 5°3.3′N 2°49.7′W | Takinta N.W. Nauli | Oil and bitumen. Mapped only. |
| 18. | African and Eastern Trade Corp. (U.A.C.) No. 1 | 5°3.7′N 2°45.0′W | Chitabini | Oil sand at 6 m (20 ft). Mapped only. Gas in sand/clay segment at 81–150 m (266–492 ft) interval. Trace fo oil at 342 m (1,122 ft). Gas in shale and sandstone at 549 m (1,801 ft). Gas in sand and shale at 732–1,006 m (2,402–3,301 ft). |
| 19. | UAC Borehole No. 2 | 5°4.2′N 2°46.9′W | N.E. Takinta | Gas in sandstone at 36 m (118 ft). Oil and gas in sandy shale at 310 m (1,017 ft). |
| 20. | SFP Borehole No. 6 | 5°6.4′N 2°50.2′W | Tano River | Gas and oil show at 152 m (499 ft). Oil in shale at 549 m (1,801 ft). Gas and oil show at 792 m (2,598 ft). |

in these wells, the possibility of finding commercial oil/gas in this basin is yet to be proved (Kesse,1978).

### Oil/Gas Possibilities in Inland Voltaian Basin

Drillers found no trace of hydrocarbon in any part of the section in the Premuase well. Electric logging confirmed this result, finding high velocities associated with the rocks in the well. The average interval velocity in the Upper Greenish-Grey Series was 4,080 m/sec (13,386 ft/sec) whereas in the Variegated Series, the average interval velocity was 4,400 m/sec

Table 6. *Result of analysis of oil and tar sand from the Tano basin (Source: Robertson Research International Ltd, 1981).*

| Oil and Tar Compositional Data | | |
|---|---|---|
| | Oil | Tar From Sand |
| Loss on topping % | 4.1 | — |
| Asphaltene content % | — | 0.5 |
| Alkanes % | 67 | 39 |
| Aromatics % | 12 | 23 |
| Resense % | 10 | 36 |
| Carbon isotope ratios | | |
| Alkanes | -26.8 | -27.4 |
| Aromatics | -26.4 | -26.4 |
| Resenes | not available | -26.2 |
| Asphaltenes | not available | -24.3 |

(14,436 ft/sec). At the contact of the Variegated Series with the Lower Greenish-Grey (1,132 m [3,714 ft]), a velocity break showed an average velocity of 4,650 m/sec (15,256 ft/sec) below this level (Watt, 1977).

The velocities in the Premuase-1 well, observed in the middle Voltaian, were high, while those in the lower Voltaian were very high considering the predominantly or entirely clastic nature of the sediments. Therefore, geologists concluded that the rocks in the well may have reached a high degree of diagenesis.

Table 7.

| Borehole No. | Depth in meters (feet) |
|---|---|
| 1 | 182.88 ( 600) |
| 2 | 60.96 ( 200) |
| 3 | 30.48 ( 100) |
| 4 | 60.96 ( 200) |
| | 335.28 (1,100) |

The results indicated deep burial and subsequent removal of the overburden. Geological conditions in the basin, therefore, were clearly adverse for preserving liquid hydrocarbons.

### Traces of hydrocarbons in the Voltaian Rocks

The Soviet Geological Survey Team encountered traces of bitumen in the 1962–1965 drilling program, giving hope that the basin, in spite of its age, might prove to be an oil province. Bitumen is usually found in dolomite intercalated in the middle Voltaian shale/sandstone sequence. On further examination of the bitumen in the Voltaian rocks, Watt (1977), who recently examined several of the core samples from the Premuase-1 well, found that bitumen in those rocks was not distributed in the natural bedding, but instead was always in fractures mostly

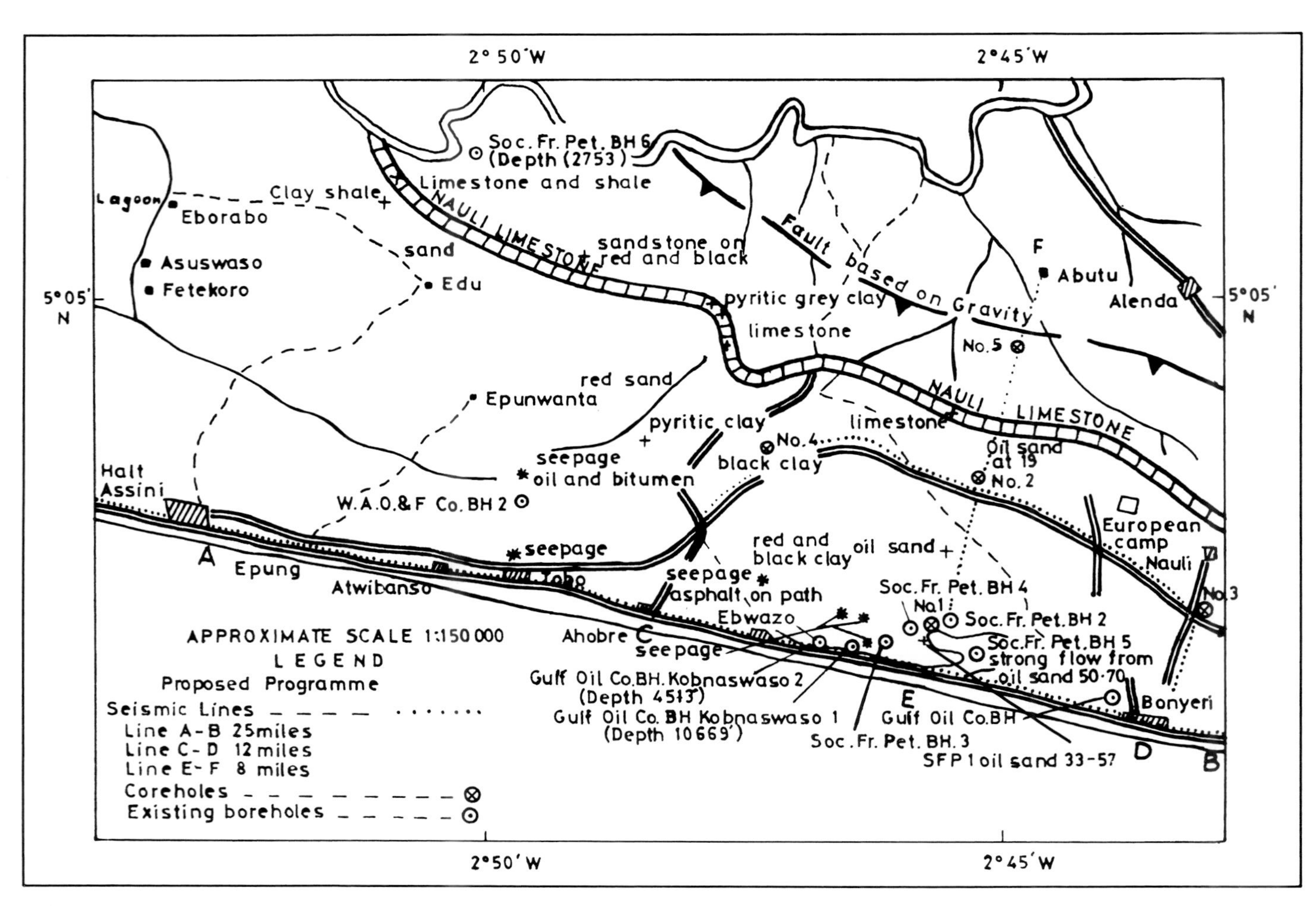


Figure 10. *Proposed seismic and drilling program for evaluating the Tano sands.*

Table 8. *Division of continental shelf.*

| Company/ Corporation | Total No. of sq km (sq mi) | Block Nos. |
|---|---|---|
| Mayflower-Volta Petroleum | 7,630 (2,946) | 1, 2, 21, & 22 |
| Mobil Oil | 4,250 (1,641) | 3, 4, & 5 |
| Texas Gas Exploration Corp. | 4,060 (1,568) | 6 & 17 |
| Texaco Oil | 5,180 (2,000) | 7, 8, 9, 11, & 12 |
| Union Carbide | 3,330 (1,286) | 18, 18A, 19, & 20 |
| Signal Oil | 1,851 ( 715) | 13 & 16 |
| Signal Oil and others | 3,903 (1,507) | 10, 14, 15, & 15A |

transverse to the bedding, and occasionally ran along the bedding planes. He thought that the bitumen had seeped downward from a younger, overlying formation that has since been eroded.

### Hydrocarbon Potential of the Voltaian Basin

The Voltaian basin is an old one with no significant structural closures. No source rocks are in juxtaposition to allow possible inward, normal, hydrocarbon migration (Watt, 1977). Geologists considered accumulation of oil in commercial quantities unlikely, although not impossible.

## THE CONTINENTAL SHELF

In 1968, the Ghanaian government invited private foreign oil companies to undertake oil prospecting work in the country. Most companies were interested in the continental shelf, so the government divided it into 24 concession blocks (Figure 11). Table 8 and Figure 12 show that by the end of the year, the entire continental shelf was under license:

These oil companies received oil prospecting licenses in December, 1968, that empowered them to prospect for oil over a three-year period beginning January, 1969. The licenses required the companies to drill one or more holes to a total depth of 3,660 m (12,000 ft) within 18 months from signing the agreement. By June, 1971, nearly all the companies had fulfilled their obligations, drilling 11 wells during 1970.

### Signal Oil Company and Oil Discovery

In June, 1970, Signal Oil Company discovered oil in a well about 14.4 km (9 mi) south of Saltpond. This well (Ghana 10-1) was positioned by seismic data in Block 10 and was drilled to a depth of 2,967 m (9,734 ft). Two producing horizons were located. The first at 2,347 m (7,700 ft), presumably in the Cretaceous, tested 1,300 BOPD of 40° API oil, while the second zone, at 2,590.9 m (8,500 ft) in the Devonian, tested 2,300 BOPD of 37° API oil.

Signal's initial tests showed that the well could produce a total of 3,600 barrels of oil a day from two horizons. The quality of the oil was good, with the sulphur content being 0.5%. Signal geologists estimated the total reserves in this area at about 7.5 million barrels.

### Evaluating The Saltpond Find

In 1971, Signal Exploration, Amoco Ghana, and Occidental Oil Companies drilled two wells near Well 10-1 of the Saltpond

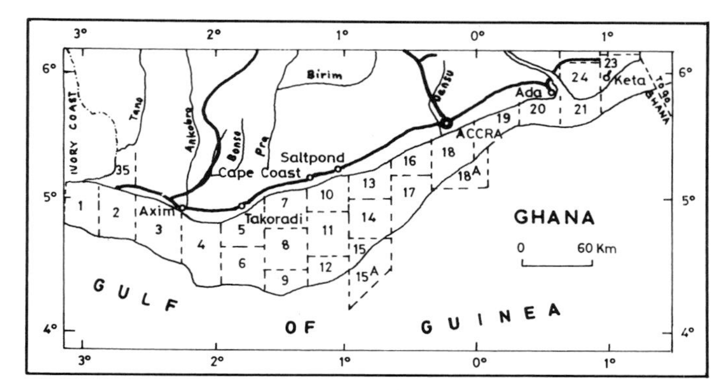


Figure 11. *Map of Ghana showing offshore concessions.*

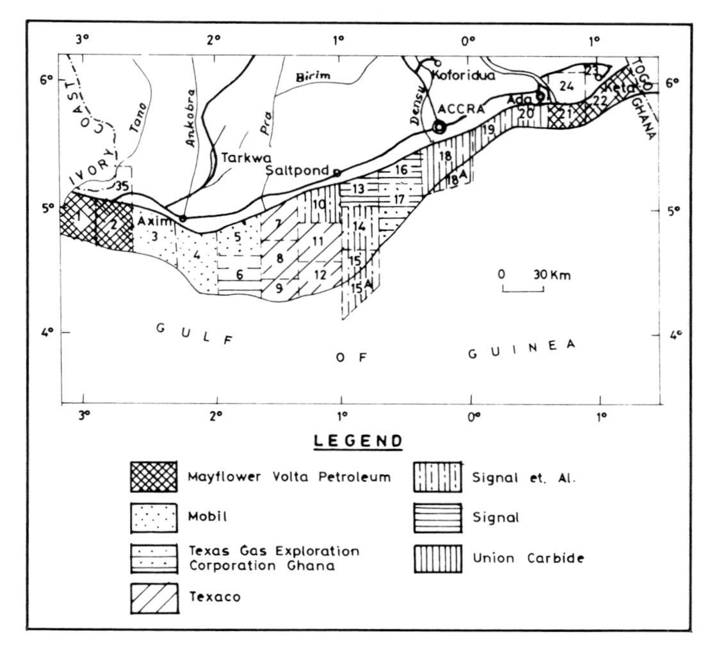


Figure 12. *Continental shelf concessions as of December, 1968.*

find. From December 23, 1974, to February 6, 1975, Amoco Ghana Exploration Company drilled another well near Well 10-1 (Kesse, 1978).

After drilling these wells, Amoco stated that the company had proven about 175 hectares (432 acres) of oil-bearing reservoir in the Saltpond area, which contains about 2.4 million barrels of recoverable oil. This amount of oil clearly did not justify the production facility, but it was possible that there were 6.34 million barrels of recoverable oil in an area of 422.01 ha (1,042 acres). An additional test well should confirm their findings.

### Oil Exploration and Exploitation, by Agri-Petco of Ghana, Inc.

On December 13, 1976, Agri-Petco International, Inc., an independent Corporation based in Tulsa, Oklahoma, U.S.A., received a Petroleum Prospecting License covering Blocks 10 and 13 in the offshore Ghanaian concessions (Figure 13). By the end of December, 1977, Agri-Petco had drilled 3 wells in Block 10, with encouraging results (Kesse, 1978).

On January 10, 1978, Agri-Petco and the Ghana government signed an Oil Winning Agreement. At that time, Agri-Petco informed the Ghana government that each of the 3 wells was capable of producing about 1,000 barrels of oil per day.

By June 6, 1978, Agri-Petco had drilled 6 wells in Block 10 and had decided to develop this field, (known as the Bonsu Field but later called Saltpond Field). On the June 12, 1978, Mr. E. David Philley, President of Agri-Petco, presented the results of their drilling activities in the Saltpond area to General I.K. Acheampong, then Head of State and Chairman of the Supreme Military Council. Mr. Philley stated that Agri-Petco was scheduled to begin production by October 1, 1978. The initial production rate was to be 5,000 to 6,000 barrels of oil per day. Table 9 shows the anticipated production from the 6 wells:

Table 9. *Projected production—late 1978 start.*

| Year | | Production (barrels) |
|---|---|---|
| 1978 | — | 519,500 |
| 1979 | — | 1,717,300 |
| 1980 | — | 1,258,100 |
| 1981 | — | 921,600 |
| 1982 | — | 672,200 |
| 1983 | — | 494,600 |
| 1984 | — | 313,700 |
| Total | | 5,897,000 |

Agri-Petco intimated that the Saltpond Field was a complex one because the 7 major producing horizons were not continuous in all the zones between the 6 wells. Recovery was further complicated by primary gas caps in three zones and established water-oil contacts in two zones. Fluid analysis of the bottom-hole samples of oil showed the oil to be near its bubble point, which meant that gas would start to break out of solution very rapidly as the bottom-hole pressure dropped. Geologists needed to use extreme care when selecting the intervals to be perforated, to avoid gas penetrating the oil zone and also water production. They planned to control initial production rates at reasonable levels. Detailed reservoir pressure, along with gas volumes, had to be carefully studied to get maximum recovery efficiency.

Current geologic mapping also indicated the potential of additional development wells. Such development has to be justified after evaluating data obtained from drilling the six wells, for example, seismic work and some production history. The plan was to drill three additional producers and two gas injection wells southeast of the first six wells. The additional drilling was to find similar reservoir conditions, along with the reinjection of produced gas (Agri-Petco of Ghana, Inc., 1978).

The ten-well scheme could have reserves of 8.85 million barrels of oil. Table 10 shows a reasonable production projection if the drilling had been completed by late 1979:

In summary, there were several reservoirs with water, oil, and gas present. Reservoir fluid analysis indicates bottom-hole pressures near the bubble point. The number of reservoirs and type of oil made this field very complex, but close production supervision of the reservoir energy should produce the maximum amount of oil from each well and the field as a whole.

### The Performance To Date of Agri-Petco of Ghana, Incorporated

Table 10 shows the quantity of oil that Agri-Petco has produced since October 1, 1978, when they started producing oil from the Saltpond Field. Note that Agri-Petco produced 2,149,204 barrels of oil between the October 1, 1978, and December 31, 1983 (Table 11). The production difficulties experienced from this field arose because the reservoir consists of seven thin pay sands with several different oil/water and gas/oil interfaces. The reservoir pressure fell below the bubble point and because of this, production declined to less than 2,000 barrels of oil per day. By the end of February, 1982, production had declined to about 1,350 barrels of oil per day (Agri-Petco of Ghana Inc., 1982).

If new techniques are not brought to bear, production from this field is expected to decline further and, in fact, to cease by 1985.

Robertson Research International Limited (1981), in their report to the Ghana government, recommended that a new reservoir engineering study reassess the reservoir in the light of its production history and, if possible, to recommend an additional development drilling or well re-completion program to prove additional recoverable reserves and prolong the productive life of the field. Robertson Research stated that such a study could cost approximately $53,000.00.

### Oil in the Ivory Coast and Its Effect on Ghana

On October 18, 1977, the Ivory Coast announced that the country had struck oil. Esso and Shell started intensive oil research work in 1971, and in 1974 they discovered promising offshore oil indications. The reserves, estimated at about 70 million tons, of which 25 million tons were commercially exploit-

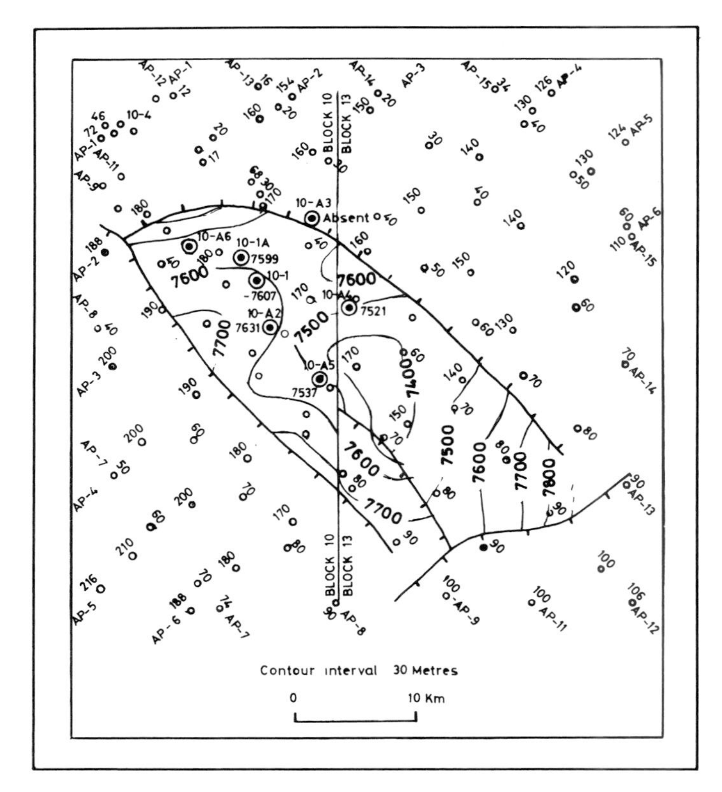


Figure 13. *Locations of Agri-Petco wells in Blocks 10 and 13, Saltpond area.*

Table 10. *Projected production—late 1979 start.*

| Year | | Production (barrels) |
|---|---|---|
| 1978 | — | 519,500 |
| 1979 | — | 1,895,000 |
| 1980 | — | 1,826,600 |
| 1981 | — | 1,461,300 |
| 1982 | — | 1,169,000 |
| 1983 | — | 935,200 |
| 1984 | — | 596,600 |
| 1985 | — | 446,800 |
| Total | | 8,850,000 |

Table 11. *Quantity of oil produced from Saltpond Field by Agri-Petco of Ghana Incorporated.*

| Period | | Quantity (in barrels) |
|---|---|---|
| 1st Oct.–31st Dec., 1978 | — | 295,988 |
| 1st Jan.–31st Dec., 1979 | — | 655,781 |
| 1st Jan.–31st Dec., 1980 | — | 664,254 |
| 1st Jan.–31st Dec., 1981 | — | 599,797 |
| 1st Jan.–31st Dec., 1982 | — | 434,750 |
| 1st Jan.–31st Dec., 1983 | — | 498,634 |
| Total | | 3,149,204 |

able, are situated some 15 km (9 mi) offshore Grand Bassam, a coastal town 45 km (28 mi) east of Abidjan.

This discovery is significant because of the Ivory Coast's basinal structure. The Ivory Coast is framed by three faults that converge on the Ghanaian shelf west of Accra (Figure 14):

1. a major Ivory Coast normal fault along the north side of the basin,
2. the Ivory Coast-Ghana ridge along the basin's southern flank, and
3. the Accra fault (Akwapim fault) trending south-southwest, crossing the coastline immediately west of Accra.

The Accra fault intersects the other two and trends in an unclear way. It may terminate in the Ivory Coast basin at its eastern end. However, west of the Accra fault, the continental shelf widens to form a marginal plateau that is up to 80 km (50 mi) wide (Robertson Research Intern. Ltd, 1981).

Sedimentation from Cape Three Points to the west end of the Ivory Coast basin seems to have been continuous from Permo-Triassic times through to the Tertiary. A dispread conglomerate of Jurassic age is 2,000 m (6,560 ft) thick in the Gulf Kobnawaso No. 1 well in the onshore Tano basin, and is also present at the west end of the Ivory Coast basin where it is 472 m (1,550 ft) thick. Maximum subsidence and deposition occurred during the Lower and Upper Cretaceous times when coarse detritus were deposited in the more positive Cape Three Points area,

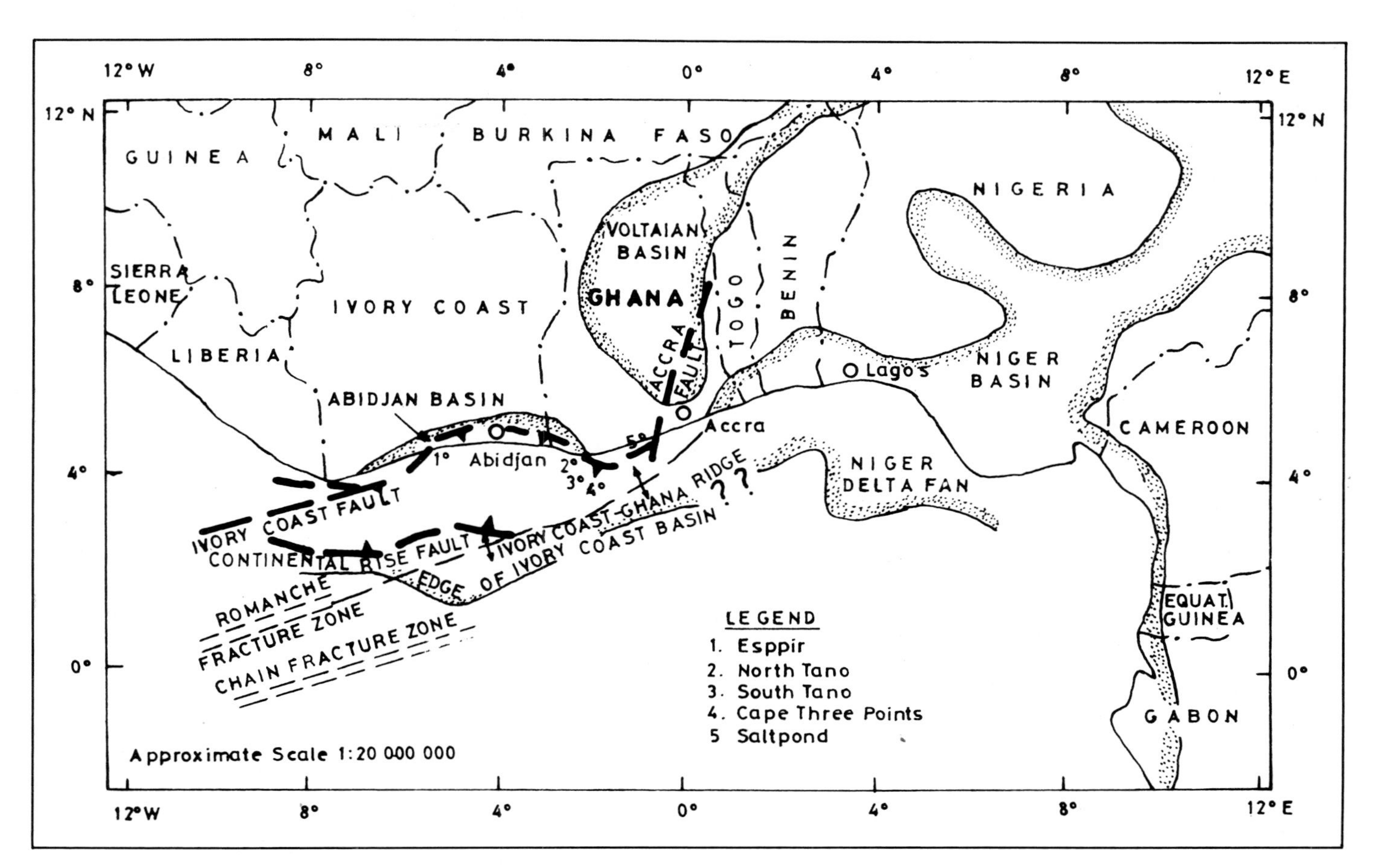


Figure 14. *Regional geological framework along the West Coast of Africa.*

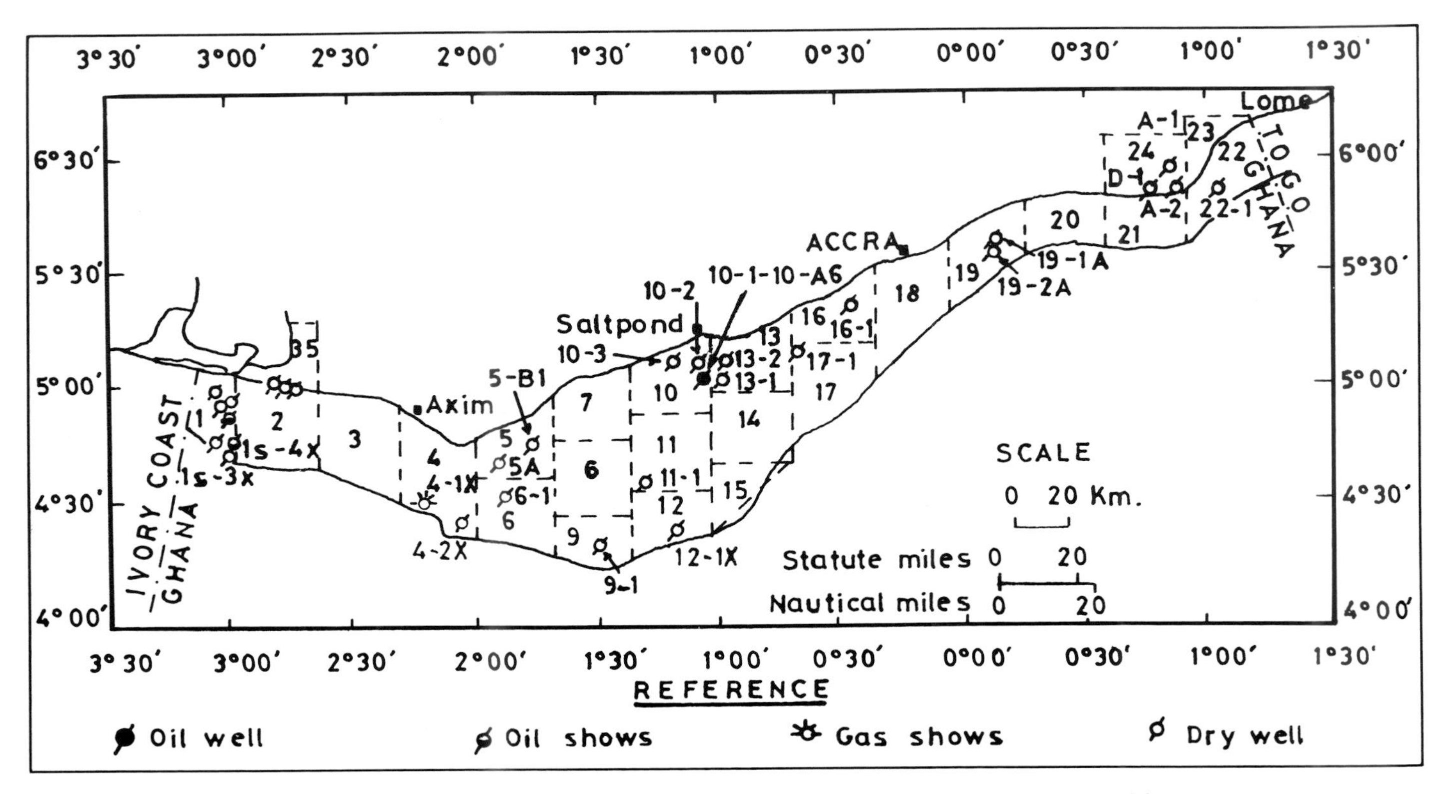


Figure 15. *Locations of oil wells drilled onshore and offshore Ghana since 1896.*

with the finer sediments deposited at the western and deeper end of the basin. A final tectonic pulse in the upper Eocene/ Oligocene formed the present-day structure of the basin and subsequent quiescence indicates the deposition of flat-lying Miocene sediments off the Ivory Coast.

### Ghana's Petroleum Potential in the Tano Basin

The discovery of oil in the Ivory Coast was of great interest to Ghana because Ghana's petroleum potential is closely associated with that of the Ivory Coast basin. This basin extends for 560 km (248 mi), along the whole of the Ivory Coast and persists eastward into Ghana for an additional 320 km (200 mi) terminating in the area immediately west of Accra.

The basin's prolific potential is only now being established, with the discovery of a giant oil field by Phillips Petroleum Company at Espoir near Abidjan. The "Ocean Oil Weekly Report" reported on March 16, 1981, that (1) the structure at Espoir could contain 3 to 8 billion barrels of crude; (2) it probably measures at least 20.8 km (13 mi) across; (3) the discovery has at least 92 m (300 ft) of perfect, clean oil sand; and (4) all 6 intervals were tested at the high end of 4,900 barrels of oil per day.

Because of these facts, Phillips contracted 5 additional drilling rigs in this area and ordered a jack-up production platform for extended production testing. They planned to start production by mid-1982 from three sub-sea completed wells.

### The Record So Far in Ghana

Currently (December, 1984), a total of 57 wells have been drilled on land and in the continental shelf of Ghana (Figure 15). Of this number, 21 have been drilled onshore while 36 have been drilled offshore.

### Seismic Activity in Ghana in 1982

In 1982, the Ministry of Fuel and Power entered into an agreement with Geophysical Service Incorporated (GSI), a subsidiary of Texas Instruments of Dallas, Texas, U.S.A., to undertake a marine seismic survey along 70% of the Continental Shelf of Ghana (Figure 16). The survey was to ensure that reliable data were available for oil companies to assess the country's oil potential.

GSI recorded approximately 7,300 km (4,500 mi) of nonexclusive seismic and magnetic data. The stacked and migrated sections are now available from GSI along with the geophysical interpretation report.

The Ministry of Fuel and Power has offered the blocks in the GSI survey area for petroleum exploration and production. At the time of writing this report, Shell Oil Company and Amoco Ghana Exploration Company have submitted bids for some of the blocks. The Ghana government has also given its approval to a new exploration and production law that is to govern exploration and production of petroleum in the country. In essence, the law makes provisions for production shaving of petroleum between the Ghana National Petroleum and the contractor.

## SUMMARY AND CONCLUSIONS

In conclusion, oil/gas possibilities onshore Ghana on the whole are yet to be proved. Offshore Ghana, oil/gas exploration should concentrate on the Ghanaian end of the Tano basin adjoining the Ivory Coast because this area is a direct continuation of and forms the eastern end of it. Even though the Ghanaian end has experienced less subsidence throughout its

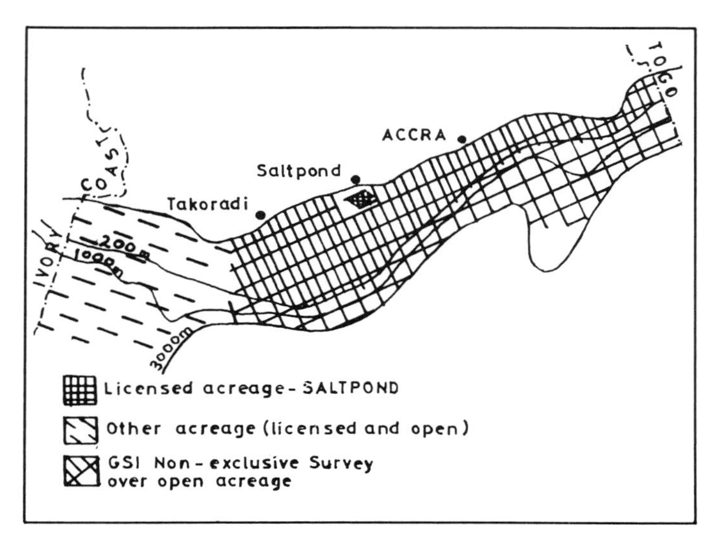


Figure 16. *Marine seismic survey area of Geophysical Services Incorporated.*

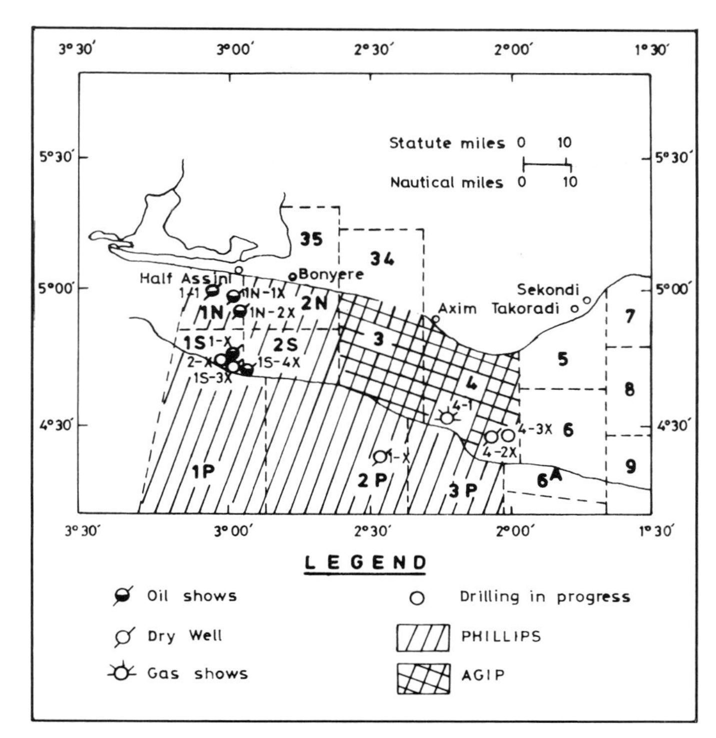


Figure 17. *Locations of oil wells drilled by Phillips Petroleum Company and Agip (Africa) Ltd since 1978.*

geological history, the sands are probably closer to their source, thicker, and more massive and dirtier, that is, not so porous or permeable as those at Espoir. The Espoir sands are cleaner, well winnowed and interbedded with thick intervals of hydrocarbon-rich marine shales.

Phillips Petroleum Company Ghana has confirmed the oil/gas possibilities in this area by drilling seven wells since July, 1978, (Figure 17).

Phillips encountered oil/gas shows at certain intervals in these wells and of late, Petro-Canada has also found oil/gas in the two wells they have drilled in this area.

## SELECTED REFERENCES

Agri-Petco of Ghana, Inc., 1978, Monthly reports to the Director, Ghana Geological Survey Department (unpublished reports).

——, 1982, Monthly reports to the Director, Ghana Geological Survey Department (unpublished reports).

Ako, J. A., and A. M. Procca, 1975, Geophysical exploration of the Voltaian basin: Annual report of the Director of Geological Survey for the year 1973–74, p. 11–13.

Amoco Ghana Exploration Company, 1975, Evaluation of the 10-1 structure, 32 p.

Anan-Yorke, R., 1971, Geology of the Voltaian basin: Special bulletin for oil exploration, Ghana Geological Survey, 29 p.

Barbu, M., et al, 1963, Gravity surveys carried out in the Keta basin and their results, 27 p.

Bates, D. A., 1956, Development of petroleum of Western Nzima: Annual report of the Director of Ghana Geological Survey for 1954–55, p. 1.

——, 1962, Geology of Ghana, *in* Agriculture and land use in Ghana, edited by J. Brian Wills, Oxford University Press, p. 51–61.

——, 1957, Prospecting for oil: Annual report of the Director of Geological Survey Ghana for 1956–57, p. 4–5.

Compagnie Generale de Geophysique, 1971, Airborne magnetometric survey, Ghana, from 8th December, 1970 to 3rd January, 1971, Ghana EP Report 42486.

——, 1976, Report of data processing of Kwame Danso prospect, Ghana EP Report 47,683,411.

Cudjoe, J. E., 1972, Report on the Nauli Limestone (unpublished report of Ghana Geological Survey Department).

Ghana Geological Survey, 1967, Exploration for petroleum and gas in Ghana, 47 p.

Hirst, T., 1948a, Oil in Apollonia: Annual report of the Director of Geological Survey for 1946–47, p. 4–5.

——, 1948b, Oil in Apollonia: Annual report of the Director of Geological Survey for 1947–48, p. 4–8.

Junner, N. R., 1939, Oil in Apollonia: Annual report of the Director of Geological Survey, 1938–39, p. 20–22.

——, and T. Hirst, 1946, The geology and hydrogeology of Voltaian basin: Gold Coast Geological Survey Memorial 8, 51 p.

Keizer, J., G. J. M. de Boer, and E. van der Bent, 1976, Results of a photogeological and geophysical study of the Voltaian basin (south part): Report EP-47500, Shell Internationale Petroleum Maatschappij Besloten Vernootschop, 14 p.

Kesse, G. O., 1975, Prospects of petroleum and other fossil energy resources discovery in Ghana: Geological Survey of Ghana Report No. 75/9, 33 p.

——, 1978, Petroleum exploration trends in Ghana, *in* G. O. Kesse and E. Jones, eds., Contributions to the Geology of Ghana: Report of Geological Survey of Ghana, Accra, Ghana, p. 24–36.

——, 1978b, The search for oil (petroleum) in Ghana: Geological Survey Report No. 78/1, 22 p.

Khan, M. H., 1970, Cretaceous and Tertiary rocks of Ghana with an historical account of oil exploration: Bulletin No. 40, Ghana Geological Survey, p. 1–43.

——, 1974a, Oil exploration laboratory: Annual report of the Director of Ghana Geological Survey for 1966–69, p. 8.

——, 1974b, Drilling for oil in the Keta basin by the Romanian Team: Annual report of the Director of Ghana Geological Survey for 1966–69, p. 17–27.

Kirton, M., et al, 1974, Gravity and magnetic survey of the Voltaian basin: Annual report of the Director of Ghana Geological Survey for 1966–67, p. 61–63.

Mitchell, J., 1958, Prospecting for oil: Annual report of the Director of Geological Survey, 1957–58, p. 12–14.

Proca, A. M., 1974, Report on a gravity Bouguer map of the Voltaian basin, (Ghana Geological Survey unpublished report).

Ricchiuto, T., R. Cavaliere, and D. Grignani, 1982, Source rock evaluation and stable isotopes study of the well Axim 4-3X (Ghana) Agip S.P.A. (unpublished report).

Robertson Research International Limited, 1981, Ghana: Preparation for a petroleum exploration project: Report No. 4609P/B, Project No. RRPS/812/B/2410. Prepared for Ministry of Fuel and Power, Accra, Ghana. p. 1–52.

Seismograph Service Limited, 1976, Field area report on a seismic reflection survey conducted in the Republic of Ghana for Shell Exploration and Production Company of Ghana Limited; Party 830, p. 1–10.

Soviet Geological Survey Team, 1965, Hydrogeological investigations in the Voltaian basin. Party No. 20: Annual report of the Director of Geological Survey for 1964–65, p. 34–35.

Watt, D. S., 1977, Premuase-1 Well résumé: Shell Exploration and Production Company of Ghana Ltd., p. 1–12.

# Natural Gas Expected in the Lakes Originating in the Rift Valley System of East Africa, and Analogous Gas in Japan

Osamu Fukuta
*Geological Survey of Japan*
*Ibaraki, Japan*

Gas, dissolved in the deep water of Lake Kivu, is geochemically similar to gases derived from the oil and gas fields situated in the Green tuff region. The water of Lake Kivu at about 400 m (1,310 ft) in depth contains 22 millimoles of methane and 77 millimoles of carbon dioxide per kilogram. Therefore, the water may emit gas of about 2.02 cu m[1]/kl at the lake surface. Thus, Lake Kivu gas may be recoverable by means of self-flowing through long pipes. Abiogenic gas resources are expected in some east African rift lakes, caldera lakes, and in some volcaniclastic rocks attributed to large-scale magmatic activities.

[1]In this paper, cu m of gas are expressed in Normal cu m, or $Nm^3$, i.e. cu m at 0° C and 1 atm.

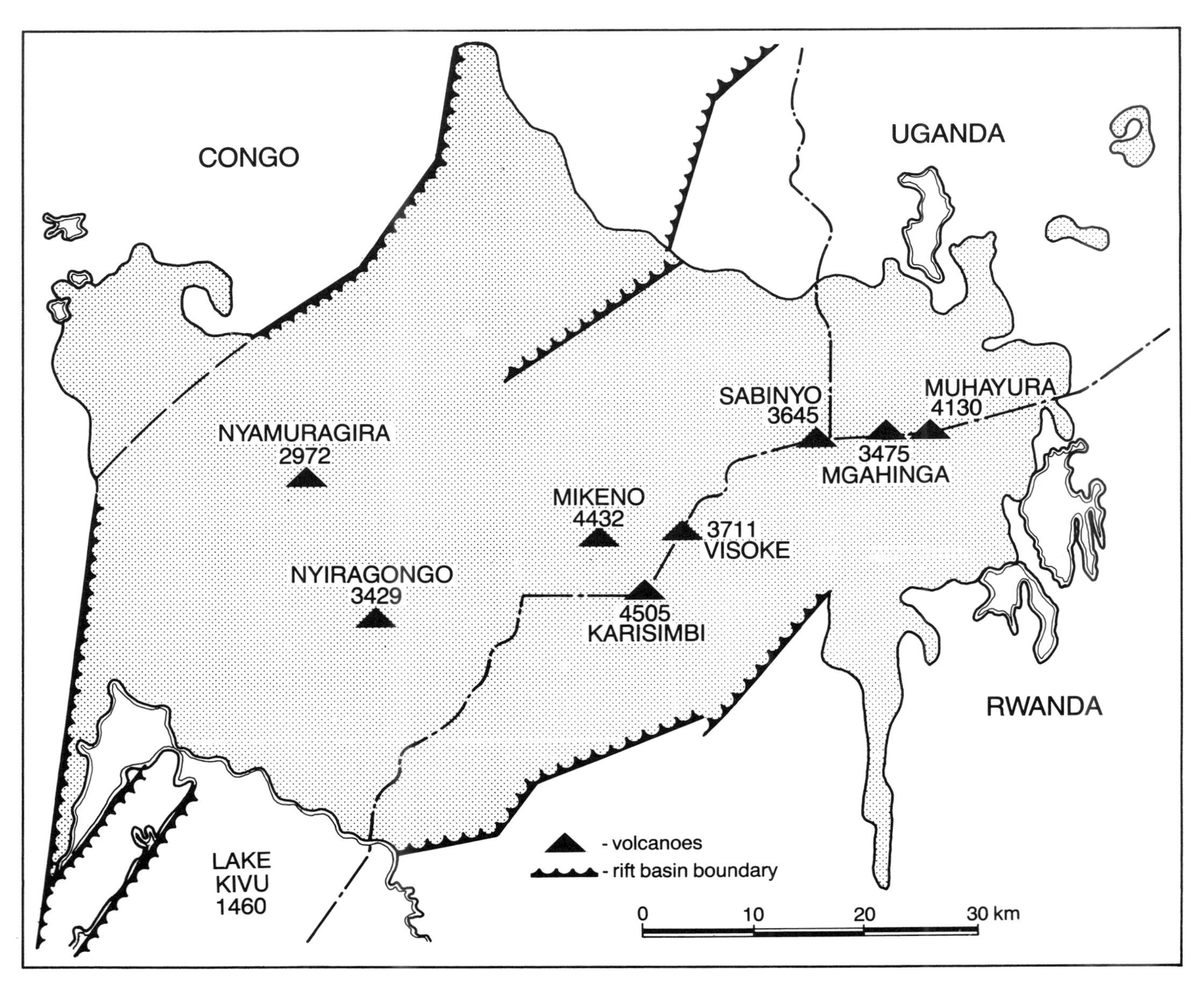


Figure 1. *Sketch map showing the positions and heights (in meters) of the eight major volcanoes of the Birunga volcanic field north of Lake Kivu.*

## INTRODUCTION

Recently, interest in abiogenic methane has increased as a result of Welham and Craig's (1979) observations of methane rising from hydrothermal vents on the East Pacific Rise and Gerlach's (1980) observations on volcanic gases from the Nyiragongo lava lake near Lake Kivu (Figure 1). Gold (1979) and Gold and Soter (1980, 1981) hypothesize that during the accretion of earth materials 4.5 billion years ago, methane and other hydrocarbons were trapped and have been migrating from their burial depth to the surface ever since. MacDonald (1983) focuses on the reaction in igneous rocks and the upward migration of primordial methane.

Wakita and Sano (1983) observe extraordinarily high $^3He/^4He$ ratios, up to $8.65 \times 10^{-6}$, in gases from wells drilled into deep reservoirs in volcaniclastic rock formations. They conclude that the formation of a natural gas reservoir in the volcaniclastic rock might be attributed to large-scale magmatic activities that occurred in the middle Miocene.

Many authors have discussed the origin of dissolved gas in Lake Kivu, an east African rift lake. This paper focuses on outward migration of primordial gas and formation of dissolved gas deposits in Lake Kivu and compares that to analogous gases in Japan.

## DISSOLVED GAS IN THE DEEP WATER OF LAKE KIVU

Lake Kivu is the westernmost and highest (1,460 m or 4,780 ft) of the rift lakes that curve through east-central Africa. The lake has a surface area of about 2,400 sq km (925 sq mi), a maxi-

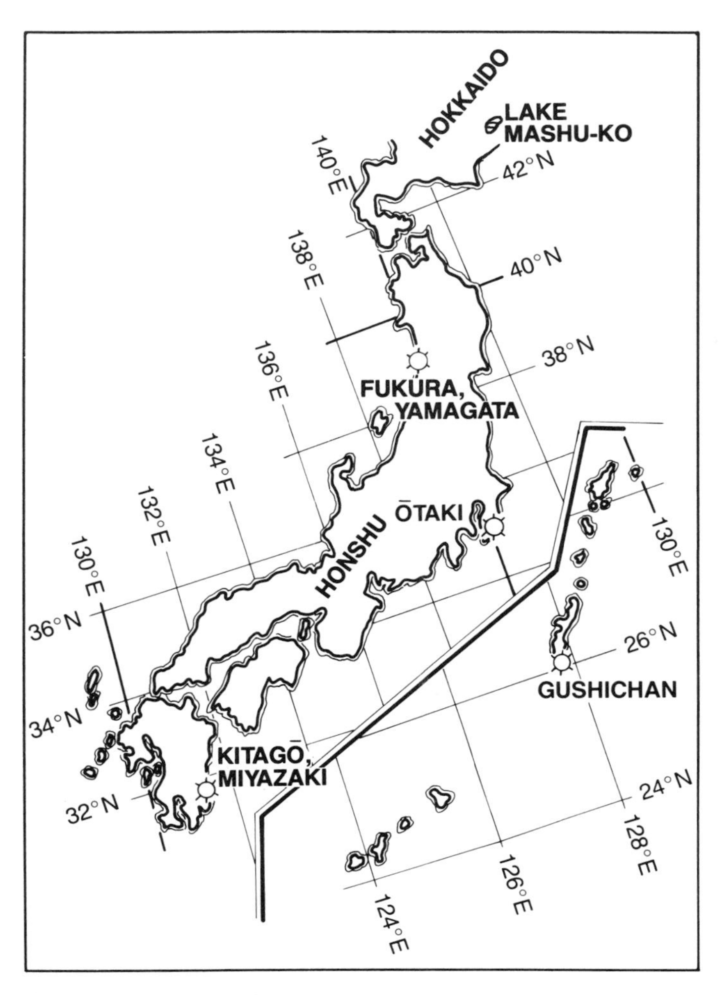


Figure 2. *Localities of some characteristic gases in Japan (three wells and one lake).*

mum depth just under 500 m (1,640 ft), and a total volume of about 580 cu km (140 cu mi). At depths of 400 m (1,310 ft), Lake Kivu contains 2,000 ml/L of gas, consisting of 22% methane and 77% carbon dioxide (Burke, 1963).

Deuser et al (1973) conclude that the methane in Lake Kivu has to be of biogenic origin, because there is no methane in the gases emanating from Nyiragongo or from nearby hot springs. Alternatively, Gerlach (1980) suggests that the methane is of volcanic origin, because the gases emanating from the volcano at high temperatures would also form methane at lower temperatures.

Gold and Soter (1980) touch upon Lake Kivu in their deep-earth-gas hypothesis. They think it probable that all waters were supplied by nonbiological methane seeping up through deep crustal fissures. Craig's discovery (Gold and Soter, 1981) that the waters of Lake Kivu contain an anomalously large amount of $^{3}He$ supports this theory.

Furthermore, MacDonald (1983) argues that the carbon and helium isotope ratios, together with their geologic settings, strongly suggest that the large quantities of methane in Lake Kivu and the gases venting along the East Pacific Rise are abiogenic.

Basically, I agree with Gold and Soter (1981) and MacDonald (1983) on Lake Kivu methane. However, Figure 1 of Deuser et al (1973) shows the concentrations of dissolved methane and inorganic carbon in Lake Kivu water. The lake's water at the 400-m (1,310-ft) depth contains 22 millimoles of methane and 77 millimoles of carbon dioxide per kilogram (see Figure 1 in Deuser et al, 1973). Lake Kivu (altitude 1,460 m; 4,790 ft) lies on nearly the same latitude as Nairobi, Kenya, (altitude 1,978 m; 6,530 ft). The average atmospheric temperature and pressure at the water level of Lake Kivu are about 19°C (66 °F) and 850 mb, respectively. The methane and carbon dioxide in water at this average temperature and pressure are estimated at 0.00283 and 0.767 cu m at 0° C and 1 atm/kl respectively. Therefore, one kl of water drawn up from 400 m (1,310 ft) in Lake Kivu may emit 2.02 cu m (70 cu ft) of gas on the lake level. Table 1 shows the calculations:

Table 1. *Calculations of gas that 1 kl of water from 400 m deep in Lake Kivu would emit at the lake level.*

| | |
|---|---|
| Methane/water: | 22 m mole/kg = 0.62 $m^3$/kl at 19°C and 850 mb |
| Carbon dioxide/water: | 77 m mole/kg = 2.17 $m^3$/kl ibid. |
| 0.62 − 0.00283 = 0.62 ($m^3$/kl) | Volume of methane at 0° C and 1 atm emitted at the lake level from 1 kl of water |
| 2.17 − 0.767 = 1.40 ($m^3$/kl) | Volume of carbon dioxide at 0° C and 1 atm emitted at the lake level from 1 kl of water |
| 0.62 + 1.40 = 2.02 ($m^3$/kl) | Volume of total gas emitted at 0° C and 1 atm at the lake level from 1 kl of water |

The water level in Kitagō-chō R 1 well, Kitagō-chō, Miyazaki, Japan (Figure 2) is 11.3 m (37 ft) below the ground surface, being 34 m (111 ft) above sea level. However, the well had been induced to gush out about 1,690 normal cu m of gas and 800 kl of brine per day by March, 1978 (Fukuta and Nagata, 1978). Its present gas and water productions are about 1,370 normal cu m and 650 kl per day respectively. The gas-water ratio has been about 2.10 normal cu m/kl. The gas from the Kitagō-chō R 1 well is mostly composed of about 3% nitrogen, 29% carbon dioxide, and 68% methane (Table 2), with carbon dioxide probably of magmatic origin.

In addition, the water level of the Gushichan R 1 well, Gushichan-son, Okinawa, Japan (Figure 2) is 41.3 m (135 ft) below the ground surface and 4.5 m (15 ft) above sea level. However, the well has gushed out about 1,190 normal cu m of gas and 360 kl of brine per day at the completion date April, 1977, with a gas-water ratio of 3.31 normal cu m/kl (Fukuta and Nagata, 1977). As Table 2 shows, the gas from Gushichan R 1 is primarily composed of methane.

Judging from the above examples, Lake Kivu gas may be recoverable by passing the water through the long pipes serving as casing pipes in some dissolved-gas wells. Methane is easily separated from carbon dioxide, because carbon dioxide's solubility in water is greater than the solubility of methane in water.

## $^3He$ RICH NATURAL GAS IN JAPAN

Wakita and Sano (1983) observed extraordinarily high $^3He/^4He$ ratios, up to $8.65 \times 10^{-6}$, in gases from wells drilled in the Green tuff region. This region is characterized by the volcaniclastic and volcanic rocks of middle Miocene age, in Japan. Wakita and Sano (1983) have thought that appreciable amounts of methane-rich gases come from the upper mantle, with the helium coming via magmatic activity.

The gas from the Fukura SK 6 well, Yuzamachi, Yamagata, Japan (Figure 2) is similar to Lake Kivu gas in its chemical composition. The gas, with a high $^3He/^4He$ ratio ($6.87 \times 10^{-6}$) consists of 29.5% methane, 67.5% carbon dioxide, 0.67% nitrogen, and 30 ppm helium (Table 2). The carbon dioxide component of the gas, therefore, originated in magmatic activity. The biogenic, methane-rich gas derived from younger sediments does not contain much helium. Therefore, the high $^3He/^4He$ ratio of the gas suggests that the methane from the Fukura Sk 6 well also largely originates in magmatic activity. The gas, therefore, indirectly supports the abiogenic origin of Lake Kivu gas.

## COSMOCHEMICAL AND PHYSICOCHEMICAL CONSIDERATIONS

The earth is believed to have originated in the solar nebula. Of the various samples we now have of the solar system, the most primitive are the carbonaceous-chondrite meteorites. Of all the various types of them, that is C1, C2, and C3 chondrites, C1 chondrites have the highest concentration of water and volatile elements such as lead, bismuth, and carbon. In fact, for all but the most volatile, element abundances in the C1 chondrites are the same as those in the sun.

Organic matter, in which carbonaceous chondrites are especially enriched, represents a complicated mixture of compounds. They include, among others, $C_{15}$–$C_{30}$ paraffins, olefins, aromatic hydrocarbons, organic acids, amino acids, nitrogenous cyclic compounds, and porphyrins (Vdovykin and Moore, 1971). These compounds may be the result of a Fischer-Tropsch type synthesis in the solar nebula. The high abundance of carbon in carbonaceous chondrites suggests that carbon may be more abundant in the earth than it is in near-surface igneous rocks.

At first, Gold and Soter (1980) attached much importance to organic compounds contained in carbonaceous chondrites. The next year, they discuss the stability regime of mantle methane in temperature-pressure space, determined by the presence of oxygen in the surrounding rock buffer (Gold and Soter, 1981). Arculus and Delano's (1981) recent measurements of the oxygen fugacity of actual mantle rocks give values as much as four orders of magnitude lower than for the quartz-fayalite-magnetite (QFM) buffer. These values are comparable instead with the iron-wüstite (IW) buffer.

Reequilibration of carbon dioxide and hydrogen of volcanic origin will form methane in certain conditions of temperature and pressure. The reaction

$$CO_2 + 4H_2 \rightleftarrows CH_4 + 2H_2O$$

proceeds to the right under increased pressure and decreased temperature. Sometimes the hydrogen in the above reaction may be generated in the temperature range of 25–270°C (77–518°F) along the active fault (Kita et al, 1982). For example, the gas from a gas well at the Otaki Junior High School, Otakimachi, Chiba, Japan (Figures 2, 3) consists mainly of 1.2% hydrogen, 2.25% nitrogen and 91.3% methane (Table 2). The biogenic methane-rich natural gas derived from the younger sedimentary formations does not contain significant hydrogen. In addition, the well is probably situated on one of many north-south faults found in the southeastern part of the southern Kanto gas field. Therefore, the hydrogen in the gas probably originated in faulting.

**Table 2.** *Chemical composition of some gases from Japan (vol. %).*

| Well | Kitagō-chō R 1 | Gushichan R 1 | Fukura SK 6 | Ōtaki Jr. H. School |
|---|---|---|---|---|
| He | 00.005 | 00.0027 | 00.0030 | 00.000087 |
| $H_2$ | 00.000 | 00.000 | — | 1.23 |
| $O_2$ | 00.56 | 00.34 | — | 00.69 |
| $N_2$ | 3.30 | 2.06 | 00.67 | 2.25 |
| Ar | 00.062 | 00.30 | — | 00.04 |
| $CO_2$ | 28.77 | 00.36 | 67.5 | 00.33 |
| $CH_4$ | 67.13 | 96.86 | 29.5 | 91.3 |
| $C_2H_6$ | 00.173 | 00.051 | — | 00.08 |
| $C_3H_8$ | 00.000 | 00.018 | — | — |
| i-$C_4H_{10}$ | — | 00.011 | — | — |
| n-$C_4H_{10}$ | — | 00.002 | — | — |
| $H_2S$ | 00.0005 -0.0022 | — | — | — |
| Analyst | Nagata | Nagata | Urabe | Yoshida |
| Reference | Fukuta & Nagata (1978) | Fukuta & Nagata (1977) | Wakita & Sano (1984) | MS |

## EXCESS BROMINE AND $^3He$

As previously stated, the methane-rich natural gas of a high $^3He/^4He$ ratio has been found in the Green tuff region in Japan. According to Horai and Uyeda (1963), Uyeda and Horai (1964), and Wakita and Sano (1983), high heat flow values occur in the Green tuff region. These include the oil and gas producing areas in the Sea of Japan side of northeast Honshu (Figure 4), where the characteristic Green Tuff Formation occupies the basal part of the continuous marine sedimentary sequence of late Cenozoic age. The formation consists mainly of altered volcaniclastic and volcanic rocks. Similar rocks and their allied ore deposits have been reported from some mid-oceanic ridges and the Red Sea (Degens and Ross, 1969; Edmond et al, 1982; Edmond and Damm, 1983). The measured values of the terrestrial heat flow in the Green tuff region are generally over 2.00 HFU (1 heat-flow unit, known as HFU, equals $10^{-6}$ cal/sq cm/sec). In contrast to these, there are areas of anomalously low heat flow, under 1.00 HFU, on the Pacific side of northeastern Japan. These areas of low heat flow, shown by the crosshatched pattern in Figure 5, seem to occupy the Kitakami and Abukuma plateaus, which are stable land masses; the eastern portion of the Kanto basin, including the southern Kanto gas field; and two areas in the Pacific off the coast of northeastern Japan. Most

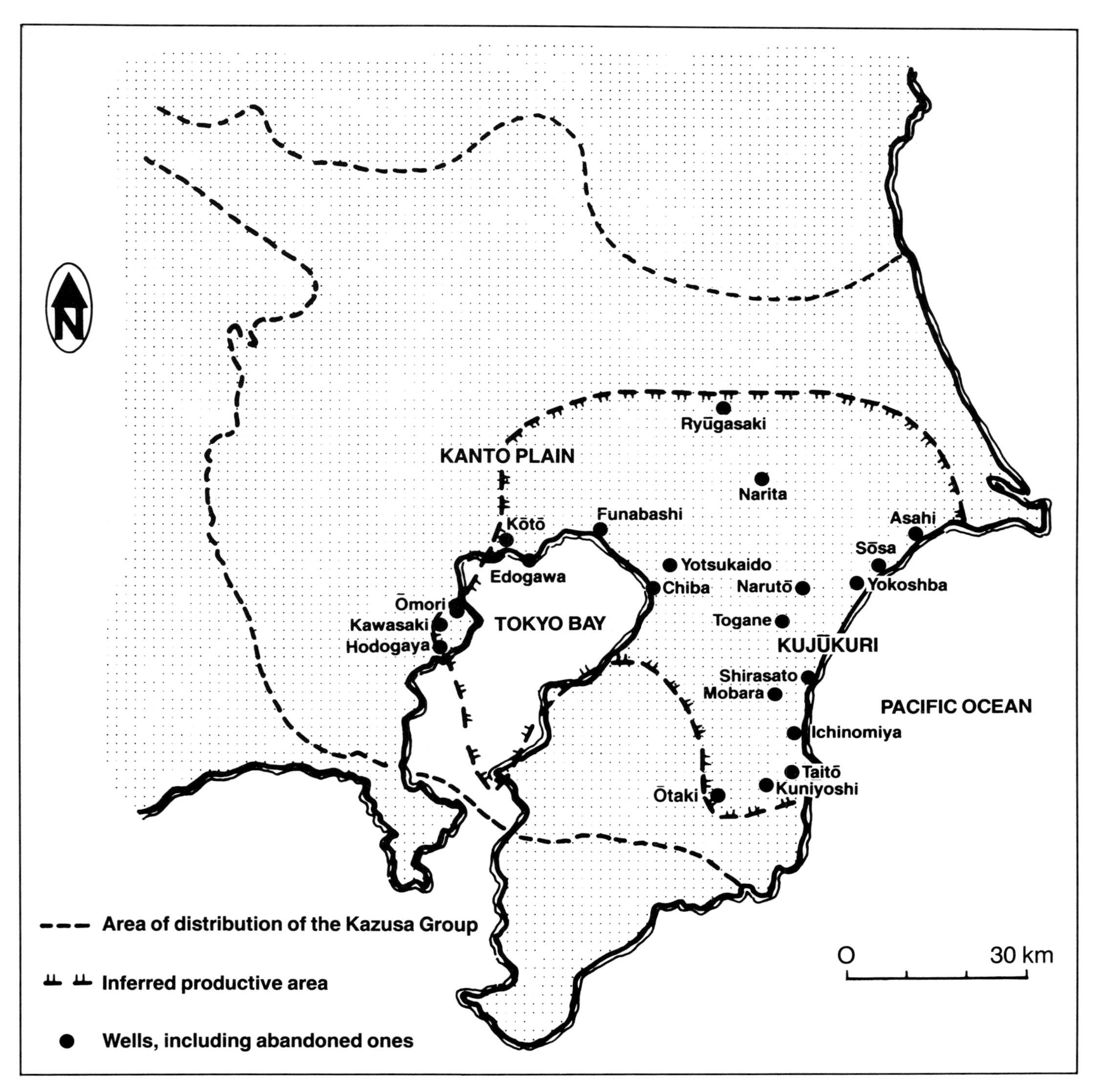


Figure 3. *Outlines of the southern Kanto gas field.*

of the heat-flow values outside the hatched area in Figure 5 are between 1 HFU and 2 HFU. The areas of moderate heat flow include the central part of Hokkaido and most of the metamorphic zones in southwestern Japan. The geothermal gradients in the upper Cenozoic formations that belong to the Green tuff region are generally over 4.00°C per 100 m (2.2°F per 100 ft). On the other hand, the gradients in the southern Kanto gas field are generally 1.50 to 1.80°C per 100 m (0.8 to 1.0°F per 100 ft).

Noguchi and Morisaki (1971) chemically analyzed 41 brine samples from the younger reservoir formations of some oil and gas fields in the Sea of Japan side of northeast Honshu (Figure 6). Table 3 summarizes their data and some values calculated by Fukuta et al (1983). Reflecting a great variety of samples, the values of $I^-/Cl^-$ and $Ca^{2+}/Mg^{2+}$ ratios are quite variable. However, the points on the $Br^-$–$I^-$ diagram (Fukuta and Nagata, 1982; Fukuta et al, 1983; Fukuta, 1985), which repre-

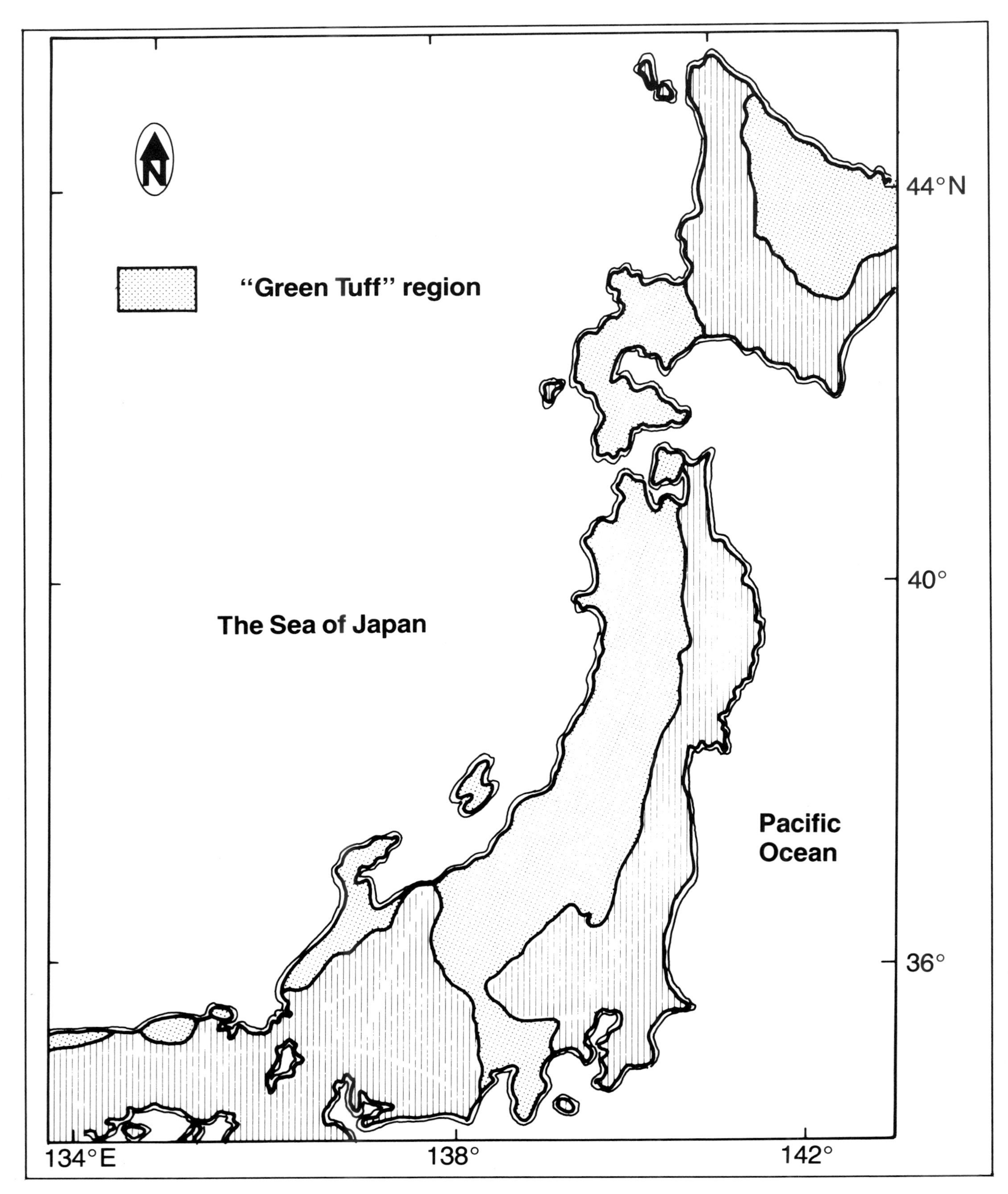


Figure 4. *Distribution of "Green Tuff" in Honshu and Hokkaido, Japan.*

**Table 3.** *Geochemical data on some brines derived from the reservoir formations of some oil and gas fields in the Sea of Japan side of northeast Honshu (contents: mg/l).*

| Field | Well | pH | $Cl^-$ | $Br^-$ | $I^-$ | $Na^+$ | $K^+$ | $Ca^{2+}$ | $Mg^{2+}$ | Fe | $Br^{-*}$ | $I^{-*}$ | $I^-/Cl^-$ | $Ca^{2+}/Mg^{2+}$ |
|---|---|---|---|---|---|---|---|---|---|---|---|---|---|---|
| Hachimori | R 45 | — | 12,736 | 76.4 | 34.7 | 4,150 | 83.0 | — | — | — | 119.2 | 54.1 | 2.72 | — |
| | R 28 | — | 12,243 | 66.6 | 27.2 | 3,730 | 74.6 | — | — | — | 108.1 | 44.1 | 2.22 | — |
| | R 24 | — | 12,741 | 70.1 | 29.7 | 4,150 | 83.0 | — | — | — | 109.3 | 46.3 | 2.33 | — |
| | R 18 | — | 12,200 | 57.5 | 16.6 | 3,920 | 78.4 | — | — | — | 93.6 | 27.0 | 1.36 | — |
| | K 8 | — | 16,951 | 112.1 | 68.5 | 3,730 | 74.6 | — | — | — | 131.4 | 80.3 | 4.04 | — |
| | C 6 | — | 12,112 | 53.0 | 14.5 | 4,800 | 96.0 | — | — | — | 86.9 | 23.8 | 1.20 | — |
| Nishiyama | R 47 | — | 11,154 | 109 | 26 | 6,450 | 65 | 23.8 | 79.6 | 2.3 | 194.1 | 46.3 | 2.33 | 0.299 |
| | R 19 | — | 11,422 | 112 | 28 | 6,730 | 44 | 35.6 | 134 | 2.4 | 194.8 | 48.7 | 2.45 | 0.266 |
| | R 77 | 7.4 | 17,661 | 170 | 49 | 8,540 | 125 | 343 | 866 | 19.9 | 191.2 | 55.1 | 2.77 | 0.396 |
| | C 55 | — | 15,742 | 129 | 38 | 8,240 | 27 | 249 | 323 | 23.5 | 162.7 | 47.9 | 2.41 | 0.771 |
| | R 78 | 6.6 | 14,179 | 125 | 43 | 7,290 | 92 | 476 | 849 | 23.0 | 175.1 | 60.2 | 3.03 | 0.561 |
| | R 87 | 7.5 | 19,678 | 128 | 55 | 9,270 | 125 | 770 | 925 | 16.7 | 129.2 | 55.5 | 2.79 | 0.832 |
| | R 6 | — | 14,179 | 136 | 37 | 7,350 | 85 | 432 | 115 | 6.8 | 190.5 | 51.8 | 2.61 | 3.757 |
| | R 13 | — | 15,675 | 145 | 42 | 7,540 | 98 | 717 | 238 | 5.7 | 183.7 | 53.2 | 2.68 | 3.013 |
| | R 11 | — | 18,569 | 128 | 15 | 8,430 | 113 | — | — | 14.3 | 155.2 | 18.2 | 0.81 | — |
| | C 26 | — | 5,440 | 44.9 | 16 | 2,890 | 33 | 175 | 427 | 12.3 | 164.0 | 58.4 | 2.94 | 0.410 |
| | C 112 | — | 9,874 | 83.4 | 29 | 5,010 | 62 | 188 | 382 | 4.1 | 167.8 | 58.3 | 2.94 | 0.492 |
| | R 28 | 8.0 | 14,816 | 143 | 38 | 7,640 | 87 | 283 | 186 | 5.4 | 191.7 | 50.9 | 2.56 | 1.522 |
| | C 1 | — | 2,516 | 20.9 | 7.5 | 1,330 | 23 | 94.5 | 150 | 12.4 | 165.0 | 59.2 | 2.98 | 0.630 |
| Higashiyama | R 21 | 7.0 | 2,293 | 13.6 | 1.3 | 2,350 | 12.8 | 29.5 | 13.7 | 6.6 | 117.8 | 11.3 | 0.52 | 2.153 |
| | C 100 | 7.4 | 2,767 | 15.3 | 0.6 | 2,880 | 14.7 | — | — | 6.0 | 109.8 | 4.3 | 0.23 | — |
| | C 66 | 7.1 | 7,340 | 50.4 | 4.4 | 7,270 | 38.5 | 12.4 | 86.5 | 4.7 | 136.4 | 11.9 | 0.60 | 0.143 |
| | C 63 | 7.0 | 2,688 | 15.4 | 1.3 | 2,660 | 14.0 | 33.1 | 15.7 | 9.7 | 113.8 | 9.6 | 0.48 | 2.108 |
| Niitsu | R 7 | — | 3,162 | 30.8 | 4.9 | 2,610 | 90 | 78.8 | 88.2 | — | 193.5 | 30.8 | 1.55 | 0.893 |
| | C 46 | 6.4 | 2,727 | 25.6 | 4.8 | 2,330 | 77 | 78.8 | 64.4 | 10.2 | 186.5 | 35.7 | 1.76 | 1.224 |
| | R 4 | 6.8 | 3,180 | 30.5 | 5.9 | 2,720 | 89 | 48.9 | — | 1.4 | 190.5 | 36.9 | 1.86 | — |
| | C 21 | 7.0 | 4,236 | 15.1 | 7.6 | 3,520 | 121 | 51.3 | 128 | 3.8 | 70.8 | 35.6 | 1.79 | 0.401 |
| | C 25 | — | 18 | 0.0 | 0.0 | 84 | 4 | — | — | 2.1 | — | — | — | — |
| | C 2 | — | 3,067 | 3.7 | 0.6 | 2,630 | 87 | 1.6 | 5.9 | — | 24.0 | 3.9 | 0.20 | 0.271 |
| | K 215 | — | 21 | 0.0 | 0.0 | 83 | 3 | 1.6 | 2.9 | — | — | — | — | 0.552 |
| | R 2 | — | 4,835 | 49.4 | 11.1 | 4,300 | 200 | 44.0 | — | 3.5 | 203.0 | 45.6 | 2.30 | — |
| Niigata | R 23 | — | 11,542 | 88.6 | 30.4 | 6,520 | 270 | 223 | 414 | 3.0 | 152.5 | 52.3 | 2.63 | 0.539 |
| | R 46 | — | 11,064 | 87.1 | 27.5 | 6,120 | 284 | 281 | 536 | 6.3 | 156.4 | 49.4 | 2.49 | 0.524 |
| | R 67 | — | 19,984 | 162.1 | 51.4 | 9,280 | 500 | 815 | 355 | 0.0 | 161.1 | 51.1 | 2.57 | 2.296 |
| | R 65 | — | 4,321 | 11.1 | 22.0 | 2,220 | 120 | 273 | 246 | 3.3 | 51.0 | 101.1 | 5.09 | 1.110 |
| | R 73 | — | 19,467 | 151.7 | 57.1 | 9,725 | 490 | 77 | 459 | 2.2 | 154.8 | 58.3 | 2.93 | 0.168 |
| | 2-3 | — | 1,959 | 15.2 | 7.0 | 1.075 | 52 | 143 | 111 | 3.6 | 154.1 | 71.0 | 3.57 | 1.288 |
| | 11B-1 | — | 18,262 | 131.5 | 32.9 | 9,130 | 500 | 426 | 689 | 2.7 | 143.0 | 35.8 | 1.71 | 0.618 |
| | 8B-1 | — | 14,516 | 103.3 | 45.8 | 7,775 | 430 | 370 | 642 | 0.12 | 141.4 | 62.7 | 3.16 | 0.576 |
| | 7 | — | 18,414 | 129.8 | 63.8 | 9,850 | 474 | 307 | 282 | 0.02 | 140.0 | 68.8 | 3.46 | 1.089 |
| | 3 | — | 621 | 9.4 | 5.9 | 810 | 18 | 31 | 38 | 0.14 | 299.2 | 187.8 | 9.50 | 0.816 |

Notes: [1]$I^-/Cl^-$: $I^-/Cl^- \times 10^3$
[2]$Br^{-*}$ and $I^{-*}$ are the $Br^-$ and $I^-$ contents calculated so that the chlorosities are equal to that in the normal seawater of 35 ‰ in salinity. (Noguchi and Morisaki, 1971; Fukuta et al, 1983).

sent the $Br^-$ and $I^-$ contents recalculated so that the chlorosities are equal to that in the normal seawater of 35 ‰ in salinity, with only one exception, are distributed on the upper side of a straight line representing the recalculated $Br^-$ and $I^-$ contents of the brine from the Hachimori oil-field (Figure 7). The $Br^-$–$I^-$ diagram (Figure 7) shows the recalculated $Br^{-*}$ and $I^{-*}$ values listed in Table 3. The regression line and its coefficient of correlation (c.c.) on the brine from the Hachimori oil field follow:

$$Br^-(mg/L) = 72.30(mg/L) + 0.7790 \times I^-(mg/L) \quad (1)$$

(c.c. = 0.9798) [Hachimori Field]

Figure 8 shows the $Br^-$–$I^-$ diagram on the brine derived from the southern Kanto gas field, the reservoir formations of which have been deposited in the neritic and upper bathyal environments in late Pliocene to early Pleistocene time. This brine is characterized by an almost stable relationship between both

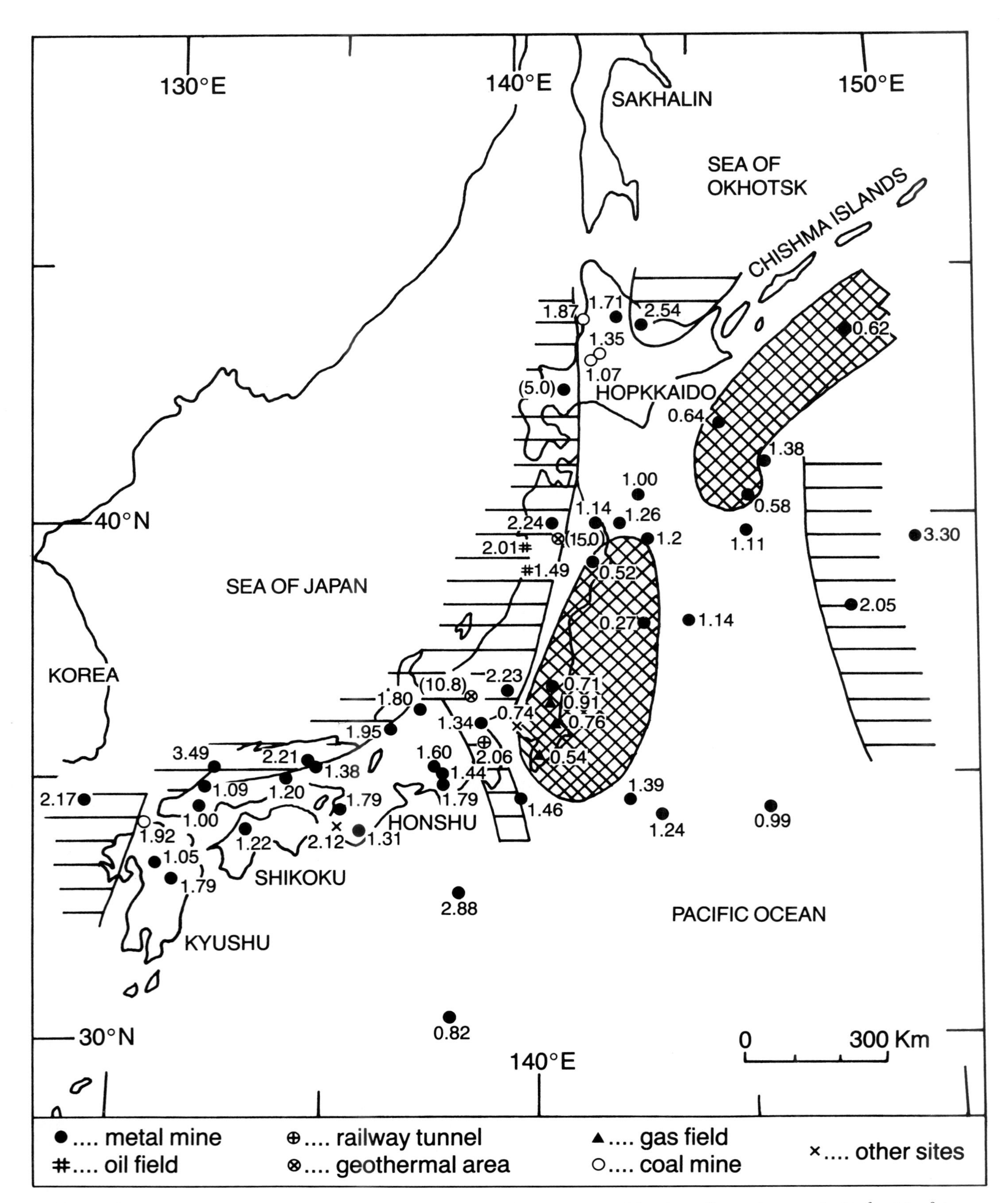


**Figure 5.** *Distribution of heat flow in and around Japan (Uyeda and Horai, 1964). Heat flow values are in* $10^{-6}$ *cal/cm²/sec.*

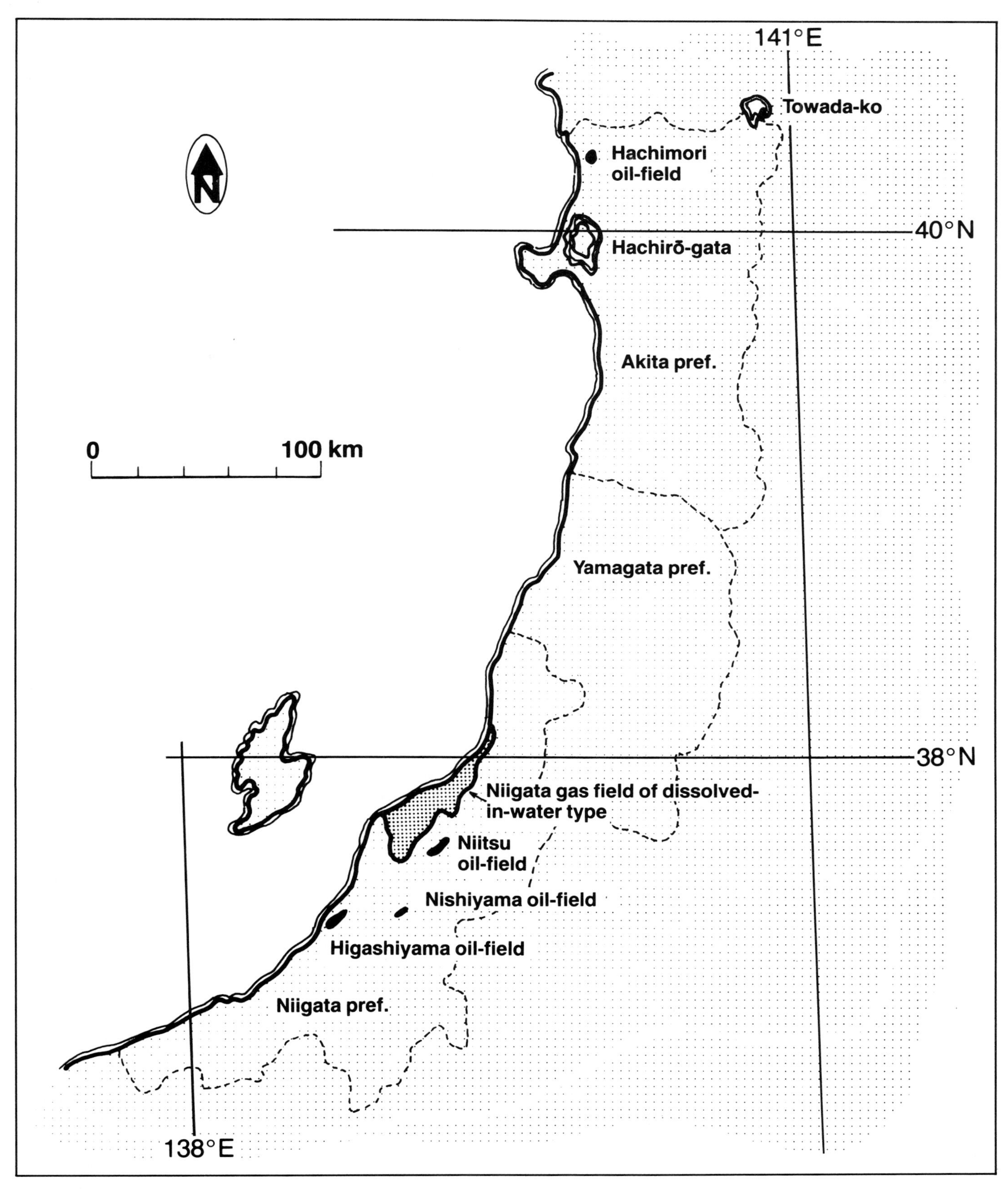


**Figure 6.** *Some oil and gas fields, sources of the brines studied by Noguchi and Morisaki (1963), in the Sea of Japan side of northeast Japan.*

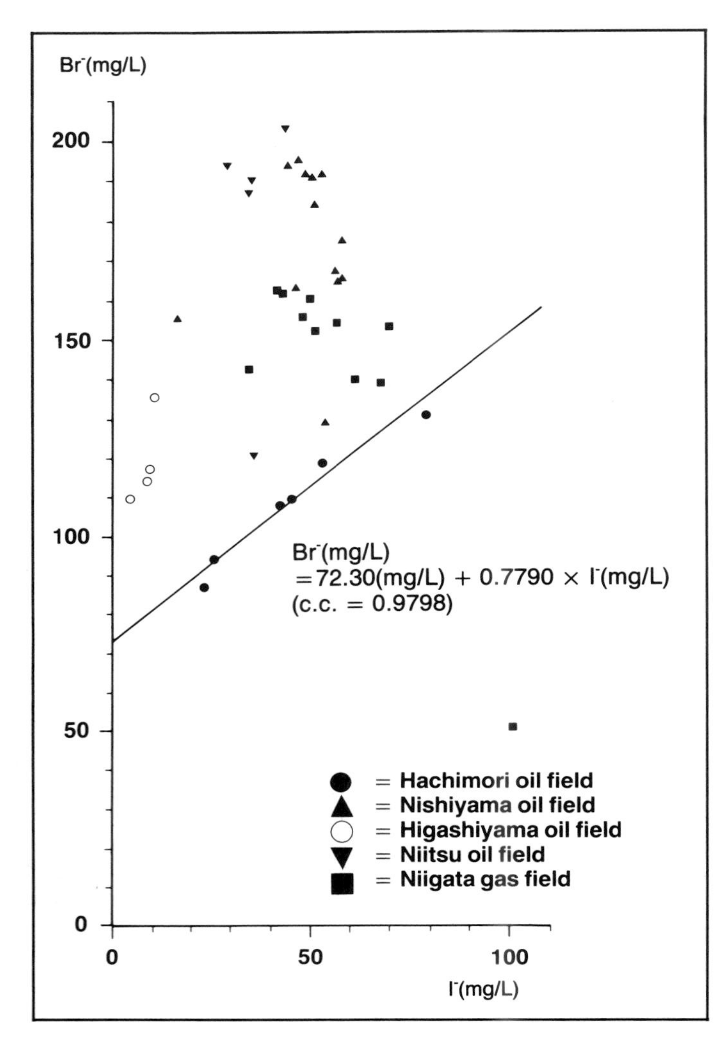


**Figure 7.** *$Br^-$–$I^-$ diagram of the subsurface brines from the Hachimori, Nishiyama, Higashiyama, and Niitsu oil fields and the Niigata gas field of dissolved-in-water type. The regression line represents recalculated $Br^-$–$I^-$ contents ($Br^{-*}$ and $I^{-*}$ in Table 3) from the Hachimori oil field.*

recalculated $Br^-$ and $I^-$ contents, a relationship that can be expressed by the following equation:

$$Br^-(mg/L) = 62.84(mg/L) + 0.6680 \times I^-(mg/L) \quad (2)$$

(c.c. = 0.9589) [Kanto field]

Normal sea water of 35‰ in salinity contains 67.3 mg/kg bromide and 0.064 mg/kg iodide (Horibe et al, 1970). Therefore, the constant 62.84 (mg/L) in equation (2) accords with the initial bromide content in the brine.

The reservoir formation of the Hachimori oil field (equation 1) is somewhat older than that of the southern Kanto gas field (equation 2). In spite of a little difference in the geologic ages of reservoir formation, the equation representing the regression line of the points on the $Br^-$–$I^-$ diagram of the brine from the Hachimori oil field (equation 1) is very close to that of the brine derived from the southern Kanto gas field (equation 2), although the constant and the coefficient of $I^-$ in the right side

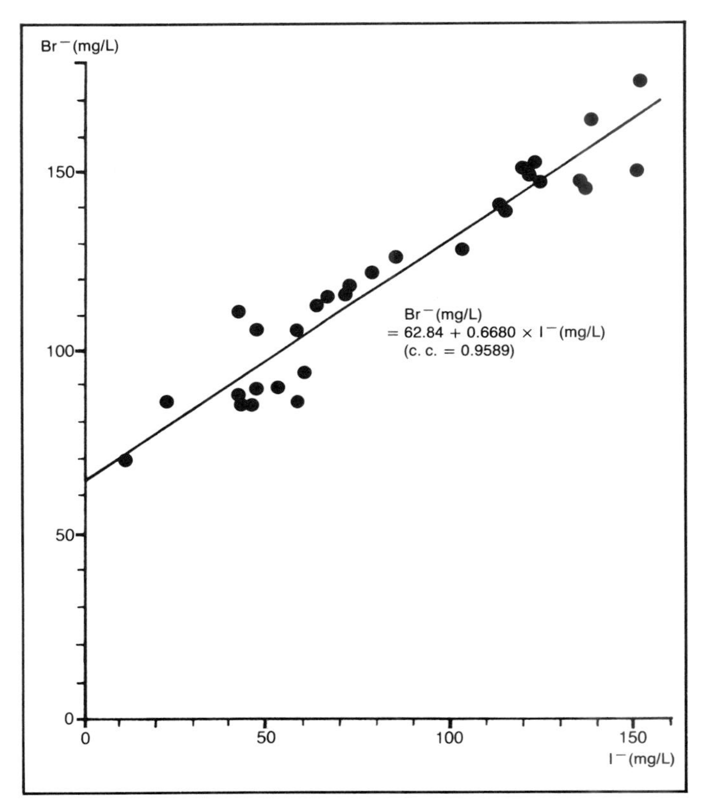


**Figure 8.** *$Br^-$–$I^-$ diagram on the brine from the southern Kanto gas field (simplified Figure 6 in Fukuta et al, 1983).*

of the equations are somewhat larger in the Hachimori oil field equation. In addition, the majority of points on the $Br^-$–$I^-$ diagram (Figure 7) are distributed on the upper side of the straight line representing equation (1).

The brines represented by many points on the $Br^-$–$I^-$ diagram (Figure 7) seem to contain some excess bromide of magmatic origin. This conclusion is also supported by the occasional occurrences of hydrogen bromide in some volcanic gases (Downs and Adams, 1973). Therefore, both the excess $^3He$ in the natural gas and the excess bromide in the subsurface brines that originated in the deeper zones of the Green tuff region are probably of magmatic origin.

## CONCLUSIONS

Some abiogenic gases may contribute to natural gas deposits. Lake Kivu gas and some gases derived from the deeper zones of the Green tuff region in Japan undoubtedly originated in magmatic activity. These gases are characterized by their excess $^3He$ of magmatic origin. The Green tuff region shows that the high values of the terrestrial heat flow are generally over 2.00 HFU, with excess bromide generally recognized in the brines derived from the region's various zones. In addition, some methane-rich natural gases have been reported from hot springs in the Kunashiri and Etorofu Islands belonging to the Chishima Islands (Kishimoto, 1982). Recently, the Kushiro Branch Office of the Japan Broadcasting Corporation (NHK) telecast a videotape recorded in the water of the Lake Machu-ko (Figure 2), a caldera lake in the eastern part of Hokkaido (Tanakadate, 1929; Katsui, 1963). The videotape shows large bubbles of natural gas that have blown out through fissures in the bottom of the lake.

## REFERENCES CITED

Arculus, R. J., and J. W. Delano, 1981, Intrinsic oxygen fugacity measurements: techniques and results for spinels from upper mantle periodtites and megacryst assemblages: Geochimica et Cosmochimica Acta, v. 45, p. 899–913.

Burke, K., 1963, Dissolved gases in East African lakes: Nature, v. 198, p. 568–569.

Degens, E. T., and D. A. Ross (eds.), 1969, Hot springs and recent heavy metal deposits in the Red Sea: Springer-Verlag, 600 p.

Deuser, W., et al, 1973, Methane in Lake Kivu: new data bearing on its origin: Science, v. 181, p. 51–53.

Downs, A. J., and C. J. Adams, 1973, Chlorine, bromine, iodine, and astatine: Comprehensive inorganic chemistry, v. 2, p. 1107–1594.

Edmond, J. M., et al, 1982, Chemistry of hot springs on the East Pacific Rise and their effluent dispersal: Nature, v. 297, p. 187–191.

Edmond, J. M., and K. V. Damm, 1983, Hot springs on the ocean floor: Scientific American, v. 248, p. 70–85.

Fukuta,O., 1985, Japanese Iodine—geology and geochemistry: Salts & Brines '85, p. 149–168.

———, and S. Nagata, 1977, Gushichan R 1 gushed out: Chishitsu (Geological) News, n. 276, p. 1–17 (in Japanese).

———, and ———, 1978, Kitagō-chō R 1 gushed out—the revival of the Nichinan gas field—: Chishitsu (Geological) News, n. 290, p. 1–22 (in Japanese).

———, and ———, 1982, Iodine deposits of dissolved-in-water type in Japan: Journal of Japanese Association of Petroleum Technology, v. 47, p. 168–185 (in Japanese).

Fukuta, O., et al, 1983, Regional characteristics recognized in the iodine deposits of dissolved-in-water type in Japan and Philippines: do., v. 48, p. 281–290 (in Japanese).

Gerlach, T. M., 1980, Chemical characteristics of the volcanic gases from Nyiragongo lava lake and the generation of $CH_4$-rich fluid inclusions in alkaline rocks: Jour. Volcanol. Geotherm. Res., v. 8, p. 177–189.

Gold, T., 1979, Terrestrial sources of carbon and earthquake outgassing: Journal of Petroleum Geology, v. 1, n. 3. p. 3–19.

———, and S. Soter, 1980, The deep-Earth-gas hypothesis: Scientific American, v. 242, p. 154–161.

———, and ———, 1981, Abiogenic methane and the origin of petroleum: Copy of the manuscript for Energy Explorat. & Exploit., 38 p.

Horai, K., and S. Uyeda, 1963, Terrestrial heat flow in Japan: Nature, v. 199, p. 364–365.

Horibe, S., et al, 1970, Chemistry of seawater: Fundamental cource of oceanography, Tokai University Press, v. 10, 381 p. (in Japanese).

Katsui, Y., 1963, Evolution and magmatic history of some Krakatoan calderas in Hokkaido, Japan: Jour. Fac. Sci., Hokkaido Univ., ser. IV (Geol. & Mineral.), v. 9, p. 631–650.

Kishimoto, F., 1982, Resources of fuel minerals in Chishima Islands: Chishitsu (Geological) News, n. 340, p. 26–29 (in Japanese).

Kita, I., et al, 1982, $H_2$ generation by reaction between $H_2O$ and crushed rock: An experimental study on $H_2$ degassing from the active fault zone: Journal of Geophysical Research, v. 87, p. 789–795.

MacDonald, G. J., 1983, The many origins of natural gas: Journal of Petroleum Geology, v. 5, p. 341–362.

Noguchi, A., and S. Morisaki, 1971, Contents and geochemical meaning of Lithium and arcenic in brine waters of oil and gas fields: Journal of the Chemical Society of Japan, v. 92, p. 145–149 (in Japanese).

Tanakadate, H., 1929, The problems of caldera in the Pacific region: Proceedings of the Fourth Pacific Scientific Congress, p. 729–744.

Uyeda, S., and K. Horai, 1964, Terrestrial heat flow in Japan: Journal of Geophysical Research, v. 69, no. 10, p. 2121–2141.

Vdovykin, G. P., and C. B. Moore, 1971, Carbon (6), Handbook of elemental abundances in meteorites *in* B. Mason, ed., p. 81–91.

Wakita, H., and Y. Sano, 1983, $^3He/^4He$ ratios in $CH_4$-rich natural gases suggested magmatic origin: Nature, v. 305, p. 792–794.

Welham, J. A., and H. Craig, 1979, Methane and hydrogen in East Pacific Rise hydrothermal fluids: Geophysical Research Letter, v. 6, p. 829–831.

## SUPPLEMENT

Recently, I have received three papers on oil and gas deposits in the Green Tuff Formation.

Komatsu, N., et al, 1984, Cenozoic volcanic rocks as potential hydrocarbon reservoirs: Proceedings of the 11th World Petroleum Congress (1983), v. 2, p. 411–420.

Sato, O., 1984, Rock facies and pore spaces on volcanic-rock reservoirs: Journal of Japanese Association of Petroleum Technology, v. 49, p. 11–19 (in Japanese).

Sekiguchi, K., et al, 1984, Geochemical study of oil and gas accumulation in Green tuff reservoir in Nagaoka to Kashiwazaki region: Journal of Japanese Association of Petroleum Technology, v. 49, p. 56–64 (in Japanese).

Sato (1984) states:

> In the Minami Nagaoka gas field, rhyolite reservoirs, having a total thickness of 1,000 m and a gas column of more than 800 m, occur at 3,800 m and deeper below the surface. The rhyolite is divided into lava, pillow breccia, and hyaloclastite in terms of rock facies, which resulted from drastic chilling and shattering in submarine volcanism. Pore spaces in the rhyolite are classified into several porosity types in terms of their origin, shape and size. The properties of porosity are closely related to each rock facies. The favorable rock facies as reservoirs are pillow breccia and lava facies. Meso- and micro-sized vugs play an important role in total porosity, while micro-sized secondary fractures affect permeability. In order to predict productive zones, it is important to understand the relevant volcanic-rock facies.

As Sato has described in his paper, rhyolite lava flows have built up a volcanic rock complex in the Minami Nagaoka gas field. On the other hand, meso-sized primary vugs and fractures in the central part of the lava dome play an important role in total porosity in the Mitsuke oil field and the Yoshii gas field.

Sekiguchi et al (1984) state,

> The formation process of hydrocarbon pools in the "Green Tuff" reservoirs has been studied from a geochemical point of view for the Mitsuke oil field, the Yoshii-Higashikashiwazaki gas field and the Minaminagaoka-Katagai gas field. The accumulated hydrocarbons originated in the shales of the Nanatani Formation and in part of the lower Teradomari Formation. The hydrocarbons were primarily expelled in oil phase from the source rocks at thermal maturity levels of 0.7 to 1% vitrinite reflectance in the low thermal levels of the reservoir. Therefore, gas-condensate was secondarily added to the reservoir from the highly matured source rocks that attained the gas-generating stage. On the other hand, hydrocarbons in the gas fields studied were primarily trapped in oil phase and subsequently cracked into gas in the reservoirs.

According to their paper, the gases from the Yoshii-Higashikashiwazaki and Minaminagaoka-Katagai gas fields consist of 82–89% methane, 9–13% gaseous hydrocarbons, and carbon dioxide and nitrogen. The carbon dioxide contents of the gases are variable, that is, 1–2% in the gas from the Yoshii-Higashikashiwazaki field and about 7% from the Minaminagaoka-Katagai field. Judging from $\delta^{13}C/^{12}C$ of methane in the gases ranging from $-32.5$ to $-34.5$ ‰, they have concluded that the gas's methane is generated when kerogen evolves. However, the diffusion speed of the heavy methane molecules is about 3% lower than that of the lighter molecules, and this difference allows a cumulative selection to take place (Gold and Soter, 1982). In addition, Sekiguchi et al (1984) believe that a substantial fraction of the slowly diffusing methane generally is lost by oxidation to carbon dioxide, chiefly in the uppermost few kilometers. The oxidation process itself may select the heavy isotope, and, in any case, it is taking place in circumstances where the number density of the heavy isotope is enhanced by 3% compared with the ratio observed in the flux. If this change results in a selection that slightly favors $^{13}C$ in the oxide, then the remaining stream of methane will become progressively depleted in $^{13}CH_4$. Therefore, the lower value of $\delta^{13}C/^{12}C$ in methane is not always supporting its thermal decomposition origin during kerogen evolution.

The gases from the above two fields are richer in carbon dioxide than the general gases derived from the sedimentary reservoirs of early Pliocene and the older ages in the non-volcanic regions. For example, the contents of carbon dioxide in the gases from the Sadowara and Joban gas fields range from 0.11 to 0.68 and 0.01 to 0.84 vol. percent, respectively. The reservoir rocks of the Sadowara and Joban gas fields are early Pliocene and Oligocene in age, respectively. Carbon dioxide in the gases from the Green Tuff Formation, therefore, is magmatic in origin and some methane in the gases makes a pair with carbon dioxide as shown in Figure 1 of Gold and Soter (1982).

Komatsu et al (1984) conclude in their paper, "The Miocene volcanic rocks which are widely distributed in the back-arc basins in the Circum-Pacific region will be a new, important target for hydrocarbon exploration." While I agree with their conclusion in general, I want to emphasize four points:

1) Some methane in the gases derived from the Green tuff reservoirs has both a magmatic and carbon dioxide origin.
2) Tectonically, the Green tuff region of Northeast Honshu is characterized by new horsts and grabens that formed during the middle Miocene. The horst and synchronous high structures of this age comprise volcanic bodies (Green Tuff Formation) and are important for hydrocarbon trapping. Hydrocarbons may have been supplied from the muddy mother rocks, which are stratigraphically higher than the Green Tuff Formation, in the graben.
3) Cathles et al (1981, 1983) present the failed rift hypothesis for the Kuroko-type massive sulfide deposits. This hypothesis can account for the general geological distribution of hydrocarbon deposits in the Green Tuff Formation and some of their geochemical characteristics.
4) The kerogen theory on the origin of hydrocarbons represented by Tissot and Welte (1978) is insufficient for some geochemical characteristics of these hydrocarbon, especially natural gas deposits.

## POSTSCRIPT

The supplement has been written on Mr. Bill St. John's advice. I would like to express my appreciation to Mr. St. John, Primary Fuels, Inc., for his kind advice. I am particularly indebted to Dr. N. Komatsu, Department of Exploration, Teikoku Oil Co.; Mr. J. Aiba, Teiseki Continental Shelf Development Co.; and Dr. Y. Fujita, Mr. O. Sato, and Mr. K. Sekiguchi, Technical Research Center, Teikoku Oil Co. for allowing me to review their new papers as well as holding valuable discussions. In addition, I am grateful to Dr. T. Gold, Center for Radiophysics and Space Research at Cornell University, for showing me his many new papers.

## Additional References

Cathles, L., et al, 1981, Failed rifts and Kuroko-type massive sulfide deposits: Mining Geology (Tokyo), v. 31, p. 50–51.

———, 1983, Kuroko-type massive sulfide deposits of Japan: products of an aborted island-arc rift: the Kuroko and related volcanogenic massive sulfide deposits: Economic Geology Monograph, 5, p. 96–114.

# The Northwest Shelf of Australia—Geologic Review of a Potential Major Petroleum Province of the Future

J. T. Forrest
*Petroleum Exploration Consultant*
*Houston, Texas*

E. L. Horstman
*Ameraust Consultants,*
*Dalkeith, Western Australia*

The Northwest Shelf of Australia consists of four major basins and a series of subbasins whose sedimentary thicknesses probably exceed 30,000 ft (9,000 m). It extends 1,500 mi (2,400 km) in a northeast to southwest direction, and averages more than 200 mi (320 km) in width. Structurally, the entire Northwest Shelf is dominated by Early to Middle Jurassic rifting. Although the tensional tectonic style predominates, compressional features are present. Proven petroleum reservoirs of the Northwest Shelf are Permian, Triassic, Jurassic, and Cretaceous sandstones. Except for the Carnarvon basin, where well density is still low, the Northwest Shelf is essentially unexplored. All the basins have most of the elements required for the generation and accumulation of petroleum. Recently announced discoveries in widely divergent areas of the shelf have created renewed interest in this large, unexplored offshore area, and may stimulate the exploration activity necessary to make the Northwest Shelf a major petroleum province of the future.

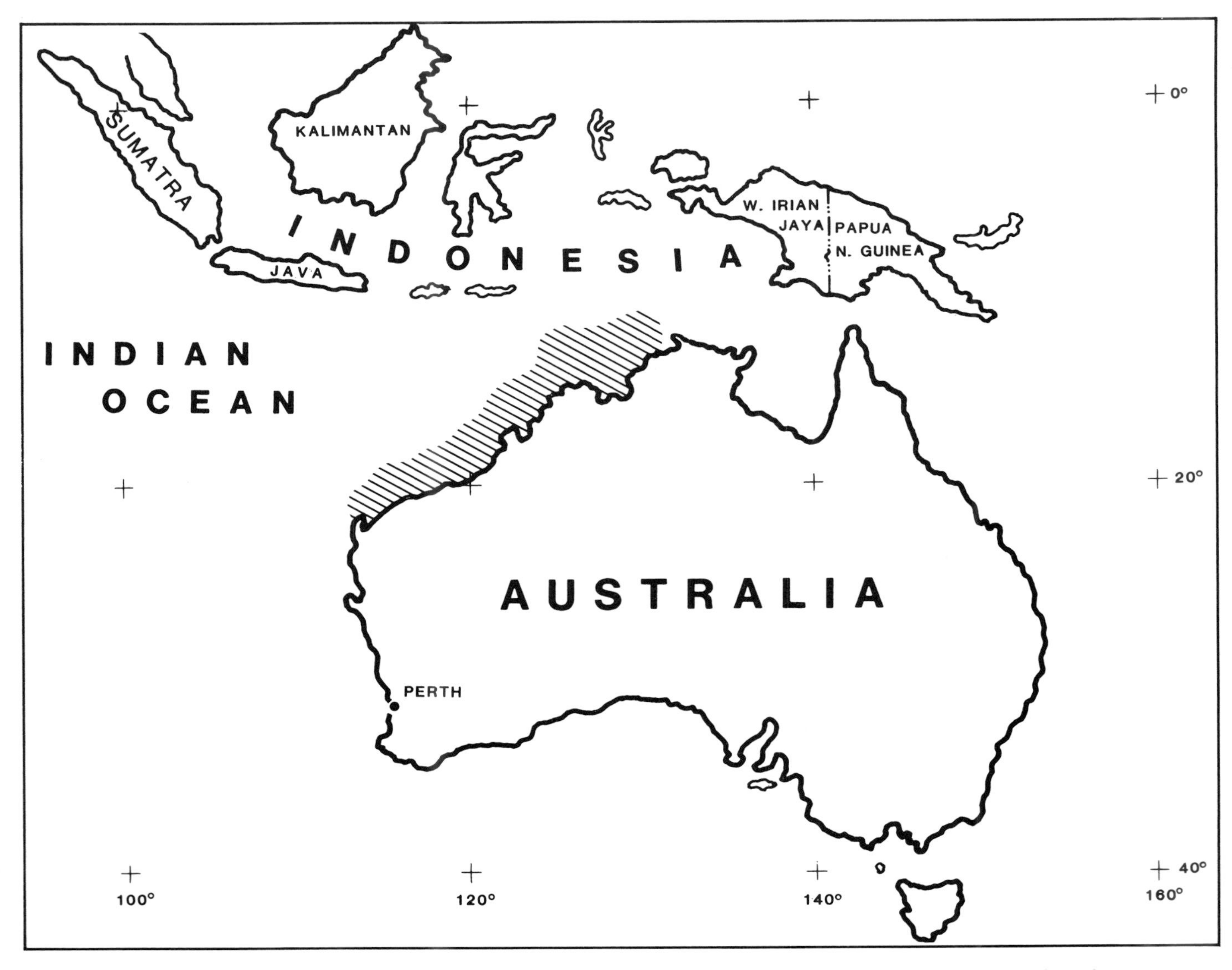


Figure 1. *Location map of the Northwest Shelf of Australia. Approximate limits of the Shelf shown in hachured pattern.*

## INTRODUCTION

"Northwest Shelf" describes a series of continental margin rift basins along the northern coast of Australia. For the purposes of this paper, the Northwest Shelf is arbitrarily limited to the area north of lat. 22° S and west of long. 131° E (Figures 1 and 2a). The prospective area contained in this broad arc varies in width from 120 to 240 mi (200 to 400 km). The length of the area of interest is approximately 1,500 mi (2,400 km). Disregarding shallow-water areas with a thin stratigraphic section over Precambrian basement, and deep-water areas with a thin sedimentary section on oceanic basement, about 300,000 sq mi (800,000 sq km) appear to be prospective for petroleum. Of this prospective area, 133,000 sq mi (345,000 sq km) are presently under lease.

The area is remote from large population centers. Perth, the largest city near the Northwest Shelf and the administrative center for much of the petroleum exploration activity, lies 650 mi (1,000 km) from the southern end of the Shelf. Smaller population centers and port facilities at Onslow, Port Hedland, Broome, and Darwin serve as supply bases for exploration activities.

Geologically, the Northwest Shelf consists of four basins. From the southwest to the northeast these are the Carnarvon basin, the Offshore Canning basin, the Browse basin, and the Bonaparte basin. Each of these can be further divided into subbasins and platform areas based on stratigraphic and structural boundaries.

## EXPLORATION HISTORY

The first oil exploration permits on the Northwest Shelf were granted to Ampol Petroleum Ltd. in 1946. These leases covered essentially all the onshore sedimentary basins in Western Australia and extended offshore to a water depth of 100 fathoms (183 m) in the Carnarvon basin and to approximately 20 fathoms (36 m) in the northern permits, an area of approximately

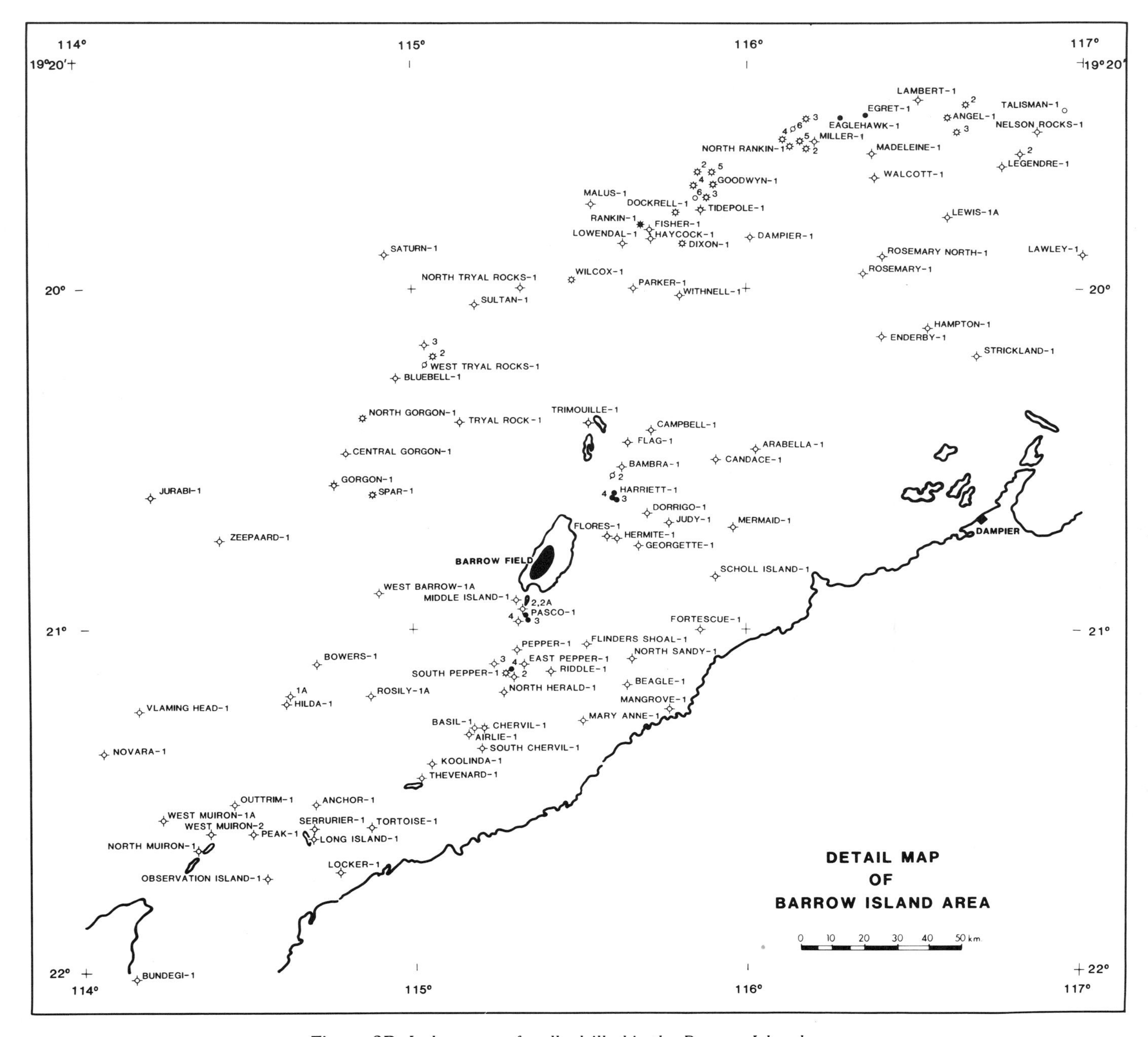


Figure 2B. *Index map of wells drilled in the Barrow Island area.*

400,000 sq mi (1,000,000 sq km). Ampol combined with Caltex in 1952 to form West Australian Petroleum Pty. Ltd. (WAPET), and initiated petroleum exploration in the coastal and offshore areas of Western Australia.

WAPET drilled their first well, the Cape Range-1, in 1953 on a surface anticline in the onshore portion of the Carnarvon basin. This well produced oil at a rate of 500 bbl per day from Cretaceous sands at 3,600 ft (1,100 m). Nine additional wells demonstrated that the initially discovered pool was extremely small, with an estimated recoverable reserve of only 85,000 bbl (Johnstone, 1979). While not an economic success, the well did give a significant boost to exploration in Australia.

The first offshore seismic work on the Northwest Shelf was carried out by WAPET in 1961. In 1964, WAPET drilled a wildcat on Barrow Island as a major step-out from previous onshore drilling. This well discovered oil in Upper Jurassic sands at 6,500 ft (2,000 m). Appraisal drilling demonstrated that the most significant reserves in the structure were contained in Cretaceous-age reservoirs at 2,300 ft (700 m) and 2,800 ft (850 m).

Prior to drilling Barrow Island, WAPET had applied for, and received in 1965, acreage west of the island to extend its holdings. Also prior to the Barrow Island discovery, Woodside (Lakes Entrance) Oil Co. and associated companies, which ultimately evolved to the present Woodside Petroleum Ltd., received leases to the offshore acreage north of and offshore from the WAPET lease interests. These leases covered an area of about 150,000 sq mi (400,000 sq km). At that time, WAPET and Woodside held petroleum leases totaling 440,000 sq mi (1.1 million sq km) on the Northwest Shelf. Later in the decade, ARCO and Aquitaine acquired the remaining open acreage in the Northern and Commonwealth Territories in the Bonaparte basin, to complete the initial leasehold picture on the Northwest Shelf.

The first success following the Barrow Island discovery occurred in 1969, with the discovery of gas in the Bonaparte basin at Petrel-1. This well was the second drilled by ARCO-Aquitaine and the fourth drilled in the offshore. Petrel-1 blew out and flowed wildly for over a year, until a relief well could be drilled. In 1968, Woodside had a noncommercial discovery of oil in its Legendre-1 well. That company's initial commercial success came in 1971, with two wells drilling simultaneously, 540 mi (875 km) apart, at Scott Reef-1 in the Browse basin and at North Rankin-1 in the Carnarvon basin; both resulted in gas-condensate discoveries. Within months, discoveries of gas-condensate were also made at Angel-1 and Goodwyn-1 near the North Rankin discovery. Two years later WAPET found gas-condensate on the North Rankin trend at West Tryal Rocks-1.

Changes in petroleum lease legislation in 1967 broke up the original, large lease blocks on the Northwest Shelf into smaller, but by world standards, still large tracts. A compulsory relinquishment of half the permit after the first 6-year term, and after each 5-year extension of the permit life, provided the opportunity for more companies to acquire exploration interests on the Northwest Shelf. From the original three groups involving nine companies, the number of permits has expanded to 32 at the present time, with approximately 90 different companies involved.

With additional companies in the Northwest Shelf exploration program, both seismic activity and drilling have increased markedly. In the latest round of drilling activity, significant gas-condensate discoveries were made by WAPET and the Woodside Group in the Carnarvon basin, while groups operated by the Occidental and Western Mining Corporation have discovered a number of small oil fields in the same basin (Forrest and Horstman, 1984). Woodside drilled a significant step-out from the 1971 Scott Reef discovery and made a smaller, but still significant, gas-condensate discovery at Brecknock-1 in the Browse basin. The most exciting recent discovery was made by Broken Hill Propriety Co. Ltd. (BHP) at Jabiru-1 in the Bonaparte basin. The discovery well tested a significant oil column on a large, faulted structure. Subsequent drilling on Jabiru has encountered both structural and stratigraphic complications preventing a reasonable evaluation of the reserve potential.

Many of the gas-condensate discoveries made to date on the Northwest Shelf are also poorly evaluated because no market exists for the gas, therefore requiring no appraisal of the reserves.

To date, essentially all the Northwest Shelf area has been investigated with a reconnaissance seismic grid and, in most of the more highly prospective areas, a detailed seismic grid. By the end of 1983, some 230 wildcat wells had been drilled on the Northwest Shelf (Figures 2A and 2B). Over half the wells have been in the Carnarvon basin. Table 1 summarizes the most significant discoveries on the Northwest Shelf. Because of inadequate appraisal drilling, reserve figures for the Shelf are uncertain, but recoverable oil and condensate discovered to date probably exceed 800 million bbl, while gas reserves probably exceed 35 trillion cubic feet (tcf).

## GEOLOGY OF THE NORTHWEST SHELF

Each of the basins collectively referred to as the Northwest Shelf is the result of a slightly asynchronous rifting event begun in the Middle Jurassic. The details of structure and stratigraphy vary from basin to basin depending on local structural configuration, structural movement, and lithologic variations in sediment source areas. The pre-Middle Jurassic sedimentary section, with its exploration objectives in the Permian, Triassic, and Lower to Middle Jurassic, on which the rift basins are superimposed, is incompletely known. Control points for the prerift unit are widely separated, and the section is rarely penetrated in wells and never in its entirety in a single well. Sufficient information is available, however, to clearly demonstrate that the Middle Jurassic rifting was superimposed on an earlier rift terrain, which, at least in part, is unrelated to the later rift movement.

With the exception of salt piercements and their resulting structures present in the Permian section of the Bonaparte basin, the principal drilling targets of the Northwest Shelf have been drape structures of postrift marine shales, over horst blocks containing prerift continental sands. As well control has increased and seismic control has improved in quality, more attention has been paid to postrift stratigraphy, and a number of small fields have been discovered associated with more recent structural activity.

Improved seismic data indicate that Tertiary strike-slip movement has generated young structures that are partially filled with currently generating and migrating petroleum.

We will discuss each of the four basins constituting the Northwest Shelf separately in the sections to follow. Time-stratigraphic and structural profiles have been prepared for each basin and are indexed in Figure 2C, while the stratigraphy of the Northwest Shelf is summarized in Figure 3.

## CARNARVON BASIN

The Carnarvon basin is the southernmost basin of the Northwest Shelf (Figure 4). The petroleum geology of this basin has been discussed by Alexander, Kagi, and Woodhouse (1980); Alexander et al (1981); Barker (1982); Crostella and Cheney (1978); Halse (1976); Moore, Hocking, and Denman (1980a, b, c); Powell (1976a, b); Thomas and Smith (1976); and Thomas, Nestvold, and Crostella (1981). The geology of the oil and gas fields within the Carnarvon basin has been discussed by Campbell and Smith (1982); Campbell, Tait, and Reiser (1984); Harris (1981); and Meath and Bird (1976). The Carnarvon basin is the most heavily explored of the Northwest Shelf basins. The only currently producing fields of the Northwest Shelf (Barrow Island and North Rankin Fields) are in the Carnarvon basin; Table 1 lists the 18 significant discoveries made there to date.

The Carnarvon basin can be divided into four subbasins with a shelf area on the east and a platform area on the west (Figures 4 through 9). The subbasins from south to north are the Exmouth subbasin, the Barrow subbasin, the Dampier subbasin, and the Beagle subbasin. The Barrow and Dampier

Table 1. *Significant discoveries of the Northwest Shelf of Australia.*

| Significant Discoveries | Recoverable Reserves tcf Gas | million bbl Condensate | million bbl Oil | Comment |
|---|---|---|---|---|
| **Carnarvon basin** | | | | |
| Barrow Island Field | — | — | 250 | |
| North Rankin Field | 7.8 | 145 | — | |
| Gorgon | ** | ** | — | 1 well, probably over 1 tcf gas |
| Central Gorgon | ** | ** | — | 1 well, probably over 1 tcf gas |
| North Gorgon | ** | ** | — | 1 well, probably over 1 tcf gas |
| West Tryal Rocks | ** | ** | — | 2 wells, probably over 1 tcf gas |
| Angel Field | 0.9 | 45 | — | 3 wells |
| Goodwyn Field area | 3.0 | 108 | — | |
| Wilcox | ** | ** | — | 1 well, probably over 1 tcf gas |
| Tidepole | 0.5 | 10 | — | 1 well |
| Egret | — | — | ** | 1 well, small oil reserve |
| Eaglehawk | — | — | ** | 1 well, small oil reserve |
| Legendre | — | — | ** | 1 well, small oil reserve |
| Harriet | — | — | ** | 6 wells, ?commercial oil field |
| South Pepper | — | — | ** | 4 wells, ?commercial oil field |
| North Herald | — | — | ** | 1 well, small oil reserve |
| Chervil | — | — | ** | 2 wells, small oil reserve |
| Talisman | — | — | ** | 1 well, small oil reserve |
| | 17.2 | 308 | 250 | |
| **Browse basin** | | | | |
| Scott Reef | 13.7 | 171 | — | 3 wells |
| Brecknock | 4.0 | 40 | — | 1 well |
| | 17.7 | 211 | 0 | |
| **Bonaparte basin** | | | | |
| Petrel | ** | ** | — | 3 wells, probably over 1 tcf gas |
| Tern | ** | ** | — | 2 wells, probably over 1 tcf gas |
| Jabiru | — | — | 35 | 3 wells ?commercial oil field |
| Sunrise | ** | — | — | 1 well, probably less than 1 tcf gas |
| Troubador | ** | — | — | 1 well, probably less than 1 tcf gas |
| | | | 35 | |

** indicates component present
—indicates component absent or presence unkown

subbasins are inseparable geologically. The original distinction was based on names used by different companies operating in the north and in the south of the area. In this paper, therefore, we have joined the two and will refer to the entire unit as the Barrow-Dampier subbasin.

The Carnarvon subbasins were a single depositional unit over most of the Mesozoic, both before and after the main period of rifting. The Exmouth subbasin is differentiated from the northern units by the southern depositional limit of Neocomian deltaic sediments. The Beagle subbasin at the northern end of the Carnarvon basin is separated from the Barrow-Dampier subbasin by a structural high, the De Grey Nose. The Beagle subbasin also lacks the thick Upper Jurassic section present to the south.

## Stratigraphy (Figures 3 and 5)

### *Permian*

The stratigraphy of the Permian section in the offshore Carnarvon basin is largely based on surface geology and on drilling in the onshore south of the area. The most recent analysis of the surface geology of the Permian section is by Moore, Hocking, and Denman (1980a, b, c). The surface section measured has a thickness of 15,000 ft (4,600 m) and is all Early Permian in age. Subsurface control has extended the stratigraphic knowledge and penetrated Late Permian age sediments.

The oldest Permian known from the Carnarvon basin is the Lyons Group of Sakmarian age, consisting of a series of fluvioglacial sands and silts interbedded with thin marine sands and shales. It is estimated to be some 7,800 ft (2,400 m) thick and to contain rafted glacial boulders. Generally, the sediments are poorly sorted and have neither good reservoir nor good source-rock potential. The unit would be too deep in most of the offshore area to be reached by current drilling techniques.

Unconformably overlying the Lyons Group is the Callytharra Formation, consisting of dark gray to black fossiliferous calcareous siltstone. The Callytharra is also of Sakmarian age and attains a maximum known thickness of 870 ft (256 m). Based on appearance only, the section is a potential source rock.

Disconformably overlying the Sakmarian sediments is a thick, entirely marine section consisting of the Wooramel, Byro, and Kennedy Groups of Artinskian-Tartarian age. The basal Wooramel Group is largely arenaceous with sandstone

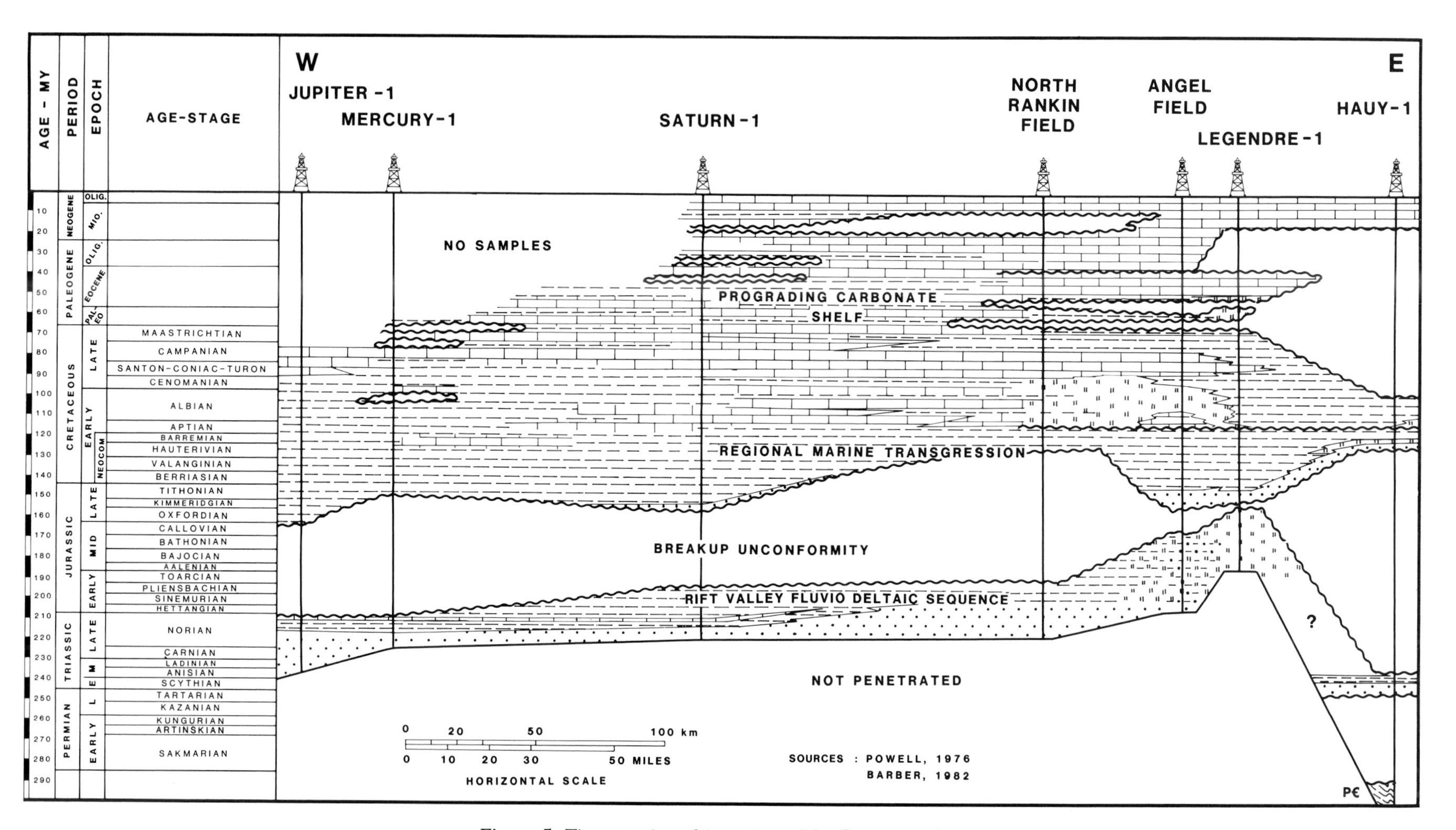


Figure 5. *Time-stratigraphic section of the Carnarvon basin.*

varying from argillaceous to clean, porous, and permeable. The unit is up to 1,640 ft (500 m) thick. The overlying Byro Group consists of interbedded siltstones and fine-grained sandstones, with occasional interbeds of black, fossiliferous shale. The shales, based on color and marine affinity, are potential source rocks. The Kennedy Range-1 well completely penetrated the Byro Group with 4,990 ft (1,520 m) of the section described by Thomas and Smith (1976). The Kennedy Group consists of laminated, fine-grained sandstones and siltstones of Kungurian to Tartarian age. The sands have been penetrated by a number of wells on the shelf areas of the Carnarvon basin and have a thickness exceeding 1,300 ft (400 m) in the Exmouth subbasin. Porous and permeable sands in these wells have not revealed any shows of gas or oil.

There is insufficient information available on any of these units to determine facies distribution and relationships. Moore, Hocking, and Denman (1980b and c) identify and describe six depositional facies in the Byro and eight in the Kennedy Group, varying from black shales deposited in a restricted environment to cross-stratified sandstones deposited above the wave base.

The paucity of modern geochemical data available from the Carnarvon basin prevents us from identifying the source-rock potential and state of maturation of the Permian section. Cook (1978) concluded that in the wells he investigated, the Permian section was possibly in the upper part of the oil-generative zone.

*Prerift Mesozoic (Triassic to Middle Jurassic)*

Triassic marine sedimentation, which succeeded the Upper Permian Kennedy Group, is represented in the Carnarvon basin by the Locker Shale. The basal Locker contains thin sandstones and appears to be conformable with the underlying Kennedy Group sands in many areas; a pronounced angular unconformity is present in other areas, demonstrating local uplift and erosion.

Well control indicates that the Locker Shale is over 3,200 ft (1,000 m) thick. The basal portion of the shale probably reflects the maximum extent of the Permian–Triassic marine transgression, with the unit becoming more regressive in character toward the top. The youngest portion of the Locker Shale is Ladinian and is typically gray to black (Crostella and Barter, 1980). Organic content varies and generally constitutes poor source rock. In the Perth basin, to the south of the Northwest Shelf, the age and lithologically equivalent unit, the Kockatea Shale is locally organic rich with high exinite content, and served as the source rock for the Dongara Field (Thomas, 1979).

The Locker Shale grades upward into the fluvial-to-deltaic Mungaroo Formation. The contact is usually placed at the point where Mungaroo sands predominate over the marine shales. The contact is diachronous, becoming younger to the northwest. The sands' source is the Precambrian Pilbara Block, to the southeast of the Carnarvon basin. The Mungaroo Formation thickens to the west from the shelf areas into the basinal environment. The thickest known development of the Mungaroo is an incomplete 10,000 ft (3,050 m) section in the Jupiter-1 well on the Exmouth Plateau. Oriented cores from wells in the North Rankin Field have established the source of the sands to be from the east of that area, and it is probable that all the Mungaroo sands prograded westward from an eastern source area.

By Late Triassic time (Norian) a marine transgression had begun in the west, reworking the older fluvio-deltaic sands and depositing a thinly interbedded sequence of calcareous sands, siltstones, and limestones. This transgressive unit, informally named the Brigadier beds by Crostella and Barter (1980), reaches a thickness of 1,080 ft (329 m) in the Saturn-1 well. On the Rankin Platform, deposition of this unit persisted into the Hettangian and on the eastern flank of the basin the fluvio-deltaic Mungaroo sands were not transgressed by the Brigadier beds until Pleinsbachian time.

Following deposition of the Brigadier beds, uplift in the source area renewed the sand supply and resulted in deposition of a marine sand unit, informally named the North Rankin beds after the North Rankin Field area where they are developed (Crostella and Barter, 1980). Deposition of the Brigadier beds concluded with a limestone bed that forms the transgressive base of the Dingo Shale.

The Dingo Shale is a thick marine sequence of dark gray siltstones and shales. Maximum known thickness of the unit is 11,463 ft (3,494 m) in the Cape Range-2 well (Geological Survey of Western Australia, 1974). The unit is as old as Sinemurian, extending, by definition, to the top of the Tithonian and encompassing almost all the Jurassic. In the Jurassic depocenters, there apparently is no significant break in deposition through the Middle Jurassic rifting period that is responsible for the Exmouth, Barrow-Dampier, and Beagle subbasins. During deposition of the Dingo Shale in the basinal areas, sands were deposited to the east along the strand line, with occasional incursions of sand into the basin. The oldest of these sand incursions is the Biggada Sand, of Callovian age. This sand has been penetrated by wells on Barrow Island and in the intervening basin between the eastern shelf areas and Barrow island. Other younger sandstones are also recognized along the eastern flank of the Carnarvon basin and are called the Learmonth Formation. Tithonian sands seen on Barrow Island intertongue with the Dingo Shale and are called the Dupuy Member of the Dingo. This member is time equivalent to the younger portion of the Learmonth Formation.

Prerift Lower-to-Middle Jurassic sediments are present over almost the entire Carnarvon basin, while postrift Middle-to-Upper Jurassic sediments are restricted to a thick section in the Exmouth subbasin, thinning northward along the length of the Barrow-Dampier subbasin. They are probably absent in the Beagle subbasin, because there is no evidence that Upper Jurassic sediments were ever deposited in areas other than where they are now present. During the rift phase, the area's uplift subjected most of the Carnarvon basin to erosion, except the subsiding graben areas in the center of the rift. The Lower-to-Middle Jurassic section has been completely removed from the tops of the Rankin Platform horsts and from the shelf areas on the east.

*Postrift Mesozoic (Late Jurassic and Cretaceous)*

As noted above, there is no significant break in sedimentation within the graben areas of the Carnarvon basin. The areas adjacent to the grabens underwent substantial uplift, however, and postrift erosion took place on these highs. The eroded materials were largely the Lower-to-Middle Jurassic marine shales. In spite of appreciable topography, there was little sand in the material transported and deposited in these areas during the Late Jurassic. The configuration of the basinal areas, bounded by the Pilbara foreland on the east and the Rankin

Platform on the west, resulted in limited water circulation and created sufficiently anoxic conditions to preserve organic matter. Geologists consider Upper Jurassic shales within the Barrow-Dampier subbasin to be the principal source rock for the oil and gas accumulations there.

A very condensed Upper Jurassic section is present on the Exmouth Plateau, 49 ft (15 m) in the Jupiter-1 and 220 ft (67 m) in the Saturn-1 wells (Barber, 1982); otherwise, Late Jurassic sedimentation was restricted to the graben areas. Some sands of Late Jurassic age were deposited along the east flank of the Carnarvon basin, as represented by the Learmonth Formation in the Exmouth subbasin and the Dupuy Member (Tithonian) of the Dingo Shale in the Barrow-Dampier subbasin. These sands' distribution indicates the presence of deltaic deposits in the Barrow Island area and in the Legendre and Angel Field areas to the north (Figure 4). Sand deposition in the Angel Field areas ended at the end of Tithonian, but in the Barrow Island area, they presaged the thick deltaic sands deposited during the Neocomian.

Marine sedimentation continued throughout the Cretaceous. In the Neocomian, a major deltaic complex developed in the southern part of the Barrow-Dampier subbasin. Explorers can locate the delta seismically by noting an obvious foreset character along the depositional front. These data indicate that the front is oriented almost east-west and is located just north of Barrow Island; deeper water turbidite sands extend out from the delta front. The Muderong shales were deposited outside the Barrow Delta area, but the more elevated of the Rankin Platform horst blocks remained above sea level until the late Neocomian or early Aptian.

Broad subsidence of the whole shelf area allowed marine sedimentation to continue through the Cretaceous. Erosion in the source area probably reduced elevations to nearly base level, and only clays and silts were contributed to the depositional areas. In Late Cretaceous, clastic sedimentation gave way to a wholly carbonate depositional regime.

#### *Tertiary*

Tertiary deposition in the Carnarvon basin consisted of a prograding, open-marine carbonate shelf. Sediments vary from shales to marl, with clastic carbonates predominating. The wedge of Tertiary sediment thickens offshore and forms the modern shelf edge. On the Exmouth Plateau, below the shelf edge, a thin, Tertiary section of deep-water carbonates is present. At the modern shelf edge, the Tertiary section attains a thickness of about 10,000 ft (3,000 m).

### Tectonics and Structure (Figures 4 and 6)

We can document fault movement in the Carnarvon basin at least as early as Late Permian. Well control suggests changes in thickness of some of the Upper Permian units across faults. On the Peedamulla and Preston shelf areas, good seismic data have established the presence of an angular unconformity at the Triassic-Permian boundary in some areas. Outside the shelf areas, the Permian is too deep and the seismic data too poor in quality to be useful in interpreting Permian structure. Powell (1976a) suggests that an elongated trough, formed in Permian time, extends the length of the northwest coast of Australia. He considers that this early block-faulting phase may have set the locus of structural movements taking place throughout the Mesozoic.

While there were minor structural movements between the Late Permian and the Middle Jurassic that affected local areas, the most important structural effect was epeiric uplift, which caused the Late Triassic regression, and is marked by the Mungaroo sand wedge. The regressive phase had already been substantially reversed by an epeiric downwarp and marine transgression when the main Callovian period of rift development began. This rifting formed the graben-bounded subbasins of the Carnarvon basin, as we know them today. The basin's sedimentary pattern has been one of transgression, since Triassic time. Minor regressive phases occurred within the Tithonian and Neocomian deltaic deposition, but these are only local phenomena. The regressive sand deposits within these deltas could be related to uplift in the sediment source area, as well as to local downwarp.

A structural disturbance in the Carnarvon basin occurred in the Miocene. This movement appears to have been strike-slip and is demonstrated in the Exmouth subbasin by documented, reverse displacement on faults at shallow depth, while normal fault displacements occur at greater depth. Seismic data indicate apparent reverse displacement on faults with predominantly normal movement elsewhere on the Northwest Shelf.

The structural traps in the Carnarvon basin consist of the large surface folds of the Cape Range area, such as the Rough Range Structure (Johnstone, 1979; Thomas and Smith, 1976) and Barrow Island (Campbell, Tait, and Reiser, 1984), and fault traps associated with the Callovian rift movement. The most prospective of these structures, from an exploration viewpoint, are the horst blocks cored by Triassic Mungaroo sands and sealed by draped, differentially compacted Neocomian Nuderong Shale (Harris, 1981; Campbell and Smith, 1982; Meath and Bird, 1976).

### Reservoir and Seal

Virtually all sandstones present in the Carnarvon basin are potential reservoirs. Porosities, even in the Permian sandstones, are generally high. The prime reservoir objectives are sands of the Upper Permian Kennedy Group, the Triassic Mungaroo Formation, the Jurassic Learmonth sands, and the time equivalent Dupuy Member of the Dingo Claystone, and the Barrow Group sands of Cretaceous age. Locally, other sands are significant reservoirs, such as the Aptian Windalia Sand at Barrow Island Field. Although it is the field's main reservoir, this sand is not widely distributed, has low permeabilities, and would not be an economic reservoir were it not on the island.

Two regional seals are present in the Carnarvon basin: the Lower Triassic Locker Shale and the Neocomian Nuderong Shale. The Albian Gearle Shale acts as a seal for the Windalia Sand at Barrow Island Field, where faulting has broken the underlying Muderong seal. All other reservoirs in that field are sealed by the Muderong Shale. Locally, the Jurassic Dingo Claystone also serves as a seal, as it does at Barrow Island for minor, Dupuy Member oil-bearing sands, and for noncommercial gas sands in the overpressured Callovian to Oxfordian Biggada Sand.

### Source and Maturation

There is little published geochemical information from the Carnarvon basin, but general concepts suggest that the potential source beds are the Permian Byro Group on the shelf areas to the east of the basin, the Triassic Locker Shale, the Jurassic

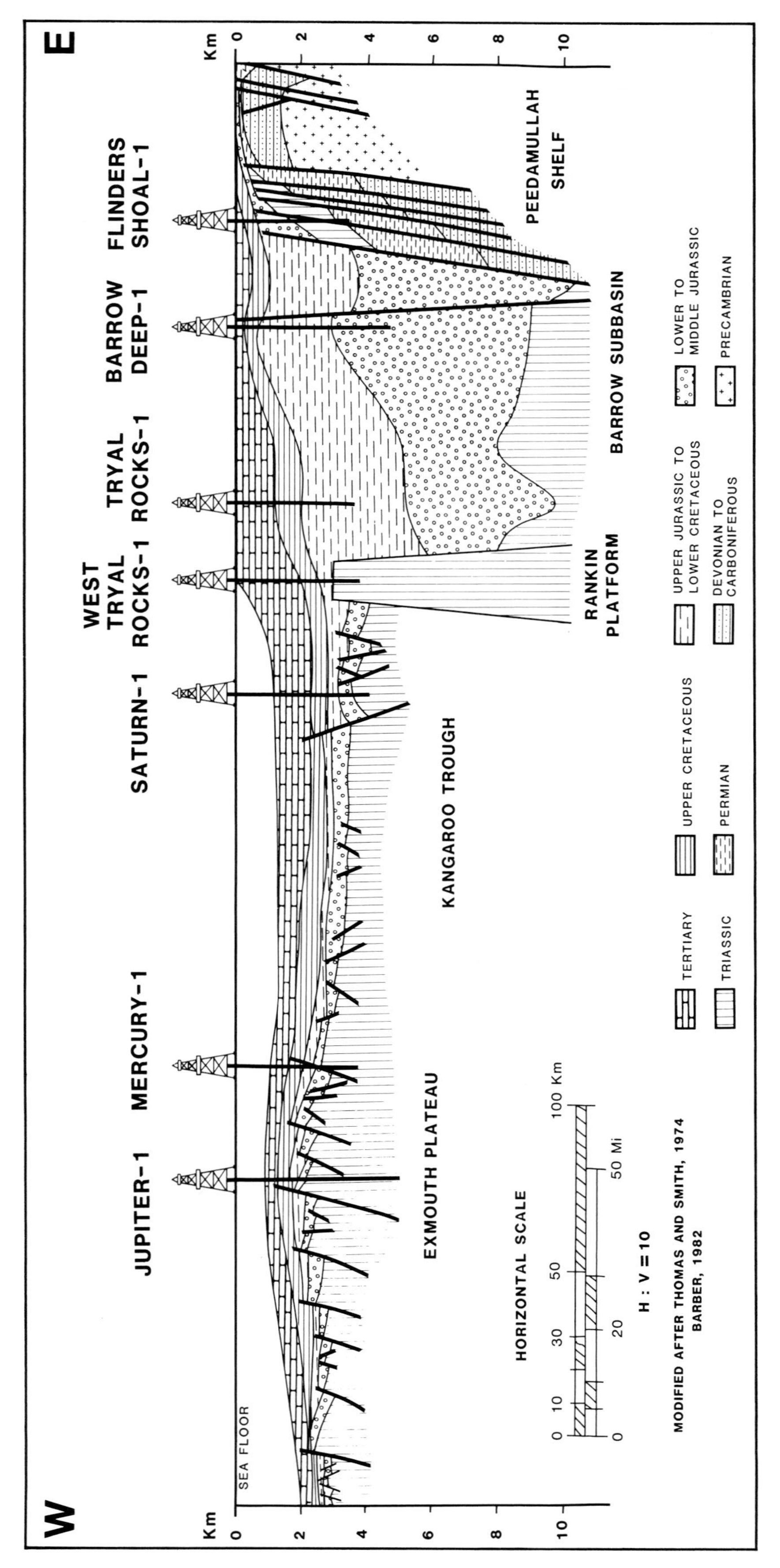


Figure 6. *Structural cross-section of the Carnarvon basin.*

Thomas, J. S., E. O. Nestvold, and A. Crostella, 1981, Exploration of part of the Northwest Shelf of Australia (A case history): Offshore South East Asia Conference, Southeast Asia Petroleum Exploration Society, February 1978.

Warris, G. J., 1976, Canning basin, Offshore, *in* R. B. Leslie, H. J. Evans, and C. L. Knight, eds., Economic Geology of Australia and Papua-New Guinea, 3 Petroleum, Australasian Institute of Mining and Metallurgy, Parkville, Victoria, Australia, 541 p.

Yeates, A. N., D. L. Gibson, R. R. Towner, and R. W. A. Crowe, 1984 Regional geology of the Onshore Canning basin *in* P. G. Purcell, ed., The Canning basin, W. A.: Perth, Australia, Proceedings of the Geological Society of Australia/Petroleum Exploration Society of Australia Symposium, 1984.

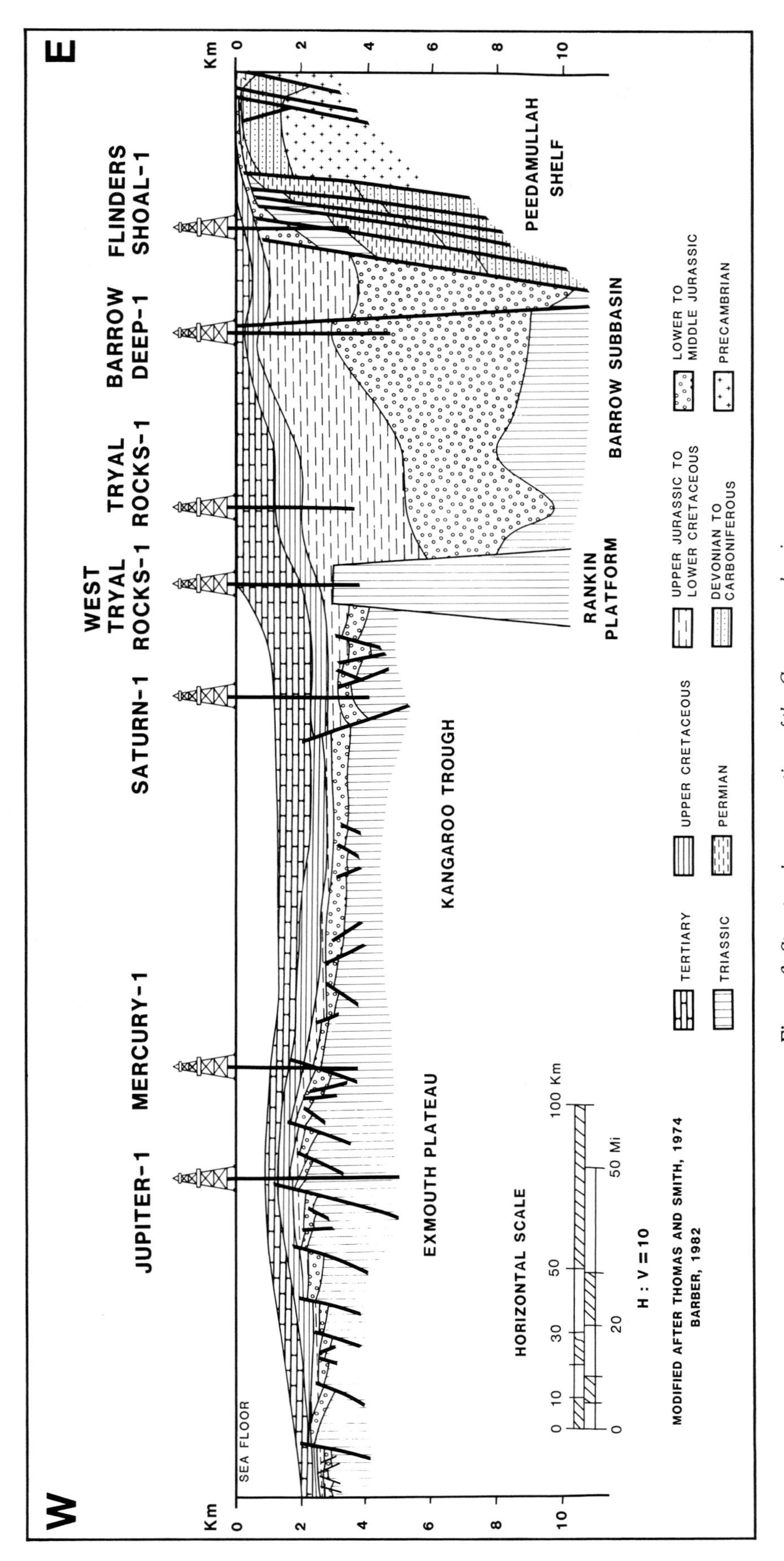


Figure 6. *Structural cross-section of the Carnarvon basin.*

Dingo Claystone, and the Cretaceous Muderong Shale. Cook and Kantsler (1980) indicate that all these units have sufficient organic carbon levels to be source rocks. Generally, vitrinite predominates over exinite, indicating a preference for a gas-prone source. They do comment, however, that the basal Locker Shale contains abundant exinite, and locally the Dingo Claystone has an exinite content varying "from rare to abundant." The Muderong Shale sequence contains abundant exinite and some alginite, suggesting a more oil-prone source rock.

Organic maceral maturity in the Carnarvon basin varies widely. Cook and Kantsler (1980) indicate that the upper level of the oil-generative window (vitrinite reflectance 0.5–1.35%) is at 5,000–6,000 ft (1,500–1,800 m) on the shelf areas, 6,000–8,000 ft (1,800–2,500 m) in the basin center, and 8,000–10,500 ft (2,500–3,200 m) on the Rankin Platform. Barber (1982) indicates that the oil window is entered on the Exmouth Plateau at 9,800–11,500 ft (3,000–3,500 m) below sea level, or roughly 6,500–8,000 ft (2,000–2,500 m) below the sea floor. This indication is consistent with the Rankin Platform province. Translating these depths to the stratigraphic section, the Triassic and Permian on the shelf areas are just within the oil-generative window, the Lower Cretaceous enters the oil window in the center of the rift basin, and the Upper Triassic is within the oil window on the Rankin Platform.

Source rock–oil correlations and oil–oil correlations in the basin suggest that all the petroleum found to date comes from the Upper Jurassic Dingo Shale (Alexander et al, 1983) with the exception of the gas on the Exmouth Plateau, where Jurassic source rock is thin and immature. Geologists consider the Locker Shale to be the source for the gas discovered on the Exmouth Plateau.

Alexander et al (1983), working on hydrocarbon weathering, suggest that the Triassic reservoirs on the Rankin Platform contained oil prior to the erosional cycle that began in Callovian time. The weathered residual oil identified by Alexander is chemically unrelated to, and earlier than, the gas-condensate now present in the reservoirs that correlates to a Jurassic source. This early petroleum accumulation must have weathered during the erosional cycle prior to the burial and maturation of the Dingo Claystone; its only logical source is the Locker Shale. No oil or gas in the Carnarvon basin has been correlated with either Permian or Cretaceous source beds.

## Field Discoveries (Figures 2A and 2B)

### *Barrow Island and Its Related Fields*

Barrow Island Field is located approximately 35 mi (55 km) off the coast of Western Australia. Campbell (1982) and Campbell, Tait, and Reiser (1984) discuss its geology. The island represents the surface manifestation of a regional anticlinal structure approximately 70 mi (112 km) in length. The Barrow Island Oil Field occupies a small area at the crest of the overall structure, where it is cut by a major northeast to southwest trending fault. The principal reservoir is the Aptian Windalia Sand, a bioturbated, very fine-grained, glauconitic sand ranging in thickness from 90 to 130 ft (27–39 m). The reservoir is at an average depth of 890 ft (550 m) and has been developed by 670 wells, including oil, gas, water-injection and water-disposal wells, and dry holes; 360 wells now produce from the Windalia Sand. The field has produced 200 million bbl to date and current production is about 22,000 bbl per day. Geologists estimate ultimate recovery to be 250 million bbl of oil.

During 1982 and 1983, a series of oil discoveries were made on both the north and south plunges of the overall Barrow Island anticline. Faulting has created a series of low-relief closures, with thin, 30–60 ft (9–20 m) oil columns, on water and under thin, free-gas caps in generally massive sands. From south to north, these discoveries are Chervil, North Herald, and South Pepper on the south plunge, and Harriet on the north plunge. Appraisal drilling has been carried out on Chervil, South Pepper, and Harriet. Currently, only the Harriet appraisal drilling has given encouragement, with four of the five appraisal wells successful. All discoveries are in shallow water, but to date it is uncertain whether any of the fields will be economically viable.

### *Rankin Platform Fields*

North Rankin Field lies some 84 mi (135 km) offshore from the iron ore port of Dampier, and is the first and only one of the Rankin Platform fields to be brought into production. Six appraisal wells were drilled to evaluate the field prior to development. A single platform has been placed in 425 ft (130 m) of water and three of seven planned development wells have been drilled. Initial production from the seven wells will be 385 million cubic feet of gas per day (MMCFGD). First deliveries to the pipeline were made in August, 1984.

The gas-condensate accumulation at North Rankin Field occurs in Upper Triassic Mungaroo sands, on a horst block flanked and sealed by Neocomian Muderong Shale (Harris, 1981). The top of the reservoir is at approximately 8,860 ft (2,700 m) with a gas/water level at 10,500 ft (3,200 m), giving a 1,640 ft (500 m) gas column. The field covers an area of 19 sq mi (50 sq km). Sand porosities average 17–21% and permeabilities range from 500 to 1500 mD. The recoverable field reserves are estimated to be 7.8 tcf of gas and 145 million bbl of condensate.

The other Rankin Platform discoveries are smaller than the North Rankin Field, and only in the case of Goodwyn Field has appreciable appraisal drilling taken place. Goodwyn, a major horst block feature similar to North Rankin, has been evaluated by five appraisal wells and a series of wildcats on subsidiary fault blocks. One of the wildcats (Tidepole-1) was successful in locating significant reserves. The gas column at Goodwyn is thinner than at North Rankin, and the several reservoir sands have different gas to water contacts. In addition, small pools of oil have been encountered in several parts of the overall area (Goodwyn-3, Goodwyn-6, and Tidepole-1). Recoverable reserves for the area (including Tidepole) are 3.5 tcf of gas and 118 million bbl of condensate and oil.

South of Goodwyn Field, discoveries have been made at Wilcox, West Tryal Rocks, Gorgon North, Gorgon Central, and Gorgon. Of these, only West Tryal Rocks has been confirmed by an appraisal well. No recoverable reserve estimates have been given for any of these discoveries, but each is expected to have over 1 tcf of gas plus condensate. West Tryal Rocks drill-stem tests recovered up to 28% inert gas (Meath and Bird, 1976), while Gorgon recorded 18% inert gas in some tests (Campbell and Smith, 1982).

North of the North Rankin Field, two small oil discoveries have been made on horst block structures similar to those in the gas fields to the south. In both cases, the seal is the Muderong Shale (Neocomian). At Egret, the accumulation is in Tithonian sand rather than in the Triassic Muderong sand. Neither discovery has been confirmed and neither area would appear to have a commercial accumulation at the present time.

To the east of the Rankin Platform, several fields have been discovered in simple anticlines or fault-closed anticlines with Upper Jurassic to Lower Cretaceous sand reservoirs. Legendre, the first offshore discovery in the Carnarvon basin, found a thin oil column in a faulted anticlinal fold. Angel Field, to the west of Legendre, is a closed structure containing relatively thin gas-condensate accumulations in Upper Jurassic sands. Two appraisal wells have been drilled in the field, and the recoverable reserves are estimated at 0.9 tcf of gas and 45 million bbl of condensate. In August 1984, an apparent discovery was made at Talisman with a thin oil column in the same sands as the Legendre accumulation. None of these fields has established sufficient reserves to be considered economic at this time.

## OFFSHORE CANNING BASIN

The geology of the Offshore Canning basin has been described by Warris (1976), while references to the area are also found in the more general Northwest Shelf papers by Powell (1976a) and Thomas, Nestvold, and Crostella (1981). The Offshore Canning basin is sparsely explored. Four of the twelve wells drilled have had Devonian carbonate objectives, well known from both surface geology and a substantial number of onshore wells. The remaining wells, with the exception of East Mermaid-1, had Permian or Triassic objectives. East Mermaid was drilled for a Jurassic section. None of the wells have encountered significant shows of oil or gas.

The Offshore Canning basin is separated from the Carnarvon basin to the south by the North Turtle Arch (Figure 4). It consists of the Bedout subbasin, a Permian-Triassic depocenter, and the Rowley subbasin, which contains a thick Jurassic section. These two subbasins are separated by a major down-to-the-west fault system. There is no clear tectonic or stratigraphic distinction between the Rowley subbasin and the Beagle subbasin to the south or between the Rowley subbasin and Browse basin to the north.

### Stratigraphy (Figures 3 and 7)

#### *Permian*

Knowledge of the Permian stratigraphy of the Offshore Canning basin is based largely on numerous onshore wells that have penetrated the section near the coast. Seismic data show that the main Paleozoic tectonic features of the onshore basin extend westward into the offshore for some distance until they are interrupted at almost right angles by the major faulting of the Middle Jurassic rift sequence.

Only two wells in the Offshore Canning basin are known to have encountered Permian sediments: Lacepede-1 and Lynher-1. Four wells have been drilled for Devonian objectives, but no information is available on these tests. Both Lacepede and Lynher are located near the major rift margin fault trend. At the time the wells were drilled, the Permian section was not considered a target section.

The onshore Permian stratigraphy, which can be projected into the offshore areas, has been presented in detail by Yeates et al (1984). A glacial phase of sedimentation began in Late Carboniferous time and extended into the Sakmarian. This fluvioglacial unit, called the Grant Group, consists of dropped clasts in a clay matrix interbedded with sandstones and shales. Locally, the sediments contain a good marine fauna, while elsewhere evidence of lagoonal and delta-plain environments is present. Following minor tectonism, the Grant Group is overlain by the Poole Sandstone, a marginal-marine to marine unit some 300–650 ft (100–200 m) thick. A marine transgression followed the Poole Sandstone, depositing the Noonkanbah Formation (Artinskian). These consisted of minor limestones, calcareous shales, and siltstones, all richly fossiliferous, that reach a thickness of 1,500 ft (460 m). A subsequent regressive unit, the Liveringa Group, consists of siltstones and sandstones deposited in a fluvial to intertidal environment. Locally, minor coal beds are present. The Liveringa reaches a thickness of 2,500 ft (760 m). Some structural movement took place in post-Liveringa time, so that the overlying Triassic section is locally disconformable with the Liveringa Group.

In the Offshore Canning basin, Lynher-1 penetrated Upper Permian sands and siltstones equivalent to the Liveringa Group. Lacepede-1 penetrated Sakmarian-age sands, silts, and shales equivalent in age to the Grant Group. In both wells, the Permian section is unconformably overlain by an Upper Triassic to Lower Jurassic section.

#### *Prerift Mesozoic (Triassic to Middle Jurassic)*

A thin basal Triassic sand developed in the onshore section, as a result of a reworking of the underlying Liveringa Group sands by the transgressive phase responsible for deposition of the Blina Shale. Onshore, the Blina Shale is described as a thinly bedded, bioturbated shale, overlain by the regressive Erskine Sand (late Scythian to early Anisian), which has fluvio-deltaic characteristics. None of the offshore wells in the public domain have penetrated this part of the Triassic known in the onshore. The Triassic section encountered in the offshore wells has been a deltaic and epicontinental sequence of interbedded sandstones and shales, younger than the onshore section described above. Seismic data indicate major block faulting in the Upper Triassic, which can be mapped on a regional basis. A time gap in the sedimentary column has been noted in wells drilled to this time-stratigraphic section. A major tectonic event in the Upper Triassic sequence is also indicated in the onshore Canning basin.

Following this Late Triassic tectonism and the resultant erosion, sedimentation recommenced, and Lower Jurassic redbeds and deltaic sediments onlap the Upper Triassic unconformity topography. The section consists of thick, fluvio-deltaic sandstones with interbedded claystones and siltstones and occasional coaly horizons.

The Middle Jurassic breakup was not a period of major structural deformation, and did not significantly disrupt sedimentation in the Offshore Canning basin. East of the major faulting that occurred at this time, a very condensed Jurassic section was deposited compared to the area west of the fault zone. In neither area is there an appreciable time gap in sedimentation such as that recognized in the Carnarvon basin.

#### *Postrift Mesozoic (Late Jurassic and Cretaceous)*

The postrift Jurassic section is relatively thin in the eastern part of the offshore basin, and thickens to the west. The immediately post-breakup section consists of marine shales and silts of Oxfordian and Callovian age. In the Bedout subbasin, some

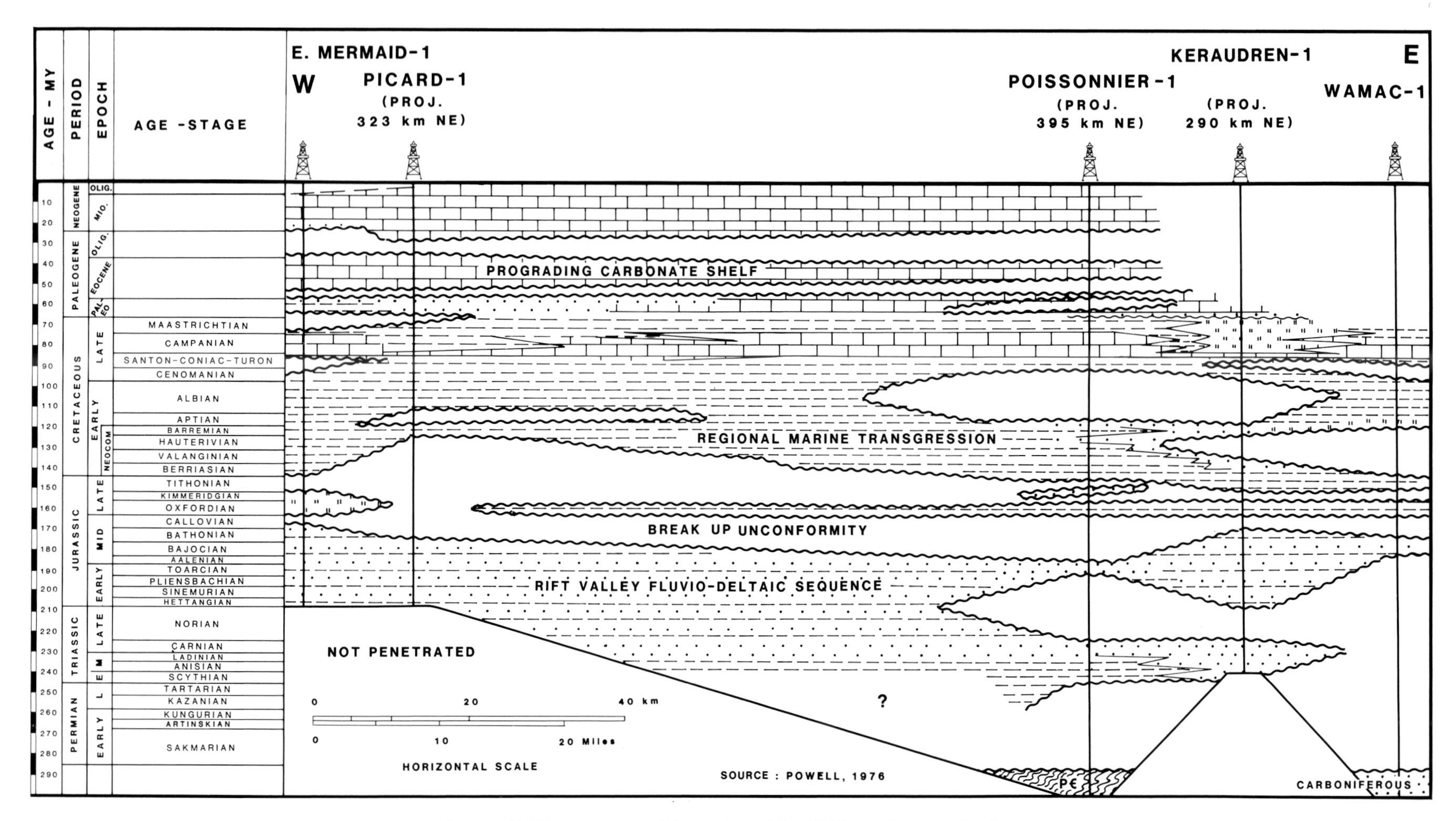


Figure 7. *Time-stratigraphic section of the Offshore Canning basin.*

areas received an appreciable thickness of Tithonian marine shale that is not present in the one well in the Rowley subbasin. This shale marks the beginning of a major marine transgression that continued through the Cretaceous. By early Neocomian time, the entire area had been inundated, and shale deposition continued until Late Cretaceous, when deposition of marls and limestones began.

*Tertiary*

During Tertiary time, broad subsidence of the shelf allowed a prograding limestone section to develop into the modern shelf edge.

### Tectonics and Structures (Figures 4 and 8)

The tectonics and structure of the Bedout and Rowley subbasins are distinctly different. The Bedout subbasin was a depocenter in Permian to Triassic time,and the structures within it are due to Triassic tectonism. The Rowley subbasin was a depocenter in the post-Middle Jurassic, and the structures in it are largely due to movements during the Middle Jurassic breakup.

The Bedout subbasin is an asymmetric, intracratonic basin increasing in depth westward to the Bedout High, a Paleozoic feature with draped Mesozoic closure. The basin has experienced some faulting, probably associated with the Late Triassic to Early Jurassic Fitzroy Movement that is well documented in the onshore basin. The Keraudren-1 well was drilled on a structure interpreted as a Triassic horst block. Other Bedout basin wells have been drilled on the Bedout High and the North Turtle Arch. Epeiric downwarp during the Mesozoic and Tertiary permitted marine sedimentation across the entire shelf area, including the Bedout subbasin.

The rifting that formed the Carnarvon basin to the south and the Browse basin to the north is documented in the Offshore Canning basin by the fault systems that make up the eastern edge of the Rowley subbasin. The Rowley subbasin appears to be open to the west, with no western flank of the rift showing on seismic data. The presence of deep water has discouraged exploration in this area.

### Reservoir and Seal

Permian, Triassic, and Jurassic sandstones are primary reservoir objectives in the Offshore Canning basin. These sands are mainly fluvio-deltaic and epicontinental in facies. Porosities and permeabilities for most of the sands generally are good.

The two regional seals common to the whole of the Northwest Shelf area are present in the Offshore Canning basin: the Lower Triassic shale (Blina Shale in the Onshore Canning basin) and the upper Tithonian to Neocomian shale resulting from the major marine transgression of that period. Locally, shales in the Lower to Middle Jurassic could provide reservoir seals. The Upper Cretacious and Tertiary carbonates have good porosity and permeability, but are not effectively sealed.

### Source and Maturation

Based on very limited well control and sparse sampling, Robertson Research (1978) reports on the source potential of the Offshore Canning basin. The paper concludes that the Triassic section contains mostly inertinitic macerals and could not source petroleum. Both prerift and postrift Jurassic sediments contain potential source rocks. However, these sediments are thermally immature over the portion of the basin sampled. The maceral content is described as humic, and is considered gas generative.

## BROWSE BASIN

The Browse basin lies offshore, northwest of the Precambrian Kimberly Block (Figure 4). Its geology has been discussed in some detail by Allen, Pearce, and Gardner (1978); Crostella (1976); Halse and Hayes (1971); Passmore (1980); and Stagg (1978). The more generalized papers on the Northwest Shelf by Powell (1976a, 1979) and Thomas et al (1978) discuss the Browse basin in a regional context.

Petroleum exploration leases over the entire Browse basin were originally held by one exploration group until 1978, and the basinal portions were held by that group until 1982. As a result of this and adverse operating conditions (extremely remote location and water depths generally in excess of 600 ft or 180 m), fewer than 20 wells have been drilled in the basin's 39,000 sq mi (100,000 sq km) area. Two discoveries, Scott Reef and Brecknock, have resulted from the drilling.

The Browse basin has not been divided into subbasins and can be considered a single depositional entity. It is bounded on the east by a Permian-floored shelf area, the Browse Shelf, and on the south by the Leveque Platform. On the west, the Scott Plateau, now a downfaulted deep-water area, provided a margin throughout much of the basin's depositional history. The north boundary of the basin is the Ashmore Block. In detail, the area south of the Barcoo-1 well is essentially continuous with the Rowley subbasin of the Offshore Canning basin, and the distinction from the Vulcan graben to the north is not clearly established.

### Stratigraphy (Figures 4 and 9)

*Permian*

The presence of a Permian section in the basin has been confirmed by three wells on the Browse Shelf: Rob Roy-1, Prudhoe-1, and Yampi-1. After penetrating some 2,200 ft (660 m) of probably Lower Permian, Rob Roy-1 drilled into Precambrian metamorphics.

The Permian section in Rob Roy-1 is silty claystone and medium- to well-sorted carbonaceous sandstones. Nonmarine sediments appear to predominate, but some marine sections also appear to be present. A 1,400 ft (425 m) Permian shale section in Prudhoe-1 is probably marine. In all three wells that penetrated it, the top of the Permian is a major unconformity. If Powell's (1976a) interpretation is correct, and a Late Permian phase of rifting took place on the Northwest Shelf, it is probably that younger marine Permian section that will be preserved in the grabens formed by tectonism.

*Prerift Mesozoic (Triassic to Middle Jurassic)*

From information in the public domain, no wells in the basin have penetrated sediments between Early Permian and Late Triassic age. Upper Triassic sections have been noted in Yampi-1, Scott Reef-1, Ashmore Reef-1, and in Lynher-1 in the northern part of Offshore Canning basin. The southern margin of the basin received fluvio-deltaic sandstones in the Late Triassic, while at Scott Reef-1 and Ashmore Reef-1, the Upper Triassic consists of interbedded shales, silts, and sandstones. Stagg

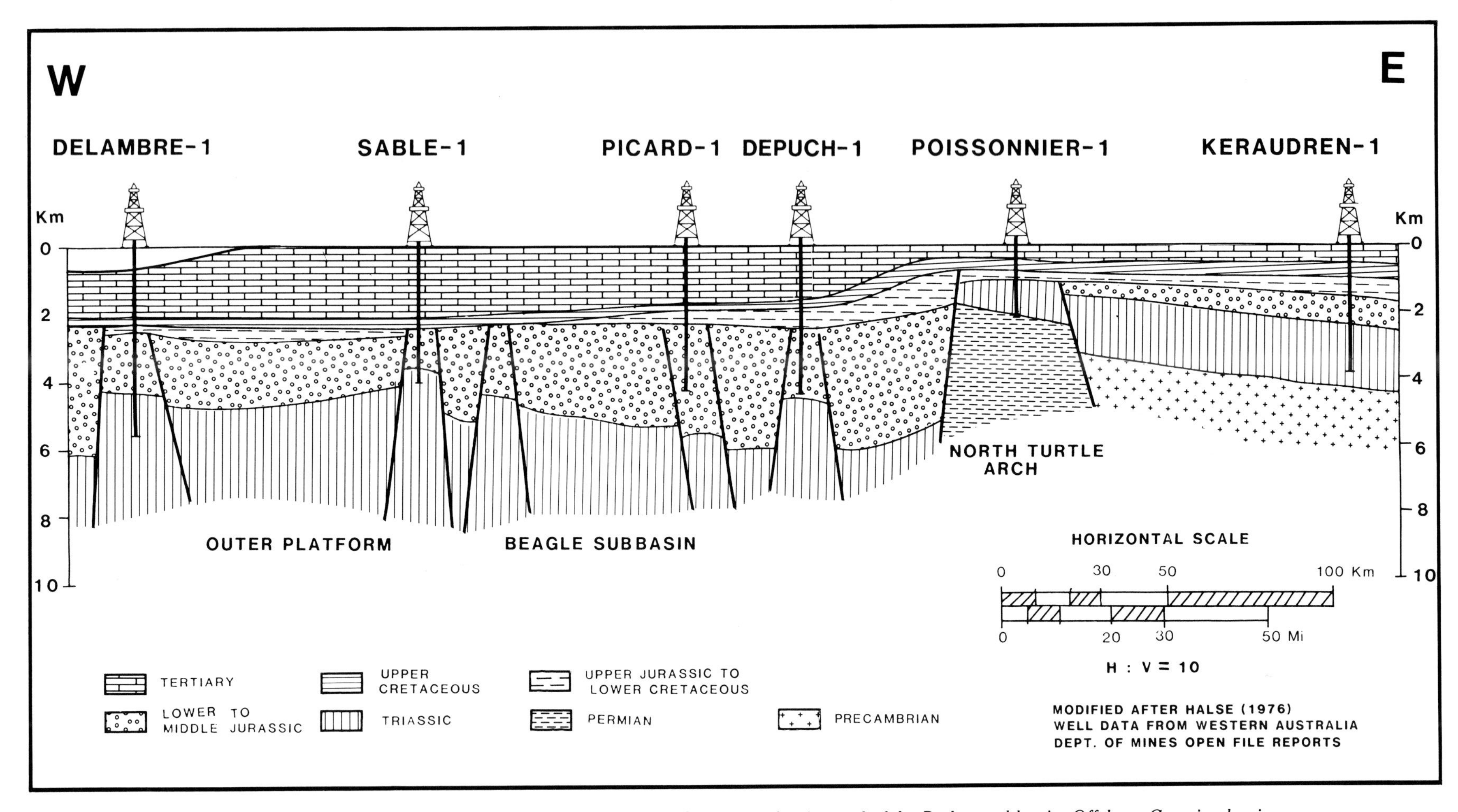


Figure 8. *Structural cross section of the Beagle subbasin, Carnarvon basin, and of the Bedout subbasin, Offshore Canning basin.*

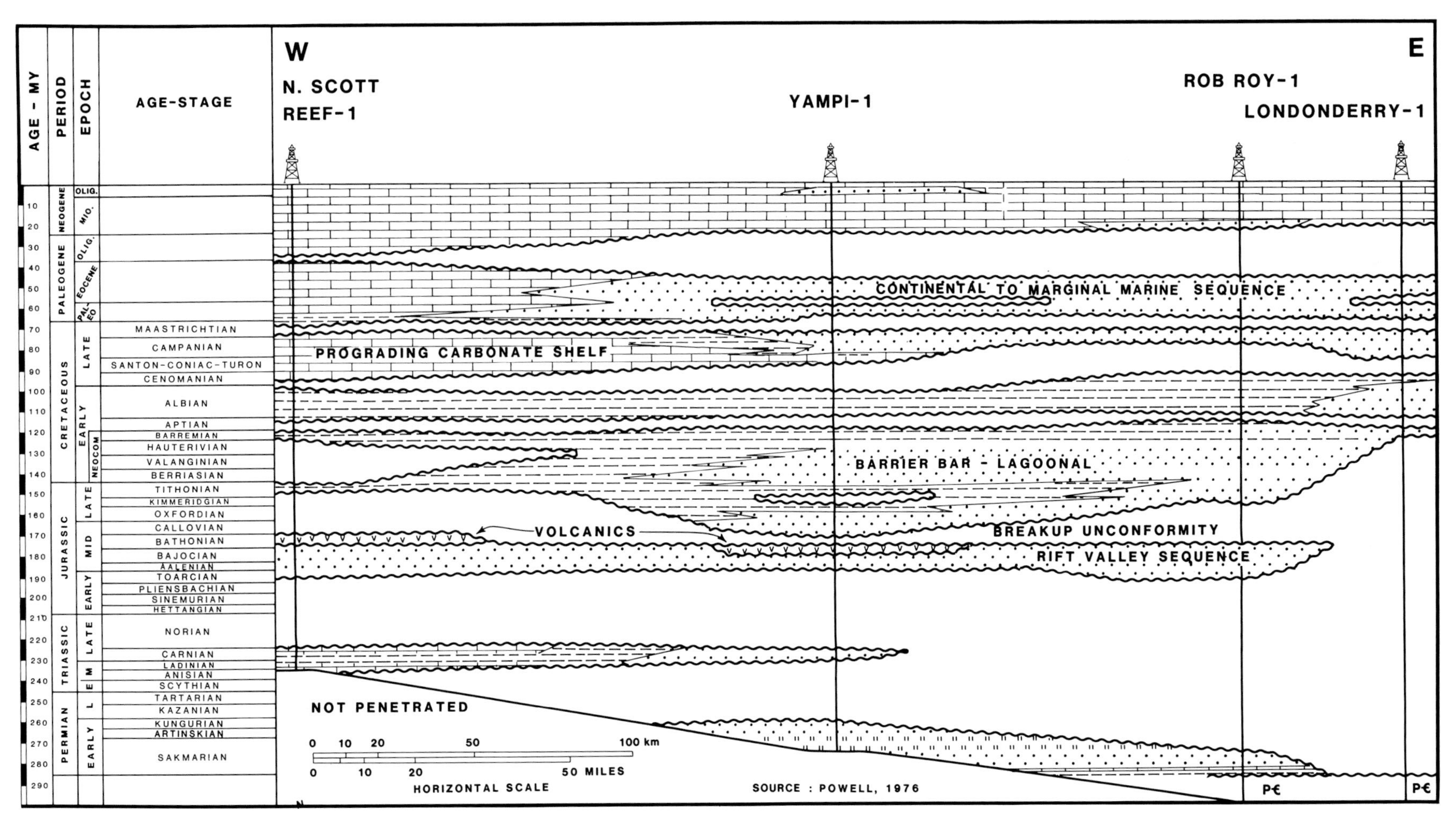


Figure 9. *Time-stratigraphic section of the Browse basin.*

(1978) believes that the Scott Plateau was emergent during the Late Triassic and may have contributed clastic material to the Browse basin from the west. Allen, Pearce, and Gardner (1978) note that dipmeter information on Scott Reef-1 supports that interpretation.

Evidence from some areas within the Browse basin suggests local structural disturbance at slightly different times during Early to Middle Jurassic. The Scott Reef-1 well shows significant stratigraphic structural thickening, indicating tectonic movement during deposition of the Lower Jurassic section. The same well also penetrated the section missing at Yampi-1. In the Ashmore Reef-1 well, the entire Lower Jurassic section is missing, while to the south at Lombardina-1, red to brown shales of that age were encountered. The Middle Jurassic sediments of the Browse basin are fluvio-deltaic over a wide area, and are probably restricted in occurrence to the area between the Scott Plateau, the Leveque Platform, the Browse Shelf, and the Ashmore Block. The Middle Jurassic depositional sequence ended with uplift and tectonism, rejuvenating movement along some of the pre-existing structural features. Locally, basic lavas were extruded. Although the igneous material is of limited extent in any one area, its occurrence is widespread, having been encountered in Lombardina-1, Yampi-1, and Scott Reef-1.

*Postrift Mesozoic (Late Jurassic and Cretaceous)*

After the rift period, the fluvio-deltaic and volcanic sequences of the Middle Jurassic were covered by a major marine transgression starting in Oxfordian time. All currently available well control, with the exception of Yampi-1, is on structurally high parts of the basin and generally reflects the missing section related to the rift. In Yampi-1, however, the time gap is short, and it is quite likely that in the basin depocenter sedimentation was continuous. The deposition of restricted marine shales proceeded outward from the basin center, onlapping the erosion surface; and by late Tithonian time, Scott Reef had been covered, as had the southern edge of the Ashmore Block, the western portion of the Browse Shelf, and the Leveque Platform. Seismic data indicate that the Scott Plateau was partially submerged at this time and received sediment deposition. The east flank of the basin received paralic sediments during the entire period of transgression.

At the end of Jurassic time and in the Early Cretaceous, estuarine and deltaic sandstones were deposited on the eastern flank of the basin. The westward extent of these sands into the basin is unknown, although wells have been drilled as far west as Caswell-2 with these sands as a principal reservoir objective. This sand incursion into the basin occurred simultaneously with the Neocomian-Tithonian delta systems' deposition in the Carnarvon basin.

Continued epeiric subsidence allowed marine deposition to move eastward during the Cretaceous, with open marine conditions established in Turonian time. The transgression reached its maximum extent during the Santonian. During this period of open marine conditions, the sediments became more calcareous in the west, while paralic-to-continental sandstones were deposited at the strandline in the east.

*Tertiary*

There is no distinct break in sedimentary pattern between the Cretaceous and Tertiary, although Maastrichtian data are lacking. Sands continued to be deposited in the east through the Eocene, while a major withdrawal of the sea occurred in the Oligocene, except in the west. During Miocene time, a major transgression took place, and the prograding wedge of Tertiary carbonates that developed over the whole of the Northwest Shelf began to form.

### Tectonics and Structure (Figures 4 and 10)

Insufficient evidence from the limited well control prevents speculating on Permian tectonism in the Browse basin. In the period from Early Permian to Late Triassic, the faulting on the eastern flank of the basin must have separated the Browse Shelf from the basin, accounting for the major time break in wells on the Browse Shelf. Rifting and graben development probably took place between the Browse Shelf and the Scott Plateau, accounting for the distribution of the Upper Triassic sediments.

Additional rifting and block faulting took place in Late Triassic to Early Jurassic time, creating the structures that are the basin's principal exploration objectives.

Epeiric structural movements dominated the Browse basin in the post-Callovian period. Existing fault rejuvenation can be documented seismically, but significant new structures do not appear to have developed. There are, however, some seismic indications of compressional displacement due to strike-slip movement along the northwest margin of the Leveque Platform, possibly related to the Miocene strike-slip movement in the Carnarvon basin.

Horst blocks occur in the Browse basin in several trends. The westernmost trend consists of a large, elongated drape anticline at the southern end of the basin where Barcoo-1 was drilled. To the north, a curved structural trend extends from the Leveque Platform through Brecknock and Scott Reef. To the northwest of the Scott Reef trend, a third major structural trend is present (Buffon-1). Allen, Pearce, and Gardner (1978) refer to these trends as the Southwest Outer Trend, the Scott Reef Trend, and the Northwest Outer Trend. Another structural trend projects northeast to southwest through the center of the basin. Allen, Pearce, and Gardner (1978) call it the Central Basin Trend. Caswell-1, Brewster-1, and Echuca-1 were drilled on this trend. On the eastern margin of the basin, a series of normal faults, downthrown to the Permian-floored platform, form a series of local horsts and grabens. The Heywood-1 well tested a structure on this trend.

### Reservoir and Seal

The principal reservoir objectives in the basin are the Triassic and Middle Jurassic fluvio-deltaic sands sealed by the Tithonian and Neocomian shales. Permian sands encountered to date on the Browse Shelf have not had good porosity and permeability. The only other potential reservoir in the basin is the upper Tithonian-lower Neocomian sand, which occurs as a wedgeout into the basinal shales on the east flank of the basin. An updip structural seal would be required for an effective trap. The sand may be present well into the basin and serve as an important hydrocarbon migration route from the generative part of the basin to the basin edge.

### Source and Maturation

The only data available on source and maturation in the Browse basin are from a report by Robertson Research (1978). The report concludes that the Permian, the Triassic, and the prerift Jurassic have no source potential. The postrift Jurassic

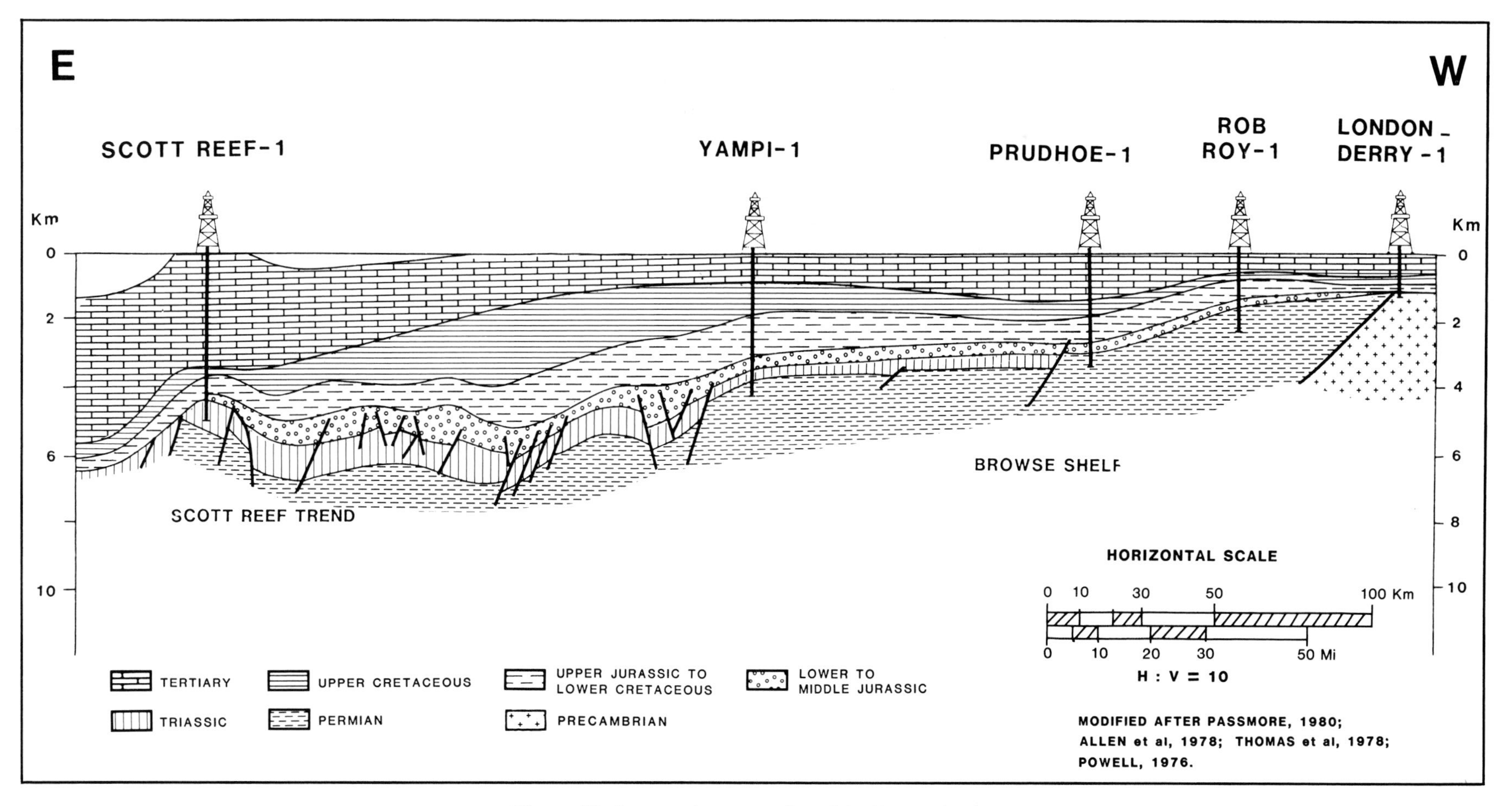


Figure 10. *Structural cross-section of the Browse basin.*

and Cretaceous sections are credited with gas source potential, but are thermally immature. These conclusions have been drawn from sparse sampling on only two wells and surely do not represent a statistically valid evaluative sampling of the basin's source potential. Oil shows in overpressured Cretaceous shales in Caswell-1 provide a more obvious demonstration that there is at least the potential for oil generation within the Browse basin.

### Field Discoveries (Figure 2A)

*Scott Reef Field*

Scott Reef Field lies 265 mi (450 km) north of Broome, the nearest port for an operations center. The discovery well was drilled in 1971, with a confirmation well in 1977. Both wells tested gas from Upper Triassic sandstones and sandy dolomites. Porosities are up to 15% and permeabilities are described as adequate. In 1982, a 12 mi (210 km) stepout well was drilled at North Scott Reef. It penetrated a much improved reservoir, raising estimated recoverable reserves for the field from 1–1.5 to 13.7 tcf of gas and 171 million bbl of condensate. While much of the field lies in very shallow water, it is separated from the mainland by depths exceeding 1,900 ft (550 m), creating engineering difficulties for a pipeline to the coast.

*Brecknock Field*

Brecknock Field lies 250 mi (400 km) north of Broome and 26 mi (42 km) southwest of Scott Reef Field, with only one well drilled on the structure. The field is estimated to have recoverable reserves of 4 tcf of gas and 40 million bbl of condensate. No other data are available on the field, but the area is on the Scott Reef Trend, and the reserves are likely to be contained in a Triassic horst similar to the Scott Reef accumulation.

## BONAPARTE BASIN

The Bonaparte basin is the northernmost basin of the Northwest Shelf (Figure 4). Petroleum geology of the basin has been discussed by Bhatia, Thomas, and Boirie (1984); Edgerly and Crist (1974); Kraus and Parker (1979); Laws and Kraus (1974); and Laws and Brown (1979). The Bonaparte basin as presently defined has a length of over 540 mi (875 km) in a northeast to southwest direction and is up to 300 mi (500 km) in width. The early discovery of gas at Petrel, and good shows of oil at Puffin, have stimulated a moderate amount of exploratory drilling in the basin. The port of Darwin on Australia's northern coast provides an operations base for the Bonaparte basin, although portions of the basin are over 500 mi (800 km) from that city. No production has been established to date in the basin.

### Stratigraphy (Figures 3, 11 and 12)

*Permian*

The onshore portion of the Petrel subbasin in Bonaparte basin contains a thick section of Cambrian to Carboniferous sediments whose geology has been reviewed by Laws (1981). Only limited information on the Lower Permian Kulshill Formation is available from onshore wells; however, a substantial number of offshore wells contribute to an understanding of the Permian section. The Kulshill Formation deposition began in Late Carboniferous time. The formation consists of a basal sandstone, a shale, and, at the top, a greywacke unit of interbedded argillaceous sand and shales. The lower sandstone member of the Kulshill is interpreted as a high-energy beach and barrier bar deposit. The sands are clean near the basin edge, becoming less porous and permeable toward the basin center. A maximum drilled thickness of 2,600 ft (800 m) of sands and minor interbedded shales has been encountered. The overlying shale member is considered to be a non-marine floodplain and lacustrine deposit. Some shale interbeds contain angular to round pebbles, probably of glacial origin. The non-marine shale member is conformably overlain by a marine section of interbedded argillaceous sands and shales with minor thin limestones.

The Kulshill Formation is conformably overlain by the Fossil Head Formation, consisting of dark carbonaceous shales and siltstones. A 100 ft (30 m) thick limestone marks the top of the Fossil Head Formation and indicates the maximum extent of marine incursion during this period.

The more regressive Hyland Bay Formation conformably overlies the Fossil Head Formation. The Hyland Bay Formation is 1,300–1,600 ft (400–500 m) thick and consists of interbedded sandstones, shales and minor coal beds and limestones. Bhatia, Thomas, and Boirie (1984) have subdivided the Hyland Bay Formation into five members. The middle member of the formation is a 1,000–1,300 ft (300–400 m) unit of interbedded sands and shales that form the reservoir section at Petrel Field. This predominantly sandstone section is separated from the upper sandstone member of the formation by another 100 ft (30 m) of limestone. The upper sandstone member hosts the gas discovered at the Tern Field. The sandstones of the Hyland Bay Formation are all considered to be deltaic in origin. Age equivalent sediments to the Hyland Bay Formation have been encountered at Osprey-1 on the Londonderry High where interbedded sands, siltstones, shales, and limestones were penetrated. Further west at Sahul Shoals-1, drillers encountered marine shales and limestones. At Troubadour-1 on the Sahul Ridge, limestones of Late Permian age were logged resting on Precambrian basement. The Hyland Bay Formation probably exceeds 3,000 ft (9,100 m) in thickness in Petrel subbasin's center.

*Prerift Mesozoic (Triassic to Middle Jurassic)*

The Mt. Goodwin Formation is a Triassic marine shale conformably overlying the Permian and marking the continuation of the marine transgression that began in Late Permian time. The Formation has been penetrated at Osprey-1, Sahul Shoals-1, and in wells within the Petrel subbasin. This massive shale and siltstone unit varies from 1,600 to 2,000 ft (500–600 m) in thickness and marks the maximum extent of Early Triassic transgression. The regressive phase that follows the Mt. Goodwin began with deposition of marine sands interbedded with shales. Continued withdrawal of the sea in the Petrel subbasin resulted in deposition of fluvio-deltaic to massive deltaic sands into Late Triassic and Early Jurassic time. A variegated shale, silt, and sandstone facies occurs in the Upper Triassic to Lower Jurassic deposits in most of this area, and as far west as the Eider-1 well. To the west of the Londonderry High, the post-Mt. Goodwin Triassic is more marine than in the Petrel subbasin, consisting largely of limestones, shales, and siltstone with only minor sandstones. Laws and Kraus (1974) indicate

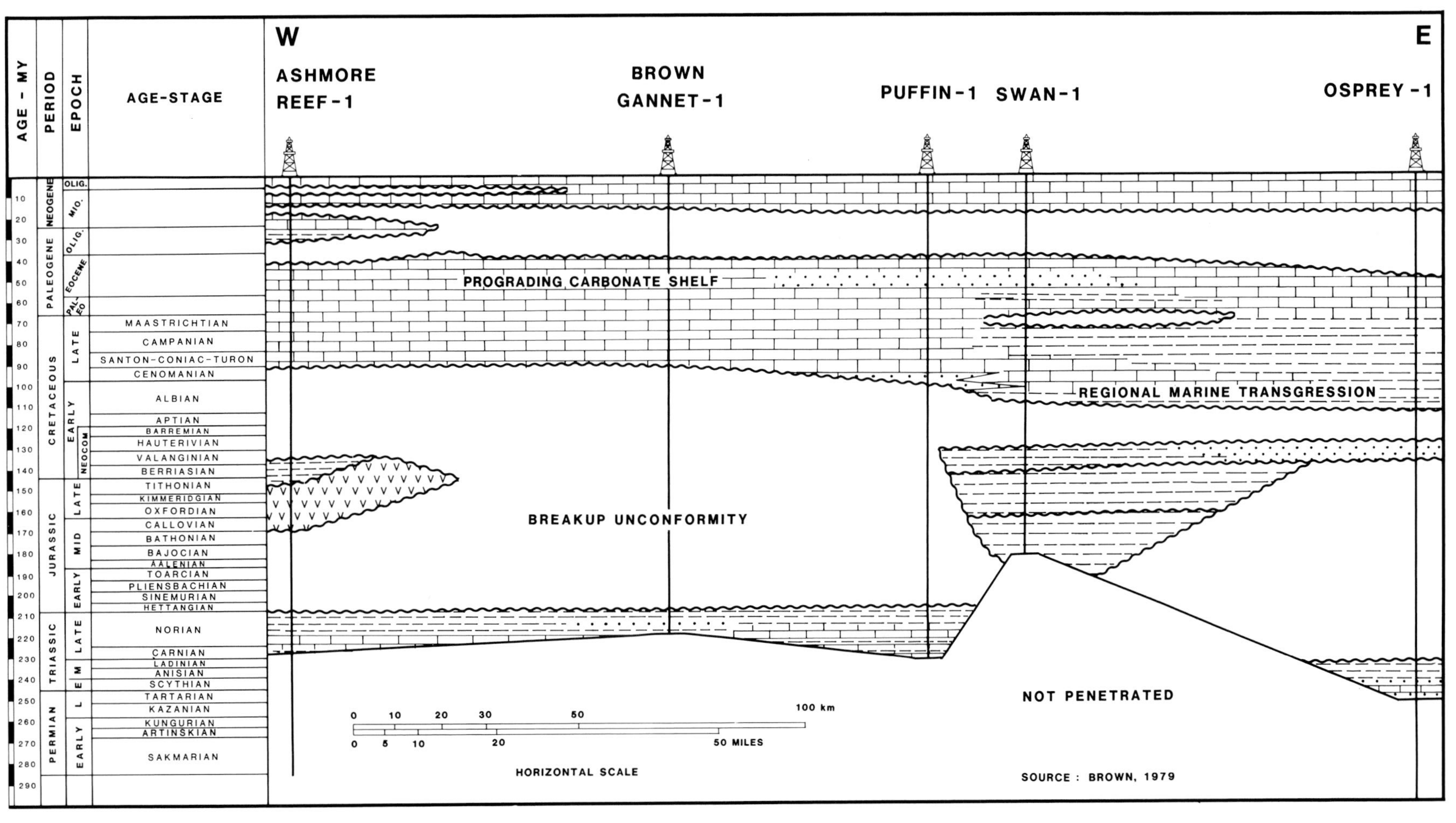


Figure 11. *Time-stratigraphic section of the Vulcan graben, Bonaparte basin.*

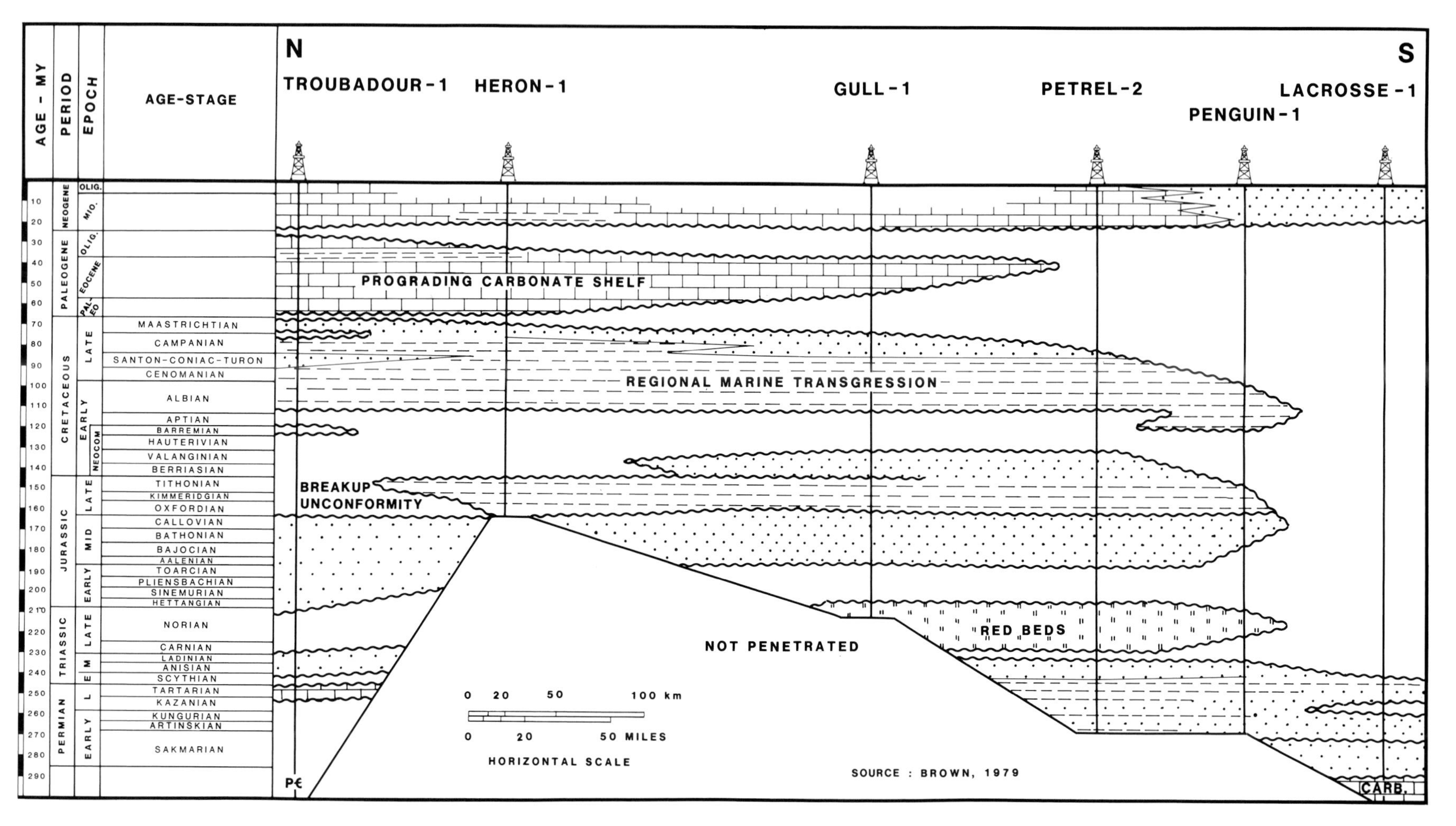


Figure 12. *Time-stratigraphic section of the Malita graben, Bonaparte basin.*

that the sandstone percentage within the Triassic increases to the northwest across the Ashmore Block, suggesting a sand source in that direction.

Early Jurassic sedimentation has not been documented in the Bonaparte basin outside the Petrel subbasin. The age equivalent of the red bed sequence encountered there either was not deposited or was eroded prior to postrift sediment deposition in other parts of the Bonaparte basin.

*Postrift Mesozoic (Late Jurassic and Cretaceous)*

Erosion in the postrift period removed a substantial section from the Ashmore Block, the Sahul Ridge, and the Londonderry High. The rifting established the Vulcan and Malita grabens as low areas, while volcanic tuffs were laid down in the vicinity of Ashmore Reef-1.

The Callovian to late Neocomian section in the Bonaparte basin is treated as a unit called the Petrel Formation. In the Petrel-2 and Gull-1 wells, there is no apparent, appreciable depositional gap in the Petrel Formation during the main Middle Jurassic breakup period, although paleontologic dating is poor in the interval. The Petrel Formation thickens northward from the southern margin of the Petrel subbasin into the Malita graben and then thins onto the Shaul Ridge. Toward the west, the formation thins over the Londonberry High, thickens into the Vulcan graben, and then thins rapidly onto the Ashmore Block, where it is absent either through nondeposition or erosion after the Aptian uplift. In the Malita graben, where the formation is thickest, (over 4,000 ft or 1,200 m), it is divided into three units. The basal unit is a fine- to coarse-grained and well-sorted sandstone, becoming more argillaceous in the center of the Malita graben. Some interbedded sandstones occur in the base of this section on the western edge of the Vulcan graben, and the unit thins to only 180 ft (55 m) on the Ashmore Block.

In the Petrel subbasin and Malita graben, the overlying Turonian to Maastrichtian section consists of interbedded shales, siltstones and sands to the south, grading to more calcareous shales in the north. The section is mainly shale over the Londonderry High. The Vulcan graben contains marls of this age, while argillaceous limestones predominate on the Ashmore Block. Between the Ashmore Block and the Vulcan graben, there is some increase in sand content.

*Tertiary*

There is no clear break in sedimentation between the Tertiary and the underlying Cretaceous, although in some areas a full age sequence cannot be established definitely. Tertiary deposition was almost entirely calcareous with only minor shales, marls, or sandstones. The section reaches 8,000 ft (2,400 m) in thickness in the Malita and, in the Vulcan graben, thins to the north over the Sahul Ridge, thickening again toward the Timor Trough. Epeiric movements from the Paleocene to Oligocene affected the nature of carbonate deposition by modifying water depths. Uplift and erosion occurred in the early Miocene; a renewed transgression took place basinwide in the late Miocene and carbonate deposition recommenced and has continued to the present.

### Tectonics and Structure (Figures 4, 13 and 14)

Initial rifting north of the Kimberley Block may have taken place as early as the Silurian, resulting in evaporite deposition. The Petrel subbasin trends northwest to southeast, parallel to the Fitzroy Trough in the Canning basin, suggesting synchronous development of the two features. The Petrel subbasin's bounding faults rejuvenated in Late Carboniferous to Permian time, allowing the thick sequence of the Kulshill, Fossil Head, and Hyland Bay Formations to be deposited. Structures that developed during this tectonic episode are generally associated with the bounding faults and consist of horsts and rotated fault blocks.

In addition to fault-induced features, the Petrel subbasin contains 24 identified salt piercement and deep-seated salt structures (Edgerley and Crist, 1974), and turtle structures formed by salt removal are also present. The salt movement probably began as early as Late Carboniferous time. Some piercements reach the Cretaceous and consequently must have had late movement. Tern Field may be located on a salt pillow structure, while Petrel Field is a turtle structure.

Following the Paleozoic movements, there is little indication of significant tectonism until Late Triassic-Early Jurassic time, when the northeast to southwest trending Malita graben was superimposed on the north end of the preexisting Petrel subbasin. The movement that opened the Malita graben displaced the Sahul Ridge to the north, leaving it as a high rim north of the graben. The Londonderry High was also uplifted at this time and became subject to erosion. The Londonderry High is not fault bounded on its east flank, and the Malita graben terminates into the flank of the high. The Ashmore Block separated from the Londonderry High is not fault bounded on its east flank, and the Malita graben terminates into the flank of the high. The Ashmore Block separated from the Londonderry High during this movement and created the Vulcan graben, which is oriented northeast to southwest, as are the southern parts of the Northwest Shelf. The Ashmore Block was also raised during the rifting and was subjected to erosion. This period of rifting developed the Bonaparte basin's subdivisions as they currently exist. Normal faulting predominated, with the formation of horst and graben structures along the graben edges.

Following this major, Late Triassic to Early Jurassic structural movement, epeiric movements during Aptian, Maastrichtian, and Oligocene time affected sedimentation, but they do not appear to have been responsible for the formation of major structural features. These movements rejuvenated existing faults and fault-controlled structures.

### Reservoir and Seal

Laws and Brown (1976) credit portions of the Permian Kulshill and Hyland Bay Formations with fair to good reservoir potential. Upper Triassic units have good reservoir potential, as do the Upper Jurassic and Lower Cretaceous sections; all potential reservoirs are sands.

Local shale seals exist within the Kulshill Formation with the oldest basinwide seal being the Lower Triassic Mt. Goodwin Formation. Locally, within the Jurassic Petrel Group, there are good shales that could seal underlying sands. The youngest basin-wide seal is provided by the lower Albian shale.

### Source and Maturation

Source rocks of the Bonaparte basin has been evaluated by Kraus and Parker (1979), Laws and Brown (1976), and Robertson Research (1978). Neither Kraus and Parker nor Laws and Brown provide detailed data upon which their conclusions

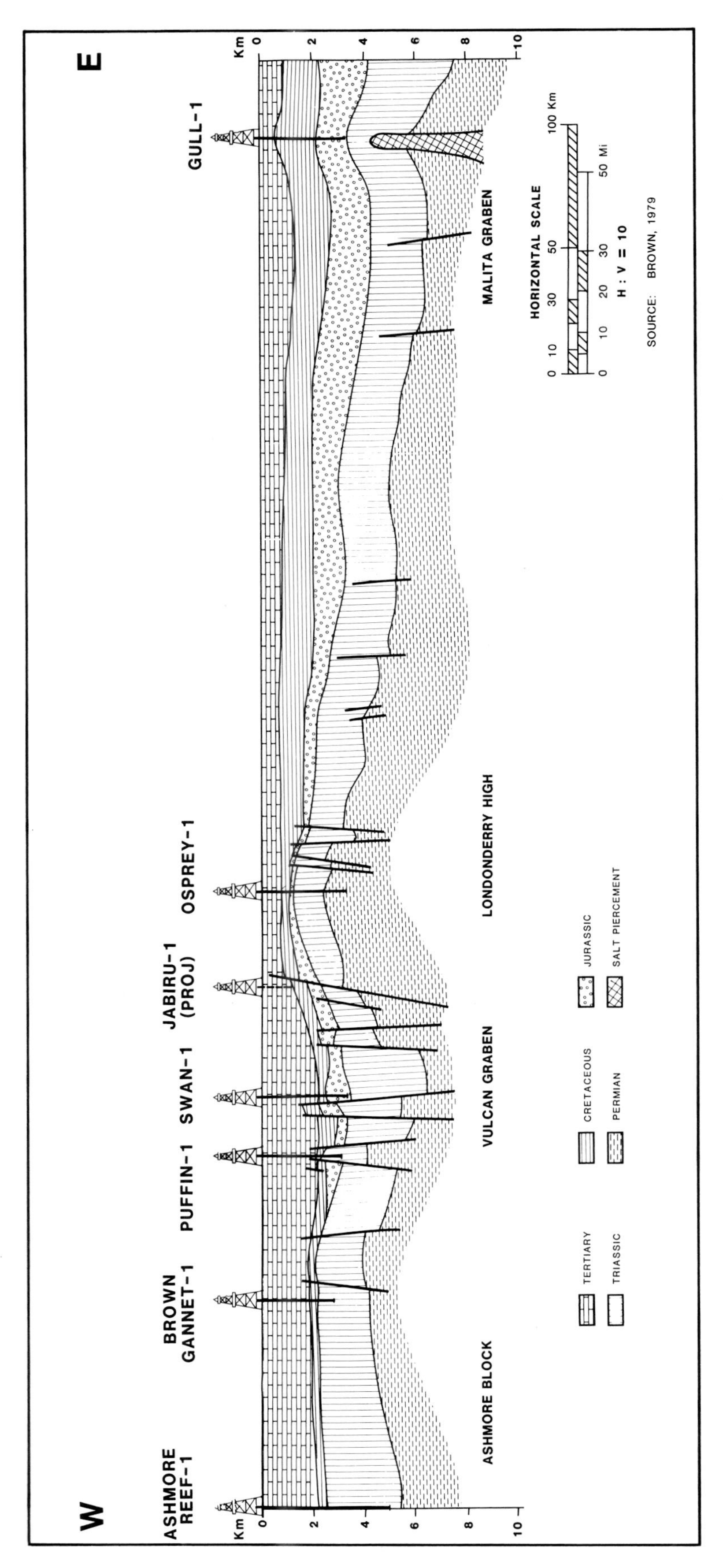


Figure 13. *Structural cross section of the Vulcan graben, Bonaparte basin.*

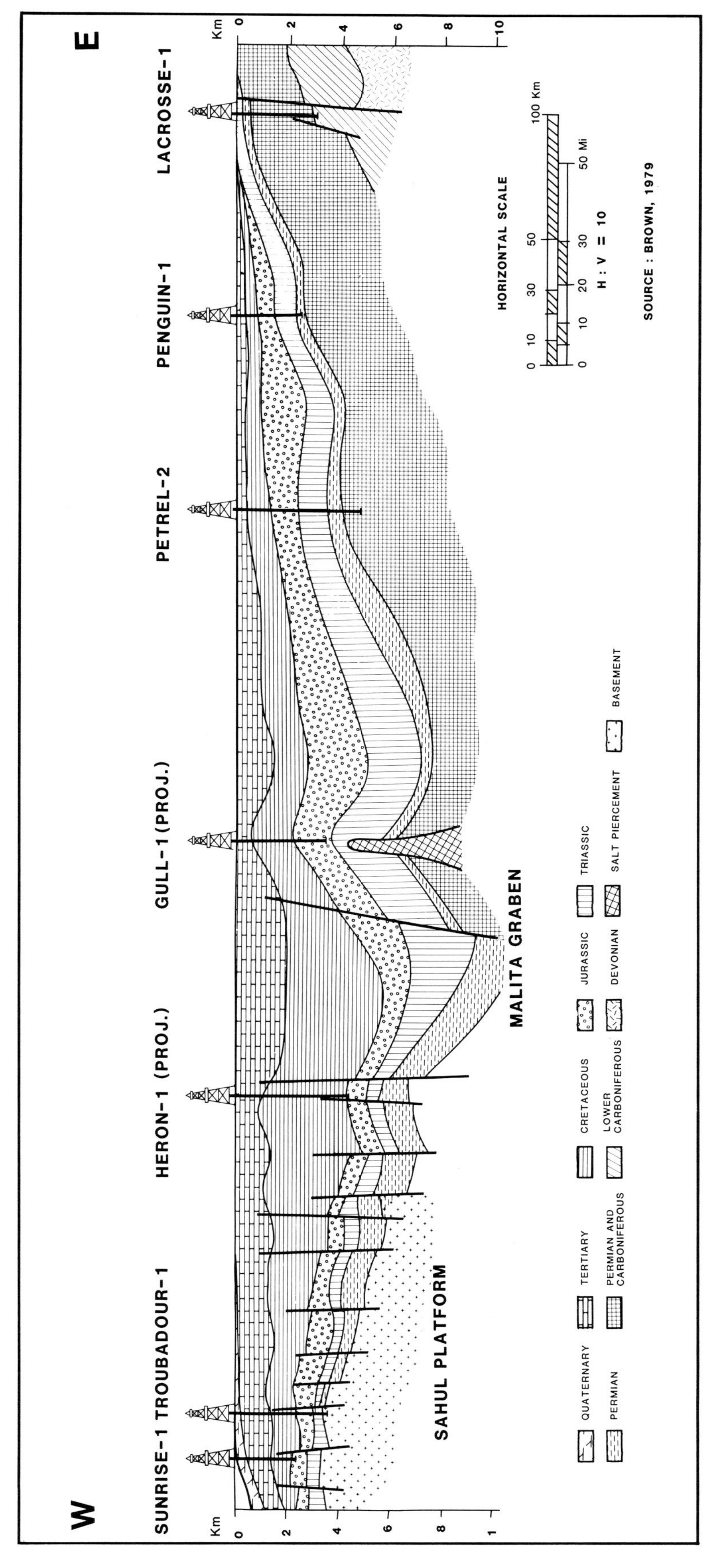


Figure 14. *Structural cross section of the Malita graben, Bonaparte basin.*

are based. Robertson Research data are presented in their report, but sampling is so scattered and so widely spaced, even within individual wells, that their conclusions must be considered statistically suspect. Of the three papers noted above, Kraus and Parker are consistently more optimistic than the other authors.

Generally, the Permian section contains a vitrinite-rich source. Kraus and Parker (1979) show that organic carbon content increases toward the depocenter, while the section is mature to postmature in the center of the basin. The Hyland Bay Formation would, therefore, appear to have the best source potential.

The Triassic section, including the Mt. Goodwin Formation shales and the overlying fluvio-deltaic section, has low organic carbon levels. The interval contains vitrinitic and inertinitic organic material and probably does not have source potential. Sedimentation in the Lower Jurassic was similar to that in the Triassic and, like that section, is not considered to have source potential.

The Upper Jurassic to Neocomian shales are locally organically rich and appear to have good source potential. Robertson Research (1978) interprets the unit to be gas prone in most areas. The top of the unit is thermally immature to marginally mature in most parts of the basin; however, within the Vulcan and Malita grabens, the unit appears to have reached the oil-generative window.

The overlying Upper Cretaceous sediments have marginal source potential and are immature except in the Heron-1 well, where the geothermal gradient is very high and the oil maturity window is entered above 5,000 ft (1,500 m).

In summary, then, source rock quality varies widely among the several argillaceous formations in the basin and also appears to vary areally within a single formation. Source potential seems to be biased toward gas production, but the basinal areas, which would contain the highest probability for a restricted depositional environment and thus oil-prone source rock, have not been sampled. No attempts to correlate the condensates from Tern and Petrel with a potential source rock have been published. The Jabiru discovery in the Vulcan graben is probably sourced from the thick, Upper Jurassic to Neocomian shales present in the graben.

### Field Discoveries (Figure 2A)

#### *Petrel Field*

The Petrel Field was discovered in 1969 when the Petrel-1 well blew out, and a relief well was drilled to kill the blowout. Two subsequent wells have demonstrated the potential for a major accumulation, although no official reserve estimates are available. The field is productive from Permian Hyland Bay Formation sandstones. The structure is most likely a turtle structure caused by differential salt withdrawal.

#### *Tern Field*

Tern Field was discovered in 1971. The field was confirmed with one well in 1982, but a third well was dry. The field is a deep-seated, salt-cored anticline, productive from the Hyland Bay Formation. No reserve figures are available.

#### *Jabiru Field*

Jabiru Field was discovered in 1983. The discovery well tested oil at rates of 7,000 bbl of oil per day, from 167-ft (51-m) oil sand. Three additional wells drilled during 1984 failed to confirm similar thicknesses of pay section, and the structure and reservoir conditions are apparently more complex than originally interpreted. The structure is reported to be a horst, containing prerift sands under an Upper Jurassic to Neocomian shale seal along the east margin of the Vulcan graben.

#### *Sahul Ridge Discoveries*

Small gas-condensate discoveries have been made at Sunrise-1 and Troubadour-1 on the Sahul Ridge. Neither of these discoveries has been confirmed, and reserves, thought to be Upper to Middle Jurassic sands, are probably small.

## CONCLUSIONS

The four basins constituting the Northwest Shelf of Australia are still in the early stages of exploration. Significant gas/condensate reserves have been established in three of the four, and significant oil reserves have been discovered in the Carnarvon basin, with a strong indication that commercial oil reserves may soon be established in the Vulcan graben of the Bonaparte basin.

Currently, almost all the basinal areas of the Northwest Shelf have been covered by at least a reconnaissance seismic grid. Well spacing is such that, except for field areas, well correlations still depend wholly on paleontology rather than on any electrical character. Well control is so sparse that stratigraphy can only be done on a broad, regional basis. None of the details required to search for potential stratigraphic traps are present. The wells, even in the most heavily drilled areas, are inadequate for accurate seismic stratigraphy interpretations.

Potential structural traps are mainly associated with tensional faulting. Salt movement in the Petrel subbasin of the Bonaparte basin, and late strike-slip movement in the Exmouth and Barrow-Dampier subbasins of the Carnarvon basin, contributed to the complexity required for structural trapping of petroleum.

Good reservoir sands and seals are provided by units of a number of different ages. Initial source rock appraisal suggests that the shelf is gas prone, but continued exploration is finding more oil. The amount of source rock data publicly available is inadequate to provide a statistically valid evaluation of the area's potential.

Current drilling has established the presence of significant volumes of petroleum, firmly proving that all elements of structure—reservoir, seal, and source—are present. The past and present levels of exploration activity on the shelf suggest it will be a long time before geologists determine the area's full exploration potential.

## REFERENCES CITED

Alexander, R., R. I. Kagi, and G. W. Woodhouse, 1980, Origin of the Windalia oil, Barrow Island, Western Australia: Australian Petroleum Exploration Association Journal, v. 20, p. 250–256.

———, ———, ———, and J. K. Volkman, 1983, The geochemistry of some biodegraded Australian oils: Australian Petroleum Exploration Association Journal, v. 21, p. 53–63.

Allen, G. A., L. G. G. Pearce, and W. E. Gardner, 1978, A regional interpretation of the Browse basin: Australian Petroleum Exploration Association Journal, v. 18, p. 23–33.

Barber, P. M., 1982, Paleotectonic evolution and hydrocarbon genesis of the central Exmouth Plateau: Australian Petroleum Exploration Association Journal, v. 22, p. 131–144.

Bhatia, M. R., M. Thomas, and J. M. Boirie, 1984, Depositional framework and diagenesis of the Late Permian gas reservoirs of the Bonaparte basin: Australian Petroleum Exploration Association Journal, v. 24, p. 299–313.

Brown, C. M., 1979, Bonaparte Gulf basin explanatory notes and stratigraphic correlations: Bureau of Mineral Resources Geology and Geophysics, Record Report, 1979/52, 19 p.

Campbell, I. R., 1982, A study of the Windalia reservoir, Barrow Island oilfield, Western Australia: Paper presented at UN Conference on Oilfield Development Techniques, Daquing, China.

———, and D. N. Smith, 1982, Gorgon 1—Southernmost Rankin Platform gas discovery: Australian Petroleum Exploration Association Journal, v. 22, p. 102–111.

———, A. M. Tait, and R. F. Reiser, 1984, Barrow Island oilfield revisited: Australian Petroleum Exploration Association Journal, v. 24, p. 289–298

Cook, A. C., 1978, Report on the organic petrology of the wells Anchor-1, Barrow-1, Barrow Deep-1, North Tryal Rocks-1, Onslkow-1, Rankin-1 and Tryal Rocks-1: Western Australian Department of Mines, Open File Report S1468(A15).

———, and A. J. Kantsler, 1980, The maturation history of the epicontinental basins of Western Australia: UN ESCAP, CCOP/SOPAC Technical Bulletin 3, p. 171–195.

Crostella, A., 1976, Browse basin, *in* R. B. Leslie, H. J. Evans, and C. L. Knight, eds., Economic geology of Australia and Papua-New Guinea, 3 Petroleum, Australasian Institute of Mining and Metallurgy, Parksville, Victoryia, Australia, 541 p.

———, and M. A. Chaney, 1978, The petroleum geology of the outer Dampier Sub-basin: Australian Petroleum Exploration Association Journal, v. 18, p. 13–22.

———, and T. Barter, 1980, Triassic-Jurassic depositional history of the Dampier and Deagle sub-basins, Northwest Shelf of Australia: Australian Petroleum Exploration Association Journal, v. 20, p. 25–33.

Edgerley, D. W., and R. P. Crist, 1974, Salt and diapiric anomalies in the southeast Bonapart Gulf basin: Australian Petroleum Exploration Association Journal, v. 14, p. 84–94.

Forrest, J. T., and E. L. Horstman, 1984, Australia's Northwest Shelf booms: AAPG Explorer, April 1984, p. 22–24.

Geological Society of Western Australia, 1974, Geology of Western Australia: Western Australia Geological Survey, Memoir 2, 541 p.

Halse, J. W., 1976, Beagle Sub-basin, *in* R. B. Leslie, H. J. Evans, and C. L. Knight, eds., Economic geology of Australia and Papua-New Guinea, 3 Petroleum, Australasian Institute of Mining and Metallurgy, Parkville, Victoria, Australia, 541 p.

———, and J. D. Haynes, 1971, The geological and structural framework of the offshore Kimberley Block (Browse basin) area, Western Australia: Australian Petroleum Exploration Association Journal, v. 11, p. 64–70.

Harris, E. J., 1981, The production geology of the North Rankin gas field, *in* Michel T. Halbouty, ed., Energy resources of the Pacific Region, AAPG Studies in Geology, n. 12, p. 273–283.

Johnstone, M. H., 1979, A case history of Rouge Range: Australian Petroleum Exploration Association Journal, v. 19, p. 1–6.

Kraus, G. P., and K. A. Parker, 1979, Geochemical evaluation of petroleum source rock in Bonaparte Gulf-Timor Sea region, northwestern Australia: AAPG Bulletin, v. 63, p. 2021–2041.

Laws, R., 1981, The petroleum geology of the onshore Bonaparte basin: Australian Petroleum Exploration Association Journal, v. 21, p. 5–15.

———, and G. P. Kraus, 1974, The regional geology of the Bonaparte Gulf-Timor Sea area: Australian Petroleum Exploration Association Journal, v. 14, p. 77–84.

———, and R. S. Brown, 1976, Bonaparte Gulf basin—southeastern part, *in* R. B. Leslie, H. J. Evans, and C. L. Knight, eds. Economic geology of Australia and Papua-New Guinea, 3 Petroleum, Australasian Institute of Mining and Metallurgy, Parkville, Victoria, Australia, 541 p.

Meath, J. R., and K. J. Bird, 1976, The geology of the West Tryal Rocks gas field: Australian Petroleum Exploration Association Journal, v. 16, p. 157–163.

Moore, P. S., R. M. Hocking, and P. D. Denman, 1980a, Modified stratigraphic nomenclature and concepts in the Palezoic sequence of the Carnarvon basin, W. A.: Western Australia Geological Survey Annual Report, 1979, p. 51–55.

———, ———, and ———, 1980b, Sedimentology of the Byro Group (Lower Permian), Carnarvon basin, Western Australia: Western Australia Geological Survey Annual Report, p. 55–64.

———, ———, and ———, 1980c, Sedimentology of the Kennedy Group (Permian), Carnarvon basin, Western Australia: Western Australia Geological Survey Annual Report, 1979, p. 65–71.

Passmore, V. L., 1980, Browse basin region explanatory notes and stratigraphic columns: Bureau of Mineral Resources, Geology and Geophysics, Record 80/42, 41 p.

Powell, D. E., 1976a, The geological evolution of the continental margin off northwest Australia: Australian Petroleum Exploration Association Journal, v. 16, p. 13–24.

———, 1976b, Dampier Sub-basin, Carnarvon basin, *in* R. B. Leslie, H. J. Evans, and C. L. Knight, eds., Economic Geology of Australia and Papua-New Guinea, 3 Petroleum, Australasian Institute of Mining and Metallurgy, Parkville, Victoria, Australia, 541 p.

Robertson Research (Australia) Pty. Ltd., 1978, Northwest continental shelf of Australia regional geologic synthesis: Western Australia Department of Mines, Open File Report S1468.

Stagg, H. M. J., 1978, The geology and evolution of the Scott Plateau: Australian Petroleum Exploration Association Journal, v. 18, p. 34–43.

Thomas, B. M., 1974, A summary of the petroleum geology of the Carnarvon basin: Australian Petroleum Exploration Association Journal, v. 14, p. 66–76.

———, 1979, Geochemical analysis of hydrocarbon occurrences in northern Perth basin, Australia: AAPG Bulletin, v. 63, p. 1092–1107.

———, and D. N. Smith, 1976, Carnarvon basin, *in* R. B. Leslie, H. J. Evans, and C. L. Knight, eds., Economic Geology of Australia and Papua-New Guinea, 3 Petroleum, Australasian Institute of Mining and Metallurgy, Parkville, Victoria, Australia, 541 p.

Thomas, J. S., E. O. Nestvold, and A. Crostella, 1981, Exploration of part of the Northwest Shelf of Australia (A case history): Offshore South East Asia Conference, Southeast Asia Petroleum Exploration Society, February 1978.

Warris, G. J., 1976, Canning basin, Offshore, *in* R. B. Leslie, H. J. Evans, and C. L. Knight, eds., Economic Geology of Australia and Papua-New Guinea, 3 Petroleum, Australasian Institute of Mining and Metallurgy, Parkville, Victoria, Australia, 541 p.

Yeates, A. N., D. L. Gibson, R. R. Towner, and R. W. A. Crowe, 1984 Regional geology of the Onshore Canning basin *in* P. G. Purcell, ed., The Canning basin, W. A.: Perth, Australia, Proceedings of the Geological Society of Australia/Petroleum Exploration Society of Australia Symposium, 1984.

# Geology and Hydrocarbon Potential of the Arafura Sea

John A. Katili
*Ministry of Mines and Energy*
*Jakarta, Indonesia*

The Arafura Sea is a continental shelf sea situated between Irian Jaya (western New Guinea) and the northern part of the Australian continent. To the south, it adjoins the stable Australian craton, and to the north, it is bordered by the Tertiary collision zone between the Australian craton and the northern Irian Jaya island arc. To the west and northwest, it is bounded by the active Banda arc collision zone, whereas to the east, it is bordered by the northern extension of the Gulf of Carpentaria that also forms the western limit of the zone of late Paleozoic granites. Shelf sediments predominate in the Arafura Sea continental shelf, ranging in age from late Paleozoic, Mesozoic to Cenozoic and underlain by granitic basement. Two tectonic styles of deformation are recognizable in the area, namely a block-faulted downwarping within stratified shelf and slope sediments of the Arafura Sea, and overthrusting of chaotic sediments from the Banda arc toward the Australian continent, in which the intensity of deformation increases from south to north. In the Malita–Calder graben and its northeastern extension, and in the Sahul Ridge, gas shows have been reported from Jurassic-to-Cretaceous fine-grained marine limestone and sandstones, while gas and condensate occur in Cretaceous sediments and Middle Jurassic, fine-grained sands.

## INTRODUCTION

Not much is known of the geology and geophysics of the Arafura Sea. The onshore geology surrounding the Arafura Sea has been described by previous authors such as Brouwer (1923); Bursch (1947); van Bemmelen (1949); Fairbridge (1951); Klompé (1957); Audley-Charles (1976); Hamilton (1974); Barber and Audley-Charles (1975); Hamilton (1979); and Jezek and Hutchison (1978).

The offshore geology along the convergent margin of the Australian–Eurasian plate has been discussed by Jongsma (1974); Curray et al (1977); Jacobson et al (1979); Bowin et al (1980); and Katili and Hartono (1983).

The geology of the region in relation to hydrocarbon prospecting along the northern Australian margin has been treated in a number of publications by Nichols (1970); Balke and Burt (1976); Balke et al (1973); Gribi (1974); Lofting, Crostella, and Halse (1975); Montecchi (1976); and Brown (1980). An unpublished report regarding the Arafura Sea Contract Area, West Irian, has been made available to the author by PERTAMINA, The Indonesian State Oil and Gas Enterprise.

Between 1972 and 1981, the bulk of the subsurface information available on the Arafura Platform area was provided by exploration as well as four early wells drilled by NNGPM between 1956 and 1959, namely the Kembelangan I, Merauke I, Aripu I, and Jaosakor I.

Since 1970, geologists have conducted seismic surveys over the entire offshore area, plus selected parts of the onshore, that provide data to construct the structural setting of the region (Figure 1).

In the southeastern part of the Arafura Sea, Champlin Indonesia Incorporated drilled three wells in 1973 and 1974, namely E-1X, E-2X, and E-3X. Phillips Petroleum Company, Indonesia, drilled ASA-1X and ASB-1X in 1971 and ASH-1X in 1974, in the offshore area north of the Aru Islands (Figure 2). In 1981, the Federal Institute of Geosciences and Natural Resources, Hannover, carried out a large, geophysical survey in the Arafura Sea.

This paper, then, compiles the geology and geophysics of the Arafura Sea based on data obtained by scientific institutions such as the Federal Institute of Geosciences and Natural Resources from West Germany. The data are substantiated by other published and unpublished materials that have been made available to the Jakarta Ministry of Mines and Energy through the courtesy of the oil industry.

Finally, the study singles out certain areas in the Arafura Sea that explorers could investigate in more detail for hydrocarbon potential.

## A REGIONAL VIEW OF THE ARAFURA SEA

The Arafura Sea is a continental shelf area situated between Irian Jaya (western New Guinea) and the northern part of the Australian continent.

To the north, it is bordered by the Tertiary collision zone between the Australian craton and the northern Irian Jaya island arc, while to the south, it adjoins the stable Australian craton. To the east, the northern extension of the Gulf of Carpentaria borders it, while to the west lies the active Banda arc collision zone (Figure 3). Australian geologists call the southwestern extension of the Arafura Shelf the "Sahul Shelf."

The bathymetry of the Arafura Shelf exhibits depths of between 50 and 80 m (160 and 260 ft), but deeper parts down to more than 600 m (1,970 ft) occur at the edges. The east-southeast trending Merauke Ridge across the Aru Islands separates the Arafura depression from a narrower "trough" in the north.

We can find in the literature an exhaustive discussion and debate on the geology of the complex resulting from the collision of the Banda Island arc system and the Australian continental crust. The debate involves the position of the subduction zone, which has far-reaching implications for interpreting the Tanimbar and Aru Troughs and the origin of the stratigraphic units occurring in the Banda, non-volcanic outer arc.

Based on recent data, such as seismic reflection, seismic refraction, gravity, and the onshore geology of Timor, the locus of subduction occurs south of Timor at the trough. Timor is underlain by Australian crust with a structural break separating the Australian shelf sediments from the accretionary wedge and structurally complicated rocks now occurring in the outer Banda arc. Audley-Charles, Carter, and Barber (1975) and Chamalaun and Grady (1978) believe that the position of subduction is between the outer and inner Banda arcs (between Timor and Wetar). Hamilton (1979) believes that the position of subduction is at the trench, outside the outer arc. Although seismic refraction and gravity data indicate that continental crust is present beneath the Timor Trough, seismic reflection data crossing the trough show that the Australian continental shelf dips under the accretionary wedge of the opposing side of the trough. As has been mentioned, the position of subduction has an implication for interpreting the geology of the non-volcanic, outer arc ridge. The position at the trough implies that the outer arc ridge is a subduction mélange, an idea held by Hamilton (1979), Katili (1974, 1975) and others. Subduction at the inter-arc position (between outer and inner arc) implies the presence of Australian material at the outer arc ridge, suggesting further interpretation on the origin of the stratigraphic units as an autochthonous, parautochthonous, and allochthonous series. Audley-Charles, Carter, and Barber (1975) believe that the outer arc ridge is essentially a zone of Pliocene collision and the site of overthrust sheets of Asian material onto Australian basement. Chamalaun and Grady (1978) believe that the overthrusting as advocated by the previous group does not occur in Timor. Their structural interpretation is that normal faulting has occurred and that whole stratigraphic units have been deposited at the Australian continental shelf.

Based on the results of seismic refraction, showing the outer-arc ridge to be underlain by Australian continental crust; and of seismic reflection that shows an observed, down-going slab under the accretionary wedge at the trough, Jacobson et al (1979) argue that subduction is at the trough and that the trough

**Figure 1.** *Marine geophysical surveys and production sharing contract areas in the Arafura Sea region (PERTAMINA and Schlüter, 1983).*

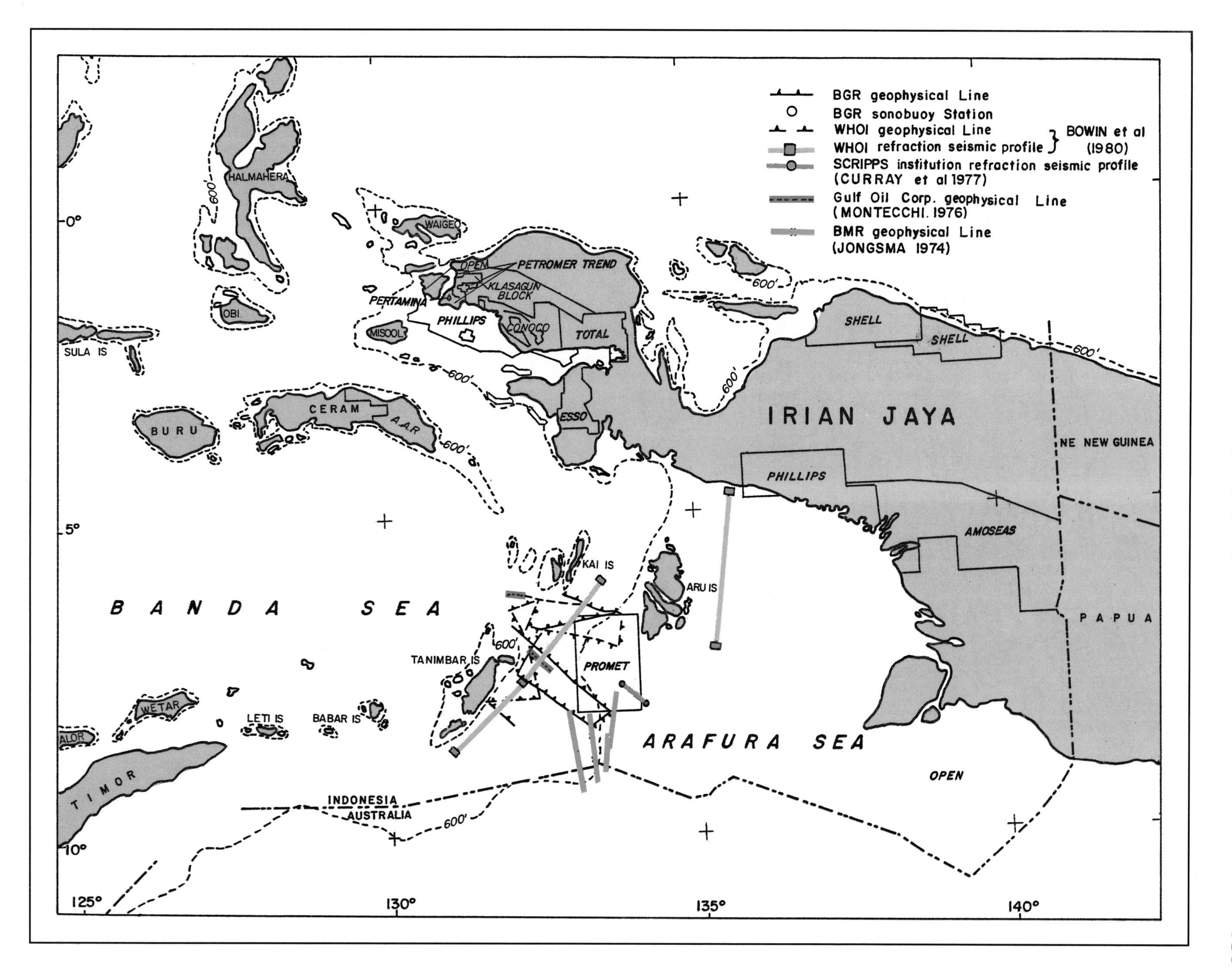
BGR geophysical Line
BGR sonobuoy Station
WHOI geophysical Line
WHOI refraction seismic profile
BOWIN et al (1980)
SCRIPPS institution refraction seismic profile (CURRAY et al 1977)
Gulf Oil Corp. geophysical Line (MONTECCHI. 1976)
BMR geophysical Line (JONGSMA 1974)
IRIAN JAYA
NE NEW GUINEA
PAPUA
SHELL
SHELL
AMOSEAS
PHILLIPS
OPEN
ARAFURA SEA
BANDA SEA
OPEN
PETROMER TREND
KLASAGUN BLOCK
CONOCO
TOTAL
ESSO
PERTAMINA
PHILLIPS
PROMET
WAIGEO
MISOOL
CERAM
A.A.R.
HALMAHERA
OBI
BURU
SULA IS
ARU IS
KAI IS
TANIMBAR IS
BABAR IS
LETI IS
WETAR
ALOR
TIMOR
INDONESIA
AUSTRALIA
600'
0°
5°
10°
125°
130°
135°
140°

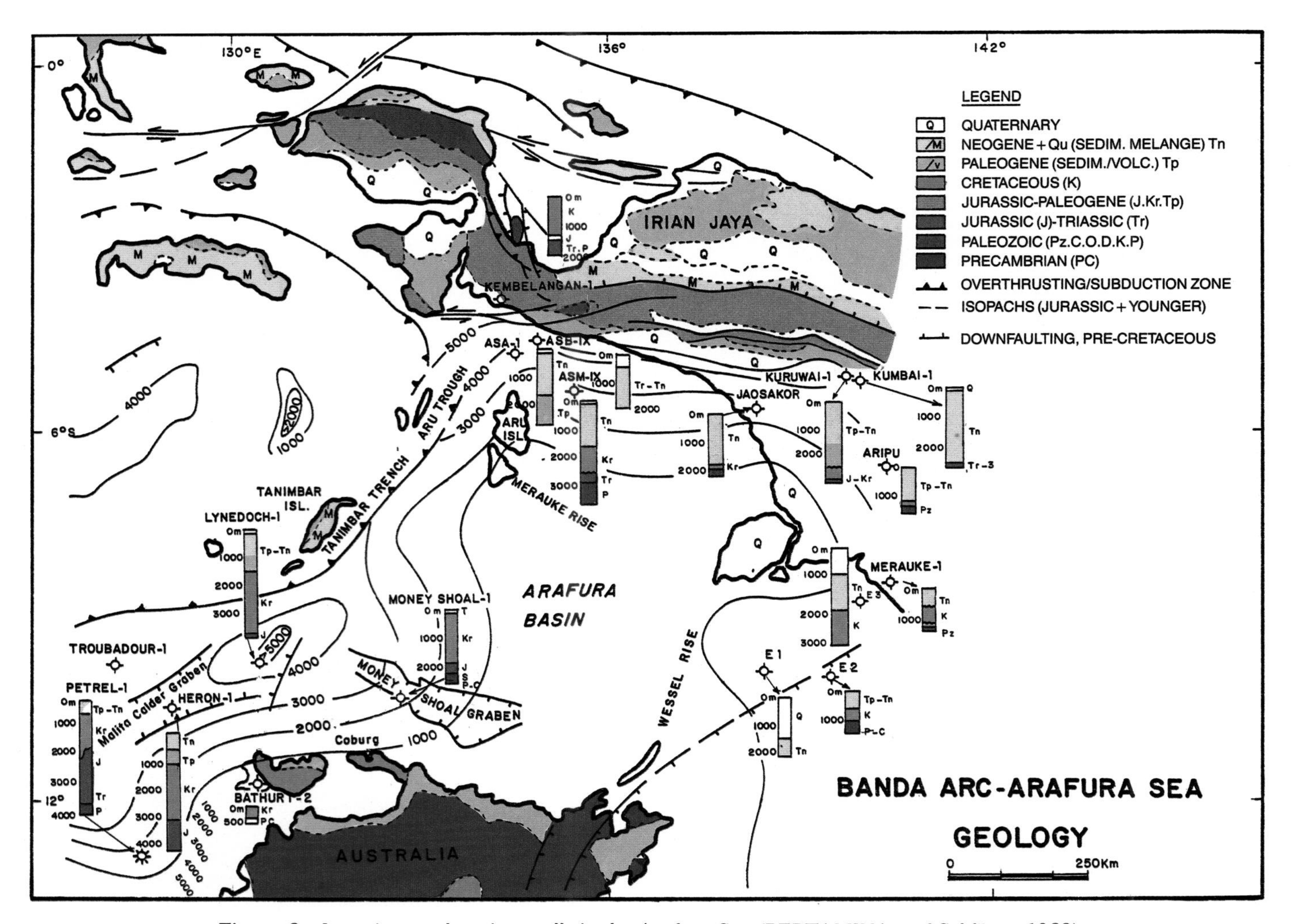


Figure 2. *Location exploration wells in the Arafura Sea (PERTAMINA and Schlüter, 1983).*

lineation represents the surface trace of the subduction zone. Later, in their interpretation of the structure of Seram, Audley-Charles, Carter, and Barber (1979) reconsider the problem of the subduction position and give two zones: A-zone at the trench and B-zone at the inter-arc position.

### Tanimbar Trench and Aru Trough

The Arafura basin is bounded on the west by two prominent features, namely the Tanimbar Trench and the Aru Trough.

Previous investigators favor the idea that these two troughs represent the eastward continuation of the Sunda-Timor subduction zone that was bent by the westward-advancing Pacific plate. The Benioff zone dips toward the Asian continent, and, consequently, establishes a chaotic wedge of imbricated sediments and crystalline basement on the northern side (Katili, 1975; Hamilton, 1979).

Cardwell and Isacks (1978) challenge this so-called one-plate model, distinguishing two separate Benioff zones. One zone extends northward beneath Timor to a depth of 600 km (370 mi) and the other, separated by the Tarera–Aiduna transform fault, extends southward beneath Ceram to a depth of about 200 km (125 m). This two-plate model could perhaps accommodate the loop shape arc without challenging the concept of rigid plate injection. However, the two-plate model for the Banda Sea is difficult to reconcile with the results of field investigations by Audley-Charles et al (1979), because their findings indicate that the stratigraphy and structure of the Ceram show remarkable resemblance to the Timor. They argue that whatever hypothesis describes the tectonic evolution of Timor would be equally applicable to Ceram, implying a preference for a one-plate model for this region.

Schlüter (1983), on the other hand, does not separate the Ceram and Tanimbar Trench along the Tarera–Aiduna fault, but instead contends that the Tanimbar Trench terminates sud-

Figure 3. *Tectonic framework of the Banda arc and Arafura Sea. (Schlüter, 1983; Circum Pacific Map and Tjia, 1977).*

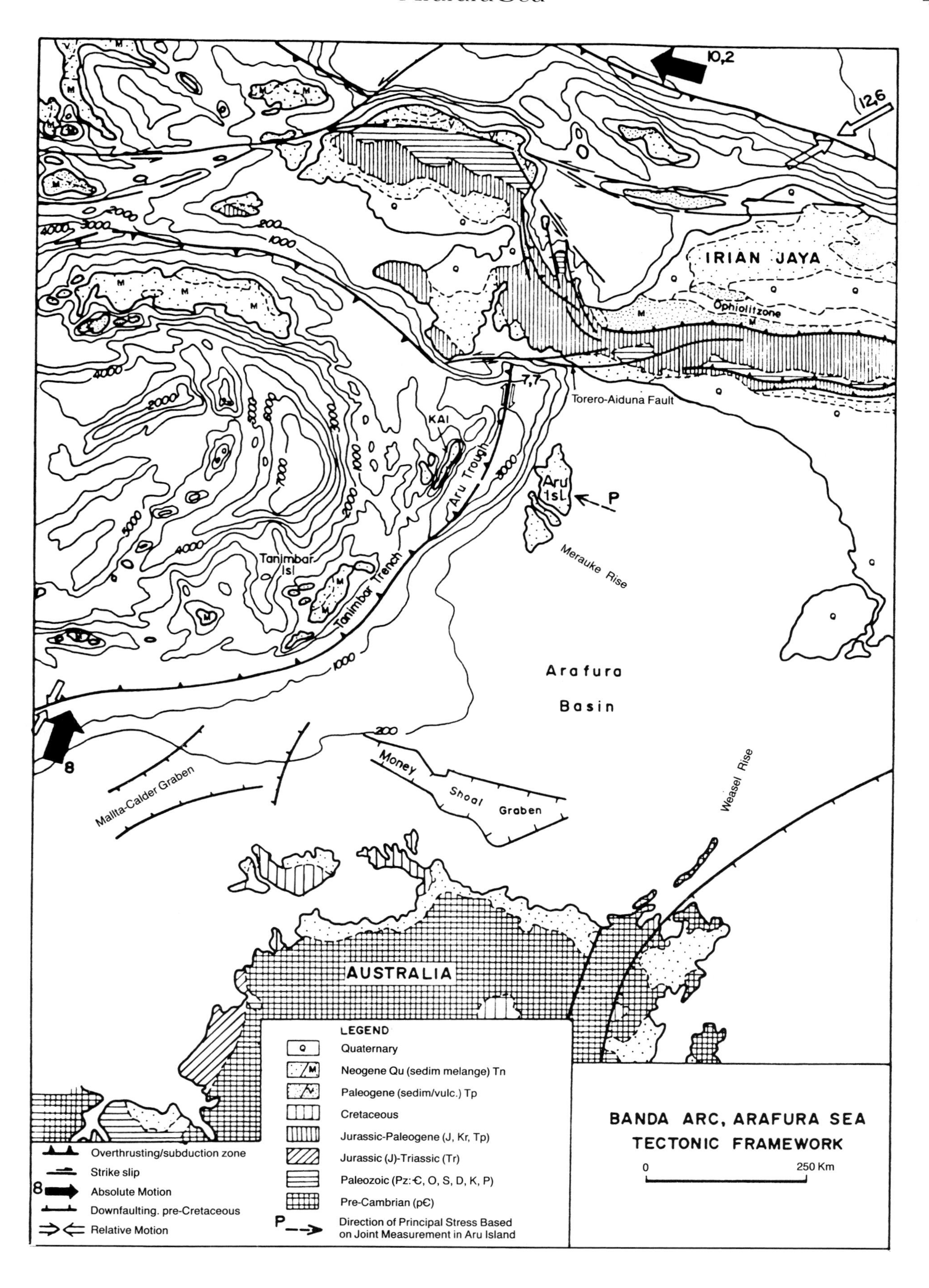
10,2
12,6
IRIAN JAYA
Ophiolitzone
7,7
Torero-Aiduna Fault
KAI
Aru Trough
Aru Isl.
P
Merauke Rise
Tanimbar Isl
Tanimbar Trench
Arafura Basin
8
Mallta-Calder Graben
Money Shoal Graben
Weasel Rise
AUSTRALIA
LEGEND
Quaternary
Neogene Qu (sedim melange) Tn
Paleogene (sedim/vulc.) Tp
Cretaceous
Jurassic-Paleogene (J, Kr, Tp)
Jurassic (J)-Triassic (Tr)
Paleozoic (Pz: Є, O, S, D, K, P)
Pre-Cambrian (pЄ)
Direction of Principal Stress Based on Joint Measurement in Aru Island
Overthrusting/subduction zone
Strike slip
Absolute Motion
Downfaulting. pre-Cretaceous
Relative Motion
BANDA ARC, ARAFURA SEA
TECTONIC FRAMEWORK
0
250 Km

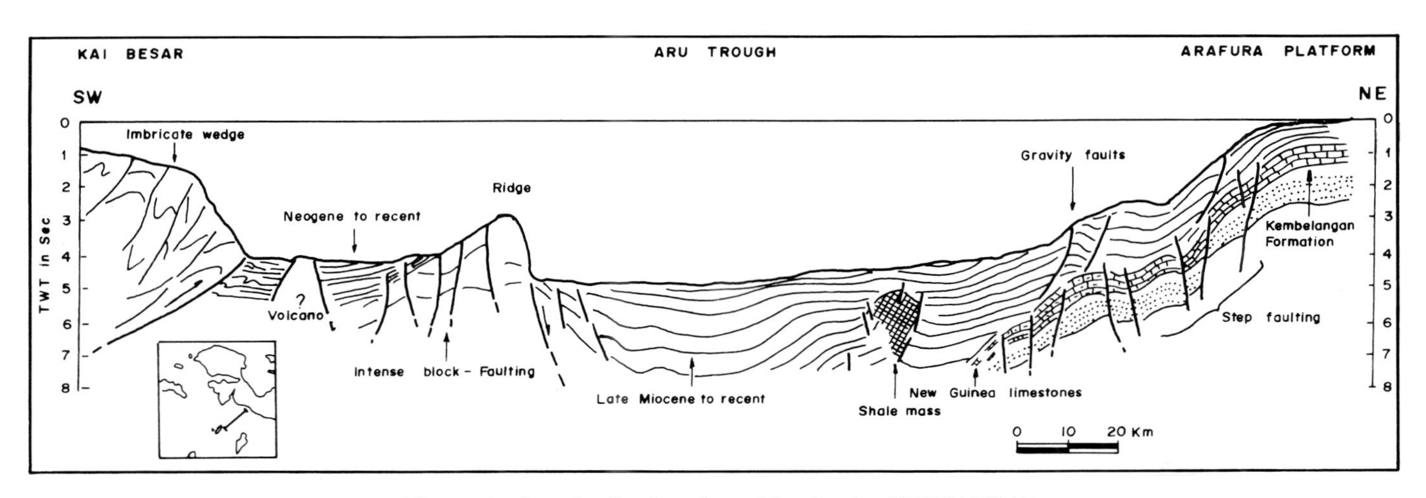


Figure 4. *Longitudinal section of Aru basin. (PERTAMINA).*

denly east of Kai Island while the Ceram Trench begins west of Kai Island. The Tanimbar Trench is considered to be the eastern extension of the Timor Trench, while to the north, it deepens and merges laterally into the 3,500 m (11,480 ft) deep Aru basin. Because the north-northeast-to-south-southwest trending contour lines turn abruptly northwestward between the Tanimbar and Kai Islands, Schlüter (1983) assumes a major fault in this area, separating the Tanimbar Trench from the Aru basin (Figure 2).

According to Bowin et al (1980), the tectonic history of this area is very complicated because the Aru Trough shows crustal extension rather than compression (Figure 4).

Schlüter (1983) contends that the most prominent difference between the Tanimbar and Kai segments of the Banda arc is that vertical tectonic movement predominates along the eastern Aru Trough rather than along the Tanimbar Trench. Large vertical offsets (up to 1.8 s) on normal faults downthrown on the basinward side occur along the eastern flank of the Aru Trough, but only small vertical offsets (up to 0.15 s) can be observed along the southeastern flank of the Tanimbar Trench. North-south step faults (Figures 5 and 6) characterize the eastern margin of the Aru Trough. More precisely, they are found offshore northwest and southwest of the main island and onshore on its western part. This type of tectonics could provide traps against faults or by draping over fault blocks.

### The Aru Archipelago

On the Aru Islands, Neogene to Quaternary marls and shallow-water limestones predominate, although there is evidence for older terrigenous material that is presumably derived from arkosic (granitic ?) outcrops (Fairbridge, 1951).

Results of previous seismic refraction investigations (Curray et al, 1977; Jacobson et al, 1979; and Bowin et al, 1980) support the idea that the Aru Archipelago is an old, peneplaned platform covered by Neogene to Recent sediments.

Southwest of Aru Island, numerous step faults are present,

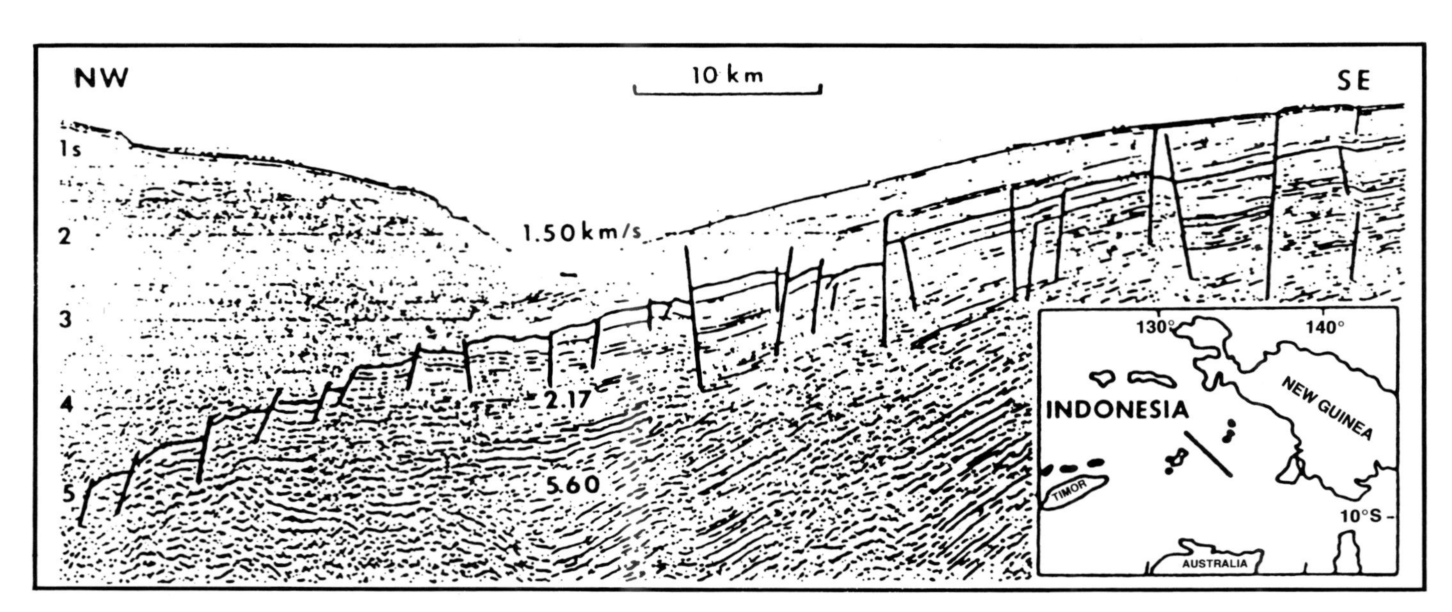


Figure 5. *Seismic sections across the Aru Trough (BGR).*

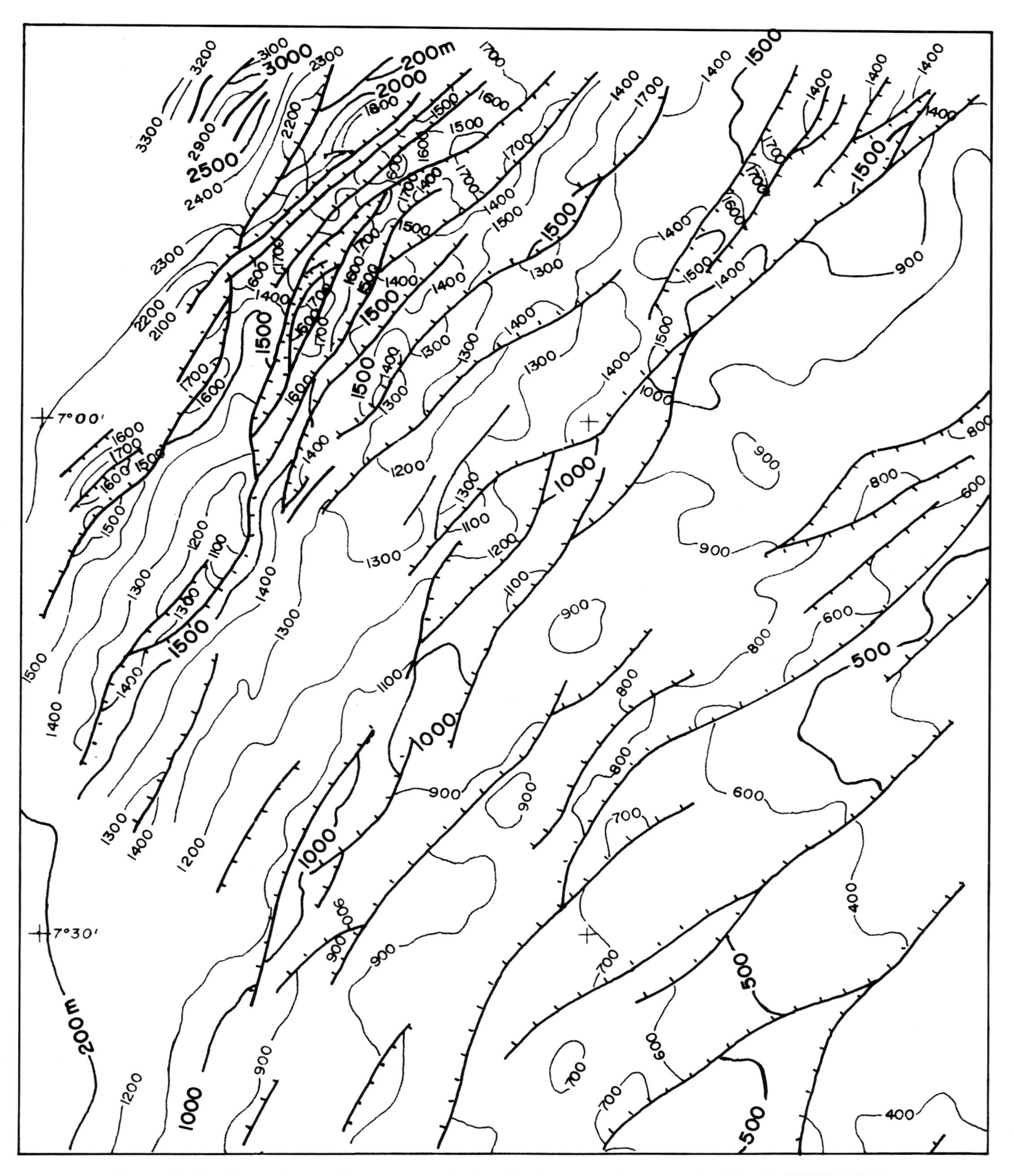

Figure 6. *Top Kembelengan Horizon map, Arafura Platform—southwestern Aru Islands.(PERTAMINA).*

delineating very narrow blocks about 2.5 km (1.6 mi) in width, with an average throw of up to 300 milliseconds (Figure 6).

The pattern both north and south of the Aru Archipelago suggests lateral strike-slip movements in addition to step faulting. The step faulting south and north of Aru is not accompanied by thickening of the Mesozoic and Tertiary sedimentation sections (Schlüter, 1983).

Tjia (1977) concludes that the N 105° E directed, maximum principle stress derived from lineaments and fracture analyses in the island of Aru appears to relate to warping of the Aru Island.

More recent investigations (Schlüter, 1981) reveal that the Aru Archipelago is an eastward-tilted block resulting from tectonic interaction of the Banda arc system and the Arafura Platform.

### The Southern Foothills Belt of the Irian Jaya

Geomorphologically, we can divide the southern foothills belt of the Irian Jaya into the foothills area, with elevations ranging from 150 to 600 m (490 to 1,970 ft) above sea level, and the alluvial plain of the southern coast of Irian Jaya. The foothills area is gently folded, with the main folding axis trending approximately east to west and gradually changing to east-southeast as it extends eastward. The geomorphologic unit consists of young Tertiary clastic sediments, called the "Boeroe Formation," that is the upper part of the 6,000 m (19,685 ft) thick sedimentary pile of Tertiary age in this area. The base of the sediments is the older Kambelangan Formation, which was located in the southern foothills of the Jaya Wijaya range. Based on seismic profiling and offshore drilling results, this sedimentary pile must wedge out southwards to the Arafura Platform.

The western margin of the Arafura Platform, all along the western part of the Aru Archipelago and its northern and southern extensions, plunges westward by numerous steps down to the Aru Trough, as identified on the seismic profiles. During the Pliocene-Pleistocene times, very thick sediments accumulated in the 3,000 m (9,840 ft) deep Aru Trough.

In 1970, Raytheon carried out a side-looking radar (SLAR) survey of the South Coastal Plain of Irian Jaya for Phillips Petroleum Company (Figure 7). The master image was produced in a scale of approximately 1:220,000, covering an area of 8,000 sq mi (20,700 sq km) from Etna Bay in the west to the foothill area of the Jaya Wijaya range in the east (geographically from long. 134° E to 138° E and lat. 4° S to 6° S). The images revealed the general trends of the area's geological structure, which mainly consists of folds and faults. Extensive alluvial fans and large taluses developed on the lower slope of the foothills area, gradually fading away to the alluvial plain. Gentle slopes dipping south and a braided river pattern characterize this zone, in contrast to the vast alluvial plain with distinct meandering river courses and estuaries in the coast.

A monoscopic model of the SLAR image interpretation recognized several types of rocks by using their textural expressions. West of Etna Bay, a very distinctive, rugged karst topography reflecting limestone occurs that the geologists interpreted as part of the Kembelengan Formation. This formation consists of clastic sediments, older in age compared to the overlying Buru Formation.

A striking east to west linear feature that exhibits very clearly in the image is a lineament extending from Etna Bay to the east with about N 85° E direction. At least five smaller lineaments are also recognized south of the main linear feature. These features might represent the west-southwest to east-northeast trending Tarera–Aiduna fault separating older rocks in the north from the Buru Formation of Tertiary age in the south. Geologists found that the Buru Formation is strongly folded in this area, with asymmetric and tight foldings between the fault lines in the Tarera–Aiduna fault zone. The folds system seems to reflect an en echelon arrangement, or dragfold, suggesting the sinistral movement of the Tarera–Aiduna fault. Approximately five to seven anticlines and synclines are apparent.

The general trend of the folding, which is west-southwest to east-northeast, changes abruptly in the area 75 km (46 mi) east of Etna Bay, exhibiting a northwest to southeast trending fold axis that we can trace for about 10 km (6 mi). Further east, the fold axis, and thus the bedding traces, change again to a dominant west-northwest to east-southeast direction. The folding here is gentle, or weak, when compared to the fold structure in Etna Bay area along the Tarera–Aiduna fault zone.

In the coastal plain, no geological structures can be observed. Along the coast, a west-northwest to east-southeast trending fold structure is suspected, based on the expression of the drainage pattern.

### The Arafura Platform

The Arafura Platform belongs to the Australian–Irian continental part of the Gondwana lithospheric plate. The platform, lying almost entirely within the limits of the 200 m bathymetric contour, is a northward, offshore extension of the Australian continent.

Schlüter (1983) distinguishes seven sequences in the Arafura Shelf, of which the top sequence is interpreted as Neogene, becoming thinner toward the shelf (Figure 8). The sequences form part of the chaotic wedge of highly disturbed rocks along the outer Banda Island arc. The underlying sequences are of Paleogene to Neogene age. They extend as much as 35 km (22 mi) beyond the Tanimbar Trench and the Aru Trough, northwestward, and mark the base of the chaotic wedge.

Based on the evidence of seismic reflection results (Balke, Crostella, and Halse, 1973; Lofting et al, 1975), the shield area can be subdivided into several intra-cratonic rises and depressions, such as the Merauke Rise, the Arafura basin, the Money Shoal graben, and the Malita Calder graben.

The intra-cratonic grabens presumably consist of thick Precambrian to Paleozoic strata that are draped by relatively thin Mesozoic–Tertiary sequences. Only the Money Shoal graben contains Jurassic to Cenozoic fluvio-deltaic deposits of more than 2 km (1.2 mi) thickness (Brown, 1980). The largely fluviatile and paralic nature of the Jurassic and Cretaceous strata in this graben, along with the presence of many small unconformi-

Figure 7. *Side-view airborne radar image interpretation of southern coastal plain of Irian Jaya, showing the readily identifiable Tarera-Aiduna strike-slip fault, which was located east of Etna Bay. The image also depicts the west-northwest to east-southeast folding along the foothill area of the Jaya Wijaya range, and the extensive covers by alluvial fans in the coastal plain.*

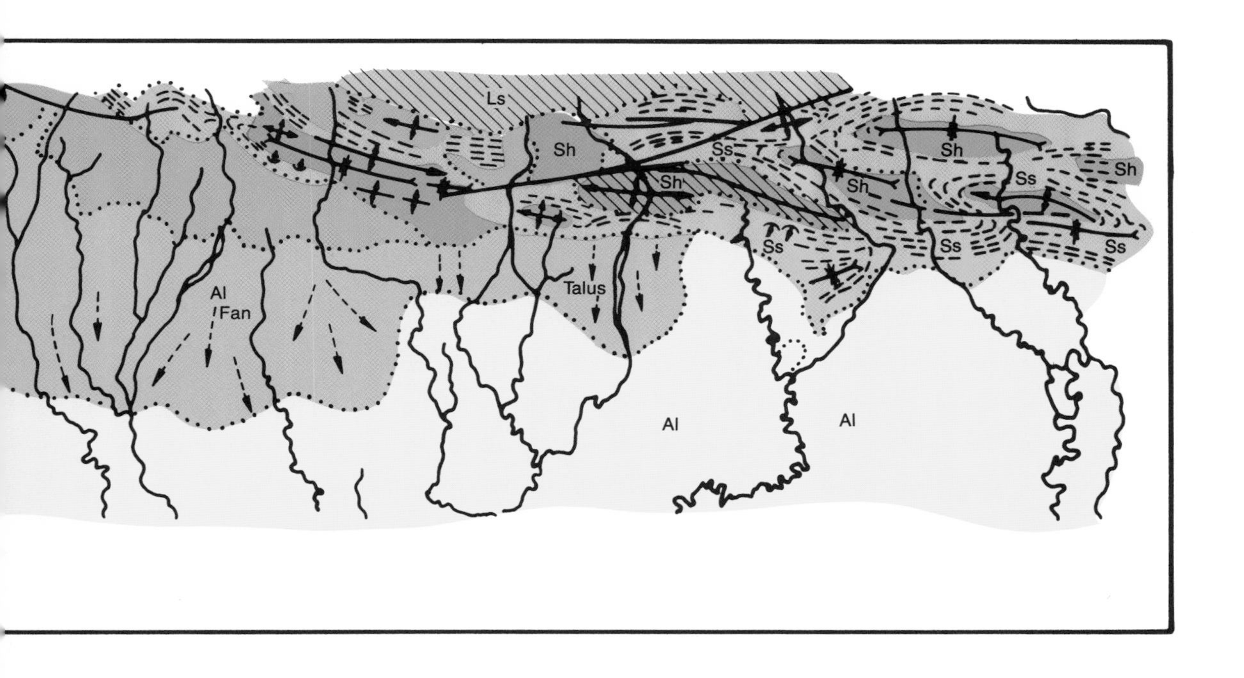

Ls
Sh
Ss
Sh
Sh
Sh
Ss
Sh
Ss
Ss
Ss
Ss
Talus
Al
Fan
Al
Al

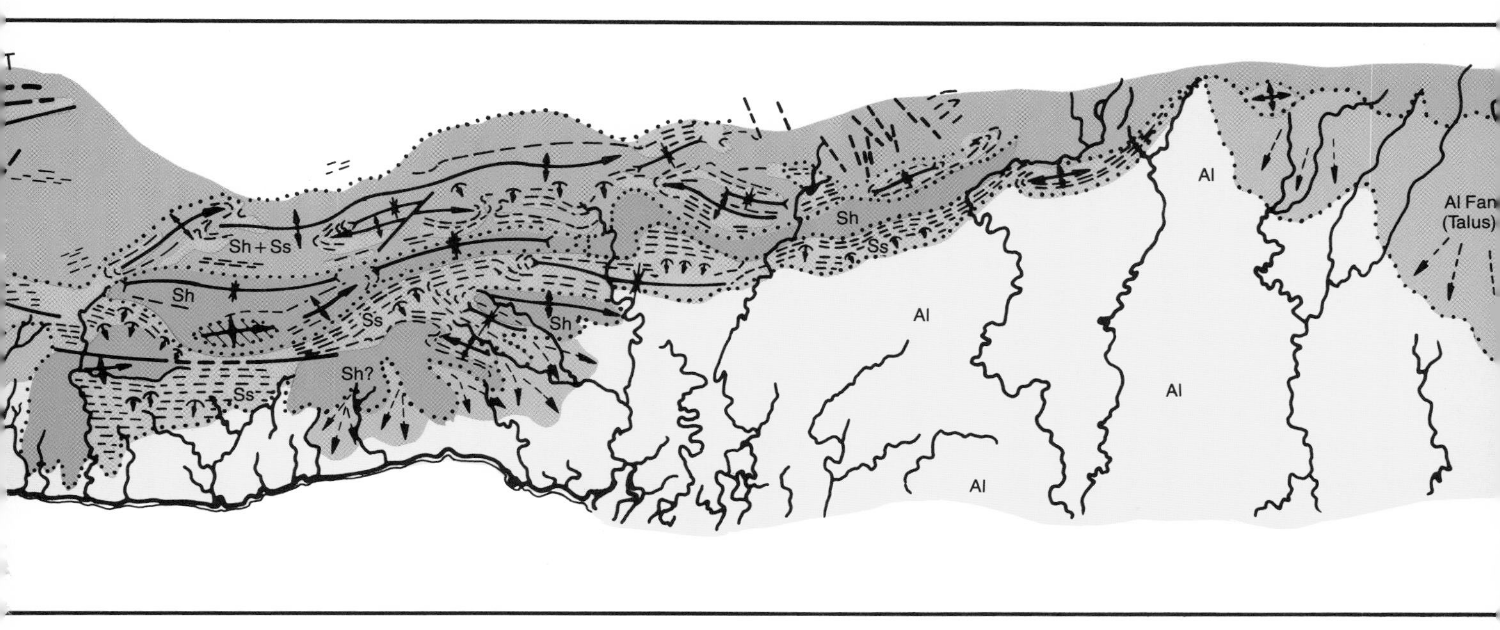


ed,
w

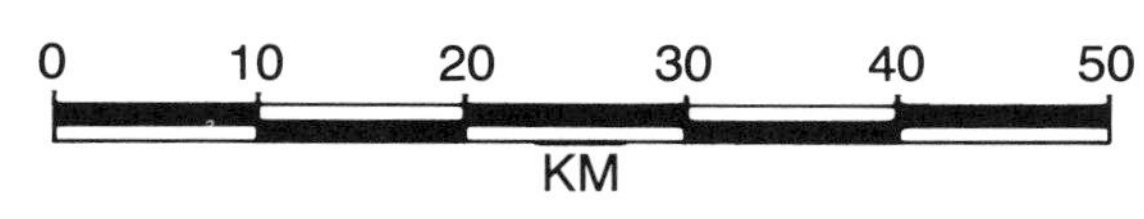


- Al fan — Alluvial Deposit
- Al — Alluvial Fan
- Ss — Sandstone
- Sh/Ss — Shale & Sandstone
- Sh — Shale
- Ls — Limestone

ation

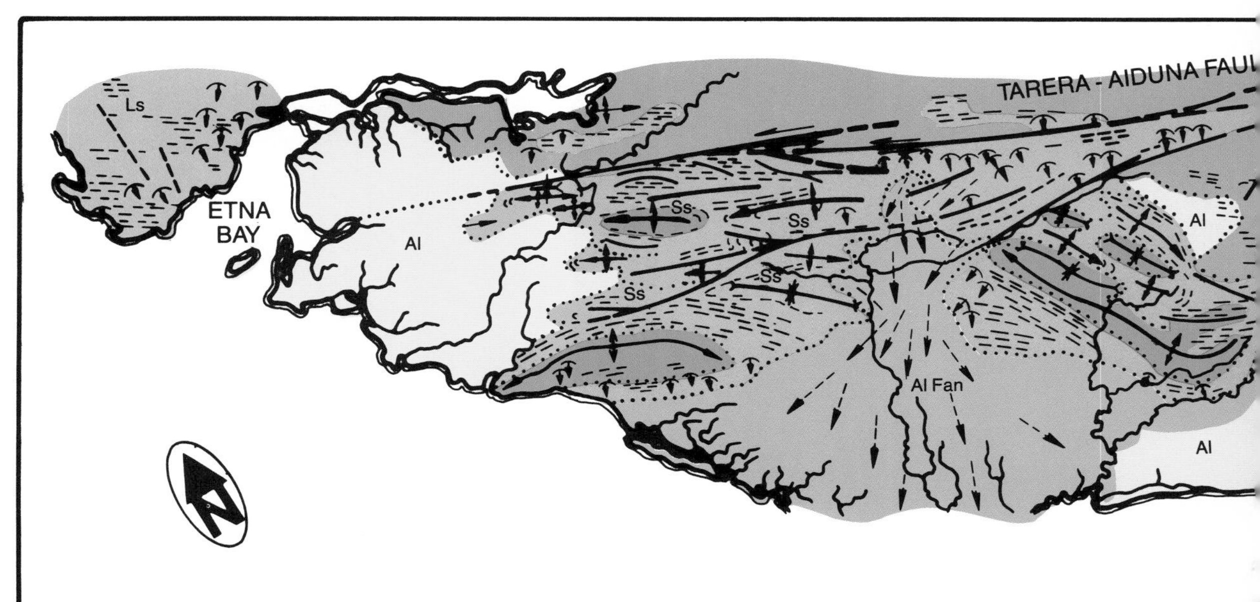
TARERA - AIDUNA FAUL
Ls
ETNA
BAY
Al
Ss
Ss
Ss
Ss
Al
Al Fan
Al

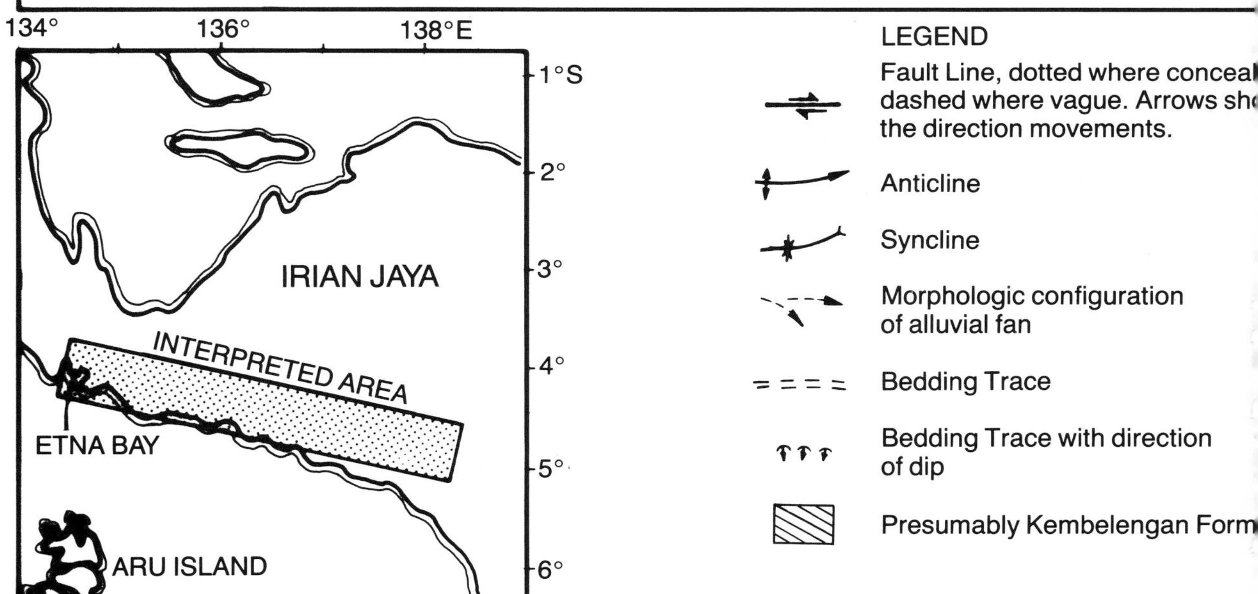
134°
136°
138°E
1°S
2°
3°
4°
5°
6°
IRIAN JAYA
INTERPRETED AREA
ETNA BAY
ARU ISLAND
LEGEND
Fault Line, dotted where conceal
dashed where vague. Arrows sh
the direction movements.
Anticline
Syncline
Morphologic configuration
of alluvial fan
Bedding Trace
Bedding Trace with direction
of dip
Presumably Kembelengan Form

| Sequence | Interpreted lithology | Tentatively interpreted age |
|---|---|---|
| Sea floor<br>A-1<br>(Consisting of subsequences A-1a and A-1b)<br>Horizon Orange | According to sampling mainly marine clay and silt(JONGSMA,1974) | Pleistocene – Early Pliocene |
| A - 2<br>Horizon Light Green<br>A - 3<br>Horizon Light Brown | Sequences A-1a to A-3 have a complex coastal downlap and coastal onlap pattern beneath the slope and probably consist of prograding deltaic sand, silt, shale and ooze | |
| A - 4<br>Horizon Yellow | Assumed platform sediments beneath shelf to middle slope with assumed seaward shelfmargin deltaic environment(clastic?) beneath lower slope. | |
| A - 5<br>Horizon Dark Brown<br>A - 6<br>Horizon Dark Green | Sequence A-5 and A-6 have mainly a parallel to subparallel pattern and probably consist of calcareous shales and argillaceous carbonates. | Late Miocene – ? Eocene ? |
| A - 7<br>Horizon Blue | Assumed shale, sand and siltstones possibly with carbonates in the upper part. | Mesozoic |

**Figure 8.** *Generalized description of sequences, interpreted lithology, and tentative ages of the Arafura Sea floor (simplified from Schlüter 1983).*

ties, indicates that a land area was present in the vicinity of the present-day Darwin Shelf shoreline during most of the Mesozoic (Balke and Burt, 1976).

Nicols (1970) contends that pre-Mesozoic structural deformation is present in the Money Shoal area, where Paleozoic prospects might exist.

Immediately south of the international border between Indonesia and Australia, the large, northeast oriented depression present is known as the Malita–Calder graben. Contrary to the Money Shoal graben, which is a pre-Mesozoic graben situated in a stable block, Balke et al (1973) place the Malita-Calder graben into the mobil zone where drape folding over a deepseated fault block, formed during Jurassic time, is predominant. It is sharply bounded on both sides and, in particular, fault hinged to the Darwin Shelf, a part of the Arafura Platform. This graben should extend into Indonesian waters, southeast of Tanimbar, and represents a Mesozoic–Tertiary sequence where thick marine sediments have accumulated in front of the descending Australian lithosphere (Nicols, 1970).

The Cenozoic and Cretaceous deposits gradually thicken from the shelf area to more basinal conditions. Simultaneously, the older Mesozoic beds abruptly thicken into the graben along a structural feature interpreted as a hinge zone. Two wells drilled in this graben encountered marine Cenozoic and Mesozoic beds. No rock older than Jurassic has been drilled into, but seismic data indicate an important thickness of Mesozoic strata below the Upper Jurassic. In one of the wells, Lower Cretaceous and Jurassic sandstones had poor porosity, but drillers found a thin gas layer in a Lower Cretaceous carbonate.

## HYDROCARBON POTENTIALS OF THE ARAFURA SEA

Major hydrocarbon occurrences are not yet known in the Arafura Sea and very little information is available regarding possible source rocks in the sea. Results from drilling in the Money Shoal graben, and in the Malita–Calder graben and its northern extension, reveal, respectively, Middle to Upper Triassic and calcareous shales, and a 2,000 m (6,560 ft) thick

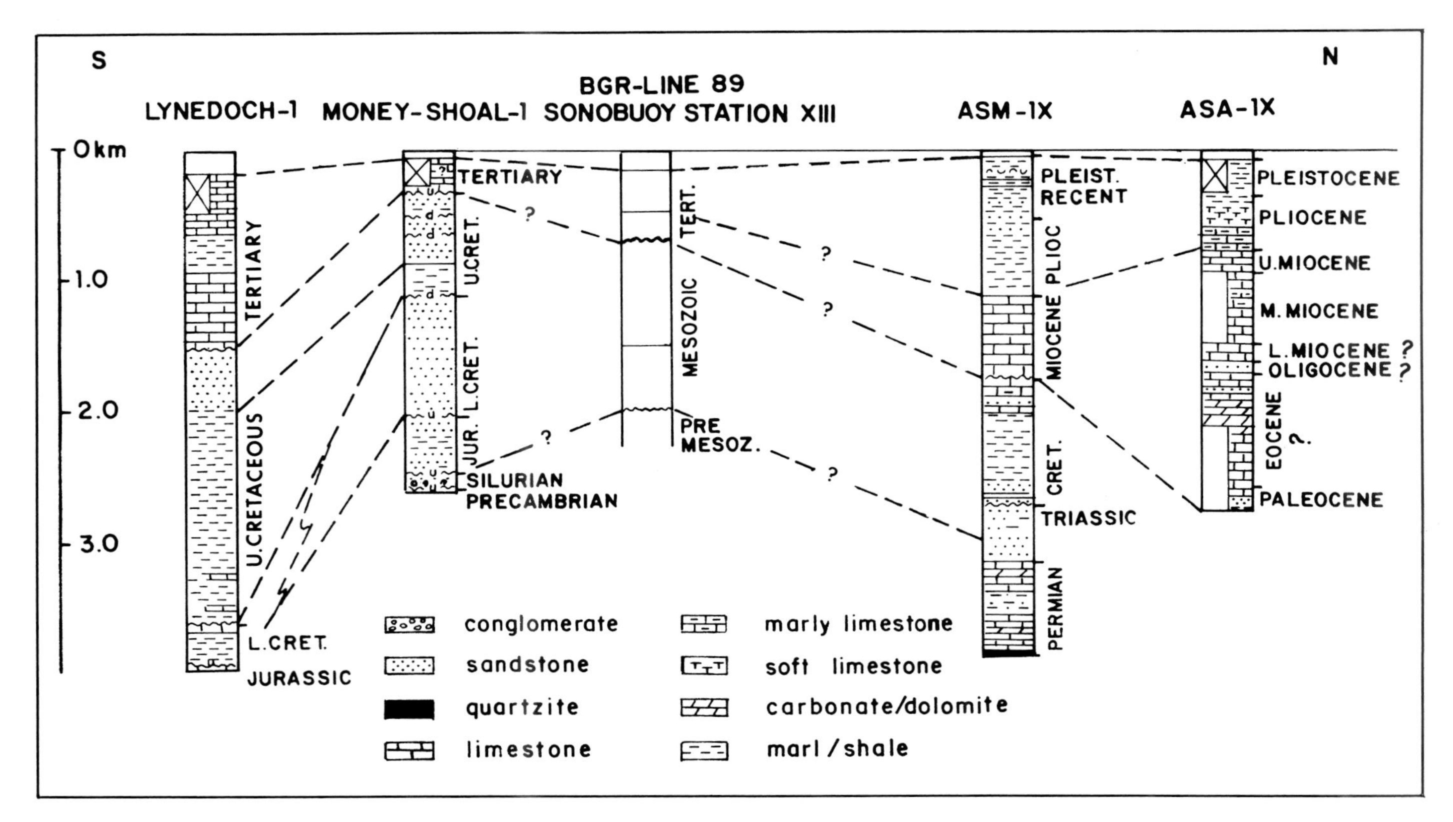


**Figure 9.** *Drill sites in the Arafura Sea chosen for stratigraphic interpretation and BGR sonobuoy station XIII (Schlüter, 1983).*

sequence of Lower to Upper Cretaceous marine shales and mudstones with intercalated sandstones (Figures 2 and 9). These are the most promising source rocks in this graben. However, as Schlüter (1983) points out, little has been published on the amount, type, and maturity of the organic matter in the sediments.

In the Malita–Calder graben and its northeastern extension, there have been gas shows within Jurassic to Cretaceous fine-grained, marine limestone and sandstone. Between the Malita–Calder graben and the Timor–Tanimbar Trench, on the Sahul Ridge, drillers found fine-grained, Jurassic sands containing gas and condensate.

A drill site east of the Aru Island (Figure 9) reveals Upper Permian to Lower Triassic sandstones and shales containing plant spores, while the overlying, Lower Cretaceous marine sandstones are rich in algae assemblages. However, the amount of organic material seems to be low in all the sequences at this drilling site (Urban and Allen, 1979).

One of the most prospective areas would be the hinge of the Arafura High, where a combination of traps and reservoir rocks presumably exists. Schlüter (1983) speculates that the hydrocarbons of the Sahul Ridge were derived from seaward lying sediments by upward migration, and suggests that this kind of process, if it has taken place, increases the hydrocarbon potential of the Aru High hinge.

Another promising area for petroleum exploration is the northern part of the Irian Jaya coastal plain. The existence of thick, sedimentary piles of the young Tertiary Buru Formation and the old Tertiary Kembelangan Formation, plus the presence of a large, active wrench fault, the Tarera–Aiduna fault, might prove that this area is a wrench-related basin. These basin styles are, by analogy, favorable for petroleum exploration. The main objective for exploring this area would be the Kembelangan Formation at an approximate depth of 1,000–1,500 m (3,287–4,900 ft) (personal communication, PERTAMINA Geologists).

Middle Jurassic to Upper Cretaceous deposits belong to the Kembelangan Formation, which can be subdivided into three members:

1) a lower transgressive member (Middle-Upper Jurassic),
2) a middle regressive member (Neocomian), and
3) an upper transgressive member (Aptian-Maestrichtian).

The Middle to Upper Jurassic member is strongly transgressive and may have covered most of Irian Jaya and its adjacent shelves, including the Arafura Platform. It was subsequently eroded from large parts of Irian Jaya during an important phase of regression, attributed to the Early Cretaceous.

The middle regressive member (0–700 m or 0–2,295 ft) of the Kembelangan Formation is mainly composed of sandstones and coarser clastics that often show cross-bedding. This member shows approximately the same geographic distribution as does the Middle to Upper Jurassic. It is well developed in the Etna Bay and Lengguru area (west of Geelvink Bay), but is also known locally from the Central Range and its southern foothills.

Kembelangan-1 is the only well to have intersected this member, 150 m (492 ft thick). The middle Kembelangan Formation is thought to represent a belt of coalescing delta complexes formed by erosional products of sediments from the Arafura Platform and the western part of the Bintuni basin.

The upper member (0–1,700 m or 0–5,575 ft) is again strongly transgressive, onlapping older rocks in large parts of the Arafura Platform. It mainly consists of dark shales and marls, intercalated with lenses of glauconitic sandstones. These sandstones become more important toward the ancient shoreline. Laterally, this member grades into a more calcareous facies.

Outcrops of the Kembelangan Formation have been mapped and described over the length of the Central Range, north of the coastal area. As has been mentioned earlier, this sedimentary pile wedges out southward to the Arafura Shelf.

Regarding the shelf deposits themselves, Schlüter (1983) argues that a sequence lies beneath the outer shelf, probably consisting of shale, sand, and siltstones, perhaps with carbonate in the upper part and of Mesozoic age. This sequence could be of interest as a potential source rock if organic material exists. Other, lower sequences, probably consisting of calcareous shales and argillaceous carbonates, might contribute to hydrocarbon generation. Schlüter further stresses that the only possibility for petroleum generation in the Arafura Sea probably exists within the Mesozoic layers.

## GEOTECTONIC CONTRAST: WESTERN AND EASTERN INDONESIA AND THE SHELF AREAS

Geologists, such as the Dutch, often consider the Arafura Shelf as the counterpart of the Sunda Shelf, which extends on the Asiatic side of the Indonesian Archipelago.

This comparison might be a sound one from a geomorphologic point of view, but the geological and structural features are totally different, which, in turn, could explain why the Sunda Shelf is richer in hydrocarbons than the Arafura Shelf.

According to Klompé (1961), the Sunda Shelf in western Indonesia is mainly the result of various Mesozoic phases of orogeny, whereas the Arafura Shelf in eastern Indonesia resulted from the Variscian orogeny. Western Indonesia is predominantly an area of Tertiary sedimentary deposits, whereas eastern Indonesia is chiefly the site of late Paleozoic and Mesozoic sedimentation.

The Sunda Shelf, previously considered to be stable, has been affected by warping and faulting as late as Quaternary (Katili and Soetadi, 1971), while some depressions in the Arafura Platform are considered to be pre-Mesozoic grabens.

The first comprehensive plate tectonic models of western Indonesia, exhibiting the various basins within an island arc system, were built by Katili (1971, 1973), who, impressed with the apparent concentricity of the belts ringing the Sundaland, shows how they could be explained by two opposing sets of oceanward-migrating Benioff zones. Pulunggono and Cameron (1984) reinforce the Malaysian geologist's point of view that the evolution of the Sundaland should partly be interpreted in terms of a complex mosaic of constantly moving fragments or microplates. A Triassic suture complex, representing a microplate boundary known as the Mutus Assemblage, is, according to both authors, the site of at least 95% of the cumulative production of the Central and South Sumatra basins because of the combination of high heat flow and the early growth of the structure.

The faulted character of the shelfal basins in western Indonesia, expressed by grabens and fault-blocks, has opened up new and additional prospects within each basin.

In relation to oil basin areas, the major tectonic features of the Sunda Shelf include the back-arc and inner-arc basins of Sumatra and Java, containing major oil and gas production because of the fortunate conjunction of favorable factors related to reservoir, source rock, and seal facies and to timely, thermal maturation. The proximity to the volcano/plutonic arc possessing favorable polarity, vis à vis the basin area, is known to produce elevated heat flow.

Intra-cratonic basins, such as the Thai and Malay basins, contain thick wedges of Tertiary clastics in which multiple reservoirs have been filled by hydrocarbons migrating from contemporaneous or older beds with favorable drainage regimes and with thermal maturation from rising isotherms in an extensional setting (Barber and Murphy, 1984).

In summary, the major tectonic elements in relation to oil basin areas in the Arafura Sea are the intra-cratonic grabens, the hinges of intra-cratonic highs, the southern foothills of Irian Jaya comprising the northern border of the platform, and, perhaps, the Arafura Platform itself.

A comparison between the outer-arc ridge of western Indonesia and that of eastern Indonesia is also of great interest when evaluating the hydrocarbon potential in the chaotic, sedimentary wedges in both areas. The arc-trench system in western Indonesia was formed by subduction of the Indian Ocean's oceanic crust beneath the Eurasian continental crust.

In the outer arc ridge of Nias Island in western Indonesia, a mid-Tertiary subduction complex developed consisting of a subduction mélange: a chaos dipping steeply to moderately, consisting of scaly clay, serpentine and peridotite, basalt and spillite, greenschist, amphibolite, and low-grade metasedimentary rocks plus interfolded and intersliced, coherent massive strata. In the tectonic mélange, mafic and ultramafic plutonic rocks, pillow basalt, and cherts form approximately 20% of the mélange inclusions. This structure suggests that at the same time that the mélange sediments were accreted, the trench fill was quite thin and pieces of oceanic crust and upper mantle were also stripped off the descending plate and accreted (Moore and Karig, 1980). No hydrocarbon occurrence has been reported in the accretionary wedge of western Indonesia.

The arc-trench system of eastern Indonesia shows an entirely different character. We can discern two distinct phases in the development of the Banda arc. In the earlier phase, the oceanic part of the Indian-Australian plate was subducted under the Banda oceanic plate, and, in the later phase, the Australian continental crust was subducted into the Banda arc subduction zone.

In the outer arc ridge of the Tanimbar Island in eastern Indonesia, the rocks consist of a highly deformed mélange of Eocene to Miocene, shallow-water limestones and marls mixed with Triassic to Jurassic clastic sediments (Hamilton, 1979). Audley-Charles and Carter (1974) assume a post-Jurassic and pre-Eocene tectonic phase.

Whereas the oceanic crust dipping in the Java Trench is covered only by relatively thin pelagic sediments, large parts of the shelf and slope sequences of the Arafura Platform are carried passively on top of the Australian lithospheric plate down into the Tanimbar Trench and Aru Trough. Schlüter (1983) believes that if these sediments, which were originally deposited in a shelf or slope environment or both, are rich in organic material, the tectonic processes in the trench and beneath the chaotic wedge, in combination with the increasing burial depth, will increase the maturity of the organic material. If reservoir rocks exist in front of the chaotic wedge, upslope migration and accumulation must be considered as possible within the faulted blocks in these rocks. Timor and Ceram's oil and gas resources suggest the presence of hydrocarbons in the outer arc ridges of Kai and Tanimbar, with their similar geological tectonic and stratigraphic conditions. Interestingly, in eastern Ceram, oil is produced also from pre-Tertiary limestone, sandstone, and siltstone. Fontaine and Manguy (1982) suggest that the Ceram oils are probably derived from Triassic source rocks.

Tamano (1984) contends that tectonic stress affecting the subduction complexes destroys the original bedding planes and sedimentary units and generates fractures. An increase in effective porosity by this brecciation raises the possibility of gas or oil traps or both, provided that the appropriate conditions for hydrocarbon generation and a seal are satisfied. Consequently, gas generated in a subduction complex may be preserved in neighboring fore-arc basins as well as in the subduction complex.

## CONCLUSIONS

In conclusion, geophysical exploration in the Australian and Indonesian waters of the little explored Arafura Sea has been successful in identifying the major tectonic elements in this region.

General stratigraphic controls have been obtained in the intra-cratonic basins and the hinges of the intra-cratonic highs as well as in the southern foothills of Irian Jaya and in the platform area. More exploring and testing will be required, especially in the structural-stratigraphic hinge zones of the Arafura High and the area of the southern foothills of Irian Jaya, including the outer arc islands of Kai and Tanimbar.

## ACKNOWLEDGMENTS

I wish to thank Dr. A. Pulunggono, from the Trisakti University Jakarta, and Messr. Soejoso and Zanial Achmad from PERTAMINA for providing me with additional data and material on the Arafura Sea and assisting me in preparing this article.

Mr. Hartono from the Marine Geological Institute, Bandung provided valuable advice on the tectonics of the Banda collision zone, while Mr. Sudradjat from the Volcanological Survey of Indonesia was kind enough to interpret the radar image of the southern foothill belt of Irian Jaya.

Finally, my greatest appreciation goes to Prof. Karl Hintz from the Federal Institute of Geosciences and Natural Resources, West Germany, for giving permission to use relevant data of the geophysical survey carried out by the SONNE-cruise in the Arafura Sea, in 1981.

## REFERENCES CITED

Audley-Charles, M. G., 1975, The Sumba fracture, a major discontinuity between eastern and western Indonesia: Tectonophysics, v. 26, p. 213–228.

——— , D. J. Carter, and A. J. Barber, 1975, Stratigraphic basis for tectonic interpretation of the outer Banda arc, eastern Indonesia: Indonesian Petroleum Association, 3rd Annual Convention, Jakarta, Proceedings, p. 25–44.

——— , D. J. Carter, A. J. Barber, M. S. Norvick, and S. Tjokrosaputro, 1979, Reinterpretation of the geology of Ceram, implications for the Banda arcs and northern Australia: Journal of the Geological Society, London, 136, p. 547–568.

Balke, B., and D. Burt, 1976, Arafura Sea area, *in* R. B. Leslie, H. J. Evans, and C. C. Knight, eds., Economic geology of Australian Institute of Mining and Metallurgy Monograph 7, p. 209–212.

——— , C. Page, R. Harrison, and G. Roussopoulos, 1973, Exploration in the Arafura Sea: Sydney, APEA Journal, v. 13, p. 9–12.

Barber, A. J., and M. G. Audley-Charles, 1976, The significance of the metamorphic rocks of Timor in the development of the Banda arcs: Tectonophysics, v. 30, p. 119–128.

——— , and Murphy, 1984, The relationship between hydrocarbon occurrences and tectonics in Southeast Asia (abs.): Taipei, Sino-French colloquium on Geodynamics of the Eurasian-Philippine Sea Plate Boundary, 22 p.

Bowin, C., G. M. Purdy, C. Johnston, G. Shor, L. Lawyer, H. M. S. Hartono, and P. Jezek, 1980, Arc-continent collision in the Banda Sea region: AAPG Bulletin, v. 64, n. 6, p. 868–915.

Brouwer, H. A., 1923, Geologische Onderzoekingen op de Tanimbar Eilanden: Algemeene Landsdrukkerij, Jaarboek van het Mijnweszen in Nederlands Oost-Indie, 50, Jaargang, 1921, p. 119–142.

Brown, C. M., 1980, Arafura and Money Shoal basins: New York, UN, ESCAP, Atlas of stratigraphy II, Mineral Resources Development Series, VII, p. 46.

Bursch, J. G., 1947, Mikropalaontologische Untersuchungen des Tertiars von Gross Kei (Molukken), Schweiz: Basel, Verlag Birkhauser, Palaentologirche Abhandlung, v. 65, n. 3, p. 1–69.

Cardwell, R. K., and B. L. Isacks, 1978, Geometry of the subducted lithosphere beneath the Banda Sea in eastern Indonesia from seismicity and fault plane solutions: Journal of Geophysical Research, v. 83 (b 6), p. 2825–2838.

Chamalaun, F. H., and A. E. Grady, 1978, The tectonic development of Timor, a new model and its implications for petroleum exploration: Journal of the Australian Petroleum Exploration Association, v. 18, n. 1, p. 102–108.

Curray, J. R., G. G. Shor, Jr., R. W. Raitt, and M. Henry, 1977, Seismic refraction and reflection studies of crustal structure of the eastern Sunda and western Banda arcs: Journal of Geophysical Research, v. 82, n. 17, p. 2479–2489.

Fairbridge, R. W., 1951, The Aroe Islands and the continental shelf north of Australia: Scope, Journ. Sci., Union University, Western Australia, v. 1, n. 6, p. 24–28.

Fontaine, H., and M. Mainguy, 1982, Don't forget Asia's Older, Petroleum News Magazine, v. 12, n. 11, p. 8–10.

Gribi, Jr., E. A., 1974, Petroleum geology of the Moluccas, eastern Indonesia, *in* L. R. Beddoes and J. S. Wonfor, eds., Proceedings of the SE-Asia Petroleum Exploration Society, v. 1, p. 23–30.

Hamilton, W., 1974, Map of the sedimentary basins of the Indonesian Region, Department of Interior, U.S. Geological Survey, Denver, Colorado.

——— , 1979, Tectonics of the Indonesian region: Washington, Geological Survey Professional Paper 1078, U.S. Department of the Interior, U.S. Government Printing Office, p. 1- 345.

Jacobson, R. S., G. G. Shor, Jr., R. M. Kieckhefer, and G. M. Purdy, 1979, Seismic refraction and reflection studies in the Timor-Aru Trough system and Australian continental shelf, *in* Watkins et al, eds., Geological and Geophysical Investigation of Continental Margins, AAPG Memoir 29, p. 209–222.

Jezek, P. A., and C. S. Hutchison, 1978, Volcanism of the Banda arc, Indonesia: International Geodynamic Conference, Tokyo, 1978.

Jongsma, D., 1974, Marine geology of the Arafura Sea: Canberra, Australian Bureau of Mineral Resources Bulletin, v. 157, p. 1–56.

Katili, J. A., 1971, A review of geotectonic theories and tectonic maps of Indonesia: Earth Science Review, v. 7, p. 143–163.

——, 1973, Geochronology of West Indonesia and its implication on plate tectonics: Tectonophysics, v. 19, p. 195–212.

——, 1974, Geological environment of the Indonesian mineral deposits, a plate tectonic approach: Publikasi Teknik Seri Geologi Ekonomi (Geological Survey of Indonesia), p. 7, Also United Nations ESCAP, CCOP Technical Bulletin, v. 9, p. 39, 1975.

——, 1975, Volcanism and plate tectonics in the Indonesian island arcs: Tectonophysics, v. 216, p. 165–188.

——, and R. Soetadi, 1971, Neotectonics and seismic zones of Indonesia: Bulletin, Proceedings, Royal Society of New Zealand, v. 9, p. 39–45.

——, and H. M. S. Hartono, 1983, Complication of Cenozoic tectonic development in eastern Indonesia, *in* Geodynamics of the Pacific Region, Geodynamic Series, v. 11, p. 387–398.

Klompé, Th. H. F., 1957, Pacific and variscian orogeny in Indonesia, a structural synthesis: Madjalah Ilmu Alam Indonesia, v. 113, p. 43–87.

Lofting, M. J. W., A. Crostella, and J. W. Halse, 1975, Exploration results and future prospects in the northern Australasian region: UK, Applied Science Publ. Ltd., 9th World Petroleum Congress Proceedings, v. 3, p. 65–81.

Moore, G. F., and D. E. Karig, 1980, Structural geology of Nias Island, Indonesia; implications for subduction zone tectonics: American Journal of Science, v. 220, p. 193–223.

Montecchi, P. A., 1976, Some shallow tectonic consequences of subduction and their meaning of the hydrocarbon explorationists, *in* M. T. Halbouty, J. C. Maker, and H. M. Lian, eds., Circum-Pacific energy and mineral resources, AAPG Memoir 25, p. 189–202.

Nicols, G. N., 1970, Exploration and geology of the Arafura Sea: Sydney, APEA Journal, p. 55–61.

PERTAMINA and Schlüter, 1983, unpublished report.

Pulunggono, A., and N. R. Cameron, 1984, Sumatran microplates, their characteristics and their role in the evaluation of the Central and South Sumatra basins: Proceedings of the 13th Annual Convention, IPA, Jakarta, v. 1, p. 121–143.

Schlüter, 1983, Geology and tectonics along the convergent Australian and Banda Sea margins from the Tanimbar Trench to the Aru Trough; results of geophysical investigations with the R/V SONNE Cruise SO-116 III in 1981: Report BGR, n. 94605, 37 p.

Tamano, T., 1984, Development of fore-arc continental margins and their potential for hydrocarbon accumulation: Petroleum Geology, p. 135–145.

Tjia, H. D., 1977, Fracture systems near Dobo, Aru Islands, Indonesia: Sains Bumi/Earth Science, Sains Malaysiana, v. 6, n. 2, p. 185–193.

Urban, L., and M. Allen, 1979, Vitrinite reflectance as an indicator of thermal alteration within Paleozoic and Mesozoic sediments from the Phillips Petroleum Corp. ASM-1X well, Arafura Sea: Singapore, Proc. Reg. Conf. Geol. Min. Res. SE-Asia, p. 103–108.

Van Bemmelen, R. W., 1949, The geology of Indonesia: The Hague, Government Printing Office, v. I, 732 p.

# Geology and Petroleum Potential of Northwestern China

Zhai Guang-Ming
*Scientific Research Institute of Petroleum Exploration and Development*
*Beijing, China*

The northwestern part of mainland China contains many large sedimentary basins such as the Zhungeer basin, the Talimu basin, the Tulufan basin, the Chaidamu basin and the basins in West Gansu Province. The deposits in these basins consist of Mesozoic-Cenozoic sediments of continental origin, and Paleozoic sediments of marine origin, totalling more than 10,000 m (32,800 ft) in thickness. Most rocks are not metamorphosed. These basins are promising targets for oil and gas prospecting. Tectonically speaking, they are widely different from basins in East China where tension forms the sedimentary basins and normal faults tend to appear from block faulting or rollover anticlines. In the northwest, however, the basins are primarily formed by compression: thrust faults and reverse faults occur, developing major structural zones and many local structures. The series of structural zones which are in rows at the piedmont of the Tianshan, Kunlunshan, Alkinshan, and Nanshan Mountains are good objectives for oil exploration.

Most of the reservoir rocks in the northwest are of Mesozoic age, with some of Cenozoic age, although geologists also consider Paleozoic rocks as exploration objectives. Within the older rocks we may find varied types of oil pools, typical structural oil pools, or large-scale stratigraphic accumulations.

Geologists have already discovered oil and gas fields in these basins. Many of the fields have started production. They may constitute an important oil production base in China.

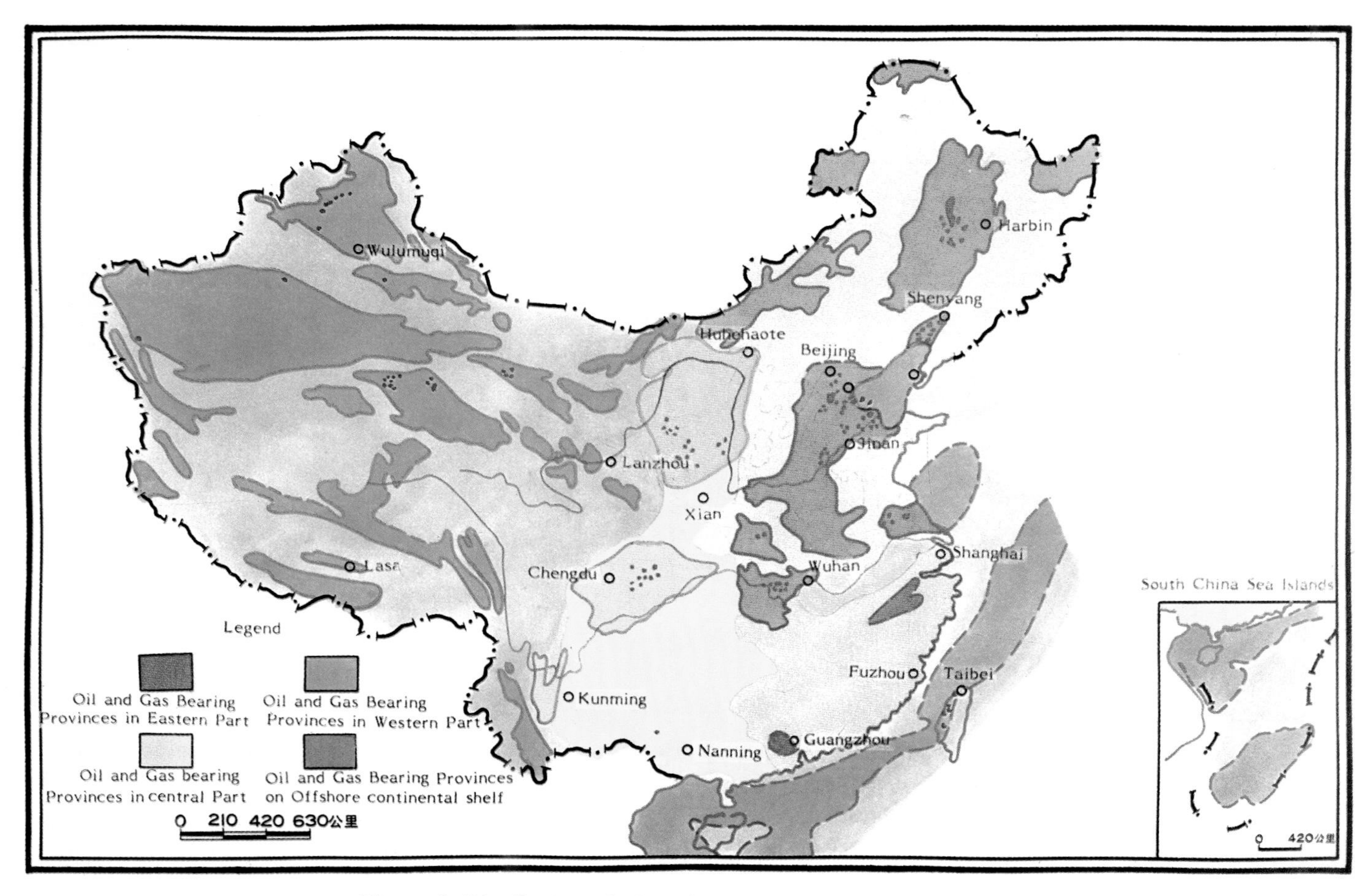


**Figure 1.** *Distribution of oil- and gas-bearing provinces in China.*

## INTRODUCTION

The northwestern part of China is the area west of the Helan and Liupan Ranges, north of the Kunlun and Aerjin Range, and covers an area of about 2 million sq km (772,260 sq mi). It is located between the Indian and Siberian plates, where its northwest and east to west trending structural framework was formed with sedimentary basins of varying size and types (Figure 1). Geologists have defined 60 Mesozoic sedimentary basins in this region, with a total area of about 1.4 million sq km (540,550 sq mi). Of these basins, 18 are larger than 10,000 sq km (3,860 sq mi). However, most of the area is not heavily explored. The limited exploration done to date occurs in only a few sedimentary basins concentrated in the Zhungeer, Jiuquan, and Chaidamu basins (Figure 2). The result of this limited activity is the discovery of 43 oil and gas fields, with reservoirs of different ages (Ordovician, Silurian, Carboniferous, Permian, Triassic, Jurassic, Cretaceous, Tertiary, and Quaternary); making this an important oil and gas potential area in China. The major discoveries include

1937: the Dushanzi oil field in the Zhungeer basin,
1938: the Laojunmiao oil field in the west Jiuquan basin,
1955: the Youshashan oil field in the Chaidamu basin,
1955: the Kelamayi oil field in the Zhungeer basin,
1958: the Yiqikelike oil field in the Talimu basin, and
1960: the Shengjinkou oil field in the Tulufan basin.

Rugged terrain in these areas has limited exploration in the past. However, improvements in exploration technology and methods and the introduction of digital seismic recording instruments into northwestern China have led to intensified research and exploration. As a result, new discoveries have been made in the Talimu and Chaidamu basins, which are the basis for conducting large-scale petroleum exploration in the northwestern part of China.

## GEOLOGIC SETTING

Over a long period of geologic time, two main structural systems developed in the northwestern part of China. Multiple tectonic movements, primarily north to south compression, led to these two systems: one is situated approximately east to west and the other trends to the northwest.

The Tianshan fold zone is a good example of the east to west trending structural system. It extends to the USSR in the west and to Mongolia in the east. It is also expressed as fault blocks of the decomposed Paleo-China Platform. With the multiple tectonic movements, these fault blocks have lifted and subsided to form the present high peaks and mountains, with Paleozoic rocks cropping out in the central part and on both the north and south flanks of the Tianshan Range.

The Zhungeer and Talimu basins lie to the north and south,

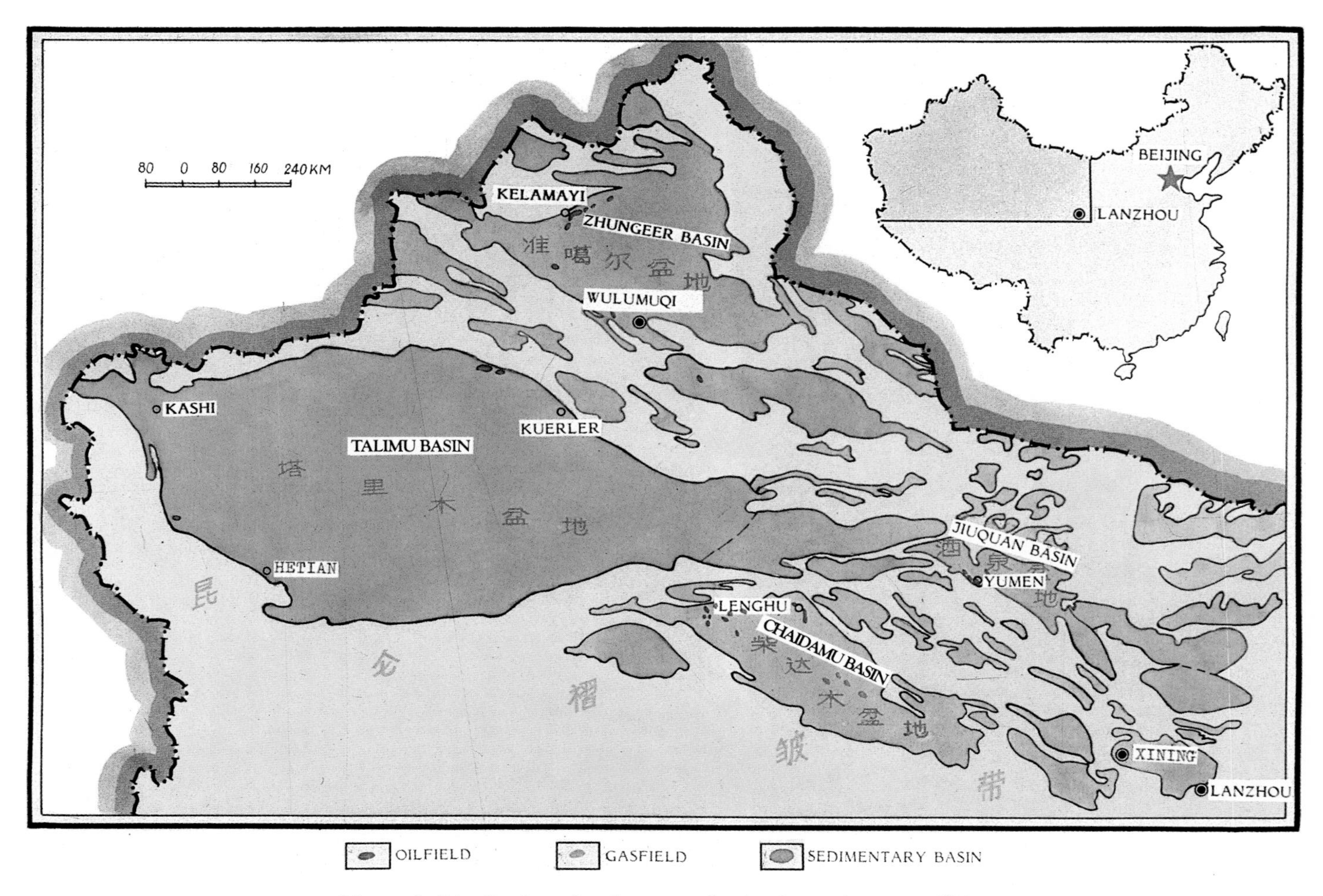


**Figure 2.** *Distribution of sedimentary basins in northwestern China.*

respectively, of the Tianshan fold zone. They are developed with the Wulumuqi mountain front depression to the north, the Kuche depression to the south, and the Tulufan-Hami intermountain depression to the east of Tianshan fold zone. The Tianshan fold zone is the principal source of sediment for the Zhungeer and Talimu basins.

The Kunlun fold zone is the product of Hercynian orogeny, mainly composed of Proterozoic and Paleozoic series. This fold zone forms the southern boundary of the Talimu basin. This folding continued until Paleozoic material was overthrust onto the Cenozoic sequence.

A Qilian fold system of Caledonian age dominates another northwest structural system. A large number of northwest trending intermountain basins have developed in the western part of Gansu. The Chaidamu basin developed south of the Qilian Range and north of the Kunlun Range. The Bachu Central structural trend, developed in the Talimu basin in Xinjiang, separating the Talimu basin into two parts, and controlled the Mesozoic-Cenozoic sedimentation on its north and south sides. The Cretaceous and Paleogene sections in the southern part are marine and the Mesozoic-Cenozoic section in the northern part is non-marine.

Many Mesozoic-Cenozoic basins in northwestern China have been formed by the Bachu Central and structural systems and the northeast trending local structure patterns. Most of these Mesozoic-Cenozoic sedimentary basins formed during late Hercynian time. Because the origin of each basin varies, the developments of late Mesozoic-Cenozoic sedimentation and structures were likewise under different controls, having a significant influence on oil and gas accumulations. The basins can be generalized into two categories:

1. Mesozoic-Cenozoic basins developed from Paleozoic massifs the Talimu, Zhungeer, Chaidamu, and Jiuquan basins.
2. Mesozoic-Cenozoic basins developed from pre-Paleozoic metamorphic rocks such as the Dunhuang and Minho basins.

The sedimentary basins in northwestern China are generally covered by deserts and gobi on the surface. Only occasional outcrops exist in the mountain front areas. Because these structural systems controlled the formation of the basins, the structural systems have also influenced the distribution of subsurface structures. Local structures occur in rows and form zones that are parallel to the front areas of the Tianshan, Kunlun, Qilian, and Aerjin mountain ranges. As a result of the multiple tectonic movements, older structures are inherited. As a result of different episodes of tectonic movement over geologic time, complicated structures exist. They comprise younger structures superimposed over older structures.

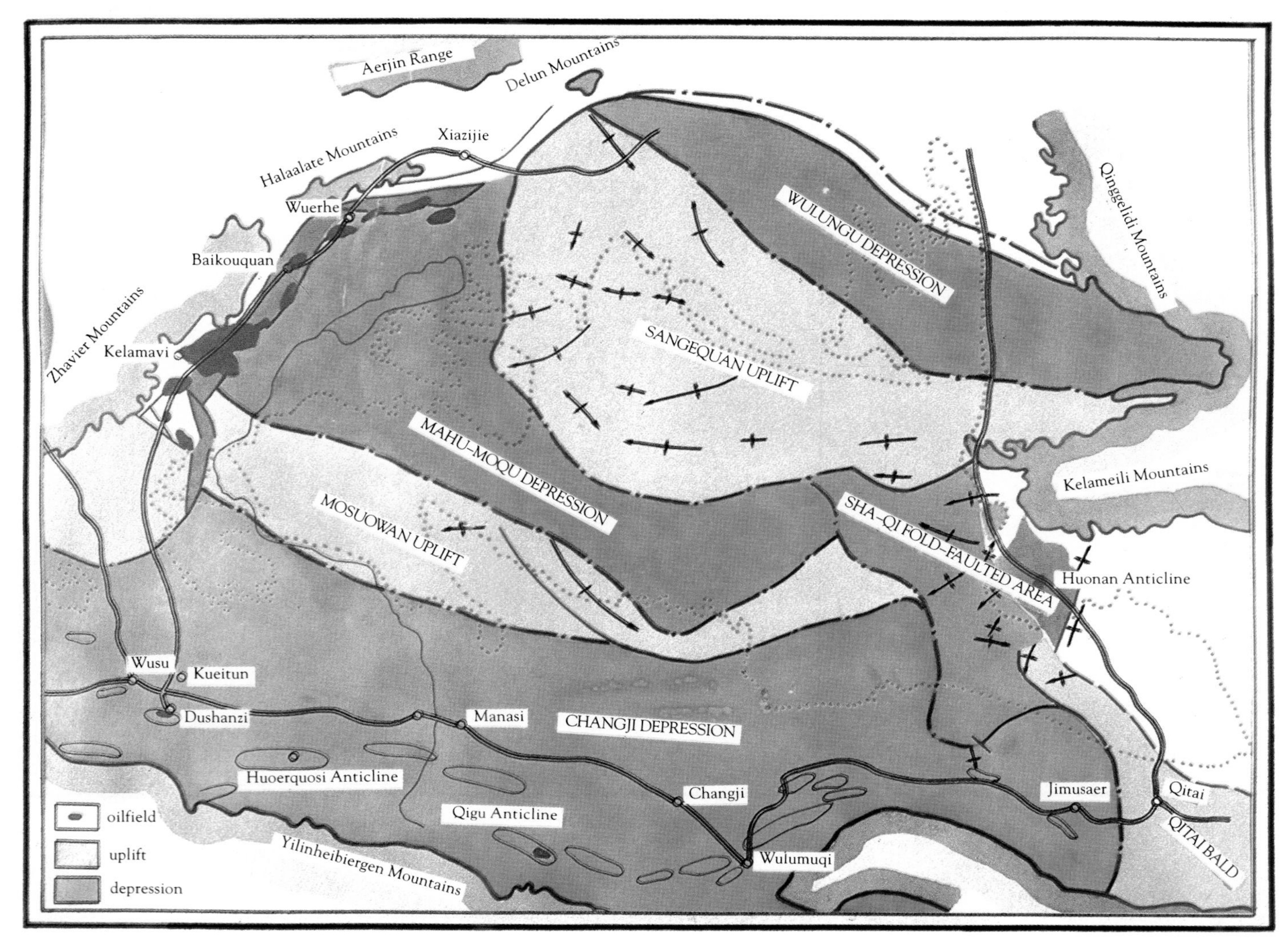


Figure 3. *Tectonic framework of Zhungeer basin.*

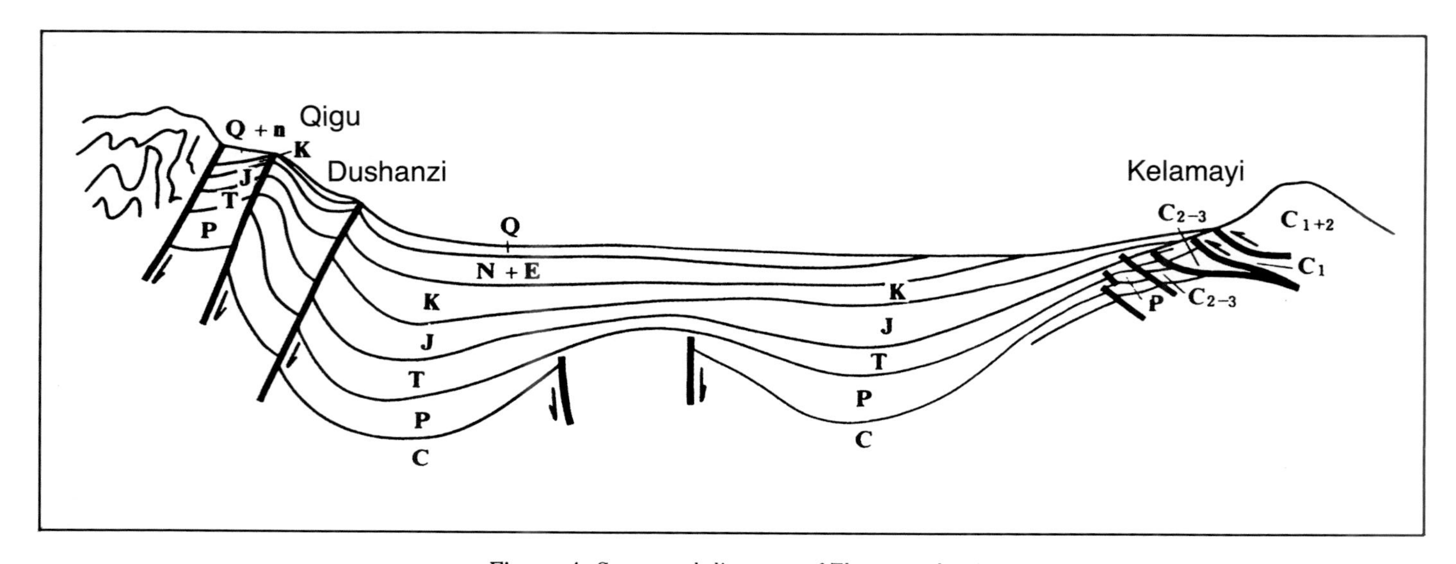


Figure 4. *Structural diagram of Zhungeer basin.*

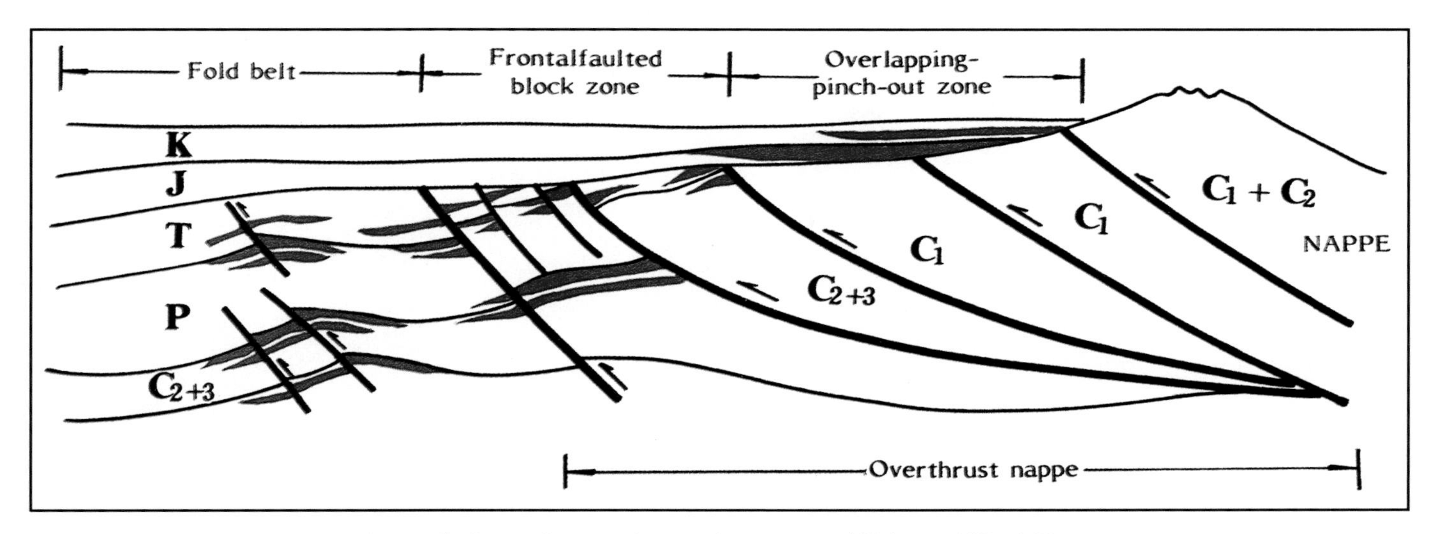


**Figure 5.** *Oil pool system in overthrust zone of Kelamayi-Xiazhijie area.*

There are two principle oil-bearing sedimentary sections in the Zhungeer, Talimu, Chaidamu, and Tulufan basins. One is dominated by marine carbonate rocks, and the other by non-marine clastics younger than Permian age (although Cretaceous and Lower Triassic marine deposits exist in some local areas). However, Mesozoic and Cenozoic sediments are overlain in nearly all 60 basins.

Since the Proterozoic Era, this region has experienced six major tectonic movements, and the basin features have changed several times. Four tectonic movements had significant influence: the Hercynian, Indo-China, Yenshan, and Himalayan movements. The basins formed through gradual evolution following each tectonic episode. The basins' common characteristics were shaped during the late Hercynian and Yenshan orogeny, and they took their present forms during the Himalayan orogenic event, with some basins still active during the Quaternary Period.

The following describes the geology of some of the main sedimentary basins.

## Zhungeer Basin

The triangular-shaped Zhungeer basin (Figure 3) is bounded by the Chengjishan Mountains on the northwest, the Aertai Mountains on the northeast, and the Tianshan Mountain Range on the south. It covers about 130,000 sq km (50,200 sq mi), with sediment 15,000 m (49,200 ft) thick in the south but thinning to 5,000 m (16,400 ft) to the north.

As a result of over 30 years' exploration, commercial oil discoveries have been made from Carboniferous, Permian, Triassic, Jurassic, Cretaceous and Tertiary reservoirs in the basin. Geologists have discovered nine oil fields in the southern and northwestern parts, with the Kelamayi oil field on the northwestern margin being the largest. Commercial oil has also been discovered in the Permian reservoir in the eastern part of the basin. While exploration in the northwest margin is more mature, all other areas in this basin are relatively unexplored.

The Kelamayi oil field is located on the northwestern margin, with oil reservoirs mainly controlled by an overthrust fault that is 250 km (155 mi) long and 20 km (12 mi) wide. Oil and gas seepages are widely found in the 5,000 sq km (1,930 sq mi) area. Carboniferous, Permian, Triassic, and Jurassic sequences dominate in the reservoirs.

The overthrust fault zone, which became active during the Early and Middle Jurassic, is a paleo-nappe structure that developed during the Carboniferous and Permian Periods. The geosynclinal zone of the Hercynian orogeny is exposed in the outcrop area west of the overthrust zone, being a northeast trending structural system. The paleocrystaline block lies in the central portion of Zhungeer basin. Based on recent seismic data (Figure 4), many northwest trending sedimentary depressions and uplifts have formed in the basin, with a very thick source-rock sequence in the depressions dominated by Carboniferous, Permian, and some Triassic and Jurassic source beds. Sedimentary cycles exist in each of the sections from middle-Upper Carboniferous to Tertiary. In the Kelamayi oil field, the main reservoirs are Carboniferous, Permian, Triassic, and Jurassic, with high production currently from the Triassic reservoirs.

Controlled by overthrust faults, oil pools in the Kelamayi oil field form in a special style (Figure 5). The middle-Lower Carboniferous is overthrust onto the middle-Upper Carboniferous, while the middle-Upper Carboniferous is overthrust onto Triassic and Permian, and the Jurassic and Cretaceous overlap the Carboniferous, Permian and Triassic. Therefore, geologists classify the oil- and gas-enriched areas into four types: (1) an overthrust nappe reservoir, (2) an overlap-pinch-out reservoir, (3) a frontal fault-block reservoir, and (4) a foldbelt reservoir. Thus, a large oil field exists over a rather extensive area, with multiple oil and gas zones and multiple, varied reservoirs. Limited by transportation, the annual production from this oil field is only about 4 million tons (28 million bbl).

The eastern part of Zhungeer basin is another potential oil- and gas-bearing area, with Kelameili Mountains to its north, and Begada Mountains to its south. Pressed by both mountain ranges, a zone trending approximately east to west forms, and a basinward extension of the base rock in the north separates the Wucaiwan from the Jimusaer depression. The Upper Permian is the main source sequence, forming very good oil- and gas-enrichment conditions.

A target for future exploration is the Jurassic sequence in the eastern part of the basin, where it forms wide outcrops with oil

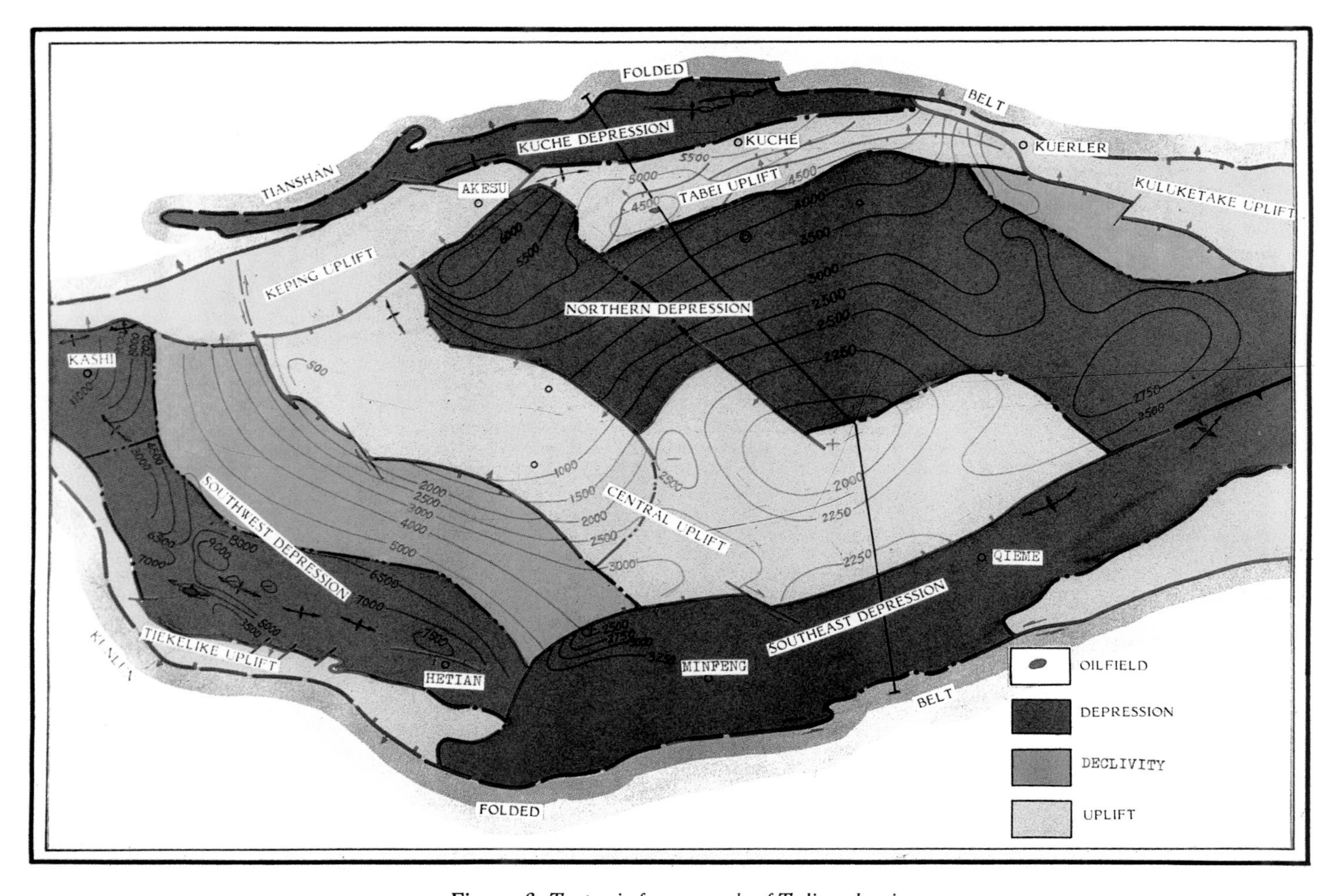


Figure 6. *Tectonic framework of Talimu basin.*

seepages and very good reservoir characteristics. A commercial discovery was made from one exploratory well drilled on Hoyanshan structure, with the Permian Pindichuan member being the pay zone. The zone, dominated by dark shale interbedded with silty fine sandstone, tested oil and gas from the interval of 1,791 to 1,863 m (5,876 to 6,112 ft).

This discovery represents a breakthrough in exploration in the eastern part of the basin. Because of the tectonic movement in the late Hercynian orogenic event, there are clear, angular unconformities and hiatuses at the top of the Paleozoic, and some local structures are not conformable with the overlying sections. Thus, in addition to the favorable traps of the local structures and uplifted horizons, unconformity and stratigraphic traps could also occur.

## Talimu Basin

The Talimu basin lies between the Tianshan and Kunlun Ranges, and is the largest inland sedimentary basin in China. The sedimentary section of Paleozoic, Mesozoic, and Cenozoic rock exceeds 10,000 m (32,800 ft), with a total sedimentary coverage of 560,000 sq km (216,200 sq mi). The lower Paleozoic consists of marine carbonates, with a thickness of over 3,000 m (9,800 ft); the Mesozoic comprises non-marine clastics of fluvial, marsh, and lacustrine facies; the Neogene is red sandy-shale; and the Paleogene is shallow-marine facies (Figures 6 and 7).

Geologists have explored the Talimu basin with a 1:1,000,000 aerial magnetic survey, a 1:500,000 to 1:1,000,000 gravity reconnaissance, a 30,000 km (18,640 mi) seismic evaluation primarily around the margin of the basin, and a geological reconnaissance survey in the outcrop areas. They found 137 surface structures, 59 subsurface structures, and over 100 oil and gas seepages. They drilled more than 50 exploratory wells, each exceeding 3,000 m (9,840 ft).

So far, geologists have established two oil and gas fields: the Kekeya oil and gas field and the Yiqikelike oil field. They have also found two structures with commercial oil (Tukermin and Shaya), and 13 structures with oil and gas shows.

Because some 70% of the basin area is desert, creating harsh surface conditions, most of the geological exploration has been done in the marginal areas. Limited exploration has occurred in the Kuche and Shache areas, while no exploration has been done in other areas including the wide desert. Exploring in this basin has intensified after highly productive oil and gas wells were drilled and the Kekeya oil and gas field was found in 1976. Since then, two G.S.I. seismic crews have contracted to do regional seismic profiles, which will provide preliminary information about the basin's structural framework.

Geologists have divided the Talimu basin into four major tectonic portions:

1. The central uplift, trending northwest with no Permian,

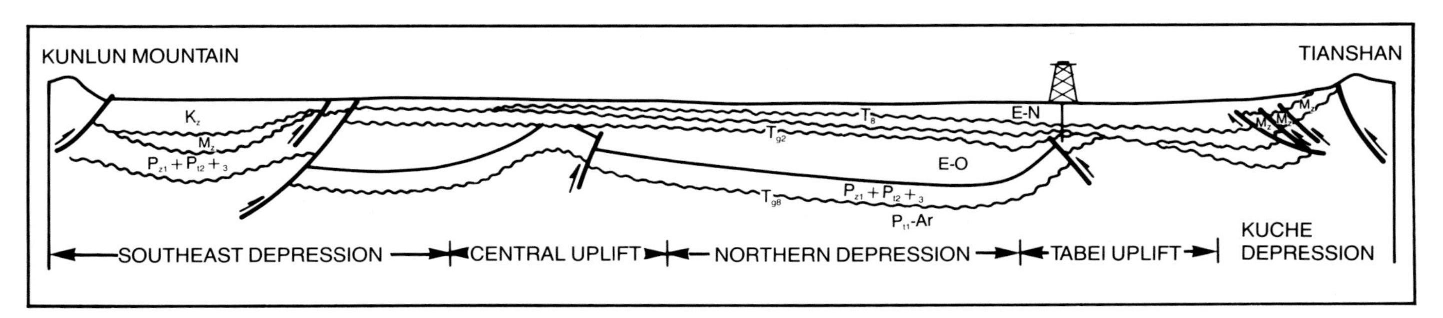


**Figure 7.** *Structural diagram of Talimu basin.*

Triassic, Jurassic, Cretaceous, or Paleogene, but with the Neogene directly overlying Carboniferous dolomite. Two stratigraphic wells have been drilled on this uplift, with oil shows in the Carboniferous interval. This large uplift between two large depressions has oil and gas potential.

2. The northern depression, covering the Tabei uplift, Kuerle slope, and Kuche depression. This depression has the most petroleum potential in the Talimu basin, with all three source sequences that exist in the basin being developed.The Kuche depression in the north has four rows of favorable east-to-west trending structural zones, and old to young sections outcrop from north to south and to the Tabei uplift further south (Cretaceous to Neogene). In the stratigraphic test drilled recently, large quantities of oil and gas were blown out from Ordovician dolomite (?) at 5,300 m (17,400 ft). This is the deepest oil pay zone ever found in China. After a seismic survey in the adjacent area to the west and east of this structure, geologists have defined a number of structures. Obviously, this is an area with great potential.

   Geologists discovered the Yiqikelike oil field in 1958, in the first structural trend in the Kuche depression. They found a Jurassic reservoir that is currently being developed of high quality oil (0.79–0.80 specific gravity).

3. The southwestern depression located between the Kunlun Range and the central uplift, includes the southwestern slope, the Yecheng depression, and the Keshi depression. In 1977, the Kekeya oil and gas field was discovered in the Yecheng depression in front of the Kunlun Range. This discovery became a major breakthrough for petroleum exploration in the Talimu basin. Of special significance is a highly productive oil well in the red bed of the Miocene Epoch. This discovery has caused petroleum exploration in the Talimu basin to be intensified. The Kekeya oil and gas field, 12 km (7.5 mi) long and 4 km (2.5 mi) wide, trending north-to-south, covers about 34 sq km (13 sq mi).The Miocene strata contain the oil and gas zones within a short, axial anticline with a steep flank in the north and gentle slope in the south. In this field, fault throws are less developed at deep horizons. Pay zones are about 4,000 m (13,100 ft) deep, with poor reservoir quality; porosity ranges from 12–16%, average permeability is 20–30 md with a few exceptions reaching 90 md. The quality of crude is good; the specific gravity is 0.79–0.83, with wax content averaging 6–9% and reaching a maximum of 16%. There are a total of five oil and gas pay zones. The Kekeya oil and gas field is a typical field, controlled by local structure, and having multiple pay zones with a net thickness of 180 m (590 ft). The five oil and gas pay zones consist of three gas condensate reservoirs and two gas condensate with oil rings.

4. The southeastern block fault area, located in front of the Aerjin Mountains, is not explored yet, and has had only two south-to-north seismic lines extended into it. Thus, the geology in this area still is not clear.

Based on stratigraphic test data, three source sections are recognized: (a) Carboniferous-Permian, consisting of limestone, dolomite interbedded with shale, and dark grey Permian shale interbedded with sandstone in the southwestern part, (b) Upper Triassic to Jurassic, comprising dark gray shale interbedded with sandstone, with wider distribution, and (c) Upper Cretaceous to lower Tertiary, consisting of marine shale and limestone.

The thicknesses of these three source sections vary, ranging from 250 to 500 m (820 to 1,640 ft), with wide distribution. Geologists consider the potential reserves to be considerable.

### Chaidamu Basin

The Chaidamu basin, bounded by the Kunlun Range in the south and the Aerjin and Qilian Mountains in the north, is a large, intermountain basin, with an area of 120,000 sq km (46,000 sq mi) (Figure 8). Since exploration began in this basin in 1954, geologists have completed a 1:1,000,000 aerial magnetic survey, a 1:200,000 gravity-magnetic survey, a geological survey of the whole area, and a semi-detailed and detailed survey of local structures. They found 170 surface and subsurface structures. The thickness of Mesozoic and Cenozoic strata in this basin exceeds 8,000–15,000 m (26,000–492,000 ft).

Based on the comprehensive study of marginal geology and geophysical data, the Chaidamu basin can be divided into three portions (Figure 9):

1. The northern part of the basin comprises a folded basement of early Hercynian age, overlain by a northern marginal block-fault zone. The entire Tertiary is a red and grayish brown section that is not source rock. However, there are block-faulted oil and gas fields of Jurassic age along its margins. From north to south, there are three rows of structural trends, dating from old to young. The Tertiary Lenghu oil field and Jurassic Yuqia oil field are in this region.
2. The western part of the basin is the folded basement of late Hercynian age, overlain by a subsided zone of predominantly Oligocene to Pliocene. This part is the most favor-

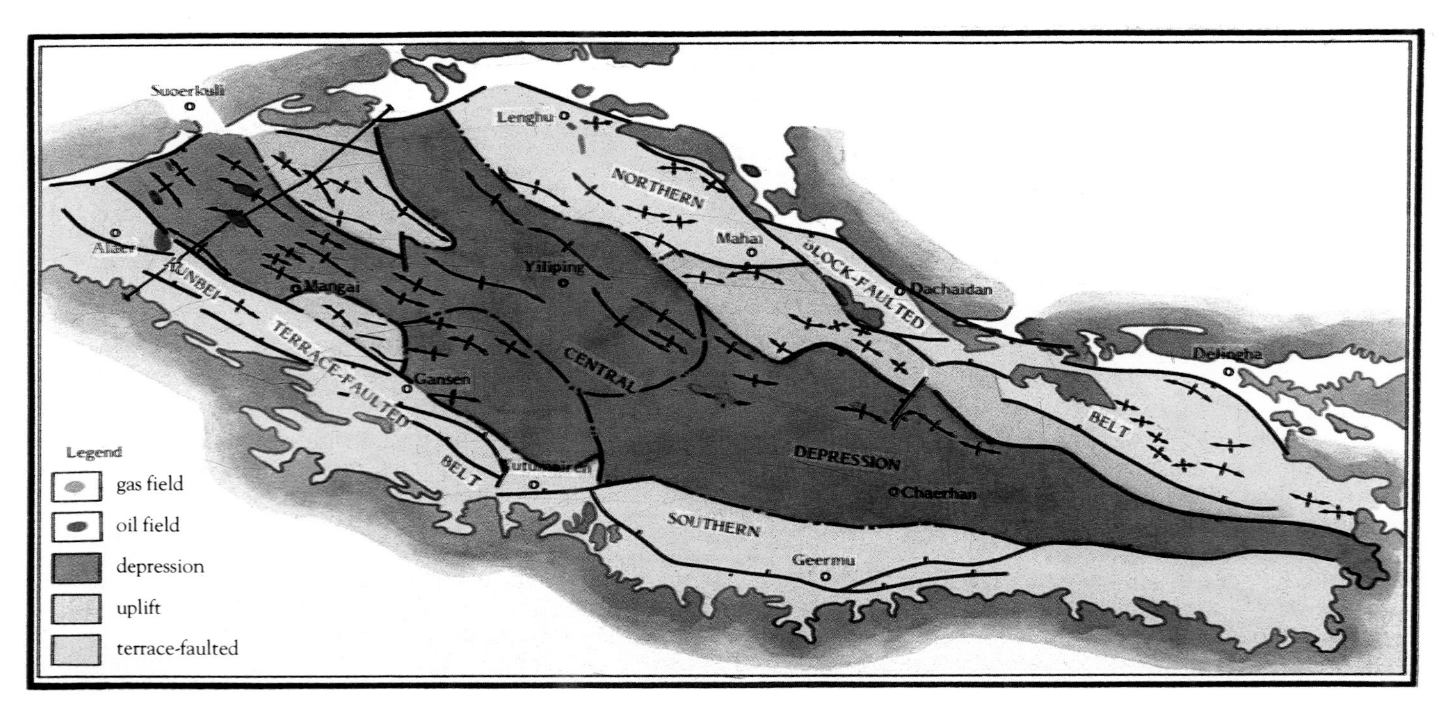


Figure 8. *Tectonic framework of Chaidamu basin.*

able oil- and gas-bearing area, with well-developed source rocks. There are four source beds ranging from Paleogene to Neogene age, with a thickness that reaches to 1,700 m (5,600 ft). Also, there are four reservoir sequences. Many surface and subsurface structures exist in this region. More significant, however, is the presence of several structures perpendicular to the northwesterly direction formed by south-to-north trending faults proven by seismic exploration in front of the Kunlun Range. These structures are very favorable for oil and gas accumulation and enrichment.

3. The eastern part of the basin is a large area with a depositional thickness exceeding 12,000 m (39,400 ft). The upper 2,000 m (6,600 ft) comprise lake and marshland Quaternary deposits. This excessive Quaternary thickness indicates new structural activity in the Chaidamu basin during that period. Four Quaternary gas fields have been discovered in this area, showing it to be relatively rich in natural gas resources. The Tertiary is also a potential objective in this area.

After exploring this basin for over 30 years, geologists have completed more than 20,000 km (12,400 mi) of seismic research, drilled 1,300 exploratory wells, and discovered 16 oil fields and five gas fields. Based on the integral geological study, there are many favorable conditions for oil and gas prospecting in this basin, including:

1. Good oil source conditions. The well data show that source rocks exist in all sequences from Jurassic to Quaternary. Based on the marginal geology of the basin, the sequence of Carboniferous to Permian is also a potential source.
2. Superior reservoir conditions. The sandstone reservoir in Guoskule oil field, at 4,000 m (13,100 ft), has 18–20% porosity and over 100 md permeability. The reservoir properties tend to be even better above 3,000 m (9,800 ft).
3. Numerous large structures. The average size of the structures is several hundred square kilometers, with some as large as 1,000 sq km (385 sq mi).
4. High formation pressure, with strong flowing capability and high well rates. Oil wells in the Guoskule oil field average 300–400 tons (2,100–2,800 bbl) per day. The Chaidamu basin is a pressure-bearing basin, which is the main factor that led to the high formation pressures. The disadvantages of the Chaidamu basin are its distance from the industrial cities, and its poor topographical conditions. Because of its climate, most of the areas in the basin are covered by gobi and alkaline earth. Although exploration is underway in this basin, progress is slow.

### Tulufan Basin

The Tulufan basin is bounded by the Begeda Mountains to the north and the Juelotake Mountains to the south. It is an inter-mountain basin of Mesozoic and Cenozoic age within the Tianshan Range folded zone, trending northwest to southeast, and covering about 40,000 sq km (15,400 sq mi). The sedimentary section is as 8,000 m (26,200 ft) thick. What little exploration the basin has had took place in the early 1950s. Exploration data for the area include a 1:200,000 gravity-magnetic reconnaissance, about 4,000 low-quality seismic lines and a few exploratory wells. Geologists discovered the Qiketai and Shengjinkou oil fields, both small fields producing from Middle Jurassic and Upper Triassic reservoirs. Since the 1960s, exploration generally has ceased because of a lack of personnel.

The general structural framework of the Tulufan basin is an

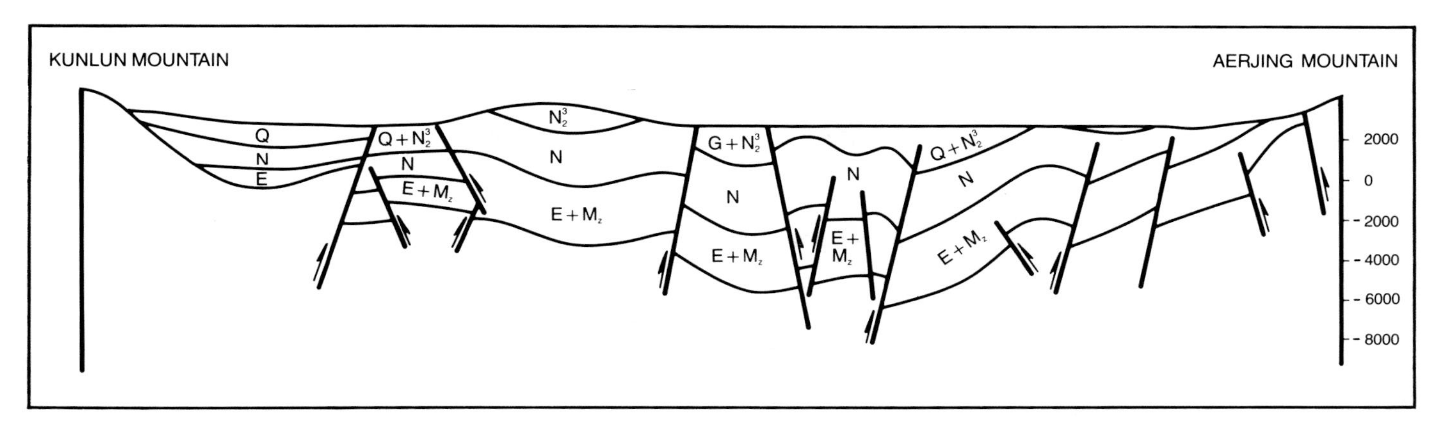


Figure 9. *Structural diagram of Chaidamu basin.*

anticlinal belt in the center of the basin that parallels the basin trend. The belt is composed of a series of east to west trending anticlines, with a large overthrust belt on the southern flank which has made the stratigraphic units on the southern flank vertical-to-overturned. The Qiketai and Shengjinkou oil fields are all found on the anticline belt, with the Middle Jurassic and the Upper Triassic being the pay zones. To the north of this anticline belt lies the northern depression covered by Quaternary gravels. It contains Triassic and Jurassic source rocks and possibly Permian source rock. According to surface geology on outcrops, the Paleozoic sequence in the north is overthrust onto the Mesozoic and Cenozoic. There are more east to west trending short axial anticlines further south, resulting in that being a favorable area for oil and gas prospecting. To the south of the anticline belt lies a south to north trending, gentle monocline covered by Quaternary deposits on the surface. The Jurassic overlaps and pinches-out to the south.

The Tulufan basin has well-developed source, reservoir, and trapping conditions, and early formed structures. These make it a favorable area for oil and gas accumulation and enrichment.

### Jiuquan Basin

The Jiuquan basin lies in front of the Qilian Mountain Range and trends northwest to southeast. It covers 2,700 sq km (1,040 sq mi) (Figure 10). Since exploration there began in 1936, geologists have found four oil fields and many oil seepages, including the Laojunmiao oil field, the Yaerxia oil field, the Baiyanghe oil field, the Shiyougou oil field, and the commercial oil flows from the deeper Cretaceous and Silurian in Yaerxia. In this basin, the principal reservoir is Tertiary. Geologists have recently found others: Cretaceous and Jurassic oil-bearing reservoirs, a Silurian metamorphosed rock reservoir, and a Carboniferous carbonate rock reservoir with oil and gas shows.

The Laojunmiao oil field is a typical, structurally controlled oil field, with a overthrust on the northern flank (Figure 11). It was found in 1938 and has been producing for 46 years, with the Baiyanghe member of Oligocene age as the pay zone. The reservoir properties are good: 20–25% porosity and 100–300 md permeability. This field has produced, to date, 19 million tons (133 million bbl) of crude and is producing at an annual rate of about 500,000 tons (3.5 million bbl) of crude.

As a result of the application of digital seismic recording equipment and an integrated geological study, geologists think that there is still potential for oil and gas. They believe that they can make discoveries in the extensional areas of both sides of the Laojunmiao structure and in the overthrust belt on the flank of the Qilian Mountains. Exploration is continuing in the Jiuquan basin.

## DISCUSSION AND SUMMARY

Being influenced mainly by the Indian and Siberian plates in the northwestern part of China, formation of most of the sedimentary basins is related to orogenic compression. Based on the overall study on the major basins, we can summarize the following characteristics:

1. Most of the basins formed in an asymmetrical shape, and developed with obvious Mesozoic and Cenozoic mountain-front depressions with a subsidence of over 10,000 m and no obvious eruptions.
2. Overthrusts control the boundaries of the basins. It is common to have several overthrust faults along basin margins, forming an overthrust zone and larger-than-normal oil accumulations.
3. Local structures in basins are repetitious and occur as various types, with either linear, or en echelon, arrangement.
4. Most of the basins formed during Hercynian and Indian-China periods, and were reworked during the Yenshanian and Himalayan orogenies. Because of multiple tectonic movements, they were formed with source, reservoir, and cap rock associations with multiple cycles, multiple ages, and multiple horizons. There are several source beds: Carboniferous, Permian, Triassic, Jurassic, Cretaceous, Paleogene, Neogene, and Quaternary. There are multiple reservoirs, such as Ordovician, Silurian, Carboniferous, Permian, Triassic, Jurassic, Cretaceous, Paleogene, Neogene and Quaternary. There are various types of structures, and many local structures, with 690 total structures currently known. These include anticlinal structures, structures in linear and en echelon shapes, flexure caused by fault-overthrust nappes, overlapping-pinch-outs, fault traps on monoclines, and salt dome diapirs. Various types of oil pools are formed because these structures are near oil-generating depressions.
5. The northward pushing of the Indian plate has formed a series of northwest and nearly east to west trending uplifts

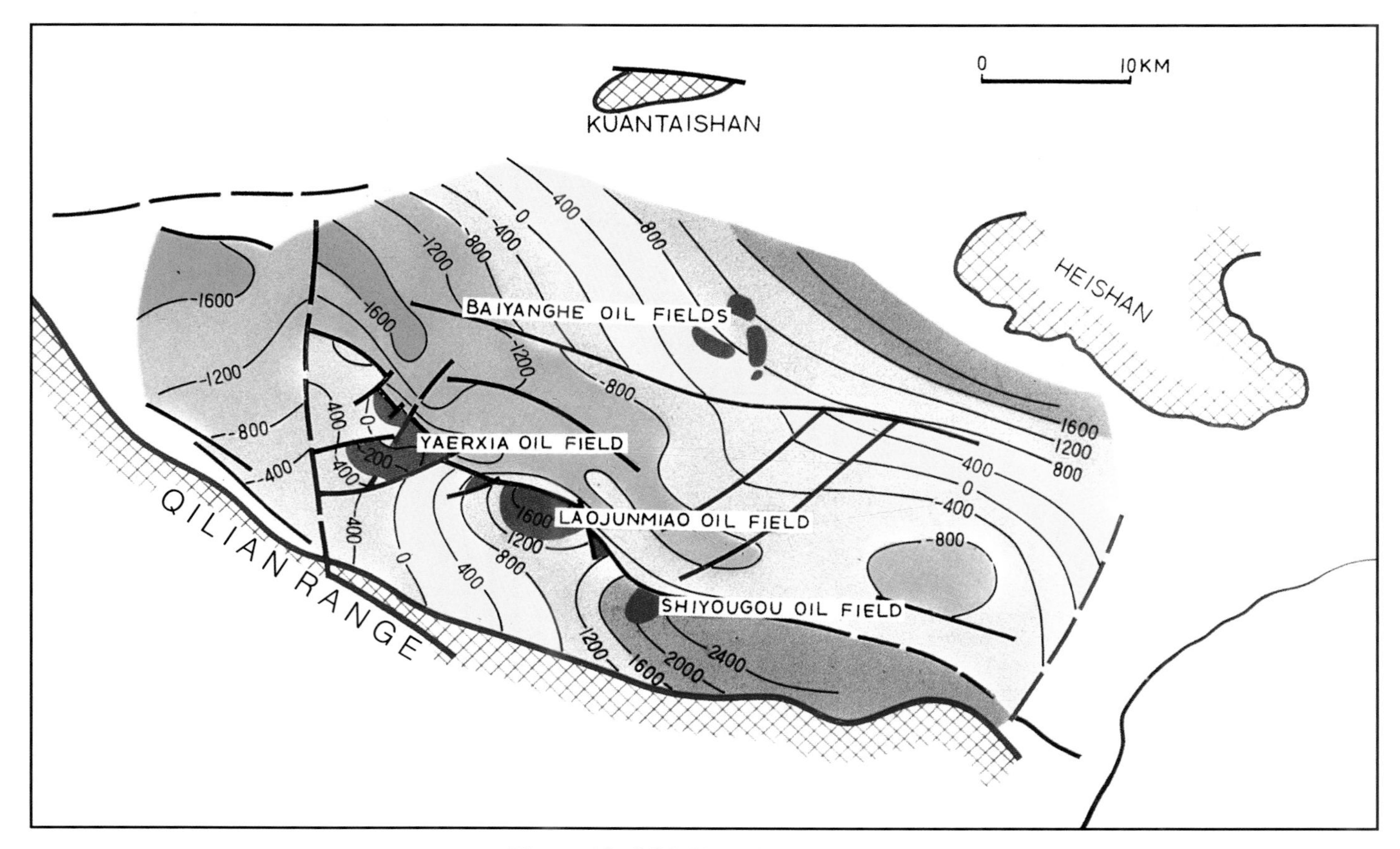


Figure 10. *Oil fields in Jiuquan basin.*

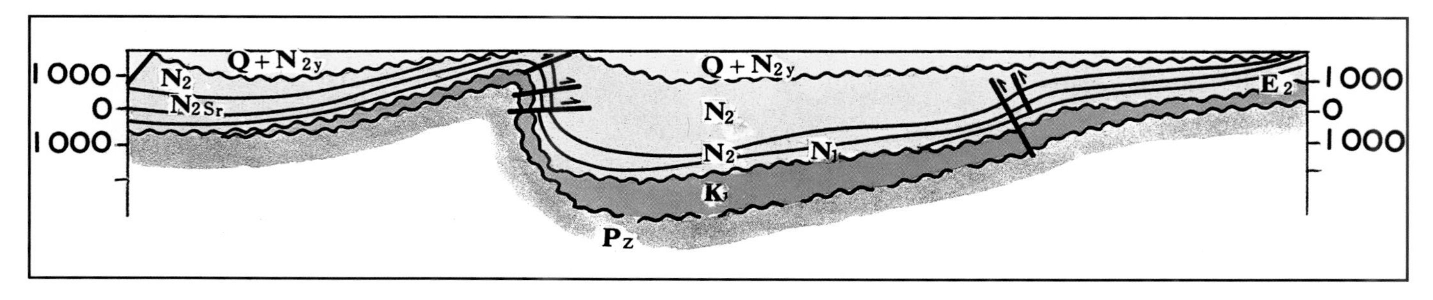


Figure 11. *Structural diagram of Jiuquan west basin.*

and subsidence zones, particularly the mountain-front depressions parallel with mountains, along the Tianshan, Kunlun, Aerjin, and Qilian ranges. These basins exceed 10,000 m (32,800 ft) in depth. Within the depressions, three to four rows of structural zones normally exist. Because these basin depressions all have oil-generating conditions, they could form oil- and gas-rich areas.

6. The crust thickness in northwestern China is greater than in the eastern part of China (Table 1). Thus, the thermal gradient in the northwestern part of China is relatively low and the threshold depth for oil generation is also relatively greater. Even at depths beyond 5,000 m (16,400 ft), not only may gas be found, but also oil, as proven by the drilling in the Zhungeer, Talimu, and Chaidamu basins. Geologists have discovered commercial oil in these basins at depths below 5,000 m (16,400 ft), revealing that further oil and gas exploration is needed for the deep horizons in the northwestern part of China.
7. In northwestern China, upper and lower structures often do not coincide. This is a result of several factors, including the hiatuses, erosion, and unconformities created by various tectonic movements, lithology differences between paleozoic marine carbonate and Mesozoic-Cenozoic nonmarine clastics, gypsum and salt rock development in local areas, and the compression of the Indian and Siberian plates. For example, sometimes the surface structure is a syncline, which may overlie the crest of an anticline subsurface. Thus, it is vital to define the deep structure clearly first, by an accurate seismic survey, in order to improve success during exploration activity.

Table 1. *Major basins of eastern and western China.*

| Name of basin | Thickness of Crust (km) | (mi) | Thermal Gradient |
|---|---|---|---|
| East Part of China | | | |
| Songliao basin | 35 | 21.7 | 3.1 - 4.8 |
| Xialiaohe basin | 34 | 21.1 | 3.1 - 3.6 |
| Bohai Gulf basin | 37 | 23.0 | 3.3 - 3.6 |
| West Part of China | | | |
| Talimu basin | 51 | 31.7 | 1.7 |
| Chaidamu basin | 53 | 32.9 | 2.8 |
| Jiuquan basin | 47 | 29.2 | 2 - 2.5 |

## CONCLUSIONS

It is to be correctly assumed that this vast northwestern region of China is yet to be adequately assessed and tested. It is felt that with more geological and geophysical studies, these basins may contribute a sizable portion of China's future petroleum needs.

# Philippine Islands: A Tectonic Railroad Siding

John J. Gallagher, Jr.
*Occidental Exploration and Production Company*
*Bakersfield, California*

In 1976, a Philippine-American consortium led by Philippines-Cities Service, Inc. discovered significant quantities of offshore oil northwest of Palawan Island. This discovery was the first commercial oil found in the Philippine Islands. New data suggest that the islands are continental and oceanic crustal fragments. These tectonic fragments have moved to their present positions along major strike-slip zones and plate boundaries. Components of each element are like railroad cars, while strike-slip and plate boundary zones are like railroad rails. Groups of components collect in certain places similar to the collection of railroad cars at railroad sidings.

Three major elements are recognized in the Philippines: the Cuyo, the Sulu, and the Luzon-Mindanao accreted terranes. The carbonate and clastic sediments of these terranes are currently deforming by strike-slip, subduction, and extension.

The tectonic history of the Philippine Island area potentially limits the size of oil accumulations. This greatly increases the risk associated with hydrocarbon exploration in the area.

## INTRODUCTION

The key to successful future exploration for oil and gas in the Philippine Islands may depend on the explorer's ability to decipher what was where, when, and how this can help locate adequate reservoirs. In other words, present knowledge may be sufficient to specify in a general way the position of each tectonic element of the very complex Philippine Island tectonic collage during each step of its history. This knowledge may have significant favorable or unfavorable implications for exploration.

A main conclusion of this paper is that the tectonic source areas for each element are fairly well identified. This paper shows, then, how the basinal areas in the Philippine Islands are associated with tectonic elements. We will reconstruct the paleo-positions of both the tectonic elements and the basins using current data and concepts about movement of the elements.

The published literature disagrees about the paleo-position of certain specific components within tectonic elements. For example, the eastern islands from Luzon southeastward to Mindanao may have moved northward (Karig, 1983), or they may have developed *in situ* (McCabe, 1984). This paper assumes that the northward movement of "railroad-car-like" components of tectonic elements along "railroad-track-like" plate boundaries explains paleotectonics, and the *in situ* interpretation explains present-day "railroad siding" tectonics.

The components of the tectonic elements on the Cuyo shelf area moved southeastward across the South China Sea and away from the South China margin. The published literature reflects this scenario, but reconstruction details differ (Ben-Avraham, 1978; Chen, Gallagher, and Kenyon, 1979; Taylor and Hayes, 1980; Holloway, 1981).

The components of the tectonic elements, including the Sulu Sea, southern Palawan Island, and the northeastern part of Borneo, moved northeastward and underwent some clockwise rotation (Karig, 1983; Taylor and Hayes, 1983).

This paper roughly defines the limits of the three tectonic elements that appear to have distinctly separate geologic histories and briefly discusses the history of each tectonic element. The focal point of the paper is the discussion of sample play concepts for basins associated with each tectonic element. The conclusion is that exploration risk is high and risked returns on investment are low because the size of the "railroad tank cars" containing oil are probably small.

## TECTONIC ELEMENT BOUNDARIES

The Philippine Islands lie along a convergent plate boundary marked by the suture zone between the Philippine Sea to the east and the South China, Sulu, and Mindanao Seas to the west (Dewey, 1977; Moore and Silver, 1983). The sedimentary basins with hydrocarbon potential are on or adjacent to the largest islands. The largest single basin in the Philippine Islands surrounds Palawan Island (Figure 1). We can distinguish three major tectonic elements based on tectonic affinities:

1. The *Cuyo* tectonic element includes several major components: the Cuyo shelf, which is the region that extends from Palawan Island north of the Ulagan Bay fault to and including a large part or all of Mindanao Island; the islands between Reed Bank and Macclesfield Bank and various other bathymetric highs in the South China Sea; and a large, prospective basinal area that surrounds the Cuyo shelf. Most of these components originated as continental crustal fragments that broke off and later drifted away from the South China continental margin (Ben-Avraham, 1978; Chen, Gallagher, and Kenyon, 1979; Kenyon, Chen, and Gallagher, 1979; Gallagher, 1981, 1984; Taylor and Hayes, 1980; Holloway, 1981, and 1982).
   Breakup may have begun in Cretaceous (Chen, Gallagher and Kenyon, 1979) and was almost certainly underway from Oligocene to Miocene (Taylor and Hayes, 1980, 1983) if the magnetic anomaly is correctly identified (Figure 2).
2. The *Sulu* tectonic element also includes several components: Palawan Island south of Ulagan Bay fault, the extensive basinal areas north and south of the island, and the Sulu Sea basin. The origin of this element is still uncertain, but it has geological affinities to Borneo (Taylor and Hayes, 1983). The Sulu element could have acted as a rigid indentor pressing into and bending the components of the Luzon-Mindanao tectonic element (McCabe, 1984). The Sulu element, therefore, would be analogous to India and the Luzon-Mindanao element would be analogous to China in the Tapponnier and Molnar (1976) tectonic analysis.
3. The *Luzon-Mindanao* tectonic element is the largest and apparently the most complete Philippine accreted terrain or "railroad siding." Its components include the major islands from Luzon to Mindanao, and the surrounding smaller islands and adjacent basins, eight of which may prove to bear hydrocarbons. This element could have been created from ocean floor by a combining suture and transform (Karig, 1983; Moore and Silver, 1983). This paper accepts that the components could have been transported great distances northward along the plate boundary (Karig, 1983, and Figure 2). Alternatively, much of the element could be arc related and not greatly displaced northward (McCabe, 1984).

Data (Figure 1) do not define well the precise limits of these three tectonic elements. However, data can support the following hypothetical tectonic histories for each tectonic element.

## CUYO TECTONIC ELEMENT

The basins within the Cuyo tectonic element extend northeastward from the Ulagan Bay fault to the region past the Cuyo shelf (Figure 1). These basins appear to have the best possibilities for oil and gas of any in the Philippine Islands. To date, however, the only commercial production comes from the basinal area northwest of Palawan Island.

The depositional history of Cuyo element basins since Late Jurassic time was shallow-water followed by deep-water and then once again shallow-water carbonates and clastics (Hatley, 1977; Beddoes, 1980). Unconformities exist in the Eocene, in the middle Miocene, and the end of the Miocene (Saldivar-Sali, Oesterle, and Brownlee, 1981). The first oil was found in fractured carbonates of lower Miocene pinnacle reefs (Brownlee

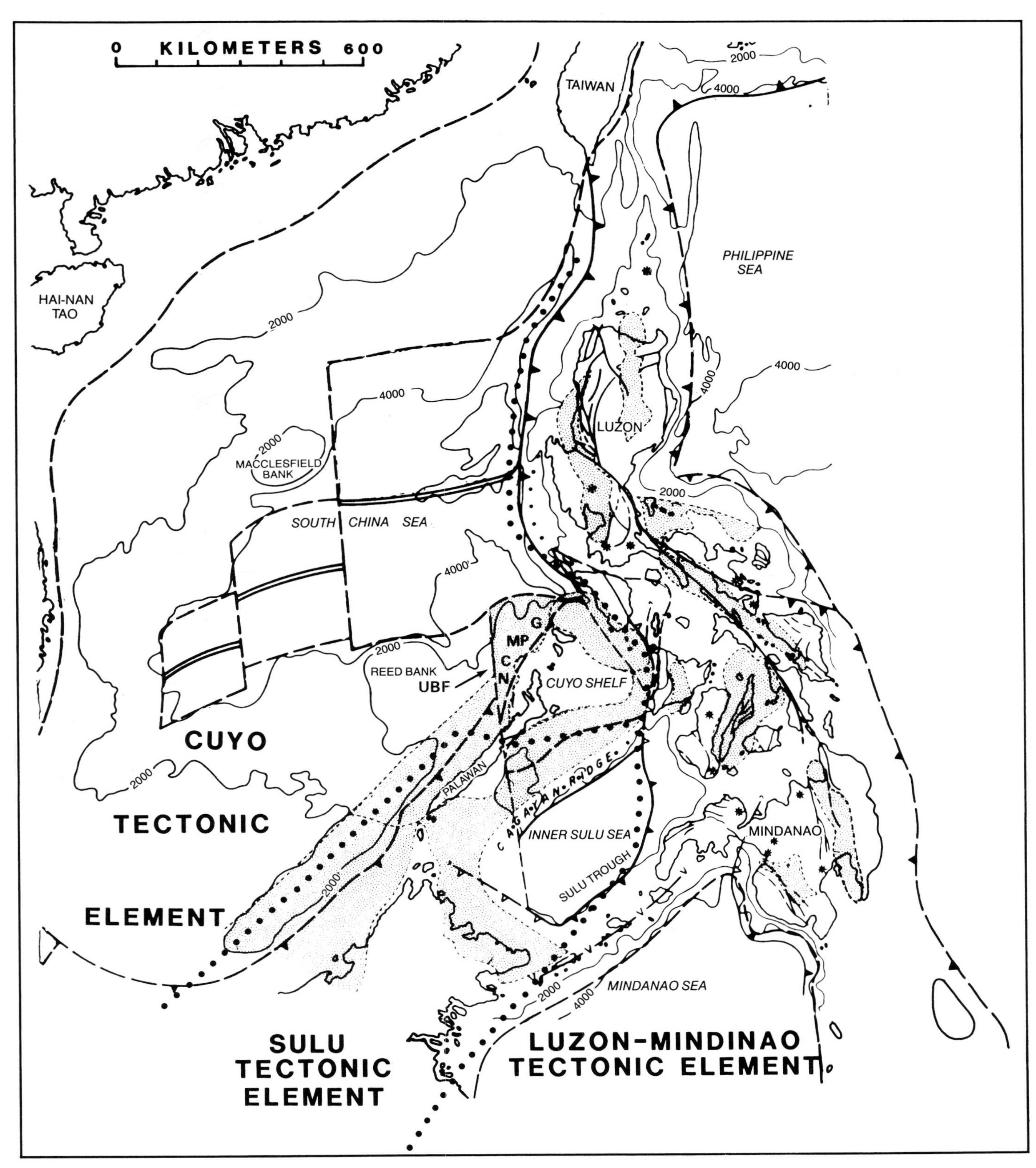


Figure 1. *Location map for the Philippine Islands showing approximate boundaries of tectonic elements (adapted from D. N. Brownlee personal communication, 1983). Basinal areas are stippled; UBF = Ulagan Bay Fault, G = Galoc-Linapacan, MP = Matinloc-Pandan, C = Cadlao-Libro, N = Nido.*

and Longman, 1981). Later, oil was found in Miocene turbidites folded into anticlines (Saldivar-Sali, Oesterle, and Brownlee, 1981; Brownlee and Longman, 1981).

The tectonic history outlined below[1] can explain the structure and stratigraphy of the Cuyo element basins and can help suggest possibilities for future play concepts. Figures 2A to 2E outline the hypothetical tectonic history of the Cuyo element. They are modeled after data and ideas in Chen, Gallagher, and Kenyon (1979), Kenyon, Chen, and Gallagher (1979), Taylor and Hayes (1980, 1983) and Karig (1983). Supporting data and concepts come from Hamilton (1979) and Holloway (1981). The plate-tectonic concepts are refinements of work by Ben-Avraham (1978). Some tectonic-stratigraphic interpretations are based on the basin classification of Murphy (1975).

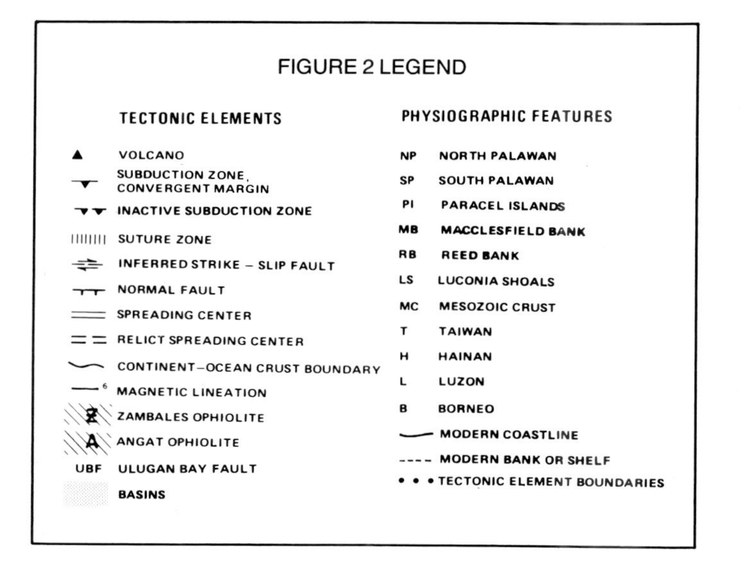


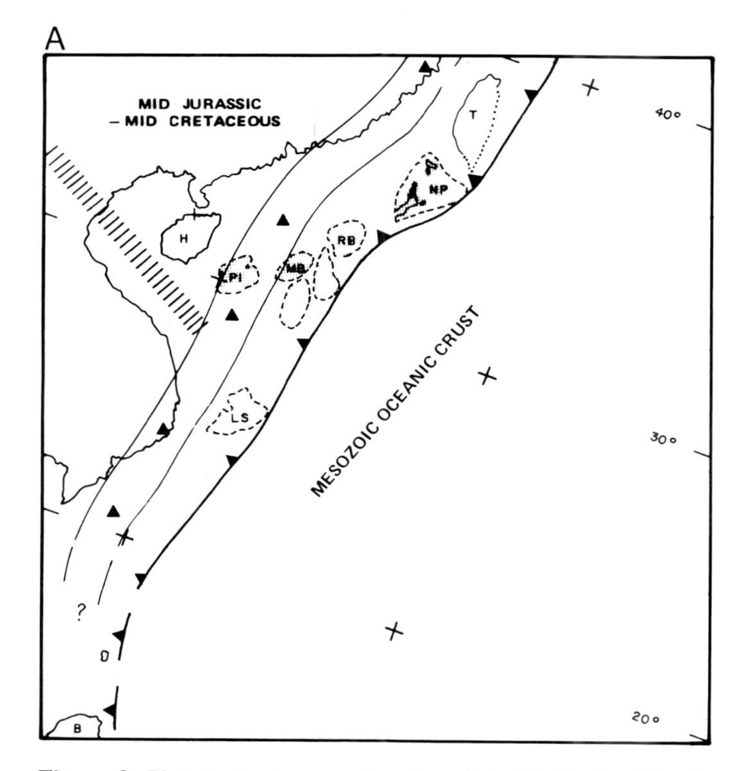


**Figure 2.** *Plate tectonic reconstruction of the Philippine Islands selectively combining data and interpretation from Ben-Avraham (1978), Chen, Gallagher, and Kenyon (1979), Kenyon, Chen, and Gallaher (1979), Karig (1983) and Hayes and Taylor (1983).*

[1]A. *Mesozoic Subduction Complex.* Figure 2A shows the presumed Mesozoic position of the Cuyo shelf around Northern Palawan Island (NP) near Taiwan (T). Reed Bank (RB), Macclesfield (MB), Paracell Islands (PI), and similar bathymetric features are components of the tectonic element in the South China Sea and consequently are likely to have histories similar to those of the Cuyo element. Chen, Gallagher, and Kenyon (1979) proposed this Mesozoic arrangement of components including the Cuyo element because of the

i. similarities between the geology in wells south of Taiwan Island and wells around Palawan Island;
ii. Eocene clastics from wells on Reed Bank with no nearby provenance that could have given rise to quartzose coarse clastics;
iii. the trend and ages of magnetic anomalies in the South China Sea (reported to Cities Service and the other members of the industrial associates at Lamont Doherty Geological Observatory during the late 1970s), which suggests separation of the Cuyo element from the South China margin;
iv. igneous and sedimentary rocks of the components that appeared to be consistent with rocks expected offshore but along trend with the three Mesozoic arc-trench complexes found onshore China.

Saldivar-Sali, Oesterle and Brownlee (1981) present data that further support the Figure 2A reconstruction. The Cadlao No. 1, Guntao No. 1 and Catalat No. 1 wells have sediments that "may be generally assigned to a continental grading to arc-trench style of deposition."

B. *Late Cretaceous to Early Eocene Rift and Strike-Slip.* Figure 2B suggests the early Tertiary (and possibly Late Cretaceous) tectonic extension of the South China Platform. Grabens were interpreted from regional seismic data for the area near the present-day China margin. Strike-slip faults were interpreted from Landsat of China and can be extrapolated offshore using seismic data (R.J. Reed personal communication, 1979).

C. *Early Oligocene to Middle Miocene Drift.* Figure 2C shows that by anomaly 11 time (approximately 32 Ma, early Oligocene), most components had separated and moved southward away from the South China Coast. By anomaly 6B time (approximately 23 Ma, early Miocene), the components were shallow areas in the middle of a sea and were approaching an arc-trench complex. Local uplift suggests initial collision because reefs were occasionally subaerially exposed (reported in Nido and Matinloc wells by Brownlee and Longman, 1981). Tectonic interpretations of stratigraphy in the Pandan and Cadlao-Libro areas further support this interpretation of Figure 2B (Saldivar-Sali, Oesterle, and Brownlee, 1981).

D. *Middle Miocene Collision.* Figure 2D shows that by mid-Miocene time, the collision occurred between the Cuyo tectonic element and the subduction complex to the south. Saldivar-Sali, Oesterle, and Brownlee (1981) present abundant stratigraphic and sedimentological data for tilting, uplift, and erosion that would be consistent with collision.

E. *Late Miocene and Younger Accretion.* Figure 2E shows the present-day geography in which the Cuyo element and the Sulu element appear to be combined. According to McCabe (1984) these tectonic elements may be welded together and may act as a rigid indentor pressuring and bending the Luzon-Mindanao terrain. If the indentor hypothesis is correct, the strike-slip faults may bound the northwest and southeast edges of the Cuyo element (? on Figure 2E). This hypothesis may explain the anticlines attributed to compression reported in the Galoc area (Saldivar-Sali, 1981). The hypothesis may also explain the Palawan trench that is generally attributed to subduction along southwestern Palawan Island (see, for example, Holloway, 1982) but which may be an extensional microcontinental edge rejuvenated as a strike-slip fault along the northwestern border of the northeastern Palawan Island.

## SULU TECTONIC ELEMENT

The basins of the Sulu tectonic element surround southern Palawan Island (southwest of Ulagan Bay fault, Figure 1) and

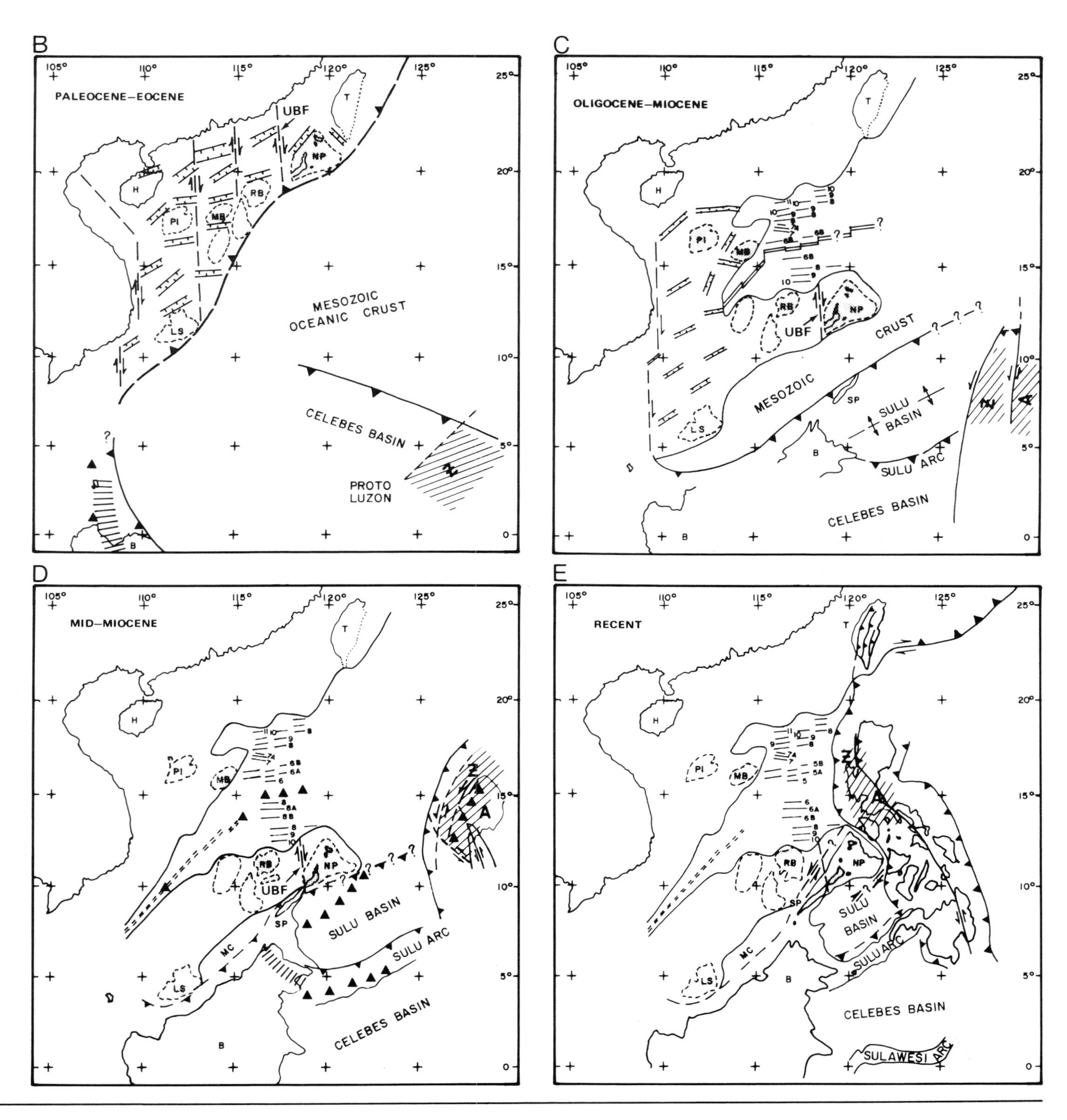


northeastern Borneo. To date, no commercial oil has been found in these basins, but the basins provide many interesting opportunities. The history of the Sulu element is not as well known as that of the Cuyo element, and most of the Sulu element basinal area is not within the present-day political boundaries of the Philippine Islands. During Mesozoic time the Sulu tectonic element may have been the southern extension of the southeast China arc-trench complexes (Figure 2A). By Paleocene time the Sulu element may have been involved in some presently undefinable way with suturing (Figure 2B) so that by Miocene time, the general geography of the Sulu element seems to have been set (Figure 2C). At mid-Miocene (Figure 2D) ophiolites were forming and may have been in place on southern Palawan Island and on Sabah (Hutchinson, 1975). The Sulu element now seems to be joined with the Cuyo tectonic element as described above (Figure 2E).

## LUZON-MINDANAO TECTONIC ELEMENT

From an oil explorer's perspective, the Luzon-Mindanao tectonic element has what are perhaps the least interesting basins, but have had the most interesting and complex tectonics. At the present time, the Mesozoic and older history of the Luzon-Mindanao element is virtually unknown (Figure 2A). Hayes and Lewis (1984) suggest that portions of Luzon formed by the end of Cretaceous in response to nearby subduction episodes. By Eocene time (Figure 2B), the Luzon and other island predecessors may have begun to develop along a proto-Pacific western plate margin. During Miocene (Figure 2C, 2D), Luzon and perhaps the other components of the Luzon-Mindanao tectonic element consolidated and had begun to move northward.

McCabe (1984) presents paleomagnetic data showing that the impact of the combined Cuyo and Sulu tectonic elements has rotated parts of the Luzon-Mindanao trend since Miocene time. The result is a bent arc composed of rigid blocks displaced along shear zones. The application of the theory of plasticity or slip-line-field theory to China can serve as an analog (Tapponnier and Molnar, 1976). This theory could lead to locating yet unknown strike-slip faults.

## PLAY CONCEPTS

The above outline of the tectonic history might prove useful in developing creative, new play concepts for exploring the Philippine Island basins. A few sample concepts follow:

1. *Basins of the Luzon-Mindanao Tectonic Element.* The Luzon-Mindanao tectonic element basins have poor reservoirs. Lack of good clastic source and cementation due to volcanic sediment diagenesis make the chances bleak for good clastic reservoirs. Carbonate reservoirs are likely, but the critical question to answer is, "Could they be large enough to be fully commercial?" At present we must answer, "NO!"

   Innovative ideas could come from more detailed restorations of the tectonic history than are shown here. Through reconstructions we may find that one or more of these basins passed by a good clastic source or were part of a substantial reef complex. Prospects in these basins would then be upgraded, but reservoir sizes would have to be large enough to justify exploration risk, and at present there are no indications that reservoirs are that large.

   Applying slip-line-field theory could help to locate strike-slip faults and construct fault-related play concepts.
2. *Basins of the Sulu Tectonic Elements.* The mostly compressional and intense tectonic history of the basins on the Sulu tectonic element could suggest that all oil has been destroyed or had leaked off. A glance at seismic lines, shows the presence of thick, sedimentary sequences with a variety of not-too-intense structures. The tectonic history further suggests that possibilities for clastic and carbonate reservoirs are good, but the present features are not attractive enough to justify exploration.

   The South China Sea basin provides north Palawan with an oil source and could also provide south Palawan and adjacent Borneo with an oil source.

   The source rock and thermal characteristics of the Sulu basin have not been adequately tested. The numerous structural trap possibilities include older compressional structures and younger, strike-slip related structures and extensional structures because of the very young continental margin subsidence.
3. *Basins of the Cuyo Tectonic Element.* The basins on the Cuyo tectonic element remain the best possibilities. Source rock does exist, as proven in offshore Palawan fields. Reservoirs, both carbonate and clastic, exist and are reported for Palawan Island. The major exploration task remaining is to identify large features attractive enough to warrant future exploration. The tectonic history may provide the key to solving this problem. Faulting is complex, as one would expect, in an area that was born in a subduction zone and later extended by rifting, jostled by drifting, uplifted and tilted by southerly directed collision, and compressed finally along strike-slip "railroad tracks" in an easterly direction. Nonetheless, a variety of structural and stratigraphic traps are predictable from tectonics and identified on seismic data.

   At A on Figure 3, for example, there is a lower Miocene carbonate buildup on the upturned edge of a tilted fault block in the Malampaya area. This fault block is one of four structures that comprise a northeasterly trend associated with a strike-slip fault zone.

   A and B on Figure 4 are carbonate buildups on a reactivated fault block in the Cadlao area. From late Oligocene to early Miocene time, during the southward drift of the Cuyo tectonic element, the feature designated by A was high. During early to middle Miocene time the feature at A was depressed relative to the feature at B, perhaps in response to collision of the Cuyo with the Sulu tectonic element.

   Strike-slip faults with very significant vertical displacements can be seen in the West Linapacan area (A on Figure 5). Smaller strike-slip splays set up attractive looking structures (A on Figure 6).

   In the Galoc area, faults below A and B on Figure 7 set up clastic reservoirs above them. At A on Figure 8 is a strike-slip fault that can be traced to the south and onshore Palawan Island. This fault separates clastics to the east from carbonates to the west, both of which are proven reservoirs.

## CONCLUSIONS

The Philippine Islands have been looked at and rejected by many major oil exploration companies. Cities Service has persisted and has been rewarded with commercial oil discoveries. However, most major companies continue to resist exploring oil basins like those of the Philippine Islands. If these companies have preconceived ideas or "knowledge" that there are no oil traps in the basins of the Philippines, then they may be making the mistake that Wallace Pratt (1952) described: ". . . our knowledge supported in some cases by elaborate and detailed studies convinced us that no petroleum resources were present in areas which subsequently became sites of important oil fields." The tectonic analysis in this paper shows that there are oil-trapping possibilities, especially in the basins of the Cuyo tectonic element. The challenge to explorers is to find "oil tank cars" or oil traps that are large enough to justify the expense of modern exploration.

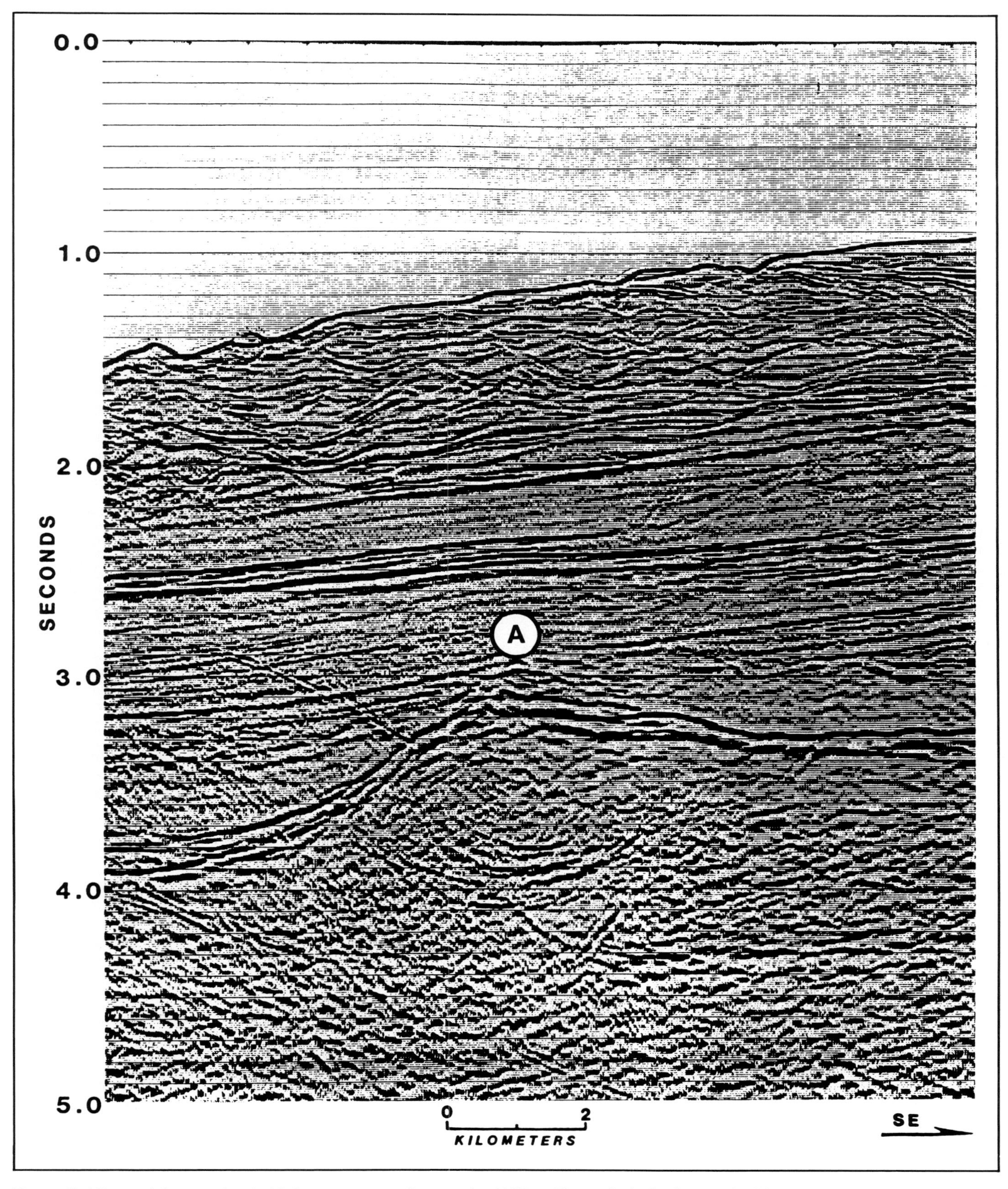


Figure 3. *Migrated time section in Malampaya area (just north of MP on Figure 1). A. Carbonate buildup on upturned basement fault block.*

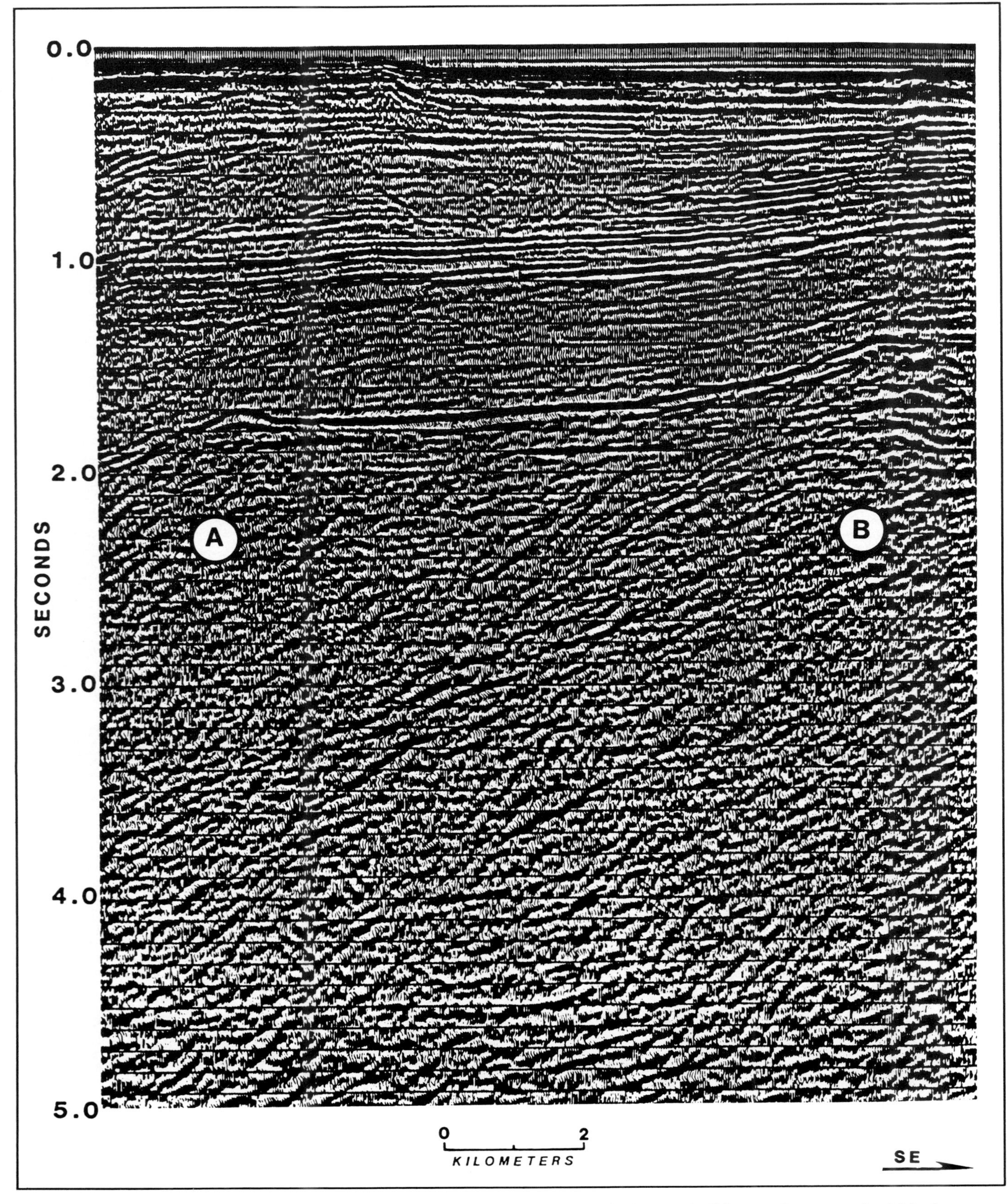


Figure 4. *Migrated time section in Cadlao area (C on Figure 1). A, B. Carbonate buildups on basement fault block.*

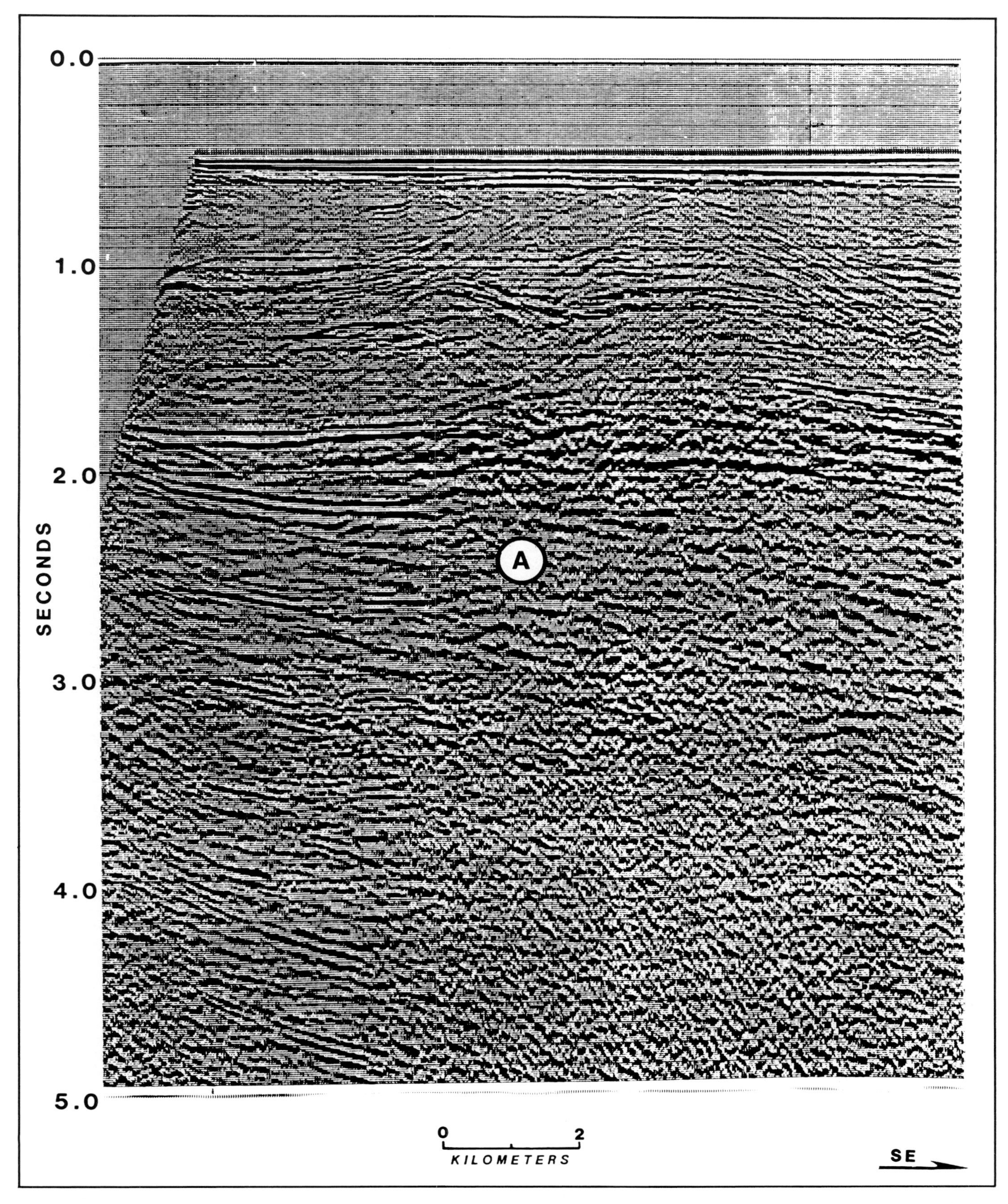


Figure 5. *Time section in West Linapacan area (west of G on Figure 1). A. Major strike-slip fault with slump west of block.*

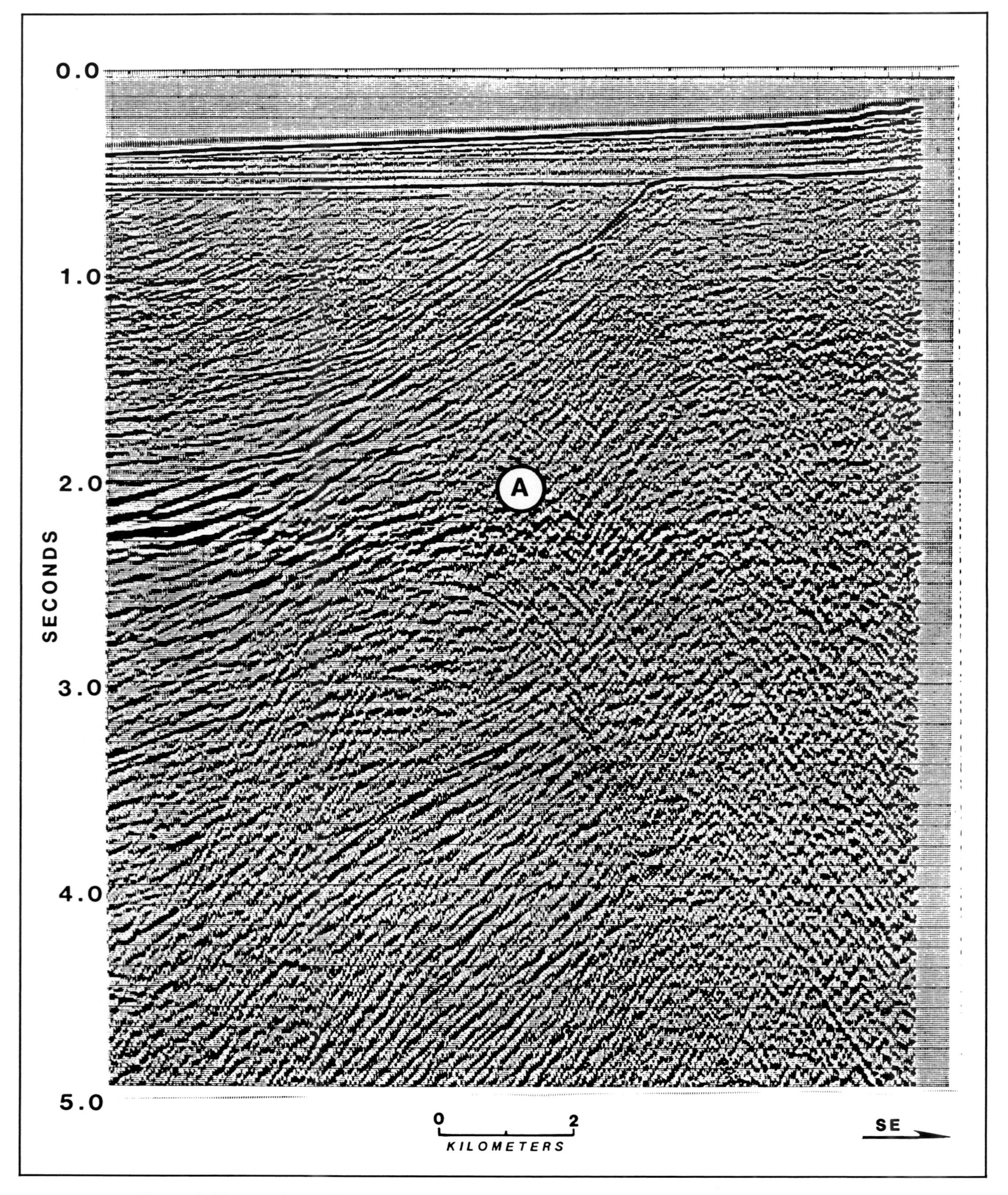


Figure 6. *Time section in West Linapacan area (South of G on Figure 1). A. Fault-bounded structure.*

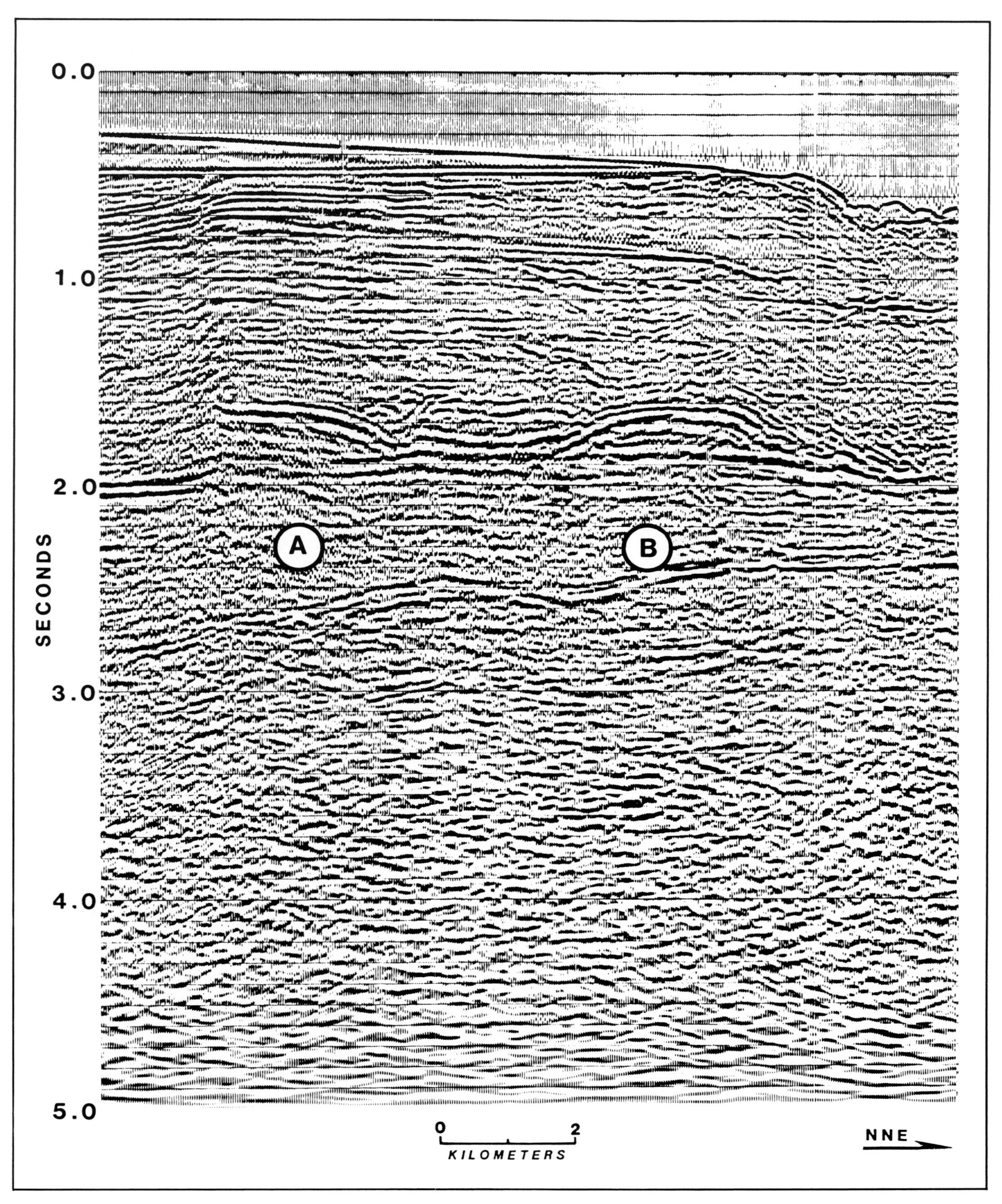


Figure 7. *Migrated time section in Galoc area (G on Figure 1). A, B. Clastic sediment accumulations over basement faults.*

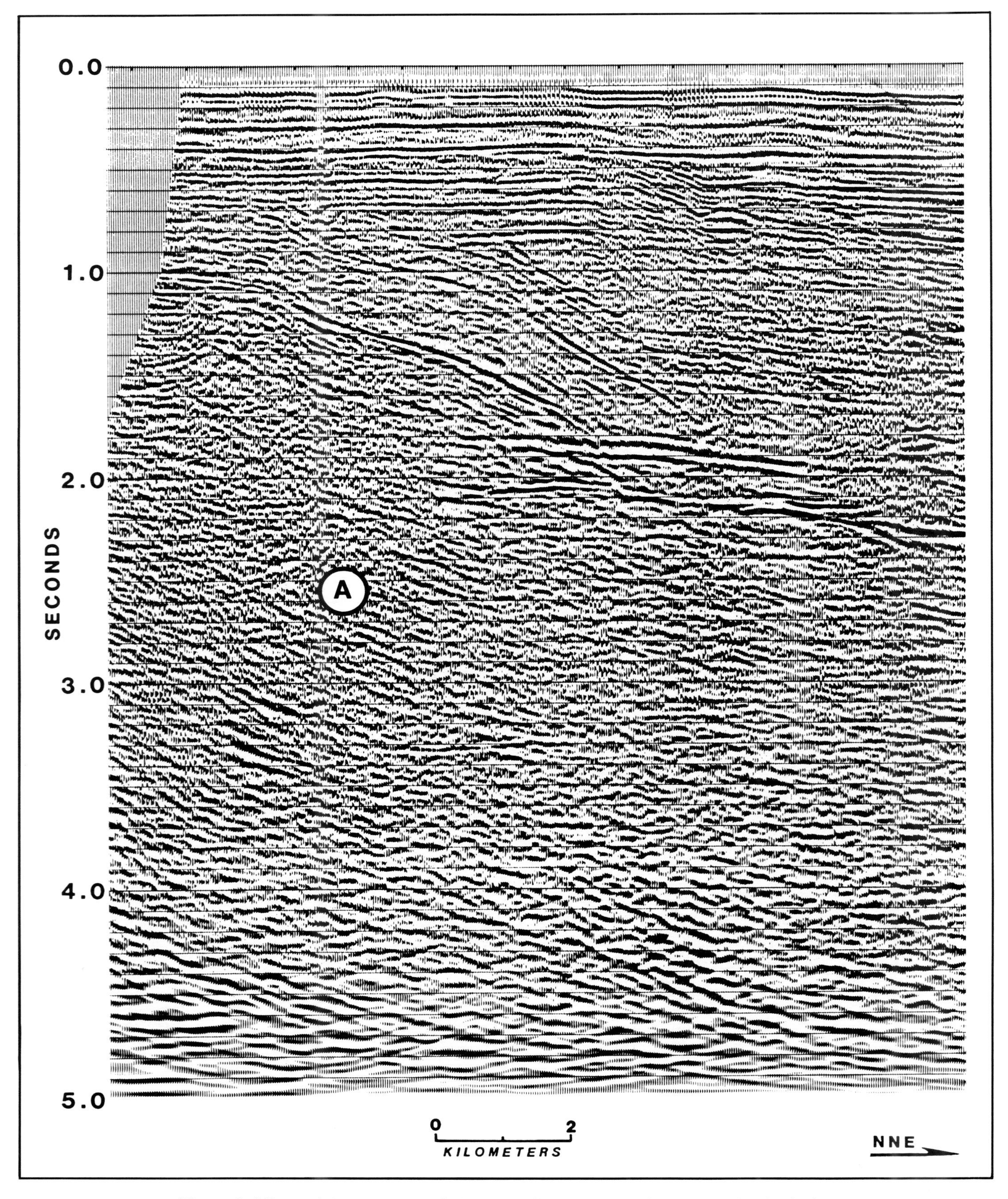


Figure 8. *Migrated time section in Galoc area (G on Figure 1). A. Major strike-slip fault.*

## REFERENCES

Beddoes, L. R., 1976, The Balabac sub-basin: Southwestern Sulu Sea, Philippines Offshore Southeast Asia Conference Southeast Asia Petroleum Exploration Society (SEAPEX) paper 15, p. 1–22.

———, 1980, Hydrocarbon plays in Tertiary basins of Southeast Asia: Offshore Southeast Asia Conference Singapore, 60 p.

Ben-Avraham, Z., 1978, The Evolution of marginal basins and adjacent shelves in East and Southeast Asia: Tectonophysics, v. 45, p. 269–288.

Brownlee, D. N., and M. W. Longman, 1981, Depositional history of a lower Miocene pinnacle reef, Nido B. oilfield: The Philippines Proceedings 4th International Coral Reef Symposium, Manila, v. 1, p. 619–625.

Chen, P. H., J. J. Gallagher, Jr., and C. S. Kenyon, 1979, Paired rift-spread systems of China: Washington, D. C., Transactions of the American Geophysical Union, v. 60, p. 398.

Dewey, J. F., 1977, Suture zone complexities: A review: Tectonophysics, v. 40, p. 53–67.

Gallagher, J. J., Jr., 1981, Tectonics of China: Continental scale cataclastic flow in mechanical behavior of crustal rocks: Washington, D. C., The Handin Volume, AGU (American Geophysical Union) Geophysical Monograph 24, p. 259–273.

———, 1984, Large-scale faults and the tectonics of China, Fourth International Conference on Basement Tectonics, Oslo, Norway p. 173–185.

Hamilton W., 1979, Tectonics of the Indonesian region, United States Geological Survey Professional Paper 1078, 345 p.

Hamburger, M. W., R. K. Cardwell, and B. L. Isacks, 1983, Seismotectonics of the Northern Philippine Island area, *in* D. E. Hayes, ed., Tectonic and Geologic Evolution of Southeast Asian Seas and Islands, Part 2: AGU (American Geophysical Union) Monograph 27, Washington, D. C., p. 1–22.

Hatley, A. G., 1977, The Nido reef oil discovery in the Philippines, its significance: Proceedings ASEAN (Association of Southeast Asian Nations) Council on Petroleum, Jakarta, Indonesia, p. 263–277.

Hayes, D. E., and S. D. Lewis, 1984, A geophysical study of the Manila Trench, Philippines, 1, crustal structure, gravity, and regional tectonic evolution: Journal of Geophysical Research, v. 89, p. 9171–9195.

Holloway, N. H., 1981, The North Palawan Block, Philippines: Its relation to the Asian mainland and its role in the evolution of the South China Sea: Bulletin Geological Society of Malaysia, v. 14, p. 19–58.

———, 1982, North Palawan Block, Philippines: Its relation to Asian mainland and role in evaluating South China Sea: AAPG Bulletin, v. 66, p. 1355–1382.

Hutchinson, C. S., 1975, Ophiolite in Southeast Asia: GSA Bulletin, v. 86, p. 797–806.

Karig, D. E., 1983, Accreted terrains in the northern part of the Philippine archipelago: Tectonics, v. 2, p. 211–236.

Kenyon, C. S., P. H. Chen, and J. J. Gallagher, Jr., 1979, Tectonics of China's basement rocks, Transactions of the American Geophysical Union, Washington, D. C., v. 60, p. 398.

McCabe, R., 1984, Implications of paleomagnetic data on the collision related bending of island arcs: Tectonics, v. 3, p. 409–428.

Moore, G. F., and E. A. Silver, 1983, Collision processes in the Northern Molucca Sea, *in* D. E. Hayes, ed., Tectonic and Geologic Evolution of Southeast Asian Seas and Islands, Part 2, AGU (American Geophysical Union) Monograph 27, Washington, D. C., p. 360–372.

Murphy, R. W., 1975, Tertiary basins of Southeast Asia: SEAPEX (Southeast Asia Petroleum Exploration Society) Proceedings, v. 2, p. 1–36.

Pratt, W. E., 1952, Toward a philosophy of oil-finding: AAPG Bulletin, v. 36, p. 2231–2236 (Reprinted AAPG Bulletin, v. 82, p. 1417–1422).

Saldivar-Sali, A., H. G. Oesterle, and D. N. Brownlee, 1981, The geology of offshore Northwest Palawan, Philippines: Proceedings Second ASCOPE (Association of Southeast Asian Nations Council on Petroleum) Conference and Exhib., Manila, Philippines, p. 99–123.

Tapponnier, P., and P. Molnar, 1976, Slip line field theory and large scale continental tectonics: Nature, v. 264, p. 319–324.

Taylor B., and D. E. Hayes, 1980, The tectonic evolution of the South China basin, *in* D. E. Hayes, ed., Tectonic and Geologic Evolution of Southeast Asian Seas and Islands: Geophysical Monograph Series, v. 2, p. 89–104, American Geophysical Union, Washington, D. C.

———, 1983, Origin and history of the South China Sea basin, *in* D. E. Hayes, ed., Tectonic and Geologic Evolution of Southeast Asian Seas and Islands, Part 2: AGU (American Geophysical Union) Monograph 27, Washington, D. C., p. 23–56.

# Wrench Faults as a Factor Controlling Petroleum Occurrences in West Siberia

Danilo A. Rigassi
*Geneva, Switzerland*

The distribution of Baykalian (late Precambrian) to Hercynian foldbelts forming the floor of the West Siberian basin suggests active wrenching throughout the Paleozoic. Strike-slip faults also control deep, sinuous Carboniferous–Early Permian basins. During the Late Permian, a complex system of narrow grabens opened in central-eastern West Siberia. Basic igneous material filled these troughs. The continental grabens of West Siberia and of central-eastern Africa are strikingly different from oceanic rifts. One difference is the large wrenching component observed along the margins of continental grabens. Depositional trends of Early–Middle Jurassic porous clastics and organic-rich shale are clearly related to strike-slip faults. In the latest Jurassic, the West Siberian basin was secluded from the open ocean due to wrench movements. Euxinic conditions that were thus created resulted in deposition of the Bazhenovo oil shale, the richest source rock in the province. Cretaceous and Tertiary wrenching produced low-amplitude drag folds that are the main hydrocarbon traps of West Siberia. A comprehensive wrench-tectonic approach to the structure of West Siberia should help locate new, vast oil and gas reserves.

## INTRODUCTION

Within less than two decades, Soviet earth scientists have turned the empty vastness of West Siberia, extending over an area of 1.2 million sq mi (3.1 million sq km), into a petroleum province that daily produces over 7 million bbl of oil and 30 billion cu ft (bcf) of gas. I estimate the still available reserves recoverable without enhanced oil recovery (EOR) methods to be 40 to 50 billion bbl of oil and close to 1,000 trillion cu ft (tcf) of gas, while cumulative production to the end of 1984 was about 20 billion bbl of oil and 60 tcf of gas. West Siberia is now first among the gas provinces of the world, and second only to the Middle East in liquid hydrocarbons. One may thus wonder whether West Siberia should be discussed in a Memoir devoted to future petroleum provinces; as a matter of fact, in spite of the above, impressive figures, West Siberia still is very far from being a mature province from the explorationist's point of view. Wildcats are much fewer than in the Middle East, both in terms of areal density and footage per areal unit. The second phase of exploration now taking place, in search of both deeper and more subtle traps, may unveil reserves surpassing those already found. Among knowledge that could lead to new discoveries, understanding the extent, nature, and ages of wrenching might play a significant role. This paper compiles the wealth of data gathered and published by Soviet geologists, and the author wishes here to thank them for their outstanding contribution to our knowledge of petroleum geology.

## ALTERNATIVE WRENCH TECTONICS

For the past dozens of years, tectonicians at large have adhered to a model in which wrench faults, forming two oblique systems of which one is dextral and the other sinistral, are the result of a compressive force. In this paper I make no assumption about the mechanism(s) that can produce wrench faults. For example, glacier crevices provide an excellent analogue of wrench faults; yet, the enclosing bedrock has no inherent tendency to compress the ice (Figure 1A). Depending upon the latitude, inertia may cause the lithosphere to glide differentially on the asthenosphere, as the Earth's velocity varies from over 1,000 miles per hour (mph) (1,600 kmph) at the Equator, to nil at the Poles (Figure 1C). Similar movements may take place between various layers—both sedimentary and non-sedimentary—of the crust. The effect of luni-solar attraction, of which the main, semi-diurnal component is greatest between approximately 10° and 40° of latitude, also should be considered (Figure 1D). On a more restricted scale, uplifting and tilting of basement blocks may cause movements within the sedimentary cover, resulting in non-compressional wrench faults and drag folds (Figure 1B).

We may thus come to newer global tectonic concepts in which the lithosphere, or certain layers, may move as do clouds and dust in the Trade Winds, with bulging taking place and moving like tidal waves. In the past, geologists have almost uniquely explained tectonical features by hypothetical, endogenous processes, the climax of this anthropomorphic approach being today's "plate tectonics." It may be time to consider the alternative view, of the rotating earth as a minute object within the field of other cosmic bodies of the universe (Rigassi, 1970).

The recognition of astroblemes and other impacts is a step in this direction. However, compared with the age and size of the Earth, such features remain negligible. Moreover, one may wonder why such phenomena are being investigated by tracing ppbs of rare elements, rather than by analyzing the whereabouts of the peculiar distribution in time of the phosphate-glauconite-oolitic iron-free silica assemblage. As Elysée Reclus (1874) stated: "Not only do the winds and the oceanic currents

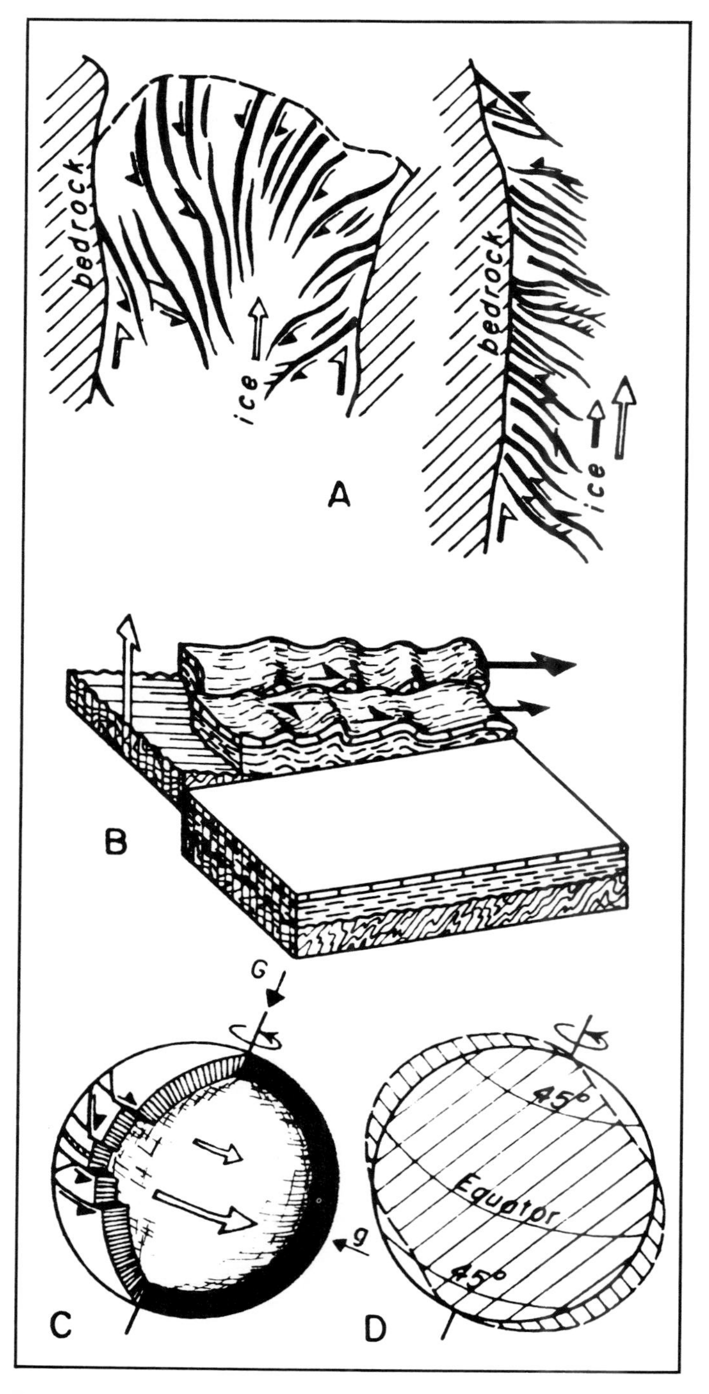


Figure 1. *Alternative wrench tectonics. A: Glacier crevices. B: Basement uplift and gliding of cover. C: Earth rotation and lithosphere inertia. D: Luni-solar attraction.*

move around the Planet, but the continents themselves with their peaks and valleys start wandering on the roundness of the Globe" (translation).

## PRE-LATE PALEOZOIC

To the east-northeast, West Siberia is limited by the platform sediments of the Tungusska syneclise, ranging in age from Riphean to Triassic. In all other directions, there are foldbelts of various Precambrian to late Paleozoic ages. Geophysical data and several wells have shown these foldbelts to form the substratum of the West Siberian basin (Figure 2). Unconformable, mainly terrestrial sediments of late Paleozoic age that originally may have been widespread have been largely eroded during the Late Permian–early Mesozoic. Depending upon the time at which the underlying series were folded, the basal part of these younger Paleozoics ranges in age from Late Devonian to Permian, being in most cases Early Permian to Late Carboniferous. The upper boundary either corresponds with the truncation by Jurassic sediments, or is transitional with Triassic clastics and volcanics within certain grabens and troughs.

The pattern of the deep-seated fold belts suggests intense wrenching during the Paleozoic. The boundaries between areas of pre- and early Baykalian and of late Baykalian (Cadomian, Pan-African) folding in eastern West Siberia along the Yenisey River indicate dextral movements along north-to-south faults and sinistral displacments on north-northwest to south-southeast faults, toward the close of the Precambrian (Surkov and Zhero, 1981, plate 3; Surkov et al, 1984). Younger sinistral movements seem to characterize the Krasnoyarsk area of southeastern West Siberia, with the northwest-to-southeast fault line separating areas of Baykalian and older folding from those affected by the Salairian event (a time equivalent of the Sardinian phase in Western Europe). On the other hand, displacement along southwest-to-northeast faults is dextral. This movement is shown by the boundaries between late Hercynian and Baykalian belts east of the Yenisey estuary (Yenisey-Khatanga Trough); by those limiting the southwest-to-northeast-trending block of Baykalian folds in the area of the Ob–Irtysh confluence; and, in southern West Siberia north of Pavlodar, by the contact between late Hercynian (Appalachian, Saalian) and early Hercynian (Wichita, Suddetic) foldbelts. In this last case, the horizontal movement is of the order of 200 km (120 mi), of which 90% is of pre-Triassic age.

## LATE PALEOZOIC

The present knowledge of West Siberian younger Paleozoics is not advanced enough for us to conclude about possible wrenching—either indigenous, or produced by the reactivation of more ancient trends—during the Late Carboniferous and Early Permian. Great sedimentary thicknesses occur

- along the Yenisey. A series of troughs, trending north-to-south and north-northwest-to-south-southeast, has 4-6 km (2.5–3.7 mi) thick, Upper Devonian–Permian, terrestrial and occasionally shallow-marine sediments. The pattern of these troughs suggests that during Hercynian times, sinistral displacment occurred along the northwest/north-northwest-to-southeast/south-southeast faults, with weaker evidence of dextral displacement of north-to-south faults. We deal with the rejuvenation of Baykalian and Salairian structural features.
- within a west-southwest-to-east-northeast trough. Across the Yenisey estuary, extending far toward the east-northeast beneath the Khatanga Trough, there are molasse-like Permian sediments up to 4 km (2.5 mi) thick. West-southwest-to-east-northeast faults, possibly with a dextral component, control the embayment.
- in southwestern West Siberia. At the Kazakh border in the Kurgan–Kustanay area, a narrow trough with Upper Carboniferous–Permian filling is bounded along its east-northeast flank by faults that seem related as "retro-charriages" or back-thrusts, to the Uralian foldbelt, there being no clear indication of a possible wrench component.

## LATEST PALEOZOIC AND TRIASSIC

The Late Permian and the Triassic were times of intense magmatic activity in West Siberia. The oldest occurrences, during Late Permian–Early Triassic, mainly consist of diabase, tholeitic basalt, and dolerite, there being in a few places ultrabasites. This ancient magmatic province extends eastward well beyond West Siberia. The province has thick traps crossed by numerous dolerites and capped by tuffaceous and volcanic-derived sediments covering the Tungusska syneclise and other parts of central Siberia, over an area in excess of 500,000 sq mi (1,300,000 sq km), and extending as far east as nearly 1,000 miles (1,600 km) from the eastern boundary of the West Siberian basin. Basalts and dolerites continued to be emplaced during the Middle and at least part of the Late Triassic, as well as being interbedded with greywackes, tuffs, and volcanic-derived sediments, and with local occurrences of acid rocks such as rhyolites. The Triassic igneous rocks are not evenly distributed across West Siberia; rather, they are limited to or more abundant within graben-like features (Figure 3). The largest such feature, the Koltogor–Urengoy graben, extends over 1,600 km (1,000 mi), with an average width of 20–50 km (12–30 mi), reaching in places up to 100 km (60 mi). From its central portion, two branches diverge, the Khuduttey graben to the north-northwest for a length of 400 km (250 mi), and the 250 km (155 mi) long Pyakipur graben to the southwest. Other grabens of limited extent occur both west and east of the southern portion of the Koltogor–Urengoy graben. In the eastern part of West Siberia, the 1,000 km (600 mi) long Khatanga–Low Yenisey graben (extending well beyond the northeastern limit of Figure 3) exhibits a west-southwest-to-east-northeast trend. It switches abruptly near the Yenisey mouth into the north-to-south-trending, 800 km (500 mi) long, Khudosey graben.

Within the above grabens, upper Paleozoic sediments either are absent or do not exhibit any overthickening compared with out-of-graben areas. The dominant trends of border faults are, in decreasing order of importance: north-to-south, northeast-to-southwest, and northwest/north-northwest-to-southeast/south-southeast. Some other grabens do contain both upper Paleozoic sediments and Triassic sediments with some igneous rocks. Here, the main trends are northeast/north-northeast-to-

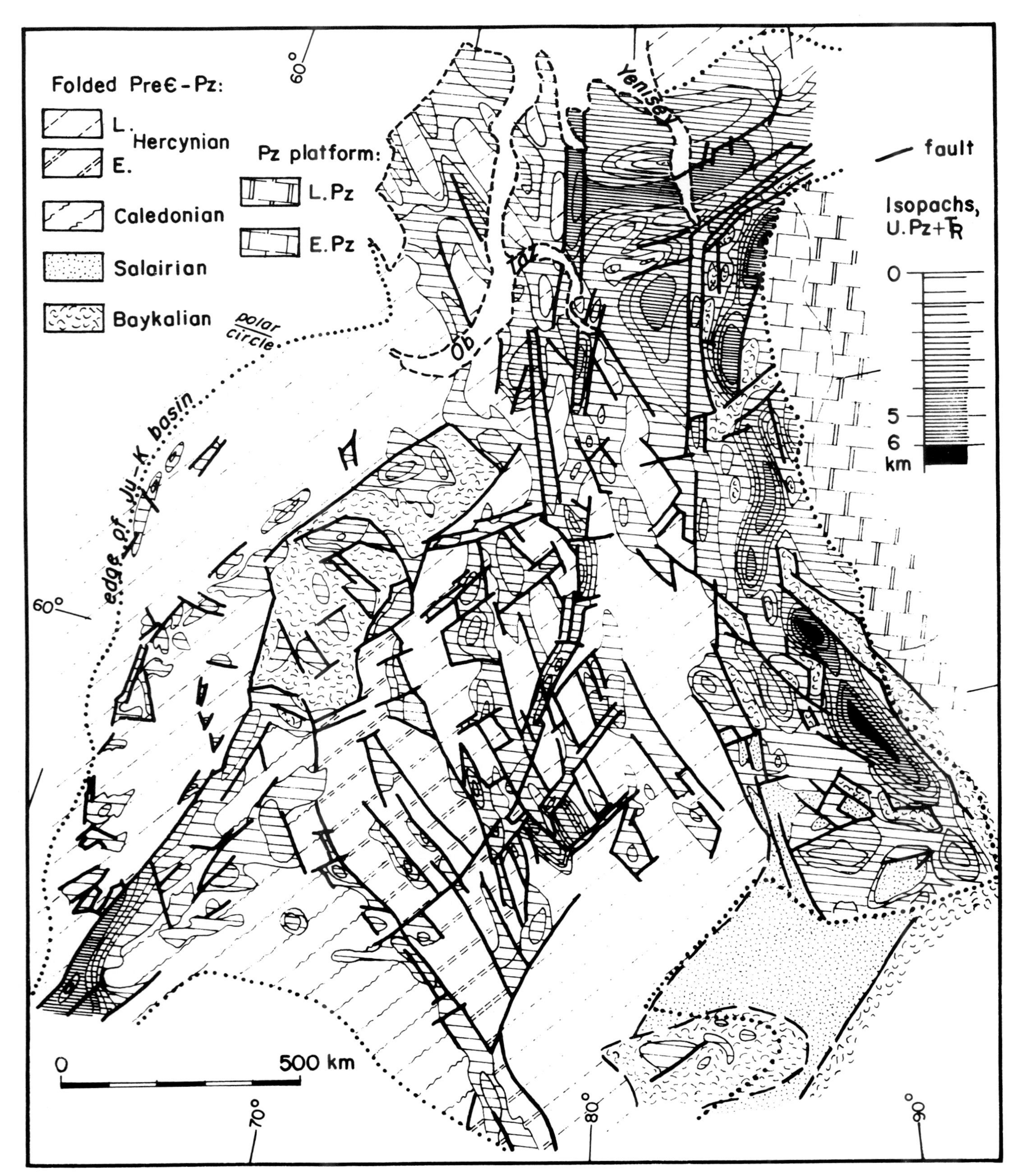


Figure 2. *West Siberia: the nature of folded Paleozoic substratum, isopachs of unfolded Paleozoic plus Triassic. (Adapted from Surkov and Zhero, 1981).*

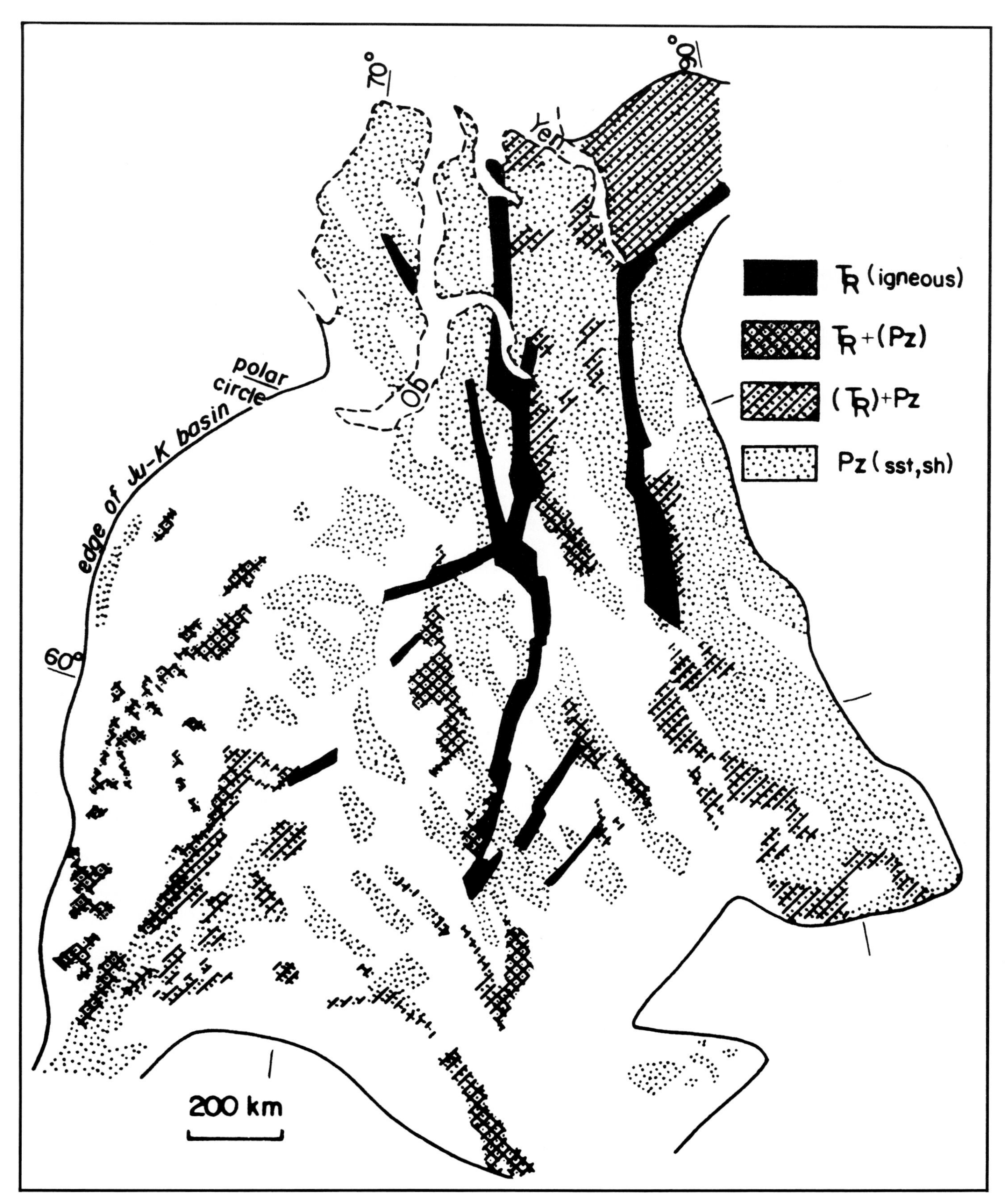


Figure 3. *West Siberia: the nature of upper Paleozoic and Triassic. (Adapted from Surkov and Zhero, 1981)*

southwest/south-southwest and northwest/north-northwest-to-southeast/south-southeast, with few north-to-south faults. Those troughs that contain only or mostly upper Paleozoic sediments are mainly oriented north-northeast-to-south-southwest and northwest/north-northwest-to-southeast/south-southeast.

The igneous-rich Triassic grabens are thus controlled by northeast-to-southwest and north-northwest/northwest-to-south-southeast/southeast faults already active during the Paleozoic, with dextral and sinistral components, respectively. Even more important are grabens along north-to-south faults, which had originated during Baykalian times. However, the Triassic wrench component seems to have been sinistral, that is, opposite to that during the late Precambrian.

In recent Soviet publications, the Triassic grabens of West Siberia are often referred to as "grabens-rifts." I prefer the purely descriptive term of "grabens" rather than "rifts" as this latter word now has definitely acquired a genetic connotation. When comparing oceanic rifts and continental grabens, one cannot avoid noticing striking differences:

1) Oceanic rifts are sinuous, linear features, only interrupted by so-called "transform" faults; features such as "triple junctions" are only known from areas with thick continental crust. Anastomosed patterns such as the junction of the Koltogor–Urengoy, Khuduttey, and Piakypur grabens of West Siberia, or those occurring in the graben system of central-eastern Africa (with every new step in petroleum exploration from Chad to Sudan unveiling new complexities) do not exist in the open ocean.
2) Along every known graben within or adjacent to a continent, a wrench component of the order of hundreds of kilometers is obvious. Even such modest features as the Rhine or Dead Sea grabens show a horizontal displacement of 100 to 200 km (60–125 mi). Evidence for a similar component has thus far not been found associated with oceanic rifts.
3) Zigzag and dog-leg abrupt changes in trend, typical of continental grabens, seem not to affect oceanic rifts. In West Siberia for example, there is a turn from the Low Yenisey to the Khudosey graben and a change in trend of the Koltogor–Urengoy graben near the Taz estuary. Elsewhere other examples include the dog-leg in the northern part of the Albert Rift, East Africa, and the passage from the Møre to the Viking and Central North Sea graben.
4) The two opposite rims of continental grabens often show considerable difference in topographic elevation, and, in many cases, possess an even greater contrast in structural elevation. In West Siberia the top of the folded Paleozoics is at −7.5 km (−4.7 mi) along the northwest rim at the Low Yenisey–Khudosey graben junction, and at only −5 to −5.5 km (−3 to −3.4 mi) along the southeast rim. At the Koltogor–Khudottey–Pyakipur "quadruple" junction, this same unconformity is at −4 km (−2.5 mi) to the west, and at −6.5 km (−4.0 mi) to the east. On the contrary, the rims of oceanic rifts only show slight differences in elevation, and should not—according to prevailing spreading models—exhibit any sizable structural conflict.
5) Similarly, both the topographic and the tectonic depth of continental grabens varies greatly along the graben trend. This is not the case with oceanic rifts.
6) Drag folds occur within and far along the flanks of continental grabens, the folding intensity greatly depending upon the thickness and nature of the sedimentary cover. However, evidence for similar features accompanying oceanic rifts is lacking.
7) Not only does the morphological and structural vertical throw vary in space along the rims of continental grabens, but there have been great changes in time, too, often reaching the inversion stage. While certain grabens are superimposed on ancestral horsts, and vice versa, nothing similar could be thought of in the case of oceanic rifts, as viewed by today's plate tectonics.
8) Some continental grabens are devoid of or poor in volcanics, both outcropping or buried (Bresse, northern Albert Rift, central-southern part of Dead Sea–Aqaba graben, and several of the newly discovered, Central African grabens). Such a setting would be absurd in an oceanic rift.

Many of these differences may result from the strong wrench component characterizing continental grabens and not oceanic rifts, suggesting different mechanisms of formation. In several continental grabens, magmatic occurrences and geophysical data show the mantle to be close to the surface; but this might be the result of graben tectonics, as opposed to rifting in the ocean caused by mantle bulging.

The beginning of the West Siberian basic (to ultrabasic) magmatic cycle during the Late Permian might be related to both the phase of easterly thrusts along the eastern margin of the Urals, and to the deposition within the Verkhoyansk belt of the flysch (Kazanian stage) followed by shallower clastics (Tatarian stage). However, the Verkhoyansk belt is over 1,000 mi (1,600 km) from West Siberia. As to the end of the West Siberian igneous activities, a relationship might exist with block faulting of early Liassic age along the Urals' eastern slope. In the Berezovo region (Low Ob), some wells have penetrated granites that are 200–210 million years old (Jurassic–Triassic boundary). In the Verkhoyansk range and adjacent areas of the Siberian Platform, regressive facies and stratigraphic gaps mark the Rhaetic.

Ever since the Early Jurassic the fault pattern of West Siberia, to a large extent inherited from more ancient structural features, has remained rather similar to what it was during the Triassic magmatic episode (Figures 13A and 13B). However, spasmodic continuation of the wrench movements already in evidence during the Triassic (and, in part, in earlier times) led to increasing complexity (Figures 13C and 13D). Figure 4 illustrates the present setting of the basin's floor.

## JURASSIC TO PALEOGENE

The depositional patterns of the Lower and Middle Jurassic sediments, grading from fluvial in south-central West Siberia to shallow-marine in the province's northern part, show sharp morphological lineaments, of which the dominant trends are north-northeast/northeast-to-south-southwest/southwest, northwest-to-southeast, and north-to-south, there being minor sub-latitudinal features in places. Between these valleys, higher areas often are sinuous, either "Z" or "S" shaped. Such a setting indicates active wrench faults, and is illustrated by the isopachs of various stages, 195 to 170 million years old (Figure 5). Displacement is clearly left-lateral along the northwesterly faults

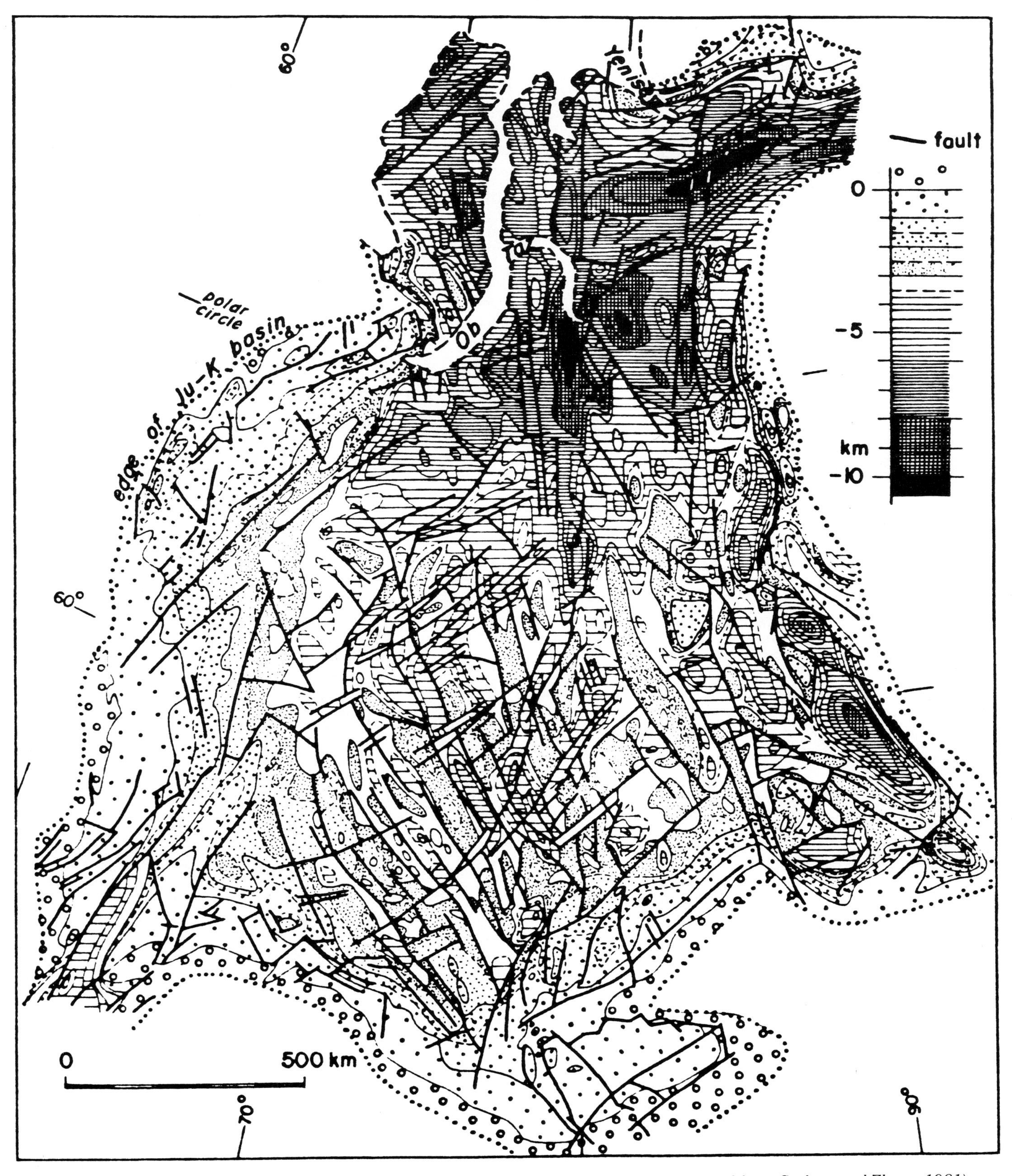


Figure 4. *West Siberia: structural map, isobaths of the top of the folded Paleozoic. (Adapted from Surkov and Zhero, 1981)*

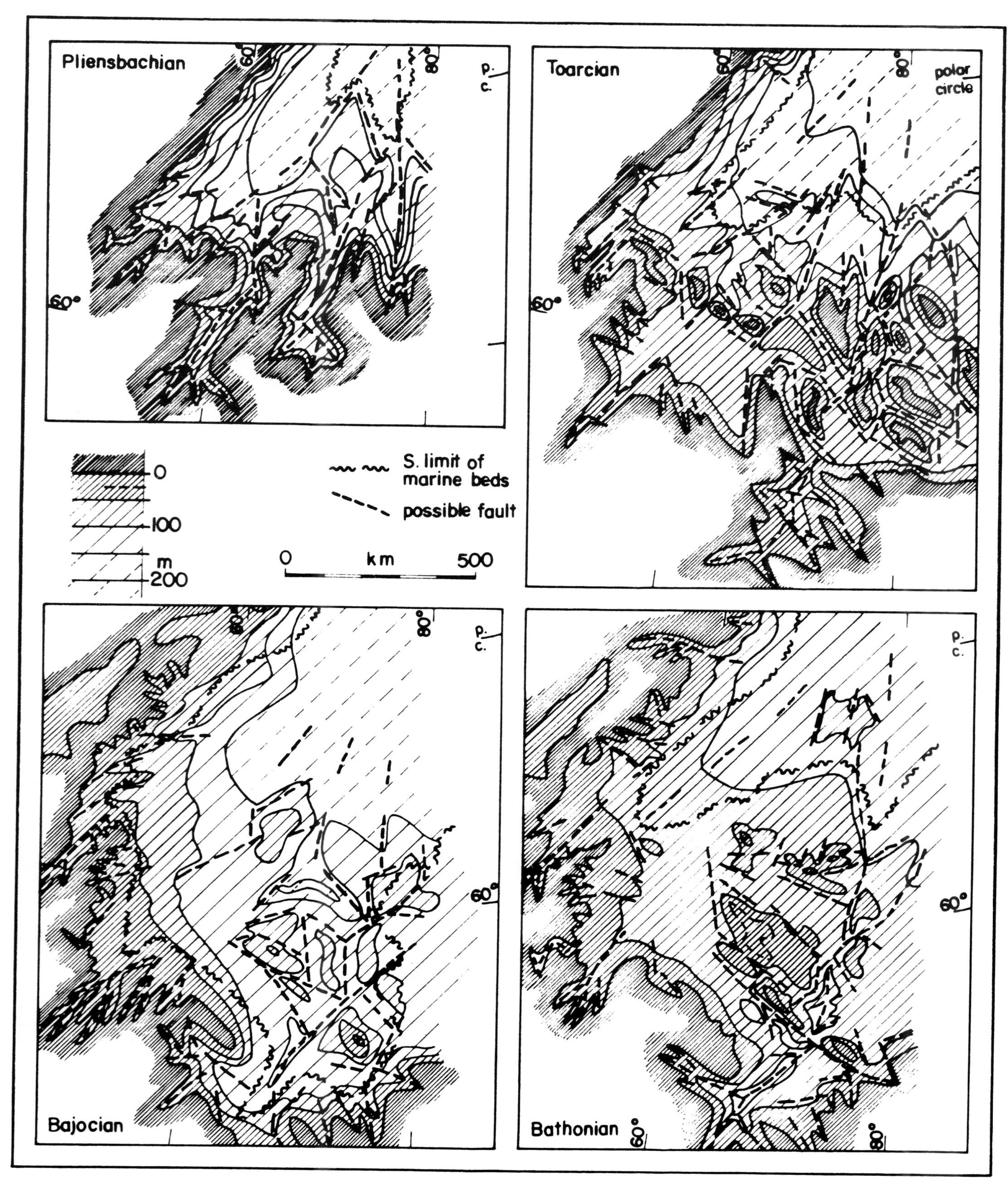


Figure 5. *Central-western West Siberia: isopachs of Liassic stages. (Adapted from Korzh, 1978).*

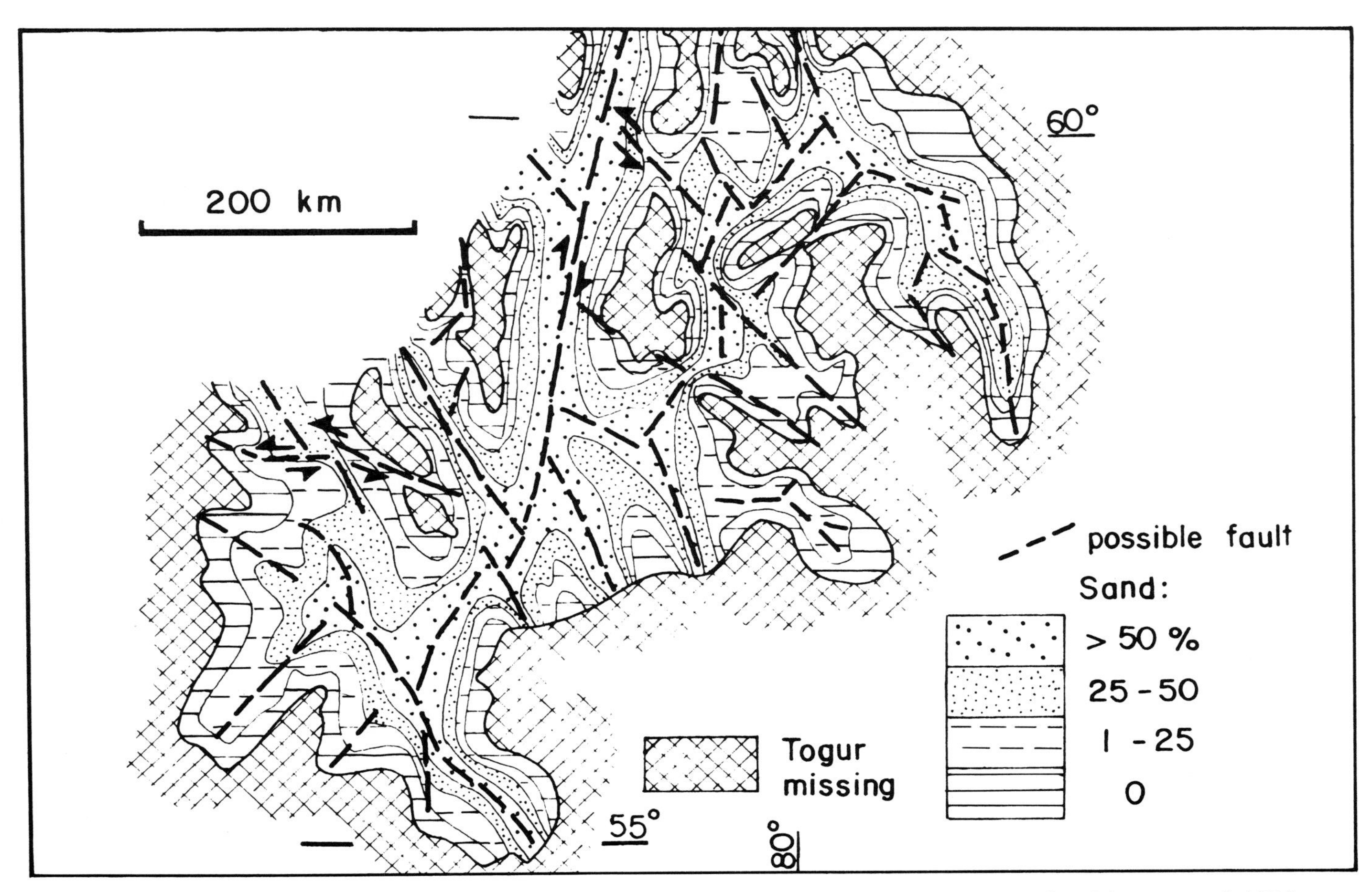


**Figure 6.** *Southeastern West Siberia: sand ratio, Togur Formation, Toarcian (Adapted from Gaydeburova et al, 1974).*

and right-lateral along the northeasterly trends, whereas we cannot conclude as to north-to-south and west-to-east lineaments. Judging from the distribution of late Liassic fluvial sands (reflecting the paleo-draining) in the southeastern part of West Siberia, there might have been local, dextral movements along north-to-south faults and sinistral wrenching along sublatitudinal trends (Figure 6). Next to having controlled the distribution of porous sands, the postulated Liassic–Dogger fault pattern allowed organic-rich shale, of both marine and lacustrine-swampy origin, to accumulate in small, rather isolated troughlets (the central part of Bajocian map, Figure 5) and small gulfs not subject to intense, coarse clastic discharge (the upper-central part of Bajocian and Bathonian maps, Figure 5). Such shale may be the source of hydrocarbons now accumulated in sands of late Liassic to Oxfordian age.

During the Late Jurassic (Volgian stage), most of West Siberia was turned into an enclosed, euxinic basin, with deposits of the Bazhenovo oil shale. Previously, marine invasions into the basin apparently came from the Lena estuary area, i.e. from the east-northeast, via the Khatanga Trough. During the Oxfordian, a barrier linking Novaya Zemlya with the Taymir peninsula foldbelt was lowered, permitting transgression from the Sea of Barents and the Kara Sea. There is evidence of a paleo-Gulf Stream already in existence, allowing the average sea temperature to be around 16°C (60°F) during the Oxfordian in the West Siberian basin, with only minor currents supplying colder water from the Lena–Khatanga embayment in the basin's northeastern part. By Volgian times, sea-water temperature had increased to 17.5° C (63°F), then abruptly decreased to 13°C (55°F) at the very end of the Jurassic (late Volgian), because currents from the Khatanga Trough gained in importance (Bogolenov, 1983). Whatever the impact of global events on the peculiar Volgian situation of West Siberia, the near-closure of the narrow strait connecting the basin with the Kara Sea and the Atlantic realm, followed by an improvement of the connection with the Khatanga Trough, undoubtedly results from sinistral wrenching along the large southwest-to-northeast faults.

This wrenching extended from the Ob estuary and the Yamal peninsula to the Khatanga Trough and the Taymir fold belt. Rapid decay of the temperature in the nearly-secluded Volgian basin produced thanatocoenoses under euxinic conditions, leading to the deposition of the Bazhenovo formation—probably the main source rock of West Siberia. Available data show that the Bazhenovo embayment, over 1,000 km (600 mi) wide at the mid-Ob latitude, narrows to 200 km (125 mi) between the Ob and Yenisey estuaries. Whether a similar oil-shale facies was deposited further north is as yet unknown. However, the overall setting of Soviet Arctic territories, as well as marine geophysical data from the Kara Sea and the Glacial Ocean, suggest the occurrence of another major northeast-to-southwest system of sinistral wrench faults. These faults extended from northern Severnaya Zemlya to northwestern

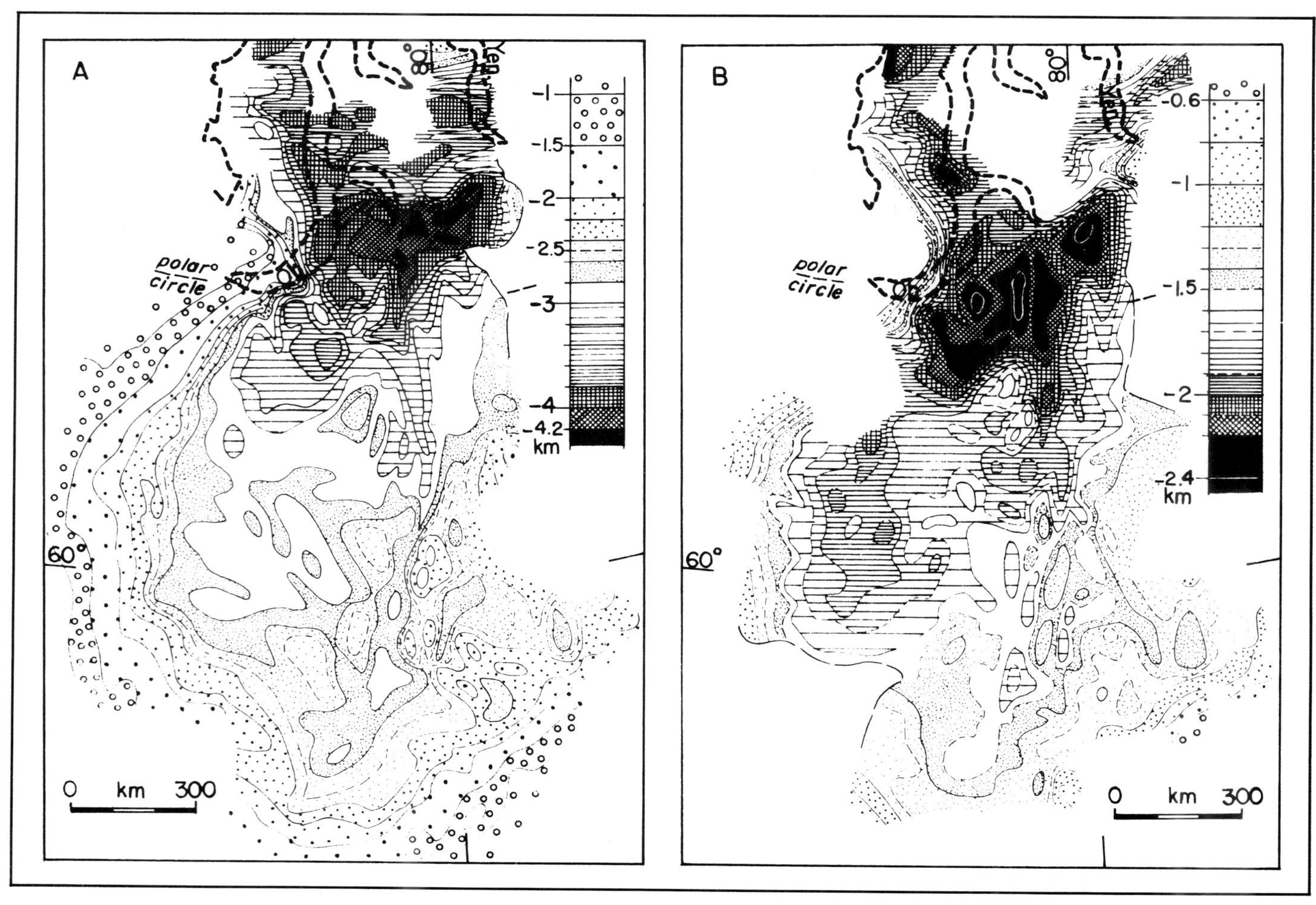


**Figure 7.** *West Siberia: structural maps. A: Isobaths, "B" reflector, base Bazhenovo formation (Volgian). B: Isobaths, "M" reflector (lower Aptian). (From Surkov and Zhero, 1981).*

Novaya Zemlya. Most of the Kara Sea, therefore, is underlain by latest Jurassic age beds of Bazhenovo facies, and might prove to be a hydrocarbon province as prolific as the land part of West Siberia. On the other hand, Volgian source rocks should not be present within the Khatanga Trough.

The present structure of the Bazhenovo Formation (Figure 7A) shows the southwest-to-northeast Ob estuary–Khatanga trends, next to the conspicuous north-to-south trends inherited from the Triassic grabens. The structure also shows the northwest-to-southeast trends now exhibiting sinistral components. When comparing this map with that of the lower Aptian (Figure 7B), and constructing an isopach map of the Volgian–Barremian series (Figure 8), several structural directions of Early Cretaceous age become obvious:

- South of latitude 65°N, sinuous troughs trending from north-northeast to north and north-northwest are not all superimposed upon Triassic grabens. Thus, Early Cretaceous movements produced some new, faint north-to-south downwarping along a few narrow belts, and resulted in Triassic grabens, by then nearly dormant, becoming more sinuous. "Z" and "S"-shaped areas of lesser deposition show the overall pattern of northeast/north-northeast-to-southwest/south-southwest sinistral and northwest-to-southeast dextral faulting to have remained active. Dextral wrenching may have occurred along north-to-south, still weak sublatitudinal fractures.
- From Latitude 65° to the Polar Circle, a similar but less obvious pattern emerges.
- Further north, the dominant features are the dextral, northeasterly Low Yenisey–Khatanga trends, and the sinistral, northwesterly Yamal peninsula trends.

Another way to apprehend paleo-movements has been proposed by Kuzin et al. (1974) (Figure 9). By comparing for each individual structure the amount of closure respectively at the top Cenomanian (91 million years old) and at the top Neocomian (124 million years old) with that at the top of the folded Paleozoic, we can construct two maps, showing, in terms of percentage, the amount of closure gained from the Neocomian to present (Figure 9A) and from Cenomanian to present (Figure 9B). Structures that had acquired most of their closure by the end of the Neocomian are restricted to up to a 300–400 km (185–250 mi) wide belt along the southern, western, and eastern margins of the basin. Many structures do not presently contain any hydrocarbons, while others (in particular in the basin's

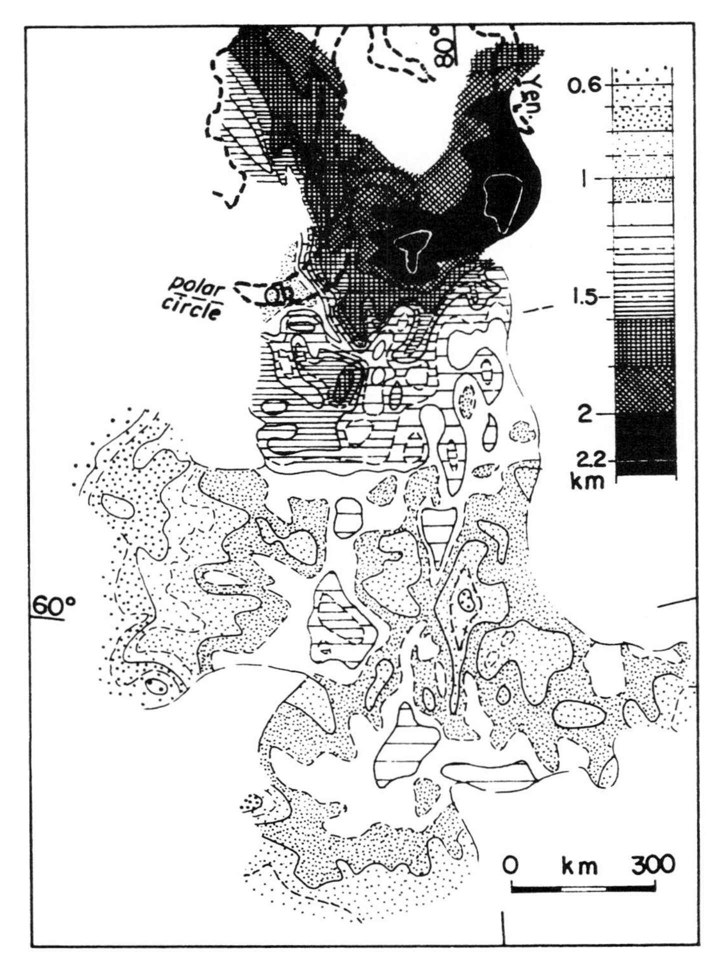


Figure 8. *West Siberia: isopachs of interval between "B" and "M" reflectors.(From Surkov and Zhero, 1981).*

southeastern part) have oil and gas pools within the Paleozoic and the Lower–Middle Jurassic. By substracting the post-Cenomanian closure from the post-Neocomian one, we can see that during the Barremian to Cenomanian most structures within the central part of the basin continued to grow, as did some closer-to-edge uplifts.

Here too, the shape of the structural pictures indicates wrenching: sinistral along north-to-south and northwest-to-southwest, dextral along northeast-to-southwest faults, uncertain on west-to-east trends. Structures born in post-Cenomanian times include a few scattered, minor uplifts near the margins of the basin, as well as some highs and culminations in its north-central portion, that is, in the mainly gas-bearing area of Urengoy–Yamburg–Medvezhe. In the maps of Figures 9A and B, the wrench-fault pattern already evident in Jurassic and older times suggests that such an approach might help us understand the structural evolution of other petroleum provinces, the North Sea, to name one.

On a local scale, West Siberian structures also allow recognition of wrench movements along faults, from the Barremian to present. Salmanov (1974) has reconstructed the closure of the BV horizon (Valanginian) of several structures in the mid-Ob area, as it was during the early Aptian [overburden about 500 m or 1,600 ft], the Turonian [overburden 1,200 m or 3,900 ft], the early Eocene [overburden 1,500–1,600 m or 4,900–5,200 ft] and at present [overburden 2,100–2,200 m or 6,900–7,200 ft] (Figure 10). Here, again, we find the usual pattern of wrench faults. However, the sub-latitudinal trends mainly exhibit left-lateral components that decrease in time. Moreover, the total amount of closure, which at Vartovsk has steadily grown from mid-Cretaceous to present, reached two maximums in the Meghion area: one in early Aptian; another in Recent times. The first situation is also found in the Samotlor area, while the second one occurs in the Pravdinsk–mid Salym region. Two types of development took place:

1) Structural growth, very likely spasmodic, took place during the entire mid-Cretaceous to present period.
2) Another that had two main phases of growth:
   a) in the late part of the Early Cretaceous.
   b) during the Neogene.

The increase in total closure at the Valanginian level from Barremian to present ranges between 50% and over 300%, being greater than it would if it were because of draping-compaction effects only.

## NEOGENE–RECENT

The present structural setting of the Eocene again shows sets of northwest/west-northwest-to-southeast/east-southeast sinistral and northeast-to-south-west dextral faults, as well as possible dextral sub-latitudinal fractures. Figure 11A gives an example for the basin's north-central part. The deformation of a lower Neogene peneplain leads to similar conclusions (Figure 12), while that of various low (Pleistocene) peneplains tells a similar story, in spite of the tilting having been very minute (maximum 0.4%, average less than 0.1%) (Figure 11B).

Space imagery, a well advanced branch of earth science in the Soviet Union, shows a wealth of major lineaments crossing West Siberia: northwest-to-southeast (clearly sinistral), northeast-to-southwest (clearly dextral), north-to-south (wrench component, if present, unclear) and west-to-east (dextral) (Bryukhanov et al, 1982; Bush, 1983; Generalov, 1983). Certain investigators see in some present sub-latitudinal morphological features the expression of newly created grabens that may still be active today (Burlakova, 1983).

## CONCLUSIONS

Under the West Siberian basin, the Moho is 34–44 km (21–27 mi) deep, compared with 46–50 km (29–31 mi) under the Tungusska syneclise and 46–53 km (29–33 mi) under the Uralian foldbelt (Karus et al, 1984). When did the crustal thinning occur? It may be related to the deepening of the late Paleozoic troughs in the eastern part of West Siberia along the Yenisey, with the Late Permian–Middle Triassic magmatic and graben formation event, with easterly thrusts along the eastern margin of the Urals (Late Permian–Rhaetic). Some major faults, south-southwest-to-north-northeast dextral, and northwest-to-southeast sinistral, do penetrate the Moho. How do we reconcile this penetration with a wandering continent? Moreover, while the crust in West Siberia as a whole is thinner, there is no

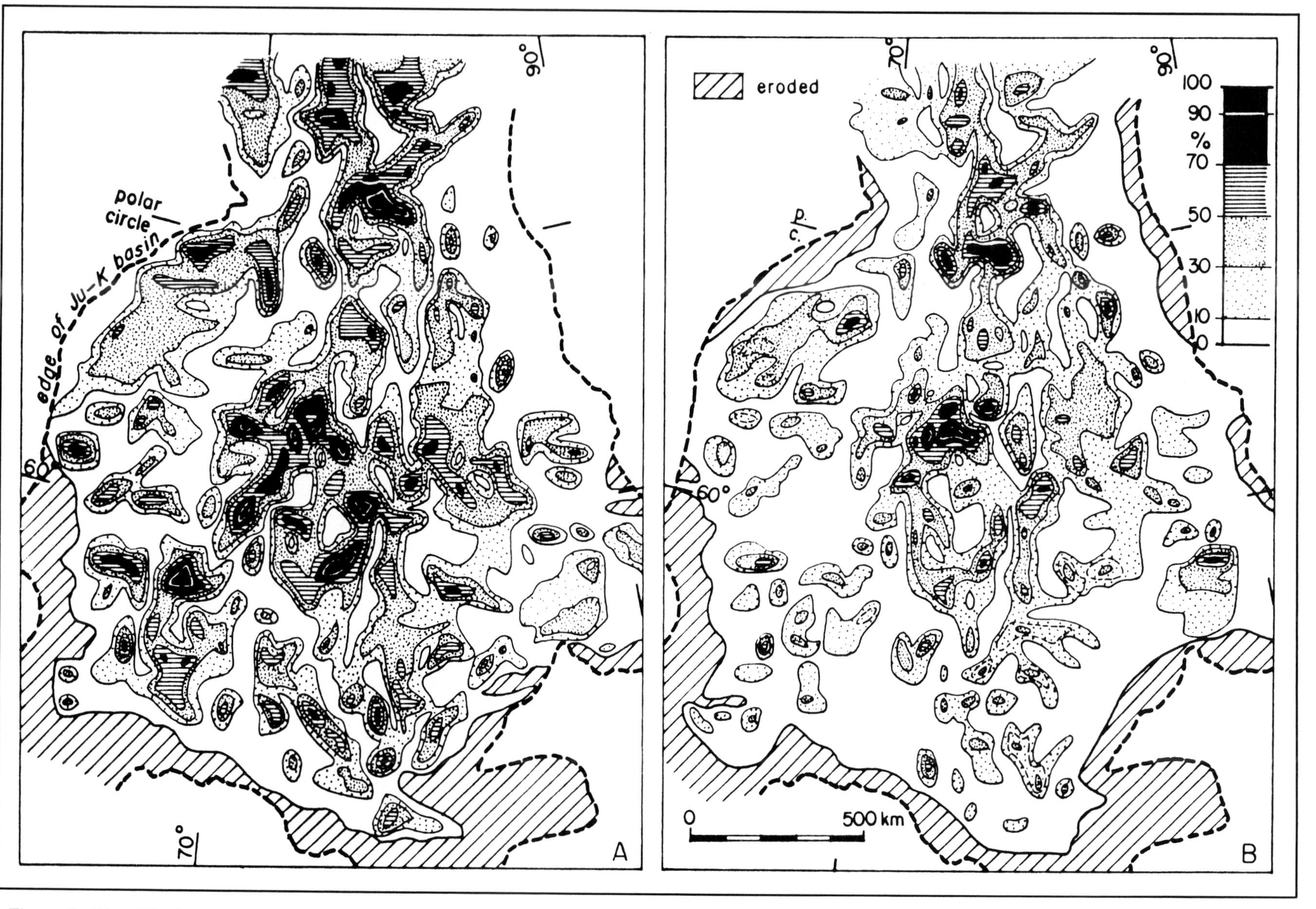


Figure 9. *West Siberia: percentage of structural closure (at top folded Paleozoic) gained since: A: Neocomian, B: Cenomanian. (Adapted from Kuzin et al, 1974).*

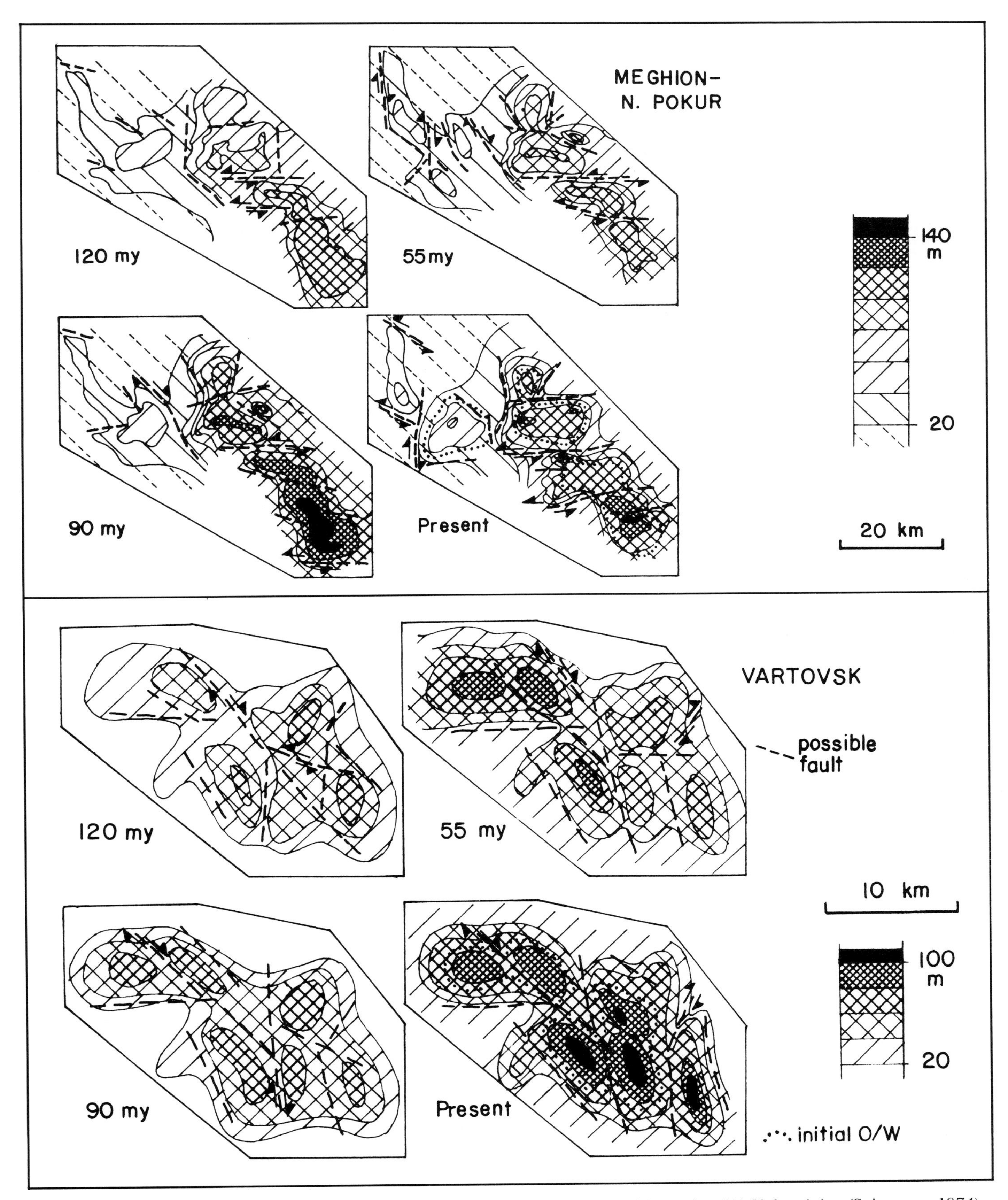


Figure 10. *Mid-Ob area: evolution in time of the structural closure at the level of formation BV, Valanginian (Salmanov, 1974).*

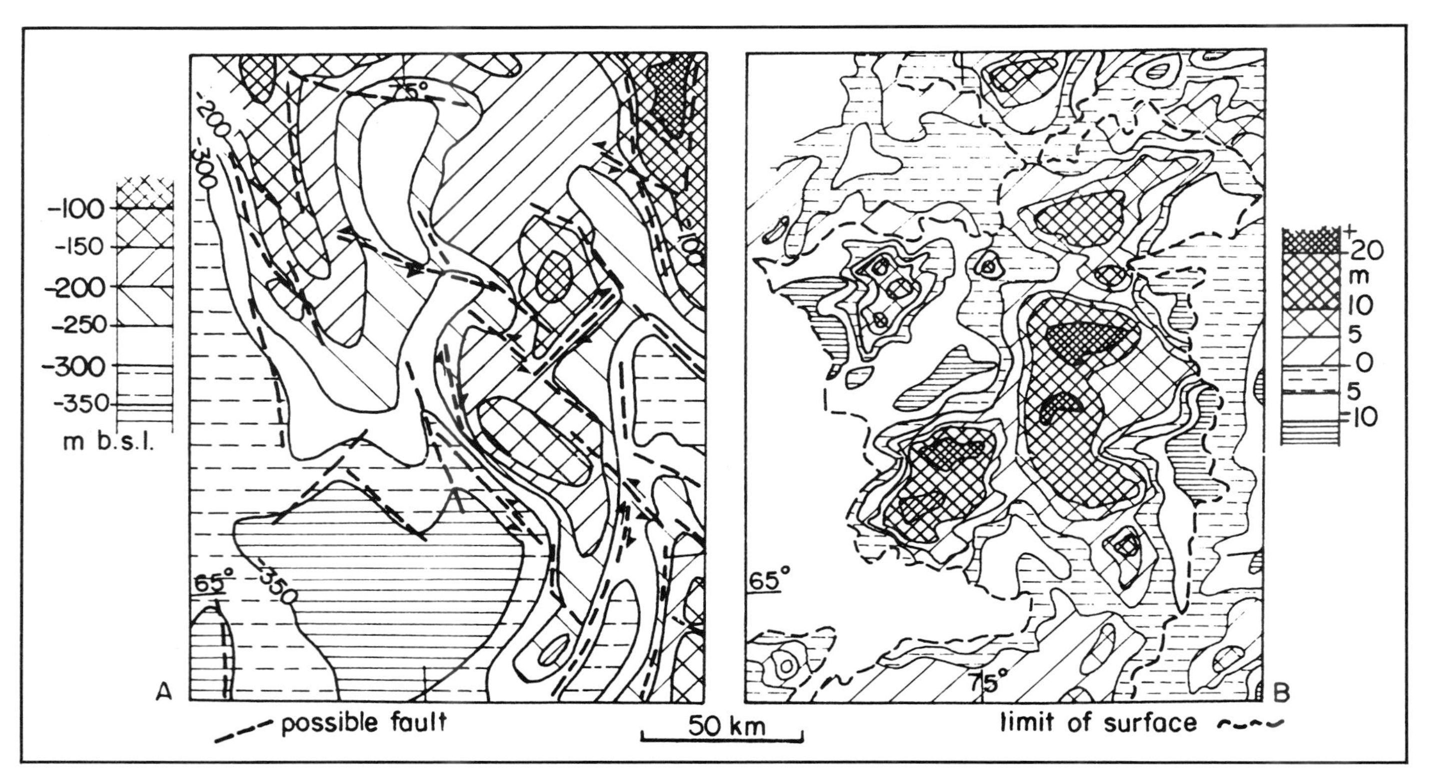


Figure 11. *North-central West Siberia (Urengoy area). A: Structural map, isobaths of top Eocene. B: deformation of peneplains. (Adapted from Reynin and Perughin, 1974).*

direct relationship with the Triassic grabens. On the contrary, the depth to the Moho is about 40 km (25 mi) in the Khuduttey graben, 37–38 km (23–23.6 mi) in the Urengoy graben and only 35 km (22 mi) in between. In the basin's southern part, all the oil and gas fields found thus far are in areas where the Moho is less than 37 km (23 mi) deep, regardless of the total sedimentary thickness, the presence or absence of unfolded Paleozoic sediments, or the thickness of the Bazhenovo source rock.

Wrench tectonic activity is evident in West Siberia, from late Precambrian to the present. A wealth of geological and geophysical data indicate this, as do the huge morphological anomalies, such as the "S" shaped change in trend of the Hercynian foldbelt from northern Novaya Zemlya to the Urals, the "S" turn of the middle and low course of the Ob River, and the "Z" turns of the Ob around Novosibirsk and of the Irtysh.

Wrench faults have guided the distribution of now-buried Paleozoic foldbelts, with different sedimentary contents and different folding intensity. They also controlled the distribution of un- and little-folded upper Paleozoic sediments. Analyzing them should help define those areas that might offer oil and gas prospects within Paleozoic reservoirs. Throughout the Mesozoic, the deposition of both sandy reservoirs and shaly source rocks again was regulated to a large extent by wrench-type movements. Mid-Cretaceous to Recent wrenching both produced the low structural closures that currently are petroleum traps, and in places segmented them or gave them a sinuous trend. One way to discover new reserves in not-too-remote areas of West Siberia might be to search for as-yet-unsuspected culminations, cut away by wrenching from other highs known to contain hydrocarbons. It would also be important to distinguish the faults of major rank that have been active since the Mesozoic and may have induced drag closure in their vicinity. They are very difficult to discriminate from the large number of minor pre-Mesozoic and Mesozoic faults, the abundance of which tends to create unwanted noise on seismic lines. These minor faults are local and are not necessarily continuations of deeper fractures. For this purpose, neotectonic studies and use of space imagery might prove to be more efficient than geophysical work.

There are in the world several analogues of West Siberia, some yet to be explored. One may think, for instance, of the "Cuvette congolaise," with its Stanleyville beds that are similar and almost contemporaneous with the West Siberian Bazhenovo. The problem here, next to accessibility, would be the presence of a thick enough post-Stanleyville cover. The Karroo of southern Africa also has many features in common with West Siberia; it has, however, much more intense diagenesis that has altered potential reservoirs. A wrench-tectonic approach of the North Sea structure may lead to as-yet-unsuspected prospecting themes and areas. The same could, we hope, be true where dealing with the Middle East, or with the large Algeria–Western Desert ensemble.

## REFERENCES CITED

Bogolenov, K. V., ed., 1983, Paleogeographiya severa SSR v Yurskom period (Jurassic paleogeography of the northern U.S.S.R.): Nauka, Novosibirsk, 190 p.

Burlakova, G. S., 1983, Koltsevye i lineyneye struktrury tsen-

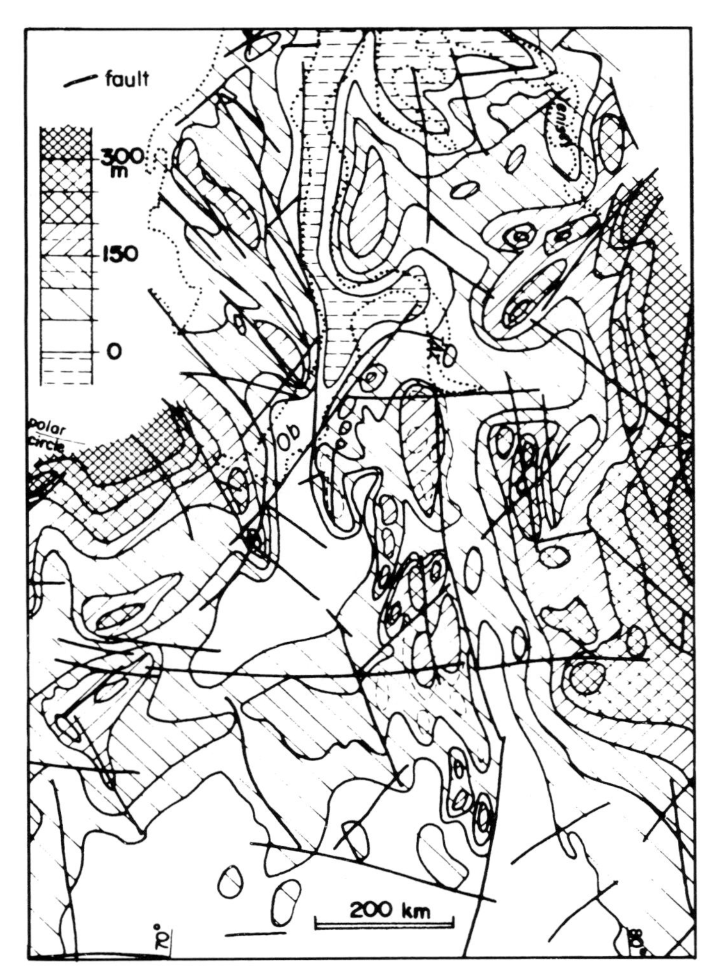


Figure 12. *Northwestern West Siberia: isohyps of Neogene (early Miocene?) peneplaned surface (from Gorelov and Rozanov, 1978).*

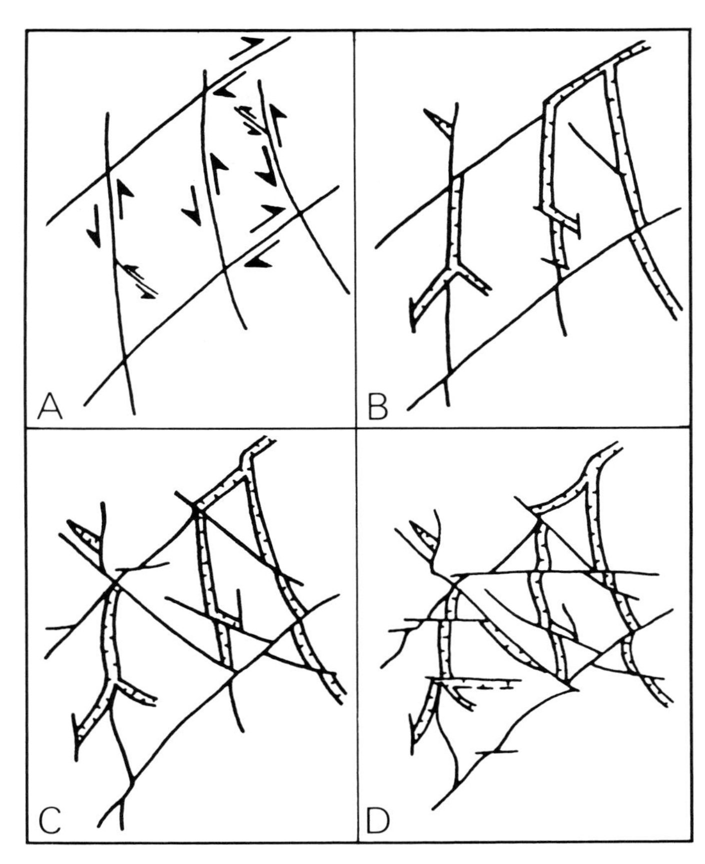


Figure 13. *Possible model of West Siberian fault evolution. A: Pre-Late Permian, B: Early Triassic, C: Approximately Middle Cretaceous, D: Present.*

tralnoy chasti Zapdno-Sibirskoy plity (Circular and linear structural features of central West Siberia: Geologiya i Razvedka, v. 8, p. 41–47.

Bryukhanov, V. N., et al, 1982, Lineynye i koltsevye struktrury (Linear and ring structures): Geotektonika, v. 1, p. 1–13.

Bush, V. A., 1983, Sistemy transkontinentalnykh lineamentov Evrasii (Systems of transcontinental faults of Eurasia): Geotektonika, v. 3, p. 15–31.

Gaydeburova, W. A., et al, 1974, Perspektivy neftegazonosnosti nizhneyurskikh otlozheniy tsentralnykh i yuzhnykh rayonov Zapadno-Sibirskoy plity (Oil and gas prospects of the Lower Jurassic series of south-central West Siberia, *in* SINNIGIMS, ed., Geologiya i neftegazonosnost Sibiri (Geology and petroleum bearing of Siberia), Novosibirsk, p. 32–36.

Generalov, P. P., 1983, O pozdekaynozoyskikh proyavleniyakh skladchatosti nagnetaniya i sdviyovykh dislokatsiy v Zapadnoy Sibiri (On the Late Cenozoic compression folding and displacment faults in West Siberia, *in* Nauka, ed., Regionalnaya neotektonika Sibiri (Regional neotectonics of Siberia), Novosibirsk, p. 15–25.

Gorelov, S. K., and L. N. Rozanov, ed., 1978, Map of the neotectonic deformation of the sediment cover of the oil and gas regions of the U.S.S.R.: VNIGRI, Moscow.

Karus, E.V., et al, 1984, Glubinnoye stroyenie Zapadnoy Sibiri (Deep pattern of West Siberia): Sovietskaya Geologiya, v. 5, p. 75–85.

Korzh, M. V., 1978, Paleogeograficheskie kriterii neftegazonosnosti Yuri Zapadnoy Sibiri (Paleogeographical criteria controlling the petroleum content of the Jurassic of West Siberia): Nauka, Moscow.

Kuzin, I. L., et al, 1974, Kaynozoyskiy etap strukturoobrazovaniya v Zapano-Sibirskoy neftegazonosnoy provintsiy (The Cenozoic stage of the structural evolution of the West Sibeia petroleum province), *in* Neftegazonosnost Zapadnoy Sibiri (Oil and gas content of West Siberia): VNIGRI, Leningrad, p. 52–68.

Reynin, I. V., and N. N. Perughin, 1974, Analiz deformatsiy poverkhnostey vyravnivaniya kak metod kolichestvennoy otsenyi tektonicheskikh struktur za etap relefoobrazovaniya (Analysis of peneplains), *in* Neftegazonosnost Zapadnoy Sibiri (Oil and gas content of West Siberia): VNIGRI, Leningrad, p. 69–75.

Reclus, E., 1874, Les Phénomènes terrestres. Les continents. Hachette, Paris, 280 p.

Rigassi, D., 1970, Tidal anomalies and regional geological structure: Science, v. 170, p. 1002–1003.

Surkov, V. C., and O. G. Zhero, 1981, Fundament i razvitie platformennogo chekhla Zapadno Sibirskoy (Basement and evo-

lution of the West Siberian platform): Nedra, Moscow, 143 p.
Surkov, V. C., et al, 1984, Evolyutsiya i glubinnaya struktura zemloy kory Sibiri (Evolution and deep structure of the earth's crust in Siberia): Sovietskaya Geologiya, v. 7, p. 41–49.
———, 1984, Tektonika fundamenta Zapdno-Sibirskoy plity i formirovanie struktur platformennogo chekhla (The basement and structural evolution of the West Siberian platform), *in* Tektonika molodykh platform (Tectonic of young platforms): Nauka, Moscow, p. 134–145.

## SUPPLEMENT REFERENCES

Evseyev, G. P., 1980, Usloviya neftegazonosnosti i osobennosti formirovaniya mestorozhdeniy nefti i gaza na Zapadno-Sibirskoy plite (Petroleum-bearing conditions and petroleum field formation in West Siberia): VNIGRI, Leningrad, 159 p.
Gladenkov, Yu. B., et al, 1984, Seismostratigraphiya i ee razvitie v Sovetskom Soyouze (Seismostratigraphy and its development in the USSR): News of the Academy of Sciences of the U.S.S.R., Geological series, 4, p. 3–20.
Kogan, A. B., ed., 1979, (Glubinnoye stroyenie vostochnykh nefteperspektivnykh territoriy SSSR po resultatam kompleksnoy interpretatsii geologo-geofizicheskykh dannykh (The deep structure of oil-prospective eastern territories of the U.S.S.R. based on complex integrated geologico-geophysical data): VNIGRI, Leningrad, 143 p.
Kontorovich, A. E., et al, 1975, Geologiya nefti i gaza Zapadnoy Sibiri (Oil and gas geology of West Siberia): Nedra, Moscow, 679 p.
Kozlovsky, E. A., ed., 1984, Moriya Sovietskoy Arktiki (The seas of the Soviet Arctic), Geologicheskoye Stroyenie SSSR (Geological structure of the U.S.S.R.), v. 9, Nedra, Leningrad, 280 p.
Krasny, L. I., 1984, Globalnaya delimost litosphery v svetc geoblokovoy kontseptsii (Global division of the lithosphere in the light of a global concept): Sovietskaya Geologiya, 7, p. 17–32.
Rudkevich, M. Ya., et al, 1984, Formirovanie neantiklinalnykh i kombinirovannykh lovushek v melovykh otlozheniyakh Zapadno-Sibirskoy plity i metodiki ikh poiskov (Formation of non-anticlinal and combined traps within the Cretaceous of the West Siberian Platform): Geologya nefti i gaza, 8, p. 17–23.
Salmanov, F. K., 1974, Zakonomernosti raspedeleniya i usloviya formirovaniya zalezhey nefti i gaza (Distribution rules and formation conditions of oil and gas reservoirs): Nedra, Moscow, 280 p..
Savostin, L. A., et al, 1984, Istoriya raskrytiya Evraziyskogo basseyna Arktiki (History of the opening of the Arctic Eurasian basins): Reports of the Academy of Sciences of the U.S.S.R., geological series, v. 275, 5, p. 1156–1161.
Solovieva, N. A., 1981, O poperechnykh narusheniyakh sredinno-okeanicheskikh khrebtov (On the transverse faults of Mid-oceanic ridges): Geotektonika, 6, p. 15–31.
Stroganov, V. P., 1984, Obosnovanie vozmozhnosti neftenakopleniya v Yurskikh depressiyakh na severa Zapadnoy Sibiri (Evidencing oil prospects in deep depressions of northern West Siberia): Sovietskaya Geologiya, 4, p. 9–19.
Vysotsky, V. N., et al, 1982, Razmeshchenie lovushek vyklinivaniya v osnovnykh neftegazonosnykh kompleksakh Zapadnoy Sibir (Distribution of pinch-out traps in the main petroleum-bearing formations of West Siberia): Geologiya nefti i gaza, 9, p. 1–6.

# Deep Mediterranean Basins and Their Oil Potential

P. F. Burollet
*Total–Compagnie Francaise Des Petroles*
*Paris, France*

Generally speaking, the Mediterranean basins have a relatively flat and uniform bottom, with water depths of 2,800 to 3,000 m (9,190 to 9,840 ft). There are several major exceptions: the Tyrrhenian Sea and the Aegean Sea are internal parts of the Alpine orogenies, with a back-arc type of stretched structure. The Ionian Sea shows gradual northward slopes, while the Eastern basin occupies a typical external position with the front of the Hellenic and Tauric nappes and the external overthrust zone known as the Mediterranean Rise.

The main petroleum exploration objectives are the Oligocene and Miocene series, deposited during and after the main orogenic phases, and covered by thick, Messinian evaporites and Pliocene–Pleistocene marine sediments. Below the Ionian Sea, the Mediterranean Rise, and the Eastern basin, Mesozoic and Paleogene series may be seen as possible targets. The zone at the shelf limit of the African Platform and the Alpine chains resembles many of the productive areas around the globe.

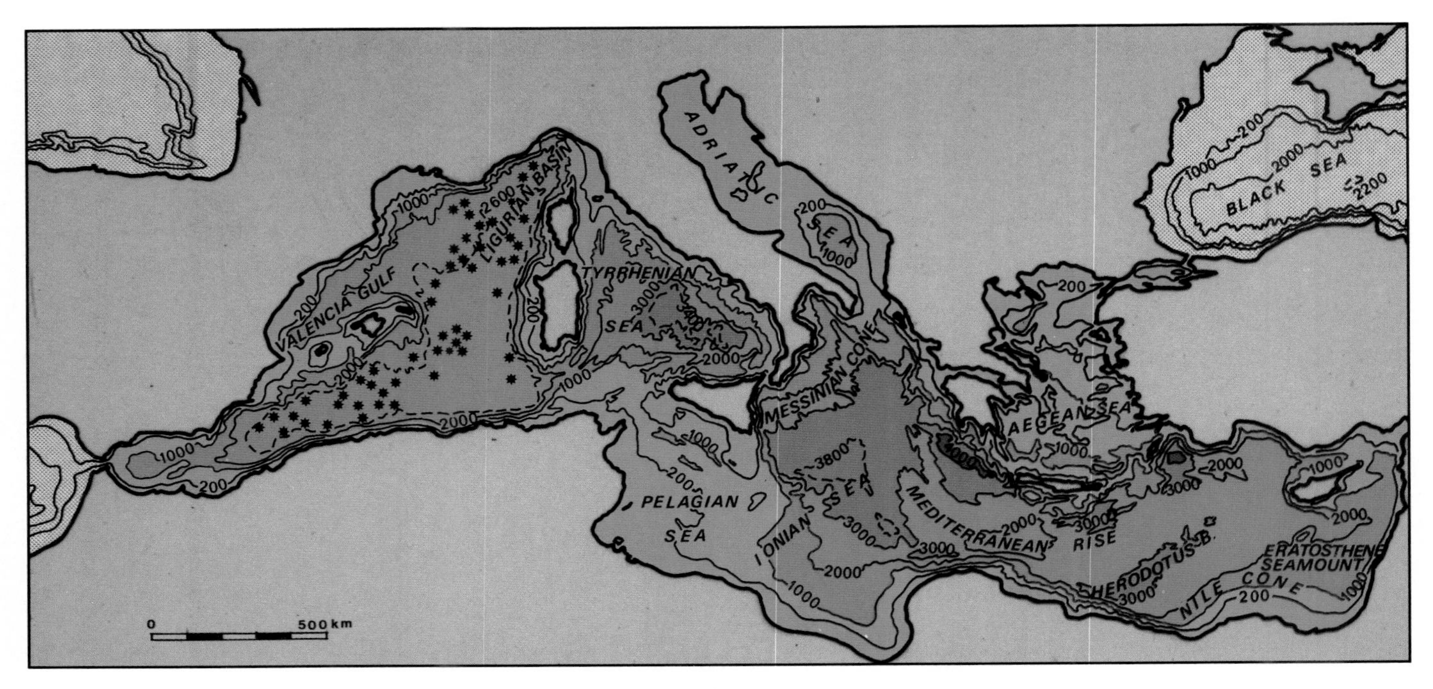


Figure 1. *Bathymetry of the Mediterranean Sea. Stars indicate the presence of salt domes.*

## INTRODUCTION

During the 1960s, hydrocarbon consumption experienced an exponential growth curve that suggested a shortage for the following 25 years. With increasing prices of crude oil, a new effort was oriented to frontier areas and especially to the deep seas. At the same time, scientific exploration of the oceans made significant progress, with increasing knowledge of the tectonics and sedimentology of the continental margins and deep basins.

Since 1970, oil companies have conducted deep-water geophysical exploration, especially in the Mediterranean. Various surveys were made by American, French, and Italian groups (Allan and Morelli, 1972; Finetti and Morelli, 1972; Giese et al, 1982; Recq et al, 1984).

During the same period, the international oil industry developed the technology for drilling in deep water, while studying ways to produce and transport hydrocarbons in these areas. At present, the maximum water depth reached by wildcats is about 2,000 m (6,560 ft).

Recently, stagnation in oil consumption and relatively low prices of crude oil have discouraged oil companies from investing in long-term projects. Most of them have concentrated their exploration budgets on areas and prospects able to provide a short-term pay out.

However, bearing in mind the political instability of the Middle East and the eventuality of an increasing worldwide demand, it is time to prepare for the prospects at the end of the century.

Lying south of industrialized Europe and near North Africa, with its tremendous demographic growth, the Mediterranean could be one of the best regions for deep-sea exploration.

Many papers have been published on Mediterranean geology. Every two years the General Assembly of CIESM (International Commission for the Scientific Exploration of the Mediterranean Sea) meets as a forum for presenting results and for discussions between geologists and geophysicists. We may underline special, collective publications gathered at certain phases of this exploration: that is, initial reports of the Deep Sea Drilling Project (DSDP)–Legs 13 (Ryan et al, 1973) and 42 (Montadert et al, 1978); and reports by Stanley et al (1972); Drooger (1973); Finetti and Morelli (1973); Nairn, Kanes, and Stehli (1977 and 1978); Biju-Duval and Montadert (1977); Burollet, Clairefond, and Winnock (1979); Wezel (1981); Burollet and Bizon (1984); Morelli (1984); and Intergovernmental Oceanographic Commission bathymetric charts (UNESCO, 1981).

Associated with the Alpine folds, the Mediterranean Sea is partially a remnant of the Western part of the Mesozoic Tethys; several basins are younger, and the origin of some of them is still a mystery for geologists and geophysicists. Nonetheless, we make progress every year with such work as geophysical surveys, bathymetric mapping, dredging and coring, and direct observation with manned submersibles. During the International Geological Congress in Moscow (August 1984), French and Russian geologists presented a palinspastic reconstitution of the Tethyan areas during Mesozoic and Cenozoic times (Dercourt et al, 1984).

## BATHYMETRY AND PHYSIOGRAPHY

The Mediterranean Sea is formed by several basins separated by straits or saddles; from west to east they are the Alboran Sea, the Algero–Ligurian basin, the Tyrrhenian Sea, the Adriatic Sea, the Ionian Sea, the Aegean Sea, and the Levantine basin. The larger basins have a relatively flat and uniform floor with water depths of 2,700 m (8,860 ft) to 3,000 m (9,840 ft). The continental slopes are relatively abrupt, and continental shelves are very narrow, except those located in front of the large rivers and those under the Adriatic and Pelagian Platforms (Figure 1).

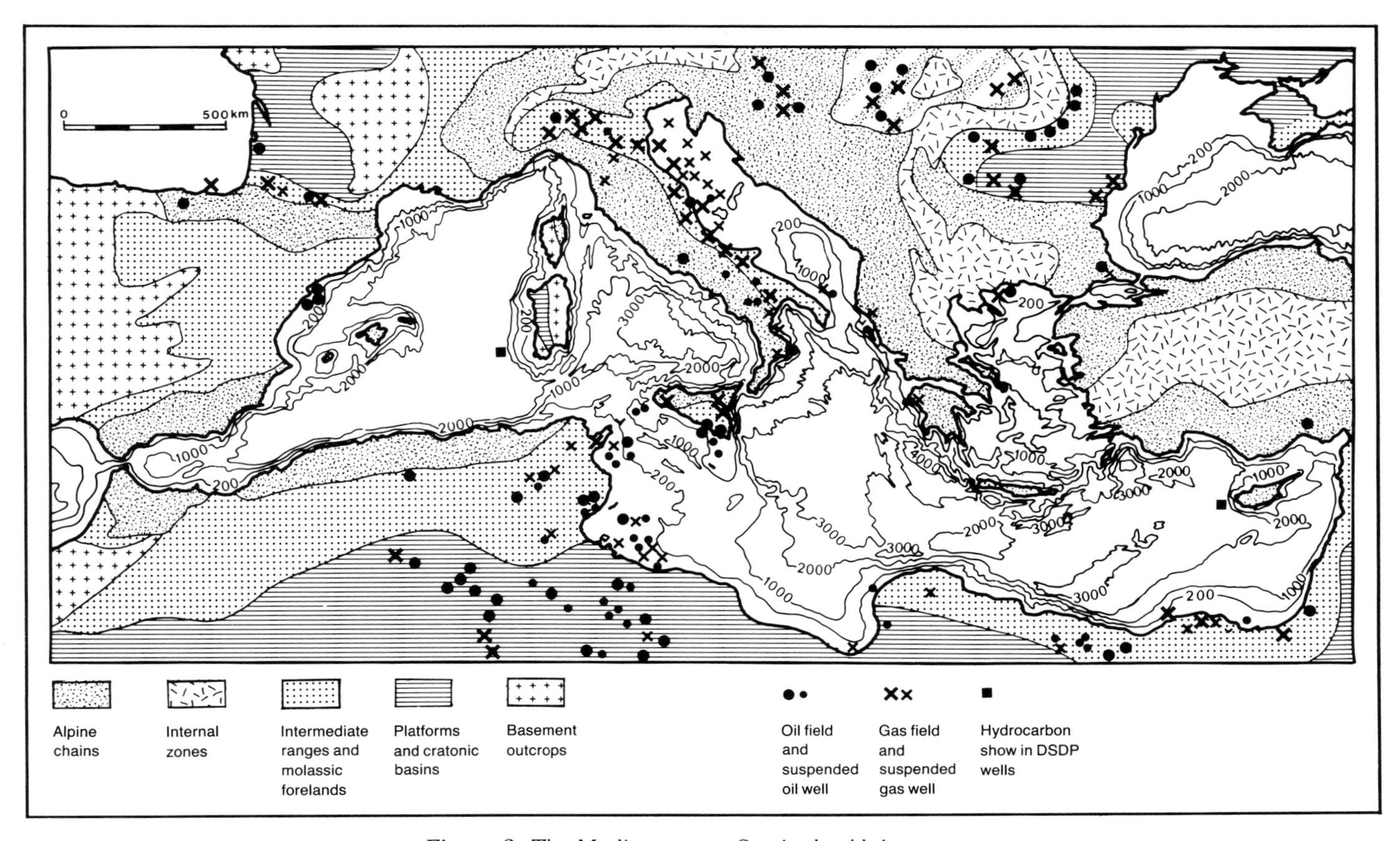


Figure 2. *The Mediterranean Sea in the Alpine system.*

Ciabatti and Marabini (1973) analyze the hypsometric curves of the Mediterranean by 500 m (1,640 ft) depth slices: the units most representative in terms of surface are the continental shelves (included in the 0–500 m or 0–1,640 ft slice), and the deep plains (2,000 to 3,000 m or 6,560 to 9,840 ft).

There are, however, several major exceptions to this scheme. The Alboran, Tyrrhenian, and Aegean Seas have a complex and uneven submarine topography with deep mounts and ridges, while the Ionian Sea shows a gradual slope from south to north. South of Greece, the western part of the Levantine basin has a hummocky ridge separated from the Hellenic arc by a deep trench.

## REGIONAL DESCRIPTIONS

Using the whole of the Alpine fold zone as a reference, one observes that each deep basin of the Mediterranean has different morphological and sedimentological characters. These differences may be explained by the basins' structural positions. The Alboran Sea and the South Balearic basins have an internal position between the diverging Betic ranges and Rif or Algerian Atlas. The rest of the western Mediterranean has an ambiguous position in relation to the break of a former Hercynian massif and the drift of Corsica and Sardinia. The Tyrrhenian and Aegean Seas are typical back-arc basins. The eastern basins and the Ionian Sea occupy an external position with regard to the Hellenic and Tauric chains, including the frontal overthrust zone called the Mediterranean Ridge or, even better, the Mediterranean Rise. The Pelagian and Adriatic Seas are cratonic platforms (Stanley, 1972; Finetti, 1976; Van Bemmelen, 1972).

The crustal nature of the basins is not simply a matter of academic and scientific curiosity. We will see that source beds and their thermic evolution depend largely on the deep structure. Even the presence and the thickness of reservoirs may be related to basement and tectonic history.

To the west, the Alboran Sea and the South Balearic basin have an internal position within the orogeny; relative to the Gibraltar arc, they may be interpreted as back-arc marginal seas. The crust is stretched and broken, and the role of shear faults is prominent. Two basins are separated by marginal plateaus and by the west-southwest to east-northeast orientated submarine range of Alboran Islands. The western basin is 1,500 m (4,920 ft) deep; the eastern one is 1,200 m (3,940 ft) deep and opens eastward to the western part of the Algero-Balearic basin (Carter et al, 1972; Durand-Delga, 1980).

The rest of the Western Mediterranean Algero-Ligurian basin has intermediate significance because of its internal position with relation to the Algerian margin, and external position everywhere else (Figure 2) (McKenzie, 1972; Caire, 1973a).

The structure results from rifting with relation to a counterclockwise, early Miocene rotation of Corsica and Sardinia, and the southward migration of the Kabylia old basement terranes during early to mid-Miocene times (Alvarez, Cocozza, and Wezel, 1974; Biju-Duval, Dercourt, and Le Pichon, 1977; Burrus, 1984; Le Douaran, Burrus, and Avedik, 1984; Rehault, Boillot, and Mauffret, 1984).

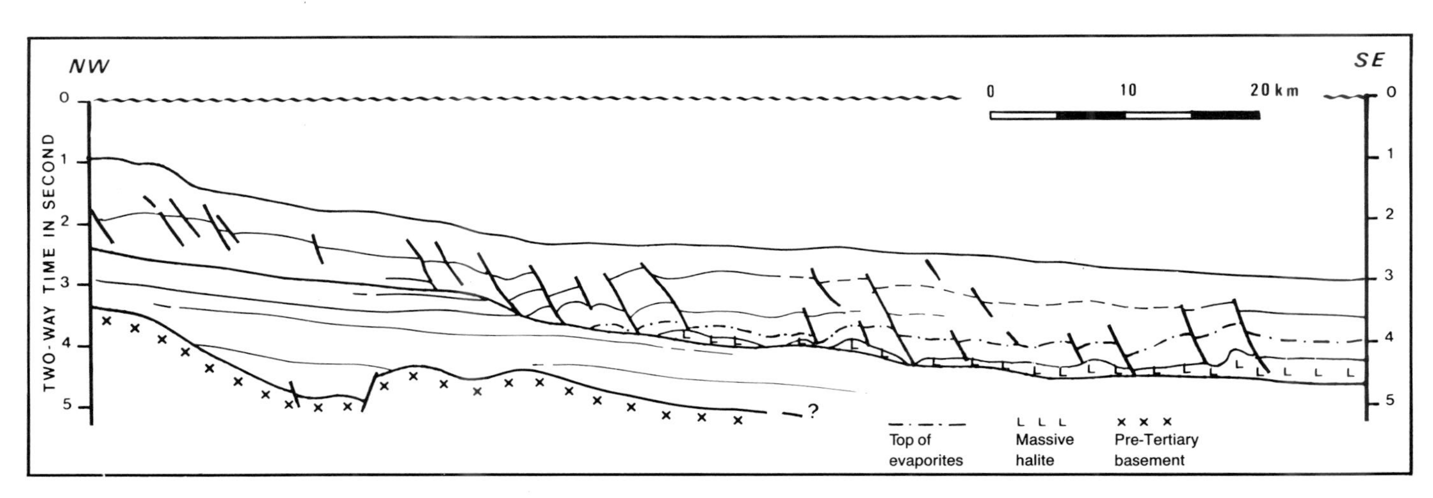


Figure 3. *Interpretation of a seismic section across the margin of the "Golfe du Lion."*

Refraction and reflection seismic surveys have defined sedimentary and crustal layers. Miocene and Pliocene–Pleistocene series were drilled during DSDP legs 13 and 42A. For safety requirements, the well was not allowed to drill through the Messinian salt. Large, positive gravity anomalies and high heat flow are in agreement with the seismic results.

The continental crust thins seaward, and the axial zone is underlain by thin, oceanic crust with a shallow Moho (± 10 km [6.2 mi] below sea level). Pre-Tertiary sediments may exist on the continental crust, as seen near the Balearic Islands and in the Gulf of Valencia. Close to the margin of Provence, or along western Corsica and Sardinia, these sediments are mainly Hercynian, more or less metamorphic, Paleozoic series (Figure 3).

The Tertiary series consists of detritic syn-rifting Oligocene sequences (inferred from seismic), Miocene shale and sand sequences with some calcareous horizons, upper Miocene evaporites including thick halite, and Pliocene–Pleistocene pelagic shales (Rehault, Boillot, and Mauffret, 1984).

The pre-evaporitic series may reach a thickness of 4 km (2.5 mi), while the evaporites are 2 km (1.2 mi) thick in the central zone and the Pliocene–Pleistocene unit is generally 1 km (0.6 mi) thick (Figure 3) (Drooger, 1973; Burollet et al, 1974; Cita et al, 1978).

In the southern part of the Mediterranean, the sea floor is very flat and regular. Off the Provencal and Ligurian coasts, numerous, recently active salt domes perturbed the Holocene sediments and may form circular submarine hills that rise above the surrounding bathyal plain, even in areas where it is formed by the distal lobes of the Rhone deep-sea fan (Figure 1).

Between the Spanish mainland and the Balearic Islands, the Valencia Trough has a southwest to northeast orientation. Water depths may be over 2,000 m (6,560 ft) in the central deep, corresponding to a channel system draining sediments from the Ebro delta and deep fan. The northwestern shelf and slope are known for their oil production.

The southern and northeastern margins of the Tyrrhenian Sea correspond to various units of the Apennine orogene. To the north are found north to south ridges and furrows corresponding to a succession of Ligurian and Tuscanian units. They disappear toward the south on the brink of the bathyal plain between 2,000 m (6,560 ft) and 3,000 m (9,840 ft). The physiography is more complex offshore from the Marches and Lucania, suggesting the shearing off of units of Tuscanian or more external zones. Along the Calabrian arc, where relative displacements were most important, the margin is narrower and rich in volcanism. North of Sicily, a succession of east to west dislocations bear the mark of the horizontal displacements of the most internal units. The present orientation of the basement and the placement of the Alpine Corsican nappes occurred at the beginning of Miocene times (Figure 4) (Boccaletti et al, 1978 and 1981; Bartole et al, 1984).

Since late Miocene, a rigid bridge of continental crust has existed along the Tunisia–Corsica–Genova axis. It acted like a puncheon, arching the Genova area and contributing to the extension of the Western Alps toward the southwest, west, and northwest. In this compression, the Corsican crust would have been thrust under the Adriatic crust of the Ligurian area (Morelli et al, 1977; Tapponier, 1977; Selli, 1974 and 1981; Fabbri et al, 1981).

The thrust of the Tyrrhenian area through the Calabrian-Sicilian arc over the Ionian crust has been mentioned above. The repartition of the main Apenninic and Sicilian nappes may be understood in light of the fact that their original area was much narrower than it is today in the transverse, east to west direction, while it was longer from north to south. The most internal zones travelled a long way toward the Calabrian arc and the northeast area of Sicily. Next to these, the other units must have been deposited in an order similar to that of their present-day succession, that is, from west to east they are as follows: Ligurids, Sicilids and Argille Scagliose complex; Oligocene–Miocene flysches tardily associated with these units (Monferrato, Reitano). Further on are units corresponding to the Tuscany nappes and autochthonous, followed by the carbonate units of the Apennines, with their associated flysch. Finally, we find the outer parautochthonous zone that underwent moderate displacement (Colantoni et al, 1981; Gasparini et al, 1982).

The abyssal plain represents the tearing apart of these zones following the gradual curving of the Calabro–Lucanian arc.

Figure 4. *Physiographic and tectonic sketch of the Tyrrhenian Sea.*

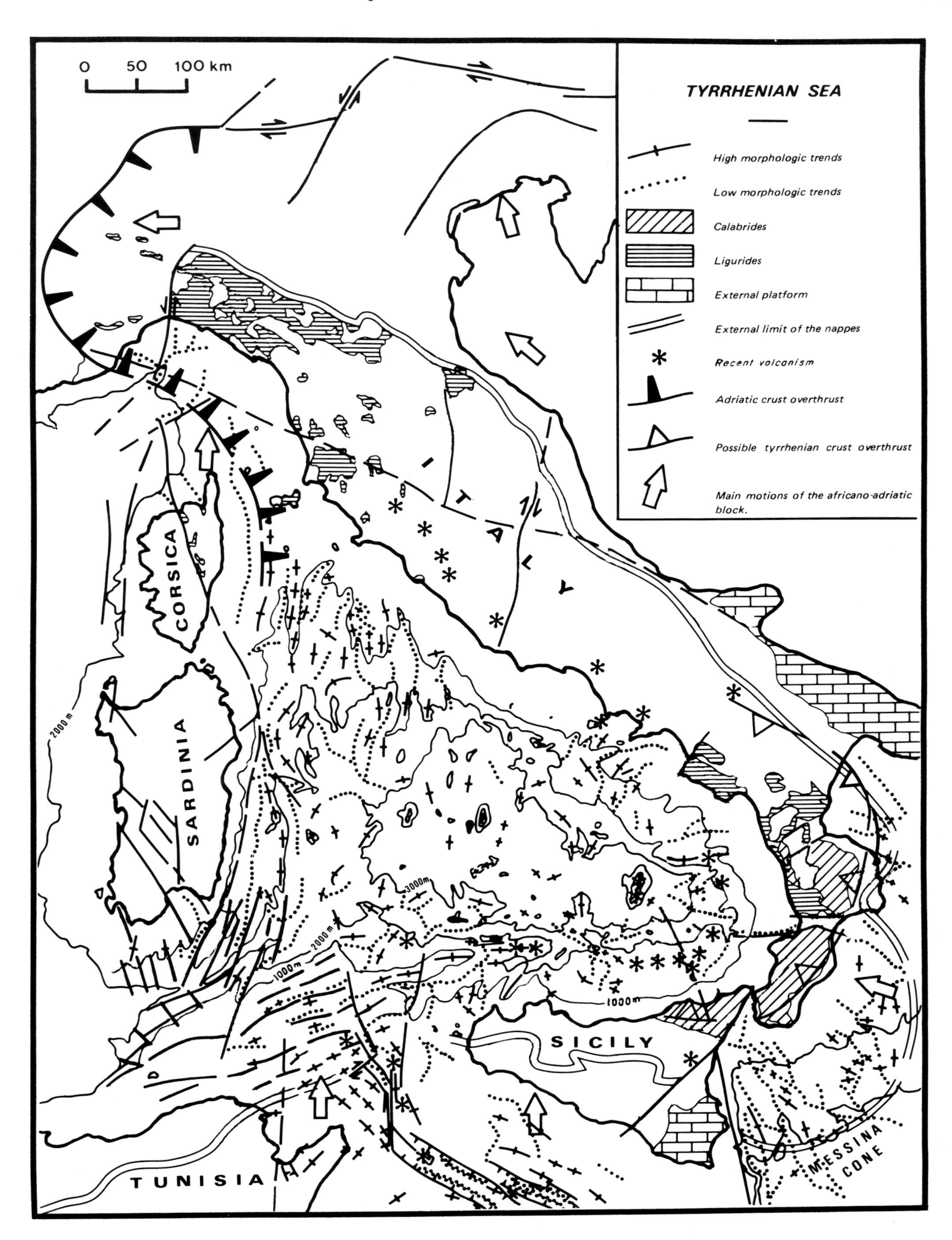

0 50 100 km
TYRRHENIAN SEA
High morphologic trends
Low morphologic trends
Calabrides
Ligurides
External platform
External limit of the nappes
Recent volcanism
Adriatic crust overthrust
Possible tyrrhenian crust overthrust
Main motions of the africano-adriatic block.
ITALY
CORSICA
SARDINIA
SICILY
TUNISIA
MESSINA CONE
2000 m
3000m
2000m
1000m
1000m

Intermediate crusts of the Alpine type were stretched and down-faulted along the margins of the basin, while in the center they disappeared completely, leaving only a thin, oceanic scar covering the mantle (Brunn and Burollet, 1979; Caire, 1973b; Della Vedova et al, 1984; Ferruci et al, 1982; Scandone, 1979).

The Adriatic Sea is mostly underlain by continental shelves; just the Southern basin has an irregular, submarine relief with a central plain where water depth reaches 1,230 m (4,035 ft). Geologically, the Adriatic Sea is a part of the Apulian Promontory surrounded by convergent alpine overthrusts of the Appennines, Southern Alps, and Yugoslavian and Albanian Dinarics (Channel and Horvath, 1977).

Several authors interpret the Adriatic promontory as a part of an Apulian plate, separated from the African cratons by a Cretaceous oceanic basin called the "Mesogea" (Biju-Duval and Montadert, 1977; Biju-Duval, Dercourt, and Le Pichon, 1977; Biju-Duval et al, 1980; Sengor and Jlmaz, 1981). Other authors, after Argand (1922), considered the Adriatic promontory as a northern appendage of the African plate, including the Pelagian and Ionian Seas (Blanchet, 1977; Horvath and Channel, 1977; Burollet, Mugniot, and Sweeney, 1978; Burollet, 1981; Giese et al, 1982).

Underlain by the thick, Mesozoic and Cenozoic series, the Adriatic Sea is a rich petroleum province, especially on the Italian side. Until recently, the presence of known oil and gas was limited to the shelf; however, new discoveries have extended the petroleum-bearing area to the southern, deeper borderland.

The Pelagian Sea, between Tunisia, Sicily, Malta, and Tripolitania, is a shallow continental shelf, cut by intermediate terraces and by the recent northwest to southeast graben-like troughs of Pantellaria, Malta, and Linosa, with water depth reaching 1,700 m (5,577.4 ft) (Burollet, Mugniot, and Sweeney, 1978; Burollet et al, 1979; Letouzey et al, 1980; Winnock, 1979) (Figure 5). Part of the African and East Tunisia craton, the Pelagian Platform, is overlain by very thick sediments including the Triassic, Jurassic, Cretaceous, and Tertiary series. Oil and gas have been found in these various formations, off Sicily, Tunisia, and Tripolitania.

The Ionian Sea is the western part of the Eastern Mediterranean. The classical sequence of morphological units as shelf, slope, rise, and the abyssal plain, is rarely to be found. On the western side, a steep escarpment brings the emergent land of Sicily and the Pelagian Platform into contact with the abyssal plain and the Cone of Messina. East of Malta, this scarp is divided in two by terraces. South of lat. 35°N, the slope gradually decreases, bringing into contact the sub-valley called the Tripolitanian Trough and the Rise of Sidra. This rise is separated from the abyssal plain by a series of submarine reliefs, trending roughly west to east, named Medina mounts, and prolonging eastward the high axes of Central Tunisia, the Kerkennah uplift, and the Medina banks. In the south, the transition between the continental shelf and the large rise of Sidra is made by a very indistinct, only slightly inclined slope scarcely noticeable in the topography. Only near Cyrenaica can a distinct slope be picked up, but here again are found the sub-sea mountains of Cyrene between the slope and the abyssal plain; as with the Medina mounts, the Cyrene mounts are made of subhorizontal Cretaceous and Tertiary beds. This composition was checked by deep diving observations and dredgings. In the Medina mounts, neritic Lower Cretaceous massive calcarenite was overlain by Upper Cretaceous pelagic limestone (Rossi and Zarudzki, 1978; Burollet, 1981; Biju-Duval et al, 1982) (Figure 5). Eastward, the narrow and deep trough of Herodotus separates the slope off Cyrenaica from the complex reliefs of the Mediterranean Rise, the southern part of which forms the vast uplift of Herodotus. Between the Ridge and the Rise of Sidra, the abyssal plain is reduced to a narrow trough at depths between 3,800 and 3,960 m (12,465 and 12,990 ft). The same is true to the northeast, where the Ridge and the Cone of Messina are separated by a valley that continues to connect with the main indentation of the Apulian slope. The limit between the Ionian Sea and Calabria, Apulia, and the Adriatic Sea is an irregular slope, very steep in places, with cirques whose unclear origins may be the result of tectonism, erosion, or solution (Biju-Duval et al, 1982; Finetti, 1981).

The crustal nature of the abyssal plain is still under discussion. The crust has a medium thickness and an intermediate velocity after seismic refraction, and there is a very strong, positive gravity anomaly (Morelli, Gantar, and Pisani, 1975a). On the other hand, heat flow has low values, too low for the normal evolution of an oceanic floor (Erickson et al, 1977). Reflection seismic shows profound and regular reflectors under the Miocene evaporites, prolonging those of the Sidra Rise and even those of the Sidra basin onshore (Finetti and Morelli, 1972; Weigel, 1974; Berckhemer, 1977; Burollet, Mugniot, and Sweeney, 1978; Makris et al, 1982) (Figure 6).

The northern border is marked by the two overthrust alpine arcs of Sicily–Calabria and of Greece, with their submarine annexes of the Cone of Messina west and the Mediterranean Rise east. Between these arcs, Apulia and the southern Adriatic Sea represent a stable block belonging to a limestone platform with the cratonic foreland of the folded belts converging toward it from two sides (Apennines and Dinarids-Hellenids).

I believe that the main part of the Ionian Sea was a vast platform, neritic until middle Cretaceous, and gradually subsiding to the north during the Late Cretaceous, a movement accentuated later by accumulation of the external, overthrust arcs on the northern part of the sea.

The Ionian Sea is surrounded by oil- and gas-producing shelves: the Adriatic Sea, the Gulf of Tarento, the Pelagian Sea, and a recent discovery northwest of Cyrenaica.

The Eastern Mediterranean or Levantine Sea is bordered on the north by the Alpine arcs of Greece, Turkey, and Cyprus. Offshore Greece, there is a deep trench and the hummocky ridge corresponds to an overthrust, intermediate or external submarine range. Going eastward these units progressively lose their complexity and their sharpness (Figure 1) (Brunn, 1976; Le Pichon et al, 1982).

To the south and east, the limits are made by moderately steep slopes cut into by canyons off Libya, Israel, and Lebanon. The main features of the Eastern Mediterranean's southern border are the large Nile delta and deep cone. Between this cone and Cyprus, the curious, circular sea mountain of Erathosthenes seems to have an anticlinal structure (Lort, 1974; Ross, 1977; Makris et al, 1984).

Researchers differ in describing the geological history and the crustal nature of the Ionian Sea basin. Researchers postulate that it was a Mesozoic ocean, remaining below the deep plain, or an intermediate foreland of the African craton that foundered progressively as the Tethyan ocean was subducted below the Greek and Tauric ranges at the end of the Cretaceous (Bein and Gvirtzman, 1977; Ben-Avraham, 1978).

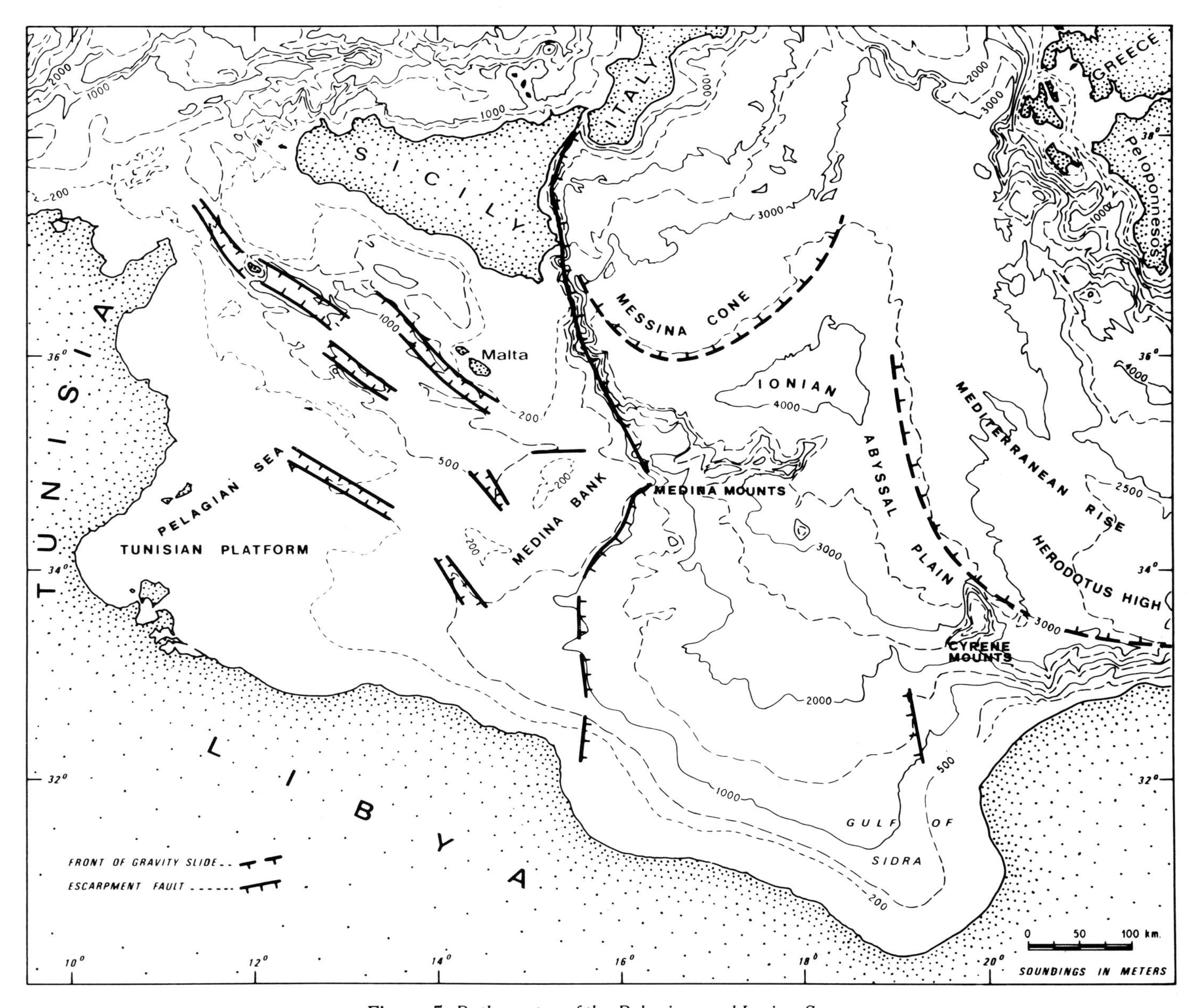


**Figure 5.** *Bathymetry of the Pelagian and Ionian Seas.*

Compressionnal features north of Egypt, offshore and on land (the Syrian arc) indicate a relative closure of the system at the end of the Cretaceous; this fact, together with the obduction of ophiolites in Cyprus and the low heat flow, suggest that the present floor is of intermediate or continental nature.

Marginal, tardi-tectonic basins exist in several parts of the arcs: the Cilicia basin north of Cyprus, the Antalaya basin northwest of Cyprus, and the Aegean Sea. These basins have a very complicated morphology, with more than 200 islands and a series of troughs. High heat flow, active volcanism and seismicity characterize a stretched back-arc marginal sea. These structures relate to a north to south distension associated with the southward curvature of the arc, caused by east to west compression between the Tauric block of Turkey and the Apulian promontory (Berckhemer, 1977; Blanchet, 1977, Le Pichon and Angelier, 1979; Angelier et al, 1982; Le Pichon et al, 1982).

Few oil and gas indications are known around the Eastern Mediterranean: the Prynos basin northwest of the Aegean Sea, oil shows in DSDP wells of Leg 42 and, in the Adana basin of Turkey. There are large discoveries, mainly of gas, off the Nile delta. In fact, the exploration is still at an early stage, partially because of the narrow shelves and local political difficulties.

## OIL AND GAS POTENTIAL

Two main kinds of source beds may be considered in the deep basins of the Mediterranean. First, the Mesozoic and Paleogene series are known on land or on the continental shelves in places where former platforms subsided. Second, Miocene and Pliocene series are contemporaneous and posterior to the formation of the post-orogenic basins.

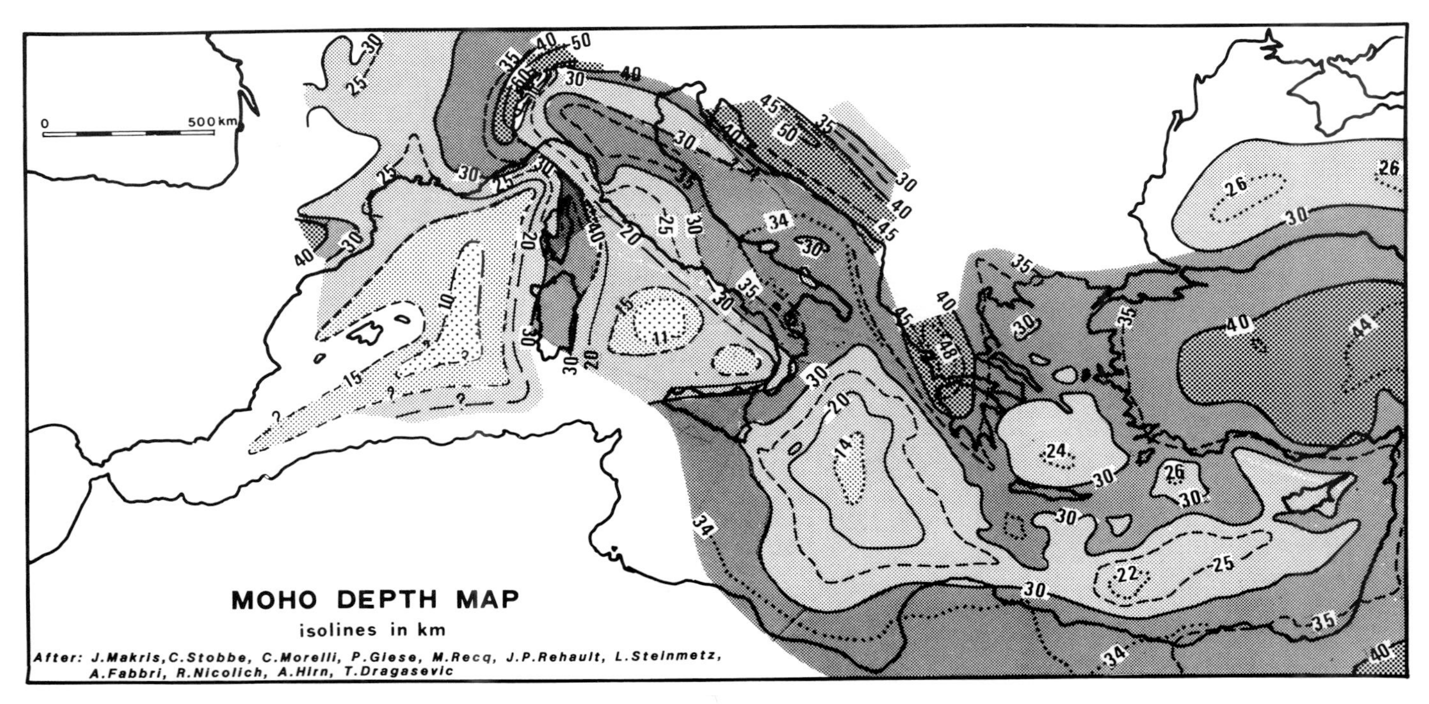


Figure 6. *Isobath map of the Moho, Mediterranean Sea.*

This second category of source beds may exist either on continental or oceanic crust, while the first type is restricted to the continental or intermediate crust areas.

## Mesozoic and Paleogene Potential Petroleum Series

### *Triassic*

Oil is known in Triassic dolomites associated with the dark, Upper Triassic and Liassic Streppenosa shales, south of Sicily in the fields of Ragusa, Gela, and others.

Oil shows are also known in Greece and Israel. Drillers encountered gas in the Jeffara plain of Tripolitania (A-1-23 well), but the source there is probably Paleozoic.

The Triassic, therefore, may represent an objective in the Pelagian Sea, in some places of the Ionian Sea, in the southern Adriatic Sea, and in some portions of the Levantine basin.

### *Jurassic*

Few shows are known in the Jurassic series around the Mediterranean, except in eastern Tunisia (Bir Ali Khalifa, for example). Oil is known in the Egyptian Western Desert; but because of the narrow shelf, no offshore exploration has occurred. A possibility may exist in the southern Levantine basin.

### *Early Cretaceous*

Several oil shows were encountered in Aptian dolomite and limestone of the Serdj Formation east of Tunisia. The Albian dark shales of the lower Fahdene Formation have a promising source potential; and recently, a discovery was reported off northwestern Cyrenaica.

The Lower Cretaceous calcarenites may be a good reservoir in the southern Adriatic Sea, the Pelagian Platform, and the Ionian Sea. In the Mediterranean Rise, if some foothills or thrust-belt-type traps were identified, the Cretaceous series would have good potential. However, until now, the hummocky surface of this zone has given very poor seismic records, even below the Herodotus uplift.

### *Late Cretaceous*

Several oil and gas fields are known in the the Upper Cretaceous reservoirs, sometimes in sands but more generally in carbonates. In the Sidra basin of Libya, source beds are Upper Cretaceous and, more often, Paleocene. In offshore Tunisia, the field of Isis is a Cenomanian reef, associated with Albian and lower Turonian black shales. In the Gulf of Valencia, off Spain, Cretaceous carbonates with karstic weathering have been supplied by the Miocene source beds. The same type of migration has been described in the Adriatic Sea.

A good potential occurs in the Levantine basin from the Cretaceous series, either in front of the Sinai or below the Nile delta and cone. The traps are very deep, but many anticlinal features are known by seismic reflection and most remain undrilled.

In the Ionian Sea, the main, foundering phase is during the Late Cretaceous and Paleocene, corresponding to the rifting in the Sidra basin. The Pelagic limestones are thin and lack porosity; however, during the Cenomanian and Turonian stages, open marine platform deposits probably existed over a large part of the Sidra rise and in the Mediterranean Rise wedges.

### *Paleocene and Eocene Series*

The Paleocene and Eocene series are prolific around the Central Mediterranean. The Sidra basin in Libya, Eastern Tunisia, and the Pelagian Sea off Tunisia and Libya have production (or undeveloped discoveries) in Paleocene reefs and in Nummulite limestone of early Eocene age. Unfortunately, below the Pelagian Sea and northward on the Sidra Rise, they grade laterally to non-porous, Globigerina limestone or shale. Series of this age may cause some interest in the Adriatic Sea, near Albania.

Between Tunisia, Sardinia, and Sicily, stratigraphic conditions may be helpful, but the tectonics show numerous overthrust units similar to those seen in Northern Tunisia or Algeria.

Around the Tyrrhenian Sea, the intermediate-depth plateaus are probably formed by an Alpine–Appeninic series of various facies. Depending on their lithologic succession and their structural complexity, some may be prospective along northwest Sicily, Sardinia, and Corsica. Along the Italian mainland, the risk of tectonic complications becomes greater, while to the southeast, the volcanic arc seems to condemn the whole area for pre-Miocene prospects.

*Neogene Series Hydrocarbon Potential*

Deep basins of the Mediterranean are partially unfilled by Neogene sediments with three main units (Drooger et al, 1973) (Figure 3):

(1) A lower unit with shale and coarser clastics that may include Oligocene sediments at the base.
(2) An evaporitic sequence with dolomite, gypsum, and halite, mainly of latest Miocene age (Messinian).
(3) A post-evaporitic, Pliocene series of pelagic shale, including some turbidites. It is capped by Pleistocene and Recent formations with pelagic muds, hemipelagites, turbidites, and sapropelic black shales.

The restricted environments of the lower unit and of some levels of the evaporite sequences are valid source rocks. Miocene hydrocarbons occur in many places around the Mediterranean, both on land and on the continental shelf: Italy, Sicily, northeastern Tunicia, Valencia/Gulf of Spain, the Prynos basin in the Greek Aegean Sea, and the Nile delta, among others. Oil shows were reported from western Algeria (Cheliff basin) and Adana Gulf (Turkey).

Pliocene formations also yield gas (in the Adriatic Sea and the Gulf of Tarento), as well as oil or gas shows on the eastern coast of the Tyrrhenian Sea.

Some analyzed samples from DSDP Legs 13 and 42 have given valuable information. At site 134 of Leg 13 on the southwestern margin of Sardinia, evaporitic beds presented in situ, immature marine organic matter, together with an enrichment in the gas and gasoline-range liquids indicating a migration from a deeper horizon where the thermal evolution was possible (McIver, 1973).

In the Messinian evaporitic levels of site 374 (Leg 42) in the Ionian Sea, organic carbon of marine origin has a good potential source rock, but it has a low level of thermal evolution (Deroo, Herbin, and Roucache, 1978).

In the Pliocene marls and nannofossil oozes of site 374, and of site 378 in the Cretan basin, as in the mid-Miocene of sites 377 (Mediterranean Rise) and 375 (Florence Rise near Cyprus) and in the Tortonian dolomitic marls of site 375, organic carbon was common and had percentages of the total weight reaching 5% in some samples. They are poorly evolved or immature material from a continental origin. The Serravallian of the Florence Rise, however, showed a more advanced stage of evolution: late diagenesis or early catagenesis (Deroo, Herbin, and Roucache, 1978).

Reservoirs can be found in sands and conglomerates that were frequently deposited after the main orogenic crises. Deep sea fans, in front of the main river mouths, may be rich in turbidites. Some Messinian series reefs are reported around the Western Mediterranean, but it is not likely that we will find them in the deep basins (Rouchy, 1981).

The evaporites, especially thick halite, represent an excellent cover or cap rock. If we take into consideration the size of the deep basins, the Miocene prospects would suggest the possibility of very large reserves.

The problem of thermal evolution is still open. Although insufficient in many places for such a young series, it may be overprinted by a locally high heat flow. Another difficulty is the poor quality of seismic reflections in the pre-evaporitic series, especially where the shallow horizons are disturbed by salt domes.

The best places for the Miocene targets are probably in the Western Mediterranean where the series is very thick. In the Tyrrhenian, the Neogene series is thin and the foundering of the main part of the basin very recent (mid-Pliocene). Thick detritic formations are reported in the northern strait, east of Corsica.

Certainly several basins of the Aegean Sea have good potential, as have the Eastern basins around Cyprus.

## CONCLUSIONS

The deep Mediterranean structures may contain major petroleum objectives. The pre-Miocene targets, for example, may be sought in the foundered or tilted basins corresponding to former platforms. These include the Tyrrhenian intermediate terraces, the South Adriatic Sea, the Eastern Pelagian Sea, the Sidra Rise and Ionian Sea, the Mediterranean Rise, and the Levantine basin, especially south and east of Cyprus, off the Nile delta and the Sinai.

Neogene prospects exist in the newly formed basins, where Messinian evaporites are thick, mainly in the Western Mediterranean. They have good potential in such local, restricted troughs as the Valencia Gulf, the North Tyrrhenian, the Strait of Sicily, the Aegean Sea, the Antalya basin, and north of Cyprus.

Exploring the deep Mediterranean will require numerous improvements in technology and in geological concepts:

(1) A better knowledge of the deep structure.
(2) Improvements in seismics to obtain images beneath the evaporites and in the overthrust areas, such as the Mediterranean Rise.
(3) Development of drilling and production technology that allows operations in 3,000 m (9,840 ft) of water with the requisite safeguards. This range would be sufficient for exploring the Western Mediterranean, the Mediterranean Rise, and several locations in the Levantine basin.

## REFERENCES CITED

Allan, T. D., and C. Morelli, 1972, A geophysical study of the Mediterranean sea: Bolletino di Geofisica Teorica ed Applicata, Trieste, v. 14, p. 291–342.

Alvarez, W., T. Cocozza, and F. Wezel, 1974, Fragmentation of the Alpine orogenic belt by microplate dispersal: Nature, v. 248 (5446), p. 309–314.

Angelier, J., N. Lyberis, X. Le Pichon, R. Barrier, and P. Huchon, 1982, The tectonic development of the Hellenic arc and the Sea of Crete, *in* X. Le Pichon, S. S. Augustithis, and J. Mascle, eds., Geodynamics of the Hellenic arc ant trench: Tectonophysics, v. 86, p. 159–196.

Argand, E., 1922, La tectonique de l'Asie: 3rd International Geological Congress Liége, v. 1, p. 171–372.

Bartole, R., D. Savelli, M. Tramontana and F. C. Wezel, 1984, Structural and sedimentary features in the Tyrrhenian off Campania, Southern Italy: Marine Geology, v. 55, n. 3–4, p. 163–180.

Bein, A., and G. Gvirtzman, 1977, A mesozoic fossil edge of the Arabian plate along the Levant coastline and its bearing on the evolution of the Eastern Mediterranean, *in* International symposium: structural history of the Mediterranean basins: Paris, Technip, p. 95–110.

Ben-Avraham, Z., 1978, The structure and tectonic setting of the Levant continental margin, Eastern Mediterranean: Tectonophysics, v. 46, p. 313–331.

Berckhemer, H., 1977, Some aspects of the evolution of marginal seas deduced from observations in the Aegean region, *in* Symposium International: Structural history of the Mediterranean basins: Paris, Technip, p. 303–314.

Biju-Duval, B., and J. Dercourt, 1980, Les bassins de la Méditerranée orientale representent–ils les restes d'un domaine océanique, la Mésogée, ouvert au Mesozoïque et distinct de la Tethys: Bulletin Société Géologique de France (7), 22: 43–60.

———, ———, and X. Le Pichon, 1977, From the Thetys ocean on the Mediterranean seas: a plate tectonic model of the evolution of the Western Alpine system, *in* Symposium International Structural history of the Mediterranean basins: Paris, Technip, p. 143–164.

Biju-Duval B., and L. Montadert, 1977a, Structural history of the Mediterranean basins. XXV Congrès CIESM (Split 1976): Paris, Technip, 448 p., 9 pl.

———, and ———, 1977b, Introduction to the history of the Mediterranean basins: Symposium International: Paris, Technip, p. 1–12.

———, et al, 1982, Données nouvelles sur les marges du bassin Ionien profond (Méditerranée orientale: résultats des campagnes Escarmed: Revue Institut Français du Petrole, v. 37, n. 6, p. 713–731.

Bizon, J. J., and P. F. Burollet, 1984, ECOMED. Ecologie des microorganismes en Méditerranée occidentale. Etude morphologique, hydrologique, et écologique du bassin entre Languedoc et Corse: Paris, Assn. Francaise des Techniciens du Pétrole, 197 p.

Blanchet, R., 1977, Bassins marginaux et Téthys alpins: de la marge continentale au domaine océanique dans les Dinarides, *in* Symposium International: Structural history of the Mediterranean basins: Paris, Technip, p. 47–72.

Boccaletti, M., and P. Manetti, 1978., The Tyrrhenian Sea and adjoining regions, *in* E. M. Nairn, W. H. Kanes, and F. G. Stehli, eds., The Ocean basins and margins, 4B: 149–200: New York: Planum Press.

———, R. Nicolich, and L. Torturici, 1984, The Calabrian arc and the Ionian Sea in the dynamic evolution of the Central Mediterranean, *in* Geological and geodynamical aspects of the Mediterranean: Marine Geology, v. 55, n. 3–4, p. 219–245.

Brunn, J. H., 1976, L'arc concave zagro–taurique et les arcs convexes taurique et égéen: collision et arcs induits. Bull. Soc. Géol. France, 18, p. 553–567.

———, and P. F. Burollet, 1979, Islands arcs and the origin of folded ranges: Geologie en Mijnbouw, 58 (2), p. 117–126.

Burollet, P. F., 1981, Structure and Petroleum potential of the Ionian Sea. Deep Offshore Technology Conference, Palma de Mallorca: Paris, ASTEO and Petrole Information Publishers, v. 1, n. 1a, p. 1–11.

———, and R. Byramjee, 1974, Réflexions sur la tectonique globale; exemples africains et méditerranéens: Paris, Notes et Mém. Compagnie Francaise des Pétroles, n° 11, p. 71–120.

Burollet, P. F., P. Clairefond, and E. Winnock, 1979, La Mer Pélagienne, Géologie Méditerranéenne: Marseille, 345 p., 6 maps.

———, J. M. Mugniot, and P. Sweeney, 1978, The geology of the Pelagian Block: the margins and basins of Southern Tunisia and Tripolitania, *in* The ocean basins and margins: v. 4B, New York, Plenum Publishing Company, p. 331–359.

Burrus, J. 1984, Contribution to a geodynamic synthesis of the Provencal basin (North–Western Mediterranean): Marine Geology, v. 55, p. 247–269.

Caire, A., 1973a, Les liaisons alpines précoces entre Afrique du Nord et Sicile et la place de la Tunisie dans L'arc tyrrhénien, *in* Livre Jubilaire M. Solignac, Annales des Mines et Géologie (Tunisie), v. 26, p. 87–110.

———, 1973b, The Calabro–Sicilian arc, *in* Gravity and Tectonics, J. Wiley and Sons, New York, p. 157–173.

Carter, T. G., et al, 1972, A new bathymetric chart and Physiography of the Mediterranean Sea, *in* the Mediterranean Sea, Stroudsburg, Dowden Hutchinson and Ross, p. 1–23.

Channel, J. E. T., and F. Horvath, 1976, The African–Adriatic promontory as a paleogeographic premise for alpine orogeny and plate movements in the Carpatho–Balkan region: Tectonophysics, v. 35, p. 71–101.

Ciabatti, M., and F. Marabini, 1973, Hypsometric researches in the Mediterranean Sea: Athénes CIESM Assemblée, 1972, v. 22, n. 2a, p. 178–180.

Cita, M. B., W. B. F. Ryan, and R. B. Kidd, 1978, Sedimentation rates en Neogene deep–sediments from the Mediterranean and geodynamic implications of their changes, *in* Initial report of the DSDP, 42, I, Washington, D.C., p. 991–1,002.

Colantoni, P., et al, 1981, Lithologic and stratigraphic map of the Italian Seas. C.N.R.: Bologna, Instituto per la Geologia Marina, Scale 1:1,500,000; Litografia Artistica Cartografica, Florence (color map).

Della Vedova, B., G. Pellis, J. P. Foucher, and J. P. Rehault, 1984, Geothermal structure of the Tyrrhenian Sea: Marine Geology, v. 55, n. 34, p. 271–289.

Dercourt, J., et al, 1984, Geologic evolution of the Tethys convergence zone from Liassic to Present between the Atlantic and Pamir: Moscow, Colloque 03, XXVII International Geological Congress, oral presentation and posters.

Deroo, G., J. P. Herbin, and J. Roucache, 1978, Organic geochemistry of some neogene cores from sites 374, 375, and 378: Leg 42 A, Eastern Mediterranean Sea, Initial Reports of the DSDP, v. 42, part 1. Washington (U.S. Gov. Printing Office), p. 465–472.

Dragasevic, T. 1969, Investigation of the structural characteristics of the Mohorovicic discontinuity in the area of Yugoslavia: Bolletino de Geofisica Teorica ed Applicata, Trieste, v. XI, n. 41–42, p. 57–65.

Drooger, C. W., 1973, Messinian events in the Mediterranean:

Amsterdam, North-Holland Publ., 272 p.
Durand-Delga, M., 1980, La Méditerranée occidentale: Etapes de sa genèse et problèmes structuraux liés à celle–ci, Paris Livre Jubilaire S.G.F. Mém. h.s. 10, 203–224.
Enay, R., J. J. Bizo, M. Mascle, Y. Morel, R. Perrier, and B. Biju-Duvall, 1982, Faune du Jurassique supérieur dans les séries pélagiques de l'escarpement de Malte (Mer Ionienne). Implications paléogéographiques: Paris, Rev. Institut Français du Pétrole, v. 37, n. 6, p. 733–757.
Erickson, A. J., G. Simmons, and W. B. F. Ryan, 1977, Review of heatflow data from the Mediterranean and Aegean Seas, *in* International Symposium: Structural History of the Mediterranean basins: Paris, Technip, p. 263–280.
Fabbri, A., and P. Curzi, 1979, The Messinian of the Tyrrhenian Sea: seismic evidence and dynamic implications. Giornale di Geologia, (2), XLIII, Fasc, p. 215–248.
——, P. Callignani, and N. Zitellini, 1981, Geologic evolution of the peri–Tyrrhenian sedimentary basins, *in* F. C. Wezel, ed., Sedimentary basins of Mediterranean margins: Bologna, Technoprint, p. 101–126.
Ferruci, F., J. Makris, R. Nicolich, M. Snoek, L. Steinmetz, and W. Weigel, 1982, Deep seismic soundings by OBS in Tyrrhenian Sea, XXVIII Congress CIESM, 2–11 December, Cannes.
Finetti, I., 1976, Mediterranean ridge: a young submerged chain associated with the Hellenic arc. Bolletino de Geofisica Teorica ed Applicata, Trieste, 69: 31–65.
——, 1981, Geophysical study on the evolution on the Ionian Sea, *in* F. C. Wezel, ed., Sedimentary basins of the Mediterranean margins: Bologna, Technoprint, p. 465–488.
——, and Morelli, 1972, Wide scale digital seismic exploration of the Mediterranean Sea: Bolletino de Geofisica Teorica ed Applicata, Trieste, v. XIV, n. 56, p. 191–342.
Gasparini, C., G. Iannoccone, P. Scandone, and R. Scarpa, 1982, Seismotectonics of the calabrian arc: Tectonophysics, v. 84, p. 267–286.
Giese, P., K. J. Reutter, V. Jacobshagen, and R. Nicolich, 1982, Explosion seismic crustal studies in the Alpine–Mediterranean region and their implications to tectonic processes, *in* Alpine–Mediterranean Geodynamics A.G.V. Geodyn. Ser. 7.
Hirn, A., 1980, Le cadre structural profond de la France: Colloque c7, Géologie de la France, 26ème Congrés Géologique International, Paris, p. 34–39.
Horvath, F., and J. E. T. Channel, 1977, Further evidence relevant of the African Adriatic promontory as a paleogeographic premise for Alpine orogeny, *in* International Symposium: Structural history of the Mediterranean basins: Paris, Technip, p. 133–142.
Intergovernmental Oceanographic Commission (UNESCO), 1981, International Bathymetric Chart of the Mediterranean (10 sheets at a scale of 1:1,000,000): Moscow, U.S.S.R., Head Department of Navigation and Oceanography.
Laubscher, H., and D. Bernoulli, 1977, Mediterranean and Tethys, *in* International Symposium Structural history of the Mediterranean basin: Paris, Technip, p. 129–132.
Le Douaran S., J. Burrus, and F. Avedik, 1984, Deep structure of the north–western Mediterranean basin: results of a two ship seismic survey: Marine Geology, v. 55, n. 3–4, p. 325–345.
Le Pichon, X., and J. Angelier, 1979, The Hellenic arc and trench system: a key to the neotectonic evolution fo the Eastern Mediterranean Sea: Tectonophysics, v. 60, p. 1–42.
Le Pichon, X., S. S. Augustithis, and J. Mascle, 1982, Geodynamics of the Hellenic arc and trench, Tectonophysics, 86, n. 1–3, 304 p.
——, N. Lyberis, J. Angelier, and V. Renard, 1982, Strain distribution over the east Mediterranean ridge: a synthesis incorporating new Sea–Beam data, *in* X. Le Pichon, S. S. Augustithis, and J. Mascle, eds., Geodynamics of the Hellenic arc and trench: Tectonophysics, v. 86, p. 243–274.
Letouzey, J., and P. Tremolieres, 1980, Paleo–Stress fields around the Mediterranean since the Mesozoic derived from icrotectonics: comparisons with plate tectonic data: 26ème Congrés Géologique International in Colloquium C5–Bureau de Recherches Géologiques et Minieres ed., Orléans, p. 261–273.
Lort, J. M., W. Q. Limond, and F. Gray, 1974, Preliminary seismic studies in the Eastern Mediterranean: Earth Planet Science Letter, v. 21, p. 355–366.
Makris, R. Nicolich, M. Snoek, and W. Weigel, 1982, Deep seismic soundings by OBS in the Ionian Sea: XXVIII Congress CIESM, 2–11 December, Cannes.
——, and C. Stobbe, 1984, Physical properties and state of the crust and upper mantle of the Eastern Mediterranean Sea deduced form geophysical data: Marine Geology, v. 55, n. 34, p. 347–363.
McIver, R. D., 1973, Geochemical significance of gas and gasoline–range hydrocarbons and other organic matter in a Miocene sample from site 134, Balearic Abyssal Plain: Initial Reports of the DSDP vol. 13, Washington, U.S. Government Printing Office, p. 813–816.
McKenzie, D. D., 1972, Active tectonics of the Mediterranean region: Geophysical Journal, Royal Astronomic Society, v. 30, p.109–185.
Montadert, L., et al, 1978, Initial reports on the D.S.D.P. 42, part: Washington, U.S. Government Printing Office, 1249 p.
Morelli, C. 1970, Physiography and Magnetism of the Tyrrhenian Sea. Bolletino di Geofisica Teorica ed Applicata, Trieste, 12 (48), p. 275–308, Trieste.
——, 1975a, The gravity map of Italy, *in* Structural Model of Italy, Maps and Explanatory Notes: Quaderno Ricerca Scientifica, v. 90, p. 427–447.
——, 1975b, Geophysics of the Mediterranean: Newsletter of the cooperative investigations in the Mediterranean, v. 7, Monaco, 29–III.
Morelli, C., 1984, Geological and geodynamical aspects on the Mediterranean. Marine Geology, 55, 3 and 4: Elsevier, 494 p.
——, in press, Promontorio Africano o microplacca Adriatica? Boll. Oceanologia Teorica ed Applicata, v. II, n. 2, p. 151–168.
——, et al, 1977, Crustal and upper Mantle structure of the Northern Apennines, the Ligurian Sea, and Corsica, derived from seismic and gravimetric data: Bolletino di Geofisica Teorica ed Applicata, Trieste, v. 75–76f, p. 199–260.
——, C. Gantar, and M. Pisani, 1975a, Bathymetriy, gravity (and magnetism) in the Strait of Sicily and in the Ionian Sea: Bolletino di Geofisica Teorica ed Applicata, Trieste, v. 17, p. 39–58.
——, M. Pisani, and C. Cantar, 1975b, Geopnhysical anomalies and tectonics in the Western Mediterranean: Bolletino di Geofisica Teorica ed Applicata, Trieste, v. 18, p. 211–249.
Nairn, A. E. M., W. H. Kanes, and F. G. Stehli, 1977, The Ocean

*ANNEX A:*

*EXPLORATORY WELLS DRILLED IN MEDITERRANEAN DEEP WATER AREAS*

*In the 35 wells drilled in Mediterranean deep water areas, 7 are oil wells and 4 were reported as gas wells.*

| Country | Name | Year | Water Depth | Total Depth | Remarks |
|---|---|---|---|---|---|
| France | GLP-1 | 1982 | 1,700 m | 3,607 m | Dry |
| | GLP-2 | 1983 | 1,250 m | 5,354 m | Dry |
| Italy | Rovesti | 1978 | 955 m | 3,347 m | Dry |
| | Federica | 1979 | 360 m | 3,712 m | Dry |
| | Aquila | 1981 | 827 m | 4,246 m | Oil |
| | Falco | 1981 | | 2,839 m | Gas |
| | | 1981 | | 2,767 m | Dry |
| | Fosca | 1981 | | 2,398 m | Dry |
| | Merlo | 1982 | | 2,307 m | Dry |
| | Grifone | 1982 | 341 m | 3,160 m | Dry |
| | Floriana | 1982 | 219 m | 2,625 m | Dry |
| Spain | Rapita | 1975 | 614 m | 9,461 m | Dry |
| | Montanazo D1 | 1977 | 470 m | 9,190 m | Oil |
| | Montanazo D2 | 1978 | 746 m | 2,741 m | Oil |
| | Montanazo C1 | 1978 | 673 m | 2,891 m | Oil |
| | Montanazo D3 | 1981 | 756 m | 2,864 m | Dry |
| Malta | Gozo 1 | 1982 | 415 m | 3,287 m | susp. |
| Greece | W. Katakolon | 1982 | | 3,218 m | Oil and gas |
| | S. Katakolon | 1982 | 220 m | 2,963 m | Oil shows |
| Yugoslavia | Juzni Jadran 3 | | 300 m | 4,000 m | Dry |
| Algeria | Habibas | 1977 | 935 m | 4,500 m | Dry |
| Libya | E1 NC 41 | 1977 | 217 m | 2,865 m | Oil |
| | A1 NC 35A | 1977 | | 4,905 m | Dry |
| | F1 NC 41 | 1978 | | 4,080 m | Gas |
| | F2 NC 41 | 1980 | 274 m | 2,789 m | Gas |
| | D1 NC 35A | 1983 | | 3,352 m | Gas |
| | E1 NC 35A | 1983 | | 3,597 m | Oil |
| | F1 NC 35A | 1984 | | | |
| Egypt | NDO 1 | 1975 | 408 m | 2,845 m | Dry |
| | NDO B1 | 1975 | 413 m | 3,810 m | Dry |
| | Nab 1 | 1978 | 528 m | 2,863 m | Gas shows |
| | GEBEL el BAHRI | 1981 | | 1,204 m | Dry |
| | G.el BAHR 1A | 1981 | | 1,219 m | Dry |


basins and margins, v. IV A. The Eastern Mediterranean: Plenum Press, N.Y., 503 p.

——, ——, and ——, 1978, The ocean basins and margins, v. IV, B. The Western Mediterranean: Plenum Press, N.Y., 447 p.

Nicolich, R., 1981, Crustal structure in the Italian Peninsula and surrounding seas: a review of DSS data, *in* F. Wezel, ed., Sedimentary basins of Mediterranean margins: Bologna, p. 19–31.

Recq, M., J. P. Rehault, L. Steinmetz, and A. Fabbri, 1984, Amincissement de la croûte et accrétion au centre du bassin tyrrhénien d'après la sismique refraction: Marine Geology, v. 55, n. 3–4, p. 411–428.

Rehault, J. P., G. Boillot, and A. Mauffret, 1984, The western Mediterranean basin geological evolution: Marine Geology, v. 55, n. 3–4, p. 447–477.

Ross, D. A., and E. Uchupi, 1977, The structure and sedimentary history of the southeastern Mediterranean Sea–Nile Cone area: AAPG Bulletin, v. 61, p. 872–902.

Rossi, S., and E. F. K. Zarudzki, 1978, Medina e Cirene: Montagne Sottomarine del Mare Jonio: Bolletino di Geofisica Teorica ed Applicata, Trieste, v. XXI, n. 77, p. 61–67.

Rouchy, J. M., 1981, La genèse des évaporites messiniennes de Méditerranée—Thèse Doctorat Univ. Paris VI, 295 p., 18 pl.

Ryan, W. B. F., D. J. Stanley, J. B. Hersey, D. A. Fahlquist, and T. D. Allan, 1970, The tectonics and geology of the Mediterranean Sea, *in* Maxwell, ed., The sea: New York, v. 4, part II, p. 387–492.

Ryan, W. B. F., et al, 1973, Initial reports of Deep Sea Drilling Project, 13: Washington, U.S. Government Printing Office, 1,447 p.
Scandone, P., 1979, Origin of the Tyrrhenian Sea and Calabrian arc: Bolletino Societa Geologica, Italy, v. 98, p. 27–34.
Selli, R. 1974, Appunti sulla geologia del Mar Tierreno: Rend. Sem. Fac. Sci. Univ. Calgiari, Suppl., v. 43; p. 327-351.
——, 1981, Thoughts on the geology of the Mediterranean region, *in* F. Wezel, ed., C.N.R. Italian Project of Oceanography: Bologna, Tecnoprint, p. 489–501.
Sengor, A. M. C., and J. Jlmaz, 1981, Tethyan evolution of Turkey: a plate tectonic approach: Tectonophysics, v. 75, p. 181–241.
Stanley, D. J., 1972, The Mediterranean Sea, a natural sedimentation laboratory: Stroudsburg, PA, Dowden, Hutchinson and Ross, 765 p. 1 map.
Steinmetz, L., F. Ferruci, A. Hirn, C. Morelli, and R. Nicolich, 1983, A 550 km long Moho traverse in the Tyrrhenian Sea from O.B.S. recorded P.N. waves: Geophysical Research Letters, v. 10, n. 6, p. 428–431.
Tapponier, P. 1977, Evolution tectonique du système alpin en Méditteranée: poinconnement et écrasement rigide–plastique: Bull. Soc. Géol. Fr., v. 19, p. 437–460.
Van Bemmelen, R. W., 1972, Driving forces of Mediterranean orogeny: Geologie en Mijnbouw, v. 51, n. 5, p. 548–573.
Weigel, W., 1974, Crustal structure under the Ionian Sea: Journal of Geophysical Research, v. 40, p. 137–140.
Wezel, F. C., 1981, Sedimentary basins of the Mediterranean margins: Bologna, Technoprint, 520 p.
Winnock, E., and F. Bea, 1979, Structure de la Mer Pélagienne, *in* P. F. Burollet, P. Clairefond, and E. Winnock, eds., La Mer Pélagienne: Marseille, Annales de l'Université de Provence, v. VI, n. 1, p. 35–40.

# Vega Field and the Potential of Ragusa Basin Offshore Sicily

M. W. Schramm, Jr.
*Seagull International Exploration, Inc.*[1]
*Houston, Texas*

G. Livraga
*Petromarine Italia S.p.A.*[1]
*Rome, Italy*

Vega, the largest oil field in the Mediterranean Sea, is located between the southern coast of Sicily and the island of Malta. Its discovery in October, 1980 was based on interpretation of very poor-quality seismic data which, nevertheless, outlined a relatively small structure. The Vega structure lies across the edge of the Late Jurassic-age "Inici" or Siracusa platform. The northeastern flank appears to be controlled mainly by facies variations between the Inici and basinal sediments, while the southwestern edge is controlled by dip.

The fractured limestone and dolomite reservoir may contain in excess of one billion barrels of oil, 30% to 50% of which may be recoverable. Heavy crude is contained within a productive area of approximately 4,000 hectares (10,000 acres). A deeper objective, the Triassic Taormina Dolomite, a major producing reservoir in the basin, has not yet been drilled in this field. The potential for discovery of other fields similar to Vega in the Ragusa basin is excellent.

[1]The above companies are wholly owned subsidiaries of Denison Mines Limited, Toronto, Canada.

## INTRODUCTION

As shown in Figure 1, the Vega field, the largest single oil field in the Mediterranean Sea, is located in Italian waters in the Malta Channel between the southeastern coast of Sicily and the island of Malta, approximately 18 km (11 mi) due south of the Sicilian coast. The field lies almost wholly within and straddles Zone C Exploratory Permits CR-76-SE and CR-80-SE, which have been partially redesignated Exploitation Concessions CC-5-ME and CC-6-IS (Figure 2).

On November 8, 1976, the Italian government granted Permit CR-76-SE to Seagull Exploration Italy, S.p.A. (now Petromarine Italia, S.p.A.); Permit CR-80-SE was granted to Seagull on September 17, 1977. Current partners in the permits and the Vega field are Montedison (Operator) 30%; Agip S.p.A., 30%; Canada Northwest Italia, 20%; Elf Italiana S.p.A., 10%; and Petromarine, 10%.

The dual permit area covers 29,792 ha. (73,616 acres) with water depths averaging 120 m (394 ft). In 1978 and again in 1980, 140 km (87 mi) of detailed reflection-seismic data, in addition to 90 km (56 mi) of reconnaissance data previously shot by Agip, revealed the presence of a drillable structure. This finding led to the drilling of Vega No. 1, the discovery well, in the Jurassic "Inici" Formation, that was completed in October 1980, on Permit CR-76-SE. Later other exploitation concessions were issued.

The quality of all seismic data on which the discovery well was drilled was poor. Consequently, to obtain better data and to better delineate the structural anomaly, an additional 201 km (125 mi) of 2D and 1,730 km (1,075 mi) of 3D data were gathered in 1981. Two outpost wells were subsequently drilled through April 1983, confirming geological prognoses of a major oil field, and development plans were begun.

The Vega field may contain 1 billion bbl of heavy crude, 30% to 50% of which is considered recoverable, within a productive area exceeding 4,000 ha. (10,000 acres).

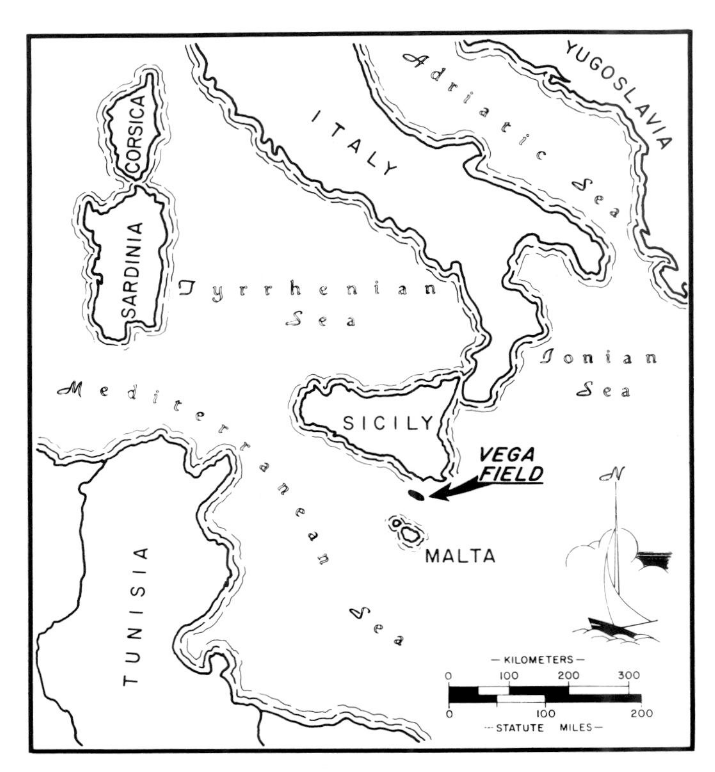


Figure 1. *Index map of the Sicily region showing the location of Vega field.*

## EXPLORATION HISTORY

The region's exploration history began when surface asphalt and gas seeps appeared on onshore Sicily, attracting local industrial exploring for hydrocarbons as early as 1901 (Vercellino and Rigo, 1970). Orthodox exploration did not begin until 1935, but only a few, non-commercial gas discoveries were made.

It was not until 1950, when the Sicilian Regional Government passed the Sicilian Hydrocarbon Law, that foreign oil companies were drawn to the area. The surface geologic setting and an associated prominent oil seep at Ragusa, in southeastern Sicily, led Gulf Oil Corporation to conduct extensive geological and geophysical studies on the prospect. Gulf spudded the Ragusa No. 1 in 1953, and encountered 234 m (767 ft) of oil-saturated vuggy, fractured, then-named Triassic Taormina Dolomite (later split into two formations and renamed the Gela and the Naftia formations; see below). As of 1980, the Ragusa field had produced 120 million bbl of 19.4° API gravity oil. Recoverable reserves had originally been estimated to be 30–40 million bbl (Figure 2).

In 1956, a second major oil field, Gela, was found 47 km (29 mi) west of Ragusa, when Agip encountered low-gravity crude in the then-named Triassic Taormina Dolomite and the overlying limestone beds of the Upper Triassic Noto Formation. Average production per well during the early development stage was 1,100 bbl/day. Cumulative production to 1980 totalled 100 million bbl of oil and 4.6 billion cu ft (bcf) of gas. Recoverable reserves of the low-gravity (8° to 14° API) crude had been calculated originally to be 90 million bbl. About a third of the field extends offshore.

Two smaller satellite fields have since been developed just east and north of Gela (Figure 2). The Ponte Dirillo oil field to the east produces from the then-named Taormina Dolomite. Cumulative production to 1980 totalled 8 million bbl of 14°–16° gravity oil. The Cammarata oil field to the north, on the other hand, produces from the Jurassic Siracusa Formation with cumulative production to 1980 totalling one million bbl.

Geologists immediately recognized the Jurassic and Triassic potential of the far offshore component of what we call the Ragusa basin, and offshore areas opened to exploration under the provisions of Petroleum Law 613 of July 21, 1967.

Offshore drilling activity started in 1973, and to date 12 exploratory wells have been drilled to explore the Triassic dolomites that proved to be so prolific onshore. Not all tests reached this objective, however, and those that did were unsuccessful. In 1978, Montedison discovered the Mila field just offshore southwest of Ragusa. Currently four wells produce from the Mila Member of the Upper Triassic Noto Formation at a depth of 3,500 m (11,480 ft). The Mila field has the capability of producing 5,000–10,000 bbl of 35° API gravity oil per day from lime-

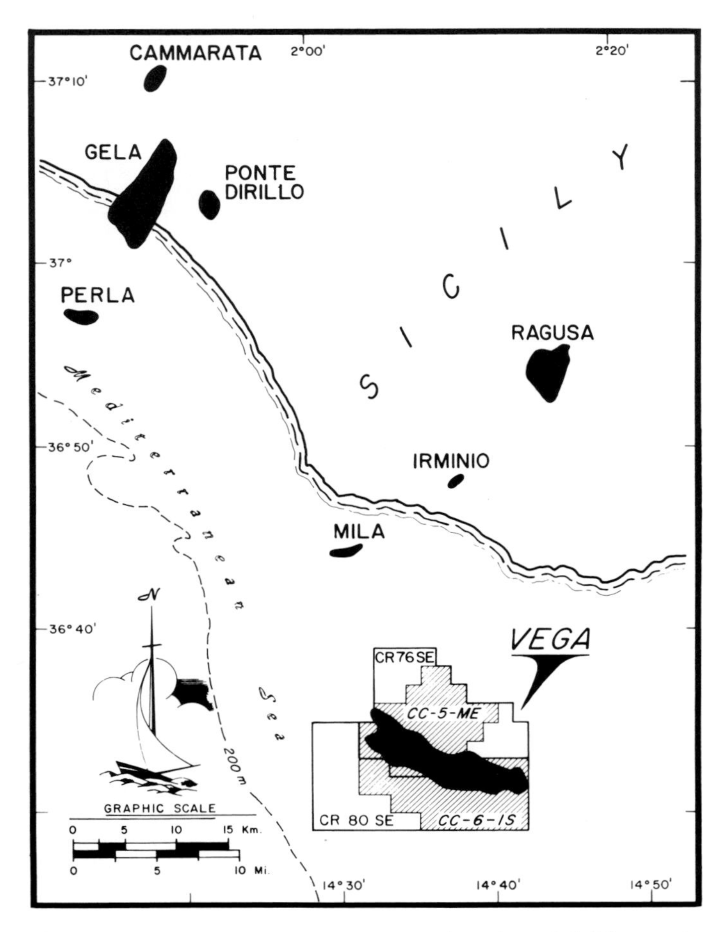


**Figure 2.** *Generalized location map of major oil fields, southeastern Sicily.*

stone whose porosity is only 5%. Recoverable reserves are estimated to be 14 million bbl.

In 1979, after having abandoned an earlier exploratory well to the Triassic dolomites on the same structure, Agip discovered the Perla field offshore, southwest of Gela. This discovery was of an Early to Late Jurassic-age shelf-edge facies of the Siracusa Formation, thus establishing an offshore counterpart of the Cammarata play. This reservoir, referred to, probably erroneously (in the Ragusa basin) as the "Inici" Formation, yields 2,000–3,000 bbl of oil/day from five wells. Estimated recoverable reserves are 7 million bbl of 15° API gravity oil.

The Vega field, the largest oil field in the region, was discovered by the Montedison group, which included Agip, Canada Northwest, Elf, and Seagull, in October 1980, on a farmout from Seagull Exploration Italy, S.p.A. (now Petromarine Italia, S.p.A.). The field is currently being developed in the Early to Late Jurassic-age, shelf-edge shoal facies ("Inici") of the Siracusa Formation at an average depth of 2,500 m (8,200 ft). A maximum gross oil column of 310 m (1,017 ft) has been drilled to date, and as much as 14,700 bbl of oil/day have been tested on a 45-day test from one of the confirmation wells. The reserves of the Vega field may be 1 billion bbl of 15.5° API gravity crude, 30 to 50% of which is considered recoverable.

## TECTONIC FRAMEWORK

The Ragusa basin, informally called the Streppenosa euxinic basin (Johnson et al, 1971) is a depositional basin that encompasses nearly the entire offshore Malta Channel and a portion of the southeasternmost part of the island of Sicily (Figure 3). We will use the name "Ragusa basin," drawing from the Ragusa field outside the town of Ragusa.

The basin's limits are defined by the presence of the Streppenosa Formation, a black shale, and the underlying Nota Formation, which may reach a maximum thickness in excess of 3,000 m (9,800 ft) just off the southeastern coast. Geologists have defined its extent where known to the north onshore, and to the east and south by the partially equivalent Siracusa Platform. Westward and eastward, the extent of the Streppenosa and Noto prism is not known because there has been no drilling.

The significance of these limits is that the Streppenosa and Noto are known only in the subsurface, and no commercial oil production from the underlying Naftia or Gela carbonates (Taormina) has been found where the Streppenosa–Noto cover is absent. Also, the Streppenosa shales are probably the primary source of rock for all accumulations in the intercalated limestone and dolomite beds and in the overlying Siracusa ("Inici").

The basin, which formed as an interior sag during Jurassic time, lies under and is confined to a portion of the extensive Tertiary-age Tunisia–Sicily Platform—the largest outcropping section covering most of southeastern Sicily. It is a large, limestone massif characterized by deep-seated horst and graben faulting. This structural region is referred to by Patacca et al (1979) as the Ragusa zone. Three periods of uplift affected the Ibleo Platform: in the Jurassic, Cretaceous, and Tertiary. Late Jurassic and Early Cretaceous sporadic movement, which plays the key role for structural accumulation of oil in the Ragusa basin, induced fault alignments that trend northwest to southeast, while Late Cretaceous and Tertiary alignments are northeast to southwest. The same alignments and structural history, noted on the Ibleo Platform, are observed on the offshore seismic data.

Figures 2 and 3 show that all fields except Perla and, notably, Vega (all of which produce from Triassic or Jurassic rocks), are oriented in a northeast to southwest direction congruent with Tertiary tectonic trends. The general northwest to southeast orientation of the Vega field, a major exception, is due to the fact that Vega produces from a Jurassic shelf-edge shoal facies that strikes in this general direction, and upon which minor Tertiary structural modification was induced.

Volcanic activity that began in Jurassic time has continued to the present, and intrusive rocks are commonly associated with the producing reservoirs in the basin.

## STRATIGRAPHY

Figure 4 shows the lithostratigraphic terminology and facies relationships used in the Vega field as they relate to other field areas in the Ragusa basin. Only the predominant lithologies are indicated symbolically. Inasmuch as there is no established stratigraphic committee in Italy to determine a common stratigraphic nomenclature, and because new terms and relation-

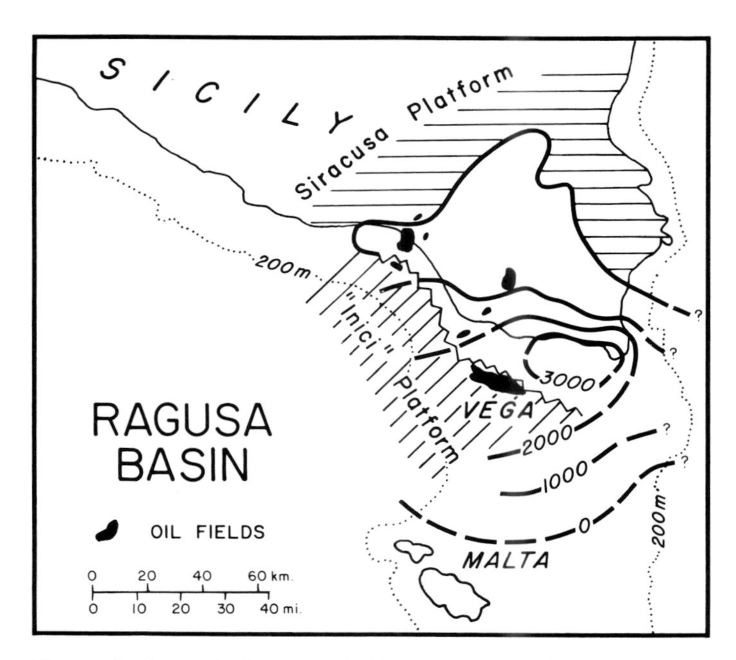


Figure 3. *Isopach (in meters), Streppenosa and Noto Formations, Ragusa basin, Sicily (Modified from Johnson et al, 1971).*

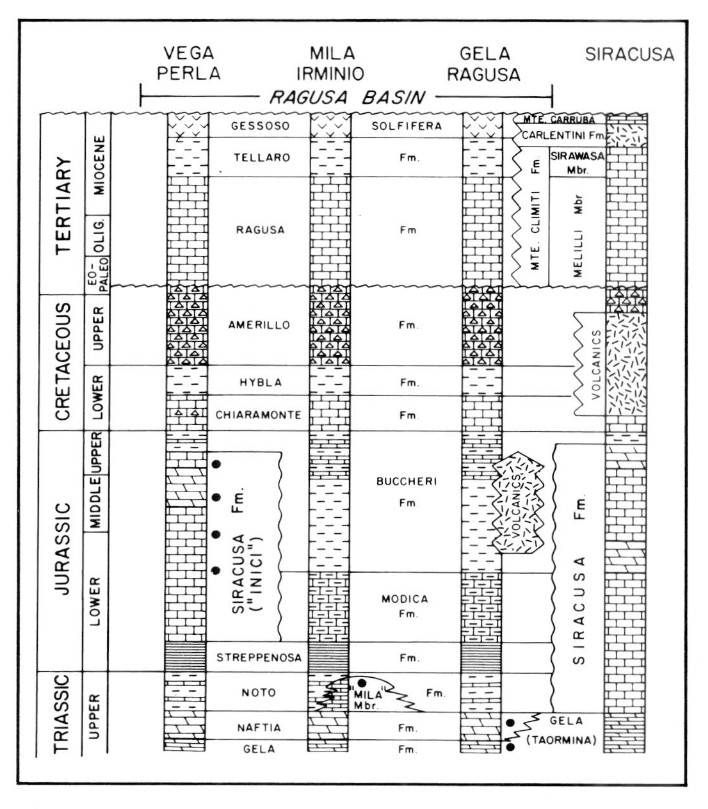


Figure 4. *Composite pre-Pliocene stratigraphic diagram, Ragusa basin, Sicily. The bullet denotes the position of oil-producing reservoirs.*

ships are evolving as the region offshore is being explored, we have elected to refer to and modify the work of Patacca et al (1979) on Mesozoic lithostratigraphy in the onshore Ragusa area, and use Tertiary lithostratigraphic nomenclature commonly employed in the region.

The deepest formation reached in the Ragusa basin is a Middle to Late Triassic-age dolomite, referred to in older literature as the Taormina Dolomite. It is the primary exploration objective in the region. This formation name was recently dropped by Patacca et al (1979) who split it into two formations and renamed it (1) the Gela Formation, a lower white, algal dolomite, overlain by (2) the Naftia Formation, a brown dolomite associated with minor evaporites.

In the Siracusa area northeast of Ragusa the Naftia facies is not recognized. Both formations produce prolifically at Ragusa, Gela, and Ponte Dirillo, but have yet to be drilled and tested in the Vega field.

The overlying Noto Formation of Late Triassic age, which formerly was considered a lower part of the Streppenosa Shale, occurs throughout the basin proper. It changes facies from intercalating dark shales and limestone beds to massive carbonate facies of the Siracusa Formation to the north, and most probably to the south toward Malta. We cannot substantiate this facies change to the west and east. In the Irminio and Mila field areas, a limestone "reefal" buildup occurs that appears to trend northeast to southwest linking the two fields. Locally, this limestone is referred to informally as the "Mila" member of the Noto Formation. Geologists are currently evaluating the potential reserves of oil in this narrow reefal trend.

The Streppenosa Formation of earliest Jurassic age directly and conformably overlies the Noto, and, together with the Noto, its limits where it changes facies to the Siracusa mark the limits of the Ragusa basin. The Streppenosa is a black, bituminous shale, and, along with the Modica and Buccheri formations, is the major source of petroleum for all the carbonate reservoirs in the basin.

The Streppenosa is normally overlain by marly limestone and shale interbeds of the Lower Jurassic Modica Formation, which is then overlain by the Lower to Upper Jurassic Buccheri basinal shales and interbedded limestones. These formations, formerly referred to as the "Villagonia" and "Giardini" formations, respectively, are also basinal facies of the Siracusa carbonate platform facies. In the southwestern part of the basin, however, the Modica and Buccheri formations change facies rather abruptly to another shelf-edge carbonate body we informally call the "Inici".

The name Inici was first applied to the Jurassic carbonate reservoir encountered by Agip in the Perla field. However, the name was drawn from Jurassic Monte Inici rocks that occur in the Palermo Gulf in northwestern Sicily in an entirely separate basin. More appropriately, this facies represents the southern counterpart of the Siracusa carbonate platform edge, but, unlike the Siracusa, it is underlain by Streppenosa bituminous shale. Nevertheless, we will follow the views of Patacca et al (1979), who think that the section defined as Inici in Perla and Vega fields is nothing but the upper part of the Siracusa Formation. The underlying source rocks extend much farther south at some terminus between the Vega area and Malta, thus extending the limits of the basin in that direction. The Siracusa "Inici" is, therefore, the main reservoir in the Vega field. Underlain by the Streppenosa shales, in lateral facies that contact the Modica and Buccheri basinal marly limestones and shales, and overlain by younger Buccheri beds, the Siracusa shelf-edge forms a

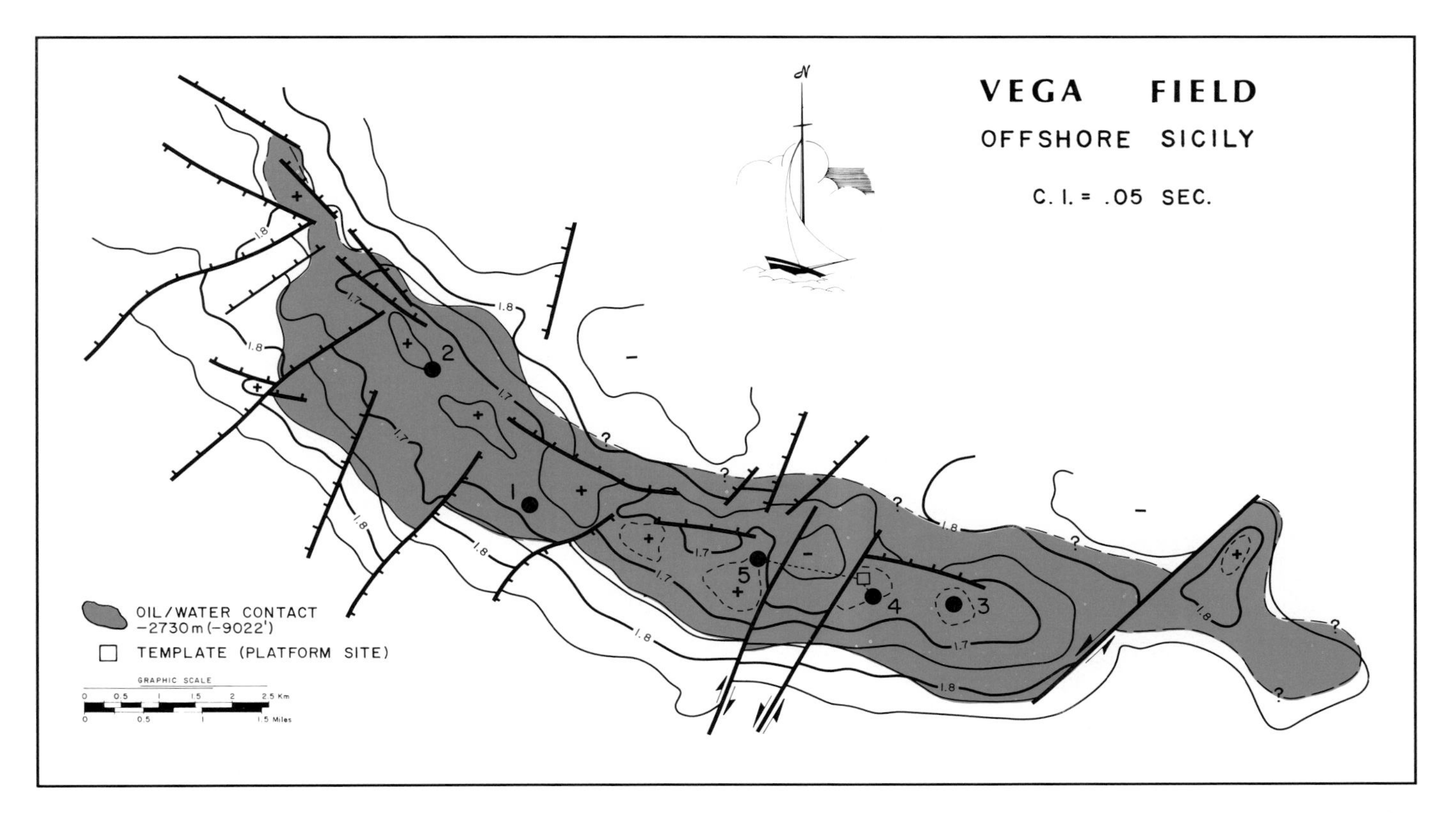


Figure 5. *Seismic time contour map of the top of Siracusa resevoir, showing the estimated limit of the Vega field based on depth conversion. Common oil/water contact occurs in wells in −2,750 m (−9,022 ft).*

belt of yet-to-be-discovered, potentially large oil fields extending from northwest to southeast across the basin (Figure 3).

Except for major and rather abrupt changes in facies and occurrence of extrusive and intrusive volcanics, with consequent changes in the nomenclature in the Siracusa and Priolo areas of eastern onshore Sicily, the Cretaceous and Tertiary nonproductive strata are relatively homogeneous regionally across the Ragusa basin. Primarily, deposition of alternating limestone and shale formations, terminating with deposition of evaporites during late Miocene, characterized the remaining stratigraphic history of the Ragusa basin.

## THE VEGA FIELD

### Structure

The Vega field, which is named after a bright star in the constellation Lyra, is a major structural–stratigraphic trap oriented in a slightly curved, elongate northwest to southeast direction (Figure 5). The field itself, which lies under a maximum water depth of about 131 m (430 ft), is approximately 14 km (8.7 mi) long and varies in width from 1 to 2 km (0.62 to 1.24 mi). The current 3D seismic studies could reveal a field length of over 17 km (11 mi). The feature in which the oil is entrapped is asymmetric in the northwestern sector and highly faulted. Northeast-to-southwest-trending normal and strike-slip faults at the Siracusa level, attributable to Late Cretaceous and Tertiary regional tectonics, dominate the feature throughout. Less abundant, northwest-to-southeast-trending, Mesozoic-age faults occur only along the northeast flank of the anomaly. One rather pronounced sinistral fault nearly disjoins the easternment segment of the field from the main body.

The faults shown in Figure 5 represent only those with sufficient throw or slip to permit mapping. The entire field area is highly faulted and fractured, which does, in fact, contribute significantly to producibility of the carbonate reservoir.

Because of the generally inferior quality of the seismic data, it is difficult to ascertain to what extent topography or lateral stratigraphic changes affect the form of the anomaly and hence the distribution of the oil (Figure 6). The fact that geologists have observed a common oil–water contact at −2,750 m (−9,022 ft) in all wells drilled to date indicates early placement of oil prior to and concurrent with major fracturing during late Mesozoic time. Readjustment may have occurred during the tectonically active Tertiary period.

The size and magnitude of the anomaly, considering its orientation, would suggest that the contrary trend of the Vega field is largely due to selective accumulation along the northwest-to-southeast-trending, porous carbonate edge of the Siracusa Platform. The extreme hummocky nature of the feature, relief, and evidence from core data suggest a prominent buildup along this part of the platform edge. There are insufficient geological and paleontological data at present to substantiate whether or not any portions of the mass are reefal.

### Reservoir Conditions

The Siracusa carbonate reservoir lies at drilled subsea depths ranging from 2,440 m (8,005 ft) to 2,650 m (8,645 ft). It com-

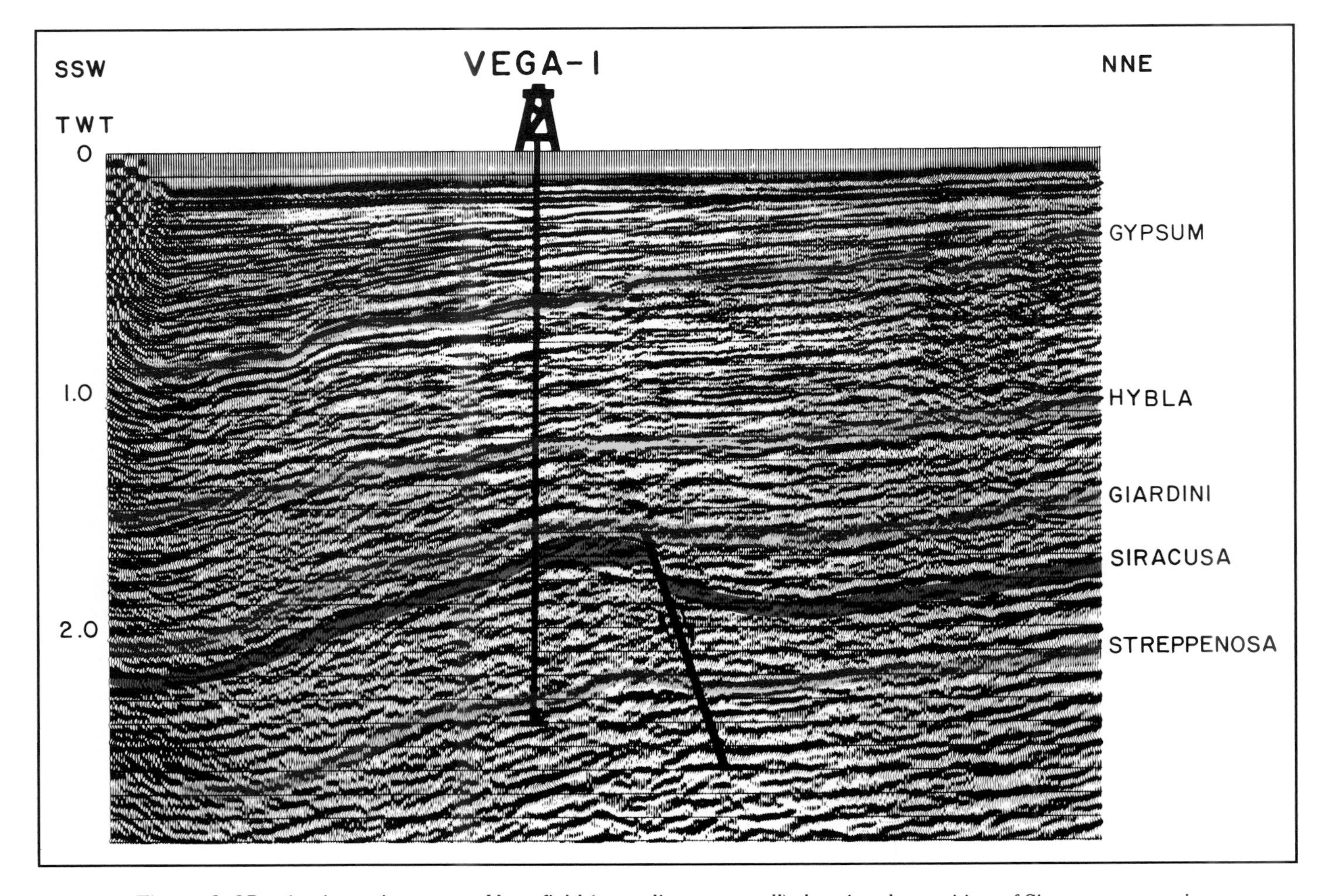


**Figure 6.** *3D seismic section across Vega field (near discovery well) showing the position of Siracusa reservoir.*

prises at least six major lithofacies, which have been identified from cores of Vega wells 1, 2, and 3. These facies are

*Facies I* - algal, intraclastic packstone and grainstone, which comprises four subfacies:
- *Facies Ia* - fenestral, wavy, laminated, algal peloidal packstone and grainstone.
- *Facies Ib* - algal, peloidal mudstone to grainstone.
- *Facies Ic* - peloidal oncoidal grainstone.
- *Facies Id* - algal boundstone.

*Facies II* - pisoid-rich (dolomitic) packstone and grainstone.
*Facies III* - dolomite and calcitic dolomite.
*Facies IV* - intraclastic grainstone, which comprises two subfacies:
- *Facies Va* - coarse graded intraclastic–bioclastic grainstone.
- *Facies Vb* - intraclastic grainstone.

*Facies VI* - lithoclastic breccia.

In general, similar facies occur within all three wells, and the sequence is interpreted to have been deposited in a broad, tidal-flat/lagoonal belt, highly dissected by tidal channels. Within this belt, linear shoals or islands were periodically emergent and underwent calcretization and karstification. The lack of well-developed calcrete and karst within Vega 2 suggests slightly deeper water sedimentation than Vega 1 and Vega 3.

The cored interval in all three wells shows an overall deepening upward, with calcretes and karstic features being most abundant in the lower part of the sequence. We interpret the sequence, therefore, to be broadly transgressive, within which periodic emergence or regression resulted in calcretization and karstification. No palynomorphs were recorded nor was stratigraphically significant microfauna recovered in any of the samples or cores studied.

The major processes to have affected the reservoir characteristics of the sequence are dolomitization and fracturing. Dolomitization is extensive in the upper part of Vega 1 and, to a lesser extent, in Vega 3. Fracturing occurs throughout the sequence but is, in general, more abundant in the upper part of all three wells. Undolomitized limestones are fully cemented by fibrous rim calcite (after submarine to supratidal aragonite) and rhombic pore-filling calcite (meteoric phreatic) and have very low porosity and permeability except where fractured.

Dolomitization created significant intercrystalline porosity, and porosity and permeability were greatly enhanced by fracturing. The dense network of fractures provided conduits for acidic formation waters that created minor intracrystalline porosity and enlarged intercrystalline and fenestral pores. The

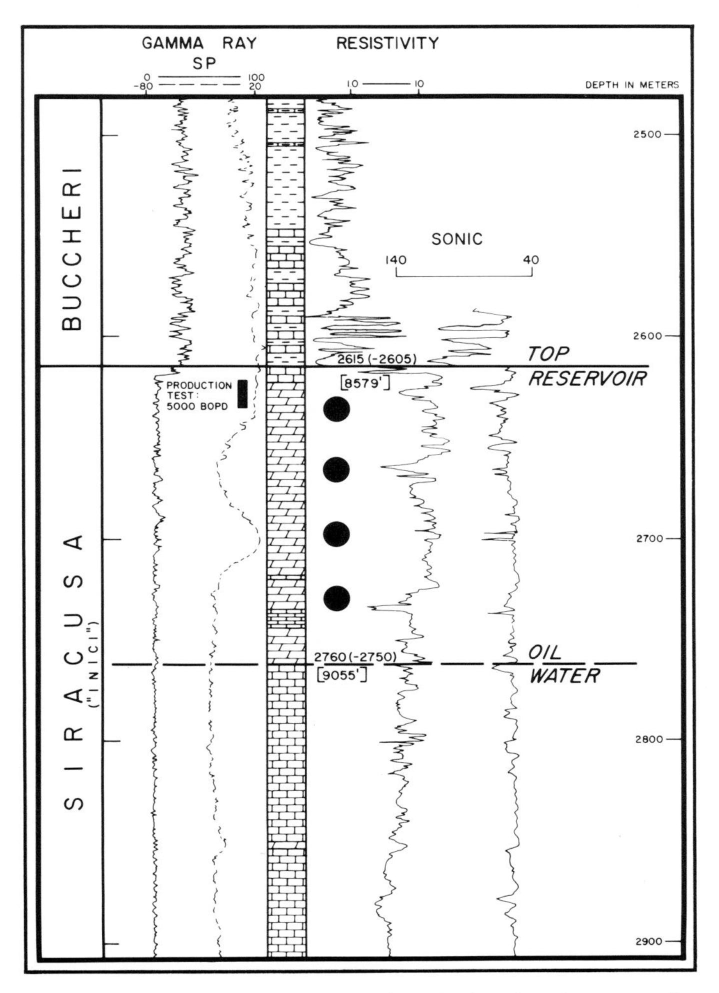


**Figure 7.** *Composite log of Vega No. 1, showing the reservoir.*

main phase of dolomitization occurred during early burial as the meteoric–marine water mixing zone moved laterally through the sequence, the extent and distribution of dolomite being controlled by position relative to the mixing zone. Dolomitization continued during later burial with the precipitation of baroque dolomite cement.

The intensity of fracturing in the upper part of the sequence may reflect an inherent lithological control; possibly the relatively deeper-water facies were not as well lithified and hence were more susceptible to fracturing. In Vega 1 and Vega 3 fracturing is more intense within the dolomite, possibly as a result of its lower ductility relative to limestone.

Pores and fractures contain 15.5° and 16.2° API gravity, respectively, intermediate-naphthenic crude throughout the reservoir. Based on analyses from Vega wells 1 and 3, the pour point of the crude is 8° to 10° C (46.4° to 50° F), and the sulphur content averages about 2.5%. Although water content is relatively low (0.2 to 0.5%), salt content is only 13 mg/l. Metal content is exceptionally low, except for nickel, which is 66 ppm. Asphaltenes range from 18.8% in Vega 1 to 22.3% in Vega 3.

Based on the five wells drilled to date, the thickness of the oil column in the reservoir ranges from 145 m (476 ft) in the Vega 1 discovery well on a flank location (Figure 7) to as much as 310 m (1,017 ft) in Vega 4. In Vega 2, the oil column is 282 m (925 ft); in Vega 3, 292 m (958 ft); and in Vega 5's deviated hole, the restored column is 270 m (886 ft). Calculations are based on a common oil–water contact at subsea, −2,750 m (−9,022 ft), throughout the field.

Average primary porosity of the reservoir determined from logs ranges from 12% to 16% in the eastern part of the field and is considerably lower in the westernmost segment.

## Production Test

A 30-day production test was run in the Vega 1 well in early 1981, to determine the productivity of the Siracusa reservoir (Figure 7). The test was run in a 9 5/8 in. cased hole open from 2,622 to 2,635 m (8,603 to 8,645 ft). Before testing, the section was treated with acid. During the test, 52,479 bbl of oil were produced. The initial flowing pressure measured 267.5 kg/sq cm absolute (3,804 psia) and remained constant throughout completion of the test. A deliverability of approximately 5,000 bbl of oil/day by diesel oil fluxing is expected from Vega 1.

In late 1982, a 45-day production test was run on the Vega 3 well in an open hole from 2,485 to 2,629 m (8,153 to 8,625 ft). A maximum production of 14,700 bbl of oil/day was reached. During the test, 182,198 net bbl of crude oil were produced using diesel injection. Average flow rate was 8,009 bbl of oil/day. The initial reservoir pressure was 256.7 kg/sq cm absolute (3,651 psia), a value that remained unchanged at the end of the test.

Vega 2 was production tested in early 1983, with 186 hours of flow, through a perforated 7-in. casing in the acidized interval of 2,495 to 2,525 m (8,186 to 8,284 ft). The initial reservoir pressure measured was 259.6 kg/sq cm absolute (3,692 psia) below the tested zone at 2,543.6 m (8,345.1 ft). Although crude recovery on previous drill-stem tests after acidization encompassing this interval reached a maximum of 5,000 bbl of oil/day, the estimated flow rate of the final, shorter tested interval was only 1,000 bbl of oil/day with diesel injection.

In late 1984, the Vega 5 deviated well was being tested; the Vega 4 well, deviated from the same platform, had not yet been tested.

## Development

The Vega discovery well in the center of the field and the two successful confirmation wells on both ends of the field were drilled and production tested from a drillship. The fourth and fifth wells were deviated from a semisubmersible, temporary platform (Figure 5) using a seabed template.

Engineering design is almost complete for the largest drilling and production platform in the Mediterranean Sea, and fabrication should begin at the end of 1984 (Figure 8). The 8,000-metric ton, eight-leg, tubular space frame structure will be 141 m (463 ft) high, have bottom dimensions of 70 m by 80 m (230 ft by 262 ft) and top dimensions of 50 m by 18 m (164 ft by 59 ft). It will be installed in 125 m (410 ft) water. The lower parts of the jacket will be designed and fabricated to allow for installing over the pre-installed seabed template.

Topside facilities include equipment for simultaneous drilling and production. Production capability will be 75,000 bbl of oil/day from 30 wells. The platform is designed for a 25-year life.

The field is expected to go on-stream from this platform in mid-1986. At least two other platforms are planned for the

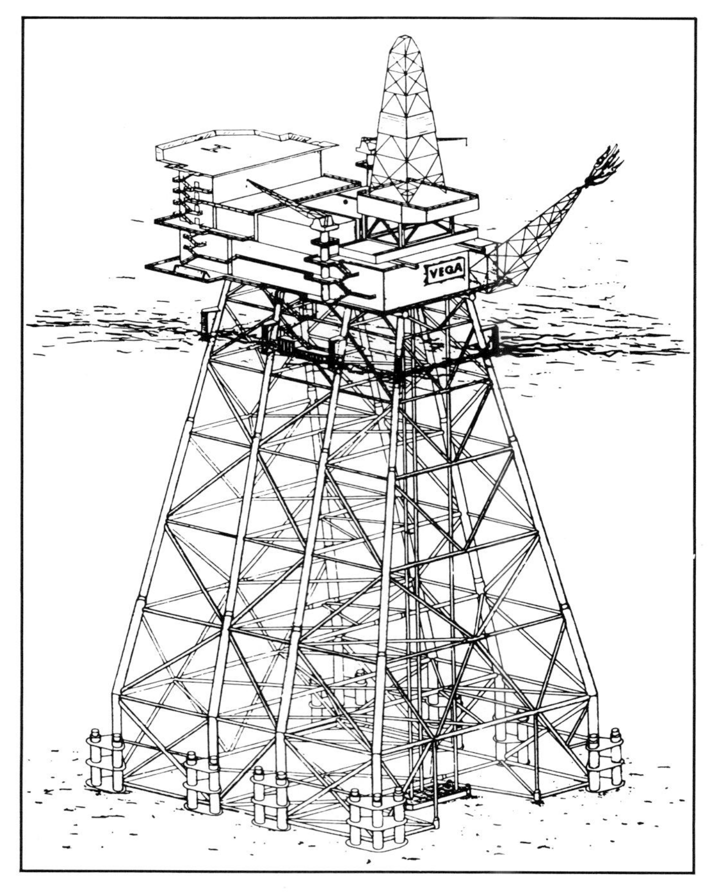


**Figure 8.** *Artist's impression of Vega Platform.*

central and northwestern segments of the field, to accommodate production of the estimated recoverable field reserves of 300 to 500 million bbl of oil.

## ACKNOWLEDGMENTS

We are grateful to Denison Mines and to the partners in Vega field, namely Montedison, Agip, Canada Northwest (CNW), Petromarine Italia and Elf for permission to present and publish this paper. We also wish to thank our colleagues for valuable ideas and contributions, in particular, A. Rigamonti and R. de Mattia.

This paper is presented not only as a tribute to Wallace Pratt, but also to the late Joseph Vercellino, geologist and founder of Seagull, without whose vision and persistence the Vega field might not have been discovered.

## REFERENCES

Johnson, M. S., B. D. Klemme, F. A. Rigo, and J. Vercellino, 1971, Geology and exploration in the Mediterranean Sea: *in* Eighth World Petroleum Congress Proceedings, v. 2, p. 85–97.

Patacca, E., P. Scandone, G. Giunta, and V. Liquori, 1979, Mesocoic paleotectonic evolution of the Ragusa zone (Southeastern Sicily): *in* Estratto da Geologica Romana, v. 18, p. 331–369.

Vercellino, J., and F. Rigo, 1970, Geology and exploration of Sicily and adjacent areas: *in* Geology of Giant Petroleum Fields, AAPG Memoir No. 14, p. 338–398.

# Petroleum Potential of the Thrust Belt and Foretroughs of Sicily

G. Sestini
G. Flores
*Consultants*
*Florence, Italy*

A review of the geology and of the known hydrocarbon occurrences, in the thrust belt of the Apennines in Sicily and its marginal troughs, points to an insufficiently tested potential for commercial petroleum discoveries. The Central basin and the northwest margin of the offshore Trapani basin hold the most promise. The Central basin is a foretrough filled by over 5,000 m (16,400 ft) of clastics, with large gravity slides that originated in the southward Miocene–Pliocene progression of subsidence and substratum thrusting and gliding. Conditions appear to have been optimal for burial and maturation of deep, Triassic potential source rocks, as well as of organic matter in the Miocene deposits beneath or within the olistostromes, or both. There are two potential plays: gas production in Oligocene and Miocene sands, and oil in deeper, Mesozoic carbonates. In the Trapani basin, oil accumulation in lower Miocene reservoirs is related to a Paleogene source-rock depocenter, which is buried under Neogene clays and a repetition of thrust sheets. Future exploration potential appears to be in the thrust belt itself, further to the northwest, rather than in the Neogene basin.

## INTRODUCTION

The southern margin of the fold and thrust Apenninic belt of Sicily, and the associated Neogene basins (Figure 1), appear to have a hydrocarbon potential that still has not been sufficiently tested. Exploration efforts have been moderate. After various attempts before the 1950s, a fair amount of exploration occurred during 1950–1960, but the oil discoveries in the southeast (e.g. Ragusa, 1953 and Gela, 1956) soon diverted attention, especially those made offshore since the early 1970s like the Nilde, Perla and Mila fields. Nevertheless, explorers have made important Tertiary discoveries in the thrust front: the gas-condensate fields of northeastern Sicily (Gagliano, Bronte, etc.), still producing some 120 million cubic feet per day (MMCFGD) since 1963, and the Nilde light oil field, in the northeast offshore (estimated at 10,000 bbl oil per day [BOPD]). A few gas finds in the center were significant, if not commercial, as well as light oil shows at the northern margin of the Central basin.

Exploration on the Adriatic side of the Apennines shows that the favorable conditions for maturation, migration, and trapping of hydrocarbons have been the syntectonic deepening of the Tertiary foretrough and enhanced burial because of the overthrusting and the accumulation of gravity slides. Examples are the olistostrome-covered Pliocene fields of San Salvo and Candela in southeast Italy (Carissimo et al, 1963; E.N.I. 1969). In Algeria in 1953, the Oued Gueterini field, 110 km (68 mi) south-southeast of Alger, produced 2 million bbl of light oil from Eocene limestones at the base of a stack of thrust sheets that in the Miocene had glided into the South Tellian Foredeep (Kieken and Winnock, 1957).

In addition, oil accumulations exist in the downflexed upper part of the foreland sequence, where the Neogene sediments buried this sequence, and where suitable reservoir facies formed, either at the paleoshelf margin or on its slope (e.g., Santa Maria Mare heavy oil field; Duvernoy and Reulet, 1977). Hydrocarbon sources are in the foretrough itself, producing mainly gas, as in the Po Valley (Mattavelli et al, 1983), or in the substratum (where the best known source is Triassic black shales), or both.

This paper reviews the geological framework and the likely hydrocarbon habitat of the Apenninic front of Sicily, to bring attention to this region as a future exploration province worthy of greater activity than it receives at present.

## REGIONAL GEOLOGICAL SETTING

The Sicilian segment of the Apennines is a thrust belt made from the imbrication of the former African (or Apulian) plate margin, overridden by the allochthonous nappes from more internal domains (Channel, Catalano, and D'Argenio, 1959). Southern and southeastern Sicily belong to the stable African foreland, which extends across the Sicilian Channel and the Pelagian Sea (Figure 1). Except for Miocene rotational effects involving the African margin during the emplacement of the nappes, and for later Pliocene–Pleistocene transcurrent movements of a regional extent (Boccaletti, Conedera, and Dainelli,1982), the foreland has changed little in the relative spatial arrangements of its parts since the Triassic. The relative transcurrent movement of Africa and Europe, which ran latitudinally north of Sicily, was sinistral, with divergence from Liassic to middle Cretaceous, and dextral, with convergence from Turonian onward (Bijou-Duval, Dercourt, and Le Pichon, 1976; Dewey et al, 1973).

In southeastern Sicily and Tunisia, the present east-northeast to northeast structural trend represents a relict of the pre-Cenozoic prevalent trend of the passive African margin, developed as normal, originally east to west, or even east-northeast faults, during the Late Triassic–middle Cretaceous phase of the Tethyan oceanic opening. This trend is evident also in the Sicilian Channel (Finetti and Morelli, 1973), in the Sardinia Straits (Auzende, 1971; Auzende, Olivet, and Bonnin, 1974), in northwestern Sicily, and in the southeast part of the Sicily Central basin. The present east to west orientation of the Sicilian chain, instead, may have been a consequence of the clockwise rotation of the northern thrust sheets relative to the substratum of the Hyblean–Sciacca foreland (Channel, Catalano, and D'Argenio, 1980).

The Calabrian massif and the Kabilies of eastern Algeria are probably fragments of the European continental margin (Iberian–Balearic, or Alboran, blocks), detached in consequence of the Oligocene–Miocene opening of the Balearic basin and the rotation of Corso–Sardinia (Alvarez, Cocozza, and Wezel,1974; Cherchi and Montadert, 1982). Calabria was then moved further east-southeast, with the stretching and formation of oceanic crust in the southern Tyrrhenian Sea.

The Calabria–Peloritani massif is a complex nappe structure with Hercynian structural units made of high- and low-grade metamorphics and granites, thrust over a marginal Mesozoic–Cenozoic sequence, the external "Limestone Chain" (Longi–Taormina zone in Sicily). Seaward of northern Tunisia, a similar Paleozoic continental basement occupies large parts of the Sardinia Straits (Auzende, Olivet, and Bonnin, 1974), and extends to some 50–100 km (30–60 mi) east and southeast of Sardinia (Quirra Seamounts, Alvarez, Cocozza, and Wezel, 1974).

Although geologists generally agree that the Apennines of northern Sicily and the Maghrebid chain of North Africa belong to the same fold system with the same time of deformation (Alvarez, 1978; Caire, 1970; Catalano and D'Argenio, 1982a; Wezel, 1974), it is not clear how the northeast to southwest structures of the Tellian Atlas of Tunisia and the Sicilian Channel relate to the east-to-west-trending ones of northern Sicily. There are indeed substantial differences in the overall structural and stratigraphic settings of Sicily and Tunisia. They especially differ in the character of the Mesozoic to Oligocene sequences, essentially because the former were closer to the African paleo-continental shelf than were the latter (Marie et al, 1984).The Tellian Atlas has no equivalent in Sicily because it is a foreland intensely deformed in the northwest, where it is affected by thrust faulting and by strong diapiric phenomena involving Triassic evaporites.

In Sicily, the little-folded Hyblean Plateau in the southeast, and the Sciacca Ridge in the southwest, which are both part of the more stable African foreland, are separated from the chain by the deep sedimentary depressions of the Central and Trapani basins. These basins are foredeeps, in the sense of strongly subsident Miocene tracts affected by extensive gravity sliding (nappes, klippen, and olistostromes). By contrast, the Neogene basins of the Medjerda Valley in Tunisia (Figure 1,C and D) are smaller and less defined, with a weak gravity expression and a thinner fill, devoid of olistostromes (Burollet and Rouvier, 1971;

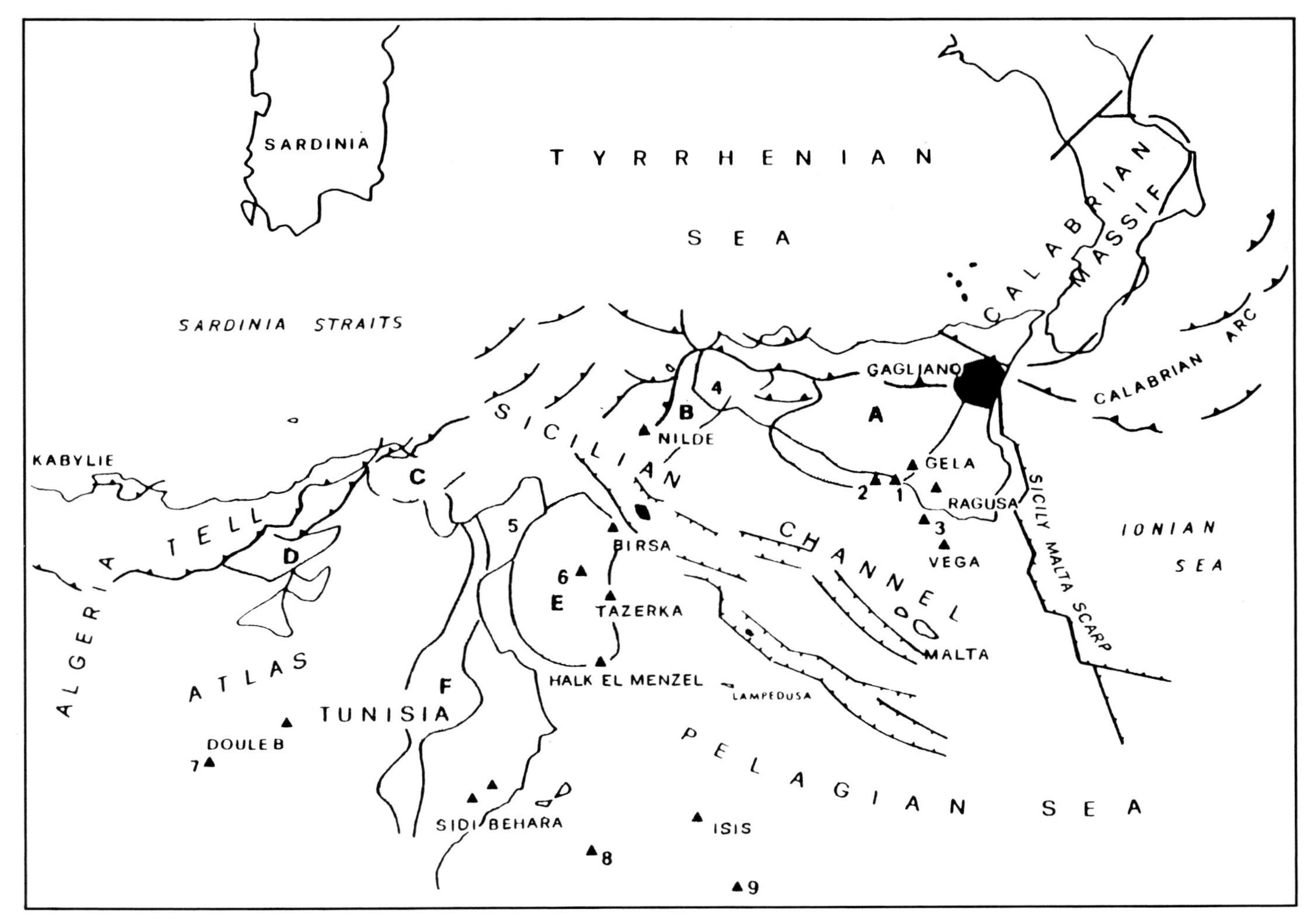


Figure 1. *Oil fields and Neogene basins of the Central Mediterranean region in relation to the Maghrebid–Apenninic thrust belt. Basins: A, Central Sicily; B, Trapani; C, Kechabta; D, Medjerda; E, Hammamet; F, Peri-Atlasic. Oil and gas fields: 1, Gela offshore; 2, Perla; 3, Mila; 4, Lippone; 5, Sidi Abdelrahman; 6, Cosmos; 7, Tamesmida-Jebel Onk; 8, Ashtart; 9, Buri.*

Rouvier, 1977). Nevertheless, both the Sicily and the North Tunisia basins rest over the more deformed margin of the foreland and have, themselves, been affected by horizontal movements.

On the other hand, the Hammamet basin and the Peri-Atlasic depression (Figure 1, F and E) are depocenters in the stable foreland of, respectively, Senonian–Miocene and late Miocene–Pleistocene age. They are unrelated to the thrust belt proper, although they border to the west and northwest on folded zones (Atlas, Cap Bon). Their eastern margins are, however, already part of the stable and more rigid foreland of the Pelagian Block, which is affected exclusively by fracturing. The north-central parts of this block (Lampedusa Plateau, Sciacca Ridge, Malta horst) are characterized by thin Mesozoic sediments, especially Neogene, owing to late Miocene and Pliocene uplift, which was contemporaneous with the northeast to southwest opening of the Pantelleria, Malta, Linosa–Medina, and Jarrafa grabens.

The Sicily–Malta Escarpment is an old dislocation, possibly initiated in Middle–Late Triassic and active in middle Cretaceous (Finetti, 1982), but displaced essentially after the Messinian (Burollet, Mugniot, and Sweeney, 1974; Finetti, 1982; Grasso and Lentini, 1984). Volcanics of various ages relate to this fault zone.

## THE STRUCTURAL FRAMEWORK OF SICILY

The highest structural units of the chain, the Sicilids, are northeast, in the Nebrodi and eastern Madonie Mountains, where they are overridden by the Longi–Taormina thrust belt of the Peloritani Mountains (Figure 2). The Sicilids include at least four nappes, with complex relations between them (Monte Soro, Reitano, Troina, and Capizzi). Lithologically, these nappes are flysch terrains, either of various sandy composition, or shaly-calcarenitic, dating from Early Cretaceous to Oligocene. These nappes originated in a complex basin to the north, and their original relations are still far from understood and a matter of debate.

Over large tracts of the southern flank of the Nebrodi and Madonie Mountains, some of the more plastic components of the Sicilids (i.e., the calcareous-argillaceous flysch) have been tectonically mixed into a fourth, more chaotic nappe ("argille

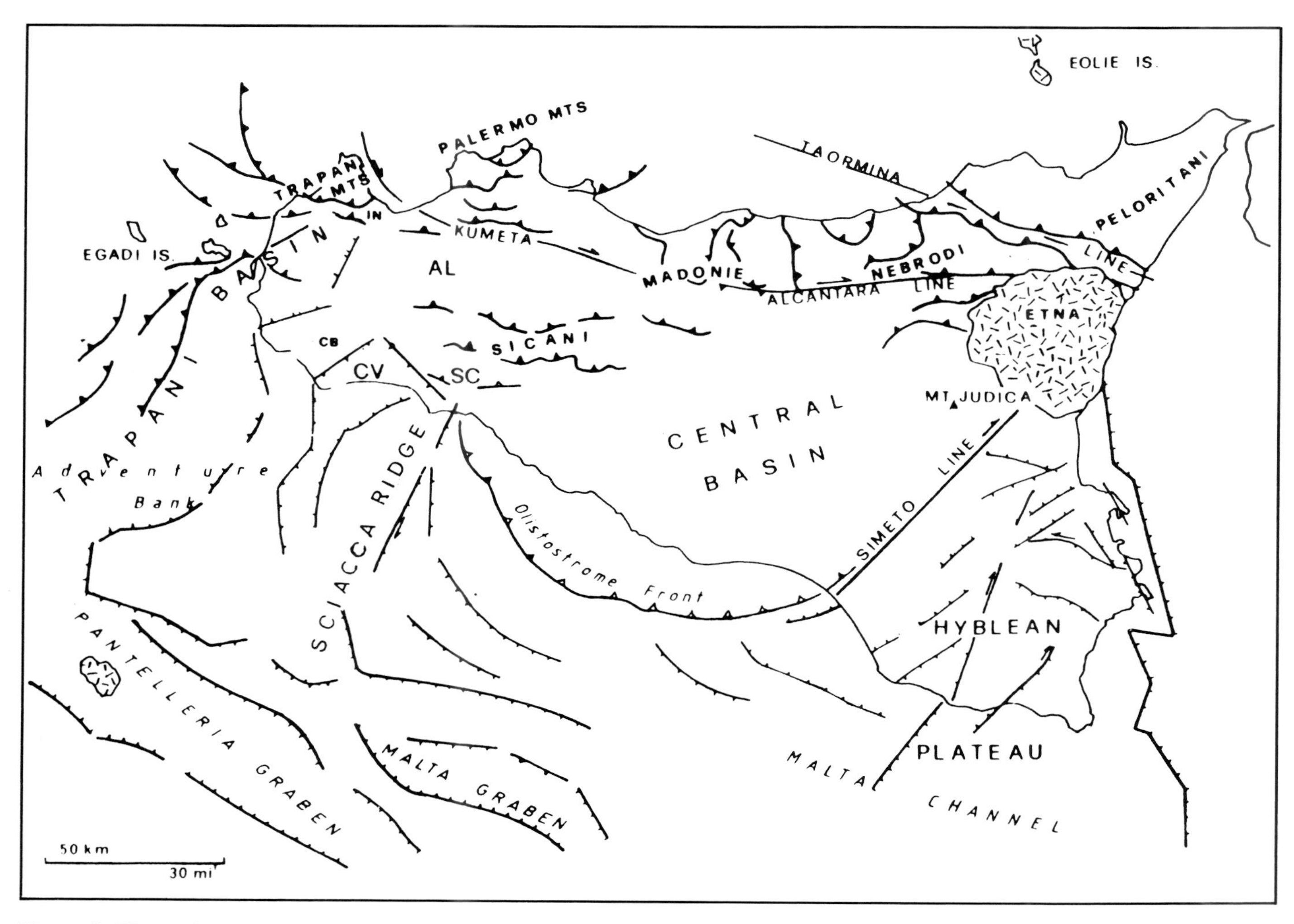


Figure 2. *The main structural units of Sicily. CV = Castelvetrano basin; CB and IN = Campobello-Inici Ridge; SC = Sciacca Zone; AL = Alcamo basin.*

scagliose" Ogniben, 1975), the forerunner of the gravity slides of the northern Central basin.

In the Madonie and eastern Palermo Mountains, the Sicilids overlie the Imerese and Panormid thrusts (Catalano and D'Argenio, 1982a; Schmidt di Friedberg, Barbieri, and Giannini, 1960). The upper member of the Imerese sequence is a thick, upper Oligocene–lower Miocene, quartzarenitic flysch, informally called "Numidian Flysch" because of its similarity to the homonymous unit of the Maghrebid Chain. The contact is usually tectonic, as the flysch is detached from the Imerese carbonates at the level of the Paleogene shales, and is itself internally broken into a number of thrust slices (Duée, 1970), with some disintegration at the southern front into the argille scagliose complex.

Northwest of the island (Figure 2), the Palermo and Trapani Mountains are essentially carbonatic (Caflisch, 1966; Catalano and D'Argenio, 1982a; Giunta and Liguori, 1972). Figure 3 shows the names and relations between the major thrust units of northern and western Sicily. In spite of the different terminologies used, there is general agreement (Broquet et al, 1984; Broquet, Caire, and Mascle, 1976; Catalano and D'Argenio, 1982a; Ogniben, 1975) that these presently east to west structural units correspond to original facies zones, or domains, that became sheared and detached more or less along their boundaries, and overlapped from north to south (Figure 3). The present differences between those in the northwest compared to those south of Palermo, are because of the disappearance of the Imerese and Sicani domains in the west and that of the Trapani domain to the east. Geologists disagree, however, about the amount of horizontal displacement, whether in the range of 10–30 km (6–19 mi), decreasing from east to west (Caflisch, 1966; Schmidt di Friedberg, Barbieri, and Giannini, 1960), or of many tens of kilometers, with a 60% shortening (Catalano and D'Argenio, 1982a).

South of the Nebrodi and Madonie Mountains, the Central basin is an area with a marked gravity low (Figure 4). The Bouguer gravity values range from +120 milligals on the Hyblean Plateau to −100 milligals in the basin center, suggesting a sediment thickness of several thousand meters. The northeast-elongated basin is about 160 km (100 mi) long and a maximum of 100 km (62 mi) wide; it covers 11,000 sq km (4,200 sq mi), 2,000 sq km (772 sq mi) of which are offshore. The sediments are Miocene–Pliocene clays, marls, and sands, with substantial intercalations of olistostromes and olistoliths (Flores 1955),

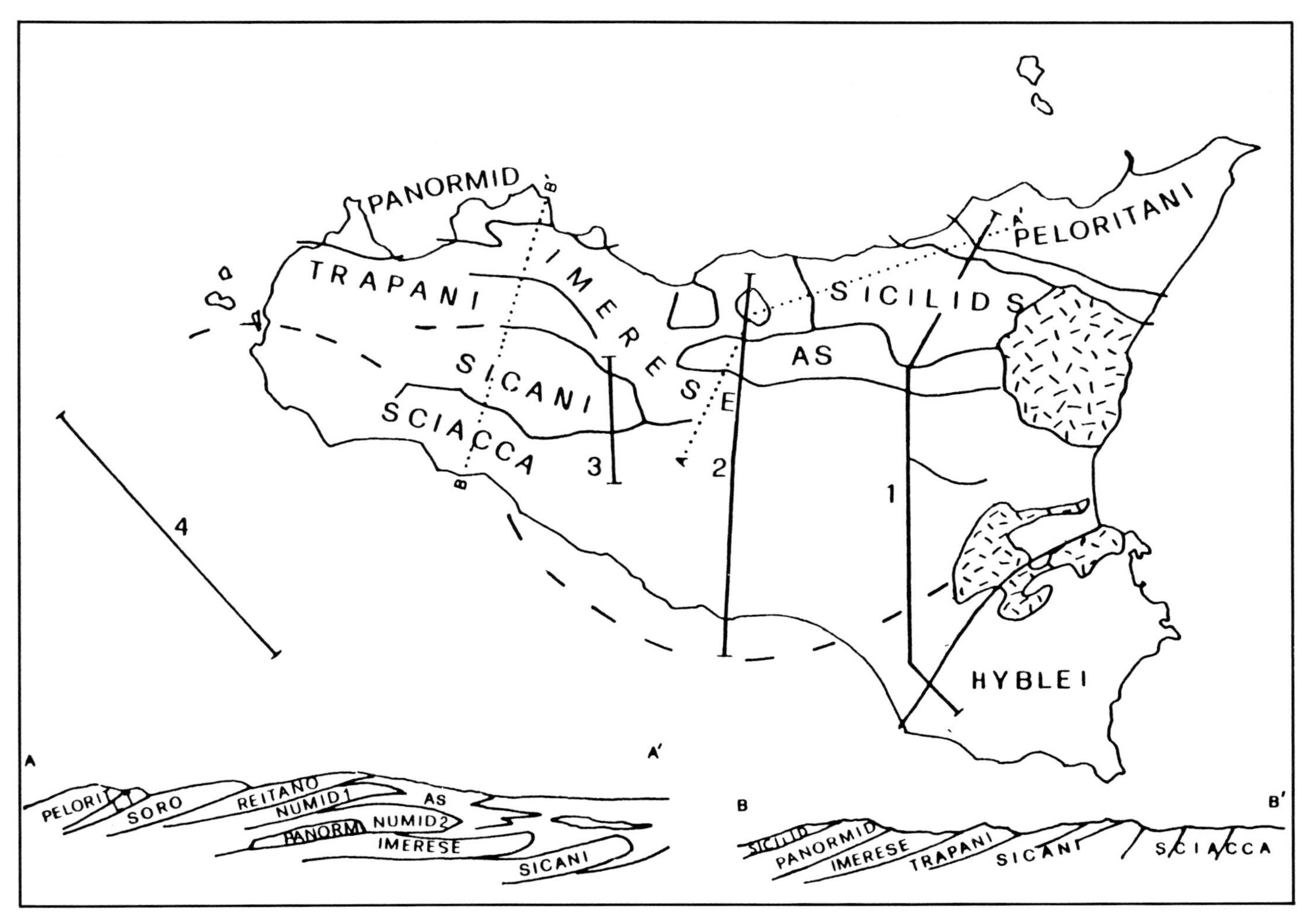


Figure 3. *Present occurrence of the sedimentary domains and structural zones of Sicily. 1, 2, 3, 4 are the cross sections of Figures 7, 8, and 16.*

some of which are wide and thick (Flores, 1959; Marchetti, 1956; Rigo, 1956). These Neogene sediments have been strongly deformed by slumping, shearing, overthrusting, and lateral movements. A series of post-lower Pliocene asymmetric folds (Figure 5) is considered to be genetically related to right-lateral, Pliocene–Pleistocene movements along the Kumeta-Alcantara and Simeto shear systems (Ghisetti and Vezzani, 1982).

The deeper structure of the Central basin is not known directly. A geoelectric-resistivity survey done in the early 1950s by GCG, on behalf of the Italian Geological Survey (Beneo, 1955; Regione Siciliana, 1961), showed a surface that is interpreted as the top of the resistive-carbonates substratum. Whereas the deep contours of this surface indicated a northeast to southwest trend, the weak resistivity contours (3 to 16Ω)of the more conductive Neogene clastics in the north, showed an east to west trend (Figure 6), agreeing with that of the Madonie and Palermo overthrusts.

We know little of the resistive substratum's nature because only basin margin wells have reached it, indicating sequences similar to those of nearby exposures in the Sicani, Madonie Mountains, Mount Judica, and the marginal belt of the Hyblean Plateau. In the two cross-sections of Figure 7, the substratum has been sketched in by extrapolation from surface structures and well data, especially from around the northern and western margins. We may assume that thrust slicing is the rule, at least in the northern half of the basin, near the chain, whereas horst blocks may be more common near the southern and southwestern edges.

The Sicani Mountains separate the Central from the Northwest Sicily basins. Though presently elevated as Pleistocene horsts (as are all the other major mountains of Sicily), they are made of a number of Triassic–Oligocene, southward-directed thrust sheets that overlap upon one another and also onto the Serravallian-Tortonian sediments of the Central basin (Broquet, 1970; Broquet, Caire, and Mascle, 1976; Mascle, 1970). Toward the north (Figure 8) there are several kilometric slabs, or klippen, that are embedded within or are entirely floating on the chaotic Miocene shales. These structures may result from shearing at the level of Permian–Triassic shales, which have become mixed with the Miocene shales (Caflisch and Schmidt di Friedberg, 1967a) and the fragmentation and sliding of the overlying rigid Mesozoic rocks into the strongly subsident Miocene basin (Broquet, 1970; Broquet, Caire, and Mascle, 1976).

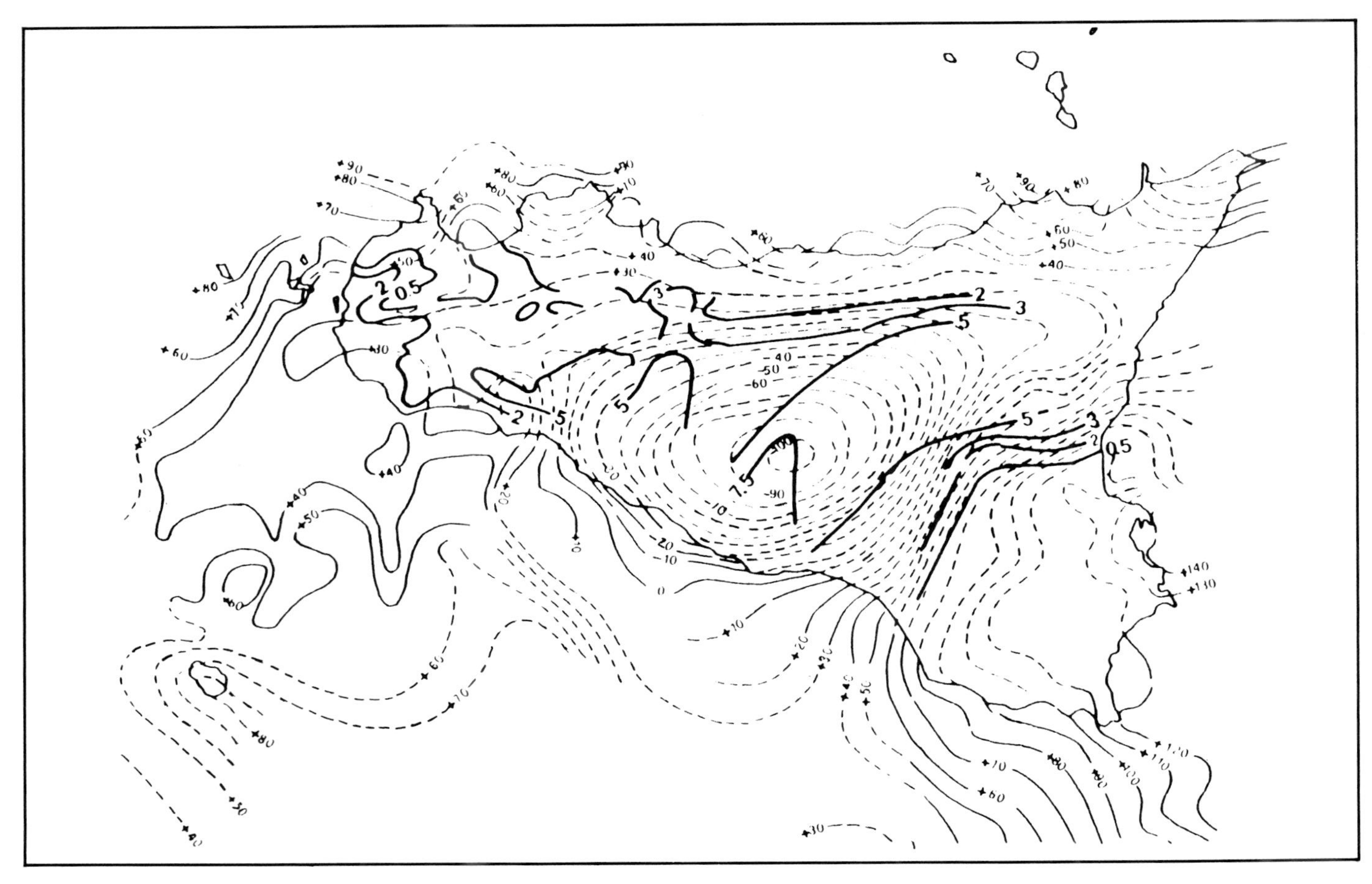


Figure 4. *Bouguer gravity anomalies of the Sicily region (based on Beneo, 1955).*

Toward the west, the Sicani are thrust directly onto the more internal parts of the Sciacca Zone.

The Miocene–Pliocene basins of northwestern Sicily stretch from west of the Palermo Mountains to the Adventure Bank in the Sicilian Channel. For convenience we have generalized the name of the Trapani basin to the whole thrust-front depression, but in fact, this area is divided into a number of more or less separate depocenters. Onshore, the Alcamo, Trapani, and Castelvetrano subbasins contain, respectively, about 1,500, 2,500 and 2,000 m (4,900; 8,200; 6,550 ft) of Neogene sediments. Offshore, there appear to be at least two subbasins, one south of the Egadi Islands, the other east of the Nilde field (this one with probably up to 3,500 m or 11,500 ft of Miocene–Pliocene deposits).

The Neogene sequence is mainly Tortonian clays and sands, with frequent olistostromes in the north (Trapani and Alcamo basins; Rigo, 1956). In the south and offshore, the sequence extends to Pliocene and Pleistocene and is devoid of gravity slides. The Castelvetrano subbasin differs from the others in being a graben in the foreland Sciacca Zone and Sciacca Ridge.

The Neogene sediments are folded with east to west trends. The substratum carbonates, on the other hand, are dissected into mainly northeast to southwest horst-graben blocks (Catalano and D'Argenio, 1982a; E.N.I., 1969).

The Hyblean Plateau is a structurally high area, sloping to the west and south toward the Malta horst, with exposures largely

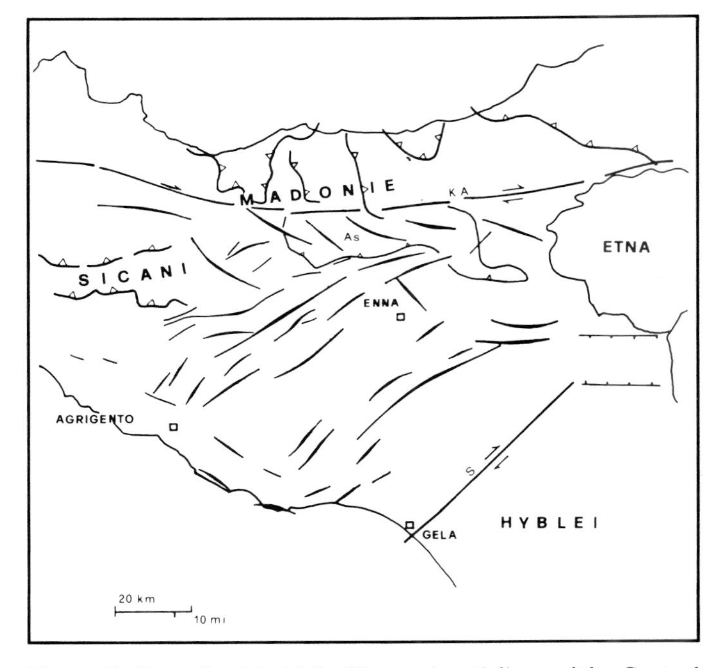


Figure 5. *Superficial (middle Pliocene) anticlines of the Central basin. KA = Kumeta-Alcantara Line; S = Simeto Line; As = Argille Scagliose nappe.*

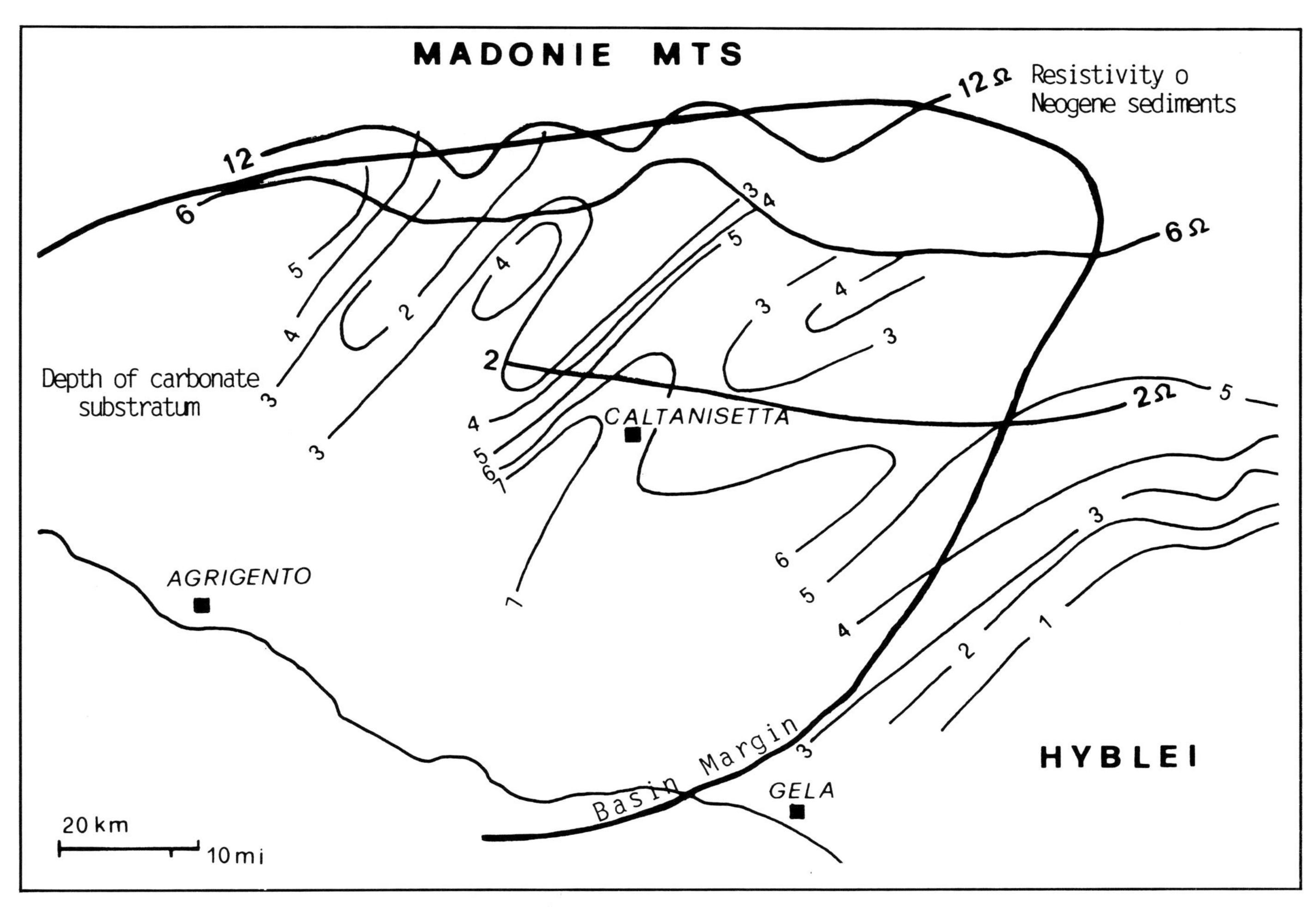


Figure 6. *Substratum configuration in the Central basin, according to depth of resistive substratum (contours in km), and the resistivity values (ohms) of the conductive overburden (re-drawn after Beneo, 1955).*

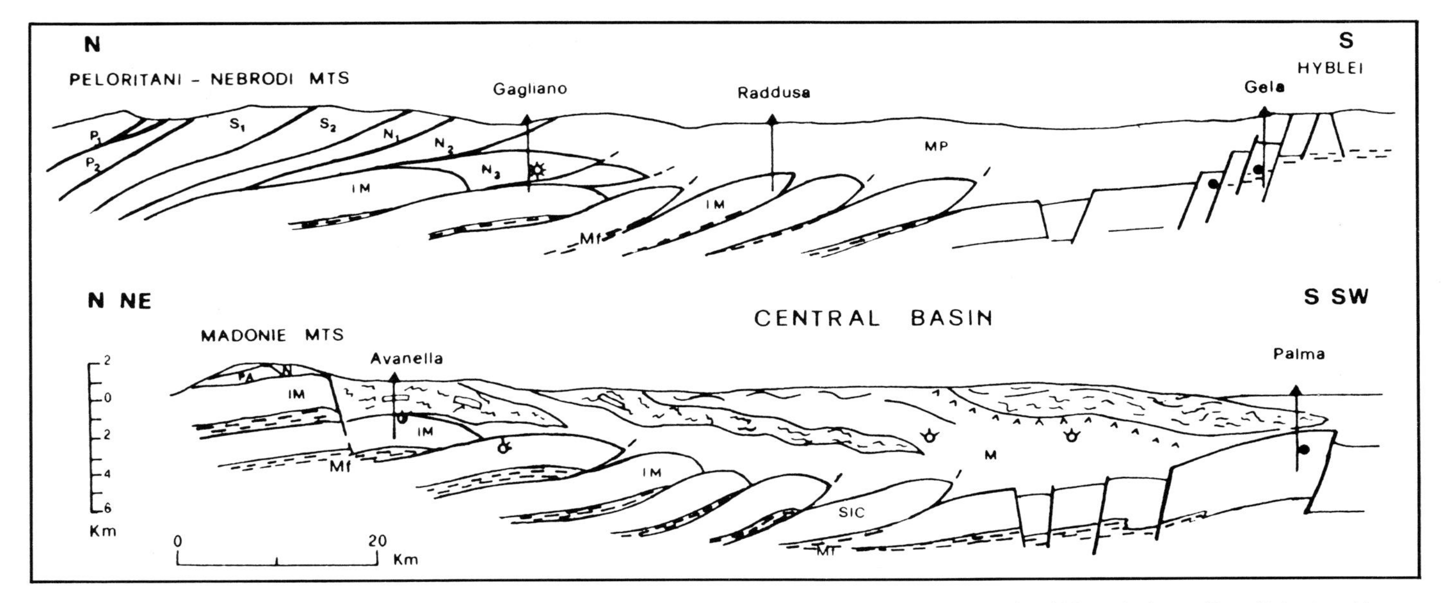


Figure 7. *Idealized structural cross sections through the Central basin (for location see Figure 3). Abbreviations: $P_1$ = Peloritani basement; $P_2$ = Limestone Chain (Longi-Taormina Zone); $S_1$, $S_2$ = Sicilid nappes; $N_1$, $N_2$, $N_3$ = different slices of Numidian Flysch; IM = Imerese Sequence; Mf = Mufara shales; MP = Miocene–Pliocene sediments, SIC = Sicani Sequence.*

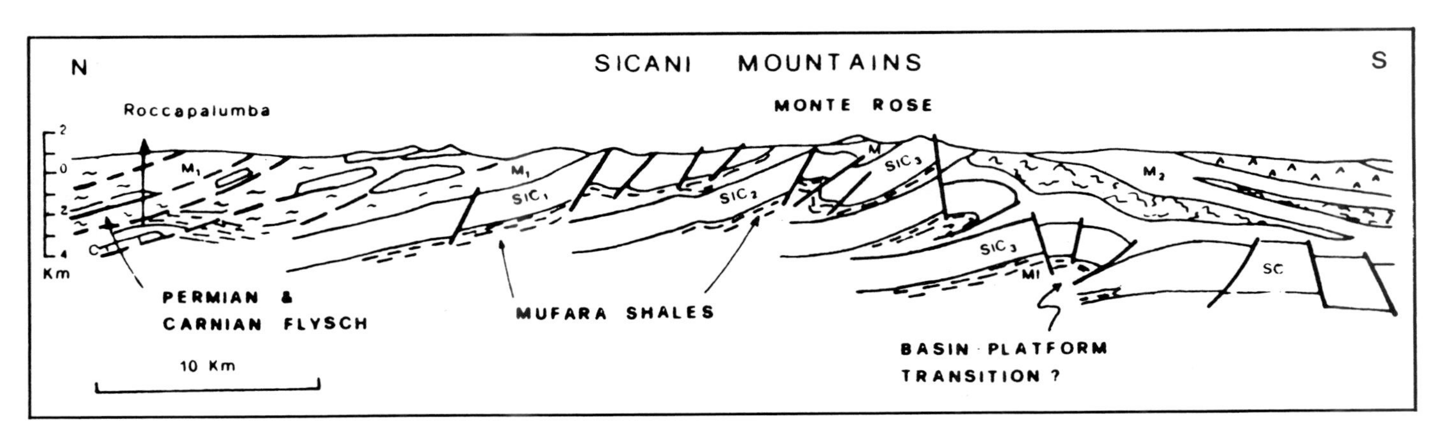


Figure 8. *Idealized cross section of the Sicani Mountains. (3 of Figure 3). Interpretation of Roccapalumba well, after Caflish and Schmidt di Friedberg, 1976a. C=Carboniferous limestones; $M_1$=Serravallian-Tortonian; $M_2$=late Tortonian; $SIC_1$=structural units of Sicani Zone; SC=Sciacca Zone.*

| | | N TUNISIA | | NW SICILY | N SICILY | SE SICILY |
|---|---|---|---|---|---|---|
| PLIO-PLEIST. | | | | RIBERA | | RIBERA |
| MIOCENE | MESSINIAN | OUED BEL KHEDIM | | GESSOSO – SOLFIFERA | | G.S. |
| | TORTONIAN | MEDJERDA | D | TERRAVECCHIA | | TELLARO |
| | SERRAVALLIAN | BEJA GRP; BEGLIA; OUM DOUIL | | | | |
| | LANGHIAN | | | | | |
| | BURDIGALIAN | AIN GRAB | C | CORLEONE BONIFATO | COLLESANO | RAGUSA |
| | AQUITANIAN | FORTUNA | | | | |
| OLIGOCENE | | KORBUS | | | | |
| EOCENE | | SOUAR EL GARIA – BOUDABBUS | B | AMERILLO | CALTAVUTURO | |
| PALEOCENE | | EL HARIA | | | | |
| CRETACEOUS LATE | SENONIAN | ABIOD | | | | AMERILLO |
| | | ALEG | | | | |
| | TURONIAN CENOMANIAN | FAHDENE | | | | |
| CRETACEOUS EARLY | ALBIAN APTIAN BARREMIAN | | | HYBLA | CRISANTI | HYBLA |
| | NEOCOMIAN | SIDI KRALIF | | BUSAMBRA | | CHIARAMONTE |
| JURASSIC | LATE | NARA | | GIARDINI | | BUCCHERI |
| | MIDDLE | | | | | |
| | EARLY | | | INICI | FANUSI SCILLATO | MODICA; STREPPENOSA; SIRACUSA |
| TRIAS | RHAETIC | | A | TAORMINA | MUFARA | NOTO |
| | NORIAN | RHEOUIS | | | | NAFTIA |
| | CARNIAN | | | | | (GELA) TAORMINA |

Figure 9. *Correlation of lithostratigraphic names used in Sicily and northern Tunisia.*

of Tertiary limestones and Miocene–Pliocene volcanics. The primarily carbonatic subsurface sequence ranges from Triassic to late Miocene age. It indicates the persistence of basin facies, especially in the Ragusa–Malta Channel area, with a basin-platform margin complex (eastern Siracusa Platform versus western Ragusa basin) in Triassic–Liassic and from Late Cretaceous to late Miocene time (Grasso and Lentini, 1982; Patacca et al, 1979; Pedley, 1981).

The present structure is characterized by intense east-northeast, northeast, and west-northwest faulting, dissecting gentle Upper Cretaceous and Pliocene folds (Ghisetti and Vezzani, 1975; Rigo and Cortesini, 1959). The faulting is mainly post-early Pliocene in age, but it probably follows a pattern established as early as Late Triassic, and re-mobilized in Middle Jurassic and middle-Late Cretaceous time (Finetti, 1982; Patacca et al, 1979; Rigo and Cortesini, 1959).

West of the Hyblean Plateau, the foreland margin slopes under the sediments and olistostromes of the Central basin (Colantoni, 1975; Finetti, 1982; Winnock, 1981). The Sciacca

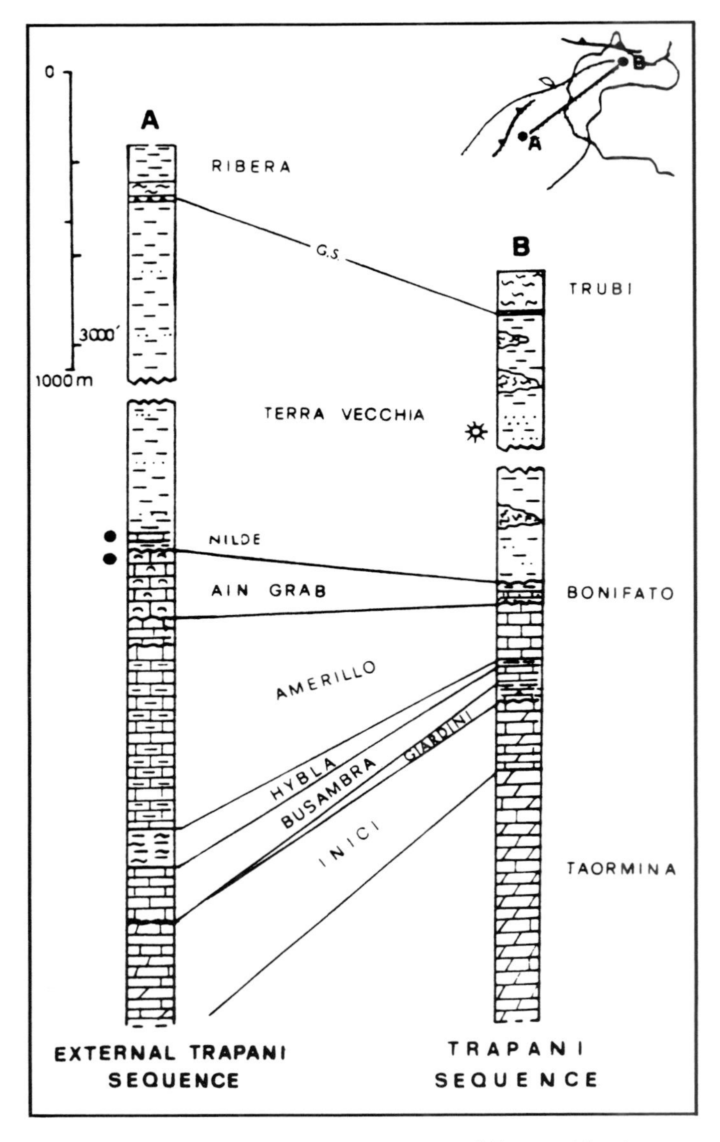


Figure 10. *Lithostratigraphic summary of Trapani basin margin.*

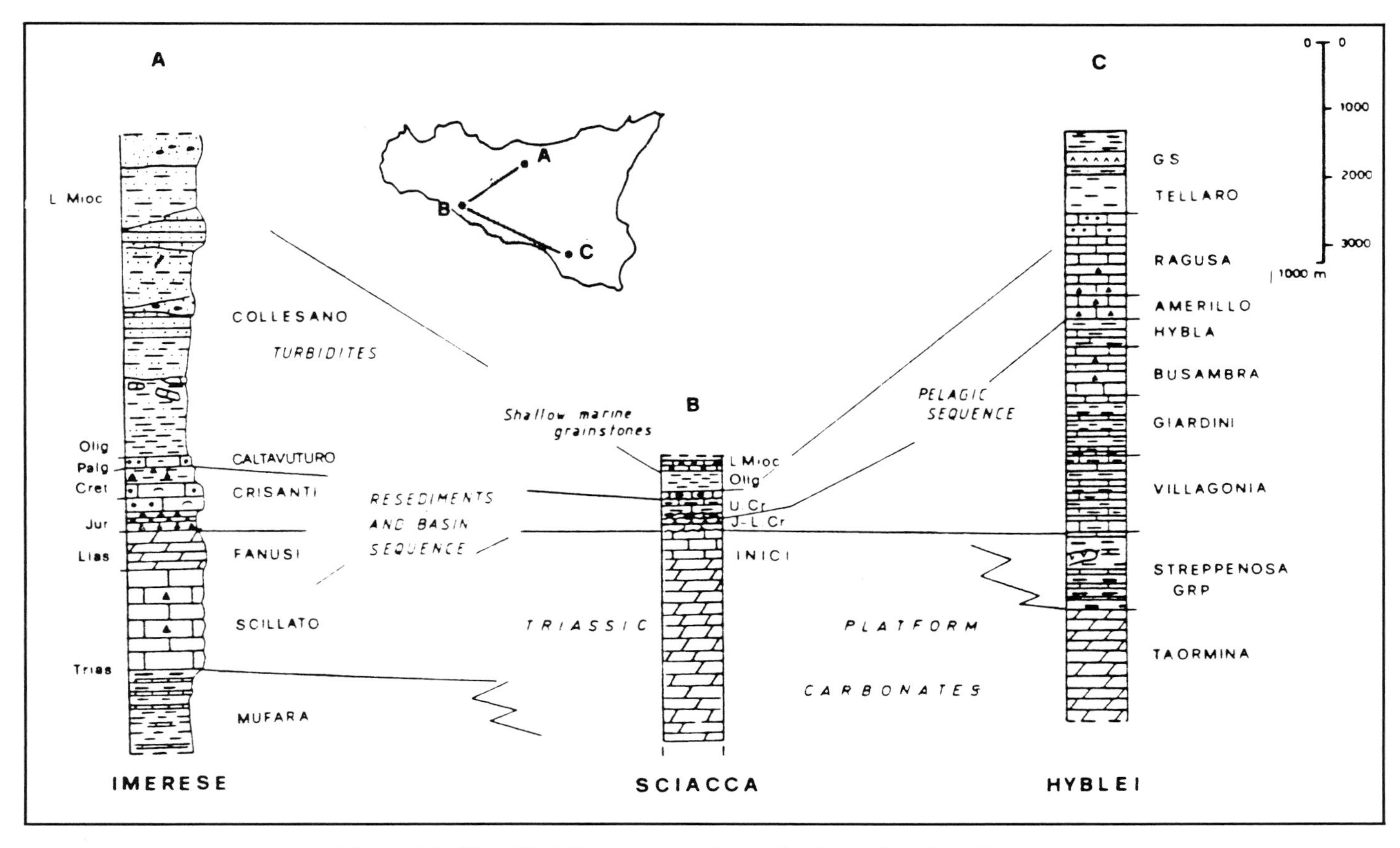


Figure 11. *Simplified lithostratigraphy of the Central basin substratum.*

Ridge is a complex horst-graben structure, with north-northeast and northeast trends, limited to the east by normal faults bearing major displacements. Uplifted in early Tertiary, it remained positive after late Miocene. Pliocene sediments (up to 500 m or 1,600 ft only) occur exclusively in the younger, narrow troughs (Winnock, 1981).

The northern margin of the foreland substratum is affected by reverse faulting at the edge of the olistostrome front of the Central basin (Figures 2 and 7), and, more intensely, in the northern part of the onshore Sciacca Zone.

## STRATIGRAPHY

From the point of view of exploration, the stratigraphic sequence of Sicily is fairly simple,[1] in spite of the complex tectonics. Of special interest are the pre-middle Miocene substratum sequence of the thrust belt margin and under the foretrough basins, as well as the Neogene stratigraphy of the latter. We can recognize four major stratigraphic subdivisions in practically all parts of Sicily (Figures 9, 10 and 11).

### Unit A: Triassic–Liassic Carbonates With Local Basin Facies

Dolomites and dolomitic limestones, sometimes thicker than 3,000 m (9,800 ft), represent lagoonal-intertidal, algal-biostromal to open-marine, intraclastic-oolitic facies (Triassic–Liassic Taormina and Liassic Inici formations), or platform margin, locally reefoidal facies (Siracusa Formation in southeastern Sicily and Malta Channel). Some evaporites are present in the northwest (Marettimo Island and a few offshore wells), probably in transition to the quantitatively more important Triassic–Liassic evaporites of Cap Bon (Tunisia). In northern Sicily (Imerese domain), the basal unit is made of cherty Halobia calcilutites (Upper Triassic Scillato Formation, 200–250 m or 650–820 ft) overlain by turbiditic dolorudites/arenites (Rhaetic Liassic Fanusi Formation, 200 m or 650 ft).

The Streppenosa Group basin facies in southeastern Sicily (600–3,000 m or 1,970–9,850 ft) includes Rhaetic black shales with laminated dolomitic limestones (tidal plain and euxinic

[1]The lithostratigraphic nomenclature of Sicily is, however, still in some state of confusion. There is not yet a reasoned system based on a critical, general review of facies variations and a consequent choice of names related to the most suitable type sections. The earlier and most applied nomenclature (Rigo and Barbieri, 1959; Schmidt di Friedberg, 1962, 1965), though established on Stratigraphic Code principles, was based on formations exposed, or in the subsurface, in different parts and in different tectonic units of the island (e.g., the Triassic–Liassic formations near Taormina, in the Longi–Taormina tectonic zone of the allochthonous Peloritani massif). New subsurface, mainly biostratigraphic, names are proposed for southeastern Sicily by Patacca et al (1979). Although these may better indicate the basin facies of the Hyblean foreland, they largely overlap with the earlier units and are not established in the proper form. Confusion also arises from the indiscriminate application of both nomenclatures for offshore wells, as well as of Tunisian sequence names (in the Sicilian Channel). Confusion further arises from the use of lithostratigraphic terms to name chronostratigraphic intervals, regardless of the facies of the latter. A systematic review and revision of Sicilian stratigraphy is long overdue and increasingly necessary.

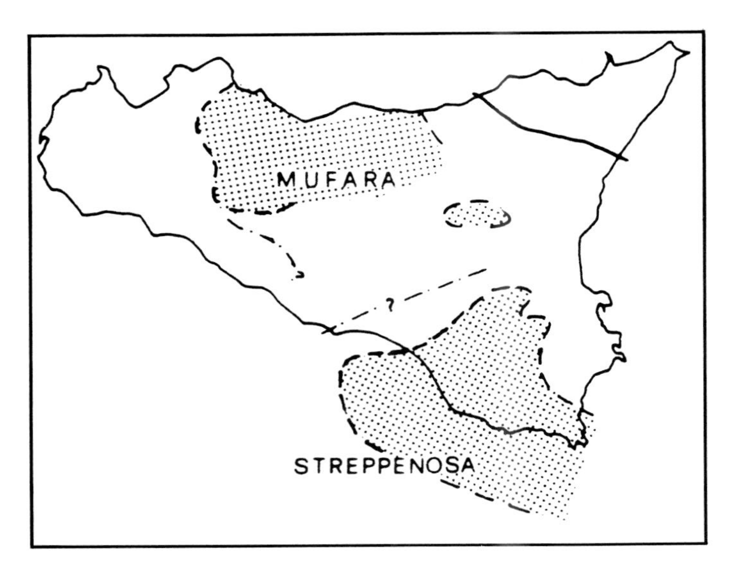


**Figure 12.** *Present distribution of Triassic–Liassic basin facies.*

lagoons, of Naftia and Noto formations; Patacca et al, 1979) and lower Liassic black marls/shales with graded calcsiltites (a deeper basin facies, Streppenosa Formation).

The Mufara basin facies (Carnian–Rhaetic, over 800 m or 2,600 ft) crops out in central-northern Sicily (Figure 12) at the base of the Sicani and Imerese sequences. The exposed upper part of the formation represents a fairly deep basin (turbiditic calcarenites, *Halobia* calcilutites, and brown, sometimes black, shales, with a characteristic occurrence of dwarf, pyritized microfauna). We do not know whether the Mufara basin started as a lagoonal depression, as did the Streppenosa, nor how far it extended, as a single or multiple depocenter.

Catalano and D'Argenio (1982b) have suggested that the Streppenosa, the Mufara, and other, similar Triassic-Liassic facies in southern Italy resulted from deposition in pull-apart basins because of transcurrent faulting in Early Jurassic. The Mufara basin could also represent the southwestern end of a marine embayment (Lagonegro basin of southern Italy; Channel, D'Argenio and Horvath, 1979; Scandone, Giunta, and Liquori, 1977), perhaps situated over an earlier, Permian, marine depression.

## Unit B: Upper Triassic To Eocene, Mostly Carbonates

The early development of the Alpine Tethys, in response to the opening of the central-southern Atlantic (Dewey et al, 1973), caused rapid crustal subsidence and drowning of the Triassic–Liassic carbonate platform. The deposition rate was drastically reduced, with deep marine characters over an irregular topography of seamounts, deep plateaus, ridges, and open basins. The Giardini Formation (cherty and crinoidal limestones, red nodular ammonitic limestones, re-sedimented bioclastic limestones, pelagic-pelecypod calcilutites, siliceous shales, and volcanics), rests disconformably on the Inici–Taormina carbonates. It is often very thin (few meters to 100 m or 330 ft) with a faunal condensation or absence, especially in northwestern Sicily (Giunta and Liguori, 1972; Jenkins, 1970; Wendt, 1969). In southeastern Sicily, the upper Liassic–Malm interval is thicker (800–1,500 m or 2,600–4,900 ft), with a fairly different, more pelitic facies (Modica and Buccheri formations; Patacca et al, 1979).

The Early Cretaceous was a tectonically quiet period in Sicily, with pelagic deposition throughout. The Busambra Formation (Tythonian–Hauterivian, 20–250 m or 66–820 ft) is pelagic white calcilutites with *Calpionella*. In southeastern Sicily, red packstones and nodular limestones (up to 300 m or 980 ft) are named the Chiaramonte Formation by Patacca et al, 1979). The Hybla Formation (Barremian–Albian) is a marly-pelitic interval, characterized in southeastern Sicily by grey-green, calcareous shales with marly limestone and marl (Rigo and Barbieri, 1959).

The Amerillo Formation (Cenomanian–Eocene) consists of compact, sometimes cherty calcilutites and marls (scaglia) with *Globotruncana* or *Globorotalia*. Thickness is variable, greater offshore (500–800 m or 1,640–2,600 ft) than onshore (200–400 m or 650–1,300 ft) in eastern Sicily. In the Sicilian Channel, the Amerillo facies is transitional to the thicker, more argillaceous and differentiated sequence of northeastern Tunisia (Figure 9).

In north-central Sicily (east Sicani Mountains, Madonie, and Mount Judica), the Crisanti Formation (Jurassic–Upper Cretaceous, about 250 m or 800 ft) comprises radiolarites, siliceous calcilutites and shales, volcanics, and Cretaceous fragmental limestones. The latter are resedimented, bioclastic material derived from the margin of the Panormid Platform. The Caltavuturo Formation (Paleocene–Eocene, 20–200 m or 65–655 ft) is made of graded nummulitic calcarenites/ rudites, with red marlstones and grey shales.

The latest Cretaceous to Eocene was a period of instability in the African foreland, related to the Eocene-Alpine deformation further north. In the Pelagian region, the instability caused reduced deposition and frequent gaps (Maestrichtian–Eocene disconformity; Marie et al, 1984; Salaj, 1977). In Sicily, an extensional event is seen by fault activity with volcanics and Maestrichtian breccias, especially thick and extensive in the Sicani external belt (Catalano and D'Argenio, 1982a; Mascle, 1970).

## Unit C: Upper Transitional Clastics

Unit C includes sedimentary sequences formed during the Eocene-Alpine tectonic stage, before the Miocene deformation of the chain. There is a sharp difference between the Oligocene-Miocene rocks of the Imerese domain, derived from a more northerly flysch basin, and the coeval sequences of the foreland, as shown by the gaps in the northern Sicily section (Figure 9).

The upper Oligocene–Aquitanian Collesano Formation. (Numidian Flysch) divides into a lower pelitic member (Alia, 300–500 m or 980–1,640 ft) and an upper member (Tavernola, 1,500 m or 4,920 ft) of fine to coarse sands, sometimes with conglomerates. In the southeast offshore of the Trapani basin Oligocene, shallow marine quartz sands, perhaps partly equivalent to the Fortuna Formation of Tunisia, are overlain by shallow marine, transgressive bioclastic limestones (the Burdigalian Ain Grab Formation, 500–650 m or 1,640–2,130 ft). Oligocene sediments are absent over large parts of the Sciacca Ridge. In northwestern Sicily, Unit C includes Oligocene–Langhian greenish marls, bioclastic limestones, and glauconitic sandstones (Bonifato and Corleone Formations, 50–200 m or 165–655 ft). In the onshore Sciacca Zone and eastern Sicani Mountains, there are thin Oligocene shales with nummulitic calcarenites.

### Unit D: Neogene Clastics of the Foretrough Basins

The Neogene sequence of the Central and Northwest Sicily basins is contemporaneous with the main phase of overthrusting in the northern chain in Langhian–Serravallian. In the Tortonian, deformation migrated south, and with the Trapanese and Imerese nappes overriding the Sicani domain. The foreland of southern Sicily was not involved, and experienced quiet, slow pelitic-calcareous deposition (Tellaro Formation, 30–100 m or 98–330 ft).

The Neogene sequence of the Central basin displays five sedimentary cycles, each one shallowing upward (Figure 13) and ending with tectonic movements or uplifts or both, at least in parts of the basin. In the Terravecchia Formation (Flores 1959), the lower, Langhian–Serravallian part is marly-shaly, becoming sandy-turbiditic upward (lower Tortonian Barbara Member). The environment was restricted, deep marine, becoming shallower and sandy-conglomeratic toward the northern basin margin and stratigraphically higher biohermal limestones.

The Terravecchia Formation's thickness varies because of the olistostrome intercalations (the Lavanche, formerly Valledolmo, olistostrome is up to 2,000 m or 6,560 ft in wells). The olistostromes are characterized by a chaotic mixture of assorted lithologic types in a shaly-sandy matrix, with mixed Mesozoic and Cenozoic faunas. In wells at the basin's northern margin, the Lavanche contains exotic masses, under and between chaotic shales, that can be easily related to parts of the Imerese and Sicilid sequences. These include blocks of Triassic carbonates and black marls, Oligocene sandstone and shales, and Eocene calcareous flysch. Toward the south-southeast the size of the olistoliths rapidly decreases, and the slump materials are more comminuted.

Mapping and stratigraphic studies (Rigo, 1956; Broquet et al, 1984) reveal a southward progradation of depositional sequences and olistostromes resulting from the continental shelf and slope advancement from the north margin of the rapidly sinking basin (Figure 13). We know of major episodes of slumping in this area in the Miocene, affecting hundreds of square kilometers, with an aggregate thickness of 2,000–3,000 m (6,560–9,840 ft).[2]

The upper Miocene sequence ends with the Evaporitic Complex or Gessoso–Solfifera Formation. Normally considered as one lithostratigraphic unit, an unconformity caused by uplift in the northern part of the basin divides it into two cycles (Decima and Wezel, 1971). At the time of deposition of the formation's lower member (Tripoli diatomite and bituminous marls), a northeast to southwest trough 400–600 m (1,310–1,970 ft) deep with a restricted circulation existed in the southwestern part of the Central basin (Broquet, Mascle, and Monnier, 1984). Although the ensuing evaporitic environment was one of shallow, restricted lagoons (Schreiber and Friedman, 1976), subsidence actively continued along the same axis (Figure 14) accumulating as much as 2,000 m (6,560 ft) of gypsum and potassium salts.

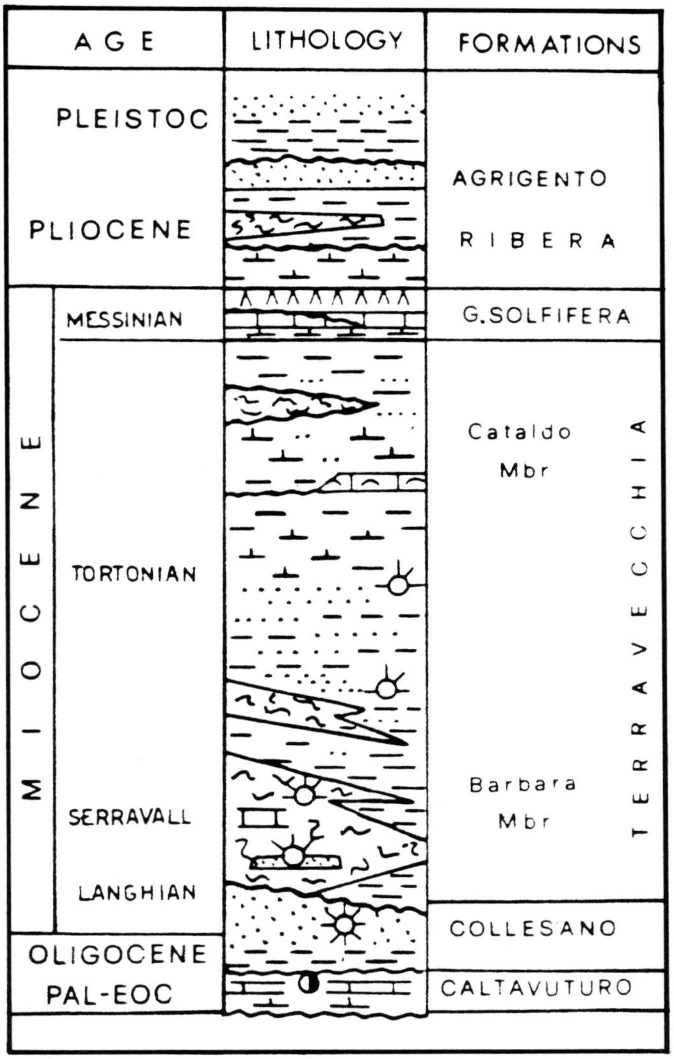

Figure 13. *Summary of the Cenozoic stratigraphy of the Central basin.*

The rapid Pliocene transgression took place over an uneven surface, characterized by highs that were quickly dismantled. The predominantly shaly Ribera Formation, in addition to remnants of the evaporitic sequence (within the lower Trubi member) also contains intervals of brecciated polygenic claystones, intercalated as submarine slumps (Narbone olistostrome, Figure 14). The slumping phenomena continued well into the Pleistocene (e.g., Gela olistostrome).

## THE PETROLEUM HABITATS OF SICILY

### Southeast Sicily Foreland

The oils of Ragusa, Gela, Perla, and Vega are heavy (6–20° API) naphthenic and naphthenic-mixed, with a high sulphur and $CO_2$ content in the associated gases (Rocco, 1959; Caflisch

[2]When first described (Flores, 1955, 1959) geologists knew of no submarine gravity slides of comparable size. Later, oceanographic research showed the existence of very large Neogene slumps and slides at the bottom of continental shelves, in front of active margins (Moore, Curray, and Emmel, 1976), and also of passive ones (Emery et al, 1975; Dingle, 1977). The Agulhas slump, along the southwest African coast, for example, has a cross-sectional area of 133.5 sq km (52 sq mi), an average thickness of 250 m (820 ft), a length of at least 700 km (435 mi), a volume of 20.33 cubic km (5 cu mi), and a total area of 7,949 sq km (3,100 sq mi) (Dingle, 1980). For comparison, the total area of Sicily is 25,708 square km (10 sq mi).

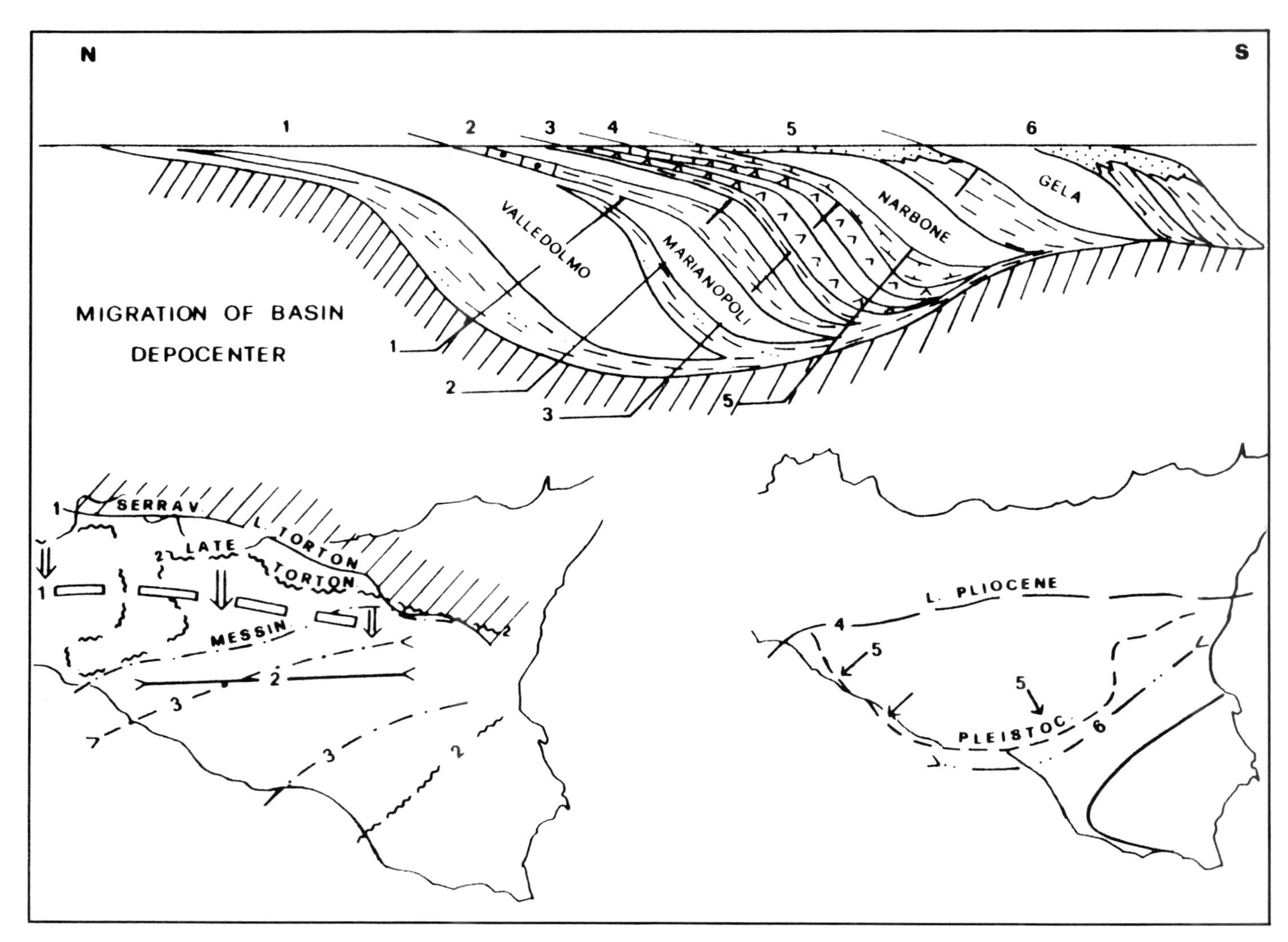


Figure 14. *Migration of depositional axes and gravity slumps in the Central basin (top, modified from Broquet et al., 1984). Stratigraphy, shorelines and depocenters: 1 = Serravallian-early Tortonian; 2 = late Tortonian; 3 = Messinian; 4 = early Pliocene; 5 = late Pliocene; 6 = Pleistocene. The arrows indicate the main direction of slumping.*

and Schmidt di Friedberg, 1967b). Exceptions are the light oils (24–39° API) of Mila, San Bartolo, and Irminio. The source is the Streppenosa shales, especially the Noto Formation (Total Organic Carbon, or TOC, 4%; Pieri and Mattavelli, in press).

In general, trapping in southeastern Sicily is in faulted anticlines or in horst traps. In the Vega and Perla fields, trapping is in the regressive reefoid carbonates of the Siracusa Formation, at the margin of the Streppenosa basin (Pacchiarotti, 1982; Schramm and Livraga, this volume). At Ragusa and Gela, the reservoir is in the vuggy-fractured Taormina dolomite under the Streppenosa shales, in a paleostructure, an east-northeast to west-southwest Middle Jurassic horst that became a drape anticline in Early Cretaceous, and was mildly re-folded later (Kafka and Kirkbride, 1960; Patacca et al, 1979; Rigo and Cortesini, 1959). Because of the faulting, the Streppenosa shales are both the source and the seal.

In general, shows in wells in the southeastern Sicily foreland are located at the top of the Siracusa–Inici carbonates, under the Buccheri (ex-Villagonia) shales. The main controlling factor is proximity to the Streppenosa source: non-fracture porosity development relates to secondary dolomitization, rather than to basin margin or high energy facies (Mattavelli, Chilingarian, and Storer, 1969).

No oil has been found in Triassic–Liassic reservoirs of the Sicilian offshore (including the Malta Horst) where the black shales are absent. Oil has not been found in the rest of Sicily where the Taormina–Inici carbonates have been repeatedly tested, but with negative results (only minor bitumen shows). Thin, stratigraphic intervals of black calcilutites, marls and shales, occur at various places in the Taormina–Inici carbonates of northwestern Sicily (e.g., in the Egadi Islands and in a few onshore and offshore wells). They probably represent local lagoons, restricted in time and place, rather than persistent basins, and are quantitatively of minor importance as potential source rocks.

### Central Basin and Northeastern Sicily Thrust Belt

The Gagliano and nearby fields (Casalini, Bronte-San Nicola, and Miraglia) produce methane and condensate (55.7° API) from the Oligocene–Miocene Collesano sandstones. The Gag-

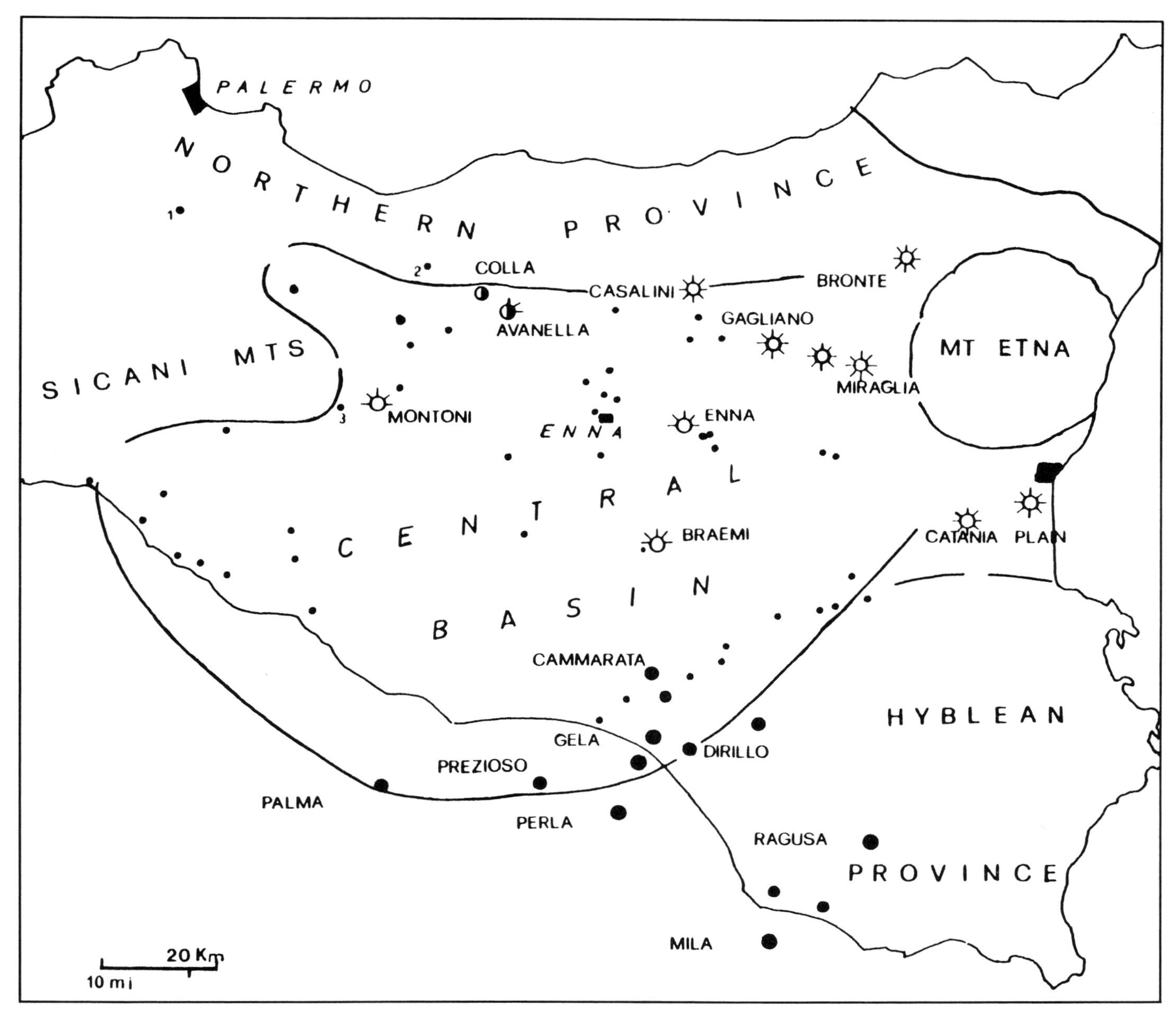


Figure 15. *Main oil and gas occurrences of the Central basin exploration province.*

liano field is located in a gentle northeast to southwest fold with a flat-topped gravity anomaly, bounded by two northeast to southwest normal faults (E.N.I. 1969; Vercellino and Rigo, 1970). The reservoir consists of several lenticular sandstone pays (porosities 6–15%) at depths of 1,500–2,740 m (4,920–8,990 ft), in a lower (autochthon?), Imerese-Numidian flysch unit, overthrust by a more internal Numidian flysch slice, which is in turn truncated by a 1,500 m (4,920 ft) thick mass of slumped argille scagliose.

The Gagliano reserves had been estimated at 700 billion cubic feet (bcf), with 20 million bbl condensate; average past production was in the range of 4–12 million cubic feet (mmcf) per well, with a daily production of about 115 mmcf and 2,600 bbl of condensate.

The surface and subsurface oil shows of the northern Central basin (Figure 15) are light (39–47° API), of paraffinic type, and low in sulphur (Caflisch and Schmidt di Friedberg, 1967b). The main oil shows at the northern margin of the Central basin were found in the Colla and Avanella wells, in fractured Caltavuturo marlstones and calcarenites and in the siliceous carbonates of the Fanusi and Crisanti formations (Flores, 1959; Schmidt di Friedberg, 1959).

The occurrence of light oil and gas in these and the Gagliano upper Miocene and lower Pliocene structures suggests late maturation and migration. At the southern margin of the Central basin, Palma-1 has light oil (37–40° API) in Inici carbonates—in a position comparable to that of Avanella.

The source of the Central basin light oils, methane gas, and condensate is unknown, although their density and chemistry suggest a deeply buried source. Apart from shales in the subsurface Caltavuturo and lower Collesano formations, whose geochemistry and potential we do not know, the only available potential source rock candidates are the shales of the Mufara Formation. Their exposed upper parts are oxidized, with 0.2% TOC, woody-herbaceous kerogene (Pieri and Mattavelli, in press).

Black shales have been reported from outcrops and from a few wells: Mount Judica, Piana Albanesi, Marineo-1, Platani-2, Cerda-1 (Caflisch 1966, Catalano and D'Argenio 1982b, ENI 1969, Regione Siciliana 1961); but to our knowledge, they have not been analyzed. The Mufara basin should have developed in a manner similar to that of the Streppenosa basin, with a restricted environment throughout.

The numerous gas shows in the basin are likely to be of Tertiary origin. In the subsurface, gas occurs both in the Terravecchia sandy intercalations (e.g., Montoni-2, estimated at 1–4 MMCFGD: and Enna-2 at 1 MMCFGD with condensate) and in the olistostromes, especially in the sandy and carbonatic olistoliths. We believe that, in addition to sources within the lower part of the Terravecchia, gravity slumping may have contributed organic matter conducive to gas generation and accumulation.

Small methane shows in Tortonian and Pliocene sands are probably organogenic, or are related to the Tripoli bituminous shales. These contain up to 2% TOC, with immature algal kerogen. Broquet, Mascle, and Monnier (1984) estimate an oil productivity of 62–114 kg/ton (0.46–0.84 bbl/ton). Unfortunately, in Sicily the Tripoli shales are exposed or are very moderately buried, but could be of interest as potential source rocks in other parts of the Central Mediterranean region below the *M* horizon evaporites, which could be considered a good seal.

### Offshore Trapani and Hammamet Basins

Light oils (36–41° API) and $CO_2$-rich gas are trapped in the porous Ain Grab calcarenites, in asymmetrical upper Miocene faulted folds (e.g. Nilde oil field) covered by the Terravecchia shales. In the nearby Hammamet basin, light oils are also located in the Ain Grab, and as in middle Miocene sands, sealed by the Oum Douil shales (e.g. Birsa, Tazerka, Oudna, Cosmos, and Halk el Menzel). The structures are Eocene–upper Miocene horst blocks, which were inverted and buried in the Pliocene (Bellaiche and Blanpied, 1979).

In both basins, there is a clear relationship between the localization of these traps and the depocenter of the source rocks located beneath them. In both areas, oils are likely to come from the Paleocene El Haria shales, and especially from the lower Eocene Bou Dabbous bituminous marly limestones (which are the source rocks of the Ashtart and the Sidi el Ytaiem oils in Tunisia; Burollet and Oudin, 1980).

In the Hammamet basin, the Albian–Cenomanian Fahdene shales could have been source rocks, also. These and the partly time-equivalent Hybla shales of Sicily draw attention as potential source rocks, corresponding to the well-known period of organic-rich sediment formation, because of widespread anoxic conditions in restricted, oceanic-opening basins (Arthur and Schlanger, 1979). The Albian black shales of onshore Tunisia are the source of the Douleb, Tamesmida, and Jebel Onk (Algeria) oils in Aptian reef carbonates (Marie et al, 1984). In Sicily, however, black organic shales occur infrequently in the Hybla and are scattered and quantitatively unimportant (e.g., Calabianca beach exposure, west of Palermo, and a few wells in the southeast).

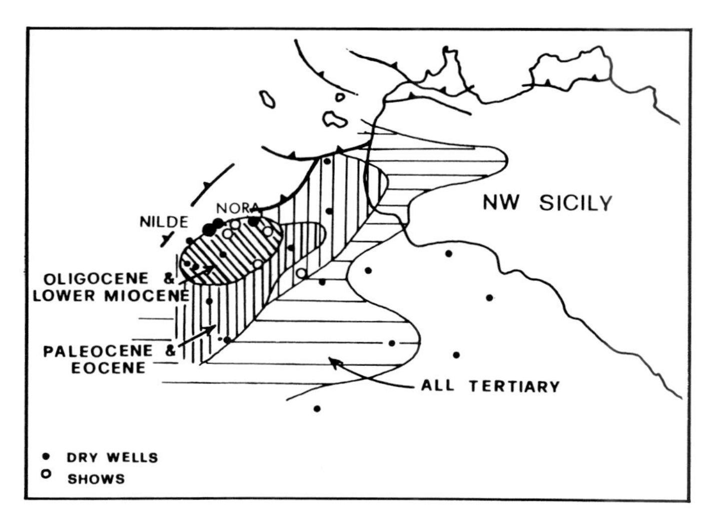


Figure 16. *Relations between oil occurrences (large black dots), Paleogene depocenters, and overall burial (all Tertiary), in the offshore Trapani basin. Small black dots indicate dry wells.*

## THE EXPLORATION POTENTIAL OF THE CENTRAL BASIN

In view of the geological complexity of the Central basin as currently known, the difficulties encountered in earlier exploration are not surprising. The first wells were drilled on surface structures or near seeps, or both, while many only penetrated the olistostromes. Gravity and seismic refraction and reflection surveys demonstrated that deeper structures existed, but their definition was vague. Reflection seismic was greatly hampered by the olistostromes, the gravity surveys by the olistoliths and the evaporite masses. Several wells, in fact, revealed a disharmony between the deep and the superficial structures and turned out to have been mislocated.

The old gravity surveys showed, nevertheless, that numerous large, positive anomalies existed. Some of these corresponded in location and trend to the surface Neogene anticlines, or to synclines; others were unrelated to either and probably derived from dense olistoliths buried in lighter shales. Some of these highs appear to be valid, particularly in areas where the evaporites are absent or reduced in thickness, and where the olistostromes are finely textured. Researchers need to re-examine all these anomalies using up-to-date gravity and (dynamite) seismic reflection surveys. Up-to-date seismic runs in the northeastern and southeastern parts of the basin have already proven that good reflections can be obtained and the deeper structures more precisely delineated. Appropriate models of structural interpretation are essential, however, in view of possible overthrust repetitions, and of the disharmony between the plastic Neogene cover and the underlying rigid carbonates that result from the superposition of the late Miocene and the middle Pliocene phases of deformation.

The main assumption behind exploring the Central Sicily basin is the source rock significance of the Mufara shales, and their having entered the oil window at the time of Miocene–

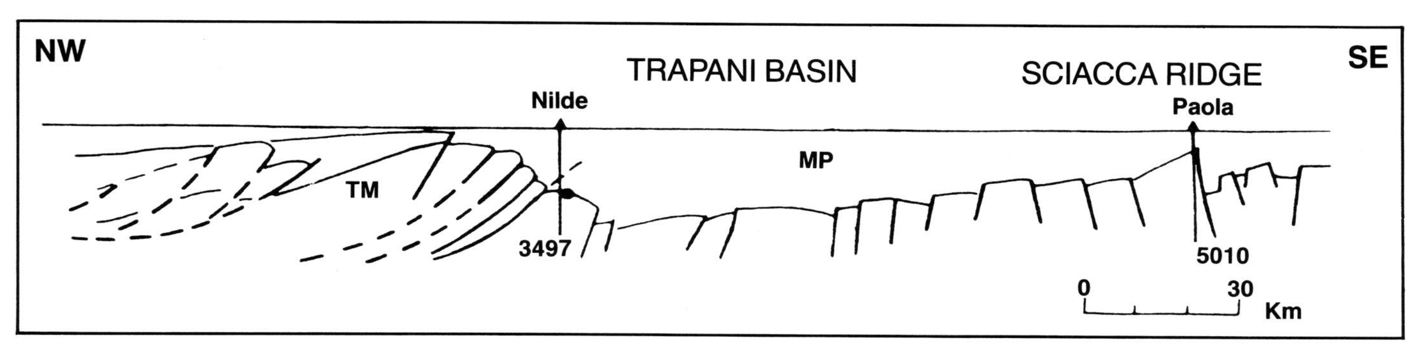


**Figure 17.** *Simplified cross section (from seismic line) of the offshore Trapani basin (profile 4, Figure 3). TM = Triassic–early Miocene; MP = middle Miocene–Recent.*

Pliocene structural development. The Mufara's extension beyond its known area of occurrence is critical (Figure 12). In the southwest, for instance, somewhere south of the Sicani Mountains (Mount Rose, Figures 8 and 11), there should be a transition to the Taormina–Inici carbonates of the basin margin in the Sciacca Zone. A similar transition should exist in the southeast toward the platform margin Siracusa dolomite that rimmed the northwestern part of the early Streppenosa basin. The stratigraphy of wells in the belt from Gela to Catania indirectly suggests this interpretation, indicating a persisting high after the drowning of the Liassic Platform. Evidence regarding the southern part of the Central basin, from Agrigento to Gela, is still indirect and conjectural. The area happens to be in the axial belt of maximum subsidence, especially evident during late Miocene–early Pliocene (Figure 14), and is oriented northeast to southwest like the foreland margin, coinciding with both the maximum gravity low (Figures 4 and 5) and the maximum depth of the resistive substratum (Figure 6). Such accentuated Neogene subsidence may have been conditioned by the shearing of plastic masses at depth, such as thick shales deposited in a basin or a graben within the Triassic interval.

The main exploration targets in the substratum are the large, asymmetric anticlines formed in the Miocene, either directly under the Terravecchia and the olistostromes, or perhaps in subthrust nappe repetitions (in the north basin margin and Sicani Mountains?). We consider the occurrence of paleostructures of the Ragusa–Gela type unlikely for two reasons:

1. The reversal of Liassic–Cretaceous normal faults to thrusts in the Miocene, with lateral deformation of the earlier horst blocks;
2. A reduced post-Liassic sedimentation, i.e., a very to relatively thin B stratigraphic interval. Normal-faulted structures are more likely to have persisted, however, under the southern and southwestern margins of the basin, especially offshore, far from the effects of the main overthrusts.

The plays envisaged in the folds are

1. Additional Collesano gas accumulations in the north.
2. Oil and gas carbonate plays associated with fractured reservoirs, either (A) at the top of the substratum (e.g., Eocene Caltavuturo calcarenites in the north and Bonifato–Ain Grab type calcarenites in the south), or in the Crisanti–Fanusi dolarenites and deeper calcarenites; or (B) in the Taormina–Inici carbonates. Target depths would be moderate in the north (3,000–4,000 m or 10,000–13,100 ft), but deep in the south (compare with the ±5,000 deep oil/gas/condensate production at Malossa and Gaggiano, in northern Italy; Pieri and Mattavelli, in press).

Basin fill objectives are secondary in importance because of the lenticularity of the sands in the non-slumped Terravecchia, their reduced thickness, porosity, and permeability toward the basin center, and their difficulty of delineation. The main gas finds have been noncommercial because of lenticularity (e.g., Enna and Montoni). Nevertheless, there are still large, untested areas. A detailed stratigraphic and geochemical study of the Terravecchia formation, including the olistostromes, would be an urgent undertaking, particularly when aimed at assessing the olistostrome potential source-rock value.

## EXPLORATION IN THE NORTHWEST SICILY BASINS

Onshore exploration in the 1950s aimed at the substratum carbonates of the prominent, gravity-delineated horsts in the Alcamo and Trapani basins (E.N.I., 1969; Regione Siciliana, 1961). Six wells tested the Trapani and Sciacca sequences, finding no hydrocarbons in the porous, thick Inici–Taormina carbonates. The Tortonian sands, however, contained gas in several wells, and there was a small, commercial discovery (Mazara-Lippone) over the Campobello ridge. The gas is certainly derived from the basin itself, because there is no obvious deep potential source rock in the Mesozoic section.

Exploration offshore since the early 1970s has produced 20 wells within the confines of the Miocene basin. As information on these wells is only partly available, we shall limit ourselves to comment generally on some of the exploration results that would be significant to future exploration.

1. Oil is found only in the lower Miocene Ain Grab and Nilde limestones. There are no significant shows in the underlying section.
2. The thickness distribution of the Cenozoic formations, including the Ain Grab, indicates depocenters at the present northwestern margin of the basin, near the folds (Figures 16 and 17). There is no well information yet on the region northwest of Nilde, but the fact that the Sciacca Ridge to the east was positive in the Paleogene suggests that the Paleogene depocenter, or basin, extended northeast beyond Nilde. This depocenter should have continued until the early Miocene (e.g., the exceptionally thick Ain Grab), indicating a pre-tectonic subsident belt in front of

the internal zones, already affected by deformation in late Oligocene and early Miocene.

3. Oil migration must have been post-Miocene, because maturation could not have occurred until burial by the Serravallian–Tortonian sediments. The post-Langhian cover of the folds, however, is relatively thin (Figure 17). Unlike the Hammamet basin, oil accumulations are located northwest of the Neogene basin. As the southeast margin of the basin was uplifted after the Messinian and the source rocks were not in that direction, migration to the folds must have taken place from the west and northwest. If this is feasible, then the main burial of the likely source rocks would not entirely have been provided by the middle-upper Miocene sediments, but especially by overthrusting. The thrusting appears to have also involved the earliest, syntectonic basin sediments, which incidentally include some bituminous shales.

## CONCLUSIONS

Encouraging conditions for commercial petroleum accumulations still exist in the Apenninic belt of central and northwest Sicily. The positive factors are the occurrence and areal distribution of likely source rocks, and their exceptional Neogene burial by thrusting and foredeep sedimentation. In the northwest, the potential area for additional discoveries is offshore in the Sicilian Channel, in the thrust zone northeast of the Nilde field, rather than in the Miocene basin to the east.

The Central basin bears considerable untapped potential, as known light-oil occurrences are matched by a Streppenosa-type basin with probable potential Triassic source rocks that would have reached maturity by the time of late Miocene–early Pliocene structuring. Future exploration targets should be the deep carbonates, as the Neogene sediments themselves bear only a moderate potential for small gas discoveries. The targets should be large, asymmetric anticlines in the northern half of the basin and at its western margin, and horsts in the southern and southeastern parts.

Exploration progress in the Central basin will depend on up-to-date techniques, such as closely spaced, highly stacked seismic lines (with high-resolution and 3D-seismic in some cases), and gravity interpretation by stripping modelling, in order to distinguish between olistoliths and deep structures, and deep from shallow structures.

Should our assumption about a deep Triassic source potential and its extension to the southwest of the basin prove reasonable, the Central basin of Sicily should have all the ingredients of an attractive exploration province, with optimal burial and source-rocks relations conducive to the development of an oil kitchen, as was the case in several other basins (Demaison, 1984), contemporaneous with the formation of suitable structural traps.

## REFERENCES CITED

Alvarez, W., 1978, A former continuation of the Alps: Bulletin of the Geological Society of America, v. 87, p. 891–996.

Alvarez, W., T. Cocozza, and F. C. Wezel, 1974, Fragmentation of the alpine orogenic belt by microplate dispersal: Nature, v. 284, p. 309–314.

Arthur, M. A., and S. C. Schlanger, 1979, Cretaceous 'oceanic anoxic events' as causal factors in development of reef-reservoired giant oil fields: AAPG Bulletin, v. 63, p. 870–885.

Auzende, J. M., 1971, La marge continentale Tunisienne: resultats d'une étude par seismique reflection, sa place dans la cadre tectonique de la Méditerrañee occidentale: Marine Geophysical Researches, v. 1, 162–177.

———, J. M. Olivet, and J. Bonnin, 1974, Le Detroit Sardano-Tunisien et la zone de fracture Nord-Tunisienne: Tectonophysics, v. 21, p. 357–374.

Bellaiche G., and C. Blanpied, 1979, Apercu neotectonique, *in* P. F. Burollet et al., eds., La Mer Pelagienne: Géologie Mediterranée, v. 6, p. 50–59.

Beneo, E., 1955, Les resultats des études pour la recherche petroliere in Sicile: 4th World Petroleum Congress, sect. 1/A/2, Paper 1.

Bijou-Duval, B., J. Dercourt, and X. Le Pichon, 1976, From the Tethys ocean to the Mediterranean seas: a plate tectonic model of the evolution of the western Alpine System, Structural History of the Mediterranean basins: Paris, Editions Technip, p. 143–164.

Boccaletti, M., C. Conedera, and P. Dainelli, 1982, The Recent (Miocene-Quaternary) rhegmatic system of the Western Mediterranean region: Journal of Petroleum Geology, v. 5, p. 31–49.

Broquet, P., 1970, The geology of the Madonie Mountains of Sicily, *in* W. Alvarez and K. Gorbandt, eds., Geology and history of Sicily: Petroleum Exploration Society of Libya, p. 210–230.

———, A. Caire, and G. Mascle, 1976, Structure et évolution de la Sicile occidentale (Madonie et Sicani): Bulletin Societé Géologique France, sr. 7, v. 18, p. 994–1013.

———, G. Mascle, and M. Monnier, 1984, La formation á tripolis du Bassin de Caltanissetta (Sicile): Revue Géologie Dynamique Géographie Physique, v. 25, p. 87–98.

———, G. Douée, G. Mascle, and R. Truillet, 1984, Evolution structurale alpine de la Sicile et sa signification géodynamique: Revue Géologie Dynamique et Géographie Physique, v. 25, p. 75–86.

Burollet, P. F., and H. Rouvier, 1971, La Tunisie, *in* Tectonique de l'Afrique: Sciences de la Terre, UNESCO, Paris, v. 6, p. 91–100.

———, and J. L. Oudin, 1980, Paléocene et Eocene en Tunisie, petrole et phosphates: Documents Bureau Recherches Géologiques Minières, v. 24, p. 205–218.

———, J. M. Mugniot, and P. Sweeney, 1974, The Geology of the Pelagian Block, the margins and basins off southern Tunisia and Tripolitania, *in* A. Nairn and W. Kanes, eds., The Western Mediterranean: 4B, The Ocean basin and Margins, p. 331–350.

Caflisch, L., 1966, La geologia dei Monti di Palermo: Rivista Italiana Paleontologia Stratigrafia, Memorie, v. 12, 108 p.

———, and P. Schmidt di Friedberg, 1967a, Un contributo delle recerche petrolifere alla conoscenza del Paleozoico in Sicilia: Bollettino Societá Geologica Italiana, v. 86, p. 537–551.

———, and ———, 1967b, L'evoluzione paleogeografica della Sicilia e sue relazioni con la tettonica e la naftogenesi: Memorie Societá Geologica Italiana, v. 6 p. 449–474.

Caire, A., 1970, Sicily in its Mediterranean setting, *in* W. Alvarez and K. Gorbandt, eds., Geology and History of Sicily: Tripoli Petroleum Exploration Society of Libya, p. 145–170.

Carissimo, L., O. D'Agostino, C. Loddo, and M. Pieri, 1963, Petroleum exploration by AGIP Mineraria and new geological information in Central and Southern Italy from the Abruzzi to the Taranto Gulf: 6th International Petroleum Congress, sr. 1, p. 267–292.

Catalano, R., and B. D'Argenio, 1982a, Guida alla geologia della Sicilia occidentale: Palermo Societá Geologica Italiana, Guide Geologiche Regionali, 155 p.

——, and ——, 1982b, Infraliassic strike-slip tectonics in Sicily and SE Apennines: Rendiconti Societá Geologica Italiana, v. 5, p. 5–10.

Channel, J. E. T., B. D'Argenio, and F. Horvath, 1979, Adria, the African promontory in Mesozoic Mediterranean paleogeography: Earth Sciences Reviews, v. 15, p. 213–292.

Channel, J. E. T., R. Catalano, and B. D'Argenio, 1980, Paleomagnetism and deformation of the Mesozoic continental margin of Sicily: Tectonophysics, v. 61, p. 391–407.

Cherchi, A., and L. Montadert, 1982, Oligo-Miocene rift of Sardinia and the early history of the Western Mediterranean basin: Nature, v. 298, p. 736–739.

Colantoni, P., 1975, Note di geologia marina sul Canale di Sicilia: Giornale de Geologia, sr. 2, v. 40, p. 181–207.

Decima, A., and C. F. Wezel, 1971, Osservazioni sulle evaporiti Messiniane della Sicilia centro-merdionale: Rivista Mineraria Siciliana, v. 22, n. 130–132, p. 172–187.

Demaison, G., 1984, The generative basin concept, *in* G. Demaison and R. J. Murris, eds., Petroleum geochemistry and basin evaluation: AAPG Memoir 35, p. 1–14.

Dewey, J. F., W. C. Pitman, III, W. B. F. Ryan, and J. Bonnin, 1973, Plate tectonics and the evolution of the Alpine system: Bulletin of the Geological Society of America, v. 84, p. 3137–3180.

Dingle, R. V., 1977, The anatomy of a large submarine slump on a sheared continental margin (SE Africa): Journal of the Geological Society of London, v. 134, pt. 3, p. 293–310.

——, 1980, Large allochthonous sediment masses and their role in the construction of the continental slope and rise off SW Africa: Marine Geology, v. 37, p. 333–354.

Duée, G., 1970, The Geology of the Nebrodi Mountains in Sicily, *in* W. Alvarez and K. Gorbandt, eds., Geology and History of Sicily: Petroleum Exploration Society of Libya, p. 187–200.

Duvernoy R., and J. Reulet, 1977, A carbonate turbidite reservoir: the formation 'carbonate scaglia', an attempt at sedimentological characterization of carbonate deposits: Elf Aquitaine, Centre Recherche Boussens et Pau, p. 255–231.

Emery, K. O., E. Uchupi, and C. O. Bowin, J. Phillips, and E. S. W. Simpson, 1975, Continental margin off western Africa Cape Saint Francis (South Africa): AAPG Bulletin, v. 59, p. 3–59.

E. N. I., 1969, Italia: Geologia e Ricerca Petrolifera, *in* C. Colombo, ed., Enciclopedia del petrolio e gas naturale, v. 6, p. 336–571.

Finetti, I., 1982, Structure, stratigraphy and evolution of Central Meditarranean: Bollettino Geofisica Teorica e Applicata, v. 24, n. 96, p. 247–312.

——, and C. Morelli, 1973, Geophysical exploration of the Mediterranean Sea: Bollettino Geofisica Teorica e Applicata, v. 15, p. 263–342.

Flores, G., 1955, Discussion: Rome, 4th World Petroleum Congress, sect. A-2, p. 120–121.

——, 1959, Evidence of slump phenomena (olistostromes) in areas of hydrocarbon exploration in Sicily: New York, 5th World Petroleum Congress, sect. 1, p. 259–275.

Ghisetti, F., and L. Vezzani, 1975, The structural features of the Hyblean Plateau and of the Monte Iudica area (SE Sicily): Bollettino Societá Geologica Italiana, v. 99, p. 57–102.

——, and ——, 1982, Il ruolo della zona di taglio M. Kumeta-Alcantara nell'evoluzione dell'arco Calabro, implicazioni e problemi: Guida Sicilia Occidentale, Societá Geologica Italiana, p. 119–123.

Giunta, G., and V. Liquori, 1972, Geologia dell'estremitá nordoccidentale della Sicilia: Rivista Mineraria Siciliana, v. 23, n. 136–168, p. 165–226.

Grasso, M., and F. Lentini, 1982, Sedimentary and tectonic evolution of the eastern Hyblean Plateau (Southeastern Sicily) during Late Cretaceous to Quaternary time: Paleogeography, Paleoclimatology, Paleoecology, v. 39, p. 261–280.

Jenkins, H. C., 1970, The Jurassic of eastern Sicily, *in* W. Alvarez and K. Gorbandt, eds., Geology and History of Sicily: Petroleum Exploration Society of Libya, p. 245–254.

Kafka, F. T., and R. K. Kirkbride, 1960, The Ragusa oilfield (Sicily): New York, 5th World Petroleum Congress, sect. I/12, p. 233–257.

Kieken, M., and Winnock E., 1957, Le champ de l'Qued Gueterini: XX Congreso Geologico Internacional, Symposio Yacimientos Petroleo y Gas, p. 24–43.

Marchetti, M. P., 1956, The occurrence of flowage materials and slides (olistostromes) in the Tertiary series of Sicily: Mexico, 20th International World Petroleum Congress, sect. 5, p. 209–225.

Mascle, G., 1970, Etude géologique des Monts Sicani: Thése, Université Paris, Rivista Italiana Paleontologia Stratigrafia, Memorie, v. 16.

Marie, J., Ph. Trouvé, G. Deforges, and Ph. Dufaure, 1984, Nouveaux elements de paléogéographie du Cretacée en Tunisie: TOTAL, Notes et Memoires, n. 19.

Mattavelli, L., G. U. Chilingarian, and D. Storer, 1969, Petrography and diagenesis of the Taormina Formation, Gela oilfield, Sicily (Italy): Sedimentary Geology, v. 3., p. 59–86.

——, T. Ricchiuto, D. Grignani, and M. Shoell, 1983, Geochemistry and habitat of natural gases in the Po basin, northern Italy: AAPG Bulletin, v. 67, p. 2239–2257.

Moore, D. G., J. R. Curray, and F. J. Emmel, 1976, Large submarine slide (olistostrome) associated with Sunda arc subduction zone, northeast Indian Ocean: Marine Geology, v. 21, p. 221–226.

Ogniben, L., 1975, Lithostratigraphic complexes and evidence for tectonic phases in Sicily and Calabria, *in* L. Ogniben, M. Parotto, A. Praturlon, eds., Structural Model of Italy: La Ricerca Scientifica, v. 90, p. 365–408.

Pacchiarotti, E., 1982, Exploration effort in the Italian offshore: Milano, Italian-Norwegian Seminar, AGIP, p. 15–28.

Patacca E., P. Scandone, P. Giunta, and V. Liquori, 1979, Mesozoic paleotectonic evolution of the Ragusa zone (SE Sicily): Geologica Romana, v. 81, p. 331–369.

Pedley, H. M., 1981, Sedimentology and paleoenvironment of the southeast Sicilian Tertiary platform carbonates: Sedimentary Geology, v. 28, p. 273–291.

Pieri, M., and L. Mattavelli, in press, The geological framework of Italian petroleum resources: AAPG Bulletin, expected 1986.

Regione Siciliana, 1961, Studied indagini per ricerche di idro-

carburi: Palermo, Assessorato Industria e Commercio, 79 p.
Rigo, F., 1956, Olistostromi neogenici in Sicilia: Bollettino Societá Geologica Italiana, v. 75, p. 185–215.
Rigo, M., and F. Barbieri, 1959, Contributo alla conoscenze strutturale della Sicilia sud-orientale: Bollettino Servizio Geologico d'Italia, v. 81, p. 349–369.
Rigo, M., and A. Cortesini, 1959, Stratigrafia pratica applicata in Sicilia: Bollettino Servizio Geologico d'Italia, v. 80, p. 1–92.
Rocco, T., 1959, Gela in Sicily: New York, 5th World Petroleum Congress, section I, p. 207–233.
Rouvier, H., 1977, Géologie de l'extreme nord Tunisien: tectoniques et paléógéographies superposées á Déxtremité orientale de la chaine nord-maghrebine: Thése Doctorat, Université Paris VI, 375 p.
Salaj, J., 1977, The geology of the Pelagian Block, *in* A. Nairn and W. Kanes, eds., The western Mediterranean: 4B, The ocean basins and margins, p. 361–416.
Scandone, P., 1975, The seaways and the Jurassic Tethys ocean in the Central Mediterranean area: Nature, v. 256, p. 117–119.
——, G. Giunta, and V. Liguori, 1977, The connection between the Apulia and Saharan continental margins in the southern Apennines and in Sicily: Memorie Societá Geologica Italiana, v. 13, p. 317–323.
Schmidt di Friedberg, P., 1959, La geologia del gruppo montuoso delle Madonie, nel quadro delle possibilitá petrolifere della Sicilia Centro-Settentrionale: Atti 2nd Convegno Internazionale 'Petrolio in Sicilia', p. 130–136.
——, 1962, Introduction á la géologie petroliére de la Sicile: Revue Institute Francais du Petrole, v. 17, p. 635–688.
——, 1965, Litostratigrafia petrolifera della Sicilia: Rivista Mineraria Siciliana, n. 91–93, p. 88–90, 198–217.
——, P. F. Barbieri, and G. Giannini, 1960, La geologia del gruppo montuoso delle Madonie (Sicilia Centro-Settentrionale): Bollettino Servizio Geologico d'Italia, v. 81, p. 73–140.
Schramm M. Jr., and G. Livraga, 1986, Vega field and potential of Ragusa basin, offshore Sicily, this volume.
Schreiber, B. C., and G. M. Friedman, 1976, Depositional environment of upper Miocene (Messinian) evaporites of Sicily, as determined from analysis of intercalated carbonates: Sedimentology, v. 23, p. 225–270.
Vercellino, J., and F. Rigo, 1970, Geology and exploration of Sicily and adjacent areas, *in* R. E. King, ed., Stratigraphic oil and gas fields, AAPG Memoir 16, p. 388–399.
Wendt, V. J., 1969, Die stratigraphische-palaeogeographische entwicklung des Jura in Westsizilien: Geologische Rundschau, v. 58, p. 735–755.
Wezel, C. F., 1974, Flysch successions and the tectonic evolution of Sicily during the Oligocene and early Miocene, *in* S. C. Squyres, ed., Geology of Italy: Petroleum Exploration Society of Libya, p. 105–127.
Winnock, E., 1980, Les depôts de l'Eocene au nord de l'Afrique, aperçu paleogeographique de l'ensemble: Documents Bureau Recherches Géologiques Minières, v. 24, p. 219–244.
——, 1981, Structure du block Pelagien, *in* F. C. Wezel, ed., Sedimentary basins of Mediterranean margins: Bologna, Technoprint, p. 445–467.

# Trapping Styles and Associated Hydrocarbon Potential in the Norwegian North Sea

R. B. Færseth
K. A. Oppebøen
A. Sæbøe
*Norsk Hydro*
*Oslo, Norway*

Exploration has been active in the Norwegian North Sea for almost 20 years, resulting in recoverable reserves that are estimated at 1.5 billion cu m (9.5 billion bbl) of oil and 2.5 trillion cu m (88.2 trillion cu ft) of gas. The 25 fields with reserves of more than 100 million bbl of oil equivalent each, together account for 98% of the total recoverable reserves. Seven fields are giant, with more than 1 billion bbl each. The distribution of oil and gas fields is intimately linked with the Mesozoic rift system of the Viking and Central grabens that, in turn, controlled the distribution and maturation of the Upper Jurassic source rock interval. In the Norwegian North Sea, hydrocarbon discoveries to date occur in four forms of traps: structural-unconformity, salt-supported, compression-related and stratigraphic traps.

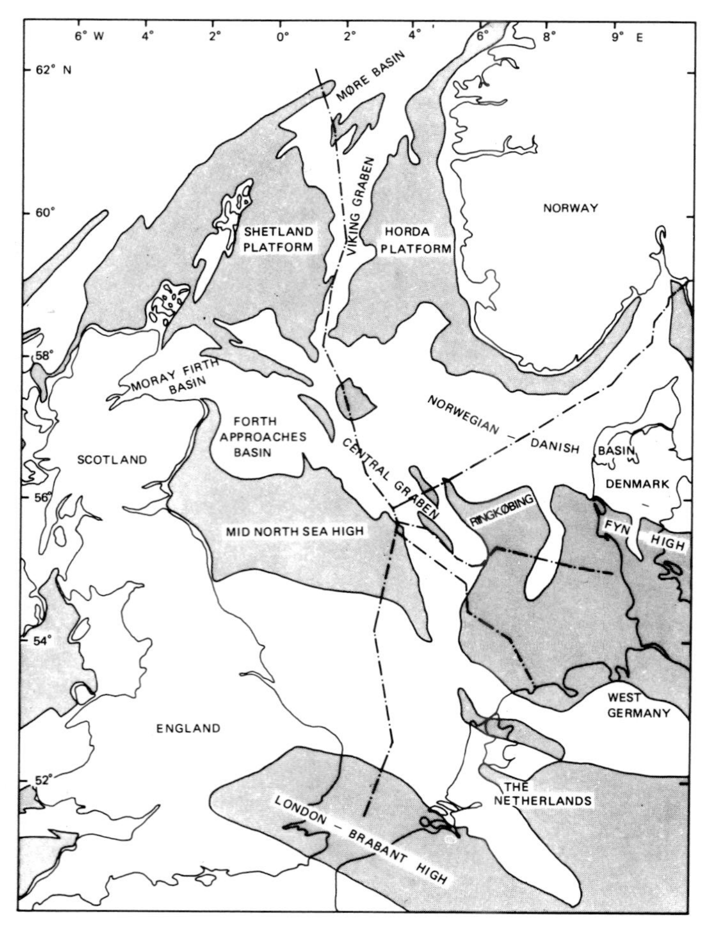


Figure 1. *Main structural elements of the North Sea.*

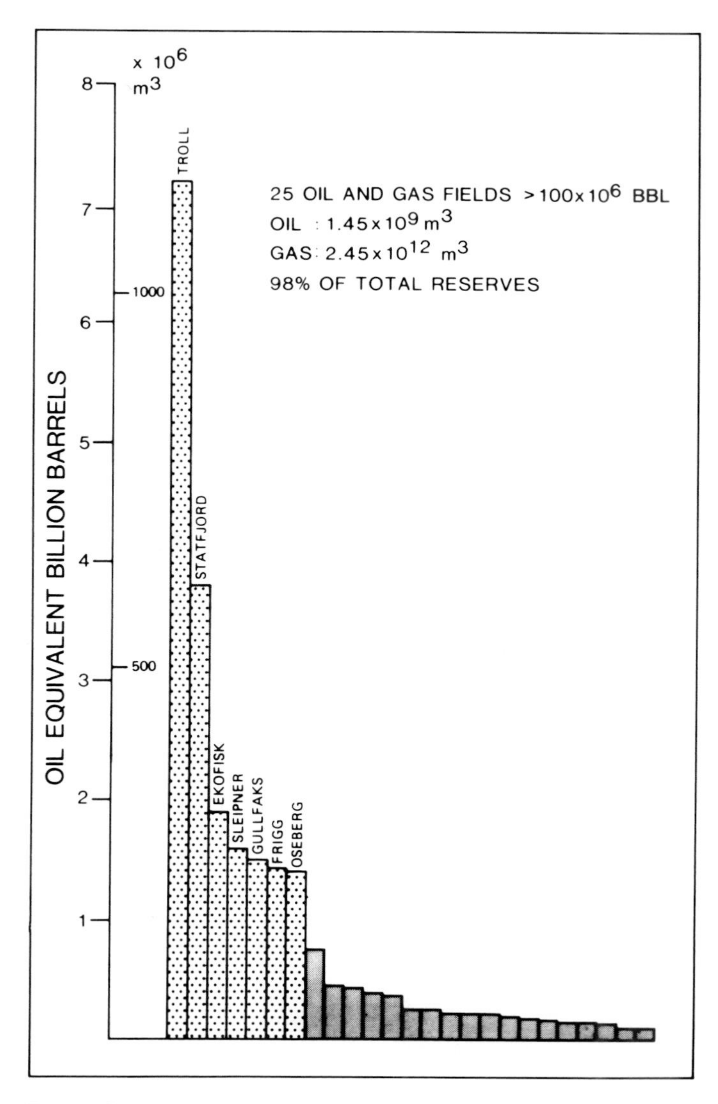


Figure 2. *Recoverable reserves of major fields in the Norwegian North Sea.*

## INTRODUCTION

The North Sea sedimentary basin, north of the Mid-North Sea-Ringkøbing Fyn High, covers an area of approximately 200,000 sq km (77,200 sq mi) (Figure 1), and is by far the most important oil province of northwestern Europe. Maximum recoverable reserves in known accumulations in Norwegian and U.K. waters are some 6.3 billion cu m (40 billion bbl) of oil equivalent.

In the Norwegian North Sea, which geologists have explored for almost 20 years, the proven recoverable reserves are estimated to be 1.5 billion cu m (9.5 billion bbl) of oil and 2.5 trillion cu m (88.2 trillion cu ft, or tcf) of gas. The total expected resources in the Norwegian sector are 5.0 billion cu m (32 billion bbl) of oil equivalent (Norwegian Petroleum Directorate, 1984). Production rates from the Norwegian sector are now at 35 million cu m (220.5 million bbl) of oil and 25 billion cu m (882 billion cu ft) of gas per year. In 1990 the production should peak at 65 million cu m (409.5 million bbl) of oil and 32 billion cu m (1.1 tcf) of gas. Through the end of 1984, approximately 0.4 billion cu m (2.5 billion bbl) of oil equivalent has been produced from the Norwegian sector, which amounts to 10% of the proven recoverable reserves. The annual expenditure on exploration drilling reached a maximum of 600 million dollars in 1982.

Current production is the result of more than 400 wells drilled to an average depth of 3,150 m (10,300 ft). Water depths range from 50 to 450 m (160 to 1,450 ft), generally less than 100 m (330 ft) in the southern half of the area. Of the fields planned for development, the Troll gas field is situated in water depths up to 350 m (1,150 ft).

The North Sea rift system is bounded to the south, east, and west by metamorphosed Paleozoic and Precambrian rocks. To the north, it grades into the Upper Cretaceous, lower Tertiary Atlantic shelves of western Norway and the Shetland Islands (Figure 1). Post-Devonian sediment thickness ranges from 2 km (1 mi) to more than 10 km (6 mi) in the grabens, which straddle the Norwegian/U.K. sector line.

The distributions of oil and gas fields are intimately linked with the Mesozoic rift system of the Viking and Central grabens (Figure 1). Although, structurally speaking, these grabens form part of the same megatectonic unit, their hydrocarbon habitat differs considerably. The main differences are age and lithology

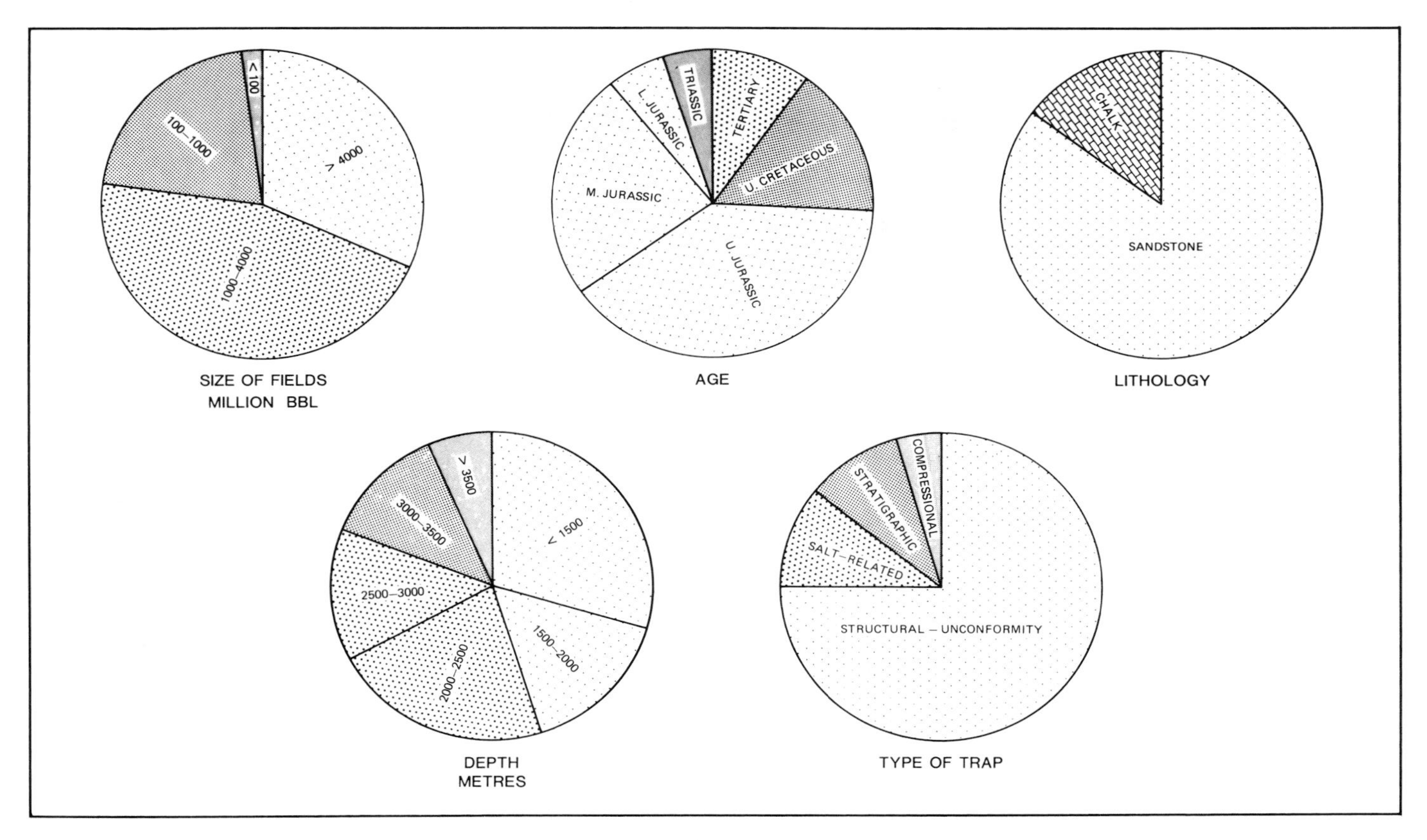


**Figure 3.** *Habitat of hydrocarbons in the Norwegian North Sea.*

of producing reservoirs, and structural style of traps containing major accumulations.

## HABITAT OF OIL

Figure 2 gives the field size distribution for the 25 fields on the Norwegian sector of the North Sea having recoverable reserves of more than 16 million cu m (100 million bbl) of oil equivalent each. The cumulative total of these fields is 3.9 billion cu m (25 billion bbl) oil equivalent, with oil being 1.45 billion cu m (9.2 billion bbl) and gas 2.45 trillion cu m. These represent 98% of the total, proven recoverable reserves.

Seven of the fields are giant, with reserves in excess of 1 billion bbl of oil equivalent each, and, together, they constitute 80% of the hydrocarbons discovered to date in the Norwegian North Sea basin.

### *Age of Reservoir*

Geologists found the oil and gas in reservoirs belonging to four geological periods—from the Tertiary to the Triassic. However, close to 70% of all, recoverable petroleum reserves and approximately half of all the oil belongs to the Jurassic period, generally occurring in reservoirs of Middle and Late Jurassic age (Figure 3). Of the seven giant fields, five have Jurassic reservoirs.

### *Lithology of Reservoir*

Figure 3 shows that 85% of the total reserves are in Tertiary, Jurassic, and Triassic sandstones. Geologists have found the remaining 15% in Upper Cretaceous-Danian chalk in the southernmost part of the Central graben.

### *Depth of Reservoir*

A majority of the fields that together account for approximately 35% of the reserves occur at depths between 2,000 and 3,000 m (6,560 and 9,850 ft) (Figure 3). Of the reserves found at a depth of less than 2,000 m (6,560 ft), the giant Troll gas field accounts for nearly 75%. Below 3,000 m (9,850 ft) and down to a maximum of 4,300 m (14,100 ft), geologists have found hydrocarbons in Upper Cretaceous chalk and Jurassic sandstones. The age of the reservoir, however, does not correlate with the depth of burial in the proven fields. The porosity of the reservoirs tends to decrease with increasing age and depth of burial. On the other hand, more than 85% of the oil and gas in sandstone reservoirs occurs at depths shallower than 3,000 m (9,850 ft), with porosities in the range of 20–35%. Fields producing from an Upper Cretaceous chalk have an average depth to the reservoir of around 3,000 m (9,850 ft). The chalk within these fields is characterized by excellent porosity preservation, in places exceeding 40%, which reflects both the sedimentary conditions and secondary microbrecciation (Kennedy, 1980).

## REGIONAL PETROLEUM GEOLOGY

The Viking and Central grabens dominate the structural framework of the Norwegian North Sea. Along these structural

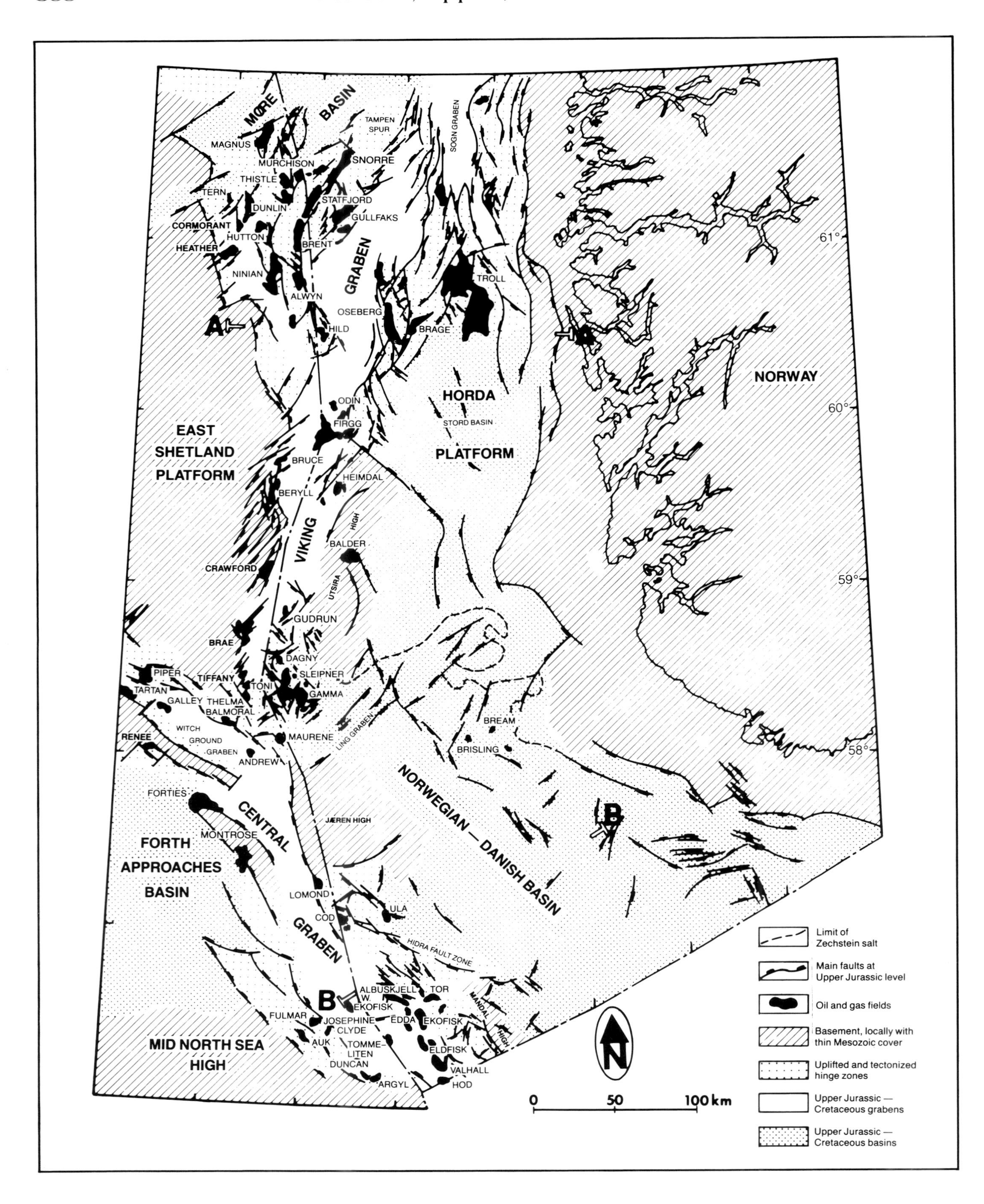

MØRE BASIN
TAMPEN SPUR
SOGN GRABEN
MAGNUS
MURCHISON
SNORRE
THISTLE
TERN
STATFJORD
DUNLIN
GULLFAKS
CORMORANT
HUTTON
HEATHER
BRENT
GRABEN
NINIAN
TROLL
ALWYN
OSEBERG
A
HILD
BRAGE
NORWAY
HORDA
ODIN
60°
STORD BASIN
EAST SHETLAND PLATFORM
FIRGG
PLATFORM
BRUCE
HEIMDAL
BERYLL
VIKING
HIGH
BALDER
CRAWFORD
59°
UTSIRA
GUDRUN
BRAE
DAGNY
PIPER
TIFFANY
SLEIPNER
TONI
TARTAN
GAMMA
GALLEY
THELMA
BALMORAL
BREAM
WITCH GROUND GRABEN
RENEE
MAURENE
LING GRABEN
BRISLING
58°
ANDREW
61°
FORTIES
CENTRAL
NORWEGIAN — DANISH BASIN
B
JÆREN HIGH
MONTROSE
FORTH APPROACHES BASIN
LOMOND
ULA
COD
GRABEN
HIDRA FAULT ZONE
ALBUSKJELL
TOR
W. EKOFISK
MANDAL HIGH
FULMAR
JOSEPHINE
EDDA
EKOFISK
CLYDE
MID NORTH SEA HIGH
AUK
TOMMELITEN
ELDFISK
DUNCAN
VALHALL
ARGYL
HOD
N
0 50 100 km
Limit of Zechstein salt
Main faults at Upper Jurassic level
Oil and gas fields
Basement, locally with thin Mesozoic cover
Uplifted and tectonized hinge zones
Upper Jurassic — Cretaceous grabens
Upper Jurassic — Cretaceous basins

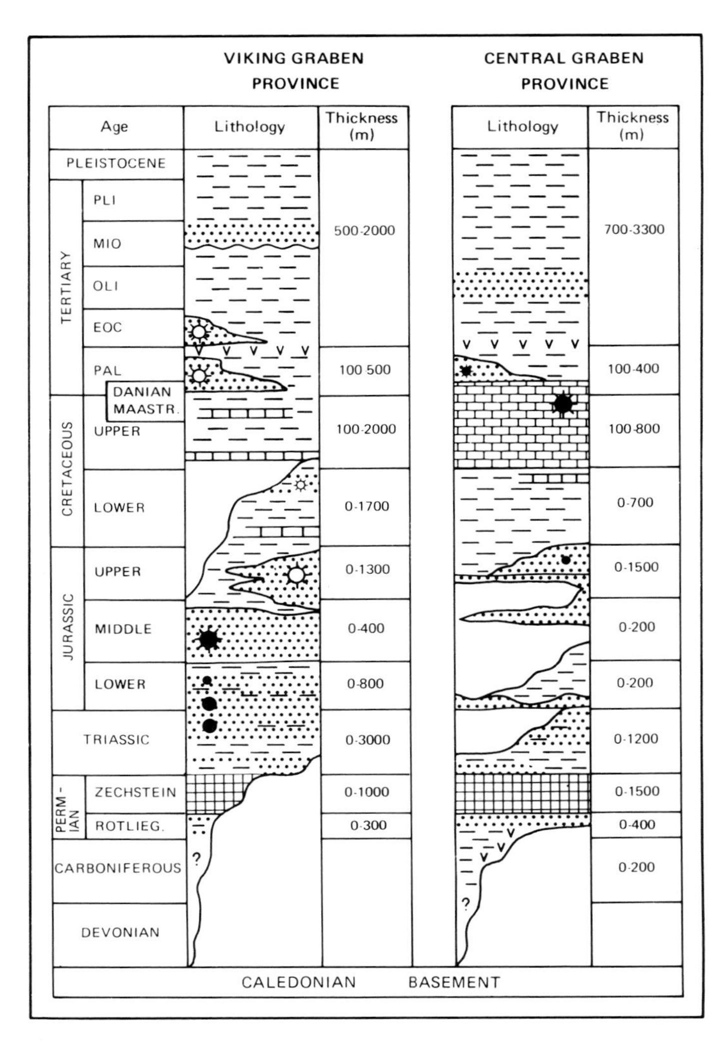


Figure 5. *Generalized stratigraphic diagrams of the Viking graben and Central graben provinces (Norwegian North Sea).*

features, the factors necessary for petroleum came together and gave rise to most of the oil and gas found in the Norwegian North Sea region (Figure 4).

The graben phase and associated faulting began in Jurassic (Callovian) time and were superimposed on a pre-graben sequence. This lasted into the Early Cretaceous, to be followed by deposition of a thick, Cretaceous and Tertiary post-graben sequence.

All three sedimentary cycles (pre-graben, graben, and post-graben) are productive in the North Sea basin. The deposition patterns of the various cycles result from the interplay of sea level changes, climatic variations, tectonic movements, and rejuvenation of relief. All have a direct bearing on the regional distribution of habitat-controlling parameters, such as source rock, reservoir, and seal.

We will now outline the structural and depositional history in pre-graben, graben, and post-graben time as encountered in the Norwegian North Sea. We will also indicate the translation of lithological units into habitat terms like source and reservoir potential.

### Pre-Graben Development

The basement underlying the North Sea sedimentary basin consolidated at the end of the Caledonian orogeny in Late Silurian time. Following this orogeny, tectonic phases affected the North Sea, which led to periodic reactivation of the basement lineaments and developed sedimentary basins of different types and orientations.

Geologists attributed a Devonian intracontinental basin that developed across the southern part of the region to major strike-slip movements along northeast to southwest oriented faults. In Carboniferous time, the Variscan tectonic activity uplifted central Europe, producing concomitant east to west oriented downwarps across the North Sea area (Ziegler, 1982). This orogeny culminated in pre-Permian time, but compression continued subtly to affect the foreland, producing the Mid-North Sea–Ringkøbing Fyn High, with adjoining basin downwarps to the north and to the south (Figure 1). Rotliegendes, mainly eolian sediments, followed by Zechstein evaporite deposits (Figures 5 and 6), cover the southern half of the Norwegian North Sea.

A rift system bisected the Variscan lineaments with an overall north to south orientation in Triassic time, and thick, sedimentary sections of red beds and conglomerates developed over major parts of the North Sea basin (Figures 5 and 6). The Triassic structural framework was consolidated in the Jurassic, allowing marine waters periodically to flow into the basin both from the north and the south. In mid-Jurassic time, a doming occurred in the central North Sea, and erosion from this high and the highs bounding the basin to the east and west fed sands into the basin. The sands, which were deposited in a broad, generally flat basin with a northerly paleoslope, represent the progradational deposits of a deltaic complex.

#### *Reservoirs*

Rotliegendes, Triassic, Lower Jurassic, and Middle Jurassic sandstones represent potential reservoirs within the pre-graben sequence (Figure 5). To date, oil and gas within the Norwegian part of this sequence occur only in Jurassic and Triassic reservoirs. The most important reservoir is the Middle Jurassic sand series, which contains 25% of the total proven reserves and 45% of all oil. A blanket distribution in the northern half of the North Sea basin characterizes the series with thicknesses generally between 100 and 300 m (330 and 985 ft), with average porosities reaching 25% to 30%. The reserves are mainly accounted for by three giant fields: Statfjord, Gullfaks, and Oseberg, all located along the northern Viking graben (Figure 4).

Other Jurassic reservoir horizons are formed by Hettangian-Pliensbachian sands with porosity values of 23% to 30%. They are generally limited to the northern part of the basin, where they contain the deeper pools of giant oil fields. The Lower Jurassic sands contain 5.5% of the total reserves.

Geologists had assumed that sands within the thick Triassic

Figure 4. *Faults, main structural elements, and location of oil and gas fields in the northern North Sea.*

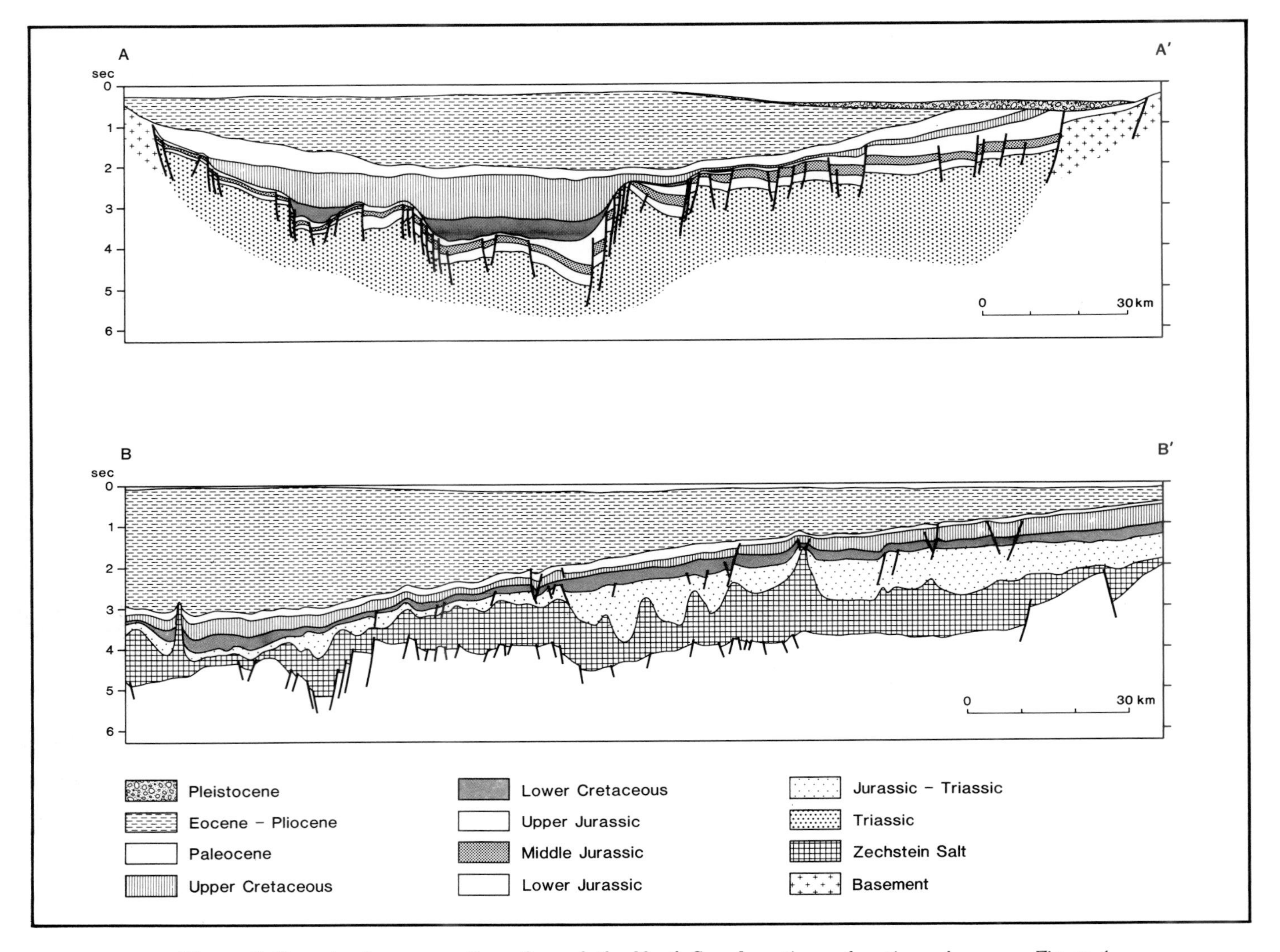


**Figure 6.** *Geoseismic cross sections through the North Sea. Locations of sections shown on Figure 4.*

sequence have a minimal petroleum reservoir potential. Over the last two years, however, they have found large oil pools along the northwestern flank of the Viking graben, in deeply eroded fault blocks, where Triassic reservoirs possess an average porosity of 20%. At the present, Triassic sandstone reservoirs account for 5% of the total proven reserves.

## Graben Development

The development of the Viking and Central grabens into pronounced axial depressions started in Callovian time (Figures 1 and 4). From this time and into the Early Cretaceous, normal faulting was exceptionally intense. The total graben system is more than 800 km (495 mi) long, running diagonally across the Mesozoic sedimentary basin. In the grabens, Upper Jurassic and Lower Cretaceous deposits developed into dominantly shaly sequences, which in places are thicker than 2,000 m (6,560 ft) (Figures 5 and 6).

Deep-water, restricted conditions produced the Kimmeridgian-Ryazanian "hot shale" source rocks. Also, the marginal basins to the east and west of the grabens subsided fairly rapidly, but the depositional environment produced a different and less prolific source-rock facies (Barnard and Cooper, 1981).

The general transgression of the seas in Late Jurassic time produced shoreline sand deposits marginal to the Viking and Central grabens, with some sands swept out to the graben edges and into the deep-water slope and ocean-basin environment.

Rotated fault blocks characterize the flanks of the Viking graben and, to a lesser extent, the Central graben, with episodic erosion over crestal parts during the graben cycle and hence reduced Upper Jurassic and Lower Cretaceous thicknesses (Figure 6). Structures located in such areas contain a major portion of the proven reserves.

### *Source Rocks*

The distribution and maturation of the Upper Jurassic organic shales closely control the position of the proven oil and gas fields. The shales occur widely, and their thickness ranges from zero to tens of meters above structural highs, increasing to hundreds of meters in grabens and parts of the side-basins (Figure 6).

The values of organic carbon vary abruptly both laterally and vertically; however, average values for a section are usually between 5% and 10%, tending to the higher values in the graben depocenters. Algal sapropel is dominant in the grabens, while inertinite dominates areas both closest to the source of sediment and where shallow-water conditions prevailed over stable highs. For these reasons, the richest source rock tends to occupy the graben areas. In such areas, the Upper Jurassic mudstones have also been buried deeply enough to realize a large part of their oil-generating capability.

In areas outside the grabens, only isolated subbasins of the Upper Jurassic shales are rich enough and have been buried deeply enough to generate sufficient oil to fill reservoir traps.

General lack of discoveries in areas where Upper Jurassic shales are immature suggests that lateral migration is not a common phenomenon in the North Sea basin.

#### *Reservoirs*

Geologists have identified more than ten individual sand formations of shallow marine origin within the North Sea basin with maximum thicknesses greater than 200 m (660 ft) (Vollset and Doŕe, 1984). Porosities average 18% to 25%, but may exceed 30%.

The Upper Jurassic reservoirs contain almost 40% of the total proven reserves, of which 140 million cu m (880 million bbl) is oil and 1.38 trillion cu m (48.7 tcf) is gas. More than 80% of the proven Upper Jurassic reserves occur in the giant Troll gas field.

### Post-Graben Development

The faulting related to the graben development shows a significantly diminishing effect up through the Lower Cretaceous sequence, indicating the transition to an interior-sag basin. The continued subsidence associated with general transgression produced a dominantly shale/carbonate depositional system through the Cretaceous period. During Late Cretaceous time, chalk accumulated in the southern North Sea basin (Figure 5), suggesting the remoteness of positive tectonic elements capable of supplying clastic materials. In this area, thick Upper Permian evaporites provided a mechanism for deforming the interior-sag till.

The Cenozoic sediments were concentrated in Mesozoic depocenters, reaching a maximum of more than 3,000 m (9,840 ft) along the Central graben (Figure 6).

Late Cretaceous, Alpine compressional stresses affected the northwest European basin system, producing continued downwarping in some areas and inversion in others (Ziegler, 1982; Pegrum, 1984). The uplifted blocks remained positive throughout the subsequent Cenozoic period. Compressional, inversion features provide trapping possibilities in the southermost parts of the Norwegian North Sea.

Continuous uplift of the Norwegian mainland through the Miocene and early Pliocene resulted in a westerly tilt and erosion of more than 1,000 m (3,280 ft) of Mesozoic and Paleogene sediments in the northeastern part of the North Sea basin (Figure 6).

Along the western flank of the basin, in the vicinity of the East Shetland Platform, Paleocene and Eocene delta development and submarine fan sedimentation produced reservoir rocks significant to the occurrence of petroleum (Lovell, 1984).

#### *Reservoir Rocks*

Reservoirs within the post-graben sedimentary cycle contain 26% of the total reserves, with Upper Cretaceous-Danian chalks as the main producers in the Norwegian part of the Central graben. These chalks form a blanket deposit over major parts of the southern North Sea. However, the ten fields producing from chalk reservoirs cluster in the southermost part of the Norwegian North Sea. The fields are mainly oil-bearing and account for 16% of total proven reserves, of which half are contained in the giant Ekofisk field.

Paleocene-Eocene sands provide reservoirs for fields located along the Viking and Central grabens. Two fields, Frigg and Sleipner Gamma (Figure 4), account for more than 75% of the total Tertiary reserves, of which 38 million cu m (240 million bbl) is oil and 340 billion cu m (12 tcf) is gas.

## TRAPPING STYLES

In terms of trap types, Figure 3 shows the dominance of structural-unconformity traps, which hold nearly three-quarters of the total reserves. The salt-supported and stratigraphic traps account for 11% of the reserves each, while 4% occur in fields with anticlinal closures related to compressional tectonics.

The amount of reserves within the structural-unconformity traps provides a measure of the importance of graben tectonics as a trap-forming agent. The hydrocarbons found within such traps occur in reservoirs belonging to both the pre-graben and graben cycles, and have accumulated under the unconformity at the base of the Cretaceous level. However, the petroleum potential of the basin, and especially that of the post-graben, interior sag cycle, is greatly improved because of closures related to halokinetic movements, Late Cretaceous-Tertiary inversion, and depositional features.

In the Norwegian North Sea, almost 85% of the proven reserves belong to the Viking graben province. Another feature is the total dominance of the structural-unconformity category of traps along the flanks of the Viking graben, while in the Central graben, the proven reserves are restricted to the three other trap categories.

### Structural-Unconformity Traps

The cumulative total of recoverable reserves found in the structural-unconformity traps are 2.9 billion cu m (18 billion bbl) oil equivalent, of which oil is 1.25 billion cu m (8 billion bbl) and gas 1.65 trillion cu m (58.2 tcf). Almost 75% of these gas reserves occur in the Troll field. Of the seven giant fields, five are located within structural-unconformity traps.

The hydrocarbon column height varies from tens of meters to more than 500 m (1,640 ft) and the extent of the fields ranges from less than 20 sq km (8 sq mi) to almost 700 sq km (270 sq mi).

#### *Regional Controlling Factors*

In general, several factors control the location of the structural-unconformity traps: Position within Upper Jurassic-Lower Cretaceous structural framework, development of seal and reservoir, and proximity to a mature source rock.

Numerous closures associated with Jurassic-Lower Cretaceous extensional tectonics and block faulting occur in the North Sea area. However, a striking feature in terms of prospectiveness is the overwhelming influence of the hinge areas represented by the uplifted shoulders of the rotated platforms

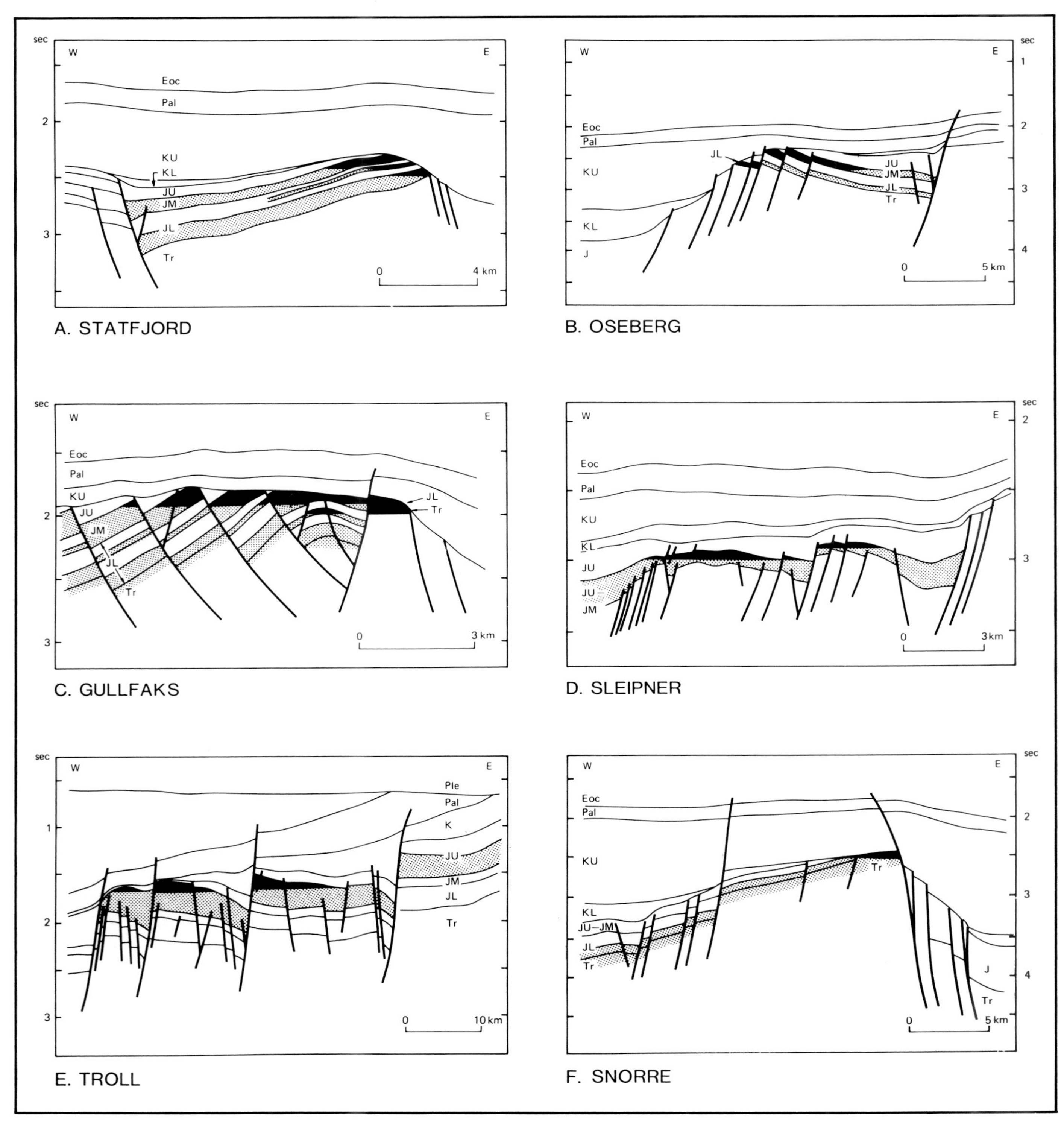


Figure 7. *Dip-oriented sections showing structural-unconformity traps in the Norwegian North Sea. Dotted intervals represent sandstone reservoirs. Main hydrocarbon accumulations in black.*

flanking the Viking graben. Jurassic and Triassic sandstones, which represent reservoir rocks in the structural-unconformity traps, occur over major parts of the Norwegian North Sea, and Upper Jurassic-Cretaceous shales and dense carbonate beds provide sealing formations. Hence, geologists ascribe two main factors for the dominance of the hinge areas. First, these tectonically complex areas provide excellent trapping possibilities. Second, closures along hinge areas are in the most favorable position for trapping oil and gas generated from mature source beds in the graben deeps.

The structural style of the Viking graben is, at the pre-Cretaceous level, dominated by tilted fault blocks and bounded by normal faults. As shown in Figure 6, the individual fault blocks exhibit an overall similar style. The bedding within the fault block generally dips away from the major boundary fault and the pre-graben and graben sequences are partially or totally truncated over the crest, producing structures characterized by a rounded morphology.

From a map view, a longitudinal north-northeast to south-southwest and an oblique northwest to southeast trend dominate the regional faults associated with the formation of the graben. In the northern part of the North Sea, the uplifted areas flanking the Viking graben form megastructures bounded on the graben side by north-northeast to south-southwest trending faults. The tilt of these structural units, away from the graben axis, give rise to numerous closures in the crestal parts and established prolific hydrocarbon trends (Figure 4).

The Statfjord field represents an example of a closure located at the crest of such a megastructure bounded by a single fault trend (Figure 4). This field is characterized by relatively gently dipping (6° to 8° west) reservoirs that are little dissected by subsidiary faults (Figure 7A). The field has two major reservoirs with separate oil-water contacts. In contrast to the previous example, the Oseberg oil and gas field, located on the opposite flank of the Viking graben (Figure 4), shows a steeper dip within the pre-graben sequence and a more complex development of the fault scarp (Figure 7B). The main reservoir is Middle Jurassic sandstones, which contain a hydrocarbon column of more than 500 m (1,640 ft). In the westernmost fault compartment, the Middle Jurassic is eroded, and a Lower Jurassic sandstone forms the reservoir. The fault between the two reservoirs is sealing.

The individual megablocks are generally segmented by another set of normal faults that are restricted to the interior of the block and do not bound the mega-unit (Figure 4). As a result, the picture becomes more complex, where variations in displacement magnitude, abrupt changes in fault orientation, termination of blocks, and the junction of different fault trends all result in closures characterized by a lack of uniqueness in structural expression.

The Gullfaks field is an example of such a complexly faulted structure (Figure 7C). This rotated, northeast to southwest oriented megastructure is dissected by sealing and non-sealing internal faults, producing from four different sand intervals. The Sleipner field, located in the southern Viking graben, is another giant confined to complex structural features. The hydrocarbons occur in shallow, marine sandstones of Callovian-early Oxfordian age, with a different oil to water contact in the two main fault compartments (Figure 7D). The Sleipner field, while classified as a structural-unconformity trap, contains Zechstein salt that extends into the southern Viking graben. The structural style in this area indicates some effect of mobile salt.

The giant Troll field, covering some 700 sq km (270 sq mi), is the only example of a field having its reserves confined to more than one megastructure (Figure 7E). The oil and gas occur in Upper Jurassic sandstones with a common oil to water contact in all fault compartments.

*Timing of Events*

An important factor in assessing the hydrocarbon potential of the structures is the sequence of events and the trap preservation through time. In the Viking graben, oil was generated from Upper Jurassic shales from mid-Cretaceous time to the Miocene, in the deepest parts of the graben, and gas from the Oligocene to Recent. Thus, the sealing of structures in Upper Cretaceous time implies that along the flanks of the graben, trap formation generally predates the onset of hydrocarbon generation. To the east of the Viking graben, the critical dip of reservoir units away from the graben axis was reduced because of the Miocene and lower Pliocene uplift of the Norwegian mainland. As shown in Figure 6, most of the megablocks in this area still maintain a gentle dip toward the basin margin. Permanence of the structural-unconformity traps and their immediate proximity to a mature source rock, therefore, contribute to the abundance of hydrocarbons in the hinge areas.

*Morphology and Closure*

Within the Viking graben province, the structural-unconformity traps exhibit individual variations in both type of closure and morphology. Principal closures are produced either by truncating the reservoir at the base Cretaceous level, or directly against the downthrown side of the faults (Figure 7). A combination of the two closure types is quite common, especially in fields producing from more than one reservoir below the base Cretaceous unconformity.

Figure 7 shows that a rounded morphology at the base Cretaceous level characterizes the structural-unconformity traps. Along the northwestern flank of the Viking graben, however, where the stratigraphically deepest erosion took place, geologists have encountered traps with a dipping, pre-graben sequence and a flat morphology at base, Cretaceous level (Figures 7C and 7F).

*Producing Horizons*

Triassic and Lower-, Middle-, and Upper-Jurassic sandstones form the reservoir horizons in the structural-unconformity traps. In general, these fields are producing from only one sandstone reservoir, which in most cases belongs to the Middle- or Late-Jurassic time series. As these sandstones either directly underlie, or intercalate with the Upper Jurassic shales, this enhances the migration of hydrocarbons expelled from the source beds into the sandstones. The hydrocarbons associated with this trapping style always tend to accumulate in the youngest sand below the base Cretaceous unconformity. This accumulation implies that in areas covered by Upper Jurassic, shallow marine sands, geologists should regard as low the chance of having hydrocarbons in Middle Jurassic or older reservoirs, even though their reservoir quality might be better.

Structural traps producing from sands of more than one generation are restricted to the megablocks located close to the graben axis, where vertical displacement occasionally exceeds 2,000 m (6,560 ft) at base Cretaceous level (Figure 6). As a result of the movements, upfaulted blocks were severely eroded, and the different Jurassic and Triassic sand intervals were placed against a mature Upper Jurassic source either on the eroded crestal parts, or along the faultscarps (Figure 7).

## Salt-Supported Traps

The fields having their reserves confined to salt-supported traps are concentrated in the southwesternmost part of the Nor-

Figure 8. *Cross-sections of salt-related traps in the Norwegian North Sea. Reservoir rocks are represented by Upper Cretaceous-Danian chalk and Upper Jurassic and Paleocene sandstones (dotted) with main hydrocarbon accumulations in black.*

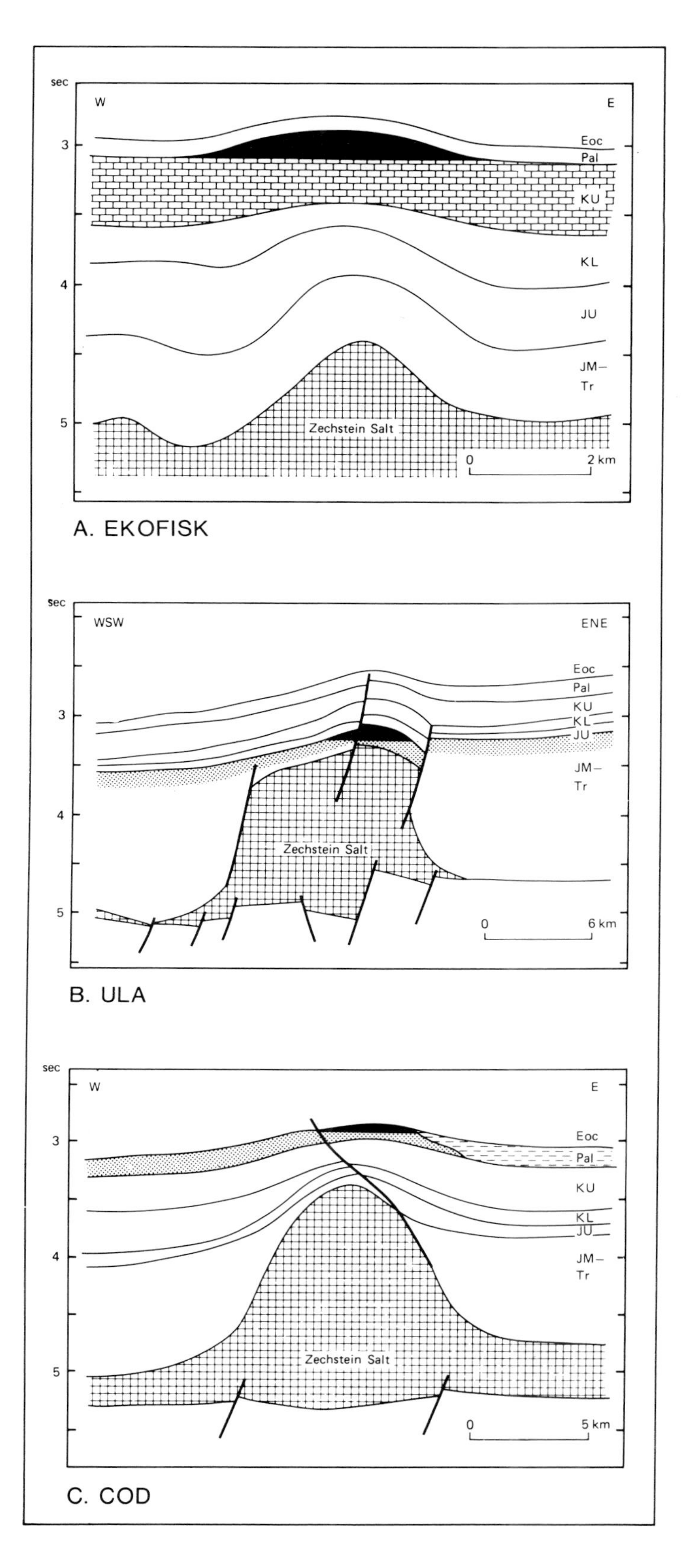


wegian sector, including the spectacular Ekofisk cluster of fields (Figure 4). The total amount of recoverable reserves within this type of trap is 200 million cu m (1.3 billion bbl) of oil and 215 billion cu m (7.6 tcf) of gas, of which more than 70% is accounted for by the Ekofisk field. Upper Cretaceous-lower Tertiary chalk contains 93% of the reserves confined to this trap category. Tertiary and Upper Jurassic sandstones provide additional reservoirs. The chalk interval forms a blanket deposit in the southern half of the Norwegian North Sea, which is basically the same area that is covered by the Zechstein salt (Figure 6). The salt basin, with thicknesses exceeding 1,000 (3,280 ft) in the southern part, has all the classical elements: pillows, piercements, turtle-backs, and rim synclines related to phases of salt flow from the Triassic into the Tertiary. Although the salt movements have created a large number of closures at top chalk level, major parts of the area remain unproductive all over the southern Norwegian North Sea basin.

The salt-related traps producing from chalk reservoirs meet four main requirements. They occur in an area

1. with abnormally high chalk porosity;
2. affected by several periods of tectonism and salt migration;
3. with a very favorable position for trapping hydrocarbons generated from mature Upper Jurassic source beds in the Central graben; and
4. characterized by a favorable timing of the sequence of events.

Within the Central graben, the chalk interval thickness exceeds 800 m (2,625 ft), but reduces to less than 200 m (660 ft) along the eastern flank of the graben, where the fields occur. In general, the chalk's primary porosity decreases from the Norwegian coast toward the west because of a deeper level of burial. The excellent porosity preservation along the Central graben, which allows the chalk to retain an average porosity of 30% to 40% at depths close to 3,000 m (9,850 ft), is probably a result of a high clay content that links to transgressive pulses in the overall regressive history and is opposed to compaction (Kennedy, 1980). Overpressure exists because of the rapid subsidence of the Central graben. The early oil migration into pores, could slow or stop the chemical compaction. Another factor that is essential for production from a chalk reservoir is microbrecciation, which improves the very low matrix permeability of this lithology. Fracturing results from tectonic stresses induced by the Tertiary growth of the structures on which the fields occur. The fracturing also allows the hydrocarbons to migrate vertically from an Upper Jurassic source through a Lower Cretaceous shale limestone sequence that in other places represents a sealing formation. The area with abnormally high average porosity coincides with a mature, Upper Jurassic source. In the deepest parts of the Central graben, oil generation from shales of this time interval started in the Paleocene, that is, shortly after the deposition of the chalk.

A map view shows that the salt-supported traps display, at top chalk level, elliptically shaped closures with their long axis oriented northwest to southeast, parallel to the Central graben

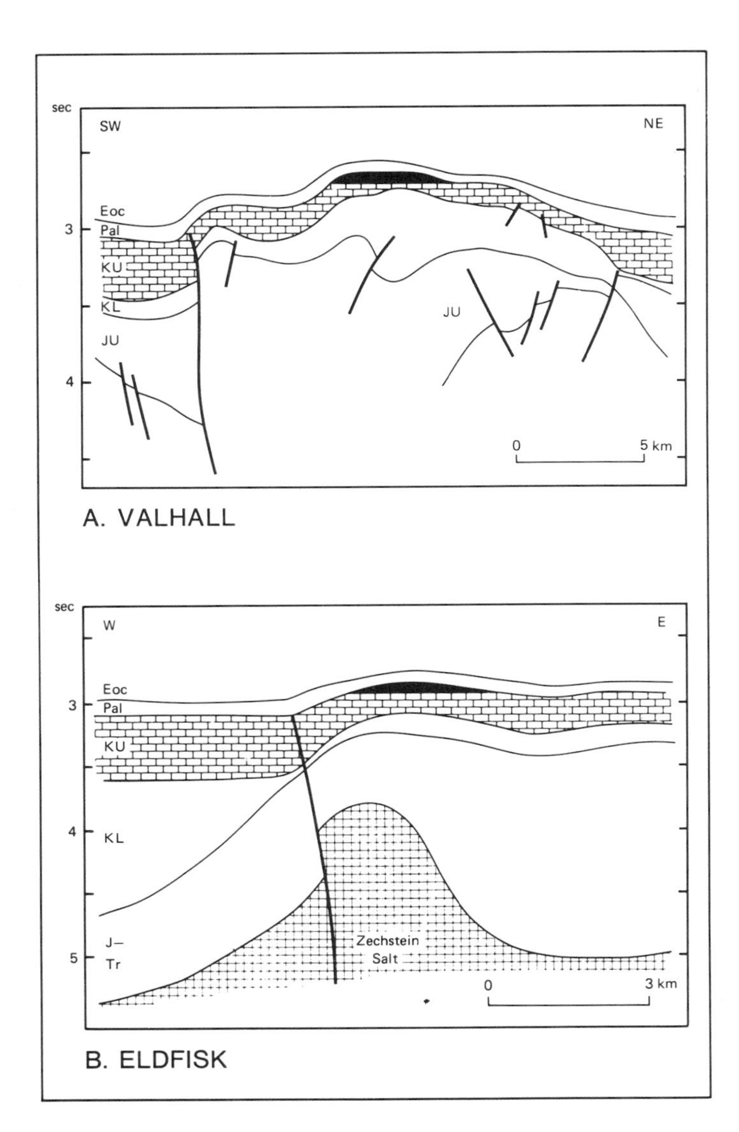


**Figure 9.** *Cross-sections of traps related to Upper Cretaceous-Paleocene inversion. Upper Cretaceous-Danian chalk forms the reservoir. Main hydrocarbon accumulations in black.*

(Figure 4). Figure 6 shows that the large-relief piercement structures, which are evidence of continued halokinetic movements, have not proved productive in the North Sea basin.

The Ekofisk field, characterized by a simple geometry, shows the top chalk level with about 240 m (780 ft) of closure over an area of 50 sq km (20 sq mi; Figure 8A). An overpressured Paleocene shale above the chalk seals a hydrocarbon column of 305 m (1,000 ft), which extends below the spill point of the reservoir. The gross reservoir units show only minor changes in thickness across the structure, indicating that main halokinesis took place in post-Danian time.

The recoverable reserves in the salt-supported traps with Upper Jurassic and Tertiary reservoirs are generally small. The Ula field, located northwest of the Ekofisk province (Figure 4), has recoverable reserves of 25 million cu m (160 million bbl) oil equivalent (Figure 8B). In this salt-supported trap, the reservoir is represented by shallow marine sands that accumulated along the flanks of the Central graben in Late Jurassic time. Their thickness is strongly affected by syn- and post-depositional halokinetic movements.

Paleocene submarine sands, building out from the East Shetland Platform, have their largest distribution in the U.K. part of the Central graben, barely reaching the Norwegian waters (Héritier et al, 1979). The Cod field on the Norwegian side of the sector line (Figure 4), which contains 11 million cu m (70 million bbl) oil equivalent, is located above a salt dome close to the pinch-out of the Paleocene reservoir sand (Figure 8C).

### Traps Related to Compression

The Hod, Valhall, Eldfisk, and Edda fields (Figure 4), with reserves that total some 96 million cu m of oil and 55 billion cu m (1.9 tcf) of gas, occur on an inversion feature within the Central graben. This structural feature forms a northwest to southeast oriented, asymmetric ridge with a steep, fault-bounded western edge and a gentle dip to the east. The fault to the west of the ridge acted as a major, normal, down-to-the-east fault during the Late Jurassic and parts of the Early Cretaceous. In Late Cretaceous time, the movement along this fault was reversed, and the sedimentary basin became a structural high with reduced thicknesses for the Upper Cretaceous-Danian chalk, which represents the reservoir (Figure 9A). The taphrogenic movement ceased in the early Tertiary, and halokinetic activity, which continued sporadically throughout the Tertiary, was intimately linked to the structural inversion and trap formation along this trend (Figure 9B). This activity gave rise to a very complex reservoir geometry. However, these movements also fractured the chalk and the underlying Lower Cretaceous shale/limestone sequence to create avenues for hydrocarbon migration from an Upper Jurassic source in the Central graben.

### Stratigraphic Traps

In general, stratigraphic trap fields occur along the southern part of the Viking graben. They are mainly gas-bearing, and the reserves occur in Paleocene and Eocene submarine sandstones with the pools typically located at depths between 1,700 and 2,100 m (5,570 and 6,890 ft). The Cod field, located on the east flank of Central graben (Figure 4), is an exception, as is a gas-discovery in Lower Cretaceous (Aptian-Albian) sandstones in the northeasternmost part of the Norwegian North Sea. The total proven reserves amount to some 40 million cu m of oil and 350 billion cu m (1,200 tcf) of gas. The Frigg Field, the only giant belonging to this category, contains more than 60% of the gas, while the Balder field contains 88% of all the oil.

The Paleocene and Eocene sediments include a large amount of clastic material that originated in the west and that, during periodic falls of eustatic sea level, was brought into the deep water of the basin by turbidity currents, creating submarine fan complexes. In general, these sands are wedging out along the eastern flank of the Central and southern Viking grabens. The hydrocarbons within the Tertiary reservoirs were, presumably, generated from a Jurassic source, and the fields occur in areas where oil and gas generation from Upper Jurassic shales started in Paleocene and Miocene time, respectively. Eocene marine shales seal the reservoir sands. The fields generally occur above the fault complex separating the Viking graben and the Utsira High (Figure 4), which was active through the Mesozoic and reactivated during Paleocene-Eocene time (Pegrum, 1984). The

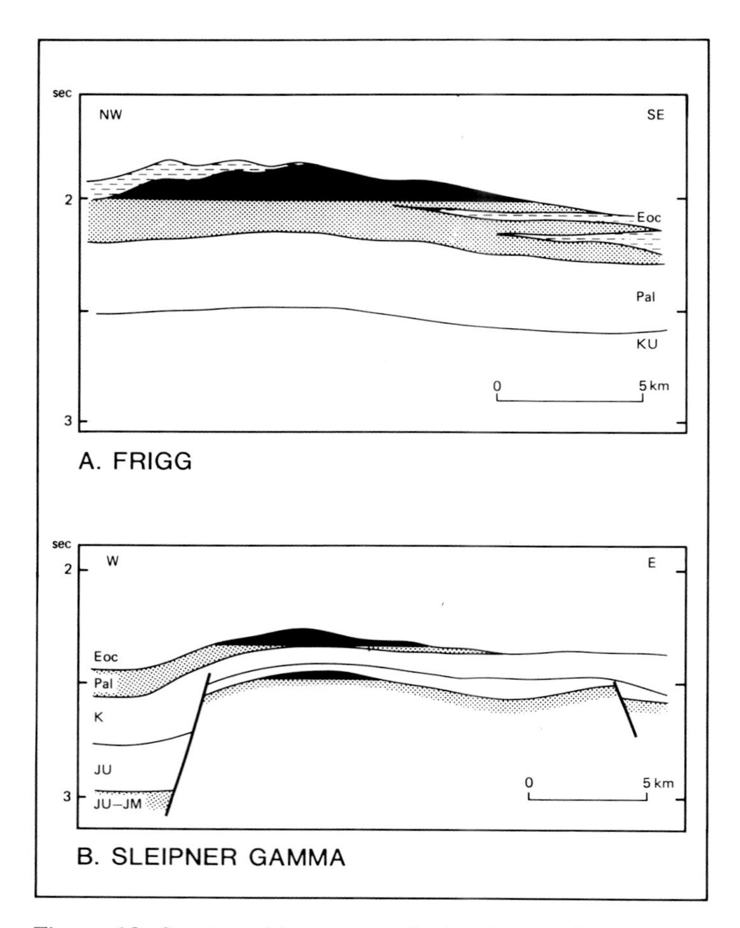


**Figure 10.** *Stratigraphic traps producing from Paleocene and Eocene sandstone reservoirs. The depositional topography is enhanced because of draping and differential compaction of sands above deep-seated structural features. Main hydrocarbon accumulations in black.*

late tectonic movements probably created fractures in the Cretaceous sequence, which is generally thin along this trend, to form avenues for the vertical migration of hydrocarbons. The Frigg field, however, is located close to the deepest part of the Viking graben, and hydrocarbons have migrated through a Cretaceous section more than 1,500 m (4,900 ft) thick, consisting primarily of shales with some limestone stringers.

Depositional topography and subsequent sub-marine erosion resulted in a mounded upper surface of the reservoir sequence. In the Frigg and Balder fields, the oil and gas are trapped stratigraphically within these mounded closures (Héritier et al, 1981; Sarg and Skjold, 1982). The Frigg field overlies a deep, complex Jurassic faulted block, and the depositional topography became enhanced by draping and differential compaction of sands. This field represents, at the top reservoir level, a closed area of 115 sq km (44 sq mi), with a vertical closure of 170 m (558 ft) (Figure 10A).

The Sleipner Gamma field (Figure 4) is located near the updip pinch-out of Paleocene sand. However, the trapping also resulted from the early Tertiary rejuvenation of the Viking graben border fault, which gave rise to a well-defined closure at this level (Figure 10B). The Cod field, located on the eastern flank of the Central graben (Figure 4), is another example of a composite trap that results from the combination of Paleocene sand pinch-out and post-Paleocene halokinetic movements (Figure 8C).

The only major discovery to date in stratigraphic traps with reservoirs other than Tertiary sands occurred in the northeastern part of the North Sea basin at the eastern flank of the Sogn graben. In this area, the reservoir is represented by massive sands of Aptian-Albian age in isolated channels capped by shales and located above a mature Upper Jurassic source rock.

## CONCLUSIONS

Exploration has established four major trapping situations in the Norwegian North Sea:

1. The structural-unconformity traps contain nearly three-quarters of the total reserves. The fields are located in hinge areas represented by the uplifted and strongly tectonized shoulders of the rotated platforms flanking the Viking graben. The hydrocarbons occur below a base Cretaceous unconformity in Triassic, Lower-, Middle- and Upper-Jurassic age sandstone reservoirs.
2. The salt-supported traps contain 11 % of the total reserves. The fields are concentrated along the eastern flank of the Central graben, including the Ekofisk cluster of fields. The hydrocarbons have been found in Upper Cretaceous-Danian chalk and Upper Jurassic and Tertiary sandstones.
3. Traps related to compression contain 4% of the total reserves. All fields are located on a northwest to southeast oriented, asymmetric ridge within the Central graben, which resulted from an Upper Cretaceous-lower Tertiary reversal of movements. Taphrogenic movements combined with halokinetic activity resulted in the closures where Upper Cretaceous-lower Tertiary chalk forms the reservoir.
4. The stratigraphic traps contain 11% of the total reserves. Main reservoirs are the Paleocene and Eocene submarine fans that originated in the west and are wedging out along the eastern flank of the Central and Viking grabens. The hydrocarbons are trapped near sand pinch-outs in closures combining depositional topography and later draping and differential compaction over deep-seated structural features.

The distributions of oil and gas fields are intimately linked with the Mesozoic rift system of the Viking and Central grabens, which, in turn, controlled the distribution and maturation of the Upper Jurassic source rock.

Almost 85% of the proven, recoverable reserves belong to the Viking graben province. The uplifted, tectonized shoulders of this graben provide many structural traps below a base Cretaceous unconformity. Permanence of the traps and their immediate proximity to a mature source contribute to the abundance of hydrocarbons along these trends.

The Central graben province is completely dominated by the salt-related, Ekofisk type of fields that clustered in the southwesternmost part of the Norwegian sector.

The concentration of fields along specific trends demonstrates that within this basin, characterized by a complex and highly varied geological history, all the factors necessary for petroleum come together only in a few areas in an optimum manner for economically significant deposits. In Norwegian waters, large areas east of the graben shoulders have remained unproductive.

The large fields influence the proven reserves. Because they are geographically large, such fields are generally confined to easily mappable features. In areas that combine the main requirements for hydrocarbons contained within structural-unconformity traps, the most spectacular closures have already been drilled. Regarding the salt-supported Ekofisk type of field, which was the first play type established in Norwegian waters, approximately 45% of the reserves have been produced, and only an insignificant percentage has been found in such traps over the last years. Hence, future exploration must be directed toward smaller and more subtle features, which will result in a significant decline in proven reserves per year.

Compared to other producing basins, exploration companies have not drilled many wells in the North Sea. However, the concentration of wells is highest in the geologically favorable areas that have proved to be hydrocarbon-bearing.

As these are the principal areas also for future resources, exploration in the North Sea has been fairly thorough, and most areas south of lat. 62°N are in an advanced stage of exploration.

## REFERENCES CITED


Barnard, P. C., and B. S. Cooper, 1981, Oils and source rocks of the North Sea area, *in* L. V. Illing and G. D. Hobson, eds., Petroleum geology of the continental shelf of North-West Europe: Heyden & Son, London, p. 169–175.

Héritier, F. E., P. Lossel and E. Wathne, 1979, Frigg field—large submarine—fan trap in Lower Eocene rocks of North Sea Viking graben: AAPG Bulletin 63, p. 1999–2020.

——, ——, and ——, 1981, The Frigg gas field, *in* L. V. Illing and G. D. Hobson, eds., Petroleum geology of the continental shelf of North-West Europe: Heyden & Sons, London, p. 169–175.

Kennedy, W. J., Chalk sedimentation in the southern Norweigian offshore, *in* The sedimentation of the North Sea reservoir rocks: Norwegian Petroleum Society, Geilo, p. 1–55.

Lovell, J. P. B., 1984, Cenozoic, *in* K. W. Glennie, ed., Introduction to the petroleum geology of the North Sea: Blackwell Scientific Publications, London, p. 151-169.

Norweigian Petroleum Directorate, 1984: Annual report, p. 1–106.

Pegrum, R. M., 1984, Structural development of the southwestern margin of the Russian-Fennescandian Platform, *in*, Petroleum geology of the North European margin: Norwegian Petroleum Society, Trondheim, p. 359–370.

Sarg, J. F., and L. J. Skjold, 1982, Stratigraphic traps in Paleocene sands in the Balder area, North Sea, *in* M. T. Halbouty, ed., The deliberate search for the subtle trap: AAPG Memoir 32, p. 197–206.

Vollset, J., and A. G. Doré, eds., 1984, A revised Triassic and Jurassic lithostratigraphic nomenclature for the Norwegian North Sea: Norwegian Petroleum Directorate, Bull. 3, p. 1–53.

Ziegler, P. A., 1982, Geological atlas of western and central Europe: Elsevier, Amsterdam, p. 1–130.

# Basin Development and Hydrocarbon Occurrence Offshore Mid-Norway

Terje Hagevang[1]
Hans Rønnevik
*Saga Petroleum*
*Høvik, Norway*

The mid-Norwegian continental margin, bordered oceanward by Tertiary oceanic crust, has a thick sedimentary sequence ranging in age from Paleozoic to present. The Jurassic contains two source rocks, one gas- and one oil-prone, and two sandstones of fair to excellent reservoir characteristics. The structures are related to Middle and Late Jurassic tectonisms, capped by Upper Jurassic and Cretaceous shales and marls. Exploration drilling started in 1980. A number of discoveries have been made, and production is expected in the 1990s.

[1]Presently with: Amerada Hess Norway, Oslo, Norway.

## INTRODUCTION

Extensive geophysical exploration mapping of the Møre–Lofoten Shelf (Figure 1) started in 1969, while exploration drilling started in 1980, with Saga Petroleum making the first gas discovery in the third well. The geophysical mapping and exploration drilling have revealed the structural and stratigraphical evolution of an area that 20 years ago was thought to consist of crystalline rocks covered by glacigene deposits.

Geologists separate the Møre–Lofoten continental margin into three bathymetric areas: the Møre Shelf, the Nordland Shelf and the Lofoten Shelf. The Møre Shelf is 60 km (37 mi) wide, with water shallower than 300 m (984 ft). The continental slope is steep, and the shelf edge is unstable with several slump scars. The continental rise ends in the Norway basin.

The Nordland Shelf is 120 km (75 mi) wide and the water is generally deeper than 300 m (984 ft). However, with several shallower banks, the continental slope is gentle and the shelf edge stable. The Vøring Plateau constitutes the western part of the margin.

The Lofoten Shelf has a water depth shallower than 200 m (656 ft) and the width of the shelf is narrower than 60 km (37 mi). A steep, unstable continental slope ends in the Lofoten basin.

The Møre–Lofoten Shelf was glaciated in the Quaternary, and most of the shelf is covered with glacigene deposits. In addition, deep, glacially eroded trenches occur close to and parallel with the coast.

## STRUCTURAL SETTING AND DEVELOPMENT

The Møre–Lofoten continental margin (Figure 2) is a passive margin, bordered oceanward by Tertiary oceanic crust that postdates anomaly 25 time (57 Ma).

Early Tertiary lava flows form several escarpments along the margin edge off Møre. The most landward escarpment is the Færøy–Shetland Escarpment, with flow basalts masking the oceanward extent of the continental crust. West of the Færøy–Shetland Escarpment there is a zone of seaward-dipping reflectors beneath the flatlying basalts. The nature of these rocks and the crust beneath is uncertain. The flatlying basalts are interpreted to be subaerial extrusions.

The transition between the continental and oceanic crust off Nordland occurs on the outer part of the Vøring Plateau. The landward boundary of the outer part of this plateau is marked by the Vøring Plateau Escarpment. This escarpment acted as a tectonic hinge zone from the middle Tertiary. Seaward of the escarpment is thin sedimentary cover over subaerial, extruded basalts.

There are three domal uplifts that geologists associate with volcanic activities. The two southern domes are of Paleocene–Eocene age (Caston, 1976), while the northern dome is of Oligocene age (Rønnevik, Jørgensen, and Motland, 1979). Seaward-dipping reflectors occur in a northeast to southwest zone on the outer Vøring Plateau.

East of the Færøy–Shetland and Vøring Plateau Escarpments, there is a zone of hummocky volcanic rocks at the Paleocene level. Geologists relate this basalt front to submarine extrusions, while Tertiary dikes and sills intrude Cretaceous strata landward of the basalts.

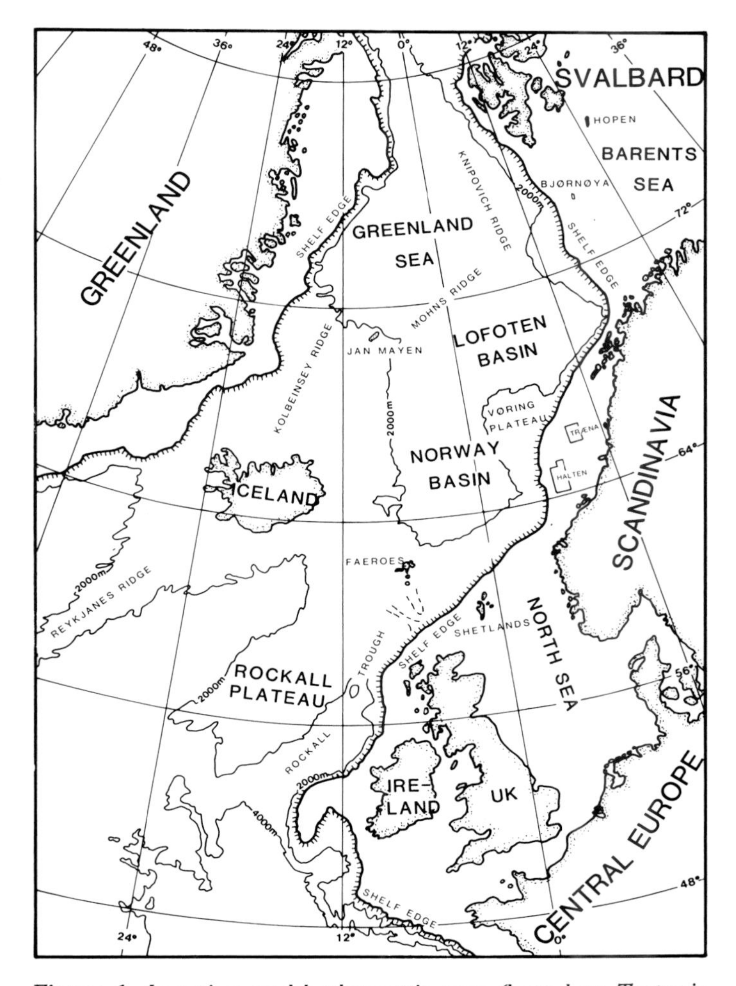


Figure 1. *Location and bathymetric map (based on Tectonic Map of Europe and Neighbouring Areas, Moscow 1979). The designated Halten and Traena tracts are situated on the Møre–Lofoten Shelf.*

Researchers have debated the nature of the crust underlying the Møre and Vøring basins. Suggestions have included Permian (Russel, 1976) or Cretaceous age (Bott, 1975; Smythe, 1983), oceanic crust or thinned continental crust (Hinz, 1972). The reflection seismic data do not support oceanic crust of Cretaceous age. Oceanic crust of Permian age cannot be disproved, but refraction seismic surveying suggests thinned continental crust. Bøen, Eggen, and Vollset (1984) have estimated a thinning factor of 3. This estimate is twice the thinning calculated for the North Sea basin.

The Møre basin is limited landward by the crystalline rock of the Norwegian coast, while the Vøring basin is limited landward by the sedimentary Trøndelag Platform.

The development of the Møre–Lofoten continental margin is dominated by the post-Jurassic subsidence of the Møre and Vøring basins. The Trøndelag Platform acted as a major, stable platform during the Cretaceous. The platform is rimmed by the Nordland Ridge in the north and Frøya High in the south. The area in between the two ridges, termed the Halten Terrace, marks the transition between the Møre basin and Trøndelag Platform. The Vestfjord basin north of the platform underwent

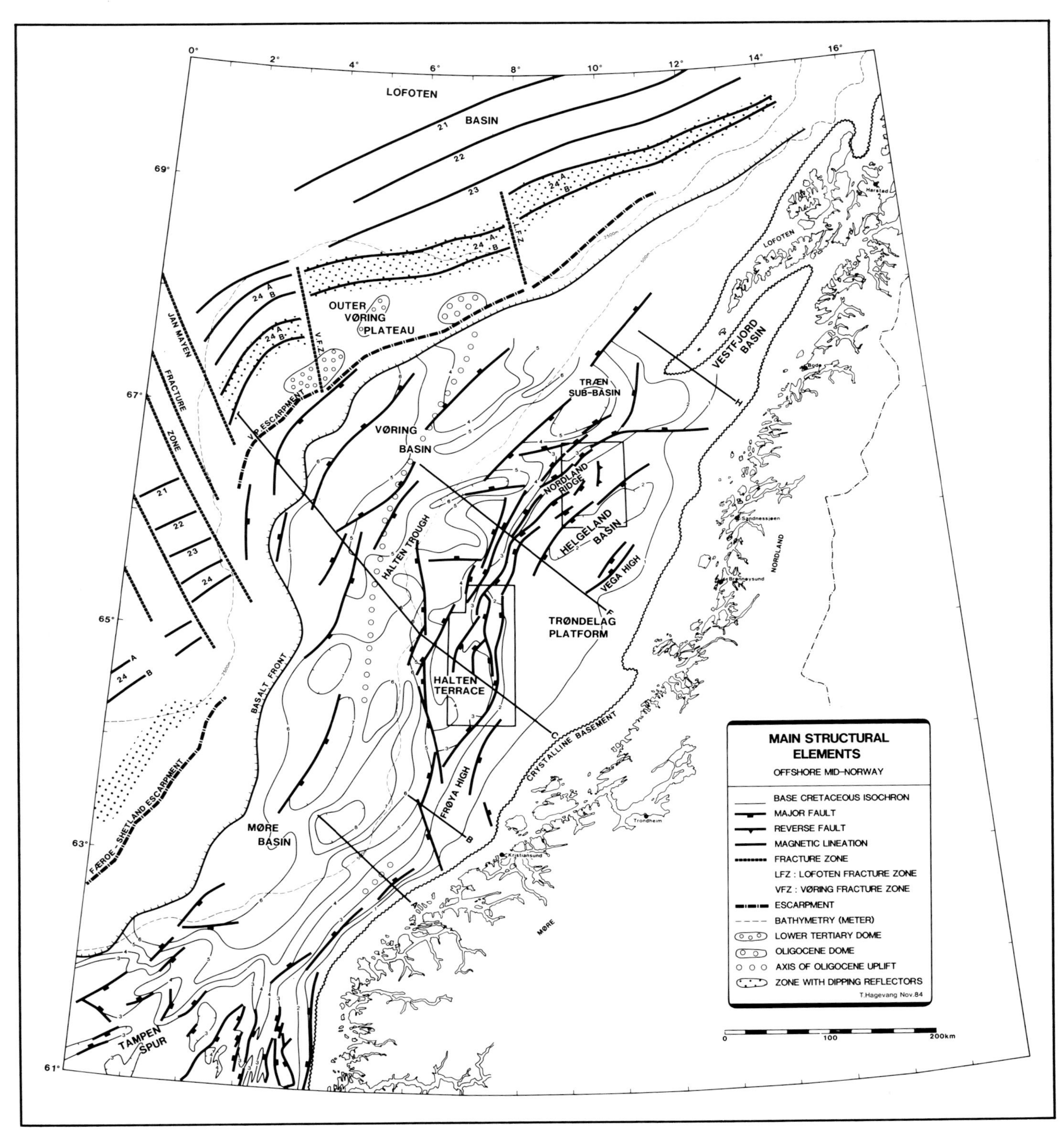


Figure 2. *Main structural elements: offshore mid-Norway. Locations of geoseismic profiles are shown. From the following sources:*
*Cretaceous structures: Saga interpretation and Bøen, Eggen and Vollset, 1984;*
*Magnetic anomalies and fracture zones: Hagevang, Eldholm, and Aalstad, 1983 and Skogseid, 1983;*
*Basalt front and Tertiary domes: Bøen, Eggen and Vollset, 1984;*
*Dipping reflectors: Modified after Talwani, Mutter and Hinz, 1983.*

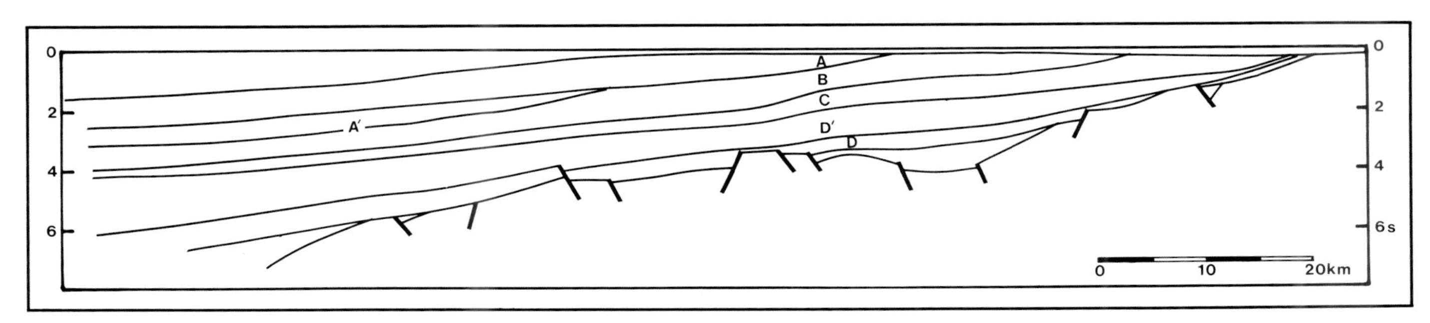


Figure 3. *Profile A. O, Pliocene; A, Lower Miocene; A', mid-Oligocene; B, top lower Eocene; C, base Tertiary; D', mid-Cretaceous; D, base Cretaceous; E, Middle Jurassic (after Rønnevik and Navrestad, 1977).*

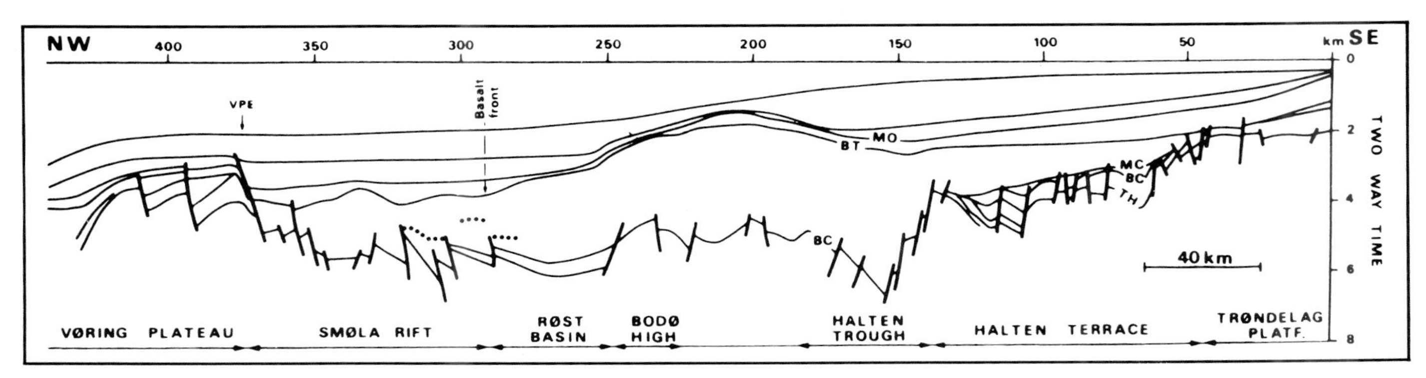


Figure 4. *Profile C. MO, mid-Oligocene; BT, base Tertiary; BC, base Cretaceous; MC, mid-Cretaceous; TH, top Hettangian; ..., volcanics; VPE, Vøring Plateau Escarpment (Bøen, Eggen, and Vollset, 1984).*

minor post-Jurassic subsidence.

The Møre basin is approximately 200 km (124.3 mi) wide. The basin axis is northeast to southwest oriented and centrally situated. The base Cretaceous reaches a depth of more than 8 sec two-way time along the axis, and the thickness of Cretaceous strata reaches more than 6 km (4 sec two-way time). The Cretaceous fill in the Møre basin has an onlap relation toward the basal sequence boundary in the southeast (Figure 3). The Norwegian mainland acted as a major provenance area during the Cretaceous.

Geologists know that Jurassic strata onlap toward the crystalline basement along the southeastern flank of the Møre basin. The present mainland and intrabasinal highs acted as provenance areas. The pre-Jurassic strata are thin in the Møre basin.

The northeastern boundary of the Møre basin is marked by a north to south fault zone with major growth during the Cretaceous age (Figure 2). The fault zone consists of north-northwest to south-southeast fault segments offset by northeast to southwest fault segments. The late Kimmerian tectonic phase affected the entire Møre basin. The northeast to southwest faults along the southeastern boundary of the Møre basin may have undergone shear movements, and the northeast to southwest offsets along the boundary fault zone in the northeast are probably of similar nature.

The Møre basin continued to grow as a separate depocenter in early Tertiary. Weak compressional structuring occurred in the coastal part of the Møre basin in the middle Tertiary.

The width of the Vøring basin is comparable to the Møre basin, and the basin is bounded landward by the Trøndelag Platform. The northeast to southwest basin axis corresponds to the Cretaceous depositional center and occurs close to the Trøndelag Platform. In the axial zone, the Cretaceous sequence reaches a thickness of more than 4.5 km (3 mi). An east-northeast to west-southwest fault zone separates the axial part into two basins: the Halten Trough and Træna subbasin.

The Vøring basin received almost no sediments during early Tertiary, so that a starved basin situation is likely. In the middle Tertiary, a major north to south basin inversion, related to the compressional tectonics, occurred across the basin (Figure 4). Yet, geologists do not fully understand the pre-Cretaceous structural setting and development of the Vøring basin.

The axial parts of the Møre basin and Vøring basin interconnect over the Halten Terrace. This terrace includes a thick, Upper Triassic–Jurassic sequence and is bounded by north to south faults that showed major growth in the Cretaceous. Growth faulting along north to south faults occurred during deposition of this sequence. The northeast to southwest faults observed on the terrace are of late Kimmerian age.

The Trøndelag Platform was a stable tectonic area during Late Triassic and most of the Jurassic periods. The separation of the Trøndelag Platform and Halten Terrace started in the lowermost Jurassic.

The late Kimmerian tectonic phase resulted in uplift of the Nordland Ridge and Frøya High. During latest Jurassic and Early Cretaceous, these highs were heavily eroded. Most of the sediments above the crystalline basement were eroded on the Frøya High, leaving only small segments of asymmetric basins (Figure 5).

The late Kimmerian tectonic phase resulted in the dome-shaped Vega High on the Trøndelag Platform (Figure 6). The

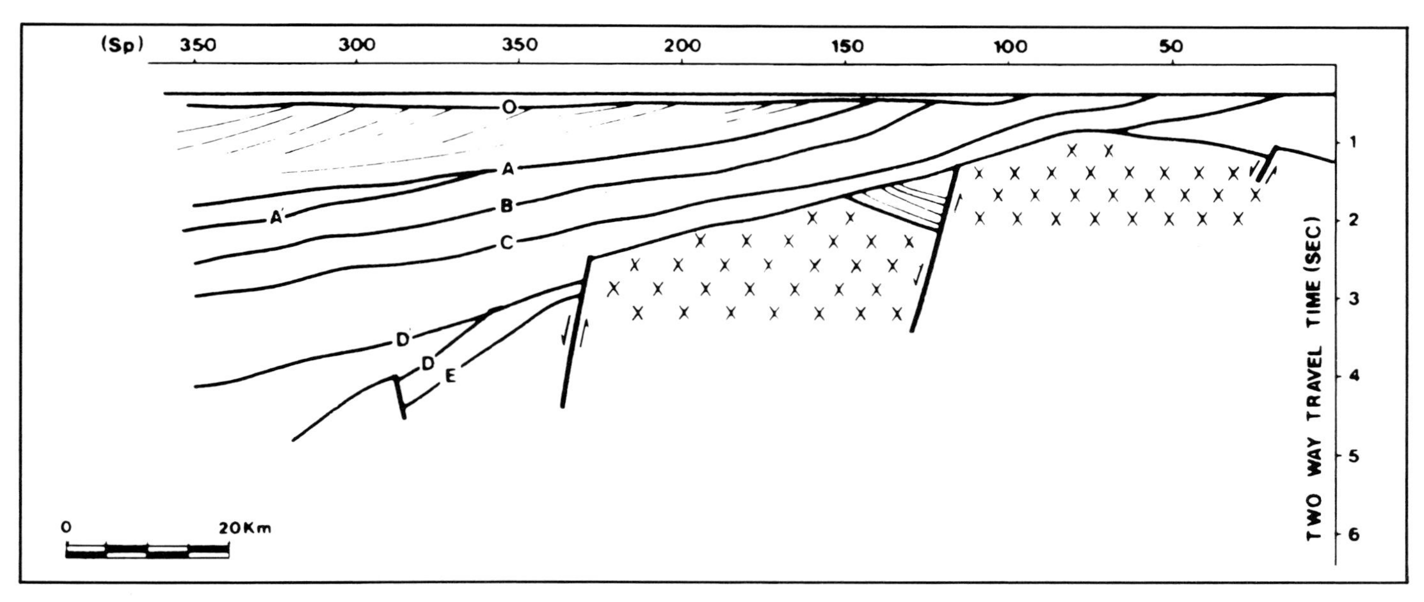


Figure 5. *Profile B. O, Pliocene; A, lower Miocene; A' , mid-Oligocene; B, top lower Eocene; C, base Tertiary; D' , mid-Cretaceous; D, base Cretaceous; E, Middle Jurassic (after Rønnevik and Navrestad, 1977).*

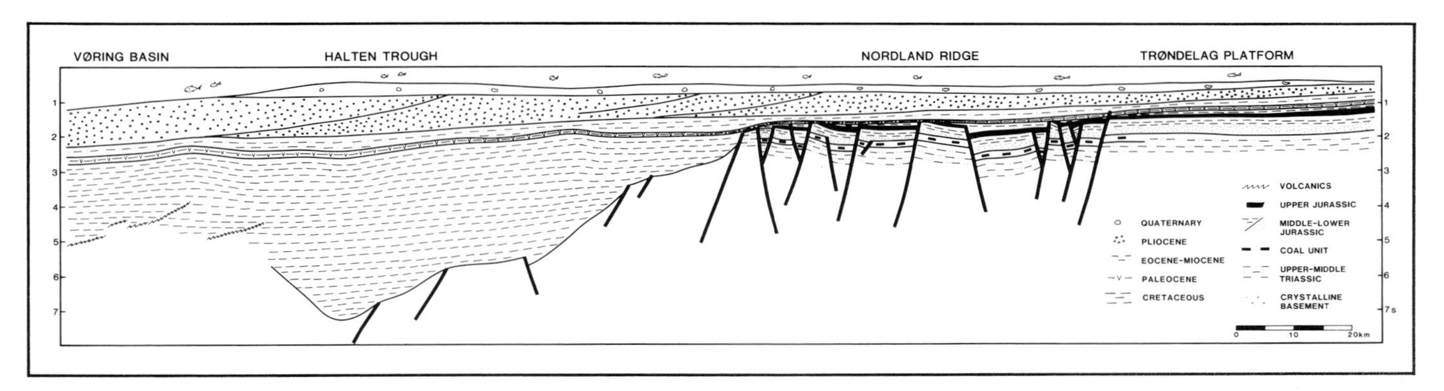


Figure 6. *Profile F.*

Cretaceous strata are thin on the platform. In the shallow Helgeland basin, a sequence thickness of 500–1,000 m (1,640–3,280 ft) occurs along the basin axis.

Thick, pre–Late Triassic strata are present on the Trøndelag Platform. However, the basin configuration and nature of the basin fill are largely unknown.

The Nordland Ridge was modified during a compressional tectonic event in the middle Tertiary. Reverse faulting along north to south shear faults occurred on the northern part of the ridge (Gabrielsen and Robinson, 1984).

The Vestfjord basin to the north of the Trøndelag Platform underwent minor subsidence during the Cretaceous and Tertiary, and, at these levels, the basin plunges into the Træna subbasin. In the Vestfjord basin there is a thick pre-Cretaceous basin fill of supposed post-Carbonifereous age (Figure 7), with the oldest sequences being deposited in asymmetric grabens. The basin was very little affected by the Kimmerian tectonic phases.

The whole shelf between Møre and Lofoten underwent rapid subsidence and progradation after the middle Tertiary compressional event. At the shelf edge, more than 1,500 m (4,920 ft) of Pliocene sediments occur.

The pre-Tertiary tectonic history of the Møre–Lofoten Shelf is similar to the tectonic development of the North Sea basin. The major difference is a larger stretching during the Kimmerian tectonic phases and a larger Cretaceous subsidence in the two major basins.

## STRATIGRAPHY AND FACIES

Geologists know the stratigraphy of the sediments that build the Møre–Lofoten Shelf, from exploration wells in the (1) southern part of the Møre basin, (2) Halten Terrace, (3) Trøndelag Platform, (4) Helgeland basin, and (5) Nordland Ridge (Figure 8). Permian rocks and a metamorphic basement were penetrated in the Træna area. Geologists, however, generally know very little about the stratigraphy of pre-Triassic rocks.

The Middle and the lower part of Upper Triassic consists of continental red bed sediments and halite. The uppermost Triassic and lowest part of the Jurassic consist of paralic sand/shale and coal sediments. The upper part of Lower Jurassic and the

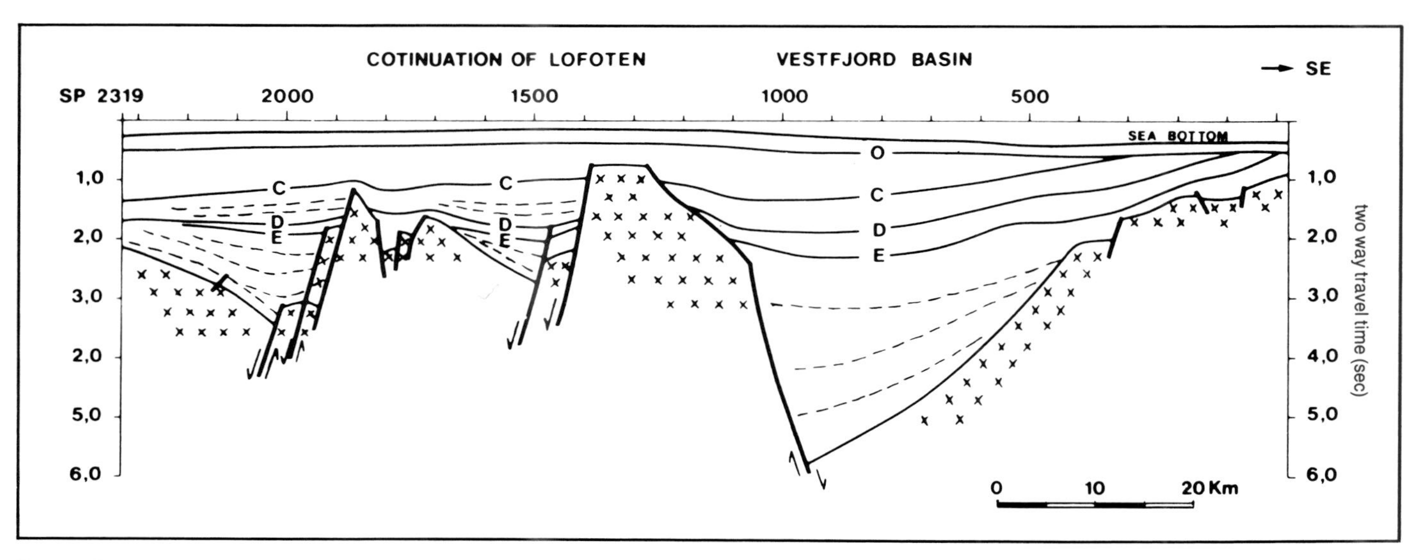


Figure 7. *Profile H. O, Pliocene; A, lower Miocene; A', mid-Oligocene; B, top lower Eocene; C, base Tertiary; D', mid-Cretaceous; D, base Cretaceous; E, Middle Jurassic (after Rønnevik and Navrestad, 1977).*

Middle Jurassic consist of marine shale and sand. The Upper Jurassic and younger sequences consist of shale in the basinal areas, while geologists know of the shallow marine sand at the Møre basin edge. Researchers know of several laterally consistent unconformities in the post-Triassic sequence.

Geologists likewise know of the post-Middle Triassic sedimentary sequence from several wells on the Halten Terrace and Trøndelag Platform (Figure 9). Figure 10 shows a generalized lithostratigraphic section for the Halten area that serves as a type section for description of the units.

### Triassic

Explorers have drilled a Triassic sequence of more than 2.5 km (2 mi), penetrating sediments down to Middle Triassic age. The lowest part of the drilled sequence consists of alluvial, continental red, silty claystones, and grey, fine-grained sandstones. These sediments are overlain by a 1,300-m (4,265-ft) thick evaporite sequence, consisting of two halite intervals, each of about 400 m (1,312 ft), separated by an anhydritic grey to brown claystone. These results suggest a sabkha-type environment.

This sequence is overlain by approximately 700 m (2,300 ft) of early Rhaetian rocks. Red-brown, silty claystones with minor thin sand beds appear, probably deposited in the distal parts of alluvial fans during an arid climate.

The upper part of the Triassic section comprises a grey bed interval consisting of fine- to medium-grained sandstones alternating with siltstones of lacustrine-fluvial origin. These findings indicate a transition to more humid conditions. The Triassic sequence observed in the Halten area continues northward on the Trøndelag Platform. The sequence is present on the Nordland Ridge except in the structurally very highest areas (Figure 8).

Generally, however, the Triassic sequence has poor to no source or reservoir rock potential.

### Jurassic

No marked lithological break occurs at the Jurassic–Triassic boundary, which lies within the lower part of a lithostratigraphic unit called the Coal Unit. This unit is close to 400 m (1,312 ft) thick in the Halten area and about half that in the Træna area. In front of growth faults on the Halten Terrace, the thickness is expected to reach 1 km (0.6 mi). The same facies is indicated along the whole shelf.

The Coal Unit is a sandstone, clay, and siltstone sequence interbedded with carbonaceous claystones, minor coal seams, and some siderite, dolomite, and limestone cemented beds. The lower part of the Coal Unit was deposited on the nonmarine part of a coastal plain that high-sinuousity river systems dissected. The cyclic sequence consists of well-sorted channel sands, overbank mudstones, and intra-channel, fresh-water swamp accumulations rich in plant material. The upper part of the Coal Unit was deposited in an estuarine to marginal-marine environment. The Coal Unit is equivalent to the Statfjord Group and the lower part of the Dunlin Group in the North Sea. The Lower Jurassic marine transgression started in Hettangian in the North Sea, in early Sinemurian in the Halten area, and in late Sinemurian in the Træna area (Figure 8).

The reservoir potential of the Coal Unit as a whole is ranked as poor because of a low sand content (less than 50%) and expected lateral discontinuity of the channel sands. Locally, the well-sorted, medium-grained channel sands can have good reservoir characteristics. However, early diagenetic siderite cementation and authigenic pore-filling kaolinite reduces the permeability (Karlsson, 1984).

A Lower Jurassic sandstone sequence with a thickness of 100–200 m (328–656 ft) overlies the Coal Unit. The lowest part of the formation consists of argillaceous siltstone with several very fine sandbeds. The sand content increases toward the top of the formation and several coarsening-upward cycles exist. The lithology of the upper part is very fine sandstones alternating with very coarse sandbeds and argillaceous siltstones.

The Lower Jurassic sandstone sequence is observed throughout the Møre–Lofoten Shelf. The sand varies in composition from quartz arenites to lithic wackes. The detrital clay is partly recrystallized to illite, and partly altered to siderite. Kaolinite is common, and a loosely packed network partially fills the pore space (Karlsson, 1984). Syndepositional formation of glauconite occurs in the upper part, indicating periods with low sediment supply. Postdepositional precipitation of pyrite associated with the organic detritus and siderite cementation occurs throughout the unit. Locally, dolomite cement has grown.

The marine transgression that began in the upper part of the Coal Unit continues into the Lower Jurassic sandstone. The boundary between the two units appears at the first occurrence of coal. Reworked carbonaceous materials are, however, found throughout the Lower Jurassic sandstone. The marine influence on the sedimentation increases upward, gradually changing the estuarine/marginal-marine environment into a shallow marine environment (Figure 11).

The fine-grained sandstones of the Lower Jurassic sandstone sequence are heavily bioturbated, giving admixed clays that destroy the reservoir quality. Early diagenetic siderite cementation associated with the clays causes local blocking of the primary porosity, effectively reducing the permeability. The uppermost part of the sequence is well-sorted marine sandstones with good reservoir characteristics (Figure 11).

A mudstone unit overlies the Lower Jurassic sandstone, and is as old as the Drake Formation in the northern North Sea. The thickness of the unit is about 100 m (328 ft) in the Halten area, and only a few meters in the Træna area and flanks of the Møre basin.

In the Halten area, the Drake Formation shows an overall upward-coarsening trend, consisting of four to five cycles. The cycles grade from claystones to siltstones in the lower part and from argillaceous siltstones to fine-grained sandstones with glauconite in the upper part. The topmost section is a tight, carbonate cemented bed rich in pelecypod shell fragments (Karlsson, 1984).

The formation represents offshore muds and sands deposited in an overall regressive phase. The fine-grained sandstones at the top of each unit have no reservoir qualities (Figure 11).

The Drake Formation is, in the Halten area, overlain by a massive, regressive sandstone complex called the Middle Jurassic sandstone, comparable with the Brent Group in the North Sea. The top of the sequence is marked by a major unconformity, referred to as the pre-Callovian or middle Kimmerian unconformity. Because of erosion, the thickness is greatly dependent on structural position, varying from 50 to 240 m (164 to 787 ft) on the Halten Terrace.

The Middle Jurassic sandstone sequence consists of three units. The lower unit consists of numerous thin-coarsening and fining-upward sequences, grading from silt or clays into poorly sorted and bioturbated fine-grained sandstones. Mica is abundant. The middle unit is a silty claystone with minor laminae of very fine sand. The upper unit is a clean, well-sorted, medium-grained sand with some coarse-grained and conglomeratic intervals (Figure 11).

The Middle Jurassic sandstone sequence was deposited in a marine environment with coastal sands prograding a stable shelf. The lower unit may be interpreted as an intertidal facies. The middle unit represents a transgressive event giving muddy deposits in a low-energy shelf with large, terrestrial input and probably restricted communication to open marine conditions. The upper unit represents the major regression, and the sands were deposited in a high-energy, wave dominated shoreface environment.

The lower sand unit is a poor to good reservoir rock depending on the degree of diagenesis. The cleaner upper sand unit has not had its good to excellent reservoir characteristics significantly reduced by the slight authigenic growth of carbonates, kaolinite and illite.

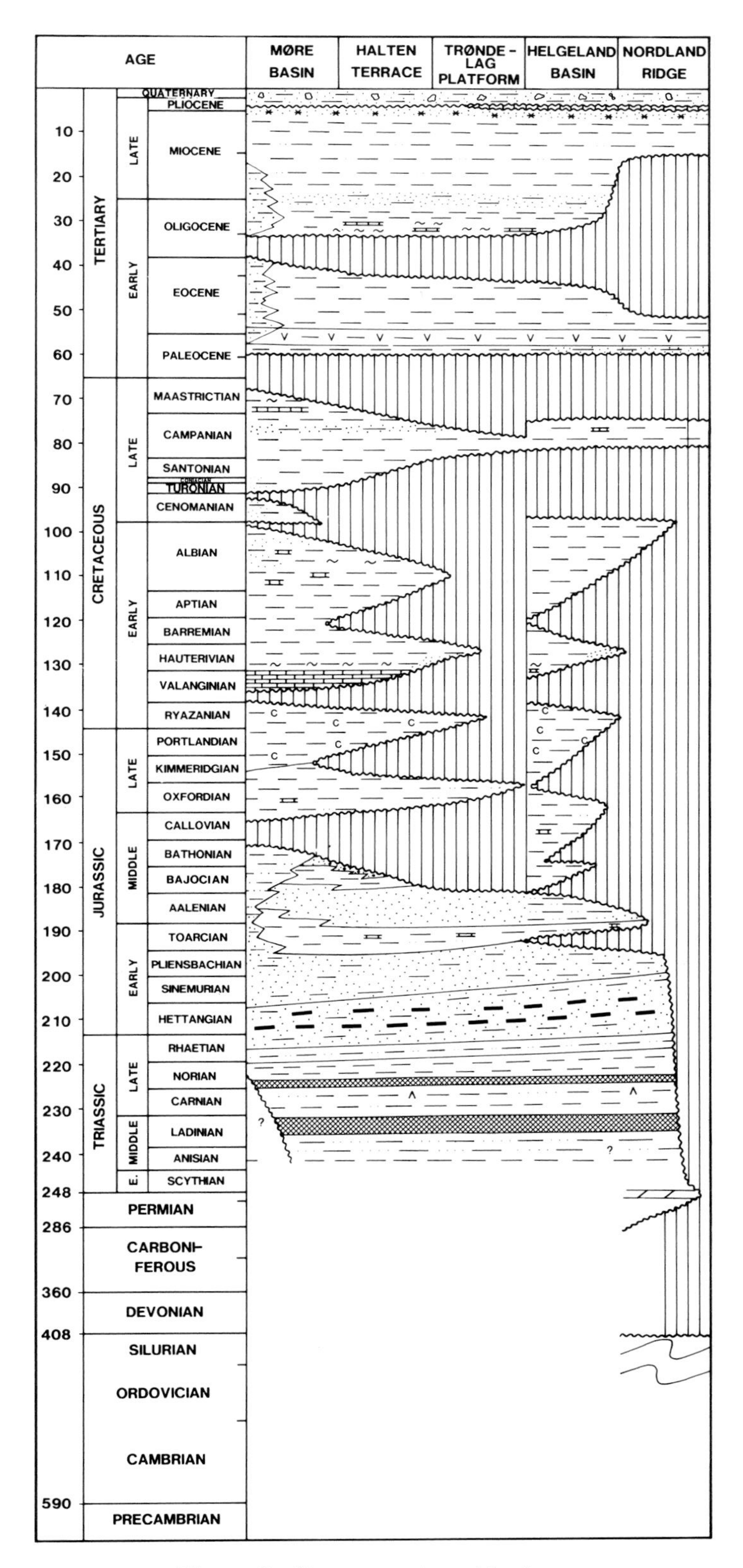


Figure 8. *Chronostratigraphic chart.*

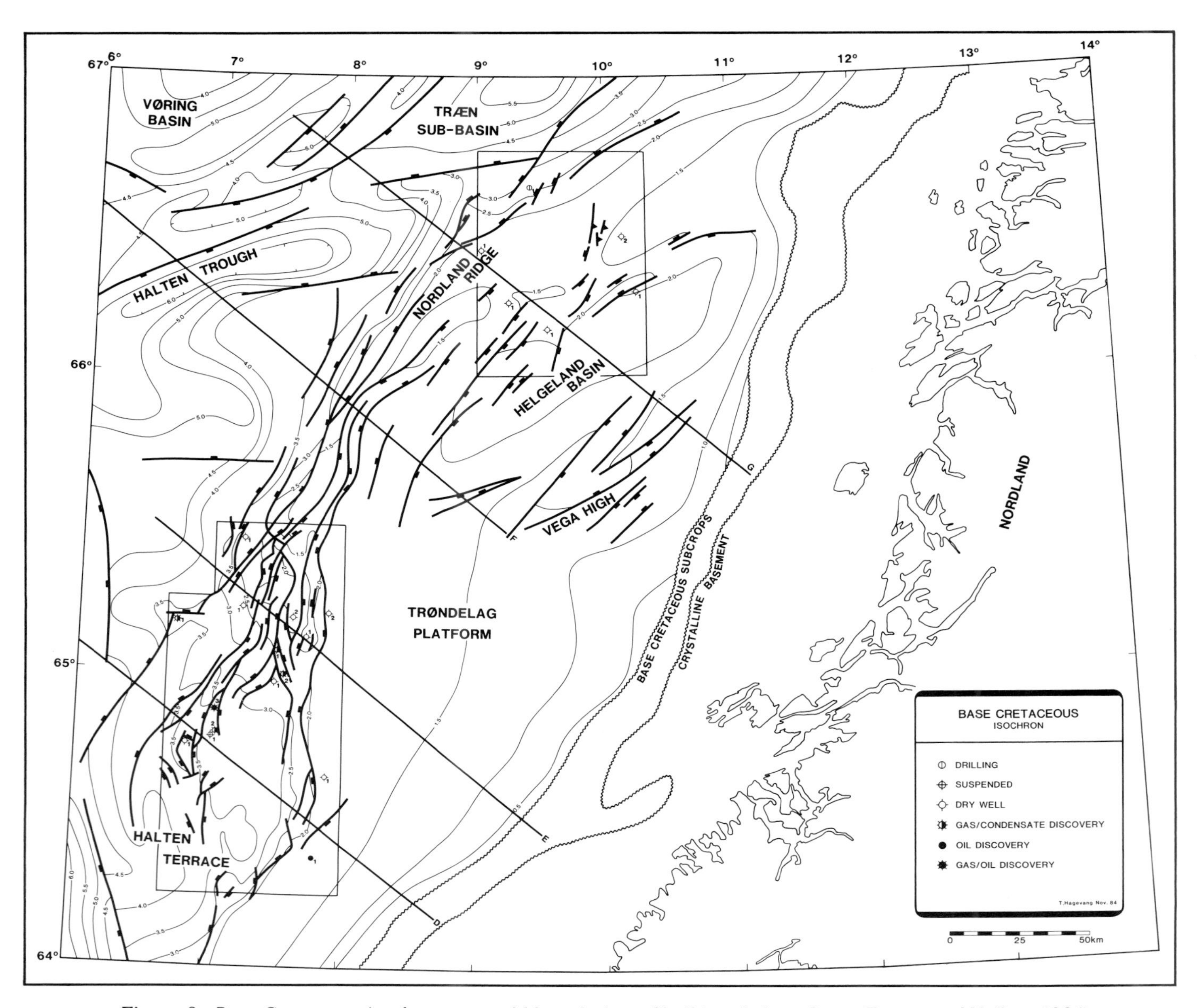


Figure 9. *Base Cretaceous isochron, central Møre–Lofoten Shelf (partly from Boen, Eggen, and Vollset, 1984).*

Generally, only the lower unit of the Middle Jurassic sandstone is preserved on the Trøndelag Platform and the major horsts of the Halten Terrace. In these cases, lying structurally high, the reservoir characteristics of the lower unit are enhanced.

The Middle Jurassic developed as claystones, in part silty and sandy in the Træna area and on the southern flank of the Møre basin (Figure 8).

The Upper Jurassic sequence rests on a regional, Middle Jurassic unconformity. The sequence ranges from Callovian to Ryazanian in age, and correlates with the Heather and Kimmeridge Clay Formations in the North Sea.

The Heather Formation consists of dark grey claystones, argillaceous siltstones, and scattered carbonate stringers or nodules, and developed as an overall, fining-upward sequence. The thickness is typically about 40 m (130 ft). The formation represents the onset of a regional transgressive period, depositing marine muds. The content of organic carbon is 0.5%–2%, and the sequence is only of moderate interest as a source rock.

The Kimmeridge Clay Formation consists of a uniform black shale/claystone, deposited in a marine basin with restricted water circulation and little clastic input. Anaerobic conditions prevented the decay of the organic material. The largest thickness of the Kimmeridge Clay Formation observed in the wells is about 60 m (200 ft), although it is anticipated to reach several hundred meters in the Møre and Vøring basins. The Kimmeridge Clay Formation is an excellent, oil-prone source rock.

The top of the Jurassic is marked by the late Kimmerian unconformity, a major unconformity only in the structurally highest areas, that also has an unconformity between the Heather and Kimmeridge Clay Formations. This intra, Upper Jurassic unconformity may be slightly older in the Helgeland

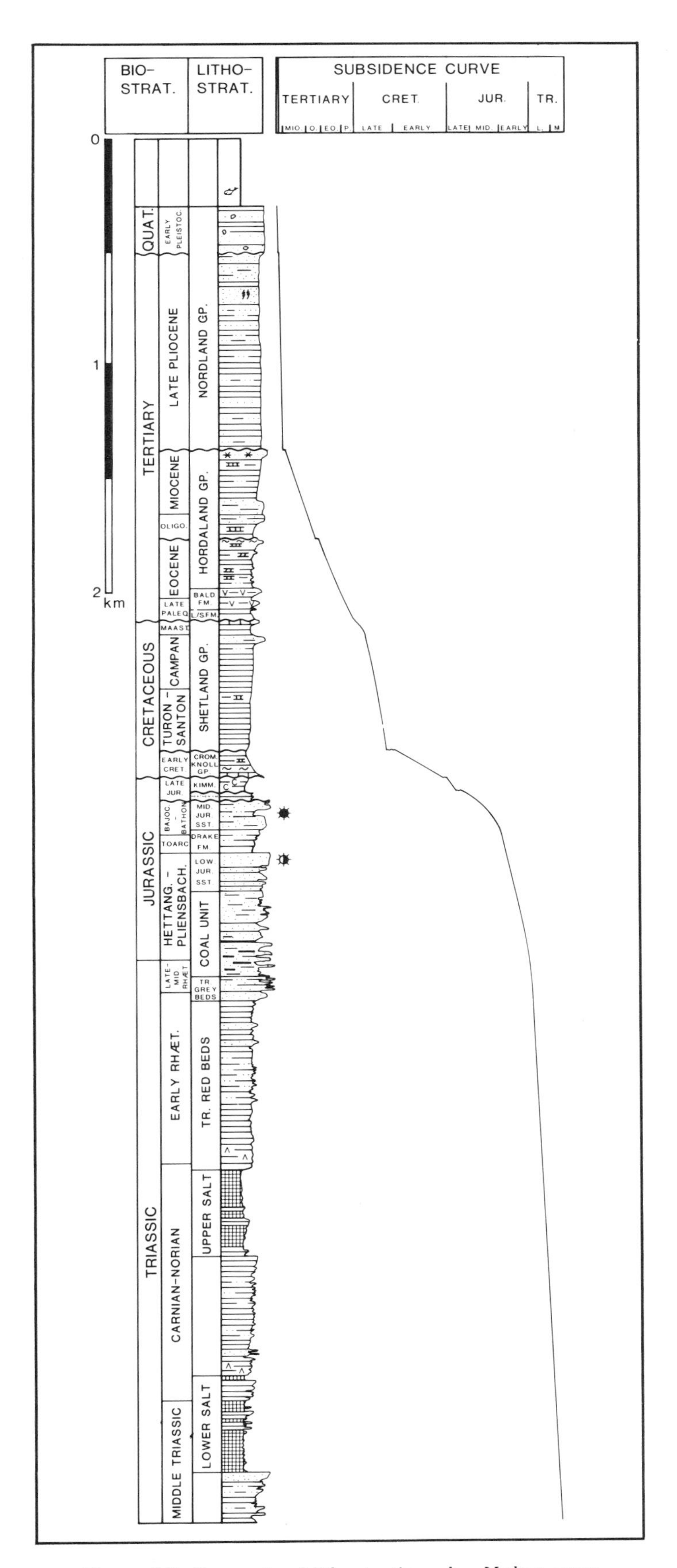


Figure 10. *Generalized lithostratigraphy; Halten area.*

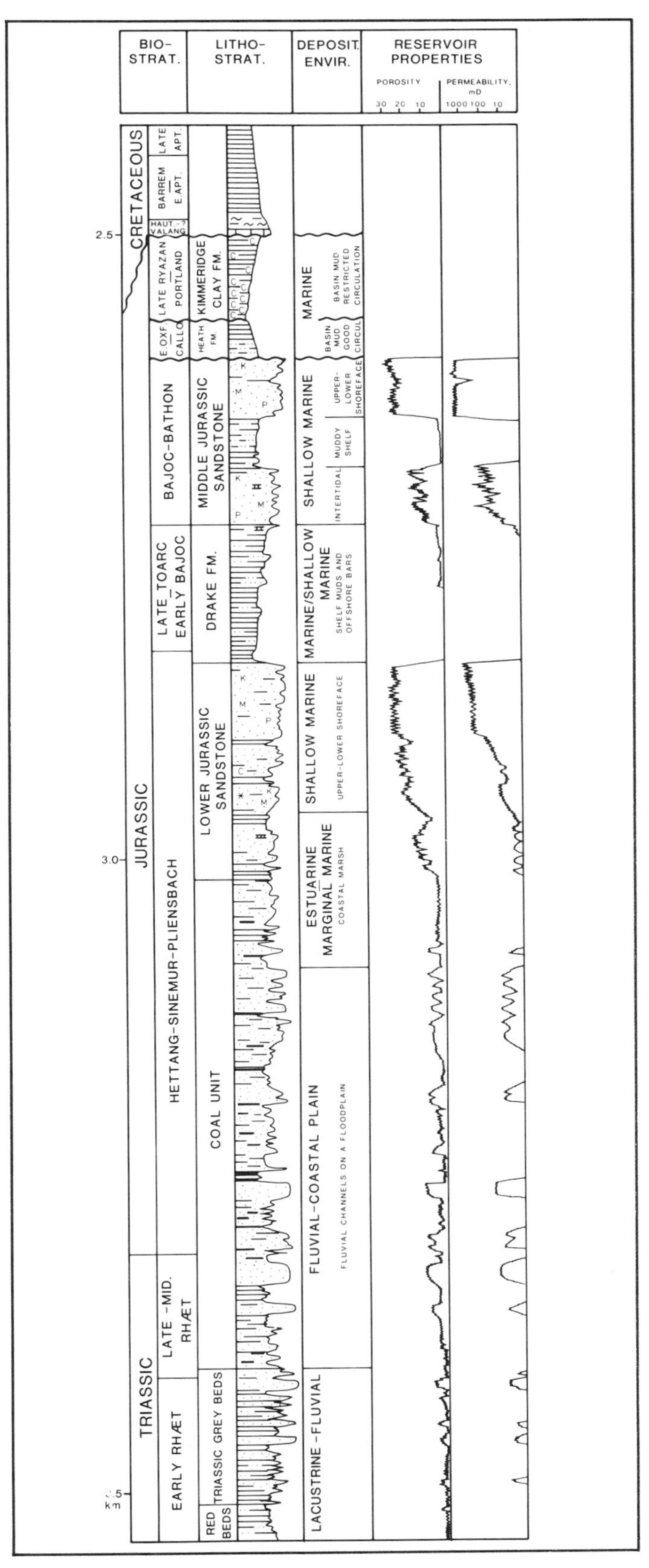


Figure 11. *Generalized Jurassic stratigraphy, facies, and reservoir properties, Halten area.*

basin than in the Halten area (Figure 8).

### Cretaceous

The Cretaceous section separates into two groups: a lower unit that is equivalent to the Cromer Knoll, and an upper unit that is equivalent to the Shetland Group in the North Sea.

The lower Cromer Knoll Group is dominated by grey-white, argillaceous limestones and red-brown marls in the lowermost part, and grey claystones with minor limestone stringers in the upper part. The age ranges from Valanginian to Albian. In the Møre basin, turbiditic sandstones appears.

The Shetland Group consists mainly of grey, uniform claystones of Cenomanian to Maastrictian age. Intercalated laminae of siltstone and very fine sandstone occur in the Campanian. The Maastrictian is very calcareous, containing a large number of limestone streaks and marly intervals (Figure 10).

Except in the structurally lowest parts, such as the Møre basin, the Cretaceous sequence is broken by several unconformities (Figure 8), and local erosion may have given coarse clastic input. The thickness of the Cretaceous section varies from 0–6 km (0 to 4 mi), depending on structural position. In the wells, geologists have observed thicknesses of from 30 to 1,300 m (98 to 4,265 ft).

The Cretaceous sediments were deposited on an open marine shelf with good to restricted bottom circulation. The sequence has no source or reservoir rock potential.

### Tertiary

The Tertiary to Quaternary sequence is generally on the order of 2 km (1 mi). At the outer part of the shelf, more than half of the sequence is of Pliocene–Pleistocene age. This high subsidence and sedimentation rate during the last 5 million years has been observed in all the wells in the area (Hollander, 1984). On the Nordland Ridge, the mid-Tertiary compressional event resulted in an unconformity that ranges in age from early Eocene to late Miocene, reducing the thickness of the lower Tertiary sequence to about 200 m (656 ft).

The lowermost Tertiary sediments geologists have observed are of late Paleocene age, and consist of a grey claystone in the Halten area, with the sequence becoming sandy in the Træna and Møre areas. These sediments are, however, of no importance as reservoir rock.

At the Paleocene–Eocene boundary, there is a tuffaceous claystone equivalent to the Balder Formation of the North Sea. This regional marker horizon reflects considerable volcanic activity associated with the Cenozoic opening of the North-Atlantic Ocean.

The Eocene-to-Miocene sequence compares to the Hordaland Group, and is predominantly a claystone sequence in the basin. The thickness ranges from 0–1 km (0.0 to 0.6 mi). The sequence is thinnest in the north because of erosion and nondeposition related to the mid-Tertiary unconformity. Smectites are the dominating clay minerals, giving a low-density and high-sonic transit time. The topmost part of the unit is a sandy interval, very rich in glauconite, indicating a marine hiatus. We can observe a sandy development on the flank of the Møre basin.

During the late Pliocene, a thick, upward-coarsening clay–sand unit was deposited, and the Quaternary consists of glacigene deposits.

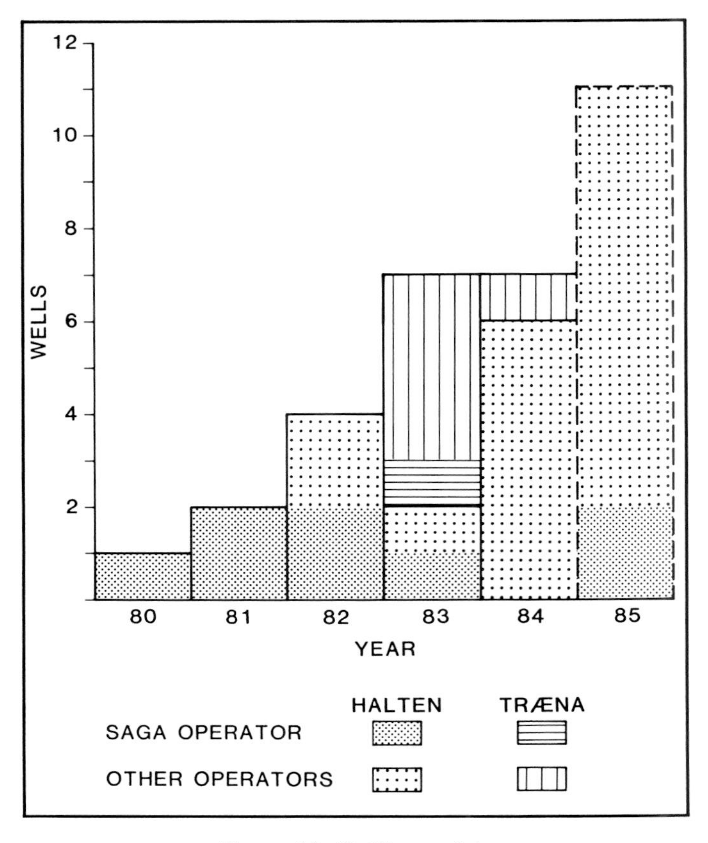


Figure 12. *Drilling activity.*

## HYDROCARBON POTENTIAL

### Discoveries

Drillers have confined offshore mid-Norway exploration to the designated Halten and Træna tracts (Figure 9). The southern part of the Møre basin, however, has been drilled in the northernmost part of the North Sea.

The activity in the Halten area started in 1980 with one well drilled. Except for a setback in 1983, drilling has increased and should peak with 11 wells in 1985. As of mid-November 1984, Saga Petroleum had drilled 6 of 14 wells (Figure 12).

Saga made the first discovery in 1981 with well 6507/11-1 (Figure 13). Gas was found in the Middle and Lower Jurassic sandstone at depths of approximately 2,350 and 2,500 m (7,710 and 8,200 ft), respectively. A production test in the upper reservoir gave 27 million cubic feet (mmcf) of gas and 870 bbl of 60° API condensate per day through a 40/64″ choke. Well 6407/2-2 was drilled in 1983 on the same prospect trend, located downflank and on a separate fault compartment (Figure 13). Gas was found in the Middle Jurassic sandstone at a depth of about 2,500 m (8,200 ft). The well flowed 35 mmcf of gas and 750 bbl of 52° API condensate per day through a 60/64″ choke.

These discoveries, named the Midgard Field, have reserves of some 3.5 trillion cubic feet (tcf) of gas and 130 million bbl of condensate, including two undrilled segments.

Wells 6407/1-2 and 6407/1-3, drilled by Statoil in 1982/83 and 1983, respectively, opened the Tyrihans Field (Figure 13). The first well, located on the southern segment, tested some 3,000 bbl of a 45° API oil and 13 mmcf of gas per day through a

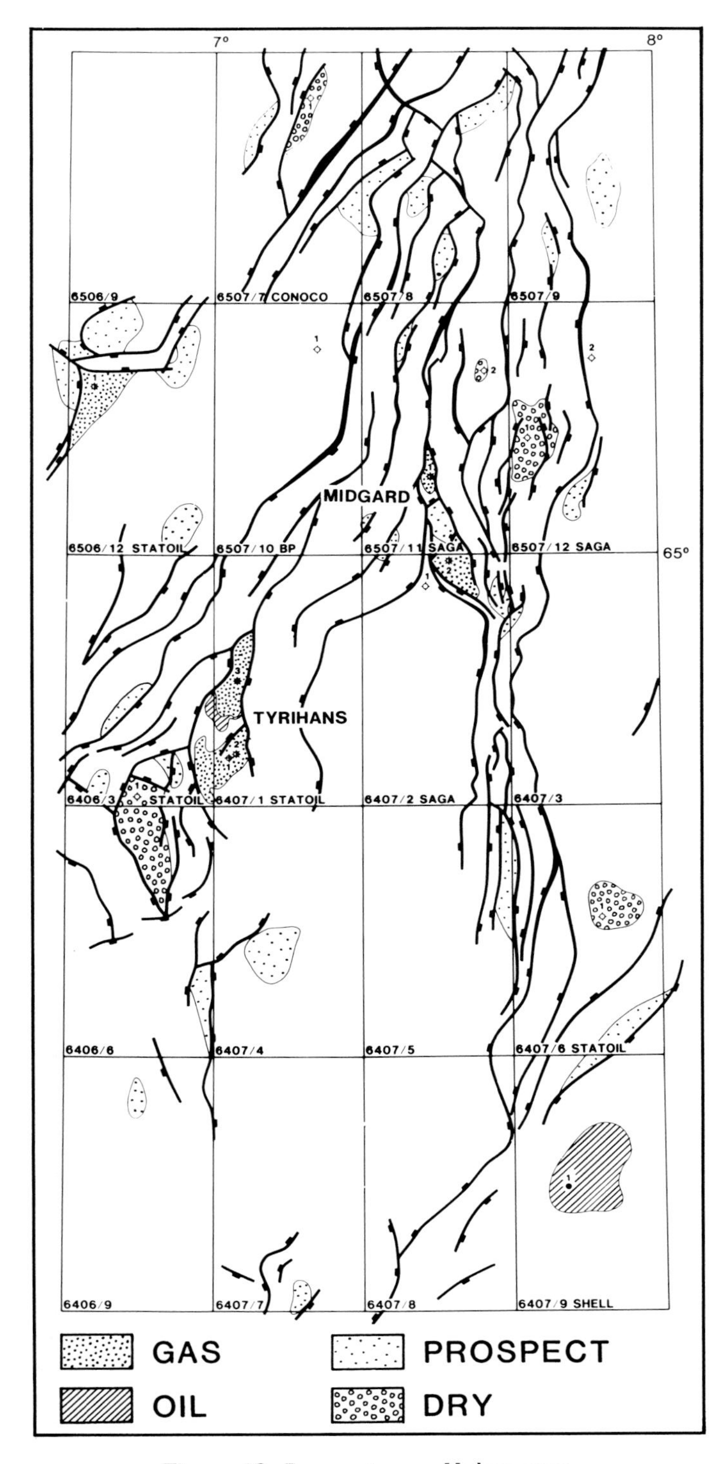


**Figure 13.** *Prospect map, Halten area.*

3/4″ choke. This well tapped the Middle Jurassic sandstone at a depth of about 3,700 m (12,140 ft) (Aasheim and Larsen, 1984). The northern segment lies some 100 m (328 ft) higher structurally than does the southern segment, and differs from it by having a thin oil zone capped by gas, as found by well 6407/1-3 (Figure 13). The well tested 5,500 bbl/day of a 31° API oil, 5 mmcf of gas through a 3/4″ choke from the oil zone, and 31 mmcf of gas and 1,600 bbl of 50° API condensate per day, through a 1″ choke, from the gas zone.

The total reserves for Tyrihans are somewhat smaller than those for Midgard. The liquid content, however, is higher. Numbers recently quoted publically are 2.5 tcf of gas and 220 million bbl of condensate/oil.

In 1984 Shell discovered oil on the Trøndelag Platform in the southeast corner of the Halten area (Figure 13). They found Jurassic sandstones, containing light oil with some condensate, at approximately 1,900 m (6,235 ft). A two-day production test gave 8,500 bbl of oil per day through a 2″ choke. The reserves, however, are uncertain. A minimum of 250 million bbl have been quoted publically.

In late 1984, Statoil made a discovery on the Halten Terrace in the northwest part of the Halten activity area. The top of the discovery is shallower than 4 km (2 mi). The drilled structure consists of several segments separated by faults (Figure 13). The well has not been completed, and both the type of hydrocarbons and the reserves are uncertain. Gas/condensate have been indicated, and a total hydrocarbon column of more than 600 m (1,968 ft) has been reported. The reserves, too, should be much larger than in the other discoveries.

The Træna tract was opened in 1983, with five wells drilled the first season. Another well was spudded in November 1984 (Figure 12). All the wells completed have been dry.

### Structures

The exploration drilling in the Halten and Træna tracts has concentrated on the Jurassic sequence. The sequence contains both reservoir and source rocks, and is sealed by Upper Jurassic and Cretaceous shales and marls. The structures are related to the middle and late Kimmerian tectonic phases, and modified by erosion and differential subsidence during the Cretaceous.

The Trøndelay Platform acted as a stable shelf, having only minor structuring (Figure 9). Locally, flat, roll-over structures or structures bounded by antithetic faults occur along the edge (Figures 13, 14, and 15).

In the Træna area, the western part of the Nordland Ridge eroded during uppermost Jurassic and Cretaceous, and only the pre-Jurassic rocks with poor reservoir potential occur below the Kimmerian unconformity (Figure 16). This condition may have produced stratigraphically sealed reservoir sands within the Cretaceous section along the western flank of the ridge. The degree of erosion on the crest and western flank of the Nordland Ridge further south is critical for prospectivity.

The Halten Terrace comprises a complex Basin and Range province in front of the southern part of the Trøndelag Platform, with the Cretaceous subsidence being intermediate (Figure 9). The base Cretaceous ranges in depth from 1.5–4 seconds (1–5 km or 1–3 mi). Drillers have tested a diversity of structural positions (Figures 14 and 15), but several undrilled prospects remain (Figure 13).

The Møre basin underwent rapid subsidence during the Cretaceous, and few Jurassic closures are observed within drillable depths (Figure 2). Local structuring does, however, occur to the south of the Frøya High. Exploration drilling in the landward, southernmost part of the basin has proven that gas occurs in stratigraphically sealed, Lower Cretaceous sands. The existence of such a play type further north in the landward part of the Møre basin seems likely.

The Vestfjord basin is unstructured and contains mainly

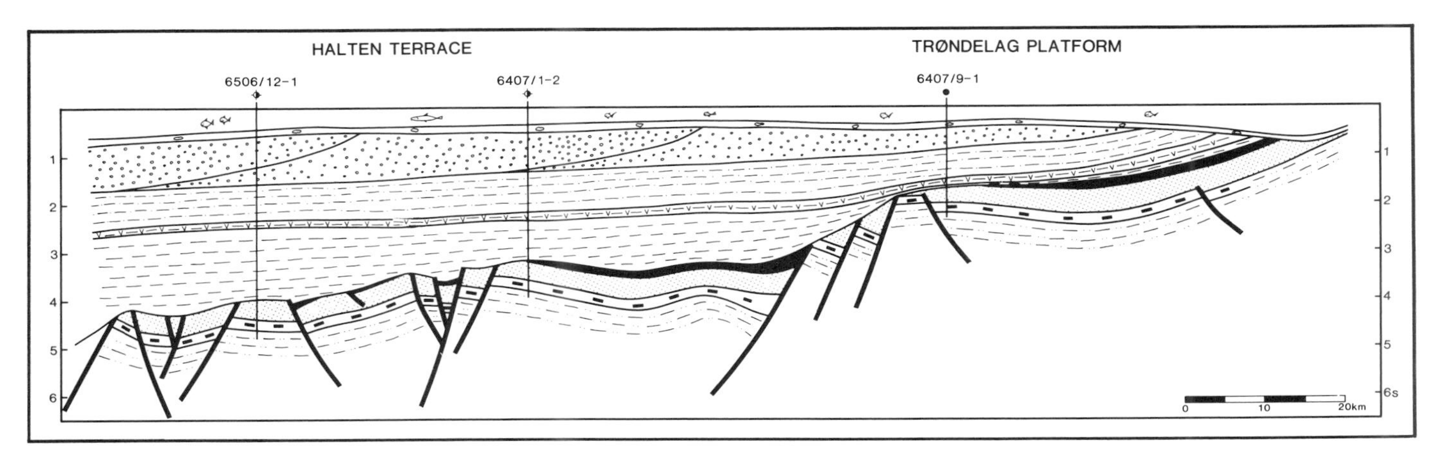


Figure 14. *Profile E.*

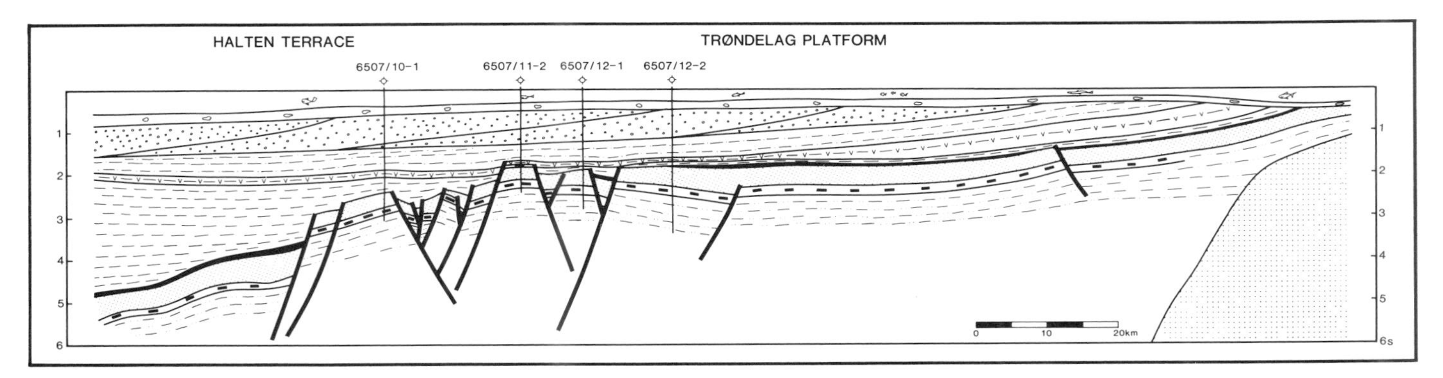


Figure 15. *Profile D.*

Paleozoic rocks. Its prospectivity is expected to be low.

The Vøring basin was structured during the Kimmerian tectonic phases, and was also affected by the middle Tertiary tectonic phase. The basin will, however, remain a white spot as far as exploration drilling is concerned in the near future because of water depth (500–1,500 m or 1,640–4,921 ft).

## Reservoir Rocks

The wells drilled in the Halten area have revealed a Middle Jurassic and a Lower Jurassic reservoir section (Figure 11). Changes in the lateral facies of the Middle Jurassic sandstone are found away from the Halten area. The Lower Jurassic sandstone, on the other hand, seems to be a persistent sand sheet throughout the area (Figure 8).

Generally, the diagenetic processes have not harmed the reservoir properties of the Jurassic sands, and geologists see a similar porosity-depth trend in the North Sea. This trend relates to the very late burial caused by the high subsidence rate during the Pliocene–Pleistocene.

The degree of post-depositional erosion during Upper Jurassic and Cretaceous is the most limiting factor on the presence of reservoir rocks.

## Source Rocks and Maturity

Geologists have identified two potential source-rock intervals: the Coal Unit of Late Triassic/Early Jurassic age, and the Kimmeridge Clay Formation of Late Jurassic age.

The individual carbonaceous claystone/coal beds of the Coal Unit are generally less than 1 m (3 ft) thick. The cumulative thickness of low density (less than 2.0 g/cu cm) lithologies, as determined from the density log, seems to be fairly constant at about 40 m (130 ft) throughout the area. The content of the total organic carbon for the entire unit averages 6–8 wt. %, the pyrolysate yield 10–12 mg/g rock, and the hydrogen indexes 150–200. The maceral composition is dominated by vitrinite over inertinite and exinite, having a potential for not only gas generation, but also heavier hydrocarbons.

Correlation of source rocks and the reservoired light gas/condensate of the Midgard Field suggests the Coal Unit as the main source rock. The gasoline range and heavier hydrocarbons were released from the source rock at moderate maturity levels during the main phase of hydrocarbon generation, while the major portion of the gas expulsion took place at a fairly advanced maturity level (Elvsborg, Hagevang, and Throndsen, in press).

Altogether, the Coal Unit can confidently be classified as a humic source rock with rich potential for gas and condensate generation at high maturity levels.

The Kimmeridge Clay Formation is a black, organically rich "hot" shale. The organic matter is dominated by amorphous, liptinitic oil-prone material. The total content of organic carbon is high, with an average of 10 wt. %. Pyrolysis shows a potential for hydrocarbon generation of 35 mg/g rock, with a correspondingly high hydrogen index of 350.

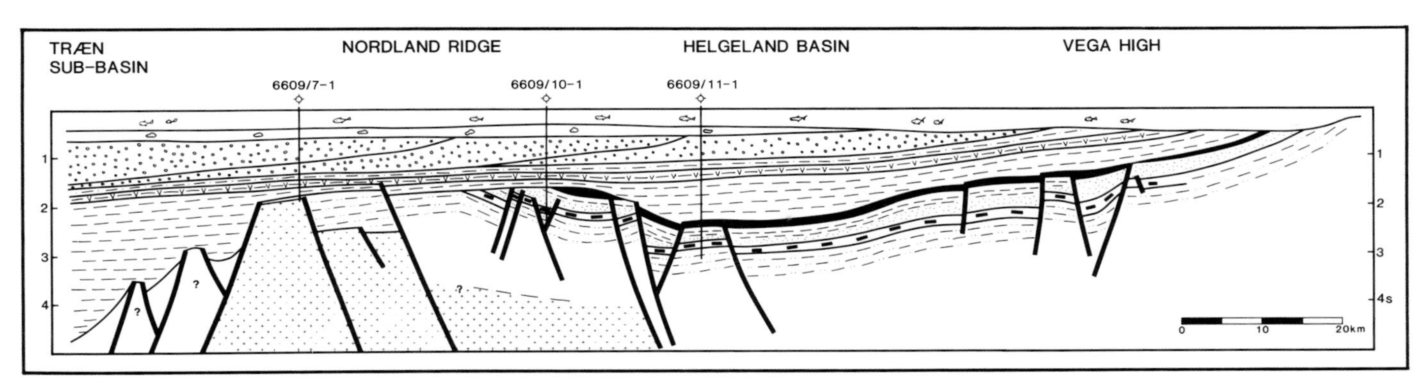


Figure 16. *Profile G.*

The maturity increases continuously with depth (Figure 17). The gradient is very steep because of the late burial; hence, the peak mature zone occurs between 4 and 4.8 km (13,000-16,000 ft). The current maturation is at equilibrium with the present temperature gradient.

Figure 18 shows a schematic hydrocarbon generation map for the two source-rock units, and also shows that the Trøndelag Platform is entirely immature. Minor expulsion of gas occurs in the very deepest part of the Helgeland basin, which explains the dry wells in the Traena area. Long-range migration from the mature basins to the west into the structures on the Nordland Ridge has not been possible because of the lack of permeable rocks across the ridge.

A wide zone with maximum gas expulsion, and a narrower zone in the peak oil window, exist in front of the Trøndelag Platform. The area between the Halten and Traena tracts will be out for bid in the Norwegian 10th concession round. The rim zone of the Trøndelag Platform in that area certainly looks attractive from a hydrocarbon generation point of view.

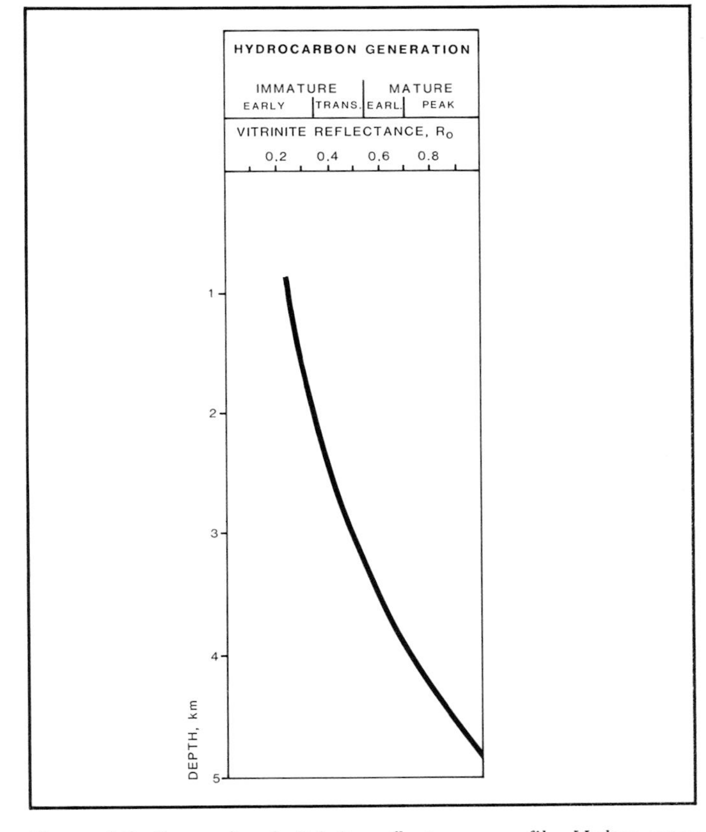


Figure 17. *Generalized vitrinite reflectance profile, Halten area.*

## SUMMARY

### Structural Development

None of the described areas show a complete structural development from late Paleozoic and onward. We can summarize the composite tectonic history for the Møre–Lofoten continental margin as follows:

(1) Late Paleozoic and Early Triassic: Deposition in fault-bounded basin.
(2) Early and Middle Triassic: Stable tectonic condition.
(3) Early and middle Kimmerian: Rifting, stretching, and doming of the Møre basin and Vøring basin. Separation of the Halten Terrace from the Trøndelag Platform.
(4) Late Kimmerian: Wrenching along the southeastern flank of the Møre basin and flank of the Trøndelag Platform results in the Frøya High, Nordland Ridge and Vega High.
(5) Cretaceous: Rapid subsidence in the Møre and Vøring basins. Moderate subsidence in the Helgeland basin.
(6) Early Tertiary: Inversions and extension in the western part of the Møre basin and Vøring basin.
(7) Mid-Tertiary: Major inversion in the Vøring basin. Compressional modification at the Nordland Ridge and minor compression in the coastal area of the Møre basin.
(8) Miocene, Pliocene and Pleistocene: Rapid subsidence and shelf progradation along the whole shelf.

### Stratigraphy

We know the Middle Triassic and younger sections from drilling on the Møre–Lofoten Shelf. The Triassic and lowest Jurassic clastic sequences were deposited under continental conditions and with a climate that changed from arid to humid. In Lower and Middle Jurassic, marine conditions were established along the entire Møre–Lofoten Shelf. A major Middle Jurassic regressive sand sequence developed in the Halten area, while we can

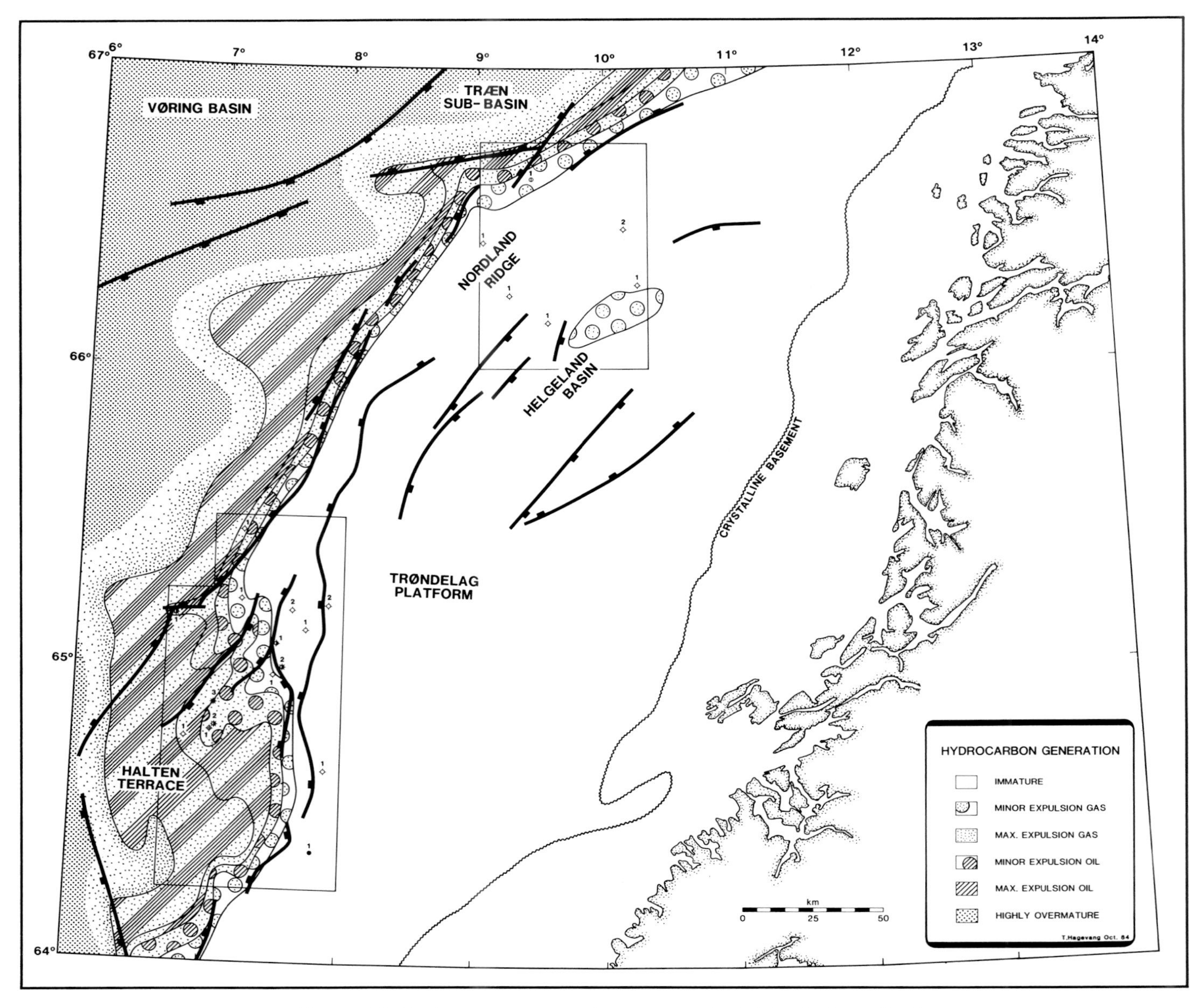


Figure 18. *Hydrocarbon generation map, central Møre–Lofoten Shelf.*

observe a shale facies north and south of this area. The post-Middle Jurassic strata that we have observed have a shaly development. Strong Upper Jurassic and Cretaceous erosion has truncated the Jurassic reservoir rock in the structurally high areas.

### Hydrocarbon Potential

The Halten area, centrally situated on the Møre–Lofoten Shelf, is a petroleum province. Four major hydrocarbon discoveries have been made, and both oil and gas are present. While the two newest fields are not delineated, more than $0.5 \times 10^9$ ton oil equivalent can be considered proven.

The established play type is Jurassic sand in fault-bounded structures, with Upper and Lower Jurassic source rocks acting as sources.

Production from the Halten area is expected in the 1990s.

## ACKNOWLEDGMENTS

The authors wish to thank Saga Petroleum for the opportunity to publish this work. Thanks are also due to Saga colleagues Werner Karlsson, Hans Augedal, Roy Gabrielsen, and Arvid Elvsborg, and former Saga employees Torbjørn Throndsen and Charlotte Robinson for their professional contribution.

Finally, we gratefully appreciate the efforts by Ingrid Crosby and Mona Pitias, who drafted the paper, and Riborg Johnsen, who typed it.

## REFERENCES

Aasheim, S. M. and V. Larsen, 1984, The Tyrihans Discovery—preliminary results from well 6407/1-2, *in* A. M. Spencer et al, eds., Petroleum geology of the North European margin:

London, Graham and Trotman Ltd., 436 p.
Bott, M. H. P., 1975, Structure and evolution of the Atlantic floor between northern Scotland and Iceland, Norges geologiske Undersøkelse, v. 316, 195–199.
Bøen, F., S. Eggen, and J. Vollset, 1984, Structures and basins of the margin from 62°N to 69°N and their development, *in* A. M. Spencer et al, eds., Petroleum geology of the North European margin: London, Graham and Trotman Ltd., 436 p.
Caston, V. N. D., 1976, Tertiary sediments of the Vøring Plateau, Norwegian Sea, recovered by leg 38 of the Deep Sea Drilling Project, *in* U. Talwani et al, eds., Initial reports of the Deep Sea Drilling Project, vol. 38: Washington, U.S. Government Printing Office, 1256 p.
Elvsborg, A., T. Hagevang, and T. Throndsen, in press, Origin of the gas/condensate of the Midgard Field at Haltenbanken, Symposium "Organic geochemistry in the exploration of the Norwegian Shelf:" Stavanger, Norwegian Petroleum Society.
Gabrielsen, R. H. and C. I. Robinson, 1984, Tectonic inhomogeneities of the Kristiansund-Bodo Fault Complex, offshore Mid-Norway, *in* A. M. Spencer et al, eds., Petroleum geology of the North European margin: London, Graham and Trotman Ltd., 436 p.
Hagevang, T., O. Eldholm, and I. Aalstad, 1983, Pre-23 magnetic anomalies between Jan Mayen and Greenland-Senja fracture zones in the Norwegian Sea, Marine Geophysical Researches, v. 5, p. 345–363.
Hinz, K., 1972, The seismic crustal structure of the Norwegian continental margin in the Vøring Plateau, in the Norwegian deep sea and on the eastern flank of the Jan Mayen Ridge between 66 and 68°N, 24th International Geological Congress, Montreal, section 8, 28–37.
Hollander, N. B., 1984, Geohistory and hydrocarbon evaluation of the Haltenbank area, *in* A. M. Spencer et al, eds., Petroleum geology of the North European margin: London, Graham and Trotman Ltd., 436 p.
Karlsson, W., 1984, Sedimentology and diagenesis of Jurassic sediments offshore mid-Norway, *in* A. M. Spencer et al, eds., Petroleum geology of the North European margin: London, Graham and Trotman Ltd., 436 p.
Rønnevik, H. C. and T. Navrestad, 1977, Geology of the Norwegian shelf between 62° and 69°N, Geojournal, v. 1, 33–46.
———, F. Jorgensen, and K. Motland, 1979, The geology of the northern part of the Vøring Plateau, *in* Proceedings of Norwegian Sea Symposium: Norwegian Petroleum Society.
Russel, M. J., 1976, A possible Lower Permian age for the onset of ocean floor spreading in the northern North Atlantic, Scottish Journal of Geology, v. 12, 315–323.
Skogseid, J., 1983, A marine geophysical study of profiles between the Vøring Plateau margin and the Jan Mayen Ridge, cand. scient. thesis: Univ. Oslo.
Smythe, D. K., 1983, Faeroe-Shetland Escarpment and continental margin of the Faeroes, *in* M. H. P. Bott et al, eds., Structure and development of the Greenland-Scotland Ridge: Plenum Publ. Corp.
Talwani, M., J. Mutter, and K. Hinz, 1983, Ocean continent boundary under the Norwegian continental margin, *in* H. M. P. Bott et al, eds., Structure and development of the Greenland-Scotland Ridge: Plenum Publ. Corp.

# Tectonic Development and Hydrocarbon Potential Offshore Troms, Northern Norway

T. Sund
O. Skarpnes
L. Nørgård Jensen
R. M. Larsen
*Statoil*
*Stavanger, Norway*

Four years of exploration activity offshore Troms has resulted in four gas discoveries, with total resources of $170 \times 10^9$ cu m ($6{,}000 \times 10^9$ cu ft) recoverable gas. Shows of heavy, waxy oil are present in most wells. Recent regional studies on structural evolution and organic geochemistry suggest that, in parts of the area where exploration activities have been concentrated so far, the oil-prone source rock has not been mature enough to expel the generated oil. In other parts, all potential source rocks became overmature 100 million years ago because of extreme subsidence during the Early Cretaceous. The regional studies clearly demonstrate the close relationship between the structural evolution and the hydrocarbon potential of the region. The different regional, structural elements are individual provinces with respect to hydrocarbon potential. The remaining hydrocarbon potential within Tromsøflaket is fair.

## INTRODUCTION

Offshore Troms is one of the new frontier areas north of lat. 62°N on the Norwegian Continental Shelf (Figure 1). Even though drilling in the Norwegian North Sea started in 1966, drilling north of lat. 62°N did not start until 1980. Limited areas on Tromsøflaket and Haltenbanken were awarded to Norwegian oil companies. They discovered minor amounts of hydrocarbons in one well in 1980, but they did not production test it.

The Tromsøflaket area is located in the southwestern part of the Barents shelf, and comprises a number of intracratonic basins and highs (Figure 2). To the west, the area is bounded by a passive continental margin, where ocean-floor spreading began during early Eocene (Talwani and Eldholm, 1977). The area open for exploration drilling so far lies between lat. 71°N and 72°N; and long. 17° 40′ E and 22° E, and covering about 1,300 sq km (502 sq mi) (Figure 2). Since the area was opened for exploration in 1980, 22 wells have been drilled.

Until September 1984, the exploration activity had resulted in the discovery of four gas fields with total resources of $170 \times 10^9$ cu m ($6,000 \times 10^9$ cu ft) recoverable gas. In addition, several minor accumulations and shows had been encountered, all in the Hammerfest basin.

Statoil's last well on Tromsøflaket, (7121/4-1, completed October 1984, Figure 3) opened a new era in exploring the offshore Troms area. Statoil found light oil (34° API) in addition to gas. We believe the oil encountered in well 7121/4-1 to have been generated from a different source than the gas, condensate, and traces of waxy oil we encountered in previous wells.

Prior to the last well, regional studies on structural development, organic geochemistry, and source rock maturation had indicated that the Hammerfest basin consists of two hydrocarbon provinces separated by a west-east trending central dome. Exploration drilling has so far been limited to the southern region (Figure 4B), and we define this region a gas province. In the area north of the central dome, computer modelling indicated the possibility of local generation and migration of oil as well as gas. Consequently, we have defined this area as a gas and oil province. Statoil's well 7121/4-1 is the only well drilled in the northern hydrocarbon province so far, and it confirms the conclusions from the regional studies.

## STRUCTURAL DEVELOPMENT OF THE TROMSØFLAKET AREA

### Pre-Kimmerian Development

The post-Caledonian development of the Tromsøflaket area probably started with deposition of terrestrial sediments in fault-bounded extensional basins. The Caledonian structural trend is northeast-southwest along the Norwegian mainland. A regional fault system follows this trend (Rønnevik and Jacobsen, 1984; Figure 2), but only a few faults related to this system have been mapped on Tromsøflaket.

A continuation of the north-south fault system seen on the Loppa High underlies the Ringvassøy-Loppa Fault Complex (RLFC) and the Tromsø basin (Figure 3). This fault system relates to the post-Caledonian rift phase found on Spitsbergen (Steel and Worsley, 1984), offshore northern Norway, and in East Greenland (Escher and Watt, 1976). After rifting, a thick salt sequence of Permian age was deposited in a paleo-Troms basin, while a paleo-Loppa High partly eroded along the eastern margin of the basin. Marginal carbonates and clastics were, therefore, deposited in the Hammerfest basin.

The fault activity in the Tromsøflaket area ceased during Middle Permian, and the whole area became part of a regional intracratonic basin until Middle Jurassic. The only indication of structural movements is an Early Triassic unconformity on the Loppa High (Figure 5).

### Post-Kimmerian Development

The major structural elements and most of the faults in the Tromsøflaket area are a result of the late Kimmerian tectonic event. Since the post-Kimmerian development of these structural elements is different, we will describe them separately.

#### *The Ringvassøy-Loppa Fault Complex and the Tromsø Basin*

The Tromsø basin is a very deep (12–15 km, or 7–9 mi), north-south trending structure bordered by the Senja Ridge to the west and the Ringvassøy-Loppa Fault Complex to the east (Figures 2 and 6). The basin development postdates the Middle Jurassic. An episode of extreme subsidence took place during Barremian and Albian, and a series of salt diapirs developed along the basin axis. The Ringvassøy-Loppa Fault Complex formed by reactivation of the post-Caledonian fault system during the subsidence of the Tromsø Basin. The fault complex is a network of north-south trending, listric faults dipping toward the basin. Faults related to the basin subsidence have prograded as far east as the Askeladden Field in the Hammerfest basin (block 7120/8, Figure 3).

Rollover anticlines formed on the downthrown side of some of the major listric faults.

#### *The Senja Ridge*

The Senja Ridge is a north-south trending structural high forming the western margin of the Tromsø basin (Figures 2 and 6). The ridge coincides with a marked positive-gravity anomaly, but falls within a quiet magnetic zone (Faleide, Gudlaugsson, and Jacquart, 1984). Two exploration wells in the central part of the ridge penetrated thick sequences of steeply dipping, Lower Cretaceous sediments below the base Paleocene unconformity. This finding seems to indicate that the ridge was formerly part of the Tromsø basin and was created by Late Cretaceous to Paleocene inversion or transpression, and also that the ridge is underlain by a non-magnetic basement high, as suggested by Syrstad, Talleraas, and Bergseth (1976).

#### *The Loppa High*

The wedge-shaped Loppa High is between the Hammerfest and Bjørønya basins (Figure 2). It was uplifted during Late Jurassic-Early Cretaceous, and became an island, with deep

Figure 1. *Location and waterdepth map. Tromsøflaket (expanded Troms I) is the most northerly area opened for exploration drilling on the Norwegian Continental Shelf. Twelve blocks have been awarded, where 22 wells have been drilled since 1980.*

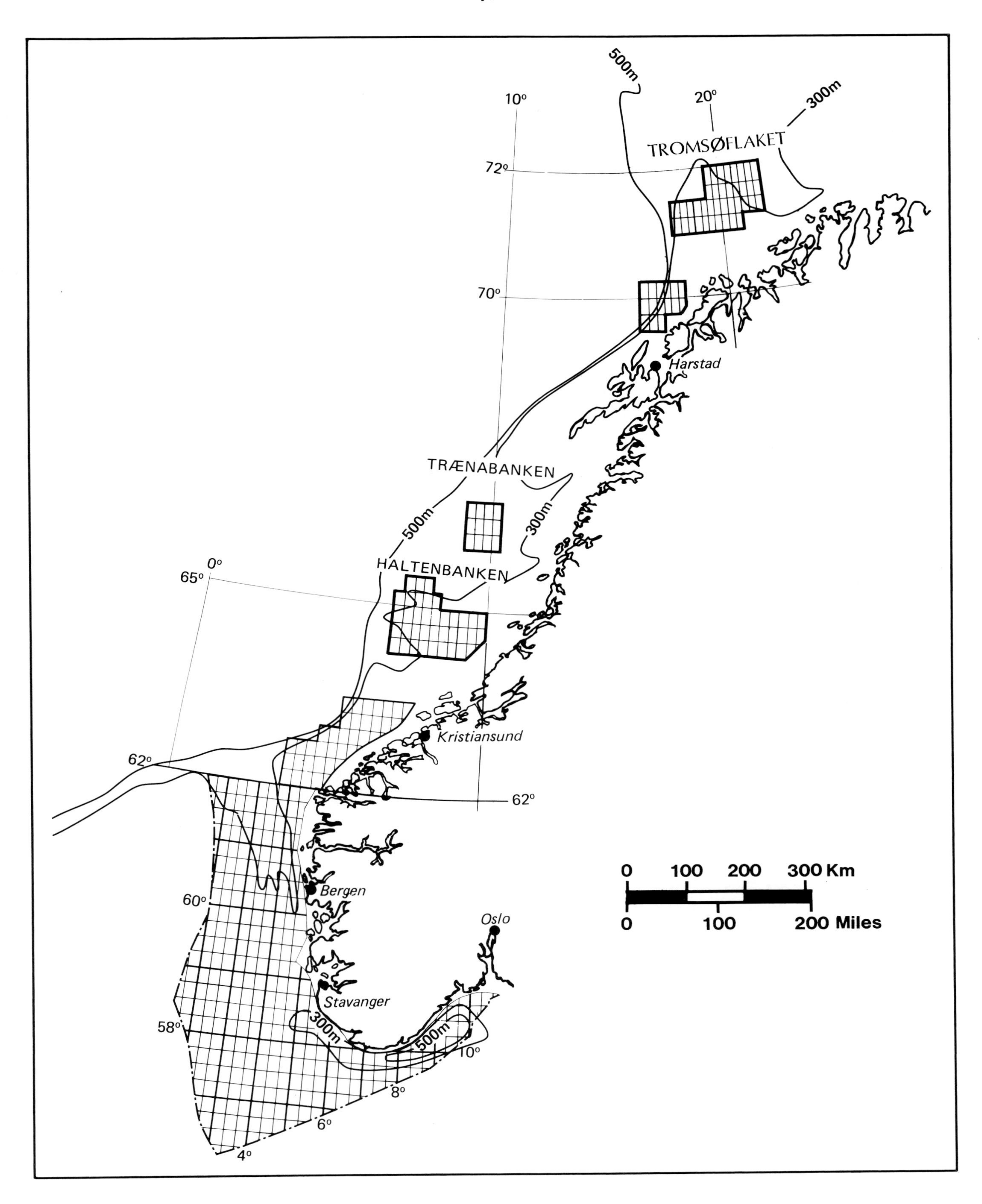
500m
10°
20°
300m
TROMSØFLAKET
72°
70°
Harstad
TRÆNABANKEN
500m
300m
HALTENBANKEN
0°
65°
Kristiansund
62°
62°
0 100 200 300 Km
0 100 200 Miles
Bergen
60°
Oslo
Stavanger
58°
300m
500m
10°
8°
6°
4°

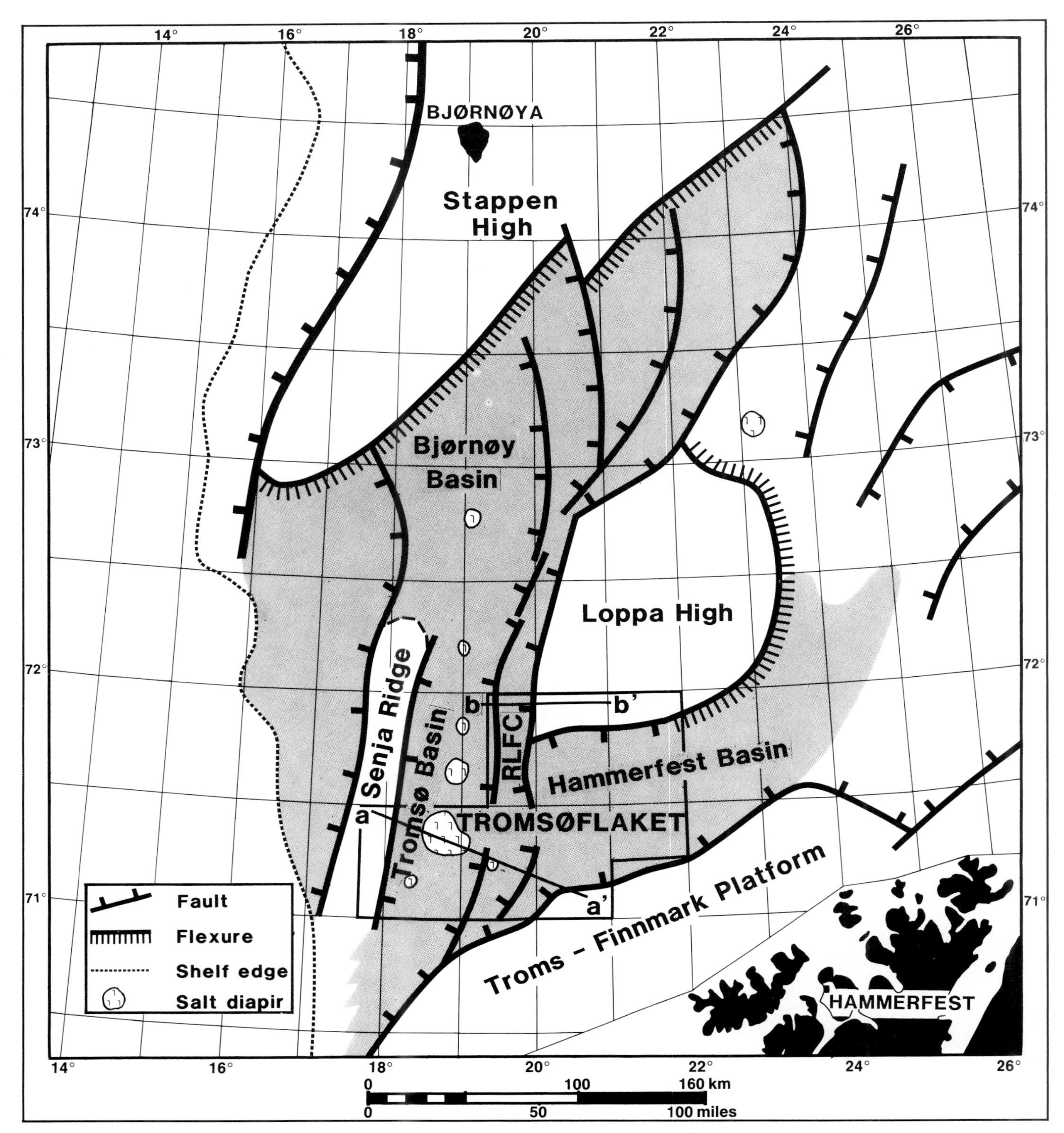


Figure 2. *Major tectonic elements in western Barents Sea. Tromsøflaket is located in the southwestern Barents Sea, where several tectonic elements meet that were mainly formed during Cretaceous. No wells have yet penetrated the Loppa High. Sections a–a′ and b–b′ are structural cross sections, shown in Figures 5 and 6 respectively. RLFC: Ringvassøy-Loppa Fault Complex.*

Figure 3. *Structural elements of eastern Tromsøflaket. Area opened for exploration activity outlined. Fault trend is N-S in Ringvassø y-Loppa Fault Complex, E-W in Hammerfest basin. Operator is shown in licensed blocks. A–A′ and B–B′ structural cross sections shown in Figures 4A and 4B respectively.*

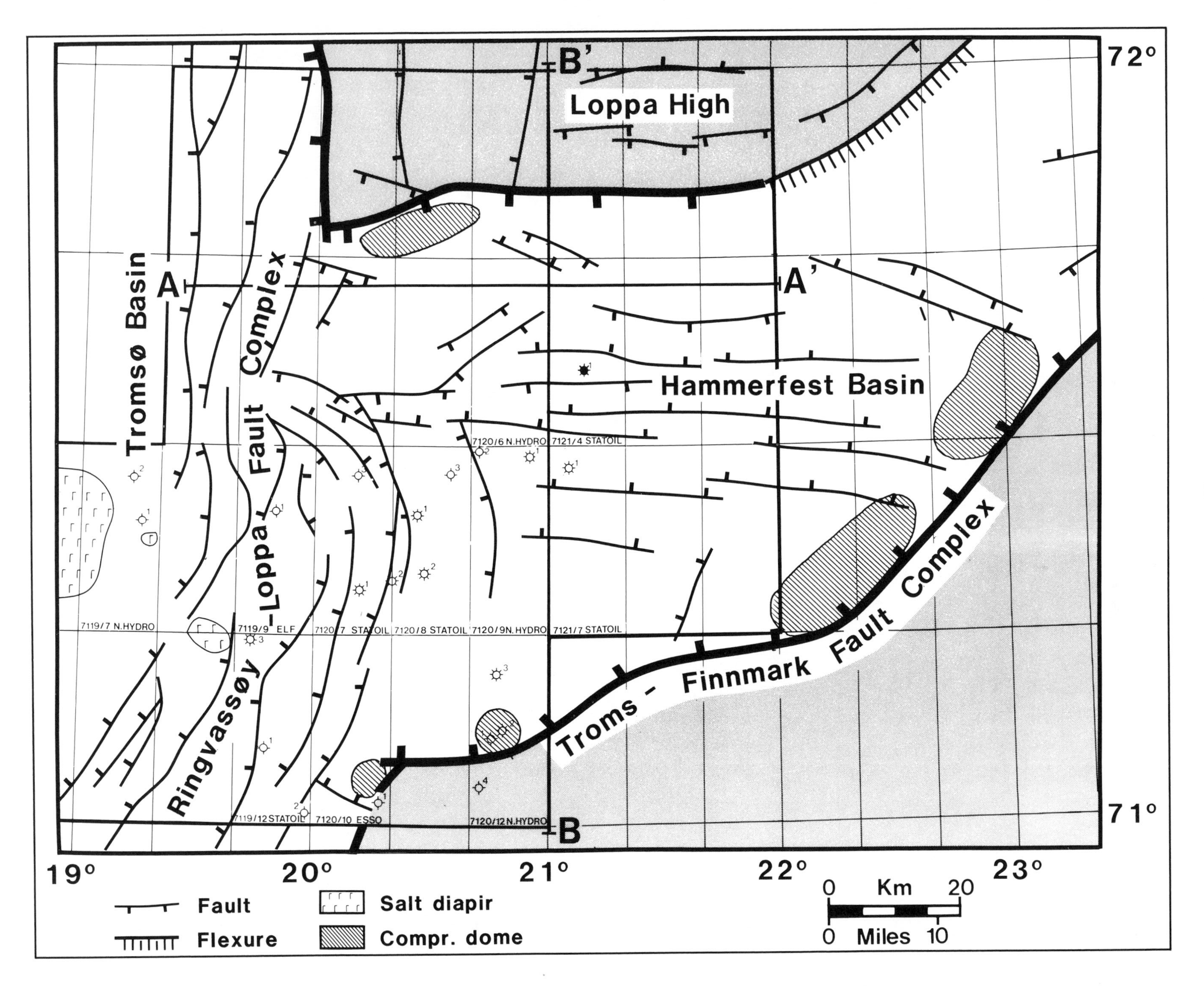
72°
71°
19°
20°
21°
22°
23°
Loppa High
Hammerfest Basin
Tromsø Basin
Ringvassøy -Loppa Fault Complex
Troms - Finnmark Fault Complex
A
A'
B
B'
7119/7 N.HYDRO
7119/9 ELF
7119/12 STATOIL
7120/6 N.HYDRO
7120/7 STATOIL
7120/8 STATOIL
7120/9N.HYDRO
7120/10 ESSO
7120/12 N.HYDRO
7121/4 STATOIL
7121/7 STATOIL
0 Km 20
0 Miles 10
Fault
Flexure
Salt diapir
Compr. dome

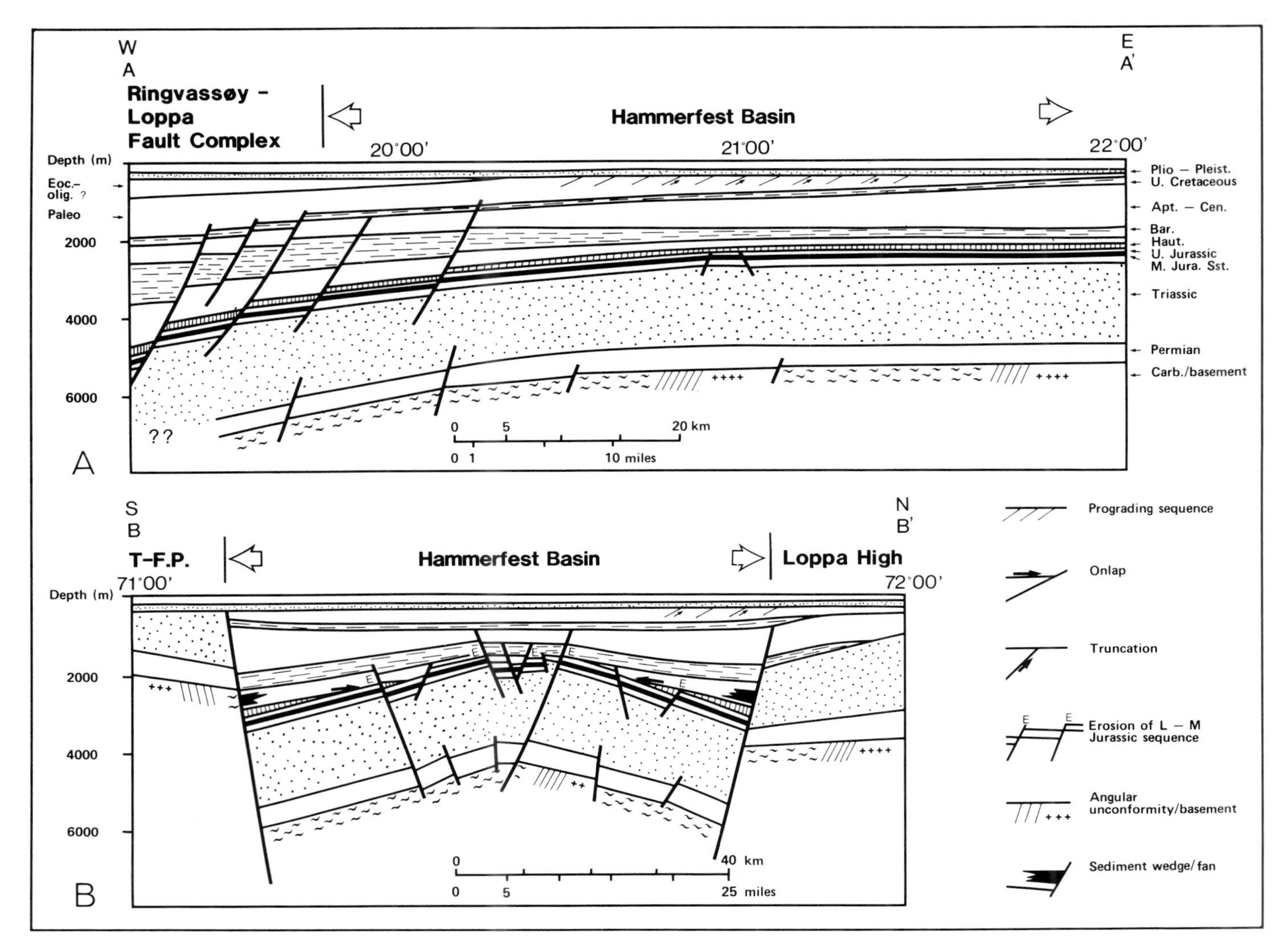


Figure 4. *Structural cross sections, Hammerfest basin. Locations shown in Figure 3, line A–A'. (A, above) E-W section demonstrating increasing fault activity into the Ringvassøy-Loppa Fault Complex. Late Tertiary unconformity is increasingly important toward the east. The Cretaceous Period demonstrates that the basin is shifting. (B, below) N-W section that demonstrates doming and symmetry of the Hammerfest basin. The central dome defines two hydrocarbon provinces as discussed in the text. T.-F.P.: Troms-Finnmark Platform. The cross sections are schematic, and the scales are only indicative.*

canyons cutting down into the Triassic sediments (Figure 5). During the Cretaceous, the island gradually subsided, finally disappearing in the early Paleocene.

*The Hammerfest Basin*

The Hammerfest basin is an east-northeast-to-west-southwest trending, late Kimmerian structure between the Loppa High to the north, and the Troms-Finnmaı k Fault complex (Figure 3) and the Troms-Finnmark Platform (Figure 4) to the south.The basin widens and deepens westward, toward the Ringvassøy-Loppa Fault Complex (Figure 4A). The basin is symmetric, with a doming of the pre-Kimmerian sequences parallel to the basin axis that was active from Middle Jurassic to early Barremian. The main subsidence of the basin occurred along the northern and southern margins (Figure 4B). The dominant east-west trending fault system in the central part of the Hammerfest basin was formed by flexural extension related to the doming. A majority of these faults dip toward the basin axis, where a small graben formed along the crest of the dome. Because of this geometry, the Hammerfest basin divides into a

Figure 6. *Structural cross section across Tromsø basin. Figure 2 shows the location (line a–a'). Faulting in Ringvassøy-Loppa Fault Complex is predominantly late Kimmerian. We can observe extreme Early Cretaceous subsidence in Tromsø basin. Computer modelling indicates the top Jurassic down to 12,000 m (39,370 ft). The Senja Ridge developed during Late Cretaceous-Paleocene. The central salt diapir is huge, as also seen on Figure 2. The cross section is schematic and scales only indicative (modified after Gloppen and Westre (1982); and Bleie, Nysæther, and Oppebøen (1982)).*

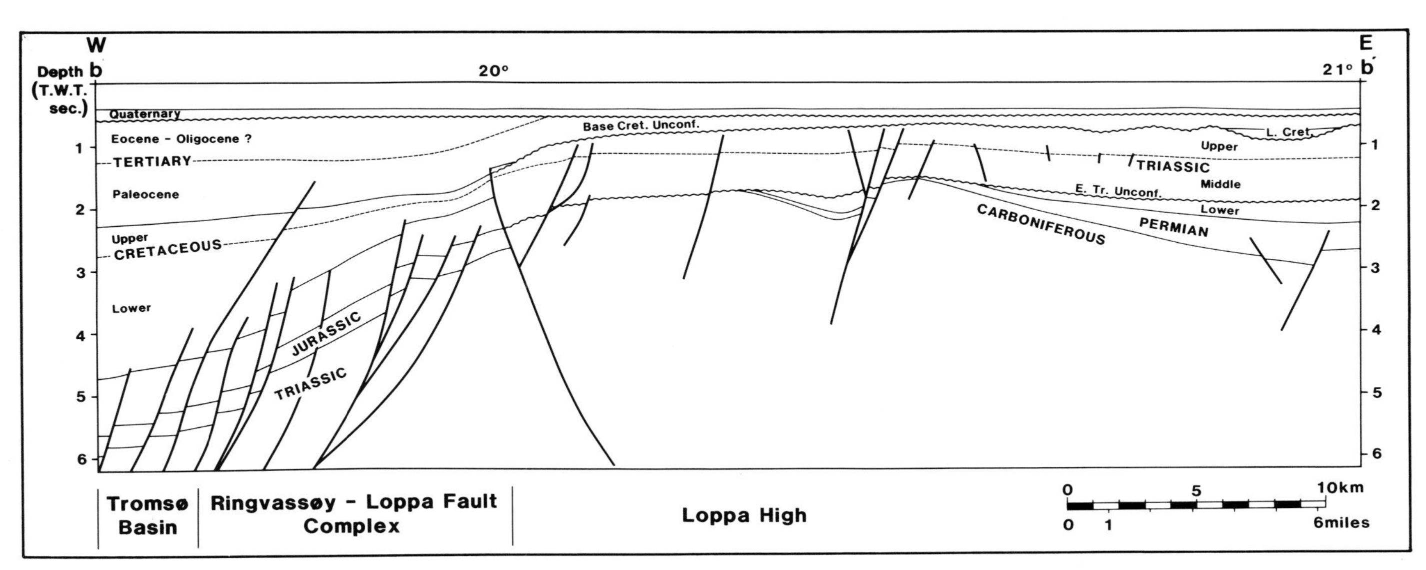


Figure 5. *Structural cross section, Loppa High. The location is shown in Figure 2 (line b–b'). The section is drawn along a seismic line and demonstrates several episodes of movements: 1) Paleozoic normal faulting; 2) Early Triassic erosion (due to uplift?); 3) late Kimmerian normal faulting and Early Cretaceous subsidence; and 4) late Tertiary erosion. In the study area the erosion is generally due to an east-west tilting of the area; on the Loppa High it is also because of inversion. In addition, Barremian lateral movements affected the area as the compressional domes indicate (Figure 3).*

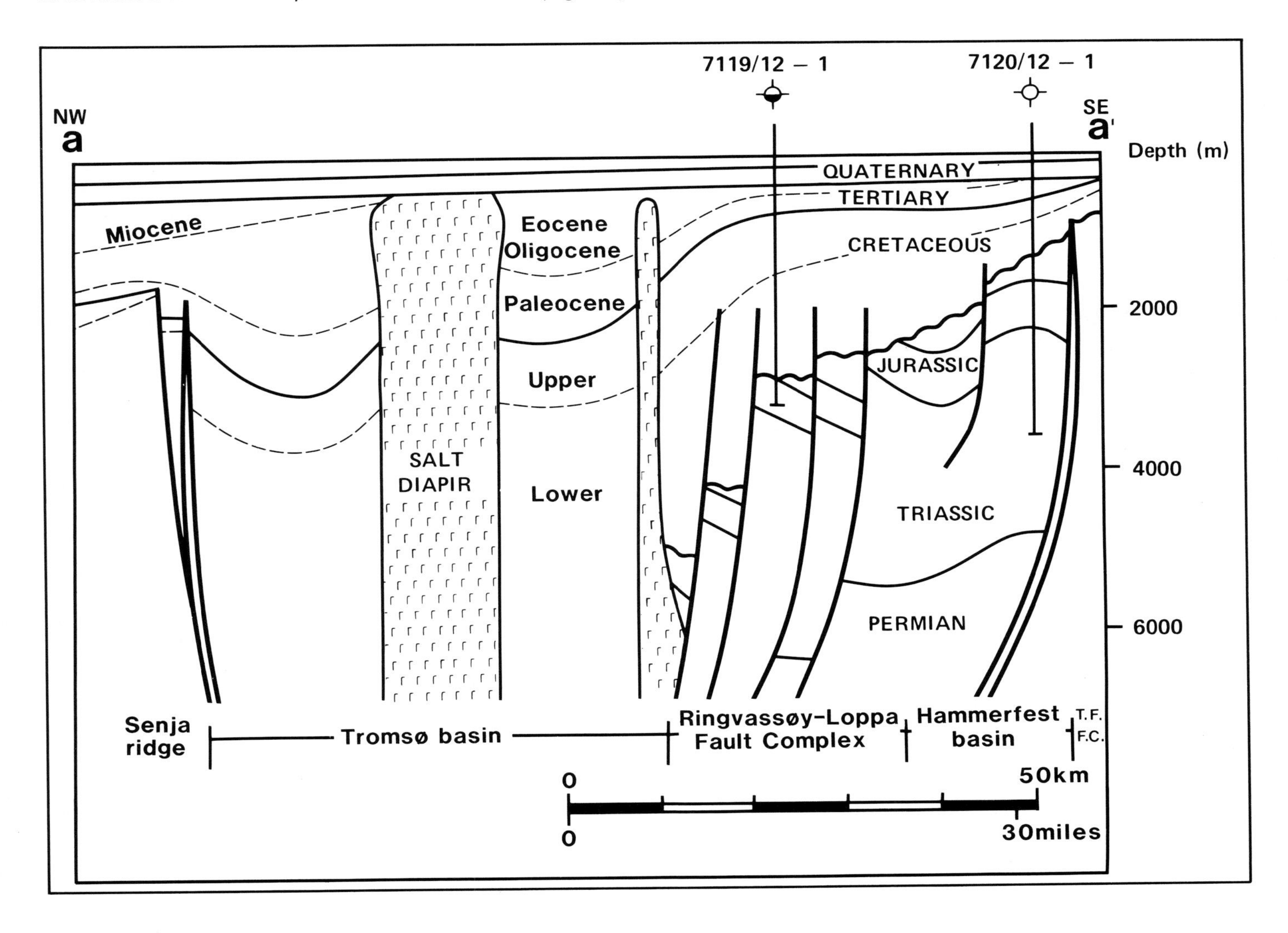

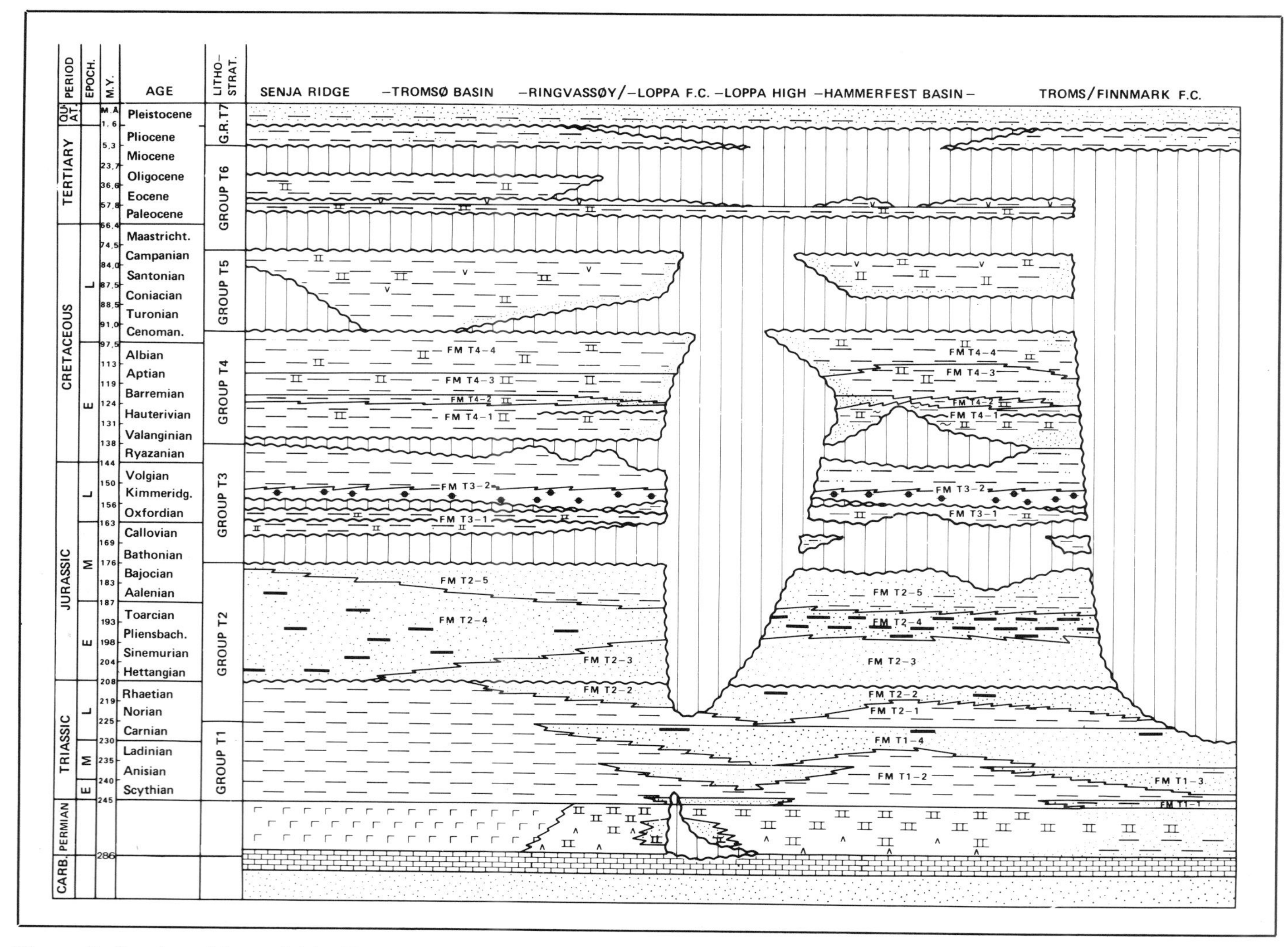


Figure 7. *Stratigraphic model for Tromsøflaket. The model shows the known or inferred stratigraphy for the different tectonic elements. The vertical lines show the missing section. The text discusses the potential reservoir and source rocks.*

northern and a southern province with respect to hydrocarbon generation and migration.

The Cretaceous subsidence of the basin was overprinted by an episode of right-lateral movements, culminating in the Barremian. During this episode, a number of compression domes and transtension grabens formed along the Loppa High and the Troms-Finnmark Fault Complex (Figure 3).

### Tertiary Development

Late Paleocene to early Eocene subsidence and sedimentation took place in the Tromsøflaket area. After the early Eocene opening of the Norwegian-Greenland Sea, uplift and erosion affected the area east of the Ringavassøy-Loppa Fault Complex, especially the Loppa High. The Loppa High was probably subjected to a second inversion phase during Cenozoic (Figure 5). This inversion resulted in a major erosion east of the Ringvassøy-Loppa Fault Complex (Figures 4, 5, and 6); and Quaternary and upper Pliocene sediments rest unconformably on Eocene (Oligocene) sediments. A more continuous Tertiary sedimentation took place in both the Tromsø basin and west of the Senja Ridge, occurring as prograding sequences (Spencer, Howe, and Berglund, 1984; Figure 6) .

## HYDROCARBON POTENTIAL

### Reservoir rocks

All discoveries and shows of hydrocarbons on Tromsøflaket have been in the Upper Triassic to Middle Jurassic sandstone sequence. In addition, Carboniferous sandstones, Upper Car-

→

Figure 8. *Thickness map (in meters) of Lower to Middle Jurassic that includes the only proven reservoir rock. The map is based on well and seismic information on 19° 40′ E, to the west on a deposit model. Details related to erosion on the fault crests (Figure 4B) are not included.*

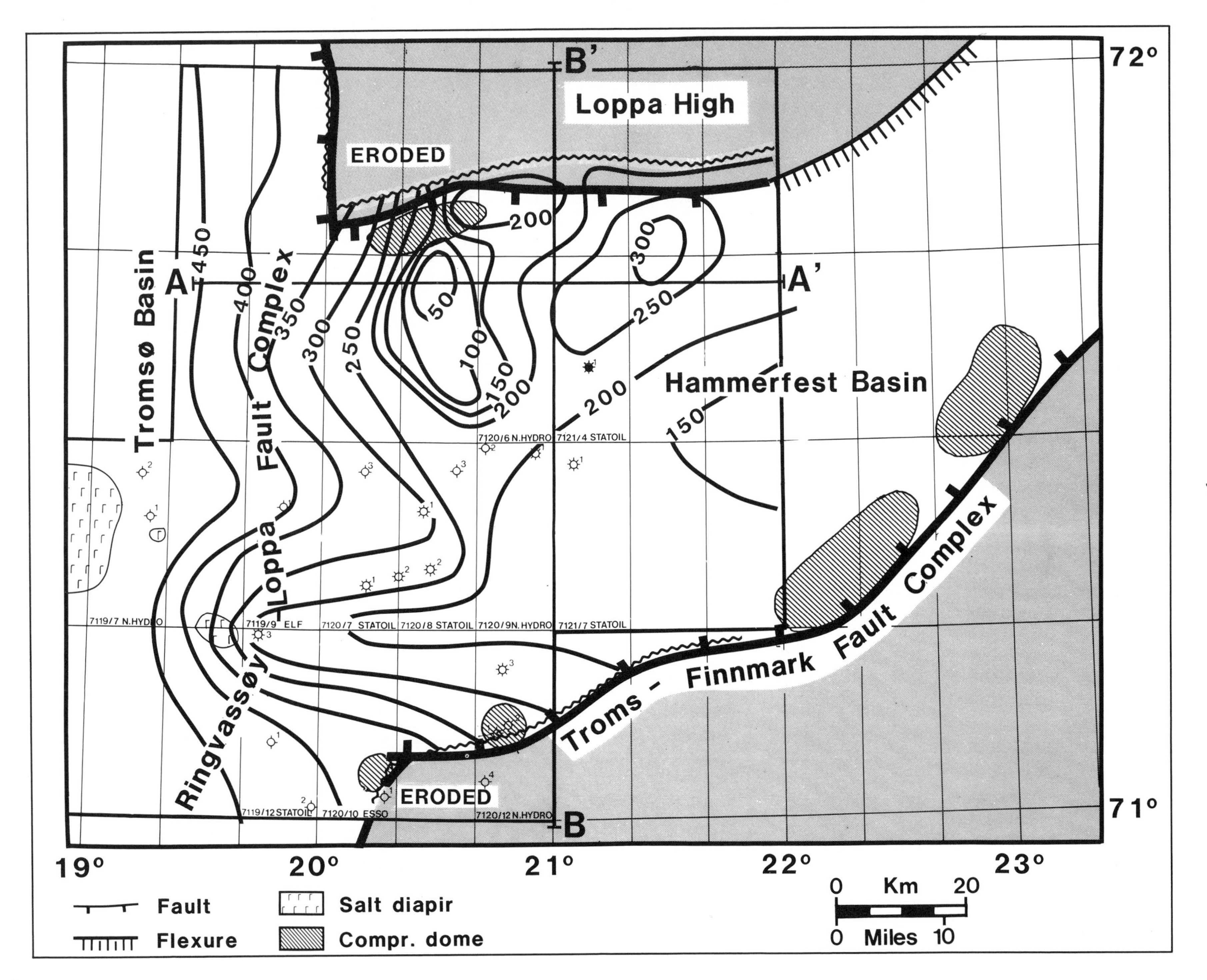
B'
Loppa High
ERODED
72°
A
A'
Tromsø Basin
Ringvassøy - Loppa Fault Complex
Hammerfest Basin
450
400
350
300
250
200
150
100
50
7120/6 N.HYDRO
7121/4 STATOIL
7119/7 N.HYDRO
7119/9 ELF
7120/7 STATOIL
7120/8 STATOIL
7120/9N.HYDRO
7121/7 STATOIL
Troms - Finnmark Fault Complex
ERODED
7119/12STATOIL
7120/10 ESSO
7120/12 N.HYDRO
B
71°
19°
20°
21°
22°
23°
Fault
Flexure
Salt diapir
Compr. dome
0 Km 20
0 Miles 10

boniferous to Lower Permian carbonates, and Cretaceous sandstones represent potential reservoir rocks (Figure 7).

The Carboniferous and Permian sequences are of particular interest on the Loppa High where the top of the carbonate sequence can be reached at approximately 1,800 m (5,900 ft) below sea level. In the Hammerfest and Tromsø basins, the Carboniferous and Permian sequences lie very deep, at more than 4,000 m (13,120 ft) and more than 12,000 m (39,370 ft) respectively.

The Upper Triassic to Middle Jurassic sandstones exhibit good reservoir properties, in particular the upper marine part of the sequence (Olaussen et al, 1984). This reservoir interval was affected by the initial structural movements of the Kimmerian phase, as demonstrated by a thickness map (Figure 8). The frequency of deposit breaks in the sequence increases upward, and locally, the late Kimmerian unconformity erodes the sequence (Figure 4B).

Argillaceous sediments dominate the post Jurassic sequence in the Tromsøflaket area. We can expect intervals with a more sandy facies along the margins of the Loppa High and the Troms-Finnmark Platform. On the Loppa High, the Early Cretaceous was a period with extensive erosion, and a system of valleys draining southward into the Hammerfest basin developed. We can see indications of fan-type prograding sequences on the seismic lines. These fans are redeposited Jurassic and Triassic sediments, but the reservoir properties and areal distribution of these Lower Cretaceous fans are uncertain. The migration path for hydrocarbons, generated in underlying source rock lithologies, also limits the hydrocarbon potential of the fan deposits.

### Source Rocks

Potential source rock lithologies are found throughout the Carboniferous-to-Tertiary sequences in the Tromsøflaket area. We find two intervals of economic interest with respect to hydrocarbon accumulations: the Upper-Triassic-to-Lower-Jurassic coal sequence and the Upper Jurassic marine shales (Figure 7).

The coal sequence consists of interbedded, continental sandstones and shales with coal beds. The organic matter is mainly land-derived type III kerogen, and, with sufficient maturity, will yield gas, condensate, and waxy oil. The interbedded sandstones and shales give a high primary migration efficiency for generated hydrocarbons. Migration and accumulation of hydrocarbons will therefore probably start immediately after generation.

The organic material of the Upper Jurassic marine shales is a type II and mixed type II/III kerogen, with Total Organic Carbon content of up to 12%. With sufficient maturation, the marine shales will yield oil and gas. The sequence is thin on structural highs, but thickens downflank, reaching maximum thickness in the Tromsø basin and in the two deep depressions in the Hammerfest basin, along the margins of the Loppa High, and the Troms-Finnmark Platform (Figure 4B). We believe that the primary migration efficiency for the Upper Jurassic shales is much lower than for the Upper Triassic to Lower Jurassic coal sequence. A higher level of maturation, therefore, is needed for the Upper Jurassic source rock before expulsion, migration, and accumulation of hydrocarbons will occur.

### Maturation, Generation, and Migration

The hydrocarbon potential of the Tromsøflaket area is directly related to the maturation history of the source rock intervals—hence to the structural evolution of the area. Figure 9 shows the current maturity on the top Jurassic level.

The Triassic and Jurassic source rocks in the Tromsø basin reached a maturation level equivalent to peak oil generation 120 million years ago, and rapid subsidence and sedimentation continued throughout the Late Cretaceous causing overmaturation of the source rocks.

In the Ringvassøy-Loppa Fault Complex, peak oil generation was reached 100 million years ago. In the deepest parts of the fault complex, the source rocks reached a stage of overmaturation 40 million years ago. The eastern part of the fault complex still generates both oil and gas. We can see several large gas leakages related to major faults on seismic lines across the fault complex. There are, however, no larger structures in a favorable setting to trap migrated hydrocarbons in the Ringvassøy-Loppa Fault Complex. Long distance migration of hydrocarbons from the fault complex into the western part of the Hammerfest basin is very complicated because of the high frequency of faults.

The Hammerfest basin has the highest hydrocarbon potential of all structural elements in the Tromsøflaket area. In this area the Triassic and Jurassic source rocks were still immature 100 million years ago. The Upper Triassic to Lower Jurassic coal sequence started to generate gas, condensate, and waxy oil 40 million years ago, both south and north of the central dome. The Upper Jurassic shales matured sufficiently for oil generation a little later. Because of the poor primary migration efficiency of the shales, expulsion and migration from the Upper Jurassic shales have taken place only where the maturation has reached the mature part of the oil window; that is, only in the northwestern part of the basin, north of the central dome (Figure 9). South of the central dome, no hydrocarbons (or very few) have been expelled from the Upper Jurassic shales.

The source rock potential on the Loppa High is poor, and the Triassic and Jurassic source rock sequences are not preserved on the High. The hydrocarbon potential on the Loppa High is, therefore, dependent on long-distance migration from the northern hydrocarbon province in the Hammerfest basin, or on generation of hydrocrabons from Paleozoic source rocks within the high.

## CONCLUSIONS

The Senja Ridge, the Tromsø basin and the Ringvassøy-Loppa Fault Complex have a very low hydrocarbon potential. In the Hammerfest basin, several gas discoveries have been

Figure 9. *Present day maturation (vitrinite reflectance) map on the top Jurassic level. The Jurassic source rocks are overlooked in the western part (see text discussion). The higher maturation is evident toward the Loppa High compared with that toward the Troms-Finnmark Fault Complex. The map is based on wells and computer modelling.*

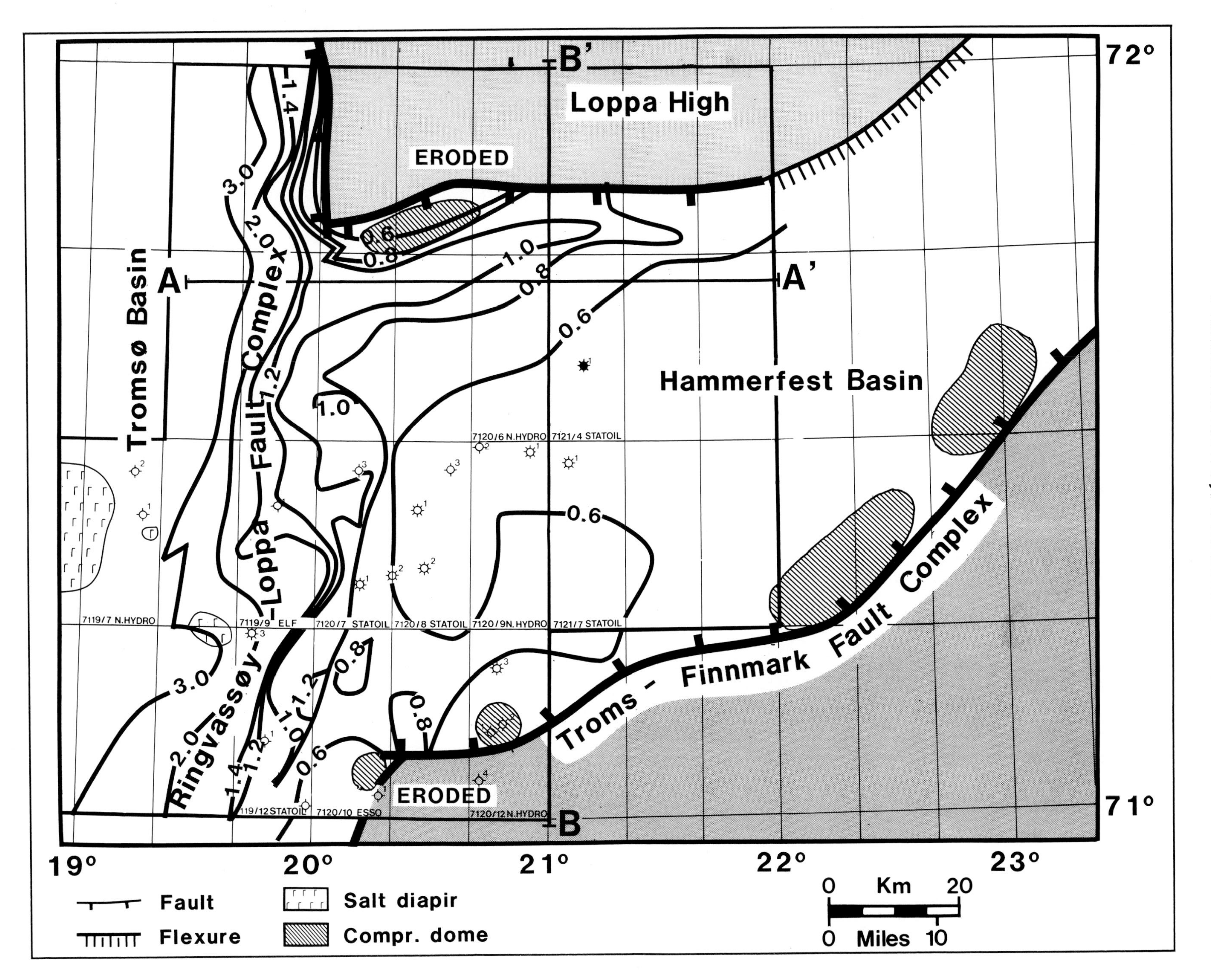
72°
71°
19°
20°
21°
22°
23°
B'
B
A
A'
Loppa High
ERODED
Hammerfest Basin
Tromsø Basin
Ringvassøy-Loppa Fault Complex
Troms - Finnmark Fault Complex
ERODED
7120/6 N.HYDRO
7121/4 STATOIL
7119/7 N.HYDRO
7119/9 ELF
7120/7 STATOIL
7120/8 STATOIL
7120/9N.HYDRO
7121/7 STATOIL
119/12 STATOIL
7120/10 ESSO
7120/12 N.HYDRO
Fault
Flexure
Salt diapir
Compr. dome
0 Km 20
0 Miles 10

made south of the central dome (Figure 10). Total discovered resources are $170 \times 10^9$ cu m ($6 \times 10^{12}$ cu ft) of recoverable gas. We foresee only a limited remaining potential in this area.

North of the central dome, one discovery of gas and light oil has been made. Future drilling in this area will prove additional oil and gas resources, and total discovered and undiscovered resources are expected to exceed $200 \times 10^6$ cu m ($7 \times 10^{12}$ cu ft) recoverable oil equivalents.

The Loppa High has a fair potential as an oil and gas province, but we are uncertain both about reservoir rocks and migration of hydrocarbons. The traps are probably dependent on long distance migration from the northern province in the Hammerfest basin, or the Bjørnøya basin. Estimates of recoverable hydrocarbons vary from negligible to $200 \times 10^6$ cu m oil equivalents.

## REFERENCES

Bleie, J., E. Nysæther, and K. A. Oppebøen, 1982, The hydrocarbon potential of the Northern Norwegian Shelf in the light of recent drilling, *in* The geological framework and hydrocarbon potential of basins in northern seas: Proceedings from Offshore Northern Seas 1982, Conference and Exhibition, part E/5, 39 p.

Escher, J., and W. S. Watt, 1976, Geology of Greenland: Geological Survey of Greenland, Copenhagen.

Faleide, J. I., S. T. Gudlaugsson, and G. Jacquart, 1984, Evolution of the western Barents Sea: Marine and Petroleum Geology v. 1, p. 123–150.

Gloppen, T. G., and S. Westre, 1982, Petroleum potential big off Northern Norway: Oil and Gas Journal, June 7, 1982, p. 114–127.

Olaussen, S., T. G. Gloppen, E. Johannessen, and A. Dalland, 1984, Depositional environment and diagenesis of Jurassic reservoir sandstones in the eastern part of Troms I area, *in* A. M. Spencer, ed., Petroleum geology of the North European Margin: Norwegian Petroleum Society, p. 61–80.

Rønnevik, H. C., and H-P. Jacobsen, 1984, Structural highs and basins in the western Barents Sea, *in* A. M. Spencer, ed., Petroleum geology of the North European Margin: Norwegian Petroleum Society, p. 19–32.

Spencer, A. M., P. C. Home, and L. T. Berglund, 1984, Tertiary structural development of Western Barents Shelf, *in* A. M. Spencer, ed., Petroleum Geology of the North European Margin: Norwegian Petroleum Society, p. 199–210.

Steel, R. J., and D. Worsley, 1984, Svalbard's post-Caledonian strata—an atlas of sedimentational patterns and paleogeographic evolution, *in* A. M. Spencer, ed., Petroleum Geology of the North European Margin: Norwegian Petroleum Society, p. 109–135.

Syrstad, E., S. Bergseth, and T. Naurestad, 1976, Gravity modelling offshore Troms, northern Norway, *in* Offshore Northern Seas 1976: Exploration Geology and Geophysics, Technology Conference and Exhibition, Stavanger 1976.

Talwani, M., and O. Eldholen, 1977, Evolution of the Norwegian-Greenland Sea: Geological Society of America Bulletin, v. 88, p. 969–999.

Figure 10. *Hydrocarbon discoveries on Tromsøflaket. All discoveries are found in the western part of the Hammerfest basin, and Snøhvit is the only discovery so far containing oil as well as gas.*

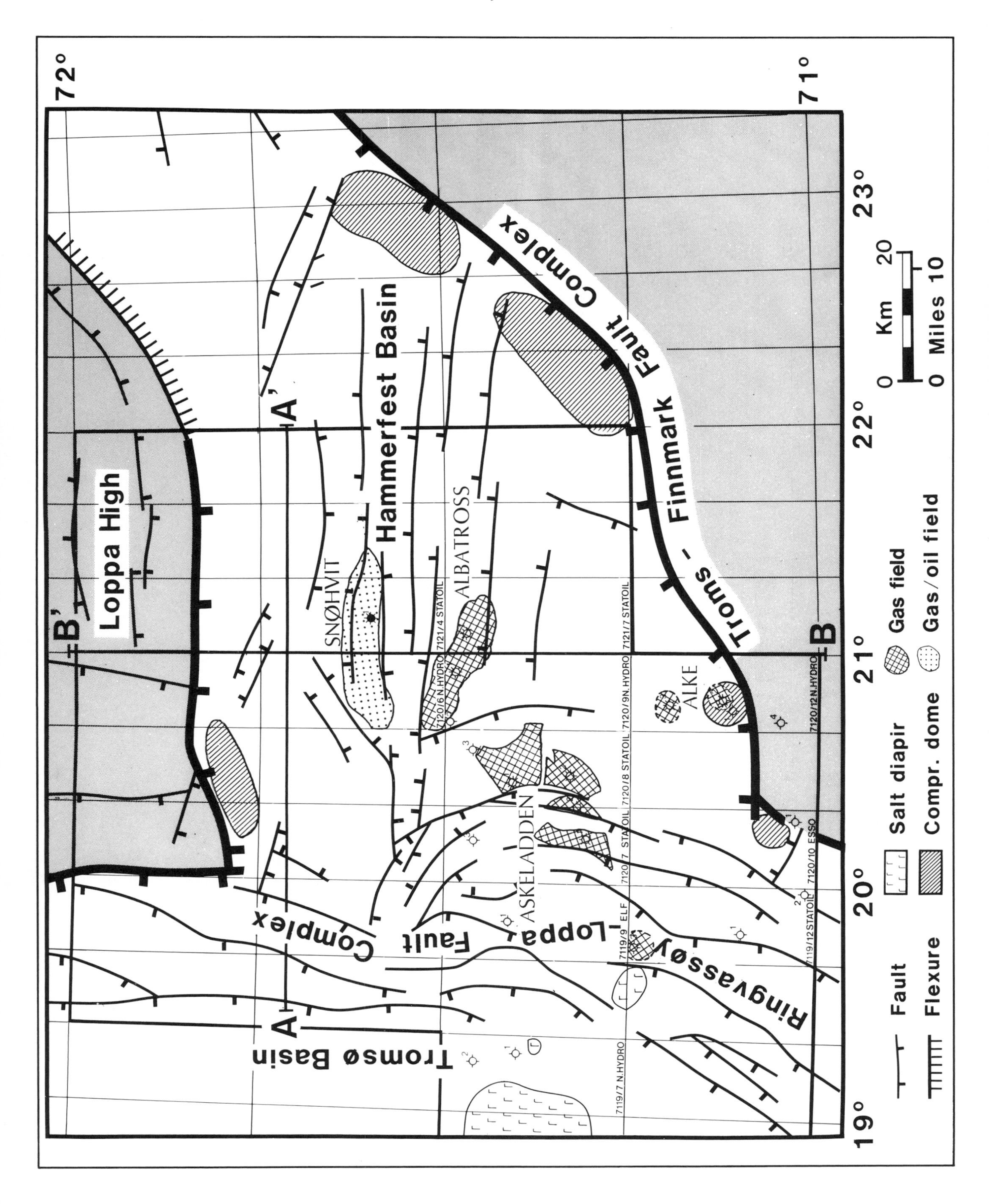
72°
71°
19°
20°
21°
22°
23°
Loppa High
Hammerfest Basin
Tromsø Basin
Troms - Finnmark Fault Complex
Ringvassøy - Loppa Fault Complex
SNØHVIT
ALBATROSS
ASKELADDEN
ALKE
A
A'
B
B'
Km
Miles
Fault
Flexure
Salt diapir
Compr. dome
Gas field
Gas/oil field

# The Permian of the Western Margin of the Greenland Sea—A Future Exploration Target

F. Surlyk
*Geological Survey of Greenland*
*Copenhagen, Denmark*

J. M. Hurst
*BP Development*
*London, England*

P. A. Scholle[1]
*Chevron Oil Field Research Co.*
*La Habra, California*

S. Piasecki
*Geological Survey of Greenland*
*Copenhagen, Denmark*

L. Stemmerik
*Geological Institute*
*University of Copenhagen*
*Copenhagen, Denmark*

F. Rolle
*BP Development*
*Copenhagen, Denmark*

E. Thomsen
*Geological Survey of Denmark*
*Copenhagen, Denmark*

The Upper Permian of central East Greenland contains a relatively thick, widely distributed, oil-prone source rock. It occurs directly adjacent to large, carbonate buildups that constitute the main potential reservoirs. It is immediately overlain by thick, coarse-grained sandstones that are also a potential reservoir. The whole sequence is draped by Upper Permian and Lower Triassic shales that may act as a seal.

This has implications for the Norwegian shelf north of lat. 62°N. There are good reasons to believe that Upper Permian rocks of the same types as in East Greenland exist, deeply buried in this region. The Upper Jurassic shales are immature over large areas of the Norwegian shelf, and the presence of deeper buried Upper Permian source rock thus adds considerably to the petroleum potential of the area.

[1]Currently Southern Methodist University, Dallas, Texas.

## INTRODUCTION

Deposits of Late Permian age occur in a belt up to 100 km (60 mi) wide and 400 km (250 mi) long in central East Greenland, from Jameson Land in the south to Wollaston Forland in the north (Figure 1) (Maync, 1942, 1961; Birkelund and Perch-Nielsen, 1976; Surlyk et al, 1984a). An isolated outlier of carbonate and sandstone occurs in the extreme northeastern corner of East Greenland (Håkansson and Stemmerik, 1984). In the present context we will restrict ourselves to describing the Upper Permian deposits in central East Greenland. They were first discovered on Wegener Halvø by Lauge Koch in 1926 and at Kap Stosch by Wordie in 1926 (Figure 1). Koch (1929) erected the Foldvik Creek Formation to describe a sequence at Kap Stosch of gypsiferous sandy shale overlain by black shale, now known to be of Late Permian age.

The Upper Permian rocks were dated as Upper Carboniferous or Lower Permian in most of the early literature. Aldinger (1935) was the first to demonstrate that the sequence contained ganoid fish fauna of Zechstein (that is, Late Permian) age. However, Rosenkrantz (1929) had already shown that the white, richly fossiliferous, reworked limestone clasts occurring in the basal Triassic at Kap Stosch were of Zechstein age. The faunal similarity to the British and German Zechstein further indicated the existence of a marine connection between the European Zechstein Sea and the Arctic Ocean (Rosenkrantz, 1929).

Maync studied all the main Permian sections north of Jameson Land. He unravelled the very complex facies relations, and his 1942 monograph forms the basis for all later studies of the Upper Permian of East Greenland. Since then, the Upper Permian has received only sporadic study, mainly in connection with regional mapping programs. Spath (1935) and Trümpy (1960, 1969) emphasized the apparently continuous nature of the Permian-Triassic boundary sequence and the alleged co-occurrence of "Permian" and "Triassic" faunal elements at this horizon. This assertion was rejected by Teichert and Kummel (1976), who demonstrated that the "Permian" elements in the basal Triassic beds were actually reworked. Thus, there is no positive evidence for continuous sedimentation across the Permian-Triassic boundary in East Greenland.

Atlantic Richfield Company (ARCO) and Nordisk Mineselskab undertook extensive field work in the sedimentary basin of East Greenland in the early 1970s. Their investigations made an initial evaluation of the hydrocarbon potential, eventually identifying areas or sequences of particular interest.

In 1980, an application for a hydrocarbon concession was submitted by the ARCO group, and a concession, signed in December 1984, covers an area of approximately 10,000 sq km (3,860 sq mi) in Jameson Land (Surlyk, Piasecki, and Rolle, 1985).

The oil section of the Geological Survey of Greenland (GGU) undertook expeditions to Jameson Land in 1982 and 1983 to evaluate the hydrocarbon potential of the area. They focused mainly on the Upper Permian sequence, which is considered the main exploration target. Other parts of the Jameson Land section were also studied with special emphasis on potential source rocks (Surlyk, 1983; Surlyk et al, 1984a, b; Surlyk, Piasecki, and Rolle, 1985). The expedition used extensive shallow-core drilling, particularly in Upper Permian black shale and carbonate units. These cores form the basis for source and reservoir rock analyses. The carbonate sequences were studied in the field by J. M. Hurst, P. A. Scholle, L. Stemmerik, and F. Surlyk, while the black shale and other potential source rocks were studied by S. Piasecki, F. Rolle, and E. Thomsen. The coarser clastic sediments and the basin configuration were studied by F. Surlyk.

It is the results of these two expeditions that are reported here. A formal lithostratigraphic scheme will be published elsewhere.

The recognition of widely distributed, potential source and reservoir rocks in Jameson Land has obvious implications for areas outside Jameson Land. The same facies types occur throughout the Upper Permian of the East Greenland basin and probably occur in the shelf areas north of Scoresby Sund, where deep Mesozoic and older pre-drift basins are known to occur (Larsen, 1984).

Furthermore, the Mesozoic sedimentary sequence in the Haltenbanken area on the Norwegian shelf is very similar to the East Greenland sequence (Jakobsen and Veen, 1984; Karlsson, 1984; Surlyk et al, 1984a; Surlyk, Piasecki, and Rolle, 1985). It is very likely, therefore, that the Norwegian shelf also contains a Permian sequence developed in the same facies associations. If this is the case, it adds considerably to the hydrocarbon potential of the Norwegian areas.

## GEOLOGIC SETTING

The most prominent geological feature in East Greenland is the Caledonian mountain belt that occupies most of the wide strip of land between the coast and the Inland Ice (Figure 2).

The fold belt is dominated by metamorphic complexes of a variety of ages. In central East Greenland, these complexes, are bordered to the east and west, and partially overlain by low-grade to non-metamorphic, middle Proterozoic through lower Paleozoic sedimentary rocks of considerable thickness (Figure 2). The main orogeny appears to have taken place in latest Ordovician to earliest Silurian time (Haller, 1971; Henriksen and Higgins, 1976).

After the main Caledonian orogenic phase, an intermontane basin formed between latitudes 71°30′–74°30′N. More than 7,000 m (22,960 ft) of Middle and Upper Devonian continental red beds were deposited. They overlie the folded Caledonian sequences with pronounced unconformity and contain acid lava flows and tuffs as well as basic dikes and sills. Sedimentation was punctuated by faulting and thrusting, resulting in conspicuous, intra-Devonian unconformities. Major faults bordered the basin.

Friend et al (1983) interpret the Devonian basins in terms of continent-continent collision and major lithospheric, transcurrent fracturing. The major fracture zones governing Devonian sedimentation in East Greenland included both vertical and

Figure 1. *Locality map of East Greenland showing the names mentioned in the text, the outcrop of the Upper Permian Foldvik Creek Group, the main Caledonian thrusts, and main late Paleozoic and Mesozoic faults. A, B, and C indicate the position of the sections shown in Figures 8, 12, and 18.*

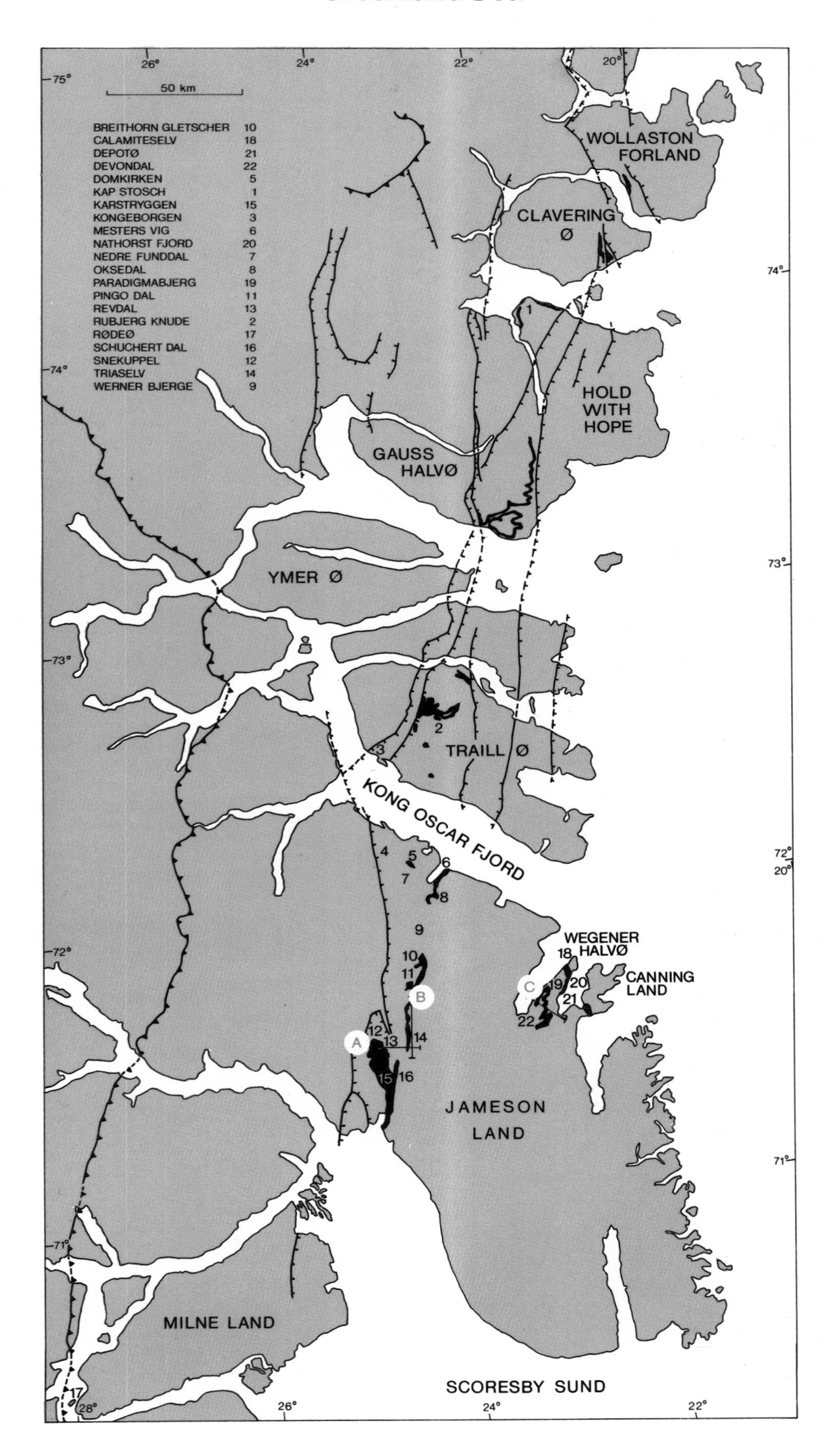
50 km
BREITHORN GLETSCHER 10
CALAMITESELV 18
DEPOTØ 21
DEVONDAL 22
DOMKIRKEN 5
KAP STOSCH 1
KARSTRYGGEN 15
KONGEBORGEN 3
MESTERS VIG 6
NATHORST FJORD 20
NEDRE FUNDDAL 7
OKSEDAL 8
PARADIGMABJERG 19
PINGO DAL 11
REVDAL 13
RUBJERG KNUDE 2
RØDEØ 17
SCHUCHERT DAL 16
SNEKUPPEL 12
TRIASELV 14
WERNER BJERGE 9
WOLLASTON FORLAND
CLAVERING Ø
HOLD WITH HOPE
GAUSS HALVØ
YMER Ø
TRAILL Ø
KONG OSCAR FJORD
WEGENER HALVØ
CANNING LAND
JAMESON LAND
MILNE LAND
SCORESBY SUND

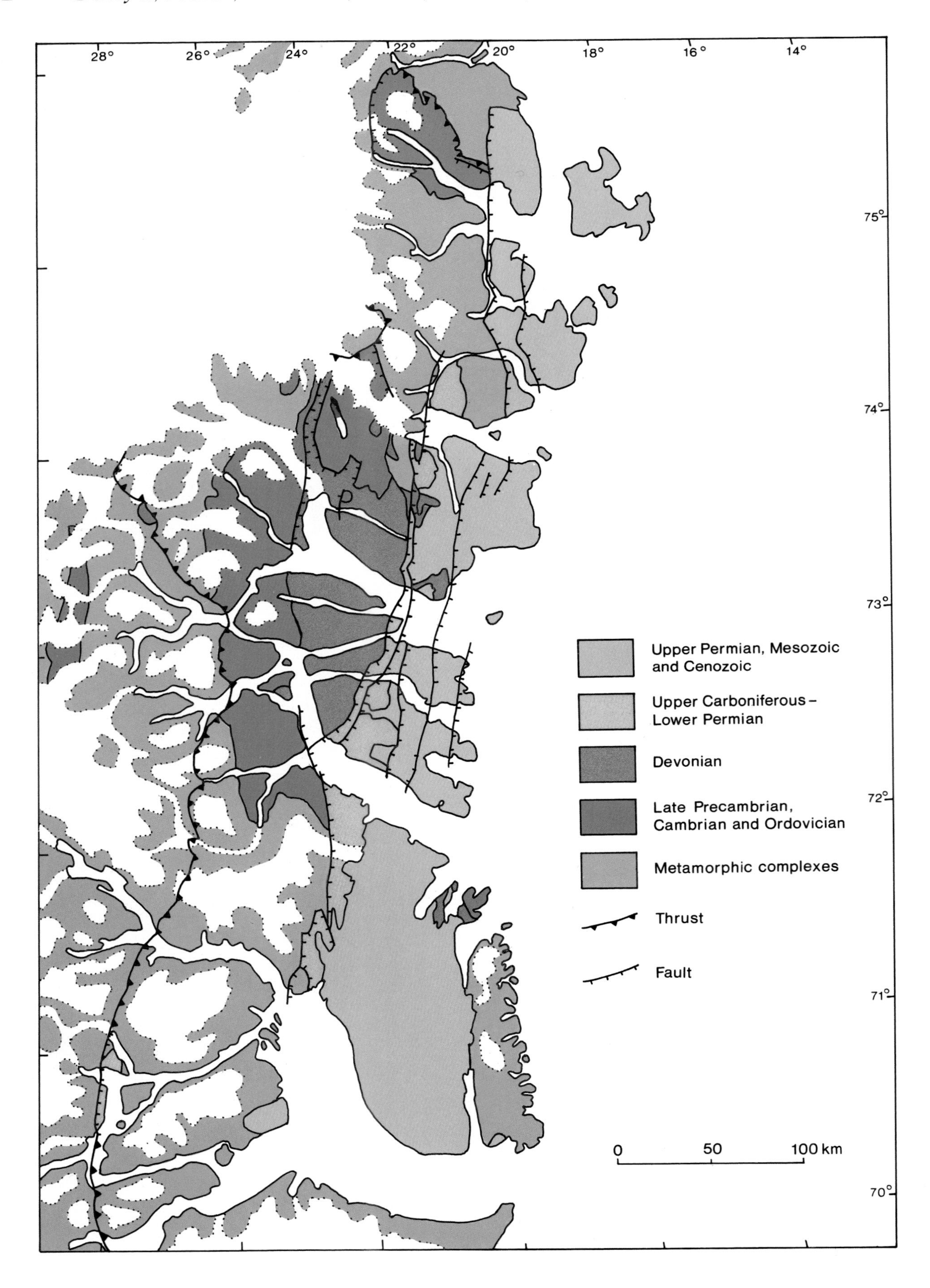

28°
26°
24°
22°
20°
18°
16°
14°
75°
74°
73°
72°
71°
70°
Upper Permian, Mesozoic and Cenozoic
Upper Carboniferous - Lower Permian
Devonian
Late Precambrian, Cambrian and Ordovician
Metamorphic complexes
Thrust
Fault
0
50
100 km

Figure 3. *Thick Lower Permian syn-rift sequence of cyclically alternating fluviatile sandstone and shale in the foreground (mountain about 800 m or 2,625 ft high). The mountains in the background consist of Caledonian basement rocks. View toward the west, northern Jameson Land.*

Figure 4. *Meander-belt sandstones separated by black shale of flood-plain origin. Lower Permian syn-rift sequence, east of Schuchert Elv.*

horizontal components of motion. Friend et al (1983) using the presence of aligned granite intrusions and other geologic lineaments, found evidence of four also roughly north-south trending zones controlling basin evolution. This hypothesis deviates in important aspects from Haller's (1970) concept of a major northwest-southeast oriented, down-warped tract receiving Devonian clastic sediments.

The Devonian sedimentation pattern appears to have continued into the Early Carboniferous when most of the basin was thrown into shallow folds and thrusts, all running roughly north-south (Bütler, 1935, 1959). The folding was followed by regional uplift causing a break in sedimentation over a long period in Early (and possibly later) Carboniferous time.

## LATE CARBONIFEROUS-EARLY PERMIAN RIFTING

A major change in the regional stress field took place some time in the Late Carboniferous. From a dominantly Devonian, left lateral, strike-slip regime (Friend et al, 1983), central East Greenland became subject to mainly east-west tension throughout its remaining geological history.

The first rifting phase preceded early Tertiary spreading by approximately 250 m.y., and the syn-rift sediments were deposited in a system of westward-tilted half-grabens (Collinson, 1972; Haller, 1971; Kempter, 1961; Surlyk et al, 1984a). The thickness of the sequence is of the order of 1.5 to 3 km (1 to 1.8 mi) and it is limited along the western basin margin by major, normal faults that were active during sedimentation. Some of these are reactivated Caledonian thrusts. The eastern border relations are not well known because the sequence is buried by younger sediments, except for a few small outliers on Wegener Halvø and Canning Land. Sedimentation seems, however, also to be controlled by faults in this area. The oldest syn-rift clastic sediments exposed in the Clavering Ø area are considered to be of early Namurian age (Halle, 1953; Bütler, 1961). The age of the bulk of the syn-rift sequence in Jameson Land is Early Permian, perhaps reaching down into the latest Carboniferous (Kempter, 1961; Piasecki, 1984). A pollen assemblage dominated by *Potonieisporites* occurs throughout the sequence, while the genus *Vittatina* had not yet appeared. This indicates deposition during a very short time interval in the basal Autunian, close to the Carboniferous-Permian boundary. The lower part of the assemblage in the Mesters Vig area suggests a possible Late Carboniferous age also in this part of the basin. The basal part of the sequence is not exposed around Mesters Vig, and it is thus very likely that syn-rift deposition began in the Late Carboniferous. Further toward the south, along the western basin margin, continental Lower Permian deposits rest directly on Caledonian basement rocks.

Deposition took place on a system of coalescing, alluvial fans along the western border fault. The fan deposits consist of conglomeratic red beds. Grain sizes rapidly decrease toward the east, where the fans gradually gave way to northward-draining river systems, and a major floodplain developed laterally to the fan systems (Figures 3 and 4). The rivers were meandering, as seen by the thick, uniform successions of alternating sandy point-bar sequences and black shale of backswamp and lake origin (Figure 4).

The boundary between the marginal alluvial fan fringe and the longitudinal fluviatile system fluctuated widely, and correlating the individual lithostratigraphic units has been difficult. A coherent lithostratigraphic scheme is being prepared.

Central East Greenland, therefore, provides good evidence for the initiation of a major rift basin in the Norwegian-

Figure 2. *Main geological divisions of the East Greenland Caledonian fold belt and the post-Caledonian formations. Modified from Henriksen and Higgins (1976) and Haller (1971).*

Figure 5. *View toward the north of tilted Lower Permian red beds overlain with angular unconformity by conglomerates of the Upper Permian Huledal Formation and by gypsum and carbonate conglomerate of the Karstryggen Formation. Note the peneplained nature of the unconformity and the erosive basis of the carbonate conglomerate (top of the hill), inner part of Revdal.*

Greenland Sea area in Late Carboniferous time. Rifting started approximately at the Lower-Upper Carboniferous boundary about 330 Ma in northern East Greenland. The rifting propagated southward and reached the Jameson Land area in the latest Carboniferous, about 290 Ma. Rift basin formation continued into the Early Permian, and the youngest syn-rift deposits are of early Autunian age: about 280 million years old.

The southward-propagating continental rift basin seems to have been filled in the opposite direction with fluviatile sediments deposited by northward flowing rivers (Surlyk et al, 1984a). This relation between rift propagation and drainage pattern has also been demonstrated for the Jurassic sequence of East Greenland (Surlyk, 1977, 1978a, b; Surlyk and Clemmensen, 1983; Surlyk, Clemmensen, and Larsen, 1981), and for the Cretaceous of West Greenland (Surlyk, 1982).

Demonstrating Late Carboniferous-Early Permian rifting in East Greenland lends support to the suggestion that the Carboniferous basins in Britain formed as a result of widespread rifting of the North Atlantic region (Russell and Smythe, 1983; Haszeldine, 1984). There is, however, no evidence for late Paleozoic spreading and formation of oceanic crust in East Greenland. The syn-rift sequence contains important Pb-Zn, barite, and fluorite mineralizations (Harpøth, personal communication, 1985). They have not been precisely dated, but they conform well with the widespread, contemporaneous rift-associated mineralization in the North Atlantic region (Russell, 1978). The Early Permian rifting and extension was also associated with intrusion of lamprophyric dikes.

## EARLY PERMIAN PENEPLANATION

The Late Carboniferous-Early Permian rifting event culminated during and after the early Autunian deposition of continental sediments.

The whole syn-rift sequence underwent further tilting, and was uplifted and eroded before the Late Permian transgression (Figure 5). The Upper Permian Foldvik Creek Group thus rests

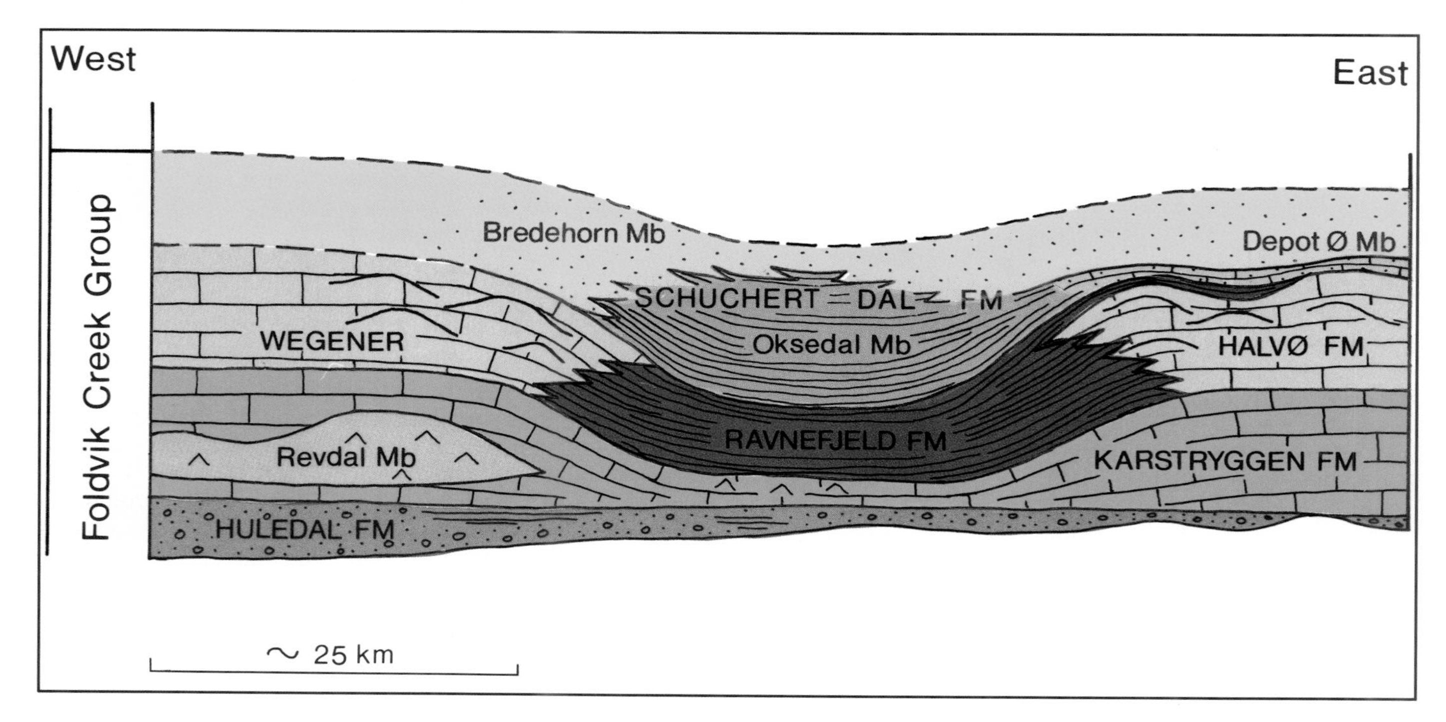


Figure 6. *Lithostratigraphic scheme of the Upper Permian Foldvik Creek Group of Jameson Land showing a highly simplified east-west cross section with the top of the basal conglomerate as baseline. The units are formally defined in a separate paper, which the authors are preparing.*

with a pronounced, angular unconformity of up to 15° on the eroded and peneplained surface of the syn-rift sediments.

The bulk of the Foldvik Creek Group is of Late Permian age, but the lower part contains a pollen assemblage dominated by *Vittatina* and appears to extend down into the uppermost Lower Permian, the Kungurian Stage.

The erosion thus occurred in a period of, at maximum, 20 million years. Two hypotheses can explain the tilting: peneplanation, and angular unconformity at the base of the Upper Permian. First, the Late Carboniferous-earliest Permian syn-rift sequence may have been tilted and eroded during a separate tectonic event some time during the Early Permian (Surlyk et al, 1984a). Second, the rifting and associated tilting may have continued without interruptions from the Early Carboniferous into the Early Permian. Regional uplift, however, caused by thermal bulging associated with the rifting, or a drop in sea-level, resulted in a lowering of erosional base level.

The pronounced half-graben geometry of the syn-rift basins, and the marked increase in the magnitude of the hiatus and in the angle of the unconformity from basinal to crestal area, lends credence to the second hypothesis. That is, rifting and associated block tilting continued through much of the Early Permian, but the area became subject to erosion rather than deposition because of a lowering of the base level. This lowering may be due to regional thermal bulging or a fall in sea-level.

The overlying, Upper Permian sequence can be considered as representing the intitial phase of basin subsidence caused by thermal contraction following the rifting event.

## LATE PERMIAN TRANSGRESSION

### Huledal Formation—The Upper Permian Basal Conglomerate

The Huledal Formation is a fluvio-marine, sandy conglomerate of basal Late Permian or slightly older age (Figure 6).

Following a long period of erosion and peneplain formation in the Early Permian, subsidence resumed. Jameson Land and surrounding areas became part of a low-lying, intermontane, graben-like basin. The relatively even and flat erosion surface became covered by sand and gravel deposited in a system of coarse-grained braid-plains. The resulting unit is a red-brown to drab colored conglomerate (Figure 7) that occurs throughout the area, except at a few localities, and is overlain by the Upper Permian transgressive marine deposits. It has a sheet-like geometry that is remarkably uniform in appearance and shows only minor lateral variation in grain-size. The thickness is typically 20–30 m (66–98 ft) with a maximum measured as 160 m (525 ft) in Hold with Hope. Locally, it reaches thicknesses of 60–70 m (197–230 ft) over the down-thrown side of tilted, Lower Permian fault blocks, but wedges out in the Wegener Halvø area. Marked thickness increases occur, especially where the lower part of the conglomerate is of local syntectonic origin and is seen to fill fault-controlled depressions in the Lower Permian substratum. These depressions may represent small, antithetic graben structures situated over rollovers adjacent to the main, synthetic faults that form the margins of the rift.

The conglomerate is poorly sorted. The matrix is sandy, and the coarse fraction is dominantly in the medium-pebble to fine-

**Figure 7.** *View toward the west in Revdal showing the red Upper Permian basal conglomerate (Huledal Formation) overlain to the left by gypsum (white) and to the center and right by algally laminated limestone (yellow), which formed a barrier limiting the evaporitic lagoon. The whole sequence is overlain by carbonate of the Karstryggen and basal Wegener Halvø Formation.*

cobble fraction. The conglomerate is petrographically immature. The clasts consist of granite, quartzite, limestone of the Proterozoic Eleonore Bay Group, Devonian igneous rocks, silt- and sandstone, and, locally, Lower Permian sandstone. The higher parts of the conglomerate are, in some cases, interbedded with or overlain by light sandstone and red and green siltstone.

The conglomerate shows mainly low-angle, highly irregular cross-bedding and scour-and-fill cross-bedding. Planar-and-trough cross-bedding formed by migrating bed forms occurs but are rare. Prominent erosional surfaces commonly overlain by crude, upward-fining units 0.5–2.0 m (1.6–6.6 ft) thick can be observed in most sections. Paleocurrents are mainly from easterly directions along the eastern basin margin in Canning Land, and from mainly westerly and northwesterly directions in Wegener Halvø and along the western basin margin. Uplifted Caledonian terrain west of the main border fault and in Canning Land were the main source areas for the conglomerates. Crestal highs of the tilted, Lower Permian sequences acted as local source areas, especially during the early phases of conglomerate deposition. Sedimentation took place in low-gradient, alluvial fans along the basin margin passing into a vast, interior, coarse-grained braid-plain. Well-defined channel systems were not developed, and sediments were probably deposited intermittently by flash floods following torrential rain storms.

The influence of the encroaching Late Permian Sea can be seen in the uppermost part of the conglomerate that contains Upper Permian marine fossils. There are no marked changes in sedimentary structures, although the bedding appears to be more regular and the environment seems to have been a wide, very shallow, protected marine bay, where deposits was dominated more by fluviatile than by marine processes. Thus, there is a shift from an intermontane braid plain to a system of coarse-grained, short-headed fan deltas.

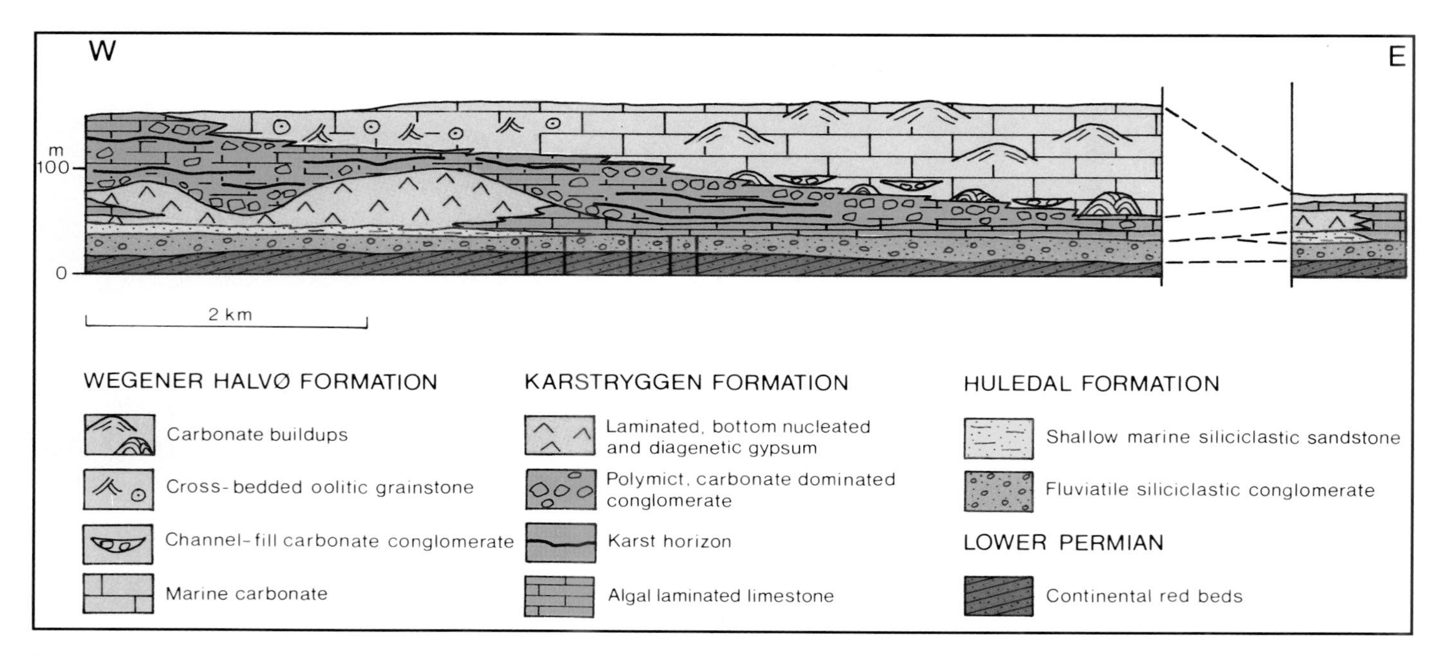


**Figure 8.** *East-west cross section (A on Figure 1) through the Foldvik Creek Group in Karstryggen at the western basin margin. The section is restored with the top of the Huledal Formation as base line. The small section to the right represents the area east of Schuchert Dal. Note the thickly developed gypsum deposits of the Karstryggen Formation located over the down-tilted part of the underlying fault block. Note also the eastward thinning of the Wegener Halvø Formation across Schuchert Dal. The vertical black lines crossing the Huledal Formation are lamprophyric dikes.*

## Karstryggen Formation—Marginal Marine Carbonate and Evaporite

The Karstryggen Formation is a marginal marine carbonate and evaporite unit of Late Permian (Kazanian) age, occurring mainly along Schuchert Dal to the west and on Wegener Halvø to the east.

### *Schuchert Dal Area*

Along the western margin of the basin, the Karstryggen Formation is dominated by limestone deposited in hypersaline, shallow marine and supratidal environments. The sequence includes algally laminated limestone, homogeneous and laminated lime mudstone, oolitic grainstone, and intraclast conglomerate. Subordinate occurrences of penecontemporaneous evaporite are interbedded with these facies. Thick sequences of laminated gypsum are found in localized areas (Figures 5, 6, 7, and 8).

The laminated limestone contains calcitized algally laminated evaporite. The lamination is usually flat, but domal forms are occasionally seen. Teepee structures, small-scale solution breccias, and thin, rip-up conglomerates are associated with the algally laminated limestone. This facies is volumetrically very important in the lowermost part of the Karstryggen Formation throughout the area (Figure 8).

The intraclast conglomerate consists of mainly, angular, unsorted pebble- to boulder-sized clasts of algally laminated limestone and penecontemporaneously dolomitized, homogeneous limestone. Toward the west on Karstryggen, however, clasts of gypsum and Caledonian basement rocks are important, and the conglomerate is associated with solution breccias related to several episodes of karsting (Figure 9). Recrystallized, reddened limestone with large vugs filled with very coarse sparry calcite occurs below undulating planar surfaces. The facies may be associated with channels 5 m (16 ft) deep and tens of meters wide, or it may occur independently. Exposure limitations preclude detailed correlation, but at least four distinct horizons are known. The channel conglomerates and the reddened limestone surfaces occur predominantly in the western side of Karstryggen, but they also extend into the eastern part of the platform. Going eastward on Karstryggen, the intraclast conglomerate becomes interbedded with algally laminated limestone and lime mudstone with a very restricted marine fauna. On the eastern side of Schuchert Dal, the intraclast conglomerate is only found filling small, erosive channels, probably of tidal origin (Stemmerik, in press).

The oolitic grainstone is cross-bedded and the grains consist almost exclusively of ooids. Rarely, algally coated gastropod shell fragments occur within this facies. Occasionally, irregular beds of diagenetic, nodular-mosaic evaporite are also associated with this facies. Oolitic grainstone in the Karstryggen Formation is restricted to localized areas along the east side of Schuchert Dal.

The lagoonal lime mudstone consists of bedded (5–100 cm or 2–39 in), non-fossiliferous, homogeneous and laminated lime mudstone. The bedding-parallel orientation of elongate silt-sized mica grains suggests that the sediments underwent little or no bioturbation. On Karstryggen, this facies is only rarely recognized, while it is widespread along the east side of Schuchert Dal. It often occurs associated with algally laminated limestone and diagenetic evaporite in cyclically developed sequences (Stemmerik, in press).

The diagenetic evaporite forms 10–50 cm (4–20 in) thick beds of nodular and nodular-mosaic evaporite (the classification follows Maiklem, Bebout, and Glaister, 1969). The evaporite was

**Figure 9.** *Polymict conglomerate from the upper part of Karstryggen Formation, westernmost Revdal. The clasts are mainly gypsum and limestone in a matrix of arkosic sand.*

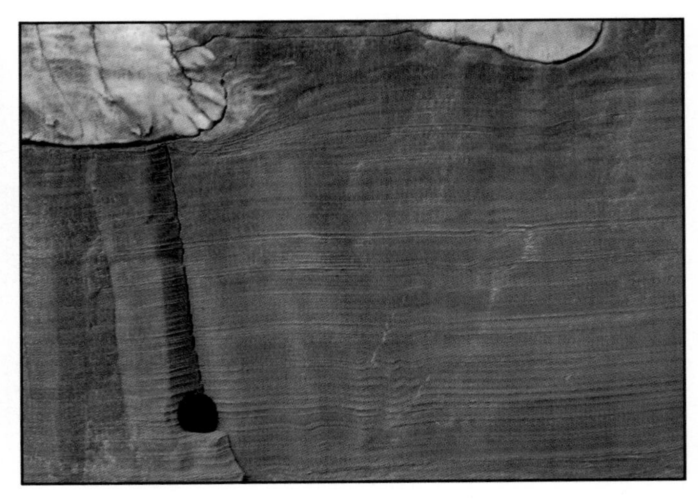

**Figure 10.** *Laminated gypsum and calcite with late diagenetic gypsum nodules (white). Karstryggen Formation, Triaselv. Lens cap for scale.*

formed within algally laminated limestone. The most common types of diagenetic evaporite are displacive and replacive horizontally elongated nodules, and enterolithically folded layers (Stemmerik, in press). Diagenetic evaporite occurs mainly on the eastern side of Schuchert Dal.

The laminated gypsum is restricted to very localized areas where evaporite precipitation dominated sedimentation. The main facies is millimeter-laminated gypsum with individual laminae being traceable over distances up to 40 m (131 ft) (Figures 10 and 11). Vertically aligned gypsum crystals, or nodules, and dome-forming, algally laminated gypsum also occur. These evaporite facies form the main part of the Revdal Member, which has only been observed in the western part of Revdal, and in two areas along the eastern side of Schuchert Dal (Figure 12). In Revdal the member varies in thickness from 0 m in the central part of Revdal to more than 100 m (330 ft) toward the west (Figures 5, 7, and 8). This variation is partly controlled by later erosion associated with the overlying conglomerates (Figure 5).

The limestone and evaporite of the Karstryggen Formation were deposited in a number of hypersaline to evaporitic, shallow marine and supratidal environments.

Algally laminated limestone and gypsum form under shallow subtidal and intertidal, hypersaline and evaporitic conditions (Logan, Hoffman, and Gebelein, 1974; Kinsmann and Park, 1976; Kushnir, 1981). These environments were widespread during the early depositional stage of Karstryggen Formation (Figure 8).

At the same time, a more than 100 m (330 ft) thick sequence mainly of laminated evaporite was deposited in a structurally controlled low area in the western part of Revdal (Figures 5, 7, and 8). These facies represent the deeper lagoonal to supratidal environments of an evaporitic lagoon (e.g., Schreiber et al, 1976, 1982; Warren, 1982; Loucks and Longman, 1982; Butler, Harris, and Kendall, 1982; Warren and Kendall, 1985). The late, diagenetic alteration of the evaporite to secondary gypsum obscures the depositional sequence in the Revdal area. A number of 10–50 m (33–164 ft) thick, shallowing-upward sequences are, however, recognizable. The intercalated laminated limestone indicates periodic lowering of the salinity of the evaporitic lagoon and migration of the surrounding carbonate facies into the area of earlier evaporite deposits.

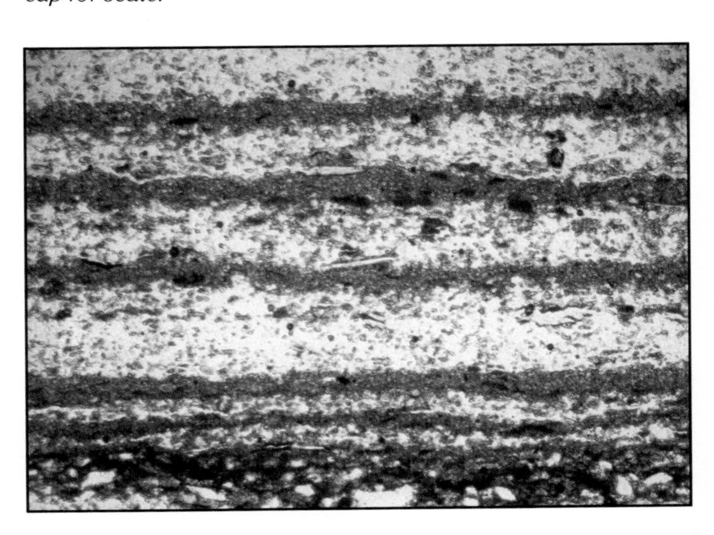

**Figure 11.** *Photomicrograph of laminated gypsum (white) and calcitic microspar (brown). Note the dark particles of organic matter in the calcite laminae. Long axis of photo = 3.2 mm (0.13 in). Revdal Member, Karstryggen Formation, Triaselv.*

The depositional environments changed considerably in the younger part of the Karstryggen Formation. Along the western margin, widespread erosion and deposits of polymict, partly intraclastic, conglomerate were associated with intense karstification (Figures 9 and 13). The karsting is most pronounced toward the west, where the conglomerate is thickest. Toward the eastern part of Karstryggen, the conglomerate is interbedded with algally laminated limestone and restricted marine lime mudstone of the Wegener Halvø Formation (Figure 8).

Further to the east, along the eastern side of Schuchert Dal, sedimentation is not affected by the input of terrigenous clastic material. The widespread sedimentation of algally laminated limestone and gypsum was gradually replaced by a more com-

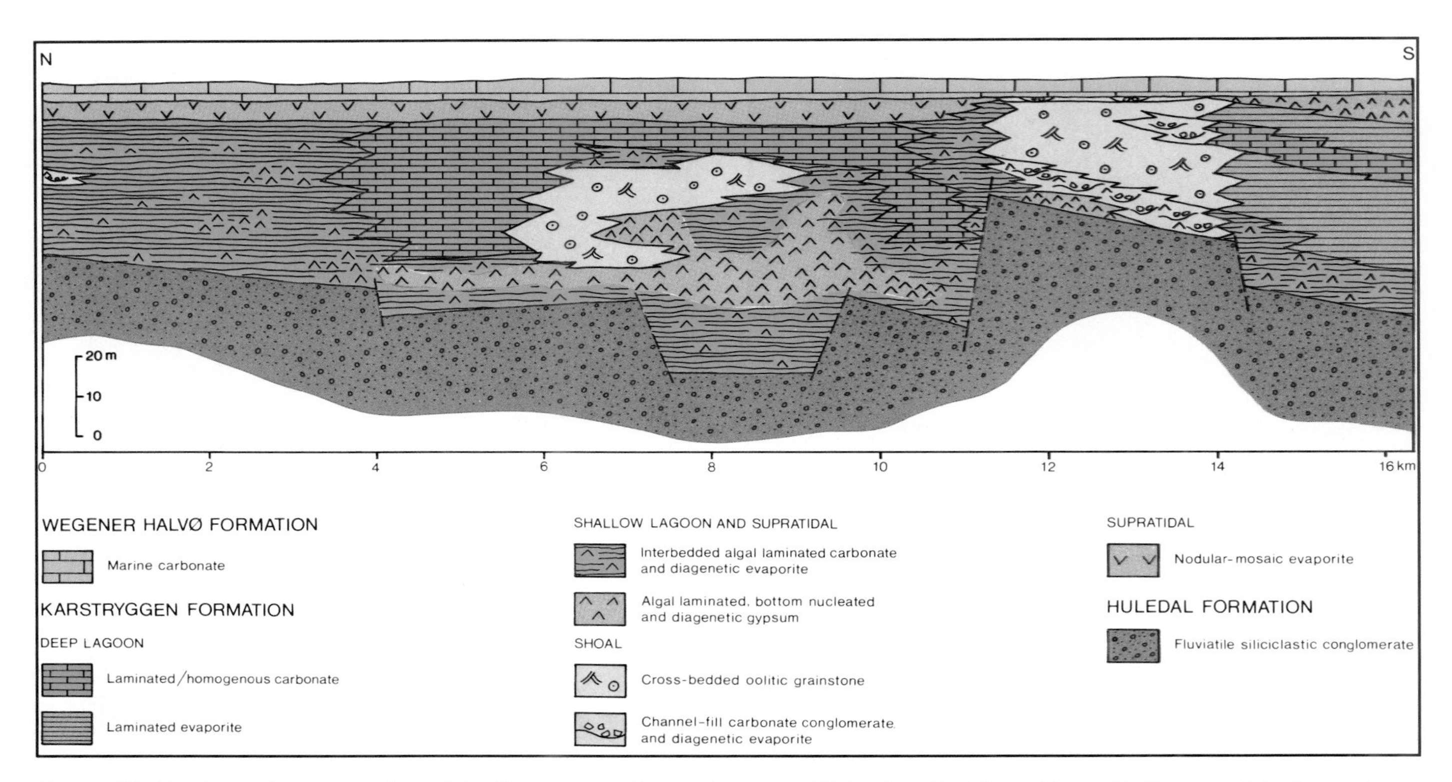


**Figure 12.** *North-south cross-section of the Karstryggen Formation east of Schuchert Dal (B on Figure 1). The top of the formation is used as a base line. Vertical scale strongly exaggerated to show the control on sedimentation exerted by small-scale faults of the Huledal Formation. Note the shoaling-upward tendency of the Karstryggen Formation.*

plex facies pattern. While the algally laminated sediments continued to form in the shallow, protected areas, lime mudstone and laminated evaporite were deposited in deeper, protected environments (Figure 14) (cf. Schreiber et al, 1976). The cyclic association of lime mudstone, algally laminated limestone, and diagenetic evaporite represents repeated shallowing-upward sequences and exposure of the margins of the lagoons.

The protected lagoonal areas were separated by high-energy shoals, where cross-bedded, oolitic grainstone and intraclast conglomerate were deposited (Figures 15, 16, and 17). The bed thicknesses, which rarely exceed 50 cm (20 in.), may reflect the water depth during deposition.

### *Wegener Halvø Area*

In the Wegener Halvø area, the Karstryggen Formation consists almost exclusively of limestone deposited under hypersaline conditions. Locally, in the easternmost parts of this area, solution breccias, calcitized evaporite nodules, and calcite pseudomorphs after gypsum indicate evaporitic conditions during the early stage of sedimentation. The remaining part of the sequence is dominated by laminated, lagoonal lime mudstone with rare intraclast conglomerate. On Canning Land, a thick section of lime mudstone with fenestrae indicates continuous deposition under peritidal conditions (cf. Shinn, 1983).

The upper part of the Karstryggen Formation is intensively karstified, with up to 50 m (164 ft) deep solution pipes. Geologists recognize at least two regional karstification events. The uppermost karst horizon forms the surface on which the marine limestone of the Wegener Halvø Formation was deposited.

**Figure 13.** *Infill in karst cave formed in carbonate conglomerate. Large, sparite crystals are to the lower left. Top of Karstryggen Formation, Revdal. Top of hammer handle for scale.*

## Wegener Halvø Formation—Marine Carbonate

The hypersaline environments of the Karstryggen Formation gradually became flooded during the ongoing, Late Permian transgression. The limestone and evaporite of the Karstryggen Formation thus gave way to marine limestone of the Wegener Halvø Formation.

The formation is characterized by limestones with a normal

Figure 14. *Laminated lime mudstone with graded, erosive limestone beds. Karstryggen Formation, east of Schuchert Dal.*

marine fauna. They are typically buff to grey colored, massively bedded, and cliff-forming. In the Karstryggen area, this formation develops as a laterally continuous platformal sequence, whereas in Wegener Halvø, it occurs mainly as local buildups surrounded and overlain by black shales of the Ravnefjeld Formation (Figure 18). There are eight main facies:

1) Laminated pelletal/micritic limestone. This facies forms discrete or laterally linked hemispherical (LLH) domes up to 5 m (16 ft) in height and more than 10 m (33 ft) across. This facies can occur with or without an associated marine fauna. It occurs commonly at the base of the formation throughout Wegener Halvø and locally at the southern end of Karstryggen (Figure 19 and 20), where it characterizes the transition from the Karstryggen to the Wegener Halvø Formation.
2) Oolitic/oncolitic grainstone. This facies is present as discrete lenticular beds (5 to 10 m or 16 to 33 ft high, up to 200 m or 656 ft wide), with small to medium-scale, high-angle cross-bedding. The oncolitic facies tends to develop in association with stromatolitic layers that locally are associated with ripped-up clasts of stromatolitic limestone. The facies occurs at the base of the formation in Wegener Halvø, at the top of carbonate buildups in Wegener Halvø, and in the westernmost exposure of the northern Karstryggen area.
3) Massive, light-grey limestone composed of fenestrate bryozoan/marine cement grainstone to wackestone (predominantly packstone). These units are poorly bedded and form discrete buildups of up to 150 m (492 ft) thick (Figures 21 and 22). They occur along the eastern side of Wegener Halvø from Devondal in the south, through Paradigmabjerg to Calamitesdal in the north.
4) Thin to thick bedded (10 cm–4 m or 4 in–13 ft), dark-grey, monospecific to polyspecific packstone to wackestone. Individual beds are internally structureless, or graded and laminated. Skeletal components consist predominantly of brachiopods (productids and spiriferids), fenestrate and trepostome bryozoans, and echinoderm debris (Figure 23). Locally, grains are extensively bored, and borings and intergranular porosity of skeletal components are filled with authigenic glauconite. The facies occurs as halos of steeply dipping beds, radially oriented with respect to facies 3. The distribution is the same as facies 3. It is presumed that a similar facies bordered the Karstryggen area, but it is now eroded or obscured.
5) Dark-grey to black, thin-bedded (< 50 cm or < 20 in), massive, graded and laminated lime mudstone. The facies forms thin deposits associated with carbonate buildups and platforms, and contains skeletal fragments and intraclasts reworked from the buildups and platforms (Figure 24). The facies occurs at east Schuchert Dal, southern Karstryggen, Traill Ø, Wegener Halvø, and in the basal part of the Depotø sequence.
6) Light-grey to tan, skeletal and pelletal wackestone. This facies forms as elliptical, lenticular, medium- to high-angle cross-bedded units. Beds dip radially away from the mounds, and there is a predominant cross-bedding direction perpendicular to the long axis of the units—that is, currents moved along the slope, not down the slope. The units are on the upper surface of a single buildup in Devondal and are the predominant facies throughout the upper part of this formation in eastern Karstryggen. They occur as dozens of isolated mounds with heights of between 10–30 m (33–98 ft) and long dimensions of tens to hundreds of meters.
7) Dark-grey, thin-bedded skeletal packstone to pelletal mudstone with abundant faunas of inarticulate brachiopods, fenestrate and trepostome bryozoans and siliceous sponge spicules. The unit is partially silicified with chert and chalcedony replacing many of the brachiopods as well as lesser numbers of bryozoans and echinoderms. Extensive compaction and stylolitization also occur. The facies directly underlies and surrounds the buildups of facies 6 in eastern Karstryggen.
8) Light-grey to yellowish tan, thin-bedded, structureless, miliolid wackestone to mudstone. The facies occurs throughout eastern Karstryggen, underlying facies 7. A major exposure horizon or conglomeratic unit separates it from the Karstryggen Formation.

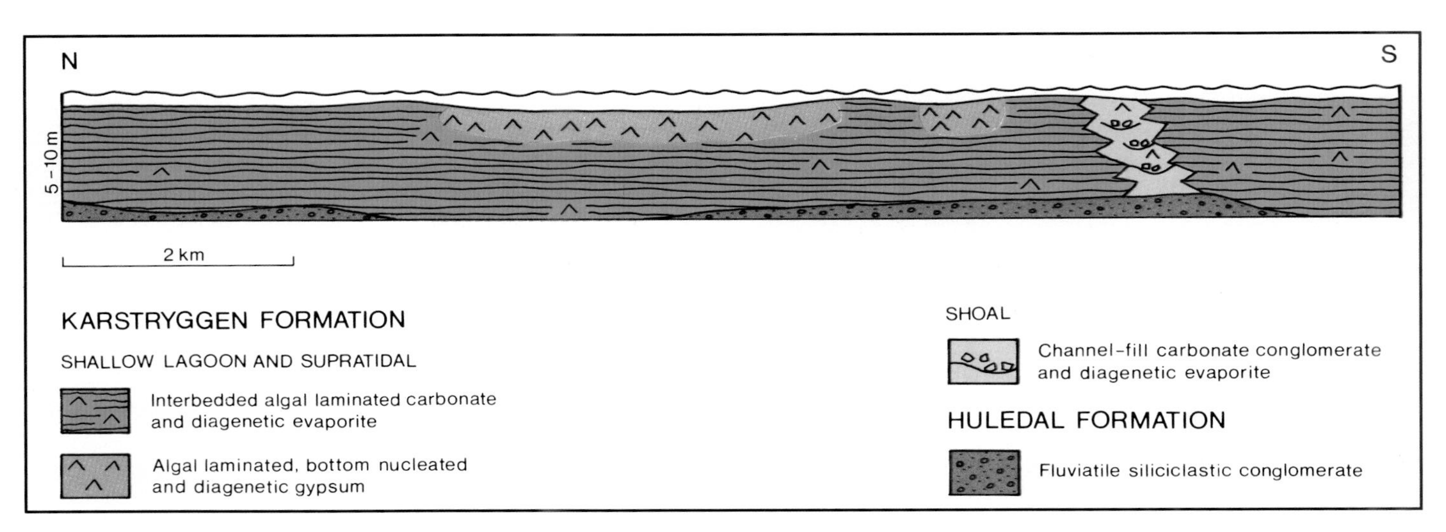


**Figure 15.** *Lateral facies variations during early Karstryggen Formation times. The section corresponds to the lower part of the Karstryggen Formation in section B (Figure 1) from about 4 to 16 km (2 to 10 mi).*

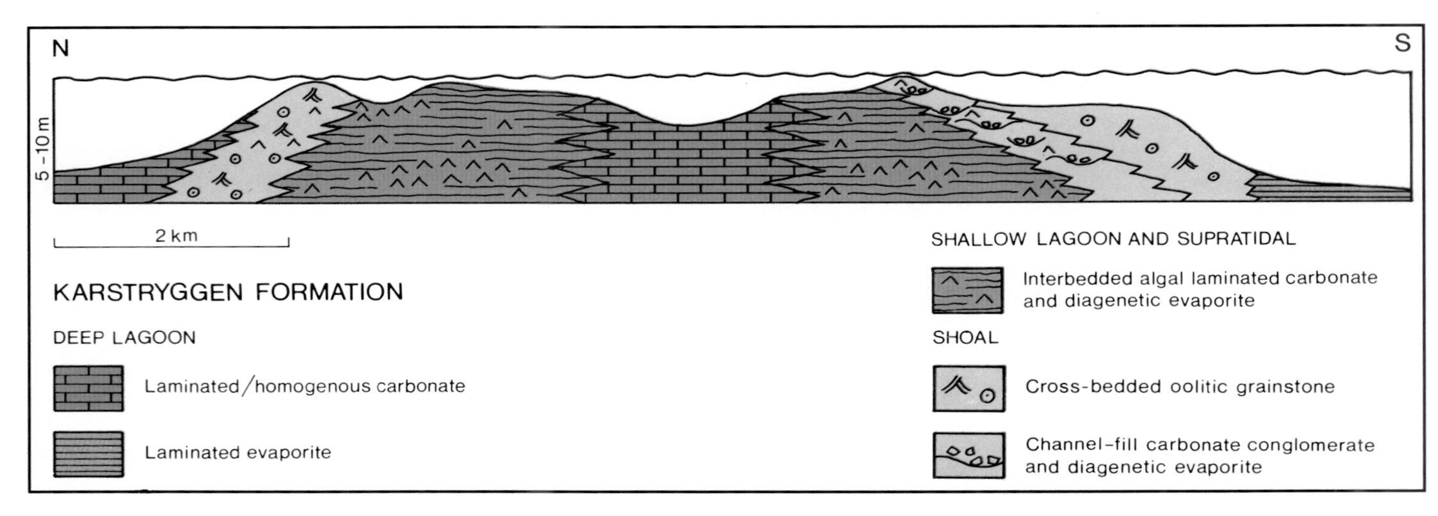


**Figure 16.** *Lateral facies variation of the Karstryggen Formation during a slightly later stage than Figure 15.*

### *Depositional Environments*

The entire Wegener Halvø Formation represents a major transgressive sequence in Jameson Land. In Wegener Halvø (Figure 18) the initial stages of the transgression are represented by small, algal stromatolite mounds (facies 1) and associated oolitic/oncolitic grainstone shoals (facies 2) that formed on topographic highs in a relatively high-energy, probably shoreline, environment. Continued transgression led to fairly rapid drowning of these incipient buildups and shoals. In a few areas along the southern margin of Wegener Halvø, deposition more-or-less kept pace with increasing water depth, forming the major buildups represented by facies 3, 4, 5, and 6 (Figure 22).

Facies 3, the buildup core, has no rigid framework. Fenestrate bryozoans are the only organisms that can be considered to be a framework builder and sediment baffling organism. Encrusting organisms that would indicate organic binding of the buildup occur only rarely in the form of thin, algally laminated beds. However, massive amounts of submarine cement provided sufficient rigidity to maintain the structural and topographic viability of the core. Flank deposits (facies 4) are compositionally similar to the core, but are almost completely lacking in early submarine cement and thus show considerable downslope reworking of sediment. In general, sediment moved as debris flows, grain flows, and turbidity currents on the steep flanks of the bioherm. With increasing transport distance from the buildup, bedding thickness and grain size decrease into facies 5 and ultimately pass gradationally into black calcareous shale of the Ravnefjeld Formation, a basinal correlative of the bioherms.

Cross-bedded shoals (facies 6) developed at the crest of the bioherms during the latest stage of bioherm formation. Although this facies may represent a later phase, relatively high-energy environment in shoaling water, the main bioherm developed in a low-energy environment. The rarity of algal borings, algal encrustations, or algal grains within the bioherm, combined with the predominantly unsorted, muddy sediments, suggests deposition below the photic zone as well as below wave base. In a shale-dominated basin, this suggests water depths in excess of 25 m (82 ft).

The larger bioherm crests appear to have stood about 100 m

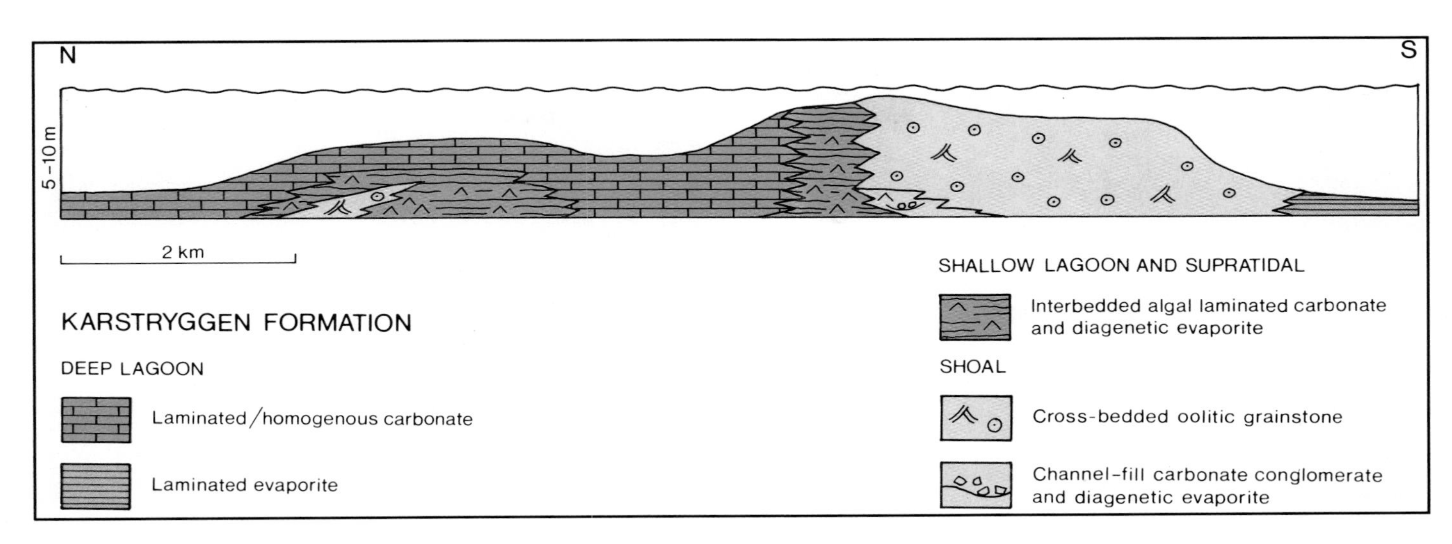


**Figure 17.** *Lateral facies variation of the Karstryggen during a slightly later stage than Figure 16, immediately before the final regression indicated by the widespread supratidal, nodular-mosaic evaporite shown on Figure 12.*

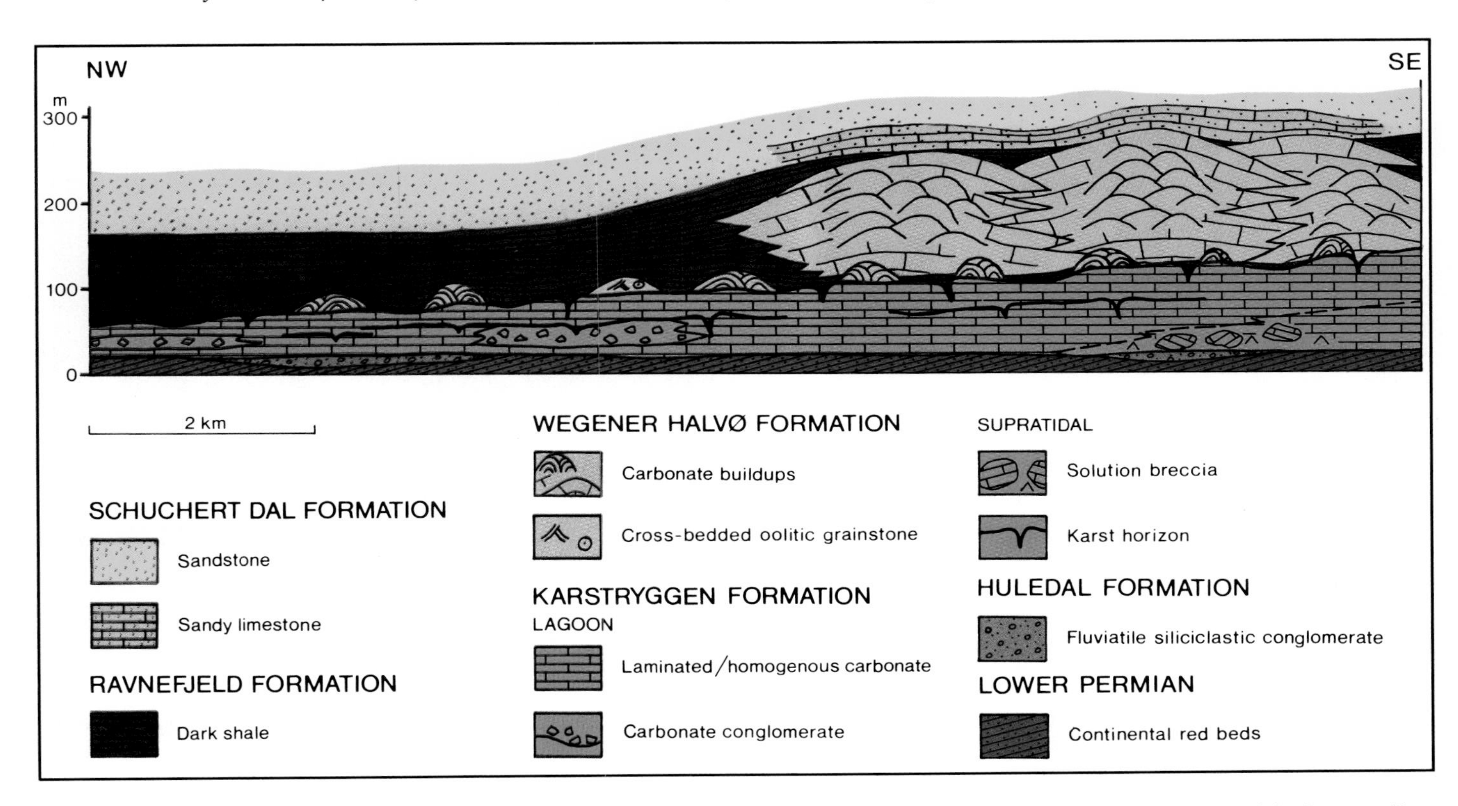


**Figure 18.** *Northwest-southeast cross section of the Wegener Halvø (C on Figure 1). Note the basinward thinning of the hypersaline Karstryggen Formation and the interfingering of the overlying carbonate buildups (Wegener Halvø Formation) and the black basinal shale (Ravnefjeld Formation). Note also the virtual absence of evaporite in this part of the Late Permian basin except for scattered occurrences associated with solution breccias (indicated with inverted V's).*

(330 ft) above the surrounding sea floor, implying total basin water depths of at least 125 m (410 ft) (Figure 25). The presence of numerous borings and the extensive glauconite formation in sediments of the largely autochthonous, proximal flank facies implies low sedimentation rates and early establishment of reducing conditions. This evidence, combined with the absence of a burrowing or autochthonous fauna in the deeper basin flank facies or in the associated shales of the Ravnefjeld Formation (Figure 25), may indicate strongly dysoxic or anoxic conditions in deeper water layers throughout much of the basin. Aborted growth for the majority of the incipient buildups (facies 1) may have resulted more from widespread distribution

**Figure 19.** *Stromatolitic buildups in the basal part of the Wegener Halvø Formation. Note the massive core and the bedded flank facies. The overhanging ledge to the right (west) represents the prograding edge of the normal marine limestone of the Wegener Halvø Formation. Southern Karstryggen.*

**Figure 20.** *Stromatolitic buildups from the basal part of the Wegener Halvø Formation overlain by thin marginal shale wedge of the Ravnefjeld Formation. Southern Karstryggen, 650 m (2,132 ft) east of Figure 19.*

of anoxic conditions than from simple drowning.

The Karstryggen area represents a broad and continuous platform at least 30 km (19 mi) long and at maximum 10 km (6 mi) wide. Lithofacies analysis indicates that a broad range of non-reef facies dominates. The Karstryggen platform is easily separated into an eastern and a western, more rapidly subsiding province both in the Wegener Halvø Formation and in the underlying Karstryggen Formation (Figure 8). This distinction is probably related to a hinge zone that also acts as a zone of weakness for intrusions.

The western half of the Karstryggen area contains a thick sequence of Karstryggen Formation, including shallow-marine to intertidal carbonates that interfinger eastward with the open marine Wegener Halvø Formation (Figure 8).

In eastern Karstryggen the lower part of the Wegener Halvø Formation consists of restricted, marine platform lime muds (facies 8) that pass upward into more open marine limestones. The uppermost part of the section expands in the northern part of the Karstryggen section and contains large, cross-bedded mounds with fully marine faunas (facies 6). These mounds formed above wave base, although the abundance of pellets and micritic matrix precludes a very high-energy depositional environment. Intermound areas (facies 7) are muddy, having large numbers of intact skeletal fragments. Both features indicate deposition in quiet, deeper water. The base of the Wegener Halvø Formation at the southern end of Karstryggen consists of stromatolitic algal domes (LLH type; facies 1) deposited in a shallow marine setting. These limestones interfinger with the western feather edge of the basinal shale of the Ravnefjeld Formation. They are normally devoid of a skeletal macrofauna with the exception of large ostracodes. The stromatolitic domes in some cases form large buildups that show a differentiation into a massive core and a bedded flank facies (Figures 19 and 20).

On the eastern side of Schuchert Dal, the Wegener Halvø Formation is extremely thin (5–15 m or 16–49 ft) and consists entirely of allochthonous (allodapic) limestones, possibly representing a distal, deeper-water facies of the Karstryggen platform (facies 5).

**Figure 21.** *Transition from massive core facies to well-bedded, flanking limestone in small buildup of Wegener Halvø Formation, Karstryggen.*

### Ravnefjeld Formation—Black, Basinal Shale

Along the western basin margin, the Ravnefjeld Formation crops out along the eastern side of Schuchert Dal, continuing toward the north to the Mesters Vig area. The outcrop continues north of Kong Oscar Fjord at the southern coast of Traill Ø over Hold with Hope to Clavering Ø (Figure 1). The other main outcrop area is at the eastern basin margin on Wegener Halvø and in small outliers on Depotø and in Canning Land. The Ravnefjeld Formation comprises most of the Posidonia Shale of Maync (1942) and the Posidonia Shale Member of Birkelund and Perch-Nielsen (1976). The lithology is well represented throughout the entire Upper Permian basin of East Greenland. In the Jameson Land area, the thickness of the formation varies

Figure 22. *Transition from core facies (right) to dipping flank beds (left) in buildup of the Wegener Halvø Formation, Devondal.*

Figure 24. *Plan view of coral rich debris flow deposit derived from the prograding normal marine carbonates of the Wegener Halvø Formation. The debris flow deposits occur in black shale of the Ravnefjeld Formation shown in Figure 20.*

Figure 23. *Graded flank beds from a buildup of the Wegener Halvø Formation. The white pebbles are crinoid ossicles. Devondal.*

Figure 25. *Small basin, between two carbonate buildups of the Wegener Halvø Formation, filled with dark shale of the Ravnefjeld Formation. Both units draped by ledge forming fossiliferous, calcareous sandstone, which is overlain by shale of the basal Triassic Wordie Creek Formation.*

from a few meters along the western margin in the southern part of Karstryggen (Figure 20) to about 60 m (200 ft) between the carbonate buildups on Wegener Halvø. The thickness in the deeply buried central part of the Jameson Land basin is not known, but is most likely of the order of 50–100 m (160–330 ft).

The Ravnefjeld Formation is a black, bituminous, laminated mudstone with a high content of muscovite and, locally, of pyrite. Calcareous concretions and concretionary beds are common. Thin, intercalated sandstones occur along the western basin margin. In the lower part of the formation, the shale interfingers with black, massive, or graded and laminated carbonate beds of the Wegener Halvø Formation (facies 5). The content of organic matter is high in contrast to the overlying shale of the Oksedal Member. The lamination of the shale is caused by the strongly parallel orientation of muscovite and the organic grains, such as woody material and sporomorphs (Figure 26). This fabric may be because of compaction. Very thin bivalve shells are also parallel to lamination (Figure 26B). The lamination, therefore, reflects a combination of texture and composition of the mud (cf. Potter, Maynard, and Pryor, 1980).

The formation is almost barren of indigenous macrofossils in the Jameson Land region except for the supposed pseudoplanktonic bivalve *Posidonia permica* Newell, which is very common at certain levels. Accumulations of disarticulated fish remains, such as scales, have been recognized at several levels in the lower part of the formation, and shark teeth are common in the westernmost outcrop at southern Karstryggen (Figure 20). A few, more complete fish fossils have been found only at Mesters Vig and on Wegener Halvø in connection with carbonate beds.

Earlier reports of indigenous macrofossils, such as fish,

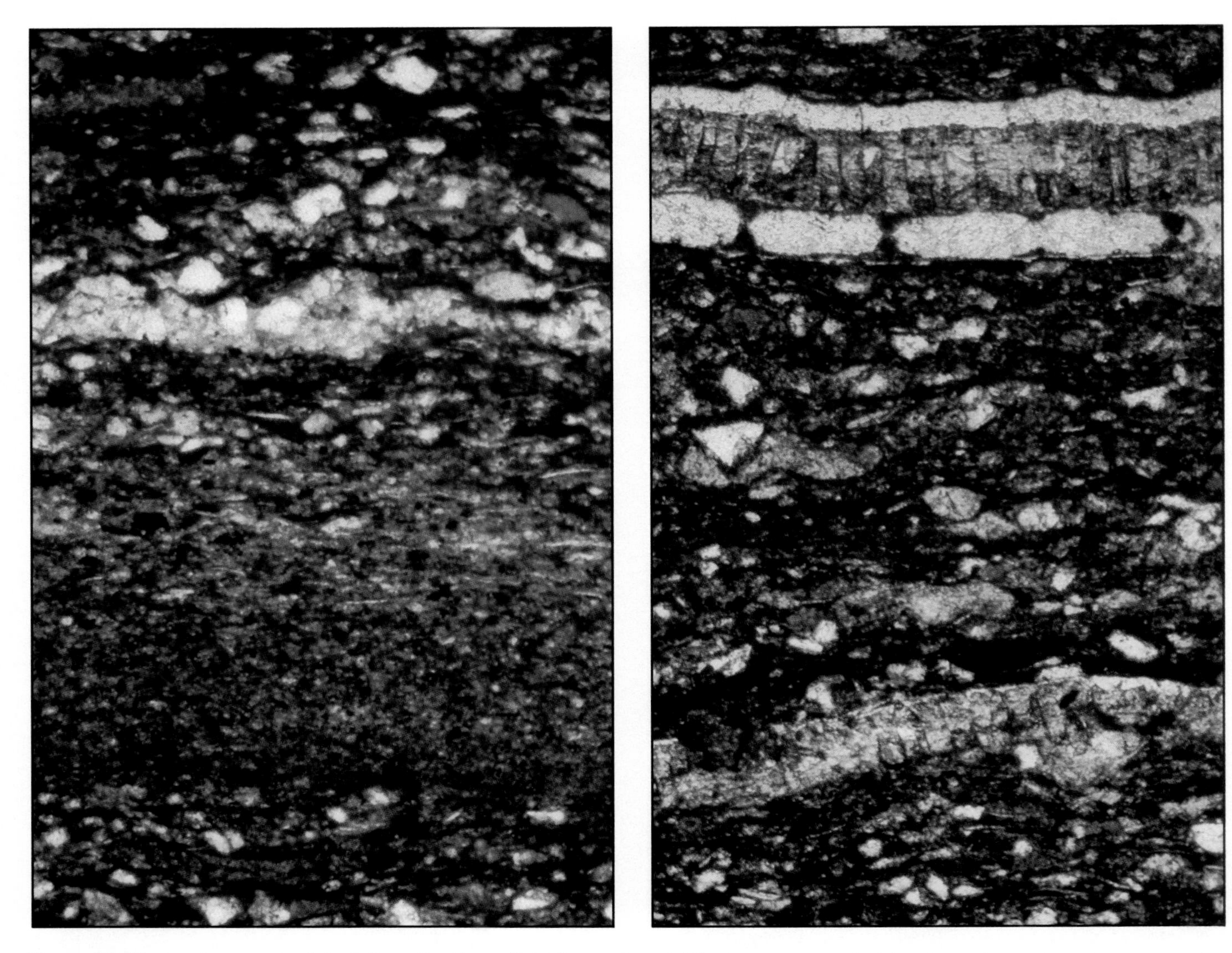

**Figure 26.** *Photomicrograph of shale from the Ravnefjeld Formation. A: The lowest interval of the photo shows detrital quartz grains in a fine-grained matrix with a high content of organic matter forming an irregular lamination. A single lamina of quartz grains forms the basis of a graded interval, with larger grains of quartz, muscovite, and organic matter confined to the lower part. Above a sharp boundary, a texture similar to the lower part of the photo reappears. The long axis of the photo corresponds to 1.7 mm (0.07 in). B: The photomicrograph shows an irregular lamination formed by the organic matter of the sediment and supported by parallel oriented muscovite grains and thin shells of the bivalve Posidonia permica. Quartz grains occur scattered throughout the fine-grained matrix.*

ammonites, belemnites, and small gastropods (Perch-Nielsen et al, 1972), cannot be assigned accurately to either the Ravnefjeld Formation or the Oksedal Member, because these two shale units were not distinguished at that time. These fossils do, however, occur in the Oksedal Member. Sporomorphs are the only microfossils that are frequent in the Ravnefjeld Formation. Tasmanites are concentrated in a few horizons, but acritarchs are absent. The earlier reported conodonts from the Posidonia Shale (Perch-Nielsen et al, 1972) were recovered only from the carbonate beds of the Wegener Halvø Formation that interfinger with the lower part of the Ravnefjeld Formation (S. Stouge 1985, personal communication). The conodonts may, therefore, not be indigenous to the Ravnefjeld Formation.

Several calcareous beds in the Ravnefjeld Formation contain brachiopods, corals, and bryozoans that were resedimented downslope from the topographically higher carbonate buildups together with carbonate debris (Figure 24).

The general absence of indigenous macrofossils in the formation in Jameson Land corresponds well with the absence of acritarchs. In contrast, acritarchs are abundant in the formation at Kap Stosch (Balme, 1979), as are macrofaunal elements (Nielsen, 1935; Bendix-Almgreen, 1976).

The Ravnefjeld Formation thus appears to represent basinal and inter-buildup mud (Figure 25) deposited below a stratified water column that was anoxic in the lower part and could be flooded by fresh water in the uppermost part. The nature of the interfingering carbonates of the Wegener Halvø Formation shows that the Ravnefjeld Formation was deposited at relatively shallow water (c. 25 m or 82 ft) along the western basin margin, while the water depth was about 125 m (410 ft) in the Wegener Halvø area.

The formation is indicated to be of Guadalupian or younger Late Permian age on the basis of conodonts; pollen and spores indicate a Kungurian-Kazanian age. The sum of the biostratigraphic evidence suggests, therefore, a latest Early Permian to earliest Late Permian age for the formation.

### Schuchert Dal Formation—Progradational Shale and Sandstone

The Schuchert Dal Formation includes four main lithologies in the Jameson Land area, but only two of these are important for the evaluation of the petroleum potential of the region (Figure 6). The fine-grained Oksedal Member was earlier included in the Posidonia Shale, and the coarse-grained Bredehorn Member was partly classified as "Complex Series," or as Martinia or Productus Limestone (e.g., Perch-Nielsen et al, 1972).

The outcrop area of the Schuchert Dal Formation is identical to the one described for the Ravnefjeld Formation. A maximum thickness of 200–220 m (660–720 ft) is recorded from western Jameson Land, while a thickness of 50 m (164 ft) is reported from Wegener Halvø, and 30–60 m (100–200 ft) are characteristic in the northern part of the Permian basin.

The Oksedal Member is a strongly bioturbated, calcareous, grey to black, micaceous mud- to siltstone with a low content of organic matter. The degree of bioturbation varies, and laminated shale units locally occur. The member has a fairly low content of autochthonous macrofossils in spite of the intense bioturbation. Ammonites, belemnites, and brachiopods have been reported (Perch-Nielsen et al, 1972), and large complete fish fossils have been collected in the Schuchert Dal area. Some sandstone beds contain abundant resedimented brachiopods and bryozoans that were transported into the environment in debris flows. Sporomorphs are frequent in the Oksedal Member in combination with acritarchs and inner linings of foraminifers.

The Bredehorn Member consists of yellow weathering, fine- to medium-grained, micaceous sandstone with conglomerate. The clast size of the conglomerates varies from 1–2 cm (0.4–0.8 in) to 20–50 cm (8–20 in) with occasional boulders larger than 1 m (3 ft). The clasts in the pebble fraction consist mainly of quartzite, while coarse cobbles and boulders consist mainly of Upper Permian limestone. The sandstone at the lower part of the member contains massive beds with wavy bedding-planes, occasionally with rounded clay clasts and coalified wood concentrated in the top of the beds. Some of the beds show slump folds, and sole marks occur at shale contacts. Conglomeratic intervals, current-produced sediment structures, and density and diversity of trace fossils increases upward through the sequence, reflecting decreasing water depth during deposition. Broadly speaking, the member comprises a coarsening-upward sequence. The top of the formation is, in most areas, represented by a significant depositional break at the Permian-Triassic boundary in combination with strong erosion in the Wegener Halvø region.

The Schuchert Dal Formation, therefore, represents the final clastic infilling phase of the Late Permian basin, reflecting a relative fall in sea-level, or a marked progradation where subsidence could not keep pace with sediment influx. Deposition took place under well-oxygenated conditions with the only exceptions being locally and periodically in the deepest part of the basin. Depositional environments were represented by coarse-grained alluvial fans along the basin margin that passed into short-headed fan deltas. The fine-grained sediments were transported further into the basin in suspension, to be deposited as the basinal mud of the Oksedal Member, while coarser sediments, representing the distal portion of the Bredehorn Member, were transported into deeper water by turbidity currents. Fossils are very uncommon in the sandy Bredehorn Member, except for reworked material, but some of this could be indigenous. Poorly preserved wood fragments occur in the turbiditic sandstones, and the member is barren of organic microfossils reflecting the coarse grain-size.

The age of the Schuchert dal Formation is indicated by fossils from the Oksedal Member. On the basis of conodonts, it is of Guadalupian or younger Late Permian age (T. Frazier, personal communication, 1985), and on the basis of pollen and spores, it is of Late Permian, Kazanian-Tartarian age. The ammonite *Cyclolobus kullingi* also indicates a Late Permian age of the Schuchert Dal Formation (see discussion in Teichert and Kummel, 1976).

The combination of basinal infill and sea-level fall ended the sedimentation in northern Jameson Land, Wegener Halvø, and further north. In the deepest part of the asymmetric basin adjacent to the western border faults, deposition apparently proceeded uninterrupted across the Permian-Triassic boundary, with similar sedimentological facies below and above the boundary.

## THE END-PERMIAN REGRESSION

A major eustatic regression was initiated in latest Permian times. This regression is, in East Greenland, reflected in the occurrence of progradational, shallow marine sandstone of the top Permian Schuchert Dal Formation (Figures 6 and 21), and in the marked unconformity at the Permo-Triassic boundary (Trümpy, 1969; Teichert and Kummel, 1976).

Birkenmajer (1977) demonstrated the presence of major, erosional channels in the top of the Foldvik Creek Group that were filled with grey, coarse sandstone-to-pebble conglomerate. One of the channels in Wegener Halvø is 35 m (115 ft) deep and more than 200 m (656 ft) wide. The channel fill shows transport direction toward the southwest with the channels being either of latest Permian or Early Triassic age.

The area was again submerged by the sea during the Early Triassic, Scythian transgression. For a relatively short period, Jameson Land was the site of a marine embayment, but continental deposition was resumed by late Scythian time (Clemmensen, 1980a, b).

The erosion accompanying the end-Permian regression was most pronounced in the Wegener Halvø area along the eastern basin margin. Toward the west in the Schuchert Dal area, the break in sedimentation is negligible, and there was perhaps continuous sedimentation across the Permo-Triassic boundary. This difference is due to the relatively greater subsidence of the western, down-tilted part of the asymmetric basin (e.g., Vischer, 1943).

## THE EARLY TRIASSIC TRANSGRESSION

Trümpy (1969) demonstrated the progressive eastward onlap of lower Scythian ammonite zones reflecting the gradual sea-level rise. The marine sediments of the early Scythian Wordie Creek Formation consist of grey and green shale with ammonites and fish-bearing calcareous concretions and subordinate grey or green glauconitic or arkosic sandstone (Perch-Nielsen et al, 1974).

In the Schuchert Dal area Late Permian sandstone and shale sedimentation continued without any pronounced break. The early Scythian shale starts to contain sandstone turbidites a few tens of meters above their base, and the formation rapidly coarsens upward to pebbly sandstone deposited from grainflows at the base of a slope to pebbly sandstone deposited by traction currents in very shallow water (Figure 27). These deposits reflect rapid deepening followed by progradation of short-headed, steep-slope fan deltas. By the end of early Scythian times, the basin was filled in, and during the remainder of the Triassic, deposition took place in continental environments, except for a brief marine interlude in the Middle Triassic.

The continental sequences mainly consist of immature, alluvial fan conglomerates along the western and eastern basin margins, and aeolian, fluviatile, and lacustrine sandstone, shale, or gypsum in the basin center (Clemmensen, 1980a, b).

Figure 27. *Coarsening-upward, fan delta sequence initiated by dark shale with thin turbidites that upward pass into channel-fill, pebbly sandstones deposited from grain flows, and well-bedded, pebbly sandstones at the top deposited from traction currents. Basal Triassic Wordie Creek Formation. Triaselv.*

## SOURCE ROCK POTENTIAL

Source rock potential in Jameson Land is restricted to several lithological units in the late Palaeozoic-Mesozoic sequence.

The Lower Permian contains excellent source rocks in the thin shale beds. These high-quality source rocks have been found only on Traill Ø. Similar beds in Jameson Land are generally of poor source rock quality and they will probably not supply hydrocarbons to the Jameson Land area because of the regional maturation there. They are not expected to occur in the central part of the Jameson Land basin because of the half graben nature of the Lower Permian basin.

The Upper Permian contains several units of grey to black shale. Only the Ravnefjeld Formation can, however, be characterized as a potentially good source rock. In addition, the Lower Triassic contains grey and green shale with a very low source-rock potential, but a thin Middle Triassic calcareous sandstone unit, the Gråklint Beds, possesses some potential.

We evaluate the Lower Permian, Upper Permian, and lowermost Triassic sediments in the following with reference to their importance for the forthcoming exploration in Jameson Land. Further, we have concentrated the source rock evaluation on the Lower Permian fine-grained shale and carbonate of floodplain and lacustrine origin, on the black basinal shale of the Ravnefjeld Formation, and on the Lower Triassic shale of the Wordie Creek Formation.

A limited number of Lower Permian shale samples have been analyzed for organic matter content. The total organic carbon (TOC) of samples from southern Traill Ø at Kongeborgen is approximately 10% in most samples, but reaches 20% in one sample. The average is 12.3% (Table 1).

The Upper Permian Ravnefjeld Formation has been thoroughly analyzed, especially in the immature area at Triaselv, east Schuchert Dal (shallow core GGU 303 102). The content of organic matter varies from more than 6% in the lowermost part to 2–4% in the thicker, upper part, with an average of 3.0% based on 77 samples (Table 1). Analyses from the overmature Ravnefjeld Formation at Wegener Halvø also indicate a relatively high TOC of 2–3%, with an average of 2.1%, despite the catagenic depletion of the organic matter. Analyses of the Oksedal Member show low organic matter content in contrast to the Ravnefjeld Formation; 38 samples of immature Oksedal Member at Triaselv show a variation of from 0.1 to 1.6% TOC, with an average of 0.6% TOC (Table 1).

The Lower Triassic Wordie Creek Formation has an even lower organic matter content. Fourteen samples of immature sediments from the Schuchert Dal area vary from less than 0.1 to 1.0% TOC, averaging 0.4% (Table 1).

The three, fine-grained Upper Permian-Lower Triassic units thus show a significant upward decrease in organic matter content reflecting progressively better oxygenated depositional environments.

### Type of Organic Matter

Three different methods have been used to analyze and describe the composition and source rock qualities of the organic matter in the Lower Permian sediments, in the Upper Permian Ravnefjeld Formation and Oksedal Member, and in the Lower Triassic Wordie Creek Formation. Optical analyses have been performed in reflected, normal light, supplemented with blue-light-induced fluorescence (coal petrography), and in transmitted light (palynology), and the same material has been subjected to Rock-Eval pyrolysis. These analyses, have been conducted on sediments with immature to marginally mature organic matter, especially from the Schuchert Dal area, unless otherwise stated. The coal petrographic analyses were carried out on polished blocks of whole rocks, while the palynological analyses were made on organic concentrates.

The description of each sediment unit starts with a coal petrographic description followed by a palynological description. The visual evaluation of the organic matter is then compared to the pyrolysis results, and the source-rock qualities are estimated.

#### *Lower Permian Shale*

The petrographic analyses of the Lower Permian deposits in

Table 1. *Summary of LECO–Rock-Eval data from East Greenland.*

| Lithological Unit | TOC | HI | OI | $S_1$ | $S_2$ | $S_1+S_2$ |
|---|---|---|---|---|---|---|
| Lithological Unit (Upper Carboniferous-Lower Permian) | 12.3 (4) | 738 (4) | 19 (4) | 1.9 (4) | 89.1 (4) | 91.0 |
| Ravnefjeld Formation (Upper Permian) | 3.0 (77) | 342 (55) | 22 (55) | 1.3 (55) | 11.6 (55) | 12.9 |
| Oksedal Member (Upper Permian) | 0.6 (38) | 23 (27) | 39 (27) | 0.1 (38) | 0.2 (38) | 0.2 |
| Wordie Creek Formation (Lower Triassic) | 0.4 (14) | 34 (14) | 81 (14) | 0.1 (14) | 0.3 (14) | 0.3 |
| Gråklint Beds (Middle Triassic) | 1.6 (3) | 177 ( (3) | 21 (3) | 0.3 (3) | 3.7 (3) | 4.0 |

(4) = Number of samples

general indicate the presence of terrestrial organic matter that is generally unfavorable in terms of oil generation. The gas potential is probably limited as well, because of the severely oxidized nature of the organic matter. A few samples from Funddal show some oil potential because of significant liptinite (i.e., bituminite) content.

Shale of flood-plain and lacustrine origin from Traill Ø represents a marked exception. High contents of organic matter have been recorded, mainly in the form of liptinite. Following the classification of Creaney (1980) and Hutton et al (1980), the liptinite is represented mainly by alginite B associated with varying amounts of alginite A and matrix bituminite. Part of the alginite A is attributed to alginite of the *Botryococcus* type. The presence of micrinite and exsudatinite indicates that liquid hydrocarbon generation may have been initiated (Teichmüller, 1974). The petrographic composition of the organic matter is characteristic for oil shales (Figure 28).

The palynological analyses of the organic matter in the Lower Permian shales show a clear variation in source-rock quality. We can distinguish and correlate to depositional environments three different types of organic matter. The organic matter from shale beds in the alluvial fan deposits along the basin margin consists of strongly carbonized, angular to rounded grains with no source-rock potential for oil and gas. The organic matter in shale from the upper part of the fluviatile cycles varies from strongly terrestrially dominated material to a dominance of amorphous kerogen. Dominance of woody material (tracheidal origin), together with varying amounts of pollen and spores, is found in overbank and flood-plain sediments. Dominance of amorphous kerogen in combination with pollen, spores, and varying content of woody material is found in backswamp and lake deposits (Figure 29A). The organic matter in most of these beds is evaluated as having good potential as a source for gas and a high proportion should have good to excellent potential as a source for oil.

The optical analyses indicate a full-scale variation in organic matter type in the Lower Permian sediments. Organic matter of type-I, oil shale, has been documented in a few beds at Traill Ø, and this interpretation is supported by Rock-Eval analysis. Hydrogen-index values exceeding 700 have been recorded associated with a high genetic potential averaging 91.0 (Table 1), classifying these deposits as excellent potential source rocks for oil.

Sediments with mixed-type organic matter, type-II, are much more common and alternate with type-III, rich sediments toward the basin margin. These sediments have not been subjected to Rock-Eval analysis.

### *Ravnefjeld Formation*

The Upper Permian, Ravnefjeld Formation shows varying organic matter content and composition. Comprehensive sampling and shallow core drilling have been undertaken at Triaselv and the material was found to be immature to early mature.

Petrographic investigations show that vitrinite and inertinite are locally important constituents of the organic matter, indicating a significant terrestrial influence. Vitrinite is present as varying amounts of detritic and structured particles, often with distinct oxidation rims. In general, the vitrinitic material appears to be unsorted, but often it shows a distinct lamination. The content of inertinite varies but appears, in general, to be lower than the content of vitrinite; however, liptinite content is locally high. Several samples are characterized by mixtures of strongly fluorescent alginite and sporinite associated with varying contents of liptodetrinite and matrix bituminite. The structured liptinitic particles are often highly degraded. In a few samples, matrix bituminite appears to be the dominant constituent. Alginite of tasmanite-type has been recorded (Figure 30). This favorable type of liptinitic matter is, however, restricted to thin horizons.

Palynological examination of material from the Ravnefjeld Formation shows that the organic matter is composed of three main fractions. Two of these fractions, pollen-spores and finely disseminated, coalified terrestrial material, form a basic content of organic matter. Granular amorphous kerogen forms the remaining, major portion of the organic matter and is responsible for the main variations in TOC (Figure 29B). Degraded tasmanites occur, concentrated in two horizons a few centimeters thick, but are otherwise absent. Acritarchs have not been recorded in this formation in Jameson Land, but the amorphous kerogen represents degraded algal material. Macrofaunal remains are absent in the palynological preparations.

The petrographic and palynological analyses indicate a mixed type of organic matter, a conclusion supported by the results of Rock-Eval analyses. The kerogen type can be classified as type-II (Figures 31 and 32) and the average genetic potential is high, classifying this formation as a good potential source rock for oil (Table 1).

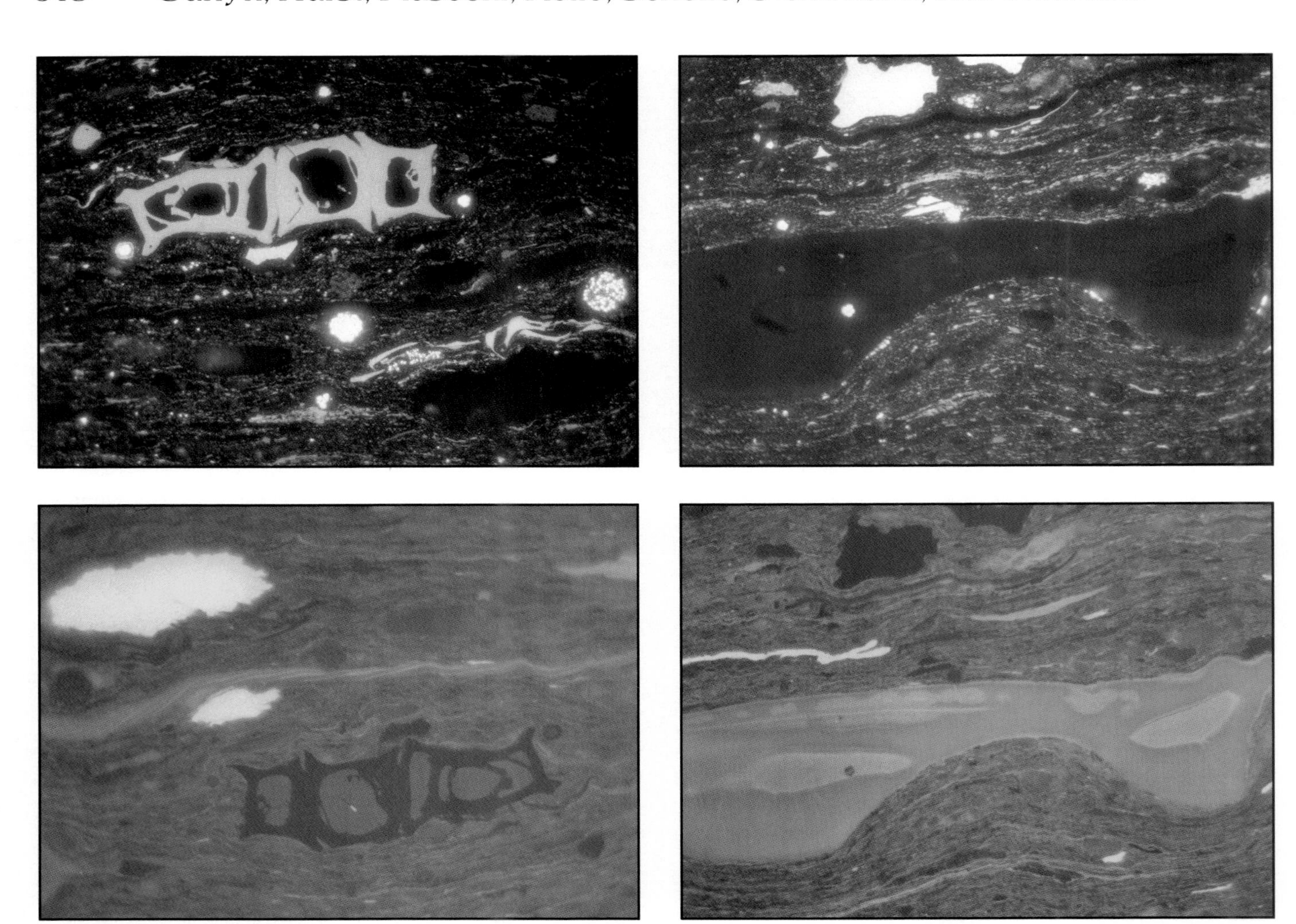

**Figure 28.** *A: Fusinite, inertodetrinite, and micrinite in a bituminous groundmass. Shale from the Upper Carboniferous or Lower Permian at Kongeborgen, Traill Ø. GGU 255024. Normal reflected light. Field width = 0.2 mm (0.008 in).*
*B: Same field as A, however, with blue-light induced flourescence. Sample dominated by lamellar type B alginite associated with structured alginite A of Botryococcus type. Exsudatinite in the cell lumina of fusinite.*
*C: Skeletal remain (ostracode?) in a bituminous groundmass with micrinite and inertodetrinite. Shale from the Upper Carboniferous-Lower Permian at Kongeborgen, Traill Ø. GGU 255024. Normal reflected light. Field width = 0.2 mm (0.008 in).*
*D: Same field as C, however, with blue-light induced fluorescence. Sample dominated by lamellar type B alginite. Exsudatinite in cavities in skeletal fragment associated with ?feeder channels.*

*Oksedal Member*

Petrographically, the organic material in the Upper Permian, Oksedal Member is dominated by oxidized vitrinite of varying grain size, associated with inertodetrinite and low contents of liptinite.

The palynological analysis of this member shows that the main difference from the Ravnefjeld Formation is the absence of amorphous kerogen. The organic matter is, therefore, dominated by pollen-spores or by strongly coalified, woody material, mainly of tracheidal origin (Figure 29C). Small amounts of amorphous kerogen occur at a few levels, but this is not characteristic for the member in the regions where it is exposed. Acritarchs form a small but significant fraction, and foraminifer test linings occur sporadically.

The organic matter is rated as unfavorable with respect to oil generation and the gas potential is also low. The occurrence of a few horizons with a limited potential for oil indicates, however, that regional variations may raise the potential of this member locally.

The Rock-Eval analyses classify the organic matter as type-III, supporting the petrographical and palynological results (Figure 34). The genetic potential of the organic matter is thus very low, classifying this unit as a poor source rock for oil (Figures 33 and 34).

*Wordie Creek Formation*

The Lower Triassic, Wordie Creek Formation is petrographically characterized by low contents of organic matter, dominated by highly oxidized vitrinite and inertinite. Only a few samples contain liptinite in significant amounts.

The organic matter composition is palynologically very simi-

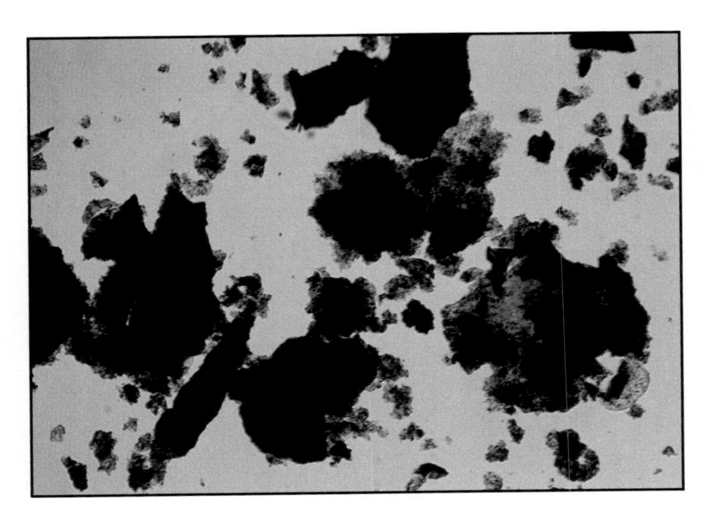

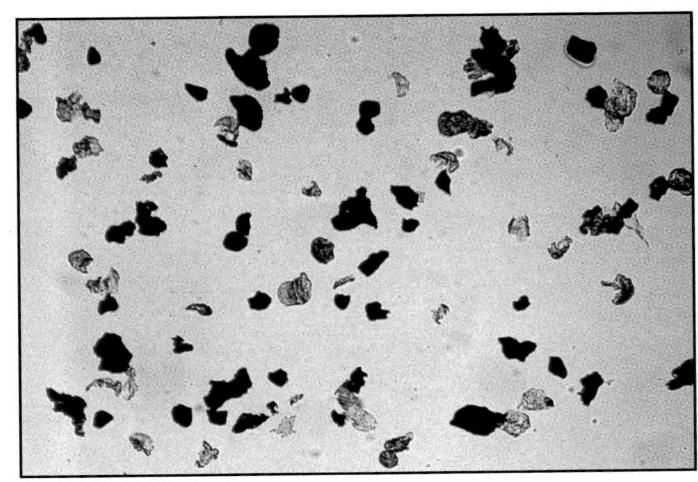

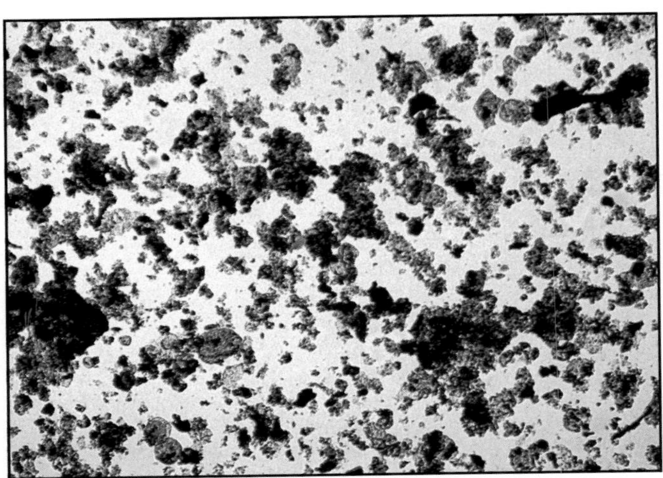

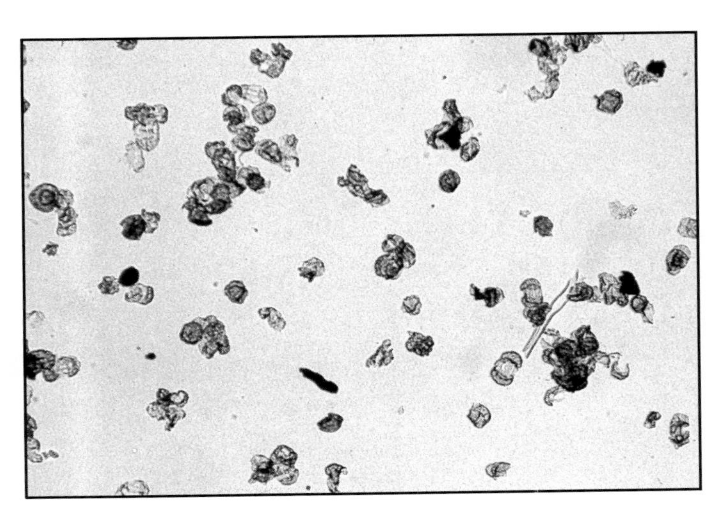

Figure 29. *Palynological preparations (concentrates) of the organic matter from:*
*A: Lower Permian shale of the Mesters Vig Formation;*
*B: Upper Permian shale of the Ravnefjeld Formation;*
*C: Upper Permian shale of the Oksedal Member, Schuchert Dal Formation;*
*D: Lower Triassic shale of the Wordie Creek Formation.*
*The sediment was treated with hydrochloric and hydrofluoric acids and the organic residues were sieved on a 20 micron mesh filter. The long axis of the photos = 0.5 mm (0.02 in).*
*A: The organic matter is completely dominated by dark, granular, coherent amorphous kerogen with scattered sporomorphs and strongly degraded algal bodies of Botryococcus-type. Other types of structured kerogen are very rare.*
*B: The organic matter is dominated by less coherent, granular, amorphous kerogen with frequent, degraded sporomorphs and carbonized grains of woody material.*
*C: The organic matter is dominated by rounded, coalified grains of woody material with a high content of degraded sporomorphs.*
*D: The organic matter is dominated by slightly degraded sporomorphs with a low content of rounded, coalified grains of woody material.*
*The 4 organic assemblages show the evolution from (A) limnic-anoxic over (B) shallow brackish/marine-anoxic environment at the western basin margin, and (C) shallow marine oxygenated to (D) deeper marine oxygenated environments.*

lar to the Oksedal Member below, reflecting a similar depositional environment. The sporomorphs, however, generally dominate over the coaly fragments. Amorphous kerogen has been found but is extremely rare (Figure 29D).

The organic matter in this formation is unfavorable for oil generation. The Rock-Eval analyses classify the organic matter as a type-III kerogen with a very low genetic potential (Figures 35 and 36).

### Generation Potential of the Ravnefjeld Formation

Estimates of the volumetric yields of a potential source rock can be useful for petroleum basin evaluation in the early stages of exploration. Besides the many pitfalls in estimating the potential yields of the source rocks, the effects of expulsion, migration, trapping, and preservation of these potential hydrocarbons also have a strong influence on hydrocarbon recovery. In the case of the Jameson Land basin, Rock-Eval analyses of

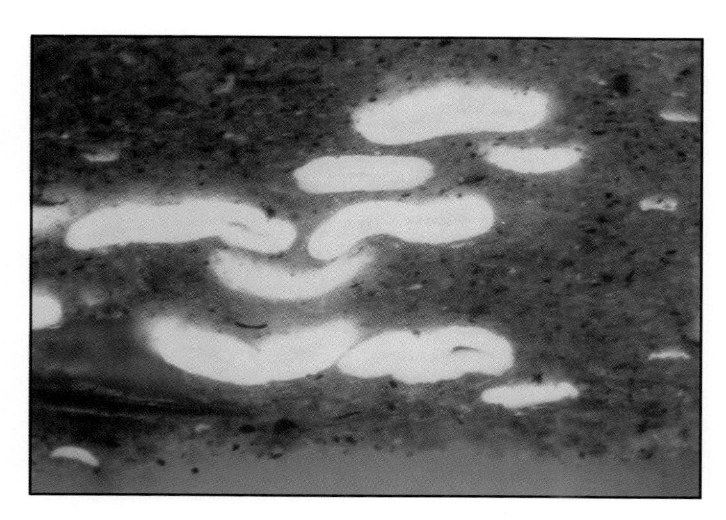

**Figure 30.** *Alginite of Tasmanites type. Shale from the Upper Permian Ravnefjeld Formation near Triaselv at Schuchert Dal. GGU 247449. Blue-light induced fluorescence. Field width = 0.5 mm (0.02 in).*

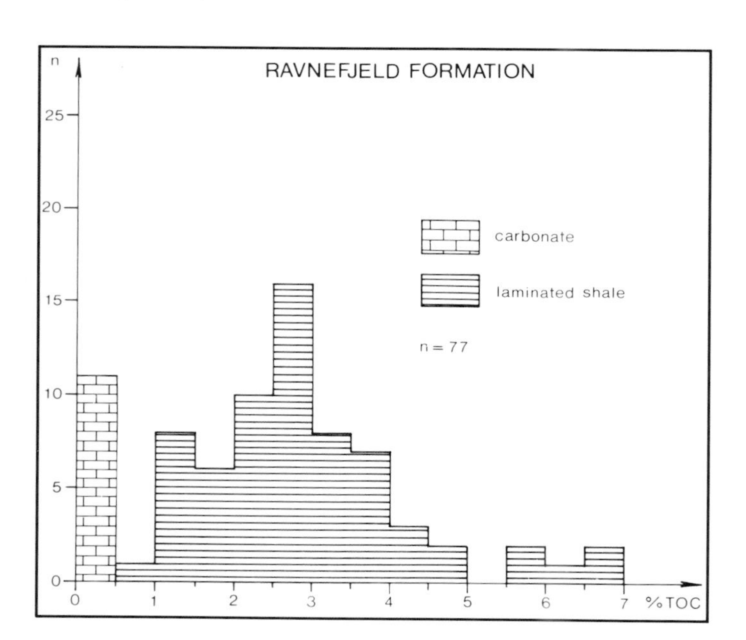


**Figure 31.** *Histogram showing the total content of organic carbon (TOC) of immature to marginally mature Ravnefjeld Formation, including carbonate beds of the Wegener Halvø Formation which alternate with the shale. Leco analysis.*

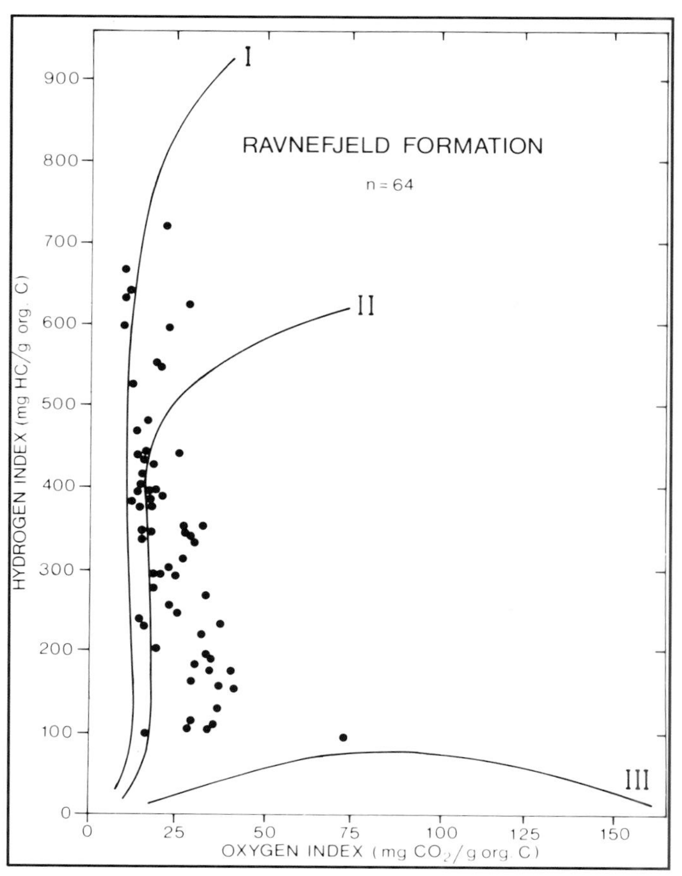


**Figure 32.** *Rock-Eval analysis of immature to marginally mature Ravnefjeld Formation. Only samples with a TOC higher than 1% are shown. Note that the organic matter is mainly of kerogen type-II.*

shallow cores of marginally mature, Ravnefjeld Formation shale at east Schuchert Dal form the basis for the estimate. The characteristics of the formation at this locality have been extended to the entire basin.

The average genetic potential from pyrolysis results ($S_1 + S_2$) based on 55 samples through the 15 m (50 ft) thick formation is 12.9 kg of hydrocarbons per ton of rock (Table 1). According to Tissot and Welte (1978), these results classify the Ravnefjeld Formation as a good source rock (more than 6 kg/t).

On the basis of time-temperature index analyses (TTI), the Ravnefjeld Formation is estimated to be post-mature within an area of at least 50 × 150 km (30 × 95 mi), and within this area, the formation should have produced the maximum amount of liquid hydrocarbons possible. The total volume of liquid hydrocarbons possibly generated within this area should to be 3,600 million ton or approximately 28,000 million barrels.

### Maturity

The regional maturity conditions were determined by analyses of the thermal alteration index (TAI) (Piasecki, 1985) and mean, random, vitrinite reflectance ($\bar{R}_0$) (Thomsen, 1985). The results are illustrated in Figure 37 for the Lower Permian, Upper Permian, and Lower Triassic deposits.

#### *Lower Permian*

In the Schuchert Dal area, three regions with different maturation levels can be distinguished (Figure 37). West of Schuchert Dal, in the Revdal-Gurreholm Dal area, a low maturity level, that is, immature-early mature, prevails. Toward the north in the Snekuppel area, higher maturity levels are reached and the shale is mature. The Snekuppel area involves a small structural high, and the sediments form a thin cover over the crystalline basement. They thus represent older strata than the southern area. Lower Permian deposits east of Schuchert Dal

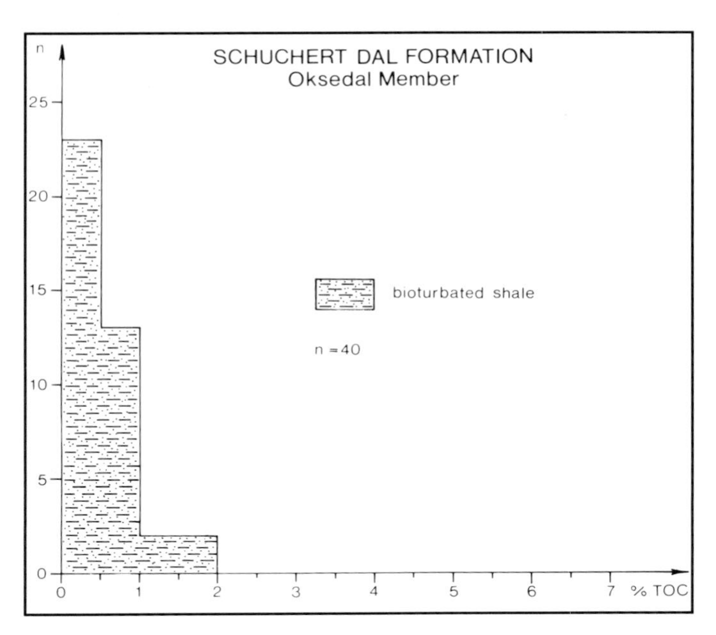


Figure 33. *Histogram showing the total content of organic carbon (TOC) of immature to marginally mature Oksedal Member, Schuchert Dal Formation. Leco analysis.*

are late mature. Toward the north, a single locality exhibits post-mature, Lower Permian deposits indicative of a northward increase in maturity similar to the conditions west of Schuchert Dal. All Lower Permian deposits are overlain by immature-early mature, Upper Permian deposits. The maturity jump across Schuchert Dal is attributed to Early Permian westward tilting and erosion of the sequence, resulting in west-to-east exposure of strata that have been progressively more deeply buried. The strongly asymmetric, half-graben nature of the Lower Permian basin suggests, however, that this maturity gradient decreases further eastward because of the eastward thinning of the sediment wedge. The apparent northward maturity increase is attributed to exposure of increasingly older strata.

In northern Jameson Land and Traill Ø, a simple maturation pattern in the Lower Permian sediments emerges from the present data. The sediments close to the western border of the basin at Skeldal and Kongeborgen are immature to early mature. Mature sediments occur in Funddal, and a post-mature level occurs in Oksedal. Upper Permian-Lower Triassic deposits in the Oksedal area are also found to be post-mature. Exposure of different levels of the sequence because of tilting can be excluded as a cause for the maturity conditions. The high maturity is, rather, caused by the proximity of the localities to the line of Tertiary intrusions extending from Werner Bjerge to southeastern Traill Ø.

Upper Carboniferous-Lower Permian sediments on Wegener Halvø and Canning Land at the eastern margin of the basin have not been analysed. They are, however, presumed to be post-mature with respect to oil generation similar to the overlying Upper Permian-Lower Triassic sediments in this region.

*Upper Permian and Lower Triassic*

In the Schuchert Dal area, all Upper Permian and Lower Triassic sections show a very uniform maturation pattern, indica-

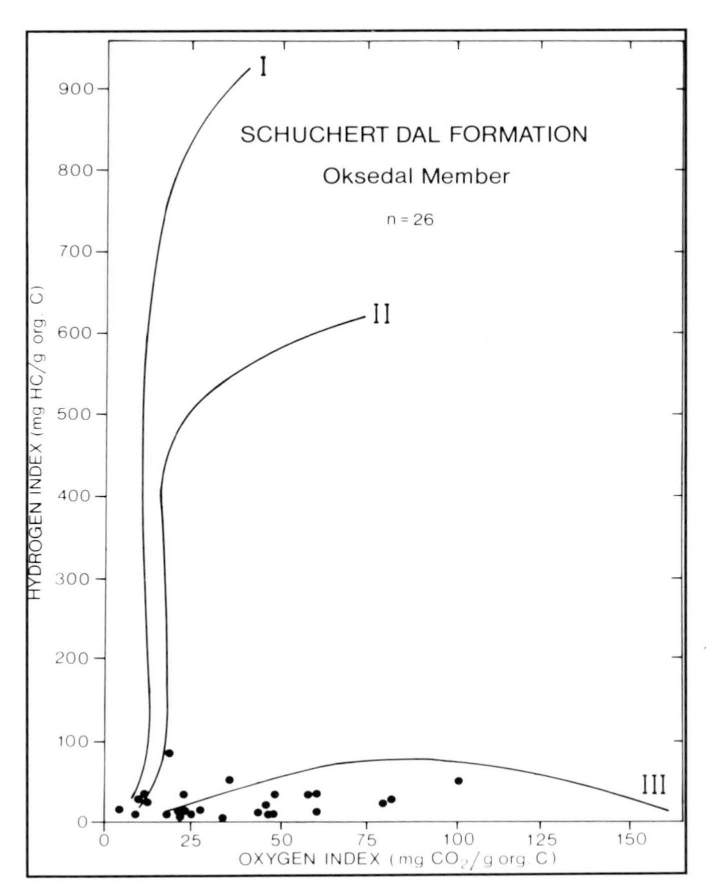


Figure 34. *Rock-Eval analysis of immature to marginally mature Oksedal Member, Schuchert Dal Formation. Note that the organic matter is typical of kerogen type-III.*

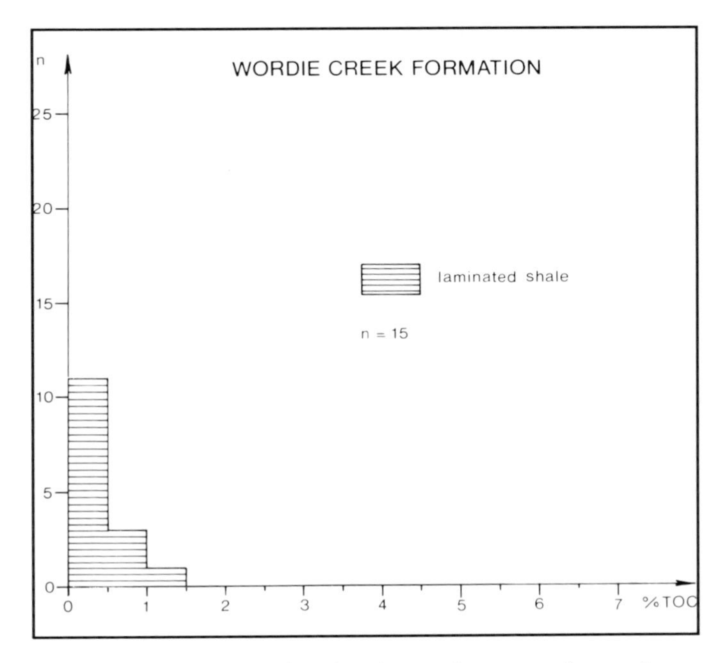


Figure 35. *Histogram showing the total content of organic carbon (TOC) of immature to marginally mature Wordie Creek Formation. Leco analysis.*

ting immature to very early mature conditions on both sides of the valley (Figure 37). Toward the north, this simple maturation pattern changes, and mature deposits are recorded at Pingo Dal, while post-mature, highly coalified deposits occur north of Pingo Dal at Breithorn Gletscher. These results are to be expected as the localities are situated close to the Tertiary Werner Bjerge intrusion. In northern Jameson Land and on Traill Ø, a maturing pattern similar to that found for the Lower Permian sequence has been recorded for the Upper Permian sequence. The Upper Permian sediments are immature to early mature at Rubjerg Knude on Traill Ø and early mature to post-mature at Domkirken. Post-mature, Upper Permian, and Lower Triassic deposits are recorded at Oksedal because of the influence of Tertiary intrusives.

The immature to early mature Upper Permian and Lower Triassic sediments along the western margin of the Jameson Land basin suggest shallow burial for that part of the basin. Deeper subsidence of these deposits in the central parts of the basin will have provided the necessary maturity levels for petroleum generation.

A large number of Upper Permian and Lower Triassic sections have been investigated on Wegener Halvø. The entire area is post-mature with respect to oil generation, but it is still within the zone of wet gas generation (Figure 37). In the southernmost part of this area at Devondal, vitrinite reflectance and TAI data give conflicting evidence of the maturity of the Upper Permian and Lower Triassic sediments. The former indicates mature and the latter post-mature conditions with respect to oil generation. Based on the present data, we cannot determine the existence of a trend of decreasing maturity toward the south from Wegener Halvø. The post-mature conditions in the Wegener Halvø area, we believe, reflect the original burial depth and thermal gradient of this region because there is no evidence of post-Permian intrusive activity. Abundant evidence exists, however, for Late Permian, pre-Triassic hydrothermal mineralization in the area. A comprehensive description of the types and timing of mineralizations in East Greenland is being prepared by O. Harpøth, B. Thomassen, and J. L. Pedersen. The region is regarded as non-prospective with respect to liquid hydrocarbons. This conclusion is also considered the case for Nathorst Fjord, between Wegener Halvø and Canning Land, because post-mature sediments occur on Depotø and in Canning Land.

### Modelling of Maturation History

The extent of the Upper Permian sequence in the Jameson Land area is only known from exposures that indicate the presence of Upper Permian deposits through the northern part of the basin. No outcrops occur south of a line extending from the base of the Wegener Halvø in the east to the south point of Karstryggen in the west (Figure 1).

In order to evaluate the hydrocarbon potential in the southern area, the maturity of potential source rocks has been modelled using TTI method described by Waples (1980). A number of assumptions have been made regarding geothermal gradients, primary and present thicknesses of sequences, and time and length of erosional periods. We have considered the subsidence history for the following stratigraphical levels: base Upper Carboniferous, base Upper Permian, base Triassic (Scythian), base Anisian, upper Norian, base Pliensbachian, base Bajocian, and upper Oxfordian. We assume non-

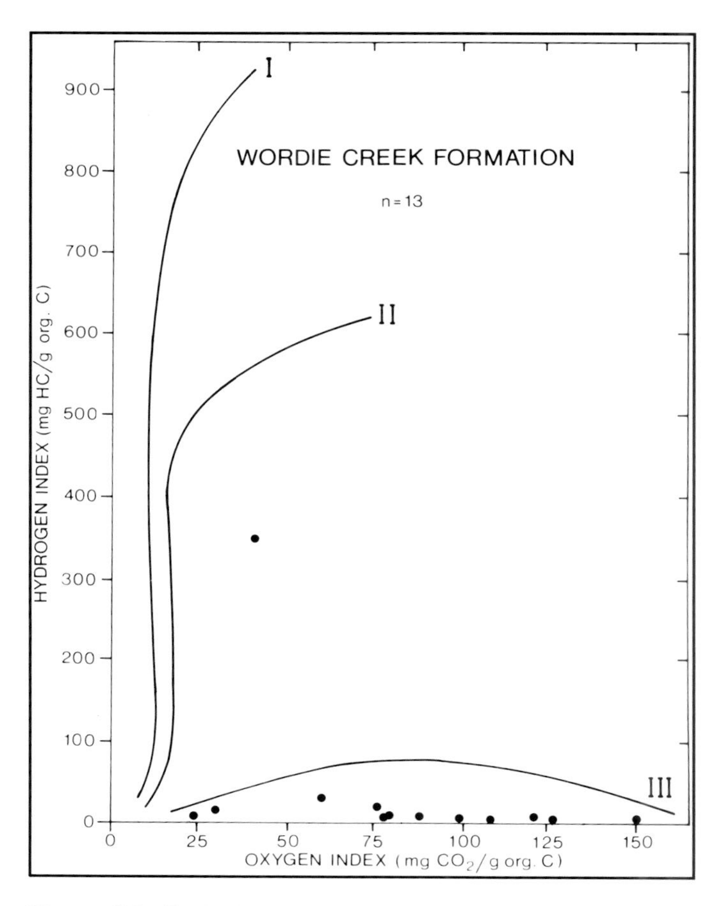


Figure 36. *Rock-Eval analysis of immature to marginally mature Wordie Creek Formation. Note that the organic matter is typical of kerogen type-III.*

deposition for the Sinemurian and the Aalenian, while considering a strong uplift for the Neogene, starting in late Oligocene. Our interpretation is based on detailed stratigraphical studies (Clemmensen, 1980b; Surlyk, 1977; and Surlyk et al, 1973) and evidence from the Werner Bjerge intrusive, which was formed as a deep-seated intrusive in the late Oligocene (Rex et al, 1979).

TTI values have been calculated assuming an average geothermal gradient of 30°C/km (1.6°F/100 ft), and a mean surface temperature through time of 20°C (68°F). Estimates of sediment thicknesses are based on measured values from outcrops. The TTI values thus derived indicate that the Upper Permian potential source rocks reached the onset of oil generation in Late Jurassic times and passed through the oil window during the Early Cretaceous. At present these deposits are within the zone of dry gas generation. Potential Middle Triassic source rocks passed through the oil window during the Early Cretaceous-Eocene time interval and are at present within the zone of wet gas generation. Potential Rhaetian-Hettangian source rocks are rated mature, reaching the onset of oil generation in the Late Cretaceous, while favorable Kimmeridgian source rocks are rated immature. This rating concurs with the results of maturity estimates in southern Jameson Land based on TAI and $\bar{R}_o$ values from Upper Jurassic surface samples. Hence, the calculated TTI values are regarded as fair estimates of the maturity conditions in the area, although we emphasize

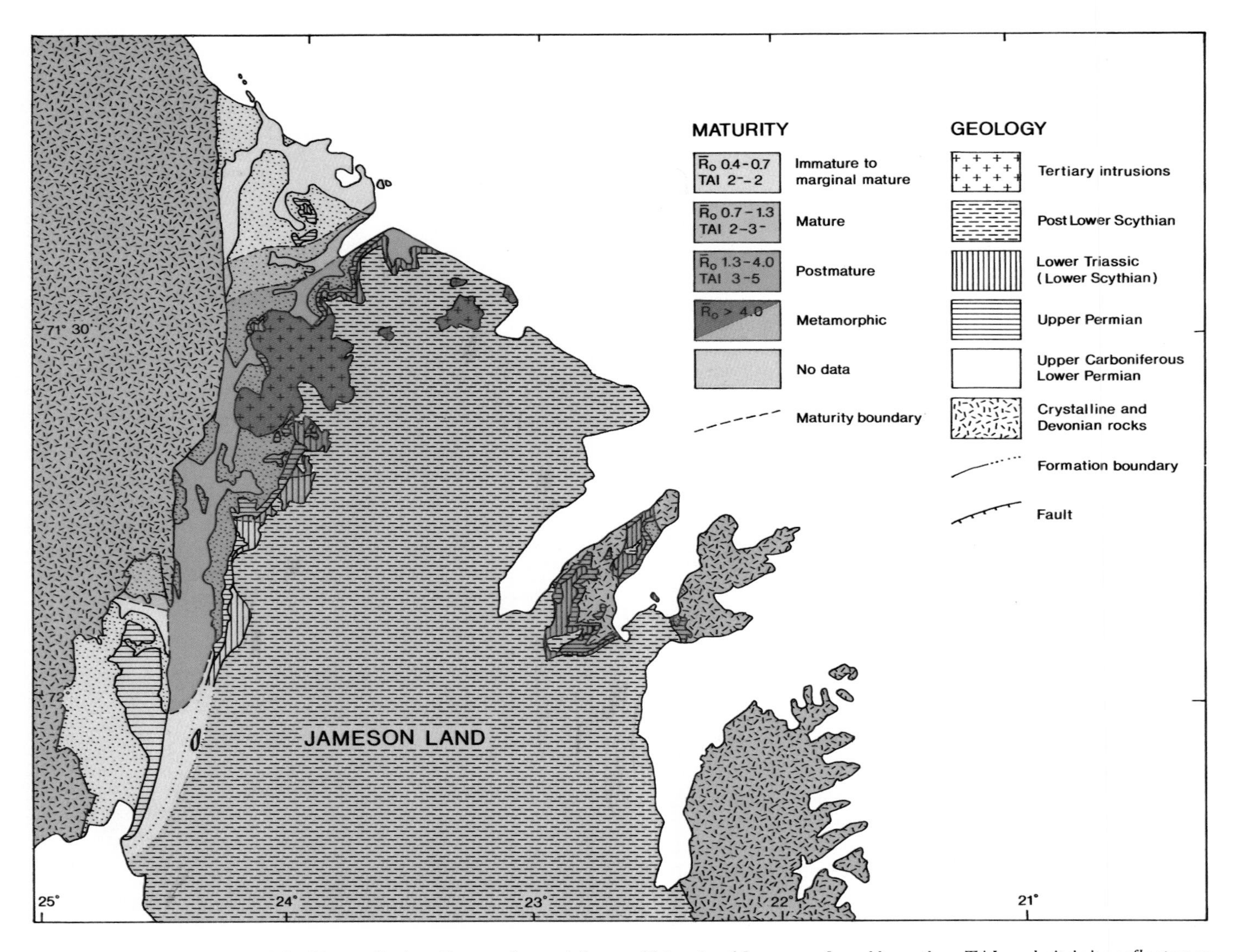


Figure 37. *Maturity map of the Upper Carboniferous through Lower Triassic of Jameson Land based on TAI and vitrinite reflectance measurements of approximately 450 samples from 40 sections through the outcrop regions. The influence of Tertiary intrusions is clearly reflected by the maturity pattern in contrast to areas where maturation variations reflect differences in burial history.*

that the basic assumption concerning geothermal gradient, surface temperatures, and sediment thicknesses is, to a large extent, based on estimates.

## RESERVOIR POTENTIAL

### Upper Permian Carbonates

#### *Karstryggen Formation*

The restricted marine and coastal carbonate of the Karstryggen Formation contains numerous units that had high initial porosity and permeability. These include some algally-laminated units, oolitic shoal deposits, and conglomeratic-intraclastic strata. However, subsequent diagenetic alteration has radically affected the reservoir potential of these units.

Early marine cementation with fibrous aragonite was responsible for extensive porosity loss in the marginal-marine grainstone. Repeated intervals of subaerial exposure apparently led to widespread leaching of aragonitic grains by fresh water. The initial stages of grain leaching led to extensive development of secondary porosity, especially in the oolitic grainstone. Locally, karstic sinkholes and cavernous porosity formed. Based on outcrop data from both Karstryggen and the Wegener Halvø, however, continued vadose and phreatic conditions generally led to virtually complete infilling of both primary and secondary porosity with blocky, non-ferroan calcite cements.

Some porosity is associated with the margins of evaporite bodies in the Karstryggen Formation in the Karstryggen area. Carbonate strata in direct proximity to these lagoonal evaporites have been completely dolomitized. These deposits consist of apparently syngenetic, medium-to finely-crystalline, subhedral dolomite, with numerous vugs, or a "rauhwacke" fabric consisting of a very porous latticework of dolomitic laminae and veins. Porosity in these bodies resulted largely from evaporite dissolution or replacement by carbonate minerals or both. In areas of interlaminated carbonates and evaporites, this porosity can be extensive, although the lateral and vertical extent of such porous zones on outcrop is only of the order of tens of meters. It is also not clear how much of this porosity development relates to outcrop weathering.

The overall reservoir potential of the Karstryggen Formation, based on present-day outcrop character, is only moderate despite evidence for widespread subaerial exposure and karstic weathering of the unit. Considerable areas may exist in which primary or secondary pores or both were not filled with freshwater cements and thus have good porosity-permeability characteristics. Furthermore, there may be areas associated with the margins of evaporite bodies with extensive porosity in altered dolomite and limestone. However, the size of such porous bodies may be quite limited, and predicting and exploring such plays would be difficult. The rarity or even absence of widespread penecontemporaneous dolomitization throughout the East Greenland Permian unfortunately precludes development of the major West Texas-style plays with very porous subtidal dolomite as reservoir and up-dip evaporite as seal. The absence of dolomite also reduces the probability of finding high primary porosity units at the depths to which the Karstryggen section has been buried (Halley and Schmoker, 1983). Finally, the fact that the Karstryggen Formation partly underlies the main source rock unit in the region (Ravnefjeld Formation) means that reservoir charging to some extent would be restricted to tilted fault blocks or to areas in which older source rocks were represent.

*Wegener Halvø Formation*

Diagenetic patterns in the Wegener Halvø Formation, much more complex than in the Karstryggen Formation, led to a broad range of porosity types and diverse reservoir potentials. Thus, the discussion of reservoir characteristics of this unit will be subdivided by area.

On Wegener Halvø the basal part of the formation is characterized by small algal buildups and associated grainstone deposits. Early subaerial exposure developed leached secondary porosity in these units, but a combination of later cement infilling and the very small size of these deposits precludes reservoir potential.

The large, bryozoan buildups, which locally developed from the stromatolitic mounds, are large enough to present locatable and economically significant targets. The core facies of these mounds contain mainly fenestrate bryozoans as their framework organisms (Figure 38). This facies had extremely high porosity for a very brief period at the time of formation, but a combination of penecontemporaneous cementation by bladed, cloudy, isopachous calcite and infiltration of pelletal and micritic sediment into remaining pores drastically reduced porosity prior to burial. Some remnant porosity (perhaps 10%) was retained and apparently persisted throughout early burial. This pore space is now filled by a late diagenetic, iron-rich, coarse, blocky calcite cement with $\delta^{18}O$ values averaging $-14‰$ (relative to PDB) (Figure 38), indicating precipitation in temperatures in excess of 100°C (212°F).

The volumetrically extensive skeletal debris of the flank facies underwent a very different diagenetic history. These strata received very little marine carbonate cementation.

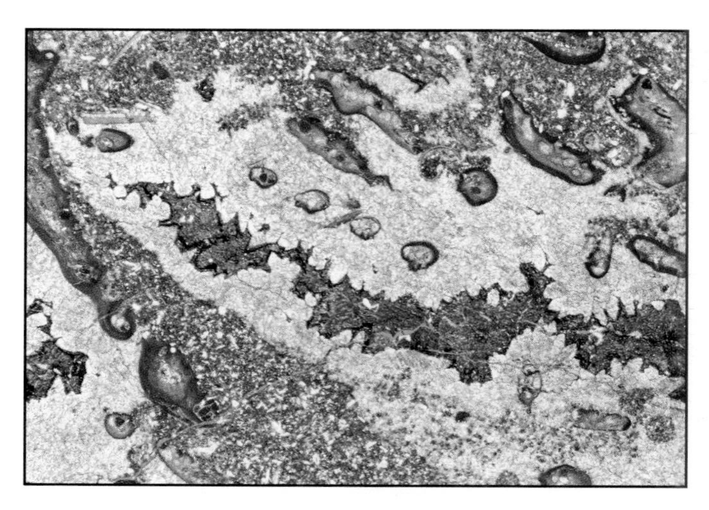

Figure 38. *Photomicrograph of a stained, thin section of core facies from a carbonate buildup of the Wegener Halvø Formation. Dark reddish grains are fenestrate bryozoans, the main framework organisms. Infiltrated silt and mud plug some porosity and are followed by massive cementation by bladed calcitic crusts (pale pink). Remnant pore space filled by "late" diagenetic iron-rich (dark blue stained) calcite. Long axis of photo = 10 mm (0.4 in) GGU 298109B. Devondal.*

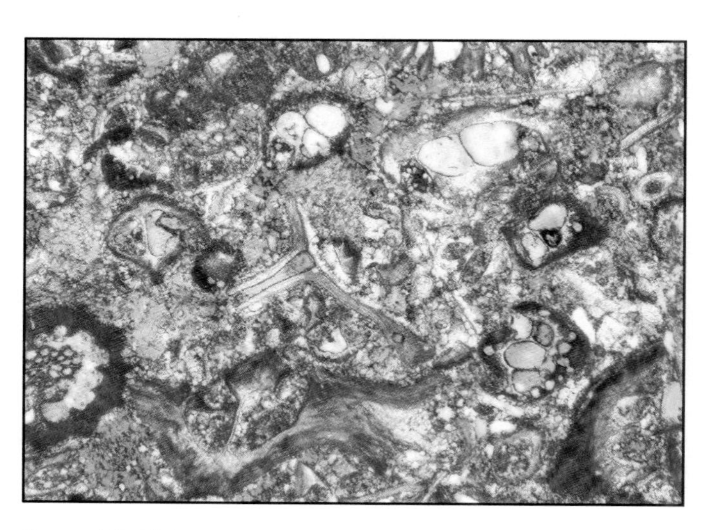

Figure 39. *Thin section photomicrograph of stained specimen of upper flank facies strata, of a carbonate buildup in the Wegener Halvø Formation. Bryozoans, brachiopods, and echinoderms are main sediment formers. Thin crusts of early marine cement line some grains and most intragranular porosity is filled iwth "early" authigenic glauconite. Large volume or remnant pore space occluded by "late" ferroan calcite cement (stained dark blue). Long axis of photo = 3 mm (0.1 in) GGU 298114. Devondal.*

Grains in the slowly deposited debris, sedimented directly adjacent to the core, were intensely bored by marine invertebrates, and both borings and intragranular primary pores were filled in with early, authigenic glauconite cement (Figures 39 and 40). Sediments deposited farther down the flanks of the bioherm by

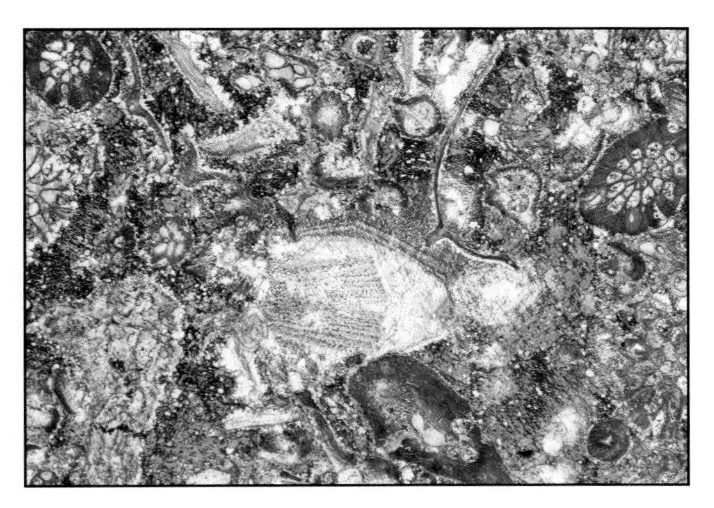

Figure 40. *As Figure 39 but showing zoned overgrowths on echinoderm fragment. Progressive increase in iron content of echinoderm overgrowths presumably indicates growth during burial, with formation waters becoming progressively more reduced through time. Long axis of photo = 5.6 mm (0.2 in). GGU 298114. Devondal.*

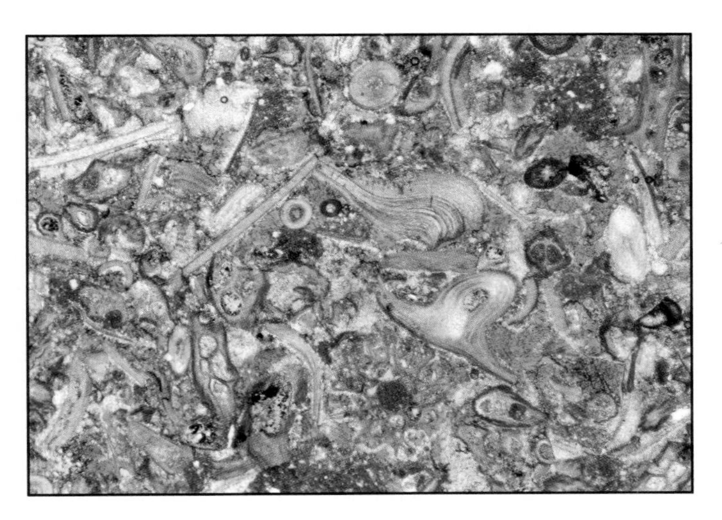

Figure 41. *Photomicrograph of stained thin section of reworked flank facies of the Wegener Halvø Formation showing virtually no early cement, some signs of compactional grain breakage and massive infill of "late" diagenetic ferroan calcite cement. Long axis = 5 mm (0.2 in). GGU 298130B. Devondal.*

grain flow and turbidity current processes were buried more rapidly, undergoing neither calcite nor glauconite cementation. Flank strata thus retained very high porosity (in excess of 40%) during early stages of burial. The lack of any cement framework, however, led to massive mechanical compacting of these strata during burial, followed by chemical compacting at later stages. Therefore, these strata exhibit extraordinary grain breakage, grain-to-grain dissolution, and late stylolitization. Although extensive cementation might be expected to result from the liberation of calcite through chemical compaction, the only intermediate-stage cements occur as zoned echinoderm overgrowths (Figure 40). These can be important porosity-destroying elements in the rare units that are rich in echinoderm debris.

Despite massive chemical and mechanical compacting of flank strata, the proximal debris units apparently reached considerable burial depths (probably several kilometers) while still retaining between 10 and 15% primary pore space. These strata show massive, late diagenetic plugging of that remanent pore space by coarse, blocky calcite (Figure 41) averaging about $-14‰$ $\delta^{18}O$ (PDB). These cements are very metal-rich, with an average of more than 15,000 ppm Fe and 11,000 ppm Mn. They were presumably precipitated from hot, ($>100°C$ or $>212°F$), reducing fluids, perhaps in association with regional barite and lead-zinc mineralization. Ferroan calcite cements also fill late fractures in these strata.

Petroleum potential of these buildups, then, depends largely on the relative timing of "late" cement formation versus petroleum generation and migration. If hydrocarbons were able to move into these structures from the adjacent, organic-rich shale of the Ravnefjeld Formation prior to the introduction of the ferroan cements, they would have encountered very porous and permeable reservoirs with excellent shale seals. Although there is no unequivocal evidence as to the relative timing of hydrocarbon migration and ferroan cement formation, a black residue is commonly visible in thin sections, and it occurs either before the earliest ferroan cement or within early stages of ferroan calcite crystals. These black to dark brown stains do not fluoresce, however, and their exact composition is currently uncertain. Ongoing organic geochemical and fluid inclusion studies may help to clarify these critical relationships.

The Wegener Halvø Formation has both a different depositional character and a different diagenetic history in its other major outcrop area. The shallow-platform carbonates of Karstryggen underwent extensive, early marine cementation and widespread, early fresh-water cementation as well. Some outcrop samples from western Karstryggen have extensive unfilled primary and secondary pore space (Figure 42), but it appears likely that, in many cases, this space results from near-surface Tertiary or Quaternary weathering. This combination of processes effectively obliterated porosity in this unit. Thus, the platform carbonate section appears to have very low reservoir potential.

In summary, the units of greatest exploration interest in the Upper Permian carbonate section are the bryozoan buildups of the type currently outcropping on the Wegener Halvø. These buildups are large enough to be locatable in seismic sections, but limited, present-day outcrop does not allow adequate prediction of whether such buildups are likely to extend in the subsurface part of the basin. The combination of porosity preservation through much of the burial history, directly adjacent source rocks, and overlying shale seals all favor reservoired hydrocarbons in these structures. Structure size may, however, be economically marginal in Arctic environments. Karstic horizons in the underlying Karstryggen Formation, especially where present on uplifted fault blocks, may also yield porous and permeable reservoirs.

### Upper Permian and Lower Triassic Sandstone

The shallow marine sandstone of the Schuchert Dal and Wordie Creek Formations consists both of well-sorted, relatively fine-grained units and coarse-grained, commonly pebbly,

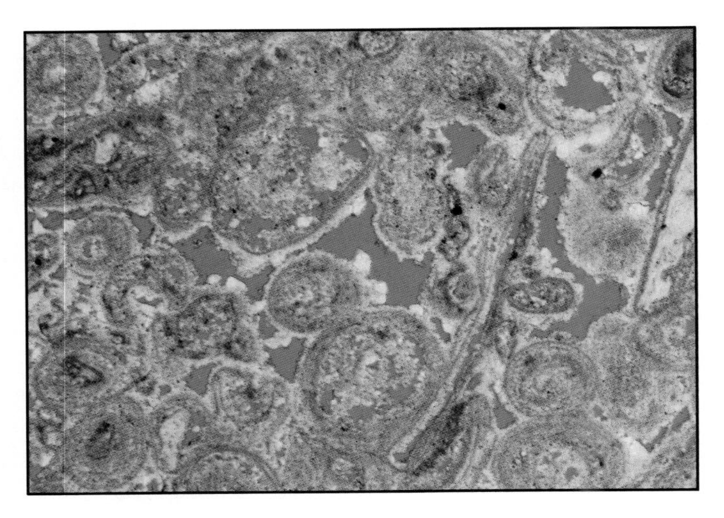

**Figure 42.** *Photomicrograph of oolitic grainstone from the Wegener Halvø Formation in the Karstryggen area showing evidence of leaching, subsequent minor fresh-water cementation by blocky calcite, and extensive, preserved primary and secondary pore space. Long axis of photo = 5.7 mm (0.2 in). GGU 292614.*

poorly sorted units. Detailed petrographic and diagenetic studies have not been done, but the initial porosity and permeability of these facies was probably high. The immature petrographic composition of some horizons suggest, however, a strong susceptibility to diagenetic alterations, in particular to neoformation of clay minerals.

The coarse-grained units are thickly developed along the basin margins, but they appear to have covered much of the basin during the end-Permian regression. They were removed by erosion during latest Permian emergence in some areas, especially along the eastern edge of the basin.

Marginal conglomerate and sandstone of the same types also occur in the lower Scythian (Figure 27), and, in the late Scythian, virtually the whole basin received continental conglomerate and sandstone (Clemmensen, 1980a).

The top Permian and overlying Triassic, therefore, contain thick, coarse-grained sequences that, under favorable diagenetic circumstances, may have a reservoir potential.

## SUMMARY AND CONCLUSIONS

The Upper Permian sequence of East Greenland is thickly developed, spatially widespread, and contains an association of facies rendering it highly attractive for petroleum exploration. Our study, which focused on the Jameson Land sequence (Figures 1, 2, and 6), generated the following points:

1. Upper Permian sediments are exposed in East Greenland over 400 km (248 mi) from central Jameson Land in the south to Wollaston Forland in the north (Figure 7).
2. The first post-Caledonian rifting phase in the Norwegian-Greenland Sea region was initiated in Late Carboniferous times and faded out during the Early Permian. The end of rifting was marked by thermal bulging and erosion leading to peneplanation, and was associated with dike intrusion and Pb-Zn, barite, and fluorite mineralization.
3. Upper Permian deposits represent the initial phase of basin subsidence caused by thermal contraction following the rifting event.
4. The sea transgressed the area during the Early to Late Permian transition after an interval of conglomerate deposition in a shallow inland braidplain (Figures 5, 6, 7, and 8). In the initial phase of transgression, shallow-water hypersaline carbonate and evaporite were deposited. They represent the Karstryggen Formation. The main facies are gypsum, algally laminated limestone, lime mudstone, oolitic grainstone, and intraclast conglomerate. The transgressive trend was punctuated by several regressive events resulting in erosion, conglomerate deposition, and karstification of the hypersaline platform sequence.
5. Continued transgression eventually led to full submerging of the area and the development of normal marine conditions (Figures 15, 16, 17, and 18). Carbonate platforms and buildups representing the Wegener Halvø Formation were formed along basin margins and probably also over structural highs within the basin. They pass laterally into black shales.
6. The main facies of the Wegener Halvø Formation are laminated pelletal/micritic limestone, oolitic/oncolitic grainstone, bryozoan/marine cement grainstone to wackestone occurring as discrete buildups, bedded mono- to polyspecific packstone to wackestone, thin-bedded lime mudstone, skeletal and pelletal wackestone, thin-bedded skeletal packstone to pelletal mudstone, and thin-bedded miliolid wackestone to mudstone. The Wegener Halvø Formation represents a major transgressive sequence. The initial stages of the transgression are represented by small, algal stromatolite mounds and associated oolitic/oncolitic grainstone shoals. Continued transgression led to drowning of these incipient buildups and shoals. In some areas deposition kept pace with increasing water depth, forming major buildups (Figures 18, 19, 20, 21, and 22).
7. The black basinal shale is laminated, has a high content of organic carbon and was deposited under anoxic conditions. It occurs throughout the Upper Permian basin as the most widespread facies.
8. With ongoing transgression, the basin became better oxygenated and the black shale gave way to light grey, bioturbated shale with a low content of organic carbon. Combined sea-level rise, subsidence, and clastic influx eventually led to the drowning of the carbonate buildups and the basin became filled with an upward-coarsening sequence topped by shallow marine sandstone (Figure 25).
9. A world-wide regression took place at the end of the Permian. In Jameson Land this regression led to the emergence of most or perhaps all the basin and was associated with erosion, especially along the eastern basin margin.
10. The area was again flooded in the earliest Triassic, and the transgression is reflected by the onlapping nature of the basal Triassic sequence along the eastern basin margin. Deposition was continuous across the Permo-Triassic boundary in the deepest part of the basin, which was situated over the western axis of the underlying half-graben. The Early Triassic marine deposition of shale

and sandstone rapidly gave way to arid, continental, red bed sedimentation, which dominated throughout the rest of the Triassic.

11. The source rock potential in Jameson Land is restricted to the thin, Upper Carboniferous-Lower Permian lacustrine shale, the Upper Permian black shale of the Ravnefjeld Formation, the thin Middle Triassic calcareous sandstone, the Gråklint Beds, the shale of the Rhaetian-Hettangian Kap Stewart Formation, the shale of the Middle Jurassic Sortehat Member, and the Upper Jurassic shale of the Hareelv Formation.
12. The Upper Carboniferous-Lower Permian shale occurs in rather narrow westwards-tilted half-graben and is thus not expected to extend throughout the Jameson Land basin. The content of organic carbon is locally high, varying between 10–20%. The sediments have a good potential as a source of gas, and some horizons have a good to excellent potential as a source of oil.
13. The Ravnefjeld Formation is locally up to 60 m (197 ft) thick in Jameson Land. Greater thicknesses may, however, occur in the deeply buried, central parts of the basin. The content of organic matter averages about 3% (Figure 31). The formation has a good potential as a source oil (Figure 32). The Middle Triassic Gråklint Beds only occur in northeastern Jameson Land, and the source potential appears to be restricted to thin limestone. The potential as hydrocarbon source seems, therefore, to be very limited. The Jurassic sequence is unfortunately exposed or occurs in a shallow subsurface position and is accordingly of a very limited potential.
14. The main reservoir potential is found in the marine carbonate buildups of the Wegener Halvø Formation. These buildups are large enough to be located seismically, but it is not known if they extend in the subsurface part of the basin. The combination of porosity preservation through much of the burial history, directly adjacent source rocks of the Ravnefjeld Formation, and overlying shale seals—all favor reservoired hydrocarbons in these structures. Karst horizons in the underlying Karstryggen Formation may also yield porous and permeable reservoirs. Fan-delta and shallow marine coarse-grained sandstone, which are widely distributed in the top of the Foldvik Creek Group and in the Wordie Creek Formation may also possess a reservoir potential, but seismic definition of this play type is difficult.
15. The maturity conditions of the basin are determined on the basis of the thermal alteration index (TAI) and mean, random vitrinite reflectance ($\bar{R}_o$). The Lower Permian is immature to early mature along the western basin margin and late mature further east of Schuchert Dal. The overlying, Upper Permian sequence is immature to very early mature in all exposures along the western basin margin, and post-mature with respect to oil generation, but still within the zone of wet gas generation on Wegener Halvø (Figure 37).
16. The depth and extent of the Upper Permian sequence in Jameson Land is not known, and the maturation stage in the subsurface can thus only be indirectly assessed. Upper Jurassic surface samples in southwest Jameson Land are, however, early mature to mature, while they are immature in central Jameson Land. The maturity of the potential source rocks has been modelled for these areas using a time-temperature index (TTI). A number of assumptions have been made regarding geothermal gradient, thicknesses of sequences, and time and length of erosional periods. The TTI values indicate that the Upper Permian potential source rocks reached onset of oil generation in Late Jurassic times and passed through the oil window during the Early Cretaceous.
17. The Upper Permian of Jameson Land, therefore, contains a relatively thick, widely distributed oil-prone source rock. It occurs directly adjacent to large carbonate buildups that constitute the main potential reservoirs. It is immediately overlain by thick, coarse-grained sandstone that also possesses a reservoir potential. The whole sequence is draped by Upper Permian and Lower Triassic shales that can act as seals.
    Maturation studies suggest that the potential Upper Permian source rocks passed through the oil window in the Late Jurassic-Early Cretaceous time interval. In southern Jameson Land, the sequence is probably overmature and in the dry gas zone, while further north it enters the wet gas zone, and, in the west-central the lower part of the oil window.
18. The presence of an extensive, potential source rock and several types of reservoir rocks throughout the Upper Permian basin of East Greenland has important implications for the Norwegian shelf north of 62°N. Very little is known of the Upper Permian in this region. There are, however, good reasons to believe that Upper Permian rocks of the same types as in East Greenland lie deeply buried. The Upper Jurassic shale is immature over large areas of the Norwegian shelf, and if Upper Permian source rocks are present, they would add considerably to the petroleum potential of the area.

## ACKNOWLEDGMENTS

The present paper presents the main results of the Oil Section of the Geological Survey of Greenland's expedition to East Greenland in 1982 and 1983. We would like to thank J. Boserup, A. Clausen, and K. Villadsen of the shallow core drilling crew for their enthusiasm and hard work. We also direct our thanks to I. Olsen, who was an extremely efficient base camp manager in both field seasons.

We discussed aspects of the geology with T. Frazier (ARCO) and O. Harpøth (Nordisk Mineselskab) in the field and in Copenhagen. D. Leythaeuser (KFA, Jülich) participated in the 1982 field season, and we acknowledge his contribution to the geochemical discussions. We thank B. Sikker Hansen for careful drafting, V. Hermansen and N. Turner for their patient work on the word processor, and J. Lautrup for photography. The paper was written while F. Surlyk and L. Stemmerik were recipients of grants from the Danish Natural Science Research Council. The paper is published with permission of the Director of the Geological Survey of Greenland.

## REFERENCES CITED

Aldinger, H., 1935, Das Alter der jungpalaeozoischen Posidonomyaschiefer: Meddr Grønland, v. 98, n. 4, p. 1–24.

Balme, B. E., 1979, Palynology of Permian-Triassic boundary

beds at Kap Stosch, East Greenland: Meddr Grønland, v. 200, n. 6, p. 1-37.

Bendix-Almgreen, S. E., 1976, Palaeovertebrate faunas of Greenland, *in* A. Escher and W. S. Watt, eds., Geology of Greenland: The Geological Survey of Greenland, p. 537-573.

Birkelund, T., and K. Perch-Nielsen, 1976, Late Palaeozoic-Mesozoic evolution of central East Greenland, *in* A. Escher, and W. S. Watt, eds., Geology of Greenland: The Geological Survey of Greenland, p. 304-339.

Birkenmajer, K., 1977, Erosional unconformity at the base of marine Lower Triassic at Wegener Halvø, central East Greenland: Rapport Grøland geologiske Undersøgelse, 85, p. 103-107.

Butler, G. P., P. M. Harris, and C. G. St. C. Kendall, 1982, Recent evaporites from the Abu Dhabi coastal flats, *in* C. R. Handford, R. G. Loucks, and G. R. Davies, eds., Depositional and diagenetic spectra of evaporites—A core workshop: SEPM Core Workshop, Calgary, 3, p. 33-64.

Bütler, H., 1935, Some new investigations of the Devonian stratigraphy and tectonics of East Greenland: Meddr Grønland, v. 103, n. 2, p. 1-35.

——, 1959, Das Old Red-Gebiet am Moskusoksefjord: Meddr Grønland, v. 160, n. 5, p. 1-188.

——, 1961, Continental Carboniferous and Lower Permian in Central East Greenland, *in* G. O. Raasch, ed., Geology of the Arctic 1: University of Toronto Press, p. 205-213.

Clemmensen, L. B., 1980a, Triassic rift sedimentation and palaeography of central East Greenland: Grønlands geologiske Undersøgelse Bulletin, v. 136, p. 1-72.

——, 1980b, Triassic lithostratigraphy of East Greenland between Scoresby Sund and Kejser Franz Josephs Fjord: Grø nlands geologiske Undersøgelse Bulletin, v. 139, p. 1-56.

Collinson, J. D., 1972, The Røde Ø conglomerate of inner Scoresby Sund and the Carboniferous(?) and Permian rocks west of the Schuchert Flod: Meddr Grønland, v. 192, n. 6, p. 1-48.

Creaney, S., 1980, The organic petrology of the Upper Cretaceous Boundary Creek Formation, Beaufort-Mackenzie basin: Bulletin of Canadian Petroleum Geology, v. 28, p. 112-119.

Friend, P. F., P. D. Alexander-Marrack, K. C. Allen, J. Nicholson, and A. K. Yeats, 1983, Devonian sediments of East Greenland VI, review of results: Meddr Grønland, v. 206, n. 6, p. 1-96.

Halle, T. G., 1953, The Carboniferous flora of East Greenland: Meddr Grønland, v. 103, n. 2, p. 1-35.

Haller, J., 1970, Tectonic map of East Greenland (1:500,000)—An account of tectonism, plutonism, and volcanism in East Greenland: Meddr. Grønland, v. 171, n. 5, p. 1-286.

——, 1971, Geology of the East Greenland Caledonides: London, Interscience Publishers, p. 1-375.

Halley, R. B., and J. W. Schmoker, 1983, High-porosity Cenozoic carbonate rocks of south Florida: progressive loss of porosity with depth: AAPG Bulletin, v. 67, p. 112-119.

Haszeldine, R. S., 1984, Carboniferous North Atlantic palaeogeography: stratigraphic evidence for rifting, not megashear or subduction: Geological Magazine, v. 121, p. 443-463.

Henriksen, N., and A. K. Higgins, 1976, East Greenland Caledonian fold belt, *in* A. Escher and W. S. Watt, eds., Geology of Greenland: The Geological Survey of Greenland, p. 183-246.

Hutton, A. C., A. J. Kantsler, A. C. Cook, and D. M. McKirdy, 1980, Organic matter in oil shales: Australian Petroleum Exploration Association, v. 1, n. 20, p. 44-67.

Jacobsen, V., and P. van Veen, 1984, The Triassic offshore Norway north of 62°N, *in* A. M. Spencer et al, eds., Petroleum geology of the North European margin: Graham and Trotman Ltd., for the Norwegian Petroleum Society, p. 317-327.

Karlsson, W., 1984, Sedimentology and diagenesis of Jurassic sediments offshore mid-Norway, *in* A. M. Spencer et al, eds., Petroleum geology of the North European margin: Graham and Trotman Ltd., for the Norwegian Petroleum Society, p. 389-396.

Kempter, E., 1961, Die jungpaläozoischen Sedimente von süd-Scoresby Land (Ostgrönland, 71$^1/_2$°N), mit besonderer Berücksichtigung der kontinentalen Sedimente: Meddr Grønland, v. 164, n. 1, p. 1-123

Kinsman, D. J. J., and R. K. Park, 1976, Algal belt and coastal sabkha evolution, Trucial coast, Persian Gulf, *in* M. R. Walter ed., Interpreting stromatolites. Developments in sedimentology 20: Amsterdam, Elsevier, p. 421-433.

Koch, L., 1929, The geology of East Greenland: Meddr Grønland, v. 73, n. 1, p. 1-204.

Kushnir, J., 1981, Formation and early diagenesis of varved evaporite sediments in a coastal hypersaline pool: Journal of Sedimentary Petrology, v. 51, p. 1193-1203.

Larsen, H. C., 1984, Geology of the East Greenland shelf, *in* A. M. Spencer et al, eds., Petroleum geology of the North European margin: Graham and Trotman Ltd., for the Norwegian Petroleum Society, p. 329-339.

Logan, B. W., P. Hoffman, and C. D. Gebelein, 1974, Algal mats, cryptalgal fabrics and structures, Hamelin Pool, Western Australia: AAPG Memoir 22, p. 140-194.

Loucks, R. G., and M. W. Longman, 1982, Lower Cretaceous Ferry lake Anhydrite, Fairway Field, East Texas, product of shallow-subtidal deposition, *in* C. R. Handford, R. G. Loucks, and G. R. Davies, eds., Depositional and diagenetic spectra of evaporites—A core workshop: SEPM Core Workshop, Calgary, 3, p. 130-173.

Maiklem, W. R., D. G. Bebout, and R. P. Glaister, 1969, Classification of anhydrite—a practical approach: Bulletin of Canadian Petroleum Geology, v. 17, p. 194-233.

Maync, W., 1942, Stratigraphie und Faziesverhältnisse der oberpermischen Ablagerungen Ostgrönlands (olim "Oberkarbon-Unterperm") zwischen Wollaston Forland und dem Kejser Franz Josephs Fjord: Meddr Grønland, v. 115, n. 2, p. 1-128.

——, 1961, The Permian of Greenland, *in* G. O. Raasch, ed., Geology of the Arctic, 1, University of Toronto Press, p. 214-223.

Nielsen, E., 1935, The Permian and Eotriassic vertebrate-bearing beds at Godthåb Gulf (East Greenland): Meddr Grønland, v. 98, n. 1, p.1-111.

Perch-Nielsen, K., R. G. Bromley, U. Asgaard, and M. Aellen, 1972, Field observations in Palaeozoic and Mesozoic sediments of Scoresby Land and north Jameson Land: Grønlands geologiske Undersøgelse Rapport, 48, p. 39-59.

——, K. Birkenmajer, T. Birkelund, and M. Aellen, 1974, Revision of Triassic Stratigraphy of the Scoresby Land and Jameson Land Region, East Greenland: Grønlands geologiske Undersøgelse Bulletin, v. 109, p.1–51.

Piasecki, S., 1984, Preliminary palynostratigraphy of the Permian-Lower Triassic sediments in Jameson Land and Scoresby Land, East Greenland: Bulletin of the Geological Society of Denmark, v. 32, p. 139–144.

———, 1985, Palynological evaluation of the regional thermal maturation of Carboniferous to Tertiary sediments in Central East Greenland: The Geological Survey of Greenland, Open file, 31.1.1985, p. 1–50.

Potter, P. E., J. B. Maynard, and W. A. Pryor, 1980, Sedimentology of shale. Study Guide and Reference Source: New York, Springer Verlag, Inc., p. 1–303.

Rex, D. C., A. R. Gledhill, C. K. Brooks, and A. Steenfeldt, 1979, Radiometric ages of Tertiary salic intrusions near Kong Oscars Fjord, East Greenland, Grønlands geologiske Undersøgelse Rapport, 95, p. 106–109.

Rosenkrantz, A., 1929, Preliminary account of the geology of the Scoresby Sound district: Meddr Grønland, v. 73, n. 2, p. 135–154.

Russell, M. Z., 1978, Mineralization in a fractured craton, *in* D. R. Bowes, and B. E. Leake, eds., Crustal evolution in northwestern Britain and adjacent regions. Geological Journal Special Issue, 10, p. 297–308.

Russell, M. J., and D. K. Smythe, 1983, Origin of the Oslo graben in relation to the Hercynian-Alleghenian orogeny and lithospheric rifting in the North Atlantic: Tectonophysics, v. 94, p. 457–472.

Schreiber, B. C., G. M. Friedman, A. Decima, and E. Schreiber, 1976, Depositional environments of upper Miocene (Messinian) evaporite deposits of the Sicilian basin: Sedimentology, v. 23, p. 729–760.

Schreiber, B. C., et al, 1982, Recognition of primary facies characteristics of evaporites and the differentiation of these from diagenetic overprints, *in* C. R. Handford, R. G. Loucks, and G. R. Davies, eds., Depositional and diagenetic spectra of evaporites—A core workshop: SEPM Core Workshop, Calgary, 3, p. 1–32.

Shinn, E. A., 1983, Birdeyes, fenestrae, shrinkage pores, and loferites: a reevaluation: Journal of Sedimentary Petrology, v. 53, p. 619–628.

Spath, L. F., 1935, Additions to the Eo-Triassic invertebrate fauna of East Greenland: Meddr Grønland, v. 98, n. 2, p. 1–115.

Stemmerik, L., in press, Cyclicity in interbedded carbonate and evaporite, Karstryggen Formation, Upper Permian, East Greenland.

Surlyk, F., 1977, Mesozoic faulting in East Greenland, *in* R. T. C. Frost, and A. J. Dikkers, eds., Fault tectonics in NW Europe: Geol. Minjbouw, v. 56, p. 311–327.

———, 1978a, Submarine fan sedimentation along fault scarps on tilted fault blocks (Jurassic-Cretaceous boundary, East Greenland): Grønlands geologiske Undersøgelse Bulletin, 128, p. 1–108.

———, 1978b, Jurassic basin evolution of East Greenland: Nature, v. 274, p. 130–133.

———, 1982, Kul på Nugssuaq, Vestgrønland: Grønlands Geologiske Undersøgelse, Danmark, p. 43–56.

———, 1983, Source rock sampling, stratigraphical studies in the upper Palaeozoic of the Jameson Land basin, East Greenland: Grønlands geologiske Undersøgelse Rapport, 115, p. 88–92.

———, and L. B. Clemmensen, 1983, Rift propagation and eustacy as controlling factors during Jurassic inshore and shelf sedimentation in northern East Greenland: Sedimentary Geology, v. 34, p. 119–143.

———, ———, and H. C. Larsen, 1981, Post-Paleozoic evolution of the East Greenland continental margin: Bulletin of the Canadian Petroleum Association Memoir 7, p. 611–645.

———, S. Piasecki, and F. Rolle, 1985, Initiation of petroleum exploration in Jameson Land, East Greenland: Grønlands geologiske Undersøgelse Rapport.

———, J. H. Callomon, R. G. Bromley, and T. Birkelund, 1973, Stratigraphy of the Jurassic-Lower Cretaceous sediments of Jameson Land and Scoresby Land, East Greenland: Grønlands geologiske Undersøgelse Bulletin, v. 105, p. 1–76.

———, S. Piasecki, F. Rolle, L. Stemmerik, E. Thomsen, and P. Wrang, 1984a, The Permian basin of East Greenland, *in* A. M. Spencer et al., eds., Petroleum Geology of the North European margin: Graham and Trotman Ltd. for the Norwegian Petroleum Society, p. 303–315.

———, J. M. Hurst, C. Marcussen, S. Piasecki, F. Rolle, P. A. Scholle, L. Stemmerik, and E. Thomsen, 1984b, Oil geological studies in the Jameson Land basin, East Greenland: Grønlands geologiske Undersøgelse Rapport, 120, p. 85–90.

———, T. Frazier, O. Harpøth, J. M. Hurst, S. Piasecki, F. Rolle, P. A. Scholle, and L. Stemmerik, in press, Lithostratigraphy of the Upper Permian Foldvik Creek Group, East Greenland. Grønlands geologiske Undersøgelse Bulletin.

Teichert, C., and B. Kummel, 1976, Permian-Triassic boundary in the Kap Stosch area, East Greenland: Meddr Grønland, v. 197, n. 5, p. 1–54.

Teichmüller, M., 1974, Über neue Macerale der Liptinit-Gruppe und die Entstehung des Micrinits: Fortschritte in der Geologie Rheinland und Westfalen, v. 24, p. 37–64.

Thomsen, E., 1985, A coalification study of Upper Palaeozoic-Mesozoic deposits from Central East Greenland: The Geological Survey of Greenland, Open file, p. 1–58.

Tissot, B. P., and D. H. Welte, 1978, Petroleum formation and occurrence: Berlin, Springer Verlag, p. 1–538.

Trümpy, R., 1960, Über die Perm-Trias Grenze in Ostgrönland und über die Problematik stratigraphischer Grenzen: Geologische Rundschau, v. 49, p. 97–103.

———, 1969, Notes on Triassic stratigraphy and paleontology of north-eastern Jameson Land (East Greenland). II Lower Triassic ammonites from Jameson Land (East Greenland): Meddr Grønland, v. 168, n. 2, p. 77–116.

Vischer, A., 1943, Die postdevonische Tektonik von Ostgrönland zwischen 74° und 75° N. Br., Kuhn Ø, Wollaston Forland, Clavering Ø und angrenzende Gebiete: Meddr Grønland, v. 133, n. 7, p. 1–195.

Waples, D. W., 1980, Time and temperature in petroleum formation: Application of Lopatin's methods to petroleum exploration: AAPG Bulletin, v. 64, p. 916–962.

Warren, J. K., 1982, The hydrological setting, occurrence, and significance of gypsum in late Quaternary salt lakes in South Australia: Sedimentology, v. 29, p. 609–637.

———, and C. G. St. C. Kendall, 1985, Comparison of sequences formed in marine sabkha (subaerial) and salina (subaqueous) settings: modern and ancient: AAPG Bulletin, v. 69, n. 6, p. 1013–1023.

# Tectonic Development of the Western Margin of the Barents Sea and Adjacent Areas

Fridtjof Riis
Jan Vollset
Morten Sand
*Norwegian Petroleum Directorate*
*Stavanger, Norway*

Interpretation of new data from the southwestern part of the Barents Sea makes it possible to refine earlier tectonic models. Following what was probably a Devonian and Early Carboniferous rifting phase, a stable platform was developed during the Carboniferous and Permian, with a north to south trending high, to the west. The north-northeast to south-southwest Senja left-lateral fault system was active in several phases through the Cretaceous and Tertiary, up to the Oligocene. This fault system caused complex deformation in the Senja Ridge and the Bjørnøya basin.

## INTRODUCTION

The Barents Sea is situated on the Norwegian and Russian continental shelves, and is bounded by the Norwegian–Greenland Sea, the islands of Svalbard and Novaja Zemlya, and the Norwegian and Russian mainland. Figure 1 shows the sea floor morphology of the undisputed Norwegian part of the area, and indicates average ice limits during summer and winter.

Because data are sparse, pre-drift plate tectonic models of the area surrounding the Arctic Sea are uncertain. However, the Barents Sea is geologically related to the Sverdrup basin in Arctic Canada. This relationship is indicated not only by plate tectonics, but also by a comparable stratigraphic development in the two basins, especially in pre-Cretaceous times. In early Tertiary times, the western margin of the Barents Sea developed as a shear margin because of the opening of the Norwegian–Greenland Sea (Myhre, Eldholm, and Sundwor, 1982; Talwani and Eldholan, 1977). The geometry changed to an ordinary passive spreading margin in the Oligocene. This plate tectonic development caused a complicated structure in the western part of the Barents Sea.

So far, 27 blocks have been opened for exploration drilling in the Norwegian part of the Barents Sea, with several gas discoveries in the Troms area. These discoveries occurred in the Hammerfest basin, where 10 exploration wells out of 18 encountered gas, and, lately, heavier components. All discoveries belong to the same play type, located on rotated fault blocks with Middle and Lower Jurassic reservoir rocks.

The Norwegian Petroleum Directorate has collected 85,000 km (52,820 mi) of multichannel reflection seismics and gravimetric-magnetometric data in the Barents Sea. A regional seismic grid, therefore, covers the area and, in large part, is semi-regional. In addition, scientific institutions have shot about 8,000 km (4,970 mi) of seismics, and industry-shot seismics cover the opened area with a detailed grid. The data vary in quality, but, in general, lines that date from 1978 and later are of good quality. Because of multiple problems, however, there are areas where the data quality is poor, such as south and northeast of Bjørnøya and east of Hopen toward Kong Karls Land.

Rønnevik, Beskow and Jacobsen, (1982) and Rønnevik and Jacobsen, (1984) have published several papers on the geology of the Barents Sea. In addition, the exposures of Devonian to Tertiary rocks on Spitsbergen and Bjørnøya have been thoroughly studied (Norsk Polarinstitutt 1971–1984; Steel and Worsley 1984; Worsley et al, 1985). Information from the Russian basins can be obtained, for example, from Nalivkin (1960). Our paper update Rønnevik's papers, with emphasis on the structural development of the southwestern part of the Barents Sea.

### Stratigraphic Correlations

Because of lack of stratigraphically deep well control, we must correlate the dating of the old reflectors with the nearby land areas. As the stratigraphy of Spitsbergen and Bjørnøya seems to fit quite well with the seismics, we will also propose a tentative dating.

In most of the Barents Sea, one can recognize a sequence of reflectors that overlie and drape over a system of block-faulted rocks (Figure 4). This is consistent with the geology on Svalbard, where main rifting took place in the Middle Devonian and in the Carboniferous. A long stable period of platform sedimentation succeeded this rifting.

In the Permo-Carboniferous sequence, there are several depositional breaks. One of them occurs in the Lower Permian (Sakmarian to Artinskian). At Bjørnøya, it is represented by an unconformity with some erosion (Worsley et al, 1985), which can be correlated to the sequence at Spitsbergen by breaks in sedimentation and change in deposition to more evaporitic-dominated environments (Worsley et al, 1985). Another break represents the boundary between clastic and carbonate sedimentation, occurring close to the top of Lower Carboniferous at Svalbard. Rønnevik, Beskow, and Jacobsen (1982) suggest that the "$F_2$" and "G" reflectors correlate with these two events.

Based on character and well data, the "$F_1$" reflector of Rønnevik, Beskow, and Jacobsen represents the top Upper Permian, characterized in Svalbard by the transition from cherty carbonates and shales to soft shales in the Triassic. This correlation also seems to be supported by preliminary results from shallow drilling made by IKU on a scientific license northeast of the Loppa High. In the Triassic sequence, the reflectors named E and E′ by Rønnevik, Beskow, and Jacobsen (1982), are tentatively correlated with top Lower Triassic and top Middle Triassic, respectively. Because of well control, dating of the Jurassic and Cretaceous reflectors is quite reliable in the southwestern part of the Barents Sea. The precise dates of the Tertiary and Upper Cretaceous strata in the western basins, however, are more uncertain. We will discuss the interpretation of the southwestern part in more detail below.

## STRUCTURAL DEVELOPMENT

### Structural trends

From the top Permian map (Figure 3), we can note two main trends. The north-to-south trend dominates in the western part, and the southwest-to-northeast trend forms the gentle basins and highs in the eastern part of the area. Composite structures occur in the junctions between these two main trends. Other east-to-west trends occur, but wrenching along the southwest-to-northeast trend often explains them. In addition, the Varanger basin represents a west-northwest to east-southeast trend parallel to the late Precambrian Timan–Kanin trend (Figure 2). Both of the two main trends are old and have been rejuvenated in several tectonic events.

### Devonian and Carboniferous Grabens

The formation of large Devonian graben structures trending north to south succeeded the Caledonian orogeny at Svalbard. Middle to Upper Devonian deformation, probably related to wrenching (Harland, 1969), caused inversion and folding of these basins. The Lower Carboniferous graben systems were developed in a separate tectonic phase. In the northwestern part of the study area, we can define large, rotated fault blocks comparable in size to the graben systems at Svalbard (cf. Figure 4). Figure 3 shows some of the most important of these faults. Geologists have mapped the faults at the deepest identifiable reflecting level, and, by correlating with Bjørnøya, believe this

→

**Figure 1.** *Barents Sea, sea-floor morphology.*

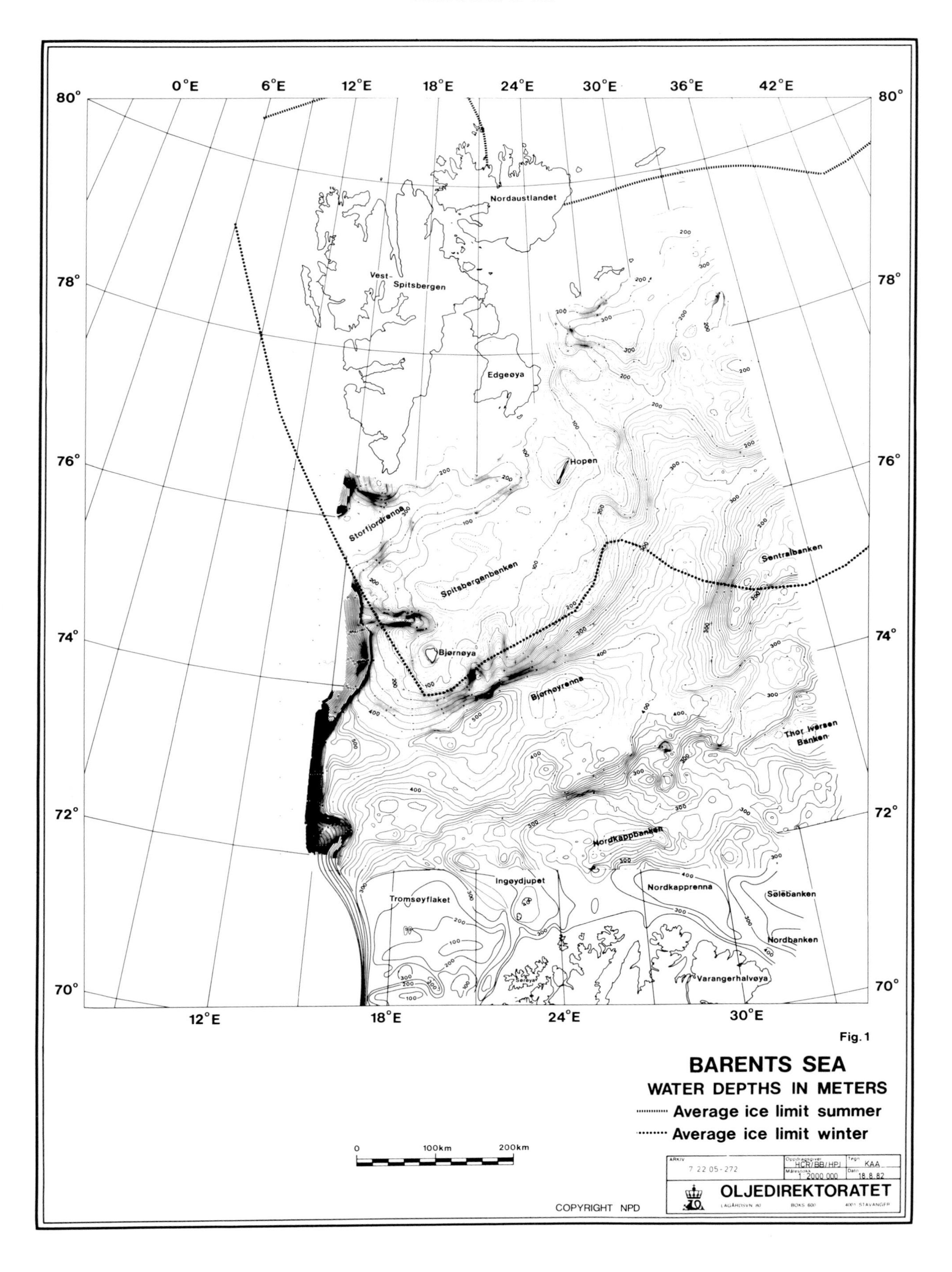
0°E
6°E
12°E
18°E
24°E
30°E
36°E
42°E
80°
78°
76°
74°
72°
70°
12°E
18°E
24°E
30°E
Nordaustlandet
Vest-
Spitsbergen
Edgeøya
Hopen
Storfjordrenna
Spitsbergenbanken
Sentralbanken
Bjørnøya
Bjørnøyrenna
Thor Iversen
Banken
Nordkappbanken
Ingøydjupet
Tromsøyflaket
Nordkapprenna
Sølebanken
Nordbanken
Varangerhalvøya
Fig. 1
BARENTS SEA
WATER DEPTHS IN METERS
Average ice limit summer
Average ice limit winter
0
100km
200km
7 22 05-272
HCR/BB/HPJ
KAA
1 2000 000
18.8.82
OLJEDIREKTORATET
COPYRIGHT NPD

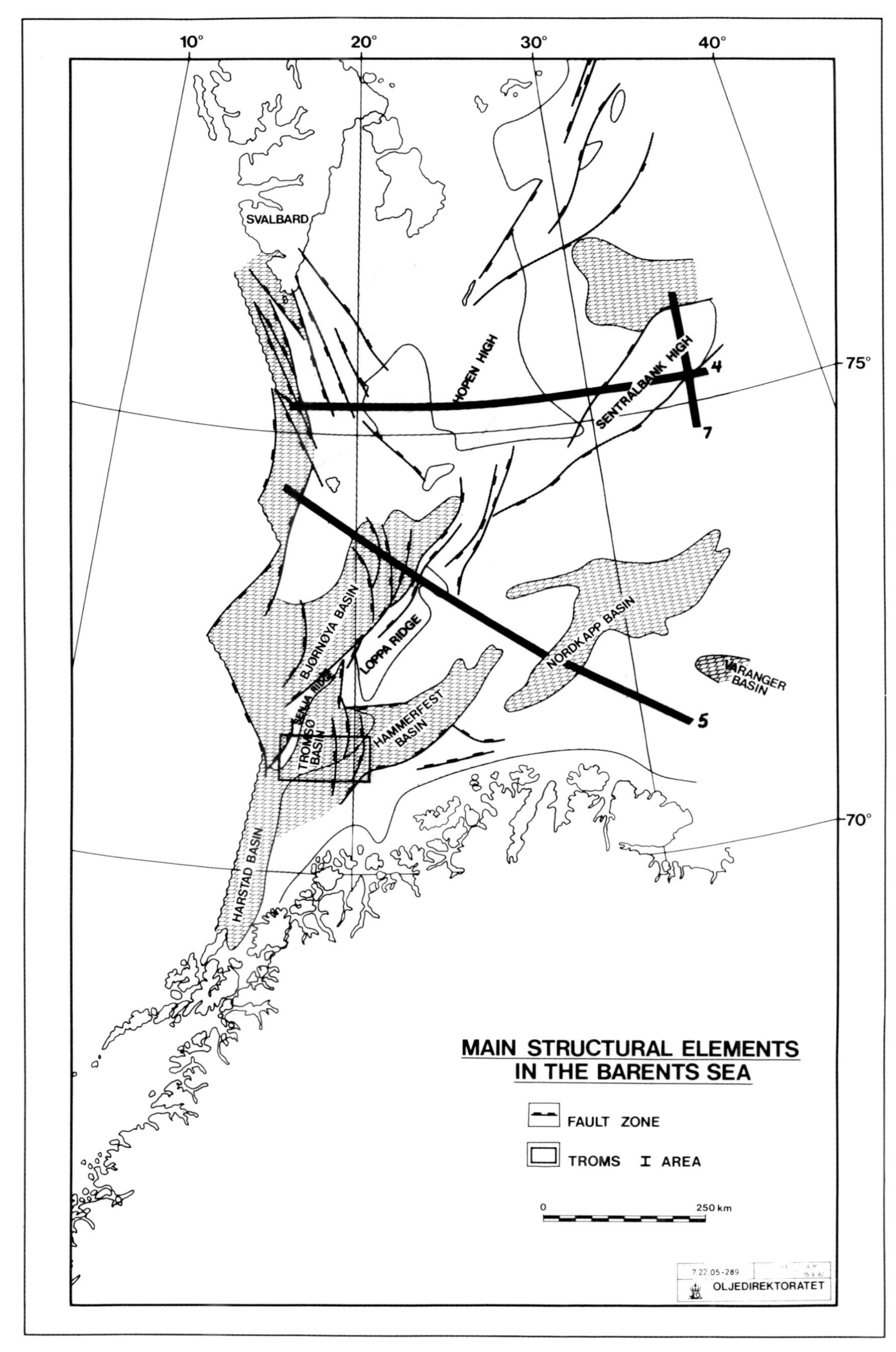


Figure 2. *Main structural elements in the Barents Sea.*

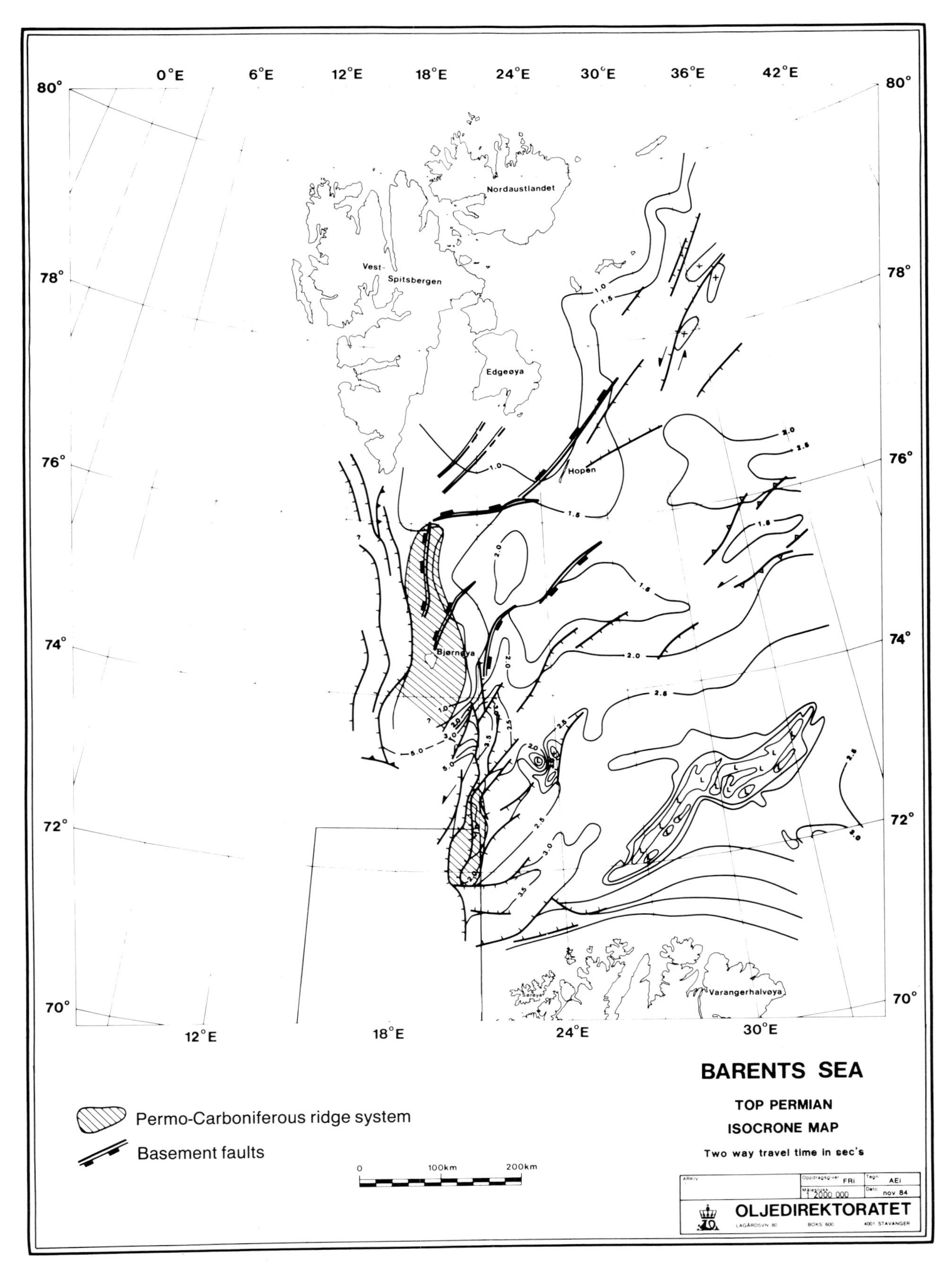


Figure 3. *Isochrone map of top Permian, Barents Sea. Frame shows location of Figure 10. L indicates salt structures.*

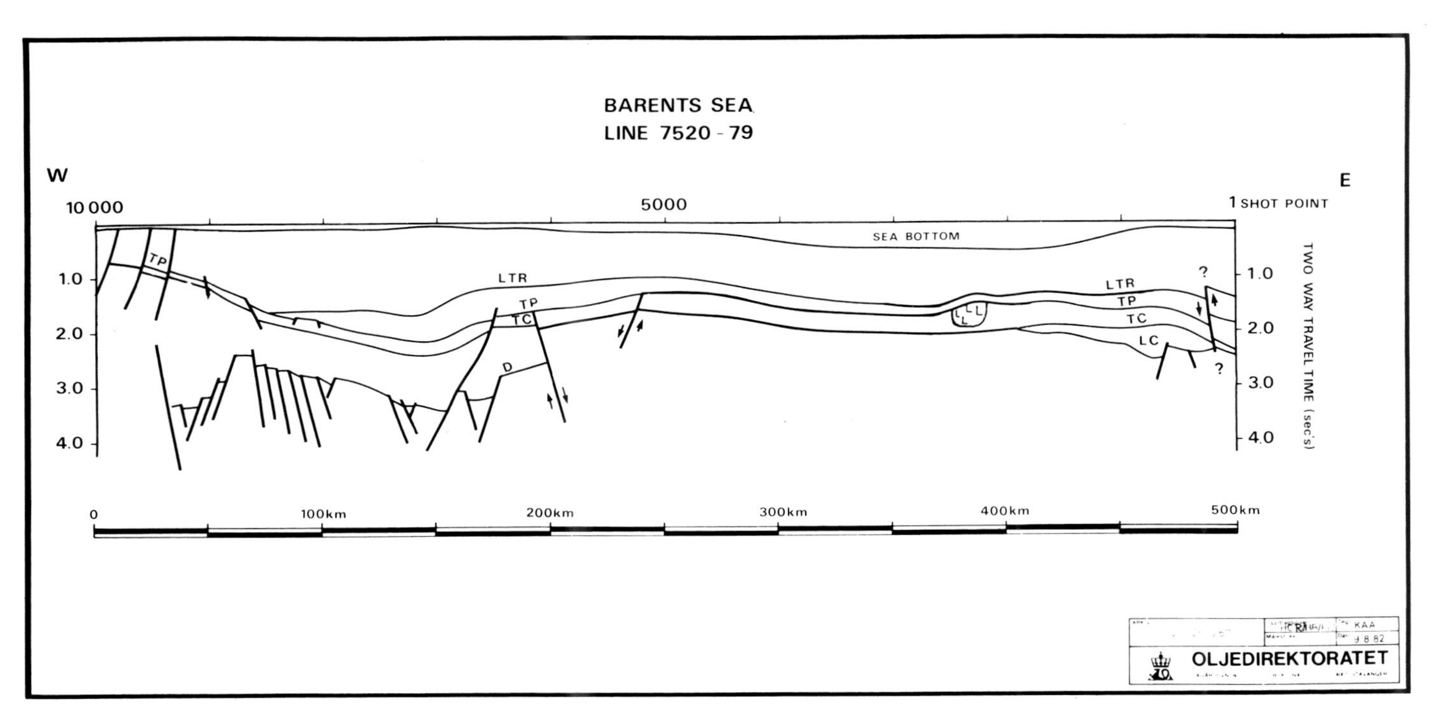


**Figure 4.** *Barents Sea Line 7520–79. The view shows an east to west profile from northern Barents Sea. LTR = Lower Triassic; TP = Top Permian; TC = Top Carboniferous; LC = Top Lower Carboniferous (approx.); D = Devonian reflector.*

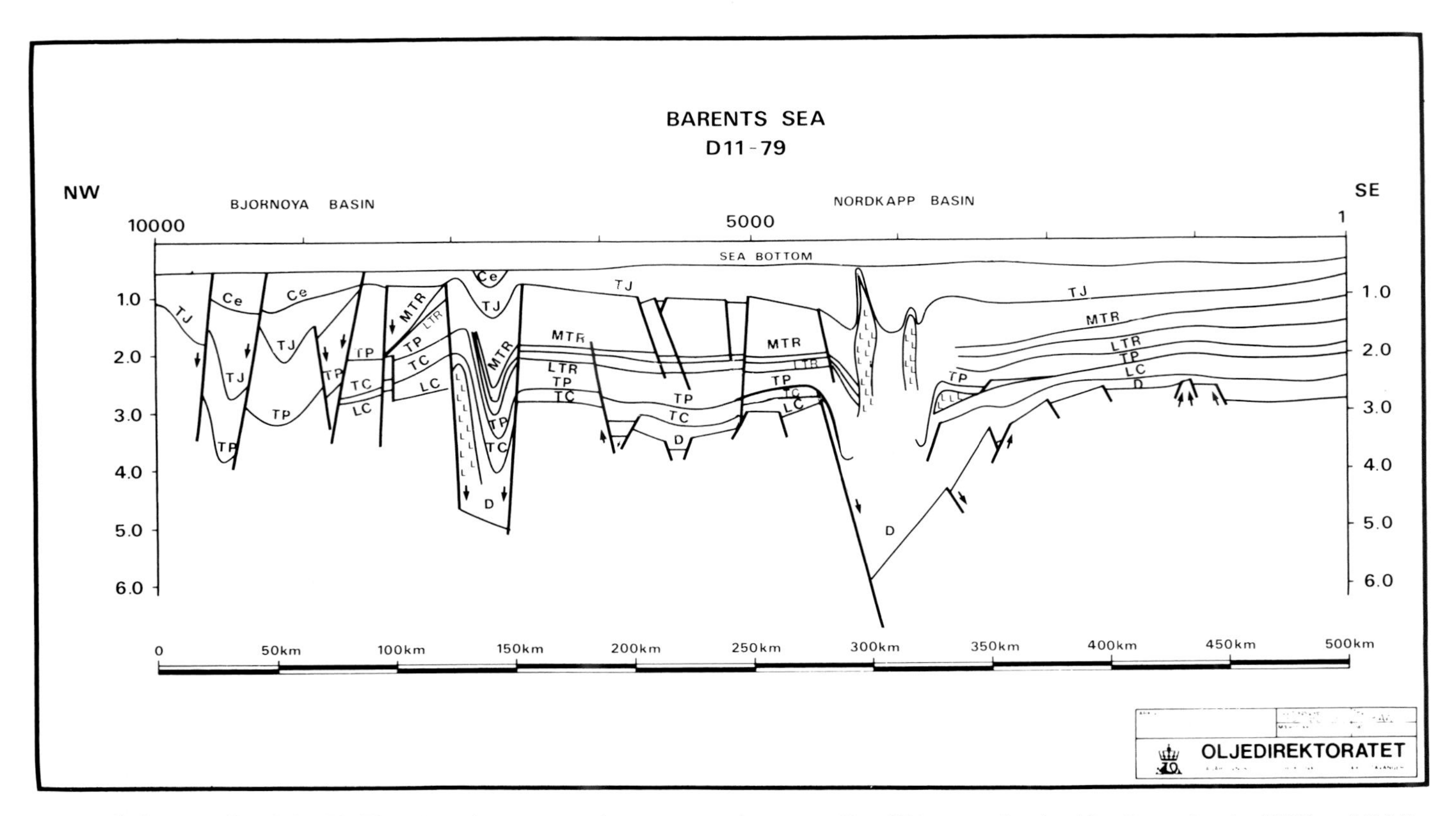


**Figure 5.** *Barents Sea D11-79. The view shows a northwest to southeast profile of Bjørnøya basin–Nordkapp basin; MTR = Middle Triassic; TJ = Top Jurassic; Ce = Cenomanian.*

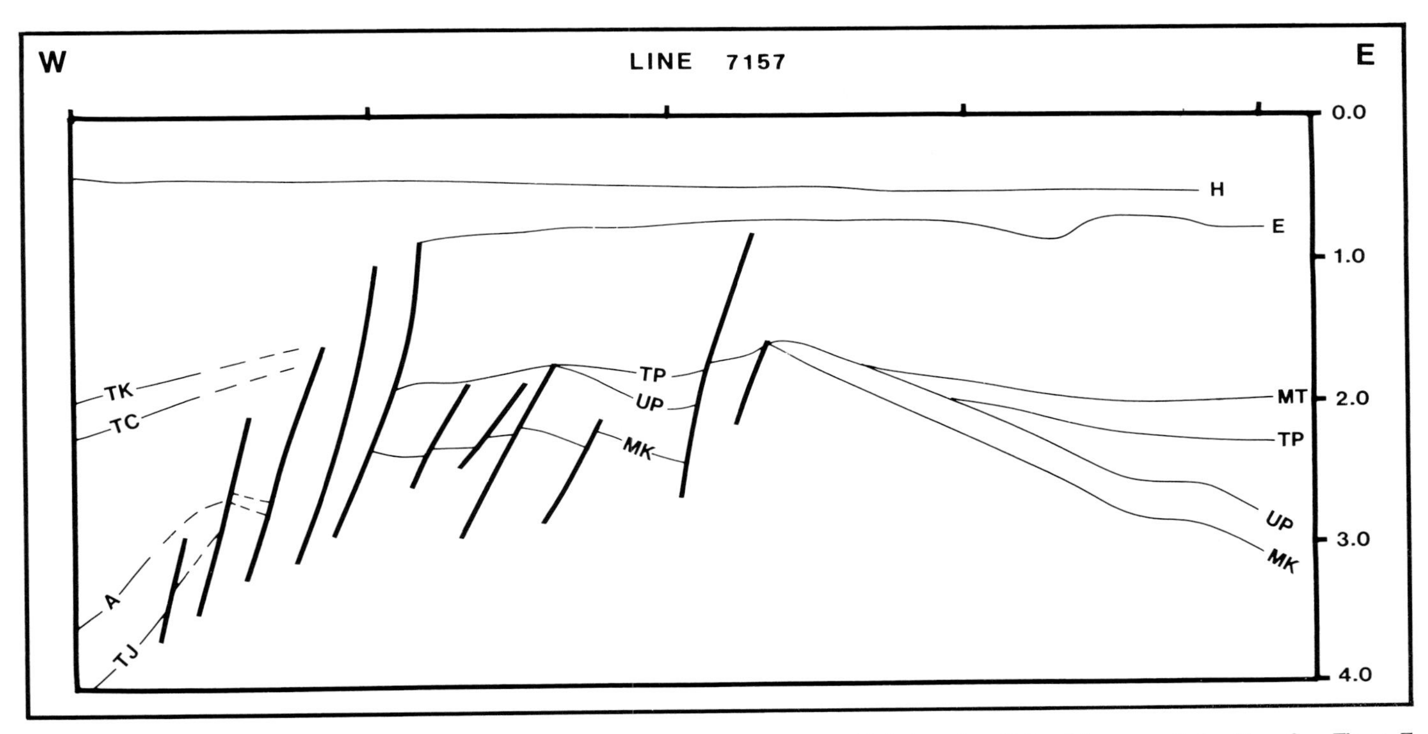


Figure 6. *Barents Sea Line 7157. The view shows an east to west profile Loppa High. Ten km (6.2 mi) between ticks. H = Sea Floor; E = Erosion Surface; TK = Base Tertiary; TC = Top Cenomanian; A = Aptian; UP = Lower Permian; MK = Top Lower Carboniferous (approx.)*

level to be the Caledonian basement. They do not know whether the mapped graben structures are of Devonian or Carboniferous age. Also, at present, they cannot map the basement reflector in other parts of the area. The basement fault blocks do not appear to have any major impact on the development of the Permo–Carboniferous sequence. However, the current topography seems to correlate with deep structural trends (Figure 3).

So far, geologists have observed only extensional features in the Devonian–Carboniferous. However, the lack of information from large areas complicates the tectonic analysis.

The Nordkapp basin salt is deeply rooted (Figure 5) and could have a Late Devonian or Early Carboniferous age (Rønnevik, Beskow, and Jacobsen 1982). Well data indicate that the Tromsø basin salt is Permian or older. Analogy suggests that the Tromsø basin developed as a salt basin in the same period as the Nordkapp basin. Seismic lines from the Loppa High indicate that the flank of such a basin could be preserved on the western part of the high.

### Middle Carboniferous to Permian

From Middle Carboniferous to Permian time most of the Barents Sea was a stable platform. Tectonic activity was limited to the western part (Figure 3; see the Permo–Carboniferous High). In this area, the north-to-south trend was active through the Carboniferous. At Bjørnøya, Sakmarian carbonates cover the rotated Carboniferous fault blocks. In the Late Permian, most of the activity had ceased at Bjørnøya.

A profile from the Loppa High (Figure 6) indicates a similar history. Triassic rocks lap onto a basement high that was tilted and eroded in Permian times. In the western part of Svalbard, the Sorkapp–Hornsund High was active in the same period (Steel and Worsley, 1984).

The existence of positive areas in the western Barents Sea may have affected the sedimentation in a fairly large area, as indicated by changes in seismic facies. In the central and eastern areas, the Permo–Carboniferous sequence thins slightly on the highs and thickens in the basins, indicating that the main tectonic features have been active for a long time. The Sentralbank High could be an exception (Rønnevik, Beskow, and Jacobsen, 1982).

### Triassic to Jurassic

The period was tectonically quiet up to the Middle/Late Jurassic. Triassic isopach maps mainly show northeast-to-southwest trends, which do not correspond to the present structural trends (Rønnevik, Beskow, and Jacobsen, 1982). Depocenters occurred in the eastern part of the study area and in basins bounding the Permo–Carboniferous Highs. The Loppa High was leveled off and buried during the Middle and Late Triassic (Figure 6).

The Jurassic sequence is now well known in the Troms area, and, apart from an insignificant discovery in the Triassic, all discoveries in the Troms area were made in Lower to Middle Jurassic (Pliensbachian–Bathonian) sandstones. The reservoir sandstones consist of two different facies regimes. Shore facies associations with several cycles of offshore-to-beach deposits dominated the upper part, while the lower part consists of coastal plain and tidal flat deposits in several upward-fining sequences (Olaussen et al, 1984).

The main parts of the reservoirs are concentrated in the

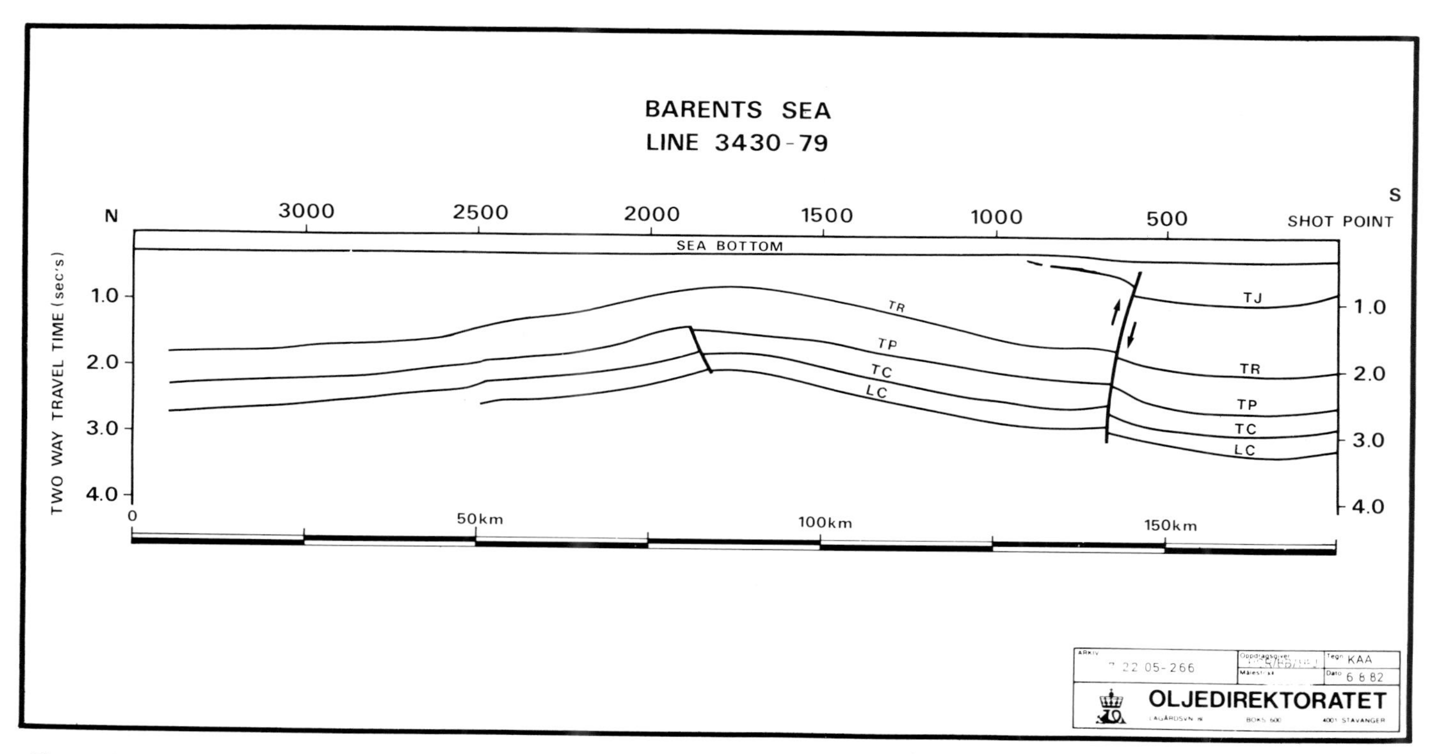


Figure 7. *Barents Sea Line 3430-79. The view shows a north to south profile across Sentralbank High. Abbreviations as Figure 4.*

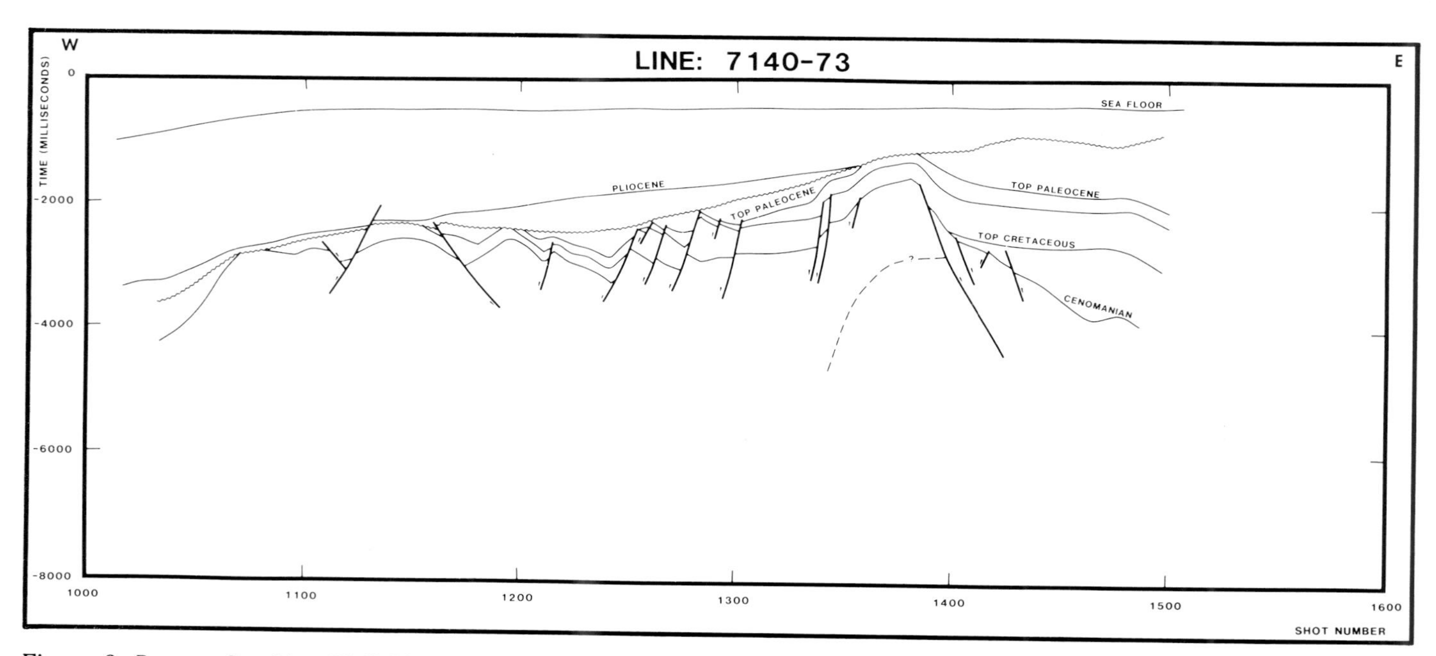


Figure 8. *Barents Sea Line 7140-73. The view shows an east-to-west profile of Senja Ridge. This view is the same profile as in Figure 11.*

upper section, called the Sto Formation (Olaussen et al, 1984). Within the Troms I area, the thickness distributions of the various sandstone beds within the Sto Formation show an overall increase towards west-northwest, except for the uppermost sandstone beds. The overlying Callovian to Oxfordian shales decrease to zero towards east-northeast. Tectonic movements in the Middle to early Late Jurassic caused nondeposition and erosion of the reservoir sandstones and the overlying shales, thereby explaining the decrease.

Geologists have debated about the direction of sediment transport into the area. Rønnevik, Beskow, and Jacobsen (1982), for example, argue that the sands originate from a western source, but have suggested an eastern source.

### Late Jurassic and Early Cretaceous

Because of erosion, precise dating of structural events younger than the Triassic is difficult in large parts of the Barents Sea. However, we can extend information from the western and southern areas, where most of the stratigraphy has been preserved.

In the Troms area, tectonic movements started in the Middle to Late Jurassic, giving rise to block faulting mainly along east to west and southwest to northeast trends. The large subsidence of the north to south trending Tromsø and Bjørnøya basins probably was initiated at the same time. The deformation continued in the Early Cretaceous with formation of the Senja Ridge and further activity along the transition zone between the Cretaceous basins to the west and the stable platform to the east. The main Cretaceous deformation phase involved faulting and folding in the Senja Ridge, and was related to wrench movements we discuss below. Activity ceased in the Late Cretaceous, and was succeeded by uplift, so that the Upper Cretaceous is condensed and is only well developed in the deep basins (Figure 8).

### Tertiary Deformation

In the early Tertiary, wrench movements continued along the same southwest to northeast trends as in the Early Cretaceous, with deformation concentrating along the main wrench zones. In the whole area, conspicuous erosion surfaces are not dated with accuracy (Figure 8), and considerable uplift probably occurred, perhaps related to the Oligocene change in spreading configuration (Talwani and Eldholm 1977; Myhre, Eldholm, and Sundvor, 1982). The deposition of a sedimentary wedge, with thicknesses up to 2.5–3 seconds two-way time succeeded the erosion (Figure 8). The wedge covers the western part of the Barents Sea south to Lofoten, but the most extensive part is situated in the western part of the Bjørnøya basin and on the Senja Ridge. The wedge is little deformed, although geologists have noted disturbances above main Tertiary faults and fold/diapir structures. Well information could indicate that the wedge as mapped in Figure 8 is as young as Pliocene/Pleistocene. This interpretation agrees with Spencer, Home, and Berglund (1984), who note that dating is difficult because of microfossil redeposition.

## TECTONIC ANALYSIS OF THE SOUTHWESTERN BARENTS SEA

The tectonic map (Figure 9) shows the main trends of the faults at the Jurassic and the mid-Cretaceous level (Figure 3 shows the location). The tectonic map defines the boundary of the Tromsø basin as the Upper Cretaceous pinch-out, which is almost coincident with the main faults on the Jurassic level. The map further describes the relations between the stable eastern areas (Loppa High and Norwegian mainland) and the subsiding western basins (Tromsø basin; Senja Ridge). Faults and domal fractures characterize the structurally complex Senja Ridge. Domal structures (stippled) are mapped at the Cenomanian level (with a few exceptions), and in general, are oriented east to west. Often, they may exhibit a complex internal structure (Figure 10), indicating a relationship with compressional tectonics rather than shale or salt flowage. However, the thick shale sequence of the Senja Ridge may have enhanced the structuring.

### Geological and Gravity Model of the Senja Ridge

A strong, positive free-air and Bouguer gravity anomaly characterizes the Senja Ridge. This gravity anomaly has a maximum relative to the Tromsø basin of well above 100 milligals.

Because drilling on the Senja Ridge has proven the existence of thick Cretaceous shaly sequences, we cannot explain this gravity anomaly. New seismics covering the northern part of the Senja Ridge indicate that the shape of the anomaly is related to the depth to the Jurassic. The anomaly's two parts correlate with the Jurassic high areas (Figure 9 and 12). Also, within the Senja Ridge, geologists can define internal horst structures that are partly offset relative to the Cenomanian axis of the ridge (Figure 9).

In order to interpret the gravity anomaly, geologists have studied free-air anomalies from observations made by the NPD 1973 survey (Figure 11). The interpretation method is based on Enmark (1981), a "2$^1/_2$-D" calculation based on the 2-D method of Talwani (1973). Modelling is done by calculating the gravity effect of bodies striking perpendicularly to the observation profile. Densities are selected relative to an average crust density of 2.67 g/cu cm. Moho appears to be at 25 km (15.5 mi) along most of the profile, decreasing to about 15 km (9.3 mi) toward the oceanic crust in the west. In addition, we assume that heavy basement is located about 1 km (0.6 mi) below the Jurassic reflector. We can draw two interesting conclusions from this model:

(1) We can account for the anomaly with the present geological model, without introducing heavy intrusions below the Senja Ridge (Figure 11). This answer indicates that the pre-Jurassic sequence is thin or faulted away.

(2) The basin at the west side of the Senja Ridge, which is little known, seems to be of a depth comparable to the Tromsø basin.

## CONCLUSIONS

### Senja Ridge Lateral Fault System

We can now explain how the Senja Ridge formed in terms of a large, southwest-to-northeast-oriented, left lateral wrench system extending from the junction between the Ridge and the continent to ocean boundary, toward the boundary between the Loppa High and the Bjørnøya basin. A further extension of the wrench system toward the Sentralbank High is possible and has been suggested by Rønnevik and Jacobsen (1984).

The observed structures fit well with such a concept. Very simple stress models would predict east-to-west-trending folds and north-to-south-trending normal faults exaggerating the depth of the large basins (Figure 13). Also, the relations between this wrench zone and the oceanic-continental boundary are predictable, taking into account that this margin had a right lateral shear in the early Tertiary. Geologists can explain the complex boundary between the Tromsø and Bjørnøya basins by such a model, which indicates that these two basins were continuous prior to wrenching.

Also, the geometry indicates a small left lateral wrench along the eastern boundary zone of the Tromsø basin, parallel to the main fault. This zone has caused the compressional features at the southwestern part of the Loppa High (Figure 14).

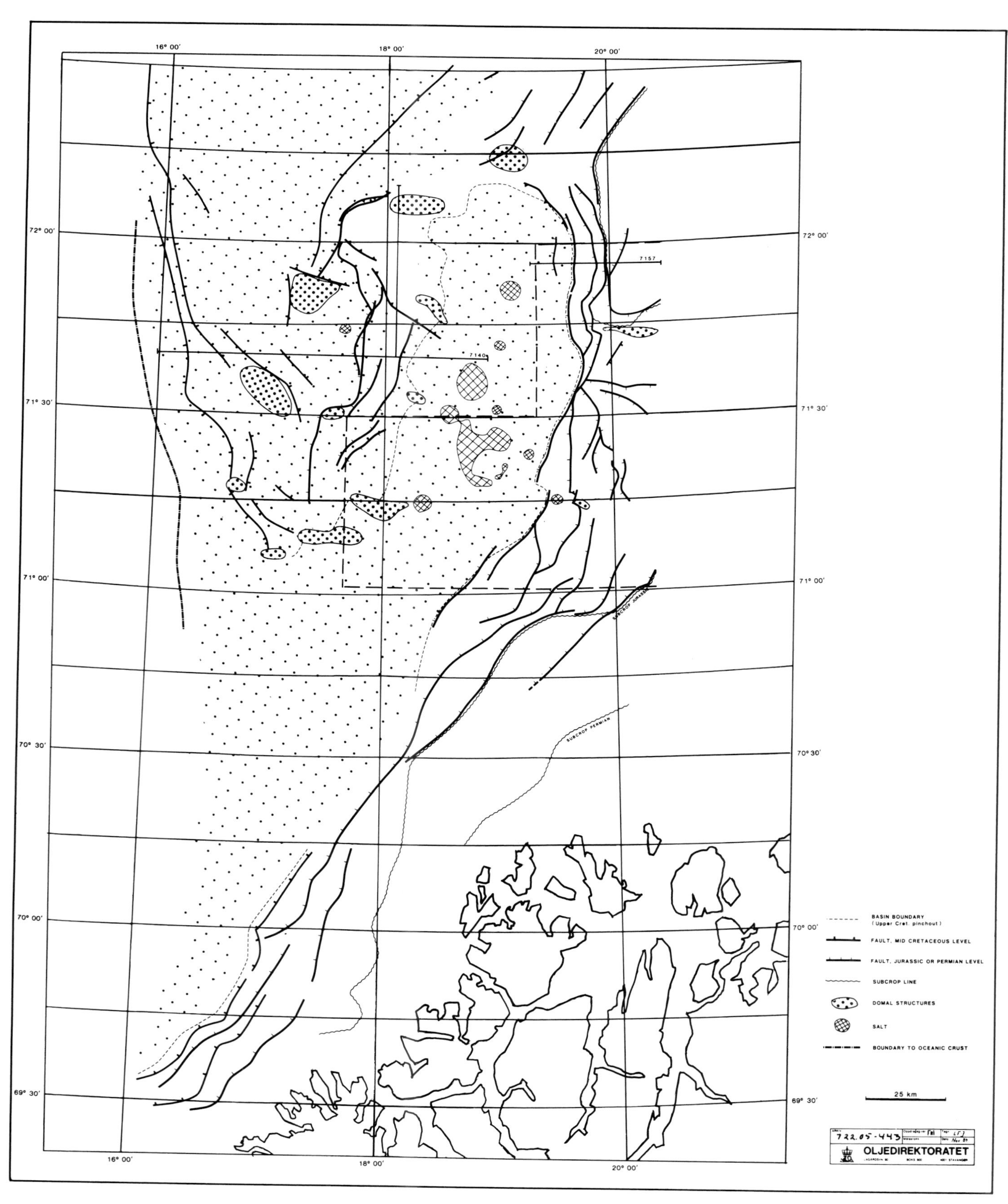


Figure 9. *Tectonic map of the Lofoten–Troms area, southwestern Barents Sea.*

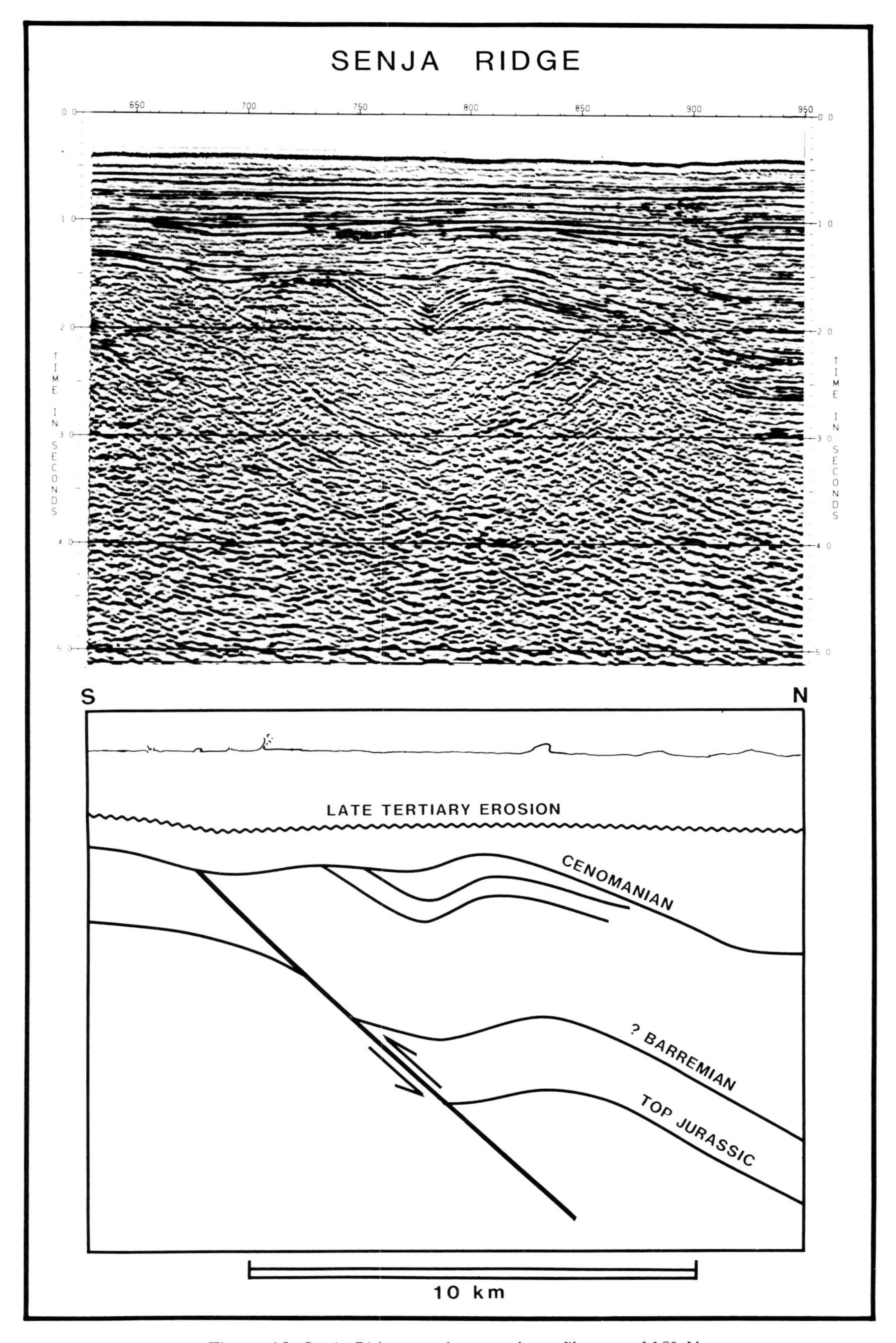


Figure 10. *Senja Ridge, north-to-south profile east of 18° N.*

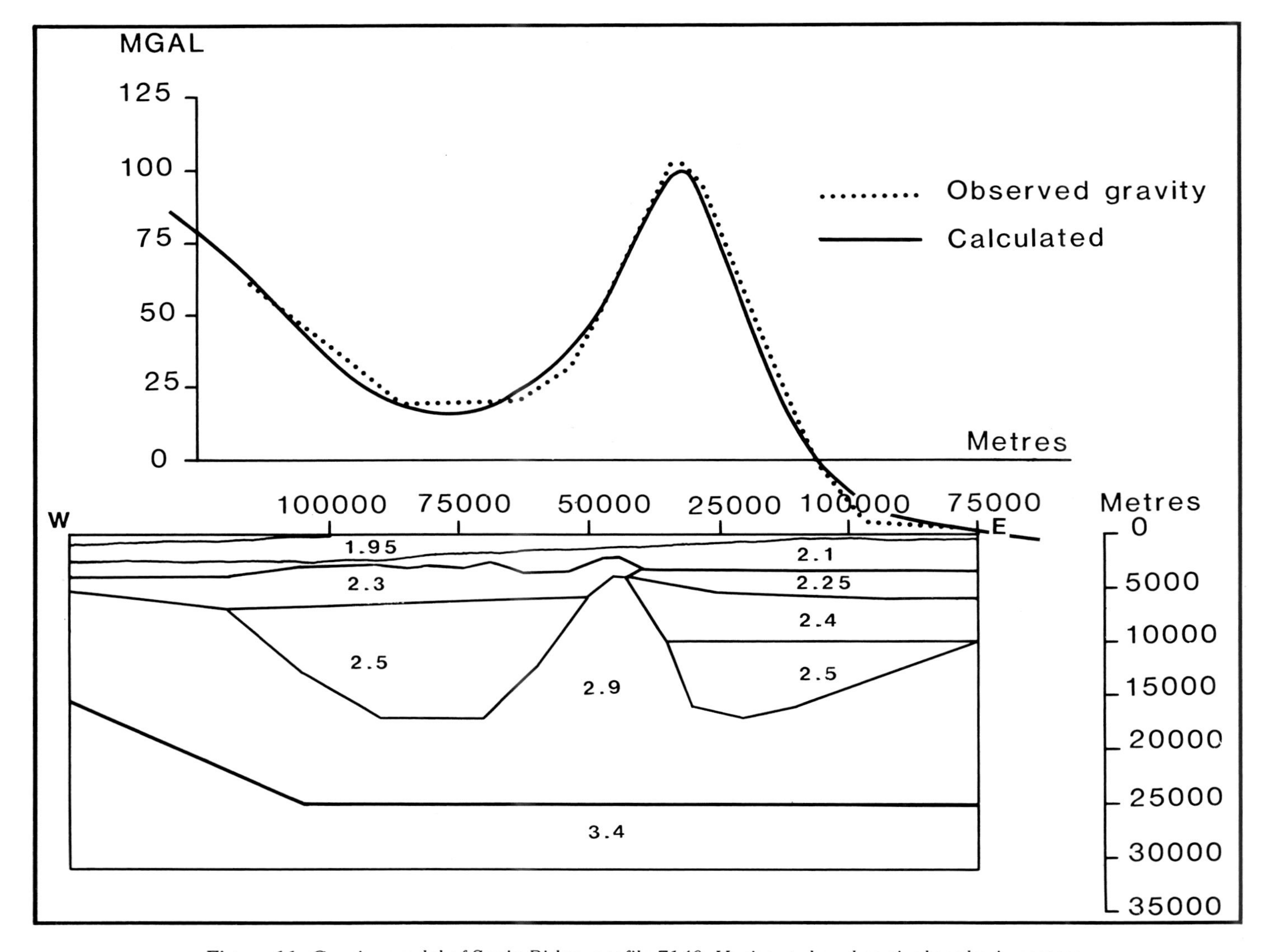


Figure 11. *Gravity model of Senja Ridge, profile 7140. Horizontal and vertical scales in meters.*

South of the junction between the main wrench system and the oceanic crust, geologists have observed no important compressional features. The important features here involve large-scale extension, and sliding and rotation of fault blocks (Figure 8).

### Timing of Wrenching

An important Tertiary phase of wrenching can be related to the opening of the Norwegian–Greenland Sea. Preceding this phase, however, important movements took place in the Early Cretaceous. We can correlate this activity with activity on the Harder Fjord Fault Zone and the Wandel Sea Strike-Slip Mobile Belt in Greenland (Håkansson and Stemmerik, 1984). In this context, it is probably also significant that the Upper Cretaceous is lacking at Svalbard.

### Regional Significance

Geologists debate how far the Cretaceous wrenching extended into the Barents Sea. The Sentralbank High was partially in post-Jurassic time (Figure 7); compare to Rønnevik, Beskow, and Jacobsen's discussion (1982). This inversion might also be the case for the areas west of the high, as erosion does not permit precise dating of the structuring. We are interpreting the relative motion along the Senja Ridge wrench zone to be at least 10–20 km (6.2–12.4 mi), which is the distance necessary for restoring the northern part of the Senja Ridge (Figure 9).

### Petroleum Geology and Exploration

We know about the Middle Jurassic play type in the Hammerfest basin with discoveries of recoverable gas resources in the

Figure 12. *Barents Sea Bouguer Anomaly map. Major highs and lows indicated.*

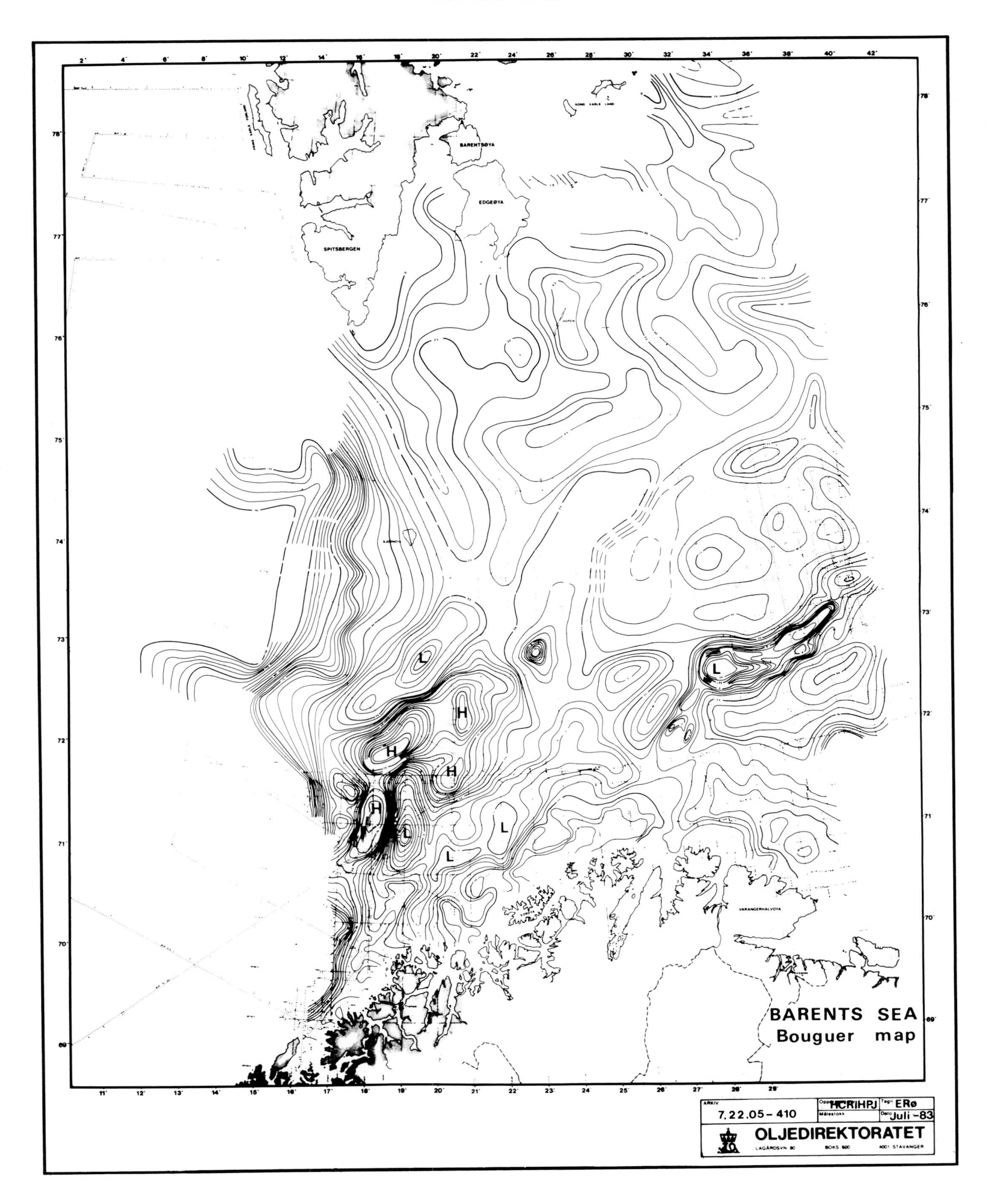
BARENTSØYA
EDGEØYA
SPITSBERGEN
HOPEN
VARANGERHALVØYA
BARENTS SEA
Bouguer map
ARKIV
7.22.05-410
HCR/HPJ
ERø
Juli -83
Målestokk
OLJEDIREKTORATET
LAGÅRDSVN 80
BOKS 600
4001 STAVANGER

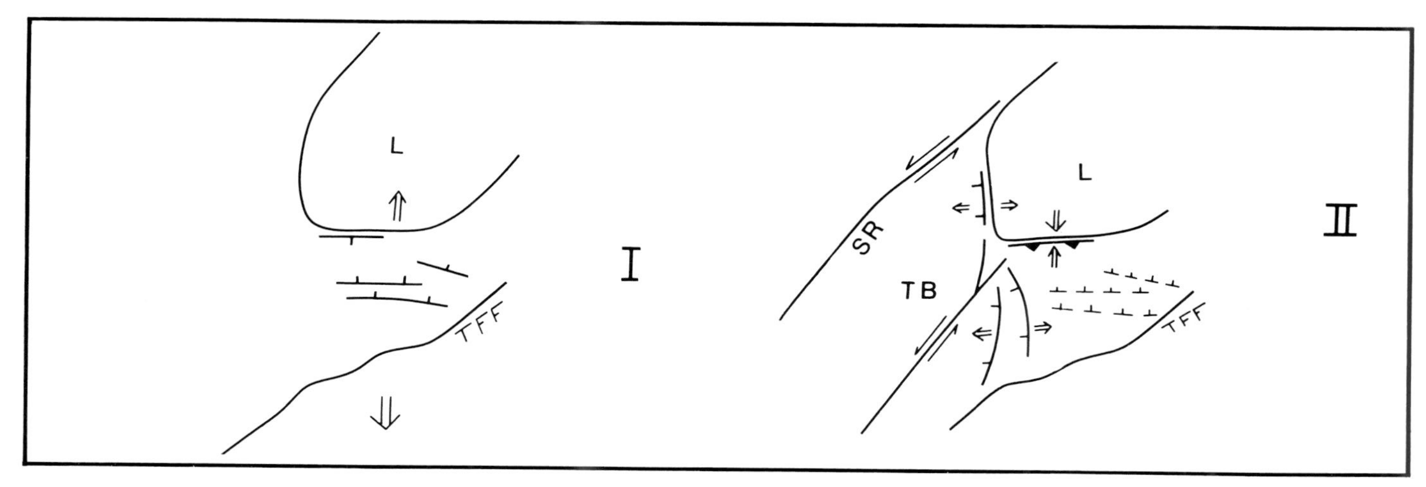


Figure 13. *Tectonic sketch showing development of Senja Ridge, Loppa High, Hammerfest basin. L = Loppa High; SR = Senja Ridge; TB = Tromsø basin; TFF = Troms–Finnmark fault.*

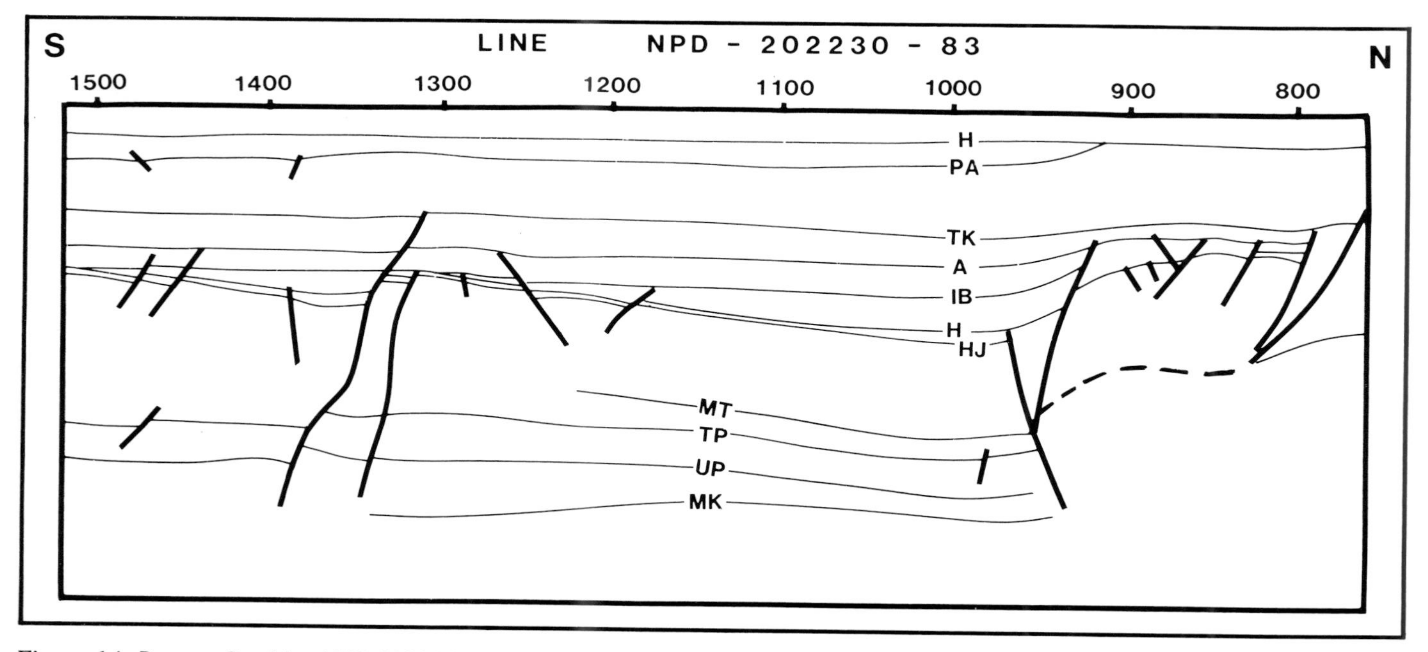


Figure 14. *Barents Sea Line NPD 202230-83. The view shows a north to south profile of Hammerfest basin–Loppa High. Abbreviations as in Figure 6. PA = Paleocene; IB = Intra-Barremian; H = Hauterivian.*

order of 200 billion cu m of gas. Source rocks for the hydrocarbons in the Troms area could be Upper Jurassic or Triassic shales. The maturation and migration history for these source rocks is not completely understood because of the complicated tectonism discussed above. Thus, for exploration purposes, timing of tectonic events could be crucial, as could a better understanding of the reservoir sands' deposition.

The 9th concession round on the Norwegian shelf offered blocks situated on the southern part of the Loppa High. Here, the Jurassic is eroded, and successful exploration will depend on Carboniferous-to-Triassic reservoir rocks. As indicated by the top Permian map (Figure 3), this condition will be the situation for large parts of the Barents Sea. Thus, results from the Loppa High will yield insights that can be very useful for further exploration.

## REFERENCES CITED

Enmark, T., 1981, A versatile computer programme for computation and automatic optimization of gravity models, Geoexploration, v. 19, p. 47–66.

Håkansson, E., and L. Stemmerik, 1984, Wandel Sea basin—The North Greenland equivalent to Svalbard and the Barents Shelf, *in* Petroleum Geology of the North European Margin: Norwegian Petroleum Society, p. 98–107.

Harland, W. B., 1969, contribution of Spitsbergen to understanding of tectonic evolution of North Atlantic region, *in* AAPG Memoir 12, p. 817–851.

Myhre, A. M., O. Eldholm, and E. Sundvor, 1982, The margin between Senja and Spitsbergen fracture zones: implications from plate tectonics: Tectonophysics, v. 89, p. 33–50.

Nalivkin, D. V., 1960, The geology of the U.S.S.R.: a short outline, Pergamon Press.

Norsk Polarinstitutt, Skr. 154 Geological map of Svalbard, 1:500,000.

Olaussen, S., A. Dalland, T. G. Gloppen, and E. Johannessen, 1984, Depositional environment and diagenesis of Jurassic reservoir sandstones in the eastern part of Troms I area, *in* Petroleum Geology of the North European Margin: Norwegian Petroleum Society, p. 98–107.

Rønnevik, H., and H. P. Jacobsen, 1984, Structural highs and basins in the western Barents Sea, *in* Petroleum Geology of the North European Margin: Norwegian Petroleum Society, p. 98–107.

Rønnevik, H., B. Beskow, and H. P. Jacobsen, 1982, Structural and stratigraphical evolution of the barents Sea, *in* A. Embry and H. R. Balkwill, eds., Arctic Geology and geophysics: Canadian Society of Petroleum Geologists Memoir 8, p. 431–441.

Spencer, A. M., P. C. Home, and L. T. Berglund, 1984, Tertiary structural development of the western Barents shelf: Troms to Svalbard, *in* Petroleum Geology of the North European Margin: Norwegian Petroleum Society, p. 98–107.

Steel, R., and D. Worsley, 1984, Svalbard's post-Caledonian strata: an atlas of sedimentational patterns and palaeogeographical evolution, *in* Petroleum Geology of the North European Margin, Norwegian Petroleum Society, p. 98–107.

Talwani, M., 1973, Computer usage in the computation of gravity anomalies, *in* B. Alder, ed., Methods of Computational Physics, v. 13, p. 343–389.

Talwani, M. Balkwill, and O. Eldholm, 1977, Evolution of Norwegian–Greenland Sea: Bulletin of the Geological Society of American, v. 88, p. 969–999.

Worsley, D., T. Agdestein, J. Gjelberg, K. Kirkemo, and R. J. Steel, 1985, Late Paleozoic basinal evolution and stabilisation of Bjørnøya—Implications for the Barents Shelf.

# Future Potential for Hydrocarbon Exploration on the United Kingdom Continental Shelf

J. R. V. Brooks
*Petroleum Energy Division*
*U.K. Department of Energy*
*London, England*

The sedimentary basins beneath the United Kingdom continental shelf are classified into exploration categories. The major areas of hydrocarbon potential are considered, and six cross-sections, each one from a different area, are presented to illustrate geological situations that may occur, and that may provide future exploration targets.

The lesser-drilled and smaller basins to the south and west of the U.K. are also discussed.

The paper gives estimates of undiscovered reserves, made by the U.K. Department of Energy, for those basins of the North Sea and the Irish Sea where geologists have carried out detailed assessments.

Considerable opportunities, it is concluded, exist for further exploration, and large quantities of hydrocarbons remain to be found.

## STATUS OF EXPLORATION ON THE UNITED KINGDOM CONTINENTAL SHELF

The Wallace Pratt Conference (December 1984) marked the 20th anniversary of the first well on the United Kingdom (U.K.) continental shelf (December 26, 1964). Since this date, exploration and development in the North Sea area of the United Kingdom continental shelf have been of considerable success, unprecedented perhaps in basins of such small areal extent.

During October 1984, there were 40 drilling installations active in the North Sea, and by the end of October 1984, a total of 2,528 wells had been drilled, of which 925 were exploratory, 449 were for appraisal, and 1,154 were developmental. The 925 exploration wells have led to the discovery of 27 oil fields and 8 gas fields currently in production; 8 oil fields and 14 gas fields are under development or under active appraisal. In addition 197 discoveries have tested hydrocarbons at significant rates. The total recoverable (proven plus probable) reserves amount to 57 trillion cubic feet (tcf) of gas and nearly 15 billion bbl of oil.

All producing offshore oil fields and those currently being considered for development lie in the sedimentary basins of the Central and Northern North Sea and the Moray Firth (Figures 1 and 6).

The gas fields currently in production are slightly more varied in their geography. Six lie in the Gas basin of the Southern North Sea, one in the Morecambe Bay area of the Irish Sea, and one across the United Kingdom and Norwegian median line. Most other dry gas discoveries are concentrated in the Southern North Sea and Irish Sea areas.

Geologically, the distribution of hydrocarbon discoveries is extremely diverse. They have been found in, although not necessarily produced from, every major geological system in the stratigraphic column from the Devonian to the Tertiary. Oil has even been found in fractured basement gneiss of Precambrian age, while shales of the Cambrian have yielded gas. Only Ordovician and Silurian formations have so far been devoid of hydrocarbons, although these lower Paleozoic systems have been drilled less frequently to date.

The distribution of geological formations in the various basins (Figure 2) may be compared with the detailed stratigraphic columns of the North Sea (Figure 3). Figure 3 also shows the geological distribution of oil and gas fields in production and under development, with gas the predominant hydrocarbon found in the Southern North Sea basin.

Oil production to date is restricted to the basins of the Northern North Sea. In the Southern Gas basin, the most prolific targets have been the Rotliegendes Sandstone of Early Permian and the sandstones of Early Triassic age. In contrast, the sandstones of the Early, Middle, and Late Jurassic are the reservoir rocks for the giant oil fields of the Viking graben within the Northern North Sea, while sands of early Tertiary age have so far been the major target for the central North Sea area. Middle and Upper Jurassic targets have also proven successful.

In addition, Tertiary (Eocene) sandstones provide the reservoir rock for the only major dry gas field in the Northern North Sea (Frigg), and Triassic sandstones are the primary reservoir rocks for gas in the Irish Sea.

While the offshore basins surrounding the U.K. have been the most prolific source of oil and gas, we should not forget that onshore exploration has continued at a very high level. Discoveries of oil and gas have been made in reservoir rocks of Early, Middle, and Late Jurassic and Triassic age in the Hampshire and Wessex basins in the southern part of England, adjacent to the English Channel. Oil has been produced from sandstones of mid-Carboniferous age since the late 1940s, and exploring these strata in central England continues; recent significant finds of oil and gas have been made in these rocks.

After 20 years of offshore exploration, the U.K. is self-sufficient in both oil and gas and is intent on remaining so. To sustain the momentum of exploration, tracts of acreage are usually put up every two years, and applications for the Ninth Round were due for submission by December 17, 1984. The Ninth Round tracts (12-minute by 10-minute blocks; Figure 4) represent some of the areas that show potential for hydrocarbon discoveries. One must always bear in mind that 90% of drilling and 90% of oil and gas production occur in the North Sea basins and only 109 exploration wells (out of a total of 925) have been drilled in the areas to the west of mainland Britain.

Let us now turn to future exploration of the U.K. continental shelf.

## EXPLORATION PHILOSOPHY

A word about the philosophy of exploration is appropriate. Much has been written about the search for the *subtle trap*. Such dialogue has reinforced the idea of exploring for unusual and unconventional reservoirs.

More than ever, the petroleum explorer needs to be positive, innovative, and open-minded. How many prospects, or basins for that matter, have been written off because "conventional geological wisdom" was against a new idea? How many wells are prematurely curtailed or never drilled to greater depths because current thinking suggests either no porosity or no reservoir/source will exist? Concepts are often discarded because the philosophy upon which they are based is not compatible with current prospect-evaluation techniques.

## STAGE OF EXPLORATION REACHED WITH SEDIMENTARY BASINS AND OTHER AREAS OF PROSPECTIVITY

Accumulated data from drilling and seismic operations in both offshore and onshore areas have made it possible to classify and rank the exploration stage of each known and potentially productive basin in the U.K. The broad assumption is that basins must be regarded as having potential until they are fully evaluated and are shown to be barren—if one can ever be certain.

Both Table 1 and Figure 5 show this classification of the most active and successful basins. Concentrating exploration in these mature (Type A) and less mature (Type B) areas may seem logical when looking for further hydrocarbons. However, this two-dimensional picture, illustrated on the map, becomes more complicated when you also consider prospects that may lie at depths below currently producing horizons.

The map of the basins (Figure 6) shows their current accepted limits at the time of their earliest infill. Essentially, they represent Late Carboniferous and Early Permian basin controls and indicate the paleogeography at that time, although this is an oversimplification to the west of the U.K. Thus, the lighter

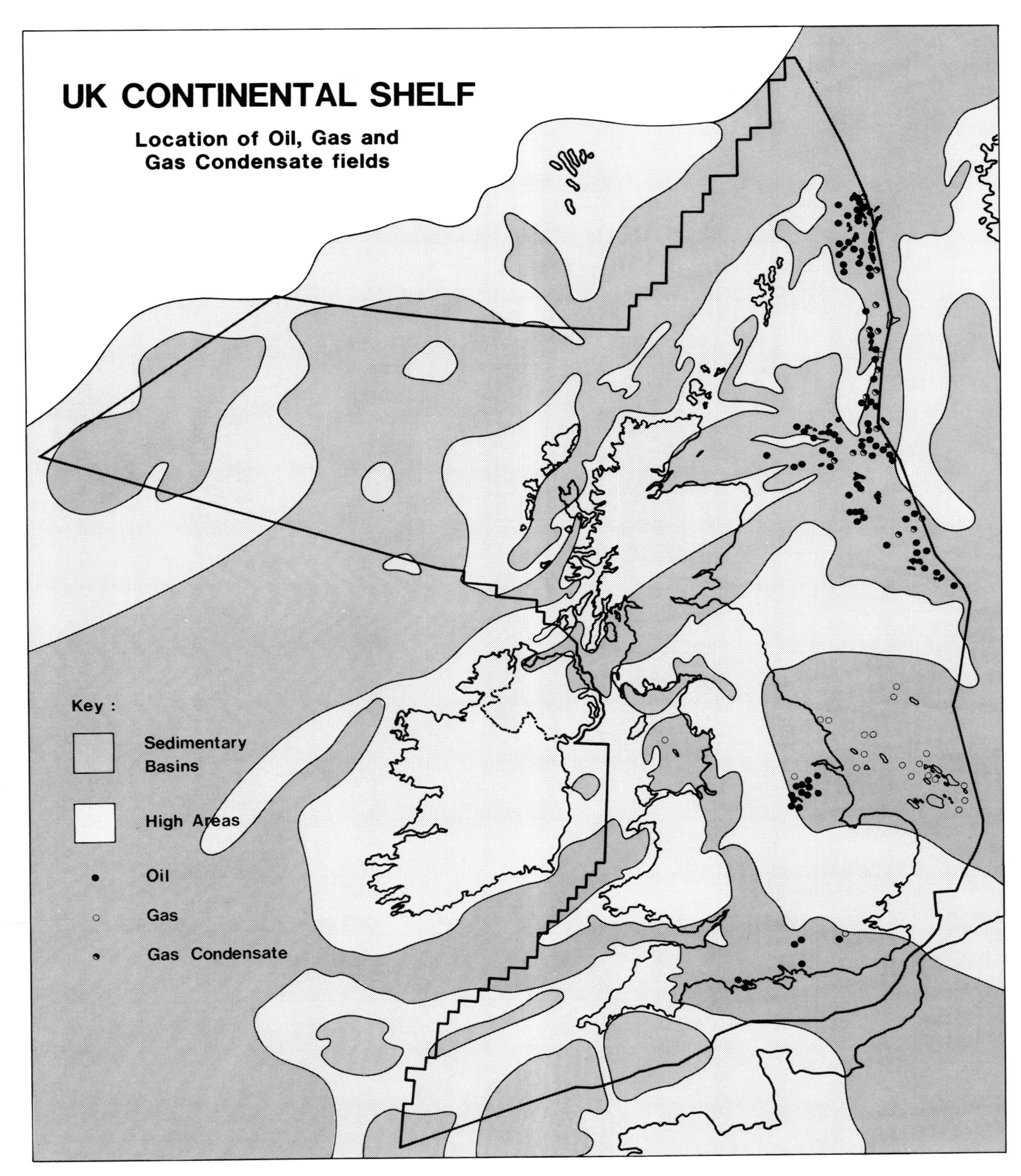


Figure 1. *Location of oil and gas fields.*

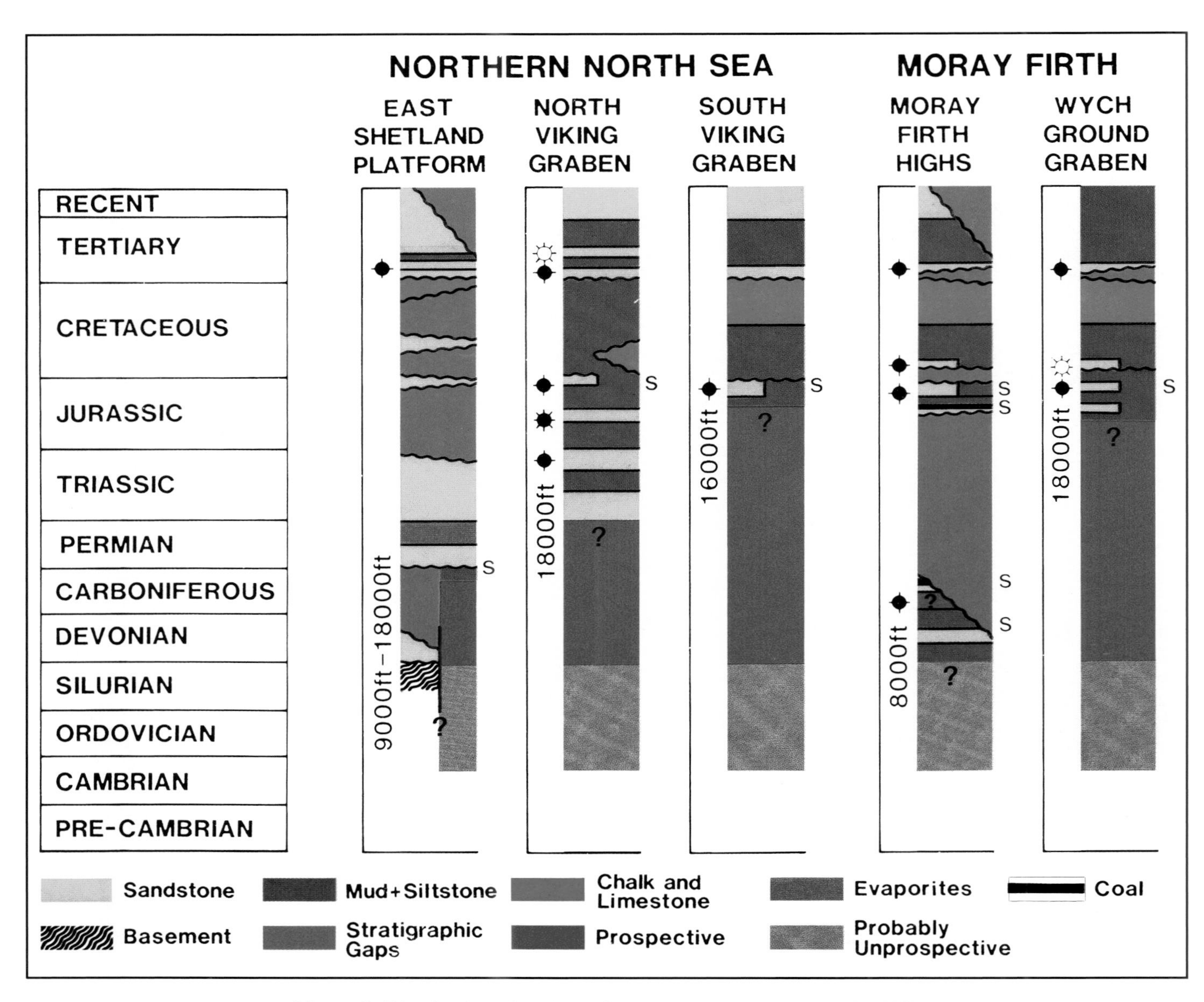


Figure 2. *Distribution of geological systems in basins around the U.K.*

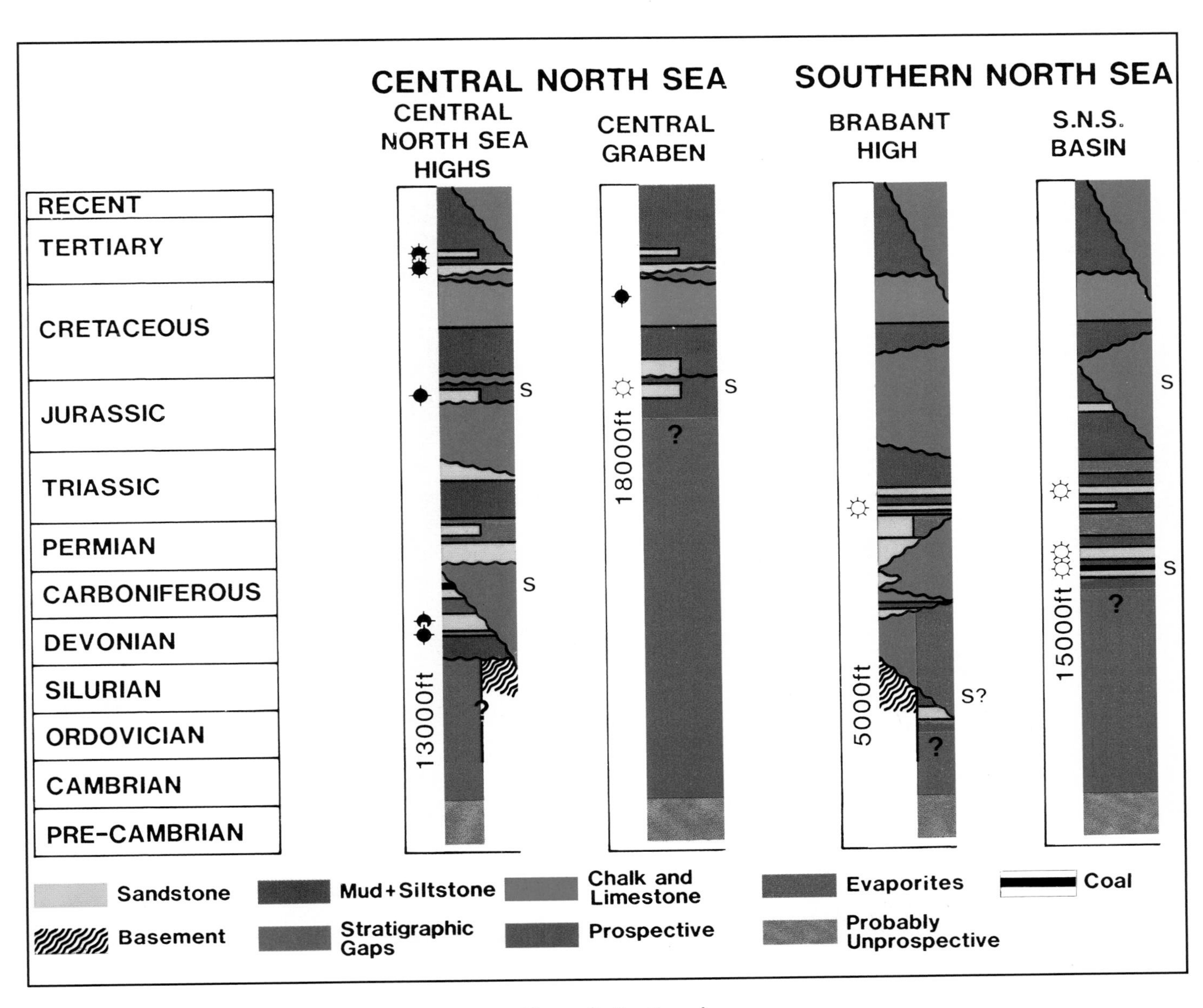


Figure 2. *Continued.*

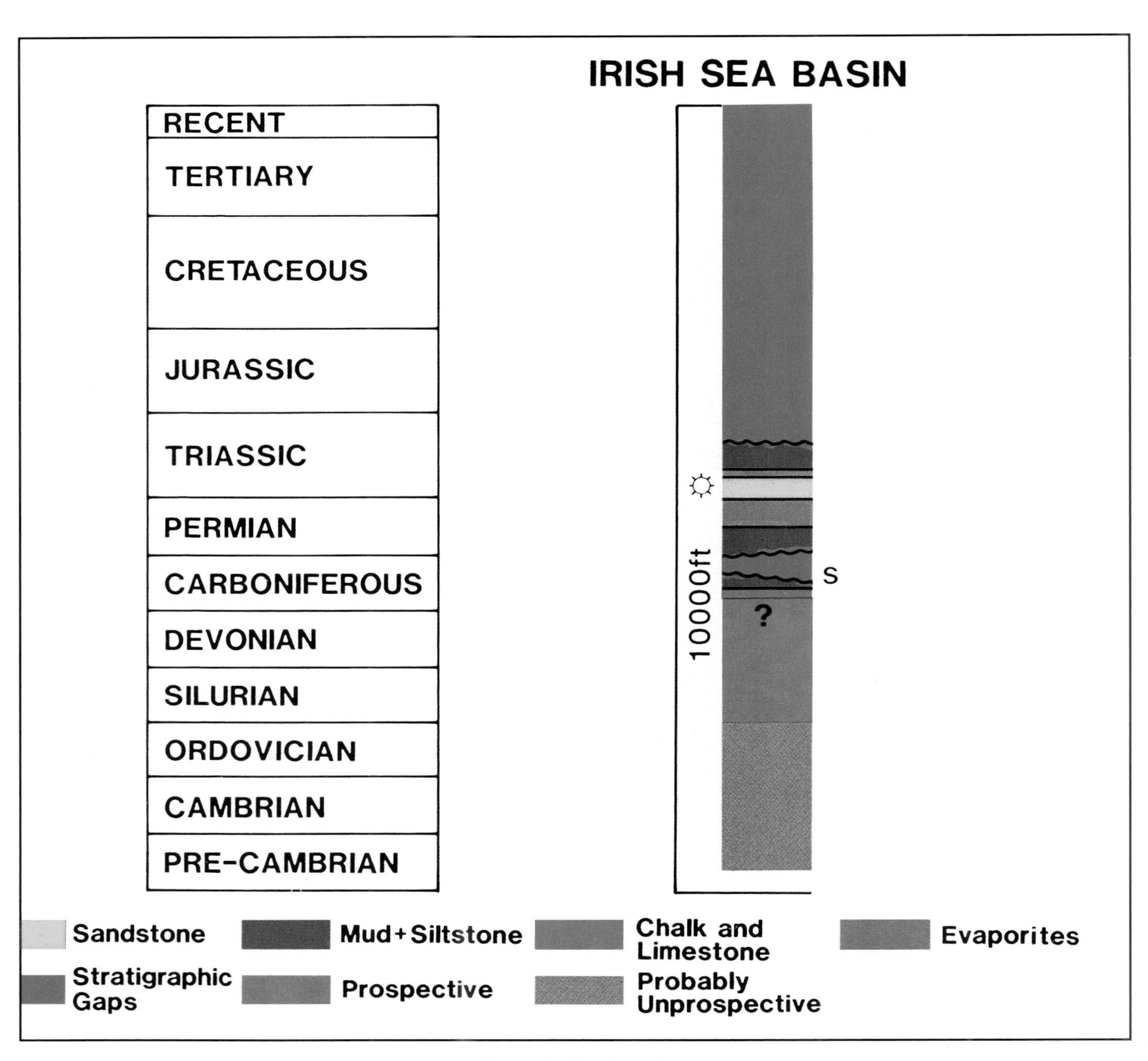


Figure 2. *Continued.*

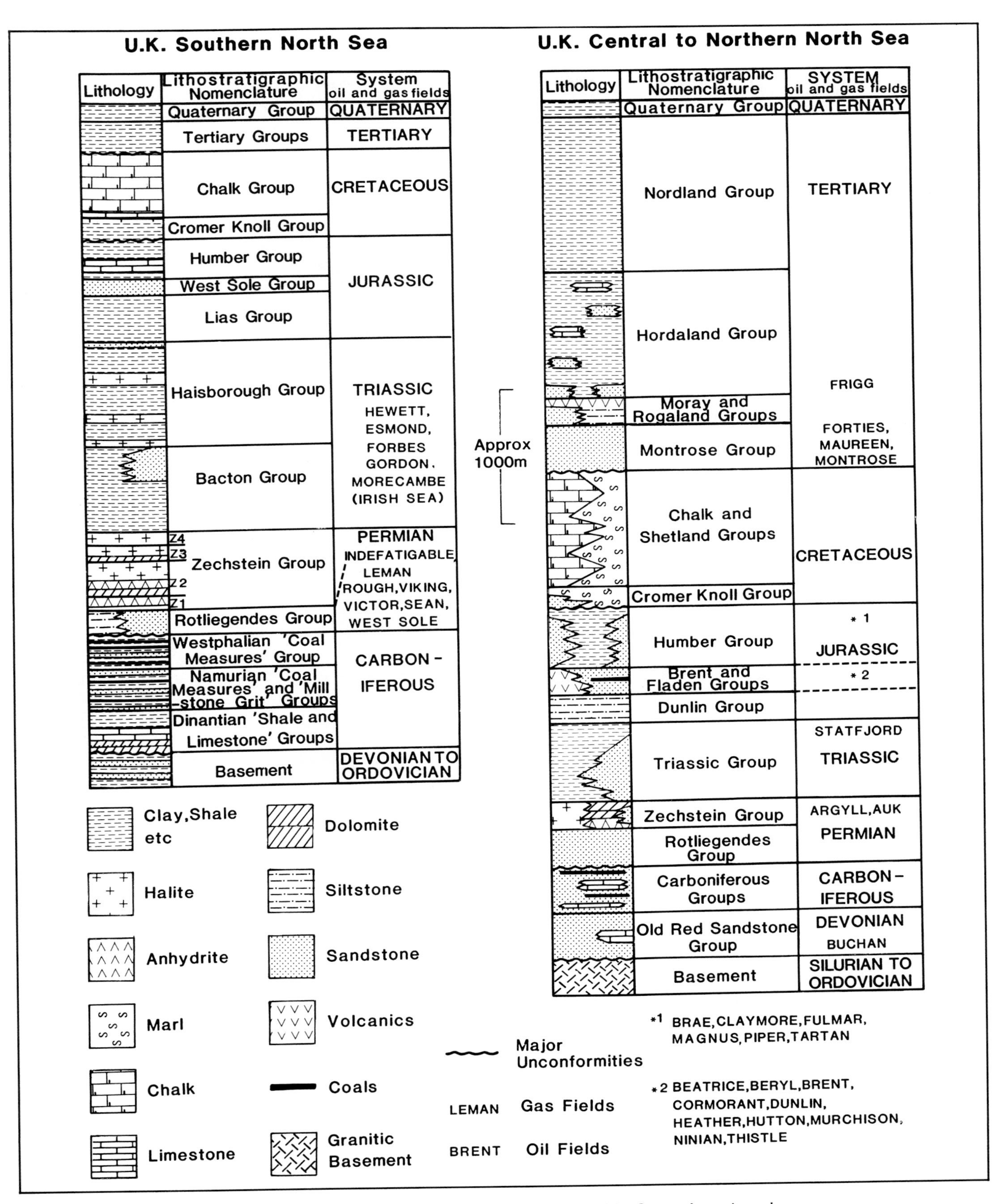


Figure 3. *Geological distribution of oil and gas fields. General stratigraphy.*

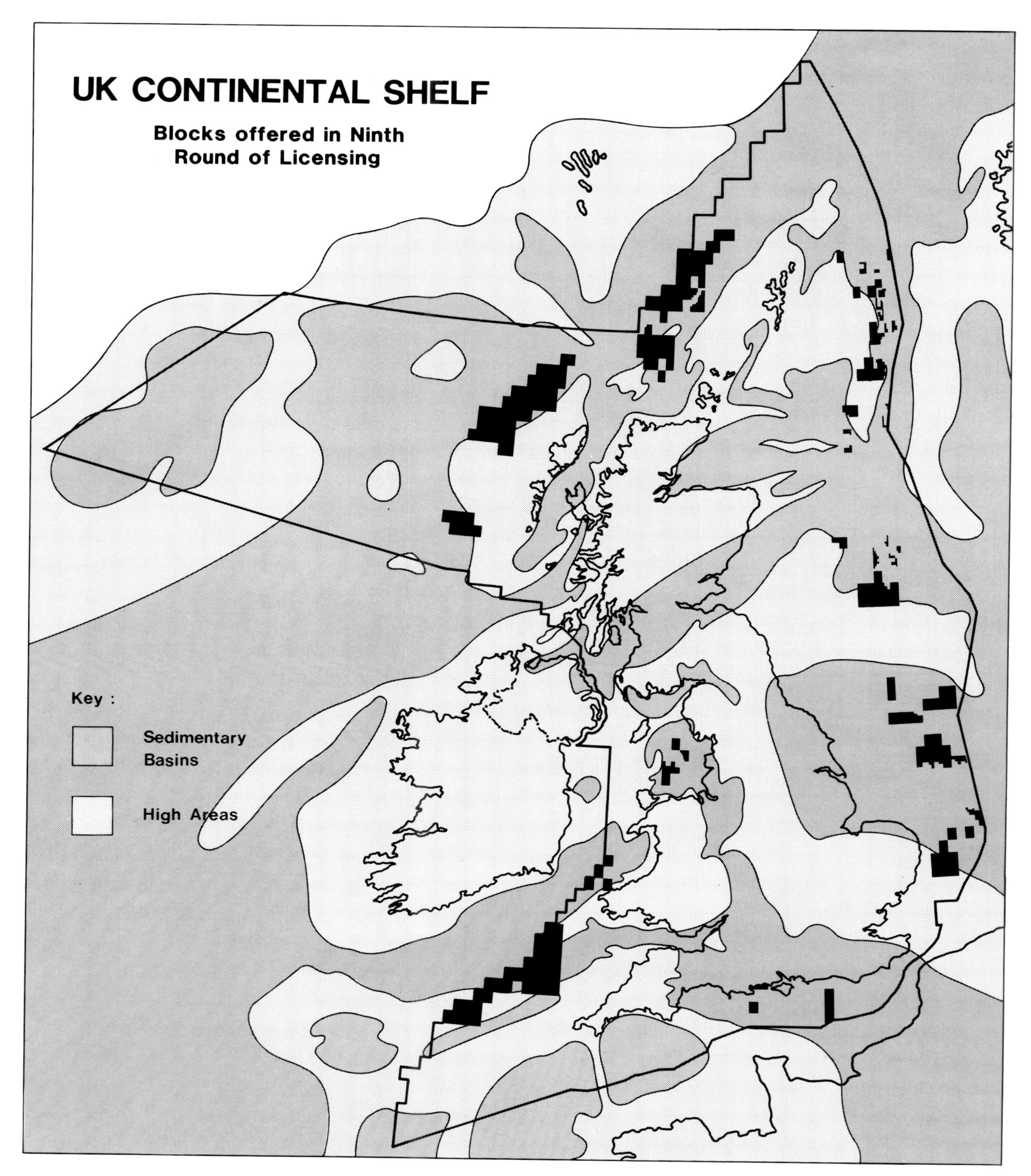


Figure 4. *Ninth Round of Licensing acreage offered.*

Table 1. *Exploration Categories of Basins and Areas*

| Category of Basin Based on Hydrocarbon Potential | | Exploration | Status | Discoveries | Examples |
|---|---|---|---|---|---|
| Hydrocarbon potential established | A | Mature. | Considerable good seismic & wells covering whole area. Major areas of current oil & gas produced scope for deeper exploration. | Small size to giant fields; e.g., Jurassic & Paleocene reservoirs (oil) & Permian reservoirs (gas). | East Shetland, Viking graben, Outer Moray Firth, Central North Sea, S. North Sea Gas basin (Rotliegendes sandstone). |
| | B | Submature. | Seismic and well coverage patchy; no deep drilling of older rocks. | Small size to giant fields.-(not as large as in A as yet) | Inner Moray Firth, North S.N. Sea, Wessex, Irish Sea, S. North Sea Gas basin (except Rotliegendes sandstone). |
| Hydrocarbon potential recognized | C | Late reconnaissance phase. | Seismic data available. Some drilling. | Early results disappointing | W. Shetlands, Unst, Cardigan Bay, SW Approaches |
| | D | Early reconnaissance phase. | Little seismic or little/no drilling | No hydrocarbon indications to date. | N. Moray Firth, E/W Orkney, Minch., Cheshire basin, English Channel |
| Hydrocarbon potential suspected | E | Frontier areas. | Includes deep-water areas of the U.K. continental shelf. Sparse seismic coverage only. | No hydrocarbon indications at present. | Margins of Mid. North Sea High; and Brabant Massif. Deep water area and deep part of most basins. |
| Hydrocarbon potential doubted | F | Cratonic areas with shallow metamorphic & igneous basement. | Some seismic. | Nil. | Rockall Bank, Scotland, Wales, Cornwall |

shaded areas (Figures 4 and 6-12) show regions of pre-Permian rocks that lie beneath shallow Mesozoic or Tertiary cover.

In some areas, these older sedimentary sequences may be present beneath the post-Carboniferous succession within drillable depths, but they may not have been explored. In other areas, the sedimentary column may be so thick that the deeper and older parts of the Mesozoic have not been reached. In the Viking graben and East Shetland basin of the Northern North Sea, wells have rarely penetrated pre-Triassic rocks, except at the basin edges; and in the southern part of the Viking graben, the top of the Middle Jurassic may not be encountered until 15,000 ft (4,570 m) subsea. In both parts of the Viking graben, even at such depths, the Mesozoic and upper Paleozoic should be attractive targets.

In contrast with the Nrthern North Sea upper Mesozoic rocks, the shallower, Carboniferous sequence lying beneath the Permian in the Southern basin of the North Sea has not been fully evaluated. No well has penetrated the full succession of Carboniferous rocks in this area, either in the U.K. sector or, it is believed, in the Dutch part.

Thus, while a classification can usefully summarize the overall status of exploration, we must consider the basins' actual geology when evaluating the total hydrocarbon potential.

The basins to the west of the U.K. are less well explored than those of the North Sea area, except for the Morecambe Bay area of the Irish Sea, between the Isle of Man and the U.K. mainland. Gas has been found there in sandstones of Triassic age, but as I will discuss later, we must also consider deeper prospects within the Carboniferous.

## POTENTIAL HYDROCARBON PROVINCES

In a paper of this nature, it is not really possible, or even desirable, to dwell on the details of every basin. I will, therefore, select six areas, each within different basins, to illustrate the "type" geology for future petroleum potential.

The six examples are shown below and in Table 2:

Table 2. *Summary of potential hydrocarbon provinces (offshore).*

| Basin | Cross-section |
|---|---|
| (A) **North Sea** | |
| *Northern North Sea* - 56°N-62°N | |
| Viking graben | (Example A–A′) |
| Moray Firth | (Example B–B′) |
| Forth Approaches and Midland Valley, mid-North Sea high | (Example E–E′) |
| *Southern North Sea* | (Example C–C′) |
| (B) **West of England and Wales** | |
| *North of Hercynian Front* | |
| Solway Firth and Morecambe Bay | (Example D–D′) |
| Cardigan Bay | |
| *South of Hercynian Front* | |
| Celtic Sea | |
| Bristol Channel | |
| South-West Approaches | |
| English Channel | |
| (C) **West of Scotland and Shetland Islands** | |
| Faroes Shetland Channel | |
| Rockall Trough | (Example F–F′) |

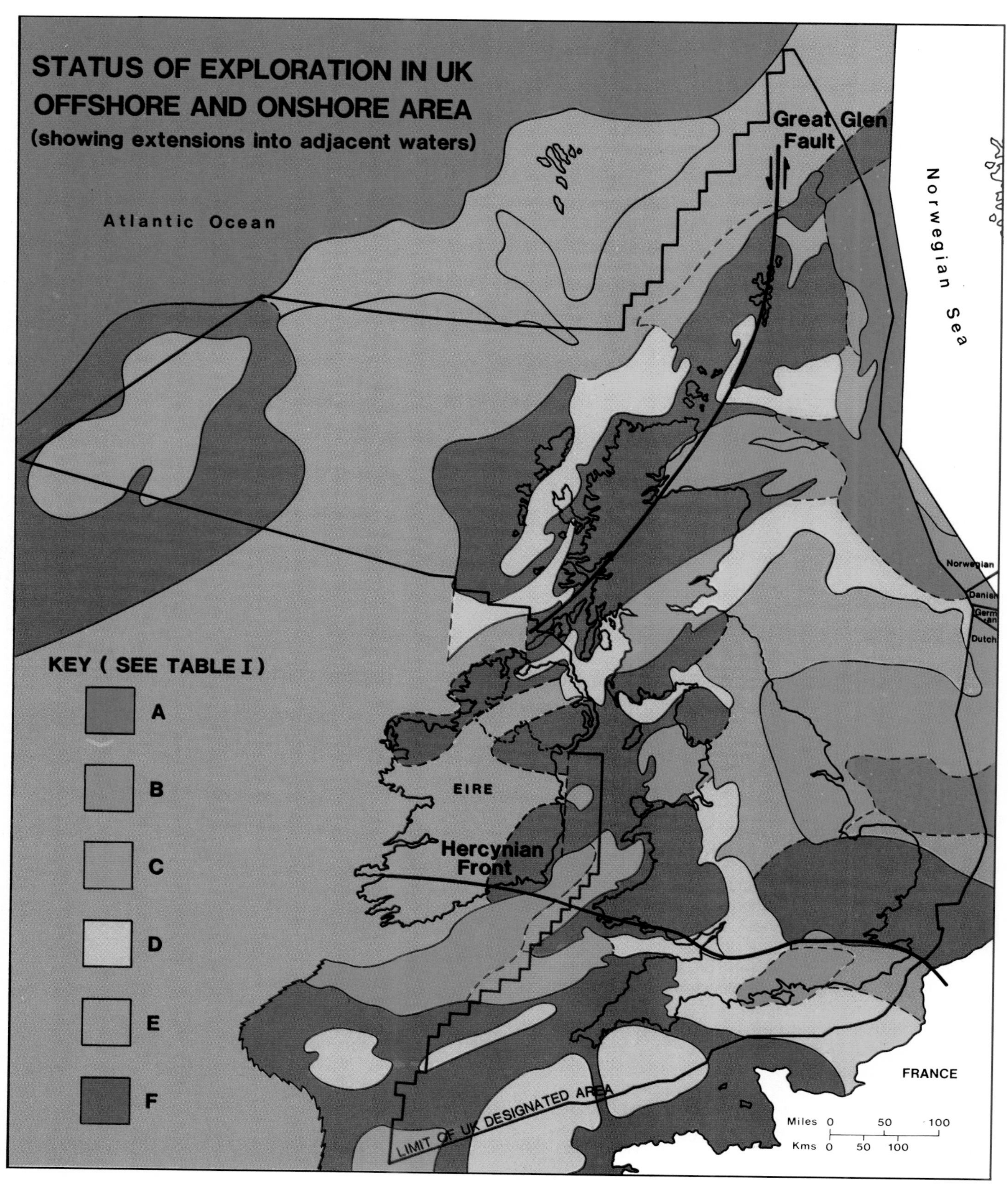


Figure 5. *Classification of sedimentary basins.*

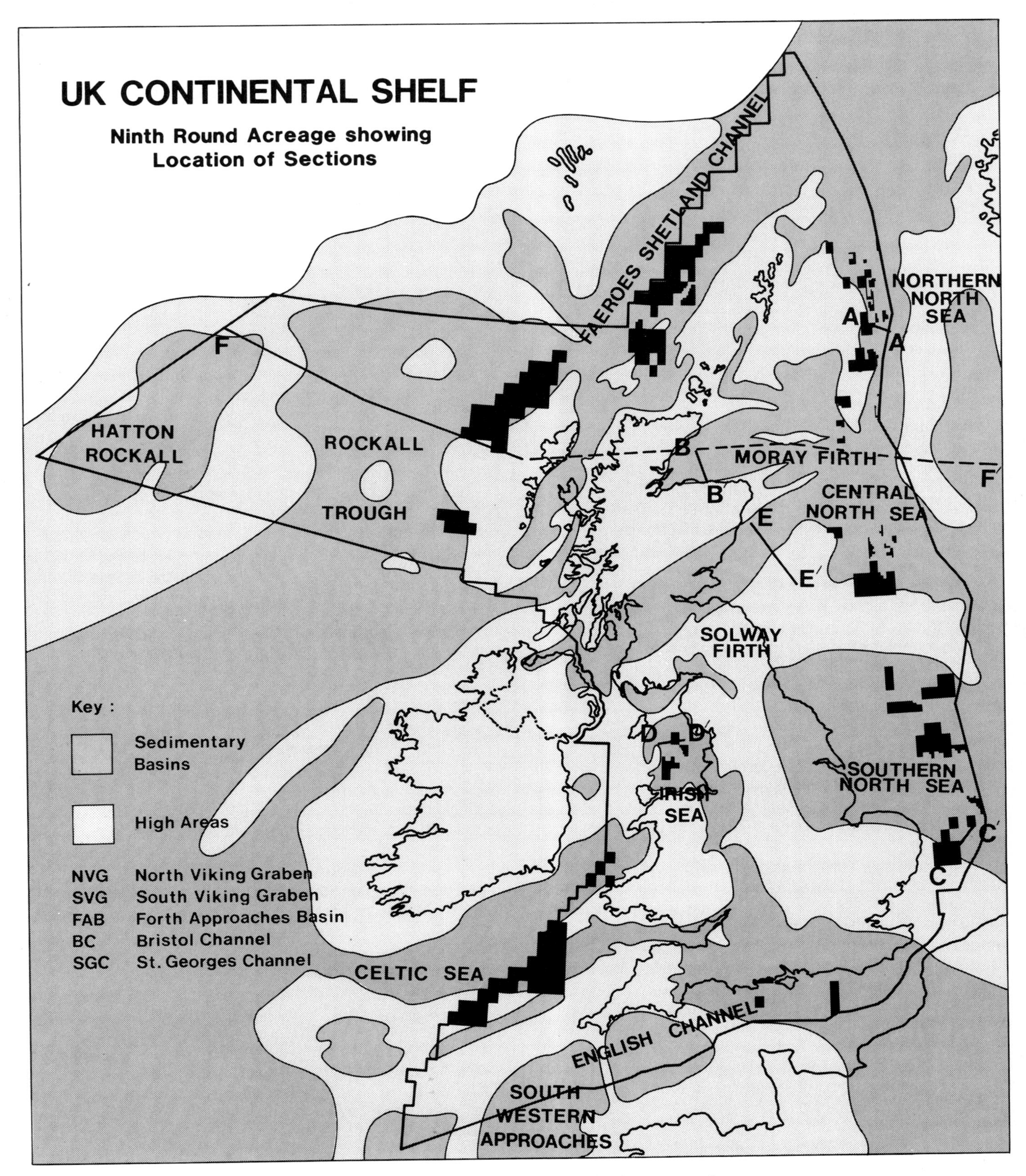


Figure 6. *Location of geological and seismic cross sections.*

Figure 6 shows the locations of these geological and seismic cross sections and the location of the major sedimentary basins and geographic areas. In addition to considering these six examples, I will review the potential of adjacent basins (see Figures 7–12).

## NORTH SEA

### Northern North Sea (56°N–62°N)

#### *Viking Graben (Quadrant 9)*

Cross-Section A–A′ (Figure 7), a west to east seismic line, extends from the East Shetland Platform in the west to the median line with Norway in the east.

Exploration in this area, although very successful, has essentially been confined to Tertiary, Cretaceous, and Jurassic horizons in the north. In the southern part of the area, where the seismic section is located, it only extends down to Upper Jurassic targets.

Listric faults within the Mesozoic and Paleozoic succession 'sole out' in a plane of decollement (or a decollement zone) at the apparent top of the Caledonide basement. On the cross-section this major fault can be seen to affect Permian and older rocks with the production of fault-controlled structures in Permian and younger sediments. This major fault system controls the westerly dipping, intra-graben fault-block structure.

Such structures have the potential for gas and possibly for oil, from sources within the deeper parts of the basin to the east. Toward the basin, the Lower Cretaceous is an eastward-dipping wedge and may provide an additional target.

Many examples of similar features are apparent from recent seismic acquisition, particularly along the edge of the graben margin adjacent to the East Shetland Platform.

#### *Inner Moray Firth (Quadrants 11 and 18)*

Together with the Viking graben and the Central graben, the outer part of the Moray Firth forms the third (northwest to southeast trending) area of a classical trilete graben system or triple junction. Mesozoic crustal extension formed these grabens and gravity measurement shows the resulting 'thin' crust beneath them.

The Inner Moray Firth basin was mainly formed by lateral movement along the Great Glen fault (shown in Figure 5) which was followed by a relatively small amount of Mesozoic extension perpendicular to the faults. Although the crust is not thinned beneath the Inner Moray Firth, the thinnest crust lies in the area of the Forties Field in the central North Sea that has had the greatest amount of basin subsidence.

There is evidence of considerable mid-Jurassic volcanicity in this area around the region of the triple junction and the Witch Ground graben. Therefore, high heat flow is recorded from this area and this anomalously high heat flow may contribute to the enhanced maturation of post-Middle Jurassic source rocks, leading to the formation of gas and gas condensate in these and younger rocks.

Oil does occur in the Middle-Jurassic of the Moray Firth, but occurrences have so far tended to be at the margin of the high heat flow areas.

The cross section B–B′ (Figure 8) from northwest to southeast in the Inner Moray Firth basin shows the half-grabens involving Cretaceous through to Old Red Sandstone (Devonian) basement rocks. The faulting beneath the Devonian in this area is considered to be Caledonian. In the center of the diagram (Figure 8), two sub-parallel reflectors, which are significant basement lineaments, dip steeply to the southeast. They are interpreted as thrusts at depth and were formed during the Caledonian orogeny. They were reactivated and controlled the location of the Mesozoic extentional faults which in turn influenced the structure of the overlying upper Paleozoic and Mesozoic strata. It is this structuring which is now the subject of interest for hydrocarbon trapping.

The larger part of the area around the Moray Firth, particularly in the north, is unexplored even by seismic and should be considered extremely prospective because much of the succession shown in Figure 3 may exist over the whole region. Oil has been found in sandstones of mid-Jurassic age (Callovian) in the Beatrice Field and hydrocarbons have been found in rocks of Devonian age. Source rocks generating both oil and gas occur, although little is known about them as yet.

The basin as a whole should be considered extremely attractive for further exploration.

#### *Forth Approaches Basin*

The northwest to southeast cross section (E–E′), in Figure 9, runs normal to the basin axis across what is essentially a northeasterly extension of the Midland Valley (of Scotland) graben.

Onshore, exploration drilling continues for Lower Carboniferous traps. Proven source and reservoir rocks occur within the Carboniferous succession and the extension of the basin offshore. Plays offshore should be enhanced by the existence of Permian salt (Zechstein Formation) acting as an effective seal to the Carboniferous and Lower Permian sandstones.

Silurian rocks are present at the southern margin of this basin where the Carboniferous rocks onlap the mid-North Sea High and the Zechstein oversteps this positive area. In parts of England and Wales, examination of Silurian shales provided geochemical evidence that they might act as source rocks given the right conditions. Under these conditions, the Silurian rocks might be productive.

The Midland Valley, with its Old Red Sandstone and Carboniferous infill, extends from the Forth Approaches graben westwards across Scotland and into Northern Ireland. Permian–Triassic sediments are locally preserved along this trend (for example, the Firth of Clyde and North Channel between Northern Ireland and Scotland), and they may offer additional targets in suitable locations.

### Southern North Sea Basin

The north to south trending Southern North Sea basin extends from the Pennine hills of England (in the west) to the Rhenish Schiefergebirge in the east. It is bounded on the north by the mid-North Sea High and the Ringkøbing-Fyn High and on the south by the London-Brabant Massif.

The main reservoir targets in this basin are anticlinal structures containing Lower Permian eolian sandstones (Rotliegendes), mainly confined to the southern half of the basin, and

Figure 7. *Cross section A–A′, East Shetland Platform to Viking graben.*

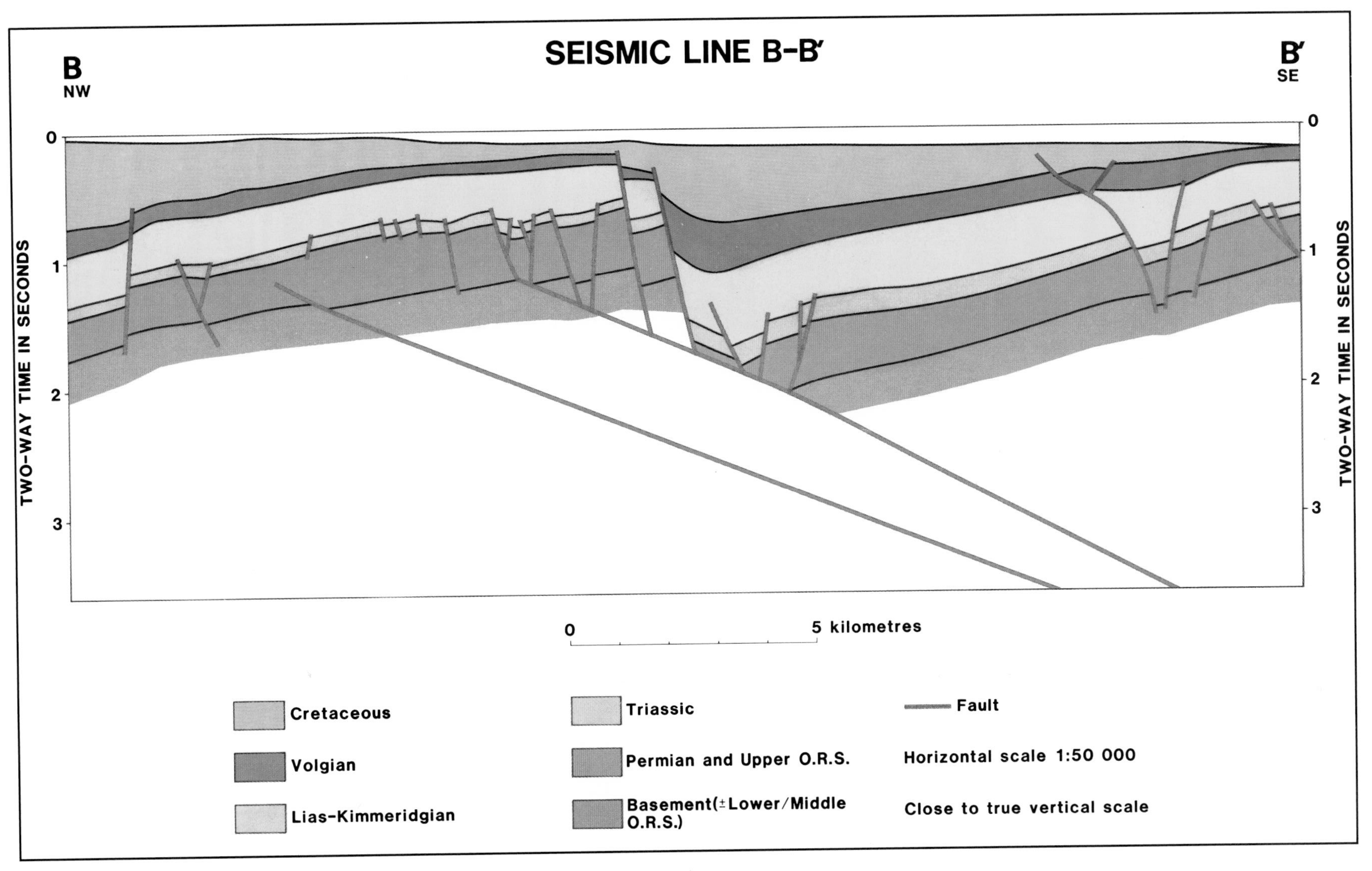


Figure 8. *Cross section B–B′, Inner Moray Firth.*

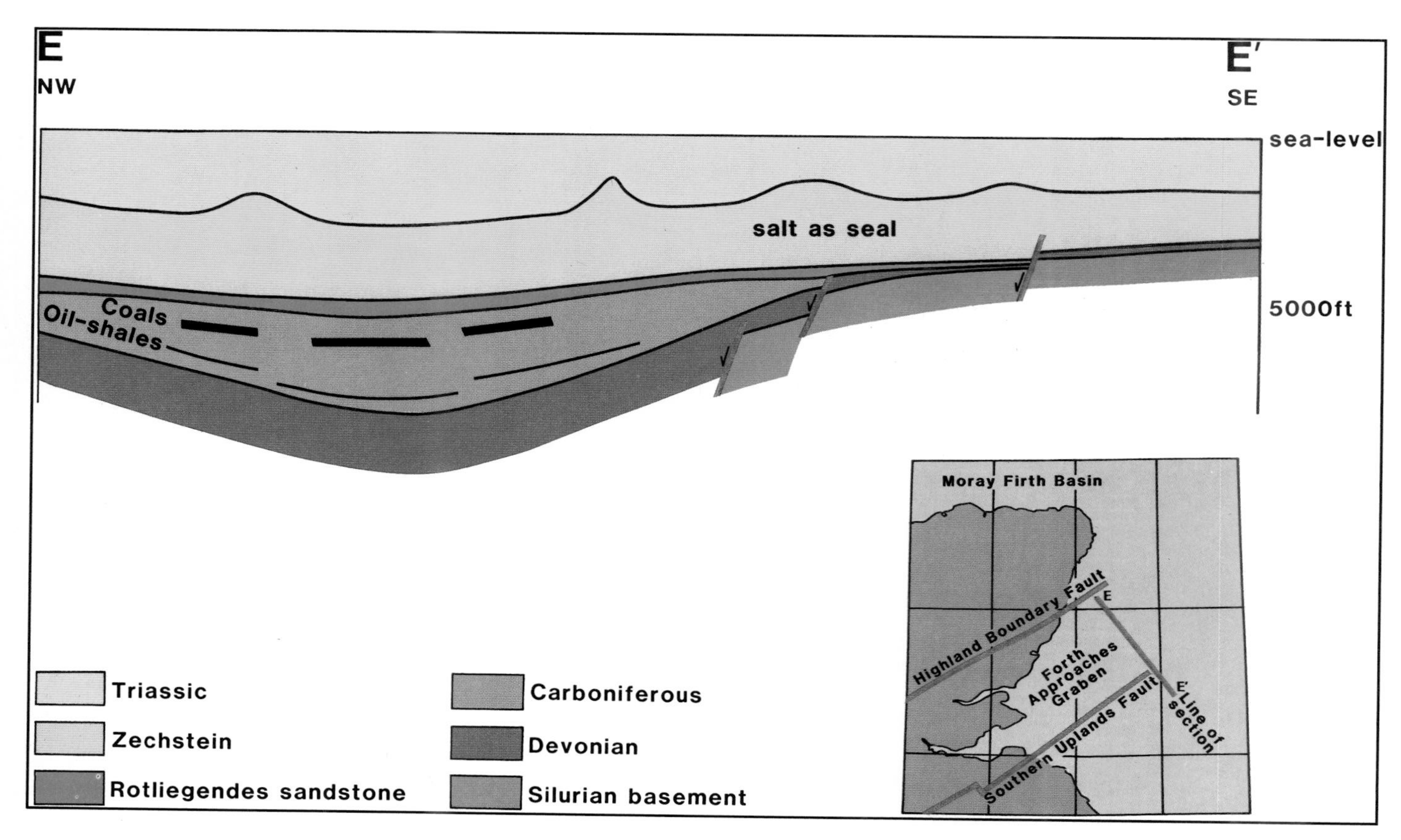


Figure 9. *Cross section E–E′, Forth Approaches basin.*

domal features containing Triassic sandstones in the north. Large accumulations of gas occur in both reservoir rocks.

The source for the gas is generally acknowledged to be coal measures of Late Carboniferous age. Seals to both reservoirs are provided by Upper Permian salt (Zechstein) and Triassic salt, respectively.

The Zechstein salt has undergone considerable movement, forming piercement pillars and walls. Movement took place well into the Tertiary. In addition, considerable erosion and inversion of parts of the basin have occurred, as shown by its complex geological history.

While a detailed geological history of this basin is beyond the scope of this paper, it is important to mention some of the complexities.

Table 1 and Figure 5 rank parts of this basin as Category B (sub-mature) because the lack of seismic resolution beneath the salt and the complexities of depth conversion procedures restrict exploration of much of the area. Even now there are parts of the basin where seismic events beneath inversion belts and the Zechstein salt cannot be resolved or even recorded. Only recently have geologists identified Carboniferous seismic horizons. There may be a considerable number of unidentified and, thus, undrilled sub-salt (Rotliegendes and Carboniferous) structures. Another reason is that exploration, which began in 1964, effectively stopped in 1970 when oil was found in the Northern North Sea, providing a better economic return than gas.

Seismic section C–C′ (Figure 10) runs southwest to northeast at right angles to the southern boundary of the basin. The section shows the thinning of the Upper Carboniferous succession onto the London-Brabant Massif and the wedge-out of seismic reflectors onto an intra-Carboniferous seismic event.

This intra-Carboniferous reflector may represent the Mississippian/Pennsylvanian (Namurian/Dinantian) boundary, which, in the subsurface of northern England, is an important structural break.

Stratigraphic traps in onlapping sediments are valid plays, and the Middle and Lower Carboniferous cannot be ruled out as potential target horizons, either here or at greater depths in the center of the basin.

In the central part of England, oil has been produced from Middle Carboniferous sandstones for nearly 40 years. The first well to discover oil in the Carboniferous was drilled in 1918.

Although few wells have been drilled deep (stratigraphically) into the Carboniferous, those that have indicate that reservoir and source rocks could be present at depth, both onshore and offshore. The results of recent drilling in both areas, in the U.K. and Dutch sectors, encourage this view.

Lithologic units beneath the Carboniferous are completely unknown. No well has penetrated the full sequence of Carboniferous rocks at any location. However, if Devonian sandstones were present on the bounding highs, in combination with

→

Figure 10. *Cross section C–C′, Southern basin.*

faulted Carboniferous, they might form an interesting reservoir. Gas has been reported from thrust-faulted Carboniferous/Devonian rocks in the Ardennes of Belgium and France.

## AREAS WEST OF ENGLAND AND WALES

West of the North Sea basins the area consists of two regimes, bounded by the "line" of the so-called "Hercynian Front" (Figure 5). The Hercynian Front is the northerly boundary of rocks affected by the Hercynian orogeny. North of the front, basement could be unmetamorphosed Carboniferous and Devonian rocks or metamorphosed Caledonian basement. South of the front, basement rocks are almost certainly metamorphosed Devonian and Carboniferous. Both the rock types and the thickness of the crust are different on either side of the Hercynian Front. Different heatflow and geothermal gradients are recorded (e.g., "hot rocks" of the Cornish Granites, south of the Front).

The above must be taken into account when considering all prospects, especially the deeper ones (Table 3).

### North of the Hercynian Front

#### *Solway Firth and Morecambe Bay (Irish Sea)*

Potential plays occur north of the Hercynian Front in the thick Carboniferous succession of the Solway basin and Irish Sea. The seismic section D–D′ (Figure 11) runs southwest to northeast across the northern part of the Morecambe Bay basin, showing events within a thick succession of Carboniferous rocks. Within this basin a large gas field (Morecambe) has been discovered and is being developed. The reservoir rock is a sandstone of Triassic age. In addition, a small oil field exists (onshore) at the southern margin of the basin and both have a Carboniferous source rock—probably the Upper Carboniferous Coal Measures.

The current, conventional view is that this source may charge reservoirs of both Triassic and Permian age. However, from analogies with outcrop and well information onshore, reservoir rocks should exist within the Carboniferous succession, which could be sourced by Carboniferous shales of perhaps Namurian age.

Marine shelf carbonates and shales are the dominant lithologies in the Lower Carboniferous (Dinantian). Marine shales within this succession have been shown to be mature since the end of Carboniferous times. Dinantian reef and shelf limestones have a poor reservoir potential, but dolomitization of these limestones, inducing secondary porosity and local fracturing near faults, may increase potential.

#### *Cardigan Bay*

This small, narrow basin, fault-bounded by the Cambrian and Precambrian rocks of Wales, is the northeasterly extension of a larger basin (the North Celtic Sea basin) that stretches to the edge of the continental margin between Eire and the Cornubian Peninsula (Cornwall). Significant hydrocarbon finds have been made in the Eire part of this basin. Gas has been discovered in Lower Cretaceous sandstones in the Kinsale Head Field (reserves reported at 1 tcf) in Eire blocks 48/20, 48/25, 49/16 and 49/21. In 1983, oil was discovered in sands of Triassic age in block 49/9. The prospects for the part of the basin within Cardigan Bay look promising despite the fact that the Hercynian Front separates Cardigan Bay from the Celtic Sea basin.

On deep seismic traverses, this "front" appears as a fault/discontinuity at considerable depth with the Celtic Sea basin straddling it. The Celtic Sea/Cardigan Bay basin has a thicker sedimentary sequence than other basins, so that this area and the nature of the Carboniferous and Devonian in this region may also be different. It seems unlikely, however, that either could be a reservoir or source rock—if the current perception of the effects of the Hercynian orogeny upon these sediments is correct.

Figure 11. *Cross section D–D′, Morecambe Bay.*

### South of the Hercynian Front

I include the potential of the lesser basins for completeness. These basins, however, are essentially too restricted in area and thickness to count as major areas of potential.

#### *South Celtic Sea Basin—Bristol Channel*

The precise relationship between the South Celtic Sea basin and Bristol Channel is not known. Drilling has taken place in both areas but with no success, resulting in the area not being licensed for some years. With encouragement from the discoveries to the northwest in Eire, companies may look at the area in a new light.

#### *South Western Approaches*

Wells have been drilled in the southwest for hydrocarbons, but none successfully. Reservoirs and seals are present, but apparently the problem is the lack of a source. It is far too early to stop exploring the basin before fully investigating the stratigraphy and source potential.

#### *English Channel*

The English Channel basin contains a similar geologic succession to that of the Wessex basin that exists onshore. A number of discoveries have been made here, mainly in sands of Middle and Late Jurassic and Triassic ages. Recently, gas was tested offshore near the Wytch Farm field, and other structures are targeted for drilling.

## WEST OF SCOTLAND AND THE SHETLAND ISLANDS

The potential for the deepwater areas west of Scotland in the Rockall Trough and west of the Shetland Islands in the Faroes Trough (Faroe-Shetland Channel) depends upon one's perception of the geological history of these two areas, especially the nature and timing of the movements of the American, Greenland, and northwest European plates. The debate about the geological history of both areas is largely based on poor seismic data and gravity information.

A considerable amount of seismic data have been acquired within U.K. designated waters in these areas. Several shallow wells have been drilled in the general area by the internationally-funded IPOD program, but only limited deep drilling has been attempted in the U.K. area.

Table 3. *Distribution of geological formations to the west of Britain.*

| Basins | Hydrocarbon Fields/Shows | Sediments Present/Inferred | Main Targets to Date | Future Targets | Notes |
|---|---|---|---|---|---|
| **North of Hercynian Front** | | | | | |
| Solway Firth | Gas in Upper Carboniferous | Permian and Triassic, Carboniferous | — | Permian and Triassic, Carboniferous | |
| Morecambe Bay/Irish Sea | Morecambe + discoveries | Thick Permian and Triassic & thick Carboniferous | Triassic | Permian and Triassic, Carboniferous reefs | |
| Cardigan Bay | Oil & gas shows | Tertiary, Jurassic, Permian and Triassic, Carboniferous | Jurassic, Permian and Triassic | Jurassic, Triassic | Carboniferous very thin |
| St George Channel | No drilling | Tertiary,? Jurassic, Permian and Triassic, ?Carboniferous | Nil | ?Jurassic, Permian and Triassic, ?Carboniferous | Basin not very deep |
| **South of Hercynian Front** | | | | | |
| Bristol Channel | Gas & oil in Eire | Tectonised Devonian-Carboniferous, Permian and Triassic, Jurassic, Cretaceous | Lower Cretaceous, Jurassic, Permian and Triassic | Jurassic, Permian and Triassic | Immature Jurassic |
| S Celtic Sea | None | Tertiary | | ?Triassic | |
| Haig Fras | No drilling | Tectonised Devonian-Carboniferous, Permian and Triassic, Lower Cretaceous, Upper Cretaceous | Nil | ? Lower Cretaceous, Permian and Triassic | Basin shallow |
| Shelf-edge Basins to Southwest | No drilling | Tectonised Devonian-Carboniferous, ?Permian and Triassic, ?Jurassic, Lower Cretaceous, Upper Cretaceous, Tertiary | Nil | Subcrop & faulted sands in Lower Cretaceous | Immature blackshales |
| SW Approaches W. English Channel | None | Tectonised Devonian-Carboniferous, Permian and Triassic, Jurassic, Carboniferous | Tertiary, Cretaceous, Jurassic, Triassic | Lower Cretaceous, Triassic | Lack of source rocks |
| E Channel/ Wessex | Oil-Wytch Farm Gas | Tectonised Devonian-Carboniferous, Permian and Triassic, Jurassic, Cretaceous, Tertiary | Lower Jurassic, Upper Triassic | Other Triassic + Jurassic | Major inversion structures |

What is clear is that both troughs are rifted, and both contain considerable thicknesses of sediment (tens of thousands of feet), although in parts of both, Tertiary lavas and intrusions obscure the full section. No magnetic stripes can be resolved for either trough, and the edge of the northwest European continental margin lies to the west of the Hatton-Rockall Bank at the eastern boundary of the Atlantic Ocean. The nature of the crust underlying these throughs is equivocal. Arguments for both continental and oceanic crust have been advanced but the evidence is not clear. It is possible that the crust here is of some intermediate or transitional type produced by extreme crustal stretching. But continental separation with the production of true oceanic crust beneath these troughts did not occur.

Figure 12 (section F–F′) shows a schematic, northwest to southeast cross section of the Rockall Trough, and illustrates the features mentioned above.

However, in terms of hydrocarbon prospectivity it is clear that significant depocenters have existed in these areas since Permian times and therefore they may well contain stratigraphic sequences, even structures similar to those in the North Sea basins.

However, if rifting of these troughs is related to the opening of

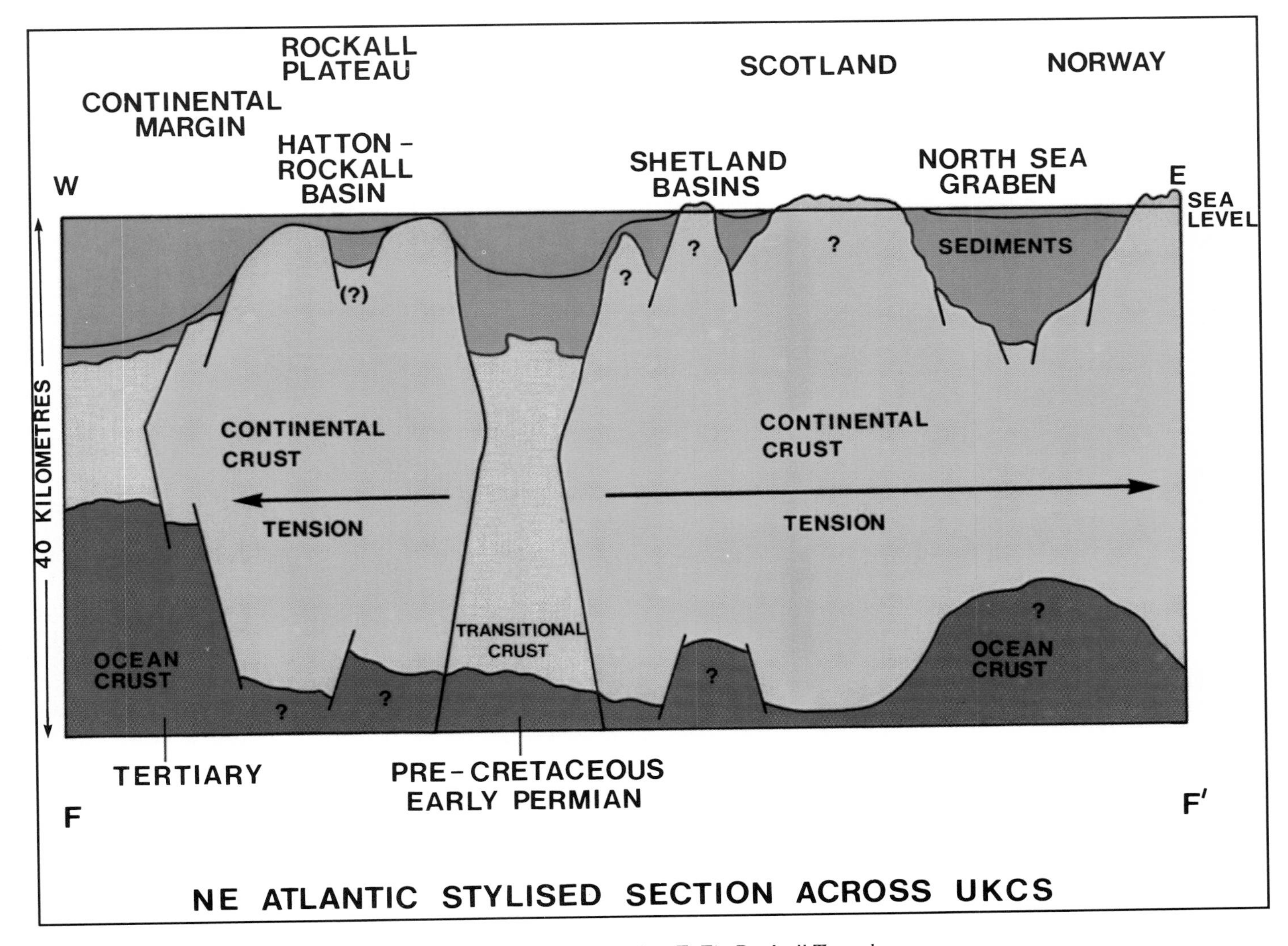


Figure 12. *Cross section F–F′, Rockall Trough.*

the Atlantic Ocean in the Early Cretaceous, the prospects for diverse sediments and structures may not exist.

The area is unknown geologically, and the wide range of uncertainty in the interpretation, together with the extreme water depths (2,000–6,000 ft [610–1,830 m]) and inhospitable environment, make these troughs the most exciting high-risk/high-reward areas currently under study.

Tracts of acreage on the eastern parts of both troughs were offered for the U.K. Ninth Round of offshore licensing by Her Majesty's Government (Figure 4).

## QUANTIFICATION OF PETROLEUM POTENTIAL

While it is not difficult to recognize the existence of petroleum potential, quantifying this potential is another matter.

The method used to quantify undiscovered hydrocarbon reserves in mature/sub-mature areas is not really applicable to areas where the stratigraphy, trap types, and timing of hydrocarbon generation is unknown. One must therefore look to a different style of basin evaluation in such areas, and perhaps use numerical models that can facilitate the large ranges of numbers and unknowns.

The U.K. has produced estimates for undiscovered resources of both gas and oil for the basins of the North Sea and Morecambe Bay. These figures are part of the U.K. estimates for ultimate "reserves" (discovered and undiscovered) of oil and gas and are included in the annual review to Parliament—*Development of the Oil and Gas Resources of the United Kingdom.*

Table 4 comes from this document and shows the range of estimates for undiscovered oil and gas in each of the major areas of the U.K. continental shelf. These estimates, together with the unquantified potential resources in subtle and stratigraphic traps and in frontier areas, suggest that prospects exist in sufficient quantities to maintain the status of the U.K. as a major explorer and producer of hydrocarbons for many years to come.

## CONCLUSIONS

Considerable potential exists for further discoveries of hydrocarbons on the United Kingdom continental shelf

— in mature areas, both in deeper plays and in stratigraphic

Table 4. *Estimates of undiscovered recoverable reserves of oil and gas on the U.K. continental shelf: reserves in future discoveries by geological area.*

| Area | Range of Estimated Reserves | |
|---|---|---|
| | Oil in million tonnes | Gas in billion cubic meters, (tcf in brackets) |
| A) North Sea 56°N - 62°N | 450 - 1900 | 15 - 130(0.5 - 4.6 ) |
| B) West of Shetland | 25 - 350 | Not Assessed |
| C) West of Scotland | 0 - 550 | Not Assessed |
| D) Southern Basin and Irish Sea | Assumed Nil | 170 - 440(6.0 - 15.5) |
| E) Remainder of U.K. continental shelf | 5 - 475 | Not Assessed |
| TOTALS | 480 - 3275 | 185 - 570(6.5 - 20.1) |

From: Development of the Oil and Gas Resources of the United Kingdom, 1984.

and subtle traps, and in undrilled, identified ('visible') structures;

— in conventional traps within lesser-drilled and undrilled basins;

— in frontier areas, such as the margins of basins, older highs and in the deep-water parts of the continental shelf westward to the continental margin.

In an attempt to encourage exploration of some of these areas, the Department of Energy offered tracts of acreage in the Ninth Round of Licensing for application by oil companies (Figure 4). The closing date for applications was December 17, 1984.

Reviewing the history of exploration on the U.K. continental shelf, the First, Fourth, and Seventh Rounds of Licensing were significant and decisive. Not only did each bring in new acreage, but also new ideas, new concepts, and new companies. The Ninth Round, which was designed to stimulate exploration into hitherto unexplored areas, may be similarly farsighted.

Since the Fifth Round in 1977, the Department has used a two-year cycle of offshore licensing. This, with 80–100 blocks awarded in each round, seems to be satisfactory for both Government and industry. It also appears to make the best use of available resources. Such a pattern of licensing, coupled with the relinquishment of acreage and the opening of new areas, will ensure continued interest in exploring the U.K. continental shelf into the 21st century.

## ACKNOWLEDGMENTS

I am extremely grateful to my colleagues, both in the Department and in the British Geological Survey, Hydrocarbons Unit for their assistance and advice, especially Chris Deegan and Jim Aithen for many hours of discussion.

My thanks are also due to Margaret Havell and her staff in the Drawing Office for the preparation of slides and diagrams and to my Secretary, Donna May, for the preparation of the manuscript.

The compilation of this paper is the work of the author. It presents his views and does not necessarily represent the official positions of the Department of Energy.

The paper is published with the permission of the Director of the Petroleum Engineering Division of the Department of Energy.

## REFERENCES CITED

Development of the oil and gas Resources of the United Kingdom, 1984. HMSO, London.

# Index

A reference is indexed according to its important, or "key" words.

Three columns are to the left of a keyword entry. The first column, a letter entry, represents the AAPG book series from which the reference originated. In this case, ME stands for AAPG Memoir Series. Every five years, AAPG will merge all its indexes together, and the letters ME will differentiate this reference from those of the Studies in Geology Series (ST) or from the AAPG Bulletin (B).

The following number is the series number. In this case, 40 represents a reference from AAPG Memoir 40. The third column lists the page number of this volume on which the reference can be found.

† = titles    * = authors